MATERIALS AT LOW TEMPERATURES

MATERIALS AT LOW TEMPERATURES

Edited by

RICHARD P. REED

and

ALAN F. CLARK

National Bureau of Standards
Boulder, Colorado

AMERICAN SOCIETY FOR METALS

Metals Park, Ohio 44073

Library of Congress Catalog Card No.: 82-73607

ISBN: 0-87170-146-4

SAN: 204-7586

Editorial and production coordination by
Carnes Publication Services, Inc.

PRINTED IN THE UNITED STATES OF AMERICA

Authors

Alan F. Clark was awarded a NAS–NRC postdoctoral research associateship at the National Bureau of Standards after receiving his Ph.D. in physics–nuclear science from the University of Michigan in 1964. Specializing in the low-temperature electrical and thermal properties of materials, Dr. Clark has published more than ninety papers and edited several conference proceedings, monographs, and texts in this field. At present he is Group Leader of the Superconductors and Magnetic Materials group in the Electromagnetic Technology Division of the National Bureau of Standards, Boulder, Colorado, and Chairman of the International Cryogenic Materials Conference Board.

Jack W. Ekin became a postdoctoral research associate at the National Bureau of Standards after obtaining his Ph.D. in physics from Cornell University in 1971. He joined the National Bureau of Standards staff in Boulder as a physicist in 1975 and is currently with the Electromagnetic Technology Division, specializing in the electromechanical properties of superconductors. He has published at least thirty-six technical papers in the area of superconductivity, twenty-one of which are about strain effects on practical superconductors.

Frederick R. Fickett is a physicist with the Electromagnetic Technology Division of the National Bureau of Standards in Boulder. He received his Ph.D. from Oregon State University in 1967. Following a postdoctoral research associateship at the National Bureau of Standards Cryogenics Division, he joined the staff in 1969. His special interest and research areas are in materials science and include studies on the electromagnetic and mechanical properties of superconductors, pure metals, magnetic materials, and polymers.

Ronald B. Goldfarb has a B.A. in electrical engineering and an M.S. in materials science, both from Rice University, and received his Ph.D. in physics from Colorado State University in 1979. He was a NAS–NRC postdoctoral research associate at the National Bureau of Standards from 1979 through 1981 and is currently a staff physicist with the Electromagnetic Technology Division at the National Bureau of Standards in Boulder. His areas of research include magnetic, amorphous, and superconducting materials and measurements of their properties.

Jerome G. Hust received his M.S. in physics from the South Dakota School of Mines and Technology in 1961. He has been employed at the National Bureau of Standards in Boulder since that time, in the Properties of Fluids group from 1961 through 1965 and then in the Properties of Solids group of the Chemical Engineering Science Division. Mr. Hust has established five Standard Reference Materials for thermal conductivity. He is now Group Leader of the Properties of Solids group.

Maurice B. Kasen received a Ph.D. from the University of Washington in 1965. He is currently a member of ASTM Committees E9 and D30 and a staff scientist in the Fracture and Deformation Division, National Bureau of Standards, Boulder, specializing in the low-temperature properties of composites. Prior affiliations include Postdoctoral Fellow, Baikow Institute of Metallurgy, Moscow, USSR (1965–1966); metallurgist, Boeing Co., Seattle, Washington (1955–1960); and metallurgist, Honeywell Corp., Minneapolis, Minnesota (1950–1955).

Hassel M. Ledbetter received a Ph.D. in metallurgy in 1969 from the University of Illinois at Urbana–Champaign. He has published approximately seventy research papers related to solid-state elastic properties; these include experiment, theory, and review. He conducts theoretical and experimental studies on solid-state physical properties — especially elastic constants, thermal expansivity, and phase transitions — in metals, alloys, polymers, and composites (both metal-matrix and nonmetal-matrix). Dr. Ledbetter is currently a metallurgist in the Fracture and Deformation Division, National Bureau of Standards, Boulder.

Harry I. McHenry is Deputy Chief of the Fracture and Deformation Division of the National Bureau of Standards in Boulder, where he conducts fracture mechanics research on cryogenic alloys and structural steels. He received a Ph.D. in metallurgy from Lehigh University. Prior to joining the National Bureau of Standards in 1974, he worked as a project structures engineer for General Dynamics, Fort Worth, Texas, where he applied fracture mechanics to the design and evaluation of aircraft structures.

John C. Moulder is a materials research engineer in the Fracture and Deformation Division of the National Bureau of Standards in Boulder. His primary technical interests are in experimental mechanics and nondestructive evaluation, especially the application of lasers and optical techniques to materials research. Prior to joining the Fracture and Deformation Division, Mr. Moulder was associated with the Thermophysical Properties Division of the National Bureau of Standards, where he studied laser-initiated metal combustion for seven years. Mr. Moulder attended the Massachusetts Institute of Technology and the University of Colorado.

David T. Read received his Ph.D. in physics from the University of Illinois at Urbana–Champaign in 1974 and has been a physicist at the National Bureau of Standards, Boulder, Colorado, since that time. Dr. Read's current research interest is applied fracture mechanics in the Fracture and Deformation Division; he has recently developed a method for direct experimental evaluation of the applied J contour integral (a crack-driving force parameter) in plastically strained metals.

Richard P. Reed obtained his Ph.D. in physical metallurgy from the University of Denver in 1966. He has published more than eighty papers on low-temperature phase transformations and mechanical properties. Dr. Reed is a founder and board member of the International Cryogenic Materials Conference and coeditor of its proceedings. He is currently Chief of the Fracture and Deformation Division, National Bureau of Standards, in Boulder, Colorado, and Gaithersburg, Maryland.

Larry Sparks received his B.S. in engineering physics from the University of Colorado in 1962. He has been employed at the National Bureau of Standards in Boulder since that time and at present is project manager in the Properties of Solids group of the Chemical Engineering Science Division. Mr. Sparks has performed research on low-temperature thermoelectric standards, thermal conductivity standards, and thermal conductivity and mechanical properties of composite materials. His work on thermoelectric thermometry resulted in the national standards for low-temperature thermocouple thermometry.

Ralph L. Tobler received his M.S. in metallurgical engineering from the University of Illinois in Chicago in 1972. His research includes mechanical properties of structural alloys at cryogenic temperatures, cryostat design, test-method development, and materials characterization. Mr. Tobler is a staff scientist in the Fracture and Deformation Division, National Bureau of Standards, Boulder, Colorado.

Preface

Research in Europe during the late 1800s led to the liquefaction of first oxygen (1877), followed by nitrogen (1883), hydrogen (1891), and, finally, helium (1908). These fluids provided scientists the opportunity to conduct research at low temperatures on subjects as basic and diverse as magnetic transitions and deformation twinning. In 1911 Onnes discovered the phenomena of superconductivity, which stimulated the fundamental studies of solid materials. For many years materials at low temperatures were found mainly in the research laboratory. In recent years, research in this field has responded to needs for materials to be used in practical systems at low temperatures.

As refrigeration techniques improved and liquid air, nitrogen, and oxygen became readily available, suitable materials were needed for storage Dewars, industrial processing equipment, and refrigerators. The need to transport and store large quantities of liquefied natural gas (LNG) led to the development and use of tough iron-nickel and aluminum alloys that are less expensive than the austenitic stainless steels traditionally used for small liquid-nitrogen Dewars. The safety and long service lifetime demanded of these large containers required an increased understanding of materials properties at low temperatures.

The space program, utilizing liquid oxygen and hydrogen fuels, led to the development and use of high-strength, lightweight materials, such as cold-drawn stainless steels and precipitation-hardened aluminum alloys, and the first use of nonmetallic composites.

The development of stabilized, practical superconductors generated a need for large, strong structures for the very high magnetic field superconducting magnets and other superconducting applications. The development of structural alloys to contain liquid helium and to withstand the large forces from superconducting magnetic fields is currently underway. In this complex application, the requirements are high strength, adequate toughness, good weldability, and minimum costs; strengthened austenitic steels are the leading candidate. Significant developments in insulating nonmetallics, structural composites, and the superconductors themselves are also the results of persistent research.

Design and test procedures for materials at low temperatures have improved along with the materials themselves. Because material properties change dramatically with temperature, new parameters take on added importance. There are many examples: The development of fracture mechanics has led to new test procedures and the opportunity for a quantitative design assessment of structural lifetime. Measurements of fracture toughness and fatigue crack growth rates are now indispensable for most cryogenic applications. Balancing the thermal expansion of components eliminates residual thermal stresses. The composite materials engineer can actually design a material to meet his needs by combining properties of constituents.

The thirteen contributors to this book formed the nucleus of the materials group of the Cryogenics Division of the National Bureau of Standards (NBS). The Division was formed in 1952 as part of the expansion of NBS to their new Boulder (Colorado) Laboratories, and our group was composed of specialists in physical and mechanical properties of solids at low temperatures. We primarily served emerging low-temperature technologies requiring cryogenic materials, such as space, LNG storage and transportation, magnetohydrodynamic energy conversion, fusion, high-energy physics, and other aspects of applied superconductivity. We evolved into the largest materials research group in the free world devoted to low-temperature research.

We have now dispersed into several disciplines, but most of our low-temperature research capabilities and interests remain. While these are still strong, we thought it appropriate to summarize our data and experience in this book. We hope our contributions will establish a firm basis for the understanding of materials behavior and property measurement to use for design and material selection at low temperatures.

We acknowledge the support and encouragement of Dave Sutter of the Office of High-Energy Physics, Department of Energy, and of Carl Henning, Dick Foley, Ed Dalder, and Don Beard of the Office of Fusion Energy, Department of Energy. Marilyn Stieg was indispensable in assisting us in editing. Expert typing was performed by Vickie Grove, Carole Montgomery, JoAnne Wilken, and Deb Wilson. We greatly appreciate the intensive reviews of specific chapters by our colleagues.

RICHARD P. REED

ALAN F. CLARK

Contents

1 Elastic Properties

H. M. LEDBETTER

1.1 INTRODUCTION

Many scientific-technological advances depend critically on solid-state elastic properties, their magnitudes, and their responses to variables like stress and temperature. Elastic constants relate to various fundamental solid-state phenomena, including interatomic potentials, equations of state, and phonon spectra. They link thermodynamically with specific heat, thermal expansivity, the Debye temperature, and Grüneisen's constant. In engineering design, they appear in calculations for load deflection, residual stress, thermoelastic stress, fracture toughness, and elastic instabilities such as buckling and jamming. Velocities of flexural, torsional, and extensional waves depend entirely on the elastic constants and the mass density. Using velocity methods, elastic-constant measurements within 0.1% become feasible; several authors claim elastic-constant-change measurements as small as one part in 10^7. This high-precision measurement capability permits using elastic constants in studying effects of temperature, pressure, mechanical stress, magnetic field, crystallographic transformation, and superconducting transition. So, elastic constants are many things to many men; they interest structural designers, materials scientists, and solid-state theorists; they connect to technological structural economics and safety, to various materials phenomena, and to fundamental interatomic forces.

This chapter emphasizes five aspects of engineering-material solid-state elastic constants: (a) general properties; (b) interrelationships; (c) relationships, especially thermodynamic, to other physical properties; (d) changes during cooling from ambient to near-zero temperature; and (e) near-zero-temperature behavior. Through these topics the author hopes to promote a better, broader familiarity with elastic and elastic-related properties, especially at cryogenic temperatures. Surveying the science and engineering literature shows that this subject, although already relatively well defined, still supports intense theoretical and experimental research. Despite the subject's maturity and activity, many main ideas remain undiffused through the engineering-materials community. This review seeks to stimulate such diffusion.

Rather than empirical observations, the author tried to collect physical interrelationships and insights. In agreement with Aristotle (4th century B.C.) who maintained that "men do not think they know a thing until they have grasped the why of it," explanations are indicated whenever possible.

Despite intense, rich thought experiments by early Greek scientists, they apparently never focused on a solid's elasticity. Galileo did. His bent-beam experiments first showed that solids respond to stress, change their volume and shape, and resume their original state upon stress removal. After Galileo, solid-state elasticity became vital in both physics and technology, a growth process reviewed by Todhunter and Pearson (1886) and by Love (1927).

1.1.1 Theoretical résumé

Explanations of elastic constants, their temperature dependence, and related phenomena usually arise from atomistic, quantum-mechanical, discrete-lattice models. A basic energy expression for solids displays Coulombic interactions among atoms and electrons. Electrostatic forces dominate all solid-state interatomic forces, attractive or repulsive. An overall interatomic potential energy, $V(r)$, that depends only on atomic coordinates, often represents these interactions. Units on $V(r)$ are energy per unit volume; sometimes it is convenient to use energy per unit mass or energy per atom. Figure 1.1 shows a typical, well-known semiempirical interatomic potential used frequently to interpret solid-state behavior. $V(r)$ is decisive; from it most fundamental physical properties follow; against it most properties should be considered. Knowing $V(r)$ permits computing various phenomena interrelated by their common source — the solid's potential energy distributed on a lattice. Among all possible crystal structures, that of lowest energy for a given $V(r)$ prevails, and equilibrium volume follows from the minimum in $V(r)$:

$$[dV(r)/dr]_{r=r_0} = 0 \tag{1.1}$$

The usual, second-order, elastic constants, C_2, relate to $V(r)$'s curvature at $r = r_0$:

$$C_2 \sim [\partial^2 V(r)/\partial r^2]_{r=r_0} \tag{1.2}$$

Many papers relating elastic constants to interatomic potentials appear in *Calculation of the Properties of Vacancies and Interstitials* (1967) and *Interatomic Potentials and Simulation of Lattice Defects* (1972).

When zero-point vibrations are neglected, at equilibrium at zero temperature, the atoms in crystalline solids occupy fixed sites. These sites form a lattice:

$$\vec{\ell} = \sum_i \alpha_i \vec{x}_i \tag{1.3}$$

where $\alpha_i = 0, 1, 2, \ldots$ and the $\vec{x}_i$ denote primitive-cell vectors. Atomic displacements from equilibrium sites, displacements small relative to r_0, determine most of a solid's interesting thermal-mechanical physical properties. For such cases $V(r)$ can be expanded about r_0 in terms of the atomic displacements $\vec{u} = \vec{r} - \vec{r_0}$:

$$V(r) = V_0 + V_1 + V_2 + V_3 + V_4 + \ldots \tag{1.4}$$

where

$$V_i = (1/i!)\,(d^iV/dr^i)u^i \tag{1.5}$$

The cohesive energy, V_0, can be taken as zero (the reference state) and independent of r. The first-order term vanishes at equilibrium according to Eq. 1.1, and for small displacements, V_2 dominates $V(r)$. A solid's potential energy is approximately a homogeneous quadratic form in the atomic displacements from equilibrium sites. For larger displacements, V_3, V_4, and higher-order terms become important. Equation 1.4 is usually truncated after V_4 because higher-order terms have very little theoretical or experimental significance at the present state of solid-state science. Nearest-neighbor interactions alone often sufficiently describe physical properties because interatomic forces usually decrease rapidly with r.

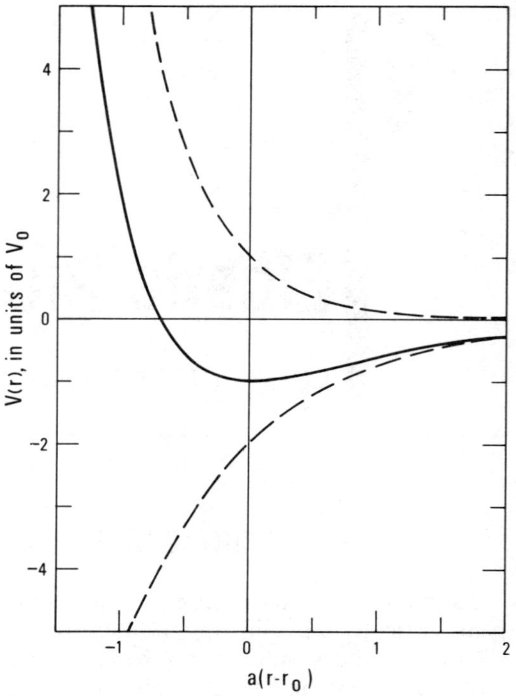

Figure 1.1 Typical example of an interatomic potential used to describe various properties of solids. Proposed originally by Morse (1929) for molecules.

A theory that predicts V_2 usually also predicts a solid's physical properties, and solids with central interatomic forces offer the simplest such opportunity. In non-central-force cases, physical-property data are used to estimate $V(r)$, which leads, in turn, to predicting and correlating other properties. Elastic constants frequently enter into $V(r)$ determinations; other properties used frequently for this purpose include phonon spectra and lattice-defect energies.

In many cases, a solid can be visualized as a set of spheres periodically placed and connected by springs that have zero tension at equilibrium. This model is characterized by a single constant, the interatomic force constant, f, often called the "spring constant":

$$V_2 = (1/2)fu^2 \tag{1.6}$$

where f relates to the second-order elastic constant according to

$$aC_2 = f \tag{1.7}$$

where a denotes interatomic spacing. The spring-constant concept applies not only to elastic constants, but to various solid-state phenomena related to lattice vibrations.

So, in this approach the quantum-mechanical task is to estimate energies and force constants, while the lattice-theory task is to compute lattice vibration frequencies and relate these to thermal-mechanical properties. As described below (§ 1.7.3, for example), this approach succeeds reasonably well; even a crude sphere-and-spring linear-chain model yields useful predictions.

Since pseudopotential theory is neither discussed nor alluded to in other parts of this review, some general remarks on this important concept are made here. A pseudopotential is a simple way of representing the interaction between a free electron and an ion consisting of a nucleus and a surrounding cloud of bound electrons. Hafner (1973) pointed out that "pseudopotential theory has emerged as one of the most simple and most powerful methods for the theoretical investigation of the properties of the solid state." Similarly, the preface to Heine's (1970) review of the pseudopotential concept describes "the ability of the method to account for quite intricate but very important physical properties with a minimum of numerical work." Cohen and Heine (1973) indicate the wide applicability of pseudopotential theory in listing properties already calculated for aluminum: "Fermi surface, pressure dependence of the de Haas–van Alphen effect, resistivity of the liquid and solid, the electron–phonon enhancement of the effective mass, phonon spectrum, compressibility, stacking fault energy, crystal structure, optical and soft x-ray spectra, superconducting transition temperature, and resistivity of alloys." Single-crystal elastic constants of aluminum have been calculated pseudopotentially by Benckert (1971) and by Suzuki (1971). Similarly, pseudopotential calculations exist for thermal expansivity (Wallace, 1968) and specific heat (Bertoni, Bortolani, and Ciunchi, 1972). So, pseudopotential theory represents a truly revolutionary concept for solid-state theorists against which observations can be compared and unified. This theory is outlined by Harrison (1966), Wallace (1972, chapter 6), and many modern solid-state-physics textbooks.

1.2 DEFINITIONS AND DESCRIPTIONS OF ELASTIC CONSTANTS

Elastic constants, which are physical properties, relate stress to strain, or force per unit area to relative length change. Either stress or strain can be induced by any elastically coupled force — mechanical, thermal, magnetic, or electrical. Since solids resist both volume change and shape change, they have at least two independent elastic constants. A low-symmetry crystal may have as many as 21 independent elastic constants. Most materials considered in this chapter are quasi-isotropic; their macroscopic elastic constants do not depend on direction. Exhibiting maximum symmetry, they are characterized elastically by two independent elastic constants. Composites and textured aggregates are anisotropic rather than quasi-isotropic. As discussed below (§ 1.9.4 and 1.9.5), they are characterized by five independent elastic constants for the transverse-isotropic case and nine constants for the orthotropic case. This section defines and describes the elastic constants used most often to characterize polycrystals.

1.2.1 Compressibility

All matter — gas, liquid, or solid — responds to pressure change and exhibits at least one elastic constant, the compressibility. It is given by

$$K_T = -(1/V)\,(\partial V/\partial P)_T \tag{1.8}$$

where P, T, and V denote pressure, temperature, and volume. Figure 1.2 illustrates the mechanical deformations with which compressibility is associated. Always positive, K has units of reciprocal pressure. Gschneidner (1964) suggested that compressibility "has been measured more than any other elastic property." Bridgman's (1949) pioneering experiments resulted in compressibility values for many solids. Some of Bridgman's experiments determined linear compressibility,

$$k_T = (A/\ell)(d\ell/dL)_T \tag{1.9}$$

where A denotes the cross-sectional area of a

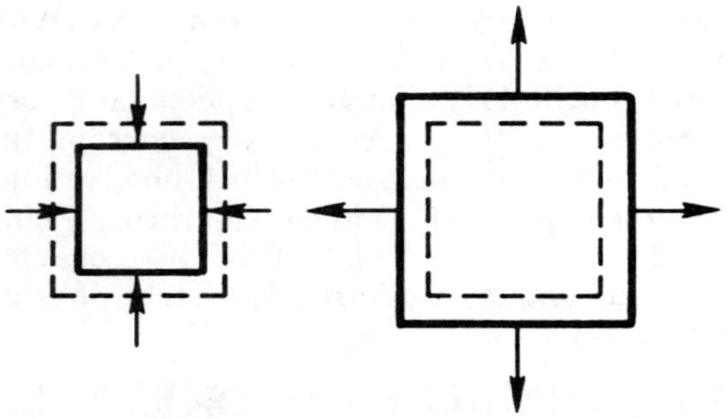

Figure 1.2 Mechanical deformations with which compressibility or bulk modulus is associated.

rod-shaped, hydrostatically loaded specimen, ℓ denotes length, and L denotes load. For isotropic solids, linear and bulk compressibilities relate according to

$$3k = K \quad (1.10)$$

The constant-entropy compressibility

$$K_S = -(1/V)(\partial V/\partial P)_S \quad (1.11)$$

where S denotes entropy, is obtained under zero-heat-flow conditions, at high frequencies, for example. Paragraph 1.5.4 describes differences between constant-temperature (isothermal) and constant-entropy (adiabatic) elastic constants.

1.2.2 Bulk modulus

Rather than compressibility, scientists often use the bulk modulus, B, to describe a solid's resistance to volume change. For isotropic materials, B and K relate reciprocally:

$$B_T = 1/K_T = -V(\partial P/\partial V)_T \quad (1.12)$$

B has pressure units, and, herein, elastic constants with pressure units are called elastic moduli. For describing solids, hydrostatic stress, σ, is often preferred over pressure, which is a negative stress. Thus,

$$B_T = V(\partial \sigma/\partial V)_T \quad (1.13)$$

Some useful relationships result from the differential thermodynamic relationships for the internal energy, U, and the Helmholtz free energy, F:

$$dU = TdS - PdV = TdS + \sigma dV \quad (1.14)$$

and

$$dF = -PdV - SdT = \sigma dV - SdT \quad (1.15)$$

Since

$$\sigma = (\partial F/\partial V)_T = (\partial U/\partial V)_{T=0} \quad (1.16)$$

then

$$B_{T=0} = V_0 (\partial^2 U/\partial V^2)_{T=0} \quad (1.17)$$

Thus, B is the curvature of the internal-energy-vs.-volume curve. All volume-dependent energy terms contribute to the bulk modulus. In terms of interatomic spacing, r, where $V_0 = (4/3)\pi r_0^3$,

$$B_{T=0} = (1/12\pi r_0)(\partial^2 U/\partial r^2)_{r=r_0} \quad (1.18)$$

Internal energy at zero temperature is often represented by a central-force pair potential energy, $\phi(r_{ij})$, such that

$$U(V_0) = V(r) \sum_{i \neq j} \phi(r_{ij})/2V_0 \quad (1.19)$$

where i and j denote specific atoms in the lattice and V_0 denotes atomic volume. Considering a face-centered-cubic (f.c.c.) lattice and summing over the 12 nearest-neighbor atoms gives, at $r = r_0$, where the derivative is evaluated at the nearest-neighbor distance,

$$B = 2(\partial^2 \phi(r)/\partial r^2)/3r_0 \quad (1.20)$$

Thus, a simple relationship exists between the bulk modulus and the interatomic pair potential. If $\phi(r)$ is known, B can be calculated simply. Similarly, any proposed $\phi(r)$ must agree with the observed B. The general relationship between elastic constants expressed as fourth-rank tensors, C_{ijkl}, and internal energy is (Wallace, 1972)

$$C_{ijkl} = (\partial^2 U/\partial \eta_{ij} \partial \eta_{kl})/V_0 \quad (1.21)$$

where η_{ij} denote components of the Lagrangean strain matrix. Thus, by using appropriate strains, all the various C_{ijkl} can be computed from $\phi(r)$. Calculating elastic constants from first principles usually involves expressing the internal energy as a function of volume, lattice parameter, or interatomic spacing. Subsequent discussion in this review frequently invokes this fundamental elastic-constant-interatomic-potential relationship. A high bulk modulus usually implies strong interatomic forces, high cohesive energy, high melting point, and high elastic-stiffness moduli.

1.2.3 Young's modulus

Engineers find Young's modulus the most familiar elastic constant, as evidenced by its many pseudonyms: extension modulus, tensile modulus, tension modulus, elastic modulus, and modulus. Despite reports to the contrary, the compressive modulus is a misnomer. Young's moduli in tension or compression are identical, theoretically and experimentally. Some difficulty on this concept arises from engineering Young's-modulus measurements using high stress. Since stress affects elastic constants and the stress–strain-curve shape, the true Young's modulus manifests itself only at very low stress.

Usually defined in terms of a uniaxially stressed rod where both stress, σ, and strain, ϵ, are measured along the rod axis, z, Young's modulus, E, becomes

$$E = \sigma_z / \epsilon_z \tag{1.22}$$

Figure 1.3 illustrates the mechanical deformations with which Young's modulus is associated.

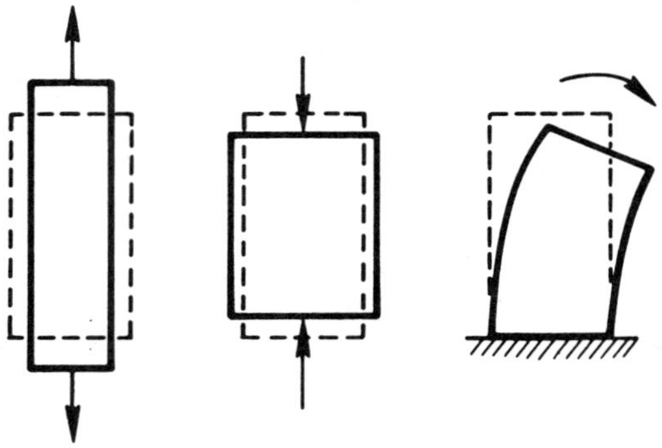

Figure 1.3 Mechanical deformations with which Young's modulus is associated.

As shown in Fig. 1.3, rod bending also involves Young's modulus. The stress along a bent rod is

$$\sigma_z = E\,\epsilon_z = E\,x/r \tag{1.23}$$

where x is a coordinate perpendicular to z, and r is the curvature radius of the neutral surface near the origin.

Young's modulus relates to the extensional sound-wave velocity in a rod by

$$v_e = (E/\rho)^{1/2} \tag{1.24}$$

where ρ denotes mass density. As described below, this relationship provides a simple and accurate method for determining E experimentally at very low stress.

1.2.4 Shear modulus

Sometimes called the torsional modulus, rigidity modulus, or transverse modulus, the shear modulus, G, relates the shear stress, τ, in simple shear to the shear strain, γ. Thus,

$$\tau = G\gamma \tag{1.25}$$

Figure 1.4 illustrates the mechanical deformations with which the shear modulus is associated. A cube sheared on one face has a shear strain related to the length change of the face diagonal by

$$\gamma = 2\,\Delta\ell/\ell \tag{1.26}$$

In torsion (see Fig. 1.4), τ is the torsional stress. For isotropic materials, simple-shear and torsional G's are identical. Both shear and torsion conserve volume but change shape. The bulk modulus, as described above, describes volume change without shape change. Stokes (1845) apparently first realized that B and G form a fundamental elastic-constant pair. Any mechanical deformation factors into a pure shear and a pure dilatation. This factorization facilitates interpretation of various elastic phenomena. For example, many-body electron–electron interactions in metals affect B but not G. Besides other examples cited below, this factorization is discussed further in § 1.5.

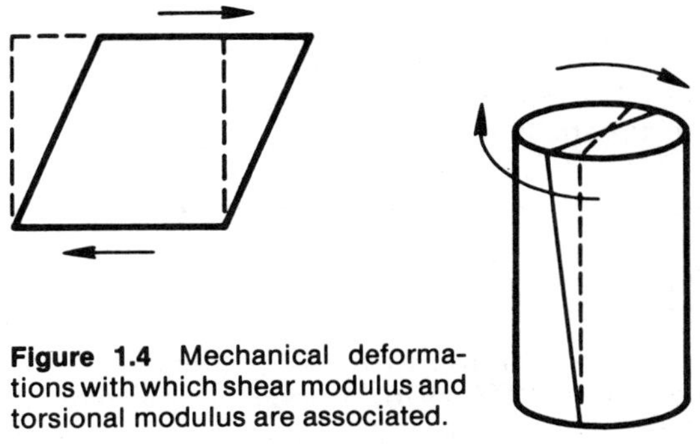

Figure 1.4 Mechanical deformations with which shear modulus and torsional modulus are associated.

The shear modulus determines the velocity of a shear-polarized (transverse) sound wave in a plate or of a torsional wave in a rod according to

$$v_t = (G/\rho)^{1/2} \tag{1.27}$$

where ρ denotes mass density. Like Young's modulus, this relationship provides a simple and accurate method for experimentally determining G at low stress.

1.2.5 Poisson's ratio

Poisson's ratio, ν, is not an elastic modulus (§ 1.2.2), but a dimensionless ratio of two elastic compliances (§ 1.5.1). Usually defined by a uniaxially stressed rod (Young's-modulus) experiment, Poisson's ratio is the negative ratio of transverse (x) and longitudinal (z) strains:

$$\nu = -\epsilon_x / \epsilon_z \tag{1.28}$$

Figure 1.5 illustrates the deformation and strains involved. Theoretically, ν is bounded by -1 and $1/2$, but negative ν values are never observed in quasi-isotropic materials. Thus, in polycrystals ν is bounded practically by 0 and $1/2$. A typical value for metals is $1/3$, 0.28 to 0.42 being the observed range for most materials. Such ν values mean that a material *tends* to maintain constant volume during uniaxial deformation. If ν is $1/2$, no volume change occurs during deformation. Liquids exhibit constant volume during deformation, and some solids (such as rubber) tend to. Köster and Franz (1961) extensively discussed ν for metals and alloys with the view that "Poisson's ratio provides more information about the character of the bonding forces than any of the other elastic coefficients."

Another useful interpretation of ν views it as a relative measure of elastic resistance to dilatation and shear since

$$\nu = (3x - 2)/2(3x + 1) \tag{1.29}$$

where x is B/G. Maximum ν occurs when x is infinite (liquids) and minimum ν when x is zero (incompressible spheres).

Many authors believe erroneously that $\nu = 0.25$ when central forces exist between atoms and that "the departure of Poisson's ratio from the value 0.25, based on the assumption that the Cauchy relations are valid, shows the extent to which central forces are involved" (Ledbetter, 1973). Ledbetter (1973, 1976) corrected this viewpoint, showing that $\nu = 0.25$ is a lower limit for central-force polycrystals and that $\nu = 0.5$ is the upper limit, which corresponds to infinite elastic anisotropy.

Various authors refer to the four elastic constants just described — B, E, G, and ν — as the bulk, engineering, macroscopic, polycrystalline, practical, or technological elastic constants. These four constants are those used most frequently to describe materials such as polycrystalline aggregates. Elastic constants in crystals have anisotropic, or tensorial, character. But in stochastically random aggregates these constants are macroscopically isotropic. Such direction-independent elastic constants are scalars; therefore, all elastic-constant relationships become much simpler.

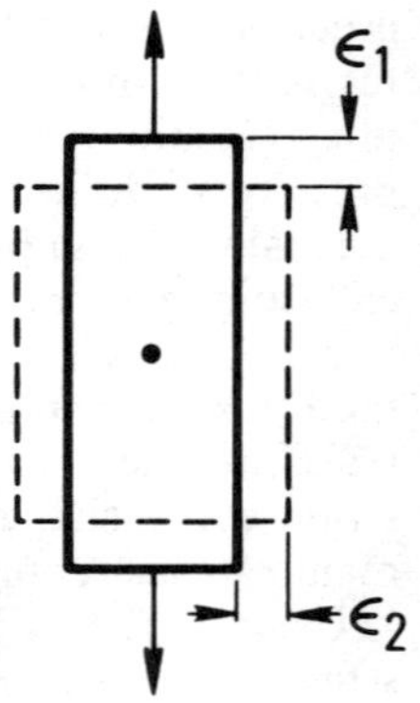

Figure 1.5 Mechanical deformation and strains with which Poisson's ratio is associated.

1.2.6 Hooke's law

Hooke's law states a linear proportionality between strain-response intensity, ϵ, and imposed stress, σ:

$$\sigma = C\epsilon \tag{1.30}$$

where C denotes an elastic-stiffness modulus. Equations 1.12 (for small strains), 1.22, and 1.25 provide examples of Hooke's law written for particular deformations: dilatation, extension, and shear. For small strains, Eq. 1.8 is an example of Hooke's law written inversely:

$$\epsilon = S\sigma \tag{1.31}$$

where the S's are elastic-compliance moduli, and for isotropic media

$$C = 1/S \tag{1.32}$$

Anisotropic media, such as single crystals, textured aggregates, or composites, require a more general form of Hooke's law:

$$\sigma_{ij} = C_{ijkl}\,\epsilon_{kl} \tag{1.33}$$

where both stress and strain are symmetrical second-rank tensors and C_{ijkl} is a symmetrical fourth-rank tensor, which has between 2 and 21 independent components depending on the material's elastic symmetry.

This chapter considers mainly Hooke's-law, or linear-elastic, solids. Nonlinear cases involve elastic constants higher than second order. The familiar Hooke's law elastic constants are the second-order constants. As Eqs. 1.2 and 1.17

show, they are second derivatives of the free energy. First-order elastic constants are mainly of theoretical interest and they vanish for a material in equilibrium.

Many elementary discussions of solid-state elastic behavior are available, for example: Feynman, Leighton, and Sands (1963, p. 38-1); Sines (1969); and Ralls, Courtney, and Wulff (1976).

1.3 ELASTIC-CONSTANT MEASUREMENT METHODS

Experimental physical-property determinations contain two elements — applied basic theory and laboratory ingenuity — and are best discussed against the theoretical background that gives them existence and validity. For brevity, only a few guiding remarks are given here. Many excellent reviews of this subject exist already: Hearmon (1946); Leibfried (1955, section 55); Hearmon (1956); Read, Wert, and Metzger (1959); Hearmon (1961); Hillier (1961); McSkimin (1961); Bradfield (1964); Alers (1966); Beyer and Letcher (1969); Krautkrämer and Krautkrämer (1969); Truell, Elbaum, and Chick (1969); Armstrong (1971); Papadakis (1973); Schreiber, Anderson, and Soga (1973); Fuller, Granato, Holder, and Naimon (1974); and Pollard (1977).

Although elastic constants are measured by a rich variety of methods, there are only four principal approaches: static stress–strain, subresonant forced vibration, resonance, and pulse. Static methods impose a stress and measure a strain response; the elastic constant is the stress–strain ratio; stress is typically 0.1 GPa. Subresonance, forced-vibration methods are used at very low frequencies (determined by the experimentalist's patience) up to typically 100 Hz. At very low stress levels (about 0.1 MPa), the stress–strain amplitude ratio is determined. Damping can be measured simultaneously as the phase lag of the strain wave behind the stress wave. Damping can also be determined by examining the logarithmic decay of the stress wave after it is switched off. Resonance methods establish standing waves by varying the stress frequency and using well-known classical vibration theory to derive elastic constants from mass density, specimen geometry, and resonance frequency; typical specimens measure a few centimeters in their largest dimension (usually one-half wavelength); resonance frequencies are typically 10 to 100 kHz, and peak stress is typically 0.1 MPa or lower. In this experiment, damping is determined from the halfwidth of the amplitude-vs.-frequency curve. Pulse methods use traveling acoustic waves with wavelengths short compared with specimen size, typically about one centimeter. Elastic constants are obtained from sound-wave velocities derived from pulse transit times in known-length specimens. Frequencies are usually near 10 MHz and peak stress is typically 10 kPa or lower. Damping is determined in a pulsed-wave propagation experiment from the pulse's attenuation (per unit length, per wavelength, or per unit time).

Static methods yield isothermal elastic constants while dynamic methods yield adiabatic constants (see § 1.5.4). Aside from this purely thermodynamic distinction, other differences arise because static methods use stresses higher by many orders of magnitude. Higher stresses mean that higher-order elastic constants and higher-order strains cannot be neglected in the stress–strain response. Although such measurements may be important for engineering applications, they do not lead to true elastic-constant measurements, which must be made at very low stress to avoid nonlinear effects.

1.4 TYPICAL ELASTIC-CONSTANT VALUES

Tables 1.1 through 1.4 give elastic constants at various temperatures for six metals — nickel, iron, copper, titanium, aluminum, and magnesium — that form the bases for most technological alloys. These data were abstracted from a compilation by Simmons and Wang (1971). Table 1.5 (Ledbetter, 1978) gives ambient-temperature elastic constants for 24 cubic elements for which reliable data exist. This table also includes mass density, ρ, the shear-wave velocity

$$v_t = (G/\rho)^{1/2} \tag{1.34}$$

the longitudinal wave velocity

$$v_\ell = \{[B + (4/3)\ G]/\rho\}^{1/2} \tag{1.35}$$

the mean wave velocity

$$v_m^{-3} = (1/3)(v_\ell^{-3} + 2v_t^{-3}) \tag{1.36}$$

and the Zener elastic anisotropy

$$A = 2C_{44}/(C_{11} - C_{12}) \tag{1.37}$$

where C_{11}, C_{12}, and C_{44} are the single-crystal Voigt elastic constants. Zener's A is useful for interpreting elastic behavior of textured aggre-

Table 1.1 Bulk modulus (reciprocal compressibility) of six metals at selected temperatures in units of GPa.

T(K)	Ni	Fe	Cu	Ti	Al	Mg
0	187.6	187.6	142.0	110.0	79.4	36.9
20	187.6	173.1	142.0	109.8	79.4	36.9
40	187.5	172.9	141.9		79.4	36.8
60	187.3	172.7	141.7		79.3	36.8
80	187.2	172.1	141.4	109.5	79.2	36.7
100	187.0	171.6	141.1		79.0	36.6
120	186.6	171.1	140.8	109.0	78.8	36.5
140	186.2	170.7	140.5		78.6	36.4
160	185.8	170.3	140.1		78.3	36.3
180	185.5	169.9	139.7	108.5	78.0	36.1
200	185.2	169.5	139.3		77.7	36.0
220	184.9	169.2	138.8	108.0	77.4	35.9
240	184.6	168.9	138.4		77.1	35.7
260	184.2	168.6	138.0		76.7	35.6
280	183.9	168.2	137.5	107.5	76.4	35.4
300	183.6	168.0	137.1	107.3	76.1	35.3

Table 1.2 Young's modulus of six metals at selected temperatures in units of GPa.

T(K)	Ni	Fe	Cu	Ti	Al	Mg
0	240.1	224.1	138.6	130.6	78.4	49.4
20	240.1	224.1	138.6	130.6	78.3	49.3
40	239.8	223.9	138.4		78.2	49.2
60	239.2	223.5	138.0		77.9	49.1
80	238.5	222.9	137.5	129.4	77.6	48.9
100	237.5	222.3	136.9		77.1	48.6
120	236.4	221.7	136.2	127.0	76.5	48.3
140	235.1	220.9	135.4		75.8	48.0
160	233.8	220.1	134.6		75.2	47.6
180	232.7	219.3	133.7	123.7	74.5	47.2
200	231.3	218.4	132.9		73.8	46.8
220	230.0	217.7	132.0	120.0	73.1	46.4
240	228.6	216.8	131.1		72.3	46.0
260	227.2	215.8	130.1		71.6	45.6
280	225.8	214.9	129.2	116.6	70.8	45.1
300	224.5	214.0	128.2	114.6	70.1	44.7

Table 1.3 Shear modulus of six metals at selected temperatures in units of GPa.

T(K)	Ni	Fe	Cu	Ti	Al	Mg
0	93.3	87.3	51.9	50.2	29.3	19.4
20	93.3	87.2	51.8	50.2	29.3	19.3
40	93.2	87.2	51.8		29.3	19.3
60	93.0	87.0	51.6		29.2	19.1
80	92.6	86.8	51.4	49.7	29.0	19.1
100	92.2	86.6	51.2		28.8	19.0
120	91.8	86.4	50.9	48.8	28.6	18.9
140	91.2	86.1	50.5		28.3	18.8
160	90.6	85.7	50.2		28.1	18.6
180	90.1	85.4	49.9	47.2	27.8	18.4
200	89.6	85.0	49.6		27.5	18.3
220	89.0	84.7	49.2	45.7	27.2	18.1
240	88.4	84.3	48.8		26.9	17.9
260	87.8	83.9	48.5		26.6	17.7
280	87.2	83.5	48.1	44.2	26.3	17.6
300	86.6	83.1	47.7	43.4	26.0	17.4

Table 1.4 Poisson's ratio of six metals at selected temperatures.

T(K)	Ni	Fe	Cu	Ti	Al	Mg
0	0.287	0.285	0.338	0.302	0.336	0.277
20	0.287	0.285	0.338	0.302	0.336	0.278
40	0.287	0.285	0.338		0.336	0.278
60	0.287	0.285	0.338		0.336	0.278
80	0.288	0.285	0.338	0.303	0.337	0.279
100	0.288	0.284	0.338		0.337	0.279
120	0.289	0.284	0.339	0.306	0.338	0.280
140	0.290	0.285	0.339		0.339	0.281
160	0.290	0.285	0.340		0.340	0.282
180	0.291	0.285	0.341	0.311	0.341	0.283
200	0.292	0.285	0.341		0.342	0.284
220	0.293	0.286	0.342	0.315	0.343	0.285
240	0.294	0.286	0.342		0.344	0.286
260	0.294	0.287	0.343		0.345	0.287
280	0.295	0.287	0.343	0.321	0.346	0.288
300	0.296	0.288	0.344	0.322	0.347	0.289

gates. Materials like aluminum, with A values near unity, exhibit nearly direction-independent elastic constants, even in highly textured aggregates. Materials like copper ($A = 3.2$) become strongly elastically anisotropic owing to slight textures. Warlimont and Hausch (1974) discuss the elastic-constant/texture relationship, which is discussed briefly in § 1.9.5. For the zero-temperature case, Table 1.6 (Ledbetter, 1978) gives data corresponding to those in Table 1.5. Tables 1.1 through 1.4 and Tables 1.5 and 1.6 represent different sources of elastic data; naturally, small differences occur between the sets of tables. In these tables, zero-temperature values represent extrapolations from 4 K; as discussed below, this introduces negligible error unless a phase transition intervenes.

Elastic moduli in Tables 1.1 through 1.6 range typically between 20 and 200 GPa. Zwikker (1954) calculated the bulk-modulus magnitude from atomic and crystal properties — electron charge, valence, dielectric constant, lattice parameter — and predicted 100 GPa, in excellent agreement with observed values shown in the tables. Zwicky (1934) also gave a simple, basic estimate of the C_{ij} elastic constants showing they should be of the order of 100 GPa.

Elastic constants of technological alloys usually do not differ significantly from those of the base metals. This is shown in a compilation of

Table 1.5 Room-temperature elastic properties of 24 cubic elements.

	ρ (g·cm^{-3})	A	B (GPa)	E (GPa)	G (GPa)	ν	v_m (cm·μs^{-1})	v_ℓ (cm·μs^{-1})	v_t (cm·μs^{-1})
Ag	10.500	2.8825	101.20	78.89	28.79	0.3701	0.1867	0.3646	0.1656
Al	2.697	1.2231	75.86	70.27	26.11	0.3456	0.3496	0.6406	0.3111
Au	19.300	2.8522	173.50	76.12	26.68	0.4269	0.1336	0.3291	0.1176
C	3.512	1.2109	442.00	1139.37	532.23	0.0704	1.3423	1.8108	1.2310
Cr	7.200	0.7144	161.87	278.38	114.72	0.2134	0.4413	0.6612	0.3992
Cu	8.930	3.2727	135.30	121.69	45.07	0.3501	0.2526	0.4678	0.2246
Fe	7.872	2.4074	166.93	205.87	79.52	0.2945	0.3548	0.5889	0.3178
Ge	5.323	1.6644	75.02	130.30	53.82	0.2105	0.3514	0.5241	0.3180
Ir	22.520	1.5645	370.37	556.74	222.79	0.2495	0.3492	0.5444	0.3145
K	0.851	6.7143	3.33	2.25	0.81	0.3874	0.1102	0.2275	0.0975
Li	0.532	8.5243	12.13	9.21	3.35	0.3734	0.2830	0.5582	0.2509
Mo	10.228	0.7140	259.77	321.83	124.40	0.2935	0.3892	0.6451	0.3487
Na	0.971	7.1624	6.51	4.81	1.74	0.3788	0.1512	0.3033	0.1340
Nb	8.570	0.4925	163.78	103.07	36.94	0.3951	0.2349	0.4986	0.2076
Ni	8.910	2.6652	185.97	212.27	81.03	0.3098	0.3373	0.5744	0.3016
Pb	11.344	4.0743	44.76	23.17	8.19	0.4136	0.0964	0.2216	0.0850
Pd	12.038	2.8096	193.06	128.23	46.15	0.3893	0.2213	0.4599	0.1958
Pt	21.500	1.5938	282.70	176.23	63.12	0.3961	0.1939	0.4131	0.1713
Rb	1.560	7.4545	2.60	1.85	0.67	0.3811	0.0740	0.1496	0.0656
Si	2.331	1.5636	97.89	161.61	65.97	0.2248	0.5889	0.8929	0.5320
Ta	16.626	1.5609	189.72	183.69	68.61	0.3386	0.2281	0.4113	0.2031
Th	11.694	3.6212	57.70	69.83	26.89	0.2983	0.1693	0.2828	0.1516
V	6.022	0.7867	155.57	129.11	47.41	0.3617	0.3160	0.6027	0.2806
W	19.257	1.0114	310.38	409.83	160.10	0.2799	0.3213	0.5216	0.2883

Table 1.6 Zero-temperature elastic properties of 24 cubic elements.

	ρ (g·cm^{-3})	A	B (GPa)	E (GPa)	G (GPa)	ν	v_m (cm·μs^{-1})	v_ℓ (cm·μs^{-1})	v_t (cm·μs^{-1})	θ (K)
Ag	10.634	2.9912	108.72	87.02	31.84	0.3666	0.1950	0.3770	0.1730	226.46
Al	2.732	1.2073	79.38	78.24	29.29	0.3357	0.3674	0.6583	0.3274	430.56
Au	19.478	2.8436	180.32	83.19	29.23	0.4231	0.1391	0.3355	0.1225	161.74
C	3.516	1.2109	442.00	1139.37	532.23	0.0704	1.3416	1.8099	1.2304	2239.62
Cr	7.231	0.6865	201.00	300.58	120.16	0.2508	0.4526	0.7068	0.4076	589.57
Cu	9.024	3.1904	142.03	132.51	49.28	0.3445	0.2625	0.4798	0.2337	344.40
Fe	7.922	2.3431	169.00	213.17	82.64	0.2898	0.3603	0.5936	0.3230	472.44
Ge	5.340	1.6651	76.52	132.92	54.90	0.2105	0.3544	0.5295	0.3207	373.36
Ir	22.659	1.5698	366.67	557.12	223.43	0.2468	0.3485	0.5416	0.3140	429.62
K	0.910	7.6267	3.66	3.10	1.14	0.3587	0.1261	0.2386	0.1120	90.52
Li	0.5375	9.2400	13.30	10.70	3.90	0.3650	0.3049	0.5868	0.2706	326.74
Mo	10.223	0.9024	265.29	335.72	130.22	0.2891	0.3481	0.6552	0.3569	474.53
Na	1.005	8.5000	7.24	5.98	2.20	0.3623	0.1664	0.3181	0.1478	147.43
Nb	8.616	0.5183	173.03	110.87	39.79	0.3932	0.2430	0.5122	0.2149	276.60
Ni	8.972	2.3859	187.60	233.52	90.33	0.2925	0.3541	0.5860	0.3173	475.98
Pb	11.593	3.8379	48.79	30.34	10.86	0.3964	0.1095	0.2336	0.0968	105.33
Pd	12.092	2.4552	195.43	134.72	48.63	0.3851	0.2266	0.4640	0.2005	275.92
Pt	21.578	1.4828	288.40	183.54	65.83	0.3939	0.1976	0.4175	0.1747	238.44
Rb	1.629	8.0657	3.21	2.39	0.87	0.3761	0.0824	0.1638	0.0730	55.31
Si	2.331	1.5640	99.22	163.21	66.57	0.2258	0.5916	0.8980	0.5344	648.87
Ta	16.754	1.6154	194.21	191.15	71.54	0.3360	0.2319	0.4158	0.2066	263.77
Th	11.888	3.4545	58.10	75.58	29.45	0.2832	0.1754	0.2862	0.1574	163.83
V	6.050	0.8130	157.04	135.23	49.85	0.3565	0.3230	0.6078	0.2870	399.19
W	19.313	0.9959	314.15	417.76	163.40	0.2784	0.3240	0.5248	0.2909	384.39

Young's moduli for forty engineering alloys, which include nickel, nickel alloys, inconels, nickel steels, stainless steels, monel, copper, copper–nickel alloys, invar, copper alloys, and aluminum alloys (Ledbetter, 1982). Except for a few cases, notably copper–nickel alloys, alloying lowers the elastic stiffness of the base metal.

1.5 ELASTIC-CONSTANT INTERRELATIONSHIPS

1.5.1 Connecting identities

Although for complete characterization isotropic solids require only two independent elastic constants, practice requires many more. This multiplicity arises because particular elastic constants best describe particular mechanical deformations. As already discussed, B, E, and G best describe dilatation, extension, and shear, respectively. Table 1.7 summarizes interrelationships among various elastic constants. These relationships can be derived simply from elastic-constant definitions and a little physics (Feynman et al., 1963, p. 38-1). Besides elastic constants B, E, G, and ν, defined above, this table contains several additional elastic constants: C_ℓ, λ, C^0_{11}, C^0_{12}, S^0_{11}, and S^0_{12}, where superscript zero denotes quasi-isotropic. Longitudinal and transverse sound velocities, v_ℓ *and* v_t, also occur. The meaning and significance of these additional constants are, briefly, as follows:

The longitudinal modulus

$$C_\ell = \rho v_\ell^2 \qquad (1.38)$$

corresponds to a longitudinally polarized elastic wave. When Poisson's ratio is zero, C_ℓ equals E. In other words, a Young's-modulus experiment using a plate instead of a rod or using a rod constrained to zero lateral displacement would measure C_ℓ rather than E.

Together with the first Lamé constant, $\mu = G$, the second Lamé constant, λ, appears in the

Table 1.7 Connecting identities among isotropic-solid elastic constants.

	B	E	G	ν	λ	C_ℓ
B, E	—	—	$\frac{3BE}{9B-E}$	$\frac{1}{2}-\frac{E}{6B}$	$3B\frac{3B-E}{9B-E}$	$3B\frac{3B+E}{9B-E}$
B, G	—	$\frac{9BG}{3B+G}$	—	$\frac{1}{2}\frac{3B-2G}{3B+G}$	$B-\frac{2}{3}G$	$B+\frac{4}{3}G$
B, ν	—	$3B(1-2\nu)$	$\frac{3B}{2}\frac{1-2\nu}{1+\nu}$	—	$\frac{3B\nu}{1+\nu}$	$3B\frac{1-\nu}{1+\nu}$
E, G	$\frac{GE}{3(3G-E)}$	—	—	$\frac{E}{2G}-1$	$G\frac{E-2G}{3G-E}$	$G\frac{4G-E}{3G-E}$
E, ν	$\frac{E}{3(1-2\nu)}$	—	$\frac{E}{2(1+\nu)}$	—	$\frac{\nu E}{(1+\nu)(1-2\nu)}$	$\frac{E(1-\nu)}{(1+\nu)(1-2\nu)}$
G, ν	$\frac{2G}{3}\frac{1+\nu}{1-2\nu}$	$2G(1+\nu)$	—	—	$\frac{2G\nu}{1-2\nu}$	$2G\frac{1-\nu}{1-2\nu}$
λ, μ	$\lambda+\frac{2}{3}\mu$	$\frac{\mu(3\lambda+2\mu)}{\lambda+\mu}$	μ	$\frac{\lambda}{2(\lambda+\mu)}$	—	$\lambda+2\mu$
C^0_{11}, C^0_{12}	$\frac{1}{3}(C^0_{11}+2C^0_{12})$	$\frac{(C^0_{11}-C^0_{12})(C^0_{11}+2C^0_{12})}{C^0_{11}+C^0_{12}}$	$\frac{1}{2}(C^0_{11}-C^0_{12})$	$\frac{C^0_{12}}{C^0_{11}+C^0_{12}}$	C^0_{12}	C^0_{11}
S^0_{11}, S^0_{12}	$\frac{1}{3}(S^0_{11}+2S^0_{12})^{-1}$	$S^{0\,-1}_{11}$	$\frac{1}{2}(S^0_{11}-S^0_{12})^{-1}$	$-\frac{S^0_{12}}{S^0_{11}}$	$\frac{-S^0_{12}}{(S^0_{11}-S^0_{12})(S^0_{11}+2S^0_{12})}$	$\frac{(S^0_{11}+S^0_{12})}{(S^0_{11}-S^0_{12})(S^0_{11}+2S^0_{12})}$
v_ℓ, v_t	$\rho(v_\ell^2-\frac{4}{3}v_t^2)$	$\frac{3\rho v_t^2(v_\ell^2-\frac{4}{3}v_t^2)}{(v_\ell^2-v_t^2)}$	ρv_t^2	$\frac{v_\ell^2-2v_t^2}{2(v_\ell^2-v_t^2)}$	$\rho(v_\ell^2-2v_t^2)$	ρv_ℓ^2

Helmholtz free-energy expansion in tensor strains:

$$F = F_0 + (1/2)\,\lambda\,\epsilon_{ii}^2 + \mu\,\epsilon_{ij}^2 \qquad (1.39)$$

for a deformed isotropic body (Landau and Lifshitz, 1959a, p. 10).

The C_{11}^0 and C_{12}^0 elastic constants follow from single-crystal notation for the Voigt elastic-stiffness constants, the C_{ij}'s. For example, cubic crystals have three independent C_{ij}'s: C_{11}, C_{12}, and C_{44}. C_{11} corresponds to a compressional–extensional deformation along a [100] direction. C_{44} corresponds to a shear on (100) in any direction contained in the (100) plane. The C_{12}-associated deformation is less physical, and $(1/2)(C_{11}-C_{12})$ is usually taken as the third independent elastic constant. It corresponds to shear on a (110) plane in a $[1\bar{1}0]$ direction. In an isotropic aggregate the macroscopic shear constants must be identical. Thus,

$$(1/2)(C_{11}^0 - C_{12}^0) = C_{44}^0 \qquad (1.40)$$

where the superscript zero indicates isotropy. As for single crystals, C_{11}^0 corresponds to a compressional–extensional deformation and $(1/2)(C_{11}^0 - C_{12}^0)$ corresponds to a shear deformation.

The S_{ij} elastic compliances parallel the C_{ij}, elastic stiffnesses (Eq. 1.32). The S_{ij}'s relate quite simply to the engineering elastic constants. For example, $1/S_{11}^0$ is Young's modulus, $-S_{12}^0/S_{11}^0$ is Poisson's ratio, and $2(S_{11}^0 - S_{12}^0)$ is the reciprocal shear modulus. Expressions for E and ν in terms of the C_{ij}^0's are less simple. Experimentally, a major distinction arises between the C_{ij}^0's and the S_{ij}^0's. Pulse experiments determine the C_{ij}'s directly, while resonance and static-stress–strain experiments determine the S_{ij}'s directly.

Figure 1.6 shows many of the relationships in Table 1.7 plotted vs. Poisson's ratio. Thus, Poisson's ratio determines the ratio of any elastic-constant pair. Lindley (1968) gives further relationships between Poisson's ratio and compressibility and other elastic constants.

1.5.2 Differential interrelationships

It is both instructive and useful to carry the connecting identities (given above) one step further and examine their differential forms. These results are summarized in Table 1.8 in terms of B and G, the fundamental elastic-constant pair; in Table 1.9 in terms of C_ℓ and $C_t = G$, the elastic moduli obtained directly in pulse experiments; and in Table 1.10 in terms of E and G, the elastic moduli obtained directly in both

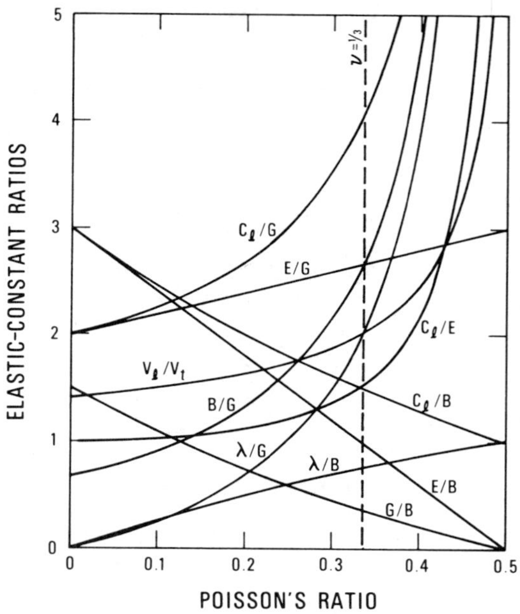

Figure 1.6 Various elastic-constant ratios vs. Poisson's ratio demonstrating relationships in Table 1.7.

resonance and static-stress–strain experiments. In all three cases, exact expressions are given together with approximate expressions that result when Poisson's ratio is assumed to be 1/3.

These tables contain the important result that E depends strongly on G but weakly on B. For most materials E and G respond similarly to changes in metallurgical variables such as composition, temperature, pressure, or magnetic field. Thus, shear moduli can be deduced reliably

Table 1.8 Differential relationships among isotropic elastic constants and the bulk and shear moduli, *B* and *G*, the fundamental elastic-constant pair.

	$\nu \simeq \frac{1}{3}$
$\frac{dC_\ell}{C_\ell} = \frac{1}{(B+\frac{4}{3}G)}\left(B\frac{dB}{B} + \frac{4G}{3}\frac{dG}{G}\right)$	$\frac{dC_\ell}{C_\ell} \simeq \frac{1}{3}\left(2\frac{dB}{B} + \frac{dG}{G}\right)$
$\frac{dE}{E} = \frac{1}{(3B+G)}\left(G\frac{dB}{B} + 3B\frac{dG}{G}\right)$	$\frac{dE}{E} \simeq \frac{1}{9}\left(\frac{dB}{B} + 8\frac{dG}{G}\right)$
$\frac{d\nu}{\nu} = \frac{9BG}{(3B+G)(3B-2G)}\left(\frac{dB}{B} - \frac{dG}{G}\right)$	$\frac{d\nu}{\nu} \simeq \frac{4}{9}\left(\frac{dB}{B} - \frac{dG}{G}\right)$
$\frac{d\lambda}{\lambda} = \frac{1}{(B-\frac{2}{3}G)}\left(B\frac{dB}{B} - \frac{2}{3}\frac{dG}{G}\right)$	$\frac{d\lambda}{\lambda} \simeq \frac{1}{3}\left(4\frac{dB}{B} - \frac{dG}{G}\right)$

Table 1.9 Differential relationships among isotropic elastic constants and the moduli C_ℓ and $C_t = G$, which are obtained directly in pulse experiments.

	$\nu \simeq \frac{1}{3}$
$\frac{dB}{B} = \frac{1}{(C_\ell - \frac{4}{3}G)}\left(C_\ell \frac{dC_\ell}{C_\ell} - 4G\frac{dG}{G}\right)$	$\frac{dB}{B} \simeq \frac{1}{2}\left(3\frac{dC_\ell}{C_\ell} - \frac{dG}{G}\right)$
$\frac{dE}{E} = \frac{1}{3(C_\ell - G)(C_\ell - \frac{4}{3}G)}\left[C_\ell G\frac{dC_\ell}{C_\ell} + (C_\ell - 2G)(3C_\ell - 2G)\frac{dG}{G}\right]$	$\frac{dE}{E} \simeq \frac{1}{6}\left(\frac{dC_\ell}{C_\ell} + 5\frac{dG}{G}\right)$
$\frac{d\nu}{\nu} = \frac{C_\ell G}{(C_\ell - G)(C_\ell - 2G)}\left(\frac{dC_\ell}{C_\ell} - \frac{dG}{G}\right)$	$\frac{d\nu}{\nu} \simeq \frac{2}{3}\left(\frac{dC_\ell}{C_\ell} - \frac{dG}{G}\right)$

Table 1.10 Differential relationships among isotropic elastic constants and the Young's and shear moduli, *E* and *G*, which are obtained directly in static and in resonance experiments.

	$\nu \simeq \frac{1}{3}$
$\frac{dB}{B} = \frac{G + 3B}{G}\frac{dE}{E} - \frac{9B - E}{E}\frac{dG}{G}$	$\frac{dB}{B} \simeq 9\frac{dE}{E} - 8\frac{dG}{G}$
$\frac{d\nu}{\nu} = \frac{E}{E - 2G}\left(\frac{dE}{E} - \frac{dG}{G}\right)$	$\frac{d\nu}{\nu} \simeq 4\left(\frac{dE}{E} - \frac{dG}{G}\right)$
$\frac{dC_\ell}{C_\ell} = \frac{1}{(3G - E)(4G - E)}\left[EG\frac{dE}{E} - (E - 2G)(6G - E)\frac{dG}{G}\right]$	$\frac{dC_\ell}{C_\ell} \simeq 6\frac{dE}{E} - 5\frac{dG}{G}$

from Young's moduli and vice versa. Any nonparallel behavior of E and G is anomalous and suggests a fundamental change in the material, for example different interatomic bonding.

It is perhaps surprising that changes in B and G affect Poisson's ratio equally but oppositely. Since B and G tend to respond in parallel to metallurgical-variable changes, ν is clearly the elastic constant least affected by changes in such variables. Conversely, small changes in ν may indicate large changes in a material's properties.

These differential equations also provide useful relationships for computing propagations of known errors. For example, B and ν determined by either resonance or static stress–strain are highly inaccurate compared with pulse-method results. In the pulse case, 1% measurement errors lead to 2% and 4/3% errors in B and ν. Similar errors in the resonance case lead to 17% and 11% errors. For determining either B or ν, the pulse method emerges clearly superior.

1.5.3 Empirical interrelationships

The previous two sections describe the functional and differential relationships between elastic constants and other elastic-constant pairs. These mathematically exact relationships emerge from elasticity theory and from the physical-geometrical meanings of the elastic constants. However useful, these relationships represent nothing more than simple, necessary functions relating a dependent variable to two independent variables.

This section considers empirical relationships between elastic constant pairs. Although only approximate, such relationships prove their usefulness in scarce-data situations. Since isotropic materials require only two independent elastic constants for characterization, any new relationship reduces the number of independent elastic constants to one.

Figure 1.7 shows a rough proportionality be-

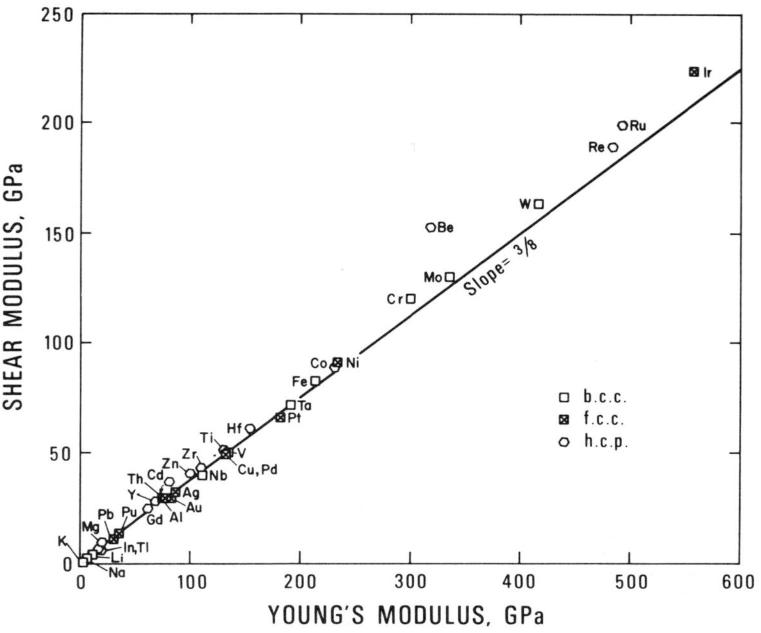

Figure 1.7 Relationship between G and E for various elements.

tween the Young's and shear moduli. From this proportionality, shear moduli, which are measured infrequently, can be estimated from Young's moduli, which are measured much more often. A slope near 0.39 fits the data better than the 3/8 slope (representing $\nu = 1/3$) shown in the figure. This empirical result is surprising because it is hard to see physically why elastic constants describing twisting and stretching relate to one another. Ledbetter (1977a) recently gave a theoretical basis to this proportionality by calculating the G/E ratio for an aggregate of crystals bound by a central-force pair potential. For the face-centered-cubic (f.c.c.) case, G/E is 0.396; and for the body-centered-cubic (b.c.c.) case G/E is 0.367 through second-neighbor interactions.

Bulk and Young's moduli relate similarly but less proportionally, as shown in Fig. 1.8. Although this proportionality is too rough for most purposes, it is useful to know that B and E are approximately proportional and that E exceeds B more often than not, especially for stiffer materials. A B-vs.-G plot shows similar dispersion, and thus a similar lack of utility. Recently, Gorecki (1979) studied extensively the G/E, G/B and E/B relationships for b.c.c., f.c.c., and c.p.h. (close-packed hexagonal) metals.

1.5.4 Adiabatic-isothermal differences

Differences between elastic constants at constant entropy and at constant temperature become important principally because dynamic measurement methods yield adiabatic values and static methods yield isothermal values. These differences have been discussed in detail by several authors, for example, Leibfried (1955, section 73), Landau and Lifshitz (1959a, p. 16), Viswanathan and Rajagopal (1961), Bhatia (1967), and Barron and Munn (1968). The main results are:

$$1/E_S = (1/E_T) - (T\beta^2/9C_P) \tag{1.41}$$

$$1/B_S = (1/B_T) - (T\beta^2/C_P) \tag{1.42}$$

$$\nu_S = (\nu_T + E_T\beta^2/9C_P) \div (1 - E_T T\beta^2/9C_P) \tag{1.43}$$

$$G_S = G_T \tag{1.44}$$

Since $ET\beta^2/C_P$ is usually small, these relationships become

$$E_S/E_T = 1 + E_S T\beta^2/9C_P \tag{1.45}$$

$$\nu_S/\nu_T = 1 + (1+\nu_T)\, E_T\beta^2/9\nu_T C_P \tag{1.46}$$

where

$$\beta = (\partial V/\partial T)_P/V \tag{1.47}$$

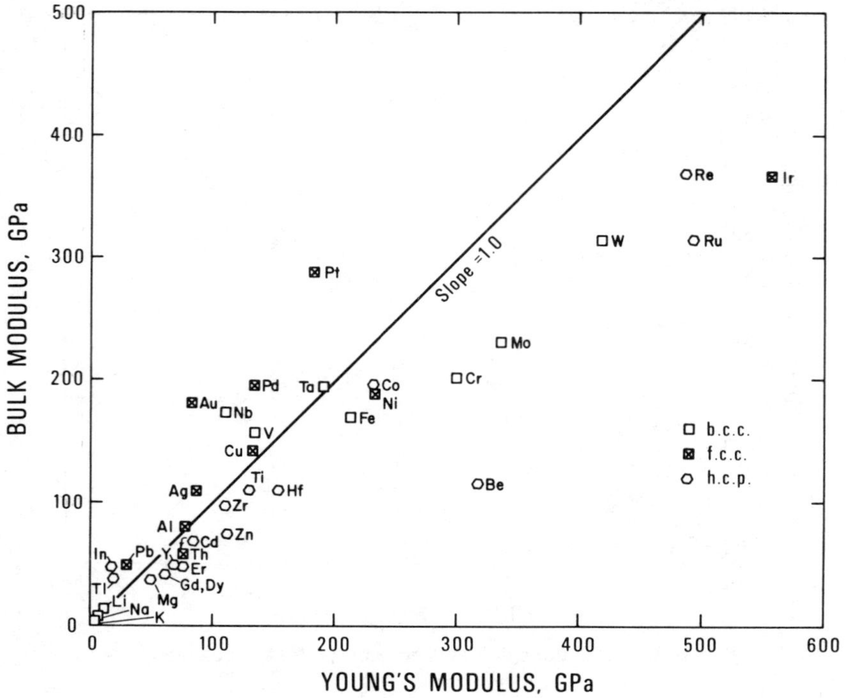

Figure 1.8 Relationship between B and E for various elements.

and C_P is the specific heat

$$C_P = T(\partial S/\partial T)_P \tag{1.48}$$

Some manipulation of these relationships gives

$$B_S/B_T = C_P/C_V \tag{1.49}$$

where

$$C_V = T(\partial S/\partial T)_V \tag{1.50}$$

Thus, in all cases the adiabatic-isothermal elastic-constant differences vanish at zero temperature. Adiabatic and isothermal shear moduli are equal at all temperatures because a shear deformation leaves volume unchanged. From the first law of thermodynamics it follows that constant-volume processes are adiabatic if they are isothermal.

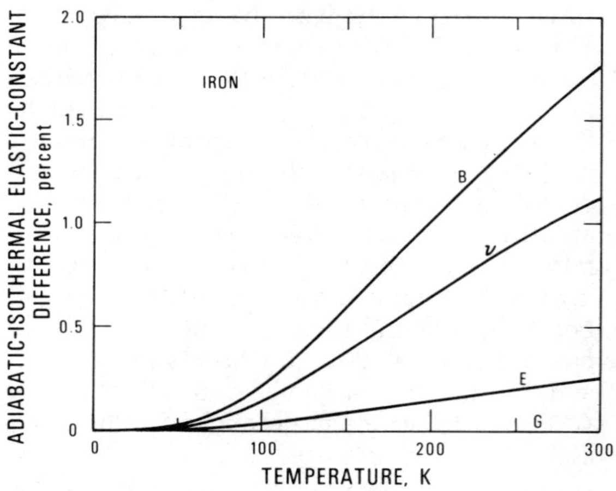

Figure 1.9 Difference between adiabatic and isothermal elastic constants as a function of temperature for the case of iron.

Figure 1.9 shows the computed adiabatic-isothermal elastic-constant difference for iron as a function of temperature between 0 and 300 K (Ledbetter, 1978). As predicted by relationships in Table 1.8, E follows G closely and shows a small difference at most temperatures. For B and ν the adiabatic-isothermal differences are negligible for most purposes up to about 300 K. However, at higher temperatures differences in B and ν can become quite large. At temperatures below about 100 K, the adiabatic-isothermal differences approach zero for all the elastic constants. Figure 1.9 is also useful for illustrating the general result that while E behaves almost like G, ν tends to behave intermediately to G and B. For stiffness-type elastic constants, such as E and B, adiabatic values always exceed isothermal values; ν behaves similarly.

Equation 1.49 is useful since C_P/C_V tabular data exist for solids at various temperatures. From such data B_S/B_T follows directly. Using relationships from Table 1.8, $\Delta E/E \simeq (1/9) \cdot (\Delta B/B)$, $\Delta\nu/\nu \simeq (4/9)(\Delta B/B)$, and of course, $\Delta G/G = 0$.

Finally, consider a simple application of Eq. 1.45. It can be shown (Wannier, 1966) that a wire stretched adiabatically undergoes a temperature change per unit strain given by

$$\Delta T/\Delta\epsilon = -(3/\beta)[(E_S/E_T) - 1] \qquad (1.51)$$

Since both β and $(E_S/E_T - 1)$ are positive, a rapidly stretched wire cools slightly during mechanical deformation. Similarly, rapid uniaxial compression causes heating since the strain changes sign. Substituting typical physical-property values into this relationship reveals that significant temperature changes can accompany adiabatic mechanical deformation.

1.6 RELATIONSHIPS BETWEEN ELASTIC CONSTANTS AND OTHER PHYSICAL PROPERTIES

Elastic and thermal properties are inseparable because they both originate in the solid's lattice-vibrational spectrum. These two types of properties differ in that the elastic constants at all temperatures relate mainly to lower-frequency (longer-wavelength) vibrations, while specific heat relates mainly to lower frequencies at lower temperatures and to higher frequencies at higher temperatures. Thus, specific heat has a much stronger temperature dependence than the elastic constants. The correspondence between elastic and thermal properties becomes especially obvious at zero temperature where the Debye characteristic temperatures determined thermally and elastically are equal (see § 1.6.3). From the first law of thermodynamics

$$dU = TdS - PdV = C_V dT - PdV \qquad (1.52)$$

and from Hooke's law for hydrostatic stress

$$P = -B(dV/V) \qquad (1.53)$$

it follows that the internal-energy change is

$$\begin{aligned} dU &= TdS + B(dV)^2/2V \\ &= dU_{th} + dU_{el} \qquad (1.54) \end{aligned}$$

which reveals a great deal concerning deformable solids. Internal energy depends on both entropy and strain. Although there are both thermal and elastic contributions to U, it cannot in turn be separated into thermal and elastic parts; once part of dU they are indistinguishable. If one considers a solid's volume to be given by an equation of state

$$dV/V = \beta dT - dP/B \qquad (1.55)$$

where B depends mainly on T, being nearly pressure independent, the thermal and elastic contributions to volume change are inseparable. As pointed out by Wannier (1966) one must "treat elastic energy as an aspect of U."

1.6.1 Thermodynamic potentials

The purpose of this section is to emphasize the important relationship between elastic constants and thermodynamic potentials: Gibbs free energy, G; Helmholtz free energy, F; enthalpy, H; and internal energy, U. Elastic constants, being thermodynamic properties, relate fundamentally to other thermodynamic properties such as specific heat, thermal expansivity, Debye's temperature, and Grüneisen's constant, γ. Interrelationships among these quantities through thermodynamic-potential derivatives are summarized in Table 1.11. Two thermal variables, S and T, and two mechanical variables, P and V, occur in the table. Strain can be substituted for volume and negative stress for pressure. From the table, C_P is clearly a thermal property, B is a mechanical property, and both β and γ are thermal-mechanical properties. B is not mechanical in the sense of plastic-deformation properties, such as yield stress. Such mechanical properties involve stress–strain variables, but they are not physical properties. Thermodynamically, in the plastic region, stress and strain are not state functions; they depend on path, history, and time. Atomistically, plastic-deformation properties differ from lattice vibrational properties by having a much larger displacement of atoms from equilibrium positions.

1.6.2 Higher-order energy terms

Before considering anharmonic effects such as the elastic-constant temperature dependence, consider the relationship between physical properties and the lattice potential energy, $V(u)$, written as a function of atomic displacement, u, from the equilibrium position:

$$\begin{aligned} V(u) &= V_0 + V_1 + V_2 + V_3 + V_4 \\ &= (1/2) fu^2 + (1/6) gu^3 \\ &\quad + (1/24) hu^4 \qquad (1.56) \end{aligned}$$

Here V_0 is a constant that can be neglected, V_1 vanishes at equilibrium, and the f, g, and h ex-

Table 1.11 Thermodynamic potentials and their derivatives.

Potential: Differential Form:	G $dG = VdP - SdT$		F $dF = -PdV - SdT$		H $dH = TdS + VdP$		U $dU = TdS - PdV$	
	Varied/Fixed	**Derivative**	**Varied/Fixed**	**Derivative**	**Varied/Fixed**	**Derivative**	**Varied/Fixed**	**Derivative**
	T/P	$-S$	T/V	$-S$	S/P	T	S/V	T
	T^2/P	$-C_P/T$	T^2/V	$-C_V/T$	S^2/P	T/C_P	S^2/V	T/C_V
	P/T	V	V/T	$-P$	P/S	V	V/S	$-P$
	P^2/T	$-V/B_T$	V^2/T	B_T/V	P^2/S	$-V/B_S$	V^2/S	B_S/V
	$P,T/T,P$	$V\beta = \gamma C_P/B_S$	$T,V/V,T$	$-\beta B_T = \gamma C_V/V$	$P,S/S,P$	$T\beta/C_P = \gamma T/B_S$	$S,V/V,S$	$-\beta B_T T/C_V = \gamma T/V$

Examples: $(\partial G/\partial T)_P = -S$, $(\partial^2 G/\partial T^2)_P = -C_P/T$

pansion coefficients are interatomic force constants. $V(u)$ is identical to F, the Helmholtz free energy of a deformed isotropic body, and it is often expanded only through second-order terms. As summarized by Barron (1955):

> The harmonic approximation corresponds to neglecting V_3, V_4, and all higher terms. Such properties therefore as the specific heat, the vibrational spectrum as determined by neutron or electron diffraction, the elastic constants (for infinitesimal strain) and the Raman and infra-red absorption frequencies all depend primarily upon V_2, and so are related to each other; whereas there are other properties which depend essentially upon the anharmonic terms V_3, V_4, etc. Some of the anharmonic properties can, however, be accounted for to a good approximation by the change of V_2 with volume (in a truly harmonic lattice V_2 is invariant under strain); these include the thermal expansion and the change of elastic constants and characteristic frequencies with volume. Others, like the anharmonic contribution to specific heat, require explicit consideration of the anharmonic terms V_3, V_4, etc.

Equation 1.56, the lattice-potential-energy/force-constant/atomic-displacement relationship has been used to describe various physical properties by several authors, including: Leibfried (1955), Peierls (1955), Dobbs and Jones (1957), Eastabrook (1957), Ludwig (1958), Leibfried and Ludwig (1961), Maradudin and Flinn (1961), Maradudin, Flinn, and Coldwell-Horsfall (1961), Foreman (1962), Barron and Klein (1962), Flinn and Maradudin (1963), Kittel (1966), Pastine (1968), and Davies (1973).

Foreman (1962) summarized many of the important previous studies:

> The anharmonic properties of crystals have been discussed in detail by numerous authors, notably by Born and his co-workers. The anharmonic terms in the crystal potential perturb the phonon frequencies when a strain is applied, thus changing the phonon contribution to the free energy. This causes thermal expansion and also a temperature-dependent contribution to the elastic constants, which would otherwise be temperature independent in the harmonic approximation. The order of magnitude of the anharmonic force constants in a solid has been estimated by Klemens (1958) in connection with the theory of phonon–phonon scattering and thermal conductivity, and in rather more detail by Van Vleck (1941) in connection with the theory of paramagnetic relaxation phenomena, but both authors discuss only the cubic terms in the Taylor series expansion of the crystal potential. A number of other authors have discussed anharmonicity, notably Born (1939, 1940, 1943), Bradburn (1943), Born and Bradburn (1943), Fürth (1944), Hooton (1955), Peierls (1956), and Stern (1961). Barron (1955) calculates the temperature dependence of thermal expansion for a face-centred cubic crystal with central forces between nearest-neighbour atoms. Sheard (1958) also calculates the thermal expansion of several metals but uses the elastic continuum model and finds the cubic anharmonicity from the measured pressure dependence of the single crystal elastic constants. . . . Hahn and Leibfried (1958) have attempted to calculate the temperature de-

pendence of the elastic constants of the alkali halides, using Coulomb forces and a short-ranged closed-shell repulsive potential between nearest-neighbour ions (Huntington 1958) and find encouraging agreement with experiment. Ziman (1960) discusses the anharmonic properties of solids on the basis of the elastic continuum model but does not include the quartic terms and does not calculate the temperature dependence of the elastic constants.

The remainder of this section extracts key results from the above studies, following Leibfried (1955), Peierls (1955), and Foreman (1962) most closely. The notation for the expansion coefficients is Leibfried's.

Statistical mechanics permits all thermodynamic properties of a system to be computed from its sum of states (*Zustandsumme*):

$$Z = \sum_i \exp(-\epsilon_i/kT) \tag{1.57}$$

Here ϵ_i is the energy of a particle in the ith state (Moelwyn-Hughes, 1964). For example

$$F = -kT \ln Z \tag{1.58}$$

and

$$U = kT^2 (d \ln Z/dT) \tag{1.59}$$

where Z depends on V and T. From formulas in Table 1.6, all the other thermodynamic properties — S, P, T, C_P, C_V, β, B_S, B_T, and γ — can be computed. In a linear-chain model (§ 1.73) the Grüneisen gamma is

$$\gamma = -ga/6f \tag{1.60}$$

Since γ is known to be always near 2 and nearly temperature independent, it follows that the ratio of the third-order elastic constant, g, to the second-order elastic constant, f/a, where a denotes interatomic spacing, should be approximately −12. This is often observed. Results collected by Fuller et al. (1974) show that C_{111}/C_{11} is -9.9 ± 5 for eight materials, including metals, semiconductors, and ionic crystals.

Peierls (1955, Eq. 2.36) gave the sum-of-states for the potential energy expressed by Eq. 1.56:

$$Z = 2\pi(m/a)^{1/2} kT \; [1 + (5g^2/24f^3 - h/8f^2)kT] \tag{1.61}$$

where m denotes atomic mass. Following Leibfried (1955, section 81) the internal energy is

$$U/kT = 1 + (g^2/12f^3 - h/8f^2)kT \tag{1.62}$$

if only the g^2 and h terms are retained. And the constant-volume specific heat is

$$C_V/k = 1 + (g^2/6f^3 - h/4f^2)kT \tag{1.63}$$

and

$$C_P/k = 1 + (5g^2/12f^3 - h/4f^2)kT \tag{1.64}$$

so that

$$(C_P - C_V)/k = (g^2/4f^3)kT = 3\gamma\alpha T \tag{1.65}$$

Neglecting zero-point energy the thermal expansion is

$$\Delta a/a = -(g/2f^2a)kT = 3kT\gamma/fa^2 \tag{1.66}$$

so that linear thermal expansivity is

$$\alpha = -gk/2f^2a = 3k\gamma/fa \tag{1.67}$$

where a denotes interatomic spacing.

The elastic constant of the monatomic linear chain, described essentially by f, varies with temperature according to

$$f(T) = f - (g^2/2f^2)kT \tag{1.68}$$

so that

$$df(T)/dT = -(g^2/2f^2)k = -18\,k\gamma^2/a^2 = -6\gamma f\alpha \tag{1.69}$$

Note the decrease with increasing temperature and the strong dependence on thermal expansivity. The difference between adiabatic and isothermal elastic constants is

$$f_S - f_T = (g^2/4f^2)kT = (3\gamma/a)^2kT \tag{1.70}$$

which can be compared with the more exact results given in Fig. 1.9.

Leibfried (1955, section 83) considered the pressure dependence of the compressibility and showed its approximate relationship to γ:

$$(dB/dP)_{P=0} = -(1/K^2)(dK/dP)_{P=0} = -ga/3f = 2\gamma \tag{1.71}$$

Thus, the predicted coefficient is approximately 4, and observed values are 4.4 for aluminum, 5.4 for copper, 5.3 for iron, 5.5 for lead, 4.2 for silicon, 4.1 for silver, and 6.0 for gold (Simmons and Wang, 1971).

Expressions for various thermodynamic properties based on a linear-chain model are summarized in Table 1.12. If the second-order, third-order, and fourth-order elastic constants, $C_2 = f/a$, $C_3 = g$, $C_4 = ha$, are known, then most of the familiar thermodynamic properties can be computed. Among the properties considered

Table 1.12 Summary of monatomic-linear-chain-model relationships for potential $V = (1/2)fu^2 + (1/6)gu^3 + (1/24)hu^4$.

Grüneisen constant	$\gamma = -ga/6f$
Internal energy	$U/kT = 1 + (g^2/3f^3 - h/2f^2)kT/4$
Specific heats	$C_P/k = 1 + (5g^2/3f^3 - h/f^2)\,kT/4$
	$C_V/k = 1 + (g^2/3f^3 - h/2f^2)\,kT/2$
	$(C_P - C_V)/k = g^2kT/4f^3 = 3\alpha\gamma T$
Lattice parameter	$a(T) - a(0) = -gkT/2f^2 = 3kT\gamma/fa$
Thermal expansivity	$\alpha = (1/a)(da/dT) = -gk/2f^2a = 3k\gamma/fa^2$
Force constants	$f(T) = f - g^2kT/2f^2 = f - 18kT\gamma^2/a^2$
	$df(T)/dT = -g^2k/2f^2 = -18k\gamma^2/a^2 = -6\gamma f\alpha$
	$f_S - f_T = g^2kT/4f^2 = (3\gamma/a)^2\,kT$
Compressibility	$(dB/dP)_{P=0} = -(dK/dP)_{P=0}/K^2 = -ga/3f = 2\gamma$

here, C_4 enters only into the specific heats. Thermal expansivity and elastic constants are determined entirely by C_2 and C_3.

The results in Table 1.12 combine in interesting and practical ways. For example, from the table it follows that

$$(dB/dT)/(dB/dP) = 9k\gamma/a^3 \tag{1.72}$$

In other words, the ratio of the temperature dependence to the pressure dependence of the elastic constants is proportional to Grüneisen's constant, which tends to be temperature independent and approximately the same for all materials. Equation 1.72 provides a possible nondestructive evaluation approach to solids. By measuring dB/dT, a much simpler measurement than dB/dP, it may be possible to study a solid's residual stress, usually considered to manifest itself as dB/dP. Salama and Ippolito (1980) reported results of preliminary studies along these lines. The general point to be made here is the interplay among various anharmonic properties via the elastic constants of second, third, and fourth order. Studies by Hiki and Granato (1966) and by Hiki, Thomas, and Granato (1967) provide other good examples of this interplay.

Anharmonic phenomena (not all independent) explained by relationships described in this section include:

a. Nonzero thermal expansivity
b. Nonzero Grüneisen constant
c. Difference between constant-pressure and constant-volume specific heats
d. Nonconstant high-temperature specific heat
e. Low-temperature specific-heat deviations from T^3 behavior
f. Temperature-dependent Debye characteristic temperature
g. Temperature-dependent elastic constants
h. Pressure-dependent elastic constants
i. Difference between isothermal and adiabatic elastic constants
j. Temperature-dependent equation of state

Other phenomena explained anharmonically but not discussed here include:

k. Thermal resistivity
l. Ultrasonic-wave attenuation
m. Phonon dispersion
n. Phonon–phonon interactions
o. Nonlinear stress–strain relationships

1.6.3 Elastic Debye temperature

Elastic and calorimetric Debye temperatures, θ_e and θ_c, are equal at zero temperature in harmonic lattices (Maradudin, Montroll, Weiss, and Ipatova, 1971). Ludwig (1958) suggested that lattice anharmonicity breaks this equality. Based on approximate, model calculations, Leibfried and Ludwig (1961) estimated that θ_e exceeds θ_c by 1 or 2%. Barron and Klein (1962) showed for a linear chain that Leibfried and Ludwig erred, that indeed $\theta_e = \theta_c$, and that this equality should apply also in three dimensions. Feldman (1964) examined the first-order quantum corrections to the elastic frequencies and to the specific heat, including zero-point energy explicitly, and concluded that the elastic frequencies equal the long-wavelength thermal frequencies, that is, $\theta_e = \theta_c$ for anharmonic crystals.

Apparently, Ludwig (1967) and most authors on the subject accepted these judgments, although Overton (1968, 1971) raised theoretical and experimental objections.

Independently and nearly simultaneously, Madelung (1910), Sutherland (1910), and Einstein (1911) suggested that a characteristic temperature θ could be computed by

$$\theta = A\ B^{1/2} m^{-1/2} V^{1/6} \tag{1.73}$$

where A is a constant that varies with lattice type, B is the bulk modulus, m is atomic mass, and V is volume. Figure 1.10 shows Eq. 1.73 for various elements. The principal deficiency of this relationship is that it emphasizes dilatational elastic waves rather than shear elastic waves, which actually dominate θ, as shown by the exact expression for the isotropic (random polycrystalline) case (Landau and Lifshitz, 1959b, sections 61, 63):

$$\theta = (\hbar/k)\ (18\pi^2/V_a)^{1/3} v_m \tag{1.74}$$

where

$$3v_m^{-3} = v_\ell^{-3} + 2v_t^{-3} \tag{1.75}$$

Where $\hbar$ and k have their usual meanings, V_a denotes atomic volume, and v_ℓ and v_t denote longitudinal and transverse sound velocities. Equation 1.74 has the same form as Eq. 1.73 since $(B/\rho)^{1/2}$ has velocity units, where ρ is the mass density. Figures 1.11 and 1.12 show that better correlations occur when either G or E is substituted for B in Eq. 1.73.

Table 1.6 gives θ's computed from elastic constants at zero temperature. Differentiation of Eq. 1.74, assuming constant volume, gives

$$d\theta/\theta = dv_m/v_m \simeq (16/17)\ dv_t/v_t + (1/17)\ dv_\ell/v_\ell \tag{1.76}$$

if $\nu \simeq 1/3$, which is a typical value. Substituting from Eqs. 1.34 and 1.35 gives

$$d\theta/\theta \simeq (49/102) dG/G + (2/102) dB/B \tag{1.77}$$

which shows that θ depends almost entirely on G to the exclusion of B.

Neglecting zero-point energy, the Debye temperature relates to the force constant, f, in Eq. 1.56 by

$$\theta = \pi(h/k)(f/m)^{1/2} \tag{1.78}$$

Figure 1.13 shows the frequency distribution for copper computed entirely from the polycrystalline elastic constants using three models: Einstein (1906), Debye (1912), and Born (1923) and Brillouin (1946). Obviously, the elastic constants offer a better method for studying the vibrational spectrum of a solid than the specific-heat method, which deals with an average of the elastic waves over all directions and over all polarizations.

For cubic metals, Mitra and Joshi (1961) discuss the relationship between the Debye θ and compressibility. Herbstein (1961) and Blackman (1955) discuss Debye temperatures obtained by various experimental methods. And Gschneidner (1964) gives the Debye temperatures of the elements. (A great need exists for a comprehensive data compilation and critical review of Debye-temperature data.)

1.6.4 Grüneisen parameter

Grüneisen's (1926) γ is the most important single parameter for describing anharmonic phenomena (§ 1.6.2), as Debye's θ is for harmonic phenomena. The two are related by

$$\begin{aligned}\gamma &= -(V/\theta)(d\theta/dV)\\ &= -(V/\nu)(d\nu/dV)\end{aligned} \tag{1.79}$$

where ν denotes frequency and γ is a dimensionless parameter that varies little with material, temperature, or pressure. As Grüneisen explained, this is because γ relates closely to the form of the repulsive potential, which changes little with material. Grüneisen also demonstrated that γ relates to other thermodynamic properties:

$$\gamma = B_T \beta V/C_V = B_S \beta V/C_P \tag{1.80}$$

where both β and C_V are anharmonic properties. Relationships in Table 1.12 show that all anharmonic phenomena can be discussed more simply by invoking Grüneisen's γ. In terms of elastic constants, the principal anharmonic phenomena are: higher-order elastic constants, temperature derivatives, and pressure derivatives. Expressions relating these to γ are given in Table 1.12 for the linear-chain model. Another useful viewpoint on γ is that it expresses the pressure change caused by heating at constant volume:

$$\gamma = V(\partial P/\partial T)_V/C_V \tag{1.81}$$

Slater (1939) discussed γ and several of its applications to solid-state thermodynamics and showed that

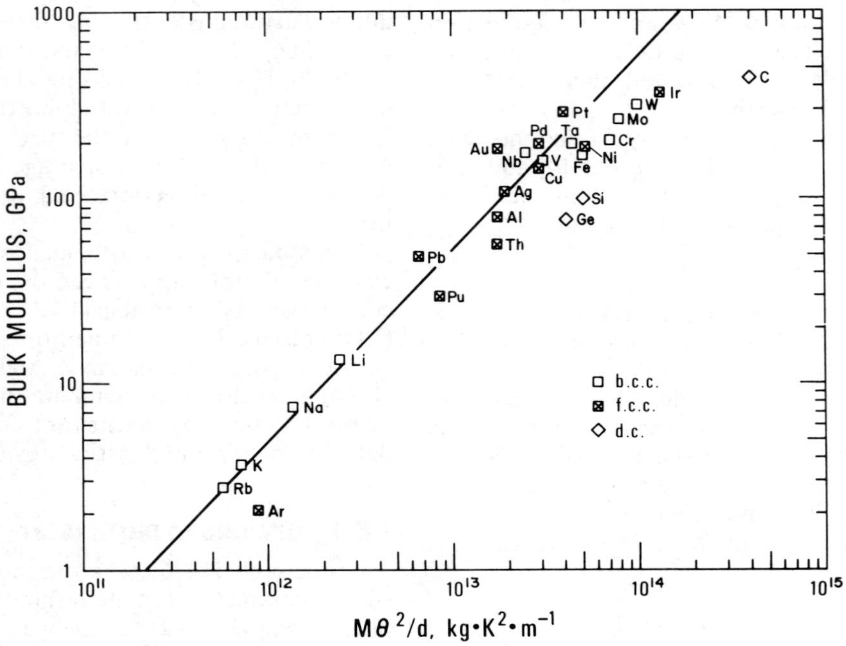

Figure 1.10 Einstein's relationship among the bulk modulus, Debye temperature, atomic mass, and interatomic distance.

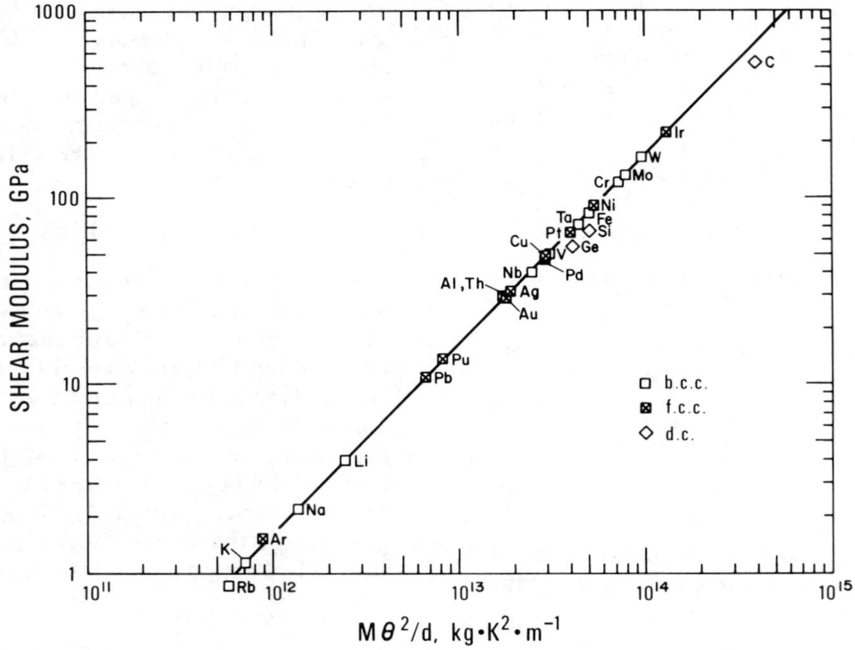

Figure 1.11 Einstein's relationship with shear modulus substituted for bulk modulus. See Fig. 1.10.

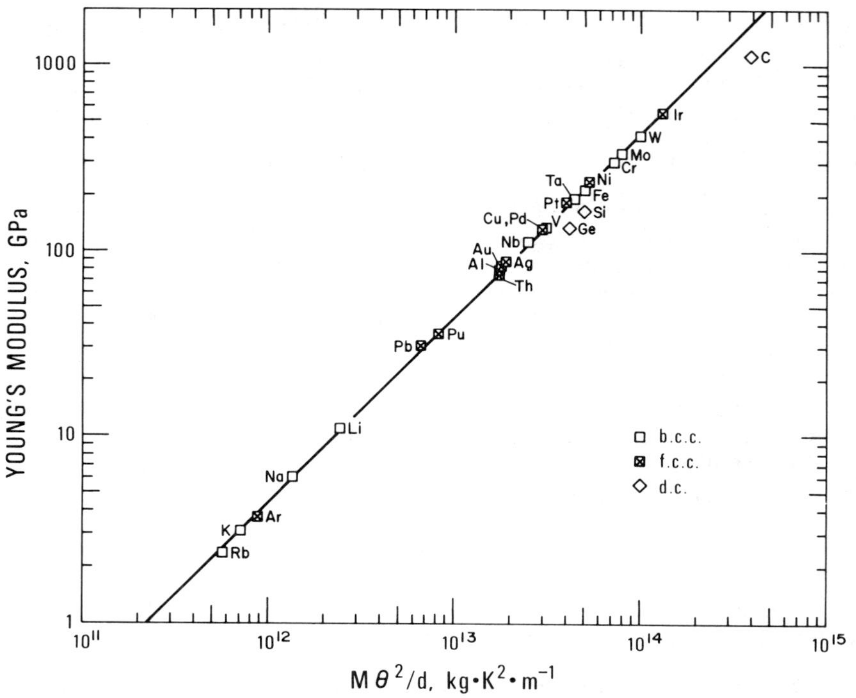

Figure 1.12 Einstein's relationship with Young's modulus substituted for bulk modulus. See Fig. 1.10.

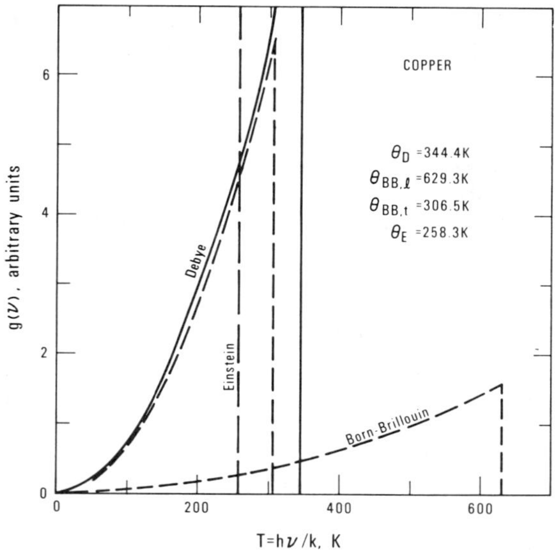

Figure 1.13 Frequency distribution of copper computed from polycrystalline elastic constants using three lattice-dynamics models: Einstein, Debye, and Born–Brillouin.

$$\gamma = -(1/6) + (V/K)(dK/dV)/2 \qquad (1.82)$$

where K is compressibility. As described in § 1.7, this relationship is important because the compressibility's temperature dependence is determined mainly by the volume change. Slater also showed that γ relates to compressibility parameters a and b by

$$\gamma = b/a^2 - 2/3 \qquad (1.83)$$

where

$$-(V - V_0)/V_0 = aP - bP^2 + \ \ldots \qquad (1.84)$$

is the equation of state in terms of volume and pressure.

Grüneisen's relationship provides insight into elastic-constant temperature dependence. If γ is temperature independent, then from Eq. 1.80

$$\begin{aligned}(1/B_T)(dB_T/dT) &= (1/C_V)(dC_V/dT) \\ &\quad - (1/\beta)(d\beta/dT) - \beta \qquad (1.85)\end{aligned}$$

Thus, knowing $C_V(T)$ and $\beta(T)$ permits computing $B_T(T)$. However, β is seldom known accurately enough for such purposes. Both B and C_V, especially B, are typically measured with higher precision than is β. The further assumption that β / C_V is also temperature independent leads to

$$(1/B_T)(dB_T/dT) \simeq -\beta \qquad (1.86)$$

Although these coupled assumptions are unreasonably crude and Eq. 1.86 is too rough for practical use, it shows the strong dependence of dB/dT on β. Zero thermal expansivity implies a small change in elastic constants with temperature. Similarly, a positive dB/dT implies a *possible* negative thermal expansivity.

Grüneisen's parameter, including elastic-constant applications, is discussed by several authors: Slater (1940); Gilvarry (1955, 1956); Vashchenko and Zubarev (1963); Pastine (1965); Anderson (1968); Knopoff and Shapiro (1969); and Holt and Ross (1970). Harris and Avrami (1972) give an elementary discussion of γ. Gerlich (1975) gives references to comparisons between elastic and thermal Grüneisen parameters.

1.6.5 Lattice dynamics and interatomic force constants

Lattice theory, being discrete, contains elastic theory, which is continuous. The essential lattice-theory parameters are the interatomic force constants

$$f_{ij}^{(\vec{k})} = -(\partial \vec{F}_i / \partial \vec{x}_j) \qquad (1.87)$$

where $\vec{F}$ denotes force and $\vec{x}$ denotes displacement. Physically, $-f_{ij}^{(\vec{k})}$ is the force on an atom at the origin in the i direction due to an atom at $\vec{k}$ being displaced in the j direction. Since $f_{ij}^{(\vec{k})} = f_{ji}^{(\vec{k})}$, there are generally six force constants for each set of neighboring atoms. Lattice symmetry usually reduces this number. It is useful to represent $f_{ij}^{(\vec{k})}$ as a 3×3 matrix, which can be diagonalized by a suitable rotation. If $\vec{x}$ is parallel to an eigenvector of $f_{ij}^{(\vec{k})}$, the force is parallel to displacement and the eigenvalue is a "spring" constant, that is $\vec{F} = -f\vec{x}$.

From interatomic force constants, the lattice vibrational spectrum (see Fig. 1.13) can be computed; this spectrum determines many important solid-state properties (Leibfried, 1955, sections 77–78). The relationship between atomic eigenfrequencies and elastic constants was established first by Einstein (1911). Models and methods for interrelating elastic constants and force constants are reviewed by several authors: Born and Huang (1954), Blackman (1955), Leibfried (1955), DeLaunay (1956), and Brillouin (1964). Simple introductory discussions are given by Ewald, Pöschl, and Prandtl (1930) and by Feynman et al. (1963, p. 39-10).

1.6.6 Specific heat

Debye (1912) established the quintessential relationships between elastic constants and specific heat, showing that specific heat derives directly from the elastic constants without difficult calculations. His elegant continuum model of a solid's acoustic spectrum has two principal properties: an upper cutoff frequency, ν_D, and a distribution function, Z, of the vibrational degrees of freedom below an upper limit ν. For an elastic body of volume V and mass density ρ, Debye found that

$$Z = (4\pi/3) V \nu_D^3 \, \rho^{3/2} [2/G^{3/2} + (1/(B + 4/3\, G)^{3/2}] \qquad (1.88)$$

where B and G are the bulk and shear moduli. The essential parameter is ν_D. Debye showed that the elastic constants determine ν_D. In terms of the bulk modulus and Poisson's ratio

$$\nu_D = (3/4\pi V_a)^{1/3} (\rho/B)^{3/2} \cdot \{2[2(1+\nu)/3(1-2\nu)]^{3/2} + [(1+\nu)/3(1-\nu)]^{3/2}\} \qquad (1.89)$$

where V_a denotes atomic volume. Thus, the acoustic spectrum of a solid can be described completely in terms of the mass density and the elastic constants, independent of the continuum frequency.

From thermodynamics (see Table 1.11), specific heat is the partial derivative of energy with respect to temperature. Planck (1906) showed that the energy of a quantum oscillator of frequency ν is

$$U = h\nu / [\exp(h\nu/kT) - 1] \qquad (1.90)$$

where h is Planck's constant, k is Boltzmann's constant, and T is temperature. The energy of a Debye solid is

$$U = (9N/\nu_D^3) \int_0^{\nu_D} u\, \nu^2 \, d\nu \qquad (1.91)$$

It follows at low temperatures that internal energy and specific heat vary as

$$U \sim T^4 / \nu_D^3 \qquad (1.92)$$

and

$$C = \partial U / \partial T \sim (T/\nu_D)^3 \qquad (1.93)$$

As discussed in § 1.6.3, calorimetric (specific-heat) and elastic Debye temperatures are identical at zero temperature. Useful independent checks on the elastic constants are therefore provided. Elastic constants can often be measured more simply and more accurately than specific heats; thus, elastic constants can contribute more to the problem of specific heats than vice versa. For example, Ledbetter (1977b) reviewed the Debye-θ data of zinc and showed that many of the calorimetric values contain large errors, but that most of the elastic values (if computed correctly) have small inaccuracies.

1.6.7 Thermal expansivity

Solids usually expand during heating because of an anharmonic interatomic potential where the potential's repulsive part exceeds its attractive part. Being an anharmonic property, thermal expansivity must relate to the elastic constants and to their temperature and pressure derivatives. For anisotropic expansivity,

$$\beta_i = S_{ij} p_j \tag{1.94}$$

where p_j is the "thermal pressure" and S_{ij} is the elastic-compliance tensor (see § 1.2.6 and 1.5.1). Thermal pressure depends on lattice properties and is strongly temperature dependent; S_{ij} is much less so. Since S_{12} is usually negative and the p_j's depend differently on temperature, negative thermal expansivity becomes possible in anisotropic materials at low temperatures (Grüneisen, 1926; Munn, 1972).

Since K and β relate volume changes to changes in pressure and temperature, respectively, one expects a relationship between these variables, and it is

$$\begin{aligned} -(\partial K/\partial T)_P &= (1/B^2)(\partial B/\partial T)_P \\ &= (\partial \beta/\partial P)_T \end{aligned} \tag{1.95}$$

Since $d\beta/dP$ is seldom studied experimentally, this relationship contributes nothing practical toward understanding elastic-constant temperature dependence. However, from the elastic-constant temperature dependence, which is measured relatively simply, $\partial\beta/\partial P$, which is difficult to measure, can be obtained. For regularly behaved materials Eq. 1.95 shows that B should decrease continuously with increasing temperature since increasing pressure decreases thermal expansivity according to Le Chatelier's principle.

The above relationships can also be expressed

$$B_T\beta = (\partial P/\partial T)_V = (\partial S/\partial V)_T \tag{1.96}$$

which is the thermal pressure (Nye, 1960).

Elastic constants completely determine thermal expansivity, as discussed in § 1.6.2 for a linear-chain model. Sheard (1958) apparently made the first computation of β from elastic-constant data. Subsequent emergence of reliable elastic constants for various materials makes such computations more feasible. Elastic constants also lead to better interatomic potentials from which thermal expansivity can be calculated.

Barker (1963) suggested an empirical relationship between β and Young's modulus:

$$\beta^2 E = \text{constant} \tag{1.97}$$

and gave it a theoretical basis. Relationships in Table 1.12 suggest that βEV, which is proportional to $k\gamma$, may be a better correlation. Correlations between Young's modulus and specific heat are less likely to succeed because specific heat depends strongly on the fourth-order term in $V(r)$. But such correlations remain unstudied.

1.6.8 Miscellaneous relationships

From definitions of β and B and some mathematics, it follows that

$$\beta B = (\partial S/\partial V)_T = (\partial P/\partial T)_V \tag{1.98}$$

The product of thermal expansivity and bulk modulus is the isothermal volume derivative of the entropy. Thus, some entropy information can be determined by simple measurement of β and B. As discussed below, entropy becomes especially important in describing low-temperature physical-property behavior. By considering equilibrium between two phases at a second-order (continuous-entropy, continuous-volume) phase transition, Eq. 1.98 leads to an Ehrenfest relationship

$$\begin{aligned} (dP/dT)_V &= (\beta_2 - \beta_1)(B_2 - B_1) \\ &= (C_{P_2} - C_{P_1})/TV(\beta_2 - \beta_1) \end{aligned} \tag{1.99}$$

for the pressure dependence of the transition temperature.

Other useful and interesting relationships can be developed from equations given in several reviews of solid-state thermodynamics: Hooton (1955), Salter (1955), Dobbs and Jones (1957), Gol'denblat (1962), Thurston (1965), and Wallace (1972).

1.7 ELASTIC-CONSTANT CHANGES WITH TEMPERATURE

Cooling from ambient to near-zero temperature changes the elastic constants of metals and alloys up to 15%. Some nonmetallics exhibit much larger changes. In all solids, stiffnesses such as Young's modulus increase, whereas elastic compliances and Poisson's ratio decrease during cooling. As described below, elastic-constant-vs.-temperature changes relate to a material's basic elasticity. Elasticity means that atoms can move about their equilibrium positions, periodically increasing and decreasing the interatomic distance. Without such lattice vibrations, physical phenomena such as thermal expansivity and phonon thermal conductivity would not occur. When atoms vibrate in an anharmonic potential well, solids expand as thermal energy increases. During heating elastic stiffness decreases mainly because volume increases but also because intrinsic interatomic forces decrease. According to Barron and Klein (1963), "the effect of the lattice vibrations on the thermodynamic and elastic properties of a crystal is determined by the harmonic frequency distribution $G(\omega)$ and its dependence on strain . . . Many properties of crystals are determined primarily by the strain dependence of $G(\omega)$, among them the thermal expansion and the vibrational contributions to the elastic constants." Similarly, Ziman (1960) pointed out that "another effect of anharmonicity is on the elastic constants, which vary with volume and with temperature. These are complicated phenomena, for which there is no elementary theory; they depend on a number of different parameters."

1.7.1 Observed regular behavior

The effect of temperature on the elastic constants of regular solids is now well established from many measurements, which are summarized by Hearmon (1946, 1956, 1966) and by Simmons and Wang (1971). Notable experimental studies of temperature effects include those of Grüneisen (1910) and Köster (1948).

By regular is meant that temperature affects only the energy of the solid's vibrational spectrum, that the form of the spectrum is unchanged by electronic, magnetic, structural, or other transitions. In metallurgical terminology, the phase is invariant with respect to temperature. Figure 1.14 shows the temperature behavior of regular materials and summarizes several important features:

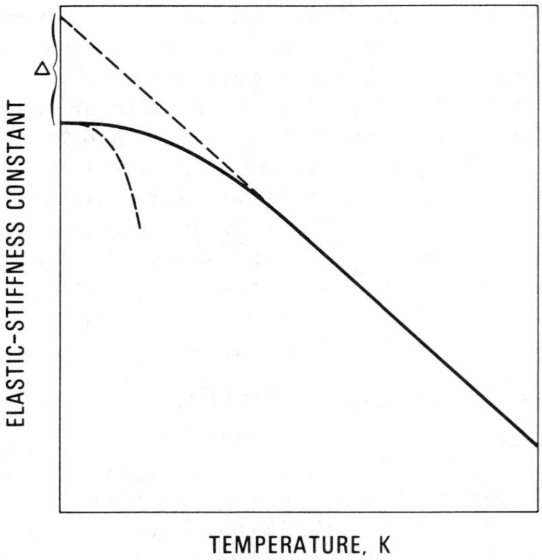

Figure 1.14 Schematic regular behavior of an elastic-stiffness constant vs. temperature.

a. Elastic stiffnesses (the C_{ij}'s) decrease continuously with increasing temperature.

b. The term dC/dT is zero at zero temperature as required by the third law of thermodynamics.

c. The term dC/dT is constant at high temperatures; that is, behavior is linear.

d. A breakaway from linearity occurs at about $\theta/2$ at low temperatures. As discussed in § 1.8, this difference between extrapolated linear behavior and actual behavior is a quantum effect caused by zero-point-energy vibrations, which contribute negatively to the elastic-stiffness constants.

e. C varies as T^4 at very low temperatures, which is predicted by many models, as discussed by Garber (1972).

f. The $C(T)$ curve's simplicity permits its description by many different models and its representation by various empirical fits, some of which are discussed in § 1.7.4.

g. Not shown in the figure, another important feature is that the elastic compliances (the S_{ij}'s) and Poisson's ratio behave inversely to the C_{ij}'s, increasing continuously with increasing temperature.

Figures 1.15 through 1.18 show how various elastic constants — bulk modulus, Young's modulus, shear modulus, and Poisson's ratio — vary with temperature between 0 and 300 K for six typical metals.

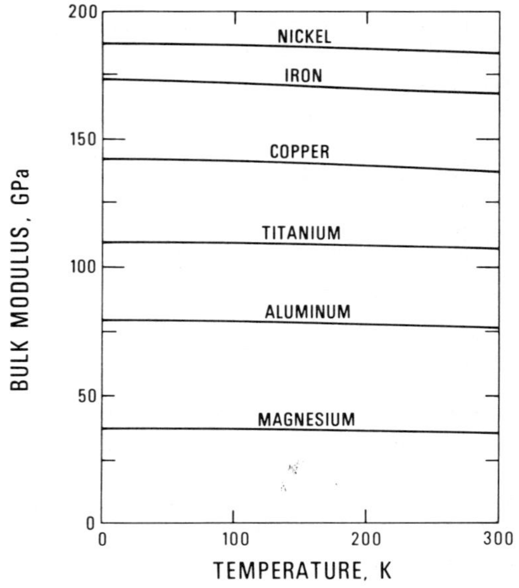

Figure 1.15 Temperature variation of bulk modulus (reciprocal compressibility) for six metals.

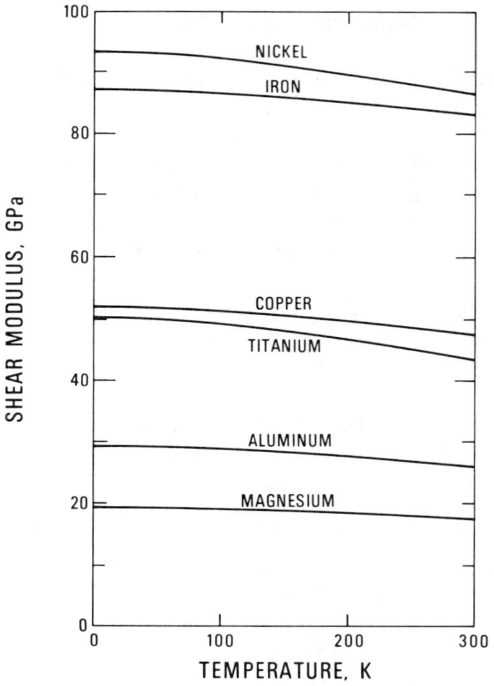

Figure 1.17 Temperature variation of shear modulus for six metals.

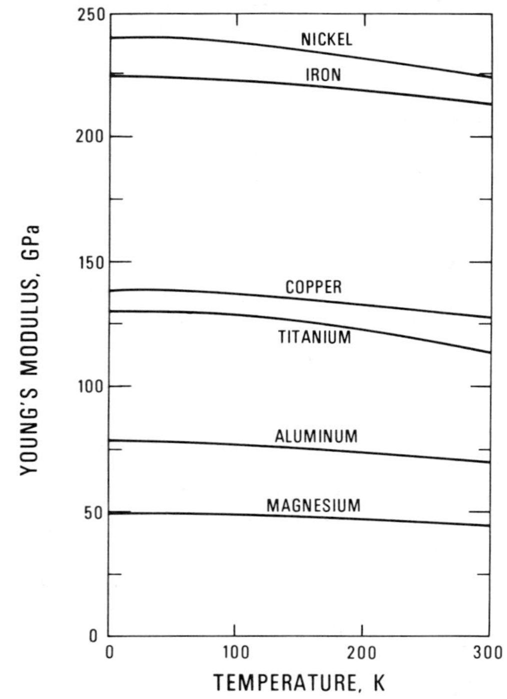

Figure 1.16 Temperature variation of Young's modulus for six metals.

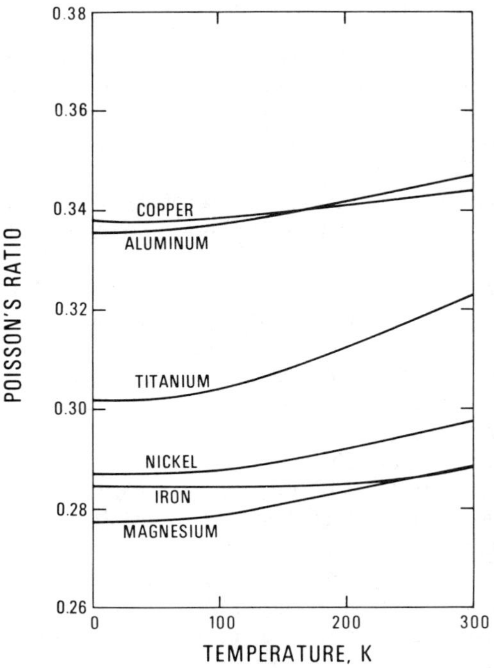

Figure 1.18 Temperature variation of Poisson's ratio for six metals.

1.7.2 Theoretical studies

As described in § 1.6.2, elastic-constant changes with temperature occur because of lattice anharmonicity. The first allowance for anharmonicity, according to Leibfried and Ludwig (1961), was made by Grüneisen and Mie in 1903 in developing their well-known equation of state. They assumed the lattice constant, a, changed with temperature so that oscillator eigenfrequencies depend also on a. This approximation, where interatomic-coupling coefficients depend on volume, is now called the quasi-harmonic approximation.

Born and his collaborators made the earliest substantive theoretical studies on elastic-constant temperature dependence: Born (1939), Fürth (1941), Born (1943), Bradburn (1943), Gow (1944), Fürth (1944), and Born and Huang (1954, chapter 2). Born's model constitutes a nearly complete theory of solid-state thermodynamics, from which (Born and Huang, 1954, p. 322) "all the normal mechanical and electrical properties of a crystal can be described in terms of its free energy . . . explicit formulae governing these properties are obtained by substituting the free energy . . . into the general results . . . temperature dependence of these formulae is in general very complex; a general discussion is possible only in the limit of very low temperatures." The model evaluates the change in lattice potential energy represented as normal-mode frequencies when the lattice is strained by various homogeneous deformations. From generalized forces between atoms, interrelationships are obtained among interatomic force constants and macroscopic elastic constants. The temperature dependence of the elastic constants arises from the anharmonic change of lattice energy. Alternatively, the temperature dependence can be considered to arise from anharmonicity-induced phonon–phonon interactions. The principal limitation of Born's model is that it considers only central-force interatomic interactions, often of a special type such as Mie–Grüneisen.

Leibfried's (1955) comprehensive review and study on the lattice theory of the mechanical and thermal properties of crystals adopts the spirit of Born's model, but more simply. In particular, Leibfried develops the solid-state physics of the linear-chain model and applies it to many physical properties, including elastic constants and their temperature derivatives. The discussion in § 1.7.3 follows largely from Leibfried's results.

Stern (1958) developed an approach similar to Born's, but with a significant difference: Stern considered energy changes under stress variations rather than strain variations. The advantage is that stress enters the energy expression directly, where its effect can be evaluated. Stern derived a proportionality between the bulk-modulus temperature derivative and specific heat that resembles the result obtained by combining Eqs. 1.80 and 1.86. The author is unaware of any theoretical or empirical followup on this perhaps-useful relationship.

Leibfried and Ludwig (1961) provided the standard statement on anharmonic effects in crystals. (See § 1.6.2 for implications of anharmonicity.) These authors use the linear-chain and quasi-harmonic models extensively, showing clearly the relationships between elastic constants (or force constants) and many other physical phenomena.

Bernstein (1963, 1965) approached the problem of elastic-constant temperature dependence from the viewpoint of the electron contributions. There are well-known calculations of elastic constants from electron energies by Wigner and Seitz (1933), Seitz (1935), Fuchs (1936), and Bardeen (1938a,b), which are reviewed by Mott and Jones (1936). Bernstein reasoned that the total electron energy is determined by electron–electron and electron–ion interactions, that these latter interactions determine the phonon frequencies, and that these frequencies determine the sound velocities, which combine with mass density to give the elastic constants. The thermal variation of the elastic constants (of metals) thus arises from two sources: the temperature dependence of the energy of the Fermi–Dirac-distributed electrons and the anharmonically induced lattice-energy change. Bernstein showed that the electrons contribute an internal energy and an elastic stiffness that both vary as T^2 at low temperatures. The elastic-stiffness contribution may be positive or negative depending on the Fermi-level density-of-states relationship. Bernstein also showed that dC/dT at low temperatures is directly analogous to $d\chi/dT$, where χ is the paramagnetic susceptibility.

Several authors reviewed the theory of anharmonic lattices, which includes the elastic-constant temperature-dependence problem. Three are discussed above: Born and Huang (1954), Leibfried (1955), and Leibfried and Ludwig (1961). Besides these, useful statements on this subject have been made by Huntington

(1958), Ludwig (1958), and Leibfried and Ludwig (1960).

The most recent lattice-theoretical review, by Ludwig (1967), emphasizes studies after Born and Huang (1954) and Leibfried (1955). These studies deal principally with details of anharmonicity, noncentral forces (especially in the shell model), more complicated crystal structures, effects of lattice defects, phonon interactions, and thermodynamic Green's functions. Ludwig devotes about twenty pages to a combined consideration of thermal expansivity and elastic-constant temperature dependence.

1.7.3. Linear-chain model

If one had a thermal equation of state, $F(V,T)$, then dC/dT would follow simply. However, such equations are complicated, even for simple lattice models.

Considering Born's (1940) thermal equation of state, Leibfried (1955, section 84), after lengthy, tedious, sometimes difficult calculations, for temperatures high compared to $\theta/2$, derived Voigt's elastic constants for a linear-chain face-centered-cubic nearest-neighbor model:

$$C_{11} = (2/\sqrt{2})f/a - (30/\sqrt{2})\gamma^2 kT/a^3 \quad (1.100)$$

$$C_{12} = (1/\sqrt{2})f/a - (21/\sqrt{2})\gamma^2 kT/a^3 \quad (1.101)$$

and

$$C_{44} = (1/\sqrt{2})f/a - (9/\sqrt{2})\gamma^2 kT/a^3 \quad (1.102)$$

so that

$$C' = (1/2)(C_{11} - C_{12}) = (1/2\sqrt{2})f/a - (9/2\sqrt{2})\gamma^2 kT/a^3 \quad (1.103)$$

and

$$B = (4/3\sqrt{2})f/a - (24/\sqrt{2})\gamma^2 kT/a^3 \quad (1.104)$$

where Grüneisen's constant $\gamma = ga/6f$, and f and g denote the force constants defined in Eq. 1.56. Using Voigt's method for averaging the C_{ij}'s, the shear modulus is

$$G = (4/5\sqrt{2})f/a - (36/5\sqrt{2})\gamma^2 kT/a^3 \quad (1.105)$$

Despite their simplicity, Leibfried's equations contain much useful information:

a. All elastic stiffnesses decrease continuously with increasing temperature.

b. The slope dC/dT does not relate to the potential energy $V(r)$ since scaling $V(r)$ does not affect the g/f ratio, which determines dC/dT. This general result for vibrational properties applies also to specific heat and thermal expansivity.

c. The relative change $(1/C)(dC/dT)$ of the elastic constants is given by g^2/f^3 and is small for strong interatomic forces (high f) and large for weak interactions; for example, covalent and van der Waals bonding, respectively.

d. All elastic constants would be temperature independent in the absence of anharmonic terms ($g = 0$).

e. Finite temperature breaks the Cauchy relationship, $C_{12} = C_{44}$, even in a central-force model.

f. The Zener anisotropy, $A = 2C_{44}/(C_{11} - C_{12})$, should be nearly temperature independent, and always near 2 for the f.c.c. case.

g. Relative changes in the bulk and shear moduli should be of the same magnitude.

h. The G/B ratio should be 0.57. As discussed in § 1.5.3, experiment shows this ratio is near 0.4. A model including also second-neighbor interactions might improve this prediction. In any case, 0.57 is not a bad first guess because this ratio depends sensitively on the model.

i. Poisson's ratio, $\nu = -S_{12}/S_{11}$ for [100]-type directions should be 1/3, and this is very close to the average observed values for metals and alloys.

1.7.4 Semiempirical models

Although relatively simple and related to basic principles, Leibfried's linear-chain model described in the previous section describes only the higher-temperature, linear region of the $C(T)$ curve. Even the most elegant and complicated basic-principle models fail to describe $C(T)$ accurately. Sacrificing fundamentals for fit leads to various so-called semiempirical models, which describe $C(T)$ quite accurately. Several such models are cited in this section, and one, devised by Varshni (1970), is discussed in more detail in the next section.

Sutherland (1891a) proposed the earliest semiempirical $C(T)$ relationship for the shear modulus known to the author:

$$G(T)/G(0) = 1 - (T/T_m)^2 \quad (1.106)$$

where T_m is the melting temperature. Sutherland (1891b) proposed another form for Young's modulus:

$$E(T)/E(0) = 1 - AT/T_m \qquad (1.107)$$

where A denotes an adjustable parameter. As implied in § 1.5.3, current knowledge predicts similar temperature dependences for G and E. Sutherland (1891c) also derived a theoretical relationship for $B(T)$ involving fundamental atomic parameters. Brillouin (1938) discussed Sutherland's $G(T)$ relationship from a general theoretical viewpoint. Siegel and Cummerov (1940) described the inadequacy of Sutherland's relationship as a melting criterion [the vanishing of $G(T)$].

Durand (1936) proposed that a general elastic-stiffness constant, C, changes with temperature according to

$$C(T)/C(0) = \exp[-AF(T/\theta)] \qquad (1.108)$$

where A denotes a material constant and θ is a characteristic temperature. For all substances, the function F is unity when T equals θ, and it is determined experimentally.

Based on a relationship used by Mack, Sutton (1953) found that the C_{44} shear modulus of aluminum could be represented within 0.5% over the 63 to 773 K temperature range by

$$C(T) = A\exp(-B\beta T)/R^2 \qquad (1.109)$$

where A and B are adjustable parameters, β denotes volume thermal expansivity, and R denotes $V(T)/V(0)$, where V is the volume. The other cubic-symmetry elastic constants, C_{11} and C_{12}, defied such a representation.

For copper, Overton and Gaffney (1955) found the C_{44} data representable within 0.2% by

$$C(T)/C(0) = \exp(-BdT)/(1 + \int_0^T \alpha \, dT)^9 \qquad (1.110)$$

where α denotes linear thermal expansivity. But, the authors emphasized the pure empiricism and lack of a theoretical basis for this relationship.

Following the advances in lattice dynamics made by Born and his collaborators (see § 1.7.2), the first suggestion of a function that describes accurately the temperature dependence over a wide range was by Wachtman, Tefft, Lam, and Apstein (1961). For Young's modulus of oxides at high temperatures they proposed:

$$E(T) = E(0) - bT\exp(-T_0/T) \qquad (1.111)$$

where the three adjustable constants are b, T_0, and $E(0)$, the Young's-modulus value at $T = 0$ K.

This relationship satisfies two theoretical criteria: $E(0)$ is approached with zero slope as T vanishes and dE/dT is linear at high temperatures where $\exp(-T_0/T)$ approaches unity. Wachtman et al. sensed that T_0 should relate to the Debye temperature and that b/E should relate to Grüneisen's constant, but these relationships were not established.

Anderson (1966) theoretically justified Wachtman's equation by considering a corresponding equation for the bulk modulus

$$B(T) = B(0) - bT\exp(-T_0/T) \qquad (1.112)$$

Anderson showed that

$$b = 3R\gamma\delta/V_0 \qquad (1.113)$$

where R denotes the gas constant, γ denotes Grüneisen's constant, V_0 denotes specific volume of average atom at zero temperature, and δ represents an important physical constant analogous to Grüneisen's constant, independent of temperature, and given by

$$\delta = -(1/\beta B)(dB/dT)_P \qquad (1.114)$$

where β denotes volume thermal expansivity. Anderson cautions that "the Wachtman empirical equation for Young's modulus apparently is valid only when the variation of Poisson's ratio with temperature is small."

Several authors use Wachtman's equation to describe the temperature dependence of either the Young's or bulk moduli for many materials, mainly nonmetallic. The author knows of only two studies that apply Wachtman's equation to metals. Varshni (1970) considered the single-crystal elastic constants of Al, Cu, Au, Ir, Fe, Pb, Mg, Mo, Ag, Ta, Th, W, MgAg, and $MgCu_2$. Nandanpawar and Rajagopalan (1978) considered the bulk moduli of Au, Ag, Cu, Mg_2Sn, and $MgCu_2$.

1.7.5 Varshni model

A semiempirical model developed by Varshni (1970) is used by the author to fit elastic-constant-vs.-temperature data for both single crystals and polycrystals between 4 and 300 K. This model is summarized by the relationship

$$C(T) = C(0) - s/[\exp(t/T) - 1] \qquad (1.115)$$

where C is an elastic-stiffness constant. There are three adjustable parameters: s, t, and $C(0)$, the value of $C(T)$ at $T = 0$. That Eq. 1.115 represents the best possible $C(T)$ relationship remains unproven. Considering three semiempirical models, Varshni showed for 51 cases of C_{ij}'s that Eq. 1.115 was superior in 30 and Eq. 1.111 in 8, although predictions of the two equations differed only slightly. Figure 1.19 shows Varshni's relationship in reduced form.

Varshni gave a theoretical basis to Eq. 1.115 by showing it follows simply from two relationships: the energy of a Planck–Einstein oscillator

$$U = (1/2)\,h\nu + h\nu/[\exp(h\nu/kT) - 1] \quad (1.116)$$

and the Leibfried and Ludwig (1961) result that

$$C(T) = C(0)(1 - D\bar{U}) \quad (1.117)$$

where $\bar{U}$ denotes the mean oscillator energy and the constant D depends on types of lattice and model. Varshni's model combines electronic and lattice effects, and it predicts a very-low-temperature $\exp(-h\nu/kT)$ dependence rather than the theoretically predicted T^4 dependence. Inspection of Eq. 1.115 shows that t is equivalent to the Einstein temperature, and $-s/t$ is the high-temperature limit of $dC(T)/dT$.

Varshni solved Eq. 1.115 by fixing t, obtaining $C(0)$ and s by least squares, determining the (minimum) t that gives lowest average percentage differences, then determining t more accurately by using smaller Δt intervals.

The author solves Eq. 1.115 differently (Naimon and Ledbetter, 1973). A residual delta is defined

$$\delta = \sum_i [C(0) - sx_i - C_i]^2 \quad (1.118)$$

where

$$x_i = [\exp(t/T_i - 1)]^{-1} \quad (1.119)$$

where C_i and T_i are the ith observed elastic constants and temperatures. Delta depends on t through $C(0)$, s, and x_i, and is minimized with respect to t using a Newton–Raphson method and $t = 200$ K as an initial estimate. By successive iterations the problem can be solved to any accuracy; one part in 10^6 typically requires less than five iterations. When t is known, $C(0)$ and s are obtained easily. For less labor, this approach yields a more accurate set of fitting parameters than obtained by Varshni's approach.

1.7.6 Recent studies

The elastic-constant temperature-dependence problem has been assailed recently from several theoretical viewpoints, most of them quite sophisticated. Indeed, these models may represent the limit of presently formulated solid-state science. They are impressive both aesthetically and in corresponding to observation.

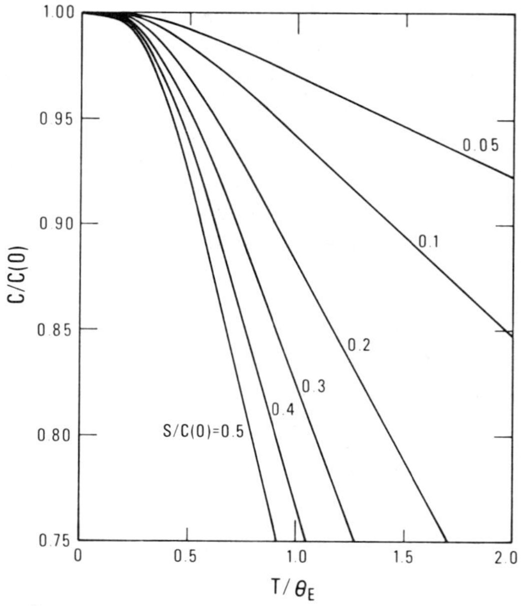

Figure 1.19 Reduced elastic-stiffness constant $C/C(0)$ vs. reduced temperature T/θ_E for various $s/C(0)$ ratios according to Varshni's relationship. θ_E denotes Einstein characteristic temperature.

Garber (1972) and Garber and Granato (1975a, 1975b) worked with a quasi-harmonic, anisotropic continuum model. Further assumptions include: no strain effects higher than first order, Debye sphere, strain dependence of frequency independent of wavelength and given by the long-wavelength limit, sinusoidal dispersion relationship outside this limit, crystalline materials, zero external stress, and lossless medium. A principal result was that elastic constants are *not* proportional to the vibrational energy as predicted by Leibfried and Ludwig (1961), whose model neglects the crystalline properties of direction and polarization. This model should work best at low temperatures, near the T^4 region. Garber and Granato (1975b) found good agreement with recent low-temperature measurements of copper's third-order and fourth-order elastic constants C_{111} and C_{1111}.

For the transition metals, a tight-binding model has been used in several related publications from Switzerland: Pasternak, Steinemann, and Peter (1968); Weinmann and Steinemann (1974a, 1974b); Walker and Peter (1977); Ashkenazi, Dacorogna, Peter, Talmor, and Walker

(1977); Talmor, Walker, and Steinemann (1977); Ashkenazi, Dacorogna, Peter, Talmor, Walker, and Steinemann (1978). In describing observation, both composition and temperature dependences, this is the most impressive model known to the author. It contains two free-electron-type free-energy parameters: the Fermi function and E_{bk}, the energy of an electron in band b with momentum k. Further details occur in the above references.

Besides agreement with observation, this tight-binding model offers a distinct advantage over previous approaches, which focus on the volume dependence of the elastic constants, C. Actually, elastic constants depend on both volume and temperature. Thus it follows thermodynamically (Callen, 1960) that

$$(dC/dT)_P = (\partial C/\partial T)_V + V\beta(\partial C/\partial V)_T \qquad (1.120)$$

where β denotes volume thermal expansivity. The left term is observed experimentally. The right terms represent, respectively, the intrinsic temperature-dependent part and the intrinsic volume-dependent part due to thermal expansion; both contain both lattice and electronic components. From the viewpoint of model calculations, the $(\partial C/\partial T)_V$ term is easiest to handle. Using Choquard's (1967) theory, Weinmann and Steinemann (1974b) evaluated the terms in Eq. 1.120 for the case of palladium. At ambient temperature for the bulk modulus they found that the $V\beta\ (\partial C/\partial T)_T$ term contributes more to dC/dT than does the $(\partial C/\partial T)_V$ term, but that the two contributions are comparable for the shear moduli. But because palladium's $C(T)$ behavior is nonregular because of electronic effects, palladium may not represent typical $C(T)$ behavior. Nevertheless, this approach separates temperature and volume effects, unlike the more familiar approaches of Born and Huang (1954) and Leibfried and Ludwig (1961).

Ruffa (1977) recently related the temperature dependence of the shear moduli of cubic metals to their thermal expansivity. In this model, the pseudopotential is approximated by a Morse potential, thermal expansivity is assumed to arise from uncoupled Morse oscillators, quantum-mechanical Morse-oscillator solutions are combined with a Debye model, and elastic shear moduli are calculated for nearest-neighbor interactions. The model's principal focus is atomic positions relative to the interatomic-potential minimum. Although agreement with observation is only approximate, Ruffa feels that including farther-neighbor interaction would improve the model's predictions.

Recent Soviet studies on elastic-constant temperature dependence are impressive. [For example, see Vaks, Zarochentsev, Kravchuk, and Saftronov (1978) and many references therein.] Using a two-parameter local pseudopotential model, Vaks et al. calculated many important thermodynamic properties of five alkali metals: (a) atomic properties at $T = 0$; (b) heat capacity and atomic mean-square displacements; (c) equation of state, pressure effects, and martensitic phase transitions; (d) thermal expansivity; (e) third-order elastic constants; and (f) temperature dependence of the second-order b.c.c. elastic constants.

1.8 ELASTIC CONSTANTS AT LOW TEMPERATURES

Unlike specific heat and thermal expansivity, which vanish at zero temperature consistent with the third law of thermodynamics, elastic constants remain finite at zero temperature, thus ensuring the solid's mechanical stability. But the third law does require vanishing temperature derivatives, dC/dT. From the third law of thermodynamics and one of Maxwell's relationships, at $T = 0$

$$(\partial S/\partial V)_T = (\partial P/\partial T)_V = 0 \qquad (1.121)$$

Pressure expanded as $P(T,V)$ must depend only on volume. Thus, the bulk modulus $-V(\partial P/\partial V)_T$ also depends only on volume, independent of T as T vanishes. Figures 1.15 through 1.18 show that experiment confirms this prediction: at low temperatures in the absence of phase transitions, elastic constants change little with temperature. Materials in metastable states may exhibit contrary behavior since the third law applies only to equilibrium.

The low-temperature T^4 dependence of elastic constants arises as follows: Low-temperature free energy in metals is given by

$$U = U_0 + ET^2 + LT^4 \qquad (1.122)$$

where the ET^2 term corresponds to electron-gas thermal excitations and the LT^4 term corresponds to lattice thermal excitations. The elastic constant is then

$$C = C_0 + (\partial^2 E/\partial \epsilon^2)T^2 + (\partial^2 L/\partial \epsilon^2)T^4 \qquad (1.123)$$

Very few reliable low-temperature elastic data

exist to test this relationship. Alers and Waldorf (1962) demonstrated a T^4 dependence for copper's C_{ij}'s between 4 and 20 K. Alers (1966) discussed this problem in detail; for the bulk modulus he estimated the magnitudes of the ET^2 and LT^4 terms.

Unpinned dislocations may lower the observed elastic stiffness by about 1% by contributing an additional strain because of their motion. Alers and Thompson (1961) demonstrated dislocation mobility in copper even at 4 K. The temperature dependence of the dislocation contribution may obscure the LT^4 term described above. Pinning the dislocations by irradiation removes the dislocation strain and restores the structure-insensitive part of the elastic constant.

Superconductivity affects elastic constants slightly, typically a few parts in 10^4 or less. Pippard (1955, 1961) gives thermodynamic developments for both shear and bulk moduli. B is discontinuous at the superconducting temperature; because of symmetry arguments G is not, but the derivative dG/dT changes dramatically. Following Pippard (1961, Eq. 8.17),

$$\Delta B/B = B\,(\partial H_c/\partial P)^2/4\pi \tag{1.124}$$

where H_c is the critical magnetic field above which superconductivity vanishes. Other aspects of the effect of the normal-superconducting phase transition on elastic constants are reviewed by Alers (1966).

Elastic constants relate directly to the most basic aspects of superconductivity. One example of this relationship is the Bardeen–Cooper–Schrieffer theoretical relationship for the superconducting critical temperature

$$T_c = (\theta/1.45)\exp(-1/NV) \tag{1.125}$$

where θ is the Debye temperature, which is determined by the elastic constants, N is the density of states at the Fermi level, and V is the net attractive potential between Cooper-pair electrons. The NV term depends on the average phonon frequency, ω, and Schrieffer (1973) suggested that a plot of T_c vs. ω might exhibit a maximum since ω relates directly to the elastic constants. DeLaunay and Dolecek (1947) published one of the earliest studies of T_c and elastic-constant relationships. Evetts and Martin (1973) point out that "the interaction parameter V . . . tends to be large in elastically soft materials. . . . if V varies significantly when a single elastic constant becomes zero the critical temperature . . . will be enhanced."

Zero-point-vibration energy and elastic constants relate intimately. Domb and Salter (1952) showed that the zero-point Debye-model internal energy for N oscillators per unit volume is

$$U_{zp} = (9/8)Nk\theta \tag{1.126}$$

where θ denotes Debye temperature, which is computable from the elastic constants, atomic weight, and atomic volume (§ 1.6.3). Zero-point energy enters the elastic constants because all energy terms contribute to them according to the second strain derivatives of the Helmholtz free energy. Salter (1954) considered this problem for the rare-gas solids and his results have several general aspects: (a) zero-point energy destroys the Cauchy relationship $C_{12} = C_{44}$ inherent in a central-force model, (b) deviation of measured elastic constants from theoretical (static-lattice) values offers a quantitative test of any zero-point energy expression; such a test relates to tests of the interatomic potential described in § 1.1.1, (c) the zero-point-energy contribution to the elastic constants is negligible for metals with high Debye temperatures. For lead ($\theta = 105$ K), Flinn and Maradudin (1963) showed using a nearest-neighbor central-force model that $\theta_{anharmonic}/\theta_{harmonic} = 0.9925$; thus the zero-temperature experimental (anharmonic) shear modulus would be about 1.5% lower than the shear modulus obtained by extrapolating the linear high-temperature (harmonic) value to zero temperature. Flinn and Maradudin also showed that at low temperatures "the largest part of . . . these anharmonic contributions arises from what is in a sense not a purely anharmonic effect, but rather a quasi-harmonic effect: the expansion of the lattice as a result of the zero-point energy." Vaks et al. (1978) used a pseudopotential model to calculate the temperature dependence of the elastic constants of five alkali metals and the zero-point-energy contribution to both the shear and bulk moduli. They found zero-point-energy contributions to the bulk modulus of 0.3 to 3.7%.

Finally, we quote some formulas for the low-temperature thermodynamic properties of a harmonic-oscillator solid (Landau and Lifshitz, 1959b, § 61). The Helmholtz free energy, the internal energy, the entropy, and the specific heat are:

$$F = E(0) - AT^4/30 \tag{1.127}$$

$$E = E(0) + AT^4/10 \tag{1.128}$$

$$S = 2AT^3/15 \quad (1.129)$$

$$C_V = 2AT^3/5 \quad (1.130)$$

where

$$A = V\pi^2 k^4 / \hbar^3 v_m^3 \quad (1.131)$$

where v_m, the mean sound velocity, follows from the elastic constants and the mass density. These relationships illustrate that all the usual thermodynamic properties are obtainable from the elastic constants and that in the low-temperature region the relationships are simple and exact.

1.9 OTHER ELASTIC-CONSTANT TOPICS

This part of the review considers miscellaneous topics that separate naturally from the main part.

1.9.1 Magnetic materials

Here, magnetic means either ferromagnetic or antiferromagnetic. Diamagnetic and paramagnetic materials are considered nonmagnetic in affecting elastic properties.

Magnetic effects can be intrinsic or extrinsic. Intrinsic effects arise from either interatomic-potential change or from a new energy term owing to magnetism onset. Any energy term that is strain dependent contributes to the elastic constants. Extrinsic effects, although not predicted energetically, alter the measured strain. For example, domain walls in ferromagnets move under stress and contribute a spurious strain that lowers the elastic stiffness below the within-domain value, which differs intrinsically from the nonmagnetic value.

The best known magnetoelastic effect is the so-called ΔE effect arising from stress-induced domain-wall-motion strain. Honda and Terada (1907) first studied this effect in detail, and it is expected in all ferromagnets where it affects only shear-type elastic constants. Besides domain walls, the effect requires magnetostrictive anisotropy, variation in the magnetostriction energy with magnetization direction. Details concerning the ΔE effect are given by Bozorth (1951), Lee (1955), and Chickazumi (1964).

Engler (1938) discovered and Döring (1938) explained another magnetic effect expected in materials with large volume magnetostriction. This effect is seen in both ferromagnets and antiferromagnets and it affects only dilatation-type elastic constants. It arises because the elastic constants at constant magnetization differ from those at constant magnetic field.

Since Heisenberg's quantum-mechanical explanation of ferromagnets using an exchange energy, scientists realize that this energy must contribute to ferromagnetic elastic constants. However, theorists have not yet provided an expression for the strain-dependent exchange energy. Recent studies have used magnetic elastic-constant measurements to deduce the nature of the exchange integral and to show that elastic-constant anomalies caused by magnetic transitions are consistent with an exchange-energy interpretation. Such effects are expected in both ferromagnets and antiferromagnets, and they affect all types of elastic constants.

Other magnetic phenomena affect elastic constants. For example, Mason (1951) described a morphic effect, a crystal-symmetry lowering traceable to the magnetostriction energy, that lowers nickel's elastic stiffness 1 to 2%.

1.9.2 Structural phase transitions

Since elastic constants are the second strain derivatives of the free energy, they must change discontinuously at both first-order and second-order phase transitions. They would be continuous through a third-order transition, but such a transition remains undiscovered. Only structural transitions are considered here; effects accompanying electrical, magnetic, order-disorder, superconducting, and other transitions are ignored.

A positive-definite elastic-strain energy leads to mechanical-stability criteria. As derived first by Born (1940) for cubic crystals, these are: $C_{11} + 2C_{12} > 0$, $C_{44} > 0$, and $C_{11} - C_{12} > 0$. Unless all three criteria are satisfied, a solid is unstable and it must transform to another crystal structure. Melcher, Guntherodt, Penney, and Holtzberg (1975) found that the bulk modulus $B = (1/3)(C_{11} + 2C_{12})$ of $Sm_{1-x}Y_xS$ goes to zero at a cubic–cubic transition induced by an f^6–f^5 d-electronic collapse accompanied by a 7% volume change. But there are very few other examples of bulk-modulus instabilities. In fact, the bulk modulus tends to be approximately conserved through phase transitions. According to Born (1939), "the violation of the condition $C_{44} > 0$, that the crystal resist shearing, is interpreted as leading to melting." Actually, this is a sufficient, but not necessary, cause for allotropic change, which is determined by the free-energy balance between competing structural phases.

Zener (1948) seems to have been first to

appreciate the relationship between low ($C_{11} - C_{12}$) values and b.c.c.–f.c.c. transitions. Zener's predictions led Barrett (1947) to discover such a low-temperature transition in lithium. This elastic instability occurs in many other materials: transition metals and alloys (Fisher and Renken, 1964; Ducastelle, 1970; Fisher and Dever, 1970; Goasdue, Ho, and Sass, 1972), 3/2 electron compounds or β-phase alloys (Nakanishi, Takamo, Hashimoto, Morimoto, and Miura, 1978), and A15 superconducting compounds, such as Nb_3Sn and V_3Si (Testardi, 1975; Bhatt and McMillan, 1976; Testardi, 1977; Kragler, 1978).

The soft-mode concept introduced by Anderson (1960) and Cochran (1959) contributed enormously to understanding solid–solid displacive phase transitions. A soft mode is a normal-vibration mode whose frequency decreases and whose amplitude increases as the transition temperature is approached. At zero frequency, there is no elastic force to restore atoms to equilibrium positions. Thus, atoms assume a new structure whose symmetry is determined by the soft-mode eigenvector (Fleury, 1971). Depending on phonon wavelength, soft-mode transitions are called either zone-center (wave vector $= k = 0$) or zone boundary ($k = \pi/a$, or similar). The ($C_{11} - C_{12}$) elastic-constant softness described above corresponds to a transverse zone-center acoustic mode.

The b.c.c.–ω transition that occurs in many materials, especially in transition metals and their alloys, provides a good example of a zone-boundary soft mode. Since these soft modes occur at $k \neq 0$, they appear, at first, unrelated to elastic-constant behavior. However, closer scrutiny alters this view. Soft-mode behavior can be computed from Born–von Karman (1912)-type lattice-dynamics models. A simple, practical model considers interatomic interactions between an atom and a limited group of neighboring atoms. The force constants, limited in number, are evaluated from physical-property data. For cubic symmetry, a three-force-constant lattice-dynamics model can proceed entirely from the elastic constants (Joshi and Rajagopal, 1968).

Nutkins's (1958) calculations provide a good simple example of the fundamental relationship between elastic constants and phase transitions. Considering the low-temperature b.c.c.–f.c.c. transitions in lithium and sodium, Nutkins computed the elastic constants of both crystal structures and from these determined the lattice-vibrational free energies, both at zero and higher temperatures. Nutkins predicted a transition temperature between 40 and 100 K, agreeing well with observation. Nutkins's study emphasizes that phase transitions occur when

$$\Delta E_b + \Delta F_{ph} = 0 \qquad (1.132)$$

where ΔE_b denotes the difference in static-lattice binding energies and ΔF_{ph} the difference in phonon free energies. Knowing the elastic constants allows prediction of the transition temperature, T_t. For example, the Debye model in the low-temperature limit predicts

$$T_t^4 = K\Delta E_b/(\theta_2^{-3} - \theta_1^{-3}) \qquad (1.133)$$

where θ denotes the Debye temperature, which can be computed from the elastic constants of phases 1 and 2 (§ 1.6.3), and K is a constant involving h, k, atomic weight, and atomic volume; ΔE_b is a decisive parameter and difficult to compute since it represents the small energy difference between two structures with large binding energies. Hafner (1976) estimated for the alkali metals, where $E_b \sim 0.4$ Ry/atom and $\Delta E_b \sim 10^{-4}$ Ry/atom, that pseudopotential calculations are almost sufficient to accurately predict T_t. Many such calculations have been reported, but the subject has not been reviewed. Typical example calculations are Saxena, Kapoor, and Bhattacharya (1969) and Vaks, Kravschuk, and Trefilov (1977).

Phase transitions, especially second-order, display many interesting interrelationships among various thermodynamic variables. Equation 1.99 for the pressure effect on the equilibrium transition temperature provides a good example. From it follows

$$\Delta B = -TVB^2(\Delta\beta)^2/\Delta C_P \qquad (1.134)$$

Theory requires (Landau and Lifshitz, 1959b, p. 434) that cooling from a high-temperature symmetrical phase to a low-temperature unsymmetrical phase always exhibits a positive ΔC_P, a positive $\Delta\beta$, and a negative $\Delta\beta$. For small volume and entropy changes, first-order transitions should exhibit similar effects. One important difference between first-order and second-order transitions, especially at low temperatures, is that the Debye-temperature change on cooling, $\Delta\theta$, is positive for first-order transitions and negative for second-order transitions. Since the shear modulus, G, relates so closely to θ, ΔG should behave similarly to $\Delta\theta$, being positive on cooling through a first-order transition and negative for a second-order case.

Several sources review the elastic-constant phase-transition relationship: Robertson (1965), Ledbetter and Reed (1974), and Warlimont, Hausch, Prasetyo, and Reynaud (1976), and Nakanishi (1980).

1.9.3 Polymers

Stephens (1968) summarized polymeric classes and properties:

> The three main divisions of polymers illustrate the differing mechanical properties which are exploited. Firstly there are elastomers, as exemplified by rubber, which in its application to tyres must absorb and damp out road vibrations, but also withstand the absorbed energy, which is converted into heat. Secondly are the thermosets, e.g. epoxy resins, which offer a great range of physical properties through their coreaction with other compounds. They are mixed in monomer form and polymerized in place to form a permanent three-dimensional cross-linked network. Thirdly are the semi-crystalline thermoplastics, of which the polyolefins are representative. As a class they offer a wide range of mechanical properties from materials of similar chemical characteristics. This results from the control which can be exerted over the polymerization process, where for example the same polymer can be produced in different forms by altering the configuration of the side groups along the main chain from regular, e.g. isotactic, to irregular (atactic). The arrangement of the side groups has an important influence on the crystallinity of the polymers, while the degree of chain branching affects the dynamic properties as in polyethylene.

Elastically, polymers differ in principle from materials described above. Polymers may exhibit either anelasticity or viscoelasticity, which are distinguished from elasticity in Fig. 1.20. Anelasticity manifests itself as a time-dependent strain equilibration. Viscoelasticity shows, additionally, only partial strain recovery upon stress removal. But these nonelastic effects may be small.

The elastic behavior of polymers falls between those of crystalline solids and liquids. Several variables determine their behavior: monomer units, molecular weight, branching, cross-linking, crystallinity, grain size, and long-range molecular order. Depending on these variables, polymeric elastic constants may exhibit strong frequency and temperature effects. Interpreting such frequency–temperature behavior can be quite complicated (Stephens, 1968, Fig. 5).

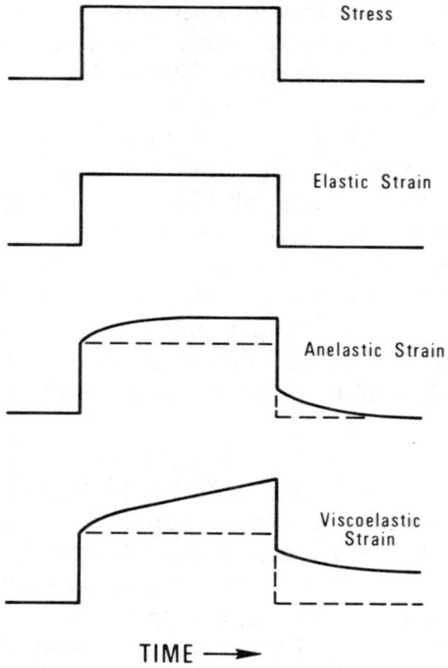

Figure 1.20 Various strain responses to an imposed stress.

Because of possible anelasticity, dynamic-modulus measurements on polymers must be interpreted carefully. Figure 1.21 shows the complex modulus, M^*, where strain lags stress by an angle, ϕ, and thus resolves into two components. A recoverable elastic strain energy occurs with the component in phase with stress. Energy associated with the component 90° out of phase with stress dissipates, usually as internal-friction heating. Experiments observe M^*, the modulus in phase with strain. Thus, both M^* and ϕ, or M_1 and ϕ, must be specified to define a polymer's dynamic elastic modulus.

Polymers differ radically from other solids by being neither chemical compounds nor systems of identical molecules. They crystallize only incompletely, and transitions involving macromolecular rearrangements always occur sluggishly. At higher temperatures, many polymers exhibit rubberlike behavior, a low elasticity modulus, and a high elongation before breaking. Staverman (1962) reviewed the thermodynamics of polymers and emphasized that all their peculiar properties result from neighboring atoms interacting with forces of very different magnitudes: strong intramolecular forces and weak intermolecular forces. Staverman emphasized also the importance of statistics and entropy in polymer-property theory. Furthermore, internal and external degrees of freedom assume definite physical meanings; thus internal and external free-energy changes accompany processes such as temperature change and mechanical deformation.

For the simplest possible polymer — isotropic, homogeneous, and having only linear terms in stress, strain, stress rate, and strain rate — Staverman (1962) enumerated seven ways to characterize a viscoelastic material:

a. $G(t)$, stress-relaxation modulus as a function of time,

b. $J(t)$, creep compliance,

c. $G_1(\omega)$, storage modulus as a function of frequency,

d. $G_2(\omega)$, loss modulus,

e. $J_1(\omega)$, storage compliance,

f. $J_2(\omega)$, loss compliance, and

g. $\phi(\omega)$, mechanical loss angle.

These experimental functions correlate with thermodynamic quantities if one invokes $g(\tau)$, the relaxation spectrum, and $f(\tau)$, the retardation spectrum, where τ denotes relaxation time.

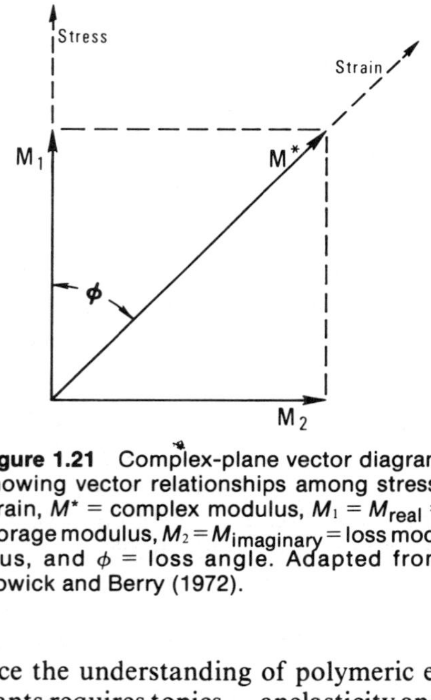

Figure 1.21 Complex-plane vector diagram showing vector relationships among stress, strain, M^* = complex modulus, $M_1 = M_{real}$ = storage modulus, $M_2 = M_{imaginary}$ = loss modulus, and ϕ = loss angle. Adapted from Nowick and Berry (1972).

Since the understanding of polymeric elastic constants requires topics — anelasticity and molecular modeling — beyond this review's scope, readers should see the books by Tobolsky (1960), Ferry (1961), McCrum, Read, and Williams (1967), and Ward (1971), and the review paper on low-temperature polymeric properties by Sauer and Saba (1970). The *Journal of Polymer Science* contains many research reports related to low-temperature polymeric elastic properties.

1.9.4 Composites

A composite's elastic constants can be important for several reaons:

a. Elastic constants can provide valuable material characterization since they are measured with low inaccuracies, typically ± 1% uncertainty.

b. Since elastic constants are anisotropic, directional properties can be studied and characterized. For example, Young's modulus along the fibers may exceed that in the perpendicular direction by a factor of 100 or more.

c. Fundamental thermodynamic properties, such as specific heat and thermal expansivity relate simply to elastic constants.

d. An elastic constant can be associated with every deformation mode, such as shear, torsion, extension, compression, and bending.

e. Composites are used mainly in their elastic regions, below their elastic limits. Many composites are elastic until failure.

f. Elastic constants provide one of the best checks on theories that relate macroscopic composite properties to microscopic constituent properties. Thus, a model valid for elastic constants can be expected to hold also for physical properties of lower tensor rank.

g. For composites, many relationships exist between elastic properties and mechanical-failure properties of a composite.

A fiber-reinforced composite's elastic constants are determined by: (a) matrix type, (b) fiber type, (c) fiber fraction, (d) fiber orientation, (e) temperature, (f) frequency, (g) fiber length, and (h) fiber distribution. Elastic constants are usually not considered dependent on fiber periodicity or fiber dimensions. For a given fiber distribution, which determines the elastic symmetry, theory shows that a composite's elastic constants are determined by fiber fraction and the elastic constants of the fiber and the matrix.

Composites with randomly spaced, uniaxially aligned filaments exhibit transverse-isotropic (or hexagonal) symmetry with five independent elastic constants. Composites with orthotropic (or orthorhombic) symmetry have nine independent elastic constants. Particle-reinforced composites are, in principle, macroisotropic with two independent elastic constants. Laminated composites exhibit orthotropic elastic symmetry, but often only four in-plane elastic constants are used to characterize laminates.

Composite elastic constants are determined experimentally by all the usual methods: load displacement, resonance (Bert and Clary, 1973; Ledbetter, 1979), and ultrasonic pulse (Tauchert and Güzelsu, 1971; Sachse, Ting, and Hemenway, 1979).

Physical properties of composites were reviewed by Hale (1976) and thermomechanical properties by Rosen (1970). What is generally known about fibrous composites was discussed by Chou and Kelly (1976) who emphasized that "the effective elastic properties of composites composed of unidirectional laminae can be predicted" but that predictive models have not been thoroughly tested because of "the lack of experimental data covering a wide range of fibre and modulus ratios."

The literature contains many predictive models for computing a composite's elastic constants from those of its constituents. Some notable papers include: Hashin and Shtrikman (1963), Hill (1963), Hashin and Rosen (1964), Rossi (1968), Rosen (1970), Behrens (1971), and Beran (1971). Although not focusing on uniaxially reinforced composites, the most complete and up-to-date review of composite elastic properties is by Watt, Davies, and O'Connell (1976).

1.9.5 Textured aggregates

Most discussion in this chapter assumes textureless aggregates. Texture, or preferred orientation, means that crystallite orientations are distributed nonrandomly. Thermal or mechanical gradients promote texture, which removes isotropy, reduces symmetry, and increases the number of independent tensor components required to specify a physical property such as elasticity. For example, a wire tends to have either five or six elastic constants, and a plate tends to have nine, unless it is cross rolled; then it has five. Several authors have reviewed this fascinating and technologically important subject: Boas and MacKenzie (1950), Hearmon (1961), Köster and Franz (1961), Green (1973), and Warlimont and Hausch (1974). Recent research papers dealing with elastic constants and texture include those of Date and Andrews (1969), Killpatrick and Young (1970), and Turley and Sines (1971a, 1971b). Although texture may occur in any given aggregate, some studies, Hopkin, Pursey, and Markham (1970) and Ledbetter, Frederick, and Austin (1980), show that texture is practically absent in many commercial alloys, especially hot-worked alloys where recrystallization occurs spontaneously at the deformation temperature.

1.9.6 Other-phenomena correlations

Because elastic constants relate directly to interatomic forces, they relate also to various chemical, mechanical, and physical phenomena that depend on and exhibit such forces. How closely these phenomena relate to fundamental forces determines the nature of these relationships, from exact first-principle interconnections to crude empirical correlations. In different ways, both types of relationships are useful. Table 1.13 lists examples of elastic-constant and other-phenomena relationships together with references that may be first-source, historical, principal, typical, review, recent, or simply convenient to the author's desk. The table's main purpose is to indicate some vital and diverse roles played by elastic constants in various scientific or technological phenomena.

1.10 CONCLUDING REMARKS

In this review the author tried to convey several basic concepts concerning polycrystalline elastic properties and their temperature dependence:

a. Elastic constants are physical, not mechanical, properties.

b. They relate thermodynamically to other physical properties such as specific heat and thermal expansivity.

c. All these properties and many others, such as the Grüneisen constant, can be derived from a general central-force interatomic potential or from the potential energy of a linear chain expanded in atomic displacements where, except for a lattice constant, the expansion coefficients are the elastic constants.

d. The Debye characteristic temperature, which relates in turn to various solid-state phenomena, is, perhaps, the most important parameter derivable from elastic constants.

e. Elastic stiffnesses in metals and alloys typically increase 5 to 15% during cooling between ambient and zero temperatures.

f. Elastic constants at low temperatures are nearly temperature independent, as required by the third law of thermodynamics.

g. Low-temperature elastic constants are important theoretically as checks on interatomic potentials, cohesion theories, and lattice specific heats.

h. Detailed theories of elastic-constant vs. temperature behavior are complicated. Although no truly satisfactory detailed theory exists, the

Table 1.13 Other phenomena that correlate with elastic constants.

Phenomenon	Sources
Atomic-vibration amplitude	James (1948) Lonsdale (1948) Tanji (1971)
Bragg intensities, Debye–Waller factor	Zener and Bilinsky (1936) James (1948, p. 215) Singh and Sharma (1971)
Characteristic frequency	Einstein (1911) Baughan (1952) Plendl and Gielisse (1965)
Cohesive energy and fusion energy	Plendl, Mitra, and Gielisse (1965) Nemilov (1968) Mulimani (1972)
Creep	Sherby (1962a) Grimvall and Sjödin (1974)
Diffusion, atomic	Zener (1951) LeClaire (1953) Aaronson and Shewmon (1967)
Dislocation properties	Gilman (1960, 1961) Fisher and Alfred (1968) Hirth (1976)
Ductility	Pugh (1954) Crutchley and Reid (1969) Stein and Shannette (1976)
Electrical conductivity	Blackman (1951) Kelly and MacDonald (1953) Meaden (1965)
Elongation	Pugh (1954)
Entropy	Lumsden (1952) Radovskii (1966)
Equation of state	Overton (1963) Swenson (1968)
Fatigue crack growth rate	Michel (1977)
Force constants and interatomic distance	Waser and Pauling (1950)
Forming and shaping	Sherby (1962b)
Fracture strength	Pugh (1954) Eshelby (1968) Hecker, Rohr, and Stein (1978)
Grain-boundary energy	Grimvall and Sjödin (1974)
Hardness	Wooster (1953) Gilman (1961) Chin (1975)
Heat of solution	Friedel (1954)
Interatomic distance	Badger (1934, 1935) Gilman (1969) Chin (1975)
Interatomic potentials	Cousins (1969) Johnson (1973) Barua and Sinha (1978)
Lattice dynamics	Leibfried (1965)
Lattice spacing and valency	Meyer, Umar, and Young (1971)
Martensite	Pace and Saunders (1971)
Melting	Lindemann (1910) Born (1939) Pines (1963)
Mössbauer emission	Visscher (1960)
Neutron scattering	Woods, Brockhouse, March, Stewart, and Bowers (1962)
Peierls force	Peierls (1940)
Phase stability	Zener (1947) Weston and Granato (1975) Cho and Puerta (1976)
Phase transitions, crystallographic	Warlimont et al. (1976)
Plastic properties of solids	Seitz and Read (1941) Zener (1942) Kelly, Tyson, and Cottrell (1967)
Point-defect energy	Granato, Hikata, and Lücke (1958) Mukherjee (1965) Hardy and Flocken (1970)
Residual stress	Zener (1949)
Solid solubility	Zener (1949)
Solution hardening	Fleischer (1963) Takeuchi (1968, 1969)
Sound velocities	Blackman (1955) Anderson (1965) Simmons and Wang (1971)
Specific heat	Debye (1912) Holm (1957) Buch (1966)
Stacking-fault energy	Weston and Granato (1975)
Strength	Garber and Gindin (1961) Kelly (1966) Tyson (1966) Milstein (1971, 1973)
Superconducting transition temperatures	DeLaunay and Dolecek (1947)
Surface energy	Grimvall and Sjödin (1974)
Thermal conductivity (lattice)	Leibfried (1955, p. 290) Peierls (1955) Klemens (1958)
Thermal expansivity	Dugdale and MacDonald (1953) Sheard (1958) Barker (1963)
Vegard's-law deviations	Friedel (1955)
Volume, atomic, in alloys and intermetallic compounds	Friedel (1954) Rudman (1965)
Wear, abrasive	Khruschov (1974)
X-ray scattering, thermal diffuse	Jahn (1942) Cartz (1958)
Yield stress	Sylwestrowicz and Gibbons (1958) Gilman (1961)
Zero-point energy	Domb and Salter (1952) Salter (1954) Barron and Klein (1963)
Others	Partington (1952) Cartz (1958) Neshpor (1959) Herbstein (1961)

subject is nevertheless quite well understood at a less fundamental level.

i. Elastic constants relate intimately to various solid-state phenomena, including: diffusion, strength, and phase transitions.

j. Empirical relationships exist also between elastic constants and many "nonelastic" phenomena. Table 1.13 lists many examples.

k. Elastic constants and their temperature derivatives are among the most accurately measured physical properties. Typically, elastic constants can be measured within 0.1% and changes, within 1 part in 10^7.

l. Differences between adiabatic and isothermal elastic constants are zero for the shear modulus at all temperatures, zero at zero temperature for all elastic constants, and typically about 2% for the bulk modulus at ambient temperature.

m. Either elastic constants or ultrasonic wave velocities are useful for detecting subtle internal transitions within a solid, for example: crystallographic transition, magnetic transition, superconducting transition, precipitation, or lattice-defect annealing.

n. Elastic constants can be used to study directional properties of materials, such as textured aggregates and composites, because they are tensors.

1.11 APPENDIX. CALCULATION OF THE BULK MODULUS

In this appendix, a complete calculation is given for the ground-state bulk modulus of an electron gas. The energy of such a system is known to be

$$U = (3\hbar^2/10m)\,(9\pi/4)^{2/3}\,r_0^{-2} \tag{A.1}$$

where $\hbar$ is Planck's constant, m is mass, and the Wigner–Seitz atomic radius, r_0, is defined by

$$(4/3)\pi r_0^3 = V/N = 1/n \tag{A.2}$$

where V is the volume and N the number of free electrons. In atomic units ($\hbar = e = m = 1$), Eq. A.1 becomes

$$U = 2.21/r_0^2 = \alpha/r_0^2 \tag{A.3}$$

where U is now in rydbergs and r_0 is in bohr radii. By defining

$$r = v^{1/3} r_0 \tag{A.4}$$

where

$$v = V/V_0 \tag{A.5}$$

where V denotes volume, then the bulk modulus becomes

$$B_s = -V\left(\frac{\partial P}{\partial V}\right)_S = -V\frac{\partial}{\partial V}\left(-\frac{\partial U}{\partial V}\right)_S = V\left(\frac{\partial^2 U}{\partial V^2}\right)_S = \frac{1}{V_0}\left(\frac{\partial^2 U}{\partial v^2}\right)_{S,\,v=1} \tag{A.6}$$

Differentiation of Eq. A.3 gives

$$\partial U/\partial v = -(2/3)(\alpha/r_0^2)v^{-5/3} \tag{A.7}$$

$$\partial^2 U/\partial v^2 = (10/9)(\alpha/r_0^2)v^{-8/3} \tag{A.8}$$

so that

$$B = \left[(4/3)\pi r_0^3\right]^{-1}(10/9)\alpha/r_0^2 \quad \text{Ry/cm}^3 \tag{A.9}$$

For sodium ($r_0 = 0.208$ nm), this predicts a bulk modulus of 9.19 GPa compared with an observed value of 7.24 GPa.

The good agreement is fortuitous. For other metals, the observed and predicted B values may differ by a factor of about 3. However, this example serves two purposes: First, for all metals, the electron gas is important and usually cannot be ignored in trying to understand their thermal and physical properties. Second, interested readers can substitute any other simple interatomic potential for Eq. A.1 — for example, Born–Mayer, Morse, Lennard-Jones — and carry through a bulk-modulus calculation. One cautionary note: because the Fermi energy varies as r^{-2}, it predicts that B varies as r^{-5}. Experimentally, one observes that B varies as r^{-4}, a dependence predicted by coulombic electron–ion interactions, $-e^2/r$, which are neglected here (Gilman, 1969).

1.12 REFERENCES

Aaronson, H. J. and Shewmon, P. G. (1967). Self-diffusion and elastic constants in the anomalous b.c.c. metals. *Acta Metall.* **15**, 385–387.

Alers, G. A. (1966). The measurement of very small sound velocity changes and their use in the study of solids. In *Physical Acoustics*, Vol. 4-A. Academic, New York, 277–297.

Alers, G. A. and Thompson, D. O. (1961). Dislocation contributions to the modulus and damping in copper at megacycle frequencies. *J. Appl. Phys.* **32**, 283–293.

Alers, G. A. and Waldorf, D. L. (1962). Variation of the elastic moduli at the superconducting transition. *IBM J. Res. Dev.* **6**, 89–93.

Anderson, O. L. (1965). Determination and some uses of isotropic elastic constants of polycrystalline aggregates using single-crystal data. In *Physical Acoustics*, Vol. III-B. Academic, New York, 43–95.

Anderson, O. L. (1966). Derivation of Wachtman's equation for the temperature dependence of elastic moduli of oxide compounds. *Phys. Rev.* **144**, 553–557.

Anderson, O. L. (1968). Some remarks on the volume dependence of the Grüneisen parameter. *J. Geophys. Res.* **73**, 5187–5194.

Anderson, P. W. (1960). *Fizika Dielectrikov*. Acad. Nauk SSSR, Moscow.

Aristotle (4th century B.C.). 993b20.

Armstrong, P. E. (1971). Measurements of elastic constants. In *Measurement of Mechanical Properties*, Vol. V, Part 2. Interscience, New York, 123–156.

Ashkenazi, J., Dacorogna, M., Peter, M., Talmor, Y., and Walker, E. (1978). The effect of band structure on elastic constants and superconductivity in Nb–Zr alloys. In *International Conference on the Physics of Transition Metals*. Institute of Physics, Bristol, England, 695–700.

Ashkenazi, J., Dacorogna, M., Peter, M., Talmor, Y., Walker, E., and Steinemann, S. (1978). Elastic constants in Nb–Zr alloys from zero temperature to the melting point. Experiment and theory. *Helv. Phys. Acta* **50**, 609 (abstract).

Badger, R. M. (1934). A relation between internuclear distances and bond force constants. *J. Chem. Phys.* **2**, 128–131.

Badger, R. M. (1935). The relation between the internuclear distances and force constants of molecules and its application to polyatomic molecules. *J. Chem. Phys.* **3**, 710–714.

Bardeen, J. (1938a). An improved calculation of the energies of metallic Li and Na. *J. Chem. Phys.* **6**, 367–371.

Bardeen, J. (1938b). Compressibilities of the alkali metals. *J. Chem. Phys.* **6**, 372–378.

Barker, R. E. (1963). An approximate relation between elastic moduli and thermal expansivities. *J. Appl. Phys.* **34**, 107–116.

Barrett, C. S. (1947). A low temperature transformation in lithium. *Phys. Rev.* **72**, 245.

Barron, T. H. K. (1955). On the thermal expansion of solids at low temperatures. *Philos. Mag.* **46**, 720–734.

Barron, T. H. K. and Klein, M. L. (1962). Thermal and elastic properties of crystals at low temperatures. *Phys. Rev.* **127**, 1997–1998.

Barron, T. H. K. and Klein, M. L. (1963). On the strain dependence of the vibrational frequency distributions of simple lattices. *Proc. Phys. Soc. Lond.* **82**, 161–173.

Barron, T. H. K. and Munn, R. W. (1968). Heat capacities of noncubic solids. *J. Phys. C* **1**, 1–10.

Barua, B. P. and Sinha, S. K. (1978). Pseudopotential of gold fitted to elastic data. *J. Appl. Phys.* **49**, 3967–3969.

Baughan, E. C. (1952). Vibration frequencies in diatomic molecules and in solid metals. *Trans. Faraday Soc.* **48**, 121–128.

Behrens. E. (1971). Elastic constants of fiber-reinforced composites with transversely isotropic constituents. *J. Appl. Mech. Trans. ASME* **93**, 1062–1065.

Benckert, S. (1971). A calculation of the second- and third-order elastic constants of aluminum. *Phys. Status Solidi b* **43**, 681–687.

Beran, M. J. (1971). Application of statistical theories to heterogeneous materials. *Phys. Status Solidi a* **6**, 365–384.

Bernstein, B. T. (1963). Electron contribution to the temperature dependence of the elastic constants of cubic metals. I. Normal metals. *Phys. Rev.* **132**, 50–57.

Bernstein, B. T. (1965). Electron contribution to the temperature dependence of the elastic constants of cubic metals. II. Superconducting metals. *Phys. Rev.* **137**, A1404–A1409.

Bert, C. W. and Clary, R. R. (1973). Evaluation of experimental methods for determining dynamic stiffness and damping of composite materials. In *Composite Materials: Testing and Design (Third Conference)*. American Society for Testing and Materials, Philadelphia, Pennsylvania, 250–265.

Bertoni, C. M., Bortolani, V., and Ciunchi, G. (1972). Thermodynamic properties of alkali metals. *J. Phys. F* **2**, 833–841.

Beyer, R. T. and Letcher, S. V. (1969). *Physical Ultrasonics*. Academic, New York.

Bhatia, A. B. (1967). *Ultrasonic Absorption*. Oxford University Press, London.

Bhatt, R. N. and McMillan, W. L. (1976). Landau theory of the martensitic transition in A-15 compounds. *Phys. Rev. B* **14**, 1007–1027.

Blackman, M. (1951). On θ values in the resistance of metals. *Proc. Phys. Soc. Lond.* **64**, 681–683.

Blackman, M. (1955). The specific heat of solids. In *Handbuch der Physik*, Vol. VII, Part 1. Springer-Verlag, Berlin, 352–382.

Boas, W. and Mackenzie, J. K. (1950) Anisotropy in metals. In *Progress in Metal Physics*, Vol. 2. Pergamon, London, 90–120.

Born, M. (1923). *Atomtheorie des festen Zustandes*. Teubner, Berlin.

Born, M. (1939). Thermodynamics of crystals and melting. *J. Chem. Phys.* **7**, 591–603.

Born, M. (1940). On the stability of crystal lattices. I. *Proc. Camb. Philos. Soc.* **36**, 160–172.

Born, M. (1943). The thermodynamics of crystal lattices, I. Discussion of the methods of calculation. *Proc. Camb. Philos. Soc.* **39**, 100–103.

Born, M. and Huang, K. (1954). *Dynamical Theory of Crystal Lattices*. Oxford University Press, London.

Born, M. and von Karman, Th. (1912). Vibrations in space lattices. *Phys. Z.* **13**, 297–309.

Bozorth, R. M. (1951). *Ferromagnetism*. Van Nostrand, New York.

Bradburn, M. (1943). The thermodynamics of crystal lattices, III. The equation of state for a face-centered cubic lattice. *Proc. Camb. Philos. Soc.* **39**, 113–127.

Bradfield, G. (1964). *Use in Industry of Elasticity Measurements in Metals with the Help of Mechanical Vibrations*, NPL Notes on Applied Science No. 30. HMSO, London.

Bridgman, P. W. (1949). *The Physics of High Pressure*. Bell, London.

Brillouin, L. (1938). On the thermal dependence of elasticity in solids. *Phys. Rev.* **54**, 916–917.

Brillouin, L. (1946). *Wave Propagation in Periodic Structures*. McGraw-Hill, New York.

Brillouin, L. (1964). *Tensors in Mechanics and Elasticity*. Academic, New York.

Buch, A. (1966). Relation between the strength characteristics of pure metals and their atomic-bond parameters. *J. Inst. Met.* **94**, 424–427.

Calculation of the Properties of Vacancies and Interstitials (1967) (ed. A. D. Franklin). National Bureau of Standards Miscellaneous Publication 287, U.S. Government Printing Office, Washington, D.C.

Callen, H. B. (1960). *Thermodynamics*. Wiley & Sons, New York.

Cartz, L. (1958). Thermal vibrations of atoms in cubic crystals. *Proc. Phys. Soc. Lond. B* **68**, 951–967.

Chickazumi, S. (1964). *Physics of Magnetism*. Wiley & Sons, New York.

Chin, G. Y. (1975). Strong and hard solids. In *Transactions of the American Crystallographic Association*, Vol. 11. American Crystallographic Association, New York, 1–13.

Cho, S-A. and Puerta, M. (1976). Debye temperature and allotropic transition of metals. *Z. Metallkd.* **67**, 546–548.

Choquard, P. F. (1967). *The Anharmonic Crystal*. Benjamin, New York.

Chou, T. W. and Kelly, A. (1976). Fibre composites. *Mater. Sci. Eng.* **25**, 35–40.

Cochran, W. (1959). Crystal stability and the theory of ferroelectricity. *Phys. Rev. Lett.* **3**, 412–414.

Cohen, M. L. and Heine, V. (1970). The fitting of pseudopotentials to experimental data and their subsequent application. In *Solid State Physics*, Vol. 24. Academic, New York, 38–248.

Cousins, C. S. G. (1969). The use of elastic shear constants to test the validity of model potential theories. *J. Phys. C* **2**, 765–767.

Crutchley, D. E. and Reid, C. N. (1969). Mechanical properties of chromium single crystals. In *High Temperature Materials, Sixth Plansee Seminar*. Springer-Verlag, Vienna, 57–66.

Date, E. H. F. and Andrews, K. W. (1969). Anisotropic and composition effects in the elastic properties of polycrystalline metals. *Br. J. Appl. Phys.* **2**, 1373–1381.

Davies, G. F. (1973). Quasi-harmonic finite strain equations of state of solids. *J. Phys. Chem. Solids* **34**, 1417–1429.

Debye, P. (1912). Zur Theorie der specifischen Wärmen. *Ann. Phys. (Leipz.)* **39**, 789–839.

DeLaunay, J. (1956). The theory of specific heats and lattice vibrations. In *Solid State Physics*, Vol. 2. Academic, New York, 219–303.

DeLaunay, J. and Dolecek, R. L. (1947). Superconductivity and the Debye characteristic temperature. *Phys. Rev.* **72**, 141–143.

Dobbs, E. R. and Jones, G. O. (1957). Theory and properties of solid argon. *Rep. Prog. Phys.* **20**, 516–563.

Domb, C. and Salter, L. (1952). The zero point energy and θ values of crystals. *Philos. Mag.* **43**, 1083–1089.

Döring, W. (1938). On the temperature dependence of the elastic modulus of ferromagnetic substances. *Ann. Phys. (Leipz.)* **32**, 465–470.

Ducastelle, F. (1970). Elastic moduli of transition metals. *J. Phys. (Paris)* **31**, 1055.

Dugdale, J. S. and MacDonald, D. K. (1953). The thermal expansion of solids. *Phys. Rev.* **89**, 832–834.

Durand, M. A. (1936). The temperature variation of the elastic moduli of NaCl, KCl and MgO. *Phys. Rev.* **50**, 449–455.

Eastabrook, J. N. (1957). Specific heat of solids at high temperatures. *Philos. Mag.* **2**, 1415–1420.

Einstein, A. (1906). Planck's theory of radiation and the theory of specific heat. *Ann. Phys. (Leipz.)* **22**, 180–190.

Einstein, A. (1911). A relationship between elastic behavior and specific heat in the case of solids with monatomic molecules. *Ann. Phys. (Leipz.)* **34**, 170–174, 590.

Engler, O. (1938). The elastic modulus of ferromagnetic substances and its dependence on temperature and magnetic field. *Ann. Phys. (Leipz.)* **31**, 145–163.

Eshelby, J. D. (1968). Stress analysis: theory of elasticity, stress analysis: fracture mechanics. In *Fracture Toughness*, ISI Publication 121. Iron and Steel Institute, London, 30–48.

Evetts, J. E. and Martin, P. J. (1973). Superconducting materials research. *Met. Sci. J.* **7**, 179–184.

Ewald, P. P., Pöschl, Th., and Prandtl, L. (1930). *The Physics of Solids and Fluids*. Blackie, London, 81.

Feldman, J. L. (1964). The relation between θ (elastic) and θ (calorimetric) near T=0° K. *Proc. Phys. Soc. Lond.* **84**, 361–369.

Ferry, J. D. (1961). *Viscoelastic Properties of Polymers*. Wiley & Sons, New York.

Feynman, R. P., Leighton, R. B., and Sands, M. (1963). *The Feynman Lectures on Physics*, Vol. 2. Addison-Wesley, Reading, Massachusetts.

Fisher, E. S. and Alfred, C. R. (1968). Effects of elastic anisotropy on dislocations in hcp metals. *Trans. Metall. Soc. AIME* **242**, 1575–1586.

Fisher, E. S. and Dever, D. (1970). Relation of the C′ elastic modulus to stability of b.c.c. transition metals. *Acta Metall.* **18**, 265–269.

Fisher, E. S. and Renken, C. J. (1964). Single-crystal elastic moduli and the hcp-bcc transformation in Ti, Zr, and Hf. *Phys. Rev.* **135**, A482–A494.

Fleischer, R. L. (1963). Substitutional solution hardening. *Acta Metall.* **11**, 203–209.

Fleury, P. A. (1971). Soft mode phase transitions I. Basic concepts. *Comments Solid State Phys.* **4**, 149–177.

Flinn, P. A. and Maradudin, A. A. (1963). Anharmonic contributions to vibrational thermodynamic properties of solids. III. The low temperature limit. *Ann. Phys. (N.Y.)* **22**, 223–238.

Foreman, A. J. E. (1962). Anharmonic specific heat of solids. *Proc. Phys. Soc. Lond.* **79**, 1124–1141.

Friedel, J. (1954). Electronic structure of primary solid solutions in metals. *Adv. Phys.* **3**, 446–507.

Friedel, J. (1955). Deviations from Vegard's law. *Philos. Mag.* **46**, 514–516.

Fuchs, K. (1936). The elastic constants and specific heats of the alkali metals. *Proc. R. Soc. Lond.* **157**, 444–450.

Fuller, E. R., Jr., Granato, A. V., Holder, J., and Naimon, E. R. (1974). Ultrasonic properties of the properties of solids. In *Methods of Experimental Physics: Solid State Physics*, Volume II. Academic, New York, 371–441.

Fürth, R. (1941). The stability of crystal lattices. V. Experimental evidence on recent theories of the equation of state and the melting of solids. *Proc. Camb. Philos. Soc.* **37**, 34–54.

Fürth, R. (1944). On the equation of state for solids. *Proc. R. Soc. Lond. A* **183**, 87–110.

Garber, J. A. (1972). The temperature dependence of second-order elastic constants in cubic crystals. Ph.D. thesis, University of Illinois, Urbana–Champaign, Illinois.

Garber, J. A. and Granato, A. V. (1975a). Theory of the temperature dependence of second-order elastic constants in cubic materials. *Phys. Rev. B* **11**, 3990–3997.

Garber, J. A. and Granato, A. V. (1975b). Fourth-order elastic constants and the temperature dependence of second-order elastic constants in cubic materials. *Phys. Rev. B* **11**, 3998–4007.

Garber, R. I. and Gindin, I. A. (1961). Physics of the strength of crystalline materials. *Sov. Phys.-Usp.* **3**, 41–77.

Gerlich, D. (1975). Elastic Grüneisen parameters for potassium metal. *Phys. Rev. B* **11**, 1365–1366.

Gilman, J. J. (1960). The plastic resistance of crystals. *Austr. J. Phys.* **13**, 327–346.

Gilman, J. J. (1961). Dislocation motions and the yield strengths of solids. In *Properties of Crystalline Solids*. American Society for Testing and Materials, Philadelphia, Pennsylvania, 69–81.

Gilman, J. J. (1969). *Micromechanics of Flow in Solids*. McGraw-Hill, New York.

Gilvarry, J. J. (1955). Grüneisen's constants for some metals. *J. Chem. Phys.* **23**, 1925–1927.

Gilvarry, J. J. (1956). Grüneisen parameter for a solid under finite strain. *Phys. Rev.* **102**, 331–340.

Goasdue, C., Ho, P. S., and Sass, S. L. (1972). Ultrasonic study of the beta–omega phase transformation in Zr–Nb alloys. *Acta Metall.* **20**, 725–733.

Gol'denblat, I. I. (1962). *Some Problems of the Mechanics of Deformable Media*. Noordhoff, Groningen, The Netherlands.

Gorecki, T. (1979). The relations between the shear modulus, the bulk modulus, and Young's modulus for polycrystalline metallic elements. *Mater. Sci. Eng.* **43**, 225–230.

Gow, M. M. (1944). The thermodynamics of crystal lattices. IV. The elastic constants of a face-centered cubic lattice with central forces. *Proc. Camb. Philos. Soc.* **40**, 151–166.

Granato, A., Hikata, A., and Lücke, K. (1958). Recovery of damping and modulus changes following plastic deformation. *Acta Metall.* **6**, 470–480.

Green, R. E., Jr. (1973). *Ultrasonic Investigation of Mechanical Properties*. Academic, New York.

Grimvall, G. and Sjödin, S. (1974). Correlation of properties of materials to Debye temperatures. *Phys. Scr.* **10**, 340–352.

Grüneisen, E. (1910). Influence of temperature on the compressibility of metals. *Ann. Phys.* (*Leipz.*) **10**, 1239–1274.

Grüneisen, E. (1926). The state of a solid body. In *Handbuch der Physik*, Vol. 10. Springer, Berlin, 1–52.

Gschneidner, K. A., Jr. (1964). Physical properties and interrelationships of metallic and semimetallic elements. In *Solid State Physics*, Vol. 16. Academic, New York, 275–426.

Hafner, J. (1973). On the influence of electronic exchange and correlation on pseudopotential calculations. I. Form factors, energy-wave number characteristics, and effective interionic pair potentials. *Phys. Status Solidi b* **56**, 579–590.

Hafner, J. (1976). Unified analysis of static, dynamic and electronic properties of the alkali metals. *Z. Phys. B* **24**, 41–52.

Hale, D. K. (1976). The physical properties of composite materials. *J. Mater. Sci.* **11**, 2105–2141.

Hardy, J. P. and Flocken, J. W. (1970). Theoretical calculations of the properties of point defects in solids. In *Critical Reviews in Solid State Sciences*. Chemical Rubber Company, Cleveland, Ohio, 605–643.

Harris, P. and Avrami, L. (1972). Some physics of the Grüneisen parameter. Picatinny Arsenal Technical Report 4423, Picatinny Arsenal, Baltimore, Maryland.

Harrison, W. A. (1966). *Pseudopotentials in the Theory of Metals*. Benjamin, New York.

Hashin, Z. and Rosen, W. B. (1964). The elastic moduli of fiber-reinforced materials. *J. Appl. Mech., Trans. ASME* **84**, 223–232.

Hashin, Z. and Shtrikman, S. (1963). A variational approach to the theory of the elastic behavior of multiphase materials. *J. Mech. Phys. Solids* **11**, 127–140.

Hearmon, R. F. S. (1946). The elastic constants of anisotropic materials. *Rev. Mod. Phys.* **18**, 409–440.

Hearmon, R. F. S. (1956). The elastic constants of anisotropic materials-II. *Adv. Phys.* **5**, 323–382.

Hearmon, R. F. S. (1961). *Applied Anisotropic Elasticity*. Oxford University Press, London.

Hearmon, R. F. S. (1966). The elastic constants of non-piezoelectric crystals. In *Landolt-Börnstein Numerical Data and Functional Relationships in Science and Technology*, Group III, Volume I. Springer-Verlag, Berlin, 1–39.

Hecker, S. S., Rohr, D. L., and Stein, D. F. (1978). Brittle fracture in iridium. *Metall. Trans.* **9A**, 481–488.

Heine, V. (1970). The pseudopotential concept. In *Solid State Physics*, Vol. 24. Academic, New York, 1–36.

Herbstein, F. H. (1961). Methods of measuring Debye temperatures and comparison of results for some cubic crystals. *Adv. Phys.* **10**, 313–355.

Hiki, Y. and Granato, A. V. (1966). Anharmonicity in noble metals; higher order elastic constants. *Phys. Rev.* **144**, 411–419.

Hiki, Y., Thomas, J. F., and Granato, A. V. (1967). Anharmonicity in noble metals: some thermal properties. *Phys. Rev.* **153**, 764–771.

Hill, R. (1963). Elastic properties of reinforced solids: some theoretical principles. *J. Mech. Phys. Solids* **11**, 357–372.

Hillier, H. W. (1961). The measurement of dynamic elastic properties. In *Progress in Solid Mechanics*, Vol. II. North-Holland, New York, 199–243.

Hirth, J. P. (1976). Anisotropic elastic theory of line defects. *Met. Sci. J.* **10**, 222–224.

Holm, M. W. (1957). Debye characteristic temperatures table and bibliography. AEC Research and Development Report, IDO–16399.*

Holt, A. C. and Ross, M. (1970). Calculation of the Grüneisen parameter for some models of the solid. *Phys. Rev. B* **1**, 2700–2705.

Honda, K. and Terada, T. (1907). On the change of elastic constants of ferromagnetic substances by magnetization. *Philos. Mag.* **13**, 36–83.

Hooton, D. J. (1955). A new treatment of anharmonicity in lattice dynamics: I, II. *Philos. Mag.* **46**, 422–432, 433–442.

Hopkin, L. M. T., Pursey, H., and Markham, M. F. (1970). *Z. Metallkd.* **61**, 535–540.

Huntington, H. B. (1958). The elastic constants of crystals. In *Solid State Physics*, Vol. 7. Academic, New York, 215–351.

Interatomic Potentials and Simulations of Lattice Defects (1972) (eds. P. C. Gehlen, J. R. Beeler, and R. I. Jaffe). Plenum Press, New York.

Jahn, H. A. (1942). Diffuse scattering of x-ray by crystals. The Faxén-Waller theory and the surfaces of isodiffusion for cubic crystals. *Proc. R. Soc. Lond.* **179**, 320–340.

James, R. W. (1948). *The Optical Principles of the Diffraction of X-ray*. Bell, London.

Johnson, R. A. (1973). Empirical potentials and their use in the calculation of energies of point defects in crystals. *J. Phys. F.* **3**, 295–321.

Joshi, S. K. and Rajagopal, A. K. (1968). Lattice dynamics of metals. In *Solid State Physics*, Vol. 22. Academic, New York, 159–312.

Kelly, A. (1966). *Strong Solids*. Oxford University Press, London.

Kelly, A., Tyson, W. R., and Cottrell, A. H. (1967). Ductile and brittle crystals. *Philos. Mag.* **15**, 567–586.

Kelly, F. M. and MacDonald, K. C. (1953). On θ values in metals. *Can. J. Phys.* **31**, 147–164.

Killpatrick, D. H. and Young, J. D. (1970). Texture and room temperature mechanical properties of dispersion strengthened Ni–Cr alloys. *Metall. Trans.* **1**, 955–961.

Kittel, C. (1966). *Introduction to Solid State Physics*. Wiley & Sons, New York.

Klemens, P. G. (1958). Thermal conductivity and lattice vibration modes. In *Solid State Physics*, Vol. 7. Academic, New York, 1–98.

Knopoff, L. and Shapiro, J. N. (1969). Comments on the interrelationships between Grüneisen's parameter and shock and isothermal equations of state. *J. Geophys. Res.* **74**, 1439–1450.

Köster, W. (1948). Temperature dependence of the elasticity modulus of pure metals. *Z. Metallkd.* **39**, 1–9.

Köster, W. and Franz, H. (1961). Poisson's ratio for metals and alloys. *Metall. Rev.* **6**, 1–55.

Kragler, R. (1978). Behavior of the elastic moduli of Nb_3Sn in the low temperature phase. *Physica B* **93**, 314–326.

Krautkrämer, J. and Krautkrämer, H. (1969). *Ultrasonic Testing of Materials*. Springer-Verlag, New York.

Kruschov, M. M. (1974). Principles of abrasive wear. *Wear* **28**, 69–88.

Landau, L. D. and Lifshitz, E. M. (1959a). *Theory of Elasticity*. Pergamon, London, 16.

Landau, L. D. and Lifshitz, E. M. (1959b). *Statistical Physics*. Pergamon, London.

LeClaire, A. D. (1953). The theory of D_0 in the Arrhenius equation for self-diffusion in cubic metals. *Acta Metall.* **1**, 438–447.

Ledbetter, H. M. (1973). Poisson's ratio for polycrystals. *J. Phys. Chem. Solids* **34**, 721–723.

Ledbetter, H. M. (1976). Poisson's ratio for central-force polycrystals. *Z. Naturforsch. A* **31**, 1539–1542.

Ledbetter, H. M. (1977a). Ratio of the shear and Young's moduli for polycrystalline metallic elements. *Mater. Sci. Eng.* **27**, 133–135.

Ledbetter, H. M. (1977b). Elastic properties of zinc: a compilation and a review. *J. Phys. Chem. Ref. Data* **6**, 1181–1203.

Ledbetter, H. M. (1978). National Bureau of Standards, Boulder, Colorado. Unpublished study.

Ledbetter, H. M. (1979). Dynamic elastic modulus and internal friction in fibrous composites. In *Nonmetallic Materials and Composites at Low Temperatures*. Plenum, New York, 267–281.

Ledbetter, H. M. (1982). Temperature behavior of Young's moduli of forty engineering alloys. *Cryogenics* **22**, 653–656.

Ledbetter, H. M., Frederick, N. F., and Austin, M. W. (1980). Elastic-constant variability in stainless steel 304. *J. Appl. Phys.* **51**, 305–309.

Ledbetter, H. M. and Reed, R. P. (1974). Elastic properties of metals and alloys, I. Iron, nickel, and iron–nickel alloys. *J. Phys. Chem. Ref. Data* **2**, 531–618.

Lee, E. W. (1955). Magnetostriction and magnetomechanical effects. *Rep. Prog. Phys.* **18**, 185–229.

Leibfried, G. (1955). Lattice theory of the mechanical and thermal properties of crystals. In *Handbuch der Physik*, Vol. 7, Part 1. Springer-Verlag, Berlin, 104–324.

Leibfried, G. (1965). The vibrational frequency spectrum of some body-centered cubic metals. In *Diffusion in Body-Centered Cubic Metals*. American Society for Metals, Metals Park, Ohio, 103–127.

Leibfried, G. and Ludwig, W. (1960). Gleichgewichtsbedingungen in der Gittertheorie. *Z. Phys.* **160**, 80–92.

Leibfried, G. and Ludwig, W. (1961). Theory of anharmonic effects in crystals. In *Solid State Physics*, Vol. 12. Academic, New York, 275–444.

Lindemann, F. A. (1910). *Phys. Z.* **11**, 609.

Lindley, P. B. (1968). Effect of Poisson's ratio on compression modulus. *J. Strain Anal.* **3**, 142–145.

Lonsdale, K. (1948). Vibration amplitudes of atoms in cubic crystals. *Acta Crystallogr.* **1**, 142–149.

Love, A. E. H. (1927). *A Treatise on the Mathematical Theory of Elasticity*. Cambridge University Press, Cambridge.

Ludwig, W. (1958). On the influence of anharmonicity on the properties of crystals. *J. Phys. Chem. Solids* **4**, 283–305.

Ludwig, W. (1967). Recent developments in lattice theory. In *Springer Tracts in Modern Physics*, Vol. 43. Springer-Verlag, Berlin.

Lumsden, J. (1952). *Thermodynamics of Alloys*. Institute of Metals, London, 88.

Madelung, E. (1910). Molecular characteristic vibrations. *Phys. Z.* **11**, 898–905.

Maradudin, A. A. and Flinn, P. A. (1961). Anharmonic contributions to vibrational thermodynamic properties of solids. Part I. General formulation and application to the linear chain. *Ann. Phys. (N. Y.)* **15**, 337–359.

Maradudin, A. A., Flinn, P. A., and Coldwell-Horsfall, R. A. (1961). Anharmonic contributions to vibrational thermodynamic properties of solids. Part II. The high temperature limit. *Ann. Phys.*

(*N. Y.*) **15**, 360–386.
Maradudin, A. A., Montroll, E. W., Weiss, G. H., and Ipatova, I. P. (1971). Theory of lattice dynamics in the harmonic approximation. In *Solid State Physics*, Supplement 3. Pergamon, New York.
Mason, W. P. (1951). A phenomenological derivation of the first- and second-order magnetostriction and morphic effects for a nickel crystal. *Phys. Rev.* **82**, 715–723.
McCrum, N. G., Read, B. E., and Williams, G. (1967). *Anelastic and Dielectric Effects in Polymeric Solids*. Wiley & Sons, New York.
McSkimin, H. J. (1961). Notes and references for the measurement of elastic moduli by means of ultrasonic waves. *J. Acoust. Soc. Am.* **33**, 606–615.
Meaden, G. T. (1965). *Electrical Resistance of Metals*. Plenum Press, New York.
Melcher, R. L., Guntherodt, G., Penney, T., and Holtzberg, F. (1975). Soft bulk modulus at the black gold phase transition in $Sm_{1-x}\ Y_xS$. In *1975 Ultrasonic Symposium Proceedings*. Institute of Electrical and Electronic Engineers, New York, 616–617.
Meyer, A., Umar, I. H., and Young, W. H. (1971). Lattice spacings and compressibilities vs. Pauling radii and valencies. *Phys. Rev. B* **4**, 3287–3291.
Michel, D. J. (1977). *Irradiation Effects on Fatigue Crack Propagation in Austenitic Stainless Steels*. NRL Memo Report 3610, Naval Research Laboratory, Washington, D.C.
Milstein, F. (1971). Theoretical strength of a perfect crystal. *Phys. Rev. B* **3**, 1130–1141.
Milstein, F. (1973). Theoretical strength of a perfect crystal with exponentially attractive and repulsive interatomic interactions. *J. Appl. Phys.* **44**, 3833–3840.
Mitra, S. S. and Joshi, S. K. (1961). Debye θ and compressibility. III. Cubic metals. *Physica* **27**, 376–380.
Moelwyn-Hughes, E. A. (1964). *Physical Chemistry*, Chapter 8, Pergamon, New York.
Morse, P. M. (1929). Diatomic molecules according to the wave mechanics. II. Vibrational levels. *Phys. Rev.* **34**, 57–64.
Mott, N. F., and Jones, H. (1936). *The Theory of the Properties of Metals and Alloys*. Oxford University Press, London, 247.
Mukherjee, K. (1965). Monovacancy formation energy and Debye temperature of close-packed metals. *Philos. Mag.* **12**, 915–918.
Mulimani, B. G. (1972). Cohesive energy and Debye temperature of close-packed metals. *Phys. Status Solidi b* **52**, K49–K50.
Munn, R. W. (1972). Role of the elastic constants in negative thermal expansion of axial solids. *J. Phys. C.* **5**, 535–542.
Naimon, E. R. and Ledbetter, H. M. (1973). Unpublished study. National Bureau of Standards, Boulder, Colorado.
Nakanishi, N. (1980). Elastic constants as they relate to lattice properties and martensite transformation. In *Progress in Materials Science*, Vol. 24. Pergamon, London, 143–265.
Nakanishi, N., Takano, M., Hashimoto, M., Morimoto, H., and Miura, S. (1978). Temperature dependence of the elastic constants above the martensitic transition temperature in Cu–15.0 at.% Sn alloy. *Scr. Metall.* **12**, 271–275.
Nandanpawar, M. L. and Rajagopalan, S. (1978). Wachtman's equation and temperature dependence of bulk moduli in solids. *J. Appl. Phys.* **49**, 3976–3979.
Nemilov, S. V. (1968). Kinetics of elementary processes in the condensed state. II. Shear relaxation and the equation of state for solids. *Russ. J. Phys. Chem.* **42**, 726–729.
Neshpor, V. S. (1959). On relations between certain thermal characteristics of solids. *Fiz. Met. Metalloved.* **7**, 559–564.
Nowick, A. S. and Berry, B. S. (1972). *Anelastic Relaxation in Crystalline Solids*. Academic, New York, 12.
Nutkins, M. A. E. (1958). Theory of phase transformations in alkali metals. *Proc. R. Soc. Lond.* **72**, 810–822.
Nye, J. F. (1960). *Physical Properties of Crystals*. Oxford University Press, London.
Overton, W. C. (1963). Relation between ultrasonically measured properties and the coefficients in the solid equation of state. *J. Chem. Phys.* **37**, 116–119.
Overton, W. C. (1968). An analysis of cubic contributions to the anharmonic frequency shift and free energy for Bravais Lattices. *J. Phys. Chem. Solids* **29**, 711–718.
Overton, W. C. (1971). Experimental evidence for a difference between thermal and elastic Debye thetas of a crystal near 0° K. *Phys. Lett. A* **37**, 287–288.
Overton, W. C. and Gaffney, J. (1955). Temperature variation of the elastic constants of cubic elements. I. Copper. *Phys. Rev.* **98**, 969–977.
Pace, N. G. and Saunders, G. A. (1971). Lattice properties in the vicinity of the martensitic transformation in TiNi. *Solid State Commun.* **9**, 331–334.
Papadakis, E. P. (1973). The measurement of small changes in ultrasonic velocity and attenuation. In *Critical Reviews in Solid State Sciences*. Chemical Rubber Company, Cleveland, Ohio, 373–418.
Partington, J. R. (1952). *An Advanced Treatise on Physical Chemistry, The Properties of Solids*, Vol. 3. Longmans and Green, London.
Pasternak, M., Steinemann, S., and Peter, M. (1968). Corrélations entre modules élastiques et structure de bande dans les métaux de transition. *Helv. Phys. Acta* **41**, 1296–1299.
Pastine, D. (1965). Formulation of the Grüneisen parameter for monatomic cubic crystals. *Phys. Rev. A* **138**, 767–770.
Pastine, D. J. (1968). A simplified theory of anharmonic contributions to the thermodynamic properties of metals. In *Behaviour of Dense Media Under High Dynamic Pressures*. Gordon and Breach, New York, 125–130.
Peierls, R. (1940). The size of a dislocation. *Proc. R. Soc. Lond.* **52**, 34–37.
Peierls, R. E. (1955). *Quantum Theory of Solids*. Oxford University Press, London.
Pines, D. (1963). *Elementary Excitations in Solids*. Benjamin, New York, 34.
Pippard, A. B. (1955). Thermodynamics of a sheared superconductor. *Philos. Mag.* **46**, 1115–1118.
Pippard, A. B. (1961). *Elements of Classical Thermodynamics*. Cambridge University Press, London.
Planck, M. (1906). Vorlesungen über die Theorie der

Wärmestrahlung. *J. Ambr. Barth.*, Nos. 149, 150, 154, 160, 166.

Plendl, J. N. and Gielisse, P.J.M. (1965). Characteristic frequencies from infrared and elastic data. *Appl. Opt.* **4**, 853–856.

Plendl, J. N., Mitra, S. S., and Gielisse, P.J.M. (1965). Compressibility, cohesive energy, and hardness of non-metallic solids. *Phys. Status Solidi* **12**, 367–374.

Pollard, H. F. (1977). *Sound Waves in Solids*. Pion, London.

Pugh, S. F. (1954). Relations between the elastic moduli and the plastic properties of polycrystalline pure metals. *Philos. Mag.* **45**, 823–843.

Radovskii, I. S. (1966). Calculation of entropy from acoustic data. *J. Appl. Mech. Tech. Phys.* (*USSR*) **7**, 175–176.

Ralls, K. M., Courtney, T. H., and Wulff, J. (1976). *Materials Science and Engineering*, Chapter 19. Wiley & Sons, New York.

Read, T. A., Wert, C. A., and Metzger, M. (1959). Elastic, anelastic, and plastic properties. In *Methods of Experimental Physics*, Vol. 6A. Academic, New York.

Robertson, W. D. (1965). Martensite transformation in terms of elastic moduli. In *Physical Properties of Martensite and Bainite*, Special Report 93. Iron and Steel Institute, London, 26–29.

Rosen, B. W. (1970). Thermomechanical properties of fibrous composites. *Proc. R. Soc. Lond. A* **319**, 79–94.

Rossi, R. C. (1968). Prediction of the elastic moduli of composites. *J. Am. Ceram. Soc.* **51**, 433–439.

Rudman, P. S. (1965). Atomic volume in Laves phases: a hemisubstitutional solid-solution elastic model. *Trans. Metall. Soc. AIME* **233**, 872–878.

Ruffa, A. R. (1977). Temperature dependence of the elastic shear moduli of the cubic metals. *Phys. Rev. B* **16**, 2504–2514.

Sachse, W., Ting, C. S., and Hemenway, A. (1979). Dispersion of elastic waves and the nondestructive testing of composite materials. In *Composite Materials Testing and Design* (*Fifth Conference*). American Society for Testing and Materials, Philadelphia, 165–183.

Salama, K. and Ippolito, R. M. (1980). The use of temperature dependence of ultrasonic velocity to evaluate internal stress. In *Proceedings of the First International Symposium on Ultrasonic Materials Characterization*. NBS Special Publication 596, National Bureau of Standards, Washington, D.C., 201–211.

Salter, L. (1954). The ideal crystal at absolute zero. *Philos. Mag.* **45**, 360–368.

Salter, L. (1955). On the thermodynamics of crystalline lattices. *Proc. R. Soc. Lond. A* **233**, 418–427.

Sauer, J. A. and Saba, R. G. (1970). Relaxation behavior of polymers at low temperatures. In *Polymers in Space Research*. Marcel Dekker, New York, 181–219.

Saxena, V. K., Kapoor, Q. S., and Bhattacharya, D. L. (1969). Binding energy and compressibility of body-centered cubic and close-packed hexagonal sodium. *Phys. Status Solidi* **34**, 145–149.

Schreiber, E., Anderson, O. L., and Soga, N. (1973). *Elastic Constants and Their Measurement*. McGraw-Hill, New York.

Schrieffer, J. R. (1973). The electron–phonon mechanism. Unpublished talk at Symposium on Superconductivity and Lattice Instabilities held at Gatlinburg, Tennessee.

Seitz, F. (1935). The theoretical constitution of metallic lithium. *Phys. Rev.* **47**, 400–412.

Seitz, F. and Read, T. A. (1941). Theory of the plastic properties of solids. *J. Appl. Phys.* **12**, 100–118.

Sheard, F. W. (1958). Calculation of the thermal expansion of solids from the third-order elastic constants. *Philos. Mag.* **3**, 1381–1390.

Sherby, O. D. (1962a). Factors affecting the high temperature strength of polycrystalline solids. *Acta Metall.* **10**, 135–147.

Sherby, O. D. (1962b). Limitations in forming and shaping. *ASM Trans. Q.* **55**, 3–13.

Siegel, S. and Cummerov, R. (1940). On the elasticity of crystals. *J. Chem. Phys.* **8**, 847.

Simmons, G. and Wang, H. (1971). *Single Crystal Elastic Constants and Calculated Aggregate Properties: A Handbook*. M.I.T. Press, Cambridge, Massachusetts.

Sines, G. (1969). *Elasticity and Strength*. Allyn and Bacon, Boston.

Singh, N. and Sharma, P. K. (1971). Debye–Waller factors of cubic metals. *Phys. Rev. B* **3**, 1141–1148.

Slater, J. C. (1939). *Introduction to Chemical Physics*. McGraw-Hill, New York.

Slater, J. C. (1940). Note of Grüneisen's constant for the incompressible metals. *Phys. Rev.* **57**, 744–746.

Staverman, A. J. (1962). Thermodynamics of polymers. In *Handbuch der Physik*, Vol. 13, Springer-Verlag, Berlin, 399–509.

Stein, D. F. and Shannette, G. W. (1976). *Relation between Electronic Structure and the Mechanical Properties of Body-Centered Cubic Metals*. ARO-10453.2-NC, Michigan Technological University, Houghton, Michigan.

Stephens, R. W. B. (1968). Sound measurements in solids and their uses. *Ultrasonics* **6**, 92–96.

Stern, E. A. (1958). Theory of the anharmonic properties of solids. *Phys. Rev.* **111**, 786–797.

Stokes (1845). Reported in Todhunter and Pearson (1886).

Sutherland, W. (1891a). V. A kinetic theory of solids with an experimental introduction. *Philos. Mag.* **32**, 31–43.

Sutherland, W. (1891b). XIX. A kinetic theory of solids with an experimental introduction. *Philos. Mag.* **32**, 215–225.

Sutherland, W. (189lc). LXIII. A kinetic theory of solids, with an experimental introduction. *Philos. Mag.* **32**, 524–553.

Sutherland, W. (1910). The mechanical vibrations of atoms. *Philos. Mag.* **20**, 657–660.

Sutton, P. M. (1953). The variation of the elastic constants of crystalline aluminum with temperature between 631K and 7731K. *Phys. Rev.* **91**, 816–821.

Suzuki, T. (1971). Second- and third-order elastic constants of aluminum and lead. *Phys. Rev. B* **3**, 4007–4014.

Swenson, C. A. (1968). Equation of state of cubic solids; some generalizations. *J. Phys. Chem. Solids* **29**, 1337–1348.

Sylwestrowicz, W. D. and Gibbons, D. F. (1958). On

the temperature dependence of yield stress for aluminum. *Philos. Mag.* **3**, 1326–1327.
Takeuchi, S. (1968). On the shear modulus parameter in the theory of solid-solution hardening. *Scr. Metall.* **2**, 481–484.
Takeuchi, S. (1969). Solid-solution strengthening in single crystals of iron alloys. *J. Phys. Soc. Jpn.* **27**, 929–940.
Talmor, Y., Walker, E., and Steinemann, S. (1977). Elastic constants of niobium up to the melting point. *Solid State Commun.* **23**, 649–651.
Tanji, Y. (1971). Debye temperature and lattice deviation of Fe–Ni (fcc) alloys. *J. Phys. Soc. Jpn.* **30**, 133–137.
Tauchert, T. R. and Güzelsu, A. N. (1971). Measurements of the elastic moduli of laminated composites using an ultrasonic technique. *J. Compos. Mater.* **5**, 549–552.
Testardi, L. R. (1975). Structural instability and superconductivity in A-15 compounds. *Rev. Mod. Phys.* **47**, 637–648.
Testardi, L. R. (1977). Structural instability of A-15 superconductors. In *Physical Acoustics*, Vol. XIII. Academic, New York, 29–47.
Thurston, R. N. (1965). Ultrasonic data and the thermodynamics of solids. *Proc. IEEE* **53**, 1320–1336.
Tobolsky, A. V. (1960). *Properties and Structure of Polymers.* Wiley & Sons, New York.
Todhunter, I. and Pearson, K. (1886). *A History of the Theory of Elasticity and of the Strength of Materials.* Cambridge University Press, Cambridge.
Truell, R., Elbaum, C., and Chick, B. B. (1969). *Ultrasonic Methods in Solid State Physics.* Academic, New York.
Turley, J. and Sines, G. (1971a). The anisotropy of Young's modulus, shear modulus, and Poisson's ratio in cubic materials. *J. Phys. D* **4**, 264–271.
Turley, J. and Sines, G. (1971b). Anisotropic behavior of the compliance and stiffness coefficients for cubic materials. *J. Phys. D.* **4**, 1731–1736.
Tyson, W. R. (1966). Theoretical strength of perfect crystals. *Philos. Mag.* **14**, 925–936.
Vaks, V. G., Zarochentsev, E. V., Kravchuk, S. P., and Saftronov, V. P. (1978). *J. Phys. F* **8**, 725–742.
Vaks, V. G., Kravchuk, S. P., and Trefilov, A. V. (1977). Equation of state and volume dependence of the thermodynamic properties of alkali metals. *Sov. Phys.—Solid State* **19**, 740–744.
Varshni, Y. P. (1970). Temperature dependence of the elastic constants. *Phys. Rev. B* **2**, 3952–3958.
Vashchenko, V. Ya. and Zubarev (1963). Concerning the Grüneisen constant. *Sov. Phys.—Solid State* **5**, 653–655.
Visscher, W. M. (1960). Study of lattice vibrations by resonance absorption of nuclear gamma rays. *Ann. Phys.* (*N.Y.*) **9**, 194–200.
Viswanathan, R. and Rajagopal, E. S. (1961). On the exact relation between C_p and C_v in crystalline media. *Physica* **27**, 1226–1228.
Wachtman, J. B., Tefft, W. E., Lam, D. G., and Apstein, C. S. (1961). Exponential temperature dependence of Young's modulus for several oxides. *Phys. Rev.* **122**, 1754–1759.
Walker, E. and Peter, M. (1977). Elastic constants of the bcc phase in niobium–zirconium alloys between 4.2 and 300 K. *J. Appl. Phys.* **48**, 2820–2826.
Wallace, D. C. (1968). Pseudopotential calculation of the thermal expansion coefficient of sodium and potassium. *Phys. Rev.* **176**, 832–837.
Wallace, D. C. (1972). *Thermodynamics of Crystals.* Wiley & Sons, New York.
Wannier, G. H. (1966). *Statistical Physics.* Wiley & Sons, New York, 27.
Ward, I. M. (1971). *Mechanical Properties of Solid Polymers.* Wiley–Interscience, London.
Warlimont, H. and Hausch, G. (1974). Elastische Anisotropic von Metallen und Legierungen. In *Mechanische Anisotropi.* Springer-Verlag, Vienna, 35–62.
Warlimont, H., Hausch, G., Prasetyo, A., and Reynand, F. (1976). Relations between elastic properties and diffusionless structural transitions. In *New Aspects of Martensitic Transformation.* Japan Institute of Metals, Tokyo, 153–158.
Waser, J. and Pauling, L. (1950). Compressibilities, force constants, and interatomic distances of the elements in the solid state. *J. Chem. Phys.* **18**, 747–753.
Watt, J. P., Davies, G. F., and O'Connell, R. J. (1976). The elastic properties of composite materials. *Rev. Geophys. Space Phys.* **14**, 541–563.
Weinmann, C. and Steinemann, S. (1974a). Temperature and volume dependence of the elastic constants of palladium. *Phys. Lett. A* **47**, 275–276.
Weinmann, C. and Steinemann, S. (1974b). Lattice and electronic contributions to the elastic constants of palladium. *Solid State Commun.* **15**, 281–285.
Weston, W. F. and Granato, A. V. (1975). Cubic and hexagonal single-crystal elastic constants of a cobalt–nickel alloy. *Phys. Rev. B* **12**, 5355–5362.
Wigner, E. and Seitz, F. (1933). On the constitution of metallic sodium. *Phys. Rev.* **43**, 804–810; **46**, 509–524.
Woods, A. B. D., Brockhouse, B. N., March, R. H., Stewart, A. T., and Bowers, R. (1962). Crystal dynamics of sodium at 90° K. *Phys. Rev.* **128**, 1112–1120.
Wooster, W. A. (1953). Physical properties and atomic arrangements in crystals. *Rep. Prog. Phys.* **16**, 62–82.
Zener, C. (1942). Theory of lattice expansion introduced by cold-work. *Trans. AIME* **147**, 361–368.
Zener, C. (1947). Contributions to the theory of beta-phase alloys. *Phys. Rev.* **71**, 846–851.
Zener, C. (1948). *Elasticity and Anelasticity of Metals.* University of Chicago Press, Chicago, Illinois, 32.
Zener, C. (1949). Relation between residual strain energy and elastic moduli. *Acta Crystallogr.* **2**, 163–166.
Zener, C. (1951). Theory of D_0 for atomic diffusion in metals. *J. Appl. Phys.* **22**, 372–375.
Zener, C. and Bilinsky, S. (1936). Theory of the effect of temperature on the reflection of x-rays by crystals. III. Allotropic crystals. *Phys. Rev.* **50**, 101–104.
Ziman, J. M. (1960). *Electrons and Phonons.* Oxford University Press, London.
Zwicky, F. (1934). On the physics of crystals, Part I. *Rev. Mod. Phys.* **6**, 193–208.
Zwikker, C. (1954). *Physical Properties of Solid Materials.* Pergamon, New York, 93.

*Available from NTIS, Springfield, Virginia.

2 Specific Heat

L. L. SPARKS

2.1 INTRODUCTION

Specific heat is a fundamental property that relates the total heat per unit mass added to a system to the resultant temperature change of the system. It has been studied for more than a century. Diverse needs have provided the impetus for this continuous study and development; specific heat experiments were used to verify Nernst's statement of the third law of thermodynamics (entropy change approaches zero as temperature approaches zero), to determine thermodynamic equilibrium for chemical reactions, and to allow calorimetric determination of the energy states of substances. Specific heat provides a thread of continuity throughout the many solid state theories; conversely these theories can be utilized to estimate the specific heat with varying degrees of success.

In addition to the scientific applications mentioned above, knowledge of the specific heat of technically important materials is often essential in the design of systems that operate at low temperatures. The increased use of cryogenic temperatures in research, aerospace, industry, and energy production and storage has created a demand for specific heat data for construction materials. Since most of the existing low-temperature specific heat data involve the elements or exotic compounds, it is frequently necessary to compute the needed values from very limited data. Calculation and estimation of specific heats are sometimes possible using the accepted models. These models are presented in this chapter with only enough solid-state and thermodynamic detail to be conceptually acceptable. Rigorous developments are left to the many appropriate texts on solid-state physics (Ziman, 1962; Rosenberg, 1963; Kittel, 1966; Gopal, 1966) and thermodynamics (Sears, 1953; Zemansky, 1968).

Specific heats are strong functions of temperature, particularly for $T \lesssim 200$ K, as shown in Fig. 2.1 (Corruccini and Gniewek, 1960 and Chang, 1977). The specific heat for the semiconductor silicon, for example, changes by more than six orders of magnitude below 200 K. Typical behavior for good conducting metals, an amorphous material (Pyrex glass), and an organic plastic (polyvinyl chloride) is illustrated in the figure.

2.1.1 Definition

Heat capacity, C, is defined as the amount of heat required to raise the temperature of a system by a unit of temperature:

$$C_x = [dQ/dT]_x \tag{2.1}$$

where x denotes any one of several constraints that may be imposed and dQ is the total heat added to the system. The most common constraints are pressure and volume; pressure is the usual constraint for solids and volume is commonly held constant in determinations involving gases.

The heat capacity per unit mass is called the specific heat or, sometimes, the specific heat capacity. Several systems of units are commonly used to express specific heat. Units found in the literature include $J \cdot g^{-1} \cdot K^{-1}$, $J \cdot kg^{-1} \cdot K^{-1}$, $cal \cdot g^{-1} \cdot K^{-1}$, $Btu \cdot lb^{-1} \cdot {}^\circ F^{-1}$, $J \cdot mol^{-1} \cdot K^{-1}$, and $cal \cdot mol^{-1} \cdot K^{-1}$. The choice of units, $J \cdot kg^{-1} \cdot K^{-1}$, was made for this

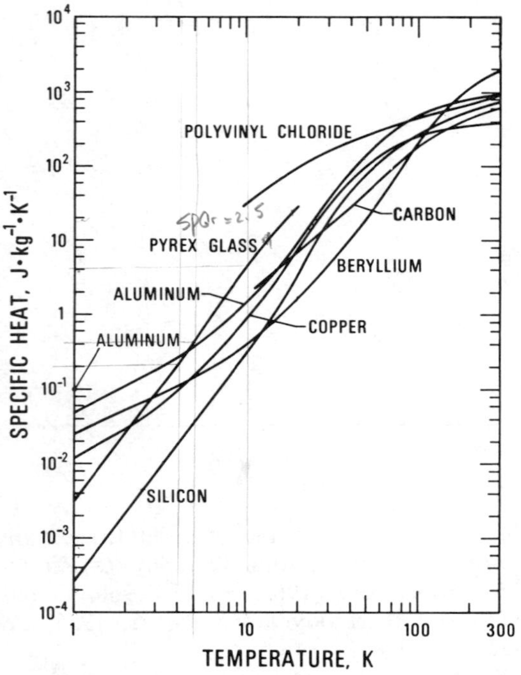

Figure 2.1 Specific heat as a function of temperature for several types of material. Typical behaviors are illustrated for metals (aluminum, beryllium, and copper), semiconductors (carbon and silicon), an amorphous inorganic (Pyrex glass) (Corruccini and Gniewek, 1960), and for an organic polymer (polyvinyl chloride) (Chang, 1977). A single point is shown for aluminum in the superconducting state at 1 K.

chapter because: (a) a mass rather than molar basis is more appropriate for discussion of alloys, and (b) these are the units specified in the International System of Units (SI). Circumstances are encountered, however, when it is necessary to convert to the molar basis. Then the unit of mass used in this chapter is the gram-mole (g-mol). It is convenient to note that 1 cal·g^{-1}·K^{-1} is equal to 1 Btu·lb^{-1}·°F^{-1}.

2.1.2 Historical background

Historical development leading to the present understanding of specific heat is particularly interesting because of the involvement of many eminent scientists, such as Boltzmann, Einstein, and Debye. The following overview of this development shows the chronology of the understanding of the concepts of lattice and electronic specific heat and the temperature dependences of the components. The models are discussed in more detail in § 2.2.2 and § 2.2.3.

Lattice specific heat. Dulong and Petit measured the specific heat of 13 solids in 1819. They found a room-temperature specific heat of approximately 25 J·mol^{-1}·K^{-1} for each material. Subsequent research verified the general constancy of this empirical value provided the measurements were well below the melting point and not too far below room temperature. Boltzmann provided theoretical justification for the Dulong–Petit value of specific heat by developing the equipartition-of-energy theorem in 1871 (Sears, 1953). Boltzmann's theorem yields a constant value of specific heat at constant volume, $C_v = 3R = 24.94$ J·mol^{-1}·K^{-1}, where R is the gas constant. As low-temperature research increased, it became clear that specific heat decreased rapidly at low temperatures. Classical theory could offer no explanation for this behavior.

Einstein (1907) extended Planck's quantum theory for electromagnetic radiation to include thermal vibrations in the lattice of solid materials. This simplified model of lattice vibration paved the way to the present understanding of many solid-state phenomena. Einstein proposed that the lattice of a solid was composed of independent oscillators and that there was only one frequency of oscillation, ν_E. The specific heat relationship derived by Einstein yields the Dulong–Petit value near room temperature and approaches zero exponentially as the temperature approaches absolute zero. Experimental values at very low temperatures indicated, however, that this exponential approach to zero was too rapid.

The next step in the evolution of specific heat theory was proposed by Debye (1912). Debye reasoned that the lattice behaved like an elastic continuum with an upper limit to the vibrational frequencies. The frequency distribution thus developed led to the well-known Debye formulation for lattice specific heats. Because of their simplicity and good agreement with experimental results, the T^3 dependence of specific heat and the Debye characteristic temperature, θ_D, are widely used in contemporary solid-state physics.

Since the introduction of Einstein's specific heat approximation, the primary effort to revise theory has involved more realistic evaluations of the frequency distributions of the phonons; the best known is the Debye quadratic distribution. Others have made use of increased knowledge of interatomic forces to represent the actual frequency distributions of the lattice oscillations. Their efforts, frequently utilizing high-speed computers to do the formidable computations, support the acceptance of this lattice theory. The expected correlation among thermal properties, elastic properties, and dielectric properties derived from this basic theory is observed.

Electronic specific heat. In about 1900 Drude, Lorenz, and others applied the kinetic theory of gases to electrons in metals by assuming that the electrons existed as free particles in the crystal lattice. The transport of heat and electrical currents by electrons was, at least qualitatively, explained using this approach. However, a major problem was created in the understanding of the specific heat of metallic conductors. If the electrons were treated as free electrons (independent oscillators) they should contribute $3R/2$ to the specific heat. This appeared to be impossible since essentially all of the observed specific heat, $3R$, was attributed to the lattice vibrations by the equipartition-of-energy theory. In 1928 Sommerfeld applied quantum statistics and Fermi–Dirac statistics to electrons in metals. This approach showed that only a small fraction of the electrons in a metal could change energy levels at ordinary temperatures and thereby contribute to the specific heat. As a result, the small electronic contribution to the specific heat at room temperature was explained. Sommerfeld's theory indicated an electronic contribution proportional to T so that it would become important, relative to the T^3 lattice term, only at very low temperatures.

2.1.3 Scope

The remainder of this chapter is devoted to expanding upon the principles mentioned in the historical sketch and presenting them in such a way that they can be applied to engineering needs. The thermodynamic and solid state relationships used by Einstein, Debye, and others are presented with emphasis on the resulting temperature-dependent relationships. The electronic contribution to the specific heat and its relative importance compared with the lattice specific heat are discussed, as are the effects of magnetic and superconducting transitions. The temperature and material dependences of the components of total specific heat are presented.

Although there are several excellent bibliographies, low-temperature data for most technically important materials do not exist. Methods of estimating these specific heats are presented. Three types of materials, graphite, polymers, and composites, are important at low temperatures but are not well described by existing specific heat theory; they are briefly discussed.

The chapter concludes with a section concerning the measurement of specific heat at low temperatures. Several rather specialized techniques have been developed to determine specific heats, but the Nernst-type adiabatic calorimeter is most frequently used for low-temperature measurements.

2.2 BASIC CONCEPTS

Practical application of specific heat data and theory often involves two branches of physics: solid state and thermodynamics. As mentioned before, the verification of very basic principles in both disciplines is often based on the observed behavior of heat capacity. Specific heat is a temperature-dependent measure of the change in the internal energy of a substance. As such, it can be affected by all temperature-dependent phenomena. For solids in particular, the lattice vibration spectrum, electron distribution, interaction of particles, and phase transitions must be considered. Sections 2.2.2 through 2.2.5 present the models used to account for these phenomena. Several thermodynamic relationships are frequently used in the discussion of these phenomena and the utilization of specific heat data. These concepts are presented, without rigorous derivation, in § 2.2.1.

2.2.1 Thermodynamic relationships

The definition of heat capacity (Eq. 2.1) involves the change in the total energy of a system of particles and therefore is governed by the laws of thermodynamics. According to the first law of thermodynamics (energy is conserved) a quantity of heat, dQ, added to a system increases the internal energy of the system and may cause external work to be done by the system. Mathematically,

$$dQ = dE + dW \tag{2.2}$$

where E represents the internal energy and W, the external work. In the case of a fluid (the simple model of a fluid can be used to develop relationships for solids with no significant loss of generality), the external work takes the form of a volume expansion, dV, against pressure, p, i.e.,

$$dW = p\,dV \tag{2.3}$$

For solids, however, not only the linear dimensions must change but the shape must also be al-

Table 2.1 The difference, $C_p - C_v$, for aluminum as calculated from Eq. 2.4.

T (K)	κ_T[a] (Pa^{-1})	β (= 3α)[b] (K^{-1})	v[c] ($m^3 \cdot kg^{-1}$)	$C_p - C_v$ ($J \cdot kg^{-1} \cdot K^{-1}$)	C_p[d] ($J \cdot kg^{-1} \cdot K^{-1}$)
0	1.260×10^{-11}	0	3.6574×10^{-4}	0	9
40	1.260×10^{-11}	6.6×10^{-6}	3.6576×10^{-4}	0.05	77.5
80	1.263×10^{-11}	27.3×10^{-6}	3.6600×10^{-4}	1.73	357
120	1.269×10^{-11}	43.8×10^{-6}	3.6653×10^{-4}	6.65	580
160	1.277×10^{-11}	53.7×10^{-6}	3.6726×10^{-4}	13.27	715
200	1.288×10^{-11}	60.0×10^{-6}	3.6810×10^{-4}	20.58	797
240	1.298×10^{-11}	64.5×10^{-6}	3.6902×10^{-4}	28.39	849
280	1.309×10^{-11}	68.1×10^{-6}	3.7000×10^{-4}	36.70	886
300	1.314×10^{-11}	69.6×10^{-6}	3.7069×10^{-4}	41.00	902

[a]Isothermal compressibility ($\kappa_T = 1/$bulk modulus) from Simmons and Wang (1971).
[b]Volume expansion from Corruccini and Gniewek (1961).
[c]Specific volume was calculated using the density at 293 K ($\rho = 2699$ kg·m^{-3}) and the thermal contraction at low temperatures from Corruccini and Gniewek (1961).
[d]Specific heat at constant pressure from Corruccini and Gniewek (1960).

tered. Practically, the specific heats of solids are measured at constant pressure, because extremely high pressures are required to keep the volume from changing with temperature.

Since the specific heat at constant volume, C_v, occurs in most thermodynamic relationships, and since the specific heat at constant pressure, C_p, results from experimental measurements on solids, a relationship between C_p and C_v is needed to determine C_v. This relationship is

$$C_p - C_v = \beta^2 v T / \kappa_T \tag{2.4}$$

in terms of temperature, T, specific volume, v, the coefficient of volume expansion, β, and the isothermal compressibility, κ_T. Viswanathan and Gopal (1961) show that this equation applies to general crystalline solids, that is, both isotropic and anisotropic. The coefficient of volume expansion and the isothermal compressibility are given by

$$\beta = (1/v)[\partial v / \partial T]_p \tag{2.5}$$

and

$$\kappa_T = (-1/v)[\partial v / \partial p]_T \tag{2.6}$$

Equation 2.4 was used to calculate $C_p - C_v$ for aluminum, as shown in Table 2.1. As seen in this table, $C_p = C_v$ at the absolute zero of temperature, and the difference $C_p - C_v$ is a slowly increasing function of temperature. The magnitude of $C_p - C_v$ is less than 10% of C_p at room temperature and always decreases as the temperature decreases. Examination of Eq. 2.4 shows that $C_p \geq C_v$ since for a mechanically stable material, $\kappa_T > 0$ and $\beta^2 \geq 0$. This can be visualized if one considers what happens when heat is added to a substance under conditions of constant pressure and then constant volume. In the constant-pressure mode, some of the added heat energy goes into expanding the substance against pressure. When the volume is kept constant, there is no external work since $dW = pdV = 0$; all of the added energy in this case must be used to increase the internal energy. The resulting temperature rise of the substance is consequently larger in the case of constant volume. A larger temperature rise for a given heat input indicates a lower heat capacity.

Equation 2.4 cannot be applied to most engineering materials owing to a lack of low-temperature data for κ_T and β. An approximation, using room-temperature values for these parameters is given by

$$C_p - C_v = A C_p^2 T \tag{2.7}$$

where $A = \beta^2 v / \kappa_T C_p^2$ (Bijl, 1957). A more approximate form may be used when room-temperature data are also lacking:

$$C_p - C_v = 0.0051\, C_p^2 T / T_m \tag{2.8}$$

where T_m is the melting-point temperature. The numerical constant has the units mol·K·J^{-1}.

Heat-balance calculations, which are frequently needed in cryogenic design, involve the thermodynamic quantity enthalpy, H, rather than the specific heat. At constant pressure

$$(\partial H / \partial T)_p = (\partial Q / \partial T)_p = C_p \tag{2.9}$$

so that

$$H_{T_2} - H_{T_1} = Q \tag{2.10}$$

and

$$H_{T_2} - H_{T_1} = \int_{T_1}^{T_2} C_p dT. \qquad (2.11)$$

Corruccini and Gniewek (1960) have compiled both C_p and H for many technical solids at low temperatures.

2.2.2 Lattice specific heat

The dominate mode of internal energy change, and therefore specific heat, for solids results from changes in the frequency of vibration of atoms or molecules about an equilibrium position in the lattice. In the case of metals, the motion of the free electrons also contributes to the specific heat but is significant only at very low temperatures. The following sections describe the lattice contribution to specific heat based on different models of lattice vibration of a solid. Emphasis is placed on Debye's development because it produces reasonable agreement with experimental results in many cases and because the Debye characteristic temperature, θ_D, is used in the theory of other properties.

Equipartition of energy. Boltzmann's equipartition-of-energy theory states that a particle in thermal equilibrium has an energy of $(1/2)kT$ for each degree of freedom it possesses (k is Boltzmann's constant). The classical picture of the crystal lattice consists of a system of atoms or molecules oscillating about fixed points so that a three-dimensional oscillator has six degrees of freedom (energy depends on both velocity and position). Each oscillator contributes an energy of $3kT$ and a mole of the substance has an energy of

$$E = 3N_A kT = 3RT \qquad (2.12)$$

where R is the universal gas constant in $J \cdot mol^{-1} \cdot K^{-1}$ and N_A is Avogadro's number. The specific heat is then

$$C_v = dE/dT = 3R \qquad (2.13)$$

This result agrees with the empirical Dulong–Petit value of 24.94 $J \cdot mol^{-1} \cdot K^{-1}$. Experimental evidence available at the time indicated that this formulation was not correct at low temperatures. The disparity between experimental and predicted results was not subtle: C_p for copper at 4 K was found, at a later date, to be only 0.02% of the $3R$ value.

Einstein's specific heat theory. In 1901 Planck developed quantum theory to describe the results of his electromagnetic radiation experiments. His conclusion was that the energy of an oscillator did not vary continuously as predicted classically, but rather changed only in steps of $h\nu$ where h is Planck's constant and ν is the frequency of the oscillation. Einstein applied this quantum concept to the thermal vibrations of atoms or molecules in a solid. He assumed that the particles of the crystal lattice oscillated independent of one another at a single frequency, ν_E. The frequency distribution resulting from this assumption is $f(\nu) = 0$ for $\nu \neq \nu_E$ and $f(\nu) \neq 0$ for $\nu = \nu_E$, as shown in Fig. 2.2a. Wave mechanics indicates that the actual allowable energy levels exist at $(m + 1/2)h\nu$, where m is zero or an integer.

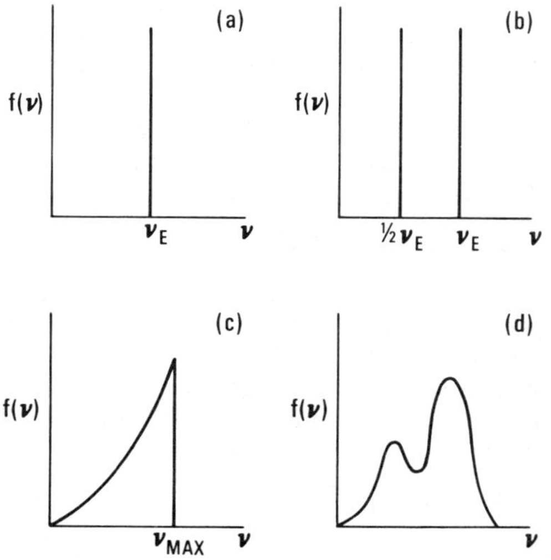

Figure 2.2 Phonon frequency distributions (a) Einstein (1907), (b) Nernst and Lindemann (Blackman, 1955), (c) Debye (1912), and (d) Born and von Karman (Blackman, 1937).

Since the frequency of atomic oscillation is approximately 10^{13} Hz, the minimum energy per oscillator is

$$E = 1/2(6.6256 \times 10^{-34} J \cdot s)(10^{13} s^{-1})$$
$$= 3.3 \times 10^{-21} J$$

Equating this minimum energy to the classical kT results in

$$T = E/k = (3.3 \times 10^{-21} J)/(1.38 \times 10^{-23} J \cdot K^{-1})$$
$$= 240 \text{ K}.$$

An oscillator with energy kT at this temperature cannot reduce its energy as the temperature is lowered because it is already at the minimum energy level. Since specific heat results from an energy change with changing temperature, there can be no contribution to the specific heat, that is, $C_v \neq 0$ when $T \gtrsim 240$ K and $C_v = 0$ when $T \lesssim 240$ K. This abrupt decrease to zero below $T \simeq 240$ K does not occur because there is a statistical probability that some particles will remain at the higher energy levels until absolute zero is reached. This probability was predicted by Boltzmann to be proportional to $e^{-h\nu/kT}$.

In the formal development of his specific heat relationship, Einstein used Maxwell–Boltzmann statistics, where the average energy of an oscillator is given by

$$\bar{E} = (1/2)h\nu + h\nu/(e^{h\nu/kT} - 1) \tag{2.14}$$

for each mode of vibration (Zemansky, 1968). The total energy of x oscillators, each with three modes of vibration, is then

$$E_{TOT} = 3x\bar{E} = 3/2\, xh\nu + 3xh\nu/(e^{h\nu/kT} - 1) \tag{2.15}$$

It follows that, at $\nu = \nu_E$,

$$C_v = dE/dT = (3rN_Ak)(h\nu_E/kT)^2 \times \exp(h\nu_E/kT)/[\exp(h\nu_E/kT) - 1]^2 \tag{2.16}$$

where the number of oscillators is the number of atoms per molecule, r, times Avogadro's number. The usual form of this, Einstein's specific heat equation, is obtained by substituting $R = N_Ak$ and

$$X_E = h\nu_E/kT \tag{2.17}$$

into Eq. 2.16. The specific heat per mole is then

$$C_v = (3rR\,X_E^2) \times \{\exp(X_E)\,[\exp(X_E) - 1]^{-2}\} \tag{2.18}$$

or

$$C_v = 3rR\,(\theta_E/T)^2 \times \{\exp(\theta_E/T)[\exp(\theta_E/T) - 1]^{-2}\} \tag{2.19}$$

when the Einstein characteristic temperature is defined as

$$\theta_E = h\nu_E/k. \tag{2.20}$$

Upon expansion (Blackman, 1955), C_v can be represented by

$$C_v \simeq 3rR[1 - 1/12(\theta_E/T)^2 + 1/240(\theta_E/T)^4] \tag{2.21}$$

when $T >> \theta_E$ and by

$$C_v \simeq 3rR(\theta_E/T)^2\exp(-\theta_E/T) \tag{2.22}$$

when $T << \theta_E$. It is clear from Eq. 2.21 that for higher temperatures ($T >> \theta_E$), C_v approaches the $3rR$ value predicted by Dulong and Pelit and by the equipartition-of-energy formulation. The exponential decrease of C_v at low temperatures ($T << \theta_E$) called for by Eq. 2.22 does not agree with experimental results (shown in Fig. 2.3). Despite the obvious shortcomings of the theory at low temperatures, it is still used as a simple approximation for calculations involving molecular vibrations at relatively high temperatures. Tabulated values of $C_v(\theta_E/T)$ are available (Gopal, 1966) and were used, along with $\theta_E = 1326$ K, to calculate the Einstein curve in Fig. 2.3.

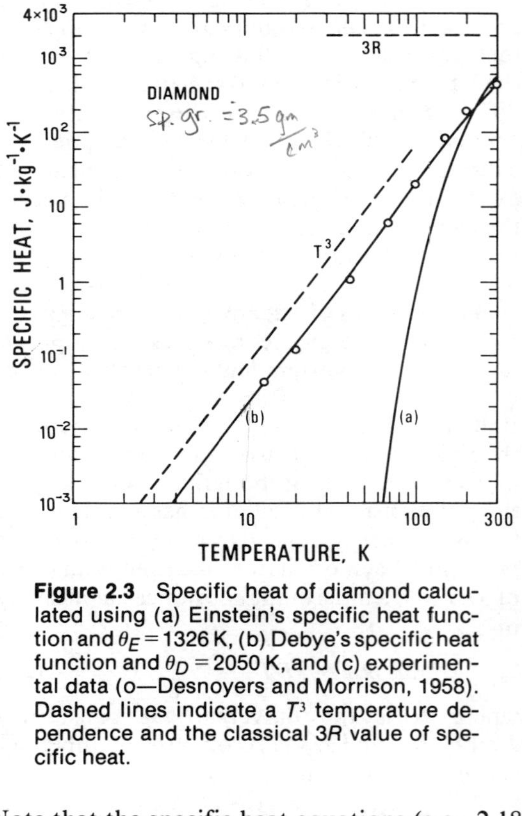

Figure 2.3 Specific heat of diamond calculated using (a) Einstein's specific heat function and $\theta_E = 1326$ K, (b) Debye's specific heat function and $\theta_D = 2050$ K, and (c) experimental data (o—Desnoyers and Morrison, 1958). Dashed lines indicate a T^3 temperature dependence and the classical $3R$ value of specific heat.

Note that the specific heat equations (e.g., 2.18, 2.19, 2.21, and 2.22) contain a term, r, to account for the number of atoms (oscillators) per molecule. For example, $r = 1$ for monatomic molecules such as diamond, $r = 2$ for diatomic molecules such as MgO, and $r = 3$ for triatomic molecules such as FeS_2. Tabulated values of the Einstein function are generally given for $r = 1$.

The problem with Einstein's theory was the assumption of a single allowable frequency of vibration. Nernst and Lindemann found that experimental data were represented better by an empirical formula involving two allowable phonon frequencies rather than one. This distribution is shown in Fig. 2.2b.

Debye's specific heat theory. The next step in the development of a representation of specific heats was provided by Debye. He assumed that the crystal lattice was a continuous isotropic medium in which only specific wavelengths could be maintained. These wavelengths were the ones that set up standing waves in the crystal. In addition, Debye established an upper limit to the frequencies (lower limit to the wavelengths) that could exist. The frequency distribution function thus established was

$$f(\nu)d\nu = 4\pi(1/v_\ell^3 + 2/v_t^3)\, V\nu^2 d\nu \qquad (2.23)$$

where v_ℓ and v_t are the wave propagation velocities in the longitudinal and transverse directions and V is the volume of the crystal. The upper limit to the frequency was established by integrating Eq. 2.23 between 0 and ν_{max} and setting the results equal to $3rN_A$ modes per mole.

$$3rN_A = 4\pi/3(1/v_\ell^3 + 2/v_t^3)v\nu_{max}^3$$

or

$$(1/v_\ell^3 + 2/v_t^3) = 9rN_A/4\pi v\ \nu_{max}^3$$

Substituting the quantity in parenthesis into Eq. 2.23 yields

$$f(\nu)d\nu = 9rN_A\nu^2 d\nu/\nu_{max}^3 \qquad (2.24)$$

The quadratic dependence and maximum allowable frequency are illustrated in Fig. 2.2c. Using the Maxwell–Boltzmann average energy, Eq. 2.14, and Eq. 2.24 for the frequency distribution, the total energy becomes

$$\begin{aligned} E_{TOT} &= \int_0^{\nu_{max}} \bar{E}f(\nu)d\nu \\ &= 9rN_A/\nu_{max}^3 \\ &\quad [h/2 \int_0^{\nu_{max}} \nu^3 d\nu + h \int_0^{\nu_{max}} \nu^3/ \\ &\quad (e^{h\nu/kT} - 1)d\nu] \end{aligned} \qquad (2.25)$$

and since $C_v = dE/dT$

$$C_v = (9rN_Ah^2)/(\nu_{max}{}^3 kT^2) \int_0^{\nu_{max}} \nu^4 e^{h\nu/kT} d\nu/(e^{h\nu/kT} - 1)^2 \qquad (2.26)$$

The Debye characteristic temperature,

$$\theta_D = h\nu_{max}/k \qquad (2.27)$$

$X = h\nu/kT$, and $X_{max} = h\nu_{max}/kT$ may be used with Eq. 2.26 to write the Debye specific heat function in the familiar form

$$C_v = 9rR(T/\theta_D)^3 \int_0^{x_{max}} e^X X^4 dX/(e^X - 1)^2 \qquad (2.28)$$

At high temperatures ($\theta_D/T \lesssim 0.1$), Eq. 2.28 may be approximated by

$$C_v \simeq 3rR(1 - X^2/20 + X^4/560 - X^6/18144) \qquad (2.29)$$

At high temperatures (small X) C_v approaches the $3rR$ Dulong–Petit value. At low temperatures the upper limit of integration in Eq. 2.28 becomes large. For $T \lesssim \theta_D/10$ it may be approximated by infinity and the integral becomes $12\pi^4/45$ so that

$$C_v \simeq 1944r(T/\theta_D)^3\ \mathrm{J{\cdot}mol^{-1}{\cdot}K^{-1}} = \beta T^3 \qquad (2.30)$$

This is the well-known Debye T^3 relationship for lattice specific heats at low temperatures. Whether it is used to calculate θ_D or β depends on what information is available. Two difficulties are encountered in applying this equation: (a) Debye theory is based on the acoustical (low temperature) mode of phonon vibration and is applicable only at very low temperatures. Practical situations, however, often necessitate using Eq. 2.30 at much higher temperatures (§ 2.3.2), and it must be recognized that the results are approximations. (b) The parameter r, number of atoms per molecule, is generally omitted from Eq. 2.30 and is seldom stated with values of θ_D or β. It must be known if these parameters are to be correctly interpreted.

Tabulated values of C_v as a function of θ_D/T are available (Gopal, 1966) and were used to calculate the Debye specific heat curve shown in Fig. 2.3. The characteristic temperature, θ_D, used in this calculation represents diamond at about $T = \theta_D/2$. As seen in Fig. 2.3, the predicted T^3 dependence exists up to about 100 K ($\simeq \theta_D/20$) owing to the very high characteristic temperature of diamond; the experimental data by Desnoyers and Morrison (1958) compare very well with the calculated values. The graphical representation of the universal Debye function is shown in Fig. 2.4. Specific heat values from this plot are for monatomic solids, that is, $r = 1$. If there is more than one atom (oscillator) per molecule, then $C_v/3R$ represents the specific heat of a fraction, $1/r$, of a mole. For example, $r = 1$ for diamond and $C_v/3R$ pertains to 1 mol of diamond; $r = 2$ for diatomic MgO and $C_v/3R$ pertains to 1/2 mol of this material; $r = 3$ for FeS_2 and $C_v/3R$ pertains to 1/3 mol of this material.

The Debye theory of specific heats was re-

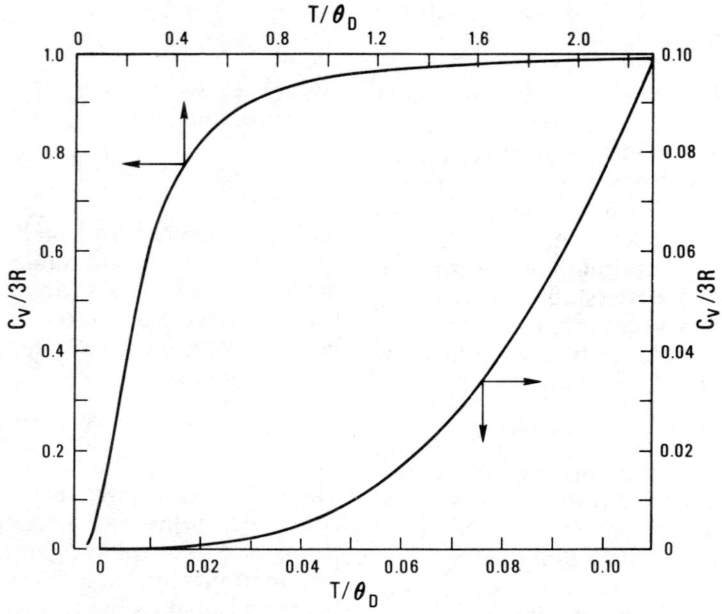

Figure 2.4 Graphical representation of the Debye function, Eq. 2.28. C_V approaches the classical value of $3R$ at high temperatures, and as required by the third law of thermodynamics, it approaches zero as the temperature approaches absolute zero.

markably successful; not only was it based on a model simple enough to be incorporated into already complex solid-state theories, but it agreed reasonably well with experimental specific heat data over a wide range of temperatures. It predicted the observed T^3 dependence of C_v when the temperature was low enough and it was based on a single parameter, θ_D. Even though Born and von Karman proposed a model (Fig. 2.2 d) for the frequency distribution of phonons in a lattice that was much more realistic than Debye's, it was essentially ignored because it was too complex. The Born–von Karman theory produced a frequency distribution that took the dispersive effect of nearest-neighbor interactions into account.

As more accurate data became available, however, it became increasingly clear that the Debye theory did have shortcomings. For instance, in Debye's theory the characteristic temperature (Eq. 2.27) is a constant for a given material. Experiment shows, however, that θ_D is actually a function of temperature. The temperature dependence of θ_D is shown in Fig. 2.5 along with the detailed phonon frequency spectrum determined from knowledge of the interatomic forces. The frequency distribution developed by Born and von Karman (Fig. 2.2d) is similar to those in Fig. 2.5. In general, the constant value of θ_D ap-

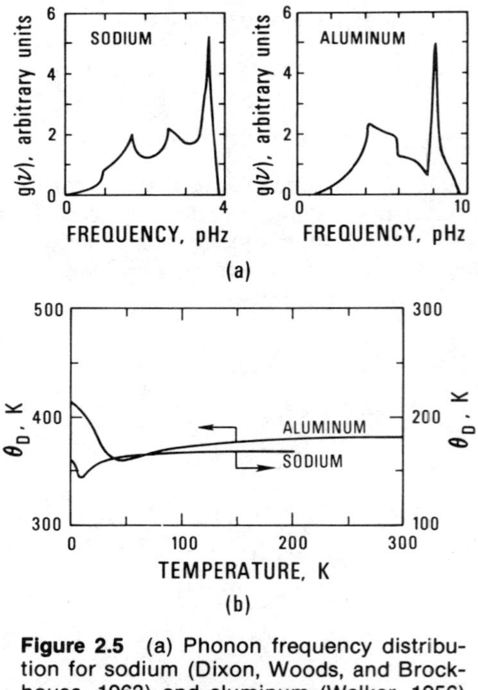

Figure 2.5 (a) Phonon frequency distribution for sodium (Dixon, Woods, and Brockhouse, 1963) and aluminum (Walker, 1956) based on phonon dispersions determined from neutron and x-ray scattering data, respectively. (b) Characteristic temperatures, $h\nu/k$, corresponding to the frequency distributions in (a).

proached at very low temperatures is different from the constant value approached at high temperatures.

Accurate representations of $f(\nu)$, based on known interatomic forces, have been developed for several materials. Experimentally observed values of θ_D correlate well with values predicted from such frequency distributions.

2.2.3 Electronic specific heat

When Sommerfeld applied Fermi–Dirac quantum statistics to free electrons in metals, the mystery of why the electronic contribution to the total specific heat was much smaller than the classically predicted $3R/2$ was solved. Sommerfeld's development (see Ziman, 1962 for a more rigorous derivation) culminates in an electronic contribution to the specific heat that is proportional to T for $T << T_F$ (T_F is the Fermi temperature). Since the lattice term is proportional to T^3, the electronic specific heat is relatively small except at very low temperatures.

The Fermi–Dirac distribution function (Kittel, 1966) gives the probability that a particular energy state will be occupied by an electron in an ideal electron gas in thermal equilibrium. This function

$$f(E) = \{\exp[(E - E_F/kT] + 1\}^{-1} \quad (2.31)$$

is plotted in Fig. 2.6. At the absolute zero of temperature, all energy states are full, $f(E) = 1$, for energies less than the Fermi energy, E_F, and empty, $f(E) = 0$, for $E > E_F$. As the temperature increases, the probability of finding a state occupied at an energy above E_F increases. The dashed line in Fig. 2.6 indicates the solution of Eq. 2.31 when $T/T_F = 0.1$. The electrons that are able to jump to a higher energy state contribute to the specific heat. Qualitatively, the electronic contribution can be easily estimated. The thermal

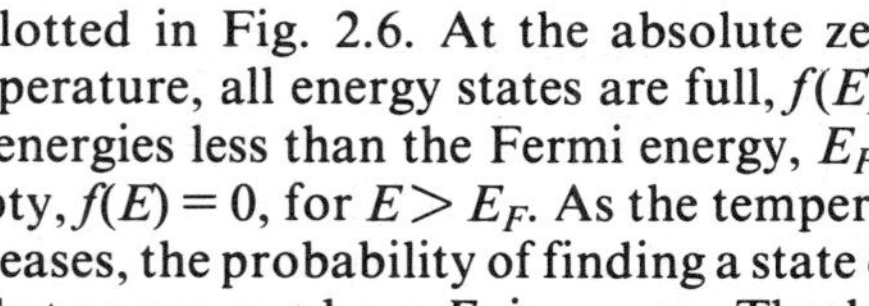

Figure 2.6 Fermi–Dirac distribution function vs. the ratio energy/Fermi energy for $T = 0$ K and $T = 0.1\ T_F$.

energy of an electron at the Fermi energy is $E_F = kT_F$ and since for most metals, $E_F \simeq 8.8 \times 10^{-19}$ J, the Fermi temperature is $T_F \simeq 6 \times 10^4$ K. The fraction of electrons that lie within kT of the Fermi energy is roughly T/T_F so that at room temperature only $300/6 \times 10^4 = 0.5\%$ of the electrons can change energy levels and, consequently, contribute to the specific heat. In a mole of a monovalent metal there would be N conduction electrons and therefore approximately NT/T_F electrons contributing to the specific heat. Each of these electrons has an energy on the order of kT so that the total energy of the electrons in question is

$$E_{el} \simeq N(T/T_F)kT \quad (2.32)$$

and at 300 K,

$$C_{el} \simeq 2NkT/T_F \simeq 0.08 \text{ J}\cdot\text{mol}^{-1}\cdot\text{K}^{-1} \quad (2.33)$$

This qualitative result indicates that C_{el} is small and proportional to the temperature.

The derivation of the electronic specific heat involves calculation of the energy of the electrons in terms of the Fermi–Dirac distribution function, Eq. 2.31, and the density of electron states, $D(E)$. Differentiating this energy with respect to T yields the electronic specific heat (Gopal, 1966).

For $T << T_F$ this results in

$$\begin{aligned} C_{el} &= 1/3\pi^2 D(E_F)k^2 T \\ &= (1.36 \times 10^{-4})\, v^{2/3} n^{1/3} T \\ &\quad \text{J}\cdot\text{mol}^{-1}\cdot\text{K}^{-1} \end{aligned} \quad (2.34)$$

where n is the number of free electrons per atom, and $D(E_F)$ includes the two allowed electronic spin orientations. The electronic contribution is generally expressed as

$$C_{el} = \gamma T \quad (2.35)$$

where $\gamma \equiv (1.36 \times 10^{-4}) v^{2/3} n^{1/3}$ J·mol^{-1}·K^{-2}. As in the previous qualitative argument, a linear temperature dependence is obtained. The temperature dependence of the electronic specific heat (Fig. 2.7) remains linear until $T/T_F \simeq 0.2$ (Stoner, 1936). At very high temperatures it approaches $3/2\ R$, the classical value for the electronic contribution per mole.

The electronic specific heat, as given in Eq. 2.35, is based on the free electron model of a metal. As in the case of Debye's elastic continuum model of the lattice, it is an oversimplification. The experimental determination of γ is discussed in § 2.2.5.

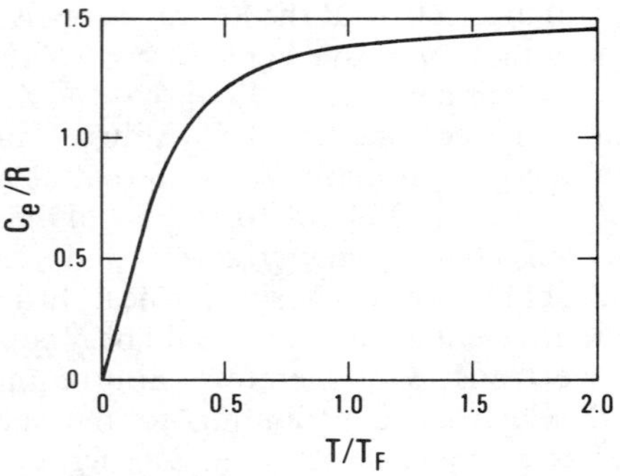

Figure 2.7 Temperature dependence of the electronic contribution to the specific heat.

2.2.4 Specific heat contributions from transitions and defects

The specific heat for well-behaved (Debye-type) materials consists of contributions from lattice vibrations and from free electrons in the case of electrical conductors. The specific heats for these materials have a characteristic temperature relationship, that is, always decreasing with decreasing temperature, as shown in Fig. 2.1. In some materials, however, experimental data reveal irregularities resulting from various transitions in the material. First-order, second-order, and λ-type transitions are experimentally observed, although theory allows for higher order transitions to exist. Each of these transitions produces a characteristic effect on the specific heat. Before discussing these effects, the nature of the transitions will be briefly reviewed. More detailed discussions of the thermodynamic basis of these transitions may be found in most texts on thermodynamics, for example, Zemansky (1968), and more on their observed behavior is found in Chapter 9.

A first-order transition is characterized by a change in entropy, S, and volume, V, at constant temperature and pressure. The Gibbs free energy, $G = H - TS$, where H is enthalpy, must not change when the temperature and pressure are constant. First derivatives of this function, however, indicate that the entropy, $-(\partial G/\partial T)_p$, and the volume, $(\partial G/\partial p)_T$, change discontinuously and $C_p = T(\partial S/\partial T)_p = \infty$. This condition is shown schematically in Fig. 2.8a. The specific heat is finite in both phases until the transition is reached, that is, there is no anticipation of the transition. Common transitions that fall into this category are melting, vaporization, and sublimation.

Some transitions occur in which S, V, T, p, and G are unchanged while the specific heat, coefficient of volume expansion, β, and isothermal compressibility, κ_T, undergo finite changes. This behavior requires that the first derivative of G be continuous and the second derivative be discontinuous. Transitions of this type are known as second-order transitions. The effect of a second-order transition on specific heat is shown in Fig. 2.8b. As was the case with first-order transitions, there is no anticipation of the transition. An example of a second-order transition is the change to superconductivity in zero magnetic field.

Another phase transition frequently observed in specific heat data is the so-called λ transition. It is similar to second-order transitions in that S, V, T, p, and G remain constant; however, λ transitions produce infinite changes in C_p, β, and κ_T. The transition is anticipated in this case, which leads to the lambda shape shown in Fig. 2.8c. This behavior in solids is observed for order–disorder transformations in alloys, magnetic and electric transformations, onset of molecular rotation, and ordering of axis orientation of molecular rotation.

Contributions to specific heat at or near these transitions may be attributed to either noncooperative or cooperative interactions. When particles do not change energy states independently, but rather function with strong interactions

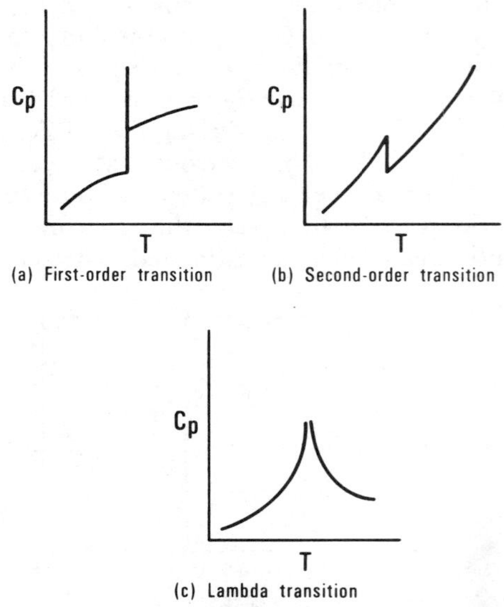

Figure 2.8 (a) Schematic of first-order phase transition. (b) Schematic of second-order phase transition. (c) Schematic of lambda-type phase transition.

among particles, cooperative effects are observed. The phenomena mentioned above in connection with second-order and λ transitions are examples of cooperative behavior. Two cooperative transitions that are particularly relevant to the specific heat of solids are magnetic-spin alignment and the normal-to-superconducting transition; noncooperative components of specific heat include the normal lattice and electronic terms and the Schottky effect.

Since specific heat depends strongly on the lattice configuration, imperfections, such as dislocations and strain, should contribute to the specific heat. Theoretical analysis of pinned dislocations and small-angle boundaries (Granato, 1958) indicates that the temperature dependence is proportional to T and T^2, respectively. Reasonable values of dislocation densities ($\sim 10^{10}$ cm^{-2}) and boundary surface areas ($\sim 10^6$ cm^2 per cubic centimeter of specimen) result in extremely small values of specific heat. Contributions from dislocations may be measurable for $T \lesssim 0.1$ K in insulators and superconducting materials where the electronic component is very small. The small-angle boundary effect is too small to be measured. Martin (1960) measured the specific heat of annealed and cold-worked 99.999% pure copper and found that the specific heat of the cold-worked specimen was about 0.15% higher than that of the annealed specimen ($20 < T < 300$K). This small difference is in the direction expected when the effect of strain on the lattice constants is considered: strain lowers the lattice constants. Collings, Jelinek, Ho, and Mathur (1977) determined the specific heat of several structural alloys with different heat treatments for $T \lesssim 20$ K, $T = 80$ K, and $T = 300$ K. The specific heat differences for similar specimens with different thermal histories were generally within the experimental uncertainty.

Magnetic contribution to specific heat. When a material capable of magnetization is considered, an additional term appears in the thermodynamic expression for the change in total energy of the system. The change is given by

$$dQ = TdS = dE + pdV + MdH \qquad (2.36)$$

where the magnetization, $M = \chi H$ is the vector sum of the magnetic dipole moments, H is the magnetic field strength, and χ is the magnetic susceptibility. Fundamental definitions for the two primary magnetic heat capacities may be derived (Gopal, 1966) and are

$$C_H = (dQ/dT)_H = T(dS/dT)_H$$

$$C_M = (dQ/dT)_M = T(dS/dT)_M \qquad (2.37)$$

The heat capacity at constant magnetization is independent of an external magnetic field, and as was the case for C_v, appears in most theoretical developments. Specific heat at constant field strength, C_H, is the quantity generally determined experimentally (corresponds to C_p in the nonmagnetic case). The difference between the specific heat contribution at low temperatures from ferro- and ferrimagnetic alignments, C_M, and C_H, analogous to Eq. 2.4, is given by

$$C_H - C_M = T(\partial M/\partial T)_H^2/(\partial M/\partial H)_T \qquad (2.38)$$

There are several types of magnetic behavior that affect the specific heat of a material: paramagnetism, ferromagnetism, antiferromagnetism, and ferrimagnetism. Diamagnetism, being independent of temperature, causes no specific heat contribution and is not considered.

A material is paramagnetic when a magnetic field tends to align the electronic magnetic moments parallel to the field. The degree of alignment is proportional to the field strength, and the alignment is destroyed by thermal agitation when the external field is removed. The susceptibility, $\chi = M/H$, is positive for materials in the paramagnetic state. The magnetic moments of ferromagnetic materials spontaneously and cooperatively assume a regular order when the temperature is low enough. The magnetization of these materials is finite even for $H = 0$; this is possible because there is a singularity in the susceptibility at the Curie temperature, T_c (Kittel, 1966). Above T_c the spontaneous parallel alignment vanishes, leaving the disordered paramagnetic state. In the presence of a magnetic field, the magnetization of ferromagnetic materials increases rapidly with increasing H and saturates at a very high level compared with materials in the paramagnetic state. A lower energy state exists for some materials when the magnetic moments or spins are aligned antiparallel instead of parallel. The spontaneous antiparallel alignment occurs through cooperative interaction for temperatures below the Néel temperature. Materials that undergo antiparallel alignment of magnetic moments are called antiferromagnetic if the moments are of the same magnitude and ferrimagnetic if they are of different magnitudes. Macroscopic behavior shows no net magnetization for antiferromagnetic materials, whereas ferrimagnetic materials behave in a manner similar to ferromagnets. A more complete discussion of magnetic properties appears in Chapter 6.

The regularity of the spin alignments mentioned above is disrupted by thermal agitation. When the disrupted spin system is Fourier-analyzed, the nonregularities appear as a set of waves. These spin waves may be quantized to yield magnons, which are the magnetic counterpart of quantized lattice vibrations (phonons). The frequency distribution of the magnons is proportional to q^2 for ferro- and ferrimagnetic materials and q for antiferromagnetic materials ($q = 2\pi$/spin-wave length). If the total energy is computed by integrating Eq. 2.14 over all values of q, then

$$C_M = dE/dT = \delta T^{3/2} \quad (2.39)$$

and for antiferromagnetic alignments,

$$C_M = dE/dT = \xi T^3 \quad (2.40)$$

The coefficients of temperature in these equations are

$$\delta = C_f N_A (k/2Js)^{3/2} \quad (2.41)$$

and

$$\xi = C_a N_A k (k/2J's)^3 \quad (2.42)$$

where J and J' are quantum exchange constants, s is the magnitude of the vector axis of precession, and C_f and C_a are constants that depend on the crystal structure. Van Kranendonk and van Vleck (1958) rigorously derived these coefficients and found that C_f = 0.113, 1/2(0.113), and 1/4(0.113) for simple cubic, body-centered cubic, and face-centered cubic lattices respectively, and that C_a = 13.7 for simple cubic lattices.

The general behavior of the specific heat near a magnetic transition is illustrated in Fig. 2.9 for

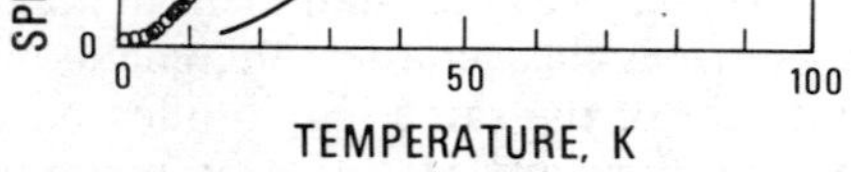

Figure 2.9 Specific heats of $MnCO_3$ and $CaCO_3$ (Kalinkina, 1963) illustrate the general effect of a magnetic transition. The Néel temperature for $MnCO_3$ is 29.4 K.

$MnCO_3$. This carbonate undergoes a para-to-antiferromagnetic transition at 24.9 K as the temperature is lowered. The difference between the $MnCO_3$ and $CaCO_3$ data is because $MnCO_3$ undergoes a transformation and $CaCO_3$ does not.

The magnetic contributions to specific heat discussed above are the result of cooperative interactions and are initiated by λ-type transitions. One more magnetic effect occurs, which is due to noncooperative changes in order. This phenomena is known as the Schottky effect and is observed in paramagnetic salts and in some ferromagnetic metals. It is a result of electronic energy degeneracies in the paramagnetic materials and nuclear energy degeneracies in the ferromagnetic materials. The observed Schottky specific heat is due to changes in internal energy that occur when nearly adjacent energy levels are occupied. The degenerate energy levels may be caused by external or internal magnetic fields.

Consider, for example, an ion having two energy levels separated by a small energy, kT': when $T >> T'$ both of the energy levels will be equally populated. In the narrow temperature range where $T \simeq T'$, many particles will jump to the higher energy level. This dE/dT over a limited temperature range gives rise to the Schottky specific heat, C_{SCH}. The actual form of the Schottky contribution to the specific heat was derived by Zemansky (1968) to be

$$C_{SCH} = R\epsilon_1^2/(k^2T^2)\,\{(g_0/g_1)e^{\epsilon_1/kT}/ [1 + (g_0/g_1)e^{\epsilon_1/kT}]^2\} \quad (2.43)$$

for a system with one degenerate energy level which is ϵ_1 above the ground state energy and has degeneracies g_0 and g_1. The more complex relationship for multiple energy levels and multiple degeneracies is given in Gopal (1966). At low temperatures, $\epsilon_1 >> kT$, and Eq. 2.43 reduces to

$$C_{SCH} = R(g_1/g_0)\,(\epsilon_1/kT)^2\, e^{\epsilon_1/kT} \quad (2.44)$$

and at high temperatures, $\epsilon_1 << kT$, it becomes

$$C_{SCH} = [Rg_0g_1/(g_0 + g_1)^2]\,(\epsilon_1/kT)^2 \quad (2.45)$$

The maximum C_{SCH} is determined by differentiating Eq. 2.43 with respect to T and setting the result equal to zero:

$$e^{-\epsilon_1/kT} = (g_0/g_1)\,(e^{\epsilon_1/kT} - 2)/ (e^{\epsilon_1/kT} + 2) \quad (2.46)$$

Values of ϵ_1/kT that satisfy this relationship for various ratios of g_1/g_0 are shown in Table 2.2 along with the corresponding maximum values

Table 2.2 The ratio ϵ_1/kT that satisfies Eq. 2.46 for the degeneracy ratios shown.

g_1/g_0	ϵ_1/kT	Max C_{SCH}[a] (J·mol^{-1}·K^{-1})
0.5	2.23	2.00
1.0	2.40	3.64
1.5	2.54	5.06
2.0	2.65	6.31
5.0	3.13	12.0

[a]The maximum Schottky specific heat is computed using these values of g_1/g_0 and ϵ_1/kT in Eq. 2.43.

of C_{SCH}. The magnitude of C_{SCH} may be several orders greater than that due to lattice and electronic contributions at liquid helium temperatures. Experimental observations of the Schottky phenomena are given in Fig. 2.10.

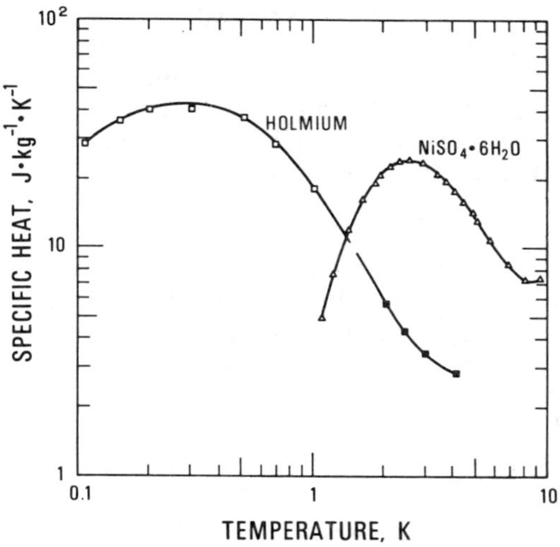

Figure 2.10 Specific heat of $NiSO_4 \cdot 6H_2O$ (△—Stout and Hadley, 1964) and holmium (■—Lounasmaa, 1962 and □—van Kemper, Miedema, and Huiskamp, 1964) as a function of temperature. The observed Schottky effect is a result of cooperative electronic spin alignment in the case of $NiSO_4 \cdot 6H_2O$ and cooperative nuclear spin alignment in the case of holmium. Below about 2 K the total specific heat of holmium is dominated by the Schottky contribution.

Specific heat in the superconducting state. Thermodynamic arguments (Zemansky, 1968) indicate that the transition from the normal to the superconducting state at zero field does not involve a latent heat and therefore must be a higher order transition. Experimental evidence indicates that it is second order. The observed specific heat of single-crystal vanadium (Fig. 2.11) at $H = 0$ shows the form of the transition (compare with the theoretically predicted curve shown in Fig. 2.8b). When the transition takes place in the presence of a magnetic field, a latent heat is observed and the transition is first order. This latent heat is given by

$$L = T(S_n - S_s) = -(VH_cT/4\pi)\, dH_c/dT \tag{2.47}$$

which gives

$$S_n - S_s = -(VH_c/4\pi)\, dH_c/dT \tag{2.48}$$

where H_c is the temperature-dependent critical field, V is the volume, and S_n and S_s are entropies in the normal and superconducting states. Since $C = T\,dS/dT$,

$$C_s - C_n = (VT/4\pi)\, d(H_c\, dH_c/dT)/dT \tag{2.49}$$

$$= (VH_cT/4\pi)\, d^2H_c/dT^2 + (VT/4\pi)(dH_c/dT)^2 \tag{2.50}$$

At $T = T_c$, $H_c = 0$ and

$$C_s - C_n = (VT_c/4\pi)\,(dH_c/dT)^2 \tag{2.51}$$

At very low temperatures the second derivative term in Eq. 2.50 dominates ($d^2H_c/dT^2 < 0$) and $C_s - C_n < 0$. The difference $(C_s - C_n)$ at T_c has been carefully determined and is 9.75, 10.6, and 41.5 mJ·mol^{-1}·K^{-1} for indium, tin, and tantalum, respectively. Corresponding values computed from Eq. 2.51 using values of dH_c/dT from Shoenberg (1952) are 9.52, 10.56, and 41.6

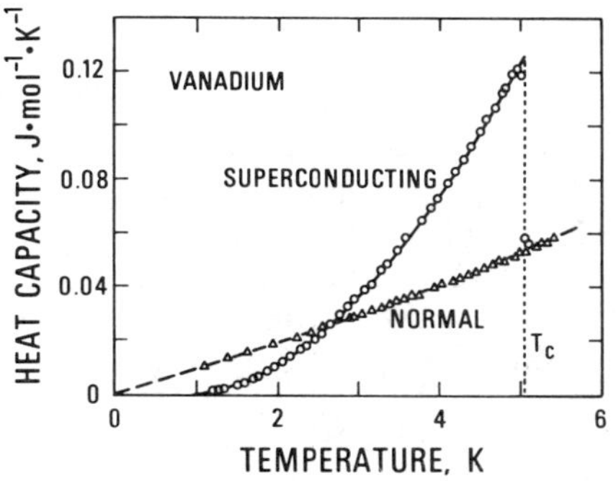

Figure 2.11 Heat capacity as a function of temperature for vanadium. The dashed line represents data taken at $H = 0$ and illustrates the second-order transition. The solid curve for normal vanadium was obtained by applying a magnetic field such that $H > H_c$ (Corak, Goodman, Satterthwaite, and Wexler, 1956).

$mJ\cdot mol^{-1}\cdot K^{-1}$, respectively (a conversion needed in this computation is $gauss\cdot cm^3 = erg/gauss$). This change in specific heat is significant, for example, $100\,(C_s - C_n)/C_n = 44\%$ for tin.

The two-fluid model of a superconducting material allows the effect of superconductivity on specific heat to be discussed in a straightforward manner. In this model electrons exist simultaneously in two forms: normal electrons (those that encounter resistance to flow) and superconducting electrons (those that experience no resistance to flow). The specific heat is given by

$$C_{\text{OBSERVED}} = C_{\text{NORMAL}} + C_{\text{SUPERCONDUCTING}}$$

These two components can be divided further:

$$C_{\text{NORMAL}} = C_{\text{ELECTRONIC, NORMAL}} + C_{\text{PHONON, NORMAL}}$$

or

$$C_N = C_{EN} + C_{GN} \tag{2.52}$$

and

$$C_{\text{SUPERCONDUCTING}} = C_{\text{ELECTRONIC, SUPERCONDUCTING}} + C_{\text{PHONON, SUPERCONDUCTING}}$$

or

$$C_S = C_{ES} + C_{GS} \tag{2.53}$$

When a material becomes superconducting, the structure of the lattice remains essentially unchanged, as evidenced by x-ray diffraction studies of the lattice and the exceedingly small changes in the elastic properties. It may be assumed, then, that $C_{GN} = C_{GS}$ so that

$$C_N - C_S = C_{EN} - C_{ES} \tag{2.54}$$

The temperature dependence of C_{ES} can be calculated by assuming a relationship for the temperature dependence of the critical field. For many superconducting materials this relationship can be approximated by the parabolic function

$$H_c = H_0[1 - (T/T_c)^2] \tag{2.55}$$

where H_0 is the critical field at the absolute zero of temperature. Using Eq. 2.55 in Eq. 2.50 yields

$$\begin{aligned} C_{EN} - C_{ES} &= (H_0^2 V/2\pi T_c^2)T - (3H_0^2 V/2\pi T_c^4)T^3 \\ &= \gamma T - 3\gamma T(T/T_c)^2 \end{aligned} \tag{2.56}$$

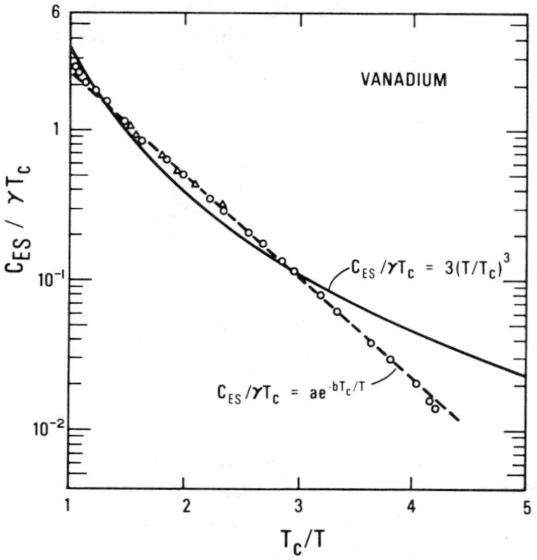

Figure 2.12 Electronic specific heat for superconducting vanadium expressed as $C_{ES}/\gamma T_c$ vs. T_c/T. Experimental data (Corak, Goodman, Satterthwaite, and Wexler, 1954) are well represented by the energy-gap formula (- - -) over the entire temperature range whereas the T^3 dependence (——) from the two fluid model fails for $T \lesssim T_c/3$.

where γ is the coefficient of normal electronic specific heat. The contribution from the superconducting electrons is

$$C_{ES}/\gamma T_c = 3(T/T_c)^3 \tag{2.57}$$

The solid line in Fig. 2.12 shows that at temperatures of $T_c/T \lesssim 3$ this T^3 dependence of C_{ES} represents the observed contribution reasonably well. The fit is not good, however, at lower temperatures. Experimental data are expressed rather well by

$$C_{ES}/\gamma T_c = ae^{-bT_c/T} \tag{2.58}$$

This relationship is shown as the dashed line in Fig. 2.12. The form of Eq. 2.58 suggests that C_{ES} is due to an energy gap. The BCS theory of superconductivity (Bardeen, Cooper, and Schrieffer, 1957) does in fact call for pairs of electrons, Cooper pairs (Cooper, 1956), to move through a crystal lattice with no exchange of energy. An energy gap on the order of kT_c exists between the normal electrons and the Cooper pairs; addition of energy to a material that is in the superconducting state causes splitting of Cooper pairs. Calculations based on the BCS quantum theory indicate that the constants in Eq. 2.58 are

$$C_{ES}/\gamma T_c = 8.5\, e^{-1.44\, T_c/T}$$
for $2.5 < T_c/T < 6$ (2.59)

and

$$C_{ES}/\gamma T_c = 26\, e^{-1.62\, T_c/T}$$
when $7 < T_c/T < 12$ (2.60)

The energy gap between the normal and superconducting electrons, $2E_0$ at absolute zero, is related to the critical temperature by the relationship

$$2E_0/kT_c = 3.52 \quad (2.61)$$

A further prediction of the BCS theory is that

$$C_{ES} = 2.43\,\gamma T_c \quad (2.62)$$

at $T = T_c$. Determinations of $C_{ES}/\gamma T_c$ for the soft superconductors, IIB, IIIA, and IVA elements, indicate that 2.43 (Eq. 2.62) is a reasonably good estimate of the observed values.

The relationships given in this section are for Type I superconductors. Many superconductors are classified as Type II superconductors; they are of great practical importance because of their capability of retaining zero electrical resistance at high values of H and T. Abrikosov (1957) developed a model that fits the observed behavior of many Type II alloys. This theory postulates that there are two critical fields with a mixed superconducting–normal state existing at field values in between. The effect of transitions on the specific heats of these alloys is still being examined. McConville and Serin (1964) found a second-order transition at the upper critical field and an apparent first-order transition at the lower critical field for a composite of niobium wires. Their data are shown in Fig. 2.13.

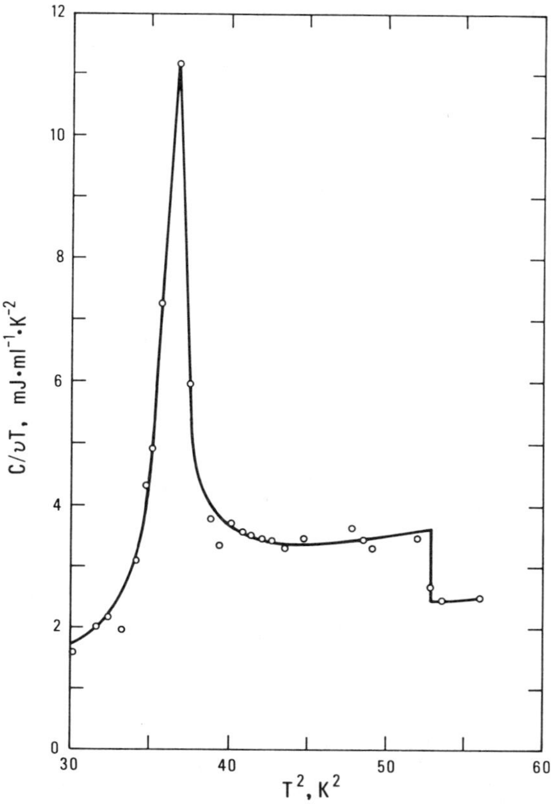

Figure 2.13 Ratio of specific heat to molar volume and temperature as a function of T^2 for a composite of niobium wires. These data include the specific heat of the addenda. The specimen was cooled prior to application of a 0.1-T field (McConville and Serin, 1964).

2.2.5 Total specific heat

The components of specific heat are assumed to be independent of one another so that the total observed specific heat can be attributed to the sum of these components:

$$C = \sum_i C_i(T) \quad (2.63)$$

The basic formula showing the temperature dependence of the lattice and electronic terms is

$$C = \beta T^3 + \gamma T \quad (2.64)$$

The lattice contribution, βT^3, comes from Eq. 2.30 and the electronic term, γT, from Eq. 2.35. When the material considered is an insulator, $\gamma = 0$ and Eq. 2.64, plotted in the form C/T vs. T^2, should produce a straight line with a slope β and an intercept at zero C/T. When the material is a conductor, that is, it has free electrons, $\gamma \neq 0$ and a similar plot produces a straight line with a slope β and an intercept γ. Data for two conductors, copper and aluminum, and one insulator, NaI, are shown in Fig. 2.14. Note that at higher temperatures the curves are not straight lines, indicating that the T^3 lattice dependence is no longer a valid approximation. The specific heat for materials that have only electronic and lattice contributions is dominated by the T^3 lattice term except at very low temperatures. When $T \lesssim 4$ K, the linear electronic term and the cubic lattice term are of the same order, as shown in Fig. 2.15. Low-temperature data are needed to allow separation of components, since the electronic term makes a relatively small contribution even for $T = \theta_D/10$.

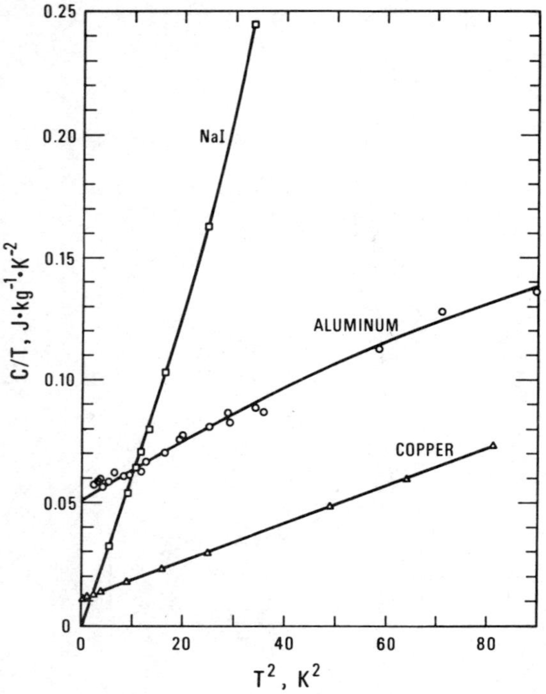

Figure 2.14 Plot of C/T vs. T^2 to yield β (slope) and γ (intercept) for the specific heat formula, Eq. 2.64. Experimental data are shown for copper (Franck, Manchester, and Martin, 1961), aluminum (Kok and Keesom, 1937), and NaI (Clayton and Marshall, 1960)

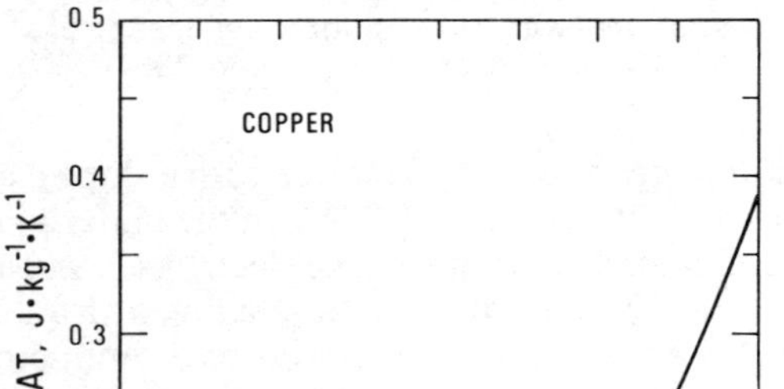

Figure 2.15 Components of total specific heat in copper. The cubic lattice term is dominate except at very low temperatures.

The spin-wave contribution from magnetic materials was shown to be $\propto T^{3/2}$ for ferro- and ferrimagnetic materials (Eq. 2.39) and $\propto T^3$ for antiferromagnetic materials (Eq. 2.40).

Since $C_{\mathrm{TOTAL}} = C_{\mathrm{ELECTRONIC}} + C_{\mathrm{PHONON}} + C_{\mathrm{SPIN\ WAVE}}$, one can write

$$C_T = \gamma T + \beta T^3 + \delta T^{3/2} \tag{2.65}$$

for ferromagnetic materials and

$$C_T = \beta T^3 + \delta T^{3/2} \tag{2.66}$$

for insulating ferrimagnetic materials. Antiferromagnetic materials occur in both metallic and nonmetallic form so that

$$C_T = \gamma T + \beta T^3 + \xi T^3 \tag{2.67}$$

for metals and

$$C_T = \beta T^3 + \xi T^3 \tag{2.68}$$

for nonmetals. The effect of the magnetic term is shown in Fig. 2.9 where $MnCO_3$ undergoes a para–antiferromagnetic transition whereas $CaCO_3$ remains in the disordered para state.

For ferromagnetic metals, the linear term dominates the total specific heat at very low temperatures and separation of the T^3 and $T^{3/2}$ terms becomes very difficult. The analytical form for ferrimagnetic materials, on the other hand, lends itself to graphical separation of components. Dividing Eq. 2.66 by $T^{3/2}$ and plotting $C_T/T^{3/2}$ vs. $T^{3/2}$ allows β, the slope, and δ, the intercept, to be determined. The separation of phonon and magnetic spin contributions is nearly impossible for antiferromagnetic materials, as indicated by the T^3 dependence of both terms in Eqs. 2.67 and 2.68.

The temperature dependences given above for the magnetic contribution to specific heat apply only at moderately low temperatures. At very low temperatures the magnetic component may decrease exponentially; at higher temperatures interaction among the spin waves appears to call for additional terms.

The analytical description of the specific heat of many alloys containing ferromagnetic iron, cobalt, or nickel requires a temperature-independent term. The cause of this anomalous behavior is generally attributed to microscopic clusters of these transition elements within a paramagnetic matrix (Hahn and Wohlfarth, 1968). The result is a two-phase magnetic system in a single-phase metallurgical system. Bennemann (1968) suggested that the effect was due to many-body interactions.

Regardless of whether superparamagnetic clusters or many-body interactions are the cause of the anomalous behavior, an empirical evaluation of data indicates that the total specific heat is given by the usual electron–phonon terms plus a constant, A.

$$C = A + \gamma T + \beta T^3 \tag{2.69}$$

At low temperatures this extra component causes an upturn in the usual C/T vs. T^2 plot, as shown in Fig. 2.16 for stainless steel AISI 310S. The magnitude of the effect is given for several alloys in Table 2.3. The references included in this table present data for other alloy systems along with more detailed discussions of the phenomenon.

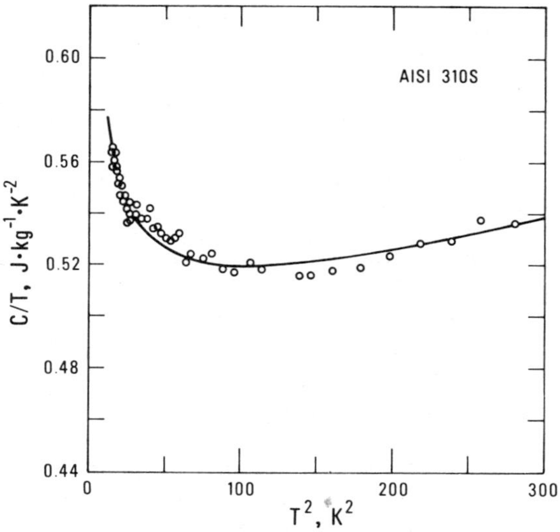

Figure 2.16 Plot of C/T vs. T^2 for stainless steel AISI 310S (Jelinek and Collings, 1975a). The rapid increase in C/T at low temperatures is a manifestation of the constant term in the specific heat formula (Eq. 2.69).

Table 2.3 Magnitude of superparamagnetic specific heat contribution for several alloys.

Alloy	A ($J \cdot mol^{-1} \cdot K^{-1}$)	Reference
74% Ni + 26% Zn	3.4×10^{-3}	Gupta, Cheng, and Beck (1964)
59.5% Cu + 40.5% Ni	5.5×10^{-3}	Schroder (1961)
72% V + 28% Fe	9.9×10^{-3}	Schroder (1961)
50% Fe + 50% Ti	9.0×10^{-3}	Schroder and Cheng (1960)
50% Fe + 50% Al	13.9×10^{-3}	Cheng, Gupta, Wei, and Beck (1964)
Stainless steel AISI 310S	21.2×10^{-3}	Jelinek and Collings (1975b)
Inconel X750	16.3×10^{-3}	Jelinek and Collings (1975b)

2.3 SPECIFIC HEAT DATA

In addition to an understanding of the analytical relationships in § 2.2, practical applications require data sources (§ 2.3.1) and methods of estimating specific heat for materials not included in published data (§ 2.3.2).

2.3.1 Data base

There is a wealth of specific heat data on the elements and nonmetallic compounds as a result of their use in crystal structure determination and chemical thermodynamics. Data on alloys are very limited, particularly below 300 K. Several excellent compilations of thermal properties were published between 1960 and 1970. Most of the significant experimental data on specific heats at low temperatures taken prior to 1970 are documented in the following references, each of which contains an extensive bibliography.

a. Specific heat and enthalpy data for several elements and other solid materials frequently used in cryogenic research are found in *A Compendium of the Properties of Materials at Low Temperatures* (1960). Comments on the critically selected data are included; the temperature range covered is $T \lesssim 300$ K.

b. Corruccini and Gniewek (1960) present tabular specific heat and enthalpy data from selected sources for several elements, alloys, other inorganic substances, and organic structures. The data are presented for $T \lesssim 300$ K.

c. Eldridge and Deem (1961) present thermal property data in graphical and tabular form for aluminum, cobalt, iron, magnesium, molybdenum, nickel, and for alloys of each. The data are compiled for $T < 4500°$ F (2755 K).

d. Hultgren, Orr, Anderson, and Kelley (1963) present tabular thermodynamic data for the metallic elements and several alloys. Discussion and phase diagrams for many binary alloy systems are included. Both cryogenic and high temperature ranges are considered.

e. Moeller, Loser, Thompson, Snyder, and Hopkins (1964) present thermodynamic properties of materials used for thermal insulation. Both graphical and tabular data are given at cryogenic and high temperatures.

f. Thermodynamic data for 31 elements and many of their borides, carbides, nitrides, and oxides for $T < 6000$ K are presented and discussed in *Thermodynamics of Certain Refractory Compounds*, Vol. 2 (1966).

g. Furukawa, Saba, and Reilly (1968) present the thermodynamic properties of copper, silver, and gold for $T \lesssim 300$ K. Detailed discussion is included along with an overview of state-of-the-art accuracies of experimental data. Critical values of θ_D and γ are also given.

h. A comprehensive compilation of thermophysical properties of matter is presented in *Specific Heat—Metallic Elements and Alloys*, Vol. 4 (1970) and *Specific Heat—Nonmetallic Solids*, Vol. 5 (1970). Experimental data are tabulated and presented graphically for cryogenic and high temperatures.

The author is not aware of any compilations published later than 1970. Data available since that time may be discovered in *Chemical Abstracts* and *Physics Abstracts*. Various computer data banks provide another source of data, usually for a fee.

2.3.2 Predicting specific heats

Specific heats of metals can be estimated by several methods. Estimates based on the relationships discussed in § 2.2 are generally adequate for engineering purposes when single elements are involved and the temperature range of interest is not near a transition. Application of the same predictive methods to alloys must be done with extreme caution. In general, the method and accuracy of the estimate depend upon the information available for the particular material.

If only ambient-temperature specific heat data are needed and no other information is available for a particular monatomic element, the Dulong–Petit value of $3R = 24.9429\ \mathrm{J \cdot mol^{-1} \cdot K^{-1}}$ may be used. Dulong–Petit values of specific heat are compared with corresponding experimentally determined data for a few elements in Table 2.4. The computed data are within about 5% of the experimental data except for beryllium and carbon. These elements were included to emphasize that this empirical rule is not universally applicable. The lighter elements with strong interatomic forces tend to deviate from the expected $3R$ value.

An important empirical extension of the Dulong–Petit rule, sometimes called the Kopp–Neumann law states: the molar heat capacity of a compound is equal to the sum of the atomic heat capacities of its constituent elements. Table 2.5 illustrates the approximate validity of the procedure at room temperature for compounds containing 2, 3, and 5 atoms per molecule. Although this rule is intended only for compounds, it is frequently used to estimate the specific heat of alloys, and the form needed is:

$$C = \sum_i C_i f_i \tag{2.70}$$

where C_i is the specific heat of the ith element and f_i is its mole fraction. Given the chemical composition and the specific heat as a function of temperature for the constituent elements, the specific heat of a material can be computed. For temperatures near 300 K the results of this type of calculation are reasonably accurate for many substances since all modes of lattice vibrations are excited and the specific heat per mole is about the same for all constituents (except for the light, hard elements mentioned above). At low temperatures specific heat depends on the lattice structure which, for an alloy, can be very different from that of the constituent elements. Although experimentally verified in a very limited number of cases, Kopp–Neumann calculations at low temperatures appear reasonably good when the elements making up the alloy have similar Debye temperatures, similar lattice structures, and no

Table 2.4 Comparison of ambient temperature specific heats for several elements as determined experimentally (reported by Corruccini and Gniewek, 1960) and the $3R = 24.94\ \mathrm{J \cdot mol^{-1} \cdot K^{-1}}$ value predicted by the empirical Dulong–Petit rule.

	Al	Be	Bi	Cr	Cu	Au	α-Fe	Pb	Ni	Si	C	Zn
Atomic wt. ($\mathrm{g\text{-}mol^{-1}}$)	26.98	9.013	209.00	52.01	63.54	197.0	55.85	207.21	58.69	28.09	12.011	65.38
Observed specific heat, C_{OBS} ($\mathrm{J \cdot mol^{-1} \cdot K^{-1}}$)[a]	24.34	17.76	25.92	23.40	24.53	25.31	24.96	26.94	26.12	20.06	8.60	25.50
Percent difference $[(C_{OBS} - 3R)/C] \times 100$	−2.5	−40.4	3.8	−6.6	−1.7	1.5	0	7.4	4.5	−0.2	−190.0	2.2

[a] C_{OBS} = observed specific heat.

Table 2.5 Comparison of ambient temperature specific heats for various compounds as determined experimentally (reported by Corruccini and Gniewek, 1960) and as computed using the Kopp-Neumann additivity principle and Dulong–Petit specific heat value of 3*R*.

	NaCl	KBr	AgCl	PbS	CuS	Ag_2S	$PbCl_2$	CaF_2	Fe_2O_3
Molecular wt. ($g{\cdot}mol^{-1}$)	58.448	119.02	143.34	239.28	95.61	247.83	278.12	78.08	159.7
Observed specific heat, C_{OBS}($J{\cdot}mol^{-1}{\cdot}K^{-1}$)	49.92	51.25	50.84	50.25	51.59	74.60	75.52	69.29	113.80
Specific heat: Kopp–Neumann, C_{KN}($J{\cdot}mol^{-1}{\cdot}K^{-1}$)	49.89	49.89	49.89	49.89	49.89	74.83	74.83	74.83	124.71
Percent difference $[(C_{OBS}-C_{KN})/C_{OBS}]\times 100$	<0.1	2.7	1.9	0.7	3.3	−0.3	0.9	−8.0	−9.6

phase changes in the temperature range of interest.

Recent experimental determinations of the specific heats of several structural alloys provide an interesting test for this method. Table 2.6 and Figs. 2.17 and 2.18 illustrate the uncertainties involved when computing the specific heats of alloys using the Kopp–Neumann additivity principle.

When the effort necessary to calculate phonon frequency distributions, $g(\nu)$, and effective characteristic temperatures, $\theta_{eff}=h\nu/k$, is considered (Fig. 2.5), it seems certain that these data will not become available for most technically important materials. Usually, not even the Debye temperature, θ_D, let alone its temperature dependence, has been determined. However, given a θ_D anywhere in the range $T \lesssim 300$ K, reasonable estimates of the lattice contribution to the specific heat can be made using tabulated values for the integral in Eq. 2.28. At low temperatures the integral is adequately represented by Eq. 2.30 and at high temperatures by Eq. 2.29. The real problem, then, in obtaining an approximate value for the lattice specific heat lies in determining θ_D. Best results are obtained when $\theta_D=f(T)$ is known so that C can be computed at specific temperatures using θ_D at corresponding temperatures. Many tables of θ_D are available in the literature and often a wide range of θ_D values are found for a given material. This is frequently due to θ_D being computed at different temperatures. Table 2.7 presents values of θ_D at $T \simeq \theta_D/2$ and $T \gtrapprox 4$ K (θ_0). The $\theta_D/2$ values favor the $100 \lesssim T \lesssim 300$ K range while θ_0 would be more applicable in the low-temperature region.

As indicated in Fig. 2.14, specific heat data at low temperatures for some materials can be used to determine the coefficient of the T^3 lattice term, β. Since

$$\beta T^3 = 1944r(T/\theta)^3 \ \mathrm{J{\cdot}mol^{-1}{\cdot}K^{-1}} \qquad (2.71)$$

the low-temperature characteristic temperature can be determined. This technique depends on being able to identify the lattice component of the total specific heat (§ 2.2.5) and knowing the number of atoms per molecule, r. If θ_D cannot be determined either because low-temperature specific heat measurements are not practical or because the nature of the results does not allow β to be determined, θ_D may be estimated using noncaloric methods. Other solid-state phenomena that involve a Debye-type characteristic temperature include the elastic constants, compressibility, melting point, expansion coefficients, and electrical resistivity (Blackman, 1955). Characteristic temperatures determined from several of these properties are shown in Table 2.7. With the exception of the specific heat and the elastic constant values at low temperatures, the characteristic temperatures derived from other properties are not equal. The difference in the thermally determined values at $\simeq \theta_D/2$ and $\simeq 4$ K is sometimes larger than the difference due to the principle involved.

Could specific heat data be estimated from very limited initial information well enough to be used in engineering design? Collings and Smith (1978a) explored the possibility of determining the specific heat of complex alloys at intermediate temperatures from low-temperature specific-heat data. Their low-temperature data ($T \leq 20$ K) for Inconel X750 were used to find the constants $A = 0.163\ \mathrm{J{\cdot}kg^{-1}{\cdot}K^{-1}}$, $\gamma = 0.23\ \mathrm{J{\cdot}kg^{-1}{\cdot}K^{-2}}$, and $\beta = 3.17 \times 10^{-4}\mathrm{J{\cdot}kg^{-1}{\cdot}K^{-4}}$ for Eq. 2.69. This value of β and a molecular weight of 0.057 $\mathrm{kg{\cdot}mol^{-1}}$ were used in Eq. 2.30 to yield $\theta_D = 476$ K. The total specific heat was estimated using Eq. 2.69. Comparison of these computed values with experimental data indicates that they are within the experimental uncertainty at 20 and 80 K, but are a factor of 20 too high at 300 K. A similar exercise for a stainless steel specimen, 310S, yields calculated values that are within the experimental uncertainty at 20 K, are 14% low at

Table 2.6 Comparison of experimentally determined specific heats, C_{EXP}, of stainless steel 310S and Invar with values computed using the Kopp–Neumann additivity rule, C_{KN}.

Alloy	Temperature (K)	C_{KN}[a] (J·kg^{-1}·K^{-1})	C_{EXP} (J·kg^{-1}·K^{-1})	$\frac{C_{EXP}-C_{KN}}{C_{EXP}}$	Composition (wt.%)
310S	4	0.34	2.19[b]	84.2	C, 0.25; Mn, 2.0;
	10	1.19	5.20[b]	77.2	P, 0.03; S, 0.03;
	20	4.51	11.68[b]	61.4	Si, 1-5; Cr, 24-26;
	80	153.3	203.0 [b]	24.5	Ni, 19-22; Fe, bal.
	300	462.4	485.0 [b]	4.7	
Invar	4	0.42	0.98[c]	57.3	Ni, 36; Fe, bal.
	7	0.86	1.88[c]	54.3	
	10	1.39	3.22[c]	56.8	
	20	4.91	12.13[c]	59.5	
	30	14.42			
	40	32.51			
	50	58.39			

[a]Specific heats as a function of temperature for the constituent elements are given by Touloukian (1966).
[b]Jelinek and Collings (1975b).
[c]Burford and Graham (1965).

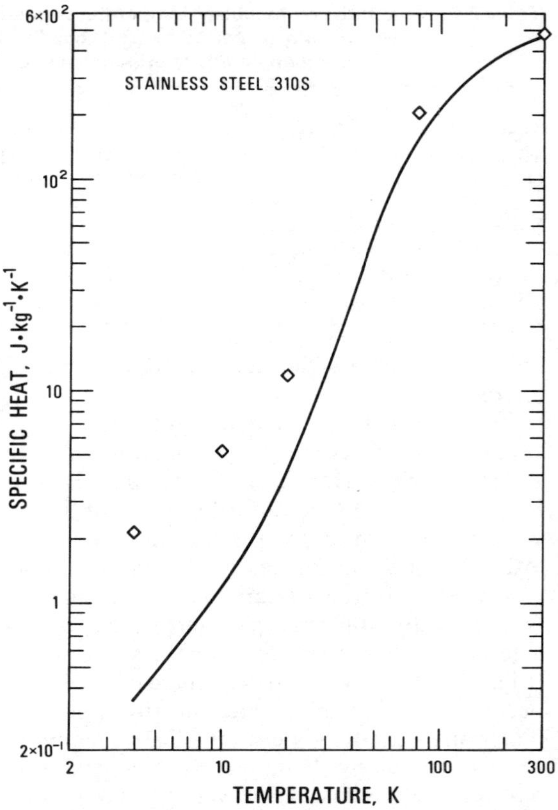

Figure 2.17 Comparison of the specific heats of stainless steel AISI 310S as determined experimentally (◇—Jelinek and Collings, 1975b) and mathematically (solid line) from the Kopp–Neumann additivity rule with values of $C_P = f(T)$ for the constituent elements from Touloukian (1966).

80 K, and are a factor of 15 too high at 300 K. It is obvious from these examples that Eq. 2.69 cannot be applied at 300 K. It can be used in the 80–K range but will result in rather large uncertainties; applicability of this equation in the intermediate range appears to be strongly material dependent.

An alternative method uses the Debye functions (tabulated values of the integral in Eq. 2.28) and θ_D. Using the θ_D, A, and γ given above for Inconel X750, $C = 7.3$, 155, and 445 J·kg^{-1}·K^{-1} for $T = 20$, 80, and 300 K, respectively. These values are within experimental uncertainty at 20 and 300 K and 15% too low at 80 K. Applying this procedure to the 310S stainless steel specimen, where $A = 0.334$ J·kg^{-1}·K^{-1}, $\gamma = 0.46$ J·kg^{-1}·K^{-2}, and $\theta_D = 508$ K, results in $C = 11.7$, 148, and 532 J·mol^{-1}·K^{-1} for $T = 20$, 80, and 300 K respectively. These values are within experimental uncertainty at 20 K, 27% low at 80 K, and 9% high at 300 K. Specific heats determined using this procedure appear to be good at $T \leq 20$ K; larger uncertainties result at intermediate and higher temperatures.

It is sometimes necessary to estimate specific heats based on much less information. For example, assume that only the specific heat at an intermediate temperature and the molecular weight are known. Again, from the data for Inconel X750 by Collings and Smith, $C_{85.46K} = 199.9$ J·kg^{-1}·K^{-1} and molecular weight $M = 57.05$ g·mol^{-1}. Using tabulated Debye functions, $C/3R = 0.457$ and $\theta_D/T = 4.32$, so that $\theta_D = 368$ K. This θ_D was found by using the total observed specific heat at 85.46 K as the lattice specific heat, since our limited data precluded accounting for the superparamagnetic and electronic terms. The discussion in § 2.2.3 and § 2.2.5, however, leads one to believe that at 80 K these terms would be small relative to the lattice term. Using this θ_D and the tabulated Debye functions to compute specific heats yields $C = 0.04$, 5.4, and 406 J·kg^{-1}·K^{-1} at $T = 4$, 20, and 300 K respectively. Comparison with Collings and Smith's data

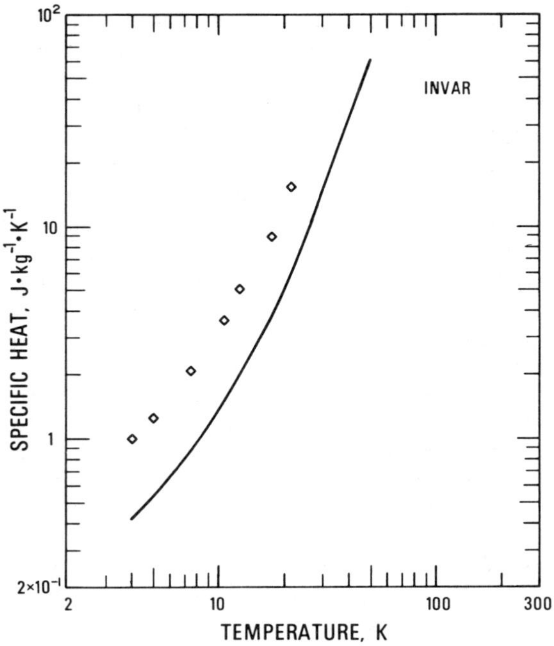

Figure 2.18 Comparison of the specific heats of Invar determined experimentally (◇—Burford and Graham, 1965) and mathematically (solid line) from the Kopp–Neumann additivity principle with values of $C_P = f(T)$ for iron and nickel from Touloukian (1966).

shows that $(C_{EXP} - C_{CALC})/C_{EXP} \times 100 = 96\%$, 26%, and 9%, respectively. This approach obviously does not work well at very low temperatures, but it begins to produce reasonable results at higher temperatures. All types of estimation must be used with caution, because phase transitions and Schottky effects are not considered.

The Debye function provides a somewhat better representation of some simple metals. The solid curves in Fig. 2.19, for instance, were calculated using the extreme values of θ_D for copper given in Table 2.7.

2.3.3. Specific heat of selected nonmetallic materials

The low-temperature specific heat of metals has been stressed in the preceding discussions. However, other materials, such as graphite, composites, and polymers, are frequently used in low-temperature applications. Much effort has been devoted to determining the temperature dependences of the observed specific heats for these materials as well as metals. The limited success of these efforts reflects the extremely complicated nature of these materials. A recently discovered material (Beall, 1972), machinable glass-ceramic, has properties that may lead to its extensive use in low-temperature apparatus. It is machinable, thermal-shock resistant, a good electrical insulator, nonmagnetic, and its thermal expansion is similar to that of metals. The specific heat of this material was determined by Lawless (1975). When $T \lesssim 10$ K, its specific heat is essentially the same as that for silicon (Fig.

Table 2.7 Characteristic temperatures as determined from specific heat, θ_D, elastic constants, θ_{el}, thermal expansion, θ_β, and electrical resistivity, θ_ρ. θ_0 data were determined at very low temperatures.

	θ_D [a]				
Material	$\theta_D/2$ (K)	θ_0 (K)	θ_{el}[e]$(T \to 0)$ (K)	θ_β[b] (K)	θ_ρ[b] (K)
Ag	220	225	226		203
Al	385	426	431		395
Au	180	164	162	180	175
Be	940	1160			
Bi	120	118[c]			
C(diamond)	2050	2200[c]	2240	1860	
Ca	230	229			
Co	440	443			
Cr	430	585	590		
Cu	310	348	344	325	333
Fe	460	464	472		
Ge	370	370[c]	373		
Hg	100	72			
K	100	91	91		
Mg	330	342			
Mn	420	450			
Mo	375	470	475		
Na	150	158	147		202
Nb	265	250	277		
Ni	440	440	476		
Pb	85	108	105		86
Pt	225	221	238	236	
Si	630		649		
Ti	355	430			
W	315	405	384		333
Zn	250	310			
AISI 304		494[d]	468[g]		
AISI 310			459[g]		
AISI 316			497[g]		
AISI 286			468[g]		
Inconel 600			465[f]		
Inconel X750		479[d]	474[f]		

[a]Gopal (1966), except as noted.
[b]Ledbetter (1982), except as noted.
[c]Blackman (1955).
[d]Corruccini and Gniewek (1960).
[e]Jelinek and Collings (1975b).
[f]Ledbetter, Weston, and Naimon (1975).
[g]Weston, Ledbetter, and Naimon (1975).

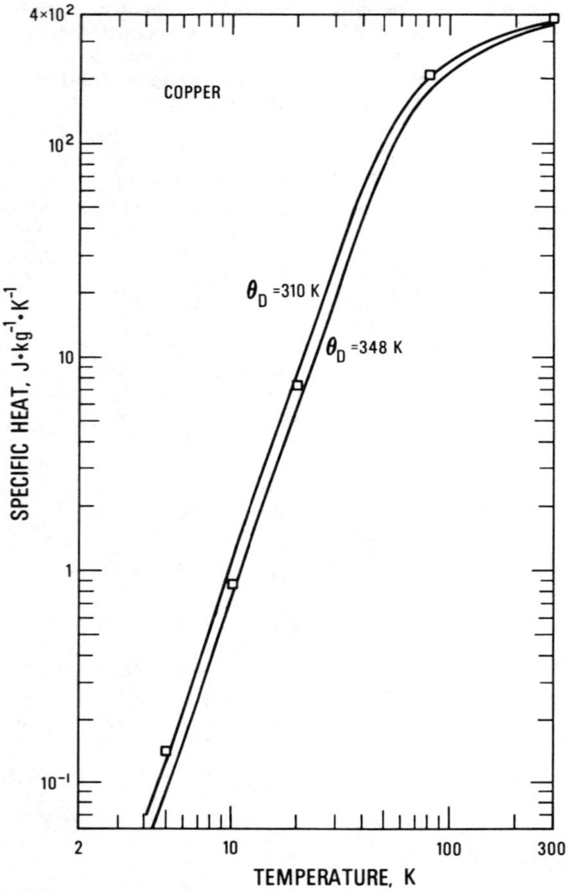

Figure 2.19 Specific heat of copper calculated using $\theta_D = 310$ K and $\theta_D = 348$ K. Experimental points (□—Sandenaw, 1959) are included for comparison. The electronic contribution (see Fig. 2.15) has been removed from the 5-K experimental point.

2.1); at $T = 19.5$ K, it is approximately 40% below the value for silicon.

Graphite. The specific heat of graphite is not well represented by the usual Debye treatment because of the highly anisotropic nature of the lattice. The observed behavior indicates that at very low temperatures, $T \lesssim 2$ K, the Debye T^3 dependence does occur. This dependence decreases in the $2 < T < 20$ K range until the lattice specific heat is proportional to T^2 for $T \gtrsim 20$ K. Data by van der Hoeven and Keesom (1963) and by DeSorbo and Nichols (1958) substantiate this general behavior. Bowman and Krumhansl (1958) suggested that the T^3 behavior was shifted to lower temperatures because of the effect of stacking faults on the apparent elastic shear constant, C_{44}. As C_{44} is reduced by increased stacking faults, more low-frequency modes of vibration become available and the specific heat is increased. Van der Hoeven and Keesom reduced the density of stacking faults by annealing, and as expected, the specific heat was reduced. The stacking fault–specific heat correlation was further corroborated by tests on four different types of specimens: natural Madagascar graphite and pile graphite (low degree of stacking faults), graphitized lampblack (high degree of stacking faults), and pyrographite (very high degree of stacking faults). The specific heat for the natural Madagascar specimens is thought to be characteristic of pure single-crystal graphite and is represented by $C = 13.8T + 27.7T^3$ in units of μJ·mol^{-1}·K^{-1} for $T \lesssim 1.0$ K. DeSorbo and Tyler (1953) determined the specific heat of a pure graphite specimen (grade GS) in the range $13 < T < 300$ K. For $13 \leq T \leq 54$ K they found $C \propto T^2$. Their results are shown in Fig. 2.1.

Polymers. Polymers are an increasingly important class of materials for low-temperature applications. This, along with scientific interest in the caloric behavior of polymers, has led to substantial theoretical and experimental attention. Genensky and Newell (1957) used the Born–von Karman model and analytical approximations to develop a frequency distribution for phonon vibrations in crystalline, chain polymers. They considered two modes of vibration perpendicular ($\perp$) and one parallel ($\parallel$) to the molecular chain. The total specific heat is then made up of contributions from these modes, as shown by

$$C_T = (2C_\perp + C_\parallel)/3 \qquad (2.72)$$

The temperature dependences of these components are predicted to be: $C_\perp$ proportional to T^3, $T^{5/2}$, and then $T^{1/2}$ as temperature increases; $C_\parallel$ proportional to T^3, going to T as temperature increases. Isaacs and Garland (1962) confirmed the T^3 dependence for crystalline polyethylene for $T < 5.3$ K. Their data indicated that $C = 1.13 \times 10^{-2} \cdot T^3$ in units of J·kg^{-1}·K^{-1}. Chang and Bestul (1973) and Chang (1977) studied the low-temperature heat capacity of two commercially important polymers: polyethylene and polyvinyl chloride (PVC). Chang's results for PVC are shown in Fig. 2.1. Chang and Bestul compared the heat capacity of linear and branched polyethylene and found a maximum difference near 30 K. The specific heat of the branched specimen was larger. Similar maximums have been reported for differences between glasses and their crystals (Guttman, 1972). This difference is attributed primarily to the volume differences of the two states, but may also be influenced by the availability of more low-frequency vibrational modes for the glassy structure.

Chang found that the specific heat of his PVC specimen was represented by the following linear relationships:

$$\begin{aligned} C &= 11.45 + 0.158T, & 140 \leq T \leq 240\ \text{K} \\ &= 10.95 + 0.162\ T, & 100 < T < 320\ \text{K} \\ &= 10 + 0.166\ T, & 80 < T < 340\ \text{K} \end{aligned} \tag{2.73}$$

where C is in J·mol^{-1}·K^{-1} and the precision is 0.1%, 0.5%, and 1%, respectively.

Composites. This broad classification includes such varied materials as metallic composites, resin-based composites, and expanded plastics. The use of these materials at low temperatures is increasing also. In general, little attempt has been made to characterize the specific heat of these materials because they tend to be very production-process dependent, they are usually anisotropic, they are affected by time and environment, and in many cases they are not homogeneous when small specimens are considered. Experimental specific heat determinations for particular specimens are of interest because they provide a qualitative view of the behavior of these composite materials.

Collings and Smith (1978b) experimentally determined the specific heat of four composites in the temperature range $3 < T < 20$ K and at ~ 80 K and ~ 300 K. The materials tested were boron–6061 aluminum, glass–resin, graphite–resin, and boron–epoxy. The results are shown in Fig. 2.20. The low-temperature data for the metallic composite were represented by an equation of the form $C = \gamma T + \beta T^3$, as expected for metals. The data for the remaining composites were fitted by polynomials of the form $\sum_n \alpha_i (T/10)^i$. A quadratic term was necessary in the temperature dependence of the nonmetallic composites. Available specific heat data for expanded plastics, foams, are based either on Kopp–Neumann additivity principle calculations (§ 2.3.2) or on calculations using

$$C = \lambda / \alpha\rho \tag{2.74}$$

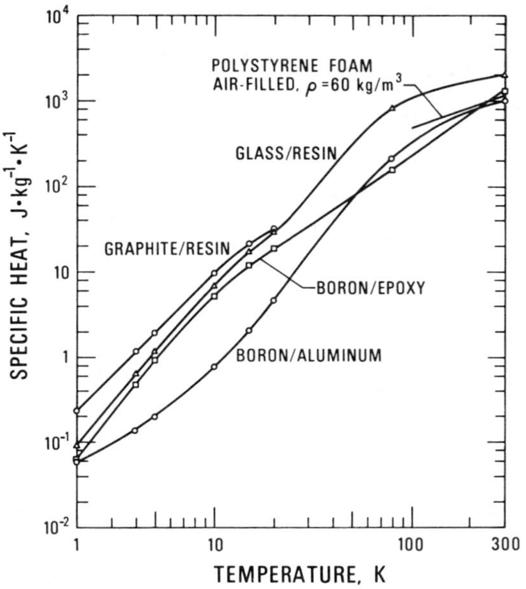

Figure 2.20 Specific heat as a function of temperature for five composite materials. The true shapes of the fiber-reinforced composite curves are somewhat uncertain because of the widely spaced data points. The calculated values for the polystyrene foam are based on 98 wt.% polystyrene and 2 wt.% air.

where λ is the thermal conductivity, α is the diffusivity, and ρ is the density. The results of such determinations are generally uncertain to about ±10%. The curve for expanded polystyrene (PS) shown in Fig. 2.20 was calculated by Ho, Desai, Wu, Havell, and Lee (1977) using the additivity principle.

2.4 METHODS OF MEASUREMENT

As discussed earlier, accurate and precise measurements of specific heats at low temperatures were needed in the first part of the twentieth century. The impetus provided first by Einstein's and then by Debye's theories for lattice specific heats led, in a very short time, to basic experimental techniques for measuring specific heats that are still used. Gaede introduced the technique of using an electrical current and a resistive wire to supply a measured amount of heat, ΔQ. Nernst and Eucken used this technique in developing a basic system of the type used today. The simplest form of the Nernst-type cryostat is shown in Fig. 2.21a. This system consists of an evacuated cryostat immersed in a cryogenic liquid. The specimen is suspended inside the cryostat by nylon or cotton threads. The heater and

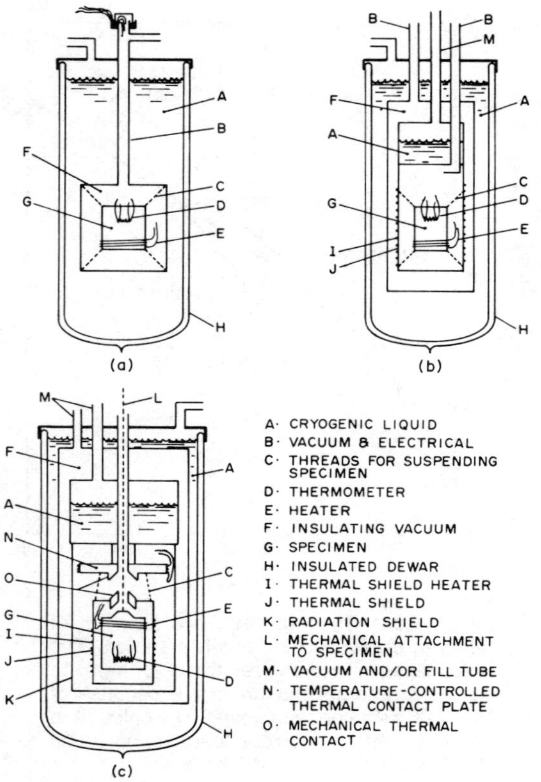

Figure 2.21 (a) Simple Nernst-type calorimeter. (b) Nernst-type calorimeter with thermal shield (isothermal or adiabatic). (c) Calorimeter with thermal shield and mechanical heat switch, which eliminates the need for exchange gas.

thermometer may be fixed directly to the specimen when a solid material is being tested. The specimen is shown as a solid in this illustration but could be in a powder or liquid form contained in a calorimeter can of known heat capacity.

Regardless of the exact cryostat configuration, several omnipresent problems are encountered in low-temperature calorimetric specific heat measurements. The main problem is thermal isolation of the specimen. Thermal contacts between the specimen and the surrounding cryostat are due to (a) solid conduction through the electrical connections and support system; (b) radiative heat transfer, normally important only when $T \gtrsim 20$ K; and (c) convective heat transfer through residual gases in the specimen chamber.

Numerous variations of the basic Nernst design have been used to reduce or account for these spurious thermal contacts. Development of semiconductor temperature transducers, such as germanium (Chapter 14) has substantially increased the precision of data when $T \lesssim 20$ K. Accuracy of data has been significantly improved as the international temperature scale (Chapter 14) used for calibration of thermometers has been made to coincide more closely with the thermodynamic temperature scale. Extension of this scale from 90 to 13.81 K (triple point of hydrogen) allows data from various sources to be referenced to a common standard. Prior to this extension, use of numerous national or laboratory temperature scales made interlaboratory comparison of data difficult. The range below 13.81 K is not yet represented by an international temperature scale.

Improvements in techniques, temperature scales, and instrumentation make it possible for contemporary specific heat measurements to be accurate to a few tenths of one percent when $4 \lesssim T \lesssim 300$ K. The precision of these measurements can be a factor of 2 better than the accuracy. The simple system shown in Fig. 2.21a can produce reasonably accurate results if the temperature difference, ΔT, between the specimen and the cryostat wall is kept small. The immersion liquid can be pumped or pressurized to achieve specific heat values over a range of temperatures. If ΔT is allowed to become large, heat losses by conduction along the electrical wires and radiation (above about 20 K) reduce the accuracy of the results. A common refinement of this basic design is to insert a thermal shield between the cryostat walls and the specimen. The temperature of the shield is controlled in one of two ways: adiabatically, that is, the shield temperature follows that of the specimen, or isothermally, which indicates that the shield is kept at a constant, known temperature while the sample temperature increases. In the ideal case, the adiabatic shield thermally isolates the specimen, that is, wires leaving the specimen are attached to the shield and therefore encounter no temperature difference and conduct no heat. Radiation is eliminated because the specimen is surrounded by the shield, which is at the same temperature. The ideal situation is never achieved in practice, of course, so measurement techniques have developed to reduce the effect of spurious heat leaks. In the isothermal mode of shield operation, a deliberate, but well-known, temperature gradient is established between the specimen and shield. The heat leaks from the specimen are then accounted for in the final results.

The next order of sophistication of a measurement system is shown schematically in Fig. 2.21b. An additional vacuum cryostat is incorporated to provide further thermal separation between the specimen and the outer cryogenic bath. The upper portion of the specimen chamber can be filled with either the same liquid used in the outer bath or with another cryogenic liquid. Pumping on the inner liquid or heating the upper part of the chamber allows specific heat data to be determined at essentially any temperature in the range $2 \lesssim T \lesssim 300$ K without large temperature differences between the specimen and its adjacent walls.

Refrigeration is supplied to the specimens in the cryostats shown in Fig. 2.21a and b by introducing helium gas into the specimen vacuum space and thus establishing a gas-conduction thermal link to the liquid bath. When the specimen reaches the desired temperature, the exchange gas is removed and the specimen-bath thermal link is destroyed. The vacuum surrounding the specimen must be on the order of 0.13 mPa (10^{-6} mm Hg) to remove convective heat conduction. At liquid-helium temperatures, helium gas adsorbs readily to the specimen and calorimeter walls. Not only does the heat of desorption affect the measurements, but the insulating vacuum may not be sufficient during the desorption.

Many ingenious variations to these designs have been utilized to isolate the specimen from exterior temperatures and reduce the long pump-out time required when an exchange gas is used. One method of cooling the specimen without the use of an exchange gas is to provide a removable mechanical thermal link (Fig. 2.21c). The specimen calorimeter is lifted by the string until it establishes a continuous thermal link with the isolating ring. The temperature of the isolating ring itself may be adjusted by an electrical heater wrapped on the ring.

The common feature found in essentially all systems is the adiabatic or isothermal shield. A comparison of experimental systems utilizing these two modes of shield control indicates that they are capable of producing equally reliable results when $T \lesssim 250$ K. Above this temperature, the rate of heat transfer in the isothermal shield systems becomes large and the necessary corrections become difficult. The adiabatic systems appear to offer advantages in terms of economy of cryogenic liquid used, faster measurement times, and simplicity of interpreting experimental data. Very complete discussions of low-temperature calorimetry along with exhaustive reference sections have been published by Westrum, Furukawa, and McCullough (1968), Stout (1968), and Hill, Martin, and Osborne (1968).

Materials for use as insulators, adhesives, heaters, and thermometers have improved significantly since the first Nernst cryostat was used. The properties of these items are discussed in Chapters 3, 4, 5, and 14.

2.5 REFERENCES

Abrikosov, A. A. (1957). On the magnetic properties of superconductors of the second group. *Sov. Phys. JETP* (Eng. Transl.) **5**, 1174–1182.

Bardeen, J., Cooper, L. N., and Schreiffer, J. R. (1957). Microscopic theory of superconductivity. *Phys. Rev.* **106**, 162–163.

Beall, G. H. (1972). Advances in nucleation and crystallization in glasses. In *American Ceramic Society Special Publication 5* (eds. L. L. Herch and S. W. Freiman). Frieman, Columbus, 251.

Bennemann, K. H. (1968). Theory of the anomalous specific heat of nickel and copper–nickel alloys at low temperatures. *Phys. Rev.* **167**, 564–572.

Bijl, D. (1957). The representation of specific heat and thermal expansion data of simple solids. In *Progress in Low Temperature Physics* (ed. C. J. Gorter). North-Holland Publishing Co., Amsterdam, 395–430.

Blackman, M. (1937). On the vibrational spectrum of a three-dimensional lattice. *Proc. Roy. Soc. A* **159**, 416–431.

Blackman, M. (1955). The specific heat of solids. In *Handbuch der Physik*, Vol. 7 (ed. S. Flugge). Springer-Verlag, Berlin, 365–382.

Bowman, J. C. and Krumhansl, J. A. (1958). The low-temperature specific heat of graphite. *J. Phys. Chem. Solids* **6**, 367–379.

Burford, J. C. and Graham, G. M. (1965). The molar heat capacity of Invar at low temperatures. *Can. J. Phys.* **43**, 1915–1917.

Chang, S. S. (1977). Heat capacity and thermodynamic properties of poly (vinyl chloride). *J. Res. Natl. Bur. Stand.* **82**, 9–17.

Chang, S. S. and Bestul, A. B. (1973). Heat capacities of polyethylene from 2 to 360 K. I. Standard samples of linear and branched polyethylene whole polymer. *J. Res. Natl. Bur. Stand.* **77A**, 395–405.

Cheng, C. H., Gupta, K. P., Wei, C. T., and Beck, P. A. (1964). Low temperature specific heat of Fe–Al alloys. *J. Phys. Chem. Solids* **25**, 759–765.

Clayton, R. N. and Marshall, B. J. (1960). Specific heat and elastic constants of sodium iodide at low temperatures. *Phys. Rev.* **120**, 332–334.

Collings, E. W., Jelinek, F. J., Ho, J. C., and Mathur, M. P. (1977). Magnetic and thermal properties of stainless steel and Inconel at cryogenic temperatures. In *Advances in Cryogenic Engineering*, Vol. 22. Plenum Press, New York, 159–173.

Collings, E. W. and Smith, R. D. (1978a). Specific heats of some cryogenic structural materials I—Fe–Ni base alloys. In *Advances in Cryogenic Engineering*, Vol. 24. Plenum Press, New York, 214–223.

Collings, E. W. and Smith, R. D. (1978b). Specific heats of some cryogenic structural materials II—composites. In *Advances in Cryogenic Engineering*, Vol. 24. Plenum Press, New York, 290–296.

A Compendium of the Properties of Materials at Low Temperatures (1960) (ed. V. J. Johnson). WADD-TR-60-56, Part 2, AD249786, 1–333.*

Cooper, L. N. (1956). Bound electron pairs in a degenerate Fermi gas. *Phys. Rev.* **104**, 1189–1190.

Corak, W. S., Goodman, B. B., Satterthwaite, C. B., and Wexler, A. (1954). Exponential temperature dependence of the electronic specific heat of superconducting vanadium. *Phys. Rev.* **96**, 1442–1444.

Corak, W. S., Goodman, B. B., Satterthwaite, C. B., and Wexler, A. (1956). Atomic heats of normal and superconducting vanadium. *Phys. Rev.* **102**, 656–661.

Corruccini, R. J. and Gniewek, J. J. (1960). *Specific Heats and Enthalpies of Technical Solids at Low Temperatures*. NBS Monograph 20, National Bureau of Standards, Boulder, Colorado, 1–20.

Corruccini, R. J. and Gniewek, J. J. (1961). *Thermal Expansion of Technical Solids at Low Temperatures*. NBS Monograph 29, National Bureau of Standards, Boulder, Colorado, 1–22.

Debye, P. (1912). Zur Theorie der spezifischen Wärmen. *Ann. Phys.* **39**, 789.

Desnoyers, J. E. and Morrison, J. A. (1958). The heat capacity of diamond between 12.8° and 277° K. *Philos. Mag.* **3**, 42–48.

DeSorbo, W. and Nichols, G. E. (1958). A calorimeter for the temperature region 1–20° K — the specific heat of some graphite specimens. *J. Phys. Chem. Solids* **6**, 352–366.

DeSorbo, W. and Tyler, W. W. (1953). The specific heat of graphite from 13° to 300° K. *J. Chem. Phys.* **21**, 1660–1663.

Dixon, A. E., Woods, A. D. B., and Brockhouse, B. N. (1963). Frequency distribution of the lattice vibrations in sodium. *Proc. Phys. Soc. Lond. A* **81**, 973–974.

Einstein, A. (1907). Die Plancksche Theorie der Strahlung und die Theorie der spezifischen Wärme. *Ann. Phys.* **22**, 180.

Eldridge, E. A. and Deem, H. W. (1961). *Report on Physical Properties of Metals and Alloys from Cryogenic to Elevated Temperatures*. ASTM-STP-296, American Society for Testing and Materials, Philadelphia, 1–206.

Franck, J. P., Manchester, F. D., and Martin, D. L. (1961). The specific heat of pure copper and of some dilute copper + iron alloys showing a minimum in the electrical resistance at low temperatures. *Proc. Roy. Soc. London A* **263**, 494–507.

Furukawa, G. T., Saba, W. G., and Reilly, M. L. (1968). *Critical Analysis of the Heat-Capacity Data of the Literature and Evaluation of Thermodynamic Properties of Copper, Silver, and Gold from 0 to 300 K*. Report No. NSRDS-NGS-19, National Bureau of Standards, Boulder, Colorado, 1–49.

Genensky, S. M. and Newell, G. F. (1957). Vibration spectrum and heat capacity of a chain polymer crystal. *J. Chem. Phys.* **26**, 486–497.

Gopal, E. S. R. (1966). *Specific Heats at Low Temperatures*. Plenum Press, New York.

Granato, A. (1958). Thermal properties of mobile defects. *Phys. Rev.* **111**, 740–746.

Gupta, K. P., Cheng, C. H., and Beck, P. A. (1964). Low-temperature specific heat of Ni-based fcc solid solutions with Cu, Zn, Al, Si, and Sb. *Phys. Rev. A* **133**, 203–206.

Guttman, C. M. (1972). Low-temperature heat-capacity differences between glasses and their crystals. *J. Chem. Phys.* **56**, 627–630.

Hahn, A. and Wohlfarth, E. P. (1968). Anomalous low temperature specific heat of alloys near the critical concentration for ferromagnetism. *Helv. Phys. Acta.* **41**, 857–868.

Hill, R. W., Martin, D. L., and Osborne, D. W. (1968). *Calorimetry Below 20° K in Experimental Thermodynamics I. Calorimetry of Non-reacting Systems* (eds. J. P. McCullough, and D. W. Scott). Page Bros., Norwich, England, 263–292.

Ho, C. Y., Desai, P. D., Wu, K. Y., Havell, T. N., and Lee, T. Y. (1977). *Thermophysical Properties of Polystyrene and Poly(vinyl) Chloride*. NBS Report GCR–77–83. CINDAS, Purdue University, Lafayette, Indiana.

Hultgren, R., Orr, R. L., Anderson, P. D., and Kelley, K. K. (1963). *Selected Values of Thermodynamic Properties of Metals and Alloys*. John Wiley & Sons, New York, 1–963.

Isaacs, L. L. and Garland, C. W. (1962). Heat capacity of crystalline polyethylene from 1.8° K to 5.3° K. *J. Phys. Chem. Solids* **23**, 311–315.

Jelinek, F. J. and Collings, E. W. (1975a). Low temperature thermal expansion and specific heat properties of structural materials. In *Materials Research for Superconducting Machinery III* (eds. R. P. Reed, A. F. Clark, and E. C. van Reuth). ARPA–2569, ADA012365.*

Jelinek, F. J. and Collings, E. W. (1975b). Low temperature thermal expansion and specific heat properties of structural materials. In *Materials Research for Superconducting Machinery IV* (eds. R. P. Reed, A. F. Clark, and E. C. van Reuth). ARPA–2569, ADA019230.*

Kalinkina, I. N. (1963). Magnetic heat capacities of antiferromagnetic Co, Ni, Mn, and Fe Carbonates. *Sov. Phys. JETP* (Engl. Transl). **16**, 1434–1438.

Kittel, C. (1966). *Introduction to Solid State Physics*. John Wiley & Sons, New York.

Kok, J. A. and Keesom, W. H. (1937). Measurements of the atomic heat of aluminum from 1.1 to 20° K. *Physica* **4**, 835–842.

Lawless, W. N. (1975). Dielectric and thermal properties of a machinable glass-ceramic at low temperatures. *Cryogenics* **15**, 273–277.

Ledbetter, H. M. (1982). Elastic properties. In *Materials at Low Temperatures*. American Society for Metals, Metals Park, Ohio (1983), 1–45.

Ledbetter, H. M., Weston, W. F., and Naimon, E. R. (1975). Low-temperature elastic properties of four

austenitic stainless steels. *J. Appl. Phys.* **46**, 3855–3860.
Lounasmaa, O. V. (1962). Specific heat of holmium metal between 0.38 and 4.2° K. *Phys. Rev.* **128**, 1136–1139.
McConville, T. and Serin, B. (1964). Specific heat of type II superconductors in a magnetic field. *Phys. Rev. Lett.* **13**, 365–367.
Martin, D. L. (1960). The specific heat of copper from 20° to 300° K. *Can. J. Phys.* **38**, 17–24.
Moeller, C. E., Loser, J. B., Thompson, M. B., Snyder, W. E., and Hopkins, V. (1964). *Thermophysical Properties of Thermal Insulating Materials.* ASD–TDR–64–5, AD601535, 1–362.*
Rosenberg, H. M. (1963). *Low Temperature Solid State Physics.* Clarendon Press, London.
Sandenaw, T. A. (1959). *Heat Capacity of Copper below 300 Degrees K, a Test of Two Calorimeter Designs.* U. S. AEC–LA–2307, 1–19.*
Schroder, K. (1961). Effect of magnetic clusters on the specific heat of Ni–Cu and Fe–V alloys. *J. Appl. Phys.* **32**, 880–882.
Schroder, K. and Cheng, C. H. (1960). Correlation of low-temperature caloric and magnetic effects in TiFe. *J. Appl. Phys.* **31**, 2154–2155.
Sears, F. W. (1953). *Thermodynamics, the Kinetic Theory of Gases, and Statistical Mechanics.* Addison-Wesley Publishing Company, Reading, Massachusetts, 296–298.
Shoenberg, D. (1952). *Superconductivity.* Cambridge University Press, Cambridge, England, 62.
Simmons, G. and Wang. H. (1971). *Single Crystal Elastic Constants and Calculated Aggregate Properties.* M.I.T. Press, Cambridge, Massachusetts, 157–158.
Specific Heat — Metallic Elements and Alloys, Vol. 4. (1970) (ed. Y. S. Touloukian). Plenum Press, New York, 1–750.
Specific Heat — Nonmetallic Solids, Vol. 5 (1970) (ed. Y. S. Touloukian). Plenum Press, New York, 1–1649.
Stoner, E. C. (1936). The temperature dependence of free electron specific heat. *Philos. Mag.* **21**, 146–160.
Stout, J. W. (1968). Low-temperature calorimetry with isothermal shield and evaluated heat leak. In *Experimental Thermodynamics I. Calorimetry of Non-reacting Systems* (eds. J. P. McCullough and D. W. Scott). Page Bros., Norwich, England, 215–261.
Stout, J. W. and Hadley, W. B. (1964). Heat capacity of α–$NiSO_4 \cdot 6H_2O$ between 1 and 20° K. Electronic energy levels of the Ni^{++} ion. *J. Chem. Phys.* **40**, 55–63.
Thermodynamics of Certain Refractory Compounds, Vol. 2 (1966) (ed. H. L. Schick). Academic Press, New York, 1–775.
Touloukian, Y. S. (1966). *Recommended Values of the Thermophysical Properties of Eight Alloys, Major Constituents and their Oxides.* Report No. CST–7590, Thermophysical Properties Research Center, Purdue University, Lafayette, Indiana, 271–343.
van der Hoeven, B. J. C., Jr. and Keesom, P. H. (1963). Specific heat of various graphites between 0.4 and 2.0° K. *Phys. Rev.* **130**, 1318–1321.
van Kemper, H., Miedema, A. R., and Huiskamp, W. J. (1964). Heat capacities of the metals terbium and holmium below 1° K. *Physica* **30**, 229–236.
van Kranendonk, J. and van Vleck, J. H. (1958). Spin waves. *Rev. Mod. Phys.* **30**, 1–23.
Viswanathan, R. and Gopal, E. S. R. (1961). On the exact relation between C_P and C_V in crystalline media. *Physica* **27**, 1226–1228.
Walker, C. B. (1956). X-ray study of lattice vibrations in aluminum. *Phys. Rev.* **103**, 547.
Weston, W. F., Ledbetter, H. M., and Naimon, E. R. (1975). Dynamic low-temperature elastic properties of two austenitic nickel–chromium–iron alloys. *Mater. Sci. Eng.* **20**, 185–194.
Westrum, E. F., Jr., Furukawa, G. T., and McCullough (1968). *Adiabatic Low-Temperature Calorimetry in Experimental Thermodynamics I. Calorimetry of Non-reacting Systems* (eds. J. P. McCullough and D. W. Scott). Page Bros., Norwich, England, 133–214.
Ziman, J. M. (1962). *Electrons and Phonons.* Clarendon Press, London.
Zemansky, M. W. (1968). *Heat and Thermodynamics*, McGraw-Hill, New York.

*Available from NTIS, Springfield, Virginia.

3 Thermal Expansion

A. F. CLARK

3.1 INTRODUCTION

Warming a solid body from absolute zero requires energy. In a free body, this energy manifests itself in two ways: an increase in temperature and a change in volume. Both of these are directly related to the additional vibrational energy of the individual atoms; the former simply because more atomic energy states are excited, the latter because the mean interatomic distance changes with energy. The ratio of the change in energy to the change in temperature is the specific heat. The ratio of the change in volume to the change in temperature is the thermal expansion.

Specific heat is discussed in Chapter 2 and the theoretical developments of the two properties go hand in hand. In fact, because the volume expansion is due to the anharmonic behavior of atomic vibrations and the specific heat is due primarily to the vibrations themselves, the theoretical development of thermal expansion has traditionally followed and has been, more or less, a refinement of specific heat theory. Measurements of the thermal expansion of materials, however, have led those of specific heat, for it is simpler to measure a length change than energy content. This has been fortunate since the thermal expansion provides a more stringent test of solid-state theories. Also, in engineering applications it is frequently a more critical design parameter.

This close relationship to specific heat is utilized in the discussion of the theory of thermal expansion, and several of the developments in Chapter 2 are referenced rather than repeated here. Following the discussion of theory, many of the measurement techniques are presented along with their advantages and disadvantages. The results of the measurements are then summarized for three classes of materials: metallics, nonmetallics, and composites. Because predicting thermal expansion values for unmeasured or new materials is often useful, one section is devoted to means of making educated guesses for low-temperature values. A short discussion of how thermal expansion data can be used is followed by a section describing where such data can be found.

3.1.1 Needs for thermal expansion data

Any structure that changes temperature during its operation undergoes a change in dimensions. This change in dimensions is a most critical design parameter, especially if dimensional stability is required or must be known. An example of required dimensional stability is the mounting structure of a large space telescope where the base structure determines the accuracy of the telescope alignment. As the orientation of the telescope changes with respect to the sun, parts of the structure will be warmed above the very low temperature of space, and the resulting expansion can distort the structure. Composite materials have been developed with extraordinarily low thermal expansions to minimize these effects (Kasen, 1979).

The dimensional change of a large liquefied natural gas (LNG) tank is an example of a situation where the dimensional changes must be known. For the custody transfer of the LNG stored in the tank shown in Fig. 3.1, the dimensional change that occurs as it is cooled while being filled can result in a large change in capacity. Even though the tank shown is lined with a low-expansion alloy, its volume can change by thirty kiloliters, which means many dollars to the buyers and sellers of LNG. Other tanks made of aluminum alloys have much larger dimensional changes.

Any structure that operates with a temperature gradient also requires knowledge of thermal expansion because dimensional changes while cooling down or during operation introduce stresses in the structure. Since these stresses are in addition to those required for operation, they must be known and designed for if safety and reliability are required. An example is the support structure of the liquid-helium Dewar for the superconducting field windings of an electric motor (Schramm and Kasen, 1978), as shown in Fig. 3.2. The narrow end of this fiberglass epoxy cone is at room temperature and the broad end is at 4 K, a temperature gradient of nearly 300 K over a distance of about 30 cm. As the cone is cooled, the colder regions contract more than the warmer, setting up a stress distribution in the structure that must be added to that caused just by the weight of the Dewar and its magnet. For this application the final stress level is important because a long fatigue life is required.

Finally, a low-temperature structure consisting of materials with different expansions will have a residual stress distribution after it is cooled. This can be a problem in such things as glass-to-metal seals or composite materials. However, widely differing expansions, if they are

Figure 3.1 Interior view of liquefied natural gas tank lined with a low thermal expansion alloy. (Photograph courtesy of McDonnell Douglas, Astronautics Laboratory and Gaz-Transport.)

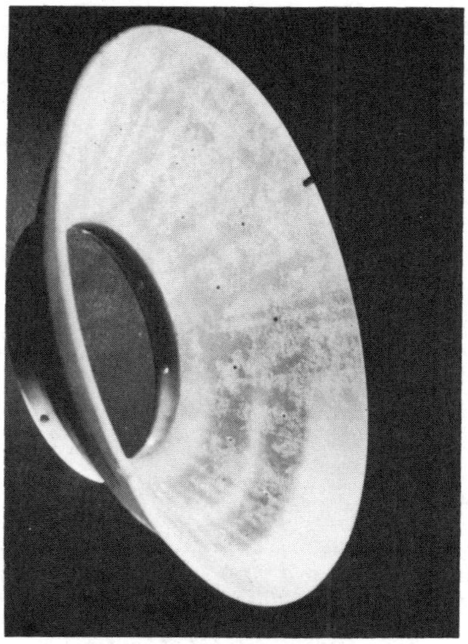

Figure 3.2 Fiberglass/epoxy support cone for a superconducting electrical motor.

known, can be utilized to the designer's advantage. For example, a multifilamentary superconducting wire consists of fine filaments of a superconductor, such as Nb_3Sn, embedded in a matrix of copper or bronze. When cooled to the operating temperature, near 4 K, the copper or bronze contracts about twice as much as the Nb_3Sn, putting it into compression. This degrades the ability of the superconductor to carry very high electrical currents (Ekin, 1981). However, when the wire is put to use in an application such as a superconducting magnet, the addition of winding, cool-down, and magnetic field stresses can return the Nb_3Sn to a relatively stress-free state where it can carry its maximum current.

Knowledge of the thermal expansion of a material is thus a prerequisite to its safe and reliable use for optimum performance. In addition, the thermal expansion is frequently used as an indicator for other materials studies, such as phase transformations (see Chapter 9). Also, as indicated earlier, it is frequently used as a test of solid-state theories.

3.1.2 Principal behavior and terminology

Most materials contract when cooled. This contraction is generally on the order of a few tenths of a percent between room temperature and absolute zero. It arises from the anisotropy about the minimum of the potential energy between two atoms as a function of the separation distance between them. In the simplest terms, the coulomb attraction and the interatomic repulsive forces are different in nature and result in a potential energy curve like that shown in Fig. 3.3. The equilibrium atomic separation at absolute zero is thus r_0, but as the energy is raised it shifts by Δr to the mean separation, r. Thus, as energy is added to a material and the temperature is raised, the mean interatomic spacing changes, resulting in a change in length.

In general, the length of a specimen is a smooth function of temperature, $L(T)$, with the normalized derivative defined as the *coefficient of linear thermal expansion,* α, or

$$\alpha(T) = \frac{1}{L}\frac{dL(T)}{dT} \qquad (3.1)$$

It will be seen later that α has the functional dependence of T^3 at low temperatures and approaches a constant at higher temperatures, generating curves for $L(T)$ and α of the shapes shown in Fig. 3.4. With few exceptions, most materials have thermal expansion curves with shapes like those shown, with only a scaling factor necessary to superimpose them. This is not true, of course, through phase transformations or for unusual bonding between atoms, such as that in some polymers.

The total change in length in a specimen between two temperatures is a function of the thermal expansion coefficient, the two temperatures, T_1 and T_2, and also the original length, $L(T_1)$. Thus, from Eq. 3.1,

$$\int_{T_1}^{T_2} \alpha(T)dt = \ln[L(T_2)/L(T_1)] \qquad (3.2)$$

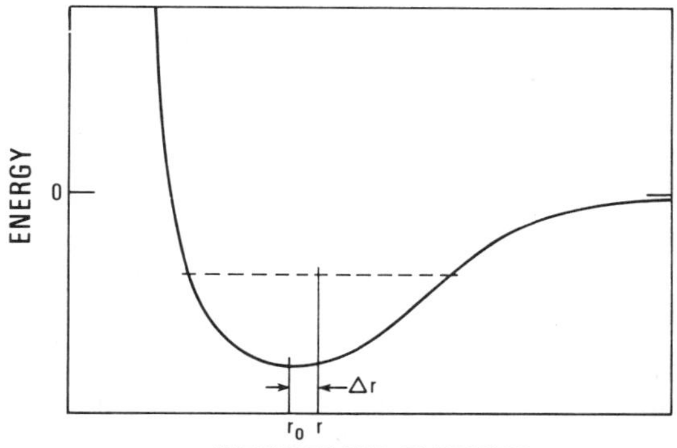

Figure 3.3 Potential energy between two atoms vs. the interatomic distance.

or

$$L(T_2) = L(T_1)\exp\left[\int_{T_1}^{T_2} \alpha(T)dT\right] \qquad (3.3)$$

If the coefficient of thermal expansion is nearly constant and a smooth function of the temperature in the temperature interval, then an average α for the temperature range can be defined as $\bar{\alpha}$. Since $\bar{\alpha}$ is small, the exponential can be expanded resulting in

$$L(T_2) \cong L(T_1)\,[1 + \bar{\alpha}\cdot(T_2 - T_1)] \qquad (3.4)$$

or

$$\frac{L(T_2) - L(T_1)}{L(T_1)} \cong \bar{\alpha}\cdot(T_2 - T_1) = \bar{\alpha}\,\Delta T \qquad (3.5)$$

In general, the *linear thermal expansion* can be defined as

$$\frac{\Delta L}{L} = \bar{\alpha}\Delta T \qquad (3.6)$$

which, at temperatures below room temperature, is occasionally referred to as thermal contraction.

The linear thermal expansion is measured in the laboratory, and the temperature for the initial length is commonly room temperature. This enables engineering calculations to have a common base for the data that is meaningful for apparatus operating above or below room temperature but has the distinct disadvantage that "room temperature" is not a very precise value. Ideally, the length at absolute zero would be best, but the actual length at 0 K is impossible to measure. The most practical compromise appears to be to choose 293 K (20° C or 68° F) as the temperature for the reference length. Thus, all of the thermal expansion data below room temperature are negative, but the thermal expansion coefficient is still positive (see Fig. 3.4).

The *Grüneisen relation* equates the linear thermal expansion coefficient to other thermodynamic parameters. This relation is the result of a straightforward calculation shown in § 3.2 and is

$$\alpha = \gamma C_V / 3BV \qquad (3.7)$$

where γ is the Grüneisen parameter; C_V, the specific heat at constant volume; B, the bulk modulus of elasticity; and V, the volume. The Grüneisen parameter is theoretically derivable and provides a good, sensitive test for fundamental atomic theories. It can be thought of as the volume dependence of the frequency of the atomic vibrations. For most materials at low temperatures its value is not a strong function of temperature, and it is frequently called the Grüneisen constant.

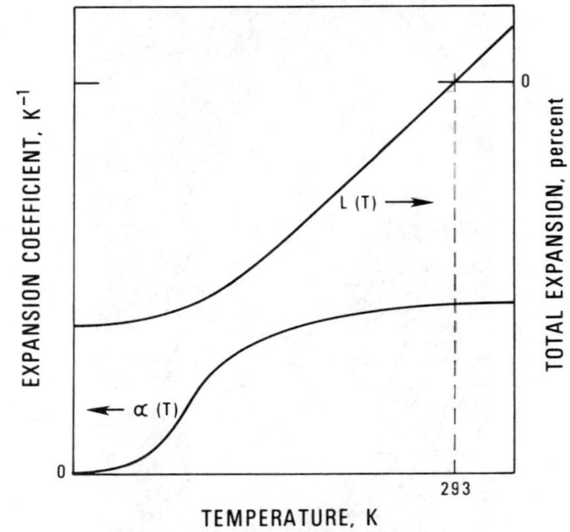

Figure 3.4 Typical curves of thermal expansion, *L(T)*, and thermal expansion coefficient, α, as a function of temperature.

One last fundamental relation is worth stating now for those who don't delve into the theory section. For all practical purposes, the *volume thermal expansion*, $\Delta V/V$, is just 3 times the linear thermal expansion or

$$\frac{\Delta V}{V} \simeq \frac{3\,\Delta L}{L} \qquad (3.8)$$

Similarly the *volume coefficient of thermal expansion*, β, is 3 times the linear coefficient of thermal expansion or

$$\beta \simeq 3\alpha \qquad (3.9)$$

These relations do not apply to anisotropic materials whose linear thermal expansion is different for different directions.

3.1.3 General material behavior

Most materials will contract when cooled, with a total length change to near absolute zero of a few tenths of a percent. In general, this contraction is smooth, monotonic, and similar in magnitude for similar materials. In Fig. 3.5 are shown broad bands of data for the general categories of glasses, metals, alloys, epoxies, and various plastics, giving an indication of the size of the linear thermal contraction and the range of data possible for each category.

Most glasses, for example, have a small con-

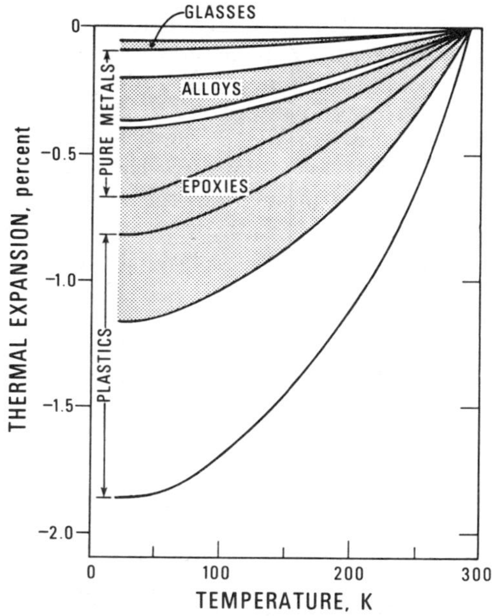

Figure 3.5 Range of thermal expansion values for various categories of materials.

traction, on the order of 0.1%, and the range is quite small (Corruccini and Gniewek, 1961). This is because the atomic bonds are fairly similar in the amorphous glassy structures, and are not strongly dependent on the kind of atoms. Some glasses and metals have been developed, however, to have expansions that match in order to make relatively stress-free glass-to-metal seals.

Metals, on the other hand, have a fairly wide range in expansion, reflecting the wide variety of atomic structure. These can vary from the tetragonal structure of indium, with a total contraction to near zero of about 0.7%, to the b.c.c. structure of chromium, with less than 0.1%. However, most alloy systems used as structural materials are based on iron, copper, aluminum, nickel, or titanium, all of which have total contractions between 0.2 and 0.4%. Since alloying constituents do not alter the basic bond behavior very much, the common alloys behave very much alike, particularly within an alloy class (Clark, 1968). One notable exception is the Fe–Ni system, which goes through the "Invar" region near 36 wt.% Ni where the phase changes from b.c.c. to f.c.c. and the thermal expansion is drastically reduced. Several other physical properties also vary widely in this region (e.g., magnetic moment, Curie temperature, specific heat). This is discussed in more detail in § 3.4.1.

There is a great variety of atomic bonding in the wide range of polymers in common use. Shown in Fig. 3.5 are the epoxies or thermosetting resins that are an example of the strongly cross-linked polymers; they range in total contraction from 0.4 to 1.2%. This disparity in thermal expansion between components in fiber-reinforced composites, such as fiberglass/epoxy, can create large internal stresses upon cool-down.

The group labeled plastics includes the more weakly cross-linked polymers, such as polytetrafluoroethylene (TFE), polyethylene terephthalate (PET), or polycarbonate. These are, in general, the more linear polymers, which are frequently used as films or fibers. Their thermal expansion can be highly anisotropic (Schramm, Clark, and Reed, 1973). In addition, the properties of these materials frequently depend strongly on the degree of crystallinity and the value of their glass transition temperature. Total contractions to 4 K vary from 0.5 to almost 2.0%.

There are some materials with either very low or very high thermal expansions. On the low end are some special glasses and some of the "Invar" alloys, with total contraction values of less than 0.05%. Some composites have been made utilizing the negative thermal expansion coefficient of graphite fibers to balance the positive contribution from an epoxy matrix, which results in a near-zero length change from room temperature to liquid-helium temperatures. On the high end of the expansion scale are the natural rubbers and elastomers, with total contractions to 4 K of 2% to 3%.

Thus, for most materials, the thermal expansion behavior below room temperature is straightforward and quite small, a few tenths percent contraction. In large structures, however, this can add up to a significant length change that must be accommodated. If not, internal strains on the order of the yield strains or, in some materials, even fracture strains result. The purpose of the remainder of this chapter is to understand and document this behavior.

3.2 THERMAL EXPANSION THEORY

Historically, the development of the theoretical basis for thermal expansion behavior has closely followed that of specific heat. This has been primarily because thermal expansion arises from the addition of anharmonic behavior of the particle vibrations to the study of their energy content. For specific heat, the vibrations are usually assumed to be harmonic. In a similar manner, the development of this chapter follows that in Chapter 2, with several additions unique to the understanding of thermal expansion, principally the addition of anharmonicity.

That the specific heat at low temperatures was found to be significantly less than the $3R$ predicted by an equipartition of energy (R is the gas constant) led to various assumptions about the frequencies of vibration allowed. Einstein was the first to account for the quantum nature of the low-temperature oscillations and assumed that all particles had the frequency predicted by Planck to be $\nu_E = kT/h$ and derived Eq. 2.16 for the specific heat (where k and h are the Boltzmann and Planck constants). Agreement with experiment was not good, however. Debye assumed a continuous frequency distribution up to a maximum allowed by the total energy content of independent oscillators and generated Eq. 2.26. Agreement with experimental data then available was remarkable, and in addition, it had the general form found empirically by Dulong and Petit of $3R$ times a function of T/θ where θ was a characteristic temperature unique to each material. This characteristic temperature was the Debye temperature, $\theta_D = h\nu_{\max}/k$, where the $\nu_{\max}$ was the maximum allowable frequency. Refinements of the theory for specific heat have generally been refinements in the treatment of the frequency distribution, such as different maximum frequencies for longitudinal and transverse vibrations (Born and von Karman, 1912; 1913a; 1913b) or more rigorous continuous distributions (Blackman, 1935a; 1935b). Accounting for anisotropies, anharmonicity, mass differences between adjacent atoms, and other complications has resulted in more complex behavior usually described by temperature-dependent values of θ_D.

To introduce anharmonicity, the potential energy curve of Fig. 3.3 must be described. The potential energy, ϕ, of a bound particle can be expressed as a function of its displacement, d, as

$$\phi(d) = pd + qd^2 + rd^3 + sd^4 + \ldots.. \quad (3.10)$$

where p, q, r, and s are constants. Since the restoring force must equal zero at zero displacement, then $p = 0$. If, to a first approximation, the potential energy is a function only of the displacement squared, that is, $\phi = qd^2$ and r, s, ... are zero, then the oscillations are "harmonic" and the bound particles are harmonic oscillators. Most of the specific heat developments have made this assumption. However, this means that any increase in energy will not result in any change in the mean displacement; that is, there will be no thermal expansion. Thus, to account for thermal expansion, higher order terms, specifically those odd in powers of d, may be included, and the coefficients must be allowed to be functions of the volume. The simplest way to handle the anharmonicity in the context of the simpler theories is to let the frequencies of vibration vary with the volume, but with the oscillations remaining harmonic. This is called the *quasi-harmonic approximation.*

In his thermodynamic development, Grüneisen (1912) noted that the dimensionless parameter

$$\gamma = \beta V / \chi_T C_V \quad (3.11)$$

was nearly a constant for all the materials then studied, where β is the volume expansion coefficient; V, the volume; χ_T, the isothermal compressibility; and C_V, the constant-volume specific heat. This Grüneisen function can be shown thermodynamically to be the logarithmic derivative of the temperature with respect to volume at constant entropy, S, or (Barron, Collins, and White, 1980)

$$\gamma = -(\delta \ln T / \delta \ln V)_S \quad (3.12)$$

In most theoretical developments it can be obtained from derivatives of the Helmholtz free energy and can therefore be thought of as the derivative of energy content with respect to volume; thus it is the measure of thermal expansion most frequently compared with theory. For the most common Einstein and Debye developments of specific heat, the energy content is directly related to the frequencies of oscillation by the Planck relation and, thus, to the earlier description of the Grüneisen gamma as a measure of the change of frequencies with volume.

Another reason that the Grüneisen gamma is so commonly utilized in the theoretical development of thermal expansion is that different contributions to the energy content usually can be separated and the volume dependence of each then calculated. Frequently separate gammas are calculated and compared with experimental values where one component of the specific heat is expected to dominate. In the theoretical description of thermal expansion that follows, the thermodynamic relations are developed first, the separate contributions for various lattice and electronic contributions are described, and finally other refinements and minor effects are mentioned.

3.2.1 Thermodynamic relationships

The volume coefficient of thermal expansion, β, is most readily related to the Helmholtz free energy, $F(V, T)$, because most experimental conditions for the study of thermal expansion permit the volume and temperature to vary under conditions of constant pressure. For these conditions one can show (Zemansky, 1968) that the entropy, S, is

$$S = -(\delta F/\delta T)_V \tag{3.13}$$

If the volume thermal expansion coefficient is defined as

$$\beta = (1/V)(\delta V/\delta T)_P \tag{3.14}$$

or the normalized change in volume with temperature at constant pressure, then the ratio of the volume thermal expansion coefficient to the isothermal bulk compressibility, χ_T, is

$$\begin{aligned} \beta/\chi_T &= (\delta S/\delta V)_T \\ \beta/\chi_T &= -\delta^2 F/\delta V \delta T \end{aligned} \tag{3.15}$$

where the isothermal compressibility is

$$\frac{1}{\chi_T} = V(\delta^2 F/\delta V^2)_T \tag{3.16}$$

The specific heat at constant volume, C_V, is commonly

$$\begin{aligned} C_V &= T(\delta S/\delta T)_V \\ C_V &= -T(\delta^2 F/\delta T^2)_V \end{aligned} \tag{3.17}$$

so that the dimensionless collection of thermodynamic parameters that Grüneisen found empirically to be nearly a constant, γ, is

$$\gamma = \beta V/\chi_T C_V \tag{3.18}$$

$$= (V/T)[(\delta^2 F/\delta V \delta T)/(\delta^2 F/\delta T^2)_V] \tag{3.19}$$

which, if the volume dependence of the energy can be expressed through one characteristic energy, E, then is (Barron et al., 1980)

$$\gamma = -\delta \ln E/\delta \ln V. \tag{3.20}$$

All of the various theoretical descriptions are refinements of how this characteristic energy is described for the solid material under consideration.

Thus, in γ there is a means of comparing these theoretical descriptions of the relationship between energy and volume with the experimentally determined quantities of volume expansion coefficient, specific heat, and bulk compressibility. The compressibility is not a strong function of temperature and can frequently be estimated, but the expansion and specific heat are both strong functions of temperature and must be measured. The theoretical description of the energy is where the choice of model allows various assumptions to be made about how the anharmonic nature of solids is introduced. Recall that this is necessary or there will be no thermal expansion. So, by changing the details of the energy description, one can calculate the effect on γ for comparison with its value determined from the experimental parameters β, χ_T, and C_V.

One of the most common assumptions about the energy is that various contributions can be treated separately and simply added. Empirically, this has been shown to be a good assumption for the specific heat (Gopal, 1966). This holds for the separation of lattice, electronic, and magnetic contributions to the total specific heat (see Eq. 2.63) as well as for the individual contributions of each discrete frequency component in a detailed description of the vibrational modes of the lattice. Therefore, from Eq. 3.20 the jth component, γ_j, is

$$\gamma_j = -\delta \ln E_j/\delta \ln V \tag{3.21}$$

and an average γ weighted by the components of specific heat is

$$\gamma = \sum_j \gamma_j C_{V_j} / \sum_j C_{V_j} \tag{3.22}$$

where the total specific heat is

$$C_V = \sum_j C_{V_j} \tag{3.23}$$

Thus, data taken and compared within a special temperature range, where only one energy mechanism is expected to contribute to the specific heat, can confirm validity of that particular model. This basic process is used to discuss various components in the remaining part of this section; a brief exposure to the more subtle refinements of the theoretical description of thermal expansion follows.

3.2.2 Lattice vibrations

The contribution of the lattice vibrations to the specific heat is, to a first approximation, from the harmonic oscillation of each atom about its equilibrium position. With harmonic oscillation there is no thermal expansion. Many frequencies are possible, $3p$, or one for each degree of freedom for each of p atoms in a unit cell. Debye established an upper limit of this discrete set of frequencies and established a theory for specific heat that had good agreement with experiment. The first three of these frequencies are considered the elastic waves or acoustic modes, where all p atoms in the unit cell move in unison. The remaining $3p$-3 modes are the optical branches and can be considered as various combinations of the sublattices moving against each other. To include the anharmonicity of the lattice, the higher order coefficients in Eq. 3.10 must be allowed to be volume dependent, and then the potential energy of displacement and the change in the frequency distribution with volume would be very complex indeed. If, for each volume, the harmonic approximation is assumed to be valid, then the discrete set of frequencies can be calculated, each of which is then permitted to change with volume. This is the quasi-harmonic approximation and the basis for most theoretical calculations of the thermal expansion.

If it is assumed that the electrons do not interact with the lattice and that the energy levels for each harmonic oscillator are quantized, then there is a free energy, f_i, associated with each normal mode of frequency

$$f_i = kT\{\tfrac{1}{2}x_i + \ln[1 - \exp(-x_i)]\} \quad (3.24)$$

where k is Boltzmann's constant and

$$x_i = \frac{h}{2\pi}\omega_i/kT \quad (3.25)$$

where h is Planck's constant and $\omega_i = 2\pi\nu_i$. (There is an implied sum over all possible wave vectors in the Brillouin zone. See, for example, Krishnan, Srinivasan, and Devanarayanan, 1979.) Then the total free energy is

$$F = \sum_{i=1}^{3p} f_i \quad (3.26)$$

Applying Eq. 3.17 to each free energy component (Eq. 3.24) one can obtain the specific heat component attributed to each mode

$$C_{V_i} = k \sum_i \sigma_i \quad (3.27)$$

where the σ_i are the Einstein specific heat functions

$$\sigma_i = x_i^2 \exp(x_i) / [1 - \exp(x_i)]^2 \quad (3.28)$$

In a similar manner through Eq. 3.19, a specific γ_i analogous to Eq. 3.20 becomes

$$\gamma_i = -\delta \ln \omega_i / \delta \ln V \quad (3.29)$$

or the logarithmic change in frequency with volume. An average $\gamma(T)$ weighted by the specific heats is then

$$\gamma(T) = \sum_i \gamma_i\sigma_i / \sum_i \sigma_i \quad (3.30)$$

which is now a function of temperature because the Einstein functions are. (Again an integration over all wave vectors in the Brillouin zone is implied.) So, if an educated guess can be made for the volume dependence of the frequencies, then a comparison can be made with the experimentally derived parameters in γ. This is usually done by calculating them from the volume dependence of the elastic wave velocities, which in turn are derived from the volume dependence of the elastic constants (see Chapter 1).

It is informative and helpful for comparing with experimental results to determine the behavior of $\gamma(T)$ at the high- and low-temperature limits. At the higher temperatures, the σ_i go to one and the $\gamma(T)$ is simply the sum of all the individual γ_i over all the normal modes and γ is a constant. At the lower temperatures, all the higher frequencies are frozen out, the optical modes do not contribute, and only the first three acoustic modes are included in the averaging. It can then be shown (Barron et al., 1980) that $\gamma(T)$ becomes

$$\gamma(0) = -\delta \ln \theta_c / \delta \ln V \quad (3.31)$$

where θ_c is the Debye characteristic temperature for specific heat as T goes to zero, which can also be calculated from the elastic constants. Reasonable agreement with experimental results has been obtained this way from even room-temperature values for the pressure dependence of the second-order elastic constants. More detailed analyses usually extend into the moments of the frequency distribution, which can be derived analytically (Barron et al., 1980). If γ is a constant, then from Eq. 3.18 it can be seen that β is proportional to C_V, which for the low-temperature limit (Eq. 2.30) is proportional to T^3. Therefore, as T goes to 0, the thermal expansion coefficient is proportional to T^3.

3.2.3 Electronic and magnetic contributions

As was shown in § 2.2.3, the electronic specific heat can be treated as additive to the lattice contribution, resulting in a term that is proportional to T. The electrons contribute to the thermal expansion directly from the volume dependence of their entropy. At low temperatures, the electronic entropy, S_e, can be expressed as

$$S_e = (2/3)\,\pi^2 k^2 T\, N(E_f) \quad (3.32)$$

where $N(E_f)$ is the molar electronic density of states at the Fermi energy, E_f, and higher order terms have been neglected (see Chapter 2). (Equations 3.17 and 3.32 lead to the expression in § 2.2.2 for C_e, Eq. 2.35.) Varley (1956) was the first to derive an electronic γ_e or

$$\gamma_e = \delta \ln N(E_f) / \delta \ln V \quad (3.33)$$

For single-band, free electrons, the density of states is proportional to $V^{2/3}$ so that $\gamma_e = 2/3$. For real metals this is complicated by the band structure, and calculations in general are not in agreement with the experimental values for different metals, which range from $+10$ to -10. Potassium with its approximately spherical Fermi surface comes closest, with a value slightly larger than 0.9.

In a like manner, the volume dependence of the entropy contribution due to magnetic interactions between spins adds to the thermal expansion. The most observable of these interactions is the onset of ordering when the spins become frozen out as the temperature is lowered. This creates anomalies in the specific heat, which also show up in the thermal expansion. Excitations involving spin reversals, or spin waves, can be quantized much as the lattice waves. A similar treatment yields a magnetic γ_m, which becomes the logarithmic derivative of the magnetic coupling energy with respect to volume or

$$\gamma_m = -\delta \ln J / \delta \ln V \quad (3.34)$$

where J is the spin exchange energy. Calculations (Meincke and Graham, 1963) suggest a term proportional to $T^{3/2}$, which, however, has not been experimentally observed (Barron et al., 1980). In general, magnetic effects on the thermal expansion are quite variable in magnitude, difficult to separate, and even more difficult to assess theoretically, especially in metals and alloys.

Rewriting Eq. 3.18, an equation for the volume expansion is obtained that can be expressed as the sum of various j contributions

$$\beta \cong (\chi_T / V) \sum_j \gamma_j C_{V_j} \quad (3.35)$$

which for the low-temperature regime can be approximated by

$$\beta = aT + bT^3 + cT^5 + ... + rT^{3/2} + ... \quad (3.36)$$

because the temperature dependence of the various contributions is known. Thus, as the temperature approaches zero, the lattice vibrations are frozen out and only the electronic contribution to thermal expansion persists, which is proportional to T. This is about as far as the theoretical prediction of the thermal expansion is carried, whereas most of the comparisons with theoretical models are done by comparing Grüneisen values for different assumptions. These can become quite complex and some of the more successful refinements are discussed below.

3.2.4 Other effects and methods

Refinements and modifications of the theoretical developments are given in this section to describe the crystal lattice more accurately or to include other energy contributions to the change in volume. These are only identified; references are given, principal conclusions are summarized, and further study is left to the reader. In general, the magnitudes of the effects described here are small and do not contribute significantly to engineering values of the thermal expansion, but they do enhance our understanding of the basic mechanisms.

In *superconductors* it is frequently assumed that the lattice contribution to the specific heat does not change through the superconducting transition, which has been supported experimentally (Barron et al., 1980) and that any differences are due to the electrons only. Munn (1969) has calculated a Grüneisen value for the superconducting electrons, γ_{es}

$$\gamma_{es} = \frac{C_{en}\gamma_{en}}{C_{es}} + \frac{C_n - C_s}{C_{es}} \frac{d \ln T_c}{d \ln V} \quad (3.37)$$

where the subscripts n and s are for normal and superconducting and T_c is the critical temperature for the superconducting transition. With the above assumption that the lattice specific heats are the same (which is the same as Eq. 2.54), it can be shown that

$$\gamma_{es} \simeq \gamma_{en} + \frac{C_{en} - C_{es}}{C_{es}} \frac{d \ln T_c}{d \ln V} \quad (3.38)$$

which is the normal electronic term plus a term from the superconducting state of the electrons. Since the electronic specific heat is usually larger in the superconducting state and the logarithmic derivative of T_c is usually positive, γ_{es} is frequently negative. The derivative in the last term can be rewritten as

$$d \ln T_c / d \ln V = \frac{1}{\chi_T T_c} \; dT_c/dP \qquad (3.39)$$

and then comparison can be made to the pressure dependence of the critical temperature (Boughten, Olsen, and Palmy, 1972).

The problems of having to use temperature or magnetic field changes to move through the superconducting transition make it difficult to compare experiments. Problems of hysteresis, trapped flux, and demagnetizing effects, have resulted in little agreement among experiments, and since the effects are small, little study of practical materials has been done. Observations of the sudden increase in specific heat, however, have been much more useful in studying the onset of superconductivity (Collings, 1983).

There are many modifications to the crystal lattice that must be accounted for in real solids: anisotropy, defects and impurities, the surface, polycrystallinity, transformations, and many others. Most theoretical developments treat these as minor perturbations on the general lattice description. In most cases, the effects on the experimental values of the thermal expansion are small and used mostly as a refined check on the theoretical description of the solid.

Many materials in the single-crystal form are highly *anisotropic*. Zinc and cadmium are the most common, with ratios of the thermal expansion in different directions of more than 5. Munn (1969) has summarized the experimental data on materials with anisotropic thermal expansion. In general, the method of treating anisotropy is to perform the thermodynamic derivatives with respect to specific directions, thereby generating thermal expansion tensors and Grüneisen relations for each direction. These, then, can be simplified knowing the crystal symmetry. The specific γ_i is then

$$\gamma_{i,mn} = -\frac{\delta \ln \omega_{i,mn}}{\delta \ln \epsilon_{mn}} \qquad (3.40)$$

where ϵ_{mn} are the strain components and the specific $\gamma_{i,mn}$ are summed using the elastic compliances (Srinivasan and Ramji, 1972).

Crystal defects and impurities can also contribute to the thermal expansion, but in general, at the low temperatures their contributions are literally frozen out. Defects, such as vacancies and interstitials, can have low-lying energy levels that can be identified with some frequencies and treated as additive to the normal frequency distribution. Impurities have an effect owing to their mass or charge difference, which changes the coupling to the rest of the lattice and thus alters the frequency distribution. Timmesfeld and Elliott (1970) showed that this effect was small, but possibly of either sign, which may explain the thermal expansion of dilute alloys. Magnetic or differently charged impurities were studied by Sheard (1969; 1972), who used measurements and predictions of γ to compare different models of impurity–lattice interactions.

The anisotropic nature of the atomic bond near the *surface* of a solid is much larger than in the bulk. Therefore, the thermal expansion could be expected to be much different (usually larger) in the last few layers near the solid boundary. This effect can be important in experiments that sample only a thin surface layer, such as low-energy electron diffraction (LEED), scanning Auger spectroscopy, or ion-scattering spectroscopy. Wallis (1975) has reviewed the theoretical and experimental results in this little-known field. Typical ratios of the surface thermal expansion coefficient to the bulk value are from 1 to 3 for the few normal metals measured and about 4 or 5 for solidified gases. Theoretically, the value is expected to rise rapidly at temperatures below 50 K.

First-order *phase transitions* are usually accompanied by an abrupt change in volume. Second-order phase transitions are accompanied by an abrupt change in the volume thermal expansion coefficient. Both can create problems in measurement and design, but both can be used to study the transitions themselves. Almost any thermodynamic text (e.g., Zemansky, 1968) will show the derivation of the Clausius–Clapeyron relation for the volume change and the Ehrenfest equation for the change in the volume thermal expansion coefficient. This latter is

$$dP/dT = \frac{1}{TV}\frac{\Delta C_P}{\Delta\beta} = \frac{\Delta\beta}{\Delta\chi_T} \qquad (3.41)$$

where dP/dT is the slope of the equilibrium curve in the P–T diagram; thus, the thermal expansion changes tend to be small. Krishnan et al. (1979) devote an entire chapter to the discussion of thermal expansion through phase transitions in crystals. The most frequent use of these studies is to mark the phase transition, but occasionally they can be used to determine the nature of the transition by the sign of the change, for example, ferroelectric vs. antiferroelectric. (See also Chapter 9.)

There are many *models and approximations*

that have been used to describe the lattice and electrons more accurately, so that thermal expansion can be used as a measure of the accuracy of the model. Barron et al. (1980) describe several of these in detail. Briefly, the anisotropic continuum model assumes that the vibrational frequencies are a function only of the possible wave vectors and thus elastic constants and their pressure derivatives can be used for calculations (Sheard, 1958; Collins, 1963). Short-range pair potentials were used by Barron (1955) to determine the changes in frequency from the off-axis restoring forces, a method good for comparing various crystal behaviors. The rigid-ion model varies the repulsive potential between two atoms and was first used by Born (1923) to predict some negative thermal expansion values. Later, Blackman (1957) used it for some good qualitative comparisons. Others describe the potential as a function of bond length and angle, consider the deformation of the ion and the electron distribution (Hardy, 1962), or try to use the pseudopotentials common to the quantum description of solids. The absence of very large anisotropies and many negative components for the thermal expansion are successfully explained by letting the internal strains be accommodated by rotation or bond relaxation (Gibbons, 1973). The description of molecular crystals including extra rotational degrees of freedom has not been very successful, probably because the vibrational and rotational frequencies are interactive.

3.3 METHODS OF MEASUREMENT

The methods of measuring thermal expansion are quite straightforward and simple. Only two parameters need measuring, the length and the temperature, to determine the function $L(T)$ from which all the rest of the data are determined. The thermal expansion coefficient is simply the normalized first derivative of $L(T)$. In practice, a change in length, ΔL, is much simpler to measure, so most measurement methods determine $\Delta L(T)$ with respect to some known material at the same temperatures. Because the magnitude of ΔL depends directly upon the original length, L_0, a reference measurement of the original sample length must also be done. Ideally, this should be measured at 0 K, but practicality dictates room temperature; it is much easier and actually a more useful reference value. Thus, 293 K (68° F) is most common; occasionally the ice point, 273 K, is used when more precision is needed.

There are many clever ways to determine a length with a great deal of accuracy. The experimental difficulties arise principally from one source, ensuring that the length change is determined with respect to a known (or constant) length. Transmitting the information to room temperature without upsetting the measurement system is not always easy but is usually straightforward. The simplest methods transmit the change in length mechanically, but electrical or optical signals have the obvious advantages of low heat loss, fewer mechanical linkages to freeze, and usually higher accuracy. Only one method, the x-ray determination of the lattice constant, is truly absolute. For discussion, the remaining methods are divided into mechanical, electrical, and optical techniques on the basis of their principal motion sensor. Table 3.1 summarizes the various methods, shows typical sensitivities and accuracies, and lists their major advantages and disadvantages.

As pointed out by Barron et al. (1980), there are some natural limits to the accuracy with which one can measure the thermal expansion of a given material. Typical values of the thermal expansion coefficient, α, near room temperature are about 10^{-5} K^{-1}. Thus, to achieve a 1% accuracy in α over a ΔT of 1 K, a resolution of 10^{-7} in length change divided by length is necessary. For a 1-cm sample this means a sensitivity of 1 nm, which is not unreasonable. At low temperatures, α has much smaller values and even more sensitivity is necessary, well beyond practical methods. The solution, of course, is to measure over a wider temperature range, which is commonly done. However, the α determined in this way is then an average over the temperature range. Note that even the best polished surfaces have a roughness of the order of nanometers, so any method relying on linking of mechanical surfaces has this limit. Similarly, it is very difficult to measure changes less than 10^{-2} or 10^{-3} of a wavelength (though not impossible) using optical methods.

A word of caution about the use of sensitivity, resolution, accuracy, and other descriptors of apparatus limits: Fantastic claims of the ability to observe a change in length that corresponds to small fractions of an atomic spacing can be absolutely true, but it takes three parameters, ΔL, L_0, and T or ΔT, to determine the thermal expansion. In the discussion below about different methods of measuring the thermal expansion, *sensitivity* refers to the minimum detectable

Table 3.1 Methods of measurement of thermal expansion at low temperatures.

Method	Sensitivity (m)	Accuracy (%)	Temperature Range (K)	Major Advantages	Major Disadvantages
Mechanical	10^{-6}	10^{-3}	4–3000	Simplicity of method Rapid Easily automated	Reference standard needed Large sample size
Electrical					
Capacitance	10^{-12}	10^{-5}	2–300	High precision Relative or absolute	Precise machining required
Differential transformer	10^{-10}	10^{-5}	2–300	Relative or absolute	Large sample size Limited sample materials
Optical					
Interferometer	10^{-12}	10^{-6}	4–600	Dynamic measurement High precision Easily automated	Temperature inaccuracies
Resonance	10^{-14}	10^{-7}	100–600	Highest precision Absolute	Precise sample preparation
Lever	10^{-10}	10^{-4}	4–300	Simplicity of method Easily automated	Relative only Mechanically sensitive
X-ray	10^{-9}	10^{-4}	4–3000	Absolute method Small sample size Study of phase transitions	Low accuracy Not dynamic

length change, *resolution* is this length change divided by the total specimen length, and *accuracy* refers to the estimate of the total uncertainty of the measurement, including all imprecision in the method and estimates of any uncontrollable biases in the calibration, reference data, temperature measurement, or reproducibility. In general, the temperature measurement is quite accurate and reproducible (see Chapter 14). Thus, a sensitivity of 10 pm in a specimen of 10 mm yields a resolution of $\Delta L/L$ of 10^{-9}, a truly exceptional measurement unless the accuracy was only 10^{-6}!

The American Society for Testing and Materials (ASTM) has established standard methods for determining the thermal expansion of a variety of materials. There are some 19 standard test methods or recommended practices that have been accepted by ASTM for a variety of special materials. Those methods that are adaptable for low-temperature use are based on either the quartz tube dilatometer or the interferometric techniques, both of which are described below.

3.3.1 Mechanical methods

The simplest and most straightforward of the means to extract the length change from a low-temperature region is a mechanical linkage. If one chooses the materials carefully so that the heat leaks are kept to a minimum and the thermal expansions of the connecting links are not large, quite accurate and reproducible data are possible. An example of this kind of apparatus is shown in Fig. 3.6, where quartz tubes are used to support the sample and to ride on top of the sample. The length change is then read at room temperature between the two quartz tubes with a simple dial indicator. This quartz tube dilatometer has been the workhorse for low-temperature measurements for many years. More data have been taken on this kind of dilatometer on more kinds of material than with any other measurement technique. (See, e.g., Arp, Wilson, Winrich, and Sikora, 1962; Rhodes, Moeller, Hopkins, and Marx, 1963; Clark, 1968.) Since most of the data discussed in the rest of this chapter were measured with devices like this, its operation and limitations are discussed in detail.

Quartz tube dilatometer. In Fig. 3.6 the sample is contained in a high-thermal-conductivity heat shield to maintain temperature uniformity, and thermocouples are attached directly to the sample. The support quartz tube is attached rigidly to the top of the cryostat, and the rider quartz tube floats on the top of the specimen against a measuring device, the simplest being a dial indicator. Linear differential transformers or other electronic readouts of motion can be used along with digital or analog voltmeter readings of the thermocouple output to provide tabular or even direct plots of length vs. temperature.

For temperature control in a tube dilatometer,

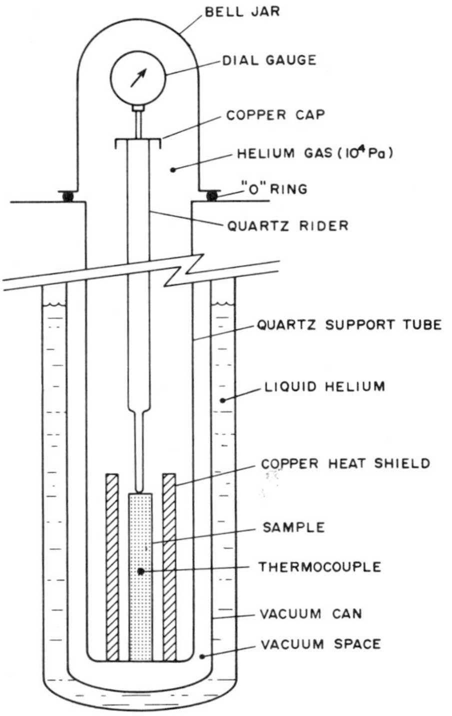

Figure 3.6 Schematic of a quartz tube dilatometer.

the sample support tube is usually isolated from a cryogen bath by a vacuum and allowed to cool slowly from room temperature to near the bath temperature. Cooling rates can be as rapid as temperature uniformity of the specimen allows; that is, the temperature gradient in the specimen due to the cooling process must be less than the accuracy limit set for the temperature measurement. For metals, this rate can be typically a few degrees per minute, so that a complete curve of length vs. temperature from room temperature to near absolute zero can be done in a few hours.

Calibration of the apparatus is accomplished by measuring a known specimen, preferably one of nearly the same thermal expansion. (Standard reference materials for thermal expansion for several different kinds of materials are available from the National Bureau of Standards in Washington, D.C. [see § 3.4.4]). A measurement of the thermal expansion of the reference material will give a value slightly different from the reference data, because the measurement is really the difference between the expansion of the specimen and that of the quartz support over the same distance. This is one of the reasons why quartz or some other low-expansion material is used for the support and rider tubes. The measured difference, then, provides the calibration of the apparatus and for most materials is only a few percent of the thermal expansion value. Specimen lengths are usually from 1 cm to 20 cm, which results in accuracy limits of about 10^{-5} in the thermal expansion and about 10^{-7} in the thermal expansion coefficient.

Other tube dilatometers. There are many other versions of the tube-type dilatometer that allow measurements to even very high temperatures. These are nicely summarized in a recent book by Valentich (1981). Using alumina or tantalum for the support tubes, for example, can extend the useful temperature range as high as 1600° C or 2200° C, respectively. In his book, Valentich describes several specific tube-type dilatometers; the various materials that can be used for specimen support and their advantages; specimen size, shape, and mounting method; calibration techniques; and much more. He concludes that accuracies of $\pm 0.5\%$ are reasonably easy to obtain with an ideal specimen size of about 1/4-in (6-mm) diameter by 2-in (50-mm) long and cooling rates of a few degrees per minute. He also describes and evaluates several different commercially available dilatometers.

The several other tube-type dilatometers used extensively for low-temperature research or data generation for a variety of materials differ only in the sensor used to determine the length change or in the cooling technique. For example, the dilatometer used by Rhodes et al. (1963) uses Fizeau plates and counts fringes and the one used by Dahlerup-Petersen (1980) to measure composite materials utilizes an inductive displacement transducer. Other commercially available systems, such as that reported by Rodrigues (1981), use the evaporating liquid helium as a source of cool gas. A balance is established between the cold cryogen and an electric heater to maintain a constant or slowly drifting temperature. In all other aspects they are still tube-type dilatometers and have the same advantages and disadvantages as the one described above, where the transmittal of the length change measurement is mechanical.

Tube-type dilatometers have another distinct advantage when the test specimens are under compressive or tensile load. The loads can be applied at room temperature, where mechanical parts work better, and can be transmitted to the specimen by the same tubes that do the measurement. This makes it possible to measure fibers, wires, or thin films, which otherwise could only be measured as rolls or stacks of material. Such a device used to measure the thermal expansion of plastic films under tension is shown in Fig. 3.7 from McConnell, Daney, and Kirgis (1970). Similarly, Fujii, Ekin, Radebaugh, and Clark (1980) modified a quartz tube dilatometer to determine the thermal expansion of several different kinds of superconducting wire.

3.3.2 Electrical methods

There are two basic methods of converting the change in length to electrical signals before the information is transmitted out of the cryostat. One method uses the specimen itself as one side of a capacitor so that a length change becomes a change in capacitance, which is detectable by a bridge circuit or a change in frequency if the capacitor is part of a resonant circuit. The other transmits the change in length to one side of a transformer such that the change in mutual inductance in the transformer is also detectable in a bridge or by a change in frequency. Both methods can be very sensitive and have the advantage that only electrical leads carry the signal to room temperature, but both are limited in application to narrow temperature ranges. In addition, the capacitance apparatus requires precise manufacture, and the inductance method is difficult to use with magnetic or superconducting materials.

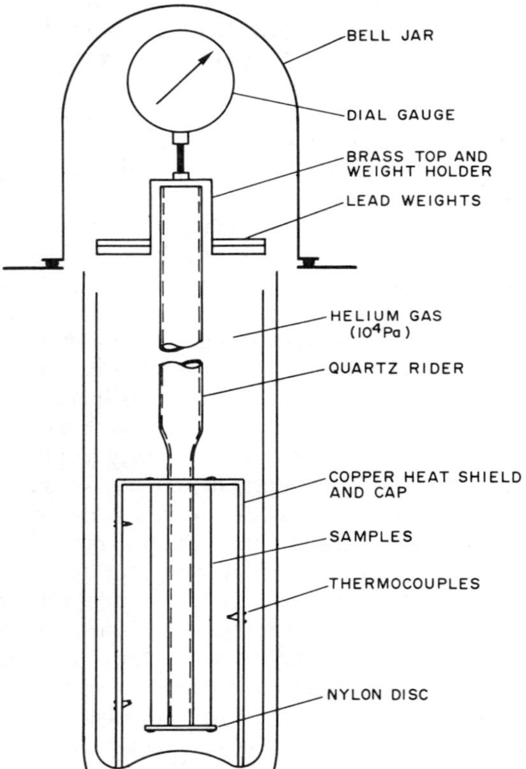

Figure 3.7 Quartz tube dilatometer for measuring films or fibers.

Capacitance. The capacitance of a parallel plate capacitor changes with either the area of or the spacing between the plates. The first is linear and the second is highly nonlinear; both have been used. Bijl and Pullan (1955) were the first to use the two-terminal capacitor to measure the thermal expansion by incorporating the specimen as one side of a parallel plate capacitor in an LC circuit and measuring the change in the resonant frequency. This resulted in a resolution in $\Delta L/L$ of 10^{-7}, an impressive improvement for the time. Because the circuits necessarily included the stray and lead capacitance and inductance, a change to the three-terminal capacitor was soon initiated by White (1961) and improved upon by Carr, McCammon, and White (1964; 1965). These and more recent devices (e.g., White and Collins, 1972; Kroeger and Swenson, 1977) have an impressive sensitivity of about one picometer.

A typical two-terminal capacitance sample cell is shown in Fig. 3.8, where the specimen forms one plate of the capacitor and the support structure of the cell, the other. Thus the measurement determines the relative thermal expansion of the specimen with respect to the cell material. This is also true in the three-terminal capacitance device also shown in Fig. 3.8, which has the advantage that the two measuring capacitor plates are shielded and the capacitance is measured with a ratio-transformer bridge circuit. Because the cell becomes the reference material, it is commonly made from oxygen-free copper whose thermal expansion is quite well known. To avoid this dependence, Carr et al. (1964) designed an absolute cell (Fig. 3.9) in which the cylindrical specimen was isolated from the reference cell by a vacuum and quartz spheres. Thus it was allowed to "float" strain free within the cylindrical capacitance plate as it drifted in temperature while heat was internally supplied to the specimen.

In all these devices, the capacitance varies inversely as the square of the spacing, so large temperature ranges, that is, large gaps, result in loss of sensitivity. On the other hand, Kroeger and Swenson (1977) inverted the geometry so the

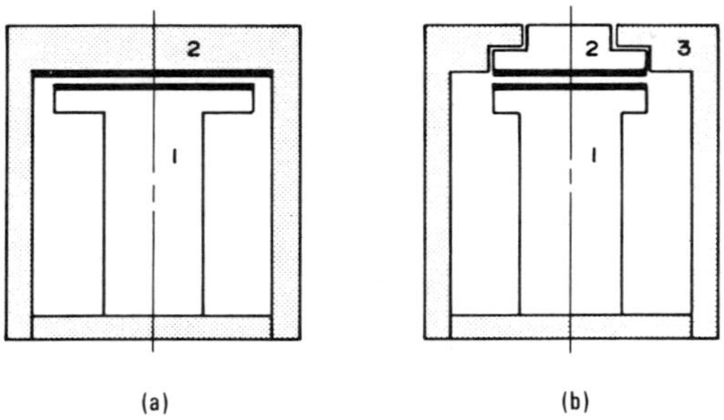

Figure 3.8 Two- (a) and three- (b) terminal capacitance cells for electrical measurement of thermal expansion.

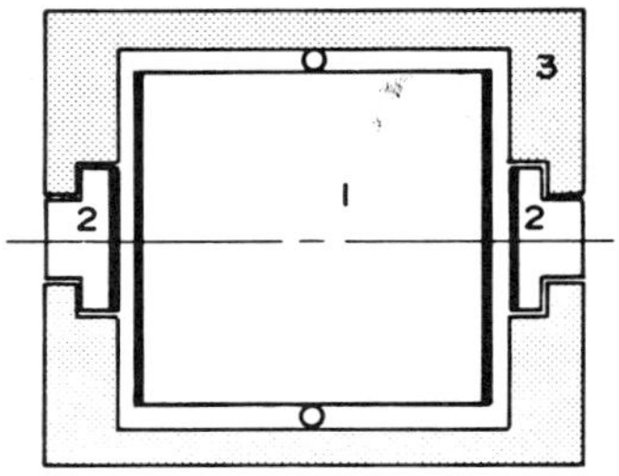

Figure 3.9 Absolute capacitance cell for electrical measurement of thermal expansion.

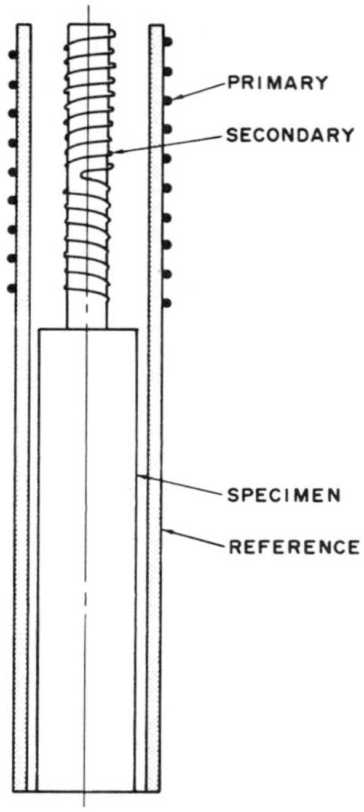

Figure 3.10 Linear variable differential transformer (LVDT) for inductive measurement of thermal expansion.

specimen contracted closer to the reference material and the sensitivity increased with the decreasing gap. Nonlinearity over a large temperature range can also be avoided by using two cylindrical capacitors that move within each other so that the capacitance varies linearly with the length change, but at the expense of a loss of sensitivity to about a nanometer (Wolfendale, 1968). Also, all of these devices have the distinct disadvantage of requiring precise design, machining, and assembly and therefore have been principally used for fundamental studies where the precision was necessary.

Inductance. The linear variable differential transformer, or LVDT, is the most common inductive device for converting displacement to an electrical signal. In this mutual inductance coil, the inner secondary is wound so that the two halves are in opposite directions and there is no mutual inductance when the secondary is centered. With a small motion, a few turns with flux linkage in one direction enter the central field region as a few turns with flux linkage in the other direction leave it (see Fig. 3.10). Most commonly this is used in the manner shown where the motion is with respect to some reference material, but it can also be used where the sample and secondary are thermally isolated from the primary.

For small motions, the change in mutual inductance is linear with displacement and easily measured with a mutual inductance bridge. An inductance sensitivity of a few nanohenrys translates into a sensitivity of a few picometers. This method was perfected by Carr and Swenson (1964), McLean, Swenson, and Case (1972), and Case and Swenson (1974) and resulted in a resolution of better than 10^{-8}. [These apparatus, where the transformer itself is at or nears the specimen temperature, are not to be confused with the apparatus of Robbins, Ohori, and Weitzel (1963) or Dahlerup-Peterson (1980), which are basically tube-type dilatometers with an LVDT sensor at room temperature.] The primary disadvantages of the low-temperature inductance sensors are their sensitivity to vibration and extraneous motion and their inability to be used with magnetic or superconducting materials.

Other electrical methods. The most frequently used technique, other than capacitance and inductance transducers, is the application of strain gauges. These devices measure the strain-induced resistance change of certain alloys in a long, thin path of the metal evaporated on a thin insulator attached directly to the specimen. Their disadvantages are relatively poor sensitivity and difficulty in calibration because of the temperature dependence of the strain resistivity (see, for example, Poore and Kesterson, 1977).

A clever technique was devised by Pritchard and Gugan (1970), who froze steel balls in solid argon and tracked their separation with inductances. Barron et al. (1980) suggest that a variation of this device using tiny magnetic particles and a SQUID should be extremely sensitive. Other devices, such as those of Pudalov and Khaikin (1969) and Mäntysalo (1965), use the change in frequency in microwave or ultrasonic cavities.

3.3.3 Optical methods

Light beams do not conduct heat, they do not transmit vibrations, and they can carry large amounts of information. Thus, they make ideal transmitters of length change information to room temperature for a dilatometer. Optical techniques vary all the way from comparing lengths through cross hairs in a telescope to resonance changes in optical cavities with resolutions in $\Delta L/L$, from about 10^{-5} to 10^{-9}. Sensitivities as low as 0.04 pm have been claimed. A monochromatic light beam reflecting off two separate surfaces interferes with itself, creating fringes of light and dark that move with a change in the relative distance between the two surfaces. Motion of less than one wavelength is easily detected. Fizeau (1864a; 1864b; 1866a; 1866b) was the first to recognize the power of interferometery and used it to measure the thermal expansion of crystalline quartz.

All optical methods have the distinct advantages listed above, can frequently be used on very small specimens, but still are limited by the mating of mechanical surfaces and associated vibrations and distortions. The advent of the narrowly monochromatic and stable wavelengths of lasers has opened up many new possibilities, and several clever techniques are pushing the state of the art.

Interferometers. All interferometers are based on the principle that two light beams from the same monochromatic source traveling two different paths will produce interference patterns when recombined. Thus, if a thermal expansion specimen is sandwiched between two mirrors, one of which is partially reflecting, one light path will be reflected at the first specimen surface, and a second will be reflected from the other. When the two light beams recombine, an interference pattern will be generated, depending upon the relative separation of the two surfaces; if the surfaces are parallel, the pattern will be rings; if they are at a slight angle, it will be lines. As the specimen changes in length, the fringe patterns move, and the number of fringes passing an index point, ΔN, is related to the length change by

$$\Delta L = (\lambda_0/2)\Delta N \tag{3.42}$$

where λ_0 is the wavelength of the light in a vacuum. Thus, one can obtain an "absolute" thermal expansion (not measured with respect to some reference material) by simply dividing by the original length or the thermal expansion coefficient from

$$\alpha = (\lambda_0/2L_0)(\Delta N/\Delta T) \tag{3.43}$$

The specimen is usually in the form of a hollow cylinder or the mirrors are separated by three identical specimen rods so that the light path is in the center. A typical schematic for this Fizeau-type interferometer is shown in Fig. 3.11 where the specimen has three points on the end of the cylinder for a reproducible and uniform mirror support. Also shown in the figure is a modification providing a reference support for the second mirror, which has the advantage of an extremely short light path so that index of refraction corrections are small (Abbe, 1889; Pulfrich, 1893), but it is then a relative type of measurement.

Basically, all interferometric systems are alike and differ only in how they create the interference pattern (Jenkins and White, 1950) and how they detect changes in it. If the light source is a pencil of monochromatic and parallel light incident on a parallel or nearly parallel pair of re-

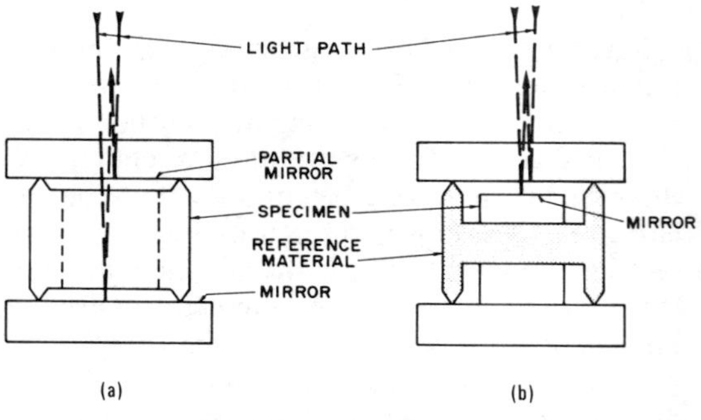

Figure 3.11 Fizeau interferometer for absolute (a) and reference (b) measurement of thermal expansion.

flecting surfaces, "fringes of equal thickness" or "Fizeau" fringes are created. If the light source is monochromatic and diverging or converging from a point source and incident on parallel surfaces, "fringes of equal inclination" or "isoclinic" fringes are formed. If the light source is monochromatic, *extended*, and parallel, such as from a point source at infinity, "Haidinger" fringes result. They are the basis of the common Fabry–Perot interferometer. A pair of Fabry–Perot plates is called an *etalon*; note that its resolution is proportional to its length. Methods of detecting the change in fringes vary from simple viewing with the eye to counting photometrically with the output of a photoelectric cell to an exotic rf sideband technique described below.

Interferometric techniques have been used to generate a large amount of the thermal expansion data, which are considered to be reference data for all kinds of materials. The Fizeau method, with a resolution of about 10^{-7} in $\Delta L/L$, was used by Ayres (1905), Dorsey (1907; 1908; 1910), and Nix and McNair (1941) for several pure metals; Buffington and Latimer (1926) for a definitive work on quartz; Rubin, Altman, and Johnston (1954) for a classical work on copper and later on other metals; Plummer and Hagy (1965) on optical materials, and for all of the standard reference materials from the National Bureau of Standards (see § 3.4.4). Fabry–Perot interferometers have been used extensively, also, and because of their exceptional resolution of 10^{-8} and better, they are extensively used for single-crystal studies (Meincke and Graham, 1965; Kinzly, 1967; Krishnan et al. 1979) and for studies of extremely low expansion materials (Berthold and Jacobs, 1976; Berthold, Jacobs, and Norton, 1977). Berthold et al. (1977) used a clever technique of applying an rf sideband to a stabilized helium–neon laser to obtain a transmission resonance in a Fabry–Perot etalon. The change in the frequency of this sideband (shown in Fig. 3.12) to maintain resonance in the cavity with a change of temperature can be very precisely measured with a frequency counter for an ultimate precision in $\Delta L/L$ of 10^{-9}, probably the best possible to date.

In other modifications of interferometric techniques, Lebesque, Van Schaik, Kruijshoop, and Blaisse (1965) and Roberts (1975) have used polarized light beams for a resolution of 10^{-8}/K. Piezoelectric oscillation of the mirror separation and measurement with a lock-in amplifier tuned to the right frequency was used by Bruins, Gar-

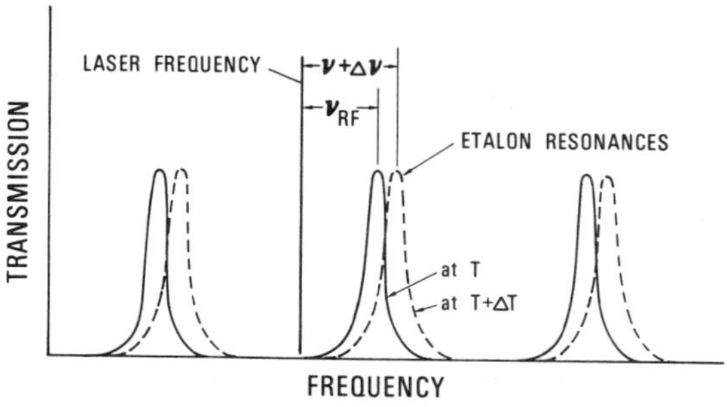

Figure 3.12 Transmission of a Fabry–Perot etalon vs. frequency showing laser frequency and change in rf sideband frequency to maintain resonance with thermal expansion.

land, and Greytak (1975) for probably the ultimate in sensitivity of 0.05 pm. All these techniques still are limited by the mating of sample and mirror surfaces, precise manufacture and alignment requirements, and a limited range of motion. But they do have the advantages of using small specimens of varied shape, the possibility of dynamic measurements, absolute or relative measurement, and lend themselves to electronic and computer control and analysis.

Optical levers, amplifiers, and comparators. The various methods of optical levers and amplifiers have been well reviewed by Jones (1961). Perhaps there are just two principles that need discussion, the use of grids and the advantage of the lever. The finest grids are made by scratching evaporated metal off of optical flats in closely spaced parallel lines. As shown in Fig. 3.13, one grid is uniform and the other has a discontinuity in the spacing so that one half of the overlap transmits light through the two grids and the other does not. With one grid fixed as a refer-

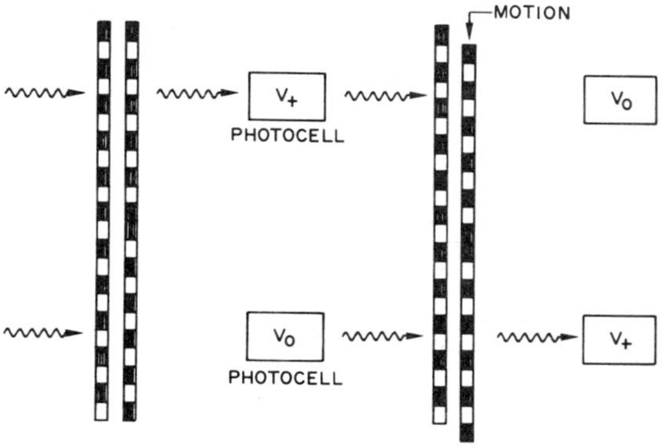

Figure 3.13 Transmission of light through offset grid.

ence and the other moving with the specimen, the differential output of the two photocells behind the grids is directly related to the relative motion. Andres (1961; 1964) used such a system with the light transmitted in and out of the Dewar by light pipes to measure the thermal expansion of several metals with the sensitivity of 0.1 nm and a claimed resolution in $\Delta L/L$ of 10^{-9}.

The obvious advantage of any lever system can also be used in optics, where a small angular motion is greatly amplified at some distance away. If a mirror is mounted on both a reference material and a specimen so that the difference in thermal expansion causes a rotation, such as in Fig. 3.14, the only limit to the amplification is the distance available and atmospheric distortion. Of course mechanical coupling, vibration, and reproducibility all limit the sensitivity as well. The method shown in the figure by Sparks (1982) uses the motion of a detector that locks onto a specific grating reflection to monitor the change in angle caused by the different thermal expansions for a resolution of $\Delta L/L$ of about 10^{-6}. Bunton and Weintroub (1968) combined both the lever and grid into a system with a resolution of 2×10^{-9}.

The optical comparator is simply two rigidly mounted microscopes with vernier filar eyepieces. With the insertion of a mirror into the optical path, the change in fiducial marks on a specimen can be compared directly with a reference scale held at constant temperature. The method is remarkably simple and direct, and if care is taken to account for distortion in the optical

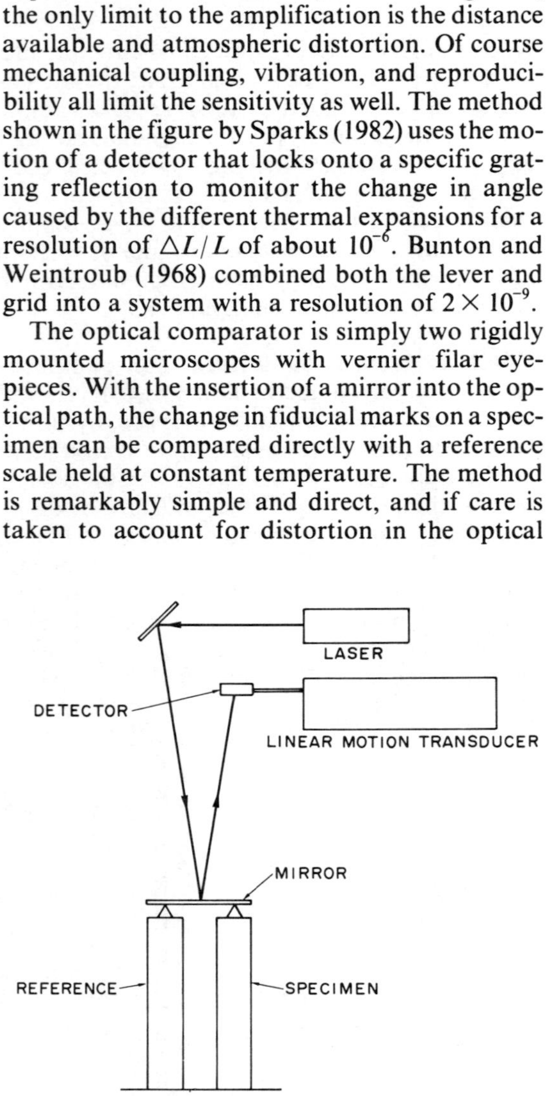

Figure 3.14 Schematic of optical lever dilatometer.

paths due to temperature gradients (Conway and Losekamp, 1965), it can be used with reasonable accuracy over wide ranges of temperature (Kirby, 1969).

3.3.4 X-ray methods

An x-ray determination of the lattice spacing of a crystal at two different temperatures will obviously determine the thermal expansion. This method has the distinct advantage of being absolute and, in addition, determines changes in lattice spacing uncomplicated by the macroscopic effects of vacancies, interstitials, impurities, or phase changes. It can also be used on very small specimens and for studying anisotropic crystals. The major disadvantages are its low resolution of about 10^{-6} in $\Delta L/L$ and that it cannot be used for dynamic measurements.

The apparatus consists of an x-ray source, a camera to record the Bragg reflections, and a temperature-controlled specimen. Techniques differ only in the way x-rays are recorded — most are recorded on film, but some ionization counters are also used. Standard methods are the single-crystal Laue reflection, Debye–Scherrer powder pattern, rotating or oscillating crystal, and rotating camera. Each has some small advantage (Krishnan et al., 1979). The basic Bragg relation for two different temperatures leads to the simple relation for the change in lattice spacing, $\Delta d/d$,

$$\Delta d/d = \Delta \sin\theta/\sin\theta \qquad (3.44)$$

where θ is the Bragg reflection angle, and a small correction due to refraction of the x-rays is ignored (Campbell, 1962). The change in lattice spacing can then be related to the distance between planes of various crystals, depending on their symmetry. For cubic crystals, the thermal expansion is

$$\Delta L/L = \Delta d/d \qquad (3.45)$$

It can be calculated for other crystal structures with more complicated expressions (Kirby, 1969). Some of the more common sources of data have used x-ray techniques: Batchelder and Simmons (1965); Simmons and Balluffi (1957); Figgins, Jones, and Riley (1956); and Bond (1960).

3.3.5 Analysis of experimental data

Nearly all dilatometers measure a length change, ΔL_i, at some temperature, T_i, with respect to some initial length, L_0, at some temperature, T_0. Thus, what is generated as the experiment progresses is increasingly larger values of ΔL_i as T_i gets farther from T_0. It is tempting (and frequently done) to subtract the adjacent ΔL values in order to determine α immediately from

$$\alpha = \frac{(\Delta L_i - \Delta L_{i-1})}{L_0\,(T_i - T_{i-1})} = \frac{(L_i - L_0) - (L_{i-1} - L_0)}{L_0\,\Delta T} \quad (3.46)$$

$$= \frac{L_i - L_{i-1}}{L\,\Delta T} = \Delta L / (L_0\,\Delta T) \quad (3.47)$$

which is really an average, $\bar{\alpha}$, over the interval ΔT. If ΔT is a small enough part of T_0, then $\bar{\alpha}$ closely approximates α (Hahn, 1970) and is then fit to some theoretically predicted equation. In reality, what is generated is the function $\Delta L(T)$ at fixed points of T_i separated by finite intervals. If normalized to L_0, then the function that is determined experimentally is $\Delta L(T)/L_0$, the thermal expansion, and this is the function that should be fit with equations and then differentiated to obtain the thermal expansion coefficient, α.

It was found empirically by Clark (1968) that a rational polynomial in temperature with the proper restraints provided the best fit to the thermal expansion, $\Delta L/L$, for most materials at low temperatures. The restraints were that the first and second derivatives must go to zero as $T \to 0$ K and $\Delta L/L$ must be proportional to T at higher temperatures. The polynomial was

$$(L_{293} - L_T)/L_T = a + T^4/(b + cT^2 + dT^3) \quad (3.48)$$

where L_0 is taken at $T = 293$ K. The thermal expansion coefficient was then determined as the derivative of the above equation at fixed points. Similarly, Hahn (1978) has used a variation of Grüneisen's equation to fit sapphire data

$$\frac{L_T - L_{293}}{L_{293}} = \frac{E(\theta_D/T)}{3a[Q_0 - kE(\theta_D/T)]} + \frac{1-a}{a} \quad (3.49)$$

where $a = L_{293}/L_0$, Q_0 and k are fitting constants, and $E(\theta_D/T)$ is a function of the ratio of the Debye temperature to the experimental temperature. In both cases, the fit to the polynomial was better than the precision of the data. For the purposes of fitting experimental data to generate reliable engineering design values, Eq. 3.48 or a simple polynomial in the temperature is usually quite adequate.

Ironically, the most stringent requirements on the experimental data are usually a comparison with theory. As Barron et al. (1980) so diligently point out, in selecting powers of the temperature for a polynomial, one must be careful to be physically meaningful in the temperature range of concern. This is especially true as one approaches $T = 0$ K where the physics is most meaningful, but the experimental errors are largest and the computer least discriminating. For metals, the thermal expansion coefficient, α, is typically fit with odd powers of T, where the T term is identified with the electronic contribution, T^3 with the long wavelength lattice contribution, and T^5 with higher order dispersive terms in the frequency spectrum. Others, such as $T^{3/2}$ and T^{-2}, can be identified with spin waves and Schottky anomalies, respectively.

At the conclusion of this section, a word about errors is most appropriate. The total uncertainty of some thermal expansion data can be estimated with a certain confidence level and consists of two parts, the standard error in the data and an estimated possible bias of the measurement system. The most common form of presentation of thermal expansion data is in a tabular form at fixed tempertures. The propagated standard error (e.g., 1×10^{-6}) for a single table value is usually less than the imprecision of a single data point in a single run (e.g., 3×10^{-6}) because each run usually consists of many data points. If several runs on identical specimens are performed, the standard error can either be larger or smaller, depending upon the reproducibility, and can be a good measure of the systematic errors introduced by the measurement system and help in estimating the bias. Added to the standard error must be an estimated possible bias for which there is no statistical formula and which is a measure of how far from the quoted value a real value could be, plus or minus its standard error. This must include estimates of inaccuracies introduced by the reference data for calibration, the calibration measurement itself, changes in experimental technique (such as cooling rate or different operators), scaling changes because the reference material has a different thermal expansion than the specimen, or possible apparatus drift or changes between calibrations. This estimated bias is usually larger than the standard error in the data.

Finally, one must assess the predictive ability of a table value quoted for a different piece of material with the same generic or trade-marked description; for example, is copper always copper? In general, large changes in composition (10 to 20%) are necessary to produce significant changes in the thermal expansion (~5%), and different heat treatments or conditions do not produce significant changes unless phase changes are involved (Clark, 1968). It was found that different metals and alloys with different heat treatments, grain sizes, or rolling directions introduced only small differences, a few percent. Thus, a generalization can be made that literature values for thermal expansion are probably good for a like material to within 5%. This is because the thermal expansion depends explicitly on the nature of the atomic bond, and only those changes that alter a large number of the bonds can affect its value.

3.4 THERMAL EXPANSION DATA

For thermal expansion there is neither a paucity of data nor a surfeit. As with most material properties, the thermal expansion is usually measured for practical materials only after it is needed for a structural design or analysis. Thus, the thermal expansion for many metals and alloys at low temperatures was measured during the period of cryogenic development for the space age. The data for nonmetallics and composites has been obtained more recently as these materials became more common in low temperature use. Nearly all the physics data as well as the usable engineering design data have been obtained and published since about 1960 when accurate measurement techniques became more common. In some cases, such as Fe–9%Ni, a material has been developed specifically for a low-temperature application, and the thermal expansion data are provided with the mechanical properties. Specialty applications frequently require a single measurement, which is often lost in the literature. With the few exceptions noted in § 3.5.3 on data sources, the data are found in scattered sources and are rarely in a uniform format.

To aid in the usefulness of this text, the principal material categories of metals and alloys, nonmetallics, and composites are discussed in a generic manner. The general thermal expansion behavior of each material category is presented, and typical values are shown in both tables and graphs. Extensive material data presentations are left to the compilations listed in § 3.5.3. Since the thermal expansion values do not change significantly with small changes in the basic material, this should suffice for most practical purposes. Specific references are discussed only if they present an unusual amount of data or contribute to the understanding of the general material category.

The format for data presentation should be explained. Graphical displays are always useful for noting general trends and for comparing similar behaviors. Graphs are not useful for generating design values because they are usually too inaccurate. Thus, tables of values of not only the thermal expansion but also of the thermal expansion coefficient are essential. Tabular values of the thermal expansion are best given in the dimensionless units of percent and referred to a zero value at room temperature, 293 K. Thermal expansion coefficient values are given in units of 10^{-6} K^{-1} at the same intervals and are, of course, referred to zero at 0 K. Because of the smoothness of the curves, precise values at intermediate temperatures are readily interpolated. In the following discussion, graphical plots may be shown for typical or representative values, but tabular data will always be for some specific material. Table 3.2 includes data for 30 of the more common or representative materials in use at low temperatures.

3.4.1 Metals and alloys

The first materials to permit precise construction of apparatus and structures were the metals and alloys. When they began to be used at temperture extremes, length changes were, not unexpectedly, noted and had to be determined and accommodated. More data exist for metallics than any other category. As can be seen in Fig. 3.15, the values for the thermal expansion of these materials are well behaved and in the middle range. The spread of values for the pure elemental metals is much larger because of the wide variety of crystal structures that occur. The range for alloys is much less, principally because they are usually made from cubic base metals and remain in the cubic structure. For the purposes of discussing like behaviors, the metals and alloys are divided into the metallic elements, metallic alloys, low-expansion materials, and miscellaneous special metals, including solders and superconductors.

Metallic elements. Because of their abundance, ease of workability, obtainable purity, and utility, particularly as base metals for alloys, the cubic metals have been studied more than any other materials. More than 160 measurements of iron and copper alone have been made

Table 3.2 Thermal expansion data for selected materials.

T (K)	$(L_{293} - L_T)/L_{293}$ (%)					
	METALS					
	Aluminum[a]	Copper[b]	Iron[a]	Nickel[a]	Niobium[a]	Titanium[a]
4	0.415[c]	0.324	0.198[c]	0.224[c]	0.143[c]	0.151[c]
20	0.415	0.324	0.198	0.224	0.143[c]	0.151
40	0.413	0.322	0.197[c]	0.223[c]	0.141[c]	0.150
60	0.405	0.314	0.195[c]	0.219[c]	0.137[c]	0.148
80	0.391	0.300	0.189[c]	0.211	0.129	0.142
100	0.370	0.282	0.181	0.201	0.121	0.134
120	0.343	0.259	0.170	0.187	0.111	0.125
140	0.312	0.234	0.156	0.171	0.099	0.113
160	0.277	0.207	0.140	0.152	0.088	0.101
180	0.240	0.178	0.122	0.132	0.076	0.087
200	0.201	0.148	0.102	0.111	0.063	0.073
220	0.160	0.117	0.082	0.088	0.050	0.058
240	0.118	0.086	0.060	0.065	0.037	0.043
260	0.075	0.054	0.038	0.041	0.023	0.027
273	0.045	0.033	0.023	0.025	0.014	0.017
293	0.000	0.000	0.000	0.000	0.000	0.000
	ALLOYS					
	Aluminum 2024-T86[d]	Aluminum 7075-T73[d]	Fe–9Ni[e]	Fe–36Ni[d]	AISI 304L[e]	AISI 633 ph[d]
4	0.396[c]	0.419[c]	0.195	—	0.306	0.193
20	0.396	0.419	0.195	0.037	0.306	0.193
40	0.394	0.417	0.193	0.040	0.303	0.192
60	0.387	0.408	0.191	0.039	0.294	0.189
80	0.372	0.392	0.187	0.038	0.281	0.183
100	0.351	0.370	0.180	0.036	0.265	0.174
120	0.325	0.343	0.170	0.033	0.245	0.162
140	0.295	0.312	0.155	0.029	0.222	0.148
160	0.262	0.278	0.137	0.025	0.195	0.132
180	0.227	0.241	0.119	0.020	0.168	0.115
200	0.190	0.202	0.100	0.016	0.140	0.096
220	0.151	0.161	0.081	0.013	0.113	0.077
240	0.111	0.119	0.060	0.010	0.083	0.057
260	0.070	0.076	0.038	0.008	0.052	0.036
273	0.043	0.047	0.023	0.006	0.031	0.023
293	0.000	0.000	0.000	0.000	0.000	0.000
	ALLOYS					
	Brass 70/30[e]	S Monel[e]	Inconel 718[d]	Ti–6Al–4V[e]	Nb–45Ti[f]	Soft Solder 50/50[a]
4	0.369[c]	0.252[c]	0.238[c]	0.173[c]	0.188	0.514[c]
20	0.369	0.252	0.238	0.173	0.187	—
40	0.366	0.250	0.236	0.171	0.184	0.510[c]
60	0.355	0.246	0.232	0.167	0.178	0.498[c]
80	0.337	0.237	0.223	0.162	0.167	0.477[c]
100	0.313	0.223	0.211	0.154	0.156	0.447
120	0.288	0.207	0.195	0.141	0.139	0.407
140	0.260	0.188	0.177	0.126	0.124	0.365
160	0.229	0.168	0.157	0.110	0.109	0.321
180	0.196	0.146	0.136	0.094	0.093	0.276
200	0.163	0.123	0.114	0.078	0.078	0.229
220	0.128	0.098	0.091	0.062	0.063	0.182
240	0.093	0.072	0.067	0.045	0.045	0.133
260	0.058	0.046	0.042	0.027	0.030	0.083
273	0.035	0.029	0.026	0.016	0.019	0.051
293	0.000	0.000	0.000	0.000	0.000	0.000

(continued)

Table 3.2 (continued)

T (K)	$(L_{293} - L_T)/L_{293}$ (%)					
	COMPOSITES					
	Epoxy[g]	**G-10CR (warp)**[h]	**G-10CR (normal)**[h]	**Cotton Phenolic (warp)**[d]	**Cotton Phenolic (normal)**[d]	**Cu/Nb–Ti Wire**[d]
4	1.16	0.241[c]	0.706[c]	0.264[c]	0.730[c]	0.265[c]
20	1.15	0.241	0.706	0.264	0.730	0.265
40	1.11	0.234	0.690	0.258	0.708	0.262
60	1.07	0.223	0.667	0.250	0.678	0.256
80	1.02	0.211	0.638	0.239	0.643	0.245
100	0.959	0.197	0.603	0.225	0.603	0.231
120	0.892	0.182	0.563	0.210	0.559	0.212
140	0.819	0.165	0.517	0.192	0.513	0.190
160	0.737	0.148	0.465	0.172	0.463	0.167
180	0.647	0.129	0.408	0.151	0.409	0.142
200	0.550	0.108	0.346	0.128	0.341	0.117
220	0.447	0.086	0.279	0.103	0.285	0.091
240	0.336	0.064	0.208	0.077	0.213	0.066
260	0.217	0.040	0.133	0.049	0.136	0.041
273	0.133	0.025	0.082	0.030	0.083	0.025
293	0.000	0.000	0.000	0.000	0.000	0.000
	MISCELLANEOUS					
	TFE[i]	**CTFE**[a]	**Nylon**[a]	**Plexiglass**[j]	**Pyrex**[j]	**Nb_3Sn**[k]
4	2.14	1.135	1.389	1.22	0.055*	—
20	2.11	1.114	1.379	1.21	0.056	—
40	2.06	1.070	1.352	1.16	0.057	—
60	2.00	1.019	1.308	1.11	0.056	—
80	1.93	0.962	1.247	1.05	0.054	0.141
100	1.85	0.900	1.172	0.99	0.050	(0.125)
120	1.76	0.834	1.088	0.93	0.046	—
140	1.66	0.763	0.996	0.86	0.042	(0.102)
160	1.54	0.686	0.896	0.78	0.037	(0.088)
180	1.40	0.604	0.789	0.69	0.032	—
200	1.24	0.517	0.673	0.59	0.027	0.067
220	1.05	0.424	0.548	0.49	0.022	0.050
240	0.855	0.324	0.412	0.37	0.016	—
260	0.645	0.214	0.265	0.24	0.010	—
273	0.50	0.134	0.161	0.15	0.006	—
293	0.000	0.000	0.000	0.00	0.000	0.000

[a] Corruccini and Gniewek (1961).
[b] Hahn (1970).
[c] Extrapolated or estimated using Grüneisen correlation.
[d] Clark (1968).
[e] Arp et al. (1962).
[f] Jelinek and Collings (1971).
[g] Dahlerup-Peterson (1979).
[h] Clark, Fujii, and Ranney (1981).
[i] Kirby (1956).
[j] Laquer and Head (1952).
[k] Schadler, Osika, Salvo, and DeCarlo (1964). Data in parentheses are from an independent source.

Tabular data are also available in the Appendix (§3.6) on: borosilicate glass, single-crystal sapphire, copper, tungsten, and fused silica.

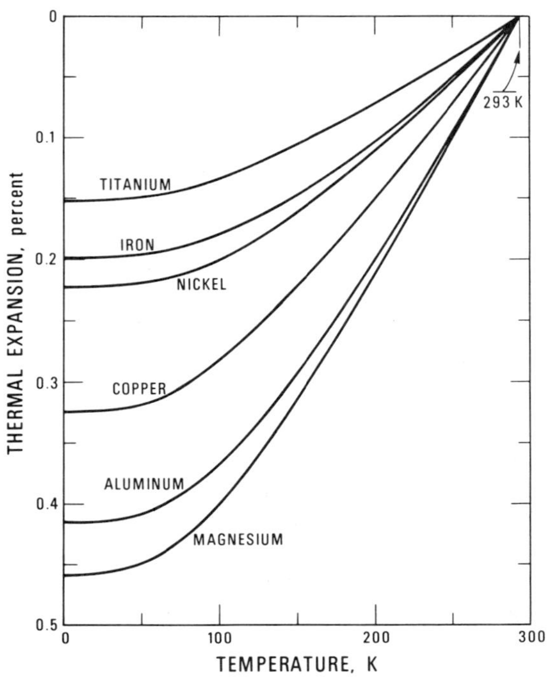

Figure 3.15 Thermal expansion vs. temperature for some of the metallic elements.

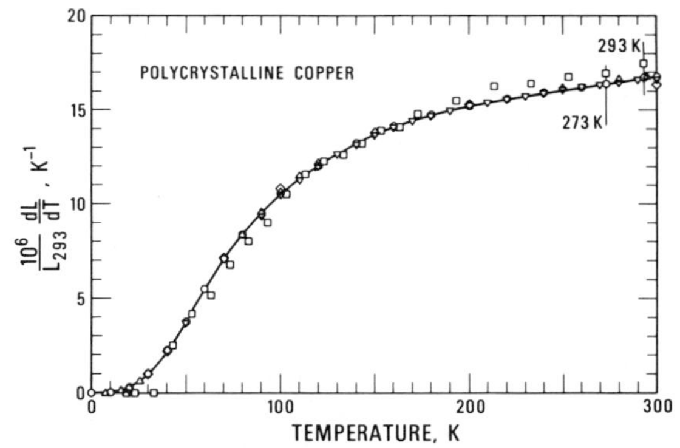

Figure 3.16 Thermal expansion coefficient vs. temperature of several different measurements of pure polycrystalline copper (Clark, 1968).

since 1900. The earliest known measurements were also made on copper and iron by van Musschenbrock in 1730 for corrections to the period of clock pendulums (Desaguliers, 1745). The thermal expansion values of cubic metals are well behaved, as seen in Fig. 3.15, and are isotropic as well. Shown as representative of the range of values for cubic metals are iron, nickel, copper, and aluminum. Iron (b.c.c.) and nickel (f.c.c.) are typical of the cubic transition metals, where the electronic component to the thermal expansion becomes significant at very low temperatures ($T < \theta_D/25$). Chromium and manganese are unique among the transition element metals in that they have antiferromagnetic transitions that are observable as discontinuities in the thermal expansion coefficient at their Néel temperatures, 311 and 95 K, respectively.

Many, many measurements have been made on copper, whose thermal expansion behavior is probably the most documented of all materials. Figure 3.16 shows representative data for the thermal expansion coefficient of copper, which has been chosen as a standard reference material (see § 3.4.4). The thermal expansion of copper (b.c.c.) is typical of the noble metals and, in fact, is almost exactly the same as that of silver and gold. These all have a strong electronic component at low temperatures. The alkali metals, such as sodium and potassium (b.c.c.), are strongly lattice dominated and have thermal expansion values several times the other cubic metals. Aluminum (f.c.c.) is representative of group IIIA metals.

The noncubic metallic elements are many and varied. Titanium (h.c.p.) is representative of a noncubic transition metal, but no low-temperature data exist for single crystals. Above room temperature it is quite isotropic and would be expected to be so below room temperature as well. Its companion element zirconium, however, does show some anisotropy (about 30%). The group II elements are all highly anisotropic and become more so as the c/a ratio deviates further from the ideal of 1.633 for these h.c.p. metals (Munn, 1969, 1972). Magnesium, which is shown in Fig. 3.15, is the least anisotropic of the II elements, with a c/a ratio of 1.623 and an $\alpha_c/\alpha_a = 1.07$. Zinc is an extreme example, with $c/a = 1.86$ and an $\alpha_c/\alpha_a = 9.5$. If zinc were shown on Fig. 3.15, the expansion parallel to the a axis would be about that of titanium, the polycrystalline average would be larger than magnesium, and that parallel to the c axis would be a long way off scale. The ratio of α_c/α_a can be quite variable with temperature. The elements aluminum, copper, iron, nickel, niobium, and titanium are included in Table 3.2.

The group V and VI elements (e.g., antimony and tellurium) are rhombohedral and hexagonal, respectively, and show anisotropy and, frequently, negative thermal expansion coefficients as well. The data are primarily of interest for comparison to theory (Barron et al., 1980). The rare earths and the actinides are nearly all noncubic and anisotropic, frequently have negative coefficients and magnetic transitions, and also provide interesting tests of our ability to model the anharmonic properties of crystals (Krishnan et al., 1979).

Alloys. A close examination of an extensive collection of alloy thermal expansion data leads to two principal conclusions (Clark, 1968). The first is that a large change in composition (10 to 20%) is necessary to produce a significant (5%) change in the thermal expansion. The second is that the thermal treatment or mechanical condition of the alloy does not have a large effect on the thermal expansion except when it produces a structure change as in the martensitic or precipitation-hardening stainless steels. That this is so could readily be concluded from the knowledge that thermal expansion depends explicitly on the nature of the atomic bond, and only those changes that alter a large number or the specific nature of the bonds will affect the thermal expansion. Specific examples of these conclusions are discussed below along with the presentation of thermal expansion data for several typical alloy systems.

The Cu–Ni system is typical of most binary alloy systems and has a reasonable amount of data for comparison. The change from one dominant alloy component to the other is smooth and linear, so much so that it can be used as a basis for predicting interpolated values with excellent precision, as is shown in § 3.5.1. Another two-component system that has been well studied is the Fe–Ni system because of its widespread use in stainless steels, but even more so because of its unusual behavior near 36% Ni alloy composition, or the "Invar" region, shown in Fig. 3.17. The severe minimum in thermal expansion near 36% Ni is frequently used in structures that must remain mechanically stable over wide temperature ranges or generate minimal residual stresses. The wide variation of expansion shown in Fig. 3.17 is used to match the expansion of nonmetals, such as in glass-to-metal seals or specific nonmetallic composites. Even in this region of rapid change of thermal expansion values, it would be reliable to predict the thermal expansion as a function of composition to better than 5%. More will be discussed in the section on low thermal expansion materials.

The Fe–Ni system can be used to emphasize another aspect of alloy thermal expansion, the effect of phase changes. This is shown dramatically in the thermal expansion of Fe–29%Ni shown in Fig. 3.18 (Clark, 1968). This alloy goes through a martensitic phase transformation upon its initial cool-down from a well-annealed state; about 95% of the material transforms to the b.c.c. martensitic phase with an accompany-

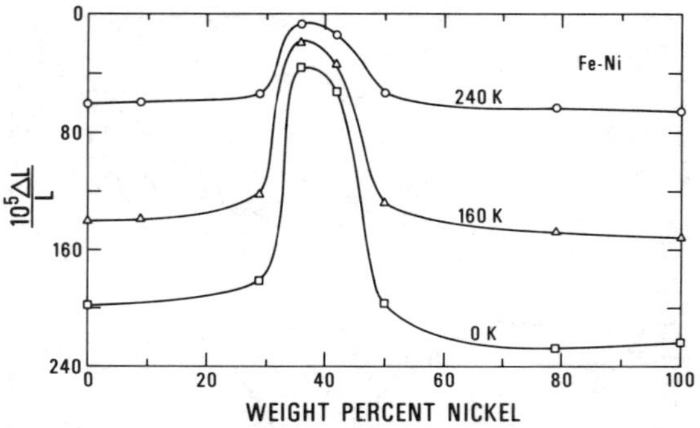

Figure 3.17 Thermal expansion at various temperatures as a function of composition for the Fe–Ni alloy system.

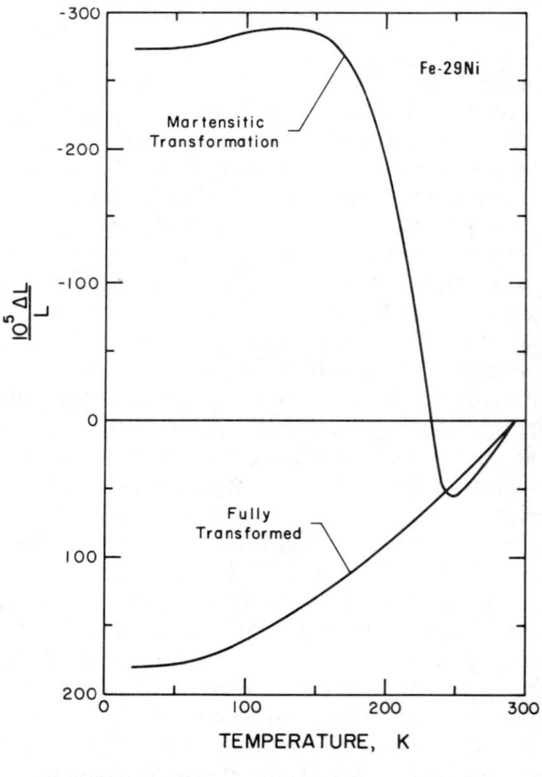

Figure 3.18 Thermal expansion of Fe–29 wt.% Ni through the first and succeeding cool-downs, showing the martensitic phase transformation on the first cool-down.

ing increase in volume. The increase is about 0.3%, significantly larger than the total length change of less than 0.2% from room temperature to near zero. The subsequent warm-up and any succeeding cool-downs follow the normal curve of the fully transformed alloy. The zero point of the transformed alloy is, of course, shifted to a new L_{293}, but it is still plotted as zero in the figure, and the initial slope shows about a 30% decrease. This effect has caused a few failures in cryogenic structures made from stainless steels.

Multicomponent alloy systems are frequently used for the high-strength alloys necessary for many low-temperature structures. Commonly used are aluminum and nickel alloys and austenitic stainless steels. Comparisons were made for more than ten aluminum alloys (Clark, 1968), and all the data fell within ±5% of that for pure aluminum. Al 2024 was selected as an alloy that would show the maximum conditional effect between the fully annealed state (O condition) and the solution heat-treated, strain-hardened, and artificially aged state (T86 condition). Less than 1.5% change was observed. Similarly, a cold-rolled and hardened alloy (Al 5083) with 4:1 grain elongation had less than 1.6% change in its thermal expansion.

If any alloy system should show compositional changes, it should be the nickel system of commercial high-strength alloys, which has large compositional variations. Shown in Fig. 3.19 are eight of these alloys (Clark, 1968), all of which have 18 to 20 wt.% Fe and varying amounts, 0 to 20 wt.%, of chromium, molybdenum, and cobalt. All the thermal expansion values fall within about ±5% of that of pure nickel, also shown in

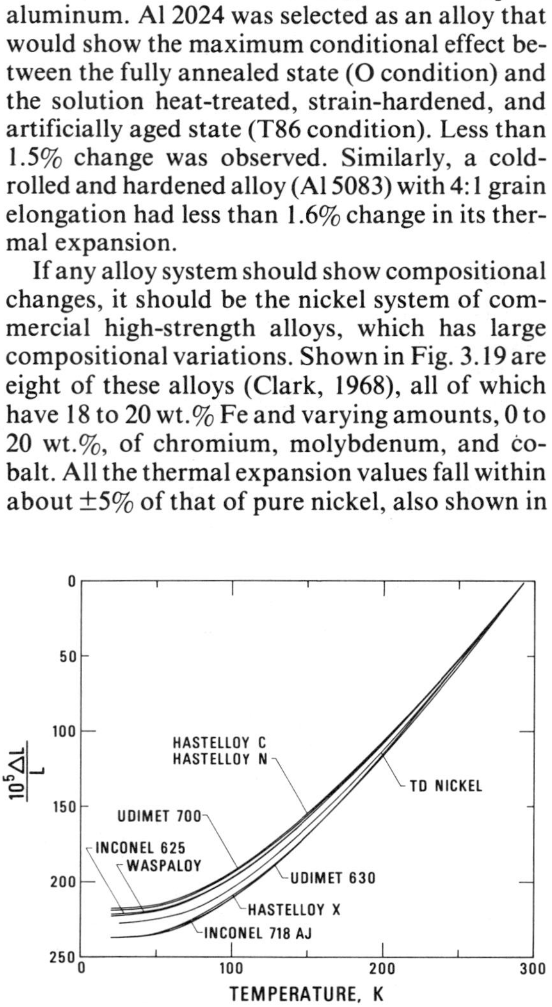

Figure 3.19 Thermal expansion vs. temperature of several nickel alloys.

the figure. It isn't until component compositions of 25 to 30 wt.% are reached, as in the Cu–Ni alloys, that the thermal expansion values become larger. Comparisons among heat treatments resulting in significantly altered grains also resulted in the conclusion that both size and shape of the grains have little effect on thermal expansion.

An alloy constant scheme has been used for the aluminum alloys where the thermal expansion of an alloy is expressed as percentage of the thermal expansion of pure aluminum (*Alcoa Aluminum Handbook*, 1962). Because of the identical shape of most thermal expansion curves and the similarity of values even for widely varying compositions or conditions, this scheme would probably work for nearly all commerical alloys. Also, because of this insensitivity to compositional changes, dilute alloy systems are studied primarily only at the very low-temperature regions for magnetic or Kondo-type interactions where the lattice contribution is quite small (White, 1972; Khan and Griffiths, 1978). Some anomalously large changes have been observed by Popov, Pervakov, and Tsibulina (1977) but with relatively low absolute values.

Invar and other low-expansion materials. As was pointed out in the previous section, there is a minimum in thermal expansion in the Fe–Ni alloy system at about 36 wt.% Ni. Guillaume (1897) was the first to notice this anomaly, and for this and subsequent studies he was awarded the Nobel prize in 1920. Because of the invariance of length with normal temperature changes, he named the alloy "Invar." It is in this region that the alloy changes phase from b.c.c. to f.c.c. and several other physical properties vary widely [e.g., magnetic susceptibility, Curie temperature, specific heat (White, 1965)]. These anomalies have been observed in other systems, such as Fe–Pt, Fe–Pd, and Zr–Zn (Wohlfarth, 1975). The low values of thermal expansion have been attributed to a balance between a positive lattice contribution and a negative magnetic one whose magnitude depends on the approximate value of the Curie temperature and is dominant at the lowest temperatures. Thus, alloys can be contrived with a variety of thermal expansion behavior, as can be seen in Fig. 3.20. Shown is a band of values for the normal range of Invar alloys plus one that is age-hardenable for increased strength and one whose zero-crossing was designed to be near room temperature.

Other materials that are frequently used because of their low thermal expansion values are also shown in Fig. 3.20. These are quartz and fused silica, which has the lowest thermal expansion for a relatively pure material, and some glasses which can also be designed to have balancing components to the thermal expansion.

Special metals. The thermal expansion data for superconductors does not generally appear in any of the compilations, so they are briefly treated here. Barron et al. (1980) do discuss several of the elemental superconductors, specifically gallium, indium, tin, lead, vanadium, niobium, and tantalum, but point out that these studies are primarily to relate the thermodynamic properties to superconductor critical properties. Thus, the differences in volume, thermal expansion, specific heat, and other properties between the normal and superconducting state are used to determine the strain dependence of the lattice–electron interactions leading to superconductivity, rather than to study thermal expansion itself.

Of the practical superconductors, only two are now in common use, the alloy Nb–Ti and the intermetallic A15 compound Nb_3Sn. Thermal expansion data for these materials are shown in Fig. 3.21. Jelinek and Collings (1975) have measured the Nb–Ti alloy system, whose data are smooth and well behaved, typical of a binary alloy system. The Nb–45 wt.% Ti, which is shown in the figure, is the most common alloy used for practical superconducting wires. The thermal expansion of a composite wire with many fine filaments of Nb–Ti in a copper matrix is also shown in the figure for comparison (Clark et al., 1981). The Nb_3Sn thermal expansion, as well as that of pure niobium from *Thermal Expansion, Metallic Elements and Alloys* (1975), is also shown. A composite copper–Nb_3Sn is also included. The composite wires have increased values of thermal expansion appropriate to the volume fraction of copper (see § 3.5.1 on predicting composite values).

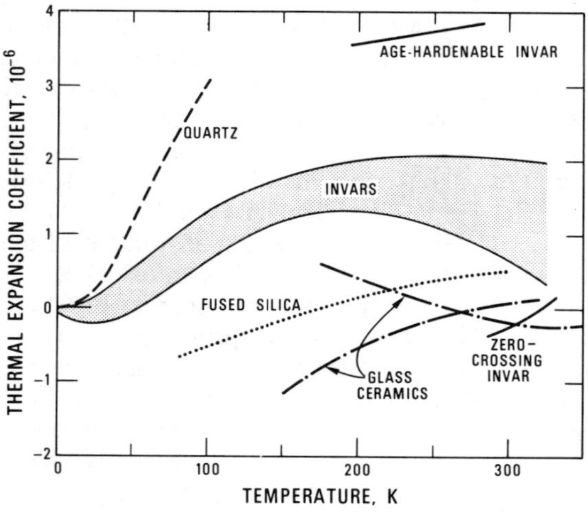

Figure 3.20 Thermal expansion coefficient vs. temperature for several low-expansion materials.

A limited amount of data exists on the other A15 compounds; there are one paper on V_3Ga —Finalyson, Gibbs, and Smith (1981) and several on V_3Si — Testardi and Bateman (1967); Fawcett (1971); Smith, Finlayson, and Shelton (1975); and Försterling and Hegenbarth (1978). Vold and Richards (1974) report on the V–Ga alloys including V_3Ga.

A final category of special alloys that is worth mentioning is solders because solders are frequently used to join dissimilar alloys for low-temperature applications. Seligmann and Sarwinski (1972) report values for several solders that are well behaved and what would be expected from their compositions.

3.4.2 Nonmetallics

The category of nonmetallics covers a very wide range of materials. Their thermal expansion values cover an even wider range, from the glasses at a total length change to 0 K of less than 0.1% to silicone rubber at more than 4%. In this grouping, crystalline solids, glasses and ceramics, and polymers are discussed. Representatives of these materials are shown in Fig. 3.22, where copper is also shown to indicate the large values for thermal expansion of some of them. Data for some are given in Table 3.2. The applications of these materials at low temperatures are just as varied as their properties: low expansion structures, seals and bearings, thermal and electrical insulations, and matrix materials for composites. Therefore, these thermal expansion data are found in widely scattered sources.

Crystalline solids. Nearly all materials solidify into ordered states, which can be identified as the well-known crystal structures. Nonmetallic materials include many of the elements, all the oxides and halides, solidified gases, innumerable salts, and many more. A recently published review text, *Thermal Expansion of Crystals* by Krishnan et al. (1979), covers in detail the ther-

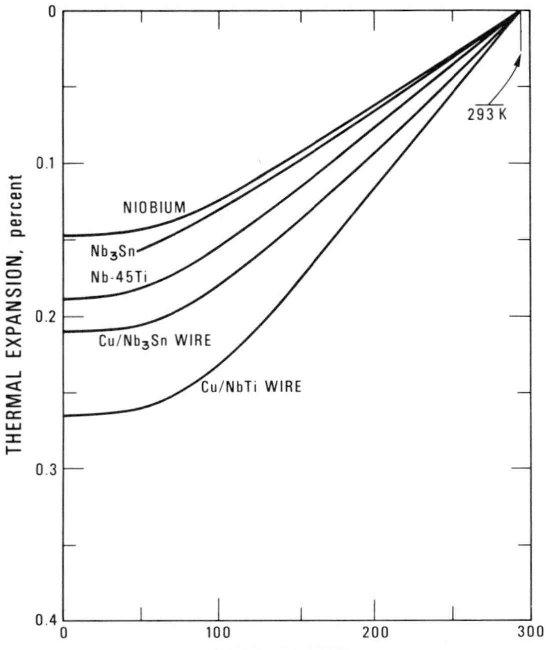

Figure 3.21 Thermal expansion vs. temperature for superconducting materials and wires.

mal expansion behavior of many different crystal structures. The review also presents data on more than 370 different crystalline materials. Because single crystals are frequently studied to compare the results with theory and to try to understand the fundamental anharmonic behavior of crystals, there is a reasonable amount of data at very low temperatures where the different contributing components to the thermal expansion can be more readily separated. Barron et al. (1980) also have an excellent review of thermal expansion data by crystal structure.

One of the first to measure the thermal expansion of several crystalline solids was Dewar (1902), who reported values for 27 different crystalline solids. Inert gas solids are frequently studied because they form in the cubic structure (f.c.c.) and are van der Waals bonded and used for comparison to theory. There are two review volumes that describe these unique solids: *Rare Gas Solids*, edited by Klein and Venables (1976; 1977). Other molecular gas solids, such as N_2, O_2, CO, CO_2, NH_3, and CH_4, have been extensively studied by Manzhelii and co-workers (Manzhellii, Tolkachev, and Voitovich, 1966; Manzhellii, Gavrilko, and Voitovich, 1967; Manzhellii, Tolkachev, and Gavrilko, 1969; Manzhellii, Tolkachev, Bagatskii, and Voitovich, 1971). Ice has been studied by Powell (1958) and Dantl (1962). Negative thermal expansion coefficients are seen frequently in superconductors, such as Si, Ge, Ga–As, Cd–Te, and In–Sb (Yates, 1972).

Most of the nonmetallic single-crystal solids that have been measured are not cubic and thus anisotropy is common. Nominally, an average thermal expansion coefficient can be obtained by simple averaging of the thermal expansion coefficients in the three axial directions (see § 3.3), but for true polycrystalline averages this becomes a much more difficult problem (see Chapter 1). This is an area where more research is needed.

Glasses and ceramics. Glasses are disordered structures, usually a mixture of oxides, that are tetrahedrally bonded in a kind of network. The thermal expansion arises from the anharmonic nature of these bonds. Both negative (SiO_2) and positive (B_2O_3) thermal expansion coefficients are observed at low temperatures (White, 1976). Insertion of impurity ions in the open spaces in the network permits the alteration of the nature of these bonds to change the coefficient. Thus, between mixing of the various oxides (e.g., SiO_2, B_2O_3, PbO, K_2O) and adding impurities (e.g., Na, K, Ca) one can create almost any thermal expansion behavior desired. Such glasses, especially ones with very low thermal expansion values, are used in matching coefficients and for optical components like mirrors and lenses in space systems. A typical glass thermal expansion curve is shown in Fig. 3.22; however, on a much larger scale, there is usually structure in the curve as the different mechanisms become dominant.

Another anomaly in the thermal expansion curve usually occurs at the glass transition temperature, T_g, above which the material is stable in the amorphous or disordered state. Rapid cooling through the transition temperature retains the "glassy" state, which is metastable at lower temperatures. When aged, glasses devitrify into a partially crystalline structure. Because the crystalline state occupies much less volume, the thermal expansion appears to go through a rapid increase and then has a smaller coefficient at the lower temperatures. Such a thermal expansion curve is shown also in Fig. 3.22 for silicone rubber, which has a T_g below room temperature. Most conventional glasses have a T_g above room temperature.

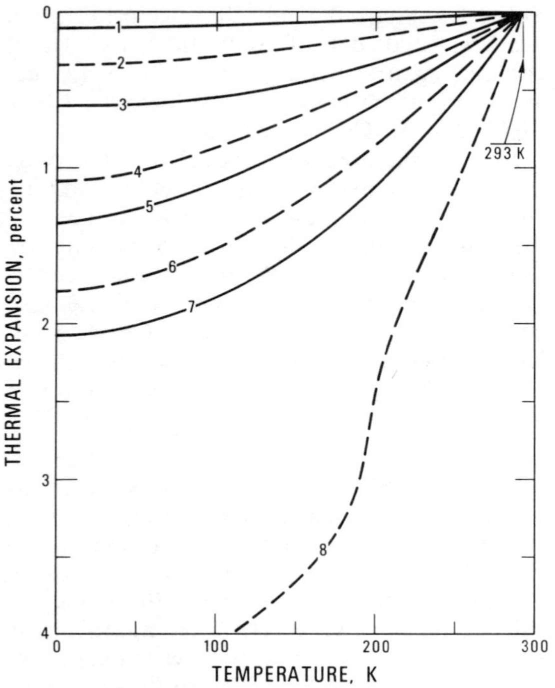

Figure 3.22 Thermal expansion vs. temperature for some typical nonmetallics: 1) glass, 2) copper, 3) polycarbonate and polychlorotrifluoroethylene, 4) polyethyleneterephthalate and polypyromellitimide, 5) rigid epoxy, 6) flexible epoxy, 7) polytetrafluoroethylene, and 8) silicone rubber.

The two compilations, Corruccini and Gniewek (1961) and *Thermal Expansion, Nonmetallic Solids* (1977), contain extensive data on glasses. More recently, White (1976) and Hortal (1978) have measured the thermal expansion of several glasses. In addition, Browder and Ballard (1977) have determined the thermal expansion of several nonglass optical materials. Few data exist on most ceramics except, of course, vitreous silica, which is used as a standard reference material (see § 3.4.4).

Two empirical correlations have been observed for optical materials that are useful in predicting thermal expansion coefficients. For the alkali halides, the coefficient is a function only of the index of refraction, n (Li, 1976),

$$\alpha = (38.8n - 20.5) \times 10^{-6} \tag{3.50}$$

The other is that the thermal expansion coefficient can be treated as additive for multioxide glasses. The average thermal expansion coefficient is just the weight percent, w_i, times the thermal expansion coefficient of the components, a_i (Morey, 1954),

$$\alpha = \sum_i a_i w_i \tag{3.51}$$

Relationships between α and the melting temperatures have also been shown (*Thermal Expansion, Nonmetallic Solids*, 1977).

Polymers. Polymers are formed by the linking together of long chains and networks of relatively small molecules called monomers, usually organic in nature. How these large molecules are joined to form a solid determines the two major categories of polymeric solids. If they are joined by a somewhat flexible molecular structure that is either linear or branched, a *thermoplastic* results. Examples of thermoplastics are the fluorocarbons, vinyls, polyimides, polyethylenes, polycarbonates, and acrylics. If the cross-linking is formed into a relatively rigid molecular structure by an irreversible chemical reaction, they are called *thermosetting plastics*. Examples of thermosetting plastics are epoxies, phenolics, polyesters, urethanes, and silicones. When the resulting cross-linking produces macroscopically flexible solids, from either category, they are sometimes called *elastomers*.

The thermal expansion behavior of polymers can be very complex because the bonds between adjacent atoms in the polymeric molecule are covalent and very strong, whereas the linking bonds are usually van der Waals in nature and very weak. Thus, in single crystal polymers, a wide variation in value as well as a high degree of anisotropy is usually seen. Barker (1967) developed a "bundle of tubes" model to describe the thermal expansion of these collections of long molecules. In polycrystalline solids, there is also an amorphous state, just as there is in glasses, and the amount of crystallization can dramatically affect the thermal expansion (see, e.g., Schramm et al., 1973). Representative values of the thermal expansion for a wide variety of polymers are shown in Fig. 3.22 where the scale is about ten times that usually presented in other figures in this chapter. The glass transition temperature, T_g, in polymers can also severely affect the thermal expansion behavior, as seen in the curve for silicone rubber and discussed in the section on glasses. Values for T_g for several of the common polymers do fall below room temperature, for example polyethylene at 140 K and polypropylene at 244 K. Usually transition temperatures are not clearly defined and depend on cooling rate, purity, stress, and other factors. They can be found in plastics handbooks (e.g., *Handbook of Plastics and Elastomers*, 1975).

The first comprehensive study of the thermal expansion in polymers was done by Laquer and Head (1952) on 27 different plastics, about all that were commercially available at the time. Two compilations have extensive data on the thermal expansion of polymers: Schramm et al. (1973), which is exclusively devoted to the low-temperature properties of six polymers, and *Thermal Expansion, Nonmetallic Solids* (1977). The six polymers in the first compilation are polytetrafluoroethylene (TFE), polychlorotrifluoroethylene (CTFE), polyethylene terephthalate (PET), polypyromellitimide (PPMI), polyparaxylylene (PPX), and polycarbonate (PC); many others are reported in the second. Not included in the above are a study on amorphous polymers at low temperatures by Roe and Simha (1974), and the measurement of the mechanical and thermal properties of nearly thirty commercial European plastics by Perrot, Blin, Bouriot, and Gilquin (1977). A study of the variation of the thermal expansion in epoxies with different catalysts and catalyst percentages by Mackowski, West, Samsel, Holderman, and Peters (1976) showed no dramatic differences due to these changes.

A few other areas of investigation are worth mentioning: McConnell et al. (1970) studied the thermoelastic expansion and creep of PET and PPMI films and fibers while under loads both below and above their elastic limits. Some unusual results, such as elongation upon cooling under load, were attributed to changes in crystallinity. Robbins et al. (1963) performed one of the few studies on elastomers. Some epoxy foams used for insulations were measured by Glaser, Black, Lindstrom, Ruccia, and Wechsler (1967). An empirical method of correlating the thermal expansion coefficient for the amorphous state with the T_g for polymers was originated by Simha and Boyer (1962) where

$$\alpha\, T_g = 0.164 \tag{3.52}$$

and a similar correlation between T_g and the difference between the thermal expansion coefficients for the amorphous and crystalline states was also determined. These results were further refined and some real limitations are discussed by Sharma, Mandelkern, and Stehling (1972).

3.4.3 Composites

One of the wonders of modern materials science is the fabrication of composites. By combining strong reinforcing fibers in a lightweight matrix, one can have the best of both properties in one material. Endless combinations of fillers, fibers, and woven cloths in both metallic and nonmetallic matrices can generate almost any desired combination of properties, truly designed materials. An entire chapter in this volume, Chapter 12, is devoted to composites and their low-temperature properties, and the details of fabrication and behavior are discussed therein. In this section, the general thermal expansion behavior of composites is discussed briefly along with a summary of a few of the more comprehensive studies. Some discussion of thermal properties of composites is also included in Chapter 12.

The thermal expansion along the reinforcing fiber in a composite tends to be dominated by the fiber, while transverse to the reinforcement the thermal expansion is dominated by the matrix. Because most matrix materials are polymers, the transverse thermal expansion is usually 2 to 5 times that in the fiber reinforcement direction. This depends strongly on the kind of reinforcement, whether it is uniaxial, crossply, cloth, or some combination, as well as on the volume fraction of the reinforcing material. Typical values for some of the more common composites are shown in Fig. 3.23. If the composite is filament

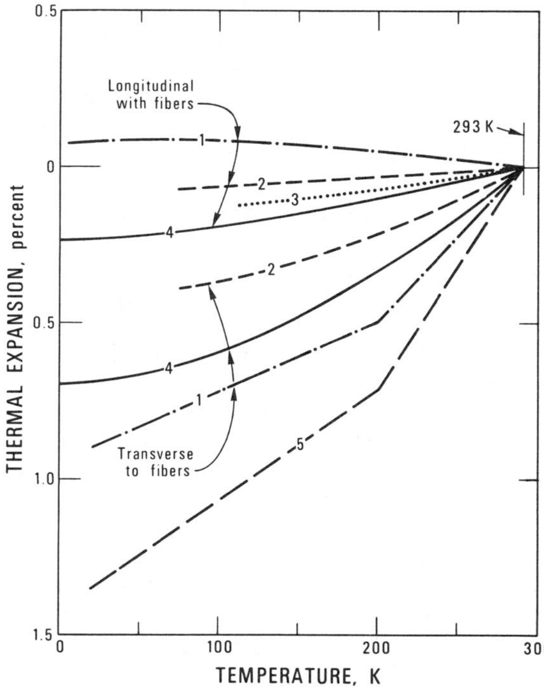

Figure 3.23 Thermal expansion vs. temperature for some typical composites: 1) polyamide/epoxy, 2) boron/aluminum, 3) concrete, 4) fiberglass/epoxy (NEMA G-10CR), and 5) fiberglass/TFE.

wound such that the reinforcing fibers are very close together, sometimes the thermal expansion is more typical of the fibers (Kasen, 1975a). This balance is discussed further in § 3.5.1, where some simple mathematical models are presented for predicting the thermal expansion of a composite.

The most common composite in use at present is fiberglass/epoxy, where the fiberglass is most typically in the form of a woven cloth. The thermal expansion behavior of these materials has been well reviewed by Kasen (1975a). The value parallel to the fibers is very close to that of the glass used, for example, about 0.1%, whereas the value perpendicular to the fibers is close to that of the matrix, varying from 0.2% to 0.8% depending upon the polymer used. More advanced composites in which boron, graphite, and polyamide fibers are used in polymer and even aluminum alloy matrices were also reviewed by Kasen (1975b). Again, the thermal expansion differs along and normal to the reinforcing fibers, but now, because graphite and the polyamide both expand slightly upon cooling, a balance can be achieved with slightly negative thermal expansion coefficients obtained. Pinheiro and Rosenberg (1978) and Pinheiro, Radcliffe, and Rosenberg (1978) have studied these materials to generate a zero-expansion composite material. Because of the strength of the aluminum alloys used, the aluminum matrix composites, although significantly strengthened by the fibers, tend to behave very much like the alloy in their thermal expansion behavior. Representative values of some of these composites are shown in Fig. 3.23 both parallel and transverse to the reinforcing fibers.

Since the two comprehensive reviews by Kasen (1975a; 1975b), he has updated the information (Kasen, 1981) by including more cloth and uniaxially reinforced composites and laminates. Also discussed are two fiberglass/epoxy composites for which the industry has minimized the variability in order to qualify them for cryogenic use, NEMA G-10CR and NEMA G-11CR. Data for these materials are included in Fig. 3.23. Dahlerup-Petersen and Perrot (1979) have reported an extensive study of European commercial composites, which includes their thermal expansion. There are also some varied composites reported in *Thermal Expansion, Nonmetallic Solids* (1977), most of which have some low-temperature data.

An aspect of the thermal expansion of composites that has not been discussed in the compilations is the theoretical calculation of behavior from that of the constituents. Three reviews of this field are: Rosen (1970), who includes other thermomechanical properties as well; Holliday and Robinson (1973) for polymer-based, filled composites; and Gulati and Plummer (1973) for laminates and structures. All conclude that these calculations are reliable with increasing accuracy, depending upon the degree of sophistication used in the model.

Filled composites are a slightly different category in that the reinforcing material does not have an extensive geometry, usually being powders or short lengths of fibers. These are frequently fabricated to alter some property other than strength, such as thermal expansion, thermal conductivity, or specific heat. Studies by Brassell and Wischmann (1974) and Shapovalenko, Tykachinskii, Kataeva, and Pyatnitskaya (1976) show that the properties (including the thermal expansion) change smoothly with the volume fraction of the filler. Extensive studies by Hartwig (1975) have included modeling their behavior (Hartwig, Weiss, and Puck, 1976) and an observation and discussion of deviations from linearity (Bauder, Hartwig, and Weiss, 1976).

One final category of composites used at low temperatures is concrete. Concrete is commonly used for storing liquefied natural gas and has several other potential applications. Very few data exist for low temperatures, but there have been extensive studies of the thermal expansion behavior of concrete near room temperature as a function of cement, sand, aggregate, and moisture content (see, e.g., Parsons and Johnson, 1944; Geymayer, 1969.) A study by Monfore and Lentz (1962) to about 110 K concluded that the thermal expansion was much less sensitive to mix and moisture content than the mechanical properties. A typical concrete is also included in Fig. 3.23.

3.4.4 Standard reference materials

The United States National Bureau of Standards (NBS) has been involved in precise determination of thermal expansion values for many years, both to develop better methods of measurement and to establish standard reference data for use by others. These data are essential because many measurement methods are comparison techniques where the thermal expansion of an unknown is compared with that of a known (see § 3.3). To fill this need, the NBS has determined precise values for five different Standard Reference Materials, ranging from the very low expansion fused silica to the often measured copper. These materials, the range of temperature for which data are available, and the range of thermal expansions and thermal expansion coefficients are shown in Table 3.3. For the temperature range up to about 1000 K, all measurements were done with a Fizeau interferometer (Hahn, 1970), and above 1000 K, with a twin microscope technique (Rothrock and Kirby, 1967). For all materials careful characterization of the material is provided, accuracies are given, values are calculated, and fits to the data described.

Because of the essential nature of these data to all thermal expansion measurements and the useful information presented with each one, the certificates for these standard reference materials are reproduced in the Appendix (§ 3.6). Barron et al. (1980) discuss other measurements of these materials and possible sources of minor discrepancies.

3.5 SOURCES AND USES OF DATA

For many materials, especially new alloys and composites that are rapidly being developed, no low-temperature thermal expansion data are available. In this final section, methods of predicting such data are outlined. There are several techniques that can generate useful engineering data from as little as a room-temperature thermal-expansion coefficient or even a specific-heat value. Surprisingly accurate projections are possible, with errors of less than a few percent. In addition, applications of thermal expansion data are briefly outlined, except for comparisons with theory, which are discussed in § 3.2. Finally, sources of data are discussed.

3.5.1 Predicting thermal expansion

As can be seen from Fig. 3.5, nearly all thermal expansion curves look alike. A single scaling factor can superimpose most of the curves, since they are all nearly linear at room temperature and approach absolute zero with zero slope and an approximate T^4 dependence. The reasons for this are outlined in § 3.2; the basic reason is that thermal expansion is determined by the nature of the atomic bond, but only as a second-order effect. In like materials the basic bonding mechanisms are the same, so unless a significant fraction of the bonds are changed, their thermal expansions are similar.

It was noted earlier that the interpretation and handling of the low-temperature thermal expansion data are greatly facilitated if they can be analytically fitted. The expansion was fitted rather than its derivative (the expansion coefficient), since all original data are taken as a length change. It was found empirically by Clark (1968) that a rational polynomial,

$$\frac{L_{293}-L_T}{L_{293}} = a + \frac{T^4}{b+cT^2+dT^3} \qquad (3.53)$$

provided the best fit to the thermal expansion data for most materials and permitted the expected thermodynamic restraints on behavior. These restraints were that the first and second temperature derivatives of $\Delta L/L$ go to zero as the temperature goes to zero and that the expansion be proportional to T^4 near zero and to T at higher temperatures. Room temperature (293 K) was chosen as the reference for convenience (and convention). The constant, a, is then the total length change from 293 to 0 K, and the room-temperature linear expansivity is approximately $1/d$. The fit of this polynomial was found to be better than the precision of the measurements for nearly all materials except the Invar-type iron alloys and other very low expansion materials.

From the room-temperature expansion coefficient. For a wide variety of metals and alloys (excluding those having low-temperature phase transformations) it was experimentally observed that the function

$$N(T) \equiv \frac{\Delta L(T)}{L} \cdot \alpha^{-1} = \frac{L_T - L_0}{(dL/dT)_T} \qquad (3.54)$$

Table 3.3 Standard reference materials.

Number	Material	Temperature (K)	Expansion	Coefficient (K^{-1})
SRM 731	Borosilicate glass	80 – 680	$-800 - 2000 \times 10^{-6}$	$2.6 - 5.3 \times 10^{-6}$
SRM 732	Single-crystal sapphire	293 – 2000	$0 - 15700 \times 10^{-6}$	$5.4 - 11.4 \times 10^{-6}$
SRM 736	Copper	20 – 800	$-3250 - 9420 \times 10^{-6}$	$0.2 - 20.5 \times 10^{-6}$
SRM 737	Tungsten	80 – 1800	$-800 - 7750 \times 10^{-6}$	$2.3 - 6.0 \times 10^{-6}$
SRM 739	Fused silica	80 – 1000	$-32 - 370 \times 10^{-6}$	$-0.7 - 0.6 \times 10^{-6}$

is nearly identical. This equation follows from the same assumptions that permit the use of the Grüneisen relation. The function $N(T)$ is simply the ratio of the total length change from absolute zero to the thermal expansion coefficient at temperature T. The value of N(293 K) for 85 metals and alloys (Clark, 1968; Corruccini and Gniewek, 1961; Arp et al., 1962) is 185.5 K with a standard deviation of 15.6 K. A sampling of this variation for a number of commercial alloys is shown in Table 3.4. (It should be noted that the standard deviation for alloys of a given class of metals is appreciably less.) Figure 3.24 shows $N(T)$ as a function of temperature for copper and aluminum. Figure 3.25 shows how this ratio, $N(T)$, for selected other metals and alloys deviates from $N(T)$ for copper by showing the percent deviation, $[(N(T)^{Cu} - N(T)^{x}) / N(T)^{Cu}] \times 100$, as a function of temperature. The phase transformation in Monel can be seen at about 250 K.

These data show that the low-temperature thermal expansion can be predicted from room-temperature properties and the function $N(T)$ with sufficient accuracy for most engineering design. Other prediction schemes, specifically for glasses and polymers, are discussed in § 3.4.2.

Table 3.4 Ratio of total contraction to room-temperature expansivity for several metallic alloys.

Material	$\frac{L_{293} - L_0}{(dL/dT)_{293}}$ (K)	Material	$\frac{L_{293} - L_0}{(dL/dT)_{293}}$ (K)
Al 2024-O	178	Waspaloy (age hardened)	173
Al 2024-T86	185	Udimet 630 (annealed)	178
Al 5083 (across)	182	Udimet 630 (age hardened)	183
Al 5083 (with)	181	Udimet 700	180
Al 7039	182	TD nickel	178
Al 7075	182	Phosphor bronze	194
Hastelloy C	171	Aluminum bronze	195
Hastelloy N	165	AISI 633 (AM350—annealed)	183
Hastelloy X	181	AISI 633 (AM350—precipitation hardened)	178
Hastelloy X (brazed)	185	AISI 430 F	174
Inconel 625	181	Fe–29 wt.% Ni	179
Inconel 718	184	Fe–50 wt.% Ni	192
Waspaloy (annealed)	182	Fe–79 wt.% Ni	186

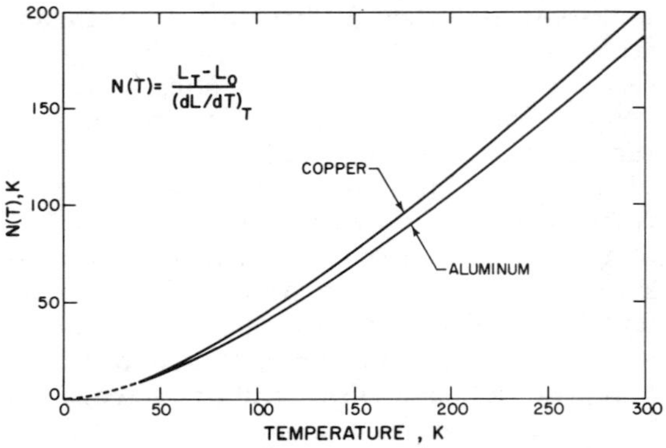

Figure 3.24 Ratio of the thermal expansion to the thermal expansion coefficient as a function of temperature for copper and aluminum.

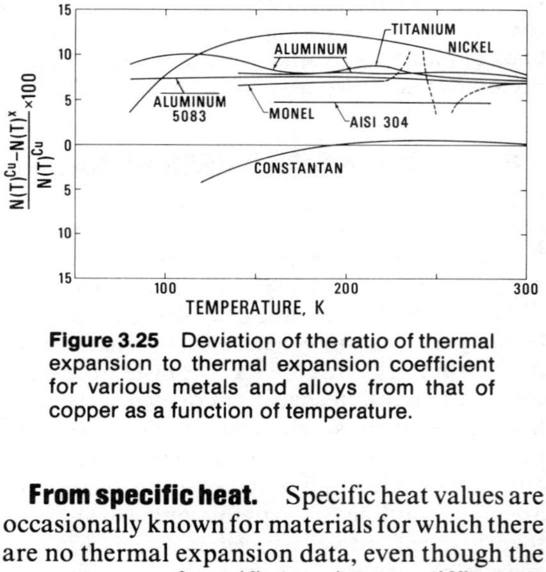

Figure 3.25 Deviation of the ratio of thermal expansion to thermal expansion coefficient for various metals and alloys from that of copper as a function of temperature.

From specific heat. Specific heat values are occasionally known for materials for which there are no thermal expansion data, even though the measurement of specific heat is more difficult. It was pointed out by Hume-Rothery (1956) that the correct (and most accurate for calculation) method of relating the expansion and specific heat is to use the appropriate experimental values in the expression (Grüneisen's first rule)

$$\frac{V_T - V_0}{V_T} = \frac{E_T}{Q_0 - kE_T} \tag{3.55}$$

where Q_0 and k are constants to be evaluated and $E_T = {}_0\int^T C_V\, dT$. For an explanation of the relation between Q_0 and γ the reader is referred to the work of Bijl and Pullan (1954) and Bijl (1957). Values of Q_0 for selected elements have been calculated by Grüneisen (1926), Hume-Rothery (1956), Simon and Vohsen (1928), and Zemansky (1951). The value of k is small and at low temperatures the term kE_T can usually be neglected. Taking the temperature derivative of the above expression and retaining the first-order terms, Simon and Vohsen used an alternate relation (Grüneisen's second rule),

$$Q_0 = \frac{C_P}{\beta} + 2kE_T \tag{3.56}$$

to calculate Q_0 for the alkali metals. In this equation $\beta = 1/V\,(dV/dT)$ is the volume expansivity. (The difference between C_P and C_V can be calculated using the volume expansivity and the bulk compressibility, but at low temperatures the difference becomes small and can often be neglected.)

Using Eq. 3.55 and available experimental data, Clark and Kropschot (1970) have calculated the temperature dependence of Q_0, shown in Fig. 3.26, for copper, nickel, and two Cu–Ni alloys between 60 and 300 K. Values below about 100 K begin to deviate from a constant Q_0 because of inaccuracies in the experimental data and, perhaps, deviations from Eq. 3.55. The values of Q_0 at 300 K are plotted in Fig. 3.27 vs. the percent nickel content to show the linear composition dependence. The composition and temperature dependence of Q_0 for a second alloy system, aluminum, Al 7075, and Al 6061 was also explored and was even more constant.

Using this method, the expansivity data for four commercial alloys were calculated using specific heat data from Touloukian (1966) and Zeigler, Mullins, and Hwa (1963). Room-temperature data were used to generate a constant, Q_0, from Eq. 3.56, and then the temperature dependence of C_P and the constant were used in Eq. 3.55 to calculate the expansion. The Q_0 constants and the results are compared with data from Corrucinni and Gniewek (1960; 1961) and Arp et al. (1962) in Table 3.5. (For easier comparison to specific heat, the volume change is referred to $T = 0$ K rather than 293 K.) The uncertainty in predicting specific heats of alloys using this technique is difficult to determine since accurate expansion and specific heat data are not available over a wide temperature range for well-characterized alloys. However, engineering estimates can be made using these techniques that will suffice for many applications.

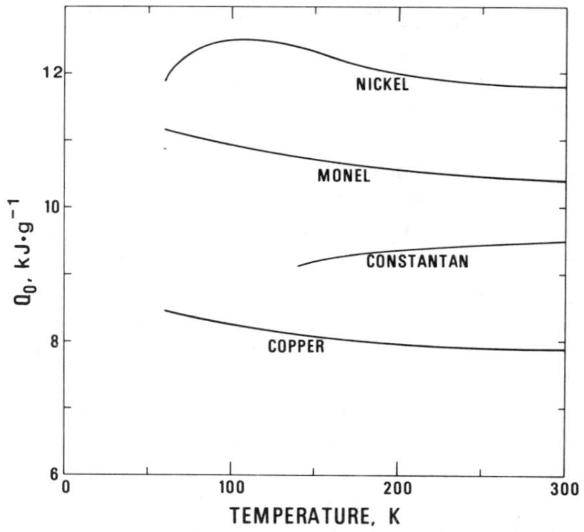

Figure 3.26 Temperature dependence of the constant, Q_0, relating specific heat to volume expansivity for the Cu–Ni system.

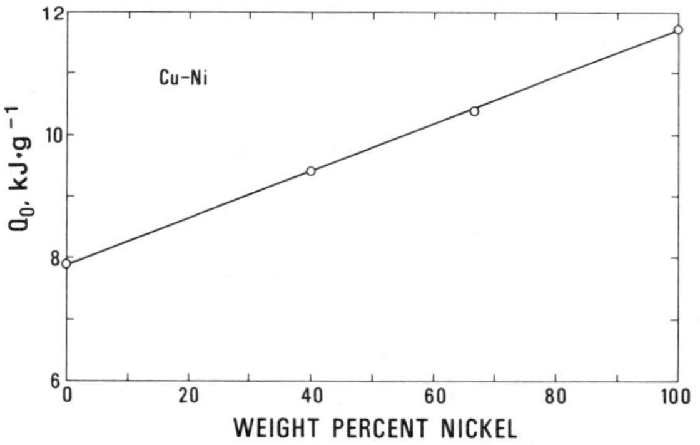

Figure 3.27 Dependence on composition of the constant Q_0 relating specific heat to volume expansivity for the Cu–Ni system.

Table 3.5 Thermal expansion as calculated from specific heat ($\triangle L/L$, %).

	Al 6061 $Q_0 = 13.0$ kJ·g^{-1}		AISI 347 $Q_0 = 9.4$ kJ·g^{-1}		Ti — C120 AV $Q_0 = 19.9$ kJ·g^{-1}		Ti — B120 VCA $Q_0 = 5.0$ kJ·g^{-1}	
T(K)	calc.	exp.	calc.	exp.	calc.	exp.	calc.	exp.
60	0.011	0.011	0.0036	0.004	0.0058	0.006	0.005	0.006
100	0.050	0.049	0.029	0.031	0.020	0.019	0.019	0.021
200	0.213	0.217	0.157	0.155	0.095	0.095	0.084	0.084
293	0.421	0.419	0.297	0.298	0.177	0.173	0.166	0.161

For composites. Frequently the thermal expansion of a composite material is unknown, although the data for its constituents are available. The thermal expansion of composite materials can be calculated by several simple methods. A series model is used when the different contracting materials act in series perpendicular to any bonding surface between them, such as perpendicular to the reinforcing fibers in a fiberglass/epoxy composite (see Fig. 3.28). A parallel model using averages weighted by the elastic modulii must be used when the contraction direction is parallel to a bond between the unlike materials, such as along a Cu/Nb–Ti multifilamentary wire. When the strain introduced by different coefficients, $\epsilon_i = \Delta T(\alpha_j - \alpha_i)$, exceeds the yield strain of a material, then a much more complex technique must be used utilizing predicted stress–strain curves (Easton, Kroeger, Specking, and Koch, 1980). Using the simplification noted above relating the thermal contraction, $\Delta L/L$, to the expansion coefficient, α, by a constant, these two simple predictive models become

$$\Delta L/L = V_f(\Delta L/L)_f + V_m(\Delta L/L)_m \quad \text{(series)} \quad (3.57)$$

and

$$\Delta L/L = \frac{E_f V_f(\Delta L/L)_f + E_m V_m(\Delta L/L)_m}{E_f V_f + E_m V_m} \quad \text{(parallel)} \quad (3.58)$$

where E is the elastic modulus and V the volume fraction of reinforcing fiber, f, and matrix, m. These equations, of course, can be generalized for more components. Now the measured values of $\Delta L/L$ can be compared directly.

The thermal expansion was measured for several fiberglass epoxy composites and their constituents by Clark et al. (1981) and calculations in both directions were compared. The fiberglass/epoxy behavior is shown in Fig. 3.29 both parallel (warp) and perpendicular (normal) to the glass-reinforcing fibers along with the predicted values for a cryogenic-grade NEMA G-11 composite using the parallel (Eq. 3.58) and series (Eq. 3.57) models, respectively. The dominance of the glass fibers in the former and the epoxy in the latter is clearly evident.

A totally different kind of composite was also studied, a superconducting wire. In this case, the matrix was copper and the reinforcing fibers were fine filaments of Nb–Ti alloy superconductor. The thermal expansion value was calculated from Nb–Ti and copper data using the parallel model; the agreement is shown in Fig. 3.30. Calculations for more complex superconducting wires (Cu/V_3Ga and Cu/Nb_3Sn) were also compared, but the agreement was not as good, probably owing to yielding of the soft copper matrix. However, modeling of this strain behavior in some simple multifilamentary A15 superconductors has been shown to work (Easton et al., 1980).

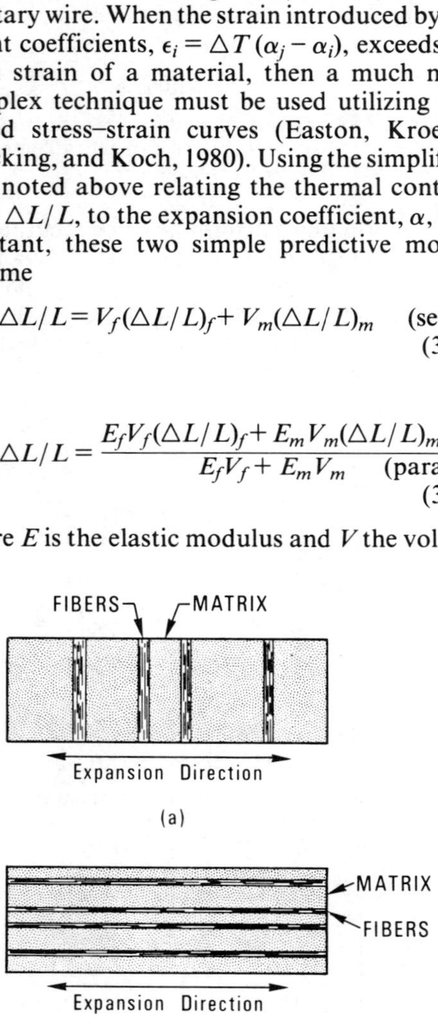

Figure 3.28 Series (a) and parallel (b) models for calculating the thermal expansion of composites.

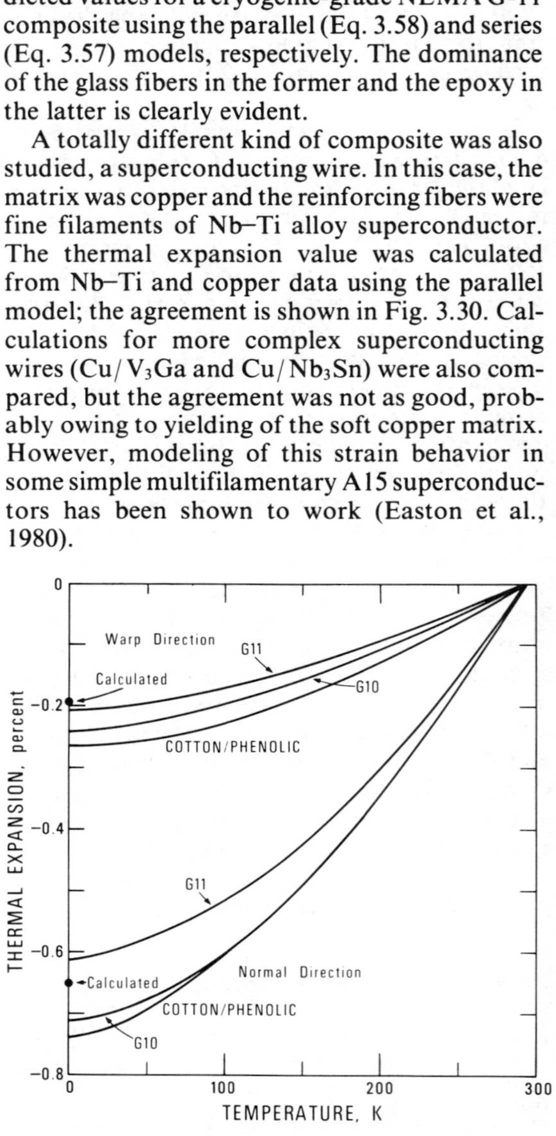

Figure 3.29 Thermal expansion vs. temperature for some composite materials compared with calculated values.

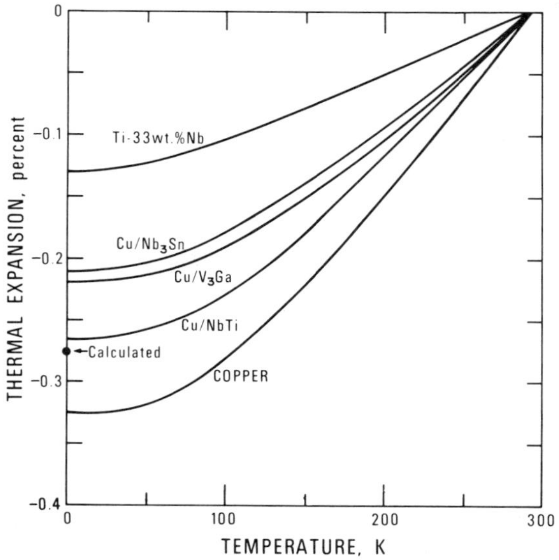

Figure 3.30 Thermal expansion vs. temperature for superconducting wire composites compared with a calculated value.

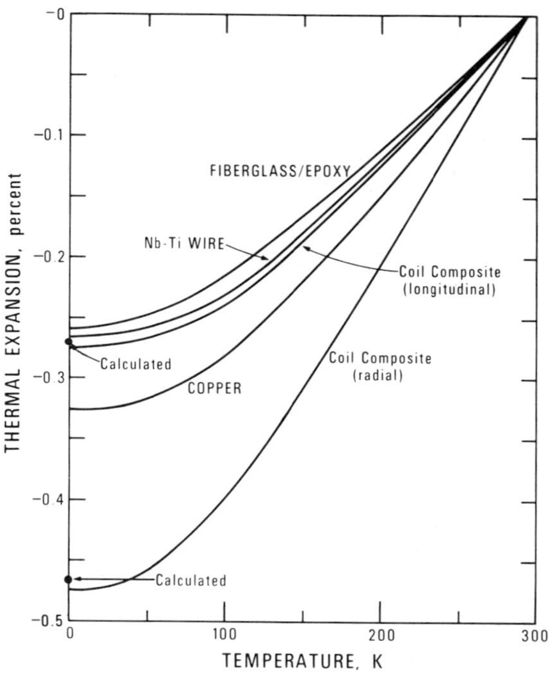

Figure 3.31 Thermal expansion vs. temperature for a superconducting coil composite and its components compared with calculated values.

Even in very complex structures, such as a superconducting coil composite studied in the work of Clark et al. (1981), these simple relationships have been shown to hold with surprising accuracy. In that composite, the reinforcing "fiber" was the superconducting wire which itself was a Cu–Nb–Ti composite. The "matrix" in the coil composite was the fiberglass epoxy composite shown above. The thermal expansion of each of these constituents can be calculated on its own. If the superconducting wire and the fiberglass/epoxy are treated as fiber and matrix even though each is a composite by itself, again, the calculated values agree closely even though the anisotropy is large (Fig. 3.31).

3.5.2 Application of thermal expansion data

The most common application of thermal expansion data is, of course, to determine a length change caused by a change of temperature. This is very straightforward and simple and is only outlined here for completeness. Volume changes are not quite as simple but are also shown. The next most common application of thermal expansion data is to determine the stresses generated in a body or structure resulting from a temperature change under two conditions. In the first, a homogeneous body constrained by external forces while undergoing a temperature change will generate internal stresses. In the second, an unconstrained body will generate internal stresses if there is a temperature distribution within it, or if there are differing thermal expansion values within the body or structure.

Length and volume changes. The exact relationship between a length of material at temperature T_1 and its new length at temperature T_2 is shown by Eq. 3.3. This equation involves an integral of the thermal expansion coefficient and is impractical for engineering use. For most purposes, the only equation needed is

$$L(T_2)-L(T_1)=\bar{\alpha}\,\Delta T\,L(T_1) \tag{3.59}$$

where $\bar{\alpha}$ is an average thermal expansion coefficient over the temperature range T_1 to T_2. Even if a precise $\bar{\alpha}$ is unavailable, a value at a nearby temperature can be adequate. For example, using the room-temperature thermal expansion coefficient for copper to calculate the length change from 200 to 220 K results in less than an 8% error. The agreement, of course, becomes worse the farther the values are from the room-temperature regime, where the thermal expansion coefficient is almost constant. Extrapolations are frequently all that is required for less than a few percent accuracy.

The most common format for the presentation of thermal expansion data is a standard table like Table 3.2. Length changes can be read directly from the table, and interpolation between even the coarsest data set is usually quite adequate. If more accuracy is required, a thermal-expansion-data fitting program on a small calculator can do quite well as can integration of coefficient data (see § 3.5.1).

Frequently, a bigger problem is that of handling anisotropy of the thermal expansion coefficient. Single crystals of many materials have different thermal expansion coefficients in different directions with respect to the crystal axes. If a polycrystalline material has a preferred crystal orientation due to rolling or grain growth during an anneal, for example, this anisotropy is reflected in the final thermal expansion properties. Proper averaging of the single-crystal properties to determine the properties of the polycrystalline material has been a major problem, especially for highly anisotropic materials (see Chapter 1). In general, the thermal expansion coefficient in an arbitrary direction that makes angles ψ and ϕ with the 3 and 1 directions, respectively (or c and a in a single crystal) is

$$\alpha=\alpha_1 \sin^2\psi \cos^2\phi+\alpha_2 \sin^2\psi \sin^2\phi+\alpha_3 \cos^2\psi \tag{3.60}$$

where α_1, α_2, and α_3 are the thermal expansion coefficients in the 1, 2, and 3 directions. The same will hold true for length changes. If the structure is axially symmetric (hexagonal, trigonal, or tetragonal in single crystals) then the thermal expansion coefficients in two directions are the same, that is,

$$\alpha_1=\alpha_2=\alpha_\perp; \quad \alpha_3=\alpha_\parallel \tag{3.61}$$

then the thermal expansion coefficient in a direction making an angle ψ with the perpendicular axis is

$$\alpha=\alpha_\perp \sin^2\psi+\alpha_\parallel \cos^2\psi \tag{3.62}$$

In a similar manner it is possible to calculate total volume changes from the volume thermal expansion coefficient where

$$V(T_2)-V(T_1)=\bar{\beta}\,\Delta T\, V(T_1) \tag{3.63}$$

where $\bar{\beta}$ is an average volume thermal expansion coefficient over the range T_1 to T_2. If the material is anisotropic, the volume thermal expansion coefficient can be calculated as

$$\beta=\alpha_1+\alpha_2+\alpha_3 \tag{3.64}$$

or, for axially symmetric materials,

$$\beta=2\alpha_\perp+\alpha_\parallel \tag{3.65}$$

or, for isotropic materials, simply

$$\beta=3\alpha. \tag{3.66}$$

Calculating thermal stresses. The basic principle for calculating thermal stresses assumes that any change in volume must be equally countered by a force that would return the material to its original shape. Thus, for a simple rod constrained at both ends, a ΔT would result in a change of length $\Delta L/L=\alpha\Delta T$. The same ΔL can be obtained by a stress, $\sigma=E\Delta L/L$, where E is Young's modulus of elasticity. (It is assumed that the stress is below the yield stress of the material and it has not deformed plastically.) Equating the two length changes yields

$$\sigma=E\alpha\Delta T \tag{3.67}$$

Note that the stress is independent of the length of the rod or the thickness of the material between constraints. This stress can build rather quickly to approach or exceed the yield strength of the material. For example, a high-strength stainless steel rod that is held fixed at both ends and cooled from room temperature to 200 K generates a tensile stress in excess of 0.25 MPa, over half of its yield strength.

Extensive texts are available outlining the development of thermal stresses in great detail. Boley and Weiner (1960) generated a rather complete thermoelastic theory based on thermodynamics, basic heat conduction, and the theory of elasticity in solids. They applied their analysis to many common problems and then extended the work into inelastic behavior. Gatewood (1957) takes a much more problematic approach and includes many very real complications of the typical classic problems. He includes such things as combinations of elastic and inelastic stresses, applied and thermal stresses, buckling and deflections, and fatigue and thermal shock. The range of these two texts should cover most problems arising from the application of materials at low temperatures. The classic problem of the bimetallic strip is outlined below to show how thermal expansion data can be applied, and then several of the more common boundary conditions and their resultant stresses are listed for easy reference.

Two metallic strips of equal thickness, d, and

differing thermal expansion coefficients, α_1 and α_2, are bonded together and of equal length, ℓ, at the reference temperature, T. When cooled to a temperature $T-\Delta T$, the composite strip bends in a uniform arc of radius, r, as shown in Fig. 3.32 for $\alpha_2 < \alpha_1$, where the α values are averaged over the ΔT. To a first approximation, the central lines in each of the strips are assumed to be neither in tension nor compression and act as free bodies, changing to lengths ℓ_1 and ℓ_2. These lengths can be expressed both in terms of the radius and the subtended angle, θ, and in terms of the thermal expansion coefficient and the temperature change so that

$$\ell_1 = r\theta; \qquad \ell_2 = (r+d)\,\theta \tag{3.68}$$

and

$$\ell_1 = \ell(1-\alpha_1\Delta T)$$
$$\ell_2 = \ell(1-\alpha_2\Delta T) \tag{3.69}$$

The ratio of the lengths then becomes

$$\ell_2/\ell_1 = 1+\frac{d}{r}$$
$$= \frac{1-\alpha_2\Delta T}{1-\alpha_1\Delta T} \cong 1-(\alpha_2-\alpha_1)\Delta T \tag{3.70}$$

so the radius of curvature is then

$$r = \frac{d}{(\alpha_1-\alpha_2)\Delta T} \tag{3.71}$$

If the ends of the bimetallic strip are constrained so no bending occurs, then the maximum bending stress, σ_b, is for an average $\alpha = (\alpha_1+\alpha_2)/2$ (Roark and Young, 1975)

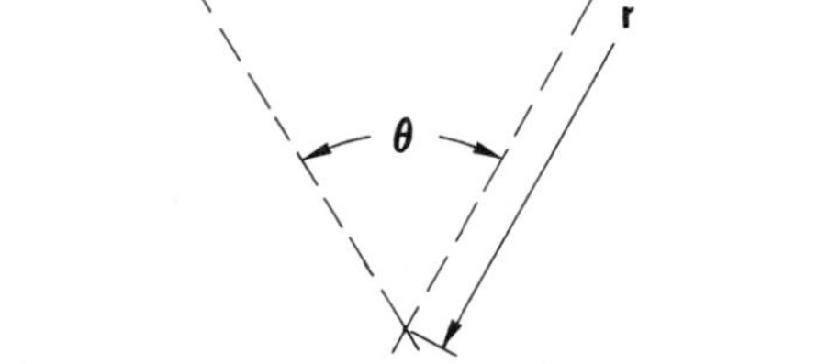

Figure 3.32 Bending of a bimetallic strip due to differing thermal expansions.

$$\sigma_b = \frac{1}{2}\,E\alpha\Delta T \tag{3.72}$$

with the warm side in tension and the cool side in compression.

The thermal stresses generated in many common shapes and subject to common constraints have been calculated several years ago (Kent, 1932; Maulbetsch, 1935; Goodier, 1937; Goodier, 1957) and have been summarized recently by Roark and Young (1975). The most useful relationships are given below, in which the values of Young's modulus, E, the thermal expansion coefficient, α, and Poisson's ratio, ν, are assumed to be averaged over the temperature range, ΔT.

a. *Externally constrained stress*
   1. A uniform straight rod fixed at the ends with a uniform temperature change (cooled) throughout has a tensile stress of

$$\sigma = E\alpha\Delta T$$

   2. A uniform flat plate held at the edges has a stress of

$$\sigma = E\alpha\Delta T/(1-\nu)$$

   3. A solid body held to the same form and volume has a stress of

$$\sigma = E\alpha\Delta T/(1-2\nu)$$

   4. A uniform rectangular bar fixed at the ends with one side at T and a uniform temperature gradient, ΔT, perpendicular to the bar has a bending stress (compression on the warm side, tension on the cold) of

$$\sigma_b = \frac{1}{2}\,E\alpha\Delta T$$

   5. Similarly, a uniform flat plate of thickness, t, held at the edges has a maximum bending stress of

$$\sigma_b = \frac{1}{2}E\alpha\Delta T/(1-\nu)$$

   [If free, its radius of spherical curvature would be $r = t/(\alpha\Delta T)$.]

b. *Internally constrained stress*
   1. A solid body whose surface is suddenly subjected to a temperature change, ΔT, has its surface under a stress of

$$\sigma = E\alpha\Delta T/(1-\nu)$$

   (compressive for a temperature rise).

2. A thin, circular disk with a small central portion of radius a raised by ΔT has radial and tangential compressive stresses in the heated portion of

$$\sigma_r = \sigma_t = \frac{1}{2} E\alpha\Delta T$$

and at a distance, r, from the radius, a, of

$$\sigma_r = -\sigma_t = \frac{1}{2} E\alpha\Delta T(a^2/r^2)$$

with the radial compressive and the tangential tensile.

3. A hollow, thin cylinder with a temperature gradient across the cylinder wall has a maximum circumferential stress of

$$\sigma = \frac{1}{2} E\alpha\Delta T/(1-\nu)$$

with compression on the warm surface, tensile on the cold.

4. A thick-walled cylinder, as above, with inner radius, b, and outer radius, c, has a stress of

$$\sigma = \frac{E\,\alpha\,\Delta T}{2(1-\nu)\log(c/b)} \times \left[1 - \frac{2b^2}{c^2-b^2}\log(c/b)\right]$$

There are many others with two transverse temperature gradients, heated surfaces, surfaces subject to ramped temperatures, hollow spheres, and almost all conceivable combinations.

The subject of transient thermal stresses is a very difficult one and very common in the application of materials at low temperatures. It is usually treated by first calculating the temperature distribution at various times from heat-transfer equations and determining the stresses at the various times as the temperature distribution comes into equilibrium. Most low-temperature structures are constrained enough so that changes in shape do not affect the heat-transfer processes and an iterative procedure is unnecessary. The reader is referred to the above references for further details.

3.5.3 Sources of data

Searching for specific values of physical properties of materials can sometimes be a very frustrating process. Thermal expansion data are no exception; in fact, surprisingly few data exist for materials commonly used at low temperatures. Outlined below, with comments on each, are sources of data for materials at low temperatures.

A frequently overlooked source is the manufacturer of the material itself. Manufacturers often supply data sheets on their specific alloys or materials, and thermal expansion is often among the data provided. Also, if only room-temperature data can be found, the methods of § 3.5.1 can be used to estimate low-temperature values.

Compilations and handbooks.

a. *Thermal Expansion, Metallic Elements and Alloys*, Vol. 12 (1975) and *Thermal Expansion, Nonmetallic Solids*, Vol. 13 (1977) of *Thermophysical Properties of Matter*. These two volumes are part of the Thermophysical Properties of Matter Data Series from the Thermophysical Properties Research Center, Purdue University and comprise probably the most extensive compilation of its kind. Most of the data are above room temperature, but because of its size, there are many low-temperature data.

b. *LNG Materials and Fluids* (1979). This handbook of materials for use with liquefied natural gas systems includes structural metals, thermal insulations, nonmetallic laminates, and the fluid components of LNG itself.

c. *Handbook on Materials for Superconducting Machinery* (1977). This collection of design data for low-temperature structural materials was assembled by the Cryogenics Division of the National Bureau of Standards and concentrates on metallic alloys; some polymers and composites are included.

d. *A Compilation of Mechanical, Thermal, and Electrical Properties of Selected Polymers* by Schramm et al. (1973). A rather complete set of low-temperature data for the six polymers polytetrafluoroethylene, polycholorotrifluoroethylene, polyethylene terephthalate, polypyromellitimide, polyparaxylylene, and polycarbonate.

e. *Cryogenic Materials Data Handbook* (1964). One of the first collections of low-temperature data for materials for use in space applications.

f. *Thermal Expansion of Technical Solids at Low Temperatures* by Corruccini and Gniewek (1961). The first extensive critical evaluation of thermal expansion data, mostly for the metallic elements and alloys.

g. *Thermal Expansion of Crystals* by Krishnan et al. (1979). This text on the subject of thermal expansion of single crystals contains extensive references to measurements at low temperatures.

h. "Thermal Expansion of Solids at Low Temperatures," an article by Barron et al. (1980). An excellent review article on the subject for the temperature range 1 to 100 K; contains many references to specific materials data.

Conference proceedings.

a. Proceedings of the International Cryogenic Materials Conferences, *Advances in Cryogenic Engineering — Materials*, Vols. 22, 24, 26, 28 (1976, 1978, 1980, 1982). This biennial conference, held in odd-numbered years since 1975, is devoted exclusively to materials at low temperatures and frequently has thermal expansion data in its proceedings.

b. Proceedings of the Cryogenic Engineering Conferences, *Advances in Cryogenic Engineering*, Vols. 1–21, 23, 25, 27, (1954–1982). A large collection of papers last indexed in Volume 20.

c. *International Symposia on Thermal Expansion*. These conferences, held approximately every other year, are devoted exclusively to thermal expansion and contain a reasonable amount of data, but they have been inconsistently published. Proceedings known to the author are:

Proceedings of the Second Symposium on Thermal Expansion (1970).

AIP Conference Proceedings No. 3, Thermal Expansion — 1971, Third Symposium (1972).

AIP Conference Proceedings No. 17, Thermal Expansion — 1973, Fourth Symposium (1974).

Thermal Expansion 6 (1978).

Journals. Listed below are selected journals that frequently contain thermal expansion data, and if one were desperate for specific information, a scan of these journals (preferably done electronically) could yield the desired values.

a. *Cryogenics*

b. *Journal of Low Temperature Physics*

c. *Journal of Applied Physics*

d. *Journal of Physics C: Solid State*

e. *Journal of Physics and Chemistry of Solids*

Specialty journals on specific materials, such as *Journal of the American Ceramic Society, Journal of Composite Materials,* and *Journal of Less-Common Metals*, would also be worth pursuing for a specific material.

Special articles. There are about a dozen articles that are frequently referenced and are extensive sources of much of the available thermal expansion data. These are listed below with a brief indication of the materials measured. None of these are compilations and all report original data.

a. Arp et al. (1962) — more than 40 engineering materials measured from 20 to 293 K, mostly metallic alloys.

b. Clark (1968) — about 30 structural alloys from 20 to 293 K.

c. Clark et al. (1981) — about 16 components of superconducting magnet composites from 20 to 293 K.

d. Laquer and Head (1952) — 20 plastics and elastomers from 20 to 300 K.

e. Mackowski et al. (1976) and Hamilton, Greene, and Davidson (1968) — epoxy resins.

f. Molby (1949) — about a dozen different kinds of glass from 80 to 300 K.

g. Rhodes et al. (1963) — about 20 metals and alloys from 20 to 573 K.

h. Robbins et al. (1963) — 26 elastomers from 76 to 300 K.

i. White (1976) and White, Birch, and Manghnani (1977) — low thermal expansion ceramics and glasses from 2 to 90 K.

Lacking success in finding the thermal expansion data desired in the above sources should not be cause for despair. There is a great deal of information available in more obscure sources. The best method to ferret this out is with the newly available electronic searching techniques on some rather extensive data bases. It may well be worthwhile to spend a few hours at your local library in an interactive search mode with such systems as:

a. Compendex. Engineering index, from 1970.

b. Conference Papers Index. Technical conference papers, from 1973.

c. Defense Technical Information Center. Department of Defense, Research and Development, from 1953.

d. Inspec. *Physics Abstracts, Electrical Engineering Abstracts,* and *Computer Control Abstracts*, from 1969.

e. Ismec. Information service in mechanical engineering, from 1973.

f. Metadex. *Metals Abstracts* and *Alloys Index*, from 1960.

g. NASA Information Bank. Aeronautics and Space, STAR, and IAA, from 1963.

h. NTIS. National Technical Information

Service, government Research and Development, from 1964.

i. Scisearch. Pure and applied sciences, from 1974.

j. SPIN. Physics, astronomy, astrophysics, geophysics, from 1975.

k. SSIE. Smithsonian Science Information Exchange, from 1974.

If none of the above yield the information needed and a predictive scheme is not adequate, the best recourse is, of course, to measure the thermal expansion on the material itself. Several laboratories in the United States have this capability at varying degrees of accuracy. The place to locate these capabilities is in the literature that has published recent measurements.

3.6 APPENDIX — STANDARD REFERENCE MATERIAL CERTIFICATES

Reproduced in this appendix are the National Bureau of Standards Certificates for five standard reference materials. The certificates include tabular data for both the thermal expansion and the thermal expansion coefficient, a description of the measurements, the statistical analysis, and references. The five Standard Reference Materials (SRMs) are:

SRM 731 — Borosilicate glass
SRM 732 — Single crystal sapphire
SRM 736 — Copper
SRM 737 — Tungsten
SRM 739 — Fused silica

National Bureau of Standards
Certificate
Standard Reference Material 731
Borosilicate Glass – Thermal Expansion

T. A. Hahn

Thermal Expansion as a Function of Temperature

T	Expansion $\Delta L/L_{293}$	Expansivity α	T	Expansion $\Delta L/L_{293}$	Expansivity α
80 K	-819×10^{-6}		293 K	0×10^{-6}	4.78×10^{-6}/K
90	−797		300	34	4.82
100	−771	2.64×10^{-6}/K	320	131	4.91
110	−744	2.86	340	230	4.99
120	−714	3.07	360	330	5.06
130	−683	3.25	380	432	5.11
140	−649	3.43	400	535	5.15
150	−614	3.58	420	638	5.19
160	−578	3.72	440	742	5.21
170	−540	3.85	460	847	5.23
180	−501	3.97	480	952	5.25
190	−460	4.08	500	1057	5.26
200	−419	4.17	520	1162	5.26
210	−377	4.26	540	1267	5.27
220	−334	4.34	560	1372	5.27
230	−290	4.41	580	1478	5.27
240	−246	4.48	600	1583	5.27
250	−201	4.54	620	1689	5.28
260	−155	4.60	640	1794	5.29
270	−109	4.66	660	1900	
280	− 62	4.71	680	2007	

This SRM is available as a square rod with 6.4 mm (1/4 in) sides; L1 is 51 mm (2 inches) long, L2 is 102 mm (4 inches) long, and L3 is 152 mm (6 inches) long. This borosilicate glass is from the same lot as SRM 717, Borosilicate Glass Viscosity Standard, with a nominal composition of 70% SiO_2, 17% B_2O_3, 8% K_2O, 3% Al_2O_3, 1% Na_2O, and 1% Li_2O. The index of refraction after fine annealing is N_D=1.4874. These rods were annealed at 789 K, cooled to 743 K at a rate of 2 degrees per hour, and then cooled to 523 K at a rate of 5 degrees per hour.

The above values of expansion and expansivity were calculated from equations based on a least squares analysis of the expansivity data from six specimens taken from various positions of the stock. A description of the experimental method, fitting procedure, and estimate of uncertainties is given in this certificate.

The overall coordination and evaluation of data leading to certification of this Standard Reference Material was performed by T. A. Hahn.

The technical and support aspects involved in the preparation, certification, and issuance of this Standard Reference Material were coordinated through the Office of Standard Reference Materials by R. E. Michaelis.

Washington, D.C. 20234
July 31, 1972

J. Paul Cali, Chief
Office of Standard Reference Materials

(over)

PROCEDURE

The apparatus used for the expansion measurements was a Fizeau interferometer with a 1-cm specimen length. Above room temperature, the measurements were made with the interferometer in a controlled atmosphere furnace using a Pt vs Pt-10% Rh thermocouple. Below room temperature, a cryostat operating with liquid nitrogen was used with a platinum resistance thermometer. The green spectral line of a mercury light source was used to produce the interference fringes. Fringe motion was measured with a filar-micrometer eyepiece. Each test specimen was made by fastening three 1-cm rods of the SRM to a Kovar ring to form a three point contact for the interferometer plates. With the uncertainties in temperature and fringe measurements, the expansivity was determined with an uncertainty of $\pm 0.03 \times 10^{-6}$/K.

Values of expansivity were calculated between equilibrium temperatures and corrections[1] were made so that α represents $\frac{1}{L_{293}} \frac{dL}{dT}$.

Tests on the six specimens indicated no appreciable differences. All of the data were pooled and the following third-order spline polynominals were obtained by the method of least squares using an Omnitab routine:

$100 \leqslant T \leqslant 293$

$$\alpha \times 10^6 = -0.8194 + 4.7611 \times 10^{-2} T - 1.4753 \times 10^{-4} T^2 + 1.7153 \times 10^{-7} T^3$$

$293 \leqslant T \leqslant 640$

$$\alpha \times 10^6 = 0.8651 + 2.3569 \times 10^{-2} T - 4.2277 \times 10^{-5} T^2 + 2.5408 \times 10^{-8} T^3$$

The standard deviation of this fit is 0.06 with 88 data points. These equations and their integrals were used to calculate the values listed in the table. A comparison of the experimental expansion data with values predicted from the equations gives a standard deviation of 8×10^{-6}. All of the data for both the expansion and expansivity were within two standard deviations of the values predicted by the equations.

[1]Hahn, T.A., Thermal Expansion of Copper from 20 to 800 K – Standard Reference Material 736, J. Appl. Phys. **41**, 5096 (1970).

National Bureau of Standards
Certificate
Standard Reference Material 732

Single Crystal Sapphire - Thermal Expansion

T. A. Hahn

Thermal Expansion and Expansivity as a Function of Temperature (IPTS-68)

T	$\frac{L-L_{293}}{L_{293}}$	$\alpha = \frac{1}{L_{293}}\frac{dL}{dT}$	T	$\frac{L-L_{293}}{L_{293}}$	$\alpha = \frac{1}{L_{293}}\frac{dL}{dT}$
293 K	0×10^{-6}	$5.38 \times 10^{-6}\ K^{-1}$	900 K	4608×10^{-6}	$8.79 \times 10^{-6}\ K^{-1}$
300	38	5.48	1000	5500	9.05
350	328	6.12	1100	6417	9.29
400	647	6.61	1200	7357	9.52
450	987	7.00	1300	8320	9.74
500	1346	7.32	1400	9306	9.97
550	1719	7.59	1500	10314	10.20
600	2104	7.82	1600	11345	10.42
650	2500	8.02	1700	12399	10.66
700	2906	8.20	1800	13477	10.89
750	3320	8.37	1900	14578	11.14
800	3743	8.52	2000	15704	11.38

This single-crystal sapphire was grown from the melt, cut into 20 cm lengths, and centerless ground to 6.4 mm diameter. Back reflection Laue photographs of each rod gave an average orientation of the c-axis of the crystal relative to the rod axis equal to 59°. Several rods were checked for variations in orientation along their lengths, but none were found. Three specimens that were thought to have the greatest difference in orientation, ~1°, were selected for the thermal expansion measurements. The expansion of these three specimens from room temperature to 1000 K, however, did not differ by more than $4 \cdot 10^{-6}$. This is well within the precision of the measurement system, and indicates that no significant difference exists between any of the rods. The tabulated values were calculated from an equation whose parameters were determined by a least squares analysis of the pooled expansion data. Three-point interpolation of the tabulated data may be used to obtain values not listed. Descriptions of the experimental methods, the equation used to fit the data, and estimates of uncertainties are given in this certificate and in reference [1].

The technical and support aspects involved in the acquisition, preparation, certification, and issuance of this Standard Reference Material were coordinated through the Office of Standard Reference Materials by R. K. Kirby.

Washington, D.C. 20234
October 3, 1977

J. Paul Cali, Chief
Office of Standard Reference Materials

(over)

This SRM is available in rods 6.4 mm (1/4 inch) in diameter and 51 mm (2 inches) in length. Inquiries for longer rods may be directed to the Office of Standard Reference Materials, National Bureau of Standards.

Procedure

In the temperature range below 1000 K, expansion measurements were made with a Fizeau interferometer [2]. Interference fringes were formed between fused-silica optical flats using the green spectral line of a mercury lamp. Fringe motion was measured with a filar-micrometer eyepiece. Each test specimen was made by grinding three 1-cm rods to fit a sample holder to form a three-point spacer for the interferometer flats. Temperatures were measured using a calibrated Pt vs. Pt-10% Rh thermocouple fastened to the sample holder. The specimens were heated in a low-pressure helium atmosphere and the expansion measured between equilibrium temperatures. Measurements on a variety of different materials with this system generally give a reproducibility on the order of $\pm 5 \times 10^{-6}$ in $\Delta L/L$ values.

In the temperature range above 1000 K, expansion measurements on one specimen were made with a twin-microscope technique [3] using a tungsten-mesh vacuum furnace. Knife edges machined into the specimen 10 cm apart defined its length. The change in length on heating between equilibrium temperatures was measured with filar-micrometer eyepieces. The temperatures were measured with W vs Re and Pt-30% Rh vs Pt-6% Rh thermocouples. Deviations in the expansion measurements result in an expected reproducibility of $\pm 15 \times 10^{-6}$ in $\Delta L/L$ values.

The tabulated values given in this certificate were calculated from a variation of Gruneisen's equation given by

$$\frac{L - L_{293}}{L_{293}} = \frac{E(\frac{\theta}{T})}{3a\left[Q_o - kE(\frac{\theta}{T})\right]} + \frac{1-a}{a}$$

where

$$E(\tfrac{\theta}{T}) = \frac{3nR\theta}{4}\left[2\,(e^{\frac{\theta}{T}}-1)^{-1} + (e^{\frac{\theta}{2T}}-1)^{-1}\right]$$

R is the gas constant (8.3143 $J \cdot K^{-1} \cdot mol^{-1}$), n the number of atoms per molecule(5), and $a = L_{293}/L_0$. The values θ=927 K, $Q_0 = 5 \times 10^6$ $J \cdot mol^{-1}$, k = 3.6, and a = 1.000674 were obtained by successive least squares approximations described by Wachtman, et al. [4] to minimize the standard deviation of the fit. The residual standard deviation of the fit is 16×10^{-6} using all 113 data points. The standard deviation of the data below 1000 K is 5×10^{-6} while above 1000 K it is 25×10^{-6}. All of the data are within approximately two standard deviations in the respective temperature intervals.

Expansivities were calculated from the derivative of the expansion equation:

$$\alpha = \frac{1}{L_{293}}\frac{dL}{dT} = \frac{Q_o C_v(\frac{\theta}{T})}{3a\left[Q_o - kE(\frac{\theta}{T})\right]^2}$$

where

$$C_v(\tfrac{\theta}{T}) = \frac{3nR}{32}\left(\frac{\theta}{T}\right)^2\left[4\,\mathrm{csch}^2\left(\frac{\theta}{2T}\right) + \mathrm{csch}^2\left(\frac{\theta}{4T}\right)\right]$$

Values of expansivity calculated from this equation were compared to values calculated from the experimental data resulting in a standard deviation of $0.04 \times 10^{-6} K^{-1}$ below 1000 K and $0.18 \times 10^{-6} K^{-1}$ above 1000 K. Examination of the deviations of the data from the above equations for both the expansion and expansivity did not indicate any systematic variations.

References

1. T. A. Hahn, Proceedings of the 1977 International Symposium on Thermal Expansion (in press).

2. T. A. Hahn, J. Appl. Phy. **41,** 5096 (1970).

3. B. D. Rothrock and R. K. Kirby, J. Res. NBS **71C,**85 (1967).

4. J. B. Wachtman, Jr., T. G. Scuderi, and G. W. Cleek, J. Am. Ceram. Soc. **45,** 319 (1962).

National Bureau of Standards Certificate

Standard Reference Material 736 Copper-Thermal Expansion

R. K. Kirby and T. A. Hahn

Thermal Expansion and Expansivity as a Function of Temperature

T	$\frac{L-L_{293}}{L_{293}}$	$\frac{1}{L_{293}}\frac{dL}{dT}$	T	$\frac{L-L_{293}}{L_{293}}$	$\frac{1}{L_{293}}\frac{dL}{dT}$
20 K	-3250×10^{-6}	0.27×10^{-6}/K	340 K	793×10^{-6}	17.07×10^{-6}/K
30	−3245	0.98	360	1135	17.22
40	−3229	2.29	380	1481	17.38
50	−3198	3.87	400	1831	17.53
60	−3151	5.48	420	2183	17.68
70	−3089	6.98	440	2538	17.82
80	−3012	8.30	460	2896	17.97
90	−2923	9.46	480	3256	18.11
100	−2823	10.46	500	3620	18.25
110	−2714	11.32	520	3986	18.39
120	−2597	12.05	540	4356	18.53
130	−2474	12.67	560	4728	18.67
140	−2344	13.20	580	5102	18.81
150	−2210	13.64	600	5480	18.95
160	−2072	14.01	620	5860	19.09
180	−1785	14.63	640	6244	19.24
200	−1487	15.14	660	6630	19.38
220	−1180	15.57	680	7019	19.53
240	− 865	15.94	700	7411	19.69
260	− 543	16.24	720	7807	19.84
280	− 215	16.50	740	8205	20.00
293	0	16.64	760	8607	20.16
300	117	16.71	780	9012	20.33
320	453	16.90	800	9420	20.51

This SRM is available as a rod 6.4 mm (1/4 inch) in diameter; SRM 736 L1 is 51 mm (2 inches) long, SRM 736 L2 is 102 mm (4 inches long), and SRM 736 is 152 mm (6 inches long). (Note: Inquiries for longer continuous rods may be directed to the Office of Standard Reference Materials, National Bureau of Standards.) The copper rods have been annealed at 811 K with no significant increase in grain size.[1] The residual resistivity ratio, R_{273K}/R_{4k}, is 62.53 which would indicate a purity of 99.99 atomic percent with about 0.012 atomic percent effective (dissolved, ionized) impurities.[2] This copper is similar to SRM 45d freezing point standard.

The above values of expansion and expansivity were calculated from equations based on a least squares analysis of the expansivity data from five specimens taken from various positions of the stock. These values differ from those in the original certificate for two reasons: (1) new high temperature measurements have been made, and (2) an improved fitting routine has been used in the analysis of the data. A description of the experimental method, fitting procedure, and estimate of uncertainties is given in this certificate.

The technical and support aspects involved in the preparation, certification, and issuance of this Standard Reference Material were coordinated through the Office of Standard Reference Materials by R. E. Michaelis.

Washington, D.C. 20234
August 5, 1975
(Revision of certificate dated November 19, 1969)

J. Paul Cali, Chief
Office of Standard Reference Materials

Procedure

The apparatus[3] used for the expansion measurements was a Fizeau interferometer with a l-cm specimen length. Above room temperature, the measurements were made with the interferometer in a controlled atmosphere furnace using a Pt vs Pt-10% Rh thermocouple. Below room temperature, a cryostat operating with both liquid nitrogen and liquid helium was used with a platinum resistance thermometer. The green spectral line of a mercury light source was used to produce the interference fringes. Fringe motion was measured with a filar-micrometer eyepiece. Each test specimen was made by fastening three l-cm rods of the SRM to an OFHC copper ring to form a three point spacer for the interferometer flats. The expansion of each specimen was measured between equliibrium temperatures. With the uncertainties in temperature and fringe measurements, the expansivity was determined with an uncertainty of $\pm 0.03 \times 10^{-6} K^{-1}$. As a result of measurements that were made on this SRM at the Oak Ridge National Laboraotry by Kollie and coworkers[4] it has been determined that at temperatures above 600 K mechanical creep had occurred during our initial measurements and the expansion values that were given in the original certificate were too low by about 0.4 percent at 800 K. The values given in this revised certificate at temperatures above 600 K are the result of new measurements that were made with an increased area of contact between the specimens and the interferometer flats. When the pressure at the contact was less than 3×10^5 Pa (40 psi) there was no evidence of mechanical creep as determined by taking measurements on both heating and cooling.

Since it was not always possible to project the measured values of expansion back to 293 K in an unambiguous manner, the calculated values of expansivity were used in the analysis of the data. Tests on the data from the five specimens indicated that no significant differences existed between them. All of the data were pooled and, as suggested by J. R. Rosenblatt of the Statistical Engineering Section of NBS, a polynomial of the form:

$$\frac{1}{L_{293}} \frac{\Delta L}{\Delta T} \times 10^6 = a_0 + a_1 T + a_2 T^2 + a_3 T^3 + b_1 (T-\theta_1)^3_+ + \ldots + b_6 (T-\theta_6)^3_+$$

where:

$$(T-\theta_i)^3_+ = (T-\theta_i)^3 \text{ if } T > \theta_i$$

$$(T-\theta_i)^3_+ = 0 \qquad \text{if } T \leqq \theta_i$$

was fit to them by the method of least squares using an Omnitab routine. The values of $\theta_1, \theta_2, \ldots$ and θ_6 were determined by the shape of the second derivative of the polynomial and a minimizing of the residual standard deviation. This procedure is considerably different from that used in the original certificate and results in a better fit to the data at all temperatures. After the initial fit to the measured values was made, corrections for finite ΔT's were applied to give the instantaneous coefficients of expansion defined by

$$\frac{1}{L_{293}} \frac{dL}{dT} = \frac{1}{L_{293}} \frac{\Delta L}{\Delta L} - \frac{1}{24} \left(\frac{1}{L_{293}} \frac{\Delta L}{\Delta T} \right)'' (\Delta T)^2$$

where values of the second derivatives, $\left(\frac{1}{L_{293}} \frac{\Delta L}{\Delta T} \right)''$, were obtained from the initial fit. A final fit to the corrected data was then made which resulted in the following values:

$a_0 = -3.6766548 \times 10^{-2}$

$a_1 = +9.4571214 \times 10^{-3}$

$a_2 = -7.6957895 \times 10^{-4}$

$a_3 = +5.3009542 \times 10^{-5}$

$b_1 = -1.2206897 \times 10^{-4}, \theta_1 = 28$ K

$b_2 = +3.9954573 \times 10^{-5}, \theta_2 = 41$ K

$b_3 = +3.0265158 \times 10^{-5}, \theta_3 = 62.5$ K

$b_4 = +1.4101666 \times 10^{-6}, \theta_4 = 90$ K

$b_5 = -2.3610759 \times 10^{-6}, \theta_5 = 178$ K

$b_6 = -2.0050557 \times 10^{-7}, \theta_6, = 350$ K

This polynomial has been decomposed into the following third-order spline polynomials which are continuous at the knots, θ_i, as are their first and second derivatives:

20 to 28 K, $$\frac{1}{L_{293}}\frac{dL}{dT} \times 10^6 = -0.037 + 9.46 \times 10^{-3}T - 7.70 \times 10^{-4}T^2 + 5.301 \times 10^{-5}T^3$$

28 to 41 K, $$\frac{1}{L_{293}}\frac{dL}{dT} \times 10^6 = 2.643 - 2.7765 \times 10^{-1}T + 9.4842 \times 10^{-3}T^2 - 6.906 \times 10^{-5}T^3$$

41 to 62.5 K, $$\frac{1}{L_{293}}\frac{dL}{dT} \times 10^6 = -0.111 - 7.616 \times 10^{-2}T + 4.5698 \times 10^{-3}T^2 - 2.9105 \times 10^{-5}T^3$$

62.5 to 90 K, $$\frac{1}{L_{293}}\frac{dL}{dT} \times 10^6 = -7.500 + 2.7851 \times 10^{-1}T - 1.1049 \times 10^{-3}T^2 + 1.160 \times 10^{-6}T^3$$

90 to 178 K, $$\frac{1}{L_{293}}\frac{dL}{dT} \times 10^6 = -8.528 + 3.12779 \times 10^{-1}T - 1.48566 \times 10^{-3}T^2 + 2.5705 \times 10^{-6}T^3$$

178 to 350 K, $$\frac{1}{L_{293}}\frac{dL}{dT} \times 10^6 = 4.788 + 8.8354 \times 10^{-2}T - 2.24845 \times 10^{-4}T^2 + 2.0939 \times 10^{-7}T^3$$

350 to 800 K, $$\frac{1}{L_{293}}\frac{dL}{dT} \times 10^6 = 13.385 + 1.4668 \times 10^{-2}T - 1.4314 \times 10^{-5}T^2 + 8.883 \times 10^{-9}T^3$$

The standard deviation of this fit is 0.059×10^{-6} with 141 data points. These equations and their integrals were used to calculate the values listed in the table. A comparison of the experimental expansion data with values predicted from the equations gives a standard deviation of 6×10^{-6}. All of the data for both the expansion and expansivity were within two standard deviations of the values predicted by the equations.

Footnotes

1 Vacuum annealing (G. E. Hicho).

2 Residual resistivity ratios (V. A. Deason and R. L. Powell).

3 Hahn, T. A., Thermal Expansion of Copper from 20 to 800 K - Standard Reference Material 736 , J. Appl. Phys. 41, 5096 (1970).

4 Kollie, T. G., McElroy, D. L. Hutton, J. T., and Ewing, W. M., A Computer Operated Fused Quartz Differential Dilatometer, AIP Conf. Proc., No. 17 - Thermal Expansion, p. 129, AIP, New York (1974).

National Bureau of Standards

Certificate

Standard Reference Material 737

Tungsten - Thermal Expansion

R. K. Kirby and T. A. Hahn

Thermal Expansion and Expansivity as a Function of Temperature (IPTS-68)

T	$\frac{L-L_{293}}{L_{293}}$	$\frac{1}{L_{293}}\frac{dL}{dT}$	T	$\frac{L-L_{293}}{L_{293}}$	$\frac{1}{L_{293}}\frac{dL}{dT}$
80 K	-814×10^{-6}	$2.30 \times 10^{-6}/K$	520 K	1040×10^{-6}	$4.70 \times 10^{-6}/K$
90	−790	2.61	560	1229	4.73
100	−762	2.88	600	1418	4.76
110	−732	3.11	650	1657	4.79
120	−700	3.30	700	1898	4.83
130	−666	3.46	750	2140	4.86
140	−631	3.59	800	2384	4.90
150	−595	3.71	850	2630	4.93
160	−557	3.81	900	2878	4.97
180	−479	3.97	950	3127	5.01
200	−398	4.10	1000	3378	5.04
220	−315	4.20	1100	3887	5.12
240	−231	4.27	1200	4404	5.21
270	101	4.36	1300	4930	5.31
293	0	4.42	1400	5467	5.43
320	120	4.47	1500	6016	5.56
360	300	4.53	1600	6578	5.70
400	483	4.59	1700	7157	5.87
440	667	4.63	1800	7754	6.07
480	853	4.66			

The above values of expansion and expansivity were calculated from equations based on a least squares analysis of the expansivity data from six specimens taken from various positions of the stock. A description of the experimental methods, fitting procedure, and estimate of uncertainties is given in this certificate.

The technical and support aspects involved in the preparation, certification, and issuance of this Standard Reference Material were coordinated through the Office of Standard Reference Materials by R. E. Michaelis.

Washington, D. C.
May 19, 1976

J. Paul Cali, Chief
Office of Standard Reference Materials

(over)

This SRM is available as a rod 6.4 mm (1/4 inch) in diameter and 51 mm (2 inches) in length. Inquiries for longer rods may be directed to the Office of Standard Reference Materials, National Bureau of Standards. This is a sintered powder metallurgical product made by compacting the starting powder at 15 tons/inch^2 and then heating to about 3400 K. The final diameter was obtained by swaging and grinding after 7 rolling operations. The density of this material is 19.23 g/cm^3. Chemical analysis indicates a purity of 99.96 wt. % while the residual electrical resistance ratio, RRR = 60, indicates a purity of 99.98 wt. %.

This SRM should not be heated above 700 K in air. Vacuum or dry nonoxidizing atmospheres (helium, argon, hydrogen, etc.) should be used when heating it to higher temperatures. Cutting and shaping of tungsten is best accomplished by electrical discharge machining.

Procedure

In the temperature range below 1120 K the expansion measurements were made with a Fizeau interferometer [1]. The green spectral line of a mercury lamp was used to produce the interference fringes. Fringe motion was measured with a filar-micrometer eyepiece. Each test specimen was made by fastening three 1-cm rods of the SRM to a tungsten ring. Temperatures above 293 K were measured with a Pt vs. Pt-10% Rh thermocouple while those below were measured with a platinum resistance thermometer. The specimens were heated and cooled in a low pressure helium atmosphere and the expansion measured between equilibrium temperatures. With the uncertainties in temperature and fringe measurements, the expansivities were determined with an uncertainty of $\pm 0.03 \times 10^{-6}$/K.

In the temperature range above 970 K the expansion measurements were made with a twin-microscope technique [2]. Knife edges machined into the specimens defined their lengths (~10 cm). The changes in length on heating between equilibrium temperatures were measured with filar micrometer eyepieces. The temperatures were measured with W/Re and Pt-30% Rh/Pt-6% Rh thermocouples. The specimens were heated in a vacuum furnace. With the uncertainties in temperature and length measurements, the expansivities were determined with an uncertainty of $\pm 0.06 \times 10^{-6}$/K.

Because it was not always possible to project the measured values of expansion back to 293 K in an unambiguous manner, the calculated values of expansivity were used in the analysis of the data. Tests on the data from the six specimens indicated that no significant differences existed between them. All of the data were pooled and a polynomial of the form

$$\frac{1}{L_{293}}\frac{\Delta L}{\Delta T} \times 10^6 = a_0 + a_1 T + a_2 T^2 + a_3 T^3 + b_1 (T-\theta_1)_+^3 + \ldots + b_4 (T-\theta_4)_+^3$$

where

$$(T-\theta_i)_+^3 = (T-\theta_i)^3 \text{ if } T > \theta_i$$

$$(T-\theta_i)_+^3 = 0 \qquad \text{if } T \leqslant \theta_i$$

was fit to them by the method of least squares using an Omnitab routine. The values of θ_1, θ_2, θ_3, and θ_4, were determined by the shape of the second derivative of the polynomial and a minimizing of the residual standard deviation. After the initial fit to the measured values was made, corrections for finite ΔT's were applied to give the expansivity defined by

$$\frac{1}{L_{293}}\frac{dL}{dT} = \frac{1}{L_{293}}\frac{\Delta L}{\Delta T} - \frac{1}{24}\left(\frac{1}{L_{293}}\frac{\Delta L}{\Delta T}\right)'' (\Delta T)^2$$

where values of the second derivative,

$$\left(\frac{1}{L_{293}}\frac{\Delta L}{\Delta T}\right)'' ,$$

were obtained from the initial fit. A final fit to the corrected data was then made which resulted in the following values:

$$\begin{aligned}
a_0 &= -2.3835532 \\
a_1 &= +8.8699747 \times 10^{-2} \\
a_2 &= -4.411438 \times 10^{-4} \\
a_3 &= +8.0325992 \times 10^{-7} \\
b_1 &= -6.3651146 \times 10^{-7},\ \theta_1 = 160\ \text{K} \\
b_2 &= -1.1648859 \times 10^{-7},\ \theta_2 = 235\ \text{K} \\
b_3 &= -4.499535 \times 10^{-8},\ \theta_3 = 330\ \text{K} \\
b_4 &= -4.8469136 \times 10^{-9},\ \theta_4 = 550\ \text{K}
\end{aligned}$$

This polynomial has been decomposed into the following third-order spline polynomials which are continuous at the knots, θ_i, as are their first and second derivatives:

$$80 \text{ to } 160\ \text{K},\ \frac{1}{L_{293}}\frac{dL}{dT} \times 10^6 = -2.383 + 8.870 \times 10^{-2}\,T - 4.4114 \times 10^{-4}\,T^2 + 8.03 \times 10^{-7}\,T^3$$

$$160 \text{ to } 235\ \text{K},\ \frac{1}{L_{293}}\frac{dL}{dT} \times 10^6 = 0.225 + 3.982 \times 10^{-2}\,T - 1.3562 \times 10^{-4}\,T^2 + 1.667 \times 10^{-7}\,T^3$$

$$235 \text{ to } 330\ \text{K},\ \frac{1}{L_{293}}\frac{dL}{dT} \times 10^6 = 1.736 + 2.052 \times 10^{-2}\,T - 5.3494 \times 10^{-5}\,T^2 + 5.03 \times 10^{-8}\,T^3$$

$$330 \text{ to } 550\ \text{K},\ \frac{1}{L_{293}}\frac{dL}{dT} \times 10^6 = 3.353 + 5.816 \times 10^{-3}\,T - 8.9484 \times 10^{-6}\,T^2 + 5.26 \times 10^{-9}\,T^3$$

$$550 \text{ to } 1800\ \text{K},\ \frac{1}{L_{293}}\frac{dL}{dT} \times 10^6 = 4.1598 + 1.4179 \times 10^{-3}\,T - 9.5104 \times 10^{-7}\,T^2 + 4.176 \times 10^{-10}\,T^3.$$

The standard deviation of this fit is 0.026×10^{-6} with 131 data points. These equations and their integrals were used to calculate the values listed in the table. A comparison of the experimental expansion data with values predicted from the equations gives a standard deviation of 5×10^{-6} in the range 80 to 1000 K and 13×10^{-6} in the range 1000 to 1800 K. All of the data for the expansivity were within three standard deviations of the predicted values.

After being heated for the first time to temperatures above 1300 K this SRM was observed to increase in length by about 40 μm/m. Subsequent heating, even to 2300 K, did not cause any further growth.

It is planned to extend the temperature range covered by this certificate to about 3000 K.

Footnotes

[1] Hahn, T. A., Thermal Expansion of Copper from 20 to 800 K - Standard Reference Material 736, J. Appl. Phys. **41,** 5096 (1970).

[2] Rothrock, B. D. and Kirby, R. K., An Apparatus for Measuring Thermal Expansion at Elevated Temperatures, J. Res. NBS **71C,** 85 (1967).

Certificate of Analysis

Standard Reference Material 739

Fused-Silica Thermal Expansion

R. K. Kirby and T. A. Hahn

Thermal Expansion as a Function of Temperature

T	Expansion $\Delta L/L_{293}$	Expansivity α	T	Expansion $\Delta L/L_{293}$	Expansivity α
80 K	-1×10^{-6}	-0.70×10^{-6}/K	320 K	$13._5 \times 10^{-6}$	$+0.53 \times 10^{-6}$/K
90	$-7._5$	−0.61	340	$24._5$	0.56
100	−13	−0.53	360	36	0.58
110	−18	−0.46	380	$47._5$	0.60
120	$-22._5$	−0.38	400	$59._5$	0.61
130	−26	−0.31	420	72	0.62
140	$-28._5$	−0.24	440	85	0.63
150	$-30._5$	−0.17	460	97	0.63
160	−32	−0.10	480	110	0.63
170	$-32._5$	−0.04	500	122	0.63
180	$-32._5$	+0.02	520	135	0.62
190	−32	0.08	560	159	0.61
200	−31	0.13	600	183	0.59
210	$-29._5$	0.19	640	206	0.56
220	$-27._5$	0.23	680	228	0.54
230	−25	0.28	720	249	0.51
240	−22	0.32	760	269	0.49
250	$-18._5$	0.36	800	288	0.47
260	$-14._5$	0.39	840	307	0.44
273	−9	0.43	880	324	0.42
280	−6	0.45	920	340	0.40
293	0	0.48	960	356	0.38
298	$+2._5$	0.49	1000	371	0.37

This SRM is available as a rod 6.4 mm (¼ inch) in diameter; L1 is 51 mm (2 inches) long, L2 is 102 mm (4 inches) long, and L3 is 152 mm (6 inches) long. (Note: Inquiries for longer continuous rods may be directed to the Office of Standard Reference Materials, National Bureau of Standards.) The rods that make up this SRM were obtained as a single drawing from an ingot of fused silica, at least 99.8 wt percent pure. These rods have been annealed by soaking at 1373 K for 7 hours and cooling to 1173 K at 12 K per hour.

The above values of expansion and expansivity were calculated from a least squares fit to expansivity measurements made on five specimens. Linear interpolation can be used between tabulated values without introducing a significant error. The expansion of this material may be altered, however, if the material is heated to temperatures above 1020 K. A description of the experimental method, fitting procedure, and estimate of uncertainties is given on the back of this certificate.

The overall coordination and evaluation of data leading to certification of this SRM was performed by R. K. Kirby and T. A. Hahn.

The technical and support aspects involved in the preparation, certification, and issuance of this Standard Reference Material were coordinated through the Office of Standard Reference Materials by R. E. Michaelis.

Washington, D. C. 20234
May 12, 1971

J. Paul Cali, Chief
Office of Standard Reference Materials

PROCEDURE

The apparatus used for the expansion measurements was a Fizeau interferometer with a 1-cm specimen length. Above room temperature, the measurements were made with the interferometer in a controlled atmosphere furnace using a Pt vs Pt-10% Rh thermocouple. Below room temperature, a cryostat operating with liquid nitrogen was used with a platinum resistance thermometer. The green spectral line of a mercury light source was used to produce the interference fringes. Fringe motion was measured with a filar-micrometer eyepiece. Each test specimen was made by fastening three 1-cm pieces of the SRM rod in holes drilled through a thin fused-silica ring. A small amount of a thin slurry of fused-silica cement was used to fasten the rods in the ring. With the uncertainties of temperature and fringe measurements, the expansivity was determined with an uncertainty of $\pm 0.03 \times 10^{-6}$/K.

Values of expansivity were calculated between equilibrium temperatures, and corrections[1] were made for the finite temperature differences. The tests on the 5 specimens indicated that no gross differences existed, and that the expansion is reversible if the temperature does not exceed 1050 K. All of the data were pooled and the following third-order polynomial was obtained by the method of least squares using an Omnitab routine:

$$\begin{aligned}\alpha \times 10^6\,\mathrm{K} = {} & -1.417 + 9.581\times10^{-3}T - 5.991\times10^{-6}T^2 \\ & -1.5642\times10^{-8}T^3 - 6.666\times10^{-6}(T-300)_+^2 \\ & +2.2128\times10^{-8}(T-300)_+^2(T+600) \\ & +1.013\times10^{-5}(T-700)_+^2 \\ & -5.37\times10^{-9}(T-700)_+^2(T+1400)\end{aligned}$$

where the notation $(\quad)_+$ means that the term is zero when the quantity in the parentheses is negative. The standard deviation of this fit is 0.026 with 169 data points. This equation and its integral were used to calculate the tabulated values of expansivity and expansion. A standard deviation of 6×10^{-6} was calculated for the expansion values.

[1] Hahn, Thomas A., Thermal Expansion of Copper from 20 to 800 K - Standard Reference Material 736, J. Appl. Phys. **41**, 5096 (1970).

3.7 REFERENCES

Abbe (1889). *Wied. Ann.* **38**, 453.

Advances in Cryogenic Engineering, Vols. 1–21, 23, 25, 27 (1954–1982). Plenum Press, New York.

Advances in Cryogenic Engineering — Materials, Vols. 22, 24, 26, 28 (1976, 1978, 1980, 1982). Plenum Press, New York.

AIP Conference Proceedings, No. 3, Thermal Expansion — 1971, Third Symposium (1972) (eds. M. G. Graham and H. E. Hagy). American Institute of Physics, New York.

AIP Conference Proceedings, No. 17, Thermal Expansion — 1973, Fourth Symposium (1974) (eds. R. E. Taylor and G. L. Denman). American Institute of Physics, New York.

Alcoa Aluminum Handbook (1962). Aluminum Company of America, Pittsburgh, Pennsylvania, 39.

Andres, K. (1961). The measurement of thermal expansion of metals at low temperatures. *Cryogenics* **2**, 93–97.

Andres, K. (1964). Thermische Ausdehnung von Metallen bei tiefen Temperaturen. *Phys. Kondens. Mater.* **2**, 294–333.

Arp, V., Wilson, J. H., Winrich, L., and Sikora, P. (1962). Thermal expansion of some engineering materials from 20 to 293 degrees K. *Cryogenics* **2**, 230–235.

Ayres, H. D. (1905). Coefficients of linear expansion at low temperatures. *Phys. Rev.* **20**, 38–51.

Barker, R. E. (1967). Grüneisen numbers for polymeric solids. *J. Appl. Phys.* **38**, 4234–4242.

Barron, T. H. K. (1955). On the thermal expansion of solids at low temperatures. *Philos. Mag.* **46**(7), 720–734.

Barron, T. H. K., Collins, J. G., and White, G. K. (1980). Thermal expansion of solids at low temperatures. *Adv. Phys.* **29**(4), 609–730.

Batchelder, D. N. and Simmons, R. O. (1965). X-ray lattice constants of crystals by a rotating camera method: Al, Ar, Au, CaF_2, Cu, Ge, Ne, Si. *J. Appl. Phys.* **36**, 2864–2868.

Bauder, P., Hartwig, G., and Weiss, W. (1976). Die thermische Ausdehnung von pulvergefüllten Epoxidharzen. Teil 2. Messung der thermischen Ausdehnung von pulvergefüllten Epoxidharzen bis zu tiefen Temperaturen. *Mater. Sci. Eng.* **22**, 265–270.

Berthold, J. W. and Jacobs, S. F. (1976). Ultraprecise thermal expansion measurements of seven low expansion materials. *Appl. Opt.* **15**(10), 2344–2347.

Berthold, J. W., Jacobs, S. F., and Norton, M. A. (1977). Dimensional stability of fused silica, Invar, and several ultra-low thermal expansion materials. *Metrologia* **13**, 9–16.

Bijl, D. and Pullan, H. (1954). Thermal expansion of simple solids at low temperatures and Grüneisen's law. *Philos. Mag.* **45**, 290–294.

Bijl, D. and Pullan, H. (1955). A new method for measuring the thermal expansion of solids at low temperatures. The thermal expansion of copper and aluminum and the Grüneisen rule. *Physica* **21**, 285–298.

Bijl, D. (1957). The representation of specific heat and thermal expansion data of simple solids. In *Progress in Low Temperature Physics*, Vol. 2 (ed. C. J. Gorter). North-Holland Publishing, Amsterdam, 395–430.

Blackman, M. (1935a). Contributions to the theory of specific heat of crystals. *Proc. R. Soc. A* **148**, 365–383.

Blackman, M. (1935b). Contributions to the theory of specific heat. *Proc. R. Soc. A* **149**, 117–125.

Blackman, M. (1957). On the thermal expansion of solids. *Proc. Phys. Soc. B* **70**, 827–832.

Boley, B. A. and Weiner, J. H. (1960). *Theory of Thermal Stresses*. John Wiley & Sons, New York, London.

Bond, W. L. (1960). Precision lattice constant determination. *Acta Crystallogr.* **13**, 814–818.

Born, M. (1923). *Atomtheorie des festen Zustandes*. Teubner, Leipzig.

Born, M. and von Karman, T. (1912). Über Schwingungen in Raumgittern. *Phys. Z.* **13**, 297.

Born, M. and von Karman, T. (1913a). Zur Theorie der spezifischen Wärme. *Phys. Z.* **14**, 15.

Born, M. and von Karman, T. (1913b). Über die Verteilung der Eigenschwingungen von Punktgittern. *Phys. Z.* **14**, 65.

Boughton, R. I., Olsen, J. L., and Palmy, C. (1970). Pressure effects in superconductors. *Prog. Low Temp. Phys.* **6**, 163–203.

Brassell, G. W. and Wischmann, K. B. (1974). Mechanical and thermal expansion properties of a particulate filled polymer. *J. Mater. Sci.* **9**, 307–314.

Browder, J. S. and Ballard, S. S. (1977). Thermal expansion data for eight optical materials from 60 K to 300 K. *Appl. Opt.* **16**(12), 3214–3217.

Bruins, D. E., Garland, C. W., and Greytak, T. J. (1975). New interferometric method for piezoelectric measurements. *Rev. Sci. Instrum.* **46**, 1167–1170.

Buffington, R. M. and Latimer, W. M. (1926). The measurement of coefficient of expansion at low temperatures: some thermodynamic applications of expansion data. *J. Am. Chem. Soc.* **48**, 2305.

Bunton, G. V. and Weintroub, S. (1968). A sensitive optical lever dilatometer for use at low temperatures and the thermal expansion of copper. *Cryogenics* **8**, 354–360.

Campbell, W. J. (1962). *Thermal Expansion of Magnesium Oxide: An Interlaboratory Study*, Bureau of Mines Report of Investigation, Bureau of Mines, Albany, Oregon, 6115.

Carr, R. H., McCammon, R. D., and White, G. K. (1964). The thermal expansion of copper at low temperatures. *Proc. R. Soc. A.* **280**, 72–84.

Carr, R. H., McCammon, R. D., and White, G. K. (1965). Thermal expansion of germanium and silicon at low temperatures. *Philos. Mag.* **12**, 157–163.

Carr, R. H. and Swenson, C. A. (1964). Application of a variable transformer to the study of low temperature thermal expansion. *Cryogenics* **4**, 76–82.

Case, C. R., II, and Swenson, C. A. (1974). Thermal expansions of high-purity and OH-doped NaCl, at temperatures below 30°K. *Phys. Rev. B* **9**, 4506–4511.

Clark, A. F. (1968). Low temperature thermal expansion of some metallic alloys. *Cryogenics* **8**(5), 282–289.

Clark, A. F., Fujii, G., and Ranney, M. A. (1981). The thermal expansion of several materials for superconducting magnets. *IEEE Trans. Magn.* **MAG-17**, 2316–2319.

Clark, A. F. and Kropschot, R. H. (1970). Low temperature specific heat and thermal expansion of alloys. *Bull. Inst. Int. Froid*, Annexe **1970-2**, 249–254.

Collings, E. W. (1983). *Applied Superconductivity, Metallurgy and Physics of Titanium Alloys*. International Cryogenic Monograph Series, Plenum Press, New York.

Collins, J. G. (1963). On the calculation of the temperature variation of the coefficients of thermal expansion for materials with cubic structure. *Philos. Mag.* **8**(8), 323–332.

Collins, J. G. and White, G. K. (1964). Thermal expansion of solids. In *Progress in Low Temperature Physics*, Vol. 4 (ed. C. J. Gorter). John Wiley & Sons, New York, 45–79.

Conway, J. B. and Losekamp, A. C. (1965). Refraction error in a comparator method for measuring thermal expansion. *Rev. Sci. Instrum.* **36**, 1245–1246.

Corruccini, R. J. and Gniewek, J. J. (1960). *Specific Heats and Enthalpies of Technical Solids at Low Temperatures; a Compilation from the Literature*. Monograph 21, National Bureau of Standards, U.S. Government Printing Office, Washington, D.C.

Corruccini, R. J. and Gniewek, J. J. (1961). *Thermal Expansion of Technical Solids at Low Temperatures*. Monograph 29, National Bureau of Standards, U.S. Government Printing Office, Washington, D.C.

Cryogenic Materials Data Handbook (1964) (eds. F. R. Schwartzberg, S. H. Osgood, R. D. Keys, and T. F. Kiefer). Technical Documentary Report No. ML-TDR-64-280, Wright-Patterson Air Force Base, Dayton, Ohio.

Dahlerup-Peterson, K. (1980). Tests of composite materials at cryogenic temperatures: facilities and results. In *Advances in Cryogenic Engineering — Materials*, Vol. 26. Plenum Press, New York, 268–279.

Dahlerup-Peterson, K. and Perrot, A. (1979). *Properties of Organic Composite Materials at Cryogenic Temperatures*. ISR-BOM/79–39, CERN, Geneva, Switzerland.

Dantl, G. (1962). Wärmeausdehnung von H_20- und D_20-Einkristallen. *Z. Phys.* **166**, 115–118.

Desaguliers, J. T. (1745). *A Course of Experimental Philosophy*. London, 436–446.

Dewar, J. (1902). Coefficients of the cubical expansion of ice, hydrated salts, solid carbonic acid and other substances at low temperatures. *Proc. R. Soc. Lond. A* **70**, 237–246.

Dorsey, H. G. (1907). Coefficient of linear expansion at low temperatures. *Phys. Rev.* **25**, 88–102.

Dorsey, H. G. (1908). Further measurements of the coefficient of linear expansion at low temperatures. *Phys. Rev.* **27**, 1–10.

Dorsey, H. G. (1910). Coefficients of linear expansion at low temperatures. *Phys. Rev.* **30**, 271–272.

Easton, D. S., Kroeger, D. M., Specking, W., and Koch, C. C. (1980). A prediction of the stress state in Nb_3Sn superconducting composites. *J. Appl. Phys.* **51**, 2748–2757.

Ekin, J. W. (1981). Mechanical properties and strain effects in superconductors. In *Superconductor Materials Science* (eds. S. Foner and B. B. Schwartz). Plenum Press, New York, 455–510.

Fawcett, E. (1971). Low-temperature thermal expansion of V_3Si. *Phys. Rev. Lett.* **26**(14), 829–833.

Figgins, B. F., Jones, G. O., and Riley, D. P. (1956). The thermal expansion of Al at low temperatures as measured by an x-ray diffraction method. *Philos. Mag.* **8**(1), 747–758.

Finlayson, T. R., Gibbs, E. E., and Smith, T. F. (1981). The low-temperature thermal expansion of polycrystalline V_3Ge. *Physica* **108B**, 1011–1012.

Fizeau, H. (1864a). Rechercher sur la dilation et la double réfraction du crystal de roche échauffé. *C. R. Acad. Sci. (Paris)* **58**, 923–932.

Fizeau, H. (1864b). *Ann. Chim. Phys.* **4**(2), 146.

Fizeau, H. (1866a). *Ann. Chim. Phys.* **8**, 335.

Fizeau, H. (1866b). Memoire sur la dilation des corps solides par la chaleur. *C. R. Acad. Sci. (Paris)* **62**, 1101.

Försterling, G. and Hegenbarth, E. (1978). X-ray study of Debye-Waller factors and lattice constants of V_3Si powder in temperature range of 25 K to 1000 K. *Krist. Tech.* **13**(6), 737–746.

Fujii, G., Ekin, J. W., Radebaugh, R., and Clark, A. F. (1980). Effect of thermal contraction of sample-holder material on critical current. In *Advances in Cryogenic Engineering — Materials*, Vol. 26. Plenum Press, New York, 589–598.

Gatewood, B. E. (1957). *Thermal Stresses with Applications to Airplanes, Missiles, Turbines, and Nuclear Reactors*. McGraw-Hill, New York.

Geymayer, H. G. (1969). *Temperature-induced Dimensional Changes in Hardened Concrete*. Miscellaneous paper C-69-10, U.S. Army Engineer Waterways Experiment Station, Corps of Engineers, Vicksburg, Mississippi.

Gibbons, T. G. (1973). Calculation of the thermal expansion for a quasi-harmonic model of tellurium. *Phys. Rev. B* **7**, 1410–1419.

Glaser, P. E., Black, I. A., Lindstrom, R. S., Ruccia, F. E., and Wechsler, A. E. (1967). *Thermal Insulation Systems, A Survey*. N67–38580, National Aeronautics and Space Administration, Washington, D.C.

Goodier, J. N. (1937). Thermal stress. *ASME J. Appl. Mech.* **4**(1), A-33–A-36.

Goodier, J. N. (1957). Thermal stress and deformation. *ASME J. Appl. Mech.* **24**(3), 467–474.

Gopal, E. S. R. (1966). *Specific Heats at Low Temperatures*. Plenum Press, New York.

Grüneisen, E. (1912). Theories des festen Zustandes einatomiger Elements. *Ann. Phys. (Leipzig)* **39**, 257–306.

Grüneisen, E. (1926). Zustand des festen Korpers (The state of a solid body). In *Handbuch der Physik*, Vol. X, 1–52. Translation available as NASA Report No. 2-18-4910 (1959) from NTIS, Springfield, Virginia.

Guillaume, C. E. (1897). Recherches sur les aciers au nickel. Dilatations aux températures élevées; résistance électrique. *Compt. Rend.* **125**, 235–238.

Gulati, S. T. and Plummer, W. A. (1973). *Influence of Poisson's Ratio on the Composite Expansion of Laminated Plates and Cylinders*. American Institute of Physics, New York, 196–206.

Hahn, T. A. (1970). Thermal expansion of copper from 20 to 800 K — standard reference material 736. *J. Appl. Phys.* **41**, 5096–5101.

Hahn, T. A. (1978). Thermal expansion of single crystal sapphire from 293 to 2000 K, standard reference material 732. In *Thermal Expansion 6* (ed. I. D. Peggs). Plenum Press, New York, 191–201.

Hamilton, W. O., Greene, D. B., and Davidson, D. E. (1968). Thermal expansion of epoxies between 2 and 300 degrees K. *Rev. Sci. Instrum.* **39**, 645–648.

Handbook of Plastics and Elastomers (1975) (ed. C. A. Harper). McGraw-Hill Book Company, New York.

Handbook on Materials for Superconducting Machinery (1977). MCIC-HB-04, Metals and Ceramics Information Center, Battelle Columbus Laboratories, Columbus, Ohio.

Hardy, J. R. (1962). Lattice dynamics of alkali halide crystals in relation to specific heat data. *Philos. Mag.* **7**, 315–336.

Hartwig, G. (1975). Low temperature properties of potting and structural materials for superconducting magnets. *IEEE Trans. Magn.* **MAG-11**(2), 536–539.

Hartwig, G., Weiss, W., and Puck, A. (1976). The thermal expansion of powder-filled epoxies. Part 1. Calculation of the expansion coefficients of powder-filled epoxies under filling factor of the thermal stresses. *Mater. Sci. Eng.* **22**(3), 261–264.

Holliday, L. and Robinson, J. (1973). Review: The thermal expansion of composites based on polymers. *J. Mater. Sci.* **8**, 301–311.

Hortal, M. (1978). Thermal expansion of some glasses at low temperatures. *Phys. Chem. Glasses* **19**(1), 5–8.

Hume-Rothery, W. (1956). On Grüneisen's equation for thermal expansion. *Proc. Phys. Soc. (Lond.)* **57**, 209–222.

Jelinek, F. J. and Collings, E. W. (1975). Low-temperature thermal expansion and specific heat properties of structural materials. In *Materials Research in Support of Superconducting Machinery—IV* (eds. A. F. Clark, R. P. Reed, and E. C. Van Reuth). Fourth Semi-Annual Technical Report, National Bureau of Standards, Boulder Colorado.

Jenkins, F. A. and White, H. E. (1950). *Fundamentals of Optics*. McGraw-Hill, New York.

Jones, R. V. (1961). Some developments and applications of the optical lever. *J. Sci. Instrum.* **38**, 37–45.

Kasen, M. B. (1975a). Mechanical and thermal properties of filamentary-reinforced structural composites at cryogenic temperatures. 1: Glass-reinforced composites. *Cryogenics* **15**(6), 327–349.

Kasen, M. B. (1975b). Mechanical and thermal properties of filamentary-reinforced structural composites at cryogenic temperatures. 2: Advanced composites. *Cryogenics* **15**(12), 701–722.

Kasen, M. B. (1979). Cryogenic applications of composite technology in the U.S.A. In *Nonmetallic Materials and Composites at Low Temperatures* (eds. A. F. Clark and R. P. Reed). Plenum Press, New York, 317–337.

Kasen, M. B. (1981). Cryogenic properties of filamentary-reinforced composites: an update. *Cryogenics* **21**, 323–340.

Kent, C. H. (1932). Thermal stresses in spheres and cylinders produced by temperatures varying with time. *Trans. ASME* **54**(18), 185–196.

Khan, J. A. and Griffiths, D. (1978). Anomalous thermal expansion of dilute and CuMn and AgMn alloys below 50 K. *J. Phys. F* **8**, 763–773.

Kinzly, R. E. (1967). A new interferometer capable of measuring small optical path differences. *Appl. Opt.* **6**, 137–140.

Kirby, R. K. (1956). Thermal expansion of polytetrafluoroethylene (teflon) from −190° to +300° C. *J. Res. Natl. Bur. Stand.* **57**, 91–94.

Kirby, R. K. (1969). *Thermal Expansion of Ceramics*. Special Publication 303, National Bureau of Standards, Boulder, Colorado, 41–59.

Krishnan, R. S., Srinivasan. R., and Devanarayanan, S. (1979). *Thermal Expansion of Crystals*. Pergamon Press, New York.

Kroeger, F. R. and Swenson, C. A. (1977). Absolute linear thermal-expansion measurements on copper and aluminum from 5 to 320 K. *J. Appl. Phys.* **48**, 853–864.

Laquer, H. L. and Head, E. L. (1952). *Low Temperature Thermal Expansion of Plastics*. AECU-2161, Technical Information Service A.E.C., Oak Ridge, Tennessee.

Lebesque, H. J. M., Van Schaik, J. W., Kruijshoop, A. W., and Blaisse, B. S. (1965). Thermal expansion coefficient of BK8 glass between 15° and 300° K. *Physica* **31**, 967–972.

Li, H. H. (1976). Refractive index of alkali halides and its wavelength and temperature derivatives. *J. Phys. Chem. Ref. Data* **5**(2), 329–528.

LNG Materials and Fluids (1979) (ed. D. B. Mann). National Bureau of Standards, Boulder, Colorado.

McConnell, P. M., Daney, D. E., and Kirgis, J. B. (1970). Thermoelastic expansion and creep of polyethylene terephthalate and polypyromellitimide film and polyethylene terephthalate fibers from 20 to 295 K. *J. Appl. Phys.* **41**(13), 5066–5070.

Mackowski, M. J., West, K. E., Samsel, D. A., Holderman, L. B., and Peters, P. N. (1976). Expansion coefficients of some epoxies in the temperature range 4–300 K. *Cryogenics* **16**, 45–47 and 736 (corrigendum).

McLean, K. O., Swenson, C. A., and Case, C. R. (1972). Thermal expansion of copper, silver and gold below 35° K. *J. Low Temp. Phys.* **7**, 77–98.

Mäntysalo, E. (1965). Determination of the thermal expansion coefficient of a lithium single crystal using ultrasonic waves. *Phys. Lett.* **16**, 17–18.

Manzhellii, V. G., Gavrilko, V. G., and Voitovich, E. I. (1967). Thermal expansion of solidified inert gases. *Soviet Phys. Solid State* **9**, 1157–1162.

Manzhellii, V. G., Tolkachev, A. M., Bagatskii, M. I., and Voitovich, E. I. (1971). Thermal expansion, heat capacity, and compressibility of solid CO_2. *Phys. Status Solidi (b)* **44**, 39–49.

Manzhellii, V. G., Tolkachev, A. M., and Gavrilko, V. G. (1969). Thermal expansion of solid CH_4 and CD_4. *J. Phys. Chem. Solids* **30**, 2759–2763.

Manzhellii, V. G., Tolkachev, A. M., and Voitovich, E. I. (1966). Thermal expansion of crystalline nitrogen, oxygen, and methane. *Phys. Status Solidi* **30**, K131.

Maulbetsch, J. L. (1935). Thermal stresses in plates.

ASME J. Appl. Mech. **2**(4), A-141–A-146.

Meincke, P. P. M. and Graham, G. M. (1963). Thermal expansion of NaCl at low temperatures. In *Proceedings of the Eighth International Conference on Low Temperature Physics London* (1962). Butterworths, London, 401–402.

Meincke, P. P. M. and Graham, G. M. (1965). The thermal expansion of alkali halides. *Can. J. Phys.* **43**, 1853–1866.

Molby, F. A. (1949). Index of refraction and coefficients of expansion of optical glasses at low temperatures. *J. Opt. Soc. Am.* **39**, 600–611.

Monfore, G. E. and Lentz, A. E. (1962). Physical properties of concrete at very low temperatures. *J. PCA Res. Dev. Lab.* **4**(2), 33–39.

Morey, G. W. (1954). *The Properties of Glass*, Second edition. Reinhold, New York.

Munn, R. W. (1969). The thermal expansion of axial metals. *Adv. Phys.* **18**, 515–543.

Munn, R. W. (1972). Role of elastic constants in negative thermal expansion of axial solids. *J. Phys. C* **5**, 535–542.

Nix, F. A. and McNair, D. (1941). The thermal expansion of pure metals: Cu, Au, Al, Ni and Fe. *Phys. Rev.* **60**, 597–605.

Parsons, W. H. and Johnson, W. H. (1944). Factors affecting the thermal expansion of concrete aggregate materials. *J. Am. Concr. Inst.* **15**(5), 457–466.

Perrot, A., Blin, M., Bouriot, P., and Gilquin, J. (1977). Proprietes mecaniques et thermiques des materiaux organiques aux temperatures cryogeniques. CERN ISR-BOM/77–8, Organisation Européenne pour la Recherche Nucleaire, Geneva, Switzerland.

Pinheiro, M. de F. F., Radcliffe, D. J., and Rosenberg, H. M. (1978). The thermal expansion and thermal and electrical conductivity of carbon and glass fibre/epoxy-resin composites from 2 to 290 K. In *Proceedings of the Seventh International Cryogenic Engineering Conference*, IPC Science and Technology Press, Guildford, Surrey, England, 494–498.

Pinheiro, M. de F. F. and Rosenberg, H. M. (1978). Carbon-fibre reinforced composites with zero expansion in the low temperature range. *Cryogenics* **18**, 373–374.

Plummer, W. A. and Hagy, H. E. (1965). Precision thermal expansion measurements on low expansion optical materials. *Appl. Opt.* **7**, 825–831.

Poore, M. W. and Kesterson, K. F. (1977). *Strain-Gage Technique for Measuring the Thermal Expansion of Solids*. Y2065, Oak Ridge National Laboratory, Oak Ridge, Tennessee.

Popov, V. P., Pervakov, V. A., and Tsibulina, Z. I. (1977). Low temperature thermal expansion of solid solution of lead in aluminum. *Phys. Met. Metallogr.* **43**(5), 202–203.

Powell, R. W. (1958). Thermal conductivities and expansion coefficients of water and ice. *Adv. Phys.* **7**, 276–297.

Pritchard, R. G. and Gugan, D. (1970). Bulk expansivity and vacancies in solid argon. *Phys. Lett. A* **32**, 124–125.

Pudalov, V. M. and Khaikin, M. S. (1969). Dilatometer with a sensitivity of 10^{-4} Angstrom. *Cryogenics* **9**, 128–130.

Pulfrich, C. (1893). Über das Abbe–Fizeausche dilatometer. *Z. Instrum.* **13**, 365, 401–437.

Rare Gas Solids, Vols. 1 and 2 (1976, 1977) (eds. M. L. Klein and J. A. Venables). Academic Press, New York.

Rhodes, B. L., Moeller, C. E., Hopkins, V., and Marx, T. I. (1963). Thermal expansion of several technical metals from 20 degrees K to 300 degrees C. In *Advances in Cryogenic Engineering*, Vol. 8. Plenum Press, New York, 278–286.

Roark, R. J. and Young, W. C. (1975). *Formulas for Stress and Strain*. McGraw-Hill, New York.

Robbins, R. F., Ohori, Y., and Weitzel, D. H. (1963). Linear thermal expansion of elastomers in the range 300° to 76° K. In *Advances in Cryogenic Engineering*, Vol. 8. Plenum Press, New York, 287–299.

Roberts, R. E. (1975). Absolute dilatometry using a polarization interferometer. *J. Phys. E.* **8**, 600–602.

Rodrigues, C. A. V. de V. (1981). French cryogenic differential dilatometer. In *Tube Type Dilatometers* (ed. J. Valentich). Instrument Society of America, North Carolina, 172.

Roe, J. M. and Simha R. (1974). Thermal expansivity and relaxational behavior of amorphous polymers at low temperatures. *Int. J. Polym. Mater.* **3**, 193–227.

Rosen, B. W. (1970). Thermomechanical properties of fibrous composites. *Proc. R. Soc. Lond. A* **319**, 79–94.

Rothrock, B. D. and Kirby, R. K. (1967). An apparatus for measuring thermal expansion at elevated temperatures. *J. Res. Natl. Bur. Stand. C* **71**, 85–92.

Rubin, T., Altman, H. W., and Johnston, H. L. (1954). Coefficients of thermal expansion of solids at low temperatures: I. Thermal expansion of copper from 15° to 300° K. *J. Am. Chem. Soc.* **76**, 5289–5292.

Schadler, H. W., Osika, L. M., Salvo, G. P., and DeCarlo, V. J. (1964). The thermal expansion of Nb_3Sn (Cb_3Sn). *Trans. Metall. Soc. AIME* **230**, 1074–1077.

Schramm, R. E., Clark, A. F., and Reed, R. P. (1973). *A Compilation and Evaluation of Mechanical, Thermal, and Electrical Properties of Selected Polymers*. NBS Monograph 132, National Bureau of Standards, Boulder, Colorado.

Schramm, R. E. and Kasen, M. B. (1978). Fiberglass-epoxy in a conical superconducting field magnet support. In *Advances in Cryogenic Engineering*, Vol. 24. Plenum Press, New York, 271–278.

Seligmann, P. F. and Sarwinski, R. E. (1972). The contraction from 295 K to 4 K of some solders, epoxies, and AGOT graphite. *Cryogenics* **12**, 239–240.

Shapovalenko, V. V., Tykachinskii, I. D., Kataeva, G. V., and Pyatnitskaya, V. V. (1976). Filled epoxy polymer with low thermal-expansion coefficient. *Instrum. Exp. Tech.* **19**(6), 1835–1837.

Sharma, S. C., Mandelkern, L., and Stehling, F. C. (1972). Relation between expansion coefficients and glass temperature. *Polym. Lett.* **10**, 345–356.

Sheard, F. W. (1958). Calculation of thermal expansion of solids from the third-order elastic constants. *Philos. Mag.* **3**(8), 1381–1390.

Sheard, F. W. (1969). Negative thermal expansion due to paramagnetic ions in cubic crystals. *Phys. Lett. A* **30**, 156–157.

Sheard, F. W. (1972). Spectroscopic applications of low temperature thermal expansion. In *AIP Conference Proceedings No. 3*. American Institute of Physics, New York, 155–168.
Simha, R. and Boyer, R. F. (1962). On a general relation involving the glass temperature coefficients of expansion of polymers. *J. Chem. Phys.* **37**(5), 1003–1007.
Simmons, R. O. and Balluffi, R. W. (1957). Low temperature thermal expansion of Cu. *Phys. Rev.* **108**, 278–280.
Simon, E. and Vohsen, E. (1928). *Z. Phys. Chem.* **133**, 165.
Smith, T. F., Finlayson, T. R., Shelton, R. N. (1975). Superconductivity and anharmonicity in V_3Si. *J. Less-Common Met.* **43**, 21–32.
Sparks, L. L. (1982). National Bureau of Standards, Boulder, Colorado. Private communication.
Srinivasan, R. and Ramji, R. R. (1972). Anharmonic properties of the hexagonal metals, magnesium, zinc, and beryllium: II. Thermal expansion. *J. Phys. Chem. Solids* **33**, 491–499.
Testardi, L. R. and Bateman, T. B. (1967). Lattice instability of high-transition-temperature superconductors. II. Single-crystal V_3Si results. *Phys. Rev.* **154**(2), 402–410.
Thermal Expansion 6 (1978) (ed. I. D. Peggs). Plenum Press, New York.
Thermal Expansion, Metallic Elements and Alloys, Thermophysical Properties of Matter, Vol. 12 (1975) (eds. Y. S. Touloukian, R. K. Kirby, R. E. Taylor, and P. D. Desai). IFI/Plenum, New York.
Thermal Expansion, Nonmetallic Solids, Thermophysical Properties of Matter, Vol. 13 (1977) (eds. Y. S. Touloukian, R. K. Kirby, R. E. Taylor, and T. Y. R. Lee). IFI/Plenum, New York.
Timmesfeld, K. H. and Elliott, R. J. (1970). Thermal expansion and volume change of crystals containing substitutional point imperfections. *Phys. Status Solidi* **42**, 859–869.
Touloukian, Y. S. (1966). *Recommended Values of the Thermophysical Properties of Eight Alloys, Major Constituents and Their Oxides*. Thermophysical Properties Research Center, Purdue University, Lafayette, Indiana.
Valentich, J. (1981). *Tube Type Dilatometers, Applications from Cryogenic to Elevated Temperatures*. Instrument Society of America, North Carolina.
Varley, J. H. O. (1956). The thermal expansion of pure metals and the possibility of negative coefficients of volume expansion. *Proc. R. Soc. Lond. A* **237**, 413–421.
Vold, C. L. and Richards, L. E. (1974). Low temperature thermal expansion of A-15 vanadium–gallium alloys. *Mater. Res. Bull.* **9**(2), 83–88.
Wallis, R. F. (1975). Effects of surfaces in lattice dynamics, In *Dynamical Properties of Solids*, Vol. 2 (eds. G. K. Horton and A. A. Maradudin). North Holland, Amsterdam, 503.
White, G. K. (1961). Measurement of thermal expansion at low temperatures. *Cryogenics* **1**, 151–158.
White, G. K. (1965). Thermal expansion of magnetic metals at low temperature. *Proc. Phys. Soc.* **86**, 159–169.
White, G. K. (1972). Expansion coefficient of coppers at 283 K and low temperatures. In *AIP Conference Proceedings No. 3*. American Institute of Physics, New York, 59–64.
White, G. K. (1976). Thermal expansion at low temperatures of glass-ceramics and glasses. *Cryogenics* **16**, 487–490 and 606 (corrigendum).
White, G. K., Birch, J. A., and Manghnani, M. H. (1977). Thermal properties of sodium silicate glasses at low temperatures. *J. Non-Cryst. Solids* **23**, 99–110.
White, G. K. and Collins, J. G. (1972). Thermal expansion of copper, silver, and gold at low temperatures. *J. Low Temp. Phys.* **7**, 43–75.
Wohlfarth, E. P. (1975). The Invar problem. *IEEE Trans. Magn.* **MAG-11**(6), 1638–1644.
Wolfendale, P. C. F. (1968). Capacitive displacement transducers with high accuracy and resolution. *J. Phys. E* **1**, 817–818.
Yates, B. (1972). *Thermal Expansion*. Plenum Press, New York–London.
Zeigler, W. T., Mullins, J. C., and Hwa, S. C. P. (1963). Specific heat and thermal conductivity of four commercial titanium alloys from 20° to 300° K. In *Advances in Cryogenic Engineering*, Vol. 8. Plenum Press, New York, 268–277.
Zemansky, H. M. (1951). *Heat and Thermodynamics*, Third edition. McGraw-Hill, New York, 272.
Zemansky, M. W. (1968). *Heat and Thermodynamics*, Fifth edition. McGraw-Hill, New York.

4 Thermal Conductivity and Thermal Diffusivity

J. G. HUST

4.1 INTRODUCTION

The increasing sophistication of modern technology has led to the development of many new materials and an increased need for in-depth knowledge of their physical properties. In particular, during the last decade there has been a massive developmental effort in low-temperature equipment. This was spurred initially by the aerospace programs. Today this technology is rapidly being transferred to the commercial sector, principally in energy-related areas such as superconducting machinery, fusion power, and low-temperature liquid fuels (liquefied natural gas and liquid hydrogen). This expansion of the number of technically important materials and low-temperature applications has created a strong demand for new and more accurate thermophysical property data. A recent survey of information requests for thermophysical data has shown a particularly strong interest in thermal conductivity data (Touloukian, 1976). It is unfortunate that these requests often must be ambiguously answered. For some technically important materials there is a total lack of data; and for others, there are numerous sets of data of widely divergent values for the "same" material. In the case of comparatively pure materials, the widely divergent low-temperature results are due mainly to slight material differences that have profound effects on such transport properties as electrical conductivity, thermal conductivity, and thermal diffusivity. In other cases, the divergent results are caused by experimental errors; 50% errors are not uncommon in the literature. This is not too surprising, because accurate thermal conductivity measurements are difficult to achieve and must be performed with the utmost care.

To satisfy the current and more stringent future demands for thermophysical properties data, two simultaneous efforts are essential. The first is to evaluate critically and correlate existing data. This step will help to identify correct and incorrect data and provide a basis for improved understanding of the fundamental mechanisms governing heat transport. The second is to measure and predict values for well-characterized technically important materials. The results of these efforts must be published in a useful and understandable format. Because transport properties are influenced by such a myriad of variables, it is virtually impossible to present "best values" for all cases. Thus it is essential that the potential causes of these variations are recognized and the best values for a particular choice of material are obtained.

4.1.1 Purpose

This chapter is intended to provide a reasonable understanding and intuitive feeling about the apparent complexities of thermal transport data. To accomplish this, salient theoretical features are presented along with practical empirical correlations and selected illustrative data. Although theoretical calculations, based on first principles, usually do not lead to accurate values of thermal conductivity, they are valuable to indicate parameter dependencies. Knowledge of these dependencies, coupled with some experimental data, is valuable in the extrapolation, interpolation, and analysis of existing data. This discussion also provides a basis for prediction of the properties of a particular material from the existing data base. No attempt is made to describe complicated rigorous theoretical developments.

Emphasis is directed toward the importance of material characterization when obtaining low-temperature thermal transport data for a specific material. At low temperatures, the generic name of the material, e.g., copper, aluminum, carbon, is generally not sufficient to characterize the material for thermal transport purposes. In other words, there is no unique answer to the question, "What is the value, or best value, or recommended value for the thermal conductivity of copper at low temperatures?" The most obvious solution to this problem is to measure the thermal transport properties of each batch of material in the condition in which it is to be used. Although in some cases this is necessary, in general it is impractical because of the expense of such measurements. Another objective of this chapter is to identify those characterization parameters that adequately distinguish the specific material within the generic material classification. To be useful, determination of these parameters must be less expensive than direct measurement of the thermal properties.

This chapter emphasizes and elaborates three points:

a. Past experimental efforts have resulted in an extensive, but by no means complete, data base of thermal transport properties of technically important materials. These data have been compiled by Touloukian, Powell, Ho, and Klemens (1970); Touloukian, Powell, Ho, and Nicolaou (1973); and Childs, Ericks, and Powell (1973) and are an excellent data base from which to interpolate and extrapolate to materials that have not been measured.

b. Inexpensive and reasonably accurate predictive methods are needed to fill in the gaps of the experimental data base. Because of the uncountable variations of materials, it is impossible to imagine that measured transport property values will ever be available for each of these. Through a combination of the results of theory, empirical relations, and the existing data base, adequate prediction of heat transport properties is often possible.

c. Compiled, recommended, or best values for materials that are defined only in generic terms rather than by more specific characterization parameters must be applied very cautiously.

4.1.2 Scope

This chapter considers the heat transport properties of solids only and emphasizes the behavior of pure metals and alloys. Temperatures below room temperature are of primary interest, but some of the results presented here are applicable to higher temperatures. Thermal conductivity and thermal diffusivity are the principal properties discussed, but because of property interrelationships, consideration must also be given to electrical resistivity, density, thermal expansion, and specific heat. The latter properties, excluding density, are discussed in other chapters of this book.

Pertinent results of theory are presented and compared with experimental data. Discussions are included on the effects of temperature, magnetic field, and metallurgical variations caused by composition, processing, and heat-treatment differences. Numerous graphs illustrate the qualitative and quantitative effects of these variables. Measurement methods and associated accuracies and pertinent empirical correlations are presented.

4.2 BASIC CONCEPTS

This section presents those basic principles and the theoretical results deemed essential to provide an intuitive grasp of heat transport in solids. First, a few elementary definitions and experimental facts are presented to illustrate the tremendous range of values and parameters that may be encountered. The examples refer specifically to thermal conductivity but are generally applicable to thermal diffusivity as well.

The definition of thermal conductivity is based on empirical observations relating heat flow and temperature gradient. For steady-state, unidirectional (or planar) heat flow through an isotropic medium, this relationship, known as the Fourier–Biot equation, is

$$\dot{Q}/A = -k\, dT/dx \tag{4.1}$$

where $\dot{Q}$ is the rate of heat flow through area A with a temperature gradient, dT/dx. The constant of proportionality, k, is known as thermal conductivity. The symbol λ is also often used in scientific literature to denote thermal conductivity, whereas the symbol k is more common to the engineering literature. The minus sign in Eq. 4.1 indicates that the direction of heat flow is opposite to the temperature gradient. It should be noted that Eq. 4.1 describes a diffusive or random energy transport process, as indicated by the presence of the temperature gradient rather than temperature difference. This is the salient difference between heat transport by conduction and other heat transport mechanisms, such as convection or radiation. The term *thermal conductivity* is not applicable without qualification to these other forms of heat transport.

If the temperature of the material is allowed to vary with time, t, at a rate $\partial T/\partial t$ and if the thermal conductivity is independent of the temperature change, one can derive

$$\partial T/\partial t = \alpha\, \partial^2 T/\partial x^2 \tag{4.2}$$

The constant of proportionality, α, is known as the thermal diffusivity and characterizes the propagation of a temperature disturbance through a material. Thermal diffusivity can be determined experimentally through Eq. 4.2 or from other property data through the derived relation

$$\alpha = k/(\rho C) \tag{4.3}$$

where ρ is mass density and C is the specific heat of the material. For most solids, the specific heat at constant pressure is normally used.

The above simple conditions were chosen only to show the definitions of k and α and the relationship between them. The definitions of these quantities under more general conditions are not considered relevant to this discussion, but the interested reader can find them elsewhere (Carslaw and Jaeger, 1959; Touloukian et al., 1973).

The following discussion is directed toward an understanding of thermal conductivity. Since thermal diffusivity can be defined in terms of thermal conductivity, specific heat, and density, and since specific heat and density are generally less sensitive to minor material variations, an understanding of thermal conductivity behavior will lead directly to an understanding of thermal diffusivity. This generality must be applied with caution in regions of anomalous behavior, such as near phase transformations, or for highly anisotropic materials.

Prior to more detailed discussions of the thermal behavior of a wide range of materials, selected data are presented. These examples illustrate the wide range of values, the general temperature dependencies, and relative intercomparisons of thermal properties for commonly used materials.

Figures 4.1 and 4.2 illustrate the thermal conductivity and thermal diffusivity of a few technically important materials spanning the range from pure metals to structural insulating materials. These figures show only a single curve for each material, as is customarily done for simplicity. This is misleading, however, for pure metals and dielectric crystals, as will be discussed in later sections. Figures 4.1 and 4.2 are used only to illustrate the appropriate range of values and are not to be used for design purposes. Note that these are logarithmic graphs spanning about five orders of magnitude in thermal conductivity and eight in thermal diffusivity. If one also considers thermal insulations, the ranges are increased by several more orders of magnitude.

Figure 4.3 shows the "smoothed" thermal conductivity curves for pure copper (better than 99.9%), one of which is depicted in Figs. 4.1 and 4.2. At room temperature the thermal properties of these copper specimens are the same within a few percent. At 10 K, however, the transport properties of these coppers span a range of nearly three orders of magnitude. Therefore, it is not meaningful to speak of the low-temperature thermal conductivity or diffusivity of copper in

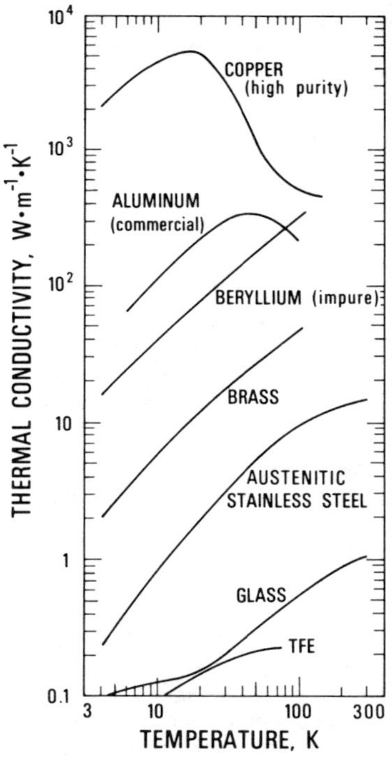

Figure 4.1 Typical thermal conductivity illustration: temperature dependencies of selected technically important materials from 3 to 300 K. Only a single curve is shown for each material, as is customary, but this is misleading for pure materials.

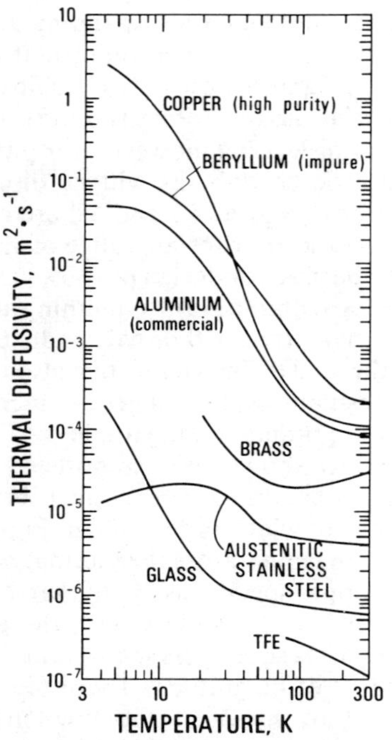

Figure 4.2 Typical thermal diffusivity illustration: temperature dependencies of selected technically important materials from 3 to 300 K. Only a single curve is shown for each material, as is customary, but this is misleading for pure materials.

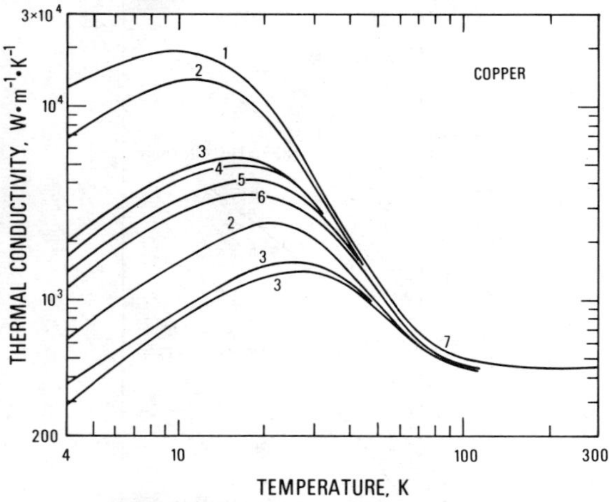

Figure 4.3 Experimentally determined curves of thermal conductivity for various pure copper specimens from 4 to 300 K. Each of these specimens is reported to contain less than 0.1% impurity. 1 — White and Tainsh (1960); 2 — Powell, Roder, and Hall (1959), 3 — White (1953); 4 — Berman and McDonald (1952); 5 — Hust and Giarratano (1974); 6 — Mendelssohn and Rosenberg (1952); 7 — Moore, McElroy, and Graves (1967).

the generic sense. Nor is it adequate to characterize the copper in terms of chemical purity, such as 99.99%, for purposes of identifying its transport properties. Similar statements are valid for other metals and nonmetals of reasonable purity.

These examples depict the problem of understanding heat transport property values for solids whose thermal properties range over eight orders of magnitude (even more if insulating materials are included). Transport property values are not only strongly dependent on temperature, but also on the chemical and physical state of the material and on other environmental conditions, such as the presence of magnetic fields.

4.2.1 Heat transport

Figure 4.4 schematically illustrates the heat transport mechanisms (conduction, convection, and radiation) that may be present in matter. Matter in the fluid state (liquid or gaseous) may exhibit all of these mechanisms. Solids may exhibit heat transport by conduction and radiation. Many technically important solids, however, are opaque; therefore, the radiation component of heat transfer is absent and only heat conduction is present. A material that is partly transparent transmits radiant energy through a process of absorption, reradiation, and

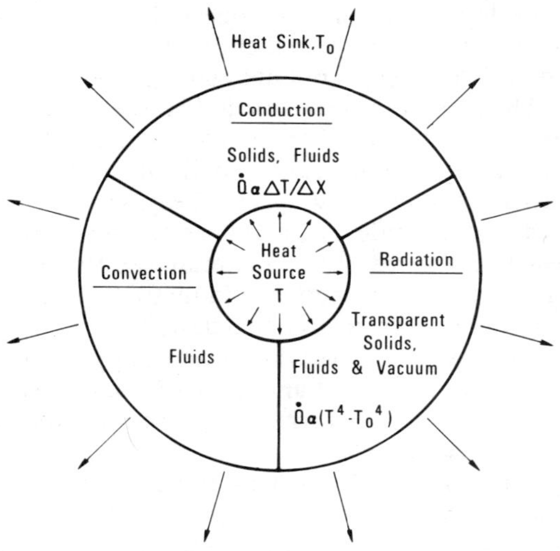

Figure 4.4 Heat transport mechanisms. Whereas relatively simple expressions can be given for heat transfer by conduction and radiation, convection is dependent on fluid specific heat, its rate of motion, temperature difference, as well as boundary layer phenomena.

scattering. If the average distance between radiation and absorption (effective mean free path) of this process is small compared with the thickness of the medium, the resultant heat transfer is a diffusion-like process and is essentially indistinguishable from conduction. The heat transfer can then be treated as if only conduction were present. If the medium is not thick compared with the mean free path, the experimentally determined thermal conductivity is dependent upon thickness, and therefore, not solely a property of the material. Conduction by electrons and lattice waves can also be size dependent, and the concomitant thermal conductivity is dependent on the size, shape, and configuration of the specimen. This will be discussed later.

4.2.2 Heat conduction and thermal diffusion

With consideration restricted essentially to heat transport by conduction in solids, the various mechanisms that determine the total conduction through a solid are considered next. Figure 4.5 illustrates the additive modes of heat conduction that may be present. In pure metals and dilute alloys, the principal conduction mechanism is electronic, i.e., the free electrons conduct most of the heat. As a metal becomes more highly alloyed, the heat conducted by the quantized lattice vibrations (phonons) becomes relatively more dominant, because the heat conducted by the free electrons is considerably more

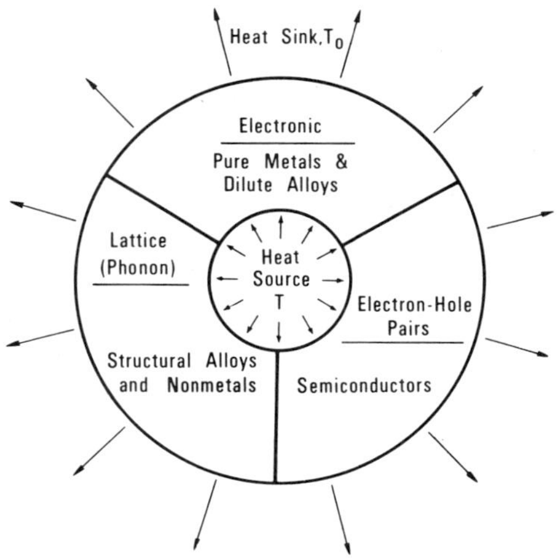

Figure 4.5 Heat conduction mechanisms. The total conduction in solids is the sum of the above components. Types of material exhibiting each mode of conduction are indicated.

reduced. In opaque nonmetals (excluding semiconductors), all of the heat is conducted by phonons. In semiconducting materials, electron–hole-pair conduction is present in addition to the electronic and phonon components. In electron–hole-pair conduction, the energy required to create the pair determines the heat transport.

These heat transport components, k_n, are normally taken to be additive, and the total heat transport is their sum, $k = \Sigma\, k_n$. In § 4.2.3 and 4.2.4, various resistive effects on these separate components and how these conduction-limiting mechanisms depend on temperature, alloying, and magnetic field are considered. It should be noted that, although the conduction components are additive, each conductive component is, in turn, limited by additive resistive mechanisms, i.e., $W_n = k_n^{-1} = \sum_m W_{nm}$. The additivity of electron and phonon conduction and selected resistance mechanisms is illustrated by the electrical analogue shown in Fig. 4.6. The temperature dependencies of various resistive mechanisms are listed in Table 4.1 (Klemens, 1958).

4.2.3 Electronic conduction

The thermal conduction by electrons is limited by interactions with phonons, lattice imperfections, other electrons, and magnetic fields. Generally, the most significant of these are scattering by phonons and lattice imperfections (chemical and physical defects). Lattice imperfections can be divided into two groups: those that are localized (point defects) and those that extend over regions that are large compared with lattice spacings. Point defects include interstitial atoms, vacancies, and substitutional impurities (different chemical and isotopic species or mass defects). Extended defects include dislocations, grain boundaries, and specimen boundaries.

Pure and dilute alloy metals. Theory indicates (see Table 4.1) that the low-temperature electron–imperfection thermal resistance decreases with temperature as T^{-1} and is proportional to the imperfection density. The electron–phonon thermal resistance increases as the square of the temperature at low temperatures and approaches a constant at high temperatures. These dependencies are shown in Fig. 4.7. High and low temperatures are defined with respect to the Debye temperature, which is between 200 and 500 K for most common metals. Table 4.2 lists Debye temperatures for selected common metals (see Chapter 1 for further details).

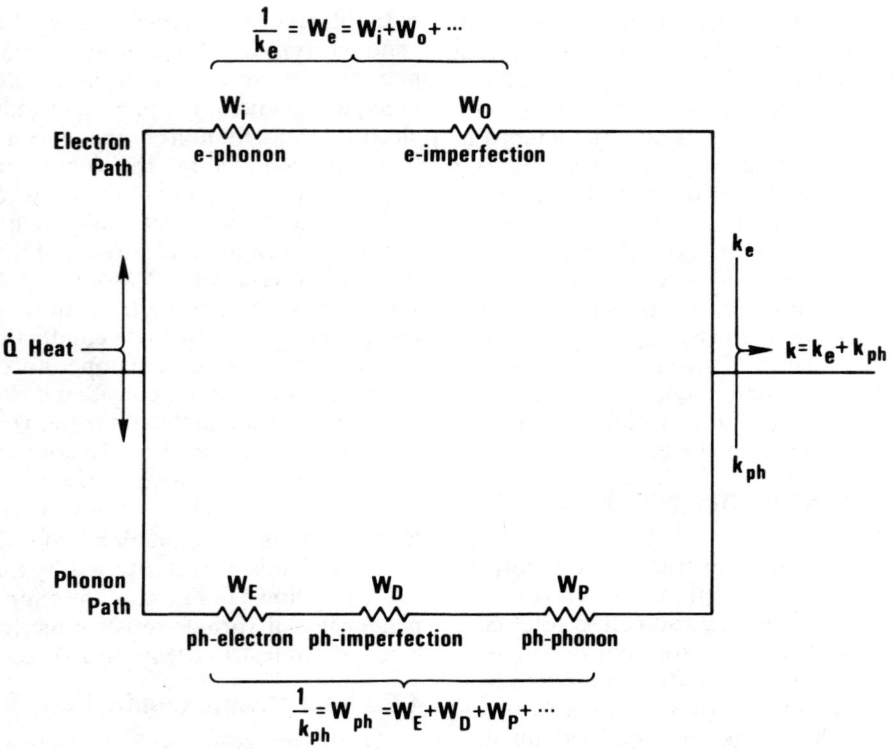

Figure 4.6 Electrical analog of thermal conduction and resistance mechanisms. The e-imperfection and ph-imperfection resistances can be further divided into resistances, such as those caused by dislocations and point defects (see Table 4.1).

Table 4.1 Variation of thermal resistance to electronic and phonon components of conduction.

Scattering Mechanism	Temperature Dependence
Low Temperatures	
Phonon–	
crystal boundary	T^{-3}
grain boundary	T^{-3}
stacking fault	T^{-1}
electrons[a]	T^{-2} to T^{-1}
dislocation	T^{-2}
point defect	T, $T^{3/2}$
phonon[b]	$T^{-3}e^{-(\theta_D/\alpha T)}$
Electron–	
phonon	T^{2}
defect	T^{-1}
High Temperatures	
Phonon–	
phonon	T
electron	constant
defect	constant
Electron–	
phonon	constant
defect	T^{-1}

[a]The temperature dependence is determined by the product of electron mean free path and phonon wave number. (T^{-1} if the product is less than unity and T^{-2} above unity.)
[b]θ_D = Debye temperature; α = constant on the order of unity

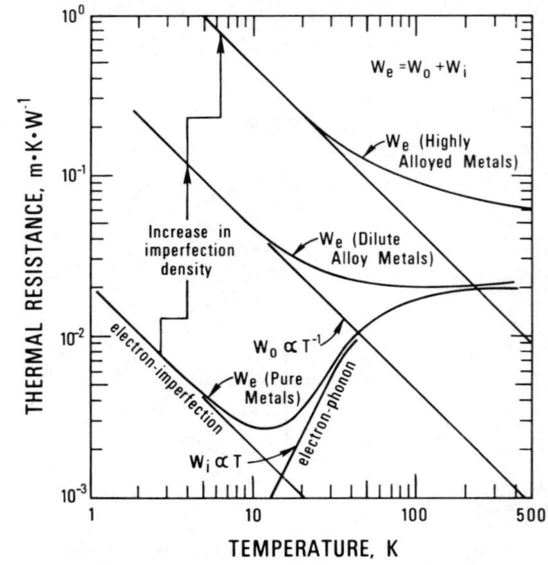

Figure 4.7 Resistance to electronic heat conduction: temperature dependencies and imperfection (defect) densities progressing from pure, annealed metals to highly alloyed metals.

Table 4.2 Debye temperatures of selected common metals (see Chapter 1).

Substance	Debye Temperature, θ_D (K)
Aluminum	430
Beryllium	1374
Chromium	590
Copper	344
Gold	162
Iron	472
Lead	105
Magnesium	386
Molybdenum	474
Nickel	476
Platinum	238
Silver	226
Tantalum	264
Tin	201
Titanium	426
Tungsten	384
Zinc	329

The electron–phonon thermal resistance is characteristic of the generic material (e.g., copper, aluminum) and is only weakly dependent on the chemical and physical imperfections. Therefore, it is an intrinsic property of the elemental (or generic) material. Conversely, the electron–imperfection thermal resistance is essentially totally dependent on the material imperfections and is most important at low temperatures. This is the reason for the large spread of low-temperature thermal transport values for pure metals and why a single value cannot be given for the thermal conductivity of, for example, "copper" at low temperatures. The thermal transport is limited by the density and type of defects and not significantly by the material lattice. One must characterize the effect of these imperfections to specify the thermal conductivity of each specific material. For pure metals and dilute alloys, excellent imperfection characterization can often be obtained with a simple, low-temperature electrical resistivity measurement, as discussed in detail later (see § 4.5)

The intrinsic and imperfection thermal resistances to electronic conduction and their sum are illustrated in Fig. 4.7, and their temperature dependencies are listed in Table 4.1. Figure 4.7 illustrates the effect of introducing imperfections (either physical defects or chemical impurities). Annealed, high-purity metals exhibit pronounced minimums in thermal resistance at low temperatures. As the imperfection density increases, the low-temperature resistance increases, and the resistance minimum gradually vanishes. The room-temperature resistance caused by defects is relatively small for pure metals and dilute alloys, and therefore, the total resistance is determined mainly by intrinsic (generic) behavior. Figure 4.8 illustrates the electronic conductivity of metals, progressing from pure metals to those with moderately high imperfection densities.

Highly alloyed metals. Higher defect concentrations and lattice disorder (as in structural alloys) further increase the resistance to both electronic and phonon heat flow. The resistance to electronic heat flow increases, however, more rapidly than the resistance to phonon heat flow. Thus, in highly alloyed metals at low temperatures, the total thermal conductivity is small and often dominated by the phonon component. Also, the defect resistance is now so large that its effect is significant at considerably higher temperatures (even at room temperature). Intrinsic (generic) behavior is, therefore, no longer exhibited by such metals. Figure 4.7 illustrates this increase in resistance to electronic conduction, whereas Fig. 4.8 depicts the suppression of the total conductivity in structural alloys.

Magnetic field effects. The effect of a magnetic field on the electronic component of thermal conduction is complicated, but a simple model may suffice to illustrate the qualitative re-

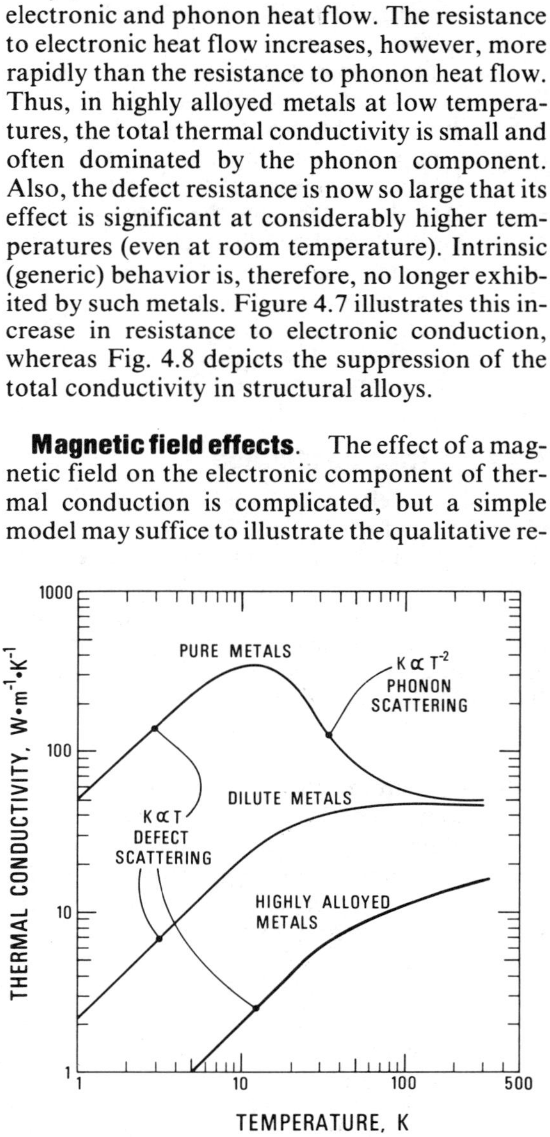

Figure 4.8 Typical electronic component of thermal conductivity for metals: temperature dependencies and imperfection (defect) densities progressing from pure, annealed metals to highly alloyed metals.

sults. An electron moving through a lattice under the influence of a magnetic field has imposed upon it a Lorentz force perpendicular to its motion and to the magnetic field direction. This force causes the electron to spiral around the field lines between collisions. In a given distance along the direction of the temperature gradient, the probability of scattering by imperfections and phonons is increased, and as a consequence, the electronic conductivity is lowered. If the electron mean free path, the average distance between collisions, is small compared with the spiral radius, the effect of the Lorentz force is small and the change in conductivity is correspondingly small, as is observed in low conductivity alloys. If, however, the mean free path is large, as it is in very pure, annealed metals at low temperatures, the increased resistance caused by the magnetic field can dominate the total resistance. Experimental data also indicate a strong correlation between thermal and electrical conductivity changes induced by magnetic fields. The quantitative decrease in conductivity caused by a magnetic field is best indicated by experimental results on thermal and electrical conductivity (see § 4.4 and Chapter 5).

4.2.4 Lattice conduction

Similar to electronic conduction, lattice (or phonon) conduction is limited by the interactions (scattering) of phonons with electrons, lattice defects, and other phonons. Again, to a first approximation, these resistive mechanisms are additive, yielding the total resistance to phonon conduction. The temperature dependencies of these resistive mechanisms are summarized in Table 4.1.

Metals. In metals at low temperatures, the most important of the phonon scattering mechanisms are phonon–electron and phonon–dislocation interactions. As the temperature increases, phonon–point defect and phonon–phonon resistances become dominant. Since in pure and dilute alloy metals, the lattice component of the total conductivity is generally insignificant compared with the electronic conductivity, very careful experiments must be performed to confirm the theoretically predicted lattice resistivity behavior, i.e., lattice resistivity decreasing as T^{-2} at low temperatures and increasing as T at intermediate and higher temperatures. The few careful experiments that have been performed on alloys to obtain such confirmation are in qualitative agreement with theory (e.g., Kemp, Klemens, and Tainsh, 1957a).

It should be noted that as the electron mean free path is further reduced in highly alloyed materials, the temperature dependence of the phonon–electron resistivity changes from T^{-2} to T^{-1}. Thus, in highly alloyed metals, both the electronic and lattice conductivities should increase as T at low temperatures, as observed experimentally.

Nonmetals. In nonmetals, excluding semiconductors, lattice (phonon) conduction is the principal energy transport mechanism. In addition, the phonon–electron scattering mechanism is absent because of the lack of conduction electrons. In crystalline dielectrics, the primary low-temperature scattering mechanisms are phonon–boundary (both crystal and grain boundary) and phonon–dislocation interactions. As temperature increases, phonon–phonon and phonon–point defect scattering become the dominant mechanisms, as in metals. Thus in relatively pure, crystalline dielectrics, the low-temperature lattice conductivity may increase as T, T^2, or T^3, depending on the magnitudes of the significant scattering mechanisms (see Table 4.1). As temperature increases, the lattice conductivity will peak and decrease as T^{-1}. As the defect concentration increases, the low-temperature conductivity of nonmetals decreases, as it does for metals, and the peak vanishes.

For nonmetals with high defect densities or amorphous materials, the usual theories of phonon conduction and scattering are no longer strictly applicable. The usual conclusion is that the mean free path of the energy carriers in materials lacking long-range order is a constant, and the thermal conductivity varies with temperature like specific heat. At low temperatures this is only qualitatively observed, indicating that the mean free path is actually a function of temperature. Detailed knowledge is clearly lacking in this area; evidently no useful predictive theory exists. Again, no attempt is made to depict anomalous behavior caused by discontinuous atomic or electronic transitions.

Although the exact temperature dependencies of the thermal conductivity of metals and nonmetals are different, the following similarities are noted:

a. The thermal conductivity of both pure metals and nonmetals rises at low temperatures with increasing temperature; at a small fraction of the

Debye temperature a peak occurs (this temperature decreases with increasing purity); above this temperature thermal conductivity decreases rapidly with increasing temperature.

b. Metals and nonmetals in the purest state have comparable maximum thermal conductivities, in spite of the fact that the heat is carried predominantly by electrons in metals and phonons in nonmetals.

c. Increased defect concentrations in either metals or nonmetals cause decreased thermal conductivity at low temperatures, and the peak, in either case, vanishes with high defect concentrations.

Magnetic field effects. At first glance, one would not expect any dependence of the phonon conductivity component on magnetic field because phonons do not directly interact with magnetic fields. Nevertheless, the phonon component can be dependent on magnetic field because of the influence of the field on the scattering source. It is possible for such magnetic effects to dominate the total thermal resistance in certain dielectric materials, but these are generally not common engineering materials (see, for example, Berman, 1976).

4.3 MEASUREMENT METHODS

There are numerous methods of measuring thermal conductivity and thermal diffusivity. The principal reason for the development of so many different methods is the large range of values for these thermal transport properties. It has already been pointed out that heat-transport properties can vary by several orders of magnitude as temperature decreases. Also, Fig. 4.1 shows a range of more than five orders of magnitude from the best to the poorest solid conductors at a given temperature. No single method is applicable to the measurement of all materials in all temperature ranges. Reported uncertainties of these various methods generally range from about 1 to 10%. True inaccuracies, however, are generally larger than those reported, and often much larger. Insufficient characterization of measured specimens also often leads the reader to suspect high inaccuracies, when in fact the measurements may be very accurate but pertinent specimen differences are unreported.

The methods of measurement can be grouped into two categories: steady state and nonsteady state. The steady-state methods are used to measure thermal conductivity directly. The nonsteady-state methods are usually used to measure thermal diffusivity directly, but they can also yield thermal conductivity and specific heat data.

An examination of the literature on experimental determinations of thermal conductivity and thermal diffusivity reveals that at temperatures below 300 K: (a) thermal diffusivity measurements by any technique are rare, and (b) steady-state thermal conductivity measurements are almost always performed using the axial (sometimes called linear, unidirectional, planar, or longitudinal) heat-flow method. Childs et al. (1973) have indicated that the axial heat-flow method is used 84% of the time. Therefore, for purposes of discussion of operational definitions and experimental techniques, this simple configuration is used. Most of this discussion, however, is also germane to other types of thermal conductivity and thermal diffusivity apparatus.

4.3.1 Operational definitions and experimental techniques

An axial-flow steady-state thermal conductivity apparatus is schematically illustrated in Fig. 4.9. At steady-state conditions the power, $\dot{Q}$, generated by an electrical heater is assumed to flow through the specimen to the temperature-

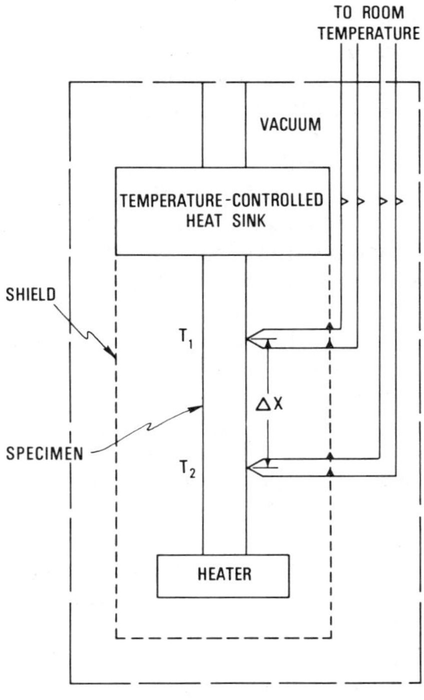

Figure 4.9 Schematic of axial heat-flow apparatus. Thermal grounding is indicated by △.

controlled heat sink. This heat is assumed to enter the bottom of the specimen and leave the top in a uniform manner so that the isothermal surfaces are planar and perpendicular to the specimen between the temperature sensors at T_1 and T_2.

Under these conditions, Eq. 4.1 can be integrated from T_1 to T_2 to yield:

$$\int_{T_1}^{T_2} k(T)dT = -(Q/A)\Delta x \tag{4.4}$$

These thermal conductivity integral values are of direct interest in some cases. Therefore, measurements are sometimes designed to determine the value of the integral Eq. 4.4 for certain common values of T_1 and T_2, such as 4 to 77 K or 20 to 77 K (the normal boiling points of common cryogenic fluids). More often, though, T_1 and T_2 are chosen to be close enough together so that, through the mean value theorem, one can write

$$k(T) \simeq \bar{k}(T) = -(\dot{Q}/A)(\Delta x/\Delta T) \tag{4.5}$$

where $\bar{k}(T)$ is the mean value of thermal conductivity between T_2 and T_1 and $\Delta T = T_2 - T_1$. To determine if $\bar{k}(T)$ is a sufficiently accurate measure of $k(T)$, one must consider the functional form of $k(T)$, the magnitude of ΔT, and the manner in which T is defined from the measured quantities T_2 and T_1. Normally, T is defined simply as the mean of T_2 and T_1, i.e., $\bar{T} = (T_2 + T_1)/2$. Frequently it is said that ΔT must be small to obtain sufficient accuracy of $\bar{k}(T)$. This statement is ambiguous because one has to ask "small compared with what?" In actuality, ΔT frequently can be quite large compared with the absolute temperature, for example. Its maximum permissible magnitude is determined by not only the measurement accuracy required, but more importantly, the functional form of $k(T)$. If $k(T)$ is independent of temperature or proportional to temperature, ΔT can be any size and $\bar{k}(T)$ will be equal to $k(T)$ at $\bar{T}$. For most materials at very low temperatures, $k(T)$ is approximately linear; in many instances this near-linear dependence extends to much higher temperatures; and in still other cases, $k(T)$ is nearly a constant over wide temperature ranges. The point is that with some prior knowledge of $k(T)$, overly restrictive and sometimes counterproductive restrictions on the size of ΔT can be relaxed. If $k(T)$ is expected to vary strongly with temperature or if no prior knowledge exists, it is best to choose ΔT so that values of $k(T)$ between T_2 and T_1 do not vary appreciably more than the allowable inaccuracy of the measurement. Further discussion of this subject along with a description of an integral measurement method (which removes the restrictions on the size of ΔT altogether) is presented by Hust and Lankford (1982). Thus Eq. 4.5 can be used to determine $k(T)$ insofar as the following important conditions are satisfied:

a. All of the energy generated by the electrical heater flows through the entire length of the specimen.

b. The heat flow is uniform, so that the isothermal surfaces are planar and perpendicular to the specimen between T_1 and T_2 (this also requires that A is a constant).

c. $k(T)$ is not a strong function of T between T_1 and T_2.

In addition to the above restrictions, one is confronted with a practical limitation on the choice of the magnitude of $\dot{Q}$ at low temperatures. This upper limit on $\dot{Q}$ is dictated by consideration of the maximum tolerable boil-off rate of the cryogenic fluid used to maintain a constant heat sink temperature. At liquid-helium temperatures, one generally limits $\dot{Q}$ to the milliwatt range.

Specimen size. Because of the above limitations on $\dot{Q}$ and ΔT, the optimum form factor, $\Delta x/A$, is directly dependent on the specimen thermal conductivity. A high-conductivity specimen must be long and thin ($\Delta x/A >> 1\ \text{cm}^{-1}$), whereas a low-conductivity specimen must be short and thick ($\Delta x/A << 1\ \text{cm}^{-1}$). Specimens of high-conductivity materials are typically rods near 0.1-cm diameter and several centimeters long, resulting in form factors of about 1000 cm^{-1}. Low-conductivity solids, such as structural alloys, may be several centimeters in diameter and about the same length with form factors near 0.5 cm^{-1}. Insulating material specimens are disks only a few centimeters thick with diameters on the order of tens of centimeters; as a result their form factors are near 0.0005 cm^{-1}.

Uncertainty. The total uncertainty in the measurement process can be conveniently divided into two components: that caused by instrumentation and calibration errors and that caused by not satisfying the above experimental conditions. The effect of instrumentation and calibration errors is readily assessed. Typical form factors for medium-conductivity materials can generally be determined to within about 1%, but are difficult to obtain to better than 0.1%. Temperature differences within the limitations discussed above can be measured to about 0.1% near room temperature, but at low temperatures (4 K), extreme care is required to do better than about 1%. The measurement of the power generated by the electrical heater is readily obtained to better than 0.1% via voltage and current determinations. One can see that by considering only the instrumentation and calibration errors, it is difficult to perform thermal conductivity measurements to better than about 1% in the range 4 to 300 K.

The second and more subtle component of uncertainty (deviations from assumed conditions) can result in much larger errors unless the utmost care is exercised. This is most likely the source of uncertainty that occasionally results in very large errors in published data. The deviation from assumed planar heat-flow and assumed zero heat loss and the effect on temperature errors are inextricably linked. For example, radial heat loss from the specimen (a) causes an error in $\dot{Q}$, (b) may cause an error in the indicated temperature of the thermometer, and (c) causes a deviation from planar heat flow. Therefore, these effects will not be considered separately.

Thermal anchoring. Consider the instrumentation leads that originate at room temperature and pass into the low-temperature cryostat as illustrated in Fig. 4.9. These leads carry heat that must be conveyed appropriately to the heat sink to avoid undesirable effects on both the specimen heat flow and other critical components in the system. The object is to anchor thermally or "ground" these leads to minimize unnecessary thermal gradients in the vicinity of the specimen and critical measuring components, such as the thermometers. Usually these electrical leads are thermally anchored to the cryogenic bath, and if a specimen shield is used, to this shield at a point that matches the temperature of the specimen at the point of connection. Such thermal anchoring is accomplished with various adhesives. Hust (1970) discusses this subject in more detail and presents equations and typical results to determine the amount of thermal tempering necessary for specific conditions. By such methods one can minimize, but not totally eliminate, heat conduction from the specimen through the leads. One must be cognizant of the fact that the heat flow from the specimen to the thermometer leads causes not only a direct heat-flow loss, but also deviations from planar heat flow and a temperature error caused by the thermal resistance between the specimen and thermometer. The latter error is extremely hard to estimate. The best approach is to achieve an isothermal region near the specimen and the best possible thermal anchoring between the thermometer and specimen. This is accomplished by thermal tempering to an appropriate component of the system and by knife-edged clamps, solders, or greased contacts to the specimen.

Radial heat loss. Next, the radial heat loss due to radiation from the specimen assembly is considered. This heat loss is generally negligible at temperatures below 100 K. Above this temperature, radiation heat loss becomes significant and increases approximately as the cube of the mean temperature for a constant specimen gradient, that is, $\dot{Q}_r \simeq$ constant $T^3 \Delta T$. This equation is valid for $\Delta T << T$, which is normally true in the measurement of thermal conductivity. (It is assumed here that the specimen gradient and the specimen-to-surrounding-shield temperature differences are comparable.) The radiation heat loss at room temperature (300 K) is, thus, about 27 times greater than at 100 K. For appropriately sized structural alloy specimens, this heat loss can be several percent of the total heat flow at 300 K even if the specimen is surrounded by a temperature-controlled shield. Normally one fills the annular space between the shield and the specimen with an insulating fiber or powder to reduce these radiation losses further.

Radial transfer of heat from the specimen assembly by conduction through the surrounding medium must also be considered. The specimen–shield assembly is normally placed in an evacuated chamber to limit the gas conduction losses. A vacuum of near 10^{-3} Pa (10^{-5} torr) is usually adequate to limit this loss to a reasonable level. White (1969) discusses this problem quantitatively. His calculations show that under typical conditions this error is about 0.1% per cm^2 of heater surface area. Typical conditions are taken to be: a vacuum of 10^{-3} Pa of helium gas, l/A equal to 30 cm^{-1}, a ratio of the temperature difference between the heater and surroundings to the temperature difference along specimen of 1, and a specimen of thermal conductivity of 1 $W \cdot m^{-1} \cdot K^{-1}$. The percentage heat loss is approximately proportional to each of the above variables, so that one can readily adjust to different conditions. The above calculation does not hold for a vacuum of less than 10^{-2} Pa, nor does it include the heat loss from the specimen or thermometers. The above result can be applied to these components to find the relative loss along the entire length of the specimen and heater by a straightforward extension. If an insulating material has been added to the annulus between the specimen and shield to limit high-temperature radiation losses, its effect on low-temperature conduction loss cannot be totally ignored. For specimens of 1 $W{\cdot}m^{-1}{\cdot}K^{-1}$ conductivity, this typically causes less than about 0.5% discrepancy, but becomes more important as specimen conductivity decreases.

As mentioned above, the deviation from planar isothermal surfaces or axial heat flow is dependent upon the existence of various radial heat losses. It also depends on the distance from the heat source and heat sink to the nearest thermometer. Clearly, the heat flow is not planar at the heat source and sink in Fig. 4.9 because of the discontinuous change in geometry. Generally, it is assumed that a distance of one or two specimen diameters from such discontinuities is required to achieve axial heat flow. It is also necessary to obtain good thermal connections from the heater to the specimen and from the specimen to the heat sink. Poor thermal connections at these locations may seriously increase the conduction and radiation errors as well as increase the deviations from axial heat flow. Soldered contacts are preferred, but adequate contacts can also be achieved with adhesive bonding or greased mechanical clamps. Wood's or Rose's metal, low-vapor-pressure greases, or expansion-matched epoxies are commonly used. It should be remembered that even a microcrack between two such surfaces is essentially equivalent to an open circuit under vacuum conditions at low temperatures.

Temperature instability of the heat sink can lead to a quasi loss of heat. In a system with a slow temperature drift, part of the heat generated by the heater goes into the heat capacity of the various components, i.e., heater, thermometer, and specimen. This quasi loss can be estimated from

$$\Delta \dot{Q} = Cm\, dT/dt \qquad (4.6)$$

where C is the specific heat, m is the mass, and dT/dt is the temperature drift rate of the component. Typically, a drift rate of a millikelvin per hour could cause a 1% discrepancy. If a thermocouple with a sensitivity of 50 μVK^{-1}, e.g., type E, is used for thermometry, one would need a voltage detection system stable to better than 0.1 μVh^{-1} to ensure that this temperature stability is achieved. The magnitude of this source of error is often underestimated or overlooked and has led to sizable errors in the published literature. Several or all of the above discrepancies can exist in any of the apparatus described in the next section.

4.3.2 Thermal conductivity apparatus

As already indicated, numerous types of apparatus are in existence to achieve specific goals and measure a wide range of substances. A few types of apparatus are briefly described here to illustrate their special features and application range.

Two rod-form, axial-flow apparatus are illustrated in Figs. 4.10 and 4.11. The basic difference is in the number of thermometers mounted along the specimen and the concomitant required specimen length. The two-thermometer type (Fig. 4.10) requires the establishment of thermal equilibrium for a single data point, and systematic errors of measurement tend to go undetected in the absence of subsidiary measurements. The type using eight thermometers results in seven data points for each thermal equilibrium. It has an additional important advantage in that some sys-

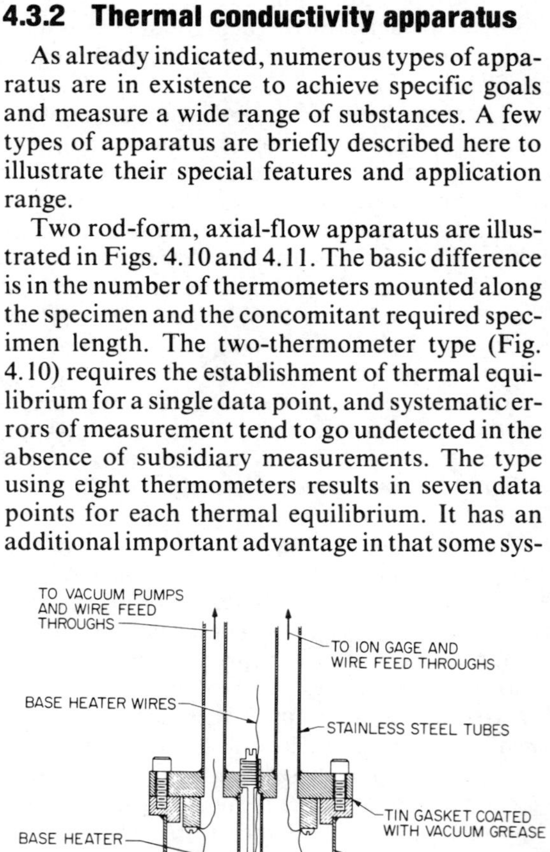

Figure 4.10 Two-thermometer type of axial-flow thermal conductivity apparatus used at Oak Ridge National Laboratory (Laubitz and McElroy, 1971).

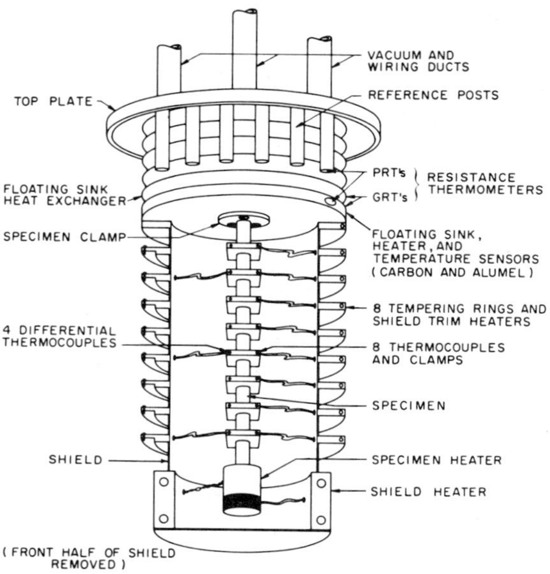

Figure 4.11 Eight-thermometer type of axial-flow thermal conductivity apparatus used at the National Bureau of Standards, Boulder, Colorado (Hust and Giarratano, 1974a).

tematic errors tend to be randomized over the specimen length. For example, if for some reason the fourth thermometer reads wrong consistently, there will be an error of one sign in the third ΔT increment and an error of the opposite sign in the fourth increment, thus randomizing this normally systematic error. Systematic errors affecting all of the thermometers will still go undetected. The eight-thermometer system suffers from additional wiring and operational complexity compared with the two-thermometer model. These types of apparatus are applicable to the measurement of (a) high-conductivity solids, such as pure metals, (b) medium-conductivity solids, such as structural alloys, and (c) low-conductivity solids, such as glasses, ceramics, and polymers, with somewhat reduced accuracy. Below this conductivity range, this form of apparatus becomes impractical.

For insulating materials near ambient temperatures, flat-plate apparatus (as illustrated in Fig. 4.12) are generally used. At lower temperatures and for very low-conductivity insulations, calorimeters (as illustrated in Fig. 4.13) are more commonly used. At cryogenic temperatures, the thermal conductivity integral between the fixed-point temperatures of common cryogenic fluids is often of more interest than thermal conductivity as a continuous function of temperature. The

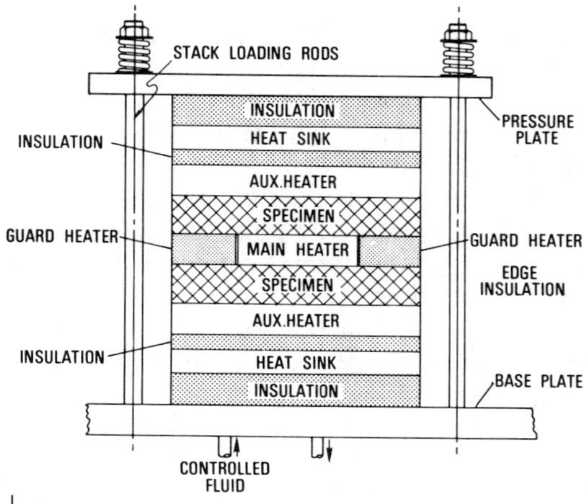

Figure 4.12 Guarded hot-plate apparatus (ASTM-C177).

cryostat illustrated in Fig. 4.13 is well suited to obtain thermal conductivity integrals. Also, such calorimeters often more closely approximate actual operating conditions of insulated cryogenic components.

4.3.3 Thermal diffusivity apparatus

Direct measurements of low-temperature thermal diffusivity are much less common than

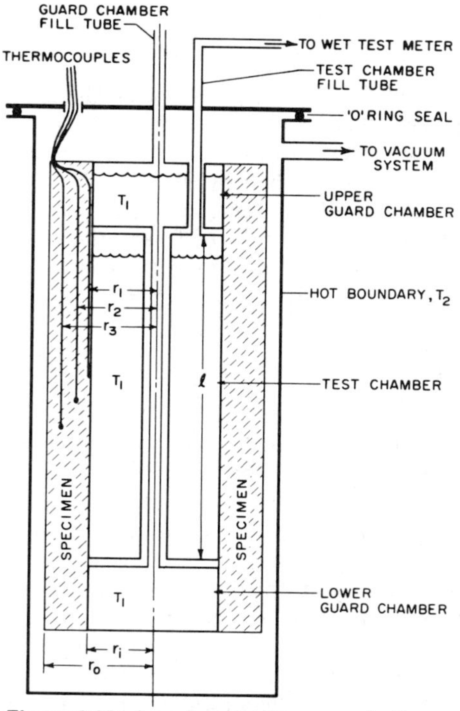

Figure 4.13 Low-temperature guarded insulation calorimeter (Kropschot, 1960).

thermal conductivity measurements. Nevertheless, for completeness, typical thermal diffusivity apparatus are briefly described here. These apparatus are divided into two categories according to the time dependence of the input energy: transient and periodic. In the transient method, a step function or energy pulse is applied to one end of the specimen, and the resultant temperature–time waveform is analyzed at one or more points along the specimen to obtain the diffusivity. In the periodic method, the applied energy is normally sinusoidal in time.

Figure 4.14 illustrates schematically the long-rod periodic apparatus. The temperatures of the two thermometers at T_1 and T_2 are plotted with an $X-Y$ recorder to yield the ellipse shown in Fig. 4.15. The parameters A_x, A_y, and ϕ are determined and yield the thermal diffusivity according to

$$\alpha = \pi\Delta x^2/[T\phi \ln(A_x/A_y)] \quad (4.7)$$

where ϕ is the phase angle between the x and y displacements and A_x/A_y is the amplitude decrement. The phase angle is determined from a measurement of the amplitudes, A_x and A_y, and the intersection distances, P_x and P_y, according to

$$\begin{aligned}\phi &= \text{arc sin } (P_x/A_x)\\ &= \text{arc sin } (P_y/A_y)\end{aligned} \quad (4.8)$$

Danielson and Sidles (1969) discuss the effects of heat loss, placement of thermocouples, guarding, and important improvements made by Abeles, Cody, and Beers (1960).

Figure 4.16 illustrates a transient (pulse) technique, using a flash tube or laser energy source. The temperature of the rear face of the specimen is recorded as in Fig. 4.17, and diffusivity is calculated from

$$\alpha = 1.37\Delta x^2/(\pi^2 t_{0.5}) \quad (4.9)$$

where $t_{0.5}$ is the time at which the rear-face temperature is 0.5 of the maximum value. The maximum value can be determined only if the specimen is well insulated. Danielson and Sidles (1969) present a detailed discussion of this method and alternate procedures.

One can expect uncertainties from diffusivity measurements that are comparable to steady-state thermal-conductivity measurements. At high temperatures, however, thermal-diffusivity measurements are becoming more popular because the results are less dependent on heat-loss errors (Danielson and Sidles, 1969). Also, diffusivity data are now of greater interest because modern applications are frequently critically dependent on transient thermal behavior.

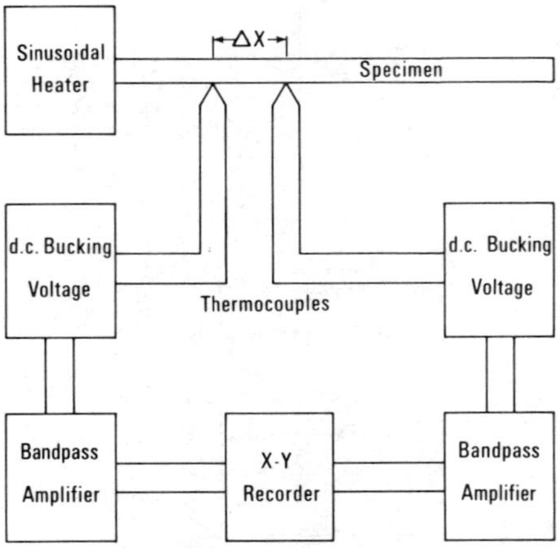

Figure 4.14 Block diagram of modified Angstrom thermal diffusivity apparatus (Danielson and Sidles, 1969).

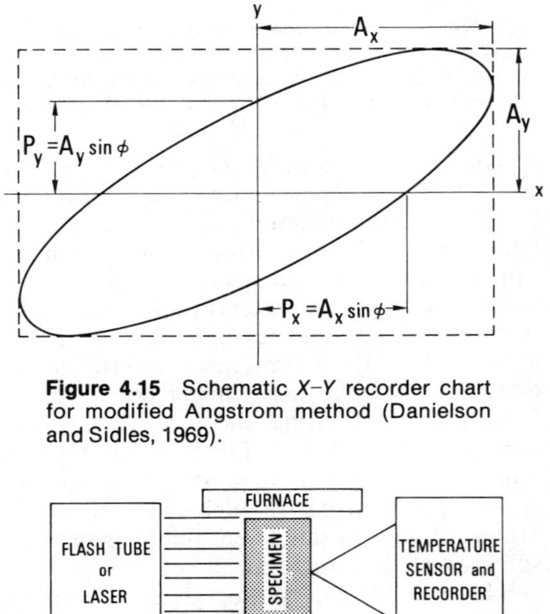

Figure 4.15 Schematic *X–Y* recorder chart for modified Angstrom method (Danielson and Sidles, 1969).

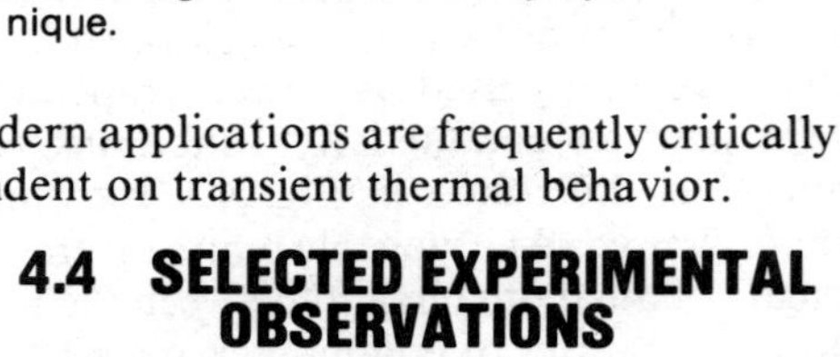

Figure 4.16 Block diagram of apparatus for measuring thermal diffusivity by flash technique.

4.4 SELECTED EXPERIMENTAL OBSERVATIONS

In § 4.2 (Basic Concepts) the results of theoretical considerations were presented to illustrate temperature dependencies and relative magnitudes predicted for various scattering mecha-

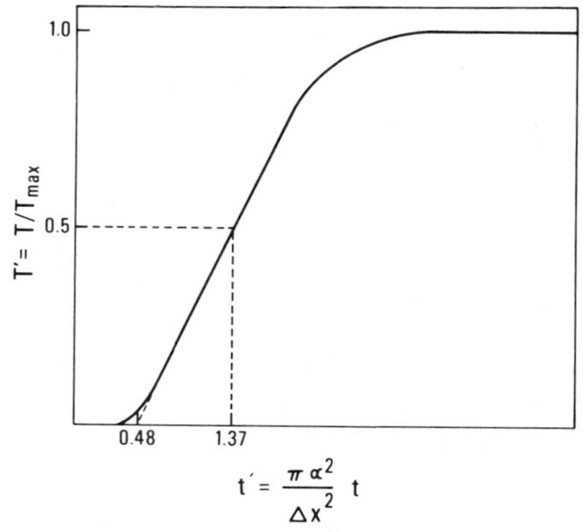

Figure 4.17 Reduced temperature of rear surface of specimen in flash diffusivity method.

nisms. It was pointed out that accurate quantitative magnitudes of these effects could not be calculated from first principles. Experimental examples are discussed here to illustrate specific magnitudes of these effects in various types of solids. The experimental results discussed are representative of the behavior of many other common materials.

4.4.1 Pure metals and dilute alloys

Experimental data for copper and aluminum are selected to illustrate the various scattering effects in pure metals and dilute alloys. Copper is chosen because more data exist for it than any other pure metal; aluminum, because the available data extend to the intermediate range between dilute alloys and highly alloyed structural materials.

Copper. Consider the error of using generic characterizations for pure metals. Figure 4.18 depicts the thermal conductivity of "oxygen-free, high-conductivity" copper as measured by two experimenters (Powers, Schwartz, and Johnston, 1950; Hust and Giarratano, 1974b). Powers et al. (1950) did not specify the residual resistivity ratio (RRR = ρ_{273K}/ρ_{4K}) of their specimen; however, from the measured thermal conductivity it can be estimated to be about 35. (See § 4.5 for further discussion of the significance of RRR values.) Hust and Giarratano (1974b) measured another specimen of copper but with a measured RRR of 214 and, therefore, a much higher thermal conductivity (Fig. 4.18). The RRR values of various specimens of such "oxygen-free" copper

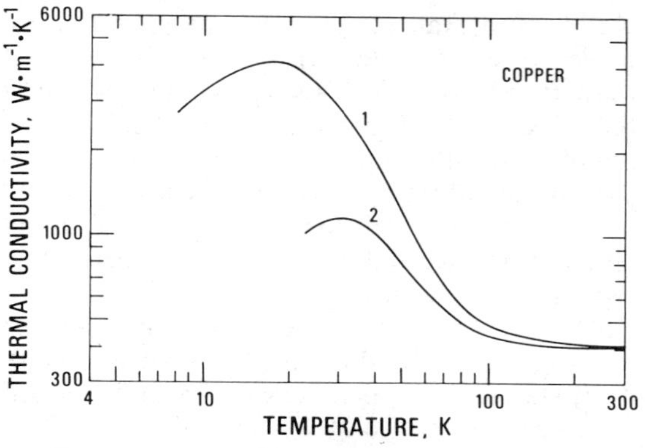

Figure 4.18 Thermal conductivity variation in oxygen-free, high-conductivity copper that is due to RRR variation, as observed by (I) Hust and Giarratano (1974b) and (2) Powers, Schwartz, and Johnston (1950).

have been measured at the National Bureau of Standards, and it was found that unless further restrictive characteristics are specified, the above wide-ranging RRR values can be expected. These differences in RRR are undoubtedly caused by small absolute variations in impurity and defect concentrations (but large relative variations).

The effect of annealing a drawn wire or, conversely, drawing an annealed wire is shown in Fig. 4.19. White (1953) measured an as-drawn wire (99.999% pure) and the same wire after annealing at 550° C for 3 h. The data by Powell,

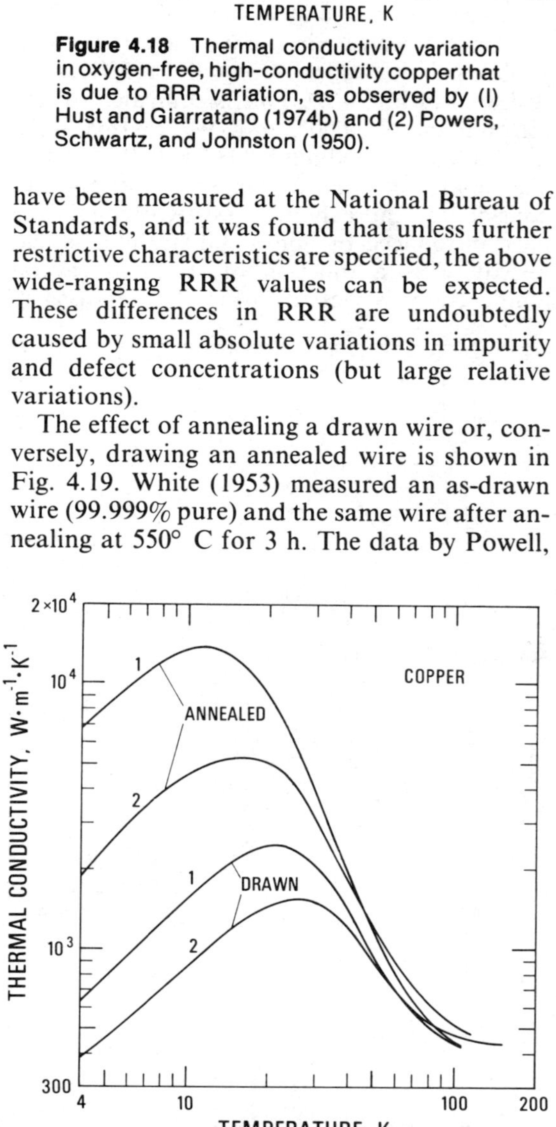

Figure 4.19 Thermal conductivity changes in pure copper induced by drawing and by annealing. 1 —Powell, Roder, and Hall (1959); 2 — White (1953).

Roder, and Hall (1959) were taken on an annealed wire (2 h at 400° C) and then after a 26.4% reduction in area by drawing. About a factor of ten reduction in low-temperature thermal conductivity resulted. In White's case, a thermal resistivity of 0.0011 m·K·W^{-1} at 4 K was removed by the anneal, and conversely, Powell et al. (1959) introduced 0.0014 m·K·W^{-1} of thermal resistivity by drawing.

The effect of strain on a copper wire was also investigated by Gladun and Holzhauser (1964). They started with an annealed wire and strained it successively up to 30.8% in elongation. This produced a decrease in conductivity from about 2750 W·m^{-1}·K^{-1} to 900 W·m^{-1}·K^{-1} at 20 K, as illustrated in Fig. 4.20. This is equivalent to the introduction of 0.00075 m·K·W^{-1} thermal resistivity at 20 K and about 0.0037 m·K·W^{-1} at 4 K. The elongation limit of copper is near 50%, so considerably more thermal resistivity could have been introduced by additional strain. Thermal resistivity as a function of strain at several temperatures is plotted in Fig. 4.21. It appears from these data that as much as 0.005 m·K·W^{-1} could be introduced by plastic deformation at 4 K.

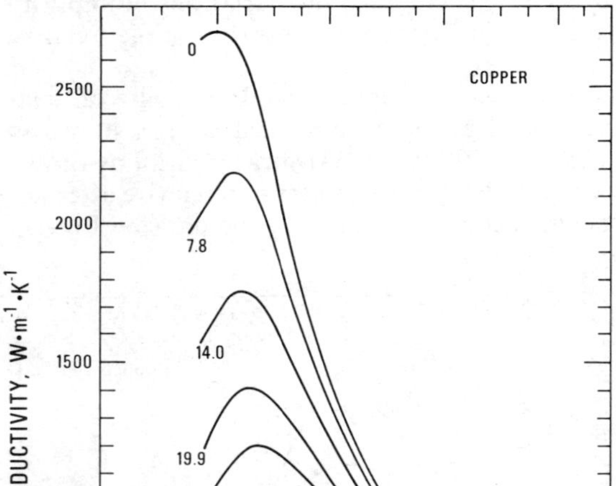

Figure 4.20 The thermal conductivity of copper as a function of temperature at constant tensile strain (Gladun and Holzhauser, 1964).

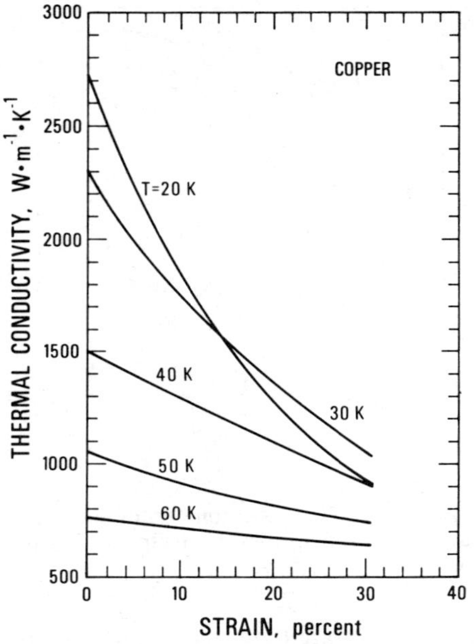

Figure 4.21 The thermal conductivity of copper as a function of tensile strain at constant temperature (Gladun and Holzhauser, 1964).

Bowman, Ziebold, and Smith (1969) investigated the effect of neutron irradiation on the thermal conductivity of copper at low temperatures. Their results are shown in Fig. 4.22. They used neutrons with energies up to about 10^7 eV and a total neutron exposure of 6.5×10^{23} nm^{-2}. The damage, as measured by increased thermal resistivity, was linear with neutron exposure. The maximum thermal resistivity introduced was 0.0038 m·K·W^{-1} at an exposure of 6.5×10^{23} nm^{-2} and a temperature of 4 K. This is nearly the same as that introduced by the 30.8% elongation of Gladun and Holzhauser (1964). Bowman et al. (1969) point out that this value would have been higher if the irradiation were done at 4 K instead of above 300 K, as some annealing occurred during irradiation and prior to measurement.

Sparks (1975) measured the effect of magnetic field up to 8 T on two specimens of copper: a high-purity specimen with an RRR of about 1500 and a lower-conductivity specimen with an RRR of about 107. His results are shown in Figs. 4.23 and 4.24. The magnetothermal resistivity introduced into the high-conductivity specimen exhibited a strong temperature dependence

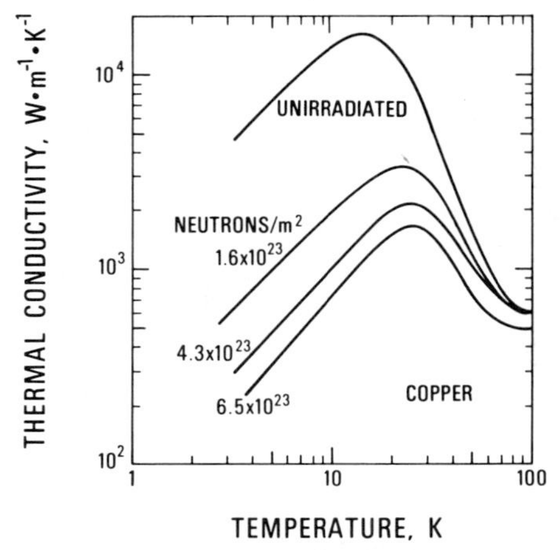

Figure 4.22 Thermal conductivity of copper as a function of neutron exposure; irradiation performed at a temperature of 300 K (Bowman et al., 1969).

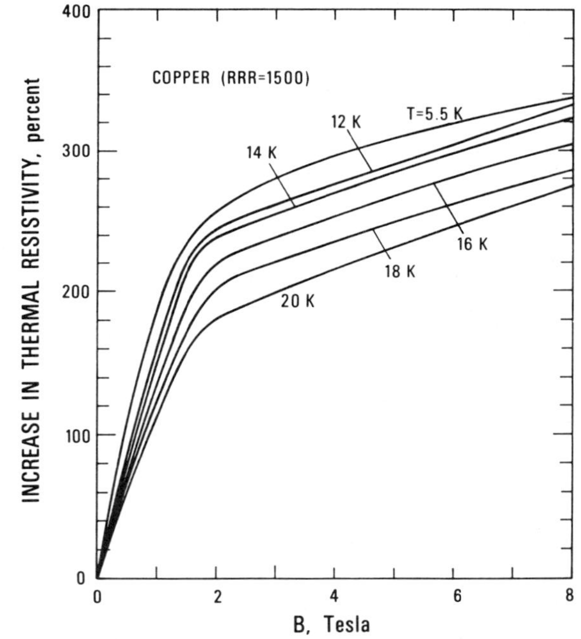

Figure 4.23 Relative change in thermal resistivity of a high-purity copper (RRR = 1500) as a function of longitudinal magnetic field at low temperatures (Sparks, 1975).

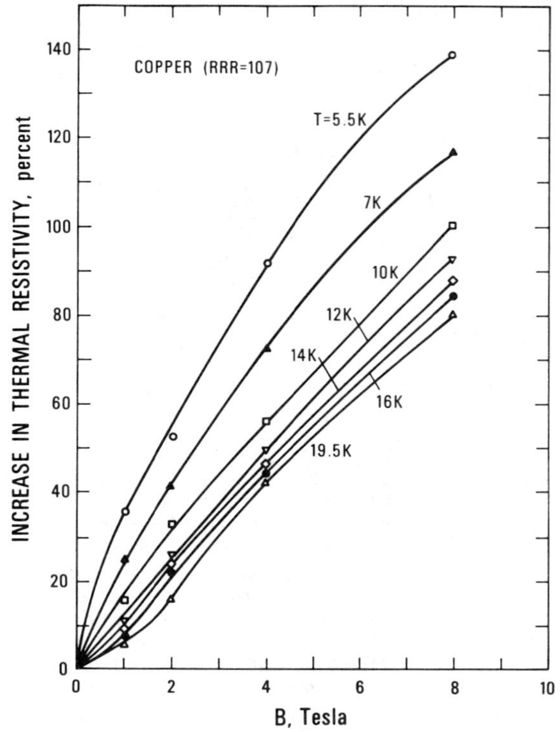

Figure 4.24 Relative change in thermal resistivity of an oxygen-free copper (RRR = 100) as a function of longitudinal magnetic field at low temperatures (Sparks, 1975).

below a field of 2 T. Above 2 T the temperature dependence was much less. The value of magnetothermal resistivity at high fields was about 0.0004 $m \cdot K \cdot W^{-1}$ near 4 K. Up to 8 T the lower-conductivity specimen exhibited a more uniform behavior, and the magnetothermal resistivity introduced near 4 K was appreciably higher, 0.002 $m \cdot K \cdot W^{-1}$. The relative change in thermal resistivity, however, is greater for the high-purity specimen at a given field. Fevrier and Morize (1973) reported measurements on two specimens of pure copper at fields up to 5 T at 4.2 K. The maximum field induced a factor-of-four increase in resistivity. Nevertheless, the ratio of the thermal and electrical conductivities was reported to be a constant for both longitudinal and transverse magnetic fields.

These numerous data on copper may create a chaotic picture, rather than an ordered one as previously indicated. To confirm that these data do indeed have a common denominator, Fig. 4.25 is presented. The data included are those for which electrical residual resistivity ratio (RRR) data are available. The RRR characterization parameter is a measure of the electronic purity of the specimen. It measures the combined effect of chemical and physical impurities and is discussed more fully in § 4.5. In Fig. 4.25 each of the experimental curves is labeled according to the measured value of RRR. The uniform progression

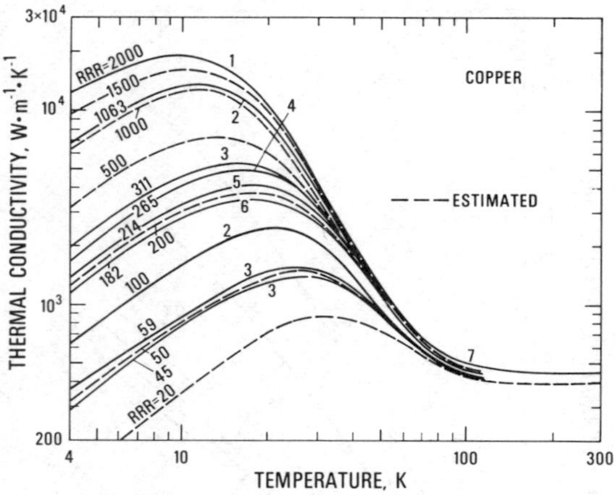

Figure 4.25 Thermal conductivity of copper with residual resistivity ratios from 20 to 2000. The estimated curves are obtained by interpolating the experimental family of curves and application of the W-F-L law (see later sections). Each of the measured specimens contain less than 0.1% impurity. 1 — White and Tainsh (1960); 2 — Powell, Roder, and Hall (1959); 3 —White (1953); 4 — Berman and McDonald (1952); 5 — Hust and Giarratano (1947); 6 — Mendelssohn and Rosenberg (1952); 7 — Moore, McElroy, and Graves (1967).

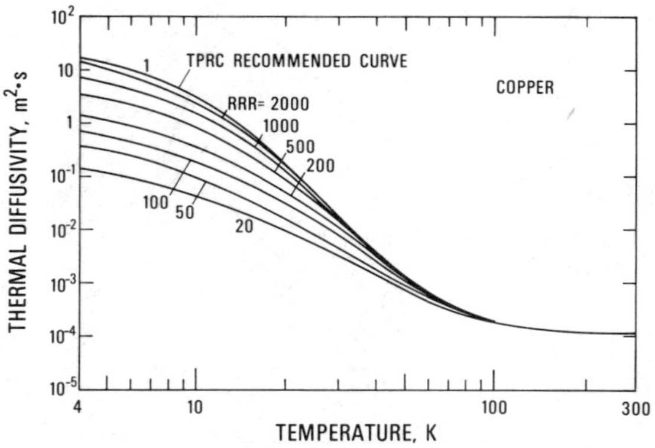

Figure 4.26 Thermal diffusivity curves derived for copper with residual resistivity ratios from 20 to 2000. The curves labeled RRR = 20 through 2000 were calculated from $\alpha = k/(\rho \cdot C)$ using values of k from Fig. 4.25 and values of C from Touloukian and Buyco (1970).

from high RRR values for the higher conductivity specimens to low RRR values for the lower conductivity specimens is obvious. Each of the measured specimens contains less than 0.1% impurity. The dashed curves are estimated on the basis of the above progression in RRR and the predictive techniques discussed in § 4.5.

The corresponding curves for the thermal diffusivity of copper, characterized by RRR, are illustrated in Fig. 4.26. In this case, very few experimental data exist. The predicted curves have been obtained through the use of the thermal conductivity curves from Fig. 4.25, values of specific heat from Touloukian and Buyco (1970), and Eq. 4.3. Note that the specific heat is nearly independent of the imperfections that cause these large variations in the thermal conductivity and thermal diffusivity. The large variations in thermal diffusivity, therefore, are a direct consequence of the large variations in thermal conductivity at low temperatures.

Aluminum. To illustrate further the order that often exists in thermal conductivity behavior, data for aluminum and its alloys are presented in Fig. 4.27. This family of curves includes

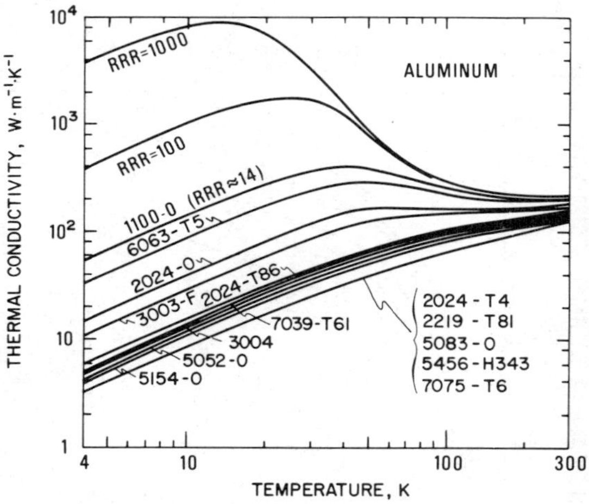

Figure 4.27 Thermal conductivity of aluminum alloys at temperatures from 4 to 300 K.

high-purity aluminum as well as some aluminum-based structural alloys. The RRR value is not shown for the structural alloys, since it loses its significance as a thermal characterization parameter at very low values. As can be seen, this class of materials forms a very consistent set of curves. The effects of material inhomogeneity and heat treatment, however, have not been extensively studied for the lower conductivity alloys.

Thermal diffusivity data can be predicted with reasonable accuracy from Eq. 4.3, the data in Fig. 4.27, and specific heat values from the literature. These predicted curves are given in Fig. 4.28.

As in pure copper, strong magnetic field effects are observed in pure aluminum. For example, Sirota, Gostishchev, and Drozd (1972) reported low-temperature thermal conductivity data at fields up to about 5 T. At zero field and a temperature of 20 K, their specimen had a thermal conductivity of 6500 $W \cdot m^{-1} \cdot K^{-1}$. At a field of 5 T the conductivity of the same specimen dropped to about 1800 $W \cdot m^{-1} \cdot K^{-1}$. Amundsen and Sovik (1970) reported measurements on two single-crystal specimens of pure aluminum at fields up to 13 T and temperatures below 4 K. A maximum increase in resistivity of more than 1.5 times was induced by the magnetic field, but the ratio of thermal-to-electrical conductivity was essentially independent of field (less than 10% variation).

Although it is clear that these selected results are not applicable to all pure metals, it is expected that they are applicable to many simple common metals. In these cases, valuable predictions of magnetothermal resistances can be made from magnetoelectrical resistance measurements. Changes in electrical resistance caused by magnetic fields are more fully discussed in Chapter 5.

4.4.2 Nondilute alloys

To illustrate the progression from pure metals to highly alloyed and disordered alloys, two additional classes of materials are discussed: (a) those that are composed principally of two metallic elements (binary alloys) and (b) those that contain more than two elements and are common engineering metals (structural alloys).

Binary alloys. Thermal conductivity data for a series of Cu–Zn alloys are illustrated in Fig. 4.29 (Lomer, 1958; Rosenberg, 1963). At 100 K, these data combined with other Cu–Zn alloy data (Kemp, Klemens, and Tainsh, 1957b) result in curve A of Fig. 4.30. If a linear law of mixtures were valid for alloys, curve B would be obtained. Nevertheless, electrical resistivity data and room-temperature thermal-conductivity data on other disordered alloys indicate that the deviation from line B, ΔW, should be of the form $Ac(1-c)$, where A is a proportionality constant and c is the relative amount of zinc in copper. Adjusting the constant A to fit ΔW at 30% Zn, curve C is obtained. The experimental values differ only slightly from this curve in the vicinity of 10% Zn. Adding curve C to curve B allows the extension of curve A to 100% Zn.

The formation of ordered lattices of copper and zinc within the alloy may have a significant effect on the transport property values. This normally decreases the scattering of energetic heat carriers and results in significantly decreased

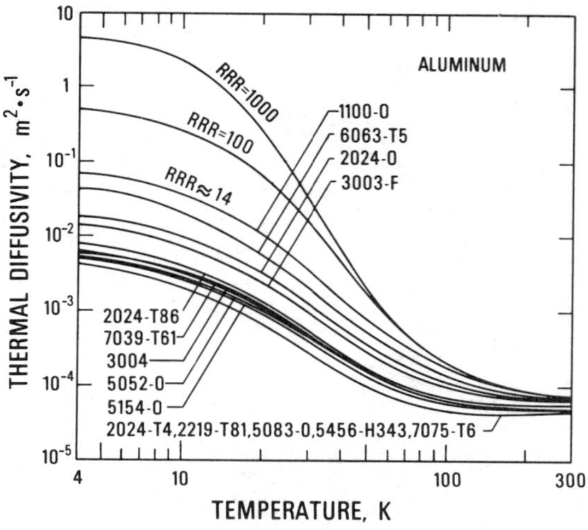

Figure 4.28 Thermal diffusivity of aluminum alloys at temperatures from 4 to 300 K.

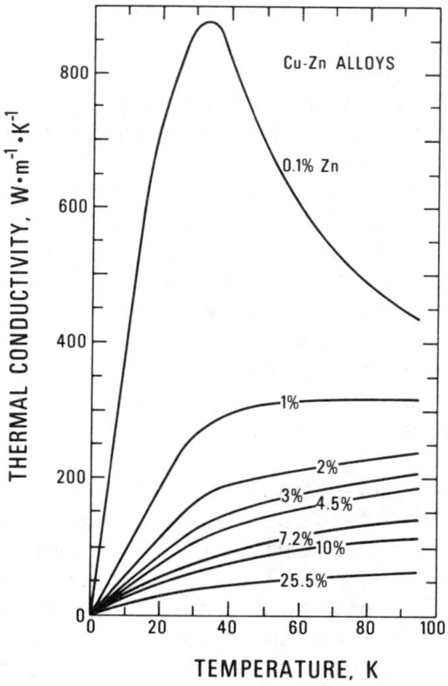

Figure 4.29 The thermal conductivity of Cu–Zn alloys (Rosenberg, 1963; Lomer, 1958).

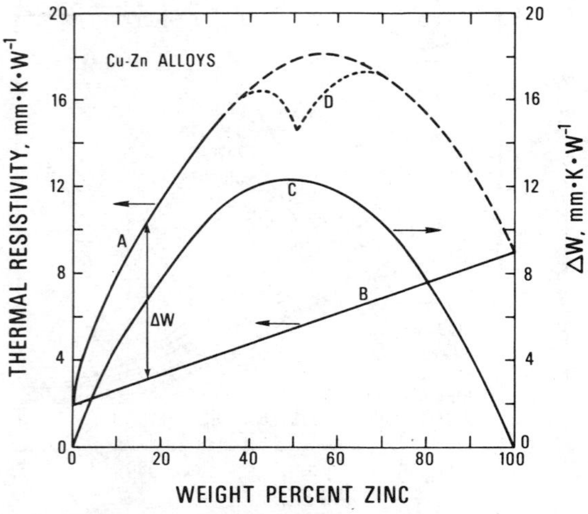

Figure 4.30 Thermal resistivity of Cu–Zn alloys at 100 K.

thermal resistivity. The dashed line D in Fig. 4.30 illustrates the typical effect of such ordering. Ordering may occur at several compositions for a given binary alloy system. Although one must be aware of the possibility of ordering in binary alloys, the conditions necessary for ordering do not commonly occur in engineering materials.

A few binary alloy systems have been investigated to determine the effects of anneal and composition. The effects of anneal are qualitatively the same as for pure metals and dilute alloys. The relative effect, however, is not as great because of the increased scattering by solute atoms and the concomitant decreased relative significance of physical defects.

Structural alloys. The behavior of some of the more common classes of structural alloys is now considered, beginning with the more well-behaved materials. Data for structural aluminum alloys have been discussed, and as noted, this class forms a very consistent set of curves.

The austenitic stainless steels, including AISI 304, 310, and 316, form another closely defined class of materials (see Fig. 4.35). In this case, the entire data set is the same to within about $\pm 10\%$ at a given temperature. Individual curves are not illustrated, since the differences in thermal properties for a given steel caused by material variations are comparable to those between different steels. Martensitic steels are more variable but are normally not considered for use at low temperatures because of their low toughness.

Brasses have been studied extensively and form a consistent family of curves similar to the aluminum alloys. This can be readily observed in the previously mentioned data compilations.

The final class of metals to be mentioned, although by no means the least important, are the high-strength superalloys. These materials, commonly used in aerospace applications, include nickel-base alloys. The thermal conductivities for nickel-base superalloys are illustrated in Fig. 4.31. As can be seen, they are low-conductivity metals and their temperature dependencies are quite at variance with one another. Also, the thermal properties of the nickel-base alloys exhibit a relatively strong dependence on heat treatment. Titanium-base alloys behave similarly.

These classes of alloys all have similar temperature dependencies (monotonically increasing with temperature and approximately linear at low temperatures). The aluminums and brasses have the highest conductivities. The nickel and titanium alloys are slightly lower, and the stainless steels fall at the bottom of this spectrum (see Fig. 4.35).

4.4.3 Nonmetals

This class of materials encompasses a vast number of materials such as elemental dielectrics, molecular crystals, ionic and valence crystals, glasses, polymers, disordered dielectrics, and minerals. One cannot hope to discuss all of these or even to give meaningful selective examples of each in this work. The examples chosen for discussion are those that vividly illustrate predicted theoretical behavior and those that are most pertinent to common engineering materials. It is important to remember that, in general, pure, crystalline nonmetals will exhibit the

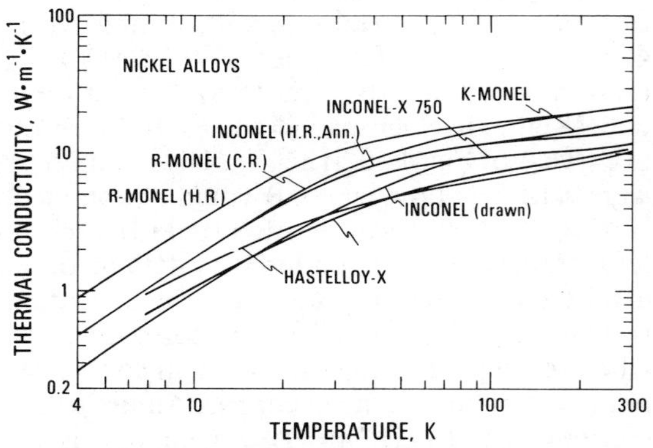

Figure 4.31 Thermal conductivity of high-temperature nickel alloys.

highest conductivities at low temperatures. Conversely, impure or disordered nonmetals will exhibit the lowest conductivities.

Crystalline dielectrics. For illustrative purposes, the most interesting dielectric solids are crystals, such as diamond, sapphire (aluminum oxide), and beryl (beryllium oxide). Although these materials have not been used extensively in cryogenic applications, they are interesting from a scientific standpoint and have been well studied. The results, too numerous to repeat here, illustrate the effects of smooth and rough specimen boundaries, grain boundaries, mass defects, irradiation damage, heat treatment, strain, and crystal orientation (anisotropy), among others (see Berman, 1976).

A single source of data that illustrate several features of phonon scattering in pure dielectric is presented by Berman (1976). These data for Al_2O_3 are shown in Fig. 4.32. They exhibit a T^3 dependence at low temperatures, indicating that the conductivity is limited by boundary scattering. The curves are grouped into two sets in this temperature region according to the surface finish. According to Berman, near the maximum the curves divide into three groups corresponding to the production batch, indicating differences in specimen defects. Above the peak, the curves decline exponentially and merge, depicting the intrinsic behavior of Al_2O_3 for this temperature range.

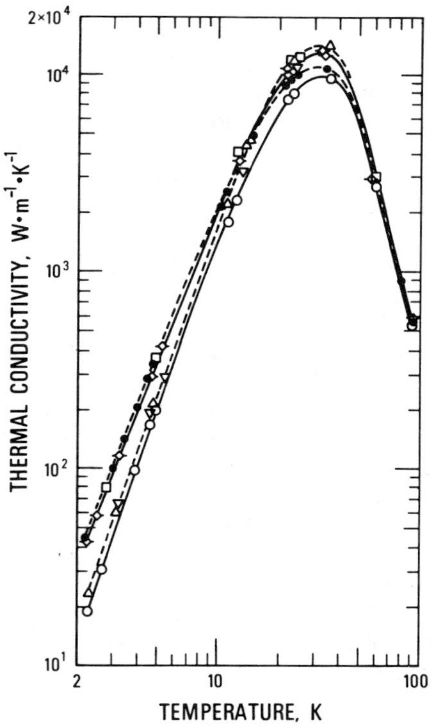

Figure 4.32 The thermal conductivity of Al_2O_3 crystals with rough and smooth surfaces. □, ●, -◇- = specimens with smooth surfaces; ○, △, ▽ = specimens with rough surfaces. ● = batch a; ○ = batch b; □, -◇-, △, ▽ = batch c (Berman, 1976).

Disordered dielectrics. Similar to highly disordered structural alloys, disordered dielectrics have low conductivities, are difficult to predict, but generally vary monotonically with temperature. Two very important classes of dielectrics are glasses and polymers. Figure 4.33, from Childs et al. (1973), illustrates the behavior of several glasses. Note that at the lowest temperature the thermal conductivity of these glasses is nearly independent of temperature, whereas at higher temperatures, it varies approximately linearly.

Figure 4.34 illustrates the behavior of several polymers. Their thermal conductivities are slightly lower at low temperatures than those of glasses but vary more rapidly with temperature and are more widely dispersed. This spread of values is caused to some extent by the degree of crystallinity. Polymers can exhibit 100% crystallinity after heat treatment. At higher temperatures, the conductivities of polymers and glasses are more similar.

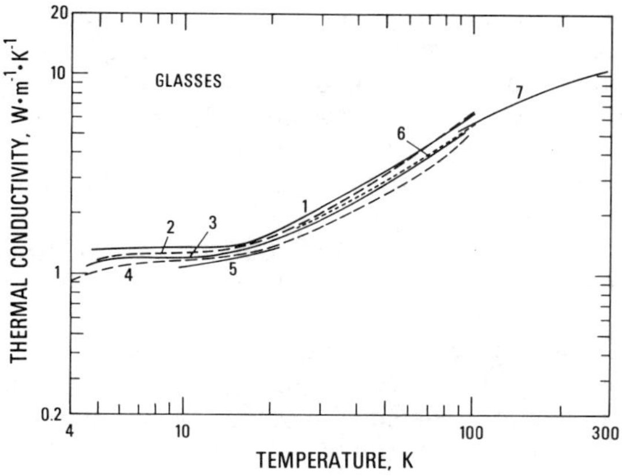

Figure 4.33 Thermal conductivity of several glasses from 4 to 300 K (Childs et al., 1973). 1 — quartz-1; 2 — quartz-2; 3 — quartz-3; 4 — Phoenix-1; 5 — Phoenix-2; 6 — glass; 7 — Pyrex.

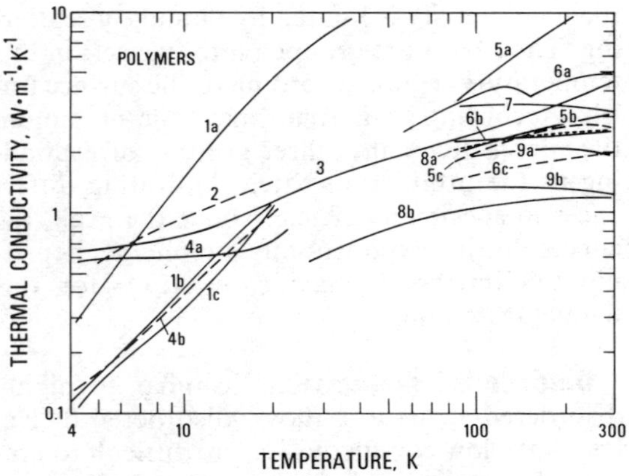

Figure 4.34 Thermal conductivity of selected polymer specimens from 4 to 300 K (Childs et al., 1973). la — polyethylene-1; lb — polyethylene-2; 2 — polytetrafluoroethylene-1; 3 — G. E. 7031 varnish; 4a — perspex plastic-1; 4b — nylon; 5a — glass-phenolic composite; 5b — glass-epoxy composite; 5c — glass-phenolic-resin composite; 6a — phenolformaldehyde; 6b — polytetrafluoroethylene-2; 6c — perspex plastic-2; 7 — polyglycol; 8a — polytetrafluoroethylene-3; 8b — polyethylene terephthalate; 9a — polyester urethane; 9b — polyisobatylene.

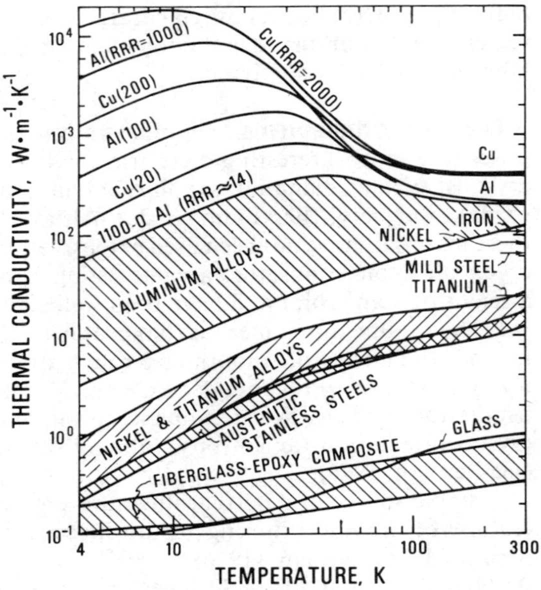

Figure 4.35 Thermal conductivity of selected classes of materials.

Structural composites. Structural composites are materials usually composed of strong fibers in a lower-strength and less expensive matrix. Generally, neither fiber nor matrix are metals, although metal-matrix and metal-fiber composites do exist. Composite applications at cryogenic temperatures are rapidly increasing. Glass-fiber polymer composites have received the most attention. Boron-fiber and graphite-fiber composites are also receiving considerable attention. The limited experimental thermal property data for composites were recently compiled by Kasen (1975). Further details and data on the thermal properties of composites are presented in Chapter 12.

In summary, Fig. 4.35 illustrates the relative magnitudes of the thermal conductivities of the materials just discussed, their temperature dependencies, and the effect of imperfections (as measured by RRR) on pure metals. The bands of values associated with certain classes of materials are also illustrated.

4.5 EMPIRICAL CORRELATIONS

Thermal transport measurements are difficult to perform accurately, time-consuming, and expensive. Therefore, from a cost-effective standpoint, it is desirable to be able to correlate thermal transport properties with other, more readily obtained characterization parameters. Measurements of these parameters allow one to check the accuracy and internal consistency of thermal measurements as well as to predict thermal properties in the absence of direct measurements. Such correlations are also useful to further our understanding of basic conduction and interaction mechanisms.

The correlations described in the following sections have been extremely useful in obtaining adequate design data. The accuracies of the various prediction techniques based on these correlations vary considerably. In some instances, thermal data can be predicted as accurately as they can be measured directly; in others, the predicted values are uncertain by more than 50%. For example, pure metal predictions based on simple electrical resistivity measurements are generally most accurate (near ± 5%). Dilute alloy predictions can be nearly as good. Predictions for highly alloyed structural metals using similar characterization techniques are not nearly as good. It should be noted that these "predictions" are based on the existing data base and, therefore, may be thought of as interpolations and extrapolations rather than first-principle predictions. Because of the absence of simple characterization parameters for nonmetals, predictions are generally more difficult.

4.5.1 Wiedemann–Franz–Lorenz law

Wiedemann and Franz (1853) observed that the thermal and electrical conductivities of metals at room temperature had a fixed ratio. Later Lorenz (1881) observed that this ratio was proportional to temperature. This relationship is referred to as the Wiedemann–Franz–Lorenz (W-F-L) law:

$$k/\sigma = k\rho = LT \qquad (4.10)$$

where σ is electrical conductivity, ρ is electrical resistivity, k is thermal conductivity, T is absolute temperature, and L is the proportionality constant called the Lorenz number.

Drude (1900) presented a theoretical derivation of the W-F-L law and obtained a value of $2.228 \times 10^{-8} V^2K^{-2}$ for L. Sommerfeld (1928) later used the more recent electron theory of metals to derive a more exact value for L. His value, commonly designated as L_0, $2.443 \times 10^{-8} V^2K^{-2}$, is generally used today. It should be pointed out that these theoretical values of L are based on electronic conduction, k_e. In this discussion L_0 denotes the particular value derived by Sommerfeld (1928), L denotes the experimental quantity $\rho k/T$, where k is total conductivity, and L_e denotes the corresponding observed electronic component, $\rho k_e/T$.

Since the original Wiedemann and Franz (1853) observations, many more metals have been measured. The values of L at room temperature for many pure metals range between 2.2 and $3.0 \times 10^{-8} V^2K^{-2}$. As the temperature is lowered, the experimentally determined values of L for pure metals decrease until temperatures near 0.1 of the Debye temperature, θ_D, are approached. At this point, L commences to rise again with further decrease in temperature. At the very lowest temperatures, experimentally observed values of L again tend toward the Sommerfeld value. The typical behavior of L for pure metals and defect-free metals is illustrated in Fig. 4.36. The depth and position of the minimum in the L vs. T/θ_D curve is determined by the effective electrical purity of the specimen as indicated by its residual resistivity. As the degree of imperfection increases in pure metals, the lattice component of thermal conductivity becomes relatively more significant and values of L tend to increase because no lattice component exists in the electrical conductivity. As the electronic component decreases in magnitude compared with the lattice component, the close relationship between k and ρ gradually becomes more nebulous. However, in some materials that are primarily phonon thermal conductors (e.g., graphite), the electrons and phonons are scattered mainly by the same defect mechanism and an empirical relationship is still observed. Figure 4.37 illustrates the Lorenz ratio progression from aluminum alloys to highly alloyed high-strength structural metals. Values of Lorenz ratio for common metals have been compiled and illustrated by Hust and Sparks (1973).

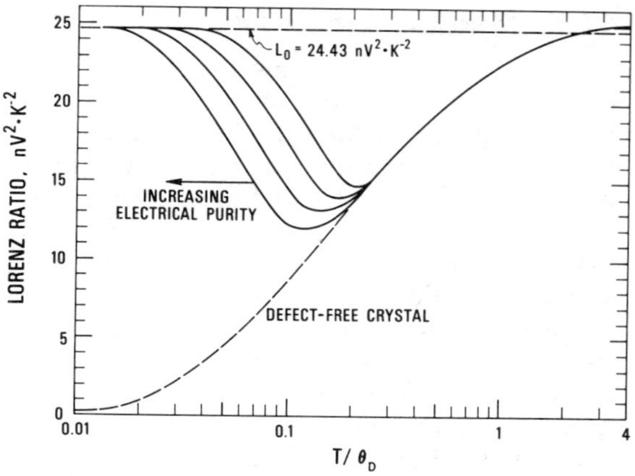

Figure 4.36 Lorenz ratio for pure metals and defect-free metals.

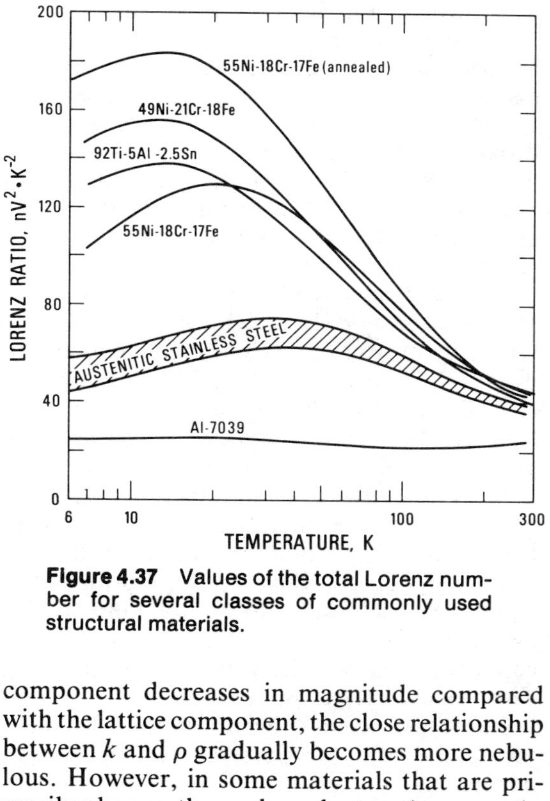

Figure 4.37 Values of the total Lorenz number for several classes of commonly used structural materials.

4.5.2 Matthiessen's rule

To apply the W-F-L law for predictions of thermal conductivity, one must know the Lorenz ratio and the electrical resistivity as functions of temperature. Here the concern is a simple method of estimating resistivity. For many pure metals and dilute alloys the resistivity can be estimated using Matthiessen's rule. This rule, expressed by Eq. 4.11, states that the total resistivity is the sum of two parts: the intrinsic resistivity, ρ_i, that characterizes the electron interactions with the phonons only, and the residual resistivity, ρ_0, that is caused by the electrons interacting with the chemical impurities and physical defects of the metal.

$$\rho(T) = \rho_i(T) + \rho_0 \tag{4.11}$$

The intrinsic resistivity is strongly temperature dependent because of the phonon dependence on temperature, whereas the residual resistivity is essentially a constant for a given specimen. The specific temperature dependence of $\rho_i(T)$ varies widely both with temperature and base material. Nevertheless, $\rho_i(T)$ is not significantly dependent upon composition changes for dilute alloys of a given base metal. Thus, if one knows $\rho_i(T)$ for a given base metal, $\rho(T)$ for other dilute compositions of that metal can be obtained from Eq. 4.11 after measuring only ρ_0. The value of ρ_0 is obtained by measuring $\rho(T)$ at low temperatures, where $\rho_i(T)$ is negligible (liquid-helium temperature, 4 K, is normally adequate). Values of $\rho_i(T)$ for several common base metals have been compiled by Meaden (1965) and Hall (1968).

A characterization parameter frequently needed to indicate the electronic purity of metals is the residual resistivity ratio (RRR). It is defined as ρ_{273K}/ρ_{4K}, where ρ_{4K} for most metals is equal to ρ_0, the residual resistivity. Since ρ_{273K} is essentially fixed for a given generic metal (e.g., copper), the value of RRR is determined by ρ_{4K}. The value of RRR is, thus, inversely proportional to the defect concentration of the specimen and directly proportional to its electronic purity.

4.5.3 Low-temperature correlation

In an earlier section (§ 4.2.3) the electronic conduction in metals was discussed and is given by

$$k_e = (aT^2 + b/T)^{-1} \tag{4.12}$$

at low temperatures. For pure metals and dilute alloys, the lattice conductivity is generally small, so k_e can be taken as total conduction. Cezairliyan and Touloukian (1965) decided to utilize this equation in reduced form as the basis for a generalized low-temperature equation:

$$k^* = 3(T^{*2} + 2/T^*)^{-1} \tag{4.13}$$

where k^* is equal to k/k_m, T^* is equal to T/T_m, and T_m is the temperature at which k is a maximum, k_m. In analogy to reduced equations of state for thermodynamic properties of fluids, they refer to this as the principle of corresponding states. The reducing parameters chosen are the temperature and thermal conductivity at the maximum in conductivity. The results of this correlation are illustrated in Fig. 4.38. The figure represents 1002 data points for a total of 22 metals (83 specimens) in varying states of impurity and imperfection content. The standard deviation of the data from Eq. 4.13 is only 3.2%. Equation 4.13 is convenient to illustrate the correctness of the form of Eq. 4.12, but it is not the most convenient form for calculating values of $k(T)$. This is because k_m and T_m are not known in terms of the best characterization parameters for metals, which is the residual resistivity, ρ_0, or residual resistivity ratio, RRR. For actual $k(T)$ calculations, it is more convenient to refer back to the original Eq. 4.12. Cezairliyan and Touloukian (1965) have also shown that a slight modification of Eq. 4.12 produces even better agreement with experimental data:

$$k_e = (\alpha' T^n + \beta/T)^{-1} \tag{4.14}$$

where $\beta = \rho_0/L$ and $\alpha' = \alpha''(\beta/n\alpha'')^{(m-n)(m+l)}$; α'', m, and n are constants for a given base metal and are best determined by least squares fit to data at temperatures below 1.5 T_m. The value of n is between 2 and 3 for most metals. For convenience Eq. 4.14 is also written in the following forms:

$$k_e = (\alpha' T^n + \rho_0/LT)^{-1} \tag{4.15}$$

$$k_e = \{\alpha' T^n + \rho_{i273}/[LT(\mathrm{RRR} - 1)]\}^{-1} \tag{4.16}$$

where ρ_{i273} is the intrinsic resistivity at 273 K.

Best values of α', m, and n as obtained by Ho, Powell, and Liley (1974) for selected elements are given in Table 4.3. Using Eq. 4.15, the constants in Table 4.3, and a specific value of ρ_0, one can calculate values of $k(T)$ at temperatures up to about 1.5 T_m as accurately as is generally required. Since T_m for many elements is about 25 K, the method is limited to temperatures below

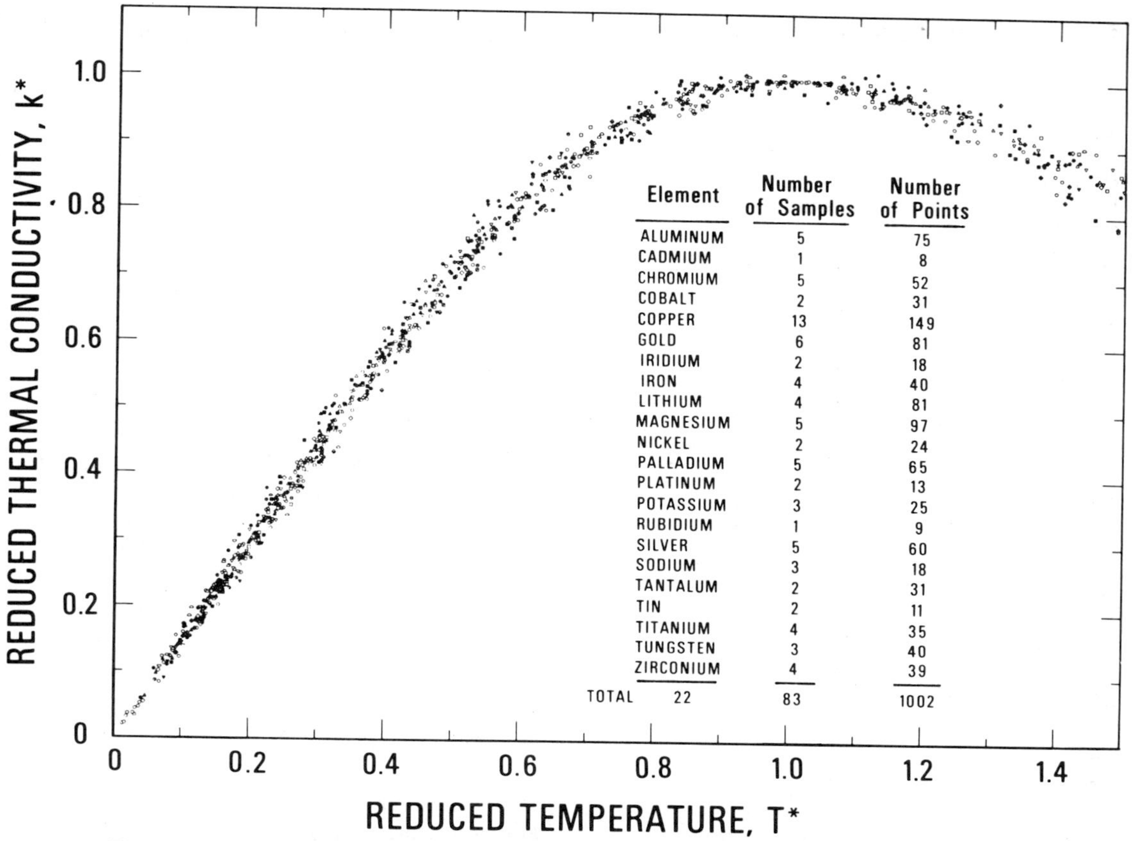

Figure 4.38 Reduced thermal conductivity vs. reduced temperature of 22 metallic elements. The k^* is k/k_m where k_m is the maximum value of k and T^* is T/T_m where T_m is the temperature corresponding to k_m (Cezairliyan and Touloukian, 1963).

about 40 K. Cezairliyan and Touloukian (1965) also presented a higher temperature correlation method based on the value of k at the Debye temperature. Using these two correlation techniques, they were able to calculate families of $k(T)$ curves for metallic elements such as those illustrated in Fig. 4.3. Such presentations for the common metallic elements would be most useful and are expected to be more commonly available soon.

4.6 PREDICTIVE TECHNIQUES

As can be seen from this review, thermal transport properties are influenced by many factors: electronic and lattice conductivity, intrinsic scattering mechanisms, physical defect and chemical impurity scattering mechanisms, boundaries, and magnetic fields. It is clear that prediction of data is, at best, a difficult task. A systematic division into temperature ranges and classes of materials makes this task more tractable. Generally the temperature ranges of interest are: $T << \theta_D$, $0.2\theta_D < T < \theta_D$, and $T > \theta_D$. The materials are classified as (a) pure metals and dilute alloys, (b) structural or highly alloyed metals, and (c) nonmetals.

4.6.1 Pure metals and dilute alloys

Pure metals and dilute alloys are generally the best known of all classes of materials, principally because more measurements have been performed on them and because of the great value of residual resistivity as a characterization parameter. To predict $k(T)$ of a material from this class, it is essential to know its residual resistivity, ρ_0, or residual resistivity ratio, RRR. Then one can make accurate predictions from either the Lorenz ratio approach (§ 5.1) or the direct calculation method (§ 5.3). If only low-temperature data (below $0.2\theta_D$) are required, the direct calculation is to be preferred. At higher temperatures, the Lorenz ratio approach is better in the absence of

Table 4.3 Constants for low-temperature thermal conductivity calculations using Eq. 4.14.

Element	m	n	$\alpha'' \times 10^4$
Aluminum	2.62	2.00	0.0479
Cadmium			
(∥ to c-axis)	5.00	4.50	0.0468
(⊥to c-axis)	5.00	4.50	0.0468
(polycrystalline)	5.00	4.50	0.0468
Chromium	2.20	2.00	0.592
Cobalt	2.20	2.10	0.540
Copper	2.63	2.21	0.0423
Gallium			
(∥ to a-axis)	2.78	2.00	2.04
(∥ to b-axis)	2.78	2.00	0.806
(∥ to c-axis)	2.78	2.00	6.57
Gold	2.46	2.00	0.460
Indium	3.00	2.00	3.50
Iridium	4.40	3.00	0.000272
Iron	2.20	2.00	0.517
Lead	3.50	3.00	4.12
Lithium	2.25	2.00	0.774
Magnesium	2.10	2.00	0.627
Molybdenum	3.20	2.60	0.00967
Nickel	2.60	2.00	0.192
Niobium	2.00	2.00	6.21
Osmium	5.80	3.00	0.00000379
Palladium	2.40	2.00	1.54
Potassium	2.10	2.00	18.0
Rhenium	3.30	2.20	0.0656
Rhodium	3.00	2.80	0.0132
Ruthenium	5.80	2.60	0.00000321
Silver	2.75	2.20	0.0730
Sodium	2.13	2.00	2.89
Tantalum	2.54	2.00	1.39
Thallium	2.80	2.00	26.2
Thorium	2.80	2.79	1.75
Titanium	2.90	2.30	0.188
Tungsten	2.80	2.40	0.0539
Zinc	3.40	3.00	0.0750
Zirconium	2.40	2.00	3.99

adequate data. If a graph such as Fig. 4.25 is available, interpolated values will usually be sufficiently accurate.

4.6.2 Structural alloys

The prediction of thermal transport data for structural alloys is considerably more difficult than for pure metals. First, because the residual resistivity is no longer as useful as a characterization parameter, and second, because fewer experimental data exist. Probably the most reliable method of prediction is by associating the unknown with a similar class of materials according to composition, heat treatment, and grain structure. For example, austenitic stainless steels are a class of materials having very similar conductivities. On the other hand, high-strength nickel alloys show considerably more spread, as illustrated in Fig. 4.31. Furthermore, the Lorenz ratios of these materials also exhibit more spread, and therefore are not as useful either. Hust (1975) observed contradictory changes in the electrical and thermal conductivities of a nickel alloy upon heat treatment: the Lorenz ratio changed more than either thermal or electrical conductivity. Fortunately, the relative variations of thermal properties in structural alloys caused by heat treatment, fabrication, microstructure, and composition variations are not nearly so great as in pure metals.

4.6.3 Nonmetals

Nonmetals, like metals, are also conveniently divided into two groups: (a) pure crystalline dielectrics and (b) noncrystalline (or amorphous) and chemically impure dielectrics. The first case is like that of pure metals in that the thermal transport properties are strongly dependent on the specific characterization of the specimen. Contrary to the case of pure metals, there is no single useful characterization parameter, such as RRR, that indicates the aggregate effect of all the specific material variations.

The noncrystalline (or amorphous) and chemically impure dielectrics are similar to structural alloys in that a single characterization parameter, other than its generic identification, does not exist. Fortunately again, the transport properties of these dielectrics are not strongly dependent upon production variations. The only practical prediction method is by comparison with a chemically similar material.

4.7 SOURCES OF DATA

Documented measurements of the thermal transport properties of solids date to the mid–nineteenth century (Wiedemann and Franz, 1853). Since that time, thousands of measurements on numerous materials have been reported. This seemingly desirable situation has led, in fact, to considerable confusion because of (a) gross experimental errors and (b) the failure of experimenters and data users to recognize the significance of specimen characterization. The latter point is especially significant at low temperatures because of the strong dependence of thermal transport properties on impurity and structural variations of materials. These facts have become increasingly clear and more widely recognized as a result of recent large-scale compilation efforts (Touloukian et al., 1970; Touloukian et al., 1973; Childs et al., 1973). These compilations represent collections of most of the experimental data (including published papers and unpublished reports and corporate literature) up to about 1970. Not only are these valuable collections of widely dispersed data, but the resulting composite graphs of classes of materials reveal and emphasize the ordered and cohesive nature of the data rather than their disordered and chaotic nature. The composite graphs, along with the previously described correlations, are valuable for extrapolation and interpolation of the existing data base to unmeasured regions.

Considerable data are available for most technically important pure metals and their dilute alloys. Unfortunately, many of these data are not characterized in terms of the RRR of the specimen measured. In spite of this lack of complete characterization, sufficient data are available in most instances to adequately define the low-temperature thermal conductivity of these materials. Although thermal diffusivity data are not as extensive, the availability of specific heat, density, and thermal conductivity data allow one to readily calculate thermal diffusivity for the same materials, with only slightly increased uncertainties. In most instances, the data required for a specific material are not directly available in the literature but, by using the methods outlined in this chapter and the compiled experimental data, generally one can adequately predict values for a properly characterized material. The most useful, recent, and comprehensive sources of compiled data are: (a) Thermal conductivity — Childs et al. (1973), Ho et al. (1974), and Touloukian et al. (1970). (b) Thermal diffusivity — Touloukian et al. (1973). (c) Electrical resistivity — Hall (1968) and Meaden (1965). (d) Specific heat —Touloukian and Buyco (1970).

Primarily because of the lack of interest in structural alloys from a scientific standpoint, these materials have been measured less extensively than pure metals. This fact makes prediction of data for structural alloys doubly difficult: first, because of the innumerable variations of this class of materials, and second, because RRR is not as pertinent a characterization parameter. Data for the most common materials of this class are available; the effect of heat treatment and microstructure variations, however, is incomplete at best. If accurate data (better than 10 to 20%) are required for an unmeasured lot of material, they often can be obtained only by direct measurement of that lot. The available data are compiled in the sources listed for pure metals and dilute alloys above.

Compiled data can also be found for specific purposes in handbooks. These data must be used with care because material characterization is often sketchy. A recent handbook, produced and sold by the Metals and Ceramics Information Center at Battelle Columbus Laboratories (Columbus, Ohio 43201) under contract to the National Bureau of Standards with funding by the Advanced Research Projects Agency, is *Handbook on Materials for Superconducting Machinery* (1974), Order No. MCIC-HB-04; the first and second supplements are dated November 1975 and January 1977. It is restricted mostly to metals and alloys, but the second supplement includes composite materials data. Another recent handbook, produced by and available from the National Bureau of Standards (Boulder, Colorado 80303), is *LNG Materials and Fluids* (1977) (ed. D. Mann). The first supplement was issued during 1978.

Older handbooks that may be useful are: *Material Properties Data Handbook* (1970), *Cryogenic Materials Data Handbook* (1966), *Aerospace Structural Metals Handbook* (1966). These are available from National Technical Information Service (NTIS), 5285 Port Royal Road, Springfield, Virginia 22161.

Because experimentalists and theoreticians are constantly adding new information to the existing data base, one cannot rely on the above-mentioned compilations for the most recent design data. The newer data are generally more valuable because of the increased awareness of the importance of complete specimen characterization. One can obtain these recent data by searching the published and unpublished literature (including corporate literature), but this can be very time-consuming, even simply to update existing compilations. A more efficient method is to contact an active data center for computer searches of their files for the more recent data only. Data centers, such as the Thermophysical Properties Research Center at Purdue University (Lafayette, Indiana), provide this service at a modest cost.

NTIS collects abstracts from U. S. government-sponsored research by 240 agencies and their contractors back to 1964. Contact: National Technical Information Service.

SSIE provides summaries of current and recently completed (United States and foreign) research projects. Contact: Smithsonian Science

Information Exchange, Inc., Room 300, 1730 M Street, Washington, D. C. 20036.

COMPENDEX, a computerized index, provides abstracts of engineering work published in 3500 journals, reports and books back to 1970. Contact: Engineering Index, Inc., United Engineering Center, 345 East 47 Street, New York, New York 10017.

A single source of information specifically on thermal properties is the proceedings of the International Conference on Thermal Conductivity. This conference is held every two years (alternating with the European Thermophysical Properties Conference, which tends to emphasize high-temperature work). The proceedings of these conferences are now published by Plenum Press. The earlier proceedings were published by the host institution. Conferences have been held since 1961.

4.8 REFERENCES

Abeles, B., Cody, G. D., and Beers, D. S. (1960). Apparatus for the measurement of the thermal diffusivity of solids at high temperatures. *J. Appl. Phys.* **31**, 1585–1592.

Aerospace Structural Metals Handbook (1966) (eds. V. Weiss and J. G. Sessler). Syracuse University Research Institute, Syracuse, New York.*

Amundsen, T. and Sovik, R. P. (1970). Measurements of the thermal magnetoresistance of aluminum. *J. Low Temp. Phys.* **2**, 121–129.

Berman, R. (1976). *Thermal Conduction in Solids.* Clarendon Press, Oxford.

Bowman, H. F., Ziebold, T. O., and Smith, J. L., Jr. (1969). Low temperature thermal conductivity of copper irradiated with reactor neutrons. In *Thermal Conductivity — Proceedings of the Eighth Conference* (eds. C. Y. Ho and R. E. Taylor). Plenum Press, New York, 175–184.

Carslaw, H. S. and Jaeger, J. C. (1959). *Conduction of Heat in Solids.* Clarendon, Oxford.

Cezairliyan, A. and Touloukian, Y. S. (1965). Correlation and prediction of thermal conductivity of metals through the application of the principle of corresponding states. In *Advances in Thermophysical Properties at Extreme Temperatures and Pressure* (ed. S. Gratch). ASME, New York, 301–313.

Childs, G. E., Ericks, L. J., and Powell, R. L. (1973). *Thermal Conductivity of Solids at Room Temperature and Below*, NBS Monograph 131, National Bureau of Standards, Boulder, Colorado.

Cryogenic Materials Data Handbook (1966) (ed. F. Schwartzberg). Air Force Materials Laboratory, Wright-Patterson Air Force Base, Ohio.*

Danielson, G. C. and Sidles, P. H. (1969). Thermal diffusivity and other non-steady-state methods. In *Thermal Conductivity*. Academic Press, New York, 149–201.

Drude, P. (1900). The electron theory of metals. *Ann. Phys.* **1**, 566–613.

Fevrier, A. and Morize, D. (1973). The effect of magnetic field on the thermal conductivity and electrical resistivity of different materials. *Cryogenics* **13**, 603–606.

Gladun, C. and Holzhauser, W. (1964). Studies of heat conductivity at low temperatures. *Monatsber. Dtsch. Akad. Wiss. Berl.* **6**, 310–313.

Hall, L. A. (1968). *Survey of Electrical Resistivity Measurements on 16 Pure Metals in the Temperature Range 0 to 273 K.* NBS Technical Note 365, National Bureau of Standards, Boulder, Colorado, 1–111.

Handbook on Materials for Superconducting Machinery (1974). Supplements Nov. 1975 and Jan. 1977. Metals and Ceramics Information Center, Battelle Columbus Laboratories, Columbus, Ohio.

Ho, C. Y., Powell, R. W., and Liley, P. E. (1974). Thermal conductivity of the elements: a comprehensive review. *J. Phys. Chem. Ref. Data* **3**, Suppl. 1, 1–796.

Hust, J. G. (1970). Thermal anchoring of wires in cryogenic apparatus. *Rev. Sci. Instrum.* **41**, 622–624.

Hust, J. G. (1975). Thermal conductivity. In *Semi-annual Technical Report on Materials Research in Support of Superconducting Machinery*. NBSIR 75–810, National Bureau of Standards, Boulder, Colorado.

Hust, J. G. and Giarratano, P. J. (1974a). Thermal conductivity. In *Semi-annual Technical Report on Materials Research in Support of Superconducting Machinery-I.* NBSIR 74–359, National Bureau of Standards, Boulder, Colorado, 29–44.

Hust, J. G. and Giarratano, P. J. (1974b) Thermal conductivity. In *Semi-annual Technical Report on Materials Research in Support of Superconducting Machinery*, NBSIR 74–393, National Bureau of Standards, Boulder, Colorado, 1–36.

Hust, J. G. and Lankford, A. R. (1982). Comments on the measurement of thermal conductivity and presentation of the thermal conductivity integral method. *Int. J. Thermal Phys.* **3** (1), 67–77.

Hust, J. G. and Sparks, L. L. (1973). *Lorenz Ratios of Technically Important Metals and Alloys.* NBS Technical Note 634, National Bureau of Standards, Boulder, Colorado.

Kasen, M. B. (1975). Properties of filamentary-reinforced composites at cryogenic temperatures. In *Composite Reliability*. ASTM STP 580, American Society for Testing and Materials, Philadelphia, 586–611.

Kemp, W. R. G., Klemens, P. G., and Tainsh, R. J. (1957a). Lattice thermal conductivity of some copper alloys. *Aust. J. Phys.* **10**, 454–461.

Kemp, W. R. G., Klemens, P. G., and Tainsh, R. J. (1957b). The electrical and thermal conductivities of some brasses at low temperatures. *Acta Metall.* **5**, 303–309.

Klemens, P. G. (1958). Thermal conductivity and lattice vibrational modes. In *Solid State Physics*, Vol. 7 (eds. F. Seitz and D. Turnbull). Academic Press, New York, 1–98.

Kropschot, R. H. (1960). Cryogenic insulation, paper presented at ASHRAE Annual Meeting, Lake Placid, New York, June 22–24.

Laubitz, M. J. and McElroy, D. L. (1971). Precise measurement of thermal conductivity at high temperatures (100–1200 K). *Metrologia* **7**, 1–15.

LNG Materials and Fluids (1977) (ed. D. Mann); Supplement 1978. National Bureau of Standards, Boulder, Colorado.

Lomer, J. N. (1958). Ph. D. Thesis, Oxford University, Oxford, England.

Lorenz, L. (1881). On the thermal and electrical conductivities of metals. *Ann. Phys.* **13**, 422–447, 582–606.

Materials Properties Data Book (1970). Materials and Processes Section, Aerojet Nuclear Systems, Sacramento, California.*

Meaden, G. T. (1965). *Electrical Resistance of Metals*. Plenum Press, New York

Powell, R. L., Roder, H. M., and Hall, W. J. (1959). Low-temperature transport properties of copper and its dilute alloys: pure copper, annealed and cold-drawn. *Phys. Rev.* **115**, 314–323.

Powers, R. W., Schwartz, D., and Johnston, H. L. (1950). *The Thermal Conductivity of Metals and Alloys at Low Temperatures I. Apparatus for Measurements between 25 and 300 K. Data on Pure Aluminum, OFHC Copper and L Nickel.* USAF TR-264-5, Columbus Cryogenic Laboratory, Ohio State University, Columbus, Ohio 1–19.

Rosenberg, H. M. (1963). *Low Temperature Solid State Physics*. Clarendon Press, Oxford.

Sirota, N. N., Gostishchev, V. I., and Drozd, A. A. (1972). Thermal conductivity of aluminum in strong magnetic fields at low temperatures. *ZhETF Pis'ma Red.* **16**, 242–245.

Sommerfeld, A. (1928). The electron theory of metals on the basis of Fermi statistics. *Z. Phys.* **47**, 1–32.

Sparks, L. L. (1975). Magnetothermal conductivity of selected pure metals and alloys. In *Semi-annual Technical Report on Materials Research in Support of Superconducting Machinery-IV*. NBSIR 75-828, National Bureau of Standards, Boulder, Colorado.

Touloukian, Y. S. (1976). Thermophysical Properties Research Center, Purdue University, Lafayette, Indiana, private communication.

Touloukian, Y. S. and Buyco, E. H. (1970). *Specific Heat*, Vols. 4 and 5 of *Thermophysical Properties of Matter*. Plenum Press, New York.

Touloukian, Y. S., Powell, R. W., Ho, C. Y., and Klemens, P. G. (1970). *Thermal Conductivity*, Vols. 1 and 2 of *Thermophysical Properties of Matter*. Plenum Press, New York.

Touloukian, Y. S., Powell, R. W., Ho, C. Y., and Nicolaou, M. C. (1973). *Thermal Diffusivity*, Vol. 10 of *Thermophysical Properties of Matter*. Plenum Press, New York.

White, G. K. (1953). The thermal and electrical conductivity of copper at low temperatures. *Aust. J. Phys.* **6**, 397–404.

White, G. K. (1969). Measurement of solid conductors at low temperatures. In *Thermal Conductivity*. Academic Press, New York, 69–109.

Wiedemann, G. and Franz, R. (1853). The thermal conductivity of metals. *Ann. Phys.* **89**, 497–531.

ACKNOWLEDGMENTS

It is a pleasure to acknowledge the assistance of several associates in the preparation of this chapter. In particular, the comments and discussions of the reviewers, F. R. Fickett and L. L. Sparks of the National Bureau of Standards, Boulder, Colorado, and D. L. McElroy of Oak Ridge National Laboratory, Oak Ridge, Tennessee, are appreciated.

*Available from NTIS, Springfield, Virginia.

5 Electrical Properties

F. R. FICKETT

5.1 INTRODUCTION

The low-temperature electrical resistivity of materials is one of the most useful of the physical properties. From it one can determine much about the chemical purity and mechanical state of a material. In more esoteric investigations, resistivity measurements provide detailed information on the electronic structure and other aspects of the basic physics of metals. In several instances, knowledge of the resistivity allows prediction of other properties, such as thermal conductivity, that are less convenient to measure. Resistivity is easily measured to reasonable accuracy for technological applications (~1%), and even very high accuracy measurements can be made with a modest investment in equipment. As one might expect, the literature on electrical resistivity is massive in quantity and, generally, of excellent quality. Excellent tutorial treatments of all aspects of the subject are available.

In this chapter a condensed treatment of topics pertaining to resistance at cryogenic temperatures is presented: its measurement, the resistive mechanisms, and the data base. The emphasis is on metals and alloys of technological importance; a topic that covers an amazingly large range of materials. Similarly, the treatment of theory and of measurement techniques is intended for those interested in the more practical aspects of the subject. In every instance, however, references are given that allow the reader to pursue the subject at any level.

Some words are appropriate here regarding what is not covered. There is no discussion of superconductors, semiconductors, or most insulators. Theory and experiment relating to single crystals or thin films are mentioned only when the information leads to a better understanding of the technologically important metals. These basic topics are well described in the referenced texts.

No discussion of resistivity at low temperatures would be complete without special mention of the book by Meaden (1965), which is the most modern general treatment available, particularly concerning pure metallic elements. The data presented there and the extensive list of references covering the period to 1964 can save one many hours of labor with the original literature.

Handbook tabulations of resistivity data, even the most recent editions of handbooks, may be of ancient origin. These data may be very accurate, but many materials, particularly pure metals, are now produced by techniques resulting in significant increases in purity. In many instances this change produces dramatic low-temperature effects and is noticeable even in the room-temperature resistivity. Some tabulations reference the sources of data, but most do not. In the latter cases, it is best to view the information with skepticism.

Regarding the pure and commercially pure metals: there is no such thing as *a* low temperature resistivity for these metals. Changes in impurity and defect content, which are minor from an industrial standpoint, may cause dramatic changes in the low-temperature resistivity. For example, recent measurements of commercial copper satisfying the CDA102 specification (99.95% Cu* + Ag) have given resistivity values at 4 K that range from 3 to 34 nΩ·cm! This does

*Unless otherwise specified, percentages are in wt. %.

not represent a failing of the manufacturer; in each instance the copper is well within the specification. The message is clear: if the low-temperature resistivity is an important parameter for the job at hand, it must be measured on the specific material to be used.

To indicate the large range of values covered by the electrical resistivity, consider Fig. 5.1, which presents typical resistivity data as a function of temperature for a number of metals, and Fig. 5.2, which indicates the resistive behavior of a variety of copper products (Fickett, 1972a). In Table 5.1, ice-point values for the resistivity of several common metals and alloys are listed. The measurements are quite recent except for some of those referenced to Hall, which are "best values" derived from an extensive literature survey. In each case the materials are very well characterized in the references. A value of the residual resistivity ratio, RRR,* is also given in the table.

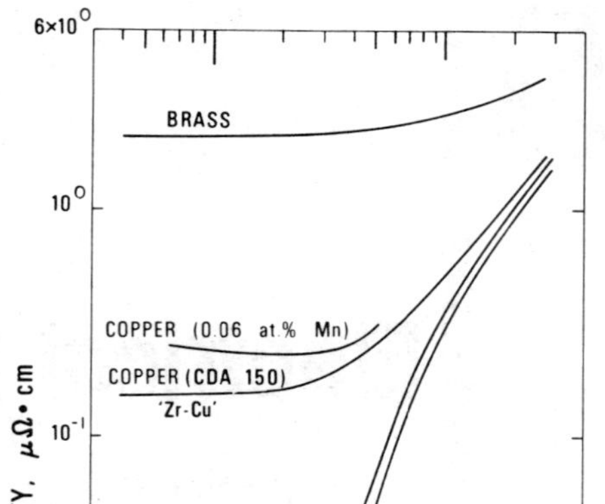

Figure 5.2 Resistive behavior of a number of "pure" coppers and copper alloys.

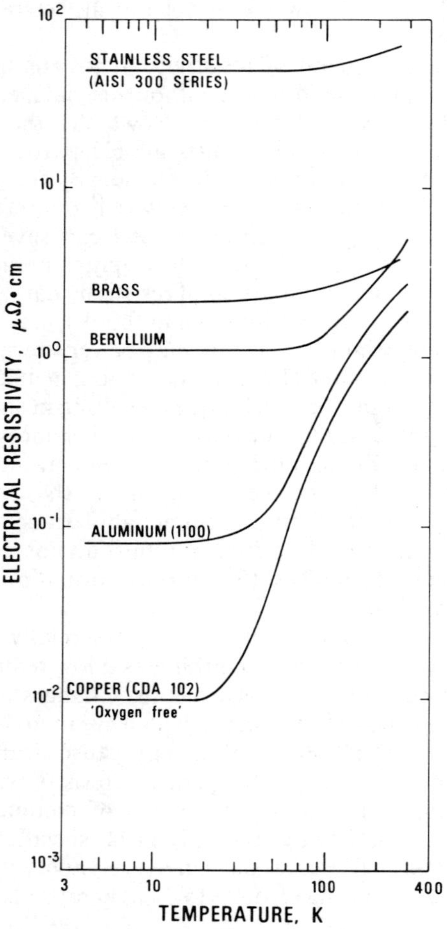

Figure 5.1 Resistivity of several common metals as a function of temperature.

Most of the figures presented here are prepared using data from the literature. Often it was necessary to extract data from small graphs, so that the exact numerical values shown here may not be very accurate. These curves are intended to indicate general behavior only; for accurate data, the referenced sources should be consulted.

*RRR — Residual Resistivity Ratio = $\rho(273\ K)/\rho(4\ K)$. In metals of commercial purity or better, the numerator depends essentially only on the thermal vibrations of the lattice and not on the impurities. The denominator depends only on the impurities for most metals, and thus the ratio is a sensitive indicator of purity; in fact, it is capable of detecting impurity levels far below those that can be seen by most analytical techniques. To first order, the ratio is independent of shape factor, i.e., it is equal to the ratio of the resistivities. In highly alloyed metals, the ratio is not as meaningful and is useful only in comparing samples of nominally similar composition.

Table 5.1 Resistivity of selected metals and alloys at the ice point.

Material	ρ ($\mu\Omega$·cm)	Reported RRR $\equiv \rho(273\ K)/\rho(4\ K)$	Reference
Pure metals			
Aluminum	2.428	40 000	Fickett (1971)
Beryllium[a]	2.70	75	Reich (1966)
Cobalt	5.24	60	Hall (1968)
Copper	1.553	12 000	Fickett (1972b)
Chromium	12.155	213	Hall (1970)
Gold	2.058[c]	300	Hust (1971)
Indium	8.00	13 000	Hall (1968)
Iron	9.07[c]	23	Hust and Giarratano (1975a)
Lead	19.2	15 000[d]	Hall (1968)
Magnesium[a]	4.12	285	Hedgcock and Muto (1964)
Manganese[b]	α 143.5	α 16	Hall (1970)
	β 91.0		
	γ 22.7		
Molybdenum	5.0	22	Hall (1968)
Nickel	6.23[c]	270	Laubitz, Matsumura, and Kelly (1976)
Niobium	13.96	1 500[d,e]	Hall (1968)
Platinum	9.60	755	Ibid.
Sodium	4.337	~ 500	Cook, Van der Meer, and Laubitz (1972)
Silver	1.48	2 000	Hall (1968)
Titanium[a]	39.4	33	Hall (1970)
Tungsten	4.94	80	Hust and Giarratano (1975b)
Tin[a]	10.1	63 000	Hall (1968)
Zinc[a]	5.45	1 400[f]	Babic, Ocko, and Rizzuto (1975)
Commercially pure metals			
Aluminum 1100	2.67	32	Clark, Childs, and Wallace (1970)
Copper CDA 102	1.56	97	Ibid.
CDA 150	1.73	11	Ibid.
Nickel TD	6.73	39	Ibid.
Commercial alloys			
Bronze, 90%	3.90	1.8	Ibid.
Naval Brass	6.32	1.6	Ibid.
Cu–Ni30	38.4	1.05	Ibid.
Stainless steels			
NBS-SRM	78.6	1.32	Hust and Giarratano (1975c)
AISI 304L	70.4	1.42	Clark et al. (1970)
AISI 316	75.0	1.39	Ibid.
AISI 321	73.9	1.38	Ibid.
AISI 410	55.9	1.46	Ibid.
Invar 36	79.5	1.58	Ibid.
Ti–6Al–4V	167.5	—	Ibid.

[a] Anisotropic resistivity observed in single crystals. Data are for polycrystalline specimens.
[b] Allotropic metal.
[c] Interpolated value.
[d] Superconducting above 4.2 K. Magnetic field applied.
[e] 4-K data from Keil, Merbold, and Diehl (1974).
[f] 4-K data from Boato, Bugo, and Rizzuto (1966).

5.2 EXPERIMENTAL TECHNIQUES AND CONSIDERATIONS

Resistivity is one of the easiest of all physical properties to measure. The basic techniques and apparatus are described in detail by Meaden (1965) and by Gerritsen (1956) as well as in many electrical measurements texts, such as those by Stout (1960), Gregory (1973), and Wolf (1973).

The Kelvin bridge method and the potentiometer method are two of the most useful classical measurement techniques. They both offer better than 1% accuracy in microohm-level measurements. Most of the textbook methods are quite time-consuming, and the approaches actually used in most laboratories are somewhat less sophisticated, since time is often of the essence and high precision is not usually required. Several of these easier techniques are described here; the more esoteric ones are left to the authors referenced above.

The equation from which the resistivity, ρ, is most often calculated is

$$\rho = RA/\ell = VA/I\ell \qquad (5.1)$$

where R is the resistance of the specimen, V is the voltage measured, I is the current, A is the cross-sectional area of the specimen, and ℓ is the length between the voltage probes. With reasonable care, both A and ℓ can be determined to 0.5% and, of course, to much higher precision in a carefully designed experiment. The usual technique of determining an average wire diameter for the calculation of A will lead to incorrect results if there are significant variations in diameter along the specimen. Most experiments are designed such that the voltages to be measured are on the order of microvolts or greater. This leads to practical specimen sizes and current levels as well as giving a voltage measurable with good accuracy with relatively inexpensive equipment. Furthermore, equipment such as this is easily adapted to the many modern computer control methods for automated operation.

5.2.1 Direct current measurements

A direct current (dc) system, which has been used successfully for routine measurements, is shown in Fig. 5.3. The current source can be anything from an automobile battery to a precision current calibrator. It should be stable over the time required for a measurement. The standard resistor is chosen primarily to have a sufficiently low resistance so that it will not suffer any significant heating at the highest currents. Allowable power dissipation limits are stated by the manufacturers — they are lower than one might expect, so it is well to check them. Reversal of the current is essential to eliminate thermoelectric voltages, a complication to be discussed shortly. The use of a commercial nanovoltmeter with its output read by a relatively inexpensive digital voltmeter provides good precision (~1%) with minimum cost and allows for greater flexibility in an environment where instruments must often be shared among several experiments. The use of a four-probe system for the leads (i.e., separate sets for current and voltage) is the best approach to low-level dc resistance measurement; two-lead systems introduce too many possible sources of error.

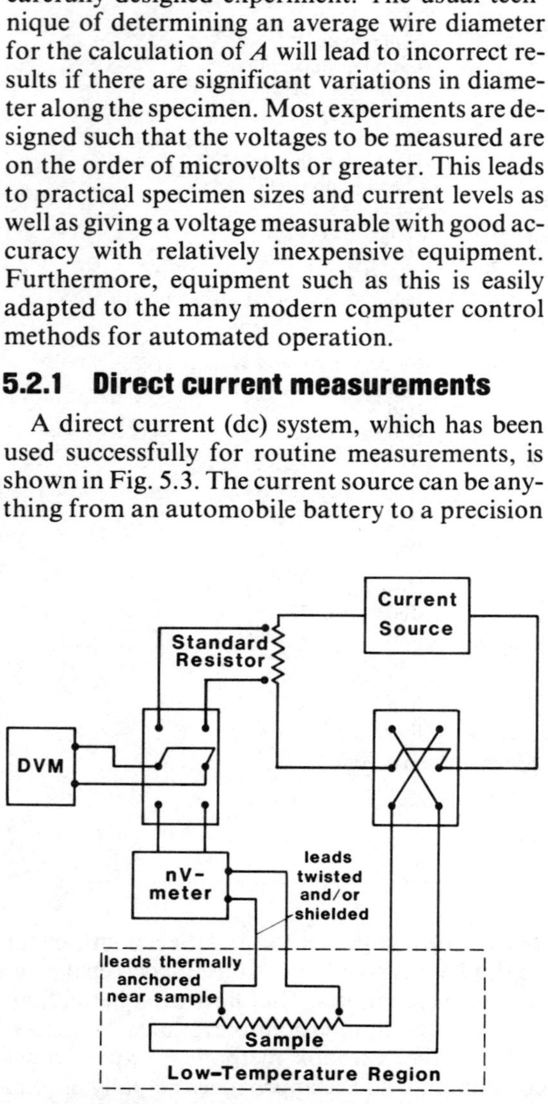

Figure 5.3 Block diagram of a system for low-temperature resistivity measurements.

5.2.2 Alternating current measurements

At first glance, alternating current (ac) techniques, particularly those involving modern lock-in amplifiers, seem to offer many advantages in low-temperature resistance measurements, particularly in the elimination of thermal voltages. In practice, however, rather heroic techniques are necessary to make accurate measurements of resistances smaller than 1 Ω (Gafert, 1976). On the other hand, a number of very nice ac bridge circuits have been developed for relatively high-resistance thermometry applications at low temperatures (Ekin and Wagner, 1970; Anderson, 1973). The only source of detailed information on lock-in measurement techniques applicable to low-temperature resistivity measurements appears to be in the manufacturers' literature. Shielding and grounding, along with the correct use of transformers, are the major problems in such an application and are discussed in detail by Morrison (1967).

5.2.3 Eddy current method

An ac technique that has proven very useful in relative resistivity determinations is the eddy current decay method. The basis for the method is described by Bean, DeBlois, and Nesbitt (1959), and detailed description of a working system is given by Clark, Deason, Hust, and Powell (1972). A schematic of the apparatus is shown in Fig. 5.4. Although not particularly accurate (1 to

5%) as a resistivity measurement, the system is extremely fast. Furthermore, it is not necessary to attach leads to the specimens, and resistivity ratios of specimens of odd shapes and sizes can be measured easily. The system is relatively expensive and parts of it must be homemade. For experiments requiring repeated low-temperature measurements on the same specimen, such as annealing recovery curves or impurity segregation experiments, it is a valuable tool. An adaptation of this technique to measurements of the anisotropic resistivity observed in some metals (§ 5.3.1) is described by Lawson (1977).

5.2.4 Low-level measurements

The relatively easy methods just described do not always suffice. The most frequent failing occurs when voltages well below the microvolt level must be measured. In fact, nanovolt signals are commonly encountered in physics experiments on pure metals. In spite of the availability of commercial nanovoltmeters and potentiometers on the market, measurements at this level are very difficult, and measurements with 1% accuracy require very specialized apparatus. A nanovolt detection system using commercial components is described by Clark and Fickett (1969). An extensive discussion of techniques using SQUID (Superconducting Quantum Interference Devices) detectors is given by Gifford, Webb, and Wheatley (1972). Other devices, which convert the low-level dc voltage to ac in the cryostat and subsequently amplify the resulting signal, have been described in the literature. Magnetic amplifiers (Foiles, 1967), Hall effect amplifiers (Chovanec, 1969), superconducting modulators (Allnut and Walton, 1972), and relay modulators (Ho, Munnings, and Fletcher, 1969) have all been used in this manner.

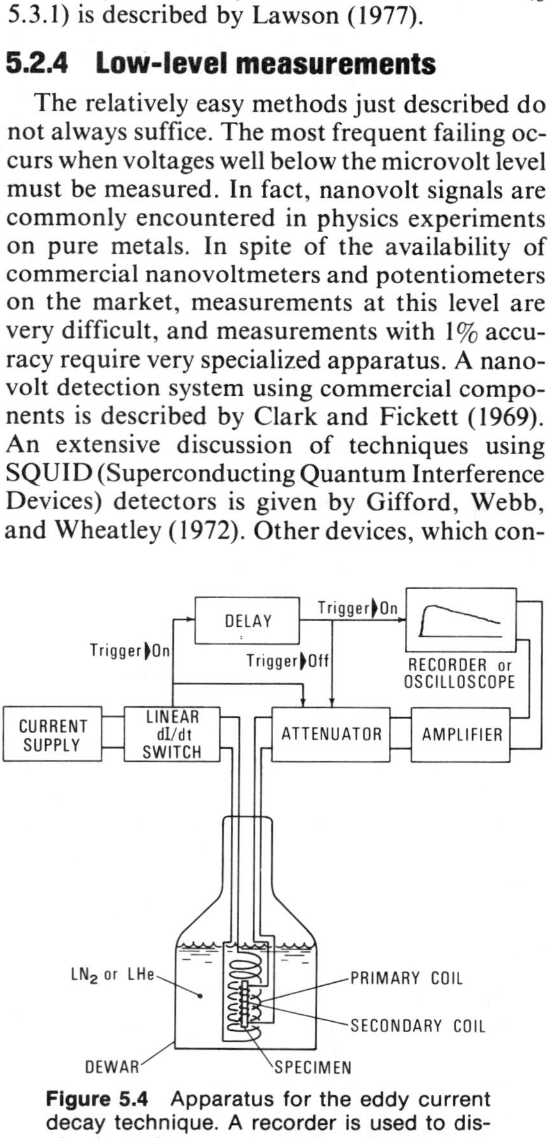

Figure 5.4 Apparatus for the eddy current decay technique. A recorder is used to display long time-constant traces (large specimens of very pure metals) and an oscilloscope for shorter time constants.

5.2.5 Additional considerations

Low-temperature measurements introduce other special problems of a technical nature that must be considered:

a. The temperature of the specimen must be accurately measured and precisely controlled.

b. At low temperatures, joining techniques for current and voltage leads are more critical than at room temperature.

c. Voltage leads traverse a large temperature gradient, leading to spurious thermal voltages and creating unwanted paths for heat conduction.

d. Uneven thermal contraction of parts of the apparatus can stress the sample.

In addition to these, there are more basic phenomena that can create havoc for the unwary, such as size effects and large resistance changes due to magnetic fields or phase transformations. These are covered elsewhere in this chapter. Only the four topics listed above are discussed in this section.

Temperature control. When resistivity measurements as a function of temperature are necessary, the experimental setup becomes much more complex. In some instances, one can take an easy way out by immersing the specimen in various baths of known temperature. The most common baths are liquefied gases and slushes (solid–liquid mixtures) of organic fluids, usually prepared by mixing with liquid nitrogen (Rondeau, 1966). Some of the temperatures attainable by this method are listed in Table 5.2. The temperatures of the liquid-gas baths are dependent on pressure and, thus, on elevation; for example, liquid helium boils at 4.0 K in Boulder, Colorado (elevation 1610 m) vs. 4.2 K at sea level. It should not need emphasis that liquid hydrogen and liquid oxygen should be used only by those familiar with their unique capabilities for total destruction of the laboratory. Some caution is also necessary with many of the organics. It is common practice to mix crushed dry ice

Table 5.2 Liquefied gas and slush bath temperatures.

Substance	Temperature[a] (K)	Bath Type
Ice water	273.15	Slush
Isobutane	262.9	Liquid B.P.
Carbon tetrachloride	250	Slush
Propane	230.8	Liquid B.P.
m-xylene	226	Slush
Trichloroethylene	200	Slush
Carbon dioxide	194.6	Solid
Methanol	175	Slush
n-pentane	142	Slush
Iso-pentane	113	Slush
Methane	111.7	Liquid B.P.
Oxygen	90.1	Liquid B.P.
Nitrogen	77.3	Liquid B.P.
Neon	27.2	Liquid B.P.
Hydrogen	20.4	Liquid B.P.
Helium (He^4)	4.2	Liquid B.P.
Helium (He^3)	3.2	Liquid B.P.

[a]All temperatures given at 0.1 MPa (1 atm).

with either acetone or alcohol to make a liquid bath at dry-ice temperature. Any cryogenic bath should be contained in an insulated container for stability and longevity. Vacuum pumping of cryogenic liquids, combined with some form of pressure control, such as a manostat system, allows attainment of ~1 K with helium, 14 K with hydrogen, 25 K with neon, and 64 K with nitrogen, but the resulting system complexity seldom justifies such an approach except for the case of helium, where only more complex options (e.g., dilution refrigerators) are available. The entire range of specialized techniques for producing and measuring temperatures below 1 K is described by Betts (1976) and is not discussed here. In all other cases, the heater temperature control techniques described below are more practical.

Continuous temperature control may be performed by a variety of techniques. White (1968) discusses several methods in general outline and includes references to the detailed literature. The technique of electrically heating a copper block in good thermal contact with the specimen but partially isolated from the coolant bath by a properly chosen heat leak is the usual choice. The electronics for temperature control are available commercially or can be built rather easily. The electrical concepts are discussed by Forgan (1974). At low temperatures, a precision of ~50 mK is easily obtained; 1 mK requires some care. Systems have been described in the literature that offer control to 50 μK (Larsen, 1968), or even 1 μK (Ries and Moore, 1970). A much less sophisticated technique is sometimes adequate: the specimen and a heat-sink block containing a thermometer are immersed in liquid helium for a 4-K measurement and then removed to the space above the liquid and allowed to drift upward in temperature. With the use of a stable X–Y recorder, this technique is a good method for looking for "glitches" in the resistivity caused by phase transformations or other factors. The temperature profiles of a particular Dewar configuration have been measured by Crooks (1969).

Bonding of specimens. In most instances, a good thermal bond between the specimen and the heat-sink block is important because thermometers are usually attached to the heat sink rather than to the specimen. It is nearly always desirable that the bonding agent be an electrical insulator. The most common materials are an electrical varnish and various greases. Thermal conductivity values for these materials have been measured (Kreitman and Callahan, 1970; McTaggart and Slack, 1969). When constructing a new apparatus, the specimen–heat-sink temperature difference should be checked, possibly with a differential thermocouple arrangement. With reasonable care, the thermal contact should be adequate for moderately precise work.

Temperature measurement. The measurement of low temperatures is discussed in many texts; two of the most useful are those by Meaden (1965) and White (1968). Rubin (1970) has published an extensive review. Thermometry materials are also discussed in Chapter 14. Low-temperature tables are now available for many standardized thermocouple types (Powell, Hall, Hyink, Sparks, Burns, Scroger, and Plumb, 1974). Table 5.3 lists some of the more common thermometers used at 4 K and above and their properties. An excellent discussion of these transducers and several others, such as gas thermometers, is given by Sparks (1974). The problems of bonding, mentioned above, and of thermal anchoring are very important.

Lead attachment. The method of attaching the potential leads and, to a lesser degree, the current leads to the resistivity specimen requires some consideration. Pressure contact arrangements are often adequate for measurements made in a liquid bath. All kinds of steel knife edges, brass screws with sharp points and fine, bare copper wires have been used. Spring load-

Table 5.3 Thermometers for the measurement of low temperatures.

Type	Temperature Range (K)	Comments
Liquid in glass:		
Toluene	180–320	
Pentane	70–300	An impure pentane.
Thermocouple:		
Type E	20–300	Useful also above room temperature.
Type KP or EP vs. Au–0.07 at.% Fe	4–20	Can be used to room temperature.[a]
Metal resistor:		
Platinum	10–300	Can be used to 1200 K—usual application is above 50 K.
Copper	20–300	Not commercially available.
Indium	4–300	Not commercially available.
Semiconductor:		
Germanium	4–100	Very sensitive to magnetic fields.
Thermistor	4–300	Very sensitive to magnetic fields.
Carbon resistor	4–100	Inexpensive, less field sensitive than the semiconductors.

[a]Au–Fe is not a standardized thermocouple type but is available commercially. Tables are found in Sparks and Powell (1972).

ing of the systems at room temperature seems to work well—almost any spring material will do. Consideration must be given to the relative thermal expansion of various parts of the sample holder. Pressure contact systems have the advantage of precise location of the probes and minimum disruption of the specimen. It has been the author's experience that pressure contacts never work well in a high magnetic field. Excessive noise always seems to occur on the voltage signal. When measurements are to be made over a temperature range or in a magnetic field, soldering, spot welding, or welding appear to be the only reasonable techniques. Solders are available with melting points from below 373 K (Bi–Pb–Sn) to about 473 K (Pb–Sn). They are nearly all superconducting, and some have critical temperatures in excess of 7 K and critical fields near 1 T (Warren and Bader, 1969). Normally this superconductivity presents no problem, but it may cause trouble in specific experiments. Soldering of potential leads generally gives a large contact area and decreases the precision with which the probe separation can be measured. A complex, but successful, technique for increasing this precision involves: coating the specimen with varnish, curing, drilling a 0.2-mm-diameter hole to the metal surface, inserting a 0.12-mm flux-coated ball of low-melting-point solder (available commercially), heating to melt the solder, and inserting a pretinned potential lead. Spot welding can be used for nearly any combination of metals, but it is easiest for higher resistivity materials. Often an intermediate metal, such as nickel, must be employed. Some arrangements, such as joining a small-diameter copper wire to one of large diameter, will tax the skill and patience of the most accomplished welder. The major problem with spot welding leads is the possibility of major damage to the specimen by mechanical deformation or arc burn. Welding, usually used for joining fine wires, is a reliable technique if a small oxy–hydrogen torch is available.

Thermal considerations for lead wires. The conduction of heat to the specimen by the leads can occasionally be troublesome. In such cases leads of high thermal resistance alloys, such as constantan, Evanohm, or manganin are sometimes used. The alloys have similarly high electrical resistances. Theoretical aspects of the problem and some practical calculations are presented by Mercouroff (1963). A more common solution is to use copper wire, but to thermally anchor a reasonable length of the lead to a heat-sink block, which approximately tracks the specimen in temperature (Hust, 1970). This technique has the added advantage of minimizing spurious thermal emf's. These thermoelectric voltages, generated in the presence of thermal gradients, arise from contacts between dissimilar metals and from inhomogeneities in the lead material. Spurious voltages in a well-designed experiment should be stable at the submicrovolt level over short time periods. All dc potential measurements should be an average of two readings taken before and after reversing the sample current.

Thermal contraction. Relative thermal contraction may cause serious problems in electrical measurements at low temperatures owing to the wide range of behavior exhibited for various classes of common materials. Clearly, a pure metal rigidly attached to a plastic substrate will be strongly compressed on cooling. The usual result of such an arrangement is that the metal tends to buckle. The obvious solution is to allow some motion of the specimen in the holder. Another is to make the entire holder out of the same type of materials as the specimen, e.g., a copper holder for pure metals, and rely on thin varnish coatings and polymer films for electrical insulation. A final possibility is that of matching the expansion coefficient of the holder material to that of the specimen. This can be accomplished by the use of filled epoxies (Seligmann and Sarwinski, 1972) or, possibly, graphite-epoxy composites of suitable composition (Fickett, 1977).

5.3 PURE METAL RESISTIVITY

Many separate mechanisms influence the electrical resistivity of pure metals and alloys. Some of these effects become particularly important at low temperatures. In the following sections, brief descriptions of the various mechanisms that are operative in producing resistance at low temperatures are presented. Clearly, not all of them are important in every metal or configuration. Solid-state physics texts contain the theoretical details; the following sections continue in a pragmatic vein.

5.3.1 Intrinsic resistivity

The intrinsic resistivity is that of a pure, perfect metal and arises from the scattering of electrons by lattice vibrations (phonons). No such perfect metal exists, of course, but at room temperature the intrinsic component is usually the dominant contributor to the resistivity of pure metals. It is also the most temperature dependent of the mechanisms.

Conceptually, experimental determination of the intrinsic resistivity is quite easy. One measures ρ vs. T and obtains a curve like that for pure copper in Fig. 5.2. The assumptions are made that the resistivity observed at the low-temperature end of the curve is due entirely to other mechanisms and is not temperature dependent. This "residual" resistivity is then subtracted from all measured values, and a new curve is created representing the intrinsic resistivity, ρ_i, which now is a true property of the metal. Often, the curve can be fit to the Bloch–Grüneisen form

$$\rho_i = \frac{K}{\theta}\left(\frac{T}{\theta}\right)^5 \int_0^{\frac{\theta}{T}} \frac{e^z z^5 dz}{(e^z - 1)^2} \tag{5.2}$$

by judicious choice of the characteristic temperature, θ. K is a constant. Meaden (1965) explains a variety of methods for determing θ, and Rogers and Powell (1958) give an extensive tabulation of values for the integral. Most common metals have θ values between 300 and 400 K (copper $\simeq$ 330 K, aluminum $\simeq$ 395 K), although some are well outside this range (beryllium $\simeq$ 1200, sodium $\simeq$ 200). At either end of the temperature range the equation reduces to a more simple temperature dependence:

$$\rho_i \propto T \text{ for } T \gtrsim \theta/2 \tag{5.3}$$

and

$$\rho_i \propto T^5 \text{ for } T \lesssim \theta/10 \tag{5.4}$$

The linear behavior is almost always observed near room temperature, as it should be, but often not at very high temperatures. The temperature coefficient of resistance, α, defined as:

$$\alpha = (R_{T_2} - R_{T_1})/R_{T_1}(T_2 - T_1) \tag{5.5}$$

is tabulated in the *Metals Reference Book* (1976) for metallic elements and for some alloys for $T_1 = 273$ K and $T_2 = 373$ K. Common pure metals tend to have values from 4 to 6×10^{-3} K^{-1} (copper and aluminum: $\alpha \simeq 4.2 \times 10^{-3}$ K^{-1}), and alloys generally $< 2 \times 10^{-3}$ K^{-1} (AISI 304 stainless steel: $\alpha \simeq 1.2 \times 10^{-3}$ K^{-1}). The low-temperature behavior expressed by Eq. 5.4 is not often observed. The experiment involves measuring the small difference between two nearly equal numbers, usually under very difficult experimental conditions. The author has tabulated the various temperature dependences reported for the resistivity of aluminum below $\theta/10$ (Fickett, 1971). They range from T^5 to T^2! Similarly, copper has been reported to have a low-temperature behavior from nearly T^6 (Clark, 1971) to T^5 (Fickett, 1972b), depending on purity. Ferromagnetic metals show even more complex behavior. In fact, it appears that there may well be no such thing as *a* temperature dependence for a real metal.

The conclusions that can be drawn from the above about the intrinsic resistivity are: (a) most metals show a linear dependence of ρ on T near room temperature, which is not greatly affected by low levels of impurities; (b) departures from linearity often occur at higher temperatures; (c) below about 20 K there is no dependable expression for the intrinsic resistivity of most metals; (d) above about 20 K, Eq. 5.2 usually provides a good estimate of ρ_i up to room temperature for well-behaved metals. One should not be too concerned about (c), since the resistivity of most useful metals is residual at low temperatures and the temperature-dependent part represents only a very small portion of the resistivity. For example, at 4 K the intrinsic resistivity of copper is $\sim 3.5 \times 10^{-7}$ $\mu\Omega$·cm (RRR $= 4.4 \times 10^{6}$) and that of aluminum is $\sim 6 \times 10^{-7}$ $\mu\Omega$·cm (RRR $= 4.0 \times 10^{6}$). These values are far above the best resistivity ratios measured experimentally (see Table 5.1).

Anisotropic metals. Metals with noncubic crystal structure show an orientation-dependent resistivity (anisotropy). Several are indicated in Table 5.1. With the exception of tin, they are all hexagonal metals. The ratio between the resistivity parallel to the *c*-axis and that normal to it, as determined from single-crystal measurements, is given in Table 5.4. Meaden (1965) lists many more in his Table III. He also discusses the manner in which one should combine the two resistivity values in order to arrive at an average resistivity, $\bar{\rho}$, for a polycrystalline material. A reasonably accurate equation is:

$$\bar{\rho} = (2\rho_{\perp} + \rho_{\parallel})/3 \tag{5.6}$$

which is slightly different from the more often quoted formula used by Hall (1968).

Anisotropy effects must be expected when rolled or drawn product forms are made from the noncubic metals and alloys. Because of preferential elongation of grains with certain orientations, the material resistivity may differ significantly from what one might expect from the above discussion.

It is difficult to make any statement about the relative temperature dependence of the directional resistivities because of the limited data. Nevertheless, it seems that for most metals, no dramatic changes in the ratio $\rho_{\parallel}$ $\rho_{\perp}$ occur on cooling to cryogenic temperatures. This was observed to be the case by the author in measurements on the highly anisotropic metal, mercury ($\rho_{\parallel}/\rho_{\perp} = 0.76$) (Fickett, 1967). On the other hand the data on zinc by Aleksandrov and d'Yakov (1963) show a variation of $\sim 20\%$ in the ratio, but it apparently occurs in a random manner.

Table 5.4 Resistivity values of anisotropic metals.

Metal	Temperature (K)	$\rho_{\parallel}/\rho_{\perp}$	Reference
Be	273	1.14	Hall (1968)
Mg	273	0.83	Ibid.
	78	0.85	Ibid.
Ti	273	1.06	Wasilewski (1962)
	77	0.90	Ibid.
Sn	273	1.45	Aleksandrov and D'Yakov (1963)
	77	1.58	Ibid.
	4.2	1.35	Ibid.
Zn	273	1.04	Hall (1970)
	77	1.14	Aleksandrov and D'Yakov (1963)
	4.2	0.92	Ibid.

Crystallographic transformations. A number of pure metals, and nearly all alloys, possess more than one stable crystallographic phase. The change from one phase to another may be caused by a variety of mechanisms, such as temperature, pressure, tensile strain, or certain impurities. The subject is one of a great deal of interest in metallurgy and materials science. There are many excellent general texts, such as the one by Christian (1965). Most such transitions occur above room temperature, but the higher temperature phases can often be obtained at room temperature and below by proper (or accidental) quenching techniques. All aspects of low-temperature phase transformations are discussed in Chapter 9. Here we present just a brief introduction to the resistive behavior.

The different phases of a metal may have vastly different resistivity values, as indicated by the data for manganese in Table 5.1, which show nearly a factor of seven variation in the ice-point resistivity. The α phase of manganese is the normal one at room temperature; the others are higher temperature phases. Transformations seldom proceed to completion and, as a result, the specimen becomes a mixture of the two phases — a situation that can lead to bizarre temperature dependences for the resistivity, since the mix of the phases is usually temperature dependent itself. To further complicate the problem, the process, although usually reversible, exhibits a large temperature hysteresis. The good news is that very few pure metals have crystallographic transformations below room temperature (Addison, 1964). Lithium transforms to hexagonal (h.c.p.) on cooling to around 77 K, sodium at ~ 36 K. Face-centered-cubic (f.c.c.) phases can be induced in these metals by deformation. Tin is tetragonal above 286 K and may have a diamond structure below that; cerium goes from f.c.c. to h.c.p. to f.c.c. in a complex series of transformations spanning the range from ~ 265 to 90 K. Gerritsen (1956) tabulates the resistivity changes that occur upon transformation for these elements and many others that undergo phase changes above room temperature. His data show that tin changes to a semiconductor, and cerium shows a decrease in resistivity as the temperature is lowered through the transition temperature. No change is shown for lithium or sodium, but a later measurement on lithium by Dugdale and Gugan (1961) shows a large cusp in the resistivity vs. temperature curve at about 70 K. The cusp is seen on lowering temperature, but is nowhere near as obvious in the increasing temperature measurement.

Magnetic transformations. Low-temperature magnetic transformations, usually from paramagnetism to antiferromagnetism to ferromagnetism as the temperature is lowered, occur primarily in the rare earth elements. These transitions result in an odd resistivity behavior with temperature (Meaden, 1971). Of the more common metals, α-manganese shows the first of these transitions over the range 70 to 90 K (Meaden and Pelloux-Gervais, 1965), as does chromium near room temperature (Meaden, Rao, Tee, and Loo, 1972). Figures 5.5 and 5.6, from the references, show the observed resistivity behavior. Similar transformations are seen in many of the ferrous alloys. Pure, soft, ferromagnetic materials, such as iron, show a strong dependence of the low-temperature resistivity on measuring current at very low current values. An excellent example is provided by the work of Takaki and Igaki (1976), which shows a variation of RRR from 500 to 170 as the measuring current density is increased to 6 $A \cdot mm^{-2}$. One should add that applying a small longitudinal field of 40 $kA \cdot m^{-1}$ (500 Oe) to the specimen *increased* the RRR to 2500! This effect most likely results from the removal of domain walls, which scatter the conduction electrons. The theory and some experimental work on the topics concerning electron scattering interactions in magnetic materials, such as electron–electron and electron–magnon scattering, are reviewed by Volkenshtein, Dyakina, and Startsev (1973).

5.3.2 The effect of impurities

The materials considered here are the dilute binary alloys, in which the impurity metals are in solid solution. In general, these alloys contain less than a few atomic percent of impurities. They may be those occurring naturally in the metal or may be introduced for experimental purposes. The effect of low levels of impurities on the resistivity of commercially pure conductors is of primary importance, particularly at low temperatures, where the impurity resistivity often represents the major contribution to the total resistivity. Most of the experimental work on the resistive contribution of various impurities was done prior to about 1970. The experiments are not as easy to carry out as one might expect. The production and proper chemical analysis of dilute alloy specimens is difficult, and particularly in the case of analysis, few laboratories have adequate facilities. Analysis below about 0.1 at.% is subject to large (20 to 50%) errors for many elements. The purity of the solvent metal is often known imprecisely. The analysis supplied with the metal is often a general analysis for that type of metal and not specific to the piece supplied. Furthermore, precipitation, compound forma-

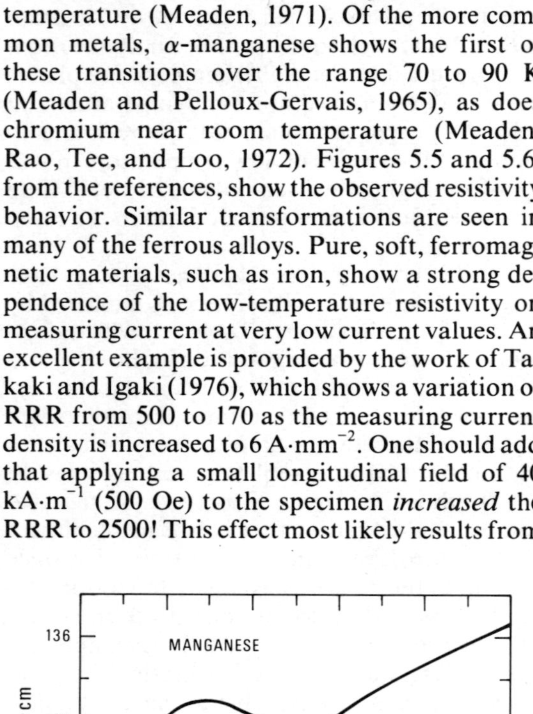

Figure 5.5 Resistive behavior of α-manganese. The metal is antiferromagnetic below ~100 K (Meaden and Pelloux-Gervais, 1965).

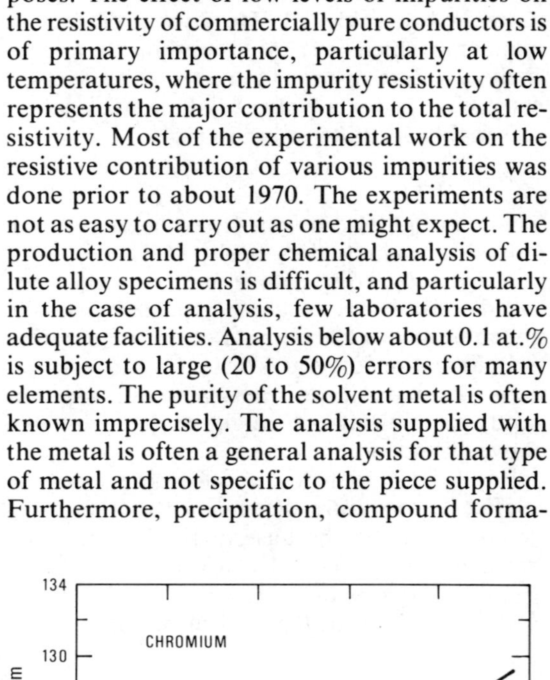

Figure 5.6 Resistive behavior of annealed chromium. The metal is antiferromagnetic below ~310 K (Meaden, Rao, Tee, and Loo, 1972).

tion, agglomeration and migration of the solute to grain boundaries can occur even at very low concentrations. These situations are not detected by conventional analysis and always result in a resistivity contribution that is below that for the true solid solution. As an example, the resistivity of iron in copper at 4 K has been observed to decrease from 1.81 $\mu\Omega$·cm/at. ppm (observed up to 100 ppm) to 1.48 $\mu\Omega$·cm/at. ppm for alloys containing 750 and 690 at. ppm Fe (Fickett, 1974a). Magnetic measurements on the higher content alloys suggest the precipitation of γ-iron.

In the following subsections, the effect of various impurities on the low-temperature resistivity of pure metals and its variation with temperature are discussed. The topic is one that has been reviewed often and well, and this treatment relies to a large extent on those reviews.

Impurity resistivity at fixed temperature. The contribution of impurity elements to the resistivity of pure metals has been studied extensively. Most of the available data are at room temperature, although a significant amount of low-temperature data exist. Table 5.5 shows aluminum data collected by the author (Fickett, 1971) and copper data from the extensive compilation by Welles (1967). Blatt (1968) presents a table covering some 40 metals as solutes in copper, silver, gold, aluminum, magnesium, lead, titanium, iron, nickel, palladium, and platinum. No indication is given of the temperature of measurement or the data sources. A table of earlier data is also presented by Gerritsen (1956). The data in the literature in general are not as convincing as one would like. In nearly every case, measurements by different experimenters result in widely different numbers, often varying by 20 to 30% for the same alloy. Thus, there is still a need for a comprehensive series of measurements on alloys that are very well characterized both chemically and metallurgically. Unfortunately, such a task could take a lifetime.

Some general statements can be made: Impurity elements in solution always cause the resistivity to increase. The increase appears to be always linear with concentration as long as the impurity remains in solution. The contributions of solute elements are additive, allowing calculation of low-temperature resistivity from a good chemical analysis. In nearly all instances, the resistive contribution of an impurity measured at room temperature is within about 5% of the value found at 4 K. The low-temperature value is usually higher.

In some instances, it is possible to increase the amount of impurity in solution far above the room-temperature solubility by very rapid quenching of the alloy. The technique is described by Ocko, Babic, Krsmik, Girt, and Leontic (1976) and in earlier work by the same group. Their studies indicate that the value of the specific impurity contribution to the resistivity is maintained to quite high percentages of solute.

Table 5.5 Resistive contributions of impurities to copper and aluminum.

Impurity Element	$\Delta\rho(\mu\Omega\cdot$cm/at.%) Copper	Aluminum
Al	0.95	–
Ag	0.6	1.0
As	6.7	0.5
Au	0.55	–
Be	0.65	0.4
Bi	5.7	1.3
Ca	0.3	0.3
Cd	0.3	0.57
Co	6.9	3.6
Cr	4.0	8.4
Cu	–	0.83
Fe	9.3	5.4
Ga	1.4	0.24
Ge	3.7	0.79
Hg	1.0	–
In	1.1	–
Ir	6.1	–
Li	–	0.86
Mg	0.8	0.46
Mn	2.9	6.9
Mo	–	7.5
Na	–	2
Ni	1.1	1.6
Pb	3.3	1.2
Pd	0.95	–
Pt	2.0	–
Po	8.5	–
Re	4.4	–
Sb	5.5	0.9
Sc	–	5
Se	10.0	–
Si	3.1	0.72
Sn	3.1	0.9
Te	8	–
Th	1.8	–
Ti	16	5.5
V	9.7	6.8
W	–	7.3
Zn	0.3	0.23
Zr	1.3	5.8

The effect of temperature. Near room temperature the dilute alloys maintain the linear dependence of resistivity on temperature observed for the pure solvent. However, the slope is seriously affected by the impurity. In particular, $d\rho/dT$ is found to decrease with increasing content of magnetic elements and to increase otherwise. This behavior is discussed by Gerritsen (1956). He presents a graph for copper alloys showing a linear dependence of the temperature coefficient of resistivity on the room-temperature electrical conductivity.

At low temperatures, the resistivity of annealed alloys is nearly always dominated by the impurity content, and usually this residual resistivity is effectively a constant below 4 K. It is generally assumed that this constant value is the value at 0 K and it is designated as ρ_0.

The topic of most interest in recent years is the observed deviations from Matthiessen's Rule. The rule, as usually seen, states that the phonon (intrinsic) resistivity of the pure solvent, ρ_p, and the impurity resistivity due to the solute, ρ_0, are additive, and that the latter contribution is independent of temperature. Therefore, the total resistivity, ρ, becomes,

$$\rho(c,T) = \rho_p(T) + \rho_0(c) \tag{5.7}$$

over the entire temperature range (c is the impurity concentration). In general, defects and surfaces also contribute to ρ_0. It has been known almost since its presentation that the rule is not exact, but when one considers that it resulted from an experiment on dirty alloys at near room temperature in the 1860s, it has had an amazing success. The way to make an exact equation is to introduce a third term, Δ, the deviation term:

$$\rho(c,T) = \rho_p(T) + \rho_0(c) + \Delta(c,T) \tag{5.8}$$

It is the study of the behavior of Δ with impurity type and content and with temperature that is the major topic of interest today. Two excellent and very thorough reviews of the subject have been published by Bass (1972) and by Cimberle, Bobel, and Rizzuto (1974). The first of these covers deviations due to various defects and size effects as well as impurities. A lengthy treatment of nickel-based alloys is presented by Yamashita and Hayakawa (1976), and a discussion of size-effect deviations is given by Boughton and Neighbor (1972). Figure 5.7, from Bass (1972), shows the various behaviors observed for Δ from 0 K to room temperature. His assessment of the available data shows that: at low temperatures,

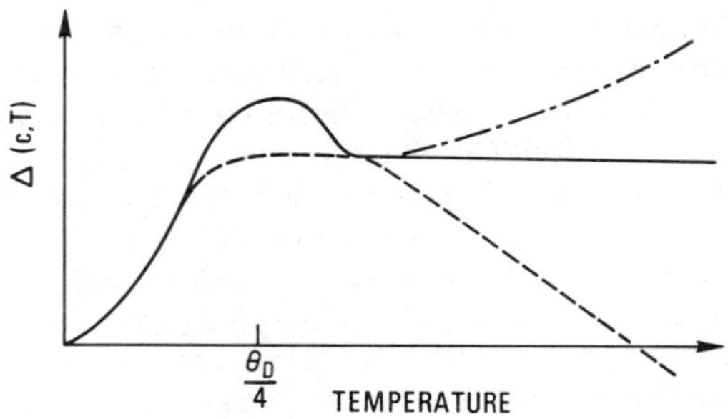

Figure 5.7 Types of behavior observed for deviations from Matthiessen's Rule in alloys as a function of temperature (Bass, 1972).

Δ rises with temperature usually more slowly than $\rho_p(T)$, but often exceeds $\rho_p(T)$ in size near $\theta_D/10$ (θ_D is the Debye temperature). The hump, which may occur at $\theta_D/4$ to $\theta_D/5$, moves to lower temperatures and increases in relative size as c decreases. For values of c greater than a few tenths of a percent, Δ normally is nonlinear with c. At high temperatures, Δ can be represented by a constant plus a linear term in T. Both of these terms are observed to vary linearly with c at constant temperature. The term linear in T may either increase or decrease with temperature. At high temperatures, Δ is usually 5 to 10% of ρ_0, but in some cases (cold-worked noble metal alloys and magnetic alloys), it has been observed to be equal to ρ_0. Table 5.6 indicates alloy systems that have been studied for deviations from Matthiessen's Rule. Most of the information in the table comes from the two reviews mentioned above (Bass, 1972; Cimberle et al., 1974).

It is important to stress that, from a technological point of view, these deviations are most often not large enough to be of concern in applications.

Kondo effect. This low-temperature phenomenon has been extensively investigated in recent years. The manifestation of this effect is a minimum in the electrical resistivity of pure metals and alloys containing dilute magnetic solutes. The classical material is Cu–Fe, but the minimum has been observed in many other systems. The resistivity minima may occur at any temperature, but most systems show it in the range 2 to 30 K. Again, the effect is usually very small, although it can be a serious impediment to the creation of the constant-resistance alloys for low-temperature applications described in § 5.4.5.

The effect arises as the result of an interaction between the magnetic moment of a localized im-

Table 5.6 Alloy systems on which studies of deviations from Matthiessen's rule have been made.

	Solvent Metal																						
	Ag	*Al*[a]	Au	*Be*	Cd	Co	Cr	*Cs*	Cu	Fe	*Ga*	*In*	*K*	Li	Mg	Ni	*Na*	Pb	Pd	*Pt*	*Sn*	*W*	*Zn*
Ag		X	X		X									X	X								X
Al	X		X						X	X					X								X
As	X								X														
Au	X	X							X	X						X				X			
Bi	X																				X		
Cd	X														X			X			X		
Ce															X								
Co		X	X			X			X	X						X							
Cr		X	X			X			X	X						X							
Cu	X	X	X													X							
Er																		X					
Fe		X	X			X			X							X							
Ga		X																					
Gd															X								
Ge		X							X	X													
Hf						X										X							
Hg														X									
In	X		X						X		X				X			X			X		
Ir						X										X							
Li															X								
Mg	X	X			X									X									
Mn		X	X			X			X	X					X	X							
Mo						X				X						X							
Nb						X	X									X							
Nd															X								
Ni			X						X	X													
Os						X				X						X							
Pb															X								
Pd	X		X							X						X							
Pt	X		X							X						X							
Re						X				X						X							
Rh			X			X										X				X			
Ru						X				X						X							
Sb	X								X						X						X		
Si									X	X													
Sn	X		X						X	X	X	X			X								
Ta		X				X	X			X						X							
Ti						X				X						X							
Tl															X								
V		X				X	X			X						X							
W						X				X						X							
Zn		X	X						X						X						X		
Zr		X	X			X										X							

[a]Data on solvents printed in italics are available with unknown solute elements in some instances.

purity and the average moment of nearby conduction electrons. The review article by Daybell and Steyert (1968) still represents the best compilation of information on the subject, although many papers have appeared since then.

Basically, the interaction is thought to occur as follows: below a critical temperature, T_k, a "quasi-bound" state forms between the moments, characterized by an antiferromagnetic exchange interaction in these alloys, i.e., the conduction electron spins shield the impurity moment. The temperature, T_k, depends exponentially on the exchange constant, J, so that very small changes in J can cause large variations in T_k. Values for T_k have been observed over five orders of magnitude for various transition elements in copper. For iron, the temperature is most likely around 30 K — its determination is not as straightforward experimentally as it is theoretically. It is interesting that the experimental effects of this quasi-bound state were known for many years before the successful theoretical treatment by Kondo (1965). The effect of this interaction on the resistivity is to cause an increase in the resistive contribution per impurity, $\Delta\rho/c$, as the quasi-bound state forms, i.e., as the temperature is lowered. This is not a sharp step, but the value of $\Delta\rho/c$ is constant above and below it. The step height is considered to be impurity independent and has a value ~0.14 nΩ·cm/at. ppm for copper-based alloys. The effect of this step on the measured (i.e., total) resistivity of the metal is to give the observed minimum in the resistivity vs. temperature curve of the alloy at near T_k. It is interesting that this Kondo minimum is seen in nearly all "pure" copper, indicative of its rather high iron content. Internal oxidation of the iron impurities in copper, as described below, removes the Kondo minimum entirely. Also, a temporary breakup of the quasi-bound state can be caused by application of a strong magnetic field.

Gases as impurities. Gases occur to a limited extent as elemental impurities in metals. Oxygen is the most common. It occurs, for example, even in "oxygen-free" copper at 10 to 20 ppm. Gas analysis is very difficult at these levels, and the content is seldom stated by the supplier. The contribution to the resistivity is of the same order as that of metallic impurities. As an example, Hess and Pawlek (1959) report a value of 5.3 μΩ·cm/at.% for oxygen in copper.

Nitrogen is added as a strengthening element in various modern stainless steels that are used extensively for low-temperature structures. Almost no data exist on the effect of this addition on the resistivity, but one would expect it to give only a small contribution to the already very high values for these metals.

Hydrogen in metals causes serious embrittlement problems. It is not a naturally occurring impurity, but often is encountered in high-temperature industrial processes. Analysis of hydrogen content is very difficult and essentially no low-temperature specific-resistivity data exist.

Thin films of alloys of alkali metals with rare gases have been made to study the physics of metal–insulator transitions (Phelps, Avci, and Flynn, 1975). At low impurity levels, the contribution to the low-temperature resistivity is normal (2.2 μΩ·cm/at.% for krypton in rubidium and 2.8 μΩ·cm/at.% for xenon in cerium). At ~40 at.% gas, the resistivity at 7 K is only ~100 μΩ·cm for both alloys, but at high percentages of the gas (>75 at.%), the resistivity exceeds 10^7 μΩ·cm.

The addition of specific gases under controlled conditions of temperature and pressure can be used to decrease the low-temperature resistivity of metals. The major application of the technique has been to effectively remove the residual iron from copper. This is accomplished by the oxygen forming compounds with the iron, primarily Fe_3O_4, which agglomerate as relatively large particles even in very dilute alloys (<4 ppm Fe). By this technique, copper single crystals with residual resistivity ratios near 65 000 and polycrystalline wires with ratios of 20 000 have been produced. These ratios are much higher than the approximately 2000 ratio of the highest purity unoxidized copper. The efficacy of the technique is illustrated in Fig. 5.8, which shows the results of 4-K resistivity measurements on a series of dilute Cu–Fe alloys (Fickett, 1976). Internal oxidation has also been used to increase the residual resistivity ratio of gold, silver, and tantalum. The techniques and a historical review have been published by the author (Fickett, 1974b).

5.3.3 Defect resistivity

Defects in the crystal structure of metals and alloys can give significant contributions to the low-temperature resistivity. In some metals, the available purity is such that defects represent the major contribution to the residual resistivity. The simple act of winding a wire around a sample holder can decrease the residual resistivity ratio by a factor of two or more. Defect resistivity is a topic dear to the heart of theorists, and many texts, such as Ziman (1960), go into great detail on the calculation of the contributions of the various defect types. Here a brief description of the defects, some comments on how they are introduced into metals, and the resistivity modifications that they cause are presented.

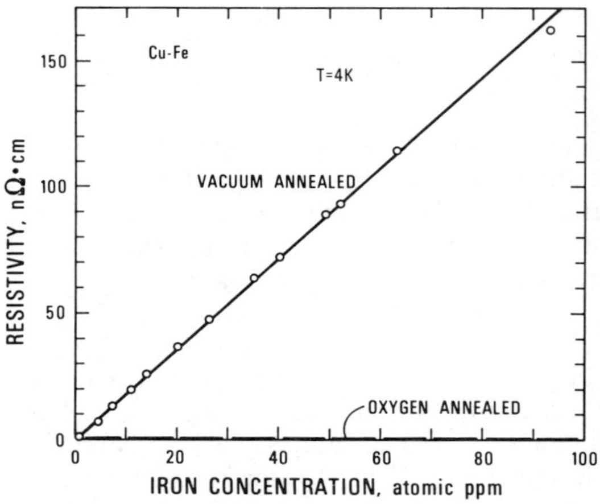

Figure 5.8 Electrical resistivity of Cu–Fe alloys before and after internal oxidation.

The resistive contribution of most defects is not strongly temperature dependent, although deviations from Matthiessen's Rule (mentioned above) do occur. Temperature still plays an important role, however. Defects formed at 4 K tend to stay put in the lattice, but as the temperature is raised, the various types start to anneal out, giving a so-called recovery curve for the defect resistivity. This recovery usually starts well below room temperature.

Defects and their production. Defects are usually grouped according to their dimensionality. The point defects are vacancies and interstitials. The term *interstitial* refers to an atom of the host lattice located off a normal lattice site. Other atoms, notably gases, may also exist in interstitial positions. Depending on the method of production, vacancies can occur alone (Schottky defects) or in vacancy–interstitial pairs (Frenkel defects). The former are found primarily in quenched specimens, and the latter in irradiated specimens. The literature on these defects is enormous. The book by Seeger, Schumacher, Shilling, and Diehl (1970) is suggested as a starting point. Direct experimental determination of defect concentrations is not possible. Typical experimental values quoted are in the range 0.01 to 0.1 at.%. The topic of radiation damage effects in metals is one of great interest, both at high temperatures, owing to fission reactor problems, and at very low temperatures, owing to the use of pure metals to protect (stabilize) superconductors in fusion devices (see *Radiation Damage in Metals*, 1976). Point defects are the first to anneal out of a metal with increasing temperature, and the annealing may start at temperatures as low as ~10 K.

Dislocations are called line defects, but their usual configuration involves loops, clusters, and tangles. Creating and measuring a known concentration of dislocations in a metal is very difficult. The units of dislocation density are lines per square centimeter, or cm^{-2}. Dislocations occur naturally in all real metals, and their concentration can be increased somewhat by quenching and a great amount by deformation. In annealed aluminum or copper, for example, the dislocation density tends to be $\sim 10^6$ lines/cm^2, although, with care, densities many orders of magnitude lower can be achieved. After severe deformation, this number may reach 10^{12} lines/cm^2. There is a massive amount of literature on this topic also. Nabarro (1967) discusses all aspects of dislocations and includes a brief section on their effect on electrical properties. Dislocations are somewhat more difficult to anneal out of metals and, in most common metals, temperatures well above room temperature are necessary for complete recovery.

The planar defects are stacking faults and grain boundaries. Stacking faults occur in conjunction with dislocations, and in fact, their existence was first proposed to account for discrepancies observed between theory and experiment on dislocation resistivity in f.c.c. metals. The units of stacking-fault (and grain-boundary) density are those of surface area per unit volume or inverse centimeters. Stacking faults are most often introduced by quenching, although deformation will also suffice. In an annealed metal, their density tends to be less than a few hundred inverse centimeters. Both Nabarro (1967) and Ziman (1960) present extensive discussions of stacking faults.

The important parameter for grain-boundary resistivity is the boundary area per unit volume or S/V. The determination of S/V from observation of a single surface of a specimen is not entirely straightforward with any of the several existing techniques. Grain sizes in metals cover the range from a few tens of micrometers to many centimeters, leading to S/V values of 10 to 1000 cm^{-1}. The introduction of controlled grain sizes in pure metals is usually performed by deformation, followed by recrystallization and subsequent grain growth. Specimen size limits the maximum grain size achievable in most pure metals. Pure aluminum wires, for example, will grow a "bamboo" structure of grains about the size of the wire diameter, whereas in copper the grains tend to be somewhat smaller and randomly located.

It should be reemphasized that the measurement of defect density is not at all simple for any defects, and errors of a factor of 2 or 3 may occur. The separate defect types never have the courtesy to appear alone, and separation of the resistive contributions is also a significant source of uncertainty. Furthermore, the presence of impurity elements may have an effect on the type and number of defects introduced by a particular procedure.

The resistive contribution of defects. From a theoretical point of view, the problem in calculating the resistivity contribution from isolated defects is the evaluation of what fraction of the resistivity is due to the core of the defect and what fraction is due to the surrounding strain field. There is very little agreement between authors. When the more realistic case of combined defects, such as dislocation loops containing stacking faults, is considered, the relative contribution of each defect type becomes a matter for debate, again with little agreement. When very large numbers of defects are introduced, such as in dislocation tangles, the contribution per unit defect is expected to decrease. One point on which all theoretical calculations agree is that the scattering from the extended defects is anisotropic and has at least some component that gives small-angle scattering (Hunter and Nabarro, 1953). Because of this, the resistivity contribution should be somewhat temperature dependent and this, indeed, appears to be the case experimentally. A further impediment to theory–experiment correlation is the refusal of the defects to occur in the same form from one experiment to the next. Linear dislocations and dislocation loops containing stacking faults are most common. The stacking faults may, however, occur as three-dimensional defects, such as multilayers or, at least in copper, as tetrahedra. Special conditions may also produce helical dislocations or other configurations. A general conclusion is that the best available theoretical calculations agree with experiment to within an order of magnitude.

The experimentally determined vacancy resistivities reported for aluminum center around 2.6 $\mu\Omega$·cm/at.% with about a 40% spread (Fickett, 1971), and Blatt (1968) suggests a value of 1.0 to 1.5 $\mu\Omega$·cm/at.% for copper. In each case the interstitial resistivity appears to be about 1/2 to 2/3 of the value for vacancies. Theoretical copper values reported by Welles (1967) are 0.4 to 1.8 $\mu\Omega$·cm/at.% for vacancies and 0.2 to 10.5 $\mu\Omega$·cm/at.% for interstitials!

Experimental dislocation resistivities in both aluminum and copper tend to be $3 \pm 1 \times 10^{-13}$ $\mu\Omega\cdot cm^3$, with the lower limit applying at 4 K. Note that multiplication by the dislocation density is necessary to obtain an actual resistivity. Again, there is a wide variation of values, ranging from 0.2 to 9×10^{-13} $\mu\Omega\cdot cm^3$, found in the literature on aluminum. Values for several other metals are given in Table 5.7.

Data on stacking fault resistivities are not as plentiful. The value for aluminum is apparently 2 to 5×10^{-7} $\mu\Omega\cdot cm^2$, and that for copper, 10^{-6} to 10^{-5} $\mu\Omega\cdot cm^2$. The lower value is usually more appropriate at 4 K, but it is not the temperature dependence that is responsible for the range of the data. Again, determination of an actual resistivity requires multiplication by the stacking fault density.

Table 5.7 Dislocation resistivity in metals.[a]

Material	Dislocation Resistivity ($\mu\Omega\cdot cm^3$)
Ag	3.4, 2.0×10^{-13}
Au	3.5×10^{-13}
Ni	9.4×10^{-13}
Fe	20×10^{-13}
K	4×10^{-13}
α-brass	18×10^{-13}

[a]Data compiled by Barbee (1965) from the literature.

Grain-boundary-resistivity values have been measured only occasionally (Kasen, 1970) and disagreement over the proper method of determining S/V for bulk specimens puts any measurement in doubt by at least a factor of 2. For copper and aluminum the contribution appears to be $\sim 10^{-6}$ $\mu\Omega\cdot cm^2$. Multiplication by S/V (in cm^{-1}) is necessary to derive a true resistivity. Room-temperature values appear to be about 20% higher than those at 4 K.

5.3.4 Size effects

The increasing application of very pure metals to cryogenic magnet construction brings the consideration of size effects very much into the realm of technology. Pure aluminum and copper conductors have electron mean free paths on the order of a millimeter, and since many conductors are of this size, better than half of the resistivity at low temperatures may be due to boundary scattering. Many new applications are being found for conducting composites both as commercial superconductors and as high-strength, high-conductivity normal metals. In these materials, the interfiber spacing may be such that scattering of the electrons from these boundaries has a significant effect on the resistance. Also, new uses are being found for classical thin films (of μm size) of aluminum in deposited circuits, very low temperature thermometry, and more exotic applications (such as dc transformers) that make use of the superconducting properties of the films. Here the discussion concentrates on the relatively large conductors.

The general topic of dc size effects in metals is the subject of an excellent review by Brandli and Olsen (1969). Three parameters appear at least implicitly, in all dc size-effect theories: the ratio

$K = d/\ell_b$, of the characteristic specimen dimension to the bulk electronic mean free path; the specularity parameter, p, which is the fraction of electrons specularly reflected from the surface; and the product $\rho_b \ell_b$ for the bulk metal, which should be a constant independent of size or temperature. This product is given in free electron theory by

$$\rho_b \ell_b = m v_f / e^2 n \tag{5.9}$$

where n is the free electron concentration, v_f is the Fermi velocity, and e and m are the electronic charge and mass, respectively. If one considers the real Fermi surface of a metal, one arrives at a more realistic (and more complex) form for the conductivity in terms of an integral over the surface of the momentum-dependent mean free path. The simpler expression above will suffice for our purposes. Also, in the following presentation p is assumed to be zero, corresponding to diffuse scattering at the surface. The exact value of p and its dependence on various parameters is a topic of great debate. The experimental literature indicates that zero is as good a value as any for calculating correction terms, particularly when one considers the uncertainties in the other parameters.

The parameter $\rho_b \ell_b$ is not truly constant with temperature for real metals, i.e., deviations from Matthiessen's rule occur (Boughton and Neighbor, 1972). Furthermore, the value determined by different types of experiments is highly variable. Brandli and Olsen (1969) list values of $\rho_b \ell_b$ for a variety of pure metals, which vary from 3.5 to 68×10^{-12} $\Omega \cdot cm^2$, with most between 6 and 20 $\times 10^{-12}$ $\Omega \cdot cm^2$. Values of 6.0×10^{-12} $\Omega \cdot cm^2$ for aluminum and 6.6×10^{-12} $\Omega \cdot cm^2$ for copper at 4 K are representative of the literature. Use of this number is really the only easy method of determining the electron mean free path.

There are many formulas available in the literature for calculation of the size-effect resistivity. The earliest relationships, which prove to be adequate in most instances, are due to a suggestion of Nordheim (1934) for wires

$$\rho / \rho_b = 1 + 1/K \tag{5.10}$$

and the Fuchs (1938) relationships for films, which, like most of their descendants, can be written simply only in the extreme limits of K $(= d/\ell_b)$. The equations, in terms of conductivity, σ, are

$$\sigma / \sigma_b \simeq 0.75\, K \ln K^{-1}, \quad K << 1 \tag{5.11}$$

$$\sigma / \sigma_b \simeq 1 - (3/8)K \quad K >> 1 \tag{5.12}$$

The specularity parameter, p, appears in the more general form of these equations, but here it has been set to zero as discussed above.

Size effects can be of importance in the world of low-temperature conductor applications. Figure 5.9 shows the measured vs. the bulk residual resistivity ratio for copper wires of quite reasonable size. One should also note that a magnetic field applied to a metal showing a significant size effect may cause a dramatic decrease in the measured low-temperature resistivity (Brandli and Olsen, 1969).

5.4 ALLOY RESISTIVITY

The alloys of concern here are the nondilute mixtures of metals. In almost all of these alloys, the lattice resistivity is significantly smaller than the temperature-independent components, and thus, there is often not much difference between the resistivity measured at room temperature and that measured near 4 K. In general, however, the variety of behavior exhibited by alloys is astonishing in its diversity.

5.4.1 Well-behaved alloys

As anyone familiar with the workings of Murphy's Law might predict, there are few alloys that fall into this category. In the most classic cases, such as Ag–Au and Cu–Au, the component metals are mutually soluble across the entire range of composition. The resistive behavior of these two alloys is shown in Fig. 5.10 (from Bardeen, 1940, reporting earlier data) and Fig. 5.11 (Johansson and Linde, 1936), respectively. Apparently no

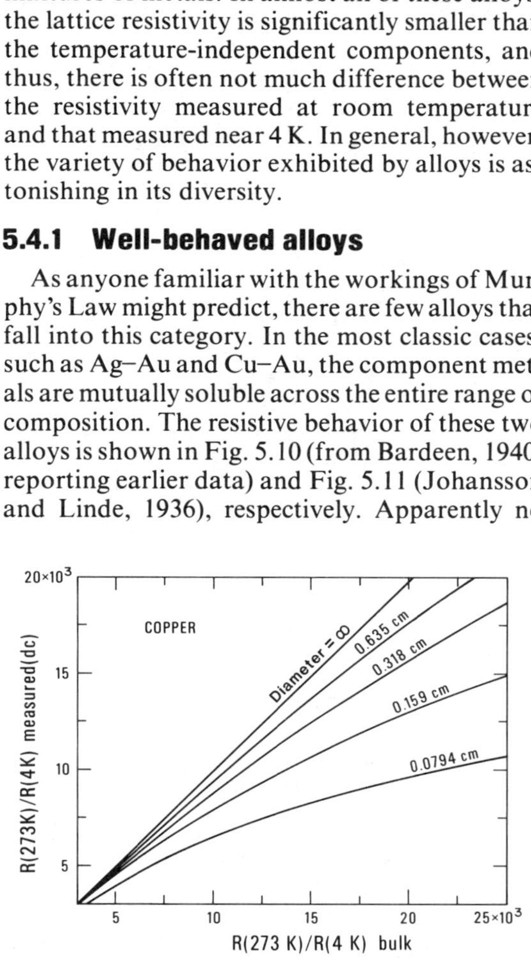

Figure 5.9 Measured vs. bulk residual resistance ratio for copper wires of various sizes. Calculation uses $(\rho\ell)_{bulk} = 0.66 \times 10^{-11}$ $\Omega \cdot cm^2$.

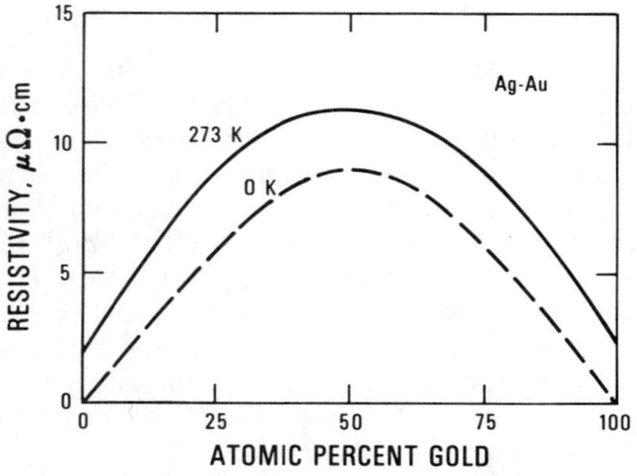

Figure 5.10 Resistivity of the well-behaved alloy system Ag–Au (Bardeen, 1940).

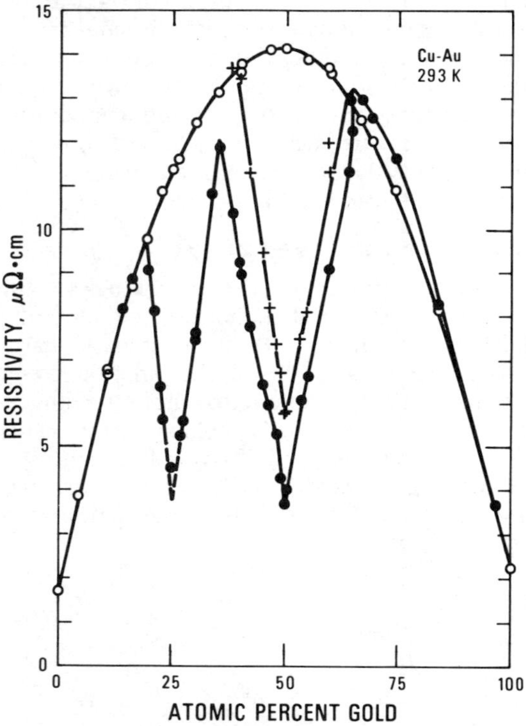

Figure 5.11 Resistivity at 293 K of the alloy system Cu–Au. o — from the well-behaved disordered alloy quenched from 650°C; • — the effect of ordering introduced by a 200°C anneal, + — the effect of a furnace cool of the alloys (Johansson and Linde, 1936).

recent measurements covering the entire composition range have been made on any binary alloy system. Bardeen also presents curves from the early literature showing similar behavior for In–Pb, Pt–Pd, and K–Rb. These alloys are homogeneous solid solutions with a random distribution of atoms. As Fig. 5.10 implies, the soluble, nonordered alloys have both a temperature-dependent and a residual contribution to their resistivity. The residual resistivity ratio over the midrange compositions is not large. A recent measurement on an Ag–25% Au alloy (Edwards, 1972) gave RRR = 1.3, and from Fig. 5.10, it looks like this value is roughly maintained over the entire midrange. The residual resistivity can be described by Nordheim's rule

$$\rho_0 \propto x(1-x) \tag{5.13}$$

where x represents the fraction of one of the constituent atoms. Note that the resistance change with composition is very large even for these simple alloys.

Other alloys that fall in the well-behaved category are those with a solid-solution phase that covers a relatively large range of composition. Of the technically important metals, the Cu–Zn alloys containing $\lesssim$35 wt.% Zn are in this class. These are the classical brasses. Figure 5.12 is from the data of Henry and Schroeder (1963) on alloys made from pure copper (RRR ~ 1300). The resistivity ratio here is also quite small, above a few percent zinc. Their Fig. 6 shows a linear temperature dependence of the resistivity for all the alloys above 40 to 50 K.

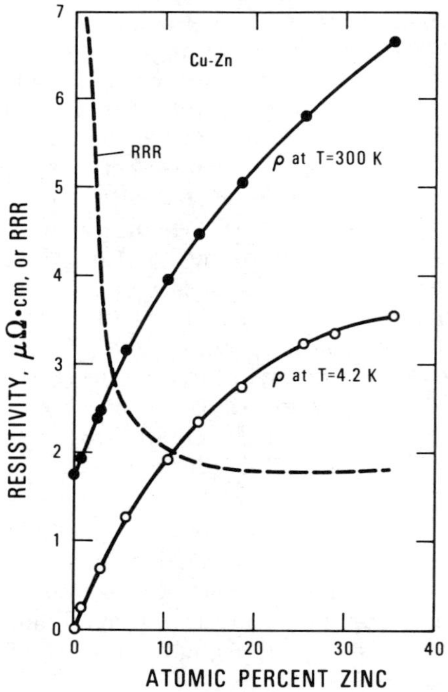

Figure 5.12 Behavior of the α-phase resistivity of brass (Henry and Schroeder, 1963).

Another system, not strictly an alloy, that exhibits uncomplicated resistive behavior is the heterogeneous mixtures of mutually insoluble metals, such as Pb–Zn or Zn–Cd, in which the two elements each form small crystallites. The resistivity is an average of values for the two pure metals and is monotonic across the composition range — it is not often linear. The temperature dependence should also, of course, be an average. The Pb–Sn solders are of this type, but the solid is more complex, consisting of a eutectic mixture (also found in Zn–Cd) and crystallites of lead with up to 25% Sn at the lead-rich end. Regardless, the observed resistive behavior is as stated above.

5.4.2 Alloys with ordering and amorphous alloys

Some disordered binary alloys can be ordered by annealing well above room temperature. When this occurs, the two types of atoms each form their own separate crystallographic structure within the alloy. Thus, the A atoms will be preferentially located in one sublattice and the B atoms in another. This ordering results in a lowering of the resistivity at all temperatures. The classical alloy system in this group is Cu–Au. The effect of ordering on the resistivity is shown in Fig. 5.11. The highest degree of ordering occurs at the compositions CuAu and Cu_3Au.* The existence of the ordered sublattices has been confirmed by x-ray techniques. The author knows of no alloys that undergo this type of ordering below room temperature — certainly none of technological importance.

As one might expect, the resistivities of ordered alloys show a greater temperature dependence than do their disordered counterparts discussed above. There is only a limited amount of information available on this topic. Cohen and Bever (1960) report RRR = 4.0 (ρ = 4.27 $\mu\Omega$·cm at 293 K) for ordered Cu_3Au and RRR = 1.2 (ρ = 11.32 $\mu\Omega$·cm at 293 K) for the disordered alloy. The latter value is also indicated by the data of Varotto and Vidoz (1970). Similar ordering behavior was observed in alloys of Ag–Mg by Gangulee and Bever (1968) with RRR values ~13 for ordered Ag–Mg. There is no listing in the literature of alloys that exhibit ordering. The most comprehensive treatment is not very recent (Nix and Shockley, 1938). The general field of order–disorder phenomena is reviewed by Guttman (1956). Low-temperature data have been reported for Ni–Fe (>65 at.% Ni), Cu–Pt, β Ag–Zn, β Ag–Cd, and β-brass. In the last three materials, the occurrence of other alloy phases tends to complicate any attempt to observe the ordering (Abriata, Cabo, and Kittl, 1968).

Alloys of the form Au_4X, where X stands for various transition elements, have been studied (Toth, Arrott, Shinozaki, Werner, and Sato, 1969; Maple and Luo, 1967). These alloys show both an atomic ordering at high temperature and a magnetic ordering, which usually occurs below room temperature. The resistivity always decreases below room temperature with a change to a steeper slope below the magnetic transition. According to Toth et al. (1969), residual resistivity ratios tend to be around 2 to 3, and the alloy resistivities at 4 K vary from ~5 $\mu\Omega$·cm for Au_4Mn to 145 $\mu\Omega$·cm for Au_4V. On the other hand, Maple and Luo (1967) report 95 $\mu\Omega$·cm at 4 K (RRR = 1.7) for the ordered alloy Au_4V. The same alloy in the disordered state shows a slowly increasing resistivity with decreasing temperature to about 40 K, followed by a slight decrease to 4 K (ρ = 217.5 $\mu\Omega$·cm at 4 K, RRR = 0.96).

Amorphous alloys are generally thin films deposited on cold substrates. Much of the existing work has been done on semiconducting or magnetic alloys, usually those containing at least one nonmetallic component. The temperature and magnetic field dependencies of the resistivity of a number of these alloys are reported by Cochrane and Strom-Olsen (1977). Measurements on the low temperature resistivity of amorphous binary metallic alloys have been made by Korn and co-workers. A review of the work on alloys of gallium, tin, lead, and bismuth with the noble metals is given by Korn, Pfiefle, and Zibold (1974). Nearly all such alloys have high resistivities (~50 to 150 $\mu\Omega$·cm at 4 K) and residual resistivity ratios close to unity. The resistivity may rise or fall with decreasing temperature depending on the alloy makeup and the specific elements involved: amorphous Cu–Sn alloys show a resistivity drop with decreasing temperature below 40% Cu, a peak at near 40 K for 40 to 60% Cu, and a rise at higher values. The residual resistivity increases with increasing copper content. A more recent paper (Korn and Pfeifle, 1976) describes measurements on amorphous Cu–Fe alloy films covering the entire composition range. Residual resistivities range from 16 $\mu\Omega$·cm for unalloyed copper to a peak of ~205 $\mu\Omega$·cm at

*These are not intermetallic compounds. They do not precipitate from the alloy, and they do not have a high resistivity.

80% Fe. The slope of the ρ–T curve is always positive and is low up to ~10% Fe but then increases sharply to a maximum at ~30% Fe.

5.4.3 Alloys with phase transitions or precipitation

A large number of alloys, both ferrous and nonferrous, undergo phase transformations below room temperature. These transformations are termed martensitic and the resulting material is called martensite. The transformations are a diffusionless shear type and the resulting martensite takes the form of plates, sheets, and needles, with an accompanying macroscopic shape change or change in surface relief. The transformation may occur spontaneously upon cooling below a characteristic temperature M_s, or it may be induced by stress below a characteristic temperature M_d. Thermally induced transformations are at least partially reversible, but often with a large hysteresis. Martensitic transformations are common in many materials of technological interest, such as stainless steels, brasses (β), titanium alloys, lithium, and sodium, to name only a few. The physics and metallurgy of the transformation are treated in many texts (see Chapter 9). A survey of the literature regarding specific low-temperature transformations is given by Reed and Breedis (1966).

With so many alloys involved, one should not expect a great deal of uniformity in the resistive behavior and, in fact, that is what is observed. Quite a few materials, however, show resistance–temperature curves like those shown in Fig. 5.13, (Hummel, Koger, and Pasupathi, 1968) for β-brass. The normal decrease of resistivity with temperature is interrupted by the transformation. The resistivity increases while the higher resistance martensite is forming, followed by a continuation of the decrease to the residual resistivity value (not reached in the experiment shown). The warming curves show the reversibility of the thermally induced martensite. The deformation martensite remains. Note that the resistivity values are quite high.

Another special group of alloys is that in which age hardening occurs owing to the slow precipitation of particles of a second phase relatively near room temperature. A number of commercial aluminum alloys are of this type. Whether the resistance initially increases or decreases upon isothermal annealing seems to depend on the temperature and, of course, on the material. Federighi and Ceresara (1964) show

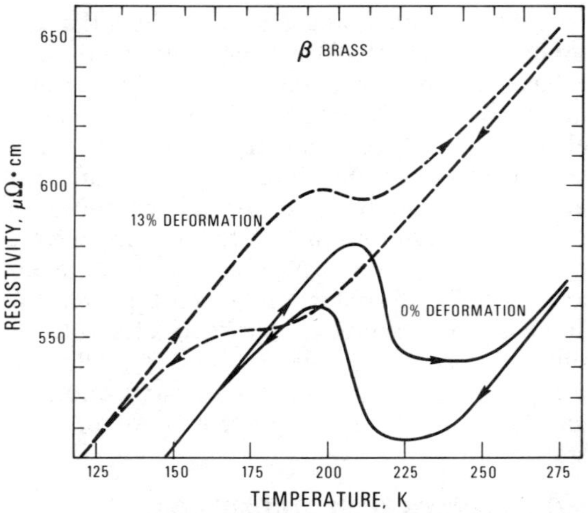

Figure 5.13 The effect of martensitic transformation of the resistivity of β-brass (Hummel, Koger, and Pasupathi, 1968).

both types of behavior for an Al–5% Mg alloy. The resistivity decreases on annealing at $T >$ 20° C and increases for $T <$ 20° C (down to −40° C). The increase in resistivity is taken as an indication of the formation of Guninier–Preston zones prior to the precipitation (Osamura, Hiraoka, and Murakami, 1973). One would expect that the resistivity at all temperatures would ultimately decrease as the precipitation goes to completion, and this is implied by the isochronal annealing experiments by Federighi and Ceresara (1964) as well as by Bardeen (1940). Low-temperature resistivity data have been reported on Al–Zn, Al–Ag, Al–Cu, Cu–Be, and Al–Mn alloys.

5.4.4 Magnetic alloys

There are two groups of interest here: alloys in which a transition element is dissolved in a noble metal and the magnetic ferrous alloys, such as the steels. The alloys of the latter group usually show only a small temperature dependence of the resistivity, most often a slight decrease with decreasing temperature.

The magnetic alloys involving a noble metal and a transition element show ferromagnetic behavior at some concentration of the magnetic element. The Curie temperature is concentration dependent and may fall well below room temperature (see Chapter 6). Three alloy systems have been investigated in some detail: Cu–Ni, Ag–Pd, and Au–Ni. Of these, the first has been the most studied. The behavior of the resistivity with alloy concentration is still quite regular at all temperatures, as shown in Fig. 5.14 and Fig. 5.15, and as expected, the resistivity ratios are not high. They are highest for the ferromagnetic compositions. Within this range, however, a wide variety of behavior is observed as the temperature is decreased. The rate at which the resistivity drops with decreasing temperature always increases below the Curie temperature. This is illustrated by the Cu–Ni data of Svensson (1936), which have been quoted by nearly everyone, and those of Clinton, Tyler, and Luo (1974) on Au–Ni, which show Curie temperatures to below 4 K. The nonmagnetic alloys in these series show maxima and minima as well as general increases in resistivity with decreasing temperature (Houghton, Sarachik, and Kouvel, 1970a). Furthermore, the behavior of a given specimen may change with time of storage at room temperature (Legvold et al., 1974) and is strongly affected by degree and type of deformation (Kawatra and Budnick, 1972). Although very interesting from the viewpoint of metal physics, these varieties of behavior usually cause only a small relative change in the resistivity — on the order of several percent.

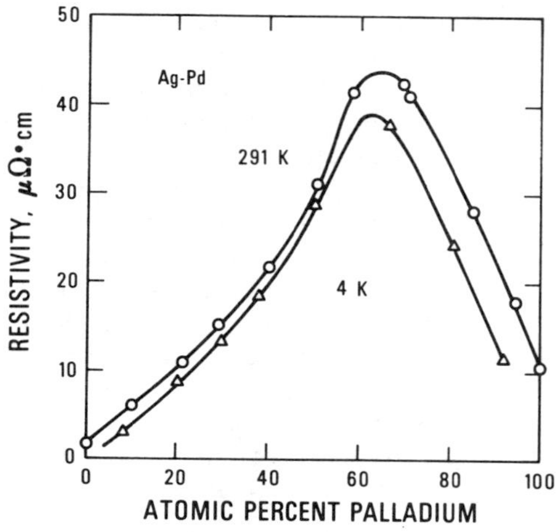

Figure 5.15 The variation of resistivity with composition for the magnetic alloy Ag–Pd. ○ — Svensson (1932), △ — Murani (1974).

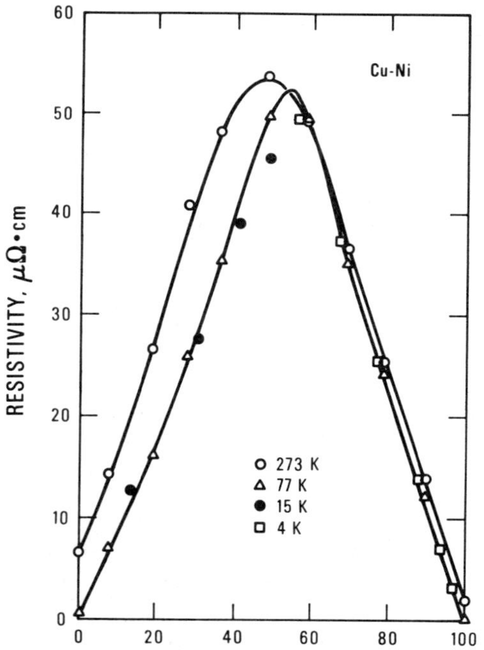

Figure 5.14 The variation of resistivity with composition for the magnetic alloy Cu–Ni, ○, △, and solid curves — Iguchi and Udagawa (1975); □ — Legvold, Peterson, Burgardt, Hofer, Lundell, Vyrostek, and Gartner (1974); • — Krupkowski and de Haas (1928). Agreement with Svensson's (1936) data is excellent.

5.4.5 Zero-temperature-coefficient alloys

A surprisingly large number of low-temperature applications require alloy wires that have been specifically designed for minimum resistivity variation over a certain range of temperature. For wires used in electrical heaters for calorimetry and thermal conductivity experiments, the entire range of temperature from 4 K to room is involved. A 1 to 5% variation in resistivity is acceptable. Wires used in temperature controllers for galvanomagnetic measurements, such as magnetoresistance, should be stable from 2 to 20 K, and should show as little variation with magnetic field strength as possible (see § 5.5.1). Variations of less than 2% are usually adequate. A wide range of new applications of the Josephson effect in voltage and current standards and comparators (Kamper, 1977) requires resistance alloys with part-per-million stability over a temperature range of a few degrees around 4 K. Magnetic field effects are frequently of concern in these devices. Finally, experiments at temperatures below 1 K often need special heaters with a low resistance coefficient. There is only limited information available in this very low temperature range. Consult the paper by Cieloszyk, Cote, Salinger, and Williams (1975) for more information.

Materials usually employed for the room-temperature to 4-K range tend to be the classical resistance alloys: constantan, manganin, Ni–Cr, and others. Some of these alloys and their properties are described in Table 5.8. These materials tend to have very high room-temperature resistivities and to show very little change in resistance with decreasing temperature to about 250 K. Below that temperature, each alloy shows a decrease in resistivity, but the amount, as shown in the table, varies greatly. The magnetoresistance is small in nearly all instances. Its actual behavior with field varies from linear, with either sign, to quadratic to very complex, but there seems to be no significant dependence on the relative orientation of the magnetic field and the wire.

The alloys used for ppm stability applications near 4 K are quite different from the classical resistance alloys just described. They tend to be not as highly alloyed and, thus, have somewhat lower resistivities, on the order of 2 to 20 $\mu\Omega$·cm. Furthermore, most of the alloys are not available commercially. Copper- and aluminum-based alloys are most often used for a variety of reasons: (a) they tend to have shallow Kondo minima in their resistivity vs. temperature curves in the region of interest owing to natural or added magnetic impurities; (b) starting metal is available with the required high purity; (c) their magnetoresistance can be made quite small by proper alloying; and (d) they are materials that tend to be both strong and reasonably ductile, thereby facilitating the production of wire.

Investigations of aluminum-based alloys have been reported by Warnecke and Kose (1977) and by Cimberle et al. (1976). The first of these describes a commercial Al–5Mg alloy showing a temperature coefficient of resistance (TCR) of $\sim 2 \times 10^{-7}$ K^{-1} between 2 and 4 K. The remainder of the alloys exhibit TCR values below 10^{-4} K^{-1} over this range with many $\sim 10^{-6}$ K^{-1}. The magnetoresistance of the Al–5Mg alloy at 4 K was $\sim 7\times$

Table 5.8 Low-temperature properties of resistance wires.

Wire	Nominal Composition (wt. %)	Nominal Resistivity at ~293 K ($\mu\Omega$·cm)	Percent Change to 4 K (%)	Magnetoresistance $\Delta\rho/\rho$ at 4 K and B (%)	References and Comments
Evanohm	75Ni–20Cr–2.5Al–2.5Cu	134	−0.9	—	Furakawa, Reilly, and Saba (1964)
		—	−0.76	—	Hust (1972)
		76.4	−0.9[a]	0 to 5 T	Lerner and Daunt (1964)
		128	—	+0.03–0.66 to 0.6 T +0.23–0.83 to 18 T	Woollam (1970); Actual $\Delta\rho/\rho$ depends on sample
Karma	76Ni–20Cr–Fe–Al	133	−0.9	—	Furakawa, Reilly, and Saba (1964)
Constantan	57Cu–43Ni	49	−6	—	Ibid.
		27.7	−7	−4.6 to 5 T	Lerner and Daunt (1964)
Manganin	84Cu–12Mn–4Ni	48	−10	—	Furakawa et al. (1964)
Ni–Cr	—	—	—	+0.002	Reported by: Cimberle, Michi, Mori, Rizzuto, Siri, and Vaccarone (1976)
PtW	92Pt–W (assumed wt.%)	66	−8	+0.024 to 9.5 T	Giauque, Lyon, Hornung, and Hopkins (1962). ρ and $\Delta\rho$ reported by Cimberle et al. (1976)
Silicon bronze	96Cu–3Si–1Zn	28.3	−14	+0.005 to 14 T	Feldman, Talley, Rojeski, Vold, and Woollam (1977). Cu–Si is not a commercial alloy.
Phosphor bronze	95Cu–5P	16.1	−23	+0.007 to 14 T	Ibid.

[a]Approximate

10^4 T^{-1}, whereas the 10^{-6} K^{-1} alloys showed $\sim 7 \times 10^{-5}$ T^{-1}, both to high fields. At very low fields, <0.1 T, the sensitivity is apparently much smaller.

Three copper-based alloys have been investigated in detail: phosphor bronze (reported by Cimberle et al., 1976), Cu–Ge (Warnecke and Kose, 1975), and silicon bronze (Sullivan, 1971). The 2 to 4–K TCR values all are 2 to 4×10^{-6} K^{-1} (the Cu–Ge and bronze TCRs are negative), and the silicon bronze magnetoresistance is $\sim 10^{-5}$ T^{-1}. Similar resistivity behavior has been reported for Cu–Ni and Cu–Pt alloy resistors by March and Thurley (1976). One of their Cu–Ni alloys shows $\Delta\rho/\rho < 1 \times 10^{-6}$ between 1.5 and 4.2 K.

5.5 EXTERNAL RESISTIVE MECHANISMS

Magnetic fields, pressure, and stress can have significant effects on the resistivity. These effects appear stronger at low temperatures, because the intrinsic (phonon) resistivity component is so small.

There are many ways in which a magnetic field can affect the transport properties of solids, and each one of them has its own name. As a group they are termed *galvanomagnetic properties.* Blatt (1968) lists 16 of these in his Table 7.7, and all solid-state physics texts discuss at least some of them in great detail. Galvanomagnetic effects in the anisotropic metals are the subject of a review by Hurd (1974). From a low-temperature electrical applications point of view, only magnetoresistance, Hall effect, and magnetothermopower are of interest, and of these, only the former is a common concern. Effects caused by time-varying magnetic fields are also important, but they are covered in § 5.6.

The effect of hydrostatic pressure on the resistivity has been investigated for a large number of conductors. It is seldom a problem in applications, but it is included here because questions sometimes arise. Stress, on the other hand, is a serious problem in nearly all conductor applications. Fatigue effects, too, must often be considered.

5.5.1 Magnetoresistance

All nonmagnetic metals show an increase in electrical resistivity with increasing magnetic field. This magnetoresistance effect has been one of the most extensively studied of the transport properties, and data exist on nearly all pure metals and a great many alloys. On copper alone there were some 60 papers published through 1972. The primary reason for the interest in this phenomenon is its ability to indicate the topology and connectivity of the Fermi surface from the transverse magnetoresistance (field applied normal to the electric current) of single crystals, and the anisotropy of the electron scattering from the longitudinal magnetoresistance. These measurements are invariably performed at liquid-helium temperatures because the magnitude of the effects scales with the product $\omega\tau$ (ω is the cyclotron frequency and is proportional to the field, H; τ is the relaxation time and scales roughly with ρ^{-1}). A low-temperature experiment is necessary to achieve the required condition $\omega\tau > 1$ with real metal samples. These topics are covered in detail in most of the texts already mentioned.

The behavior of the magnetoresistance of polycrystalline metals used in various technical applications at low temperatures is the concern here. These metals are used primarily as conductors of electricity; the highly alloyed metals used in structural applications usually show a negligible magnetoresistance. It should be stressed that magnetoresistance is a phenomenon that is important only at low temperatures and only for pure metals. A measurable effect, even at 4 K, with the magnetic fields usually available, requires resistivity ratios on the order of 10 to 100 for the common metals.

The observed behavior of many polycrystalline metals can be described by an empirical rule developed by Kohler (1938). The mathematical form of the rule is:

$$\Delta\rho/\rho_0 = f(H/\rho_0) \qquad (5.14)$$

where H is the applied field and ρ_0 is the resistivity at zero field. The function f for polycrystalline metals is single valued, increases monotonically, and depends only on the metal and the relative orientation of the field and current. Note that the magnetoresistance, therefore, does not depend on impurity content or temperature, except as they affect ρ_0. Kohler diagrams are log–log plots of observed data. Usually the plot is of $\Delta\rho/\rho_0$ vs. $H \cdot RR(T)$ where $RR(T)$ is the resistivity ratio $\rho(273\ K)/\rho(T)$. When a number of metals are compared, a reduced Kohler plot is often used in which $\rho(273\ K)$ is replaced by the resistivity at the Debye temperature. One such plot, which has appeared in nearly every review (see Fig. 46 of Meaden, 1965), presents data for a large number of metals. The data are quite old and a reevaluation is needed, but the general behavior is as predicted.

Copper shows an amazing degree of agreement with Kohler's rule. A plot made from the author's data (Fickett, 1972b) is shown as Fig. 5.16. The 200 data points used represent temperatures from 4 to 20 K, fields to 9 T, and residual resistivity ratios from 10 to 10 000. There is some evidence that extended defects, with their temperature-dependent resistivity, may cause deviations from this plot. Also, radiation damage studies have shown deviations, especially at low fields (Williams, Klabunde, Redman, Coltman, and Chaplin, 1979). Note that field-induced resistance increases by a factor of 100 are possible in very pure copper!

Aluminum shows no inclination to obey Kohler's rule (Krevet and Schauer, 1976) for a wide range of parameters. The resistivity rises and then nearly saturates as the field is increased. At a given field, $\Delta\rho/\rho_0$ shows a peak with increasing temperature at near 20 K. Maximum values of $\Delta\rho/\rho_0$ run from 0.5 to 5, far below those observed for copper.

Some general comments: Nearly any polycrystalline metal will show some agreement with Kohler's rule over a small purity range and at fixed temperature, and the Kohler plot will be linear. Very few metals show agreement when wide ranges of temperature, purity, defect concentration, and field are used. The longitudinal magnetoresistance of a given sample is always lower than the transverse, usually by a factor of 2 or more. The longitudinal magnetoresistance always saturates with increasing field. Samples

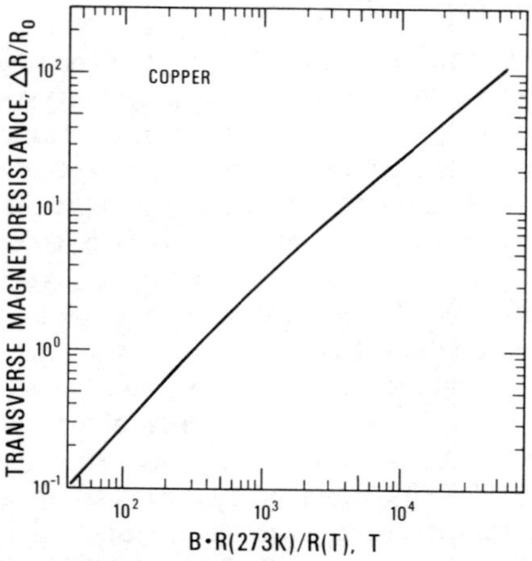

Figure 5.16 Kohler plot for the transverse magnetoresistance of copper.

whose dimensions are on the order of the electron mean free path or smaller will show a decrease in resistance with applied magnetic field. Ferromagnetic metals and alloys frequently show a decrease in magnetoresistance with increasing field. For alloys, the drop is usually small compared with the residual resistivity; for example, Houghton, Sarachik, and Kouvel (1970b) report $\Delta\rho \sim 0.25\ \mu\Omega\cdot$cm for a Cu–42 at.% Ni specimen with a residual resistivity of 47 $\mu\Omega\cdot$cm. Pure ferromagnetic metals, on the other hand, may show very large changes. Takaki and Igaki (1976) report a drop in ρ_0 from ~40 n$\Omega\cdot$cm to ~3 n$\Omega\cdot$cm for a polycrystalline iron specimen with RRR = 2500.

It is usually the resistivity at a given field and temperature that is important in applications. Figure 5.17 shows the behavior of this quantity for aluminum and copper as the purity of the metal increases. The message is clear: beyond a certain value it makes no sense to increase the purity, since the increased magnetoresistance compensates for the decrease in zero field resistivity.

5.5.2 Hall effect

When a magnetic field is applied normal to a conductor carrying current, an electric field, E_H, appears across the conductor normal to both the field and the current owing to the Lorenz force on the current carriers. This is the Hall field. The Hall coefficient, R, is defined by the equation:

$$E_H = RJH \tag{5.15}$$

where H is the applied field and J the current density. This is a somewhat simplistic description; R actually is made up of off-diagonal elements of the tensor resistivity. The Hall field is an odd function of H. Sometimes voltages appear on the Hall-field measuring probes that are even functions of the field (transverse-even and planar-Hall voltages). Thus, a measurement of R almost always requires reversal of both H and J.

The Hall voltages that arise in pure metals and alloys may be of either sign and tend to be very small. However, because R does not decrease greatly with decreasing temperature, the component of the Hall voltage that appears on misaligned voltage probes attached to a pure metal sample at 4 K can be very large compared with the resistive voltage. Typical data on the temperature dependence of R for simple metals are shown in Fig. 5.18, adapted from Alderson et al. (1968).

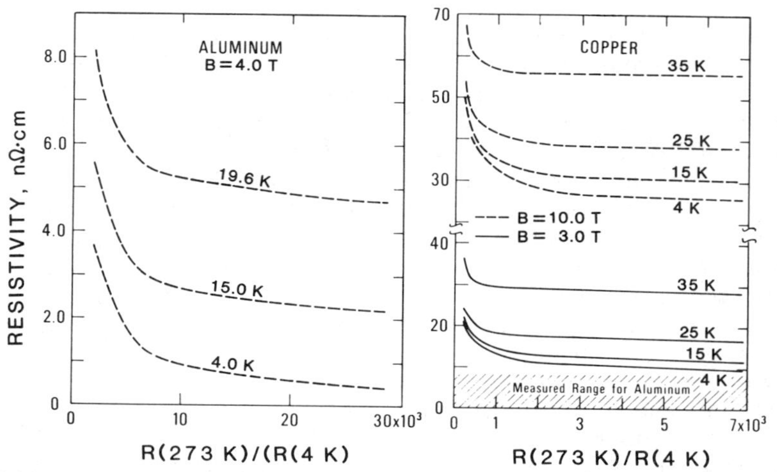

Figure 5.17 Resistivity as a function of purity as measured in the transverse configuration at several magnetic fields for aluminum and copper.

In simple theory, R is proportional to the inverse of the number of conduction electrons. Its detailed behavior in various field regions (R really is not a constant at all) tells a great deal about the Fermi surface and the scattering interactions that occur on it. Furthermore, the Hall effect in semiconductors and semimetals (which have very low free electron concentrations) is very large (cf., $R \simeq 5 \times 10^{-11}$ m^3/A·s for copper and $R \simeq +10^4$ m^3/A·s for indium-doped germanium) and usually increases with decreasing temperature. Thus, these materials are often employed in the measurement of magnetic fields.

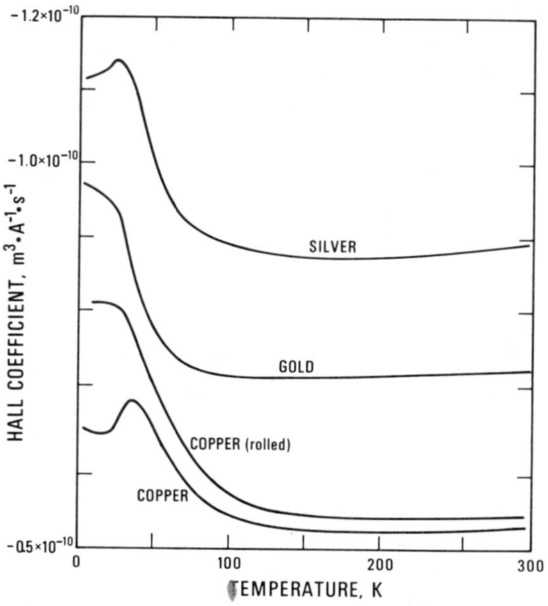

Figure 5.18 Temperature dependence of the Hall coefficient as measured by Alderson, Farrell, and Hurd (1968) for very pure ("six nines") polycrystalline noble metals at 1.206 MA·m^{-1} (15.16 kOe).

There is an enormous amount of data available on the Hall effect in nearly any pure metal and a surprising number of alloys. The comprehensive book by Hurd (1972) covers the physics of the effect and reviews all the data.

In magnetic metals, an anomalous Hall effect occurs in which the Hall voltage rises linearly with applied field to a certain field value and then changes to a linear rise with a much lower slope. This behavior is seen not only in ferromagnetic metals, but also in other metals with strong localized moments. In fact, it may present a means of observing magnetic transitions in alloys, such as the paramagnetic-to-antiferromagnetic transitions that occur at low temperatures in the stainless steels. Apparently no Hall effect data exist on the stainless steels.

5.5.3 Pressure effect

It is not often that electrical conductors are subject to hydrostatic pressure in actual applications. Meaden (1965) and Gerritsen (1956) describe the experimental situation adequately. The early very comprehensive work of Bridgman on the resistivity variation of nearly all metals with pressure is still the standard. That work, and some later experimental results have been compiled and reviewed by Lawson (1956). Bundy and Strong (1962) provide a further updating and a metal-by-metal review of the data.

Bridgman noted that there were 31 normal metals for which the resistivity decreased with increasing pressure. This group includes all the commonly used metals. The changes induced in the room-temperature resistance of most of these metals tend to be relatively small, varying from 1 to 5% of the resistivity at atmospheric pressure when measured at pressures to $\sim 10^8$ kg·m^2. The percentage changes were observed to be somewhat larger at lower temperatures. At higher pressures, $d\rho/dP$ becomes smaller. The total resistance decrease for copper at 10^9 kg·m^{-2} was observed to be ~13% vs. 1.9% at 10^8 kg·m^{-2}.

The second group of metals, which includes the alkalis, semimetals, and rare earths, shows a wide range of behavior, often with very large changes in resistivity. Nearly all the work that has been reported in the literature since the reviews mentioned above has been on these materials.

To return briefly to the case of copper, a particularly detailed experiment was performed by Dugdale and Gugan (1957) on two commercial wire samples and one of copper containing 0.056 at.% Fe. They indicate that the rate of decrease of the *intrinsic* resistivity with pressure is temperature dependent and about 4 times higher at 4 K than at room temperature. On the other hand, the *residual* resistivity increases with pressure at temperatures up to 80 K, the rate of increase becoming smaller as the temperature increases.

5.5.4 Stress and fatigue

Application of either constant or variable stress to a conductor nearly always causes the resistivity to increase. At stress levels in the elastic range of the metal, the increase is reversible on removal of the stress. This effect is the basis of the alloy strain gauges. Their properties and low-temperature behavior are discussed in Chapter 14. As metals are stressed into the plastic region, irreversible changes occur as defects are formed. In real applications, the defects described earlier never occur alone, and seldom can the relative amounts of each be determined. The problem in conductor applications is to determine how the electrical resistivity changes as the metal is subject to the stresses and temperatures encountered in operation. The actual resistivity increase of a metal under stress is a strong function of many parameters, including:

a. *Temperature*. The lower the temperature, the greater the increase in resistivity for a given deformation.

b. *Starting condition*. Prior cold work may increase the initial strain dependence of the resistivity.

c. *Alloying elements*. In general, a larger strain sensitivity is observed with increased alloying content. Some ordering alloys, however, may show a decrease in resistivity on straining.

d. *Type of stress applied*. Rolling, drawing, simple tension, shear, and compression all apparently give different results for the same amount of deformation, although the data are very limited.

One of the most important of these features from a low-temperature application viewpoint is the effect of temperature of deformation on the resistivity, as illustrated in Fig. 5.19 for copper (Broom, 1952). The tendency of the curves toward saturation is generally observed at large deformations. At a given low temperature, the curves are fit by the empirical equation

$$\Delta\rho = K\epsilon^n \tag{5.16}$$

where K is a constant and $n \sim 1.0$ to 1.3 for aluminum (Arp, Kasen, and Reed, 1969) and 1.4 to 1.8 for copper (Welles, 1967). Typical additional resistivities induced by 1 to 30% strains are in the range 1 to 200 μΩ·cm in these two metals.

Metals that are severely deformed or irradiated at low temperatures show large increases in the residual resistivity. The resistivity of the annealed state is at least partially regained as the temperature is raised. The so-called recovery

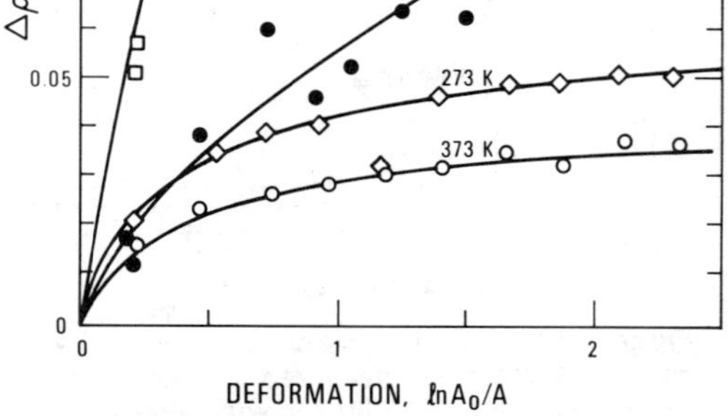

Figure 5.19 Effect of temperature on the resistivity increase due to deformation of copper by drawing (Broom, 1952)

curve shows that the residual resistivity decreases in a number of stages that occur in rather well-defined temperature ranges. There are a total of five main stages, which may, in turn, show sub-stage structure. The stages, in order of increasing temperature, are: I. Recombination of nearby vacancies and interstitials, most often seen in irradiated specimens. Many substages are observed. II. Recombination of diffusing vacancies and interstitials and migration to dislocations. III. Clustering of vacancies and migration of clusters. IV. and V. Annealing of dislocations in plastically deformed specimens. A schematic view of aluminum data compiled by Reed (1972) is shown in Fig. 5.20. For a given material or test, some stages are not always well defined, and the mechanism operative in every case is not totally agreed upon. Often, one or more of the stages are not observed at all. The exact amount of recovery observed for each stage and the range of temperature over which it occurs depend on many factors, such as: (a) the specific metal and its chemical purity, (b) how the initial state of the metal came about, i.e., by irradiation, slight deformation, or large deformation, and (c) the method of deformation or the type of particle used for irradiation.

Fatigue at low temperatures, even at relatively low strain values, can cause significant increases

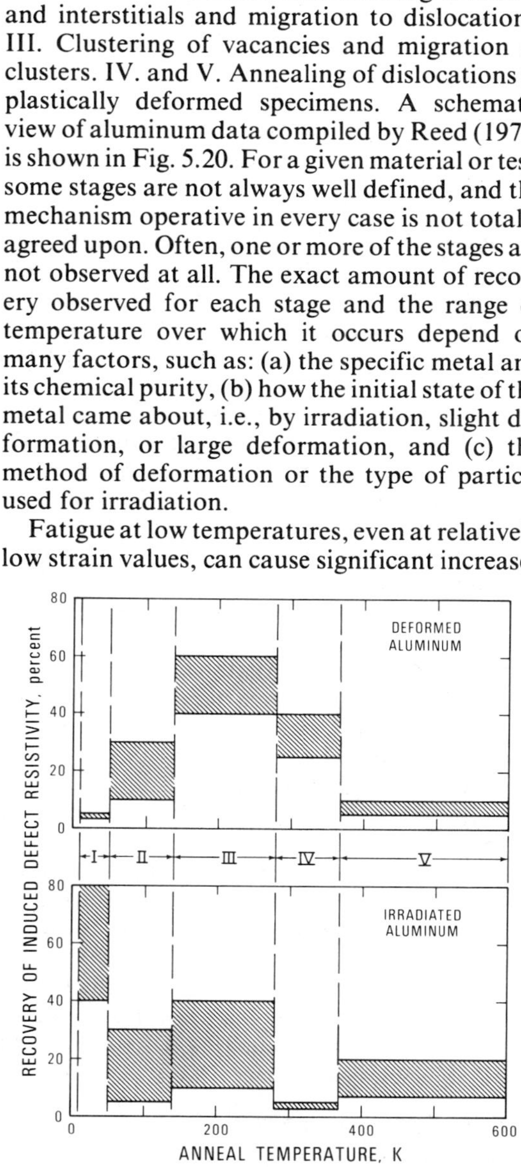

Figure 5.20 The recovery of aluminum as it is annealed through the five stages described in the text. The shaded ranges indicate the wide variations in recovery that may be observed at a given stage, depending on the detailed parameters of the experiment.

in the residual resistivity. The details, of course, depend on the amount of strain, the temperature, and the starting material. This is a topic of current interest, but not much is available yet in the way of data. The limited data that exist indicate that the resistivity rises rather rapidly in the first few cycles and then more slowly, much like the curves of Fig. 5.19 if the abscissa were in cycles. Typical residual resistivities induced in pure copper by fatigue (0.2% constant strain) to several hundred cycles range from about 2 nΩ·cm for RRR = 2200 starting material to 11 nΩ·cm for RRR = 223 (Fisher, Kim, and Turner, 1976). Recent measurements on copper as a function of strain (to 0.88%) and large numbers of cycles (to 10^5) show the resistivity increase to be exponential over the entire range (Ekin, 1978).

5.6 AC RESISTIVITY

The electrical resistivity of conductors can be significantly higher for alternating current than for direct current, even at quite low frequencies. This is especially true for the pure metals at low temperatures where the electronic mean free paths may be quite long. Furthermore, under ac excitation, power losses due to induced eddy currents occur even in the structural alloys. Although these losses are usually negligible at room temperature, the very low specific heat of all materials near 4 K can lead to unacceptable temperature excursions with very moderate power inputs by room-temperature standards. In conductors, the losses show up as undesirable increased impedances and produce the added heat loads. Ferromagnetic materials suffer additional complex losses because of their hysteretic behavior and temperature- and field-dependent permeabilities.

Dielectrics are most often associated with ac applications and, thus, a brief introduction to their low-temperature electrical behavior is also included in this section.

5.6.1 The skin effect — classical and anomalous

Nearly all texts on electromagnetism (e.g., Jackson, 1962) treat the case of the penetration of an electromagnetic wave into a conductor. The field energy is nearly entirely magnetic in nature and the wave is exponentially damped, decreasing to $1/e \simeq 37\%$ of its surface value at a distance, δ, into the metal. This distance is termed the *classical skin depth*, or *penetration depth*, and for good conductors, is given by:

$$\delta = (2\rho/\omega\mu)^{1/2} \qquad (5.17)$$

where ω is the applied angular frequency and μ is the permeability in henrys per meter. Note that this equation is in SI units and thus ρ must be expressed in ohm-meters and δ has the units of meters. The implication is that alternating currents may be concentrated quite near the surface of a conductor, and at very high frequencies (e.g., ~100 MHz where $\delta \simeq 0.007$ mm for copper at room temperature), they may be described as "flowing entirely on the surface." On the other hand, at powerline frequencies the conductor size is usually less than a few skin depths and the current is *not* confined to the surface. Values of δ are shown in Table 5.9 for metals with RRR = 1000 and a 1.0–Hz current, a rather modest and realistic set of values. Clearly, temperature has a dramatic effect on the ac behavior! Because the current distribution changes, the inductance of the metal also decreases as the frequency increases. Calculation of the ac resistance, R_{ac}, of a given conductor configuration is a complex problem in differential equations. A variety of these calculations are presented by Stoll (1974) and Lammeraner and Stafl (1966). The case of a circular wire of radius, r, is one of the easier to treat. The resulting equations are:

$$R_{ac}/R_{dc} = 1/4 + r/2\delta + 3\delta/32r$$
for $r >> \delta$ (5.18)

$$R_{ac}/R_{dc} = 1 + r^4/48\delta^4$$
for $r < \delta$ (5.19)

These relationships are represented graphically in Fig. 5.21. To this scale, the highest order terms are negligible. This graph, in conjunction with Table 5.9 shows that a wire that has essentially dc characteristics at room temperature ($r < \delta$) may exhibit a much higher resistance at 4 K where $r >> \delta$.

Resistance increases can also occur due to the

Table 5.9 Skin depths for several metals at 1.0 Hz (RRR = 1000).

Material	Temperature (K)	Skin Depth, δ (mm)
Copper	293	66
	4	2.1
Aluminum	293	82
	4	2.6
Iron	293	16
(μ = 100)	4	0.5
Iron	293	5.0
(μ = 1000)	4	0.16

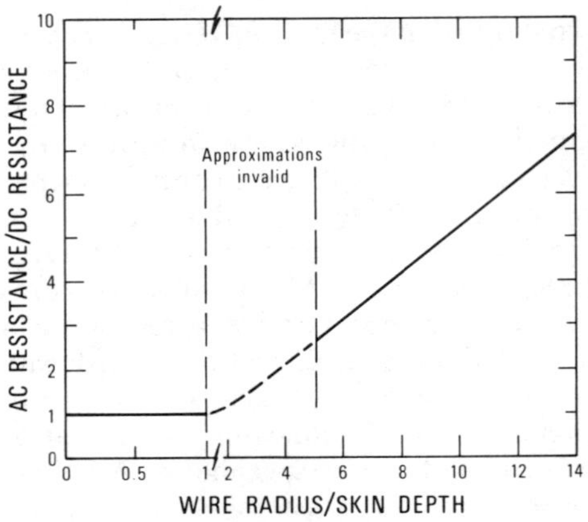

Figure 5.21 The effect of skin depth on the ac resistance of a circular wire.

presence of other active or passive conductors. These can give a "reflected" resistance of significant size (~15% of R_{ac}) at small separations. The details, again, depend strongly on geometry and the interested reader is referred to the texts just mentioned.

In the situations described above, the skin depth is assumed to be large compared with the electron mean free path. In nearly all practical situations this is justified, but at low temperatures it is not always the case and the anomalous skin effect must be considered. The theory is complex (see: Rosenberg, 1963; Ziman, 1960), but in essence it says that the only electrons that contribute to the conductivity are those travelling in such a direction that they do not leave the skin depth during the time required to travel one mean free path. Clearly, as the skin depth gets smaller, the directions satisfying this condition become fewer, approaching grazing incidence. The number of electrons available to carry the transport current become fewer, and the resistance increases. A detailed calculation, under a number of assumptions, shows the surface resistance then becomes independent of the purity of the metal and significantly larger than the value predicted by the classical skin effect. As an example, the anomalous skin effect becomes significant in pure aluminum (RRR = 10 000) at 4 K for frequencies as low as ~10 Hz.

5.6.2 Eddy currents and magnetic diffusivity

The basic mechanisms responsible for eddy currents in conductors are just those alternating field effects described above. Nevertheless, eddy currents are normally considered to be separate dissipative mechanisms caused by time-varying magnetic fields. With a few exceptions, such as their use in determining the resistivity of odd-shaped specimens (see § 5.2.3) and some shielding applications, they are quite undesirable in low-temperature systems. They introduce unwanted heat into the system. They create magnetic fields that then interact with the inducing fields to give unexpected forces, which can be quite large. Many of these problems become more severe at low temperatures. The effect of lowered specific heat has already been mentioned. The much lower resistivity of even commercially pure metals allows a changing field to induce currents that are orders of magnitude larger than would occur at room temperature. These, in turn, greatly increase the various reaction forces.

The easiest problem to analyze is a nonferromagnetic semi-infinite plate normal to the z-axis with a sinusoidal field, H_x, parallel to its surface

$$H_x = H_s \cos \omega t \tag{5.20}$$

Maxwell's equations applied to this system lead to a diffusion equation:

$$\partial^2 H_x / \partial z^2 = (\mu / \delta)\, \partial H / \partial t \tag{5.21}$$

with the solution

$$H_x = H_s e^{z/\delta} \cos(\omega t + z/\delta) \tag{5.22}$$

where δ is the skin depth as defined above. Thus, the field decreases exponentially with increasing distance below the surface (the direction of $-z$), giving the classical skin depth. The peak field below the surface occurs $z/\omega\delta$ seconds after the surface peak, representing a diffusion of flux into the metal. The diffusion constant, or magnetic diffusivity, from Eq. 5.21 is therefore

$$D_{\text{mag}} = \rho / \mu \tag{5.23}$$

The magnetic diffusivity at 4 K tends to be on the order of 1 cm^2/s for pure metals and several hundred cm^2/s for alloys. This number is frequently compared with the thermal diffusivity (see Chapter 4) to determine whether eddy current heating can be removed by conduction as rapidly as it is generated, i.e., is $D_{\text{th}} > D_{\text{mag}}$? This aspect of eddy current heating is particularly important in superconducting magnet systems; it and other topics unique to those materials are treated by Brechna (1973) and discussed in Chapter 13 here.

The calculation of the power dissipated by eddy currents is strongly related to the geometry of the structure and most often requires complex mathematical techniques for a general solution. Stoll (1974) gives many useful techniques and examples. For a flat plate with two sides (reasonably conventional) and of thickness a, located in a parallel uniform field, the eddy current losses per unit area, P_{ec}, in the limiting cases are given by

$$P_{ec} \simeq H_s^2 \rho / \delta, \qquad a >> \delta \tag{5.24}$$

and

$$P_{ec} \simeq H_s^2 (\rho/\delta)(a^3/6\delta^3), \qquad a << \delta \tag{5.25}$$

The SI units of power loss are watts per square meter. It is important that the proper equation be used. The frequency dependence of the losses is quite different in the two cases, varying as $\omega^{1/2}$ in Eq. 5.24 and as ω^2 in Eq. 5.25. In the first case, the eddy currents are said to be inductance limited, and in the second, resistance limited. As long as the simple inequalities $a > \delta$ and $a < \delta$ are not violated, results accurate to ~10% can be obtained from the above formulas.

Since the skin depth drops as $\rho^{1/2}$ as the temperature is lowered, it is quite possible that a switch from one inequality to the other may occur somewhere along the line for pure metals. An aluminum plate 1-cm thick with RRR = 10 000 and in a 1-Hz field has δ = 82 mm at room temperature and δ = 0.82 mm at 4 K. If the peak field is 3 T (H = 2.4 MA/m), the losses at room temperature will be 560 W/m^2 (by Eq. 5.25) and 1866 W/m^2 (by Eq. 5.24) at 4 K! By the way, if one had persisted (wrongly) in using Eq. 5.25 at 4 K, the result would have been more than seven orders of magnitude larger!

In the calculations outlined here, the possibly large effect of magnetoresistance in some very pure metals, notably copper, has been neglected. The increased resistance in very high fields will always result in decreased eddy current losses, but exact calculations require detailed information on the specific material.

In the construction of high-power transmission lines, the conductor is a cable made up of smaller wires, both to minimize eddy current losses and to ensure that a nearly uniform current distribution exists throughout each separate wire (i.e., negligible skin effect) so as to make optimum use of the metal. Further, to ensure that each wire carries the same current, the separate wires are transposed.

The ac losses that occur in ferromagnetic materials are even more complex. The various analyses outlined above for eddy current effects usually hold, but the permeability, μ, is no longer independent of H, and this fact introduces nearly insurmountable problems for the various loss calculations. The only new low-temperature complication comes about because the value of μ is also temperature dependent, as described in Chapter 6. For some metals the resistivity may be a strong and complex function of field, which complicates the calculations, but this is only likely in the case of pure ferromagnetic metals. The result of these effects is that the skin depth and magnetic diffusivity may become strongly dependent on the instantaneous value of the applied magnetic field.

Hysteresis losses often exceed the eddy current losses in magnetic steels, and other losses occur under ac excitation related to such phenomena as domain wall motion. For the most part, these effects are not strongly temperature dependent, and their description is best left to the many excellent treatments of room-temperature behavior, such as the book by Heck (1974). The temperature-dependent phenomena are treated in Chapter 6.

5.6.3 Dielectrics

Polymers and other solid dielectric materials are extensively employed at low temperatures as electrical insulators, thermal insulators, and in the case of fiber-reinforced materials, as structural members. In their electrical insulator role, it is most often the ac properties that are of concern, although there are also problems in dc applications. Most of the relatively small amount of available information on the ac electrical properties of dielectrics at low temperatures has been accumulated over the last decade as a result of the interest in cryogenic or superconducting powerlines. Detailed discussions of the materials problems associated with superconducting powerlines are given in the book by Rechowicz (1975) and the review by Bogner (1974).

Application of laboratory data on pure dielectric materials to the design of actual structures is difficult because of a host of geometrical and environmental effects, which often prove to be more severely limiting than the bulk properties. Another difficulty is that these materials are seldom well characterized or uniform in composition. The degree of crystallinity of polymers and the porosity of commercial fiberglass epoxy materials are two examples of properties that may vary over wide ranges and that can have significant effects on the low-temperature electrical properties. Existing electrical properties data are reviewed in the texts just mentioned and in more detail by Fallou (1975). A compilation of low-temperature electrical and mechanical properties data for six polymers is also available (Schramm, Clark, and Reed, 1973). A recent paper (Meyer, 1976) describes the 4-K electrical properties at gigahertz frequencies for a number of materials. A few other papers on specific polymers are listed by Johnson (1977).

The properties of dielectrics are strongly influenced by the degree of charge separation created by an applied electric field. Polar dielectrics show a large effect, whereas little change is observed for the nonpolar materials. The polar or nonpolar nature of a particular material is best determined from the literature on that material. Some examples of nonpolar dielectrics are polyethylene, polypropylene, and polytetrafluoroethylene (TFE); polar dielectrics are paper, polyethylene terephthalate (PET), and polypyromellitimide (PPMI).

The properties of interest in electrical applications are:

a. Breakdown strength (dc, ac, and impulse)
b. Dielectric constant (ϵ)
c. Loss tangent (tan σ)

The breakdown strengths for polymers tend to be tens of kilovolts per millimeter of thickness and are not greatly affected by temperature. Some polar dielectrics show increases as large as 10 to 20% on cooling to 4 K. Knowledge of this property is almost always of no value except as a means of material comparison. In real situations, the breakdown usually occurs at much lower electric fields due to surface flashover or interactions between the dielectric and the cryogen or gas in which it is located. Many specific cases are discussed in the references given above. Another property, the volume resistivity, is sometimes measured. This parameter gives an indication of the number of available free carriers, and is usually measured in conjunction with radiation damage studies.

The dielectric constant determines the capacitance of a system. Generally, small values are desirable since high capacitances result both in high charging currents and in high ac losses. It is hard to generalize, but some comments can be made. Most often the dielectric constant of nonpolar dielectrics is not greatly temperature dependent. That of the polar dielectrics tends to decrease with decreasing temperature. The dielectric constant is significantly higher, often by more than an order of magnitude, for polar materials than for nonpolar ones, even at low temperatures. Some typical temperature dependences measured at low frequency are shown in Fig. 5.22. Frequency has little effect on the shape of the temperature-dependence curve, although an overall decrease (< 10%) in magnitude is often seen for the polar materials with increasing frequency to the low gigahertz range.

The loss tangent expresses the sum of a number of loss mechanisms that occur because dielectric materials are not perfect and molecules and free charges exist that can be moved around in various ways by the applied electric field, creating heat at the expense of electrical energy. In the case of ac power transmission, the power loss is given by

$$P = \omega C V^2 \tan \delta \tag{5.26}$$

where $1/\omega C$ is the capacitive reactance and V is the applied ac voltage. Typical values of $\tan \delta$ for polar materials at 4 K are 10^{-3} to 10^{-4}, whereas the nonpolar materials have values below 10^{-5}. Some typical curves as a function of temperature are shown in Fig. 5.22. Many peaks and valleys occur, but the loss tangent generally decreases with decreasing temperature and most of the erratic behavior is over at 4 K. This is consistent with a picture in which many of the dissipative molecular motions are "frozen" at low temperatures. The effect of frequency on the loss tangent is widely variable and no generalizations can be made.

When all the uncertainties are combined with the large thermal contraction and rapid embrittlement of most of these materials, their use at low temperatures becomes an exercise in frustration, and it is clear that no totally acceptable dielectric material exists at present. Future directions appear to be toward chemically tailoring some of the more promising polymers, evaluating mixtures of polymers and polymers filled with low-loss powders, and evaluating a few promising nonpolymeric materials.

5.7 LOW-TEMPERATURE ELECTRICAL PROPERTIES OF SPECIFIC METALS AND ALLOYS

The aim of this section is to provide information that will help the reader to find and interpret the existing resistivity data on metals and alloys and to use those data in conjunction with his own measurements to determine resistance behavior over a wide temperature range. The metals considered are primarily those with technological applications at low temperatures.

Few sources of compiled data on the electrical

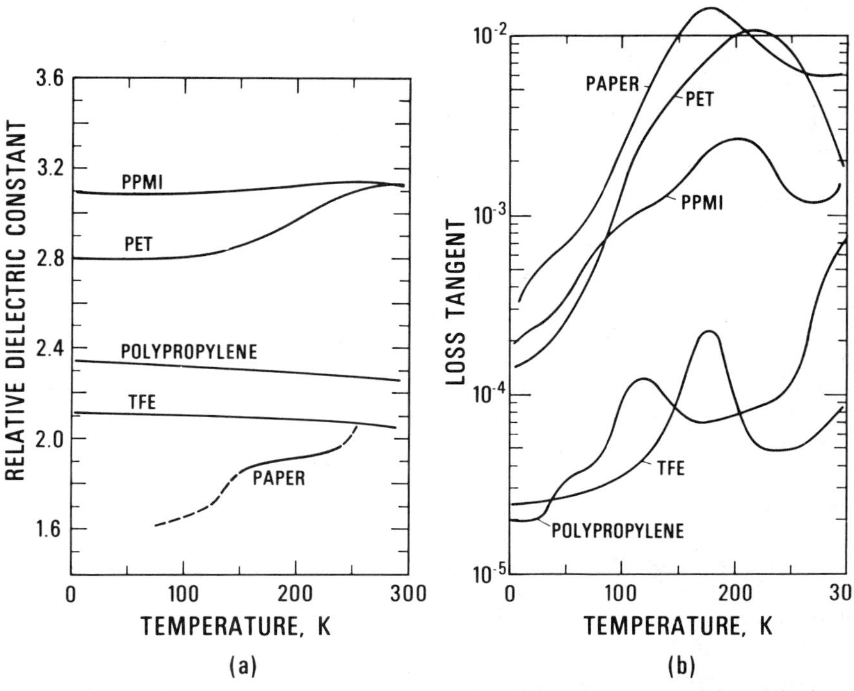

Figure 5.22 The effect of temperature on the relative dielectric constant (a) and the loss tangent (b) of several common dielectrics at near powerline frequencies. All data except those on the paper dielectric are from Chant (1967) and were measured at 75 Hz. The data on paper are from Allen and Kuffel (1968) as measured at 47 Hz.

resistivity as a function of temperature exist. The documents by Stewart and Johnson (1961) and Hall (1968, 1970) and the tabulation by Meaden (1965) cover a number of pure metals. Copper is treated in detail by Welles (1967) and aluminum by Milek and Welles (1969). Data on both pure metals and structural alloys are compiled in the *Handbook on Materials for Superconducting Machinery* (1974). A large amount of the alloy data is from Clark et al. (1970). Other sources of alloy resistivity data at low temperatures are two handbooks dealing primarily with mechanical properties: the *Materials Properties Data Book* (1970) and the *Aerospace Structural Metals Handbook* (1977).

Often the handbooks present curves of resistivity as a function of temperature without showing the actual data points. For a few metals so many data exist that this is the only practical approach, but usually the situation is quite the opposite. Thus, this practice frequently masks the fact that the measurements were made at only three or four temperatures and a solid curve on the graph does not guarantee that the resistivity is actually known over the entire range. This is especially a problem in the range 10 to 77 K where many metals are showing the first significant deviations of the resistivity from residual behavior with increasing temperature.

Room-temperature resistivity data on 31 commercial alloys are given in the *Handbook of Chemistry and Physics* (1974) and data on 129 noncommercial alloys are presented in the *Metals Reference Book* (1976).

5.7.1 Conductors

All the metals that have been seriously proposed for conductor applications below room temperature have been at least of commercial purity (RRR $\gtrsim$ 50) or better. Because of the effects of handling and the problems of manufacture, it seems unlikely that materials with RRR $>$ 10 000 will ever find very wide use. The metals of most interest are polycrystalline copper, aluminum, nickel, and beryllium. Sodium has been used for power transmission applications at room temperature, but its use has been discontinued because of a variety of problems. It has been discussed as a low-temperature conductor, but the problems become even more formidable at cryogenic temperatures (Bogner, 1974).

Figures 5.23 and 5.24 show plots of the intrinsic resistivity for copper and aluminum as determined by the Bloch–Grüneisen formula (Eq. 5.2) and the measured room-temperature resistivity.

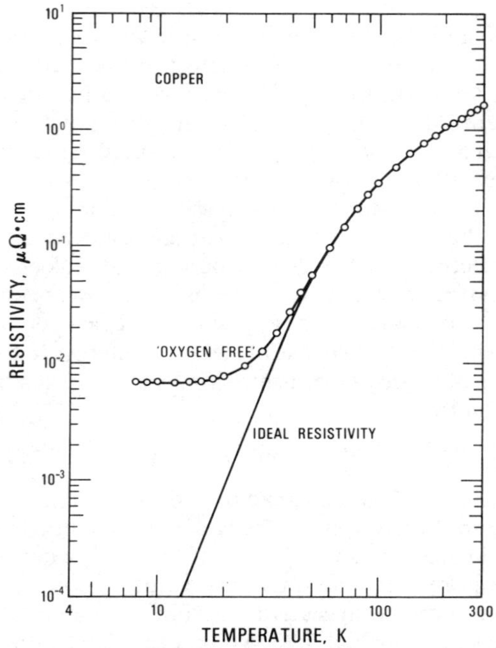

Figure 5.23 The intrinsic resistivity of copper calculated from the Bloch–Grüneisen formula using $\theta = 330$ K and $\rho(273\text{ K}) = 1.553$ $\mu\Omega$·cm. Data for "oxygen free" copper, as measured by Hust and Giarratano (1974), are also shown.

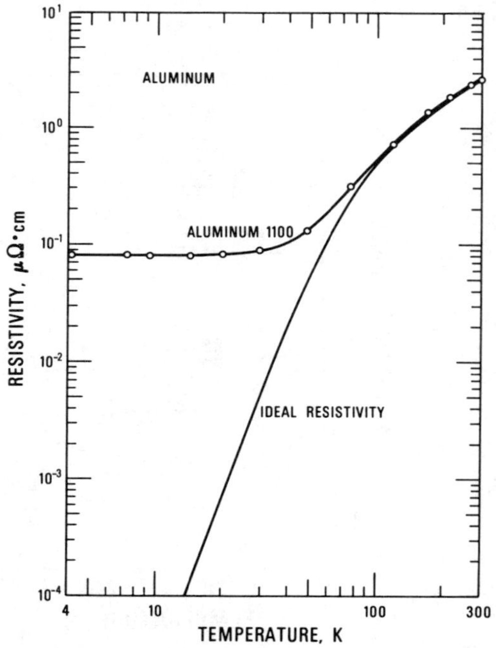

Figure 5.24 The intrinsic resistivity of aluminum calculated from the Bloch–Grüneisen formula using $\theta = 400$ K and $\rho(273\text{ K}) = 2.428$ $\mu\Omega$·cm. Data for the 1100 alloy, as measured by Clark and Sparks (1973), are also shown.

A very good estimate of the resistivity of a given material at any temperature can be obtained by measuring the residual resistivity and adding that value to those given in the figures. The Bloch–Grüneisen formula does not describe the very low temperature resistivity for very pure metals with great accuracy, but it represents the most reasonable approach and will result in usable data, at least for purposes of estimation. If better data are needed, the only option is to actually make the measurement.

The figures also show actual data points from one measurement to indicate how the technique outlined above may be applied. The serious user will need to make a larger graph; Table 5.10 provides the data. Calculated values for nickel and beryllium are also included in the table.

5.7.2 Structural alloys

The data on these alloys are quite limited, particularly if accurate values are required at closely spaced temperatures. Fortunately, from the data that do exist, it appears that most of these alloys show regular behavior with decreasing temperature. In nearly all cases, the resistivity decreases quite linearly to about liquid-nitrogen temperature, and then tends to flatten out to a residual value, usually around 20 K. In some nickel alloys a slight minimum is observed, usually above 20 K. In general, the more highly alloyed metals show a higher resistivity at all temperatures, but the curves often tend to be quite parallel for all members of an alloy series, as shown in Fig. 5.25 (from Clark et al., 1970) for aluminum alloys. For most alloys, the residual resistivity ratio is 2 or less and is, thus, not a very meaningful quantity. Table 5.11 from Clark et al. (1970), lists resistivity values for a large number of alloys that

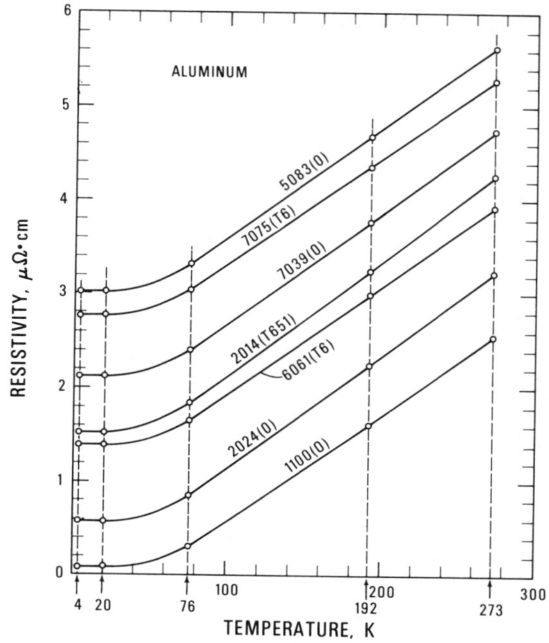

Figure 5.25 The electrical resistivity of a number of aluminum alloys, as measured by Clark et al. (1970).

Table 5.10 Intrinsic resistivity values for several metals.

Temperature (K)	Resistivity ($\mu\Omega$ cm) Copper $\theta = 330$ K	Aluminum $\theta = 400$ K	Nickel $\theta = 400$ K	Beryllium $\theta = 1300$ K
300	1.728	2.720	6.976	3.460
295	1.696	2.666	6.837	3.313
273	1.553	2.428	6.229	2.70
250	1.400	2.174	5.581	2.095
225	1.232	1.896	4.867	1.510
195 (CO_2)	1.028	1.556	3.995	0.9203
175	0.8901	1.326	3.407	0.6100
150	0.7142	1.035	2.656	0.3226
125	0.5353	0.7432	1.903	0.1417
100	0.3556	0.4604	1.182	4.846×10^{-2}
90	0.2852	0.3549	0.9109	2.882×10^{-2}
77.3 (LN_2)	0.1997	0.2325	0.5969	1.349×10^{-2}
70	0.1541	0.1712	0.4393	8.289×10^{-3}
60	9.819×10^{-2}	0.1007	0.2586	4.687×10^{-3}
50	5.307×10^{-2}	4.951×10^{-2}	0.1271	1.530×10^{-3}
40	2.213×10^{-2}	1.861×10^{-2}	4.776×10^{-2}	5.047×10^{-4}
27.2 (LNe)	3.782×10^{-3}	2.889×10^{-3}	7.416×10^{-3}	7.312×10^{-5}
20.4 (LH_2)	9.114×10^{-4}	6.904×10^{-4}	1.773×10^{-3}	1.734×10^{-5}
10	2.608×10^{-5}	1.942×10^{-5}	4.987×10^{-5}	4.893×10^{-7}
4.2 (LHe)	3.402×10^{-7}	2.538×10^{-7}	6.516×10^{-7}	6.399×10^{-9}

Table 5.11 Electrical resistivity of alloys at low temperatures (from Clark, Childs, and Wallace, 1970).

Material	Electrical Resistivity, $\mu\Omega\cdot$cm				
	273 K	192.4 K	75.75 K	19.65 K	4.0 K
Al 1100 (O)	2.67	1.617	0.310	0.084	0.082
Al 2014 (T651)	4.27	3.26	1.839	1.535	1.531
Al 2024 (O)	3.24	2.25	0.854	0.586	0.583
Al 2024 (T4)	4.35	3.39	2.02	1.745	1.742
Al 2024 (T6)	4.00	3.02	1.624	1.341	1.337
Al 2024 (T86)	4.21	3.23	1.866	1.581	1.578
Al 5083 (H113) (across)	5.71	4.74	3.36	3.07	3.07
Al 5083 (H113) (with)	5.68	4.72	3.34	3.05	3.05
Al 5083 (O)	5.66	4.69	3.32	3.03	3.03
Al 6061 (T6)	3.94	3.00	1.656	1.385	1.381
Al 7039 (T61)	4.37	3.40	2.02	1.741	1.738
Al 7039 (O)	4.75	3.78	2.40	2.12	2.12
Al 7075 (T73)	3.77	2.77	1.375	1.095	1.092
Al 7075 (T6)	5.29	4.36	3.04	2.77	2.76
Hastelloy C	126.8	125.8	124.0	123.1	123.3
Hastelloy N	120.6	120.0	119.5	119.4	120.1
Hastelloy X (annealed)	113.8	111.6	108.3	107.2	107.6
Hastelloy X (brazed)	115.0	112.9	109.4	108.1	108.7
Inconel 625	127.7	126.6	124.8	123.9	123.4
Inconel 718 (a)	114.5	112.1	108.3	107.7	108.1
Inconel 718 (b) (aged)	106.6	103.6	99.2	98.4	98.8
Inconel 718 (b) (annealed)	122.5	121.4	118.5	118.3	118.8
Inconel X (aged)	122.9	121.0	118.3	117.3	117.6
Inconel X (annealed)	124.5	123.1	121.0	120.4	120.9
René 41	126.3	123.9	120.0	118.3	118.1
TD Nickel	6.73	3.87	0.741	0.185	0.173
Udimet 630	112.2	109.4	105.1	104.1	104.5
Udimet 700	131.5	127.5	120.7	117.4	116.5
Waspaloy (aged)	119.1	116.0	110.8	108.7	108.2
Waspaloy (annealed)	124.3	121.9	117.9	116.0	115.4
OFHC Cu	1.559	1.013	0.200	0.017	0.016
OFHC Cu (60% CD)	1.581	1.034	0.222	0.032	0.031
Amzirc	1.730	1.177	0.364	0.163	0.163
Berylco 25 (annealed)	10.51	9.68	8.49	8.18	8.21
Berylco 25 (precipitation-hardened)	9.21	8.40	7.24	6.92	6.92
Aluminum Bronze D	16.2	15.3	14.0	13.8	13.9
Ni–Al Bronze	18.35	17.39	15.97	15.66	15.74
Phosphor Bronze A	10.48	9.83	8.88	8.58	8.59
Cartridge Brass, 70%	6.65	5.85	4.65	4.22	4.22
Commercial Bronze, 90%	3.90	3.29	2.38	2.10	2.11
Admiralty Arsenical	6.93	6.15	4.97	4.56	4.56

have been used at low temperatures. This table is the source of most of the alloy data found in modern compilations. This information is particularly valuable in estimating the eddy current heating in large structures.

5.7.3 Other materials

There are, obviously, many metallic elements and alloys that have not been mentioned here. The complete task would take several volumes and each material would be of very limited interest. A search of the files of the National Bureau of Standards Cryogenic Data Center yielded approximately 1400 references on resistivity alone. It seems that the resistivity of nearly every alloy imaginable has been measured at some low temperature or other. For information on a specific material, a search of the many computer systems now available is suggested. The Cryogenic Data Center at the National Bureau of Standards has been closed. The Metals and Ceramics Information Center (MCIC) at the Battelle Columbus Laboratories, Columbus, Ohio is an excellent source of bibliographies. A relatively recent publication (Johnson, 1977) on low-temperature materials properties gives an up-to-date bibliography that covers electrical properties of many of the materials discussed here, including some nonmetallics.

5.8 REFERENCES

Abriata, J. P., Cabo, A., and Kittl, J. (1968). Critical temperature for ordering in beta-AgZn and beta-AgCd. *Scr. Metall.* **2**, 41–46.

Table 5.11 (continued).

Material	Electrical Resistivity, μΩ·cm 273 K	192.4 K	75.75 K	19.65 K	4.0K
Naval Brass	6.32	5.41	3.89	3.28	3.25
Red Brass, 85%	4.66	4.00	3.02	2.72	2.74
CuNi 10	18.39	17.75	16.86	16.72	16.65
CuNi 30	38.4	37.8	36.7	36.5	36.4
CuNiSi (aged)	3.87	3.30	2.45	2.21	2.21
CuNiSi (annealed)	7.11	6.53	5.66	5.41	5.43
CuSi–A	23.2	22.5	21.6	21.4	21.3
Ti–5Al–2.5Sn	160.4	152.2	139.0	135.4	135.7
Ti–6Al–4V	167.5	161.0	150.1	146.9	—
Ti–6Al–4V (ELI)	156.3	149.0	136.6	133.1	—
Ti–13V–11Cr-3Al	149.2	150.9	155.1	157.4	156.7
AISI 670 (Co Haynes 25)	94.6	91.7	87.0	84.8	83.9
AISI 302	71.0	63.6	51.9	49.5	49.6
AISI 303	70.7	63.1	51.1	49.1	49.3
AISI 304L	70.4	63.0	51.4	49.4	49.6
AISI 310	87.3	81.7	72.4	68.8	68.5
AISI 316 (b) (cold drawn, annealed)	76.5	69.7	58.5	55.3	55.3
AISI 316 (c) (cold drawn, annealed)	77.6	70.7	59.5	56.3	56.3
AISI 316 (d) (annealed)	75.0	67.9	56.6	53.8	53.9
AISI 316 (e) (cold drawn, annealed)	71.8	64.3	52.3	50.2	50.3
AISI 321	73.9	66.8	55.6	53.4	53.6
AISI 347	70.8	63.4	51.9	50.0	50.1
AISI 410	55.9	48.7	39.5	38.2	38.2
AISI 416	66.2	60.4	53.4	52.3	52.3
AISI 430F	58.9	51.0	40.9	39.4	39.4
AISI 440C (hardened)	61.0	55.6	49.0	47.9	47.8
AISI 440C (annealed)	61.0	55.7	49.1	47.9	47.8
AISI 4340	31.0	25.8	19.27	18.16	18.15
AISI 630 (17-4 PH)	101.7	99.7	97.1	96.8	96.9
AISI 631 (17-7 PH) (annealed)	101.5	99.5	96.9	96.7	96.8
AISI 631 (17-7 PH) (precipitation hardened)	70.9	65.8	57.8	57.6	57.5
AISI 632 (PH 15-7 Mo)	80.0	73.5	63.2	61.5	61.6
PH 13-8 Mo	100.1	98.1	95.3	94.9	95.0
AISI 633 (annealed)	64.4	57.7	44.2	43.2	43.3
AISI 633 (precipitation hardened)	74.5	68.9	60.7	59.5	59.5
Fe–29 Ni	35.4	28.4	22.4	21.1	20.9
Invar 36	79.5	69.0	54.4	50.5	50.2
LR 35	71.9	59.5	43.8	40.2	40.0
Fe–50 Ni	39.2	29.2	19.5	17.6	17.4
Fe–79 Ni	20.6	16.3	12.2	11.5	11.4

Addison, W. E. (1964). *The Allotropy of the Elements.* Oldbourne Press, London.

Aerospace Structural Metals Handbook (1977). Five volumes. Mechanical Properties Data Center, Traverse City, Michigan.

Alderson, J. E. A., Farrell, T., and Hurd, C. M. (1968). Hall coefficients of Cu, Ag, and Au in the range 4.2–300° K. *Phys. Rev.* **174**, 729–736.

Aleksandrov, B. N. and D'Yakov, I. G. (1963). Variations of the electrical resistance of pure metals with decrease of temperature. *Sov. Phys.–JETP* **16**, 286–294.

Allan, R. N. and Kuffel, E. (1968). Dielectric losses in solids at cryogenic temperatures. *Proc. IEEE* **115**, 432–440.

Allnut, J. G. and Walton, A. J. (1972). A low power, high stability, superconducting modulator. *J. Phys. E* **5**, 131–132.

Anderson, A. C. (1973). Low-noise ac bridge for resistance thermometry at low temperatures. *Rev. Sci. Instrum.* **44**, 1475–1477.

Arp, V. D., Kasen, M. B., and Reed, R. P. (1969). Magnetic energy storage and cryogenic aluminum magnets. Technical Report to AFAPL. Report No. AFAPL–TR–68–87.*

Babic, E., Ocko, M., and Rizzuto, C. (1975). The impurity resistivity of ZnFe alloys. *J. Low Temp. Phys.* **21**, 243–255.

Barbee, T. W., Jr. (1965). Some aspects of dislocation dynamics in metals and the influence of dislocation distribution on low temperature electrical transport properties. Tech. Report 17, ONR contract Nonr–225(34). AD 625 023.*

Bardeen, J. (1940). Electrical conductivity of metals. *J. Appl. Phys.* **11**, 88–111.

Bass, J. (1972). Deviations from Matthiessen's rule. *Adv. Phys.* **21**, 431–604.

Bean, C. P., DeBlois, R. W., and Nesbitt, L. B. (1959). Eddy current method for measuring the resistivity of metals. *J. Appl. Phys.* **30**, 1976–1980.

Betts, D. S. (1976). *Refrigeration and Thermometry below One Kelvin*. Sussex University Press/Crane, Russak & Co., New York.

Blatt, F. J. (1968). *Physics of Electronic Conduction in Solids*. McGraw-Hill Book Co., New York.

Boato, G., Bugo, M., and Rizzuto, C. (1966). The effect of transition-metal impurities on the residual resistivity of Al, Zn, In, and Sn. *Nuovo Cimento* **45B**, 227–240.

Bogner, G. (1974). Transmission of electrical energy by superconducting cables. In *Superconducting Machines and Devices: Large Systems Applications* (eds. S. Foner and B. B. Schwartz). Plenum Press, New York, 401–548.

Boughton, R. I. and Neighbor, J. E. (1972). Ordinary size effects and deviations from Matthiessen's rule in the resistance of fine wires. *J. Low Temp. Phys.* **7**, 241–270.

Brandli, G. and Olsen, J. L. (1969). Size effects in electron transport in metals. *Mater. Sci. Eng.* **4**, 61–83.

Brechna, H. (1973). *Superconducting Magnet Systems*. Springer-Verlag, New York.

Broom, T. (1952). The effect of deformation on the electrical resistivity of cold-worked metals and alloys. *Proc. Phys. Soc. Lond. B* **65**, 871–881.

Bundy, F. P. and Strong, H. M. (1962). Behavior of metals at high temperatures and pressures. In *Solid State Physics*, Vol. 13 (eds. F. Seitz and D. Turnbull). Academic Press, New York. 81–146.

Chant, M. J. (1967). Dielectric properties of some insulating materials over the temperature range 4.2–300° K. *Cryogenics* **7**, 351–354.

Chovanec, F. (1969). The low temperature Hall effect amplifier as a zero indicator at low temperature measurements. *Cryogenics* **9**, 475–476.

Christian, J. W. (1965). *The Theory of Transformations in Metals and Alloys*. Pergamon Press, New York.

Cieloszyk, G. S., Cote, P. J., Salinger, G. L., and Williams, J. C. (1975). Thermal and electrical properties of Evanohm and other resistance alloys below 4 K. *Rev. Sci. Instrum.* **46**, 1182–1185.

Cimberle, M. R., Bobel, G., and Rizzuto, C. (1974). Deviations from Matthiessen's rule at low temperatures: an experimental comparison between various metallic alloy systems. *Adv. Phys.* **23**, 639–671.

Cimberle, M. R., Michi, V., Mori, F., Rizzuto, C., Siri, A., and Vaccarone, R. (1976). A search for low temperature standard resistance alloys. In *Proceedings of the 6th International Cryogenic Engineering Conference*. IPC Science and Technology Press, Guildford, Surrey, England, 190–193.

Clark, A. F. (1971). The low temperature electrical resistivity of various purities of copper. *Bull. Am. Phys. Soc.* **16**, 344.

Clark, A. F., Childs, G. E., and Wallace, G. H. (1970). Electrical resistivity of some engineering alloys at low temperatures. *Cryogenics* **10**, 295–305.

Clark, A. F., Deason, V. A., Hust, J. G., and Powell, R. L. (1972). *The Eddy Current Decay Method for Resistivity Characterization of High Purity Metals*. NBS Special Publication 260–39, National Bureau of Standards, Boulder, Colorado.

Clark, A. F. and Fickett, F. R. (1969). A nonsuperconducting detection system for low level dc voltages. *Rev. Sci. Instrum.* **40**, 465–468.

Clark, A. F. and Sparks, L. L. (1973). The low temperature electrical resistivity of various purities of aluminum. *Bull. Am. Phys. Soc.* **18**, 310. Abstract only, data obtained from the authors.

Clinton, J. R., Tyler, E. H., and Luo, H. L. (1974). Electrical and magnetic properties of Au–Ni alloys. *J. Phys. F* **4**, 1162–1169.

Cochrane, R. W. and Strom-Olsen, J. O. (1977). Temperature and magnetic field dependences of the resistivity of amorphous alloys. *J. Phys. F*, 1799–1810.

Cohen, J. B. and Bever, M. B. (1960). The effects of cold work on the alloy Cu_3Au. *Trans. AIME* **218**, 155–163.

Cook, J. G., Van der Meer, M. P., and Laubitz, M. J. (1972). Thermal and electrical conductivities of sodium from 40 to 360 K. *Can. J. Phys.* **50**, 1386–1401.

Crooks, M. J. (1969). Temperature profiles in experimental Dewars. *Cryogenics*, **9**, 32–35.

Daybell, M. D. and Steyert, W. A. (1968). Localized magnetic impurity states in metals: some experimental relationships. *Rev. Mod. Phys.* **40**, 380–389.

Dugdale, J. S. and Gugan, D. (1957). The effect of pressure on the electrical resistance of copper at low temperatures. *Proc. R. Soc.* **241**, 397–407.

Dugdale, J. S. and Gugan, D. (1961). The effect of the martensitic transformation on the electrical resistance of lithium and dilute lithium–magnesium alloys. *Cryogenics* **1**, 103–107.

Edwards, L. R. (1972). Effect of pressure on the resistivity of AgAu alloys. *Phys. Status Solidi B* **51**, 537–545.

Ekin, J. W. (1978). Fatigue and stress effects in NbTi and Nb_3Sn multifilamentary superconductors. In *Advances in Cryogenic Engineering*, Vol. 24. Plenum Press, New York, 306–316.

Ekin, J. W. and Wagner, D. K. (1970). A simple ac bridge circuit for use in four-terminal resistance thermometry. *Rev. Sci. Instrum.* **41**, 1109–1110.

Fallou, B. (1975). A review of the main properties of electrical insulating materials used at cryogenic temperatures. *Proceedings of the 5th International Conference on Magnet Technology*, Laboratori Nazionali del CNEN, Frascati, Italy, 644–658.

Federighi, T. and Ceresara, S. (1964). Resistometric investigation of preprecipitation in the Al–Mg5% alloy. Final Report 13.177/6401.104 to European Research Office, U.S. Army.*

Feldman, R., Talley, L., Rojeski, M., Vold, T., and Woollam, J. A. (1977). Upper limit for magnetoresistance in silicon bronze and phosphor bronze wire. *Cryogenics* **17**, 31–32.

Fickett, F. R. (1967). *Longitudinal Magnetoresistance of Single-Crystal Mercury*. Available from University Microfilms, Inc., Ann Arbor, Michigan, Order No. 68–6003.

Fickett, F. R. (1971). Aluminum 1. A review of resistive mechanisms in aluminum. *Cryogenics* **11**, 349–367.

Fickett, F. R. (1972a). Properties of nonsuperconducting technical solids at low temperatures. In *Proceedings of the 4th International Conference on Magnet Technology*, USAEC CONF–720908, 498–516.*

Fickett, F. R. (1972b). A preliminary investigation of the behavior of high purity copper in high magnetic fields. *INCRA Report 186*, International Copper

Research Association, New York.
Fickett, F. R. (1974a). A preliminary investigation of the behavior of high purity copper in high magnetic fields. *INCRA Report 186B*, International Copper Research Association, New York.
Fickett, F. R. (1974b). Oxygen annealing of copper: a review. *Mater. Sci. Eng.* **14**, 199–210.
Fickett, F. R. (1976). A preliminary investigation of the behavior of high purity copper in high magnetic fields and a final summary of project 186. *INCRA Report 186C*, International Copper Research Association, New York.
Fickett, F. R. (1977). Structural materials for cryogenic applications. In *Proceeding of the 6th International Cryogenics Engineering Conference*. IPC Science and Technology Press, Guildford, Surrey, England, 20–33.
Fisher, E. S., Kim, S. H., and Turner, A. P. L. (1976). Cyclic stress effects on transport properties of superconducting composite materials. In *Proceedings of the 6th International Cryogenics Conference*. IPC Science and Technology Press, Guildford, Surrey, England, pp. 390–394.
Foiles, C. L. (1967). Magnetic amplifier for use at 4.2 degrees K. *Rev. Sci. Instrum.* **6**, 731–733.
Forgan, E. M. (1974). On the use of temperature controllers in cryogenics. *Cryogenics* **14**, 207–214.
Fuchs, K. (1938). The conductivity of thin metallic films according to the electron theory of metals. *Proc. Camb. Philos. Soc.* **34**, 100–108.
Furukawa, G. T., Reilly, M. L., and Saba, W. G. (1964). Electrical resistances of wires of low temperature coefficient of resistance useful in calorimetry (10°–380° K). *Rev. Sci. Instrum.* **35**, 113–114.
Gafert, U. (1976). An ac resistance bridge for low resistance measurements. *Phys. Scr.* **14**, 257–259.
Gangulee, A. and Bever, M. B. (1968). The silver-rich solid solutions in the system silver–magnesium: II long range order. *Trans. Metall. Soc. AIME* **242**, 278–283.
Gerritsen, A. N. (1956). Metallic conductivity, experimental part. In *Handbuch der Physik*, Vol. 19 (ed. S. Flugge). Springer-Verlag, Berlin, 137–226.
Giauque, W. F., Lyon, D. N., Hornung, E. W., and Hopkins, T. E. (1962). Calorimetric determination of isothermal entropy changes in high magnetic fields at low temperatures. $CoSo_4{\cdot}7H_20$. *J. Chem. Phys.* **37**, 1446–1452.
Gifford, R. P., Webb, R. A., and Wheatley, J. C. (1972). Principles and methods of low-frequency electric and magnetic measurements using an rf-biased point contact superconducting device. *J. Low Temp. Phys.* **6**, 533–610.
Gregory, B. A. (1973). *An Introduction to Electrical Instrumentation*. Macmillan Press, London.
Guttman, L. (1956). Order–disorder phenomena in metals. In *Solid State Physics*, Vol. 3 (eds. F. Seitz and D. Turnbull). Academic Press, New York, 145–223.
Hall, L. A. (1968). *Survey of Electrical Resistivity Measurements on 16 Pure Metals in the Temperature Range 0 to 273° K*. NBS Tech. Note 365, National Bureau of Standards, Boulder, Colorado.
Hall, L. A. (1970). *Survey of Electrical Resistivity Measurements on 8 Additional Pure Metals in the Temperature Range 0 to 273 K*. NBS Tech. Note 365–1, National Bureau of Standards, Boulder, Colorado.
Handbook of Chemistry and Physics, 54th edition (1974) (ed. R. C. Weast). CRC Press, Cleveland, Ohio.
Handbook on Materials for Superconducting Machinery (1974). No. MCIC–HB–04, Metals and Ceramics Information Center, Battelle, Columbus Laboratories, Columbus, Ohio. Supplements were issued November 1975 and January 1977.
Heck, C. (1974). *Magnetic Materials and Their Applications*. Crane, Russak and Co., New York.
Hedgcock, F. T. and Muto, Y. (1964). Low-temperature magnetoresistance in magnesium and aluminum containing small concentrations of manganese or iron. *Phys. Rev.* **134**, 1593–1599.
Henry, W. G. and Schroeder, P. A. (1963). The low-temperature resistivities, and thermopowers of α-phase copper–zinc alloys. *Can. J. Phys.* **41**, 1076–1093.
Hess, E. G. and Pawlek, F. (1959). Electrical conductivity of pure copper and its changes through impurities. *Z. Metallkd.* **50**, 57–70.
Ho, N. S., Munnings, R. H., and Fletcher, R. (1969). An improved nanovolt relay modulator. *Cryogenics* **9**, 380–381.
Houghton, R. W., Sarachik, M. P., and Kouvel, J. S. (1970a). Anomalous electrical resistivity and the existence of giant magnetic moments in Ni–Cu alloys. *Phys. Rev. Lett.* **25**, 238–239.
Houghton, R. W., Sarachik, M. P., and Kouvel, J. S. (1970b). Electrical resistivity and magnetoresistance of paramagnetic Ni–Cu giant moment alloys. *Solid State Commun.* **8**, 943–945.
Hummel, R. E., Koger, J. W., and Pasupathi, V. (1968). The effect of deformation on the martensitic transformation of beta brass. *Trans. AIME* **242**, 249–253.
eHunter, S. C. and Nabarro, F. R. N. (1953). The propagation of electrons in a strained metal lattice. *Proc. R. Soc. A* **220**, 542–561.
Hurd, C. M. (1972). *The Hall Effect in Metals and Alloys*. Plenum Press, New York.
Hurd, C. M. (1974). Galvanomagnetic effects in anisotropic metals. *Adv. Phys.* **23**, 315–433.
Hust, J. G. (1970). Thermal anchoring of wires in cryogenic apparatus. *Rev. Sci. Instrum.* **41**, 622–624.
Hust, J. G. (1971). *Thermal Conductivity Standard Reference Materials from 4 to 300 K: IV. AGARD Gold*. Unpublished NBS Report, National Bureau of Standards, Boulder, Colorado.
Hust, J. G. (1972). Electrical resistance ratios of Evanohm heater wire at low temperatures. *Rev. Sci. Instrum.* **43**, 1387–1388.
Hust, J. G. and Giarratano, P. J. (1974). Thermal conductivity. In *Semiannual Report on Materials Research in Support of Superconducting Machinery*. NBSIR 74–393, National Bureau of Standards, Boulder, Colorado, 1–36.
Hust, J. G. and Giarratano, P. J. (1975a). *Thermal Conductivity and Electrical Resistivity Standard Reference Materials: Electrolytic Iron SRM's 734 and 797, from 4 to 1000 K*. NBS Special Publication 260–50, National Bureau of Standards, Boulder, Colorado.
Hust, J. G. and Giarratano, P. J. (1975b). *Thermal*

Conductivity and Electrical Resistivity Standard Reference Materials: Tungsten SRM's 730 and 799, from 4 to 3000 K. NBS Special Publication 260–52, National Bureau of Standards, Boulder, Colorado.

Hust, J. G. and Giarratano, P. J. (1975c). *Thermal Conductivity and Electrical Resistivity Standard Reference Materials: Austenitic Stainless Steel, SRM's 735 and 798, from 4 to 1200 K.* NBS Special Publication 260-46, National Bureau of Standards, Boulder, Colorado.

Iguchi, E. and Udagawa, U. (1975). Analysis of Young's modulus in the polycrystalline Cu–Ni alloy with the theory for Nordheim's rule. *J. Phys. F* **5**, 214–226.

Jackson, J. D. (1962). *Classical Electrodynamics.* John Wiley & Sons, New York.

Johansson, C. H. and Linde, J. O. (1936). X-ray and electrical investigation of the CuAu system. *Ann. Phys.* **25**, 1–48. (In German).

Johnson, D. M. (1977). *Low Temperature Properties of Selected Materials — A Bibliography with Descriptors.* MCIC–77–31, Metals and Ceramics Information Center, Columbus, Ohio.

Kamper, R. A. (1977). Superconducting devices for metrology and standards. In *Superconductor Applications: SQUIDS and Machines* (eds. B. B. Schwartz and S. Foner). Plenum Press, New York, 189–247.

Kasen, M. B. (1970). Grain boundary resistivity of aluminum. *Philos. Mag.* **21**, 599–610.

Kawatra, M. P. and Budnick, J. I. (1972). Transport properties of ferromagnetic systems near the critical point: electrical resistivity. In *Dynamical Aspects of Critical Phenomena* (eds. J. I. Budnick and M. P. Kawatra). Gordon and Breach, New York, 257–291.

Keil, D., Merbold, U., and Diehl, J. (1974). Influence of annealing treatments on residual resistivity and the field dependence of the electrical resistance of niobium at 4.2 K. *Appl. Phys.* **3**, 217–221. (In German).

Kohler, M. (1938). The magnetic resistance of pure metals. *Ann. Phys.* **32**, 211–218. (In German).

Kondo, J. (1965). s–d scattering at low temperatures. *Progr. Theor. Phys.* (Kyoto) **34**, 204–209.

Korn, D. and Pfeifle, H. (1976). Electrical resistivity of metastable copper–iron solid solutions. *Z. Phys.* **B23**, 23–26.

Korn, D., Pfeifle, H., and Zibold, G. (1974). Electrical resistivity of amorphous alloys. *Z. Phys.* **270**, 195–202.

Kreitman, M. M. and Callahan, J. T. (1970). Thermal conductivity of apiezon N grease at liquid helium temperatures. *Cryogenics* **10**, 155–159.

Krevet, B. and Schauer, W. (1976). Transverse magnetoresistance and its temperature dependence for high-purity polycrystalline aluminum. *J. Appl. Phys.* **47**, 3656–3669.

Krupowski, A. and DeHaas, W. J. (1928). The properties of the Ni–Cu alloys at low temperatures. *Commun. Leiden* **194a**, 1–12.

Lammeraner, J. and Stafl, M. (1966). *Eddy Currents.* Iliffe Books, London and The Chemical Rubber Co., Cleveland, Ohio.

Larsen, N. T. (1968). 50 microdegree temperature controller. *Rev. Sci. Instrum.* **39**, 1–12.

Laubitz, M. J., Matsumura, T., and Kelly, P. J. (1976). Transport properties of the ferromagnetic metals. II. Nickel. *Can. J. Phys.* **54**, 92–102.

Lawson, A. W. (1956). The effect of hydrostatic pressure on the electrical resistivity of metals. In *Progress in Metal Physics*, Vol. 6 (eds. B. Chalmers and R. King). Pergamon Press, New York, 1–44.

Lawson, N. S. (1977). Measurement of the electrical resistivity of anisotropic metals by the eddy current method. *J. Phys. E* **10**, 786–789.

Legvold, S., Peterson, D. T., Burgardt, P., Hofer, R. J., Lundell, B., Vyrostek, T. A., and Gartner, H. (1974). Residual resistivity and aging-clustering effects of Cu-rich Cu–Ni alloys. *Phys. Rev. B* **9**, 2386–2389.

Lerner, E. and Daunt, J. G. (1964). Magnetoresistance effect in advance and Evanohm wires at low temperatures in fields up to 50 kG. *Rev. Sci. Instrum.* **35**, 1069–1070.

Maple, M. B. and Luo, H. L. (1967). The electrical resistivity of Au_4V. *Phys. Lett. A* **25**, 121–122.

March, J. F. and Thurley, F. (1976). Copper-based alloys for low temperature measurement by resistivity. *PTB–Mitt.* **86**, 430–435.

Materials Properties Data Book (1970). Report 2275, NERVA Program. Three volumes. Prepared by Aerojet General Corp., Sacramento, California.

McTaggart, J. H. and Slack, G. A. (1969). Thermal conductivity of General Electric No. 7031 varnish. *Cryogenics* **9**, 384–385.

Meaden, G. T. (1965). *Electrical Resistance of Metals.* Plenum Press, New York.

Meaden, G. T. (1971). Conduction electron scattering and the resistance of the magnetic elements. *Contemp. Phys.* **12**, 313–337.

Meaden, G. T. and Pelloux-Gervais (1965). The electrical resistivity of α-manganese between 2 and 325° K. *Cryogenics* **5**, 227–228.

Meaden, G. T., Rao, K. V., Tee, K. T., and Loo. H. Y. (1972). Manifestations of magnetic transitions in the thermal conductivities and Lorenz functions of metals and alloys. In *Dynamical Aspects of Critical Phenomena* (eds. J. I. Budnick and M. P. Kawatra). Gordon and Breach, New York, 349–358.

Mercouroff, W. (1963). Minimization of thermal losses due to electrical connections in cryostats. *Cryogenics* **3**, 171–173.

Metals Reference Book, fifth edition (1976) (eds. C. J. Smithells and E. A. Brandes). Butterworths, London.

Meyer, W. (1976). Dielectric properties of polymeric materials at low temperatures and high frequencies. In *Proceedings of the 6th International Cryogenics Conference.* IPC Science and Technology Press, Guildford, Surrey, England, 367–371.

Milek, J. T. and Welles, S. J. (1969). *Chemical Composition and Electrical Resistivity of Aluminum Alloys.* Data Sheet DS–161, Electronic Properties Information Center, Hughes Aircraft Co., Culver City, California.

Morrison, R. (1967). *Grounding and Shielding Techniques in Instrumentation.* John Wiley & Sons, New York.

Murani, A. P. (1974). Localized enhancement effects in Pd–Ag alloys. *Phys. Rev. Lett.* **33**, 91–94.

Nabarro, F. R. N. (1967). *Theory of Crystal Dislocations.* Clarendon Press, Oxford.

Nix, F. C. and Shockley, W. (1938). Order–disorder transformations in alloys. *Rev. Mod. Phys.* **10**,

1–70.
Nordheim, L. (1934). The theory of the thermoelectric effect. *Actual. Sci. Ind.* **131**, 1–23. (In German).
Ocko, M., Babic, E., Krsmik, R., Girt, E., and Leontic, B. (1976). Some properties of AlSc solid solutions. *J. Phys. F* **6**, 703–707.
Osamura, K., Hiraoka, Y., and Murakami, Y. (1973). The resistivity maximum during Guinier–Preston zone formation in Al–Zn alloys. *Philos. Mag.* **28**, 809–825.
Phelps, D. J., Avci, R., and Flynn, C. P. (1975). Metal–insulator transition in metal–rare gas alloys. *Phys. Rev. Lett.* **34**, 23–26.
Powell, R. L., Hall, W. J., Hyink, Jr., C. H., Sparks, L. L., Burns, G. W., Scroger, M. G., and Plumb, H. H. (1974). *Thermocouple Reference Tables Based on the IPTS–68.* NBS Monograph 125, National Bureau of Standards, Boulder, Colorado.
Radiation Damage in Metals (1976) (eds. N. L. Peterson and S. D. Harkness). American Society for Metals, Metals Park, Ohio.
Rechowicz, M. (1975). *Electric Power at Low Temperatures.* Clarendon Press, Oxford.
Reed, R. P. and Breedis, J. F. (1966). Low-temperature phase transformations. In *Behavior of Metals at Cryogenic Temperatures.* ASTM STP 387, American Society for Testing and Materials, Philadelphia, 60–132.
Reed, R. P. (1972). Aluminum 2. A review of deformation properties of high purity aluminum and dilute aluminum alloys. *Cryogenics* **12**, 259–291.
Reich, R. (1966). Electric and superconducting properties of metals with different purities. III. *Mem. Sci. Rev. Metall.* **63**, 21–58.
Ries, R. P. and Moore, B. K. (1970). An ac resistance bridge and temperature controller. *Rev. Sci. Instrum.* **41**, 996–1003.
Rogers, W. M. and Powell, R. L. (1958). *Tables of Transport Integrals.* National Bureau of Standards Circular 595, National Bureau of Standards, Boulder, Colorado.
Rondeau, R. E. (1966). Slush baths. *J. Chem. Eng. Data* **11**, 124.
Rosenberg, H. M. (1963). *Low Temperature Solid State Physics, Some Selected Topics.* Clarendon Press, Oxford.
Rubin, L. G. (1970). Cryogenic thermometry: a review of recent progress. *Cryogenics* **10**, 14–22.
Schramm, R. E., Clark, A. F., and Reed, R. P. (1973). *A Compilation of Mechanical, Thermal and Electrical Properties of Selected Polymers.* NBS Monograph 132, National Bureau of Standards, Boulder, Colorado.
Seeger, A., Schumacher, D., Shilling, W., and Diehl, J. (1970). *Vacancies and Interstitials in Metals.* John Wiley & Sons, New York.
Seligmann, P. F. and Sarwinski, R. E. (1972). The contraction from 295 K to 4 K of some solders, epoxies, and AGOT graphite. *Cryogenics* **12**, 239.
Sparks, L. L. (1974). *Low Temperature Measurement, ASRADI Oxygen Technology Survey*, Vol. IV. NASA SP–3073.*
Sparks, L. L. and Powell, R. L. (1972). Low temperature thermocouples: KP "normal" silver, and copper versus *Au*–0.02 at%Fe and *Au*–0.07 at%Fe. *J. Res. NBS*, **76A**, 263–283.
Stewart, R. B. and Johnson, V. J. (1961). *A Compendium of the Properties of Materials at Low Temperatures (Phase II).* WADD Tech. Rept. 60–65 Part IV.*
Stoll, R. L. (1974). *The Analysis of Eddy Currents.* Clarendon Press, Oxford.
Stout, M. B. (1960). *Basic Electrical Measurements*, second edition. Prentice-Hall, Englewood Cliffs, New Jersey.
Sullivan, D. B. (1971). Resistance of a silicon bronze at low temperatures. *Rev. Sci. Instrum.* **42**, 612–13.
Svensson, B. (1932). Magnetic susceptibility and electrical resistivity of the polycrystalline (alloy) series Pd–Ag and Pd–Cu. *Ann. Phys.* **14**, 699–711. (In German.)
Svensson, B. (1936). Ferromagnetic resistivity increase of copper–nickel alloys. *Ann. Phys.* **25**, 263–271. (In German).
Takaki, S. and Igaki, K. (1976). Electrical resistivity of high-purity iron at 4.2 K. *Trans. Jpn. Inst. Met.* **17**, 353–359.
Toth, R. S., Arrott, A., Shinozaki, S. S., Werner, S. A., and Sato, H. (1969). Studies and Au_4X-ordered alloys: electron and neutron diffraction, resistivity and specific heat. *J. Appl. Phys.* **40**, 1373–1375.
Varoto, C. F. and Vidoz, A. E. (1970). On the effect of structure on the electrical resistivity of some ordering solid solutions. *Phys. Status Solidi a* **3**, 697–706.
Volkenshtein, N. V., Dyakina, V. P., and Startsev, V. E. (1973). Scattering mechanisms of conduction electrons in transition metals at low temperatures. *Phys. Status Solidi b* **57**, 9–42.
Warnecke, P. and Kose, V. (1975). Resistance alloy for low temperatures. *Rev. Sci. Instrum.* **46**, 1108–1109.
Warnecke, P. and Kose, V. (1977). Aluminum based resistance alloys for low temperatures. *Cryogenics* **17**, 635–636.
Warren, W. H., Jr., and Bader, W. G. (1969). Superconductivity measurements in solders commonly used for low temperature research. *Rev. Sci. Instrum.* **40**, 180–182.
Wasilewski, R. J. (1962). Observations on electrical resistivity of titanium. *Trans. Metall. Soc. AIME* **224**, 5–8.
Welles, S. J. (1967). *Copper, Data Sheet DS 156.* Electronic Properties Information Center, Hughes Aircraft Co., Culver City, California.
White, G. K. (1968). *Experimental Techniques in Low-Temperature Physics.* Oxford, Clarendon Press.
Williams, J. M., Klabunde, C. E., Redman, J. K., Coltman, R. R., Jr., and Chaplin, R. L. (1979). The effects of irradiation on the copper normal metal of a composite superconductor. *IEEE Trans. Magn.* **MAG–15**, 731–734.
Wolf, S. (1973). *Guide to Electronic Measurements and Laboratory Practice.* Prentice-Hall, Englewood Cliffs, New Jersey.
Woollam, J. A. (1970). Evanohm resistance to 18 T (180 kG) at 4.2 K. *Rev. Sci. Instrum.* **41**, 284–285.
Yamashita, J. and Hayakawa, H. (1976). Deviations from Matthiessen's rule in electrical resistivity of nickel-based alloys. *Prog. Theor. Phys.* **56**, 361–374.
Ziman, J. M. (1960). *Electrons and Phonons.* Clarendon Press, Oxford.

*Available from NTIS, Springfield, Virginia.

6 Magnetic Properties

F. R. FICKETT and R. B. GOLDFARB

6.1 INTRODUCTION

All materials respond to an applied magnetic field in one way or another. The response is usually easily measured and often gives detailed information about the atomic, molecular, or crystalline structure of the material. Because of this and because of the many uses of magnetic materials, the field of magnetic properties has an enormous range. This chapter discusses a relatively small portion of the field — the low-temperature behavior of materials of technological importance. Included in this group are the structural metals and alloys, soft magnetic materials, permanent magnet materials, ferrites, and amorphous materials. Basic physics is presented when it is necessary for background. Most of the specifically nuclear, atomic, or molecular effects, such as nuclear magnetic resonance, are not discussed. With few exceptions, films, magnetic bubbles, powders, and single crystals are not considered. The purpose of this chapter is to provide a relatively uncomplicated view of magnetism in materials used at low temperatures for the user of materials who occasionally needs to take magnetic properties into account.

The chapter provides a simple exposition of the origins and effects of magnetism in the materials of interest. Particular attention is paid to the systems of units, since in this field they can be very confusing. The effect of magnetic behavior on other physical and mechanical properties is discussed. Experimental techniques are described only briefly, but comprehensive references are provided. Each of the specific classes of materials is then considered in more detail. This is followed by a discussion of the existing low-temperature data and a guide to the literature.

The number of textbooks and treatises on the subject of magnetism is large. The reader will find nearly any topic covered in at least one of them. Three excellent general references are Cullity (1972), Chikazumi (1964), and Morrish (1980). The solid-state physics of magnetism is described in an introductory fashion by Kittel (1967) and in detail by Martin (1967) and Mattis (1981). The multivolume series, *Magnetism: A Treatise on Modern Theory and Materials* (1963–1973), covers a number of advanced topics.

The magnetic properties of many materials are described in the classic text by Bozorth (1951) and in *Ferromagnetic Materials* (1980). Magnetic materials, both hard and soft, used by the electrical industry are treated by Heck (1974). Electrical engineering applications are covered by Watson (1980). Parker and Studders (1962) cover permanent magnets. Soft magnetic materials are treated in the volume by Chen (1977), and the more exotic materials such as ferrites and garnets, as well as alloys, by Tebble and Craik (1969). Amorphous materials are discussed in the volumes *Amorphous Magnetism* (1973) and *Amorphous Magnetism II* (1977). Metallurgical aspects of magnetism are covered in the two volumes of *Magnetism and Metallurgy* (1969), and topics specific to rocks and minerals, in the text by Stacey and Banerjee (1974).

6.2 PROPERTIES OF MAGNETIC MATERIALS

6.2.1 Terms and units

In this section, the concepts, definitions, and systems of units that are unique to the study of magnetic properties are discussed. The exposition is intended to provide an understanding of the terminology and to review a subject that was learned at one time or other, but in which the sharp edge may have faded with time.

There is probably no other field in the physical sciences that is more confusing in terms of nomenclature and units than magnetism. Each author seems to have his own terminology. Units, when stated at all, tend to be handled in a cavalier fashion where the system (e.g., emu) is stated instead of the actual unit, leading to the possibility of bizarre combinations such as "emu^2." A survey of units used by the authors of papers in the 1974 Magnetism and Magnetic Materials Conference (Bennett, Page, and Swartzendruber, 1976) gave the following combinations for the units of "susceptibility": $emu{\cdot}g^{-1}$, $emu{\cdot}cm^{-3}$, $emu{\cdot}mole^{-1}$, $emu{\cdot}g^{-1}{\cdot}kOe^{-1}$, $emu{\cdot}g^{-1}{\cdot}At{\cdot}V^{-1}$, and $emu{\cdot}Oe^{-1}{\cdot}mole^{-1}$.

The SI system of units, based on a rationalized mks system, has been much less accepted in this field than in others. There are a number of valid reasons for this, which are discussed in the paper just mentioned. Furthermore, almost without exception, handbook data are expressed in cgs emu. The few brave souls who have attempted to convert these data to SI for use in their own texts have occasionally been tricked by the complexity of the conversions. For example, mass susceptibility may be expressed in emu as $cm^3{\cdot}g^{-1}$. The conversion factor to obtain $m^3{\cdot}kg^{-1}$ in SI is $4\pi \times 10^{-3}$, not 10^{-3}. The set of units used herein, and described in the reference above, is consistent with SI, but the data are also accompanied by the original units, usually in parentheses. The best advice regarding systems of units is quite pragmatic — choose one and stick with it.

The units, their symbols and the conversion factors are given in Table 6.1. Note that, in the case of susceptibility, different symbols are used for the SI and emu quantity. This choice is made to avoid confusion since the volume susceptibility is dimensionless in both systems, but the numerical values are *not* the same.

The equations expressing the relationships between the various quantities are:

in cgs emu	in SI	
$B = H + 4\pi M$	$B = \mu_0(H + M)$	(6.1)
$\chi = M/H$	$\kappa = M/H$	(6.2)
$F = \chi VH\nabla H$	$F = \mu_0\kappa VH\nabla H$	(6.3)
$w = BH/8\pi$	$w = BH/2$	(6.4)

Here F is the force on a specimen of volume V, w is the volume energy density, and μ_0 is the permeability of free space. The conversion of equations written in one set of units to the other is not straightforward. Table 6.2 shows the required substitutions.

Working only in SI, a few more quantities and equations that appear frequently in the literature can be derived. The first is the permeability, μ, defined by

$$B = \mu H \tag{6.5}$$

From Eqs. 6.1, 6.2, and 6.5, it is clear that

$$\mu = \mu_0(1 + \kappa). \tag{6.6}$$

The relative permeability, μ_r, given by

$$\mu_r = \mu/\mu_0 = 1 + \kappa \tag{6.7}$$

is frequently encountered in discussions of ferromagnetic materials. The relative permeability (SI) happens to be numerically equal to the permeability in cgs (Table 6.1).

Note that m, μ, and κ are properties of materials, but H is not; it is the value of the applied magnetic field that exists in the absence of any material. The flux density, B, is mixed in this respect, but to avoid confusion it should probably be used only to describe material behavior. Thus, the very satisfactory phrase "a ten-tesla magnet" should really be "a $7.96\text{-}MA{\cdot}m^{-1}$ magnet." A significant rush to change the convention is not expected. Note also that M, H, and B are vectors.

A seeming contradiction arises when one attempts to evaluate H inside a magnetized body. The applied field, H_0, induces magnetic poles on the specimen and some of the return flux from these poles is inside the specimen, which leads to a lower effective magnetizing field, H, for the material. The effect is a geometric one and, for specimens that approximate spheroids uniformly magnetized along a major axis, can be expressed by use of a *demagnetizing factor*, D, as

$$H = H_0 - DM \tag{6.8}$$

in either system of units. In general, D is a tensor. Some values of D for particular cases are given in Table 6.3.* An extensive graphical presentation and tabulations for spheroids and cylinders are given by Bozorth (1951), Brown (1960), and Crabtree (1977). This effect is only significant for highly magnetic materials or in very precise

* The sum of D for three orthogonal axes is equal to one (SI).

Table 6.1 Units for magnetic properties.

Property	cgs emu Symbol	cgs emu Units	Conversion Factor, C [a]	mks SI Symbol	mks SI Units
Flux density, magnetic induction	B	gauss (G)	10^{-4}	B	tesla (T) [b]
Flux	ϕ	maxwell (Mx)	10^{-8}	ϕ	weber (Wb)
Magnetic field strength, magnetizing force	H	oersted (Oe) [c]	$10^3/4\pi$	H	$A{\cdot}m^{-1}$ [d]
(Volume) magnetization	M	$emu{\cdot}cm^{-3}$ [e]	10^3	M [f]	$A{\cdot}m^{-1}$
(Volume) magnetization	$4\pi M$	gauss (G)	$10^3/4\pi$	M	$A{\cdot}m^{-1}$
(Mass) magnetization	σ, M	$emu{\cdot}g^{-1}$	1	σ, M	$A{\cdot}m^2{\cdot}kg^{-1}$
(Volume) susceptibility	χ	dimensionless $emu{\cdot}cm^{-3}{\cdot}Oe^{-1}$ $emu{\cdot}cm^{-3}$	4π	κ	dimensionless
(Mass) susceptibility	χ_ρ	$cm^3{\cdot}g^{-1}$ $emu{\cdot}g^{-1}{\cdot}Oe^{-1}$ $emu{\cdot}g^{-1}$	$4\pi \times 10^{-3}$	κ_ρ	$m^3{\cdot}kg^{-1}$
(Molar) susceptibility	χ_{mol}	$cm^3{\cdot}mol^{-1}$ $emu{\cdot}g^{-1}{\cdot}mol^{-1}{\cdot}Oe^{-1}$ $emu{\cdot}mol^{-1}$	$4\pi \times 10^{-6}$	κ_{mol}	$m^3{\cdot}mol^{-1}$
Magnetic moment	m	$erg{\cdot}Oe^{-1}$, $erg{\cdot}G^{-1}$	10^{-3}	m	$A{\cdot}m^2$, $J{\cdot}T^{-1}$
Permeability	μ	dimensionless	$4\pi \times 10^{-7}$	μ	$henry{\cdot}m^{-1}$ $(H{\cdot}m^{-1})$ [g]
Relative permeability	—	—	1	μ_r	dimensionless

[a] Multiply a number in cgs emu system by C to convert it to mks SI.
[b] $T = Wb{\cdot}m^{-2} = V{\cdot}s{\cdot}m^{-2}$.
[c] Dimensionally, 1 Oe = 1 G.
[d] Often expressed as ampere-turn·m^{-1}.
[e] The designation "emu" is not a unit. Sometimes substitution of "erg·Oe^{-1}" or "erg·G^{-1}" will yield correct units.
[f] Often called intensity of magnetization, I, or magnetic polarization, J. It is the magnetic moment per unit volume.
[g] $H{\cdot}m^{-1} = Wb{\cdot}A^{-1}{\cdot}m^{-1}$.

work, and in these cases it is of prime importance. A case of particular interest is the long cylinder. Its evaluation depends on the permeability, and for very high values of μ, length-to-diameter ratios of approximately 1000 may be required to ensure $D = 0$ to the required precision. D is dimensionless in both sets of units but requires the conversion

$$D \text{ (cgs)} = 4\pi D \text{ (SI)} \qquad (6.9)$$

Table 6.2 Conversion of equations.[a]

Quantity	Symbol in cgs Equation	Replace by This Symbol for SI Equation
Flux density	B	$(4\pi/\mu_0)^{1/2}B$
Magnetic field strength	H	$(4\pi\ \mu_0)^{1/2}H$
Magnetization	M	$(\mu_0/4\pi)^{1/2}M$
Volume susceptibility	χ	$(1/4\pi)\kappa$
Magnetic moment	m	$(\mu_0/4\pi)^{1/2}m$

[a] Bennett et al. (1976).

The quantum mechanical description of atomic and electronic magnetic moments leads to another natural unit for magnetism, the *Bohr magneton*, defined by

$$\mu_B = eh/4\pi m_e = 9.274 \times 10^{-24}\ A{\cdot}m^2 \qquad (6.10)$$

where h is Planck's constant, and m_e and e are the electronic mass and charge. Orbital and spin

Table 6.3 Demagnetizing factors.

Specimen Shape	D (cgs)	D (SI)
Sphere	$4\pi/3$	1/3
Long needle,[a] $H\perp$	2π	1/2
Long needle,[a] $H\parallel$	0	0
Thin film, $H\perp$	4π	1
Thin film, $H\parallel$	0	0
Toroid, H around torus	0	0

[a] These values also can be used for long cylinders and square rods, but only as rough approximations.

moments of electrons are simple multiples of this basic moment. If there were no other complications, atomic moments would turn out to be integral multiples of μ_B. In reality, overlapping energy bands lead to nonintegral values, but the concept is still frequently used in relating experimental results to atomic structure.

6.2.2 Types of magnetic behavior

The magnetic behavior of solids can be divided into five major classes: diamagnetism and paramagnetism are effects due to isolated atoms (ions) and conduction electrons; ferromagnetism, antiferromagnetism, and ferrimagnetism are effects arising from a collective behavior of the atoms in a solid. Some of the major features of these classes of materials are described in Tables 6.4 and 6.5. In addition to these major categories, there are many subcategories, such as the superparamagnetism exhibited by collections of small magnetic particles and the spin-glass behavior of amorphous and atomically disordered solids.

Diamagnetism. Diamagnetism arises from shielding currents generated when a field is applied. These induced currents must be the type that persist even when the field is no longer changing. This occurs in atoms (because of electronic motion around the nucleus) and in superconductors. The induced currents give a moment that opposes the applied field and is proportional to it, and thus the susceptibility is constant, negative, and small ($\kappa \sim -10^{-6}$) (Fig. 6.1). Diamagnetism occurs in all atoms, but is usually overridden by a positive moment arising from the spins in unfilled shells that results in a net paramagnetism. In certain superconducting materials below the critical field, the susceptibility is very large ($\kappa = -1$), representing perfect diamagnetism. This effect is due entirely to persistent electron currents in the material. The value of the diamagnetic susceptibility of noncubic materials depends on the direction of the measuring field with respect to the crystal axes. This anisotropy can be very large: $\kappa_{||}/\kappa_{\perp} = 1.4$ for zinc and 2.4 for indium. These numbers are taken from data tabulated and cited by Tebble and Craik (1969). In inhomogeneous fields, note that diamagnetic materials tend to move away from the high-field region (Eq. 6.3).

Copper is representative of the diamagnetic metals of technical importance at low temperatures; its behavior is typical, and the results of several measurements are shown in Fig. 6.2. Copper always contains iron impurities. These impurities can impart a strongly temperature-dependent paramagnetism, so that at low temperatures, most specimens show a significant decrease in their diamagnetism, as indicated by the less negative susceptibility in the figure. Note the rather large variation in the susceptibility of the same material ("five-nines" or 99.999% pure

Table 6.4 Major categories of magnetic behavior.

Type	Dependence of M on H	Dependence of κ or μ on T	Magnitude of κ	Magnetic Remanence	Example of Material
Diamagnetism	Linear	Independent[a]	Small (<0)	None	Cu, organic compounds
			Large (<0)	None	Superconductors below critical temperature
Paramagnetism	Linear	Independent[a]	Small	None	Al
	Linear	Increases with decreasing T	Large	None	Some transition metals
	Linear	$\kappa = C/T$	Small to moderate	None	Ionic crystals
	Linear	$\kappa = C/(T-\theta)$	Moderate	None	Ferromagnets & antiferromagnets above critical temperature
Ferromagnetism	Saturates	Increases to saturation with decreasing T	Variable with H; large	Large (hard) Small (soft)	Permanent magnet alloys Fe, Ni, Co
Ferrimagnetism	Saturates	Complex — usually increases to saturation with decreasing T	Variable with H; large	Moderate	Fe_3O_4, rare earth garnets
Antiferromagnetism	Increases	Peaks at critical temperature	Variable with H; moderate	None to small Small	Some stainless steels Ni–O, Mn, Cr

[a]Data often indicate a moderate increase in magnitude with decreasing temperature.

Table 6.5 Magnetic properties of representative materials.

Type	Material	At $T \cong 295$ K[a]	At $T = 4.2$ K[a]	References and Comments
Diamagnetic	Cu	$\kappa = -9.3 \times 10^{-6}$	$\kappa = -9.4 \times 10^{-6}$	Fickett (1976); very pure Cu
	Nb	Paramagnetic	$\kappa = -1$	Superconductor below 9 K
Paramagnetic	Al	$\kappa = 2.0 \times 10^{-5}$	$\kappa = 2.5 \times 10^{-5}$	Fickett (1976)
	Ce	$\kappa = 1.6 \times 10^{-3}$	$\kappa = 23 \times 10^{-3}$	Edelstein (1968)
	Potassium–chrome–alum[b]	$\kappa = 1.6 \times 10^{-3}$	$\kappa = 32 \times 10^{-4}$ [c]	DeHaas and Gorter (1930)
	Fe–22Cr–13Ni–5Mn stainless steel	$\kappa = 2.1 \times 10^{-3}$	Antiferromagnetic	Ledbetter and Collings (1979)
Ferromagnetic	Low-carbon steel	$\mu_r = 5250$ (max.)	$\mu_r = 4900$ (max.)	McInturff and Claus (1970a)
	Si steel (1.8% Si)	$\mu_r = 6250$ (max.)	$\mu_r = 9700$ (max.)	McInturff and Claus (1970a)
Ferrimagnetic	Fe_3O_4	$\mu_r \geq 5.3$	$\mu_r \geq 5.5$	Jacobs (1959) – 4 K; Nagata (1961) – RT Assumes saturation at 0.11 MA·m^{-1} (1.4 kOe)
Antiferromagnetic	Fe–22Cr–13Ni–5Mn stainless steel	Paramagnetic	$\kappa = 6.0 \times 10^{-3}$	Ledbetter and Collings (1979)

[a]Si units
[b]$Cr_2(SO_4)_3 \cdot K_2SO_4 \cdot 24H_2O$
[c]$T = 14$ K

copper). This possibly indicates some contamination during the wire drawing process.

In most applications at low temperatures, diamagnetism is not a serious concern because the susceptibility is so small. For some sensitive measurements, however, the magnetic properties of the apparatus must be taken into account. For example, an interesting situation has been observed in orbital flights of superconducting magnets where diamagnetic liquid helium is forced away from the high-field regions of the magnet, no doubt much to the consternation of the experimenters.

Paramagnetism. Often the atomic structure or band effects result in numbers of unpaired electron spin moments in a material. These moments are aligned in the direction of an applied magnetic field and result in an increased magnetization in the material. This paramagnetic magnetization is linear with field and thus gives a positive susceptibility (Fig. 6.3). As always, exceptions exist. One of the more important is that at low temperatures and very high fields the magnetization may approach saturation with increasing field, and thus κ becomes field dependent. By Eq. 6.3, a paramagnetic material is attracted to the high-field region of an inhomogeneous field.

Paramagnetism is observed in many metals, in substances containing metallic ions, and in fer-

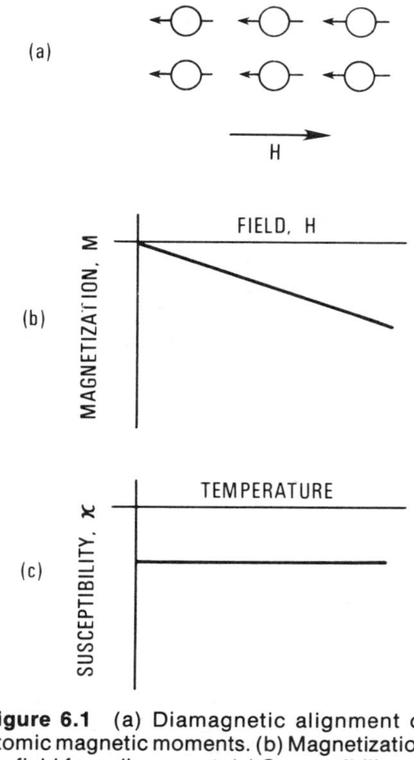

Figure 6.1 (a) Diamagnetic alignment of atomic magnetic moments. (b) Magnetization vs. field for a diamagnet. (c) Susceptibility vs. temperature for a diamagnet.

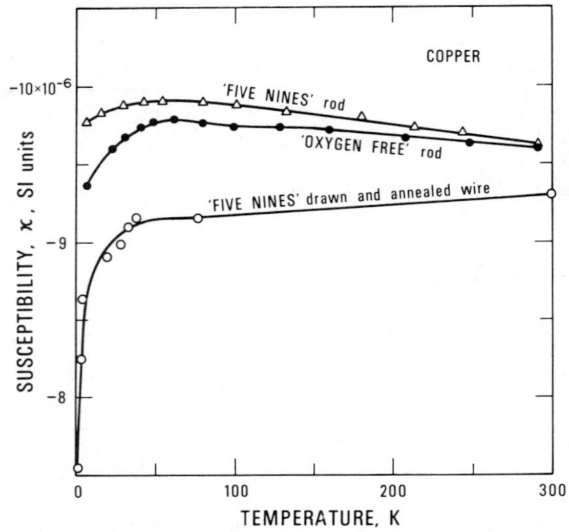

Figure 6.2 Magnetic susceptibility of diamagnetic copper as a function of temperature. △, ● — Hurd (1966); ○ — Bowers (1956).

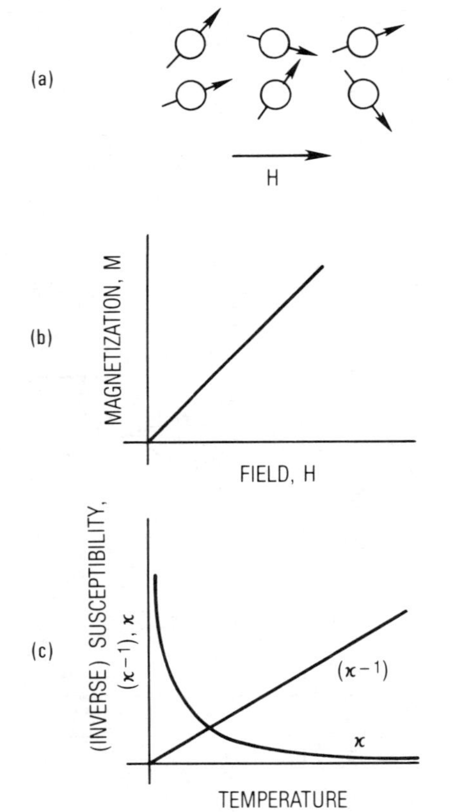

Figure 6.3 (a) Paramagnetic alignment of atomic magnetic moments. (b) Magnetization vs. field for a paramagnet. (c) Susceptibility and reciprocal susceptibility vs. temperature for a paramagnet.

romagnetic and antiferromagnetic materials above their transition temperatures. A material of interest at low temperatures that fits none of these categories is molecular oxygen. The liquid is highly paramagnetic, $\kappa = 3.4 \times 10^{-3}$ at the boiling point (90.1 K). The other common cryogenic fluids are diamagnetic, although oxygen contamination in liquid nitrogen usually results in a large paramagnetism for the liquid.*

The various classes of paramagnetic materials have quite different temperature dependences for the magnetic properties:

a. Nonferromagnetic metals, in which the paramagnetism arises primarily from the conduction electrons (Pauli paramagnetism), show a small susceptibility that is nearly temperature independent. This behavior is shown by aluminum and to a lesser extent by Cu–Ni alloys with nickel content below the critical one for ferromagnetism (~40%). Data for these materials are plotted in Fig. 6.4.

Orbital diamagnetism is always in competition with this paramagnetism and usually is of the same order of magnitude. Whether a metal is diamagnetic or paramagnetic depends on the balance of these two effects. In general, metals with electrons of higher effective mass are paramagnetic.

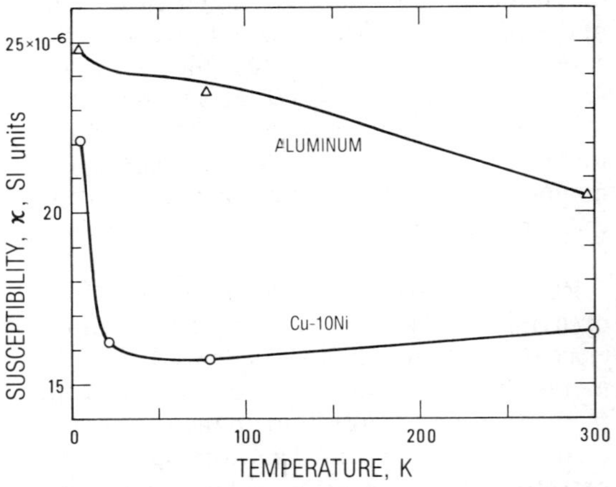

Figure 6.4 Magnetic susceptibility of relatively well-behaved paramagnetic metals. △ — Fickett (1976); ○ — Pugh and Ryan (1958). Solid curve for aluminum representing about 30 points from graphical data of Cooper and Miljak (1976).

*More than once, experimenters measuring susceptibility of materials in liquid nitrogen have found an unexpected large diamagnetism that later proved to be due to the displacement from the measuring coil of nitrogen contaminated with oxygen.

b. The transition metals, because of their unfilled inner shells, complex band structure, and high electron state density, behave differently. Their susceptibility often is relatively large and is observed to increase significantly, but not uniformly, with decreasing temperature. For example, Nagasawa's (1970) measurements on palladium show an increase in κ from about 7.2×10^{-4} at room temperature to about 12×10^{-4} at 4 K.

c. Noninteracting paramagnetic metal ions have a susceptibility that is strongly temperature dependent and obey the *Curie law*

$$\kappa = C/T \tag{6.11}$$

where C is a constant determined by the ionic density and magnetic moment and T is the absolute temperature. In terms of other parameters,

$$C = nm_p^2/3k_B \tag{6.12}$$

where n is the number of atoms per unit volume, k_B is Boltzmann's constant, and m_p is the average *paramagnetic moment* per atom, regardless of the alignment of individual moments. This *Langevin* paramagnetism is most often observed in metallic salts. The same behavior is also shown by the ferromagnetic and antiferromagnetic materials above the temperature at which their spontaneous magnetization breaks down, leaving, again, noninteracting ions. The equation that expresses this behavior is the *Curie–Weiss* law,

$$\kappa = C/(T - \theta) \qquad (T > \theta) \tag{6.13}$$

where the Weiss constant, θ, is the appropriate critical temperature. In very dilute alloys of iron in copper and other similar systems, κ also follows a Curie–Weiss law, but θ is negative (Ekström and Myers, 1972).

Four standard paramagnetic reference materials (aluminum, platinum, palladium, and manganese fluoride) covering the range of susceptibility $\kappa = 2 \times 10^{-5}$ to 6×10^{-3} are sold by the National Bureau of Standards and described in *NBS Standard Reference Materials Catalog* (1981). These materials, available in a variety of forms, are helpful in the calibration of experimental apparatus.

Ferromagnetism. In certain instances, the atomic spin moments interact with each other to produce ferromagnetism (or antiferromagnetism, or ferrimagnetism). Ferromagnetic materials (Fig. 6.5) are usually characterized by a large field-dependent permeability and, often, by a magnetization that persists after application and removal of a magnetic field. The materials possess a characteristic temperature, the *Curie temperature*, T_C, above which they are paramagnetic, as described earlier. Curie temperatures are higher than room temperature for most ferromagnets of interest here (iron, 1043 K; nickel, 631 K; cobalt, 1394 K; Fe_3O_4, 858 K), but a large number of materials have Curie temperatures well below room temperature. The literature guide edited by Connolly and Copenhaver (1972) tabulates Curie temperature values with references to the original literature. The ferro-

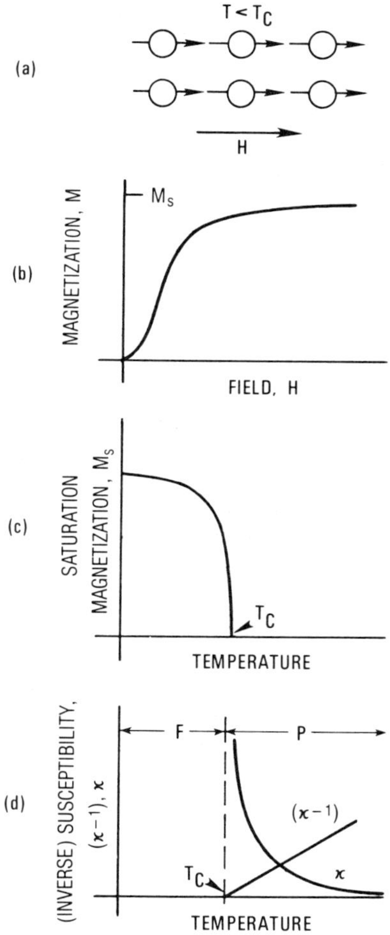

Figure 6.5 (a) Ferromagnetic alignment of atomic magnetic moments. (b) Magnetization vs. temperature showing saturation magnetization, M_S. (c) Saturation magnetization vs. temperature below the Curie temperature, T_C. (d) Susceptibility and reciprocal susceptibility vs. temperature above the Curie temperature, T_C, showing ferromagnetic (F) and paramagnetic (P) intervals.

magnetic metals of technical interest are usually alloys containing iron, nickel, or cobalt, or some of the rare-earth metals, such as gadolinium. A few alloys with entirely nonferromagnetic components exhibit ferromagnetism. These are the manganese-based Heusler alloys, which are of the form Cu_2MnX where X may be aluminum, indium, tin, or bismuth.

Some of the iron-based alloys are not ferromagnetic. The large class of austenitic stainless steels is an example. These metals are usually antiferromagnetic or spin glass at low temperatures and paramagnetic at room temperature and above. In general, ferrous alloys with the body-centered cubic (b.c.c.) structure are ferromagnetic, whereas the face-centered cubic (f.c.c.) ones are not. Thus, ferromagnetic alloys tend to be somewhat more brittle than similar nonferromagnetic alloys. A good example is the use of magnetic (martensitic) stainless steel (b.c.c.) for the blades of table knives where a good edge is desired and the use of more ductile austenitic alloys (f.c.c.) for the more elaborately shaped handles.

The major features of ferromagnets from an applications viewpoint are best illustrated by a plot of magnetization vs. applied field strength for a typical material, as shown in Fig. 6.6a, and the more common hysteresis loop, a plot of B vs. H, shown in Fig. 6.6b. The two plots are related by Eqs. 6.1 and 6.5. The plot of Fig. 6.6a represents the behavior of the material alone; parameters derived from it are often termed *intrinsic* parameters. The dashed lines indicate the first magnetization of a demagnetized, or virgin, material. The slope of this line near the origin in Fig. 6.6a is the *initial susceptibility*, κ_i (in Fig. 6.6b it is the *initial permeability*, μ_i). The magnetization rises rapidly with increasing field to a saturation value, the *saturation magnetization*,

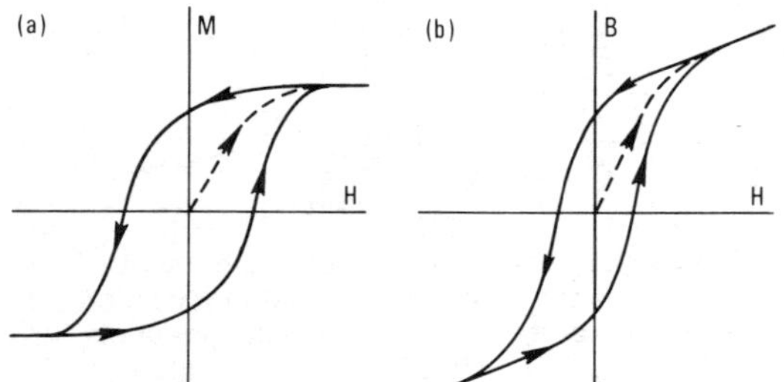

Figure 6.6 Behavior of the magnetization (a) and the flux density (b) for typical hysteretic ferromagnetic materials. The dashed line is the initial magnetization curve for virgin material.

M_s. The *ferromagnetic moment* per atom may be calculated from $m_f = M_s/n$, where n is the number of atoms per unit volume. As opposed to m_p, the paramagnetic moment, m_f depends on the degree of ferromagnetic alignment of the moments. The value of H at saturation is termed the *saturation field strength, H_s*. If the applied field is then decreased, the magnetization does not usually return to zero with H. The magnetization at this point (H = 0) is termed the *remanent* (or *residual*) magnetization or just *remanence*, M_r. The remanent flux density, B_r, is just $\mu_0 M_r$. By reversing and increasing the applied field, the magnetization can be brought to zero at a point termed the *intrinsic coercivity*. The flux density, B, goes to zero at a smaller field point, called the *coercivity, H_c*. The remainder of the loop is a reflection of the first two quadrants. When a magnetic material is driven around the hysteresis loop, by an applied alternating field for example, an energy, W, proportional to the area of the loop is converted to heat in a unit volume of the material where

$$W = \mu_0 \oint H dM = \oint H dB \tag{6.14}$$

The hysteresis loop of Fig. 6.6b can be used to illustrate another confusing aspect of the technology — the existence of a variety of permeabilities. (It should also be mentioned that permeability is actually a tensor, a fact that will henceforth be ignored.) The *initial* permeability was defined above. The most general definition of permeability is that given by Eq. 6.5, that is, the slope of the line from the origin to a point on the $B-H$ curve, but clearly the value varies depending on the location of the end point on the hysteresis loop. The *differential* permeability is the slope of the tangent to the curve at a point. The *maximum* permeability is the largest value of μ for the initial magnetization curve; note that it may be smaller than the differential permeability. A large number of other permeabilities used in the industry are described by Heck (1974). In each case, the permeability has the SI units of $H \cdot m^{-1}$, but the usual approach is to convert to a *relative* permeability by dividing by μ_0.

Of the properties described, only the Curie temperature and the saturation magnetization are insensitive to the material structure. All other properties are significantly affected by the physical and metallurgical state of the substance.

A few words are in order regarding the physical basis for ferromagnetism. It can be shown using quantum theory that the energy of interaction, U, of atoms with spins $\vec{S}_i$ and $\vec{S}_j$ can be written as

$$U = -2\, J_{ex} \vec{S}_i \cdot \vec{S}_j \qquad (6.15)$$

where J_{ex} is the exchange integral. Evaluation of J_{ex} is a complex process related to the overlap of the atomic charge distributions. Ferromagnetic materials are unique in that they give a positive value for J_{ex}, indicating that the lowest energy state is one in which the spins are aligned parallel to each other. There are other processes, such as thermal agitation, that act to oppose this alignment. Furthermore, alignment does not occur uniformly throughout a macroscopic specimen; for example, a piece of iron in zero field does not exhibit a moment consistent with complete saturation. This has been explained and experimentally verified by the concept of domains. The domains are small regions within a ferromagnetic material that are individually saturated, but are positioned with respect to each other such that the net external moment is small. When a magnetizing field is applied, favorably oriented domains grow at the expense of the others, and finally the magnetization rotates toward the direction of the applied field. Kittel (1967) gives a particularly good presentation of these effects, illustrated with photographs of actual domains.

Magnetic anisotropy is seen in single crystals of ferromagnetic materials. The magnetization saturates quickly for some crystal axes: [100] in iron, [111] in nickel, and along [0001] (the c-axis) in cobalt; these are termed *easy* axes. For *hard* axes, the value of H required to achieve saturation is much higher. The magnetocrystalline energy, that is, the energy needed to change the direction of magnetization, can be written in terms of direction cosines and a set of anisotropy constants. The constants are simply related to the field necessary to perform the motion, the *anisotropy field*. The constants and fields are strong functions of temperature (Carr, 1969). Magnetic anisotropy occurs in rolled polycrystalline steels because of preferential grain orientations.

In terms of applications, the ferromagnets can be divided into two groups: *soft ferromagnets*, such as pure iron, which are easily magnetized and demagnetized, and *hard ferromagnets*, such as the many permanent magnet materials of which the inverse is true.

Soft materials are used in electromagnets, transformer cores, shielding of magnetic fields, inductors for ac circuits, overlay patterns for the manipulation of magnetic bubbles, and ferrite cores. A complete treatment of the physical and metallurgical aspects of these materials is given by Chen (1977). The majority of soft ferromagnets are metals. They usually have very narrow hysteresis loops, that is, low values of coercive force and high permeabilities (e.g., for pure iron, $H_c \simeq 0.8\ \text{A·m}^{-1}$, $\mu_{max} \sim 2 \times 10^5$). Included in this group are the pure ferromagnetic elements, Si–Fe Permalloys, Mumetal, Perminvars, and the recently developed amorphous alloys. In low-frequency ac applications, the metals, which are good conductors, are usually laminated to reduce eddy current losses. At higher frequencies, powdered materials held together with nonmetallic binders are used.

Hard ferromagnets are used primarily in relatively small-scale applications, such as in loudspeakers, meters, and electronic instruments. Some large-scale applications are in motors, generators, and ore separators. The materials as a group are mechanically hard and usually brittle. The history of magnetism began with these materials — the lodestones of ancient times. A brief, but very interesting history of permanent magnets is given by Andrade (1958). A good permanent magnet material requires stable (well-pinned), oriented regions, each with a large moment. This gives a large hysteresis loop (H_c values to $0.4\ \text{MA·m}^{-1}$ in some cases). The conventional display used for material evaluation is the second quadrant of the loop, shown in Fig. 6.7, often called the *demagnetizing curve*. The nearly universal figure of merit for general applications is the maximum BH product, which is

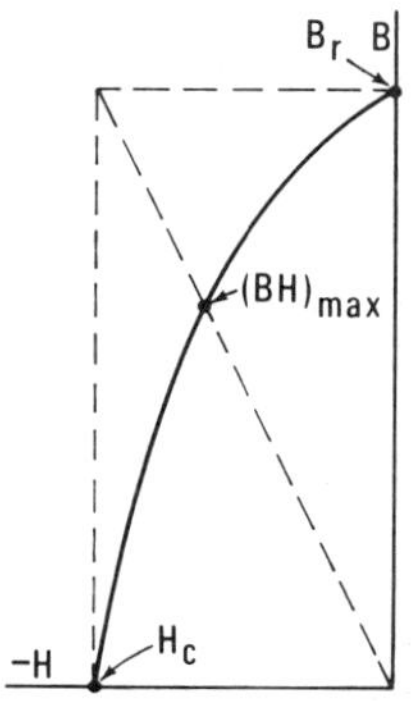

Figure 6.7 Demagnetization curve for a permanent ferromagnet showing the $(BH)_{max}$ "figure of merit." The construction determining $(BH)_{max}$ is approximate.

related to the maximum magnetic energy that can be established in an air gap by the material. The graphical construction shown in the figure allows a good estimate of the $(BH)_{max}$ point. In actual engineering practice, there are other considerations that occasionally result in a lower $(BH)_{max}$ material being preferred for a particular application (Parker and Studders, 1962). Typical values for $(BH)_{max}$ are 1 to 10 kJ·m^{-3}, with a few materials approaching 100 kJ·m^{-3}. Heck (1974) lists seven classes of permanent magnet materials. The classes and a typical example from each are: (a) steels (quench hardened, martensitic steels, Fe–C); (b) precipitation alloys (Fe–Al–Ni–Co); (c) cold-worked alloys (Fe–Co–V); (d) superstructure alloys (ordered precipitated phase, Pt–Co); (e) dust magnets ($SmCo_5$); (f) ceramic magnets (barium ferrite); (g) exchange anisotropy materials (oxide-coated cobalt dust). The special qualities of each of these classes and the details of material preparation are discussed in the reference.

The behavior of the magnetization of any ferromagnetic material with decreasing temperature is generally as shown in Fig. 6.5c. This is due to decreasing thermal agitation of the spin system. The theoretical derivation of the curve is relatively easy. The exact shape is dependent on the total electronic angular momentum quantum number, J, for the material. Tabular values of relative magnetization vs. reduced temperature for different J values are available (Darby, 1967). The curve for $J = 1/2$ agrees well with the data for iron, nickel, and cobalt (Chen, 1977), metals that, unfortunately, have actual J values much higher than 1/2. Since most materials of interest have rather high Curie temperatures, the change between room temperature and 4 K is seldom more than 10% for metallurgically well-behaved materials. On the other hand, the structure-sensitive properties, such as coercive force, remanence, and permeability, may vary greatly with decreasing temperature. The largest of these variations occurs when the anisotropy of the magnetization changes with temperature.

Antiferromagnetism. Antiferromagnetic materials have an antiparallel arrangement of aligned moments that leads to a zero net magnetic moment for the material (Fig. 6.8). The exchange interaction, J_{ex} (Eq. 6.15), is negative. Materials in this category tend to be oxides and halides of the transition elements. Chromium, manganese, and some "nonmagnetic" stainless steels are also antiferromagnetic. The physics of the exchange coupling is often quite different from that in ferromagnets. In oxide antiferromagnets, for example, it is indirect and labeled *superexchange* (Chen, 1977). Antiferromagnetism is frequently considered to be due to two distinct sublattices, on each of which the moments are aligned, but in a direction opposite to that of the other sublattice. One should not, however, take this picture as the only possibility. Neutron-scattering studies on stainless steels have been interpreted in terms of a four-sublattice model. A description of long-range antiferromagnetism in disordered Fe–Ni–Mn alloys is given by Kouvel and Kasper (1963).

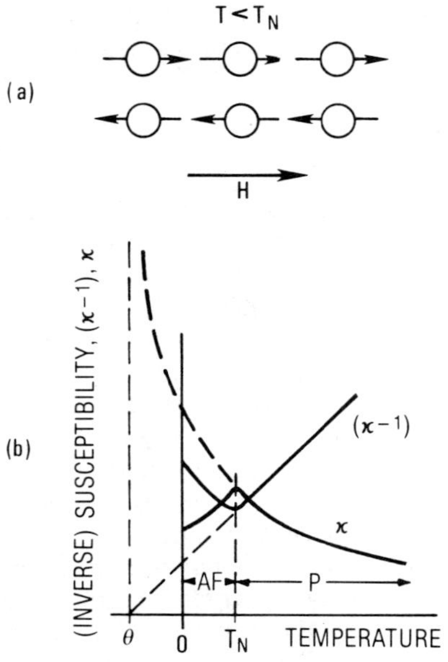

Figure 6.8 (a) Antiferromagnetic alignment of atomic magnetic moments. (b) Susceptibility and reciprocal susceptibility vs. temperature showing antiferromagnetic (AF) and paramagnetic (P) intervals.

The antiferromagnetic state only exists below a critical temperature, the *Néel temperature, T_N*. Above this temperature the material is paramagnetic (Fig. 6.8b) and obeys the Curie–Weiss law (Eq. 6.13) with a negative critical temperature. A domain structure is seen in antiferromagnetic substances, but the observation is experimentally difficult. Figure 6.9 shows the temperature dependence of the susceptibility of an austenitic stainless steel. The peaked susceptibility curve is usually taken as the indication of antiferromagnetism, but the transition can also be detected by electrical resistivity and elastic constant measurements. Many cubic antiferromagnetic materials undergo a slight lattice distortion on cooling through the Néel temperature.

Néel temperatures of various materials range from well below 4 K to far above room temperature. Connolly and Copenhaver (1972) have tabulated Néel temperatures for many materials.

Ferrimagnetism. Ferrimagnets are more complex materials than those discussed above. They typically contain two types of magnetic ions on separate sublattices. Each sublattice is ferromagnetically ordered, but antiparallel to the other. They are sometimes called uncompensated antiferromagnets. Thus, the ferrimagnets have a net magnetization, but it is significantly smaller than one would calculate from knowledge of the total number of magnetic ions (Fig. 6.10). The "standard" material in this class is the magnetic mineral magnetite, Fe_3O_4. The formula can be written $Fe_2O_3 \cdot FeO$. The Fe^{3+} ions are antiferromagnetically aligned, and the Fe^{2+} ions are ferromagnetically coupled. Other materials have this structure with the FeO replaced by XO, where X is one of a large number of divalent metals. They are usually called ferrites. As expected, the ferrimagnetic materials have a Curie temperature (sometimes called the ferrimagnetic Néel temperature) and occur both with hard and soft magnetism below that temperature. They tend to have high electrical resistivity and are relatively easy to manufacture; they are used as permanent magnets in high-frequency electronics and in computer applications.

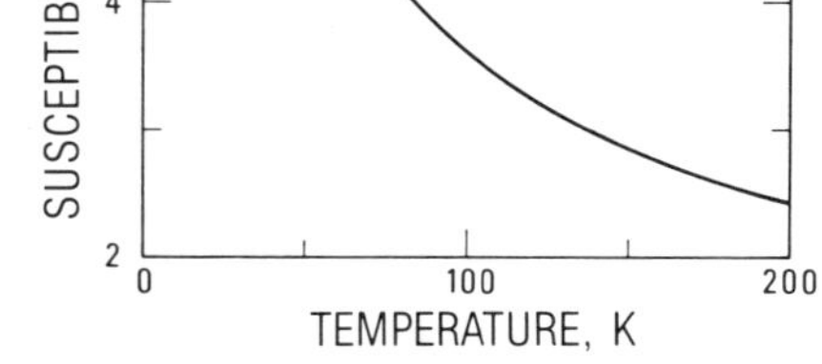

Figure 6.9 Temperature dependence of the susceptibility of an antiferromagnetic commercial stainless steel (Ledbetter and Collings, 1979).

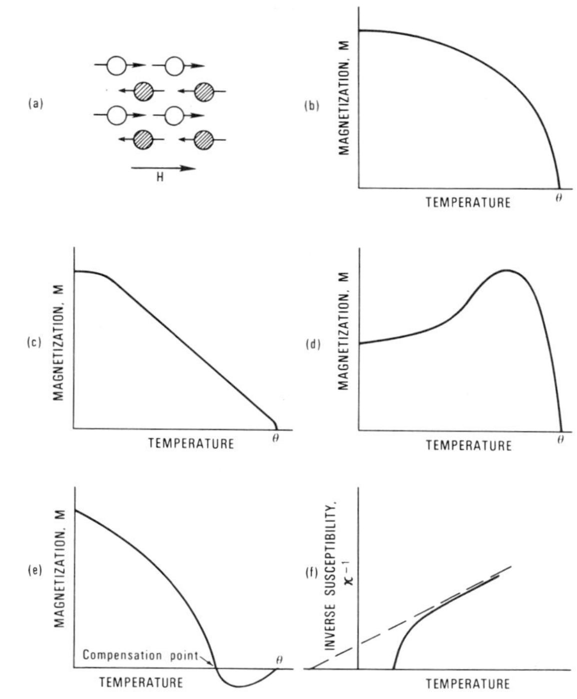

Figure 6.10 (a) Ferrimagnetic alignment of atomic magnetic moments. (b–e) Saturation magnetization vs. temperature for ferrimagnets. (f) Reciprocal susceptibility vs. temperature for a ferrimagnet.

Above the Curie temperature, ferrimagnets do not follow the Curie–Weiss law (Eq. 6.12), but exhibit a somewhat more complex temperature dependence given by

$$1/\kappa = T/C + 1/\kappa_0 = S/(T-\theta) \qquad (6.16)$$

where C, κ_0, S, and θ are constants. This behavior is plotted in Fig. 6.10f. Typical Curie temperatures are 600 to 700 K, but many exceptions exist. A very few materials have critical temperatures below 100 K. Curie temperatures are tabulated by Connolly and Copenhaver (1972) for ferrimagnetic materials, but no distinction is made between these and ferromagnets.

Because of the complex nature of these substances, the magnetization vs. temperature may take a number of forms for ferrimagnets. The magnetization of each sublattice depends on temperature in a different way. Some of the possibilities suggested by Néel (1948) are shown in Fig. 6.10. Apparently all these types of behavior have been observed. The type shown in Fig. 6.10e is particularly interesting, in that the magnetization goes to zero below the Curie temperature.

This is termed a *compensation point*. Most ferrimagnetic materials of technological interest are more well behaved, following curves like those in parts b or c of the figure. Saturation magnetization values for ferrites tend to be about 300 kA·m^{-1} ($B_r \sim 0.4$ T), and initial relative permeabilities run from ten to several hundred.

Single-domain grains. The number of magnetic domains that form in a ferromagnetic material is that number which minimizes the total of the magnetostatic and domain-wall energies. When a ferro- or ferrimagnetic material is divided into increasingly finer particles, the coercivity is observed to rise, often dramatically, until it peaks and falls very rapidly to zero, as shown in Fig. 6.11. The peak of the curve corresponds to the creation of particles too small to sustain more than one domain. Thus, any change in the saturated magnetization must be accomplished by rotation of the magnetic moment with respect to the structure — a more difficult operation energetically than the growth of preferentially aligned domains. The particle size at which this occurs is dependent both on the material and the shape of the grain but tends to be on the order of several micrometers. These materials are extensively used in magnetic recording and in the dust-type permanent magnets mentioned earlier. A good example of the increase in coercivity is provided by pure iron, which has a coercivity of 0.8 A·m^{-1} in bulk and 0.24 MA·m^{-1} in whisker form. The precipitate particles formed in the precipitation-alloy permanent magnets tend to be nearly single domain also. A complete treatment of the theory of fine magnetic particles is given by Kneller (1969). A certain amount of experimental data is also presented.

Superparamagnetism. If single-domain particles are further reduced in size, the coercivity vanishes (Fig. 6.11) because the thermal fluctuations randomize the moments fairly quickly. (Actually, these materials can exhibit an initial magnetization, but it decays exponentially with time after removal of the inducing field.) The most popular of several names for this state is superparamagnetism (Fig. 6.12).

For common ferromagnetic materials, parti-

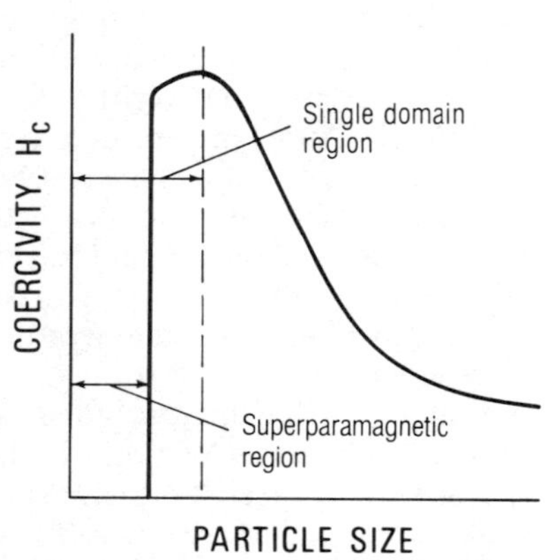

Figure 6.11 The behavior of the coercivity of finely divided magnetic material as a function of particle size.

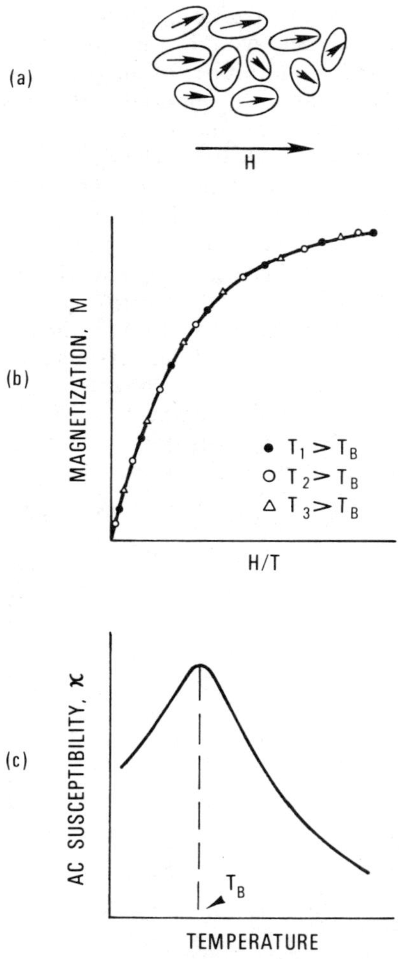

Figure 6.12 (a) Cluster magnetic moment alignment for a superparamagnetic ensemble. (b) Magnetization, *M*, vs. field, *H*, scaled by temperature, *T*. The data all fall on a single curve characterized by the Langevin function. (c) Susceptibility vs. temperature for a superparamagnetic ensemble illustrating blocking.

cles with sizes below about 15 nm are in this category. A collection of such particles will exhibit superparamagnetic behavior (Bean and Livingston, 1959). Such collections may be packed powders, precipitates in nonmagnetic matrices, and perhaps materials containing local ferromagnetic clusters. Superparamagnetism has been suggested as the low-temperature magnetic state of several structural alloys.

The magnetization process in multidomain ferromagnetic materials consists initially of the enlargement of favorably oriented domains via domain-wall motion. In single-domain particles, an applied field causes only rotation of magnetic moments and, thus, there is no hysteresis. In a collection of these particles, each acts as a single supermoment. The overall behavior resembles paramagnetism. As opposed to paramagnetism, however, saturation can be achieved in laboratory fields. As a function of field, the magnetization, M, follows the classical Langevin equation

$$M = Nm\ L(x) = Nm\ (\coth x - 1/x) \quad (6.17)$$

where $x = mH/k_BT$, N is the number of particles per unit volume, m is the net moment per particle, Nm is equal to the saturation magnetization, and H is the magnetic field. In the limit of small fields or high temperature, $L(x)$ approaches $x/3$ and

$$M = Nm^2\ H/3k_BT \quad (6.18)$$

which is the familiar Curie law. Naturally, m is many μ_B. When magnetization curves at different temperatures are plotted as a function of H/T, it is found that the data superimpose on a single Langevin curve (Fig. 6.12b).

At low temperatures, superparamagnetic particles undergo a blocking process in which the particle moments become viscous (Néel, 1955). If a magnetic field is applied, hysteresis is evident. The temperature at which this occurs depends on the particle volume, V, and the crystalline, shape, or other anisotropy constant, K. In the absence of an external field, the relaxation time, τ, for the particle moment to decay is

$$\tau = \tau_0 \exp (KV/k_BT) \quad (6.19)$$

where τ_0 is a characteristic time equal to 10^8 to 10^9 s. When τ becomes large with respect to measurement times, blocking is said to occur.

If ac susceptibility is measured as a function of temperature, a falloff is seen below the blocking temperature, T_B. Above T_B, a Curie–Weiss law is followed (Fig. 6.12c). Blocking effects are important in paleomagnetism where blocking in igneous rocks at high temperature can be used as a record of the direction of the earth's magnetic field in earlier times (Stacey and Banerjee, 1974).

Spin glasses and mictomagnetism. Experimental and theoretical research on spin glasses comprises a sizable portion of the current research literature on magnetics. The annual Conference on Magnetism and Magnetic Materials, and the triennial International Conference on Magnetism and International Conference on Low Temperature Physics devote a large number of sessions to spin glasses. Perhaps this interest will decay as a consensus is reached on definitive experiments and theoretical models. The ubiquitous nature of spin-glass effects at low temperature justifies a familiarity with them.

The terms *magnetic glass* and *spin glass* describe a state of randomly separated magnetic moments (spins) frozen in random directions with no long-range order. Indirect exchange interactions (Anderson, 1970; 1973) or spin-orbit interactions (Préjean, Joliclerc, and Monod, 1980; Levy and Fert, 1981) are believed to dominate the behavior. It may occur in both amorphous and atomically disordered crystalline alloys. Another designation for this state, due to Beck (1971), is *mictomagnetism,* which emphasizes the mixed ferro- and antiferromagnetic interactions that characterize these alloys. In practice, *spin glass* has come to refer to dilute magnetic constituents in nonmagnetic host matrices, and *mictomagnetism* refers to more concentrated alloys that exhibit magnetic clustering. Many austenitic stainless steels possibly fall into this latter category (Collings and Hart, 1979).

Spin glasses are classically characterized by a peak or cusp in magnetic susceptibility, κ, as a function of temperature, which does not correspond to an antiferromagnetic phase transition (Fig. 6.13). Such peaks are sharp in dilute spin glasses and broader in mictomagnets. The temperature at which the peak occurs is often designated as T_f, the freezing temperature. Above T_f, κ follows a Curie–Weiss law, as in the case of antiferromagnets. Unlike antiferromagnets, however, θ is often zero or positive. Neutron diffraction does not show evidence of an antiferromagnetic phase transition at T_f, nor are there singularities in the specific heat or electrical resistivity. It is evident that the spin-glass transition is not a classical phase transition.

The apparent T_f is often dependent upon the

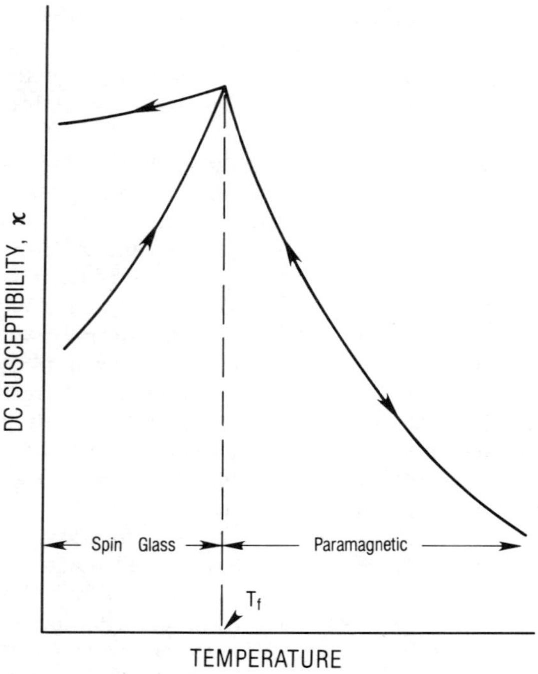

Figure 6.13 The dc susceptibility, κ, vs. temperature, T, upon warming and cooling, showing spin-glass and paramagnetic intervals.

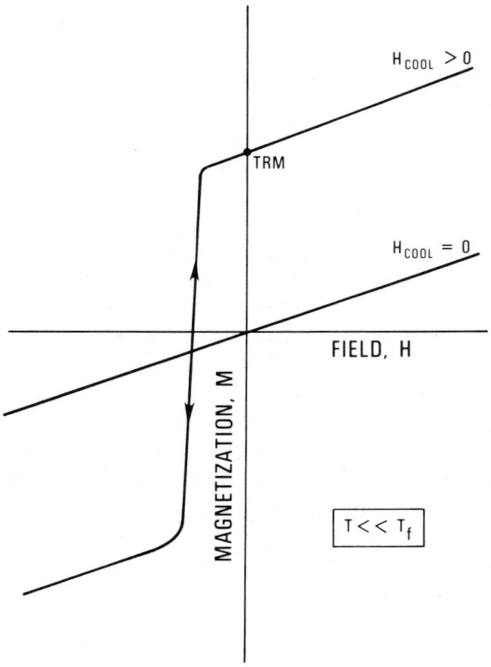

Figure 6.14 Magnetization, M, vs. field, H, obtained after cooling to below T_f in the presence and absence of a field. For the case of field cooling, the thermoremanent magnetization, TRM, is shown.

magnitude of the magnetic field used to measure κ and, in the case of ac methods, its frequency. Below T_f, time-dependent effects appear, unlike the case of antiferromagnetism. In dc magnetic fields and temperatures below T_f, κ increases as a function of time to values no greater than the maximum κ at T_f, tending toward a presumed equilibrium state. This same state may be obtained by field cooling to temperatures below T_f or by measuring κ as a function of decreasing temperature (Fig. 6.13). The thermoremanent magnetization (TRM) thus obtained at temperatures much below T_f is usually extremely stable in mictomagnets. Fields below some large threshold are unable to dislodge the TRM, giving rise to the shifted hysteresis loops (Fig. 6.14) first discovered by Kouvel, Graham, and Jacobs (1959). Large fields applied to a zero-field cooled alloy at low temperature will induce an isothermal remanent magnetization (IRM). Even in large laboratory fields (40 T), the magnetization of spin glasses does not saturate at low temperatures.

For certain alloys (usually amorphous) and when a ferromagnetic constituent is more concentrated, there appears an intermediate ferromagnetic-like state for temperatures between the Curie temperature, T_C, and the freezing temperature, T_f (Fig. 6.15). The freezing temperature in these alloys may be identified by a peak in the imaginary (quadrature) component of low-frequency ac susceptibility, κ'' (Goldfarb, Fickett, Rao, and Chen, 1982). The real (in-phase) component, κ', exhibits a clear Curie transition at the temperature at which κ'' first appears upon cooling.

The physics of the freezing process from a clusters approach is discussed by Tholence and Tournier (1974); Kouvel (1961); Knitter, Kouvel, and Claus (1977); Beck (1978); and Goldfarb and Patton (1981). The collective phase transition approach is treated by Edwards and Anderson (1975) and Sherrington and Kirkpatrick (1975). Further details of spin-glass theories and experiments can be found in the reviews by Heber (1976), Mydosh (1975; 1978), and Blandin (1978).

Other types of behavior. It is clear from the brief survey just presented that the magnetic behavior of materials is an extremely complex sub-

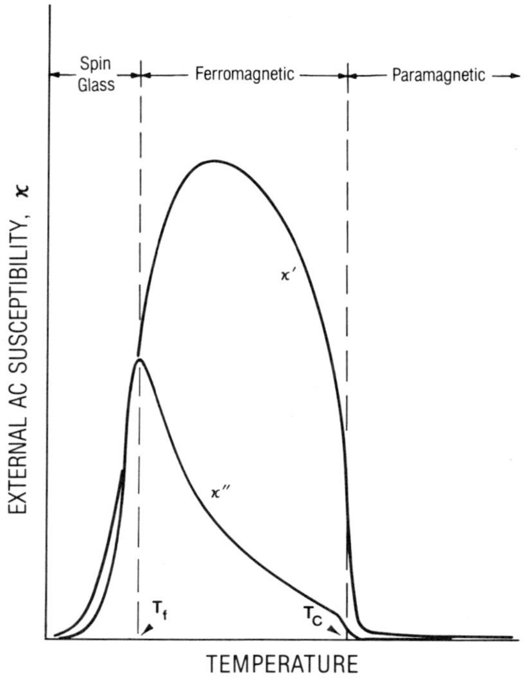

Figure 6.15 External ac susceptibility vs. temperature for a spin glass with a ferromagnetic-like intervening state. The real (κ') and imaginary (κ'') components of susceptibility are shown.

ject. Thus, it is not surprising that effects exist that cannot be conveniently categorized under the major headings above. These effects are not usually of concern, but it is good to be aware of their existence.

a. Magnetic impurities. As previously mentioned, many dia- and paramagnetic metals contain iron as a natural impurity. The available information indicates that clustering of iron impurities occurs at very low impurity levels (~100 ppm) and precipitation of relatively large iron particles can occur at higher levels. Furthermore, oxides of the transition elements can form during annealing procedures, and these ferrimagnetic materials can give a low-level remanent magnetism to nonmagnetic metals. In all instances the effects are enhanced significantly at low temperatures.

b. Parasitic ferromagnetism and weak ferromagnetism. These two terms appear to be equally popular. The effect arises from a lack of exact antiparallelism of the spins in some antiferromagnetic materials that gives rise to a spontaneous magnetization. The magnetization tends to be very small ($\sim 10^{-2}\ \mu_B \cdot mol^{-1}$). The substances tend to be nonmetallics. Typical is the mineral hematite, α-Fe_2O_3, which is a perfect antiferromagnet below 263 K but exhibits weak ferromagnetism above that (Stacey and Banerjee, 1974).

c. Metamagnetism. Some antiferromagnetic substances become ferromagnetic in high applied fields. These materials are said to be metamagnetic. An example is $MnAu_2$. At some critical field, the moments of the antiferromagnet, which are aligned against the field, rotate into the field direction. The critical field is only slightly temperature dependent.

d. Superconductivity. This phenomenon is the subject of Chapter 13, and thus it is only mentioned briefly here. In the simplest cases, superconductors below their transition surface are perfectly diamagnetic; all magnetic flux is excluded. Practical superconductors, however, interact with magnetic flux in various ways, which can primarily be described by a picture that has flux vortices flowing across the material under the influence of the applied fields (see Chapter 13).

6.2.3 Effects of the magnetic transformations

The transformations that magnetic materials undergo at the Curie or Néel temperatures nearly always have a noticeable effect on a number of physical and mechanical properties. The Curie and Néel temperatures of most magnetic materials are well above room temperature, so that the effects described are not common in low-temperature applications. Nevertheless, they sometimes do occur well into the cryogenic temperature range; stainless steel is a good example. The behavior of the magnetic susceptibility at the transition was discussed above.

Specific heat. The contribution of the magnetic state to the specific heat of materials below the transition temperature is additive with the lattice and electronic components. The spin waves provide an additional mechanism for absorbing energy. Both ferromagnetic and ferrimagnetic materials have a magnetic specific heat that varies as $T^{3/2}$. This is not easily seen in metals because of the electronic term ($\propto T$), but it is dominant in the nonconducting ferrites. Antiferromagnets show a T^3 dependence for the magnetic term, which can be 10 to 20 times larger than the lattice contribution in nonmetallic substances.

Very near the transition temperature, the specific heat of most magnetic materials rises very rapidly, approaching a logarithmic singularity. A typical graph of the total specific heat is shown in Fig. 6.16a. The graph is characteristic of a metal with a high magnetic transition temperature (e.g., nickel). For materials with lower transition temperatures, the behavior is similar, except that above the transition the specific heat rises more steeply, and the peak represents a larger fraction of the total value.

Electrical resistivity. The electrical resistivity of conductors, both ferromagnetic and antiferromagnetic, is affected by the ordering transition. All the ferromagnetic metals behave roughly as shown in Fig. 6.16b. The magnetic term appears to contribute most strongly to the resistivity when the temperature variation of the magnetization is largest, that is, below but near the Curie temperture. The curve is always steepest just below T_C. This behavior results in a sharp peak in the derivative (the temperature coefficient of resistance) at the transition. The behavior of antiferromagnetic metals is illustrated by the other curve shown in Fig. 6.16b. The exact shape of the curve is dependent on the specifics of the specimen preparation.

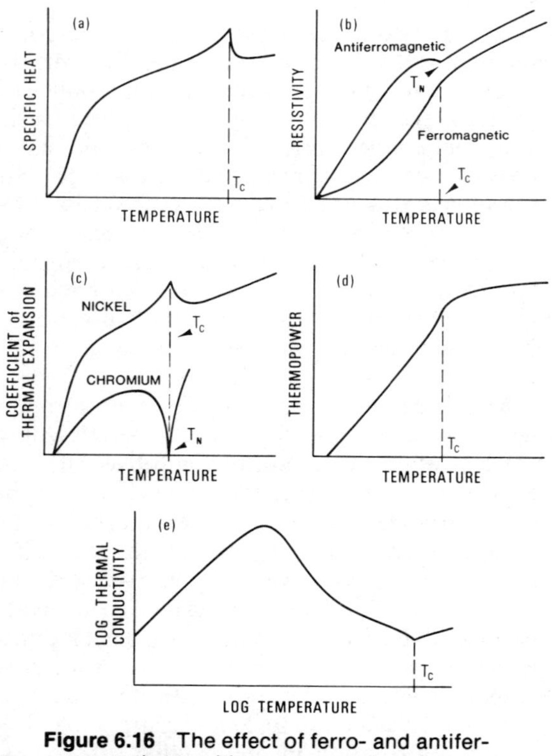

Figure 6.16 The effect of ferro- and antiferromagnetic transitions on a number of material properties.

The nonmetallic ferrites have very high resistivity values that decrease exponentially with increasing temperature both above and below the Curie temperature. The value of the exponential constant, however, often decreases significantly at T_C. Measurements on ferrites are usually made at rf frequencies. The resistivity is strongly frequency dependent in many cases.

Thermal expansion. The thermal expansion of some magnetic metals is greatly affected by the magnetic interaction. The effect may be opposite to that of conventional thermal expansion; this behavior is used in the formulation of the low-expansion alloys, primarily Fe–Ni ("Invar") alloys. The transition through the critical temperature does not greatly affect the thermal expansion, $\Delta\ell/\ell$, and is not noticeable on most data plots. On the other hand, the thermal expansion coefficient, $\alpha=(1/\ell)d\ell/dT$, measured in a very precise experiment can show a significant change. Typical behavior is shown schematically in Fig. 6.16c for nickel (ferromagnetic) and chromium (antiferromagnetic). The temperature axis is normalized to the critical temperature. The similarity of the nickel thermal expansion coefficient data to those observed for the specific heat is to be expected from the thermodynamic relationship between them.

Elastic constants. The elastic constants and sound velocities show rapid variations at the critical temperature, and their measurement is one of the most sensitive and useful methods for detecting these transformations in the cryogenic temperature range (Ledbetter and Collings, 1979). Theoretical treatments have been quite successful in explaining the observed elastic constant anomalies by means of an exchange-energy calculation.

Thermopower. The thermopower of the magnetic metals behaves somewhat like the resistivity (see Fig. 6.16d). The slope change at the transition causes a sharp peak in the first derivative of the thermopower. The behavior shown is typical of nickel; other metals show the effect to a greater or lesser extent, and chromium, as usual, shows a different variation, somewhat like that seen for the resistivity. A detailed treatment and a tabulation of the effect in various metals is given by Blatt, Schroeder, Foiles, and Greig (1976).

Thermal conductivity. Slight irregularities in the thermal conductivity-vs.-temperature curves occur for the common ferromagnetic and antiferromagnetic metals at the transition temperature. In all cases, the form appears to be as shown in Fig. 6.16e, although few data exist. One might expect a more dramatic effect for transitions at lower temperatures, but there are almost no useful data on common materials with very low transition temperatures. In the uncommon metals, Boys and Legvold (1968) have observed significant changes in the conductivity of erbium and dysprosium crystals, both at the Curie temperature and the Néel temperature. Many complex theoretical studies have been made on thermal transport and electron and phonon scattering by spin waves in magnetic materials. The compilation by Touloukian, Powell, Ho, and Klemens (1970) cites a number of pertinent papers and presents the limited data mentioned above.

6.2.4 Unique properties of magnetic materials

The order found in magnetic materials gives them some unique properties at temperatures below their transition tempertures. Unlike the properties described in the preceding section, these phenomena do not occur to any extent in nonmagnetic substances.

Magnetostriction. Many ferromagnetic materials undergo a change in dimension upon application of a magnetic field. This magnetostrictive effect (earlier termed the Joule effect) has applications in transducers of various sorts, from phonograph needles to sonar transmitters. The effect is complex; its magnitude and sign depend on the material, its thermal history, the crystal orientation with respect to the applied field, the magnitude of the field, the direction of measurement with respect to the applied field, and the temperature. Its origin is in the rotation of the magnetization of fully saturated domains into alignment with the applied field. The new equilibrium of the magnetic forces and the elastic forces between atoms causes a contraction or expansion of the lattice. The magnetostriction is thus tied to the anisotropy constants discussed in § 6.2.2. There is an inverse effect, the Villari effect, in which the permeability of a ferromagnetic material is changed in the direction of an applied stress. These and other piezomagnetic effects are of concern in geophysical studies of rock magnetism (Nagata, 1961). Magnetostriction is not always desirable; a mild, but familiar and annoying, manifestation is the ac hum from large transformers. The effect is usually small. At saturation in polycrystalline metals and upon application of a field, $\Delta \ell / \ell$ varies from −3.5 parts per million (ppm) for iron to −50 ppm for cobalt. Fe–Ni alloys with less than 80% Ni have values ranging from 0 to +30 ppm, depending on composition. The amorphous alloys based on Fe–Ni and cobalt offer the prospect of an adjustable magnetostriction ranging from negative to positive. Rare-earth–iron alloys have recently been shown to have a very large magnetostriction, +1753 ppm in $TbFe_2$ (Clark, 1974). The ferrites all show negative magnetostriction except for Fe_3O_4. Values to −540 ppm have been observed in cobalt ferrite, but numbers near 20 ppm are more common.

As the temperature is lowered, the magnitude of the saturation magnetostriction of polycrystalline specimens increases for all materials that have been measured (not very many). For metals, the increase appears to scale with the square of the saturation magnetization (Dietz and Burkhard, 1966), whereas for ferrites, the behavior is similar but slightly more complex. Data for several materials are shown in Fig. 6.17. A few magnetostriction measurements on metals have been made at 4 K, primarily on the paramagnetic transition metals. Two papers by Fawcett (1970a; 1970b) present data on many of these metals and alloys. The magnetostriction values tend to be very low, 0.1 to 0.0001 ppm.

In addition to the spontaneous magnetostriction, there is a *forced magnetostriction*, which continues to increase even after saturation, and the *form effect*, which gives a shape-dependent term. Usually these other terms are significantly smaller than the spontaneous one, but not always. The interested reader is referred to the texts mentioned earlier.

A final effect in this group is the so-called ΔE effect, which is seen in unsaturated ferromagnetic materials. Young's modulus, E, appears to be strain dependent, because the applied stress causes domain-wall motion and, at high stress, a change in orientation of the domains. The accompanying magnetostriction appears as an anomaly in the modulus. There is also a frequency dependence of E due to domain-wall viscosity.

Thermomagnetic effects. The temperature of ferromagnetic materials is observed to rise upon sudden application of a magnetic field.

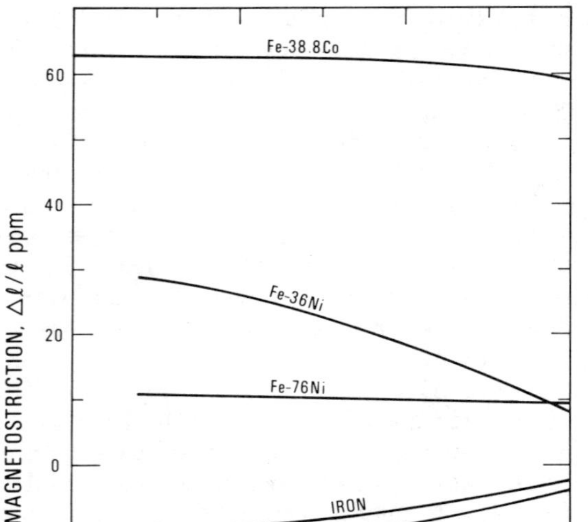

Figure 6.17 The temperature dependence of the saturation magnetostriction for selected polycrystalline alloys and a commercial cubic ferrite (Dietz and Burkhard, 1966). Compositions are assumed to be in weight percent.

This reversible temperature increase is called the *magnetocaloric effect*. It is largest near the Curie temperature and amounts to a maximum of about two degrees for iron and slightly more than one degree for nickel. The thermodynamic treatment (Carr, 1969) shows the magnitude of the effect to be proportional to the temperature derivative of the magnetization, and hence, the maximum is near the Curie temperture.

A similar effect in certain highly paramagnetic salts, notably cerium magnesium nitrate (CMN), is used to provide cooling in the very low temperature region below about 0.1 K and down to about 1 mK by *adiabatic demagnetization*. Materials, such as CMN, have an anomalously high specific heat (Schottky anomaly) at these temperatures. A magnetic field of several hundred kiloamperes per meter is applied to the salt, which is in contact with a heat sink. The heat induced by aligning the magnetic dipoles is transferred to the lattice and then to the sink. The salt is then isolated from the sink and the field removed. To resume their random positions, the dipoles extract energy from the lattice and the salt cools. Its high specific heat prevents rapid warming, and in instances where κ obeys a Curie–Weiss law, its temperature can be determined from a susceptibility measurement. The choice of a salt for cooling is discussed by Rosenberg (1963), and magnetic thermometry, by Betts (1976). Demagnetization of *nuclear* spin systems using much higher magnetic fields (~ 10 $MA \cdot m^{-1}$) on copper has been used to produce much lower temperatures, near 1 μK. Still lower temperatures have been obtained with more complex materials (Betts, 1976).

Magnetoresistance and elastoresistance. The electrical resistivity, ρ, of ferromagnetic materials under the influence of an applied magnetic field is complex. Near room temperature, polycrystalline, pure ferromagnetic metals behave as shown in Fig. 6.18, which presents Bates's (1946) data for cobalt. If the field is parallel to the current (longitudinal magnetoresistance), a more or less normal rise in $\Delta\rho/\rho$ is followed by a significant decrease. When the field is perpendicular to the current (transverse magnetoresistance), $\Delta\rho/\rho$ may rise slightly and then become negative. The changes, which are due largely to domain orientation effects, seldom exceed a few percent at room temperture. Hysteresis of a few tenths of a percent is often observed in the magnetoresistance of these metals. The detailed behavior of the magnetoresistance in alloys is even more complex and depends strongly on composition, heat treatment, temperature, field magnitude, and orientation. Results for a number of specific alloys of technological importance are shown by Bozorth (1951) and for some permanent magnet alloys, by Bates (1946). Like that of nonmagnetic metals, the magnetoresistance of most pure ferromagnetic metals and alloys increases significantly as the temperature is lowered. The transverse–longitudinal anisotropy remains, but $\Delta\rho/\rho$ becomes positive. A Kohler plot (see Chapter 5) can be made in which measurements of a given specimen at various temperatures all fall on a line. The field used for the plot should be the flux density, B, in the metal. Metals with different purity do not fit on the same line, however (Smit, 1951).

The behavior of alloys at low temperatures is very dependent on their composition. In general, alloys of two magnetic metals, such as Co–Ni, or alloys with primarily one magnetic component, like Ni–10% Cu show a large increase in $\Delta\rho/\rho$

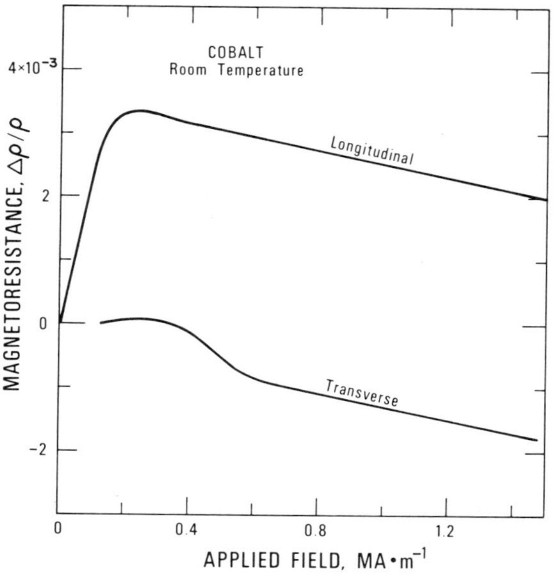

Figure 6.18 Magnetoresistance of polycrystalline cobalt (Bates, 1946). The behavior is typical of the ferromagnetic pure metals and some alloys.

with decreasing temperature (Bozorth, 1951), whereas alloys with a dilute magnetic component have a relatively small negative $\Delta\rho/\rho$ that increases with decreasing temperature. An example is provided by the work of Jacobs and Schmitt (1959) on Cu–Co alloys to 1.7 K. A number of theoretical treatments exist, but none completely explains all of the observations. The references quoted above may seem a bit dated, but with the exception of a few rare-earth alloys, there apparently has been no recent experimental work on the low-temperture magnetoresistance of ferromagnetic materials.

When unidirectional stress is applied to magnetostrictive materials, the electrical resistivity measured in the direction of the stress undergoes a change that, plotted as $\Delta\rho/\rho$ vs. stress, looks much like the magnetoresistance in both shape and magnitude. It is positive for metals with a positive magnetostriction and negative if the magnetostriction is negative. This effect is due to the preferential orientation of domains along the stress axis (positive magnetostriction) or normal to it (negative magnetostriction). The effect has been termed *elastoresistance* and is the basis for the construction of most strain gauges.

Hall effect. The Hall effect is usually manifested as a voltage across a plate-like specimen transverse to an electrical current, I, when a magnetic field, H, is applied normal to the plate. In this ideal situation where the voltage, current, and field are mutually orthogonal, the Hall voltage, V_H, is given by

$$V_H = RIB/t \tag{6.20}$$

where R is the Hall constant, B is the flux density $\mu_0 H$, and t is the plate thickness. The units (SI) of R are $m^3 \cdot A^{-1} \cdot s^{-1}$ (or $\Omega \cdot m \cdot T^{-1}$). In nonmagnetic materials, R is a measure of the effective number of carriers and may be positive or negative, depending on whether holes or electrons predominate. In most nonferromagnetic metals, R is small ($\sim 10^{-11}$ $m^3 \cdot A^{-1} \cdot s^{-1}$) and not strongly dependent on temperature down to about 100 K; below that, rapid changes on the order of 20% may occur, particularly in pure metals. Another common quantity is the Hall resistivity, ρ_H, defined by

$$\rho_H = E_H/J = RB \tag{6.21}$$

where E_H is the electric field associated with the Hall voltage, and J is the current density. Note that the units of ρ_H are $\Omega \cdot m$. In magnetic materials, the magnetization strongly affects the Hall resistivity and a new equation is used

$$\rho_H = \mu_0(R_0 H_0 + R_s M) \tag{6.22}$$

where H_0 is the applied field, R_0 is the *ordinary* Hall coefficient, and R_s is termed the *spontaneous* Hall coefficient. Note that the Hall resistivity can exist even in the absence of an applied field.

The exact form of this equation varies from author to author. It is discussed in detail in the comprehensive book on the Hall effect by Hurd (1972), which also contains much of the existing Hall-effect data on metallic elements and alloys. The value of R_s is often very large compared with that of R_0 and is extremely temperature sensitive, as seen in Fig. 6.19, which compares data on several metals from the work of Volkenshtein and Fedorov (1960). R_s is also very sensitive to specimen preparation and history. In many cases, the Hall effect provides an excellent means of detecting Néel transitions from the antiferromagnetic to the paramagnetic state (Rhyne, 1968).

Magneto-optical and photomagnetic effects. The magneto-optical effects are those in which the plane of polarization or the refractive index of electromagnetic radiation incident upon a solid are affected by the magnetization of the material. Applications of these effects have been mainly in computer memories, laser-driven optical recorders, and microwave electronics. Three effects, Faraday, Kerr, and Cotton–Mouton are the most commonly used. They are all strongly dependent on field orientation, temperature, and the wavelength of the radiation.

The *Faraday effect* is observed in transmission. It is seen as a rotation of the plane of polarization of electromagnetic radiation when it passes through a magnetized substance. The amount of rotation is proportional to the path length and the magnetization. Ferromagnetism is not required; the effect is observed in glass, for example, but its magnitude is significantly greater in ferro- and ferrimagnets. For metals, thin films are obviously required for optical observation, whereas relatively large specimens of ferrite can be used in the microwave region and rare-earth materials in the infrared. All of the general references contain some information on the Faraday effect (and the other effects too). Low-temperature data are scarce. Low-temperature magneto-optical recording is discussed by Suits (1972).

The *Kerr effect* is similar to the Faraday effect, but is observed on reflection. The incident plane-polarized light is reflected with the plane of polarization rotated by a small amount that depends on the magnetization.

The *Cotton–Mouton effect* is a birefringence phenomenon in which the index of refraction for incident radiation varies depending on whether the plane of polarization is normal or perpendicular to the magnetization. The effect is observed in transmission.

In contrast to the above effects, *photomagnetic* behavior is characterized by a change in the magnetic properties of a substance that is caused by incident photons. The effect is observed only at cryogenic temperatures and usually in rather exotic materials, such as doped yttrium iron garnet in which illumination produces changes in initial permeability and coercive force. This effect is reviewed by Enz and Teale (1970).

Hysteresis. The general features of hysteresis loops and losses were discussed above. Hysteresis effects in even mildly ferromagnetic and ferrimagnetic materials can present serious problems at low tempertures because most materials have very low specific heats and, thus, significant heating can occur with little energy input.

There are many different types of hysteresis loops; their shape depends on the material and its preparation (Heck, 1974). In Fig. 6.20 is shown a

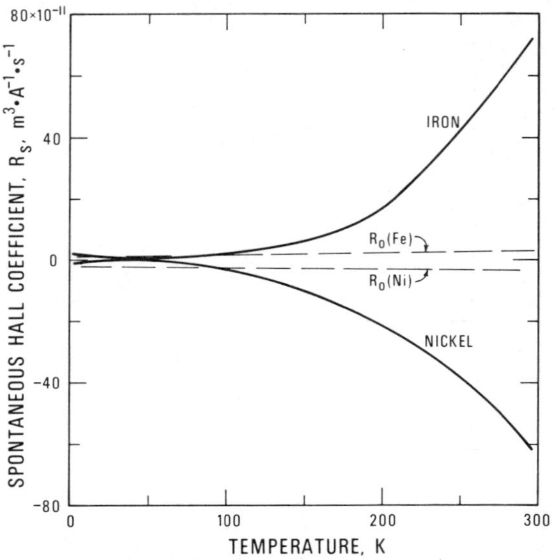

Figure 6.19 The normal and spontaneous Hall coefficient as a function of temperature for iron and nickel (Volkenshtein and Fedorov, 1960).

schematic diagram of some of the more common loops. The material characteristics are given in Table 6.6. A distinguishing parameter is the remanence ratio, the ratio of the remanent magnetization, M_r, to the saturation magnetization, M_s. The loops labeled "Perminvar" and "Isoperm" have come to be designated by these material trade names, although they are characteristic of other materials as well.

The effect of low temperatures on the hysteresis loops of materials has not been investigated in much detail. Measurements on magnet steels at 4 K by McInturff and Claus (1970a) in fact show a slight increase in M_s and a significant increase in the coercivity, H_c, from 80 $A \cdot m^{-1}$ at room temperature to 120 $A \cdot m^{-1}$ at 4 K. This demonstrates that M_s is an intrinsic property of the material, whereas H_c is structure sensitive. (Cooling to 4 K can cause significant structure modifications.) In Fig. 6.21, data from Gniewek and Powell (1962) show a significant increase in the hysteresis losses with decreasing temperature. The curve is an average for a variety of silicon steels used in electrical machinery. Their experiment did not determine the actual values of M_s and H_c, but it is reasonable to suspect that here, also, the increased losses arise from an increased H_c.

In metals and alloys, hysteresis losses are often accompanied by electrical losses from eddy cur-

Table 6.6 Hysteresis loop types (Fig. 6.20).

Designation	Remanence Ratio	Production and Typical Materials	Typical Applications	Comments
Normal	0.5–0.8	Most common "untreated" polycrystalline magnetic materials.	Electrical machinery Transformers Electronics	
Rectangular	0.8–1.0	Induced in many metals by stress and thermomagnetic treatment; spontaneous in some Mg–Mn ferrites.	Switching Computer storage	
Perminvar	0	Intrinsic to a wide range of Fe–Ni–Co alloys; some ferrites (Mg–Fe ferrite)	Inductance cores Transformers	Constant μ and low losses at low field. Exact form of loop varies widely.
Isoperm	<0.4	Mechanical working of Fe–Ni alloys (sometimes with added Al or Cu); Mn–Zn ferrites	Inductance cores Transformers	Narrow loop, high μ.

rents induced by the field variation. Eddy current effects (treated in detail in Chapter 5) are particularly serious for pure metals at low temperatures. Experimentally, eddy current losses can be separated from hysteresis losses by the fact that the eddy current loss per cycle is frequency dependent, whereas the hysteresis loss is independent of frequency.

6.3 MEASUREMENT TECHNIQUES

This section provides a brief description of some of the techniques and devices used for determining magnetic properties. Detailed descriptions can be found in the references. Measurement techniques for other properties are covered elsewhere in this book and, therefore, are described only cursorily.

The magnetic properties of interest in commercial applications are covered by specific tests and methods outlined by ASTM in their *Annual Book of Standards* (1981). No low-temperature measurements are covered. The adaptation of room-temperature techniques for magnetic properties measurement to low temperatures usually requires only the addition of a low-temperature environment in the form of a cryogenic liquid or a temperature-controlled cryostat. Both of these options are discussed in Chapter 5.

Some ac techniques applied to pure metals are bothered by greatly increased eddy current effects at low temperatures. Also, the low specific heat of most materials can result in unexpectedly large temperature excursions with relatively small energy inputs. Both of these effects can be minimized by careful choice of technique and clever equipment design.

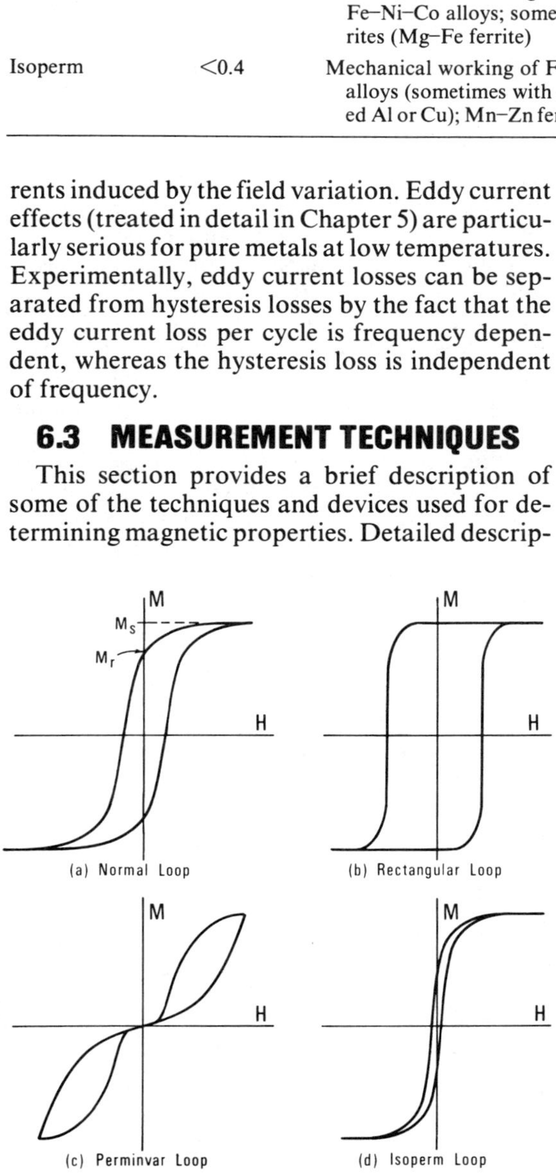

Figure 6.20 Several commonly observed types of hysteresis loops for ferromagnetic and ferrimagnetic materials.

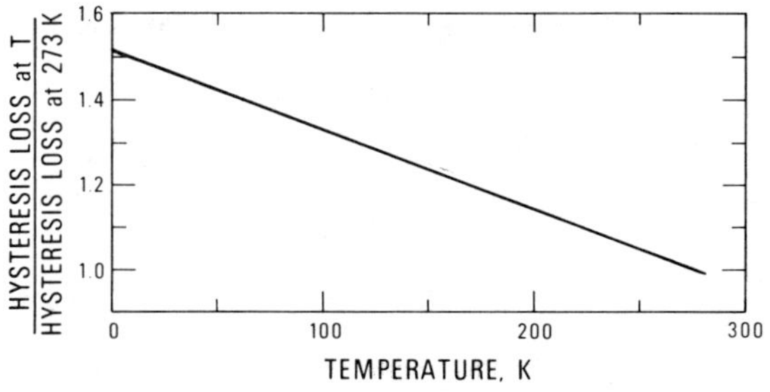

Figure 6.21 The effect of temperature on hysteresis losses. The curve averages values for several commercial Fe–Si alloy steels (Gniewek and Powell, 1962).

6.3.1 Production and measurement of magnetic fields

With few exceptions, measurement of the magnetic properties of materials requires the application of a magnetic field different from that of the earth. In many instances, field gradients are also required. Occasionally, fields must be excluded from an experimental region.

For most applications, fields and field gradients are produced by current-carrying coils, which may or may not be associated with iron pole tips and yokes (Zijlstra, 1967). In recent years, superconducting coils have produced high magnetic fields in large volumes with very small and uncomplicated power sources. Brechna (1973) discusses the design of many such systems. Permanent magnets are often used in systems where a small volume and fixed field are required. Exclusion of flux and the creation of nearly field-free regions is done with superconducting shields or with highly permeable soft magnetic alloys for small- and medium-sized applications. Larger systems use carefully constructed air-core dc coils to compensate for the earth's field. Recent research in magnetocardiography and magnetoencephalography with very sensitive superconducting flux detectors (SQUIDs) requires large volumes in which the ambient field is allowed to fluctuate only very slowly (Williamson, Kaufman, and Brenner, 1977). For these experiments, entire rooms are shielded with thick plates of pure metal, usually aluminum. Figure 6.22 shows schematically the range of magnetic fields available and the more common techniques for achieving them.

The measurement of magnetic fields involves a wide range of instruments employing proton precession, nuclear magnetic resonance, rotating coils, the Hall effect, magnetoresistance, magneto-optical effects, and permeability variation (flux-gate devices). As usual, the higher the required precision, the more time-consuming the measurement. Nuclear-magnetic-resonance, rotating-coil, and Hall-effect devices are described elsewhere in this book. An excellent review of many general measurement techniques is given by Weiss (1967). Techniques for very high fields are reviewed by Grivet (1962).

Measurements of dc fields are seldom made with the sensor itself at low temperature. However, some Hall-effect devices using semimetals, semiconductors, and metallic magnetoresistors have been used at low temperatures for moderate and high fields. Two papers (Sample and Rubin, 1976; Rubin, Nelson, and Sample, 1975) report evaluations of commercial Hall effect devices to quite high fields. SQUID devices are used for very low field determinations (Clarke, 1977). In most low-temperature experiments, however, the field is measured at near room temperature before the cryostat is inserted (for copper-wound magnets) or in an insert Dewar (for superconducting magnets). Most often, the magnet is calibrated once and a plot of field vs. magnet current is made. Later measurements then use the measured coil current and this plot to determine the field. This can lead to some problems with superconducting magnets at low field values since frozen-in flux can cause field errors as high as several tens of kiloamperes per meter. The ac-field measurement techniques used at low temperatures are essentially identical to those described in the literature for room temperature.

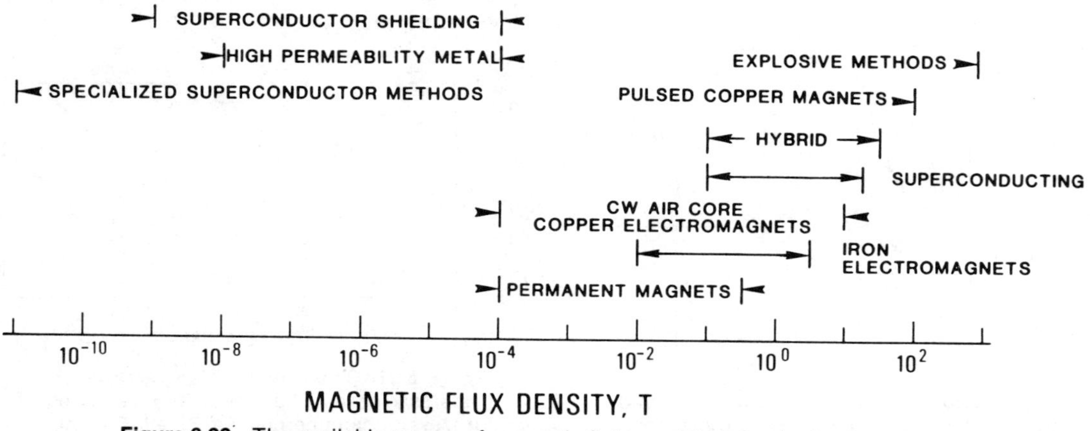

Figure 6.22 The available ranges of magnetic fields and their methods of production.

The only real concern arises because resistance of the coils involved can be much lower, and this can affect some of the circuit parameters.

6.3.2 Magnetic measurements

This section considers the measurement of the strictly magnetic properties, those associated with atomic or molecular magnetism and the exchange interaction. For simplicity of discussion, the phenomena have been divided into several groups. This division is somewhat artificial. Measurement techniques for one group often can be applied to another. Many common measurement techniques and commercial instruments are described in *Magnetism and Metallurgy* (1969) and Bozorth (1951). In addition, the texts used for university electricity and magnetism laboratory courses often have excellent descriptions of particular techniques.

Hysteresis loops, *B-H* and *M-H* curves. The complete hysteresis loop of a ferro- or ferrimagnetic material can be measured by various electromagnetic induction techniques, both dc and ac. The dc methods use the deflection of a ballistic galvanometer upon reversal of the magnetizing current as a measure of B, or movement of the specimen between counterwound pickup coils in a fixed field as a measure of M. Various clever automated systems have been devised to eliminate the point-by-point nature of this method. More modern systems replace the galvanometer with an integrating digital voltmeter. Alternating current devices, called induction curve tracers, drive the specimen continuously around the hysteresis loop while detecting B with a pickup coil on the specimen. This type of measurement is obviously quite rapid but, until the introduction of modern electronic techniques, was often only semiquantitative. The techniques described above are most useful to moderate field strengths ($\sim$100 kA·m^{-1}); the ac methods are particularly valuable at low fields. High-field measurements require more elaborate devices, such as the vibrating sample magnetometer described below.

The ideal specimen shape for these measurements is a torus onto which the magnetizing and pickup coils are wound — a geometry that is not easy to produce. Rod-shaped samples and open coils can be used, but very large corrections to the measured magnetization may be required because of the demagnetization effect. Note that DM, the demagnetizing field, is dependent not only on the shape of a specimen, but also on its magnetization. Heck (1974) gives an excellent discussion. Numerous devices have been used to avoid having to determine this correction. Often an attempt is made to complete the magnetic circuit by clamping the specimen onto the open end of a U-shaped piece of iron or steel, thereby creating an approximation of a toroidal system. Such devices are called yokes and permeameters.

Low-temperature measurements of hysteresis loops of steels at 77 K and 4.2 K using a two-coil induction technique are described by McInturff and Claus (1970b). A temperature variable device offering either a ballistic or an integrating fluxmeter mode of operation is described by Winter, Rothwarf, Leupold, and Breslin (1978).

Hysteresis losses. Because of the importance of minimizing hysteresis losses in the electrical industry, many commercial devices exist for their determination. Three techniques are commonly used. The first uses the methods just described to plot the hysteresis loop; then the loop area is determined, which gives the loss per cycle. The wattmeter method multiplies voltages proportional to H and B, as determined by coils around a (toroidal) specimen. A Hall-effect device is often used as the multiplication element. The calorimeter method measures the heat generated by repeatedly cycling the material around the loop. Each technique has its own problems and achievement of an accuracy better than about 5% is difficult.

Low-temperature measurements of losses are not common. Gniewek and Powell (1962) describe a wattmeter type of measurement on four commercial Fe–Si alloys from 4 K to room temperature. Their results are presented in Fig. 6.21. Work on ac losses in superconducting materials involves various experimental devices, which should be usable almost without modification for magnetic materials measurements. These techniques and instruments are described in Chapter 13.

Permeability, coercivity, and remanence. It is possible, of course, to determine the permeability, coercivity, and remanence of ferromagnetic and ferrimagnetic substances from the measured hysteresis loop, but this is seldom done since other methods yield these parameters directly.

High relative permeabilities ($\mu_r \sim$ 10 to 5000) are usually measured by the induction techniques used for determination of the complete loop. The specimen must be completely demagnetized prior to measurement, usually by heating it above the Curie temperature or by ac demagnetization. Lower relative permeability values ($\mu_r = 10$ to 1.01) require the same techniques used for susceptibility measurements described below. Ansorge, Billan, Gourber, Haren, Henrichsen, and Verdier (1978) describe two devices for low-temperature measurement (4 to 300 K), that together cover the total range of permeability given above. The high permeability measurement uses ac induction with computer processing and control. The low permeability measurement uses a ballistic technique with a single coil and electronics that give a direct digital display of the result. Both devices are claimed to have an accuracy of 1% or better. Three other low-temperature systems using dc induction, ac induction, and deflection of a magnet (astatic magnetometer), are described in an earlier paper by Dralle and Moore (1965).

Coercivity has been measured directly by a force balance technique (discussed below) in which the applied field is varied in magnitude while the force is monitored. The force will go to zero at $H = \pm H_c$ and, of course, at $H = 0$. The intersection of the parabolic curves with the $F = 0$ axis provides a measure of H_c (Nagata, 1961). Another device, a coercimeter, makes use of the fact that, when a magnet is placed in a uniform field, the field lines are distorted near the poles except when the applied field, H_a, equals H_c. A Hall probe with its plane parallel to the applied field is situated near one pole of the magnetized material. The applied field is increased until the Hall probe gives a zero reading (Parker and Studders, 1962). Apparently neither technique has been used specifically for low-temperature coercivity determinations, although there is no reason to suspect that there would be any problems.

Remanence is one of the few magnetic properties that does not require the presence of a magnetic field during the measurement. Many types of magnetometers are suitable for this measurement. The three mentioned here cover the entire normal range of variation of magnetic moment from about 10^{-12} to 10^{2} A·m^2. Note that these are given in the units of magnetic moment, whereas most data are presented as remanent magnetization (moment per unit volume, M_r) or as remanent flux density ($B_r = \mu_0 M_r$). To further complicate the situation, data are often presented in cgs emu units. Table 6.1 enables the truly energetic reader to perform all the required conversions.

One device for remanent moment measurement is the spinner magnetometer, which is most suitable for large specimens with large moments, although the device is capable of detection down to about 10 nA·m^2 with special electronics. The specimen spins near a stationary detector, usually a flux-gate device or a coil, and the flux is detected by the induced ac signal. Two measurements, with different sample mountings, allow determination of both the magnitude and direction of the remanence. The vibrating sample magnetometer (VSM) is similar to the spinner in operation and range, except that it is usually more sensitive. The sample size required tends to be smaller than for the spinner. As the name implies, the specimen is vibrated relative to a coil and the flux change is measured. Synchronous detection schemes allow very high sensitivity. Cryogenic cooling of the detection coils and other methods (Foner, 1975) can be used to give even higher sensitivity.

For the measurement of very small moments ($\geqslant$ 1 pA·m^2), cryogenic instruments using superconducting quantum interference devices (SQUIDs) are often used. In concept, there is no upper limit to the moment that can be measured by a SQUID system, but realistically 0.1 mA·m^2 is about the maximum.

All three devices are available commercially and have been used for low-temperature measurements. The SQUID and VSM systems are sold with a variable temperature capability ranging from 4 K to room temperature. A relatively simple homemade SQUID magnetometer system is described by Fickett and Sullivan (1974), and a more complex device with higher sensitivity, by Philo and Fairbank (1977).

Magnetic susceptibility. Measurement of the moment induced in a nonferromagnetic material by an applied external field gives the magnetic susceptibility. This quantity is of great interest in fundamental studies of materials. It is, of course, related to permeability. Typical values of the SI susceptibility, κ, range from 10^{-5} to 10^{-2} and may be negative or positive (see Table 6.5). Both VSM and SQUID systems are used to measure susceptibility. They contain a magnet that provides a variable uniform field at the specimen, which may go as high as 4 MA·m^{-1}.

Many other devices have been employed to measure susceptibility. An excellent descriptive and historical review of them is given by Bates (1961). Historically, the most significant one is probably the force balance, which is still used in many laboratories. In its simplest form, the balance measures the force exerted on the specimen by a field gradient (Eq. 6.3). Modern systems use electronic balances and, often, elaborate optical feedback systems to allow direct reading of the susceptibility. Force balance systems are used nearly exclusively for the measurement of magnetic properties at high temperatures or in oxidizing or reducing atmospheres. A number of commercial units are available.

All susceptibility instruments require calibration. This is usually performed by measuring a material of known susceptibility. As mentioned in § 6.2.2, the National Bureau of Standards, Office of Standard Reference Materials sells several susceptibility standards covering the range (SI) 2×10^{-5} to 6×10^{-3}. A technique for performing relatively rough calibrations over a higher range ($\sim 10^{-4}$ to 10^{-1}) using powdered Fe_2O_3/glass mixtures has been described (Rosenbaum, Larson, Hoblitt, and Fickett, 1979). Carefully constructed coils carrying a known current and thus having a known moment are sometimes used also. The measurement of para- or diamagnetic substances is often plagued by the presence of ferromagnetic impurities in the specimen. Correction techniques for force balance measurements are outlined by Bates (1961). These techniques are based on the fact that the dia- or paramagnetic substance experiences a force proportional to $H\,dh/dx$, whereas the force on a saturated ferromagnetic particle is proportional to dH/dx.

Many low-temperature magnetometer systems are described in the literature. An elaborate VSM for use at very low temperatures (0.3 to 4.2 K) is described by Mangum and Thornton (1970). It uses a piezoelectric vibrator system. A force balance device adapted for measurements from 4.2 K, or below, to room temperature is described by Hurd (1966) and Dellby and Ekstrom (1971). A device for low-temperature measurements that uses a horizontal force configuration (a pendulum magnetometer) and produces an output voltage proportional to the force on the specimen has been built by Yamauchi, Sato, and Watanabe (1972). This paper also cites several earlier apparatus of similar design. Salinger and Wheatley (1961) describe a ballistic galvanometer system for susceptibility measurements in the 1.6- to 4.2-K range.

Magnetic particle concentration. Several devices have been developed to determine the relative amount of ferromagnetic material in a nonmagnetic matrix. The most common application is the determination of the ferrite content of welds in nonmagnetic austenitic stainless steels or in base metal castings. They also can be used to indicate the presence of the martensitic phase in austenitic stainless steels that occurs on cooling or mechanical deformation of the metal. Clearly, many of the devices already described could be adapted to this purpose; however, the advantages of those made specifically for this application are speed and ease of measurement. One commercial device uses a clock spring and a scale (torsion balance) to measure the force necessary to lift a calibrated permanent magnet from the surface of the metal. Its use and calibration are described by Reed and Mikesell (1960). A device built specifically for detection of stainless steel martensite in liquid-oxygen tanks is described by Fickett (1971). Experiments with induction instruments designed to actually monitor the transformation to martensite in an austenitic specimen while under stress are outlined by Kelha and Niinikoski (1971).

Nuclear, atomic, and molecular effects. The experiments to detect local moments and their formation are, for the most part, beyond the scope of this presentation. Magnetic susceptibility measurements made as a function of temperature and low-temperature electron-paramagnetic-resonance methods are often used. Neutron-scattering experiments allow the direct determination of the microscopic distribution of magnetization within a magnetic material. Nuclear techniques involving the Mössbauer effect, γ–γ angular correlation measurements, and nuclear magnetic resonance are all used to determine the properties of the magnetic state of impurities and the host lattice. An excellent introduction to this extensive field is available in the series *Magnetism* (1963-1973).

6.3.3 Combined effect measurements

The measurement of specific heat, thermal conductivity, and electrical resistivity at low temperature is described in other chapters. Calorimeter systems necessary for magnetocaloric work are much the same as those used in specific heat experiments. Thermoelectric effect measurements at low temperatures are discussed by Blatt, Schroeder, Foiles, and Greig (1976). Magnetostriction and thermal expansion measurement techniques are often the same; the latter are described elsewhere in this text. A strain gauge apparatus for low-temperature magnetostriction measurements in high magnetic fields is described by Ricodeau, Melville, and Lee (1972). Optical techniques for low-temperature experiments are not significantly different from those used at room temperature. The cryostat construction, however, may become exceedingly complex when windows leading into cold, high-field regions are required. White (1968) gives an excellent discussion of the entire field of low-temperature apparatus.

6.4 EXPERIMENTAL RESULTS

Little research has been performed on the low-temperature magnetic properties of technical materials. Even less has been done in which a continuum of temperatures is used. Usually measurements are made only at a few fixed points and, thus, tend to miss much of the detailed structure.

When faced with the need to find low-temperature data on the magnetic properties of a material, there really are few alternatives to going directly to the literature via one of the abstract journals. There is a National Bureau of Standards Special Publication (Carr, Gavert, Moore, Wawrousek, and Westbrook, 1976) that lists various sources of information on magnetic properties, including handbooks, trade publications, and information centers. It describes the data base for each source in some detail. Most of the sources have few low-temperature data. A useful bibliography of low-temperature properties of materials, including magnetic properties, has been published by the Metals and Ceramics Information Center (Johnson, 1977). Handbooks of low-temperature materials properties, such as the *Handbook on Materials for Superconducting Machinery* (1974-1977) and *Materials Handbook for Fusion Energy Systems* (1980), contain some data on structural materials.

6.4.1 Well-behaved, highly magnetic metals

This group contains most of the magnetic elements, ferromagnetic Cu–Ni alloys, many steels, nickel-based alloys, and the highly permeable soft ferromagnetic alloys. In all cases, lowering the temperature increases the magnetization at saturation (as would be expected from the discussion in § 6.2.2) but usually only very slightly, since the Curie temperatures are well above room temperature. Furthermore, the hysteresis loops of the hard ferromagnets are significantly enlarged because the coercivity increases as well as the remanence. This effect can lead to moderate increases in hysteresis losses. Data presented by Gniewek and Ploge (1965) on several magnetic alloys illustrate this well. In steels, the maximum permeability usually increases with decreasing temperature, whereas the lower-field permeability is not greatly affected (McInturff and Claus, 1970b). For the highly permeable, soft ferromagnetic materials, cooling to 4 K has little effect on the saturation magnetization, but often causes a large increase in the coercive force. The permeability tends to decrease with decreasing temperature (Ackermann, Klawitter, and Drautman, 1971), which in turn decreases the effectiveness of these materials as magnetic shields (Martin and Snowdon, 1975). The effect may well be due to stress resulting from thermal contraction of the constrained metal, which destroys the careful softening anneal used for these shielding materials.

6.4.2 Well-behaved, slightly magnetic metals

This class includes the "nonmagnetic" stainless steels, the nickel-based superalloys, the f.c.c. Fe–Ni alloys, and other alloys that contain a ferromagnetic constituent metal, but which are not strongly ferromagnetic. Many of the alloys in this group transform as the temperature is lowered, such as the paramagnetic-to-antiferromagnetic shift of the stainless steels. The "well-behaved" designation still applies, because the metals have quite uniform and predictable properties at the temperatures at which they are normally used, even after repeated thermal cycling. Nearly all the metals in this group have a room-temperature relative permeability of 1.02 or less. Permeability is not often meaningful, even when the measurement field is given, since the slope of the curve changes constantly. The low-temperature magnetic properties of these materials are defined by the behavior of their magnetization-vs.-field curves.

Data taken at 4 K for two alloys, the relatively stable stainless steel AISI 316 and the superalloy Inconel 718, are shown in Fig. 6.23 from the

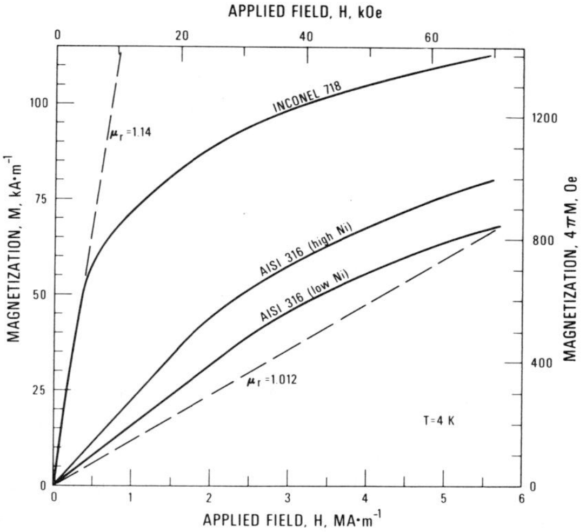

Figure 6.23 The magnetization of several weakly magnetic alloys vs. applied field as measured by Efferson and Leonard (1967). The dashed lines show the limiting permeability values.

work of Efferson and Leonard (1967). These curves are typical of the materials in this class. A series of measurements by Collings (1979) has produced a large amount of low-temperature data on the magnetization as a function of field and temperature for a number of 300-series stainless steels, Inconel 625, and AWS 330 weld metal. The exact type of magnetism exhibited by the stainless steel alloys is still debated. Suggestions have ranged from superparamagnetism to spin glass to a ferro–antiferromagnetic mixed system.

A study of a large number of Fe–Ni–Cr alloys at low temperatures in which Néel temperatures were determined is described by Warnes and King (1976). This work builds on earlier magnetic properties studies of alloys by Larbalestier and King (1970; 1973). A discussion of the stability of commercial stainless steels for low-temperature service is presented by Geyari (1976). The low-temperature behavior of the permeability of a variety of alloys used in spacecraft construction has been reported by Dralle and Moore (1965) and Droll and Iufer (1967).

6.4.3 Metals with magnetic inclusions

Strongly magnetic inclusions in either nonmagnetic or weakly magnetic materials occur for a variety of reasons. Whether they are troublesome depends, of course, on the particular application. Once again, little information is available regarding the low-temperature magnetic properties, and yet, they might provide one of the most reliable methods of nondestructive evaluation. Three important kinds of inclusions are: martensite in stainless steel, ferrite in welds, and metallic and nonmetallic magnetic impurities.

Martensite in stainless steel. Most of the austenitic stainless steels will transform partially to a b.c.c. martensitic phase when stressed at low temperature. Occasionally, some of the less stable ones will transform as a result of the thermal contraction alone. This transformation occurs below a temperature that is highly dependent on the exact composition of the steel, but it is usually below room temperature. The martensitic transformation effects of many alloys have been compiled by Reed and Breedis (1966), and a detailed discussion of phase transformations is given in Chapter 9. In austenitic stainless steels, stability against transformation is enhanced by increased nitrogen, carbon, nickel, manganese, and chromium. Adding significant amounts of nickel and chromium greatly increases the cost of the steel. A list of transformation temperatures for commercial stainless steels and some similar alloys is given by Williams, Williams, and Capellaro (1976). The region of stability of Fe–Cr–Ni alloys is shown in Fig. 6.24 from the work of Schlump (1975).

The magnetic properties of the martensite are not particularly noteworthy. It is a hard ferromagnetic phase and, most likely, behaves much like the ferromagnetic alloys. Apparently no detailed low-temperature studies of the magnetic behavior have been made. It is mentioned here because the possibility of the martensite forma-

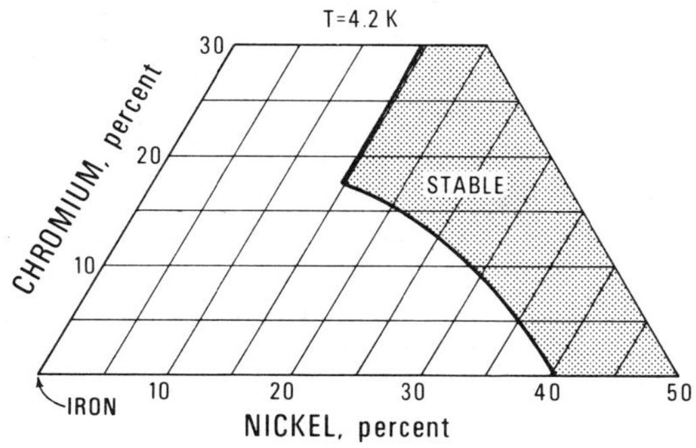

Figure 6.24 The stability of f.c.c. Fe–Cr–Ni alloys at 4 K (Schlump, 1975). Addition of nitrogen or carbon can increase the size of the region significantly.

tion in stainless steels at low temperatures frequently is not considered by people making material choices, and in magnetic applications, this phase can cause severe local field perturbations and contribute to the surface distortion of the metal and decreased ductility.

Ferrite in welds and castings. Some cast "nonmagnetic" stainless steels will attract a small bar magnet, as will most stainless steel welds. Both of these effects are due to the precipitation of ferrite particles when the molten metal cools. These particles, which are of variable composition, behave as relatively soft ferromagnets. It is possible to create welds with no ferrite, but they are considerably more susceptible to cracking. No comprehensive low-temperature studies of the magnetization properties are available, but preliminary work by the authors on cast stainless steel indicates that at 4 K, the (small) coercivity increases by about 40% and the saturation magnetization by about 20% over their room temperature values.

Magnetic impurities. Both metallic and nonmetallic materials that are ostensibly nonmagnetic may contain amounts of magnetic impurities, which can be significant in low-temperature applications. These impurities may be isolated, highly paramagnetic atoms, which have a strong temperature-dependent susceptibility that can be quite a bit higher at low temperatures than at room temperature. The effect of very low levels of the natural impurity, iron, on the susceptibility of pure copper at 4 K is shown in Fig. 6.25 from measurements by Fickett (1974). At this temperature very little iron is required to change the susceptibility from diamagnetic to paramagnetic. All the samples measured have about the same susceptibility at room temperature. A complete review of the theory of this behavior is given by Grüner and Zawadowski (1974).

Another disturbing effect can occur in materials containing small amounts of magnetic elements. Either at the surface (at room temperature) or throughout a sample (during heat treatments), hard magnetic compounds, such as Fe_3O_4, may form and give the sample a small, permanent moment. Sometimes clustering of ferromagnetic impurity atoms occurs during heat treatment, resulting in a soft ferromagnetic behavior for some metals, such as copper with natural iron content above about 100 ppm. The presence of hard impurities, primarily iron oxides, in epoxies and other organic systems occasionally causes difficulties when these materials are used in the construction of very sensitive magnetometers. Cabrera (1975) lists the remanent magnetization of some common construction materials at 4.2 K.

6.4.4 Metallic glasses

Amorphous alloys have properties of both solids and liquids, much like common glasses. Thus, amorphous alloys are often referred to as metallic glasses when they are prepared by rapid solidification of the liquid (Chen, 1980). They are noncrystalline, supercooled liquids in a metastable solid state. In addition to rapid quenching from the melt, amorphous alloys may be prepared by vapor deposition, sputtering, plasma spraying, laser melting, electrolytic and electroless deposition, ion implantation, and neutron irradiation.

The metallic glasses of commercial interest fall into the general class of metal–metalloid alloys. They typically combine various amounts of iron, nickel, and cobalt with 15 to 30% of elements from the group boron, silicon, carbon, and phosphorus. Other elements are sometimes added to enhance a desired property; for example, molybdenum is used to increase the temperature stability of some alloys. Within this group there are three subgroups; all are soft ferromagnets. The iron-based materials such as $Fe_{80}B_{20}$ are inexpensive, their saturation magnetization and soft magnetic properties are worse than those of the other groups, and their proper-

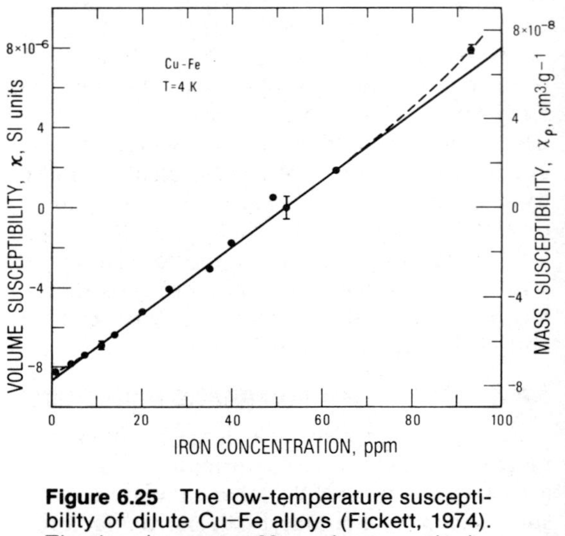

Figure 6.25 The low-temperature susceptibility of dilute Cu–Fe alloys (Fickett, 1974). The slope increase at 93 ppm is assumed to be due to the beginnings of a ferromagnetic interaction.

ties are sensitive to stress. However, they (and the others) have very low losses in ac applications. The Fe–Ni-based alloys offer better properties at higher cost. The cobalt-based alloys cost even more and offer the best soft magnetic properties, stress insensitivity, and, in contrast with the others, very low magnetostriction. Many reviews of the properties of these materials and techniques for their fabrication are available: Chaudhari, Giessen, and Turnbull (1980); *Metallic Glasses* (1978); *Ultrarapid Quenching of Liquid Alloys* (1981); *Glassy Metals I* (1981); *Amorphous Magnetism* (1973); *Amorphous Magnetism II* (1977); Chen (1980); Chen, Leamy, and Miller (1980); Warlimont and Boll (1982); Masumoto and Egami (1981); and *Fourth International Conference on Liquid and Amorphous Metals* (1980).

Most of the commercial alloys are prepared by rapid quenching at rates of 10^5 to 10^6 degrees per second. Production lines can generate 100-mm-wide amorphous ribbons at rates of 100 km/h. This is usually achieved by induction melting the alloy and forcing the liquid against a polished rotating heat sink, such as a copper wheel. The resulting product is typically 20 to 50 μm thick, although thicknesses to 100 μm have been achieved. A comprehensive study of production methods is given by Jones (1981).

X-ray diffraction patterns of amorphous alloys lack the Bragg peaks associated with crystalline materials. Instead, a broad low-intensity peak with smaller peaks of higher order is seen, characteristic of the average spacing of atoms (radial distribution function). Many amorphous alloys are at or near eutectic compositions in the equilibrium phase diagram, where the melting point is minimized. Just like liquids, amorphous alloys are homogeneous; they do not segregate into different intermetallic compounds.

Metallic glass ribbons are very hard, typically 1000 kg/mm^2 Vickers. Electrical resistivity is about three times that of the crystalline alloys of the same composition. Crystallization of amorphous alloys may be achieved by annealing at relatively low temperatures. Crystallization occurs at many nucleation sites, yielding very fine-grained alloys which themselves have significant commercial applications, especially in high-frequency electronics (Krause and Werner, 1981).

Amorphous alloys are chemically homogeneous, have no grains or grain boundaries, and no phase separation. Although their breaking strengths are comparable to most steels, yield strengths are two to three times larger than those of steel. Toughness, measured as the energy required to tear, is larger than that of steel. Some amorphous alloys containing chromium and potassium are significantly more resistant to corrosion than stainless steel (Gilman, 1980).

Proposed and extant applications of amorphous materials are discussed in the reviews by Boll and Warlimont (1981), Luborsky and Johnson (1980), and Mort (1980). They have been suggested for high-strength fibers in composites, hydrogen storage, catalysts, razor blades, protective coatings, delay lines, and superconductors, to name a few of the interesting uses. By far the most promising applications are in magnetics: transformers, recording heads, motors, magnetic shielding, and piezomagnetic sensors. Highly efficient power distribution transformers have been made whose core losses are much less than those of conventional transformers. The absence of magnetocrystalline anisotropy and the high mobility of domain walls due to the absence of grain boundaries and impurity pinning sites lead to very soft magnetic behavior with almost no hysteresis at low frequencies. Coercive forces of 4 A·m^{-1} (0.05 Oe) and lower are typical.

Eddy currents are minimized by the smaller thickness and higher resistivity of the amorphous ribbons compared with those of transformer-steel laminations. At frequencies above 1 kHz, ac losses are primarily due to so-called anomalous eddy current losses or eddy current damping of domain wall motion. This may be decreased by increasing the number of domain walls and decreasing their velocity. Partial recrystallization to form precipitates is one method for achieving this (Hasegawa, Ramanan, and Fish, 1982).

Magnetic tape heads with high permeability, saturation magnetization, resistivity, hardness, wear resistance, and small magnetostriction have been made. Their efficiency is comparable to conventional recording head materials.

As mentioned above, metallic glasses of technological interest are soft ferromagnets at and below room temperature. It is notable that long-range ferromagnetic order can exist in a structurally disordered material. Most commercial alloys achieve their optimum soft magnetic qualities after heat treatment in a magnetic field at temperatures below their crystallization temperatures. This relieves stresses and also results in an induced anisotropy and in changes in Curie temperature. The Curie temperature in some amorphous alloys is above the crystallization temperature, but always well below the Curie temperature of the crystalline material. Many amorphous alloys exhibit spin-glass behavior at cryogenic temperatures.

Saturation magnetizations are similar in amorphous and crystalline materials of the same atomic composition provided that the crystalline alloy remains single phase. Note, however, that these saturation magnetization values are considerably less than those of *pure* transition-metal alloys because of dilution by the glass-forming metalloids. Additionally, the electronic contribution of the metalloids reduces the moments of the metal ions, perhaps by filling the 3*d* bands.

Many of the envisioned applications of metallic glasses are at ambient temperature (Johnson, 1980). They also have the potential for replacing more conventional practical superconductors, many of which require elaborate processing. As yet, however, amorphous materials cannot match the critical temperatures and currents available in conventional superconductors. These parameters are often improved after crystallization, which removes the alloys from the realm of the glasses. Nevertheless, the initial amorphous approach may well prove to be a major advance in production technique.

6.4.5 Permanent magnet materials

Permanent magnet applications at cryogenic temperatures have been very few, and as a result, little effort has been made to evaluate the effect of lowering temperature on the various properties. There is a large body of literature on a number of very specific alloys, compounds, and mixtures, but usually the data are only to liquid-nitrogen temperature. Parker and Studders (1962) give some of the data on commercial permanent magnets, including the Alnico series, barium ferrite, and Fe–Co. The saturation magnetization of most permanent magnets increases with decreasing temperature, but the remanence may either increase or decrease depending on the specimen shape. The coercivity either increases or decreases as the temperature is lowered, or it may have a minimum at some temperature. These variations are large, frequently a change of 50% or more to the lowest temperatures. The changes are not always reversible; the text just cited lists the effect on the magnetization of cooling to about 80 K and rewarming a number of commercial permanent magnet materials. Losses as high as 10% are observed, depending on the shape and composition.

Low-temperature research on the newer rare-earth–cobalt materials, such as $SmCo_5$, has begun. Nagel, Klein, and Menth (1976) report on experiments with misch-metal–cobalt magnets that show the remanence and saturation magnetization increasing significantly as the temperature is lowered. The coercive force variation is a strong function of misch-metal composition. Parker, Misroch, and Oesterreicher (1975) showed that substituting nickel for some of the cobalt in the samarium–cobalt materials results in a very large coercive force at 4.2 K, about 4 $MA{\cdot}m^{-1}$! Room-temperature values are about 0.4 $MA{\cdot}m^{-1}$, nearly independent of the nickel content, but the coercive force at 4.2 K for zero nickel content is only 0.04 $MA{\cdot}^{-1}$. These parameters could lead to a number of low-temperature applications for these materials, perhaps even replacing superconducting magnets in some instances.

6.4.6 Nonmagnetic materials

The only magnetic property of common interest for nonmagnetic materials is, of course, the susceptibility. The successful operation of many low-temperature magnetic experiments depends on the careful choice of the materials from which the apparatus is to be made. Even if one is fortunate enough to have materials that have none of the inclusions or impurities mentioned above, it is often helpful to know the susceptibility at the temperature of operation. It may be significantly different from its room-temperature value. The room-temperature values from different sources are not always consistent. Many handbook data are from experiments performed early in this century on materials that were of considerably lower purity than are now available. For room-temperature values for elements, both Heck (1974) and Tebble and Craik (1969) provide tabulations which, unfortunately, do not always agree. The only systematic study of the susceptibility of many different materials at low temperatures appears to be that performed by Salinger and Wheatley (1961). They made measurements at 4.2 K and 1.6 K to determine the temperature variation of the susceptibility of a large array of materials such as epoxies, thread, cotton, paper, polymers, and metals. They do not report room-temperature values. A few materials which show changes that occur on cooling are listed in Table 6.5. Jacobs, Rodbell, and Roth (1962) have reviewed the data on high-field susceptibility of a large number of nonmetallics, primarily ferrites, at 4.2 K. Low-temperature data on Cu–Sn alloys and nonmagnetic Cu–Ni alloys are given by Fickett (1982).

Finally, several magnetic effects of interest in

the cryogenic temperature range involve the inert gases that are most often used in their liquid state for cooling experiments. Sample and Rubin (1978) have calculated and measured the effect of a high magnetic field (15 T) on a column of liquid cryogen that extends out of the field. Forces that are due to the interaction of the pure liquids with the field gradient cause temperature changes of tens of millikelvins because of modification of the hydrostatic head correction. However, the "normal" contamination by small amounts of highly paramagnetic oxygen can cause significant changes in this effect. The gases considered are nitrogen, argon, and helium. A report by Bozler, Bartolac, Luey, and Thompson (1978) indicates that a very thin layer of ^{3}He on dense exfoliated graphite has a susceptibility behavior that extrapolates to a ferromagnetic transition at 0.38 mK. Clearly, there is still much to be learned about magnetic behavior at low temperatures.

6.5 REFERENCES

Ackermann, F. W., Klawitter, W. A., and Drautman, J. J. (1971). Magnetic properties of commercial soft magnetic alloys at cryogenic temperatures. In *Advances in Cryogenic Engineering,* Vol. 16. Plenum Press, New York, 46–50.

Amorphous Magnetism (1973) (eds. H. O. Hooper and A. M. de Graaf). Plenum Press, New York.

Amorphous Magnetism II (1977) (eds. R. A. Levy and R. Hasegawa). Plenum Press, New York.

Anderson, P. W. (1970). Localisation theory and the Cu–Mn problem: spin glasses. *Mater. Res. Bull.* **5**, 549–554.

Anderson, P. W. (1973). Topics in spin glasses. In *Amorphous Magnetism*, (eds. H. O. Hooper and A. M. de Graaf). Plenum Press, New York, 1–14.

Andrade, E. N. da C. (1958). The early history of the permanent magnet. *Endeavour* **17**, 22–30.

Annual Book of ASTM Standards (1981), Part 44. American Society for Testing and Materials, Philadelphia, Pennsylvania.

Ansorge, W., Billan, J., Gourber, J. P., Haren, P., Henrichsen, K. N., and Verdier, A. (1978). Magnetic permeability measurements at cryogenic temperatures. In *Proceedings, Sixth International Conference on Magnet Technology*. ALFA, Bratislava, Czechoslovakia, 1066–1070.

Bates, L. F. (1946). The magneto-resistance of high coercivity alloys. *Proc. Phys. Soc. (Lond.)* **58**, 153–164.

Bates, L. F. (1961). *Modern Magnetism*. Cambridge University Press, New York.

Bean, C. P. and Livingston, J. D. (1959). Superparamagnetism. *J. Appl. Phys.* **30**, 120S–129S.

Beck, P. A. (1971). Some recent results on magnetism in alloys. *Metall. Trans.* **2**, 2015–2024.

Beck, P. A. (1978). Properties of mictomagnets (spin-glasses). *Prog. Mater. Sci.* **23**, 1–49.

Bennett, L. H., Page, C. H., and Swartzendruber, L. J. (1976). Comments on units in magnetism. In *Magnetism and Magnetic Materials — 1975*. AIP Conference Proceedings No. 29, American Institute of Physics, New York.

Betts, D. S. (1976). *Refrigeration and Thermometry Below One Kelvin.* Crane–Russak Co., New York.

Blandin, A. (1978). Theories versus experiments in the spin glass systems. *J. Phys. (Paris) Colloq.* **39**, C6, 1499–1516.

Blatt, F. J., Schroeder, P. A., Foiles, C. L., and Greig, D. (1976). *Thermo-electric Power of Metals.* Plenum Press, New York.

Boll, R. and Warlimont, H. (1981). Applications of amorphous magnetic materials in electronics. *IEEE Trans. Magn.* **MAG–17**, 3053–3058.

Bowers, R. (1956). Magnetic susceptibility of copper metal at low temperatures. *Phys. Rev.* **102**, 1486–1488.

Boys, D. W. and Legvold, S. (1968). Thermal conductivities and Lorenz functions of Dy, Er, and Lu single crystals. *Phys. Rev.* **174**, 377–384.

Bozler, H. M., Bartolac, T., Luey, K., and Thompson, A. L. (1978). Magnetic susceptibility of ^{3}He in grafoil at ultralow temperatures. *Phys. Rev. Lett.* **41**, 490–493.

Bozorth, R. M. (1951). *Ferromagnetism*. D. Van Nostrand Co., Princeton, New Jersey.

Brechna, H. (1973). *Superconducting Magnet Systems*. Springer-Verlag, New York.

Brown, W. F., Jr. (1960). Single-domain particles: New uses of old theorems. *Am. J. Phys.* **28**, 542–551.

Cabrera, B. (1975). The use of superconducting shields for generating ultra-low magnetic field regions and several related experiments. Ph.D. Thesis, Dept. of Physics, Stanford University (unpublished), 157–165.

Carr, M. J., Gavert, R. B., Moore, R. L., Wawrousek, H. W., and Westbrook, J. H. (1976). *Critical Surveys of Data Sources: Electrical and Magnetic Properties of Metals.* NBS Special Publication 396-4, National Bureau of Standards, Washington, D.C.

Carr, W. J., Jr. (1969). Principles of ferromagnetic behavior. In *Magnetism and Metallurgy,* Vol. 1 (eds. A. E. Berkowitz and E. Kneller). Academic Press, New York, 45–92.

Chaudhari, P., Giessen, B. C., and Turnbull, D. (1980). Metallic glasses. *Sci. Amer.* **242**, 98–117.

Chen, C. W. (1977). *Magnetism and Metallurgy of Soft Magnetic Materials.* North-Holland, New York.

Chen, H. S. (1980). Glassy Metals. *Rep. Prog. Phys.* **43**, 353–432.

Chen. H. S., Leamy, H. J., and Miller, C. E. (1980). Preparation of glassy metals. *Ann. Rev. Mater. Sci.* **10**, 363–391.

Chikazumi, S. (1964). *Physics of Magnetism.* John Wiley & Sons, New York.

Clark, A. E. (1974). Magnetic and magnetoelastic properties of highly magnetostrictive rare earth–iron laves phase compounds. In *AIP Conference Proceedings* No. 18, Part 2. American Institute of Physics, New York, 1015–1029.

Clarke, J. (1977). Superconducting quantum interference devices for low frequency measurements. In *Superconductor Applications: SQUIDS and Machines* (eds. B. B. Schwartz and S. Foner). Plenum Press, New York, 67–124.

Collings, E. W. (1979). Magnetic susceptibility and magnetization studies of some commercial austenitic stainless steels. In *Materials Studies for Magnetic Fusion Energy Applications at Low Temperatures — II* (eds. F. R. Fickett and R. P. Reed). NBSIR 79–1609, National Bureau of Standards, Boulder, Colorado, 203–280.

Collings, E. W. and Hart, S. C. (1979). Low temperature magnetic susceptibility and magnetization studies of some commercial austenitic stainless steels. *Cryogenics* **19**, 521–530.

Connolly, T. F. and Copenhaver, E. D. (1972). *Bibliography of Magnetic Materials and Tabulation of Magnetic Transition Temperatures.* Plenum Press, New York.

Cooper, J. R. and Miljak, M. (1976). Single impurity behavior and interaction effects in the magnetic susceptibility of AlMn and AlCr alloys. *J. Phys. F* **6**, 2151–2164.

Crabtree, G. W. (1977). Demagnetizing fields in the de Haas–van Alphen effect. *Phys. Rev. B.* **16**, 1117–1125.

Cullity, B. D. (1972). *Introduction to Magnetic Materials.* Addison–Wesley, Reading, Massachusetts.

Darby, M. I. (1967). Tables of the Brillouin function and of the related function for the spontaneous magnetization. *Br. J. Appl. Phys.* **18**, 1415–1417.

DeHaas, W. J. and Gorter, C. J. (1930). Determination of the susceptibility of potassium–chromium–alum. *Proc. Acad. Sci. (Amsterdam)* **33**, 676–679.

Dellby, B. and Ekstrom, H. E. (1971). A magnetic susceptibility balance for use in the temperature range 1.6 K – 300 K. *J. Phys. E* **4**, 342–345.

Dietz, G. and Burkhard, H. (1966). Measurement of the saturation magnetostriction of polycrystalline ferromagnetic alloys and ferrites between −140° C and 160° C (in German). *Z. Angew. Phys.* **21**, 386–388.

Dralle, A. V. and Moore, J. B. (1965). Magnetic properties of materials. In *The Goddard Summer Workshop Program in Analysis of Space Data and Measurement of Space Environments.* NASA N66-23443.*

Droll, P. W. and Iufer, E. J. (1967). *Magnetic Properties of Selected Spacecraft Materials.* NASA Ames Research Center Tech. Memo. X-60414 N68-34165.*

Edelstein, A. S. (1968). Spin compensation in an element, cerium. *Phys. Rev. Lett.* **20**, 1348–1351.

Edwards, S. F. and Anderson, P. W. (1975). Theory of spin glasses. *J. Phys. F: Metal Phys.* **5**, 965–974.

Efferson, K. R. and Leonard, W. J. (1967). *Magnetic Properties of Some Structural Materials Used in Cryogenic Applications.* Oak Ridge National Laboratory Report ORNL-4150, 126–132.*

Ekström, H. E. and Myers, H. P. (1972). The magnetic susceptibility of dilute CuFe alloys in the temperature range 1.7–300 K. *Phys. Kondens. Mater.* **14**, 265–274.

Enz. U. and Teale, R. W. (1970). Photomagnetic effects. *Philips Tech. Rev.* **31**, 33–39.

Fawcett, E. (1970a). Magnetostriction of paramagnetic transition metals. I. Group 4 — Ti and Zr; group 5 — V, Nb and Ta; group 6 — Mo and W. *Phys. Rev. B* **2**, 1604–1613.

Fawcett, E. (1970b). Magnetostriction of paramagnetic transition metals. II. Group VIII metals Ru, Rh, Pd, Ir, Pt, and their alloys. *Phys. Rev. B* **2**, 3887–3890.

Ferromagnetic materials, Vols. 1 and 2 (1980) (ed. E. P. Wohlfarth). North-Holland, Amsterdam.

Fickett, F. R. (1971). *Martensite Transformation Detection in Cryogenic Steels (Magnetometer Development).* NBS Technical Note 613, National Bureau of Standards, Boulder, Colorado.

Fickett, F. R. (1974). *A Preliminary Investigation of the Behavior of High Purity Copper in High Magnetic Fields.* INCRA Report 186B, International Copper Research Association, New York.

Fickett, F. R. (1976). *A Preliminary Investigation of the Behavior of High Purity Copper in High Magnetic Fields.* INCRA Report 186C, International Copper Research Association, New York.

Fickett, F. R. (1982). Electric and magnetic properties of CuSn and CuNi alloys at 4 K. *Cryogenics* **22**, 135–137.

Fickett, F. R. and Sullivan, D. B. (1974). Magnetic studies of oxidized impurities in pure copper using a SQUID system. *J. Phys. F* **4**, 900–905.

Foner, S. (1975). Further improvements in vibrating sample magnetometer sensitivity. *Rev. Sci. Instrum.* **46**, 1425–1426.

Fourth International Conference on Liquid and Amorphous Metals (1980) (ed. F. Cyrot-Lackmann). *J. Phys. (Paris) Colloq.* **41**, C8.

Geyari, C. (1976). Design considerations in the use of stainless steel for vacuum and cryogenic equipment. *Vacuum* **26**, 287–297.

Gilman, J. J. (1980). Overview of the technology and significance of metallic glasses. *J. Phys. (Paris) Colloq.* **41**, C8, 811–819.

Glassy Metals I (1981) (eds. H.–J. Guntherodt and H. Beck). *Topics in Applied Physics,* Vol. 46. Springer-Verlag, Berlin.

Gniewek, J. J. and Ploge, E. (1965). Cryogenic behavior of selected magnetic materials. *J. Res. Natl. Bur. Stand.* **69C**, 225–236.

Gniewek, J. J. and Powell, R. L. (1962). Temperature dependence of magnetic losses. In *Advances in Cryogenic Engineering,* Vol. 7. Plenum Press, New York, 303–310.

Goldfarb, R. B., Fickett, F. R., Rao, K. V., and Chen, H. S. (1982). Spin-freezing below the ferromagnetic transition determined by the imaginary component of ac magnetic susceptibility. *J. Appl. Phys.* **53**, 7687–7689.

Goldfarb, R. B. and Patton, C. E. (1981). Superparamagnetism and spin-glass freezing in nickel–manganese alloys. *Phys. Rev. B* **24**, 1360–1373.

Grivet, P. (1962). High field magnetometry. In *High Magnetic Fields* (eds. H. Kolm, B. Lax, F. Bitter, and R. Mills). M.I.T. Press, Cambridge, Massachusetts and John Wiley & Sons, New York, 54–84.

Grüner, G. and Zawadowski, A. (1974). Magnetic impurities in non-magnetic metals. *Rep. Prog. Phys.* **37**, 1497–1583.

Handbook On Materials for Superconducting Machinery (1974-1977). Report MCIC-HB-04, ADA 002698, plus two supplements, ADA 023228 and ADA 035936. Metals and Ceramics Information Center, Battelle Columbus Laboratories, Columbus, Ohio.

Hasegawa, R., Ramanan, V. R. V., and Fish, G. E. (1982). Effects of crystalline precipitates on the soft magnetic properties of metallic glasses. *J. Appl. Phys.* **53**, 2276–2278.

Heber, G. (1976). Spin glasses models and typical properties. *Appl. Phys.* **10**, 101–110.

Heck, C. (1974). *Magnetic Materials and Their Applications.* Crane–Russak Co., New York.

Hurd, C. M. (1966). A magnetic susceptibility apparatus for weakly magnetic metals, and the susceptibility of pure copper in the range 6° K–300° K. *Cryogenics* **6**, 264–269.

Hurd, C. M. (1972). *The Hall Effect in Metals and Alloys.* Plenum Press, New York.

Jacobs, I. S. (1959). Evidence for triangular moment arrangements in $MO \cdot Mn_2O_3$. *J. Phys. Chem. Solids* **11**, 1–11.

Jacobs, I. S., Rodbell, D. S., and Roth, W. L. (1962). *Research on Spontaneous Magnetization in Solid Bodies.* General Electric Research Laboratory Report ASD-TR-61-630, AD 277380.*

Jacobs, I. S. and Schmitt, R. W. (1959). Low-temperature electrical and magnetic behavior of dilute alloys: Mn in Cu and Co in Cu. *Phys. Rev.* **113**, 459–463.

Johnson, D. M. (1977). *Low-Temperature Properties of Selected Materials — A Bibliography with Descriptors.* Report MCIC-77-31, Metals and Ceramics Information Center, Battelle Columbus Laboratories, Columbus, Ohio.

Johnson, W. L. (1980). Superconductivity in metallic glasses. *J. Phys. (Paris) Colloq.* **41**, C8, 731–741.

Jones, H. (1981). Experimental methods in rapid quenching from the melt. *Treatise Mater. Sci. Technol.* **20**, 1–72.

Kelha, V. and Niinikoski, T. (1971). Susceptibility measuring device for the following of a martensitic transformation during tensile testing. *Acta Polytech. Scand., Phys. Incl. Nucleon. Ser.*, **Ph 76**, 1–16.

Kittel, C. (1967). *Introduction to Solid State Physics,* 3rd edition. John Wiley & Sons, New York.

Kneller, E. (1969). Fine particle theory. In *Magnetism and Metallurgy,* Vol. 1 (eds. A. E. Berkowitz and E. Kneller). Academic Press, New York, 365–471.

Knitter, R. W., Kouvel, J. S., and Claus, H. (1977). Magnetic irreversibility and metastability in a Cu–Mn alloy. *J. Magn. Magn. Mater.* **5**, 356–359.

Kouvel, J. S. (1961). The ferromagnetic–antiferromagnetic properties of copper–manganese and silver–manganese alloys. *J. Phys. Chem. Solids* **21**, 57–70.

Kouvel, J. S., Graham, C. D., Jr., and Jacobs, I. S. (1959). Ferromagnetism and antiferromagnetism in disordered Ni–Mn alloys. *J. Phys. Radium* **20**, 198–202.

Kouvel, J. S. and Kasper, J. S. (1963). Long-range antiferromagnetism in disordered Fe–Ni–Mn alloys. *J. Phys. Chem. Solids* **24**, 529–536.

Krause, R. F. and Werner, F. E. (1981). High frequency magnetic properties of some amorphous alloys. *IEEE Trans. Magn.* **MAG–17**, 2686–2691.

Larbalestier, D. C. and King, H. W. (1970). Problems with stainless steel formers for superconducting magnets. *Cryogenics* **10**, 410–412.

Larbalestier, D. C. and King H. W. (1973). Austenitic stainless steels at cryogenic temperatures. 1 — Structural stability and magnetic properties. *Cryogenics* **13**, 160–168.

Ledbetter, H. M. and Collings, E. W. (1979). Low-temperatures magnetic and elastic-constant anomalies in three manganese stainless steels. In *The Metal Science of Stainless Steels* (eds. E. W. Collings and H. W. King). AIME, New York, 22–40.

Levy, P. M. and Fert, A. (1981). Origin of magnetic anisotropy in transition metal spin-glass alloys. *J. Appl. Phys.* **52**, 1718–1719.

Luborsky, F. E. and Johnson, L. A. (1980). Applications of magnetic amorphous alloys. *J. Phys. (Paris) Colloq.* **41**, C8, 820–826.

Magnetism and Metallurgy, Vols. I and II (1969) (eds. A. E. Berkowitz and E. Kneller). Academic Press, New York.

Magnetism: A treatise on Modern Theory and Materials, Vols. I–V (1963–1973) (eds. G. T. Rado and H. Suhl). Academic Press, New York.

Mangum, B. W. and Thornton, D. D. (1970). Vibrating sample magnetometer for use at very low temperatures and in high magnetic fields. *Rev. Sci. Instrum.* **41**, 1764–1766.

Martin, D. H. (1967). *Magnetism in Solids.* M.I.T. Press, Cambridge, Massachusetts.

Martin, D. L. and Snowdon, R. L. (1975). Effect of temperature on magnetic shielding with Co-netic. *Rev. Sci. Instrum.* **46**, 523–524.

Masumoto, T. and Egami, T. (1981). Designing the composition and heat treatment of magnetic amorphous alloys. *Mater. Sci. Eng.* **48**, 147–165.

Materials Handbook for Fusion Energy Systems (1980). Report No. DoE/TIC-10122. Hanford Engineering Development Laboratory, Richland, Washington.

Mattis, D. C. (1981). *The Theory of Magnetism I.* Springer-Verlag, New York.

McInturff, A. D. and Claus, J. (1970a). *μ Tables for Steels as a Function of Temperature.* Report BNL 14630, Brookhaven National Laboratory, Upton, New York.

McInturff, A. and Claus, J. (1970b). *Low Temperature Iron Measurements.* Report BNL 14556, Brookhaven National Laboratory, Upton, New York.

Metallic Glasses (1978) (eds. J. J. Gilman and H. J. Leamy). American Society for Metals, Metals Park, Ohio.

Morrish, A. H. (1980). *The Physical Principles of Magnetism.* Krieger, Huntington, New York.

Mort, J. (1980). Applications of amorphous materials. *Phys. Technol.* **11**, 134–141.

Mydosh, J. A. (1975). Spin glasses and mictomagnets. In *Magnetism and Magnetic Materials — 1974.* AIP Conference Proceedings No. 24, American Institute of Physics, New York, 131–137.

Mydosh, J. A. (1978). Spin glasses — recent experiments and systems. *J. Magn. Magn. Mater.* **7**, 237–248.

NBS Standard Reference Materials Catalog (1981). NBS Special Publication 260, Office of Standard Reference Materials, National Bureau of Standards, Washington, D.C.

Nagasawa, H. (1970). Resistance minimum phenomena in exchange enhanced Pd and Pt alloys. *J. Phys. Soc. Jap.* **28**, 1171–1180.

Nagata, T. (1961). *Rock Magnetism.* Plenum Press, New York.

Nagel, H., Klein, H. P., and Menth, A. (1976). Mischmetal–cobalt as a permanent magnetic material. *J. Appl. Phys.* **47**, 3312–3320.

Néel, L. (1948). Magnetic properties of ferrites: ferrimagnetism and antiferromagnetism. *Ann. Phys. (Paris)* **3**, 137–198.

Néel, L. (1955). Some theoretical aspects of rock magnetism. *Adv. Phys.* **4**, 191–243.

Parker, F., Misroch, M., and Oesterreicher, H. (1975). Magnetic hardness in $Sm_{0.167}Co_{0.833-x}Ni_x$. *Mater. Res. Bull.* **10**, 1075–1078.

Parker, R. J. and Studders, R. J. (1962). *Permanent Magnets and Their Applications.* John Wiley & Sons, New York.

Philo, J. S. and Fairbank, W. M. (1977). High-sensitivity magnetic susceptometer employing superconducting technology. *Rev. Sci. Instrum.* **48**, 1529–1536.

Préjean, J. J., Joliclerc, M. J., and Monod, P. (1980). Hysteresis in *Cu*Mn: The effect of spin orbit scattering on the anisotropy in the spin glass state. *J. Phys. (Paris)* **41**, 427–435.

Pugh, E. W. and Ryan, F. M. (1958). Magnetic susceptibility of copper–nickel and silver–palladium alloys at low temperatures. *Phys. Rev.* **111**, 1038–1042.

Reed, R. P. and Breedis, J. F. (1966). Low-temperature phase transformations. In *Behavior of Materials at Cryogenic Temperatures.* ASTM Special Technical Publication No. 387, American Society for Testing and Materials, Philadelphia, 60–132.

Reed, R. P. and Mikesell, R. P. (1960). The stability of austenitic stainless steels at low temperatures as determined by magnetic measurements. In *Advances in Cryogenic Engineering,* Vol. 4. Plenum Press, New York.

Rhyne, J. J. (1968). Anomalous Hall effect in single-crystal dysprosium. *Phys. Rev.* **172**, 523–529.

Ricodeau, J. A., Melville, D., and Lee, E. W. (1972). Apparatus for the measurement of magnetostriction in pulsed magnetic fields. *J. Phys. E* **5**, 472–474.

Rosenbaum, J., Larson, E., Hoblitt, R., and Fickett, F. R. (1979). A convenient standard for susceptibility apparatus calibration. *Rev. Sci. Instrum.* **50**, 1027–1029.

Rosenberg, H. M. (1963). *Low Temperature Solid State Physics.* Clarendon Press, Oxford.

Rubin, L. G., Nelson, D. R., and Sample, H. H. (1975). Characterizations of two commercially available Hall effect sensors for high magnetic fields and low temperatures. *Rev. Sci. Instrum.* **46**, 1624–1631.

Salinger, G. L. and Wheatley, J. C. (1961). Magnetic susceptibility of materials commonly used in construction of cryogenic apparatus. *Rev. Sci. Instrum.* **32**, 872–874.

Sample, H. H. and Rubin, L. G. (1976). Characterization of three commercially available Hall effect sensors for low temperatures and magnetic fields to 23 T. *IEEE Trans. Magn.* **MAG–12**, 810–812.

Sample, H. H. and Rubin, L. G. (1978). Magnetic field induced temperature changes in cryogenic liquids: N_2, Ar, and He^4. *Cryogenics* **18**, 223–229.

Schlump, W. (1975). Mechanical and magnetic properties of austenitic chrome–nickel steels at low temperatures (in German). *Tech. Mitt. Drupp* **33**, 93–99.

Sherrington, D. and Kirkpatrick, S. (1975). Solvable model of a spin glass. *Phys. Rev. Lett.* **35**, 1792–1796.

Smit, J. (1951). Magnetoresistance of ferromagnetic metals and alloys at low temperatures. *Physica* **16**, 612–627.

Stacey, F. D. and Banerjee, S. K. (1974). *The Physical Principles of Rock Magnetism.* Elsevier, New York.

Suits, J. C. (1972). A review of some recent work in magneto-optical recording. *IEEE Trans. Magn.* **MAG–8**, 421–425.

Tebble, R. S. and Craik, D. J. (1969). *Magnetic Materials.* Wiley–Interscience, New York.

Tholence, J. L. and Tournier, R. (1974). Susceptibility and remanent magnetization of a spin glass. *J. Phys. (Paris) Colloq.* **35**, C4, 229–235.

Touloukian, Y. S., Powell, R. W., Ho, C. Y., and Klemens, P. G. (1970). *Thermal Conductivity: Metallic Elements and Alloys.* Plenum Press, New York.

Ultrarapid Quenching of Liquid Alloys (1981) (ed. H. Herman) *Treatise on Materials Science and Technology,* Vol. 20. Academic Press, New York.

Volkenshtein, N. V. and Fedorov, G. V. (1960). Temperature dependence of the Hall effect of pure ferromagnets. *Sov. Phys. — JETP* **11**, 48–50.

Warlimont, H. and Boll, R (1982). Applications of amorphous soft magnetic materials. *J. Magn. Magn. Mater.* **26**, 97–105.

Warnes, L. A. A. and King, H. W. (1976). The low temperature magnetic properties of austenitic Fe–Cr–Ni alloys. 2. The prediction of Néel temperatures and maximum susceptibilities. *Cryogenics* **16**, 659–667.

Watson, J. K. (1980). *Applications of Magnetism.* John Wiley & Sons, New York.

Weiss, H. (1967). Measurement of magnetic fields. In *Handbuch der Physik,* Vol. 23 (ed. S. Flugge). Springer-Verlag, New York, 198–216.

White, G. K. (1968). *Experimental Techniques in Low Temperature Physics.* Clarendon Press, Oxford.

Williams, I., Williams, R. G., and Capellaro, R. C. (1976). Stability of austenitic stainless steels between 4 K and 373 K. In *Proceedings of the Sixth International Cryogenic Engineering Conference.* IPC Science and Technology Press, Guildford, Surrey, England, 337–341.

Williamson, S. J., Kaufman, L., and Brenner, D. (1977). Biomagnetism. In *Superconductor Applications: SQUID's and Machines* (eds. B. B. Schwartz and S. Foner). Plenum Press, New York, 355–402.

Winter, J. J., Rothwarf, F., Leupold, H. A., and Breslin, J. T. (1978). Variable temperature multimode magnetometer. *Rev. Sci. Instrum.* **49**, 845–849.

Yamauchi, H., Sato, Y., and Watanabe, H. (1972). Automatically recording magnetometer for use at low temperatures. *Jap. J. Appl. Phys.* **11**, 269–270.

Zijlstra, H. (1967). *Experimental Methods in Magnetism*, Vol. IX. North-Holland, Amsterdam.

---

*Available from NTIS, Springfield, Virginia.

7 Mechanical Properties

D. T. READ

7.1 INTRODUCTION

The mechanical properties of a material describe the relations between the stresses acting on the material and its resulting deformations. Uses of mechanical property measurements range from studies of the flow of atoms past each other to selection of the most durable materials for structural supports for superconducting magnets. Stresses capable of producing permanent deformations, which remain after the stresses are removed, are considered in this chapter. Yield strength, ultimate strength, elongation to fracture, and reduction of area are important mechanical properties that describe the response of a material to stresses. The yield strength is the stress at which significant permanent deformation is first produced; the ultimate strength is the maximum stress sustained before fracture; the elongation to fracture is the extent of deformation before fracture; and the reduction of area is the difference between the initial cross-sectional area and the area at the fracture surface. The various mechanical properties are discussed thoroughly in the texts by Dieter (1976), Nadai (1950), and Tegart (1966). The effects of cryogenic temperatures on the mechanical properties of metals and alloys are reviewed in this chapter; the effects on polymers and glasses are discussed briefly. Only macroscopically homogeneous materials are considered — composite materials are discussed in Chapter 12. Semiconductors and materials with covalent and ionic bonds are omitted because they are not currently used in structures for cryogenic service. The fundamental mechanisms controlling temperature-dependent mechanical behavior, phenomena encountered in low-temperature testing, and the mechanical properties of some representative engineering metals and alloys are described. Modifications of test procedures for low temperatures and sources of data are also included. Related topics are discussed in Chapter 1, Elastic Properties, and Chapter 8, Fracture Mechanics.

7.2 TEMPERATURE DEPENDENCE OF STRENGTHENING AND FLOW

General aspects of temperature-dependent mechanical behavior may be discussed using the deformation behavior maps shown in Figs. 7.1a and 7.1b (Ashby, 1972). The axes of these graphs are normalized temperature and normalized stress. Temperature, T, is normalized to the melting temperature, T_m; stress, σ, is normalized to the shear modulus, G. Boundaries between different types of behavior are shown. These maps correspond to behavior that is observed in constant-tensile-strain-rate experiments at ordinary strain rates. At the lowest stresses the specimen deforms elastically, i.e., it returns to its original shape when the stress is removed. Increasing stress eventually brings the specimen to the end of the elastic region where plastic strain (permanent deformation) or brittle fracture occurs. Plastic deformations that are very much larger than the concurrent elastic deformations may occur in this region. The applied stress necessary to continue the deformation process is increased by specimen strain hardening until failure.

In Fig. 7.1a, the behavior characteristic of a pure, annealed face-centered cubic (f.c.c.) mate-

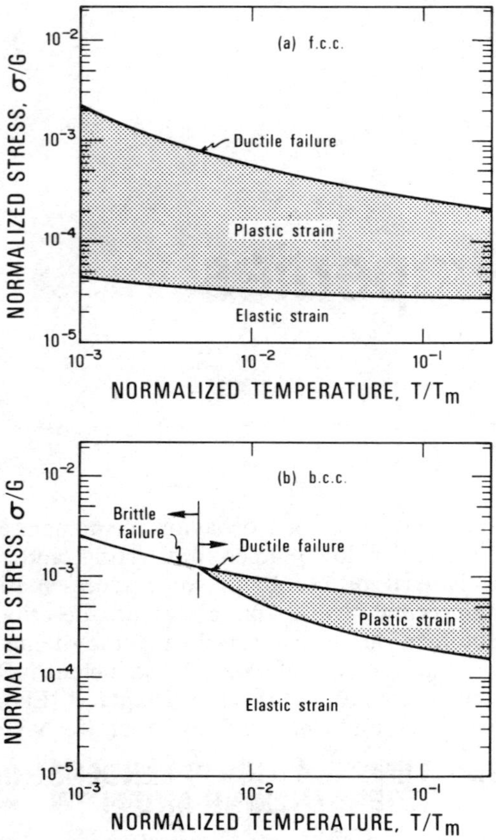

Figure 7.1 Simplified deformation behavior maps (Ashby, 1972): (a) f.c.c. material; (b) b.c.c. material.

rial is shown. The small increase of the yield strength that occurs even when the normalized temperature is decreased by two orders of magnitude is characteristic of f.c.c. behavior. The ultimate strength of such a material, which is shown as the ductile failure line, increases much more than the yield strength on cooling. Note that no region of brittle behavior is shown for f.c.c. materials. The large increase in ultimate strength, together with the relatively small yield strength increase in f.c.c. materials, results from ductile, rather than brittle, failure.

Figure 7.1b illustrates the classic body-centered cubic (b.c.c.) behavior. Note the large temperature dependence of the yield strength, the smaller temperature dependence of the ultimate strength, and a region where the specimen fails before any significant plastic deformation occurs. This brittle region can be extended to higher temperatures by introducing sharp cracks or other stress concentrators or by stressing more rapidly.

7.2.1 Basic mechanism of plastic flow in crystalline solids

Plastic flow in crystalline solids at low temperatures occurs by dislocation motion. Dislocations are one-dimensional regions in which the perfect crystalline arrangement of the atoms has been broken in specific ways (Fig. 7.2). Figure 7.2a shows one partial plane of a crystal with no dislocations. The circles indicate atomic positions. This atomic arrangement is imagined to extend infinitely in the plane of the figure and to be replicated infinitely in both directions perpendicular to the paper. The crystal is cut along the dotted line and its extension to infinity on the left, and the two half-planes above the cut are translated one atomic spacing to the right and joined to the half-planes below the cut, Fig. 7.2b.

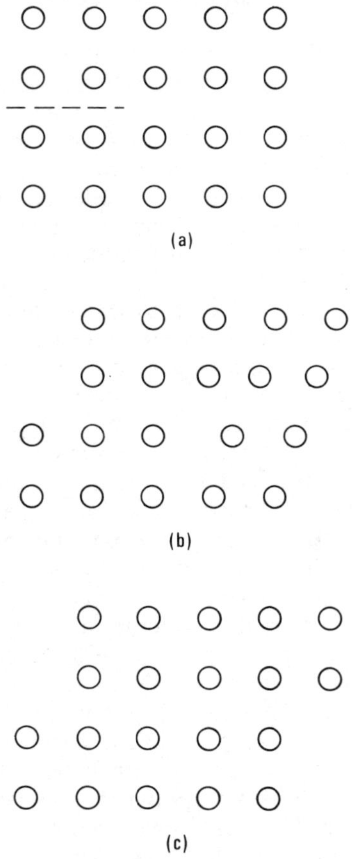

Figure 7.2 Formation of a dislocation in a simplified lattice: (a) the atomic bonds are cut along the dotted line; (b) after translation one atomic spacing to the left, the atoms above the cut are rebonded to those below; (c) when the edge dislocation has traversed the whole crystal, an increment of strain is produced.

The atoms are shifted to new equilibrium positions by interatomic forces. The edge of the excess half-plane is left dangling. Far away from the unmatched edge, the crystal lattice has its original structure with no large distortions. A region of high stresses surrounds the unmatched edge, extending to infinity in both directions perpendicular to the paper. This essentially one-dimensional region of atomic mismatch accompanied by high stresses is known as an edge dislocation. Figure 7.2c shows that when the edge dislocation has traversed the whole crystal, an increment of strain has been produced. Another dislocation type is the screw dislocation. A screw dislocation would be produced if the atomic plane above the cut in Fig. 7.2a were translated one atomic spacing perpendicular to the plane of the figure. Edge and screw dislocations both produce strain by their motion through the crystal lattice.

Dislocations are present in all structural metals. They may be absent in noncrystalline materials and in some materials produced under very special conditions, such as metal whiskers. Dislocations are generated in the production of metal during solidification and forming and in processing steps, such as machining. Furthermore, dislocations are capable of multiplication under the action of applied stress. Thus, elimination of dislocations from structural materials is impractical.

Dislocations move more readily in some crystal planes and directions than in others. These planes and directions of dislocation motion are referred to as slip planes and slip directions. The combination of a slip plane and a slip direction defines a slip system. In f.c.c. materials, dislocations move in the $\{111\}$ planes in $\langle 110 \rangle$ directions. In h.c.p. materials, they move in the basal planes in any of the six close-packed directions. In b.c.c. materials, the motion is often in $\{110\}$ planes in $\langle 110 \rangle$ directions.

Dislocations experience forces owing to stress fields produced in a material by loads applied at its surfaces. The force per unit length on a dislocation line is proportional to the shear stress across the slip plane in the slip direction. Such stresses exist even when the specimen as a whole is subjected to a simple tensile stress. The resolved shear stress in the slip plane in the slip direction $\tau(\phi, \lambda)$ is given by

$$\tau = T \cos\phi \, \cos\lambda \tag{7.1}$$

where T is the tensile stress, ϕ is the angle between the slip plane normal and the tensile axis, and λ is the angle between the slip direction and the tensile axis. From this equation, it can be seen that the maximum shear stress in a slip direction occurs in slip planes at 45° to the tensile stress axis in slip directions at 45° to the tensile axis, and that the value of the resolved shear stress for this case is one-half the axial tensile stress.

Detailed treatments of dislocations and their relation to plastic strain are given by Cottrell (1956), Friedel (1967), Nabarro (1967), and Read (1953). Dislocations were postulated by theorists in the 1920s and were used in the 1930s to explain why the strengths of crystalline metals were found to be far below the expected values for perfect crystal lattices (Cottrell, 1956). The expected value of the shear stress required to produce plastic strain quoted by Cottrell was 1/30 of the shear modulus. Pure metal single crystals are invariably found to deform plastically at stresses far below the theoretical value, because of the presence of dislocations.

Studies of the resistance of single crystals to plastic deformation commonly employ measurements of the shear stress required to produce various amounts of shear strain. A graph of shear strain (on the x-axis) against resolved shear stress (on the y-axis) is called a *flow curve* because it displays the resolved shear stress required to produce further plastic flow at a given strain level. The stress required at some strain level for further plastic strain is commonly referred to as the *flow stress*; the flow stress at which plastic strain is first detected is called the *critical resolved shear strength*. Because the flow stress depends on the strain, it cannot be used as a material property unless the strain level is specified or understood. However, the same mechanisms that control the critical resolved shear strength also contribute to the flow stress. Thus, strengthening mechanisms may be studied using the flow stress as the variable studied, and discussions of strengthening mechanisms often consist of descriptions of various effects on the flow stress.

The suitability of a material for structural purposes is evaluated using (among other properties) its yield strength, mentioned earlier as the tensile stress required to produce significant plastic deformation. The yield strength is analogous to the critical resolved shear strength defined above. The two important differences between the yield strength and the critical resolved shear strength are: the yield strength refers to polycrystalline aggregates rather than single crystals and the yield strength is a tensile strength rather than a resolved shear strength. The random orientation of the grains in a polycrystal produces a range of resolved shear stresses for a given tensile stress. The grains with the highest resolved shear stresses will, of course, yield first, and the others later. The behavior of a polycrystal is an average over all its individual crystallites. The problems of calculating the effects of random orientations of crystallites in a polycrystalline aggregate on its yield strength and of the effects of the contact between neighboring grains on the stress and strain patterns have been investigated (McLean, 1962). The grain boundaries have significant effects on the stresses necessary to initiate and increase plastic strain. The strengthening by grain boundaries adds to that produced by other mechanisms. The yield strength of a polycrystalline aggregate is a result of all the mechanisms that contribute to the critical resolved shear strength of its crystallites and the additional effects of the grain boundaries and the varying orientations of the individual grains (see § 7.2.6).

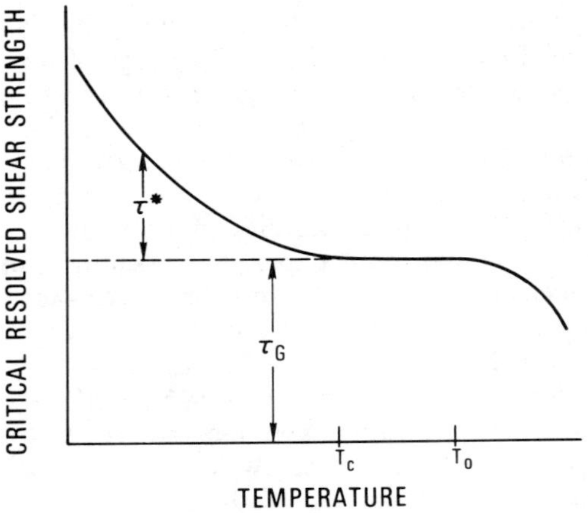

Figure 7.3 Typical form of the temperature dependence of critical resolved shear strength.

7.2.2 Thermal and athermal contributions to strength

Engineering metals yield at stresses higher than most pure single crystals because of several types of obstacles to dislocation motion. This dependence of the strength properties on the defects of the crystal lattice, rather than on the properties of the lattice itself, produces strong variations of yield strength with temperature, chemical composition, and thermal and mechanical treatment.

The typical form of the temperature dependence of the critical resolved shear strength is shown in Fig. 7.3. Strongly temperature-dependent contributions, called the thermal part and denoted τ^*, are produced by short-range obstacles, that is, lattice defects that produce significant stress fields that extend only over a few atomic distances. These defects include interstitials, vacancies, solute atoms, and other dislocations perpendicular to the slip plane. In some cases the dislocation interacts with the lattice itself. The stress exerted on the dislocation by the lattice, the Peierls stress, opposes the passage of the dislocation past each row of atoms, somewhat as a bumpy road opposes the motion of a wheel. Weakly temperature-dependent contributions to the flow stresses, designated τ_G and referred to as athermal, result from lattice defects that produce stress fields extending over many atomic distances. These defects include precipitates, parallel dislocations, and dislocation pile-ups. For this discussion, grain boundaries are considered to behave as long-range obstacles; flow by grain boundary motion generally does not occur in engineering materials in the temperature range 0 to 300 K. Possible exceptions may be found among low-melting-point metals, such as indium.

The thermal part of the critical resolved shear stress, τ^*, varies from its full value at 0 K to nil at some higher temperature (T_c in Fig. 7.3). For most engineering materials, T_c is between 200 and 400 K. At some still higher temperature, T_0, new slip systems may operate, resulting in a strength decrease as the temperature is increased. Such processes are not relevant to the low-temperature behavior of conventional structural materials. In these, τ^* is much less at room temperature than its full 0-K value. If such a material gains part of its room-temperature strength from short-range obstacles to dislocation motion, its critical resolved shear strength may be expected to rise considerably on cooling to liquid-helium temperature.

The source of the temperature dependence of τ^* is thermal activation, thermal energy assisting dislocations to overcome their obstacles. The thermal activation of dislocations overcoming obstacles may be modeled by considering a point particle free to move in one dimension placed in a potential energy minimum. The particle has a certain average thermal energy, kT, where k is Boltzmann's constant and T is the absolute temperature in Kelvin. From the Boltzmann distribution formula, the probability that the particle will have a certain energy value at a given time can be calculated. The probability exists that the particle momentarily could have considerably more energy than the average value — enough energy to escape its potential well. Once the magnitude and shape of the potential well are specified, the probability that the particle will escape from the potential well within a given time can be calculated as a function of temperature. The probable escape rate is zero at zero temperature (in the absence of quantum effects) and rises with temperature. Application of this model to plastic deformation is made by identifying the dislocation moving in its slip plane in the slip direction with the particle in the well. The strength and shape of the obstacle to the motion of the dislocation and the force on the dislocation due to externally applied stresses are used to calculate the potential well energy profile. The macroscopic strain rate is calculated from the total rate of motion of dislocations past their obstacles, which is, in the model, the escape rate of particles from the potential well. Complications prevent using such a model to predict strain rates of real materials accurately as a function of stress and temperature, because the dislocation is actually an extended lattice defect moving through a series of obstacles with a distribution of magnitudes and energy profiles. A recent summary of results from thermal activation theory was given by Kochs, Argon, and Ashby (1975). Dorn and Rajnak (1964), Li (1968), Smidt and Bement (1968), and others have fitted the temperature dependences of yield and flow strengths of several materials to models based on thermal activation theory, showing that the concept of thermal activation can be used to explain satisfactorily the thermal part of the critical resolved shear strength of crystalline solids.

The athermal part of the critical resolved shear strength occurs because of long-range stress fields set up in the lattice by dislocation pile-ups, precipitates, grain boundaries, and similar defects. These stress fields tend to oppose the dislocation motion that produces plastic flow and so decrease the net effect of the externally applied stress. These effects are incorporated into the thermal activation model outlined above by simply subtracting the athermal part of the critical resolved shear strength from the externally applied stress. Because the stress fields of dislocations and precipitates are generally proportional to the shear modulus, their contribution to the strength has the temperature dependence of the shear modulus, that is, an increase of about 15% or less on cooling from room temperature to 4 K for typical engineering materials.

Usual structural materials for cryogenic use are face-centered cubic polycrystalline alloys, which often contain significant amounts of many elements and are formed by rolling, drawing, or forging. Many strengthening mechanisms operate in these materials: strain hardening, grain boundary hardening, solution hardening, precipitation hardening, and others. These may combine in complicated ways. In the following sections, the various strengthening mechanisms are considered individually. Many of the studies cited have employed tensile tests of single-crystal specimens strengthened by only one mechanism, for experimental simplicity. The strengthening mechanisms studied, however, affect all mechanical properties and are applicable to polycrystals as well as to single crystals. Where the effects of a given strengthening mechanism on fatigue, toughness, or other properties have been studied or found to be important, the fact will be noted.

7.2.3 Crystal structure effects

Most engineering materials have face-centered cubic (f.c.c.), body-centered cubic (b.c.c.), or hexagonal close-packed (h.c.p.) crystal structures. The temperature dependences of the mechanical properties of pure, annealed f.c.c. materials, such as copper, aluminum, nickel, and lead, are quite distinct from those of b.c.c. materials, such as iron and niobium. The properties of pure, annealed h.c.p. materials, such as titanium and zinc, are generally in between those of f.c.c. and b.c.c. materials. Figure 7.4 shows the behavior trends. Table 7.1 gives a short list of metallic elements by crystal structure.

The room-temperature values of the yield strengths of pure, annealed f.c.c. materials are generally quite low. Values of 5 MPa or lower are not unusual. Furthermore, in pure f.c.c. materials the critical resolved shear strength is weakly temperature dependent. The values of the changes of the critical resolved shear strengths of pure, well-annealed single crystals of the f.c.c. metals on cooling from 300 to 4 K are of the order of 5 MPa or less, a small amount compared

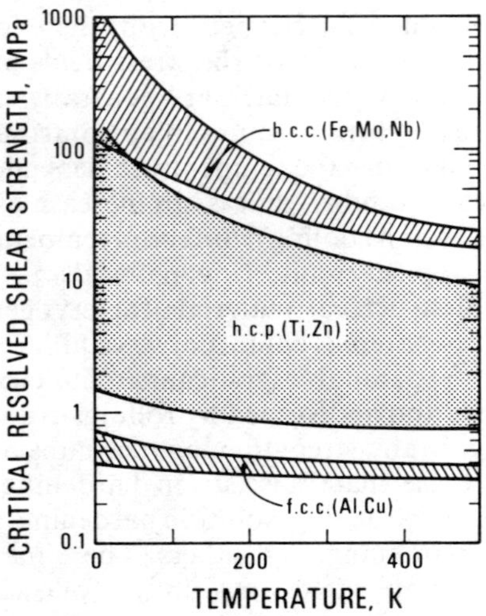

Figure 7.4 Critical resolved shear strengths of f.c.c., h.c.p., and b.c.c. materials.

Table 7.1 Commonly used metallic elements and alloys of the f.c.c., b.c.c., and h.c.p. crystal structures.

Elements	Alloys
F.c.c.	
Copper	Austenitic stainless steels
Aluminum	
Nickel	Brass (< 30 at.% Zn)
Lead	
Silver	
Gold	
Platinum	
B.c.c.	
Iron	Ferritic stainless steels[a]
Molybdenum	Carbon steel[a]
Tungsten	Nickel steels[a] (< 10 at.% Ni)
Niobium	Nb–Ti (< 40 at.% Ti)
Vanadium	
Chromium	
Lithium	
Sodium	
Potassium	
H.c.p.	
Zinc	
Cadmium	
Titanium	
Cobalt	
Magnesium	
Beryllium	

[a]Other phases may be present.

with useful yield strengths in structural materials. These changes may not be small, however, compared with the value of the critical resolved shear strength of the pure, annealed material at 300 K. As the material is deformed, the flow strength increases as a result of strain hardening, which is discussed in detail later. The maximum flow-stress value is reached at or just prior to fracture. The analogous property for polycrystalline materials is the ultimate strength, mentioned earlier as the maximum stress observed before fracture. Maximum flow stresses of 100 to 200 MPa or higher are commonly observed in pure f.c.c. materials. The maximum flow stress of f.c.c. materials increases strongly with decreasing temperature, doubling or more on cooling from 300 to 4 K. Thus two distinctive features of f.c.c. metal behavior are a weakly temperature-dependent initial flow stress and a strongly temperature-dependent maximum flow stress.

The yield strengths and critical resolved shear strengths of pure b.c.c. materials are consistently greater than those of f.c.c. metals even at room temperature, and rise dramatically with decreasing temperature. At room temperature, yield strength values of pure iron at room temperature as low as 25 MPa are observed; at 80 K, the values are about 275 MPa and continue to increase with decreasing temperature. The cause of this behavior is still not well understood. The view that a short-range interaction between dislocation and lattice strongly enhances the thermal part of the yield strength in b.c.c. materials is becoming prevalent (Sato and Meshii, 1975). Dorn and Rajnak (1964) have used such a model to explain the temperature dependence of the yield strengths of common b.c.c. metals. Wigley (1971) clearly states alternative viewpoints and gives references to the literature.

The behavior of h.c.p. materials is generally intermediate, between that of b.c.c. and f.c.c. materials. This can be understood by considering the slip behavior of f.c.c., b.c.c., and h.c.p. materials. In contrast to the four primary slip planes of f.c.c. materials, h.c.p. materials have only one primary slip plane, the basal plane. Strong b.c.c. materials like iron have no planes on which slip may easily occur because of the stronger interaction of dislocations with the host lattice.

7.2.4 Strain hardening

Strain hardening (or work hardening) refers to the increase of the flow stress with increasing strain. The hardening effect arises from the dislocations created during plastic flow. The interaction of moving dislocations with the stress fields from parallel dislocations contributes to the athermal part of the flow stress; perpendicular dislocations contribute to the thermal part.

The phenomenon of strain hardening has generally been demonstrated through measurements of flow of single crystals. Face-centered cubic materials have been the most thoroughly studied, although considerable data on b.c.c. materials have been accumulated in the past several years. Flow curves of pure, annealed f.c.c. single crystals generally have the form shown in Fig. 7.5 (Seeger, 1957). The complete single-crystal stress–strain curve can be regarded as consisting of three stages, as indicated in the figure. After elastic deformation, f.c.c. single crystals usually display a region with a low flow stress and little strain hardening, referred to as stage I or easy glide. The intersection of the extrapolation of stage I of the stress–strain curve with the stress axis at zero strain is termed the critical resolved shear strength and denoted τ_0. The slope of the flow curve in any plastic region is called the strain-hardening coefficient. The reason for the low strain-hardening coefficient in stage I, according to Dieter (1961), is that the moving dislocations traverse the slip plane to the crystal surface and escape. During easy glide, slip always occurs on only one slip system, so stage I is sometimes termed single slip.

Stage II is a nearly linear part of the flow curve during which the flow stress increases rapidly because of dislocation interactions. Dislocations are moving on intersecting slip planes (multiple slip), and their intersection produces a pronounced strengthening effect. Stage II is also called the region of "linear hardening." The strain-hardening coefficient in stage II, θ_{II}, remains relatively constant with decreasing temperature in pure f.c.c. materials. Since stage II results from multiple slip, it may begin prematurely if stage I is abbreviated or eliminated by experimental arrangements not sufficiently favorable to slip on only one set of slip planes. The crystal orientation and stressing arrangement must be carefully selected to obtain single slip. Decreasing the temperature delays the onset of stage II and increases its extent.

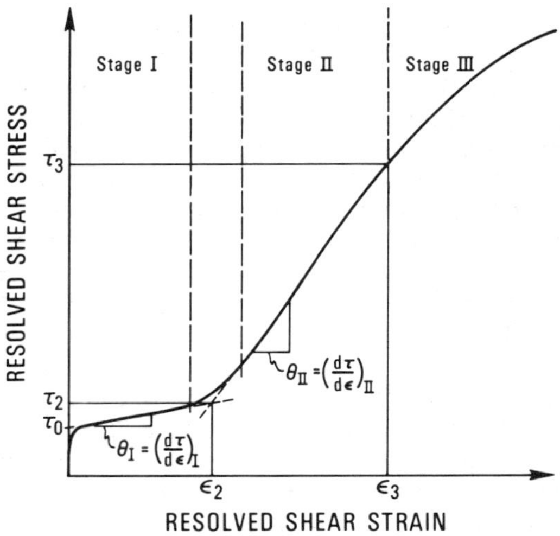

Figure 7.5 Representative stress–strain curve for f.c.c. metal single crystals.

Stage III is called dynamic recovery — dynamic because the process occurs during deformation, and recovery because the thermal effects reduce the work-hardening coefficient relative to that in stage II (Seeger, 1957). It is theorized that the strain-hardening coefficient decreases in stage III because the high stresses and thermal effects enable the dislocations to circumvent the obstacles formed during stage II, perhaps by changing slip planes (cross slip).

The stress–strain curve of a polycrystal is different from the flow curve of a single crystal. In stressing polycrystals, a distribution of resolved shear stresses is applied because the orientations of the individual grains vary; multiple slip is required in the individual grains (for compatibility of deformation from grain to grain); and the grain boundaries act as barriers to dislocation motion. These effects produce strain hardening in polycrystals even below the strain at which the yield strength is measured, typically 0.2%. The polycrystal curve is rounded, without the sharp transition to easy glide shown in Fig. 7.5, and the stresses required to reach equivalent strain levels are much higher, 3 to 15 times higher according to McLean (1962).

The orientation, temperature, and composition dependence of the flow stress of f.c.c. single crystals have been studied quite extensively, and considerable data are available on materials of other crystal structures as well. The curve shown in Fig. 7.5 is a diagram of idealized behavior, but deviations are quite common. For example, high temperatures and high purities tend to suppress stage I. Stage II is sometimes absent altogether. Stage III may be terminated prematurely because the crystal finds another way to deform at those high stresses, e.g., twinning (Blewitt, Coltman, and Redman, 1957). For a detailed discussion of the flow curves of single crystals, the reader is referred to Clarebrough and Hargreaves (1959).

The large difference between the values of the

flow stress at very small and very large strains in Fig. 7.5 demonstrates the capacity of the f.c.c. metals for strain hardening. Decreasing the temperature does not increase the critical resolved shear strengths of most pure f.c.c. metals very much, but their ultimate strengths increase considerably on cooling. Increased elongation with decreasing temperature is also characteristic of pure f.c.c. material behavior; this is sometimes observed in engineering materials. The strain-hardening behavior of f.c.c. metals is exemplified by Fig. 7.6 for copper (from Carreker and Hibbard, 1953).

The strain-hardening behavior of b.c.c. materials (Fig. 7.7) differs from that of f.c.c. materials. Body-centered cubic materials harden by a nearly constant amount with decreasing temperature. Thus, as the temperature is lowered, the ratio of ultimate strength to yield strength decreases and the b.c.c. materials tend to become less ductile. The weaker strain hardening of b.c.c. materials relative to f.c.c. materials is due to the absence or very limited extent of stage II of the flow curve, a region of significant strain hardening in f.c.c. materials. Cross slip, which is the stage III slip mode, occurs at stresses hardly above the high critical resolved shear strength. The weakness of some dislocation intersections and the ease of slip on nonprimary slip planes in b.c.c. materials relative to f.c.c. materials are cited by McLean (1962) as additional reasons for the weak strain hardening of b.c.c. materials.

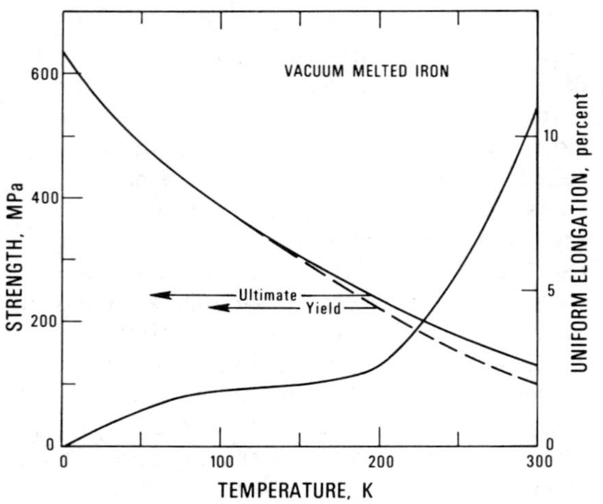

Figure 7.7 Tensile behavior of vacuum melted iron, grain size ASTM No. 8, total carbon, nitrogen, and oxygen content less than 70 ppm, metallic impurities less than 1000 ppm; yield and ultimate strengths and elongation to fracture (Smith and Rutherford, 1957).

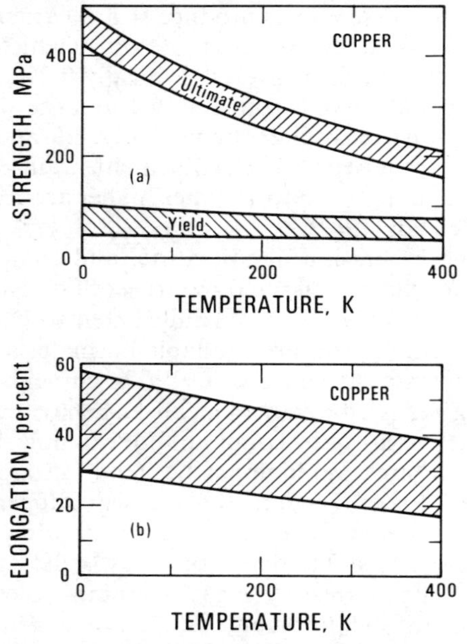

Figure 7.6 Strain-hardening behavior of copper: yield and ultimate strengths and elongation vs. temperature (Carreker and Hibbard, 1953).

7.2.5 Solution strengthening

Solid-solution strengthening, that is, increasing strength by mixing metallic elements to form a solid solution, is found in essentially all practical structural metals. In the f.c.c. materials, solid-solution strengthening is a convenient way to produce practical alloys with better strength. In b.c.c. materials, solution additions are sometimes used to increase strength, but often are used to increase toughness, that is, resistance to brittle fracture at cracks, notches, or other stress concentrations. In some rare cases, adding solute material decreases the strength of b.c.c. materials at low temperatures, e.g., Ta–Re (Haasen, 1968), but in f.c.c. materials, only increases in strength have been obtained through additions of solute. A very complete study of this latter effect using zinc additions to copper was made by Mitchell and Thornton (1963). Figure 7.8, taken from their work, shows the effect of temperature on the critical resolved shear strength of Cu–Zn single crystals. The effects of the zinc solute are visible at all temperatures, but become more pronounced as the temperature decreases.

Solution strengthening was reviewed by Evans (1974). Solute atoms provide obstacles for the motion of dislocations through the lattice, thus

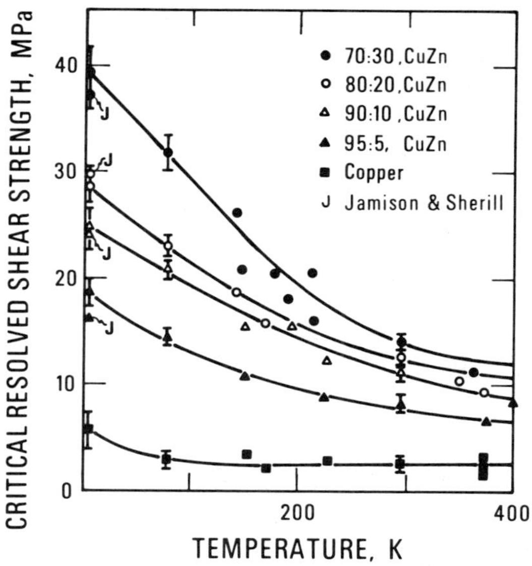

Figure 7.8 Temperature dependence of the initial flow stress of copper and Cu–Zn single crystals (Mitchell and Thornton, 1963).

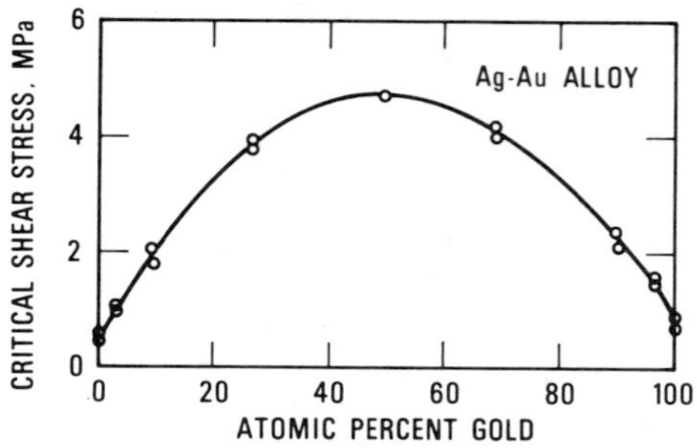

Figure 7.9 Critical resolved shear stress as a function of composition in Ag–Au alloy single crystals (Sachs and Weerts, 1930).

increasing the stress necessary for flow. Dislocations and solutes attract (or repel) one another with a short-range interaction, the strength of which depends on the difference in size between the solute and solvent atoms. Valence or chemical bond differences between solute and solvent atoms may also contribute to the strength of the dislocation–solute interaction. The solute atoms are relatively immobile at the strain rates used in tensile testing in the temperature range 0 to 300 K. Moving dislocations must pass the solute atoms; the interaction between the moving dislocation and the stress field (or electric field, in some materials) of the solute atoms hinders the motion of the dislocation, raising the stress necessary to push it through the lattice and so raising the flow stress. The strengthening effect of the solutes is largest at low temperatures and decreases with increasing temperature, as shown in Fig. 7.8.

Data, such as those shown in Fig. 7.8, have been obtained on a variety of solvent and solute elements in single and polycrystals. The general behavior as a function of solute concentration, as exemplified by Ag–Au single crystals, is shown in Fig. 7.9. In the low-solute concentration range, a strength contribution proportional to the square root of the solute concentration is expected for solution hardening.

The effect of solutes in increasing the maximum flow stress is less than their effect on the critical resolved shear strength, but of course, the critical resolved shear strength is quite low in the absence of solutes.

The effect of temperature on the ultimate tensile strength of a solution-strengthened alloy is greater than its effect on the yield strength. This strong temperature dependence of the ultimate tensile strengths in solution-strengthened alloys is similar to the behavior of pure f.c.c. materials.

The low-temperature strengths of b.c.c. metals are not improved as significantly by alloying as those of f.c.c. metals. Small amounts of the interstitial carbon are particularly effective in increasing the low-temperature strength of iron (Stein, 1966). Alloying usually raises the yield strengths of b.c.c. materials, but does not increase their temperature dependences significantly. Generally, the ultimate tensile strength has approximately the same temperature dependence as the yield strength.

7.2.6 Athermal strengthening by grain boundaries and precipitates

Grain-boundary strengthening has been found to obey the Hall–Petch relation, which states that the grain-boundary contribution to the yield strength or flow stress is inversely proportional to the square root of the grain diameter. That is, the smaller the grains, the higher the strength. The Hall–Petch relation for grain boundary strengthening applies at low temperatures as well as at ambient temperatures; the grain-boundary strengthening coefficient is not a strong function of temperature.

To illustrate the effects of temperature on precipitation hardening, the temperature dependences of the yield strengths of Al–Cu single crystals with three different heat treatments are

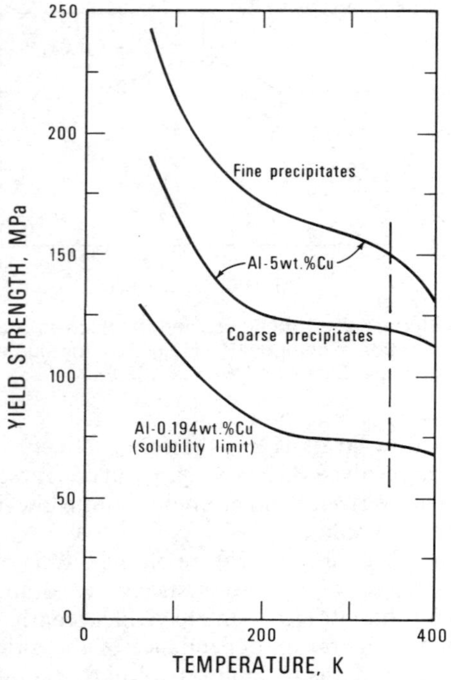

Figure 7.10 Yield strength as a function of temperature for an Al–Cu alloy in three conditions: fine precipitates, coarse precipitates, and solid solution (Starr, Shaw, and Dorn, 1954).

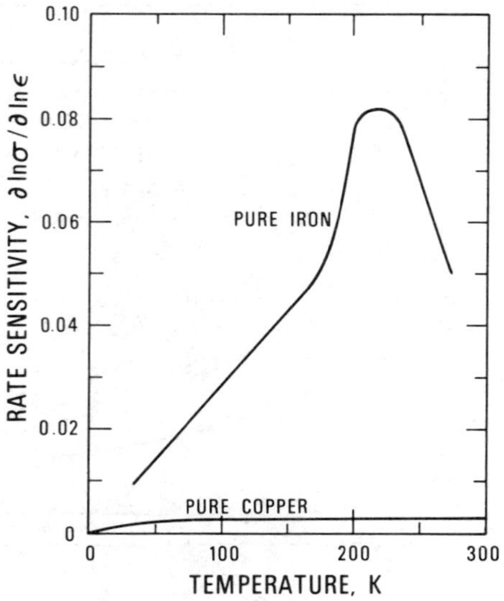

Figure 7.11 Sensitivity of flow stress to strain rate change as a function of temperature for pure copper and pure iron (Basinski, 1959; Basinski and Christian, 1960).

compared in Fig. 7.10 (Starr, Shaw, and Dorn, 1954). All crystals contained the solubility limit of copper in solution, giving rise to a thermal contribution to the yield strength from solution hardening. Precipitates were formed in the two crystals with copper contents in excess of the solubility limit. As a result of precipitation hardening, the yield strengths of both alloys containing precipitates were increased well above the crystals with only solid-solution hardening, as can be seen in the figure. (At temperatures to the right of the dashed line above 300 K, additional high-temperature flow mechanisms are activated, which lower the strength.)

Grain boundaries and precipitates both extend over many atomic distances. Their stress fields are long range, and dislocations cannot pass through them by thermal activation. The temperature dependence of the stress fields of grain boundaries and precipitates is like that of the shear modulus (typically a 15% or smaller increase between room temperature and absolute zero); their contributions to the strength have this same temperature dependence.

7.2.7 Strain-rate effects

In standard tensile tests, at ordinary strain rates (10^{-5} to 10^{-2} s^{-1}), the effect of the strain rate on mechanical behavior is generally insignificant compared with the effect of temperature. Special tests using explosive loading can produce effects comparable to those produced by lowering the temperature to the liquid helium range from room temperature. Strain-rate effects tend to be greatest in materials that have the strongest temperature dependences of strength. Solution-hardened f.c.c. materials have stronger strain-rate effects than pure f.c.c. materials, and b.c.c. materials have the strongest effects (Conrad, 1965). The increase of flow stress owing to strain rate is shown in Fig. 7.11.

Strain-rate effects in tensile tests are usually measured to learn about the mechanisms of plastic deformation; deformation kinetics theoretically relates strain rates to other effects on mechanical properties. A recent study in which strain-rate data were used through deformation kinetics to study deformation mechanisms is reported by de Meester, Doner, and Conrad (1975).

Material toughness is affected by strain rate through its effect on the flow stress, and it is in dynamic fracture toughness tests that the practical consequences of strain-rate effects are shown.

This subject is discussed in more detail in Chapter 8.

7.3 TENSILE PROPERTIES

The usual tensile properties, yield strength, ultimate tensile strength, elongation, and reduction of area, are widely used for engineering evaluation of materials for low-temperature structural service. Yield strength and ultimate tensile strength are determined using an engineering stress–strain curve (Fig. 7.12). The engineering stress, σ, is the applied tensile load, P, divided by the original cross-sectional area, A_0.

$$\sigma = P/A_0 \quad (7.2)$$

The engineering strain, e, is the change in length, $\Delta\ell$, of the uniformly strained region (the gauge length) divided by its original length, ℓ_0.

$$e = \Delta\ell/\ell_0 \quad (7.3)$$

The engineering stress–strain curve is analogous to the flow curve described earlier. With reference to the engineering stress–strain curve of Fig. 7.12, the yield strength, σ_y, is the stress at 0.2% offset strain; the ultimate strength, σ_u, is the maximum stress obtained (calculated using the maximum load, P_{max}, and the initial specimen area):

$$\sigma_u = P_{max}/A_0 \quad (7.4)$$

The elongation is the engineering strain at fracture, e_f, expressed in percent:

$$\%\ \text{elongation} = e_f = (\ell_f - \ell_0)/\ell_0 \quad (7.5)$$

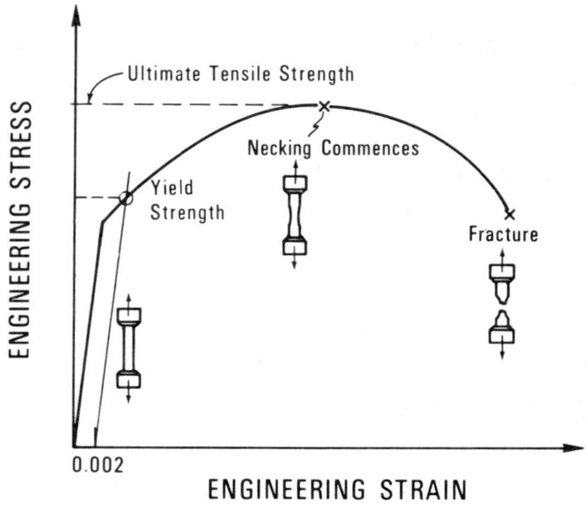

Figure 7.12 Usual form of the engineering-stress–strain curve for an engineering material.

where ℓ_f is the gauge length at fracture. Reduction of area, q, is the difference between the original specimen cross-sectional area, A_0, and the final cross-sectional area, A_f, divided by the original area, expressed as a percentage:

$$q = (A_0 - A_f)/A_0 \quad (7.6)$$

The texts by Tegart (1966) and Dieter (1976) provide more detailed treatments.

Usual procedures, instrumentation, and terminology for ambient-temperature tensile-property measurements are usually applicable to low-temperature testing. The measurement of specimen strain in the cryogenic environment requires special design and construction of extensometers to endure thermal cycling without debonding of strain gauges, breaking of wires, or other problems. Strain gauges appropriate for cryogenic temperatures must be used, and their current sometimes must be limited to prevent bubbling of the cryogen, which produces electrical noise in the strain signal. Calibration of extensometers in the appropriate cryogenic environment is a necessity. With care, extensometers for cryogenic service, using electrical resistance strain gauges and standard electronic amplification and read-out apparatus, can be used to measure reproducibly changes in gauge length around 2 μm over times of a few minutes. In a special set of measurements using capacitive sensors for displacement and special extensometer fixtures and electronic apparatus, a noise level of the order of 10^{-6} in the strain was reached by Reed and Durcholz (1970).

Other special considerations for low-temperature tensile testing are: (a) visual monitoring of specimen strain is possible but usually cumbersome and, therefore, should be avoided, (b) miniature and subsize specimens often are used to conserve cryogens, and (c) load fixtures carefully designed for safe, efficient low-temperature operation are required. Equipment and procedures for low-temperature tensile testing are described by Reed (1961).

Uncertainties in the yield strength of $\leq 2\%$ can be achieved, even in cryogenic environments. This is quite adequate for design purposes, and in fact, specimen-to-specimen variation is often of this size. The exception to this is the determination of the upper yield point. This quantity and the uncertainties involved are discussed in § 7.3.5.

Many phenomena often occur during low-temperature testing that are less commonly, or perhaps never, encountered in room- and high-temperature measurements. The ordinary room-temperature tensile stress–strain curve (Fig. 7.12) has a well-defined elastic region and smooth behavior through the transition into the plastic region, the maximum in the stress, and the remaining stress–strain curve all the way to failure. Such curves are commonly obtained at ambient temperature, but at low temperatures distinctive behaviors are often observed. These effects are discussed below.

7.3.1 Serrated stress-strain curves

Tensile stress–strain curves having the form shown in Fig. 7.13 are described as serrated, having many load drops, or exhibiting discontinuous yielding. Such curves are a low-temperature phenomenon, except for some materials that undergo strain aging during testing at ambient temperatures. The discontinuous yielding process consists of an initiation event (which is not visible in the stress–strain curve), a sudden drop in the load accompanied by sudden specimen extension, an elastic increase in the load, possibly some plastic deformation, another initiation event, and another sudden drop in the load. Adiabatic deformation has been advanced as a probable explanation for the serrations (Basinski, 1957, 1960). A nucleating deformation, perhaps at a stress concentration within the specimen and aided by a thermal fluctuation, occurs that produces enough heat to drive the sample into an unstable condition. The heat produced by the deformation raises the temperature of the sample (adiabatic as opposed to isothermal conditions) and lowers the flow stress enough for the deformation to continue at the applied stress level, which produces more heat, continuing the process. The low specific heat and thermal conductivity of impure metals at low temperatures make such a process quite plausible, although the details of the nucleating deformation are not well understood. Recent measurements on Fe–Cr–Ni alloy by Read and Reed (1980) showed that the amount of mechanical energy dissipated in each load drop was approximately equal to the amount of thermal energy required to heat the portion of the specimen that was strained during the load drop to its measured maximum temperature. It was also shown that the maximum temperature reached during load drops could be calculated using the specimen specific heat over the experimental temperature range, the temperature dependence of the flow stress, and the specimen volume strained. Serrated yielding is pro-

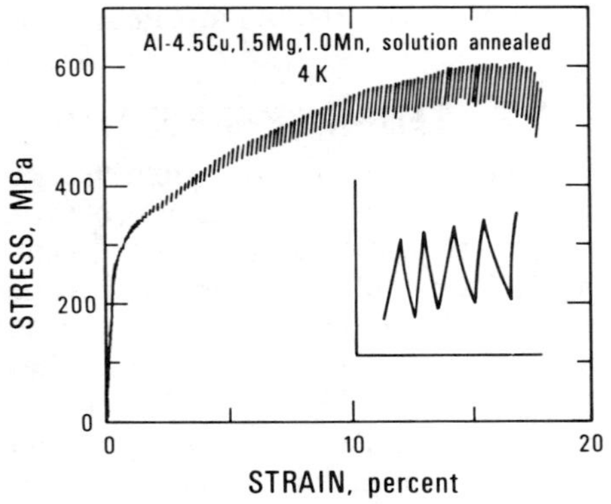

Figure 7.13 Serrated stress–strain curve obtained at very low temperature.

moted by high strain rates because smooth plastic deformation, the alternative to the serrated yielding process, has less time to occur. Very low temperatures (the behavior usually changes drastically between 4 and 20 K), mechanical disturbances (such as tapping on the cryostat), and elastically stiff load fixtures promote serrated yielding. Some materials that exhibit serrated yielding are the austenitic stainless steels (AISI 304, 310, 316) (Tobler, 1976a), brasses, bronzes, and commercially pure coppers (Reed and Mikesell, 1967), and some aluminum alloys (Basinski, 1957, 1960).

The strain-rate and temperature dependences of serrated yielding are considered to be evidence in favor of the adiabatic heating explanation. The fact that discontinuous yielding occurs in materials that do not exhibit the martensitic transformation is generally taken as evidence that the mechanism for the load drops cannot be martensitic transformation. The crystallographically closely related phenomenon of twinning was reported to produce serrated yielding in some experiments (Blewitt et al., 1957).

Discontinuous yielding with many load drops in a material undergoing a tensile test at 4 K is an indication that the material has significant capacity for plastic deformation. In fact, the toughest known structural steels at 4 K, the Fe–18Cr–8Ni alloys, exhibit considerable serrated yielding during tensile tests at 4 K.

7.3.2 Twinning

In twinning, the atoms inside a finite volume of the parent phase are realigned as a new crystal with the same structure as the parent phase but with a different orientation. The twinned portion is a mirror image of the parent crystal. The deformation is a simple shear distributed over the volume of the twin. The lattices of twin and parent adjoin at a plane of symmetry called the twinning plane.

Low temperatures promote deformation twinning in some materials by elevating the initial stress required for slip higher than that required for twinning or by allowing work hardening by slip to reach the stress level where twinning occurs.

Churchman and Cottrell (1951) found that in single crystals of iron with 0.02% C, deformation by twinning occurred at 83 K, whereas at 196 and 295 K, yield points occurred without twinning. The twinning was reported to be accompanied by audible clicks and large load drops. In this case, deformation by slip occurred at the higher temperatures but was absent at the lower. The stress required for deformation twinning at the lower temperatures was less than that necessary for slip, so twinning occurred.

Erickson and Low (1957) confirmed the effect of decreasing temperature in promoting twinning in a commercial low-carbon (0.04%) polycrystalline steel. They reported that at 20 K twinning occurred without a yield point, at 77 K twinning occurred in deformed regions after the yield point was observed, and above 125 K no twins were found, even in deformed material. Sleeswyk and Helle (1963) observed twins in pure iron specimens after plastic deformation between 77 and about 160 K. Twins were not found in specimens deformed at higher temperatures. It was concluded that the yield strength at and above 77 K was controlled by obstacles to dislocation motion rather than by twinning phenomena.

Copper strain-hardens to a progressively higher stress level as the temperature is lowered. At sufficiently low temperatures, the stress required for twinning becomes available. Convincing evidence of deformation twinning in copper single crystals deformed at 4 and 76 K was reported by Blewitt et al. (1957). Since then, many reports of deformation twinning in f.c.c. metals and alloys have appeared.

For discussions of the general aspects of twinning, see Reed-Hill (1964), Hall (1954), Cahn (1954), and the conference proceedings *Deformation Twinning* (1964).

7.3.3 Martensitic transformations

Strain-induced martensitic transformations have been observed during tensile tests at cryogenic temperatures. Iron–chromium–nickel alloys and related systems have been studied extensively. Hexagonal close-packed and body-centered cubic martensites were identified in 18Cr–9Ni stainless steel deformed at low temperatures by Guntner and Reed (1962). The transformation martensite promotes the ductility of this stainless steel.

The relationships between amounts of martensite produced and stress, strain, temperature, and chemical composition are known only imprecisely, especially for structural alloys. Difficulties in studying this problem include the absence of a general method for determining martensite content that is quick, precise, and functions *in situ* during the tensile test and the complex dependence of martensitic transformations on chemical composition and temperature. Magnetic methods that meet the above criteria exist, but many martensites of interest are not magnetic.

Effects of martensitic transformations on the flow stress or yield stress have been studied by several investigators. A sharp drop in the flow stress is often observed at the temperature where strain-induced martensitic transformation is initiated. The strain-induced martensitic transformation, although in general enhanced by lowering the temperature, may be suppressed at 4 K.

7.3.4 Anomalous yield-strength temperature dependence

A maximum in the temperature dependence of the flow stress below 50 K has been reported in copper alloyed with zinc, silver, nickel, germanium, and silicon (Schwarz, Isaac, and Granato, 1977). A yield strength maximum in the same temperature range for dilute alloys based on copper, silver, and gold was reported by Basinski, Foxall, and Pascual (1972). Increased plasticity, that is, decreased flow stress, of metallic specimens in the superconducting state relative to the normal state is a related anomalous low-temperature behavior; Pustovalov, Startsev, and Fomenko (1970) and Kostorz (1973) have reviewed the experimental facts and proposed explanations. In their more recent discussion, Schwarz et al. (1977) explain the observed anomalous temperature dependence of the yield strength by reference to a theoretical model developed by Granato (1971a, 1971b) to explain the increased plasticity in superconductors. The model considered the temperature and magnetic field dependences of the frictional drag interaction between moving dislocations and conduction electrons and crystal lattice thermal vibrations (phonons). The interaction can be described in classical terms, although quantum mechanical calculations are necessary for an accurate quantitative description. The conduction electrons scatter off the dislocation because of the deviations from perfect crystallinity in the neighborhood of the dislocation. For a moving dislocation, the average impact velocity is different at the leading and trailing edges. This difference results in a net force on the dislocation opposing its motion. The net force is proportional to the dislocation velocity, so it can be handled as a frictional drag hindering the motion of the dislocation. The dislocation–phonon interaction can be described similarly.

According to Granato's model, lowered drag between moving dislocations and stationary (on the average) electrons, which occurs at sufficiently low temperatures or in the superconducting state, allows each dislocation to oscillate about its pinning sites, its frequency governed by its mass per unit length and line tension. The oscillation allows the dislocation to exert an extra increment of force (the inertial effect) on the pinning site, and thus reduces the externally applied stress necessary for plastic flow. A simpler explanation of the superconducting–normal flow stress difference based on the viscous drag equation was not adequate because it did not explain the absence of an effect of strain rate on the flow-stress decrease (Alers, Buck, and Tittmann, 1969).

This drop in the flow stress was observed by Kojima and Suzuki (1968) by straining a lead specimen at 4 K in a variable magnetic field. When the field was on, the specimen was in its normal state; when the field was turned off, the specimen entered the superconducting state. The specimen flow stress in the superconducting state was found to be about 0.1 MPa lower than in the normal state. Galligan, Lin, and Pang (1977) claimed that an observed increase of the flow stress of copper (not a superconductor) with magnetic field is also a dislocation drag effect.

The effect of dislocation drag on flow stress was used to explain flow stress decreases in both superconductors and nonsuperconductors. In superconductors, the transition to the superconducting state turns off the electron–dislocation interaction. The absence of electron drag allows plastic flow at lowered stresses. In nonsuperconductors, the temperature dependence of the total dislocation drag affects the temperature dependence of the flow stress. Frictional drag on moving dislocations is produced by phonons (thermal lattice vibrations) as well as by electrons. At room temperature, the phonon drag is much larger than the electron drag. As the temperature is lowered, the phonon drag decreases in a highly nonlinear way, with a big drop to near zero between 100 and 20 K. The phonon drag drops off much more rapidly than the yield strength increases over this temperature range. This drop in the phonon drag causes a rapid decrease in the component of the yield strength that is produced by the phonon drag mechanism. Over the narrow temperature range where the yield strength decrease due to the decrease in the dislocation–phonon drag outweighs the usual yield strength increase with decreasing temperature, the total yield strength decreases with decreasing temperature. Where the temperature region in which yield strength increases with decreasing temperature meets the region where it decreases with decreasing temperature, a maximum in the temperature dependence of the yield strength is found.

Yield strengths of commercial brass alloys that are lower at 4 than at 20 K have also been reported (Reed and Mikesell, 1967), indicating that this anomalous behavior may not be restricted to single crystals or very soft materials. The review by Verkin (1978) references work on titanium, aluminum, and Fe–Cr–Ni alloys that also showed yield strength maxima at temperatures near 20 K.

Quantum-mechanical effects on the yield strength have been proposed to explain anomalous yield strength–temperature behavior. Consider the particle-in-a-potential-well model (§ 7.2.2). An escape rate for a particle in a potential well can be calculated on the basis of quantum-mechanical tunneling. The escape rate due to tunneling is temperature independent; therefore at sufficiently low temperatures, the only operative escape mechanism is tunneling. Application of this concept to dislocations escaping from obstacles implies that the thermal part of the flow stress should cease increasing with decreasing temperature at sufficiently low temperatures. Most estimates, however, of the temperature required for observable quantum mechanical tunneling effects (Arko and Weertman, 1969) place it within a few hundredths of a degree or less of absolute zero, which implies that the proposed quantum mechanical effects have not actually been observed.

7.3.5 Yield points

In some materials, notably mild steel at room temperature, the initial part of the stress–strain curve has the form shown in Fig. 7.14, i.e., as the

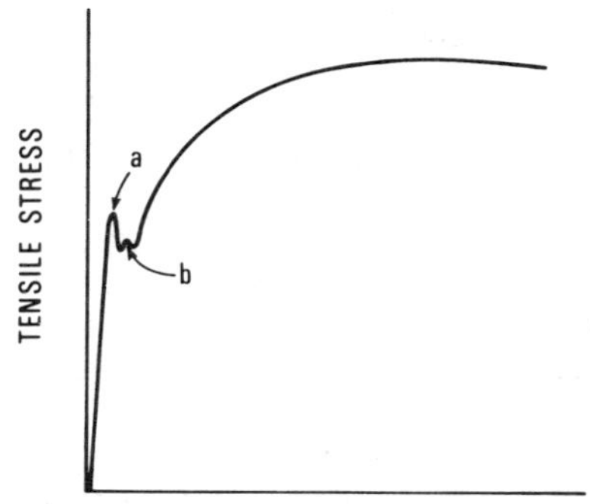

Figure 7.14 Yield points in stress–strain curve: (a) upper yield stress; (b) lower yield stress.

specimen is stretched, the stress suddenly decreases. This phenomenon is referred to as a yield point, load drop, or sometimes sharp yield point to emphasize the suddenness of the change of slope of the stress–strain curve. The maximum value of the stress at the yield point is called the upper yield stress, and the minimum value of the stress immediately following the yield point is the lower yield stress. Surface markings are associated with the yield-point phenomenon. A band of deformed metal is formed at the upper yield point at a stress concentration on the specimen. The load drop to the lower yield point is coincident with the band formation. These bands, called Lüders bands, gradually propagate along the length of the specimen as the specimen is extended. The stress does not increase with strain until the Lüders bands have covered the entire test section. The value of the upper yield stress is difficult to measure reproducibly because the stress distribution in the specimen before yield is rarely homogenous. McLean (1962) discusses apparatus effects and concludes that whereas the commonly observed ratios of upper to lower yield stresses are about 1.1, values as high as 2 may be observed with special experimental techniques to eliminate nonaxial loading.

Low temperatures and high impurity concentrations promote the appearance of a yield point in stress–strain curves; the temperature effect is shown in Fig. 7.15. In metals that have a ductile–brittle transition, lowering the temperature below the transition usually produces brittle fracture immediately after yield so that the lower yield point is not observed. Yield points are observed in materials of each of the three main metal and alloy crystal structures and are generally believed to be caused by the sudden unlocking of dislocations from impurity atoms. The yield-point phenomenon has been studied intensively in steels, often in connection with the process of strain aging, which does not occur at cryogenic temperatures in ordinary materials.

7.3.6 Cottrell-Stokes Law

The Cottrell–Stokes Law states that the ratio of the flow stresses at two temperatures is a constant independent of strain after the first few percent of plastic deformation. Cottrell and Stokes (1955) discovered this behavior in aluminum.

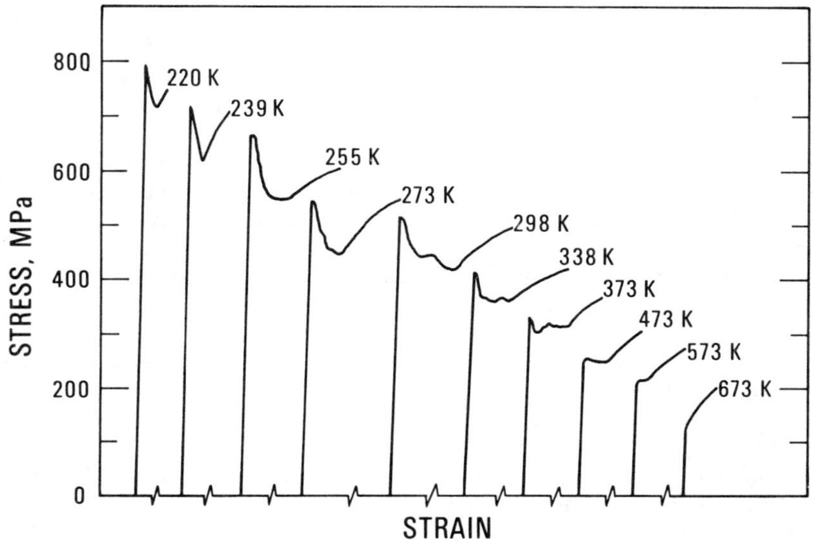

Figure 7.15 Effect of temperature on the yield point of molybdenum (Hahn, 1962).

Subsequent investigations have shown that this law is obeyed for many pure f.c.c. metals, both single- and polycrystalline, and that it applies for changes in strain rate as well as temperature. The Cottrell–Stokes Law is not applicable to b.c.c. materials or to f.c.c. alloys that are strain or precipitation hardened; mixed results have been obtained with h.c.p. materials. Since this law is concerned with the thermal part of the flow stress, it properly applies only when at least one of the two temperatures considered is low enough so that the thermal contribution to the flow stress is significant.

7.3.7 Work softening

Work softening is the name given by Stokes and Cottrell (1954) to an effect that involves straining a specimen first at a low temperature and then at a higher temperature. At the higher temperature, their specimens exhibited a yield point, then a decrease in the load required to continue the deformation, and then continuous deformation at loads equal to those for specimens strained only at the higher temperature. The yield-point effect was not due to diffusion of solutes to dislocations and so must have been an effect of the dislocation structure introduced during the low-temperature deformation. A similar effect was observed by Haasen and Kelly (1957): a small yield-point effect was introduced into a metal sample deformed at a relatively low temperature by fully or partially unloading the specimen and then reloading.

7.4 TOUGHNESS, FATIGUE, AND CREEP

Low-temperature toughness and resistance to creep and fatigue depend on the same basic mechanisms of plastic flow and resistance to dislocation motion as tensile properties, but with added complications. The behaviors of these properties are not as well understood on a microscopic level as tensile property behavior. Most studies have been oriented toward evaluation of practical materials for structural use rather than toward fundamental understanding of the basic mechanisms using pure and single-crystal specimens. The treatment of toughness, fatigue, and creep in this section has the same orientation. Fatigue and creep are discussed as they apply to unnotched specimens. Fatigue in notched and cracked specimens and current practical applications of the concept of toughness are discussed in Chapter 8, Fracture Mechanics.

7.4.1 Toughness

Toughness represents the ability of a material to resist fracture in the presence of large stresses in the neighborhood of a notch, crack, or other stress concentrator and is measurable in special fracture toughness tests. Toughness is an important low-temperature property because the decrease in toughness on cooling that occurs in ferritic steels is the most deleterious effect of cryogenic environments on commonly used structural materials. Toughness is controlled by the basic mechanisms of strength and deformation described in § 7.2, but the relationship between the measured property and the fundamental deformation processes is complex, in contrast to that for tensile properties. The complexity is due to the spatially nonuniform stress and strain fields and multiaxial stresses that occur in toughness testing owing to the use of notched and cracked specimens. A quantitative theoretical treatment of toughness would produce, in contemporary terminology, a calculation of the temperature dependence of the plane strain critical stress intensity, K_{Ic}. This quantity describes the resistance of a material under stress to fractures that begin at a stress concentrator. Such a calculation is currently not available.

The Charpy impact test and other toughness measurements. The Charpy impact test has been used widely in the United States as a low-temperature toughness measurement method for ferritic and stainless steels and their welds (*Annual Book of ASTM Standards*, 1976). In this technique, a heavy pendulum fitted with a special impact fixture called a tup is made to swing into and through a notched bar of specimen material. The energy absorbed by the specimen is related directly to the difference between the height to which the pendulum swings without impacting a specimen and the height reached after fracturing a specimen. Measurements of the deformation produced by the test are commonly made on the fractured specimen. Commercially manufactured Charpy impact testing machines capable of delivering a variety of energies at impact are available. Specimen temperature is controlled by cooling the specimen to the desired temperature by immersion in an appropriate cryogen, such as liquid nitrogen, and then rapidly inserting the specimen into the machine and immediately impacting it. Lower temperatures are reached by using a flimsy, disposable container filled with cryogen around the specimen. Internal specimen heating from the high strain rates during fracture limits the effectiveness of these techniques.

Charpy impact energy is plotted as a function of temperature, as shown in Fig. 7.16. Significant scatter is commonly observed in the values of energy absorbed in Charpy tests, but the results of many tests are usually represented by a smooth curve like the one in Fig. 7.16. The transition between high and low values of absorbed energy generally occurs over a fairly narrow temperature range in low-strength (less than about 700 MPa or 100 ksi) ferritic steels. The absorbed energy often changes much more gradually in other materials. Over the last decade, Charpy specimen lateral expansion opposite the notch has replaced the impact energy in some codes and standards. Lateral expansion, measured opposite the notch, is believed to represent more consistently the fracture resistance of materials with a wide range of yield strengths. In some United States codes, a lateral expansion of at least 0.38 mm has been required (see Chapter 8). In Fig. 7.16, lateral expansion is plotted against temperature along with the impact energy. Other quantities observed in a Charpy test that may be plotted against temperature are the percent fibrous (shear) fracture visible on the fracture surface and the lateral contraction at the notch root. Neither of these has been accepted by code-writing groups as a standard test of material quality.

The most severe limitation of the Charpy test is that the results are a qualitative indication of toughness and cannot be used directly in structural design calculations. Other limitations are the scatter often observed in the data, its inability to show the toughness transition in materials that fail by low-energy tearing instead of cleavage, and its inapplicability to nonferrous materials and specimens less than 5-mm thick.

Notch tensile testing is another technique for comparative toughness evaluation. In this method, a notch is machined around the circumference of a cylindrical specimen, or at the edges of a plate specimen, which is then loaded in tension. The notch strength or notch tensile strength is the maximum load sustained divided by the remaining cross-sectional area of the specimen at the notch. The ratio of the notch strength to the ultimate tensile strength or to the tensile yield strength (measured on an unnotched specimen) is an index of the specimen toughness. The notch–yield ratio, that is, the ratio of notch strength to yield strength, is recommended as a more meaningful parameter in a proposed test method (*Annual Book of ASTM Standards*, 1976). Like the Charpy impact technique, the notch tensile method is a qualitative measure of toughness.

A variety of notch depths and root radii have been used by various investigators for notch tensile testing, although the proposed test method referenced above specifies fixed notch dimensions. The stress at the notch root is higher than the average stress far from the notch. The stress concentration factor, K_t, gives the ratio of the

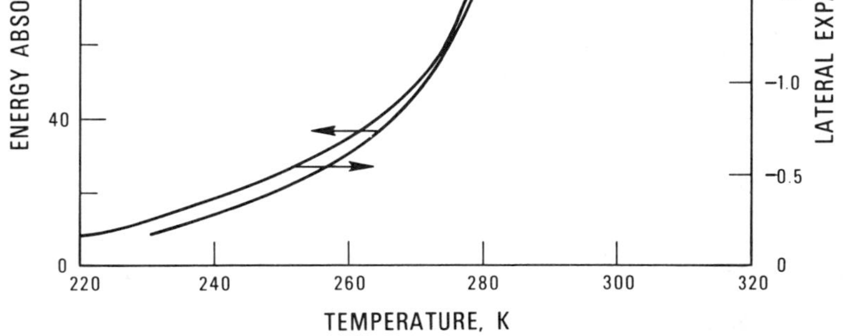

Figure 7.16 Temperature dependence of energy absorbed and lateral expansion in Charpy impact toughness test for A-533-B steel (Gross, 1970).

stress at the notch to the average stress far from the notch. Derivations of stress concentration factors for several notch configurations are given in the classic work by Neuber (1946). A more recent compilation by Peterson (1973) gives stress concentration factor values for a wide variety of notch configurations encountered in structural design.

Ductile-to-brittle transition in ferritic steels. A much-used concept related to toughness is the ductile–brittle transition. Some materials, in practice mainly b.c.c. iron alloys, are capable of ductile behavior at high temperatures but become brittle at low temperatures. The ductile–brittle transition temperature marks this transition. Many structural materials show a gradual transition between brittle and ductile behavior and so do not have a genuine ductile–brittle transition temperature. One theory of the ductile–brittle transition is that the mechanisms of failure by brittle fracture and failure by ductile fracture following yielding are potentially operative at all temperatures (Knott, 1973). These two failure modes are viewed as alternate possibilities; the one that requires lower stress is realized. At high temperatures, the ductile–failure stress is lower; at low temperatures, the brittle–failure stress is lower. The ductile–brittle transition temperature, T_{DB}, is the temperature at which the ductile and brittle fracture stresses are equal. Below T_{DB} brittle fracture occurs, and above T_{DB} ductile fracture occurs. Increases in the strain rate or the presence of notches raises the local effective yield strength, the ductile fracture stress, and T_{DB} (Fig. 7.17). This model rationalizes the

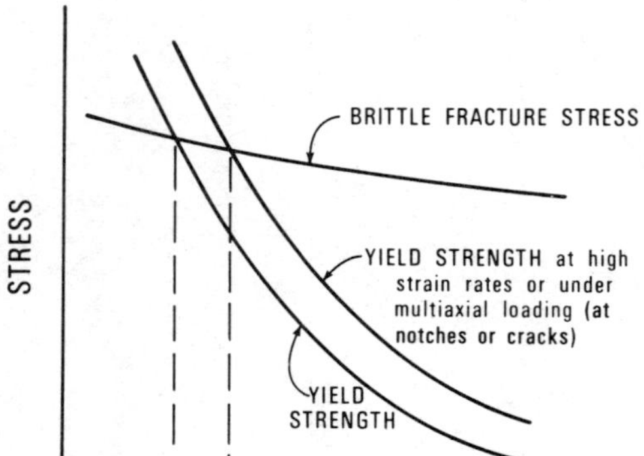

Figure 7.17 Postulated temperature dependences of ductile and brittle fracture stresses showing effects of high strain rate and stress concentration.

variability of the ductile–brittle transition temperature with material, strain rate, and other factors, but it cannot be verified experimentally, because under any set of experimental conditions only one of the two hypothesized fracture modes can be observed.

Ductile–brittle transitions occur in practically all b.c.c. materials, including iron, and in some h.c.p. materials. Face-centered cubic metals and alloys usually do not become brittle at low temperatures and have been considered less susceptible to brittle fracture than b.c.c. materials. Face-centered cubic metals that are precipitation hardened to very high strength levels may exhibit brittle behavior. Solution strengthening usually produces lower strength levels and more ductile behavior. Evidence for the toughness of copper and aluminum alloys has been reported by Reed and Mikesell (1967) and Kaufman and Johnson (1962). Details on the suitability of copper and aluminum alloys for use in toughness–sensitive applications may be found in Chapter 11, Structural Alloys. Fracture toughness has been found to vary inversely with yield strength within groups of closely related materials in several experimental studies (Wanhill, 1978; Hirth and Froes, 1977; Hahn and Rosenfield, 1975; Stout and Gerberich, 1978; Pellini, Goode, Puzak, Lange, and Huber, 1965; and Read and Reed, 1979, 1981). No quantitative theoretical explanation for this behavior has been found, although attempts have been made.

7.4.2 Fatigue

When materials are stressed repeatedly at loads below their ultimate tensile strengths, failure due to fatigue may eventually occur. Fatigue is discussed by Forrest (1966), Frost, Marsh, and Pook (1974), Kennedy (1963), and Sandor (1972). In standard texts, fatigue at low temperatures is treated briefly because fatigue strengths generally increase as the temperature is decreased and because of the large volume of room-temperature data available. Important exceptions are brittle materials, ductile or intermediate materials behaving in a brittle manner because of notches or stress concentrators, and ductile materials fatigued at constant strain amplitudes at high strains (low-cycle fatigue) (Nachtigall, Klima, and Freche, 1968). Fatigue testing of structural materials at low temperatures is done occasionally in cases where the adequacy of the fatigue resistance of a critical material is in question. Rotating superconducting machinery and pulsed superconducting magnets are examples of devices that require low-temperature fatigue resistance.

Fatigue testing. The two common methods of applying fatigue stresses for testing are direct stress and rotating–bending. In direct stress, axial loads are repeatedly applied. In rotating–bending, rotation of the specimen or of a test fixture causes a bending moment to be applied repeatedly. Few experiments involving rotating–bending have been performed at cryogenic temperatures. Specimens for direct-stress fatigue testing closely resemble those used for tensile testing, i.e., smooth round bars with grips on the ends. Hourglass-like shapes are sometimes used to prevent buckling in compression. The testing machine for fatigue experiments must be capable of rapid cycling to accomplish tests in a reasonable amount of time. Two basic variations are used in controlling the deformation: load control and strain control. In load-controlled fatigue testing, the minimum and maximum loads on the specimen are held constant during cycling, regardless of the strain values obtained; conversely, in strain-controlled fatigue testing, the minimum and maximum strain values are controlled, regardless of the load values required. The distinction between these two types of control is significant when stresses above the yield strength are used because the relationship between stress and strain becomes nonlinear.

Results of a series of fatigue tests may be expressed as an S–N curve, in which the number of cycles to failure is plotted on a logarithmic x-axis, and the stress range is plotted on a linear y-axis. The endurance limit is the stress at which this curve becomes horizontal, if it does. If the curve does not become horizontal, the endurance limit is defined as the stress required for failure at a specified arbitrary number of cycles, such as 10^8.

Fatigue at low temperatures. Fatigue of pure metals at low temperatures was studied by McCammon and Rosenberg (1957). They found that in copper, silver, gold, aluminum, cadmium, and magnesium, the fatigue strength (cyclic-stress amplitude required to fracture the specimen in 10^5 cycles) increased significantly with decreasing temperature, roughly as the ultimate tensile strength. In zinc and iron, which are subject to a ductile–brittle transition, the fatigue strength was difficult to measure below the transition, because at low stresses the material appeared to last indefinitely, whereas at high stresses, the material fractured in a brittle manner. The limited number of fatigue failures that were obtained in the midrange of stresses showed clearly that the fatigue strengths of these materials were decreased at temperatures below their ductile–brittle transitions. It was observed in this study that the increases observed in fatigue strength on cooling were due to increases in the resistance of the material to both crack initiation and crack growth.

Some fatigue data are available on structural materials at 4 K. Aluminum alloys were studied under load control by Schwartzberg, Kiefer, and Keys (1964). Aluminum and titanium alloys and steels were tested under strain control by Nachtigall et al. (1968) and Nachtigall (1975). At high strain levels fatigue resistance generally decreased with decreasing temperature; at low strain levels it increased.

The effect of the frequency of the applied stress on fatigue life is a difficult problem to study experimentally, although some work has been done at room temperature. The general conclusion from these studies is that increasing the frequency of stressing into the kilohertz range generally increases the endurance limit, with a maximum and subsequent decrease sometimes being observed. Very few data are available on this subject, which is not surprising in view of the experimental difficulties: accurate stress measurement over a wide frequency range requires unconventional techniques; heating from hysteresis increases with frequency; effects from corrosive environments may decrease with frequency. Changes in frequency, even well below the kilohertz range, can have pronounced effects on the rate of corrosion fatigue. In such situations, the fatigue life may decrease greatly with decreasing frequency. This can be a perplexing design problem because time for complete testing at low frequencies may not be available.

Cyclic strain at an amplitude of 1 to 2% and frequencies of about 0.3 Hz was found to produce heating in an AISI 304 specimen immersed in liquid helium (Shepic and Schwartzberg, 1978). This heating decreased at lower frequencies. It was implied that this frequency dependence of the specimen temperature, the cyclic stress–strain curve, and, presumably, the specimen fatigue resistance must be accounted for in 4-K fatigue testing.

7.4.3 Creep

Creep, that is, plastic deformation with time under constant load, is not significant in the design of most components for service at room and cryogenic temperatures. It could conceivably become important in the design of high-precision components for long service lives at high stresses (Wigley, 1971) or in components made from low-melting-point metals, such as indium O-rings and gaskets, and in polymers at and below room temperature. An investigation of creep in mercury and cadmium single crystals at liquid-helium temperatures was recently reported by Arko and Weertman (1969). It was established that creep is possible at liquid-helium temperatures, but that, contrary to the high-temperature behavior, stresses near the yield strength are required and creep rates decrease rapidly after loading.

7.5 ENGINEERING METALS AND ALLOYS

The behaviors of common engineering metals and alloys at low temperatures are described briefly to indicate the range of strength values produced by the various strengthening mechanisms in alloys based on several metallic elements. Because the mechanical properties of a given lot of material are determined by its own chemistry and processing, the data given below can only be exemplary. A detailed discussion of the relative advantages of candidate structural materials is presented in Chapter 11.

7.5.1 Copper and copper alloys

Room-temperature critical resolved shear strengths equivalent to a tensile yield strength of about 0.4 MPa (60 psi) have been reported for pure, well-annealed copper single crystals with very low dislocation densities (Young, 1968). This low value illustrates the fact that crystal-lattice defects in f.c.c. metals provide their strength; a material with practically no yield strength can be made by removing all dislocations, impurities, grain boundaries, and other strengthening lattice defects. Engineering materials are considerably stronger. More usual yield strength values for commercial purity copper are 20 to 60 MPa. Tensile yield and ultimate strengths as functions of temperature are shown in Fig. 7.18 for commercially pure (99.97% Cu) annealed copper; cold-drawn copper; admiralty brass, an alloy of copper with about 30 wt.% Zn, and a precipitation-hardened Cu–Be alloy. Behaviors typical of relatively pure, annealed f.c.c. metals as illustrated by 99.95% pure, annealed copper are: the low value of the yield strength at all temperatures, the weak dependence of yield strength on temperature, ultimate strength significantly larger than yield strength, and temperature dependence of the ultimate strength stronger than that of the yield strength. The large capacity of relatively pure f.c.c. metals for work hardening may be illustrated by a comparison of the behaviors of annealed and cold-drawn copper, also shown in Fig. 7.18. Cold drawing increases the yield strength to about 400 MPa and slightly increases the ultimate strength. The temperature dependence of the yield strength in the cold-drawn condition is weaker than in the annealed condition because work hardening increases the athermal component of the yield strength more than the thermal part in this material.

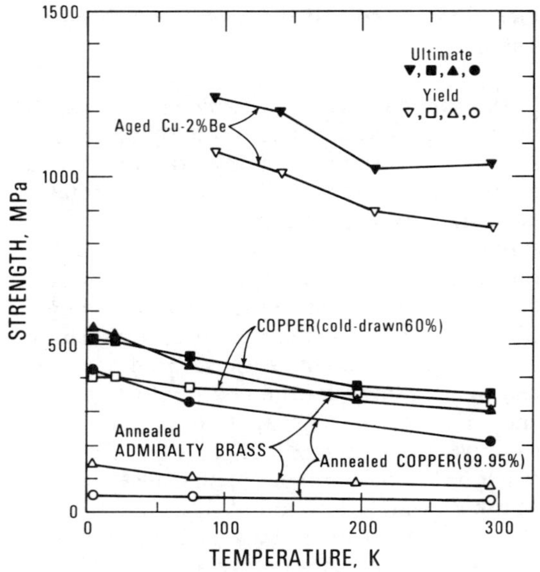

Figure 7.18 Yield and ultimate tensile strengths as a function of temperature for commercially pure, annealed, and cold-drawn copper (polycrystalline); annealed admiralty brass; and aged beryllium copper. Open symbols are yield strengths; closed symbols are ultimate tensile strengths (*Handbook on Materials for Superconducting Machinery*, 1977).

Up to about 33% of zinc is substitutionally soluble in copper. Above 33%, precipitates of the beta phase form, and ductility decreases at low temperatures. The f.c.c. solid solutions of zinc in copper are called the alpha brasses, the most useful class at low temperatures. Brass has been widely used for tubing and small machined parts in cryogenic equipment. The mechanical properties of a high-zinc alpha brass, admiralty brass, are compared with copper in Fig. 7.18. The brass has a higher yield strength by about a factor of 2 and a higher ultimate strength by about 40% over the range 0 to 300 K. The admiralty brass has a yield strength several times that shown in Fig. 7.8 for single crystal 70Cu–30Zn brass because of strengthening by grain boundaries, residual cold work, and additional chemical components. The temperature dependence of the yield strength of the commercial material is weaker than that of the single crystal because the additional strengthening mechanisms have raised the athermal part of the yield stress more than the thermal part. Most alpha brasses have sufficient toughness to prevent brittle fracture in low-temperature applications.

Alloys of copper with beryllium may be precipitation hardened. The strength properties of a typical alloy are also illustrated in Fig. 7.18. This graph shows the effectiveness of precipitation hardening in increasing strength. Usually, such dramatic increases in strength are accompanied by decreases in toughness, which limits the usefulness of some very strong materials in applications where toughness is important.

The low-temperature mechanical properties of copper and many copper alloys were compiled by Reed and Mikesell (1967).

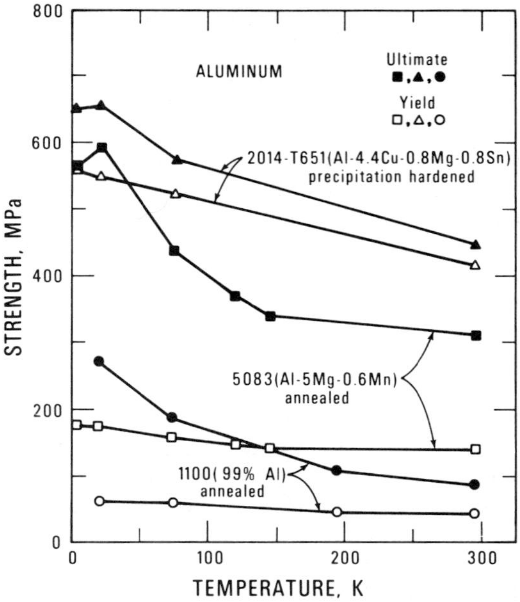

Figure 7.19 Yield and ultimate tensile strengths as a function of temperature for annealed, commercially pure, and solution-hardened aluminum and a precipitation-hardened alloy. Open symbols are yield strengths; closed symbols are ultimate tensile strengths (*Handbook on Materials for Superconducting Machinery*, 1977).

7.5.2 Aluminum and aluminum alloys

Aluminum and its alloys exhibit much the same behavior as copper and its alloys. Observed yield strengths range from 0.5 MPa (70 psi) or less for the purest research material (99.999% Al single crystal) to well over 500 MPa for precipitation-hardened materials. The yield and ultimate strengths of three commercial alloys are compared in Fig. 7.19. Annealed commercial purity aluminum, alloy designation 1100-0, has a nearly constant yield strength of approximately 60 MPa over the range 0 to 300 K, and significant work hardening is shown by the considerable difference between yield and ultimate strengths. A common solution-strengthened aluminum alloy in the annealed condition, aluminum 5083-0, has yield and ultimate strengths over twice that of the 1100-0 alloy. The yield strength of a precipitation-strengthened aluminum alloy, aluminum 2014, is about 10 times that of the commercially pure material, and its ultimate strength is close to its yield strength. For the annealed solid-solution alloys 1100-0 and 5083-0, the yield strengths have weak temperature dependences and the ultimate strengths have strong temperature dependences, behavior typical of f.c.c. materials. In contrast, the precipitation-hardened alloy has decreased capacity for work hardening. The difference of about a factor of 2 in the strength scales used for copper and aluminum alloys is explained by the difference of about the same factor in the shear moduli of these materials. Mechanical behavior of aluminum and dilute aluminum alloys has been reviewed by Reed (1972).

7.5.3 Austenitic stainless steels

Austenitic stainless steels are alloys of iron with chromium and enough nickel or manganese, or both, to ensure the stability of the f.c.c. (austenitic) crystal structure at room temperature. Other constituents, especially carbon and nitrogen, significantly affect the behavior of these steels and the stability of the austenite at low temperatures. The austenitic phase is metastable in most of these alloys and martensitic transformations may occur on cooling to cryogenic temperatures or during deformation at low temperatures. The transformation products influence the strength, toughness, and deformation behavior, as discussed in Chapters 9 and 11. Austenitic stainless steels are discussed extensively in the *Handbook of Stainless Steels* (1977).

The strength of annealed austenitic stainless steels is produced primarily by solution hardening. These steels work-harden significantly; therefore thermal and mechanical treatments significantly affect their mechanical properties.

Nitrogen additions significantly increase both the room- and low-temperature strengths of austenitic stainless steels. For example, increasing the nitrogen content 0.1% can double the 4-K yield strength of 19Cr–9Ni stainless steel (Read and Reed, 1979). Norstrom (1977) has shown that the nitrogen appears to have two separate effects. It provides solution hardening, which is especially effective in raising the low-temperature strength, and it significantly enhances the hardening effect of the grain boundaries, providing strength increases at moderately high as well as at low temperatures. This is an example of the complexity of strengthening mechanisms in practical materials.

Strengths of several versions of 19Cr–9Ni stainless steel (AISI type 304) are plotted against temperature in Fig. 7.20. It can be seen that substantial increases in yield strengths are obtainable by work hardening and by adding nitrogen. Stainless steel AISI 304N contains intentionally added nitrogen (0.12 wt.% for the material of Fig. 7.20), which greatly increases the low-temperature yield strength. The general level of strengths available is higher than for aluminum or copper alloys because of the higher shear modulus of stainless steel.

The strain-hardening behavior of austenitic stainless steels at low temperatures is influenced strongly by deformation-included martensitic transformations. The 4-K stress–strain curves of Guntner and Reed (1962) for AISI 304, which forms deformation-induced martensites at 4 K, and AISI 310, which does not, show that the strain-hardening coefficient of the AISI 304 is more than 5 times larger than that of AISI 310 at strains of 15 to 30%.

Fatigue test results for AISI 304L (low carbon) and 304N stainless steel are shown in Fig. 7.21 (Nachtigall, 1975; Shepic and Schwartzberg, 1978). Note that the fatigue resistance at liquid-nitrogen and liquid-helium temperatures is equal or slightly better than that at room temperature. A slight extrapolation of Nachtigall's results to 10^6 cycles indicates that the cyclic stress range required for failure at 10^6 cycles is of

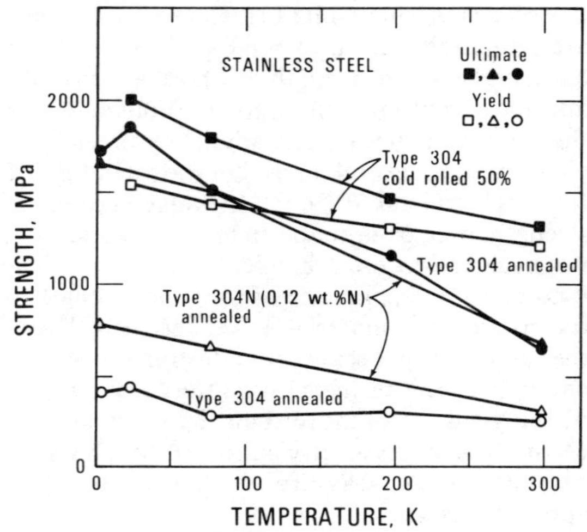

Figure 7.20 Yield and ultimate tensile strengths as a function of temperature for annealed, nitrogen-strengthened, and cold-drawn stainless steel 304. Open symbols indicate yield strengths; closed symbols indicate ultimate tensile strengths (*Handbook on Materials for Superconducting Machinery*, 1977; Read and Reed, 1979).

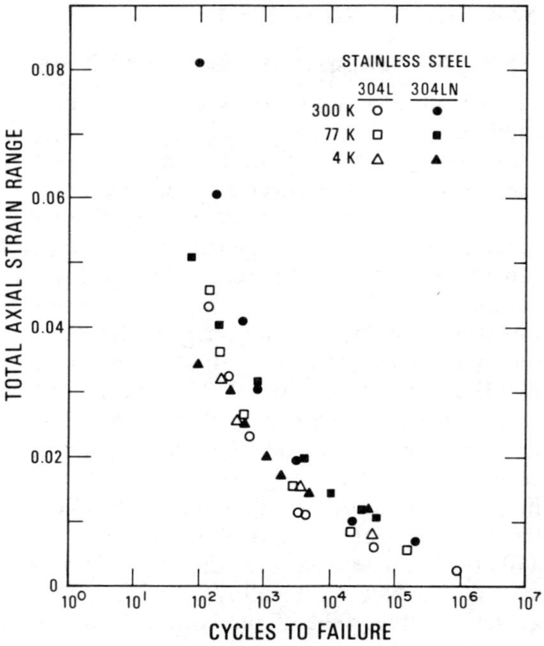

Figure 7.21 Fatigue test results for stainless steels 304L and 304LN at 300, 78, and 4 K (Nachtigall, 1975; Shepic and Schwartzberg, 1978).

the order of half the ultimate tensile stress, for fatigue at zero mean stress. Total stress ranges of around 800 MPa (116 ksi) were required to produce failure of Nachtigall's stainless steel 304L specimens after 10^6 cycles.

7.5.4 Titanium and titanium alloys

Titanium alloys, used in aerospace applications because of their low density and high strength at high temperatures, are also useful at low temperatures because of their high strength and low thermal conductivity. The toughness of some titanium alloys decreases severely at temperatures in the neighborhood of 77 K and below, limiting their usefulness at lower temperatures. Strengths of some commercial titanium alloys are shown in Fig. 7.22.

The crystal structure of titanium alloys is variable through their chemical compositions and thermal and mechanical treatments. Titanium-based materials potentially useful for cryogenic structures include unalloyed titanium, Ti–5Al–2.5Sn, and Ti–6Al–4V. Unalloyed titanium and Ti–5Al–2.5Sn have the h.c.p. crystal structure (referred to as the α phase in this context), and Ti–6Al–4V contains precipitates of the b.c.c. structure (Ti–6Al–4V is said to contain α and β phases). Interstitial carbon, oxygen, and nitrogen in small amounts can severely embrittle titanium alloys at low temperatures (Shannon and Brown, 1968; Tobler, 1976b). The annealed commercial titanium alloys have only slight work-hardening capacity compared with annealed austenitic stainless steels. Titanium alloy ultimate strengths range from about 1.25 times the yield strength for commercially pure titanium at room temperature to 1.03 times the yield strength for precipitation-hardened titanium alloy (Ti–6Al–4V) at 4 K; their ultimate strengths have slightly weaker temperature dependences than the yield strengths. The high, strongly temperature-dependent yield strengths of these materials and the relatively small and constant differences between their yield and ultimate strengths are due to their mainly h.c.p. crystal structure. More highly alloyed materials, with a higher β phase (b.c.c.) content, exhibit these same behaviors.

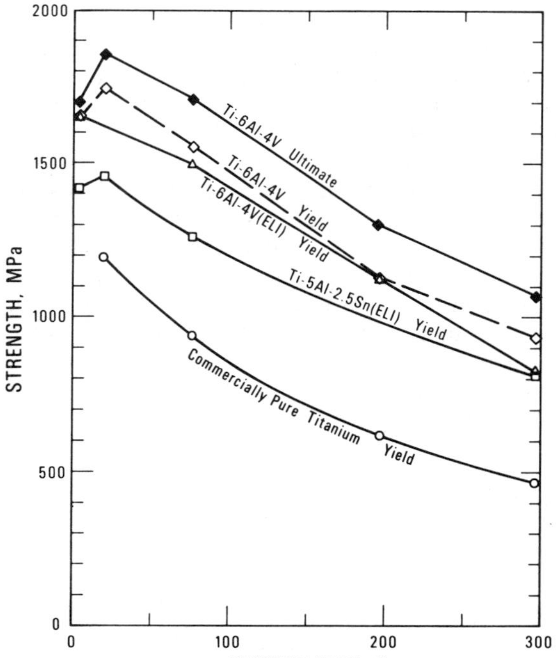

Figure 7.22 Yield strengths as a function of temperature for titanium, Ti-5Al-2.5Sn (ELI), Ti-6Al-4V (ELI), and Ti-6Al-4V, and ultimate strength of Ti-6Al-4V (*Handbook on Materials for Superconducting Machinery*, 1977, and Tobler, 1976a).

7.5.5 Iron and b.c.c. iron alloys

Iron is by far the most widely used metallic element in structural metals and alloys, but the low-temperature ductile–brittle transition (see § 7.4.1) of many ferrous alloys limits their usefulness at low temperatures. Typical b.c.c. behavior is demonstrated most clearly by the pure iron and carbon steels. The yield strength is high even in pure iron and has a strong temperature dependence (Fig. 7.23). The ultimate strength is only slightly above the yield strength at room temperature and approaches the yield strength with decreasing temperature. The tensile properties of the carbon steel give fair warning of its low toughness at low temperatures: the yield and ultimate strengths are practically identical and the elongation and reduction of area are low. Addition of nickel lowers the ductile–brittle temperature and increases the elongation at low temperatures (Fig. 7.24), making Fe–9Ni, for example, useful at temperatures as low as 77 K, if it is properly processed. The addition of nickel (f.c.c.) to iron (b.c.c.) increases the austenite stability, providing some retention of austenite at lower temperatures. The presence of the austenite at lower temperatures is thought to contribute to greater ductility, effectively serving to lower the ductile–brittle transition temperature.

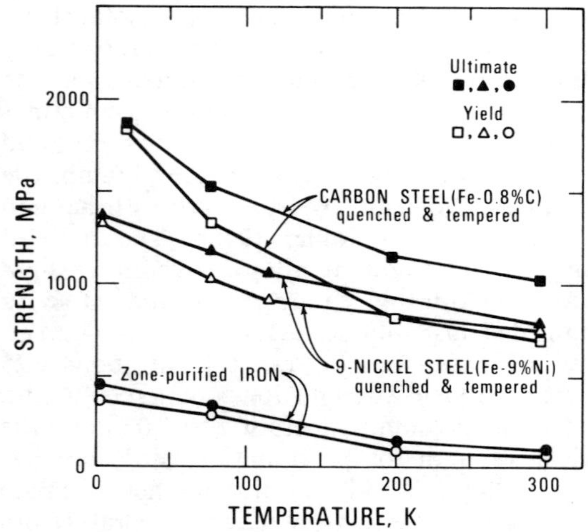

Figure 7.23 Yield and ultimate tensile strengths as a function of temperature for zone-purified iron, 9-nickel steel, and carbon steel. Open symbols indicate yield strengths; closed symbols indicate ultimate strengths (Smith and Rutherford, 1957; Tobler, 1976a; Warren and Reed, 1963).

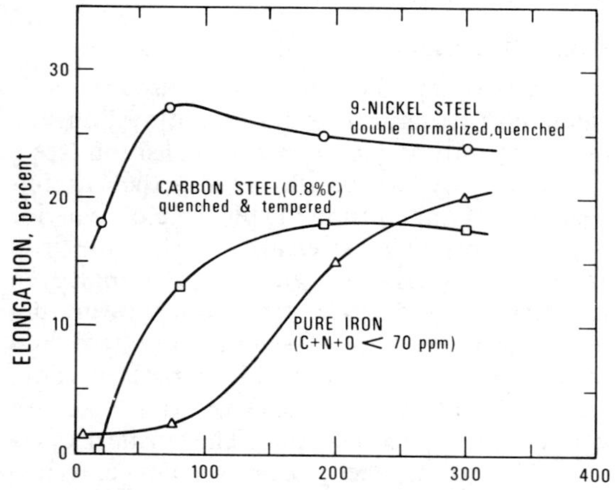

Figure 7.24 Elongation to fracture for zone-purified iron (elongation only is shown), 9-nickel steel, and carbon steel plotted against temperature (Smith and Rutherford, 1957; Tobler, 1976a; Warren and Reed, 1963).

7.6 POLYMERS AND GLASSES

7.6.1 Polymers

Polymers exhibit strikingly varied behavior depending on the temperature and the time scale of the observation. A classic simple polymer observed for common experimental times of the order of 1 s to 1 h has a glassy region, a rubbery region, and a viscofluid region as the temperature is raised. These behaviors are demonstrated by loading a polymer specimen to a given stress, measuring the magnitude of the deformation produced after a fixed interval of time, and plotting deformation as a function of temperature, as shown in Fig. 7.25 for an amorphous polymer. The term *glassy* indicates the similarity observed between the properties of polymers at low temperatures and those of glass. As the temperature is raised, the polymer becomes less brittle, and its strength and moduli begin to drop steadily and significantly. Between the glassy and rubbery regions, the polymer passes through a leathery region. Throughout the leathery region, the polymer exhibits some ductility and toughness, along with rapidly decreasing strength and moduli. Eventually the behavior stabilizes in the rubbery region, where the polymer can be stretched severely and will eventually return almost to its original size. The deformation in the rubbery region is strongly suppressed in a crystalline polymer (discussed below).

As the temperature is raised further, the polymer eventually becomes a viscous liquid, then less viscous, and then melts completely.

The glass transition temperature, T_g, is the approximate upper limit of the glassy region. It is usually determined experimentally by a thermal expansion technique. Commonly observed values of T_g are in the range 220 to 370 K. Glassy mechanical behavior is not restricted to temperatures below T_g. High values of the elastic shear modulus can be observed at temperatures above T_g if stresses and strains are applied and measured within short times. Significant plastic deformation can be obtained in tensile stress-strain experiments below T_g, even at conventional rates of deformation (see "Amorphous and crystalline polymers," below).

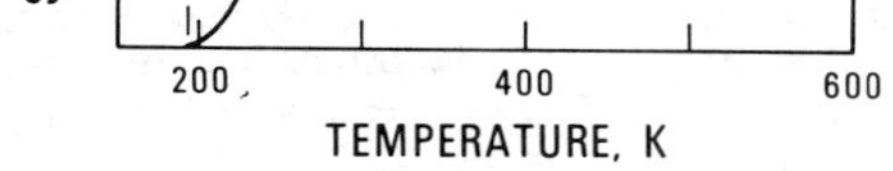

Figure 7.25 Thermomechanical curve of a simple polymer.

In polymers, a distinction must be made between nonlinear deformation, that is, lack of a linear relationship between stress and strain; viscoelastic behavior, in which strain increases with time under a constant stress; and irreversible plastic flow. Nonlinear viscoelastic strains can be almost completely recovered in the rubbery behavior region. But in the glassy region, a nearly linear portion of the stress–strain curve is followed by nonrecoverable strain at nearly constant load. Just below T_g, strains of 300 to 1000% may be obtained (Ward, 1971). Well below T_g, rupture occurs at low strains (~10%).

Recoverable deformation in polymers occurs by the mechanism of thermally activated reorientation of molecular chain segments toward the stress axis (Meares, 1967). Nonrecoverable deformation has contributions from this mechanism and additional contributions from molecular chains sliding past one another. This sliding is made possible by the severing of some molecular links.

An extensive compilation of the mechanical properties of some polymers including polytetrafluoroethylene (TFE), polyethylene terephthalate (PET), polycarbonate (PC), and four others, has been prepared by Schramm, Clark, and Reed (1973). The low-temperature strengths of polymers are well below those of the usual structural metals. The temperature dependences of their tensile strength and elongation are usually significant, as in the b.c.c. metals. The tensile strengths of polychlorotrifluoroethylene (CTFE), PET, PC, TFE, and an epoxy resin (NASA Resin 2) are shown in Fig. 7.26.

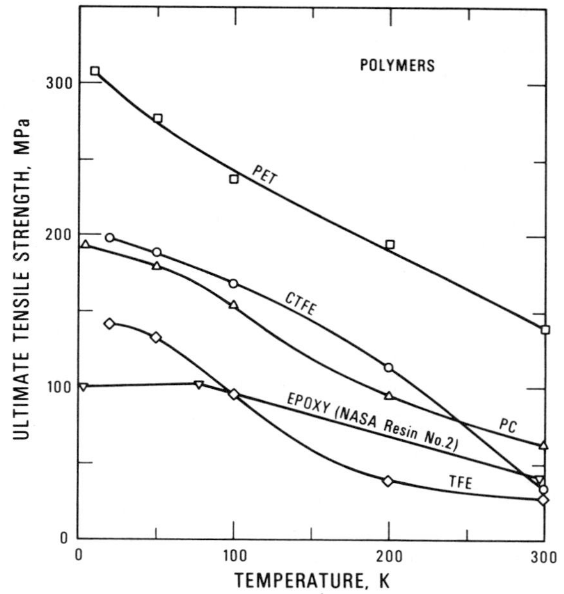

Figure 7.26 Ultimate tensile strengths as a function of temperature for CTFE, PET, PC (Schramm et al., 1973), and NASA Resin 2 (Soffer and Molho, 1967).

Amorphous and crystalline polymers. Polymers exist in amorphous and crystalline states. The crystalline state in a polymer means that the molecules occupy positions of three-dimensional order. The degree of crystallinity of a specimen of a crystallizable polymer is determined by the time–temperature path followed on cooling from the molten state. In some cases, minute single crystals have been obtained by crystallization from very dilute solutions. In PET, stretching an amorphous specimen produces small crystallites oriented along the stress axis.

Crystallinity in polymers has profound effects on their mechanical properties. Deformability in the glassy temperature region is suppressed to such a degree that they may become quite brittle below T_g, in contrast to amorphous polymers, which generally can sustain some forced deformation at and slightly below T_g. The strain produced by a given stress level in the leathery and rubbery temperature regions is likely to be much lower in the crystalline state than in the amorphous state. Thus, increased crystallinity in a polymer increases its strength, but also promotes brittle behavior at temperatures below T_g.

A dramatic instance of the effect of increased crystallinity on strength has been reported by Armeniades, Kuriyama, Roe, and Baer (1968). They found that the tensile strength of 54% crystalline PET film was more than double that of amorphous PET film over the temperature range 110 to 300 K. Even larger differences were reported below 110 K. Differences in Young's modulus of more than a factor of 10 between natural rubber specimens of about 5 and 20% crystallinity are described by Meares (1967).

Crazing. Crazes, which are opaque striations in planes normal to the tensile stress, form in some amorphous polymers, notably polymethylmethacrylate and polystyrene during plastic deformation in the brittle temperature range (Ward, 1971). Crazes can also be formed by chemical action. Research into the structure of crazes has shown them to be regions of lower refractive index and lower density than the surrounding material (Berry, 1972). These property changes are due to microscopic voids. It was found that crazes in PMMA contained 50 to 55% by volume of free space, presumably in the form of interconnecting voids. Crazes are sharply bounded; their thickness is much less than their other dimensions. They have significant load-bearing capacity, although they do not have the strength of the surrounding material.

Crazes and the submicroscopic flaws from which they grow control the tensile strength of polymers in the range of brittle behavior, which is a large portion of the cryogenic temperature range for most polymers. The existence of these intrinsic, so far undetectable, flaws is demonstrated by measurements of the tensile strength of specimens with externally introduced flaws of varying sizes. The data fit the Griffith theory of brittle fracture (Chapter 8) for externally introduced flaws down to a certain size, a_0. Smaller, externally introduced flaws have no effect on the strength, indicating the existence of intrinsic flaws of size a_0, which control the tensile strength of the polymer (Berry, 1972). At sufficiently high stresses, these invisible intrinsic flaws become visible crazes; fracture then follows, initiating at one of the crazes. In a study on polymethylmethacrylate (PMMA), the crazing was found to be much more severe and the effective intrinsic flaw size over twice as large at liquid-nitrogen temperature than at room temperature. Data on the relationship between crazing, flaw size, and tensile strength for other polymers are less conclusive, but similar effects to those observed in PMMA cannot be ruled out (Berry, 1972).

7.6.2 Glass

"Glass can be defined as an inorganic product of fusion which has been cooled to a rigid condition without crystallization" (Jastrzebski, 1959). Glasses have an amorphous structure characteristic of a supercooled liquid made up of a three-dimensional network of primary units, which have local order but no long-range order (Zarzycki, 1968). The lack of long-range crystalline order in glasses precludes the formation and motion of dislocations; therefore plastic deformation by dislocation motion is absent. At high temperatures (800 K and above), ordinary glasses behave as viscous liquids, flowing gradually under applied stress (Babcock, 1977). At room temperature and below, however, the viscosity of glass is so high that it behaves as a brittle solid, because no mechanism of plastic deformation on a practical time scale is available.

The theoretical strengths of glasses are of the order of 10 GPa (Phillips, 1972). These strengths have been achieved in practice in small samples under carefully controlled laboratory conditions. In most standard tests of a group of glass specimens, the failure stresses have been found to be widely distributed, with the minimum found to be of the order of 35 MPa (Hillig, 1961). This is due to the presence of stress concentrators on the surfaces of virtually all bulk glasses. The inability of glass to deform plastically prevents relief of the stresses around such flaws, and brittle fracture results. Glass is quite strong in compression, and the cause of specimen failure in compression tests is believed to be a tensile component of the stress that could not be eliminated by the experimental arrangement.

Kropschot and Mikesell (1957) and Hillig (1961) report increases in the strength of glasses on cooling from room to liquid-nitrogen temperature, but no further improvement was observed in tests conducted at 20 K. Like the room-temperature strength, the practical low-temperature strength of glass is far lower than the theoretical value because of fracture initiation at surface flaws.

Fatigue damage in the conventional sense does not occur in glass, but the application of a constant stress for a long time may result in failure. This is known as static fatigue and is believed to be caused by the growth of preexisting flaws to critical dimensions under the action of a steady stress. Fluctuating stress may also produce flaw growth at about the same rate as an equivalent static stress.

Glass may be fractured by thermal stresses because of its low thermal conductivity, which results in nonuniform contraction and, therefore, internal stresses under rapid cooling. This liability can be mitigated by specifically designing the chemical composition of the glass to produce low thermal contraction.

7.7 SOURCES OF DATA

Many compilations and handbooks of low-temperature mechanical properties exist. Some examples are the *Handbook of Materials for Superconducting Machinery* (1977) maintained by the Metals and Ceramics Information Center, Battelle Columbus Laboratories, and the handbook entitled *LNG Materials and Fluids* (1977) by the National Bureau of Standards. *Mechanical Properties of Materials at Low Temperatures* by Wigley (1971) contains property information, guidance on material selection, and many references.

Proceedings of topical conferences on cryogenic materials are useful data sources. These include the biennial publication *Advances in Cryogenic Engineering*, now in its twenty-eighth volume, which contains proceedings from the Cryogenic Engineering Conference and the International Cryogenic Materials Conference. The International Cryogenic Materials Conference has recently started a new series of conference proceedings, the Cryogenic Materials series, beginning with *Nonmetallic Materials and Composites at Low Temperatures* (1979). The American Society for Testing and Materials in Philadelphia, Pennsylvania, has sponsored recent conferences on cryogenic materials and published the proceedings as ASTM special technical publications, *Fatigue and Fracture Toughness—Cryogenic Behavior* (1974) and *Properties of Materials for Liquefied Natural Gas Tankage* (1975).

The *Metals Abstracts* indexes data on the mechanical properties of metals and alloys from journals and reports. Wohlbier's annual mechanical properties volumes (Wohlbier, 1974; 1975; 1976) provide useful summaries of current mechanical property literature. A critical survey of data sources on mechanical properties of metals has been compiled by Gavert, Moore, and Westbrook (1974).

References that provide background material on and more detailed discussions of topics covered in this chapter are listed in Table 7.2.

7.8 AUTHOR'S AFTERWORD

The mechanical behaviors of metals at low temperatures are generally well understood on physical grounds, measurement techniques are adequate to determine the necessary material properties, and a substantial amount of test data on practical materials has been accumulated. Opportunities for new knowledge and practical advantages remain. Solution softening in b.c.c. materials and low-temperature anomalies of the temperature dependence of the flow strengths of some f.c.c. materials are not well understood. The often-observed inverse correlation between strength and toughness lacks a quantitative explanation. Stronger and cheaper welds with adequate toughness for low-temperature service are needed. More convenient and economical low-temperature test techniques, especially for toughness, would be useful.

Table 7.2 References on background material for low-temperature mechanical properties of materials.

§ 7.2	Temperature dependence of strengthening and flow
	Arsenault (1975)
	Conrad (1960, 1964)
	Cottrell (1956)
	Li (1968)
	McLean (1962)
	Nix and Menezes (1970)
	Peckner (1964)
	Reed-Hill (1964)
	Tegart (1966)
	Tien and Ansell (1966)
	Weertman and Weertman (1965)
§ 7.2.1	Basic mechanism of plastic flow in crystalline solids
	Friedel (1967)
	Gilman and Johnston (1967)
	Johnston and Gilman (1959)
	Nabarro (1967)
	Read (1953)
§ 7.2.3	Crystal structure effects
	Christian (1970)
	Hirsch (1968)
	Soo and Galligan (1969)
	Stein and Low (1966)
§ 7.2.5	Solution strengthening
	Haasen (1965)
	Haasen (1968)
	Fleischer and Hibbard (1963)
§ 7.2.6	Athermal strengthening by grain boundaries and precipitates
	Brown and Ham (1971)
§ 7.3.3	Martensitic transformations
	Reed (1962)
§ 7.3.5	Yield points
	Hahn (1962)
§ 7.4.2	Fatigue
	Forrest (1966)
§ 7.6.1	Polymers
	Argon (1973)
	Billmeyer (1962)
	Bueche (1962)
	Meares (1967)

The austenitic stainless steels have been used for years where strength and toughness at low temperatures are required. These materials have been chosen recently for use in the Mirror Fusion Test Facility at Lawrence Livermore Laboratory and in two of the three United States–manufactured magnets for the Large Coil Project at Oak Ridge National Laboratory. The strength of the austenitic stainless steels can be significantly increased by adding nitrogen, although this decreases the toughness; thus an alloy can be custom-made for an application requiring a specific strength level. Possible improvements in the characteristics of these alloys through manganese additions are currently under study. A serviceable material for most low-temperature structural applications can be found among the austenitic stainless steels.

Because cryogenic engineering on the structural size scale of the current fusion magnets is a new field, a service history for austenitic stainless steels needs to be established. Construction materials should be carefully characterized; structures should be thoroughly instrumented; and detailed performance records should be kept and made generally available to designers. Only through such efforts will the durability and economy of future cryogenic structures be ensured.

7.9 REFERENCES

Alers, G. A., Buck, O., and Tittmann, B. R. (1969). Measurements of plastic flow in superconductors and the electron–dislocation interaction. *Phys. Rev. Lett.* **23**, 290–293.

Annual Book of ASTM Standards (1976). American Society for Testing and Materials, Philadelphia, Pennsylvania, 765–776.

Argon, A. S. (1973). A theory for the low-temperature plastic deformation of glassy polymers. *Philos. Mag.* **28**, 839–865.

Arko, A. C. and Weertman, J. (1969). Creep deformation of Cd and Hg at liquid helium temperatures. *Acta Metall.* **17**, 687–699.

Armeniades, C. D., Kuriyama, I., Rowe, J. M., and Baer, E. (1968). Mechanical behavior of poly(ethylene terephthalate) at cryogenic temperatures. In *Cryogenic Properties of Polymers* (eds. T. T. Serafini and J. L. Koenigh). Dekker, New York, 155–169.

Arsenault, R. J. (1975). Low temperature deformation of bcc metals and their solid-solution alloys. In *Treatise on Materials Science and Technology*, Vol. 6 (ed. R. J. Arsenault). Academic Press, New York, 1–99.

Ashby, M. F. (1972). A first report on deformation-mechanism maps. *Acta Metall.* **20**, 887–897.

Babcock, C. L. (1977). *Silicate Glass Technology Methods.* John Wiley & Sons, New York.

Basinski, Z. S. (1957). The instability of plastic flow of metals at very low temperatures. *Proc. R. Soc. Lond. A* **240**, 229–242.

Basinski, Z. S. (1959). Thermally activated glide in face-centered cubic metals and its application to the theory of strain hardening. *Philos. Mag.* **4**, 393–432.

Basinski, Z. S. (1960). *Aust. J. Phys.* **13**, 354.

Basinski, Z. S. and Christian, J. W. (1960). The influence of temperature and strain rate on the flow stress of annealed and decarburized iron at subatmospheric temperatures. *Aust. J. Phys.* **13**, 299–308.

Basinski, Z. S., Foxall, R. A., and Pascual, R. (1972). Stress equivalence of solution hardening. *Scr. Metall.* **6**, 807–814.

Berry, J. P. (1972). Fracture of polymeric glasses. In *Fracture*, Vol. 7 (ed. H. Liebowitz). Academic Press, New York, 37–92.

Billmeyer, F. W., Jr. (1962). *Textbook of Polymer Science.* Interscience, New York.

Blewitt, T. H., Coltman, R. R., and Redman, J. K. (1957). Low temperature deformation of copper single crystals. *J. Appl. Phys.* **28**, 651–660.

Brown, L. M. and Ham, R. K. (1971). Dislocation-particle interactions. In *Strengthening Mechanisms in Crystals* (eds. A. Kelly and R. B. Nicholson). Halstead, Essex, England, 9–135.

Bueche, F. (1962). *Physical Properties of Polymers.* Interscience, New York, 110.

Cahn, R. W. (1954). Twinned crystals. *Adv. Phys.* **3**, 363–445.

Carreker, R. P., Jr. and Hibbard, W. R., Jr. (1953). Tensile deformation of high-purity copper as a function of temperature, strain rate, and grain size. *Acta Metall.* 1, 654–663.

Christian, J. W. (1970). Plastic deformation of bcc metals. In *Second International Conference on the Strength of Metals and Alloys*, Vol. I. American Society for Metals, Metals Park, Ohio, 31–70.

Churchman, A. T. and Cottrell, A. H. (1951). Yield phenomenon and twinning in α-iron. *Nature* **167**, 943–945.

Clarebrough, L. M. and Hargreaves, M. E. (1959). Work hardening of metals. In *Progress in Metal Physics*, Vol. 8 (eds. B. Chalmers and R. King). Pergamon, New York, 1–103.

Conrad, H. (1960). Effect of temperature on yield and flow stress of bcc metals. *Philos. Mag.* **5**, 745–751.

Conrad, H. (1964). Thermally activated deformation of metals. *J. Met.* **16**, 582–588.

Conrad, H. (1965). Cryogenic properties of metals. In *High Strength Materials* (ed. V. F. Zackay). John Wiley & Sons, New York, 436–505.

Cottrell, A. H. (1956). *Dislocations and Plastic Flow in Crystals.* Clarendon, Oxford, England, 9–10.

Cottrell, A. H. and Stokes, R. J. (1955). Effects of temperature on the plastic properties of aluminum crystals. *Proc. R. Soc. Lond. A.* **233**, 17–34.

Deformation Twinning (1964) (eds. R. E. Reed-Hill, J. P. Hirth, and H. C. Rogers). Gordon & Breach, New York.

de Meester, B., Doner, M., and Conrad, H. (1975). Deformation kinetics of the Ti–6Al–4V alloy at low temperatures. *Metall. Trans. A,* **6**, 65–75.

Dieter, G. E., Jr. (1961). *Mechanical Metallurgy.* McGraw-Hill, New York.

Dieter, G. E., Jr. (1976). *Mechanical Metallurgy*, 2nd edition. McGraw-Hill, New York, 195.

Dorn, J. E. and Rajnak, S. (1964). Nucleation of kink pairs and the Peierls mechanism of plastic deformation. *Trans. Metall. Soc. AIME* **230**, 1052–1064.

Erickson, J. S. and Low, J. R. (1957). The yield-stress temperature relation for iron at low temperature. *Acta Metall.* **5**, 405–406.

Evans, K. R. (1974). Solid solution strengthening of face-centered cubic alloys. In *Treatise on Materials Science and Technology*, Vol. 4 (ed. H. Herman). Academic Press, New York.

Fatigue and Fracture Toughness — Cryogenic Behavior (1974) (eds. C. F. Hickey and R. G. Broadwell). ASTM STP 556, American Society for Testing and Materials, Philadelphia, Pennsylvania.

Fleischer, R. L., and Hibbard, W. R., Jr. (1963). Solution hardening. In *Proceedings of the Conference on the Relation between the Structure and Mechanical Properties of Metals*, Vol. I. Her Majesty's Stationery Office, London, 262–297.

Forrest, P. G. (1966). *Fatigue of Metals*. Pergamon, Oxford, England.

Friedel, J. (1967). *Dislocations*. Pergamon, Oxford, England.

Frost, N. E., Marsh, K. J., and Pook, L. P. (1974). *Metal Fatigue*. Clarendon, Oxford, England.

Galligan, J. M., Lin, T. H., and Pang, C. S. (1977). Electron–dislocation interaction in copper. *Phys. Rev. Lett.* **38**, 405–407.

Gavert, R. B., Moore, R. L., and Westbrook, J. H. (1974). *Critical Surveys of Data Sources: Mechanical Properties of Metals*. Special Publication 396–1, National Bureau of Standards, Washington, D.C.

Gilman, J. J. and Johnston, W. G. (1957). The origin and growth of glide bands in lithium fluoride crystals. In *Dislocations and Mechanical Properties of Crystals* (eds. J. C. Fisher, W. G. Johnston, R. Thomson, and T. Vreeland, Jr.). John Wiley & Sons, New York, 116–163.

Granato, A. V. (1971a). Dislocation inertial model for the increased plasticity of the superconducting state. *Phys. Rev. Lett.* **27**, 660–664.

Granato, A. V. (1971b). Dislocation inertial effects in the plasticity of superconductors. *Phys. Rev. B* **4**, 2196–2201.

Gross, J. G. (1970). Effect of strength and thickness on notch ductility. In *Impact Testing of Materials*, ASTM STP 466, American Society for Testing and Materials, Philadelphia, Pennsylvania, 21–52.

Guntner, C. J. and Reed, R. P. (1962). The effect of experimental variables including the martensitic transformation on the low-temperature mechanical properties of austenitic stainless steels. *ASM Trans. Q.* **55**, 399–419.

Haasen, P. (1964). Mechanical properties of solid solutions and intermetallic compounds. In *Physical Metallurgy* (ed. R. W. Cahn). North-Holland, Amsterdam, 821–885.

Haasen, P. (1968). Our present understanding of the strength of solid solutions. In *Proceedings of the International Conference on the Strength of Metals and Alloys*, supplement to *Trans. Jap. Inst. Met.* **9**, XL.

Haasen, P. and Kelley, A. (1957). A yield phenomenon in face-centered-cubic single crystals. *Acta Metall.* **5**, 192–199.

Hahn, G. T. (1962). A model for yielding with special reference to the yield-point phenomena of iron and related bcc metals. *Acta Metall.* **10**, 727–738.

Hahn, G. T. and Rosenfield, A. R. (1975). Metallurgical factors affecting fracture toughness of aluminum alloys. *Metall. Trans.* **6A**, 653–670.

Hall, E. O. (1954). *Twinning and Diffusionless Transformations in Metals*. Butterworths, London.

Handbook of Stainless Steels (1977) (eds. D. Peckner and I. M. Bernstein). McGraw-Hill, New York.

Handbook on Materials for Superconducting Machinery (1977). Metals and Ceramics Information Center, Battelle Columbus Laboratories, Columbus, Ohio.

Hillig, W. B. (1961). Strength of bulk fused quartz. *J. Appl. Phys.* **32**, 741.

Hirsch, P. B. (1968). Some aspects of the deformation of body-centered cubic metals. In *Proceedings International Conference on Strength of Metals and Alloys*, published as a supplement to *Trans. Jap. Inst. Met.* **9**, XXX–XXXIX.

Hirth, J. P. and Froes, F. H. (1977). Interactions between fracture toughness and other mechanical properties in titanium alloys. *Metall. Trans.* **8A**, 1165–1176.

Jastrzebski, Z. D. (1959). *Nature and Properties of Engineering Materials*. John Wiley & Sons, New York, 307.

Johnston, W. G. and Gilman, J. J. (1959). Dislocation velocities, dislocation densities, and plastic flow in lithium fluoride crystals. *J. Appl. Phys.* **30**, 129–144.

Kaufman, J. G. and Johnson, E. W. (1962). Notch sensitivity of aluminum alloy sheet and plate at −320° F based upon notch-yield ratio. In *Advances in Cryogenic Engineering*, Vol. 8. Plenum Press, New York, 678–685.

Kennedy, A. J. (1963). *Processes of Creep and Fatigue of Metals*. John Wiley & Sons, New York.

Knott, J. F. (1973). *Fundamentals of Fracture Mechanics*. Butterworths, London.

Kochs, V. F., Argon, A. S., and Ashby, M. F. (1975). *Thermodynamics and Kinetics of Slip, Progress in Materials Science*, Vol. 19 (eds. B. Chalmers, J. W. Christian, and T. B. Massalski). Pergamon, New York.

Kojima, H. and Suzuki, T. (1968). Electron drag and flow stress in niobium and lead at 4.2 K. *Phys. Rev. Lett.* **21**, 896–898.

Kostorz, G. (1973). The influence of the superconducting phase transition on the plastic properties of metals and alloys, *Phys. Status Solidi B* **58**, 9–42.

Kropschot, R. H. and Mikesell, R. P. (1957). Strength and fatigue of glass at very low temperatures. *J. Appl. Phys.* **28**, 610–614.

Li, J. C. M. (1968). Kinetics and dynamics in dislocation plasticity. In *Dislocation Dynamics* (eds. A. R. Rosenfield, G. T. Hahn, A. L. Bement, and R. I. Jaffee). McGraw-Hill, New York, 87–116.

LNG Materials and Fluids (1977). Thermophysical Properties Division, National Bureau of Standards, Boulder, Colorado.

McCammon, R. D. and Rosenberg, H. M. (1957). The fatigue and ultimate tensile strengths of metals between 4.2 and 293 K. *Proc. R. Soc. Lond. A* **242**, 203–211.

McLean, D. (1962). *Mechanical Properties of Metals*. John Wiley & Sons, New York.

Meares, P. (1967). *Polymers: Structure and Bulk Properties*. Van Nostrand, London.

Mitchell, T. E. and Thornton, P. R. (1963). The work-hardening characteristics of Cu and α-brass single crystals between 4.2 and 500 K. *Philos. Mag.* **8**, 1127.

Nabarro, F. R. N. (1967). *Theory of Crystal Dislocations*. Clarendon, Oxford, England.

Nachtigall, A. J. (1975). Strain-cycling fatigue behavior of ten structural materials tested at liquid helium, liquid nitrogen, and ambient air. In *Properties of Materials for Liquefied Natural Gas Tankage*. ASTM STP 579, American Society for Testing and Materials, Philadelphia, Pennsylvania.

Nachtigall, A. J., Klima, S. J., and Freche, J. C.

(1968). Fatigue behavior of rocket engine materials to −452° F (4 K). *J. Mater.* **3**, 425–443.

Nadai, A. (1950). *Theory of Flow and Fracture of Solids*. McGraw-Hill, New York.

Neuber, H. (1946). *Theory of Notch Stresses*. J. W. Edwards, Ann Arbor, Michigan.

Nix, W. D. and Menezes, R. A. (1970). Physics of strengthening mechanisms in crystalline solids. In *Annual Review of Materials Science* (ed. R. A. Huggins). Annual Reviews, Palo Alto, California, 313–346.

Nonmetallic Materials and Composites at Low Temperatures (1979) (eds. A. F. Clark and R. P. Reed). Plenum Press, New York.

Norstrom, L. A. (1977). The influence of nitrogen and grain size on yield strength in type AISI 316L austenitic stainless steel. *Met. Sci. J.* **11**, 208–212.

Peckner, D. (1964). *The Strengthening of Metals*. Reinhold, New York.

Pellini, W. S., Goode, R. J., Puzak, P. P., Lange, E. A., and Huber, R. W. (1965). *Review of Concepts and Status of Procedures for Fracture-Safe Design of Complex Welded Structures Involving Metals of Low to Ultra-high Strength Levels*. U.S. Naval Research Laboratory Report 6300, Washington, D.C.

Peterson, R. E. (1973). *Stress Concentration Design Factors*. John Wiley & Sons, New York.

Phillips, C. J. (1972). Fracture of glass. In *Fracture*, Vol. 7 (ed. H. Liebowitz). Academic Press, New York, 1–35.

Properties of Materials for Liquefied Natural Gas Tankage (1975) (ed. J. G. Kaufman). ASTM STP 579, American Society for Testing and Materials, Philadelphia, Pennsylvania.

Pustovalov, V. V., Startsev, V. I., and Fomenko, V. S. (1970). Plastic deformation of lead in the normal and superconducting states, *Phys. Status Solidi* **37**, 413–423.

Read, D. T. and Reed, R. P. (1979). Toughness, fatigue crack growth, and tensile properties of three nitrogen-strengthened stainless steels. In *The Metal Science of Stainless Steels* (eds. E. W. Collings and H. W. King). American Institute of Mining, Metallurgical, and Petroleum Engineers, New York, 92–121.

Read, D. T. and Reed, R. P. (1980). Heating effects during tensile tests of AISI 304L stainless steel at 4 K. In *Advances in Cryogenic Engineering*, Vol. 26. Plenum Press, New York, 91–101.

Read, D. T. and Reed, R. P. (1981). Fracture and strength properties of selected austenitic stainless steels at cryogenic temperatures. *Cryogenics* **21**, 415–417.

Read, W. T. (1953). *Dislocations in Crystals*. McGraw-Hill, New York.

Reed, R. P. (1961). A cryostat for tensile tests in the temperature range 300 to 4 K. In *Advances in Cryogenic Engineering*, Vol. 7. Plenum Press, New York, 448–454.

Reed, R. P. (1962). The spontaneous martensitic transformations in 18% Cr, 8% Ni steels. *Acta Metall.* **10**, 865–877.

Reed, R. P. (1972). Aluminum 2. A review of deformation properties of high purity aluminum and dilute aluminum alloys. *Cryogenics* **12**, 259–291.

Reed, R. P. and Durcholz, R. L. (1970). Cryostat and strain measurement for tensile tests to 1.5 K. In *Advances in Cryogenic Engineering*, Vol. 15. Plenum Press, New York, 448–454.

Reed, R. P. and Mikesell, R. P. (1967). Low temperature (295 to 4 K) mechanical properties of selected copper alloys. *J. Mater.* **2**, 370–392.

Reed-Hill, R. E. (1964). *Physical Metallurgy Principles*. Van Nostrand, New York.

Sachs, G. and Weerts, J. (1930). Zugversuche an Gold-Silber Kristallen. *Z. Phys.* **62**, 473.

Sandor, B. I. (1972). *Fundamentals of Cyclic Stress and Strain*. University of Wisconsin Press, Madison, Wisconsin.

Sato, A. and Meshii, M. (1975). Microyielding and effect of electron irradiation in high-purity iron single crystals. *Phys. Status Solidi A* **28**, 561–569.

Schramm, R. E., Clark, A. F., and Reed, R. P. (1973). *A Compilation and Evaluation of Mechanical, Thermal, and Electrical Properties of Selected Polymers*. U.S. Government Printing Office, Washington, D.C.

Schwartzberg, F. R., Kiefer, T. F., and Keys, R. D. (1964). Fatigue behavior of aluminum and titanium sheet materials down to −423° F. In *Advances in Cryogenic Engineering*, Vol. 10. Plenum Press, New York, 1–13.

Schwarz, R. B., Isaac, R. D., and Granato, A. V. (1977). Dislocation inertial effects in the plastic deformation of dilute alloys of lead and copper. *Phys. Rev. Lett.* **38**, 554–557.

Seeger, A. (1957). The mechanism of glide and work hardening in face-centered cubic and hexagonal close-packed metals. In *Dislocations and Mechanical Properties of Crystals* (eds. J. C. Fisher, W. G. Johnston, R. Thomson, and T. Vreeland, Jr.). John Wiley & Sons, New York, 243–329.

Shannon, J. L., Jr. and Brown, W. F., Jr. (1968). A review of factors influencing the crack tolerance of titanium alloys. In *Applications Related Phenomena in Titanium Alloys*. ASTM STP 432, American Society for Testing and Materials, Philadelphia, Pennsylvania, 33–63.

Shepic, J. A. and Schwartzberg, F. R. (1978). Fatigue testing of stainless steel. In *Materials Studies for Magnetic Fusion Energy Applications at Low Temperatures I* (eds. F. R. Fickett and R. P. Reed). NBSIR 78–884, National Bureau of Standards, Boulder, Colorado.

Sleeswyk, A. W. and Helle, J. N. (1963). Ductile cleavage fracture, yielding and twinning in α-iron. *Acta Metall.* **11**, 187–194.

Smidt, F. A., Jr. and Bement, A. L. (1968). Thermally activated dislocation motion and its application to the study of radiation damage. In *Dislocation Dynamics* (eds. A. R. Rosenfield, G. T. Hahn, A. L. Bement, and R. I. Jaffee). McGraw-Hill, New York, 409–429.

Smith, R. L. and Rutherford, J. L. (1957). Tensile properties of zone refined iron in the temperature range from 298 K to 4.2 K. *J. Met.* **9**, 857–864.

Soffer, L. M. and Molho, R. (1967). *Cryogenic Resins for Glass-Filament-Wound Composites*. NASA Report N67–25076.*

Soo, P. and Galligan, J. M. (1969). On the temperature dependence of the flow stress in some body-centered cubic metals. *Scr. Metall.* **3**, 153–156.

Starr, C. D., Shaw, R. B., and Dorn, J. E. (1954). Effect of dispersions of $CuAl_2$ on the elevated temperature tensile properties of aluminum–copper alloys. *Trans. Am. Soc. Met.* **46**, 1075–1088.

Stein, D. F. (1966). The effect of carbon on the strain-rate sensitivity of iron single crystals. *Acta Metall.* **14**, 99–104.

Stein, D. F. and Low, J. R. (1966). Effects of orientation and carbon on the mechanical properties of iron single crystals. *Acta Metall.* **14**, 1183–1194.

Stokes, R. J. and Cottrell, A. H. (1954). Work-softening in aluminum crystals. *Acta Metall.* **2**, 341–342.

Stout, M. G. and Gerberich, W. W. (1978). Structure/property/continuum synthesis of ductile fracture in inconel alloy 718. *Metall. Trans.* **9A**, 649–658.

Tegart, W. J. M. (1966). *Elements of Mechanical Metallurgy*. Macmillan, New York.

Tien, J. K. and Ansell, G. S. (1966). *Alloy and Microstructural Design*. Academic Press, New York.

Tobler, R. L. (1976a). Cryogenic effects on the fracture mechanics parameters of ferritic nickel alloy steels. In *Materials Research for Superconducting Machinery*, Vol. 6 (eds. R. P. Reed, H. M. Ledbetter, and E. C. Van Reuth). ADA036919.*

Tobler, R. L. (1976b). Fatigue crack growth and J-integral fracture parameters of Ti–6Al–4V at ambient and cryogenic temperatures. In *Cracks and Fracture*. ASTM STP 607, American Society for Testing and Materials, Philadelphia, Pennsylvania, 346–370.

Verkin, B. I. (1978). The development of cryogenic materials science in the USSR. In *Advances in Cryogenic Engineering*, Vol. 24. Plenum Press, New York, 37–51.

Wanhill, R. J. H. (1978). Microstructural influences on fatigue and fracture resistance in high strength structural materials. *Engr. Fract. Mech.* **10**, 337–357.

Ward, I. M. (1971). *Mechanical Properties of Solid Polymers*. Interscience, New York.

Warren, K. A. and Reed, R. P. (1963). *Tensile and Impact Properties of Selected Materials from 20 to 300° K*. National Bureau of Standards Monograph 63, U. S. Government Printing Office, Washington, D.C.

Weertman, J. and Weertman, J. R. (1965). Mechanical properties strongly temperature dependent. In *Physical Metallurgy* (ed. R. W. Cahn). Interscience, New York, 793–819.

Wigley, D. A. (1971). *Mechanical Properties of Materials at Low Temperatures*. Plenum Press, New York.

Wohlbier, F. H. (1974, 1975, 1976). *Mechanical Properties*, Vols. 1–3. Trans Tech Publications, Cleveland, Ohio.

Young, F. W. (1968). Dislocation configurations in copper. In *Dislocation Dynamics* (eds. A. R. Rosenfield, G. T. Hahn, A. L. Bement, and R. I. Jaffee). McGraw-Hill, New York, 313–336.

Zarzycki, J. (1968). Glasses—their salient chemical and physical properties. In *Chemical and Mechanical Behavior of Inorganic Materials* (eds. A. W. Searcy, D. V. Ragone, and U. Colombo). Interscience, New York.

* Available from NTIS, Springfield, Virginia.

8 Fracture Mechanics

R. L. TOBLER and H. I. McHENRY

8.1 INTRODUCTION

Fracture may be defined as the mechanical separation of a solid owing to the application of stress. Fractures of engineering materials are broadly categorized as ductile or brittle, and fracture toughness is related to the amount of energy required to create fracture surfaces. In ideally brittle materials such as glass, the energy required for fracture is simply the intrinsic surface energy of the material, as demonstrated by Griffith (1920). For structural alloys at room temperature, considerably more energy is required for fracture because plastic deformation accompanies the fracture process.

Many common structural materials, notably carbon and low-alloy steels, become brittle at low temperatures. Therefore, many cryogenic structures require the use of conservative design procedures to ensure adequate structural safety. The prevention of brittle fracture is an important consideration because of the sudden and uncontrolled nature of such failures. Experience gained from service failures indicates that most brittle fractures are initiated by small flaws that grow as cracks to a critical size during operation. Accordingly, there are three elements to the fracture problem: the initial flaw, crack growth, and fracture toughness (i.e., crack tolerance). To account for these factors, a design approach based on fracture mechanics has been developed and applied in varying degrees to the design of nuclear pressure vessels, bridges, aircraft, and cryogenic missiles structures. A test approach based on fracture mechanics is frequently used to characterize materials even when a thorough structural analysis is not conducted.

Fracture mechanics is the study of the influence of loading, crack size, and structural geometry on the fracture resistance of materials containing natural flaws and cracks. When applied to design, the objective of the fracture mechanics analysis is to limit operating stresses so that a preexisting flaw of assumed initial size will not grow to critical size during the desired service life of the structure. Service life is calculated on the basis of probable initial flaw sizes limited by inspection, a stress analysis of the structure, and experimental data relating crack growth and fracture to fracture mechanics parameters.

In this chapter, the concepts of fracture mechanics and their application to materials evaluation and the design of cryogenic structures are reviewed. Emphasis is placed on an explanation of the technology, a review of fracture mechanics testing methods, and a discussion of the many factors contributing to the fracture behavior of materials at cryogenic temperatures.

8.2 LINEAR ELASTIC FRACTURE MECHANICS

This section states the fundamental ideas underlying the foundation of fracture mechanics. The physical basis of fracture mechanics stems from the work of Griffith (1920), who demonstrated that the strain energy released upon crack extension is the driving force for fracture in a cracked material under linear elastic conditions. The elastic strain energy, V, is the work done by a load, P, causing a displacement, Δ:

$$V = P\Delta/2 = CP^2/2 \tag{8.1}$$

where $C = \Delta/P$, the elastic compliance.

The loss of elastic potential energy with crack extension of unit area, A, is defined as the strain-energy release rate, G. For a crack extending at constant deflection or at constant load,

$$G = dV/dA = (P^2/2)\, dC/dA \tag{8.2}$$

Irwin and Kies (1954) used this relationship to characterize the fracture resistance of structural materials by defining a critical strain-energy release rate, G_c, at the critical load, P_c, when fracture occurs in a specimen with a known compliance function, dC/dA.

8.2.1 The stress intensity factor

Irwin (1957) determined the stress distribution near the tip of a crack located in a linear elastic body and developed a stress analysis basis for fracture mechanics. The magnitude of the crack tip stress field, σ_{ij}, was found to be proportional to a single parameter, K, the stress intensity factor:

$$\sigma_{ij} = K(2\pi r)^{-1/2} f_{ij}(\theta) = K{\cdot}f\,(\text{position}) \tag{8.3}$$

where r and θ are cylindrical position coordinates, $r = 0$ at the crack tip, and $\theta = 0$ in the crack plane. K is a function of the applied stress, σ, a is the crack length, and $Y(a)$ is a factor dependent on structural geometry:

$$K = Y(a)\sigma(\pi a)^{1/2} \tag{8.4}$$

Tada, Paris, and Irwin (1973) and Sih (1973) have published handbooks of K formulas for various geometries and boundary conditions.

Irwin (1957) demonstrated that the strain-energy release rate and stress intensity approaches are related:

$$K^2 = E'G \tag{8.5}$$

where for plane stress,* $E' = E$, the elastic modulus; for plane strain,* $E^{-1} = E/(1-\nu^2)$, where ν is Poisson's ratio. Thus, it is equivalent to attribute the driving force for fracture to the crack tip stress field, which is proportional to K or to the elastic strain-energy release rate, G. The stress intensity, K, is used more commonly than G, because K can be computed for different structural geometries using stress analysis techniques.

Fracture occurs when the crack tip stress field reaches a critical magnitude, that is, when K reaches K_c, the fracture toughness of the material. K_c is a mechanical property that is a function of temperature, loading rate, and microstructure, much the same as yield strength is; K_c is also a function of the extent of plastic strain at the crack tip relative to the other specimen or structure dimensions. If the plastic zone is small compared with the specimen dimensions and the crack size, then K_c approaches a constant minimum value defined as the plane strain fracture toughness, K_{Ic}. Measurement methods, representative data, and factors influencing fracture toughness are discussed in § 8.4.

8.2.2 Crack tip plasticity

Applicability of the linear elastic analysis has been extended to conditions approaching net section yielding by correcting for the zone of plasticity that exists at the crack tip (Irwin, 1960a). The assumption is that the plastic material at the crack tip strains without carrying the incremental load; therefore, the crack behaves as if it were a slightly longer crack in a linear elastic material. The adjustment is made by adding the radius of the plastic zone, r_y, to the crack length, a, such that the expression for the stress intensity factor becomes:

$$K = Y(a + r_y)\, \sigma[\pi(a + r_y)]^{1/2} \tag{8.6}$$

where

$$r_y = 1/2\pi\, (K/\sigma_y)^2 \tag{8.7}$$

where σ_y is the yield strength at the crack tip. The r_y correction modifies the crack tip stress field to account for the elastic stress redistribution that is due to the localized plasticity.

8.2.3 Size effect

The extent of crack tip plasticity is influenced by specimen dimensions, particularly thickness. Thin specimens are subject to plane stress conditions (i.e., one of the principal stresses, the stress in the thickness direction, is zero), and thick specimens are subject to plane strain (e.g., the strain in the thickness direction is zero). As specimen thickness, B, increases, σ_y increases from σ_{ys} (the engineering yield stress of the material at 0.2% strain) to $(3\sigma_{ys})^{1/2}$ because of a geometric constraint to plastic deformation associated with a transition from plane stress to plane strain conditions (Irwin, 1960b). The maximum value of σ_y is reached when the plastic zone size is limited to about 5% of the thickness. Thus, in a given material, the plastic zone size as computed by Eq. 8.7 may vary with thickness by a factor of 3 — leading to a strong dependence of K_c on thickness, as shown in Fig. 8.1. The inflection point of

---

*See § 8.2.3 for definition.

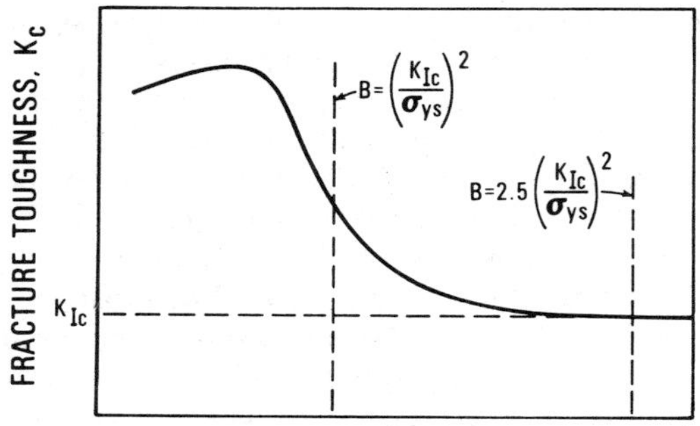

Figure 8.1 Fracture toughness transition in structural alloys.

the curve in Fig. 8.1 occurs at approximately $(K_{Ic}/\sigma_{ys})^2$. Therefore, for maximum toughness:

$$B < (K_{Ic}/\sigma_{ys})^2 \text{ or } K_{Ic} > \sigma_{ys}(B)^{1/2} \tag{8.8}$$

Equation 8.8 is useful in material selection (Rolfe, 1970).

Of great importance is the fact that the curve in Fig. 8.1 approaches an asymptote at a thickness of $B \geq 2.5(K_{Ic}/\sigma_{ys})^2$. At this point the fracture toughness value, K_{Ic}, is a material constant, independent of further increase of thickness.

8.2.4 Summary

The crack tip stress field is the driving force for fracture, and the magnitude of this stress field is proportional to the stress intensity factor, K. K is a function of crack size, applied stress, and structural geometry and can be computed using stress analysis methods. The resistance to fracture is a material property defined as the fracture toughness, K_c; fracture occurs when $K = K_c$. Thus, for a given structural geometry and material, critical crack sizes can be determined as a function of applied stress and vice versa:

$$a_c = [K_c/\sigma_y(a)]^2 \tag{8.9}$$

This approach is applicable to conditions of localized crack tip plasticity, that is, where $\sigma \leq 0.8\sigma_{ys}$ or $r_y \leq 0.3a$, when the r_y correction is used to account for crack tip plasticity (Paris and Sih, 1965).

8.3 ELASTIC-PLASTIC FRACTURE MECHANICS

High-toughness structural materials undergo extensive plastic deformation prior to fracture. Therefore, the concepts of linear elastic fracture mechanics are being extended to account for elastic–plastic behavior. Three approaches currently under development include the crack opening displacement, the J-integral, and the R-curve methods. The methods of elastic–plastic fracture mechanics have been reviewed by Turner and Burdekin (1974) and by Paris (1977).

8.3.1 The crack opening displacement

The crack opening displacement (COD) concept is a crack tip strain criterion for fracture, stemming primarily from the work of Wells (1964). For a crack in an elastic body, the crack opening displacement, v, at a distance r from the crack tip is given by Westergaard's (1939) displacement equation:

$$v = 2K/\pi E' \, (2\pi r)^{1/2} \tag{8.10}$$

Under conditions of small-scale yielding, the displacement at the crack tip, δ, can be calculated by assuming the effective crack tip ($a_{\text{effective}} = a + r_y$) is at a distance, r_y, from the actual crack tip:

$$\begin{aligned} \delta &= 2v = 4K/\pi E' \, (2\pi r_y)^{1/2} \\ &= 4K^2/\pi E'\sigma_y \end{aligned} \tag{8.11}$$

Theoretically, fracture occurs when $\delta = \delta_c$, the critical COD. In practice, a characteristic value for δ exists only for the crack initiation event; significantly more scatter exists for δ measured at maximum load or final fracture.

The COD approach is limited by the analytical and experimental uncertainties of the crack tip region. Analytically, δ is defined as the COD at the interface of the elastic–plastic boundary and the crack surface. Experimentally, δ is calculated from displacement measurements taken remotely from the crack tip; direct physical measurements of δ have been abandoned because they are not precise. Further uncertainty is introduced by the term σ_y in Eq. 8.11, which may vary by 75%, depending upon the degree of elastic constraint — a crack tip characteristic that cannot be measured directly.

Despite these shortcomings, the COD approach is being widely used because it offers a significant improvement over linear elastic methods in the plastic range. A useful empirical correlation, known as a design curve, has been developed relating COD, crack size, and applied strain for a wide range of structural and material combinations (Turner and Burdekin, 1974). Test methods for measuring δ_c have been standardized, and the measured δ_c data are used with the design curves to evaluate the critical flaw size and to compare material toughnesses [*Methods for*

Crack Opening Displacement (COD) Testing, 1979].

8.3.2 *J*-integral

The *J*-integral developed by Rice (1968) is a characterization of the elastic–plastic field in the vicinity of the crack tip. The use of the *J*-integral as a fracture criterion was suggested and experimentally demonstrated by Begley and Landes (1972). J is defined as the line integral:

$$J = \int_{\Gamma} [w dy - \vec{T}(\partial \vec{u} / \partial x) ds] \qquad (8.12)$$

where Γ is any contour surrounding the crack tip, w is the strain-energy density, $\vec{T}$ is the force vector normal to Γ, $\vec{u}$ is the displacement vector, and s is the arc length along Γ. The *J*-integral is path independent for linear and nonlinear elastic materials (Rice, 1968) and nearly so for most structural materials (elastic–plastic) under monotonic loading conditions (Hayes, 1970). Thus, J can be computed with numerical methods by analyzing loads and displacements along a contour away from the crack tip, that is, in a region where the analysis methods are quite accurate. This eliminates the uncertainties of the crack tip region — a problem that seriously limits the usefulness of the COD method.

An equivalent interpretation is that J is equal to the change of the pseudopotential energy (the area under the load–displacement curve), U, upon an increment of crack extension of unit area, A:

$$J = dU / dA \qquad (8.13)$$

For the linear elastic case, the potential energy equals the strain energy ($U = V$), and therefore, Eq. 8.13 is the same as Eq. 8.2 and $J = G$. Thus, J appears to be a logical extension of linear elastic fracture mechanics into the elastic–plastic range. Because of the irreversibility of plastic deformation, the energy interpretation of the *J*-integral does not apply to the process of crack extension, and J is not equal to the energy available for crack extension in elastic–plastic materials as G is for elastic materials. J is simply an analytically convenient, measurable parameter that is a characteristic of the elastic–plastic field at the crack tip. The contribution of Begley and Landes (1972) was to demonstrate that crack initiation under elastic–plastic conditions occurs at a characteristic value of J, called J_{Ic}; J_{Ic} is related to the linear elastic plane strain toughness, K_{Ic}, in the same way G is related to K in Eq. 8.5. Thus, *J*-integral methods can be used to determine K_{Ic} in specimens significantly smaller than the size requirements for linear elastic response.

In the past, the *J*-integral concept has been considered only applicable to crack initiation; nevertheless, Paris, Tada, Zahoor, and Ernest (1977); Hutchinson, Needleman, and Shih (1978); and others now indicate that the *J*-integral is applicable to crack propagation. For most materials that fail in the elastic–plastic range, significant fracture resistance exists after crack initiation. Therefore, in some cases the *J*-integral may be unduly conservative as a fracture criterion. Another factor currently limiting the routine application of the *J*-integral is the analytical complexity in determining the applied J at a given load level. The analysis must properly account for the stress–strain characteristics of the material. Thus, for a given structural configuration, *J*-integral solutions are required for each distinct material instead of the single solution needed for K analyses.

8.3.3 *R*-curves

The *R*-curve concept, introduced by Krafft, Sullivan, and Boyle (1961), is a characterization of the increase in fracture resistance accompanying the slow crack extension that precedes unstable fracture. The *R*-curve is constructed by plotting crack extension, Δa, as a function of the driving force for fracture expressed in terms of G, K, J, or δ. The level of driving force required to extend the crack is defined as the resistance, R, of the material. An *R*-curve is shown in Fig. 8.2, where R is expressed in terms of K and is denoted by K_R. The *R*-curve may be used as a fracture

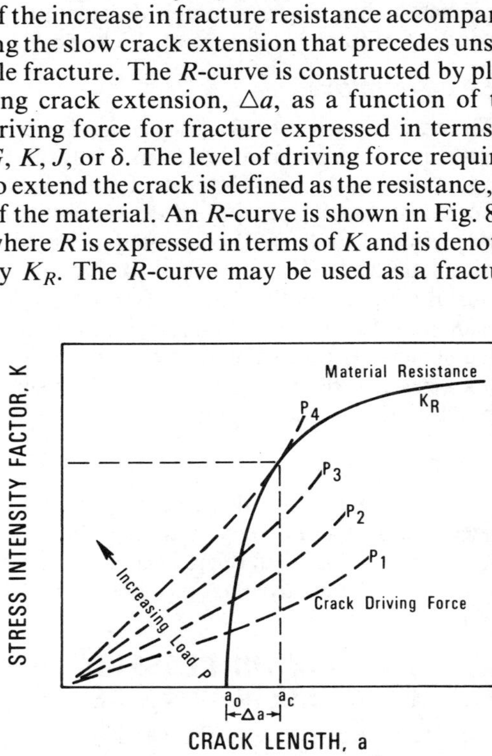

Figure 8.2 Crack growth resistance curve and crack driving force curves in *R*-curve format (McCabe and Heyer, 1973).

criterion when crack-driving-force curves expressed in terms of K vs. a at constant load in Fig. 8.2 are shown on the same plot. Fracture is predicted when the following conditions are met (Srawley and Brown, 1965):

$$K = K_R \text{ and } \partial K/\partial a = \partial K_R/\partial a \qquad (8.14)$$

K_R and K are computed for the test specimen and structure, respectively, using the appropriate value of $Y(a + r_y)$ in Eq. 8.6.

In applying the R-curve concept as a fracture criterion, it is assumed that the R-curve is a property of the material for a given thickness and temperature, and the influence of planar geometry on the predicted instability point is considered in the calculation of K for the driving-force curves. The first assumption implies that the R-curve is independent of specimen size and configuration and the initial crack size. This assumption has not been clearly established because results that both agree (McCabe and Heyer, 1973) and disagree (Sulllivan, Freed, and Stoop, 1973) with this hypothesis have been reported. The validity of the second assumption is limited to the case of small-scale yielding where the linear elastic analysis is appropriate. These uncertainties, coupled with the complexity of the proposed test method (*Annual Book of ASTM Standards*, 1981a), have limited the applicability of the R-curve approach.

The R-curve approach is used as a measure of fracture toughness for plate thicknesses where valid K_{Ic} data cannot be obtained because of the size requirements for linear elasticity. For example, the nickel steels have been evaluated at temperatures down to 76 K using the R-curve approach (Sarno, Bruner, and Kampschaefer, 1974). The results can be misleading if one compares materials on the basis of fracture toughness; that is, an initiation criterion, such as K_{Ic}, indicates a substantially lower toughness than an instability criterion, such as the R-curve.

8.3.4 Summary

The COD and J-integral concepts are efforts to define single-parameter characterizations of the fracture process that are applicable to linear elastic and general yielding fracture. In the COD concept, attention is focused at the crack tip; fracture theoretically occurs when a critical displacement develops at the crack tip. The J-integral concept examines the stress–strain conditions along an arbitrary contour away from the crack tip; fracture theoretically occurs when the potential energy available for crack extension reaches a critical value defined by J. For the linear elastic case, δ_c and J_{Ic} are consistent with the linear elastic fracture criteria:

$$G_c = K_{Ic}^2/E' = J_{Ic} = \sigma_y \delta_c \qquad (8.15)$$

The principal limitation in extending these criteria to general yielding fracture is that both δ_c and J_{Ic} are initiation parameters. Only the R-curve approach accounts for the significant increase in fracture resistance that accompanies the slow crack growth preceding unstable fracture. The application of the COD, J-integral, and R-curve approaches to the design of cryogenic structures to date has been limited to characterizing the fracture toughness of the materials.

8.4 FRACTURE TOUGHNESS TESTING

Fracture toughness is a single-parameter characterization of the fracture resistance of a material containing a crack. The single parameter depends on the fracture criterion chosen and varies as a function of temperature, loading rate, and microstructure. Several fracture criteria were discussed in § 8.2 and § 8.3: G, K, K_{Ic}, J_{Ic}, K_R, and δ_c. For the purposes of this section, K_{Ic} is used as the linear elastic fracture criterion and J_{Ic} is used as the fracture criterion for the elastic–plastic and fully plastic cases. These criteria are commonly used for cryogenic testing, and many experimental K_{Ic} and J_{Ic} data are now available at 4 K. Load–displacement records representing the three fracture cases are shown in Fig. 8.3.

Although the J_{Ic} value can be used to obtain an estimate of K_{Ic}, denoted $K_{Ic}(J)$, it should be understood that K_{Ic} and $K_{Ic}(J)$ may represent significantly different fracture behavior. K_{Ic} is the critical K level for the linear elastic case, at which significant measurable extension of the crack occurs, often triggering unstable fracture. For valid measurements of K_{Ic}, the critical K level must be reached prior to significant plastic deformation, that is, the plastic zone size is negligible, less than 2% of the crack length and the thickness. In contrast, $K_{Ic}(J)$ for elastic–plastic or fully plastic fracture indicates an estimate of K_{Ic} calculated from J_{Ic}, the J-integral value at which the first measurable extension of the crack occurs. Since plastic deformation prior to the onset of cracking does not invalidate the J_{Ic} measurement, smaller specimens can be used. Either: (a) significant amount of stable crack extension under rising load is displayed before final fracture occurs or (b) the elastic–plastic behavior culminates in

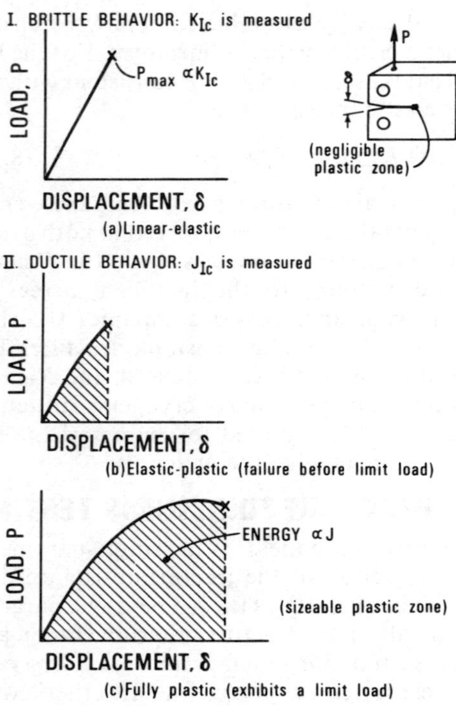

Figure 8.3 Load–displacement behavior observed in fracture toughness tests.

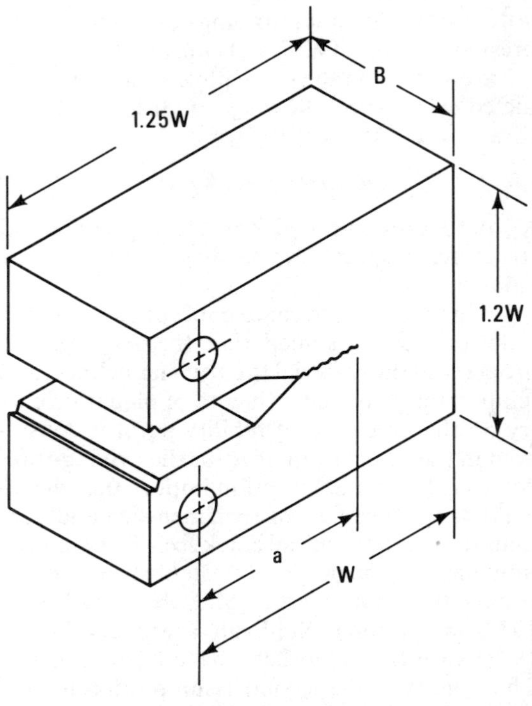

Figure 8.4 ASTM E 399 compact specimen for fracture toughness testing.

plastic instability, as discussed by Paris et al. (1977). The differences in K_{Ic} and $K_{Ic}(J)$ may be clarified by consideration of their test methods.

8.4.1 Linear elastic fracture toughness

A standard test method, ASTM E 399 (*Annual Book of ASTM Standards*, 1981b), is used to measure plane strain fracture toughness, K_{Ic}. The standard is designed to ensure that linear elastic conditions prevail throughout the test. This is achieved by requiring a sufficiently large specimen for the particular toughness and yield strength of the material being tested. Planar dimensions are sized to ensure elastic response of the specimen, and the thickness is sized to ensure sufficient through-thickness constraint. The dimensional criteria in ASTM E 399 are:

$$B, a \geq 2.5(K_{Ic}/\sigma_{ys})^2 \qquad (8.16)$$

where B and a are defined in Fig. 8.4 for the compact specimen, which is generally used for cryogenic testing.

The specimens are precracked by fatigue cycling to an initial relative crack length of $a/W \simeq 0.5$. Precracking loads are limited to low values to keep the plastic zone size at the crack tip small; the fatigue loads must be such that the maximum K level during fatigue is less than 0.6 K_{Ic}. (Changes in yield strength must be taken into account if precracking is performed at room temperature and testing at cryogenic temperatures.) Subsequently, the specimens are monotonically loaded to failure. During loading, load, P, and displacement, Δ, are measured and a P–Δ curve is recorded. The critical load, P_Q, as defined in the ASTM E 399 test method, is either the maximum load or the load at a 5% secant offset from the linear part of the P–Δ test record. If the maximum load in the test exceeds 1.10 P_Q, the test is invalid. A trial value of fracture toughness, K_Q, is calculated from the critical load, the measured crack length, the specimen dimensions, and the specimen calibration function $Y(a/W)$, as follows:

$$K_Q = P_Q Y(a/W)/BW^{1/2} \qquad (8.17)$$

If all the conditions of ASTM E 399 are met, such as precracking procedures, load–displacement record, and specimen dimensions, then $K_Q = K_{Ic}$.

The K_{Ic} test method was standardized by ASTM in 1970. Uncertainties in K_{Ic} measurements obtained by this method lie between 4 and 10% (Antolovich and Chanai, 1975). Three tests per material per temperature are considered sufficient to demonstrate reproducibility.

8.4.2 *J*-integral fracture toughness

Various methods of measuring J_{Ic} have been proposed. A standard for J_{Ic} measurement was published in 1981 (*Annual Book of ASTM Standards*, 1981c) based largely on the method proposed by Landes and Begley (1974), as shown schematically in Fig. 8.5. Typically, a series of deeply notched compact specimens are precracked to $a/W \simeq 0.6$, and each specimen is loaded to a *J*-level in the region where crack extension is anticipated. The load–displacement curve is recorded on an X–Y recorder, with displacement being measured at the load line. Then the specimen is unloaded and heated to tint the region of crack extension. The specimen is fractured, and the crack extension, Δa, is measured from the exposed fracture surface. *J* is calculated from the load–displacement record and specimen dimensions using the equation:

$$J = (A/Bb)f(a_0/W) \tag{8.18}$$

where A is the area under the load–displacement curve, $b = (W - a)$ is the uncracked ligament, and $f(a_0/W)$ is a function of crack length.

The results of a test series are plotted as J vs. Δa. On the same graph, the line defined by Eq. 8.19 is drawn:

$$J = 2\bar{\sigma}\Delta a \tag{8.19}$$

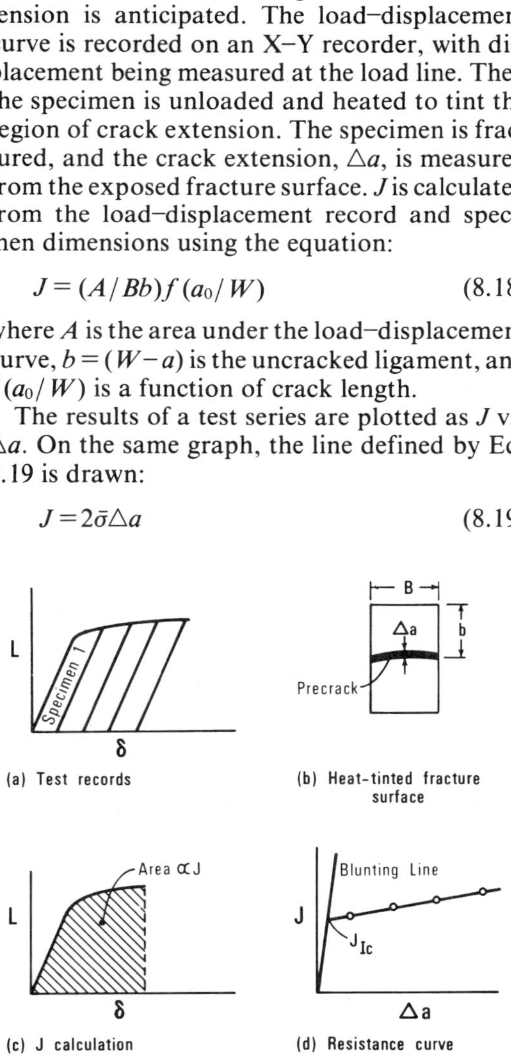

Figure 8.5 Schematic of the *J*-resistance-curve test method for J_{Ic} determination.

where $\bar{\sigma}$ is the flow stress (the average of the yield and ultimate strengths). The intersection of the J–Δa plot and the $J/2\bar{\sigma}$ line is defined as J_{Ic}, the value of *J* at the onset of crack extension. Apparent crack extension at Δa values of less than $J/2\bar{\sigma}$ is attributed to deformation at the crack tip instead of material separation. Begley and Landes (1972) have proposed that the linear elastic plane strain fracture toughness can be estimated from J_{Ic} as follows:

$$K_{Ic}(J) = [J_{Ic}E/(1 - \nu^2)]^{1/2} \tag{8.20}$$

where E = Young's modulus and ν = Poisson's ratio.

Considerable research is currently in progress to improve the J_{Ic} measurement method. Uncertainties associated with J_{Ic} measurements by the method of Fig. 8.5 are about ±20% (Carlsson and Markstrom, 1977). A single-specimen method has been developed in which the specimen is periodically unloaded by approximately 10% and then reloaded to a higher *J* level. The slope of the unloading curve is measured and related to crack extension by the compliance method, and *J* is calculated from the area under the curve prior to unloading. Thus, it is possible to construct a J–Δa curve using one specimen instead of five, a particular advantage in tests at liquid-helium temperature.

8.4.3 Other test methods

Many of the reported fracture toughness data have been obtained using test methods other than the currently favored K_{Ic} and J_{Ic} methods. Care must be taken in comparing these data with K_{Ic} and J_{Ic} values and in applying the data to materials selection or design. Difficulties arise owing to differences in measurement criteria and fracture criteria.

In both K_{Ic} and J_{Ic} testing, the measurement point (i.e., P_Q for K_{Ic} and the $J/2\bar{\sigma}$ intersection for J_{Ic}) is near the onset of crack extension. Fracture toughness data are frequently reported where the maximum load values are used to calculate fracture toughness, and the results are reported in terms of K or J values — often with a change of subscript (e.g., K_{max}, K_{IE}, K_C, J_C). For example, Kaufman, Nelson, and Wygonik (1974) estimated the cryogenic fracture toughness of 5083-O aluminum from stress intensity calculations using maximum load criteria; Lake (1975) estimated this value from *J*-integral calculations also using maximum load criteria. These results indicate a higher toughness than that reported for 5083-O aluminum by Tobler and Reed (1977a), who used the J-integral test procedure described in § 8.4.2.

Similarly, significant differences in toughness may be attributed to differences in the fracture criteria. Data obtained by the COD or R-curve methods evaluate the fracture toughness of ductile materials after significant crack extension and plasticity have occurred. Consequently, the toughness values for the same material are higher than those obtained by the K_{Ic} or J_{Ic} methods. Fracture-toughness data obtained for 5% Ni steel at cryogenic temperatures by the R-curve approach and the J_{Ic} approach are compared by McHenry and Reed (1977). At 111 K, the base metal and heat-affected zone toughness values obtained by the J_{Ic} procedure were 30 and 50% lower, respectively, than values obtained for the same plate of material by Sarno et al. (1974) using the R-curve method.

8.4.4 Specimens

There are many fracture specimen types and sizes, each offering specific advantages and disadvantages. Some of these are listed in Table 8.1. The choice of a particular specimen geometry depends on technical or scientific purposes, as well as economic considerations. The three-point bend and compact specimens (standard specimens of the ASTM E 399 and E 813 methods) are much preferred for general laboratory materials evaluation, because K and J calibrations for these specimens are accurately known and relatively low loads are required during testing. Data obtained with nonstandard specimens must be evaluated carefully to ensure that the same frac-

Table 8.1 Fracture specimens and their advantages and disadvantages.

Specimen Type	Advantages	Disadvantages
Compact specimen	– High K_{Ic} measurement capability for size – Standard specimen (ASTM E 399-81) – Low loads required	– Expense of machining
Three-point bend	– Standard specimen (ASTM E 399-81) – Low loads required – Suitable for wide range of orientations	– A long span transverse to loading direction, which may be a disadvantage for some cryostats
Center-cracked tension	– Pure tensile loading	– High loads and large material requirements
Double-edge notched	– Pure tensile loading	– High loads and large material requirements
Single-edge notched	– Easy notch preparation	– High loads and material requirements
Double cantilever beam	– Tapered specimens, can be designed such that the value of K is independent of crack length	– Long span transverse to loading direction – Side grooving may be necessary to guide cracking direction – Machining expenses
Surface-flawed specimen (part-through crack)	– Simulates a flaw type commonly found in service	– Size requirements are difficult to establish – K solution not precisely known – High loads are required
C-shaped specimen	– Special geometry suitable for bar stock	– Limited applicability
Wedge-opening loading specimen	– Larger width than compact specimen – May be bolt loaded at one end	– Expense of machining

ture criteria are used, such as the onset of cracking. The compact specimen is particularly attractive for cryogenic testing, because relatively small testing apparatus and a minimum of cryogens are needed (Fowlkes and Tobler, 1976).

8.4.5 Apparatus

Cryostats are chambers for applying loads to test specimens in low-temperature environments. Cryostat designs vary according to function and ingenuity, but the goal is always to achieve a mechanically and thermally efficient design. A number of cryostats and associated instrumentation for mechanical testing at cryogenic temperatures are described in the literature: Reed (1962), Hartwig and Wuchner (1975), Novikov (1977), Fowlkes and Tobler (1976), and Horiuchi, Shimada, Fukutsuka, and Tokuda (1977). Fowlkes' design is shown in Fig. 8.6.

The cooling media most frequently used for tests at fixed temperatures are alcohol and dry ice at 195 K, liquid nitrogen at 77 K, liquid hydrogen at 20 K, and liquid helium at 4 K. Tests at intermediate temperatures may use other cooling baths, or evaporated cryogens and temperature-regulating apparatus (Tobler, Mikesell, Durcholz, and Reed, 1975).

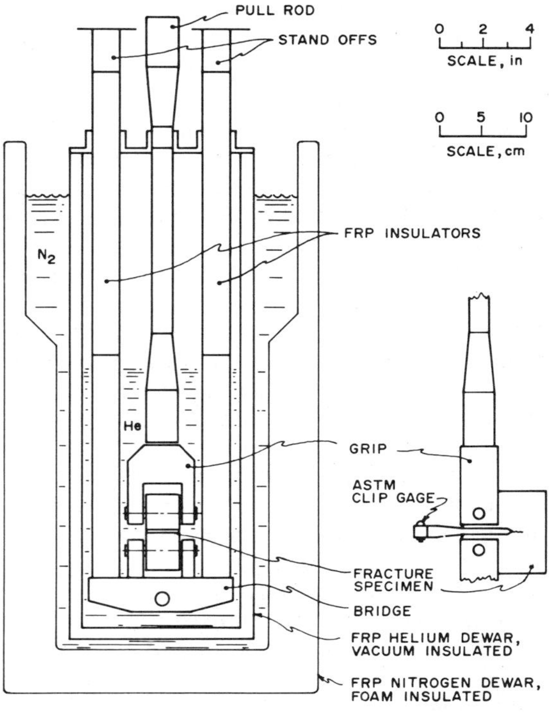

Figure 8.6 Cryostat for fracture testing in liquid helium.

Testing at 77 K or above is relatively easy because nitrogen is inert and less expensive than other cryogens. Nitrogen-bath cryostats can be readily constructed using minimal insulation, and only minor modifications may be necessary for adaptation to conventional mechanical test machines. On the other hand, testing at 20 or 4 K is more complicated and more expensive. Vacuum-insulated Dewars and thermally efficient cryostats are mandatory because of the high volatility and cost of hydrogen and helium. Formerly, liquid-hydrogen testing was frequently conducted in support of aerospace programs, but hydrogen is a hazardous material. Today, liquid-helium testing is more common.

The factors involved in apparatus design for fatigue and fracture testing at extreme temperatures down to 4 K are discussed by Fowlkes and Tobler (1976). To reduce the costs of long-term fatigue tests, a fatigue cryostat should have low thermal conductivity, low specific heat, and low density to minimize cool-down requirements and steady-state heat gain. Apart from this, the overall design resembles that for static tensile or fracture tests. Conventional double-beam ASTM E 399-type clip gauges with commercially available strain gauges suitable for 4-K service have been designed to be directly attached to the specimens. With such a gauge submerged in liquid helium, a low excitation voltage of 1.5 Vdc is used to minimize the electrical noise that is due to local boiling at the strain gauge films.

8.4.6 Specimen orientation

Most structural alloys are anisotropic — their fracture toughness varies with direction. Fracture anisotropy is caused by microstructural inhomogeneities, such as irregular grain structure, chemical segregation, crystallographic texturing, or inclusion morphology. Specimen orientation is important since it determines the direction of crack propagation through the microstructure.

Goode's (1972) notation for fracture specimen orientation has become common. This notation uses two letters; the first designates the direction of load application, and the second designates the direction of crack propagation. These letter designations are illustrated in Fig. 8.7.

In some alloys, nearly equiaxial microstructures are achieved. Then, fracture properties may be nearly isotropic. Thus, data for a recrystallization-annealed Ti–6Al–4V* alloy tested in six

*Chemical composition is in weight percent throughout this chapter.

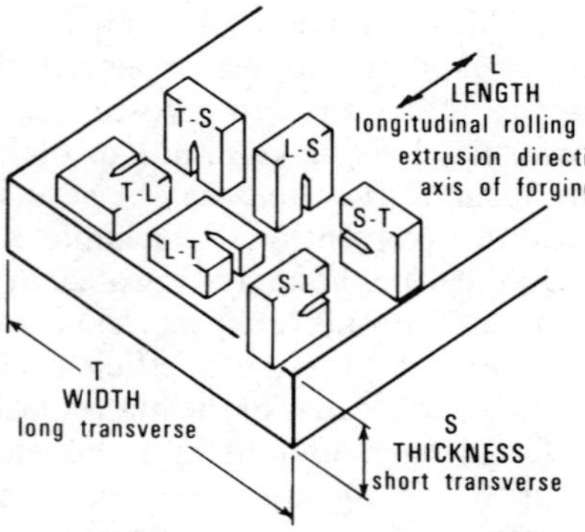

Figure 8.7 Standard notation for specimen orientation.

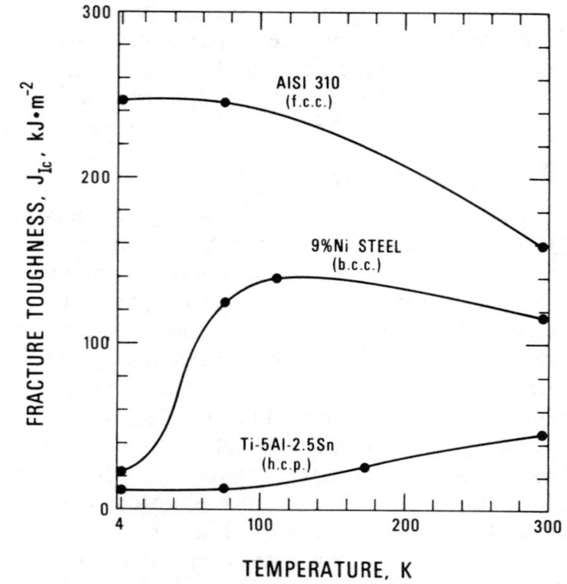

Figure 8.8 Temperature dependence of fracture toughness for alloys, illustrating characteristic behavior for three different crystal structures.

different orientations at 76 K show no measurable variation in K_{Ic} (Tobler, 1978). The fracture resistance of most alloys, however, is quite dependent on orientation, more so than some tensile properties. Rolled plates having elongated grains may exhibit J_{Ic} variations of up to 2:1 with the two most common specimen orientations ranking as follows: LT > TL. This knowledge of anisotropy can be used to advantage by orienting components judiciously in relation to the maximum service stress.

8.4.7 Loading rate

Plastic deformation is a time-dependent process that can be suppressed at high loading rates. High loading rates produce higher yield strengths and reduce the toughness of some alloys, particularly the body-centered cubic (b.c.c.) alloys. Dynamic tests are typically performed at high rates between 10^4 and 10^6 MPa·m$^{1/2}$s^{-1}, whereas static K_{Ic} tests are conducted at stress-intensity-factor rates of about 1 MPa·m$^{1/2}$s^{-1}. As a rule, alloys that show rate-sensitive tensile behavior also show rate-sensitive fracture behavior. If the tensile yield strength of an alloy is raised significantly while its ductility is lowered at high strain rates, reduced fracture toughness values may be expected for dynamic loading. This is typically the case for low-strength ferritic steels (Hahn, Hoaglund, Rosenfield, 1971; Shoemaker and Rolfe, 1971). In tests of a precipitation-hardened austenitic nickel-based superalloy, however, rate effects on fracture toughness at 77 K were found to be negligible (Wells, Kossowsky, Logsdon, and Daniel, 1976).

8.5 THERMAL AND METALLURGICAL EFFECTS ON TOUGHNESS

As shown in Fig. 8.8, fracture toughness may increase or decrease as temperature is lowered, depending on metallurgical factors. Porter (1970) has reviewed the influence of metallurgical factors on the toughness of steels, and many of his ideas are used in the following discussion.

To be suitable for cryogenic applications, structural alloys should fracture in a ductile manner at all service temperatures. Ductile fracture is caused by the formation and growth of the voids that eventually comprise the fracture surface. Thus, the toughness of ductile metals is related to the factors that influence the nucleation and growth of voids. Voids nucleate most readily at second-phase particles, such as inclusions and precipitates (1 to 10 μm in size), as a result of interfacial separation, fracture of the particle, or matrix separation caused by strain concentration near the particle. Voids grow and coalesce by ductile tearing of the matrix. Ductile tearing resistance is a function of the strength and ductility of the matrix. As matrix strength increases, less energy is dissipated by plastic deformation during tearing, and toughness is reduced. Increased matrix strength also tends to activate additional void nucleation sites. Consequently, yield strength is inversely proportional to fracture toughness, as shown in Figs. 8.9 and 8.10.

Brittle fracture requires less energy for surface formation than ductile fracture. Lower energy fracture modes include cleavage and intercrystalline (grain-boundary) fracture. Cleavage is a

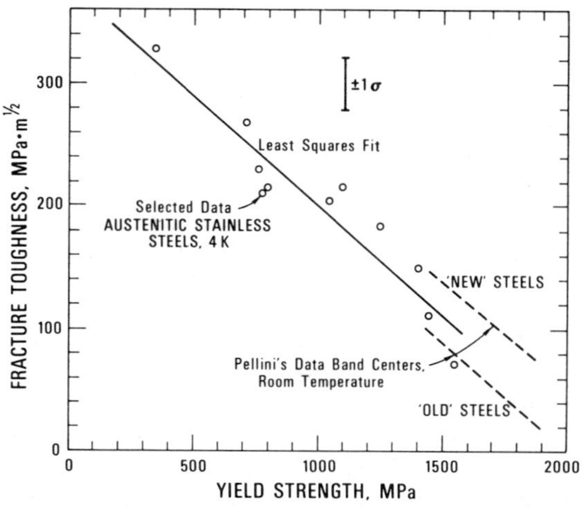

Figure 8.9 Strength vs. toughness trend line for austenitic stainless steels at 4 K.

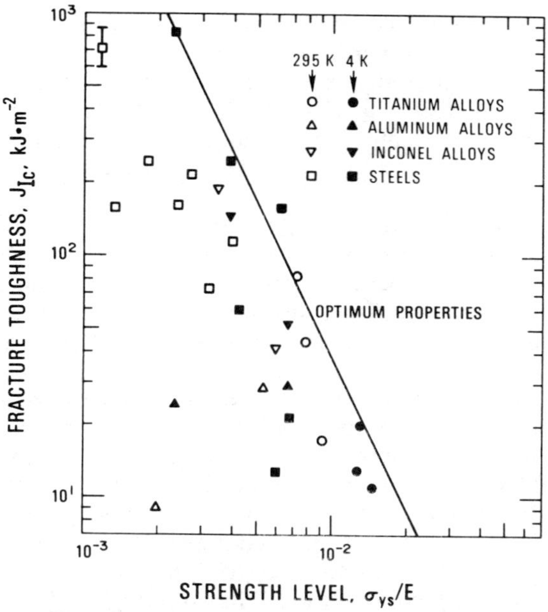

Figure 8.10 Fracture toughness vs. strength level for a wide variety of commercial alloys tested at 295 and 4 K.

fracture mode in which material separation proceeds along preferred crystallographic planes without sizable plastic deformation prior to fracture. Metals subject to cleavage usually have a large increase in yield strength as temperature is decreased. Cleavage occurs when the cleavage fracture stress is reached before the energy required for void formation is exceeded. Intercrystalline fracture occurs when the cohesive strength of the grain boundary is exceeded before cleavage or ductile fracture occurs. As matrix strength increases with decreasing temperature, intergranular failure may occur more readily in a susceptible alloy, such as the Fe–Cr–Ni–Mn–N alloys tested by Read and Reed (1978).

In the following subsections, the influence of selected metallurgical factors on toughness at low temperatures is discussed.

8.5.1 Crystal structure

Representative behaviors for alloys having the three common crystal structures are shown in Fig. 8.8. As indicated by these data, crystal structure is a reliable guide for qualitative prediction of temperature dependence: face-centered cubic alloys typically exhibit high toughness throughout the ambient-to-cryogenic range, body-centered cubic alloys exhibit precipitous decreases in fracture toughness at critical transition temperatures, and hexagonal close-packed alloys are noted for comparatively low toughness at all temperatures. Metallurgical factors (composition, purity, and processing) can strongly modify the fracture behavior, and in every crystal structure class there are exceptional alloys.

Face-centered cubic alloys. Annealed f.c.c. alloys have low strength and high fracture toughness. Included in this group are copper alloys, aluminum alloys, austenitic stainless steels, and nickel-based superalloys. Ductile tearing is typical, even in thick sections. Fracture toughness values for these alloys usually increase between 295 and 4 K, often showing a broad maximum at temperatures near 77 K. Tests on AISI 310S and Fe–23Ni–15Cr–9Mn stainless steels and 5083-O and Al–6Mg aluminum alloys have demonstrated J_{Ic} increases ranging up to 175% between 295 and 4 K.

Precipitation-hardened f.c.c. alloys are favored for some cryogenic applications because their strengths are significantly greater than those of annealed alloys. Generally, these alloys retain adequate toughness, although they may exhibit linear elastic fracture in practical section thicknesses. The variation of fracture toughness in precipitation-hardened f.c.c. alloys between 295 and 4 K is usually not greater than ±30%.

Metastable austenitic stainless steels transform partially to b.c.c., h.c.p., or both martensitic phases when cooled and stressed at cryogenic temperatures, as discussed in Chapters 9 and 11. These transformations may improve fracture resistance, if the transformation absorbs energy that would otherwise be available for crack extension (Zackay and Parker, 1976). But adverse effects are also possible since the martensitic transformation products may be brittle. The most commonly used stainless steels, AISI types 304 and 316, transform during toughness tests at cryogenic temperatures, but the fracture toughness of these alloys remains at high levels.

Body-centered cubic alloys. Because of their ductile-to-brittle transitions, alloys having the b.c.c. structure are of limited use at cryogenic temperatures. The transitions are associated with a change in fracture mode from void coalescence to cleavage as temperature is reduced. The transition temperature range is a function of the metallurgical and mechanical variables that alter matrix strength. Factors tending to raise the transition temperature for a given alloy and heat-treatment combination include increasing grain sizes, thicker sections, and higher loading rates. Nickel alloying decreases the transition temperatures of ferritic steels, as shown in Fig. 8.11.

Hexagonal close-packed alloys. The fracture toughness of h.c.p. metals and alloys is usually quite low, and many exhibit transitional behavior such as that observed in b.c.c. alloys. Beryllium, for example, exhibits room temperature K_{Ic} values of about 7 to 23 MPa·m$^{1/2}$, and a 30% decrease occurs as temperature is reduced to 77 K (Harrod, Hengstenberg, and Manjoine, 1969). Low-temperature ductile-to-brittle transitions have been observed in zinc, beryllium, and magnesium, but not in cadmium. Under certain conditions, titanium alloys have exhibited abrupt toughness reductions (Tobler, 1978).

8.5.2 Composition

The chemical composition of an alloy determines its crystal structure, phase balance, and potential strengthening and deformation mechanisms. Thus, composition has the primary influence on material behavior, including fracture toughness. Chemical composition includes impurities dissolved in the matrix or present as precipitates and inclusions. There have been few systematic investigations to determine the influence of controlled alloy additions and impurity levels on fracture toughness at cryogenic temperatures. There are many noteworthy examples, however, of the influence of composition on fracture properties; some of these are discussed below.

Alloy additions increase the stability of the f.c.c. phase in austenitic stainless steels. Fracture tests by Read and Reed (1978) on nitrogen-strengthened Fe–Cr–Ni–Mn stainless steels indicate that toughness at 76 and 4 K may increase with increasing austenite stability. However, the influence of austenite stability on low-temperature toughness is not clear in the Fe–Cr–Ni stainless steels such as AISI 304, 310, and 316.

Phase balance can have a marked effect on toughness at cryogenic temperatures. Important examples are the stainless steel weld metals, such as AWS grades 308L and 316L, which are formulated to provide 5 to 10% ferrite, a b.c.c. phase, and the balance austenite. The ferrite is needed to prevent hot cracking, but at cryogenic temperatures the ferrite is brittle and lowers the toughness of the weld metal (Szumakowski and Reid, 1978; Read, McHenry, Steinmeyer, and Thomas, 1980). Other examples of phase balance are the 5, 6, and 9% Ni steels, where the nickel content results in a 5 to 10% retained austenite after the alloys have been properly heat-treated. The retained austenite is thought to contribute to improved toughness of these alloys at temperatures from 300 to 76 K. Nickel may also serve to reduce the ductile-to-brittle transition temperature by increasing the ease of cross-slip in the ferrite matrix (Jolley, 1968). The benefits of nickel are shown in Fig. 8.11.

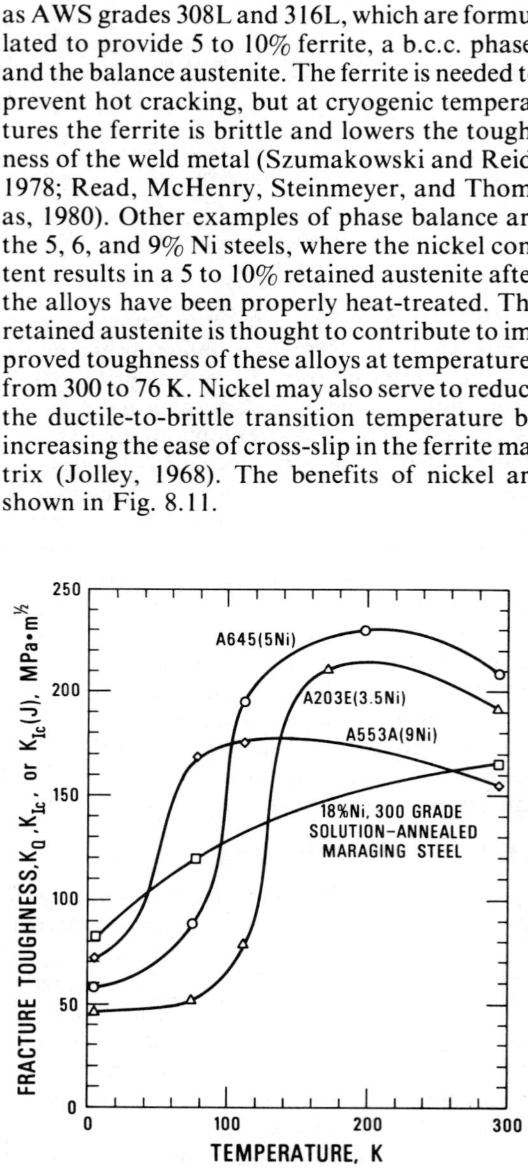

Figure 8.11 Temperature dependence of fracture toughness for various nonaustenitic steels, showing the beneficial effects of increased nickel content on transition temperature and subtransition range toughness.

Potential strengthening mechanisms are a function of alloy content. Solid-solution strengthening of austenitic stainless steels by interstitial nitrogen provides a large increase in yield strength at cryogenic temperatures and a corresponding decrease in fracture toughness, as shown in Fig. 8.9 (Read and Reed, 1978). By comparison, solid-solution strengthening by substitutional elements, such as magnesium in aluminum, has a relatively small effect on the temperature dependence of strength and toughness. For precipitation-hardened alloys having an f.c.c. matrix, changes in strength and toughness at low temperatures are also small.

In addition to intentional alloying elements, commercial alloys inevitably contain impurity elements that cannot be economically removed during processing. In most cases, impurities either dissolve interstitially, thereby reducing matrix toughness, or precipitate in the solidifying metal because they are less soluble upon cooling. Precipitates formed in this way increase the ease of void formation and reduce toughness. Occasionally impurities segregate to grain boundaries causing severe toughness losses that are due to intergranular embrittlement. A classic example is temper brittleness in steels where small concentrations of impurities segregate to the grain boundaries and form a continuous intergranular fracture path.

8.5.3 Processing

Processing includes producing, refining, and casting of the alloy; working the ingot into a suitable product form; and heat-treating the final product. Within the limits imposed by chemical composition, processing controls the alloy microstructure and, consequently, the properties. Production, refining, and casting operations determine the cleanliness and homogeneity of the alloy. Working the ingot into a suitable product form influences anisotropy, grain size, homogeneity, and cold working. Heat treatment provides the final microstructural control. All of these processes influence toughness at all temperatures. Special consideration is given below to two effects that take on particular importance at cryogenic temperatures: grain size and grain-boundary precipitation.

Processing to minimize grain size is desirable because matrix strength is increased with a minimum change in toughness. Grain refinement is particularly beneficial to alloys that undergo a ductile-to-brittle transition, such as the ferritic steels, because it lowers the transition temperature. Tetelman and McEvily (1967) attribute this to the energy consumed in nucleating cleavage cracks in adjacent grains having different orientations. Stonesifer and Armstrong (1977) evaluated the grain-size dependence of fracture toughness for a ferritic steel with a transition temperature range of 275 to 475 K. The results, shown in Fig. 8.12, indicate a large grain-size dependence at 300 K, where some ductile tearing can occur, and essentially no dependence at 4 K, where cleavage cracks are easiest to nucleate.

Porter and Dabkowski (1970) used grain refinement to reduce the fracture-mode transition temperature of a martensitic steel to approximately 76 K. Jin, Morris, and Zackay (1975) incorporated this principle in their development of an ultrafine-grained ferritic Fe–12Ni alloy that remains ductile and tough at temperatures approaching absolute zero.

Processing treatments that cause grain-boundary precipitation of intermetallic compounds, usually carbides, can reduce toughness at cryogenic temperatures. Read and Reed (1978) attributed significant toughness reductions in Fe–22Cr–13Ni–5Mn stainless steel at 4 K to grain-boundary carbide precipitation; solution of the precipitates by a 1000° C anneal increased toughness by 40%. Wells et al. (1976) found that grain-boundary precipitation in AISI 310S stainless steel reduced the J_{Ic} fracture toughness at 4 K

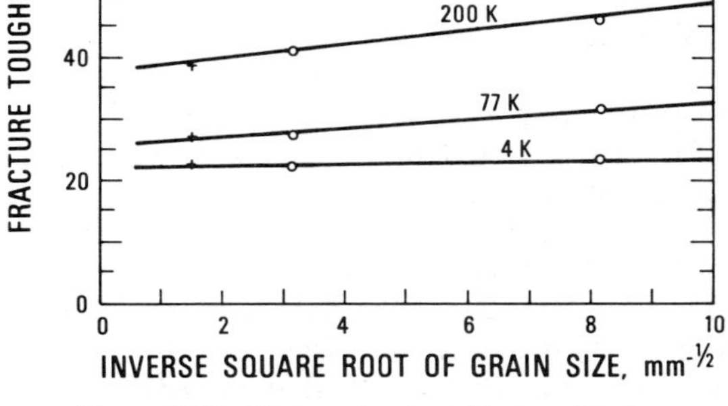

Figure 8.12 The grain-size effect on fracture toughness of a ferritic steel at various temperatures in the brittle, subtransition temperature range.

by 50%. Kossowsky (1977) also demonstrated that the low toughness at room temperature and 4 K of one heat of Inconel 750 evaluated by Logsdon (1977) was caused by grain-boundary precipitation of carbides. Embrittlement that is due to grain-boundary precipitates only occurs when the matrix strength exceeds the grain-boundary strength; that is, it may occur at cryogenic temperatures but not necessarily at room temperature.

8.6 FATIGUE CRACK GROWTH

Fatigue progresses in three stages: crack initiation, crack growth, and fracture on the final cycle. Therefore, the fatigue lifetime of a component is determined by the number of cycles required to initiate and propagate a crack to critical proportions. In complex structures, cracks already exist as a result of manufacturing or fabrication or are assumed to exist because inspection methods are not sensitive enough to verify their absence. In low-cycle fatigue applications, crack initiation occurs quickly and may account for little of the total fatigue life. Clark (1974) describes how fatigue crack initiation and growth properties affect material selection and design criteria.

In reviews of extensive crack propagation data, Paris (1964) observed that a log–log plot of the crack growth rate, da/dN, vs. the stress intensity range, ΔK, is a straight line, obeying an equation of the form:

$$da/dN = c(\Delta K)^n \quad (8.21)$$

where c and n are empirical constants and $\Delta K = K_{max} - K_{min}$ (K_{max} and K_{min} are the maximum and minimum stress intensity factors of the fatigue cycle). This relation implies that the cyclic crack tip stress field described by ΔK is the driving force for fatigue crack extension. Cyclic profile, frequency, mean load, and stress state are of secondary importance. The principal limitation of Eq. 8.21 is the failure to account for environmentally enhanced fatigue crack growth (Wei, 1970). Environmental factors, including temperature, increase the importance of the secondary variables, but Eq. 8.21 still provides a useful basis for the empirical analysis of crack growth data. Accordingly, most data are presented as log–log plots of da/dN vs. ΔK. Recent research (Crooker and Lange, 1970) demonstrates that when fatigue crack growth rates are measured over a more complete range (10^{-8} to 1 mm/cycle), the curve has three distinct regions, as shown in Fig. 8.13. At low growth rates, the curve approaches a threshold value of ΔK, denoted ΔK_{TH}, below which fatigue crack growth does not occur. At intermediate growth rates, the curve is linear on a log–log scale and conforms to Eq. 8.21. Finally, accelerated crack growth occurs when K_{max} approaches K_c. Most of the existing fatigue crack growth data at cryogenic temperatures fall in the intermediate region and are characterized by Eq. 8.21.

8.6.1 Test methods

Fatigue crack growth rate measurements are generally conducted in fatigue test machines capable of applying a constant load amplitude cycle at frequencies on the order of 10 Hz. The standard method for constant amplitude fatigue crack growth rate measurements above 10^8 m/cycle is designated ASTM E 647 (*Annual Book of ASTM Standards*, 1981d). For the reasons discussed in § 8.4.4, the compact specimen is generally used for cryogenic testing. The specimens are precracked by fatigue cycling at a load amplitude equal to or less than the test-load am-

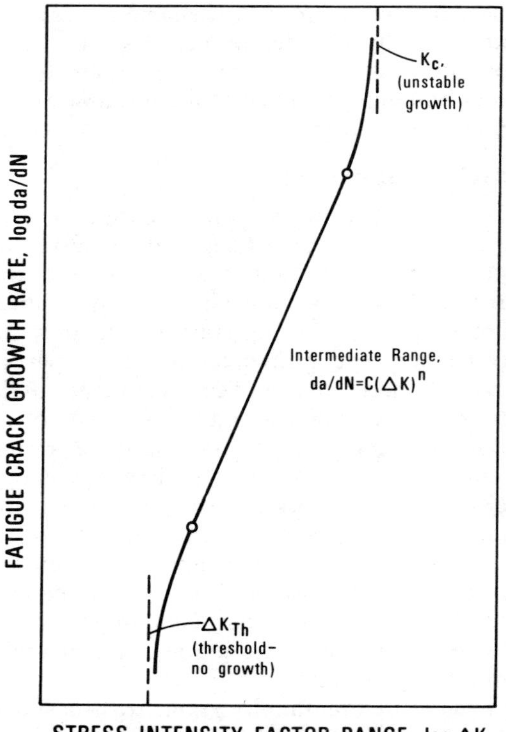

Figure 8.13 Schematic plot of fatigue crack growth rate data trends, illustrating the sigmoidal curve of da/dN vs. ΔK.

plitude. During the test, crack length is measured as a function of number of cycles and the data are plotted and reduced in terms of da/dN vs. ΔK, as shown in Fig. 8.13. The growth rate, da/dN, is the slope of the a vs. N curve at a given value of a, and the stress intensity range is the ΔK level at that value of a calculated for the specific specimen configuration. For the case of the compact specimens:

$$\Delta K = (P_{max} - P_{min})\ Y(a/W)/B(W)^{1/2} \quad (8.22)$$

where P_{max} and P_{min} are the maximum and minimum values of the constant-amplitude load cycle.

The basic data in fatigue crack propagation tests are the cycle number and crack length. The cycle number is readily monitored by electronic or mechanical counters. Several methods of crack length determination have been successfully used in room-temperature studies: direct visual measurement, ultrasonic sensing, electrical potential measurements, and the compliance method.

The compliance method is the most suitable and commonly used method for cryogenic testing. It is an indirect technique, applicable at all temperatures even if the specimen is enclosed in a cryostat and inaccessible. It is based on the correlation between specimen compliance (displacement per unit load) and crack length: compliance increases as crack length increases. A number of variations on the basic approach are described in the literature (McHenry, 1971; Fowlkes and Tobler, 1976).

The basic procedures for calibrating the compliance method are illustrated schematically in Fig. 8.14. Crack-front striations (beach marks) on specimen fracture surfaces are created by changes in the minimum fatigue load. The crack lengths are then measured, averaged, and plotted against their compliance values. The resultant curve is then fit using a polynomial expression that is used to infer crack lengths from compliance data recorded during fatigue crack growth tests.

The compliance method offers advantages for testing thick specimens, because the crack-front curvature is accounted for in the average crack length derived from the compliance value. In contrast, the visual method measures the crack length at the specimen surface. Otherwise, the accuracies of the visual and compliance methods are comparable.

Variability is an important consideration in fa-

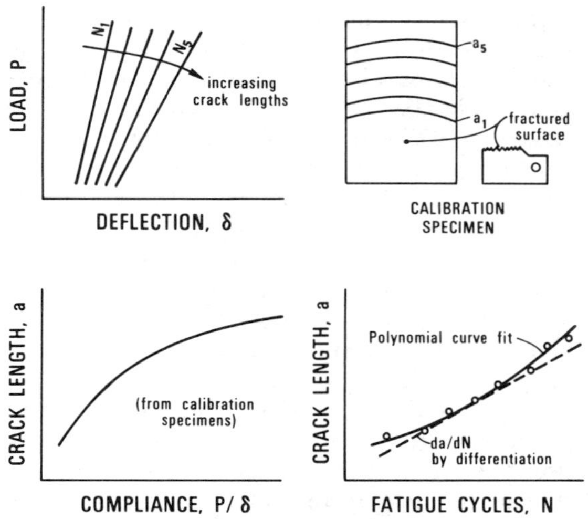

Figure 8.14 Compliance method for fatigue crack growth rate measurements.

tigue crack growth rate testing. In an evaluation of round-robin tests by 15 laboratories on a single heat of steel, Clark and Hudak (1975) concluded that interlaboratory variability amounted to a factor of about 3:1 in terms of the value of da/dN at a given value of ΔK. Intralaboratory variability was as high as 2:1. The amount of scatter must be considered in any data evaluation because it may mask the subtle effects of material or test variables.

8.6.2 Mechanical test variables

The data trends illustrated in Fig. 8.13 should hold as long as ΔK is a valid descriptive parameter for the crack tip stress field. In cases of extreme plasticity, ΔJ is a better parameter for correlating fatigue crack growth (Dowling and Begley, 1976). Nevertheless, empirical results for a broad range of materials and test conditions indicate that ΔK is generally applicable. For example, consider the data comparison (James, 1976) for low-strength AISI 304 stainless steel alloys tested at room temperature. This data comparison represents results from 15 different laboratories, using 10 different specimen geometry–thickness combinations. The reported scatter factor of 3.8 is not significantly higher than the factor of 3 reported by Clark and Hudak (1975) for a single heat of steel tested under more controlled procedures.

Mean stress is proportional to the stress intensity ratio, $R = K_{min}/K_{max}$, and can significantly influence fatigue crack growth rates. In general, for constant ΔK values, da/dN often increases as

R increases. A typical data trend is illustrated in Fig. 8.15 for an Fe–21Cr–6Ni–9Mn austenitic stainless steel (Tobler, unpublished results). Here, a change in R from 0.1 to 0.5 at $\Delta K = 30$ causes da/dN to increase by factors of 2.7 or 4.4, depending on the test temperature. The effect at 4 K is greater, being associated with a more brittle cracking mode at this temperature.

The stress intensity ratio is usually held constant at a value between 0 and 0.1 for laboratory tests. But in service applications, R may be higher or variable, and this must be considered in fatigue crack growth predictions. Forman, Kearney, and Engle (1967) and James (1976) proposed formulas to predict da/dN as a function of stress ratio with success for some materials at room temperature. The same formulas might be applicable at cryogenic temperatures.

Cyclic frequency and waveform are important variables for tests at elevated temperatures and in corrosive environments because creep and corrosion are time-dependent processes. However, at cryogenic temperatures, cyclic frequency

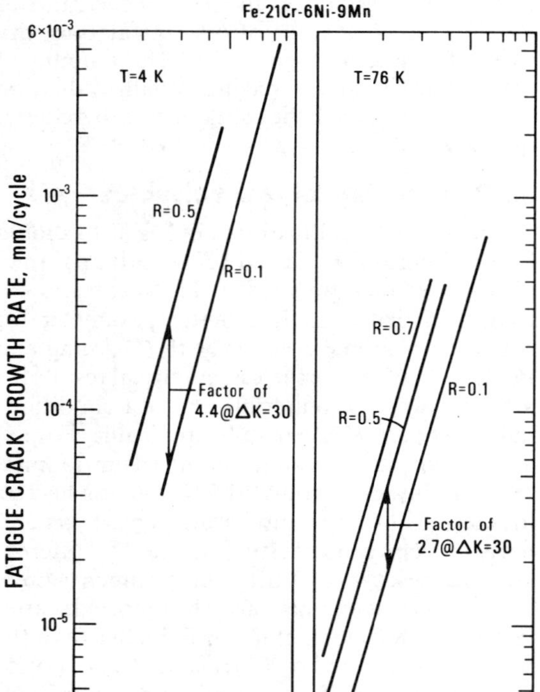

Figure 8.15 Fatigue crack growth rate data for an Fe–21Cr–6Ni–9Mn steel, illustrating stress ratio effects at two test temperatures.

and waveform are expected to have little influence on fatigue crack growth rates. One exception relates to adiabatic heating during tests at 4 K. Since the specific heats of alloys at 4 K are low, fatigue cycling causes heating owing to plastic strain energy dissipation. The heat generated should increase with cycling rate and ΔK. Cursory tests in the authors' laboratory using 3.8-cm-thick AISI 304 stainless steel specimens revealed a moderate temperature rise above the 4-K test temperature, but there was no measurable effect on fatigue crack growth rates. The possible effects of adiabatic heating at extreme cryogenic temperatures need to be investigated further, in alloys that exhibit a strong sensitivity to temperatures between 77 and 4 K.

8.6.3 Test environment

The room-temperature fatigue properties of some alloys are more sensitive to the chemical environment than the static tensile properties or fracture toughness. Fortunately, most cryogenic environments are inert or their chemical reactivity is abated, so that few problems with structural alloys are encountered. However, cryogenic oxygen and fluorine may attack some materials, and liquid nitrogen at 76 K causes crazing in some polymers (Hiltner and Baer, 1974). Some room-temperature environments, normally considered benign, may actually cause accelerated cracking. For example, unconditioned room-temperature air contains enough moisture to accelerate the fatigue crack growth rates of a wide variety of steel and aluminum alloys (Wei, 1970). This moisture effect must be recognized when comparing data for inert cryogenic environments with data for unconditioned laboratory air at room temperature. Unless data for dehumidified air are available, it may be difficult to differentiate between temperature effects and chemical reaction.

8.6.4 Specimen orientation

Crack growth rates in the intermediate range are relatively insensitive to specimen orientation. At high ΔK values, however, orientation effects become obvious and are usually associated with inhomogeneities due to grain structure, second-phase particles, or inclusion distribution and morphology.

For an alloy such as recrystallization-annealed Ti–6Al–4V (ELI), which has a nearly symmetric grain structure, it is not surprising that the fatigue crack growth rates in three orthogonal directions were found to be equivalent (Tobler,

1978). Some wrought alloys having elongated grain structures also exhibit nearly isotropic fatigue crack growth resistance. An example is rolled and annealed 5083-O aluminum alloy (Tobler and Reed, 1977a). Despite obvious microstructural inhomogeneity, there was no measurable difference in the crack growth rates for TL and LT orientations, but there was a 37% variation in the J_{Ic} values at 76 K. Similarly, results for annealed Fe–13Cr–19Mn–0.2N alloy demonstrate a 40% difference in J_{Ic} for the TL and LT orientations at 76 K, but there is no difference in crack growth rates for these two orientations (Tobler et al., 1978). Apparently, there is no simple correlation between the monotonic and cyclic cracking resistances for these materials.

8.6.5 Fatigue life calculations

Equation 8.21 is frequently used to estimate the life of a cracked structure subject to fatigue. For a crack assumed to exist at a selected location in a structure, the relationship between K and crack length is determined by stress analysis or from handbook values. Experimental data for the material relate the fatigue crack growth rate and the critical crack size to ΔK and K_c, respectively. Life is calculated by integrating Eq. 8.21 between the limits set by the initial flaw size and the final size, based on fracture toughness data:

$$N = \int_{a_i}^{a_c} da/[c(\Delta K)^n]$$

$$= 1/[c(\Delta\sigma)^n] \int_{a_i}^{a_c} da/[f(a)]^n \qquad (8.23)$$

For complex load histories, $\Delta\sigma$ takes on many values as a function of time, and Eq. 8.23 must be integrated sequentially using numerical methods. Several computerized techniques have been developed to evaluate Eq. 8.23, including some to account for load interaction effects. A load interaction effect is the beneficial effect of peak loads on subsequent low-load growth rates. For example, da/dN is frequently reduced below expected values during the low-stress amplitude cycles that follow high-stress amplitude cycles.

8.7 TEMPERATURE DEPENDENCE OF FATIGUE CRACK GROWTH

The mechanisms of fatigue crack growth are not clearly understood. The exponential dependence of growth rate on ΔK implies that cyclic crack tip plasticity, the extent of which is proportional to ΔK, plays an important role. In addition, certain crack-tip processes, such as chemical reactions, accelerate growth rates at a constant ΔK. At cryogenic temperatures, fatigue crack growth rates may be less than at room temperature owing to elevation of the yield strength, which reduces cyclic plasticity. Exceptions occur when the low temperatures cause phase transformations or change the micromode of fracture. Accelerated growth caused by chemical reactions does not occur at cryogenic temperatures owing to insufficient thermal activation. In the following subsections, the fatigue crack growth behavior of structural alloys at cryogenic temperatures is reviewed for four groups of alloys: stable f.c.c. alloys, austenitic stainless steels, ferritic steels, and titanium alloys. Consideration of metallurgical factors influencing fatigue crack growth rates, as in § 8.5, is difficult because of the limited data available.

8.7.1 Stable f.c.c. alloys

Stable alloys usually exhibit improved fatigue crack growth resistance at cryogenic temperatures. Typical are the data for AISI 310S stainless steel, shown in Fig. 8.16. The fatigue crack

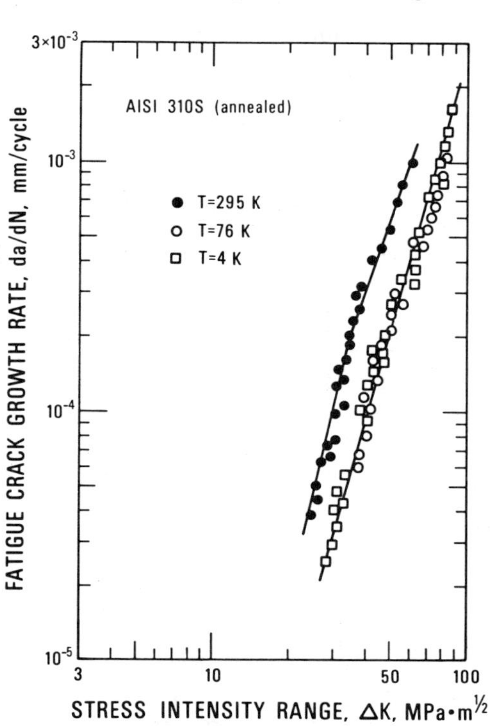

Figure 8.16 Effect of temperature on the fatigue crack growth rates of AISI 310S, an f.c.c. alloy.

growth rates of this alloy at a given ΔK value are lower at 76 and 4 K than at room temperature, but there is no difference in rates at the two cryogenic temperatures. Other stable f.c.c. alloys (such as 5083-O aluminum, Fe–25Ni–20Cr and Fe–25Ni–14Cr stainless steels, and nickel-based superalloys) show similar behavior (Tobler and Reed, 1979).

In Fig. 8.17 the 4 K fatigue crack growth rates of a number of common cryogenic alloys are compared. When plotted as da/dN vs. ΔK, the fatigue crack growth rates show broad differences between the alloys, with the higher modulus alloys exhibiting lower rates at a given ΔK value. When the effect of modulus is normalized by plotting da/dN vs. $\Delta K/E$, it appears that alloys having stable f.c.c. crystal structures are generally preferable to those having b.c.c. or h.c.p. crystal structures (Tobler and Reed, 1977b).

Metallurgical treatments that reduce the toughness of f.c.c. alloys, as discussed in § 8.5, have less effect on fatigue crack growth. For example, Logsdon, Wells, and Kossowsky (1976) and Logsdon (1977) report on the basis of tests at 4 K that:

a. Cold-worked Cr–Ni–Mn stainless steel had a 70% reduction in toughness but improved fatigue crack growth resistance, compared with the annealed condition.

b. Grain-boundary carbides in 310S stainless steel reduced toughness 50% but did not affect fatigue crack growth.

c. Precipitation heat treatment of Inconel X-750 reduced toughness 40% but did not affect fatigue crack growth.

In the same investigation, grain-boundary carbide precipitation in Inconel X-750 had a deleterious effect on both toughness and fatigue crack growth. These findings indicate that a simple correlation between fracture toughness and fatigue crack growth at intermediate ΔK values is not forthcoming.

8.7.2 Austenitic stainless steels

Unstable austenitic stainless steels exhibit martensitic transformations to h.c.p. and b.c.c. phases when cooled to or stressed at cryogenic temperatures. Since these low-temperature transformations occur locally within the plastic zone, crack growth resistance can be significantly affected; adverse or favorable effects may be observed, depending on alloy composition. Results for the Fe–Cr–Ni stainless steels AISI 304 and AISI 316 reveal no measurable difference in fatigue crack propagation rates at 295, 76, and 4 K, although extensive transformation occurs at the cryogenic temperatures. Results for AISI 304L alloy show reduced cracking rates in association

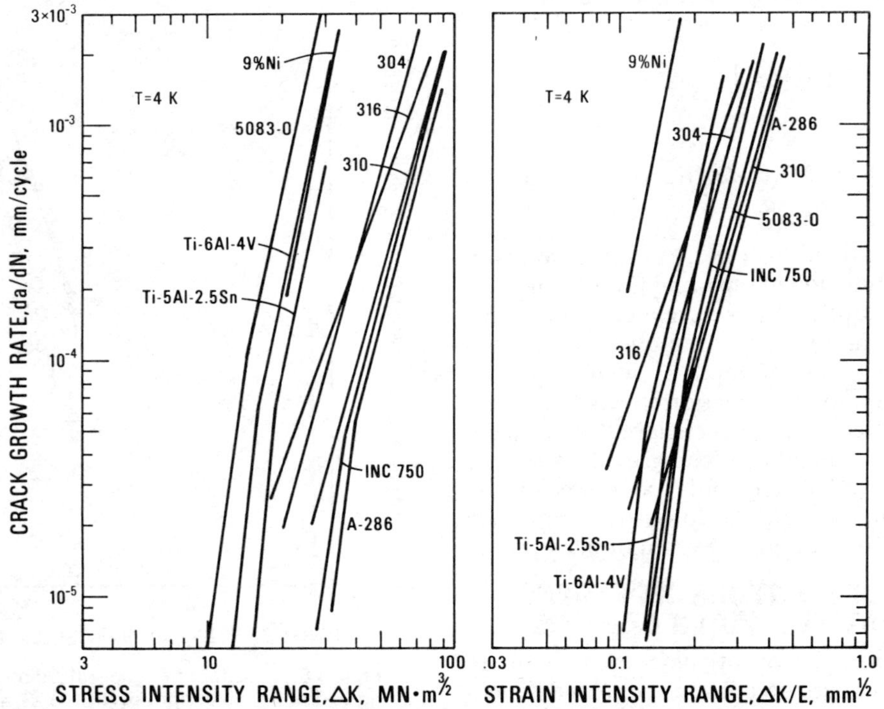

Figure 8.17 Normalized fatigue crack growth rate data for structural alloys.

with transformations at 76 and 4 K, whereas results for AISI 304N show the opposite trend (Tobler and Reed, 1979).

Austenite stability may have a significant influence on fatigue crack growth in nitrogen-strengthened stainless steels. In Fe–Cr–Ni alloys, AISI 304N and 304LN exhibit 4 to 5 times greater growth rates at 4 K than AISI 304 or 304L. The growth rates of Fe–Cr–Ni–Mn alloys at 4 K appear to decrease with increasing austenite stability in the order: Fe–18Cr–3Ni–9Mn, Fe–21Cr–6Ni–9Mn, and Fe–22Cr–13Ni–6Mn. Austenite stability is discussed further in Chapter 9.

8.7.3 Ferritic steels

The fatigue crack growth behaviors of ferritic steels show dramatic changes at low temperatures, which correlate with their fracture toughness behavior. In general, reduced temperatures have insignificant effects or are slightly beneficial as long as ductile cracking modes are operative. The onset of cleavage, however, ultimately causes a dramatic acceleration of fatigue crack growth at the lowest temperatures. The behavior of 9% Ni steel, shown in Fig. 8.18, is typical: as the temperature is reduced, there is a slight improvement between 295 and 111 K followed by a rapid loss of fatigue crack growth resistance between 111 and 4 K. The temperature dependence of fatigue crack growth resistance in this case closely parallels the temperature dependence of fracture toughness (cf. Figs. 8.18 and 8.8).

8.7.4 Titanium alloys

Fatigue data available for h.c.p. alloys are limited to the results for two titanium alloys: Ti–5Al–2.5Sn (single phase, h.c.p.) and Ti–6Al–4V (two phase, h.c.p. and some b.c.c.). Current data indicate that, within measurement abilities, these alloys have nearly temperature-independent fatigue crack growth behavior in the intermediate ΔK range. Data at temperatures between 295 and 4 K are shown in Fig. 8.19 for a low-interstitial Ti–5Al–2.5Sn alloy. Other observations for Ti–6Al–4V alloys suggest that the temperature independence extends from 4 K to at least as high as 563 K (Tobler, 1978; Wei and Ritter, 1972).

8.8 FRACTURE MECHANICS BEHAVIOR OF NONMETALS

Attempts have been made to characterize the failure of polymers, ceramics, and composites us-

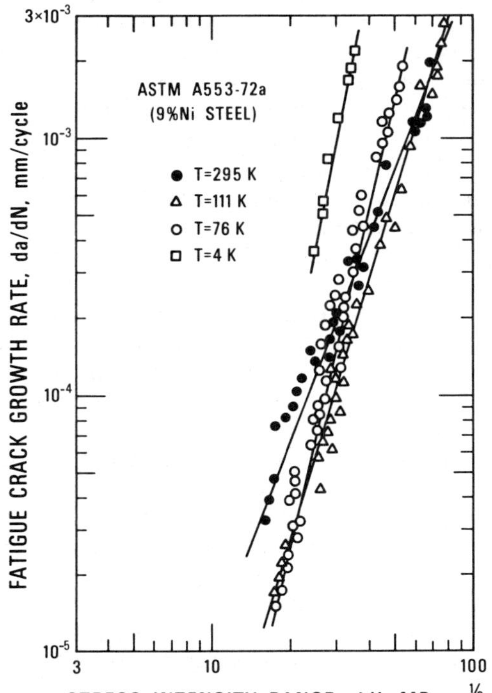

Figure 8.18 Effect of temperature on the fatigue crack growth rates of Fe–9Ni steel, a b.c.c. alloy.

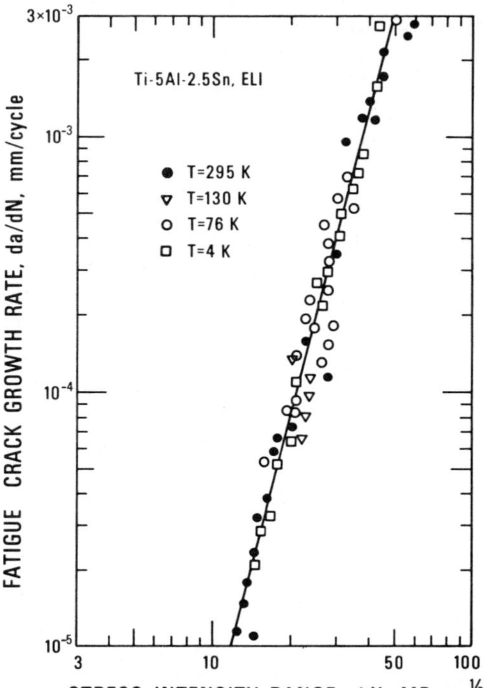

Figure 8.19 Effect of temperature on the fatigue crack growth rates of Ti–5Al–2.5Sn, an h.c.p. alloy.

ing established fracture mechanics parameters. For nonmetals, cryogenic data are quite rare, and there are no standard fracture toughness test procedures. Thus, although the concepts of fracture mechanics have been shown to be applicable to polymers and ceramics, the data generated are primarily of scientific interest. There are some cryogenic data for foams (Fowlkes, 1974), but none for glasses.

8.8.1 Polymers

Linear elastic fracture toughness tests have been applied to some polymers that experience only small amounts of plastic deformation prior to fracture. A review of extensive data for polymethylmethacrylate (Marshall and Williams, 1973) indicates the critical stress intensity factor is a valid fracture parameter for this strain-rate-sensitive material, providing the effects of crack velocity are considered. Most work in this area pertains to room-temperature conditions, but low-temperature studies on thermosetting epoxies (Kneifel, 1979) and structural thermoplastics, such as polycarbonate (Martin and Gerberich, 1976), are available. The K_{Ic} values for plastics are usually less than 10 MPa$\cdot$m$^{1/2}$ at low temperatures. Martin and Gerberich's data for polycarbonate between 325 and 225 K show a decrease from 4.5 to 2 MPa$\cdot$m$^{1/2}$, but at still lower temperatures ($\sim$100 K) K_{Ic} increases again to 5 MPa$\cdot$m$^{1/2}$. Changes in fatigue crack growth rates, as described by the Paris equation, correlated well with K_{Ic} trends at all temperatures.

8.8.2 Ceramics

Ceramics, such as alumina (Al_2O_3) and other hot-pressed or sintered polycrystalline oxides, nitrides, and carbides, exhibit low, temperature-insensitive fracture toughness values. For example, commercial grade 96% alumina shows a linear elastic toughness value of 4.3 MPa$\cdot$m$^{1/2}$ at 300 K, increasing to 5.2 MPa$\cdot$m$^{1/2}$ at 77 K (Kirchner, Gruver, and Sotter, 1976).

8.8.3 Composites

Notwithstanding limited success under simplified conditions, the application of fracture mechanics to composites, such as laminated plastics, generally has been unsuccessful because of theoretical and experimental difficulties.

Composites are distinctly inhomogeneous and anisotropic. They do not fail in the way metals do, that is, with fracture resulting from the propagation of a single, well-defined crack. Even under controlled laboratory conditions, it is extremely difficult to produce a uniform crack in these materials, and some results for brittle materials show that physical fracture mechanisms are not fundamentally changed by the introduction of a machined notch (Reifsnider, 1978).

In an angle-ply composite or any other complex layup, there are both opening and sliding fracture modes operative under pure tensile loading, in addition to numerous energy-dissipating mechanisms, such as interface debonding, delamination, and fiber breakage, that produce nonlinearities. No general fracture mechanism with universal applicability has yet been devised, and it appears theoretically impossible to relate the remote load to a conventional stress intensity factor when the damage processes are complex and combined. Other failure criteria for composites are being researched (Beaubien, Clifford, Mast, Mulville, Sutton, Thomas, Tirosh, Wolock, and Hunston, 1978).

8.9 FRACTURE PREVENTION

Many factors contribute to brittle fracture and its prevention. Consider, for example, Eqs. 8.9 and 8.22 for determining critical crack size and cycle life. These relations require an understanding of the material properties, the service load history, probable initial flaw sizes, and the K vs. a function at the crack location. A complete understanding of all these aspects obviously is not possible for a complex structure. Thus, to ensure safety, it is necessary to limit the uncertainties with a fracture control plan, which consists of the following elements:

a. Establishment of design criteria and analysis assumptions.

b. Design for damage tolerance.

c. Implementation of the controls necessary to ensure the validity of design assumptions.

The design criteria and analysis assumptions need to be clearly defined in the procurement specification or agreed upon early in the design phase. Criteria relevant to fracture control include the design service life, the anticipated service-load history, and the thermal and chemical environment. The principal analysis assumptions are the initial flaw size and location, the fracture criterion (plane strain or plane stress), and the load used in critical-flaw-size calculations.

Material and structural technologies must be integrated into the design of safe, functional, and economical structures. The design should incorporate the results of the structural analyses for strength, rigidity, fatigue life, and fracture resistance. Careful attention to design detail is essential because fatigue cracks initiating at stress concentrations are the most likely source of failures.

A variety of controls are necessary to ensure that the fabricated structure is adequately represented by the design and analyses. Material procurement and processing are controlled by specifications, which ensure that the materials used have the fracture properties required by the design. Inspection of the raw material and the final assembly ensures that the flaws in the fabricated structure are smaller than those assumed in the crack growth analysis. For fracture-critical parts, records provide traceability of product quality from the raw material heat identification, through all manufacturing operations, to the final assembly. These records are needed to limit the scope of a retrofit program when service failure is due to inferior material quality.

As cryogenic technology advances, there is an increased tendency to use higher strength materials, higher operating stresses, and thick sections to build larger and more complex structures. These trends increase the hazard of brittle fracture and sometimes justify the use of fracture mechanics methods. In the following subsections, the potential uses of fracture mechanics in conventional structures are proposed and sophisticated applications of fracture mechanics to the design of liquefied-natural-gas (LNG) ships and aerospace pressure vessels are reviewed.

8.9.1 Storage tanks and piping systems

For cryogenic structures, such as storage tanks and piping systems, there has been considerable service experience. The design practices, the fabrication methods, and the type and extent of inspection have proven satisfactory. The existing design codes published by the American Society of Mechanical Engineers (ASME), the American Petroleum Institute (API), and the American National Standards Institute (ANSI) fulfill the need for a fracture control plan. There is, however, a trend in engineering design toward increased structural efficiency through the rational determination of structural loads, working stresses, and material properties. This trend has created pressure for changing certain aspects of the design codes in the interest of cost reduction. Fracture mechanics can be used to assess the impact of these changes on structural safety.

In most cases, the allowable design stress levels for materials used in cryogenic structures are based on their strength at room temperature. Because the strength increases at cryogenic temperatures, a higher allowable stress level may be acceptable. Fracture mechanics can provide a quantitative assessment of the effect of increased stress levels on the flaw tolerance of the material at low temperatures and, therefore, can provide a rational basis for modifying allowable stress.

The qualification of materials and welding procedures for fabricating structures for liquid-helium service in accordance with the ASME Pressure Vessel Code of the ANSI Piping Code is difficult. Charpy impact tests are not appropriate at temperatures below 76 K because of testing difficulties and adiabatic heating of the test specimen. Standards have not been established for other tests, such as the notched tension test. Fracture mechanics tests are a bit cumbersome for qualification testing, but may provide a useful basis for comparison with other tests. Examples of this approach are: the correlations of K_{Ic} with unit propagation energy in the tear test for aluminum alloys (Kaufman and Holt, 1965) and correlations of K_{Ic} with the impact energy in Charpy tests for steel alloys (Barsom and Rolfe, 1970) as functions of fracture toughness.

8.9.2 Shipboard LNG tanks

Several types of containment systems have been devised for the marine transport of LNG (Thomas and Schwendtner, 1972). Generally, the containment system consists of a primary barrier to contain the cargo for the life of the ship and a secondary barrier to contain the cargo for 15 days in case the primary barrier fails. Spherical free-standing tank designs do not require a secondary barrier, if fracture mechanics analysis indicates satisfactory "leak-before-break" behavior.

The leak-before-break criterion requires demonstration by test and analysis that a subcritical crack will grow through the thickness of the tank wall and cause a detectable amount of cargo leakage for a safe period of operation prior to failure. The minimum period of safe operation is defined as 15 days at sea under storm conditions. Fracture mechanics analysis is used to compute the minimum flaw size that will allow leakage of sufficient gas to be sensed by the gas detectors (Tenge and Solli, 1973). The period of safe operation is determined by integrating Eq. 8.23 between the limits set by this minimum flaw size and the critical flaw size determined from fracture toughness measurements; crack growth data for the tank material and the appropriate cyclic stress history for storm conditions are used.

8.9.3 Aerospace pressure vessels

Cryogenic vessels used by NASA for launch vehicles and spacecraft gas containers are designed, fabricated, and tested in accordance with an overall fracture control plan [*NASA Space Vehicle Design Criteria* (*Structures*), 1970]. One of the principal elements of this plan is the proof pressure test. The proof test concept is shown schematically in Fig. 8.20. The proof test is conducted at a pressure and a stress level, σ_{pr}, that exceeds the maximum service pressure and stress level, σ_s; thus, the critical crack size during proof test, a_{pt}, is smaller than the critical crack size during service, a_s. The flaw size that could exist in the pressure vessel following proof test is conservatively estimated to be equal to a_{pt}; that is, any flaw larger than a_{pt} would have caused failure during proof test. During service, there is a period of safe operation until the initial flaw grows to the critical size, a_s. This period of safe operation is determined from empirical data or Eq. 8.23 (relating crack length to service time, t) with all the appropriate loads and environments simulated in the test or analysis.

8.9.4 Future applications

In the future, fracture mechanics will be used to establish a more rational basis for design codes and to improve standards for material acceptance, welding procedure qualification, and nondestructive inspection. In addition, many new design ideas will be developed for constructing superconducting devices and hydrogen fuel equipment. Superconducting systems for energy-storage and fusion magnets will involve enormous structures; capital investment considerations will justify the use of fracture mechanics methods. Public safety considerations probably will lead to the application of fracture mechanics to the design of equipment for storing, handling, and transporting liquid-hydrogen fuel.

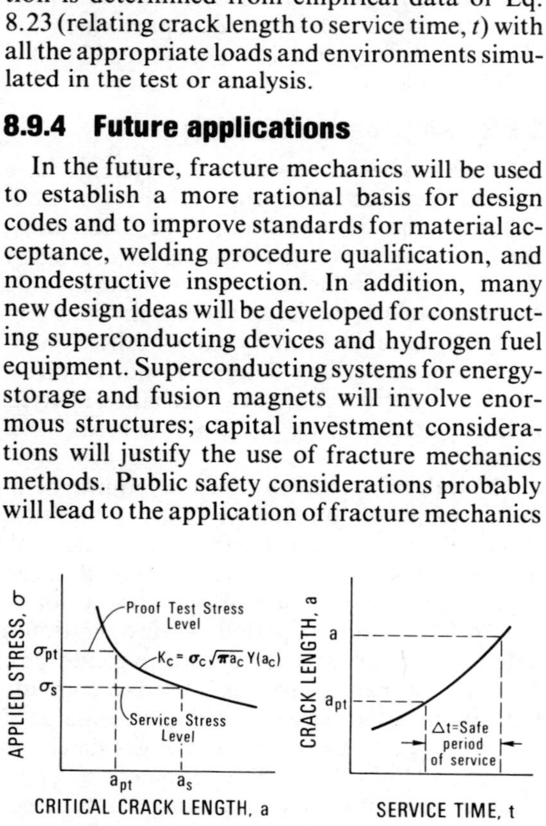

Figure 8.20 The proof test concept.

8.10 SOURCES OF INFORMATION

An abundance of information on fracture mechanics is available. Textbooks have been written by Knott (1973), Broek (1974), Lawn and Wilshaw (1975), and Rolfe and Barsom (1977). Extensive surveys of specific applications for aircraft structures (Liebowitz, 1974) and space vehicles (Liu, Ehret, and Malcahy, 1974) have been prepared. Analysis handbooks that include compilations of stress intensity factor solutions have been written by Tada et al. (1973) and Sih (1973). Material property handbooks are beginning to include fracture mechanics test data, and the *Damage Tolerant Design Handbook* (1973) is specifically limited to the compilation of fracture mechanics data on aerospace materials. Experimental methods in fracture mechanics are reviewed in a training manual written by Hickerson (1973) and a monograph, *Experimental Techniques in Fracture Mechanics*, edited by Kobayashi (1973). Fracture mechanics concepts are applied to failure analysis in a book by Colangelo and Heiser (1974).

Current research in fracture mechanics is published quarterly by two journals: *Engineering Fracture Mechanics* and *International Journal of Fracture*. The American Society for Testing and Materials (ASTM) regularly publishes the proceedings of symposia on fracture mechanics topics, including the papers presented at the annual National Symposium on Fracture Mechanics. Work in progress is presented at the semiannual meetings of ASTM committee E-24 on Fracture Testing of Metals.

8.11 REFERENCES

Annual Book of ASTM Standards (1981a). Standard practice for R-curve determination. E561-81, American Society for Testing and Materials, Philadelphia, Pennsylvania, 673–692.

Annual Book of ASTM Standards (1981b). Standard test method for plane-strain fracture toughness of metallic materials. E399-81, American Society for Testing and Materials, Philadelphia, Pennsylvania, 588–618.

Annual Book of ASTM Standards (1981c). Standard test for J_{Ic}, a measure of fracture toughness: E813-81, American Society for Testing and Materials, Philadelphia, Pennsylvania, 810–828.

Annual Book of ASTM Standards (1981d). Standard test method for constant amplitude fatigue crack growth rates above 10^{-8} m/cycle. E647-81, American Society for Testing and Materials, Philadelphia, Pennsylvania, 765–783.

Antolovich, S. D. and Chanai, G. R. (1975). Errors associated with fracture toughness testing. In *Third International Conference on Fracture*. Paper II-242, Verein Deutscher Eisenhüttenleute, Dusseldorf, West Germany, 1–5.

Barsom, J. M. and Rolfe, S. T., (1970). Correlations between K_{Ic} and Charpy V-notch test results in the transition-temperature range. In *Impact Testing of Metals*. ASTM STP 466, American Society for Testing and Materials, Philadelphia, Pennsylvania, 281–302.

Beaubien, L. A., Clifford, M. F., Mast, P. W., Mulville, D. R., Sutton, S. A., Thomas, R. W., Tirosh, J., Wolock, I., and Hunston, D. L. (1978). Failure criteria for composite structures. In *High Performance Composites and Adhesives for V/STOL Aircraft* (eds. W. D. Bascom and L. B. Lockhard, Jr.). NRL Memorandum Report 3721, Naval Research Laboratory, Washington, D.C., 1–148.

Begley, J. A. and Landes, J. D. (1972). The *J*-integral as a fracture criterion. In *Fracture Toughness*. ASTM STP 514, American Society for Testing and Materials, Philadelphia, Pennsylvania, 1–20.

Broek, D. (1974). *Elementary Engineering Fracture Mechanics*. Noordhoff International, Leiden, Netherlands.

Carlsson, A. J. and Markstrom, K. M. (1977). Some aspects of nonlinear fracture mechanics. In *Fracture 1977*, Vol. 1 (ed. D. M. R. Taplin). University of Waterloo Press, Waterloo, Ontario, Canada, 683–694.

Clark, W. G., Jr. (1974). How fatigue crack initiation and growth properties affect material selection and design criteria. *Met. Eng. Q.* **14**, 16–22.

Clark, W. G., Jr. and Hudak, S. J., Jr. (1975). Variability in fatigue crack growth rate testing. *J. Test. Eval.* **3**, 454–476.

Colangelo, V. J. and Heiser, F. A. (1974). *Analysis of Metallurgical Failures*. John Wiley & Sons, New York.

Crooker, T. W. and Lange, E. A. (1970). How yield strength and fracture toughness considerations can influence fatigue design procedures for structural steels. *Weld. J. Res. Suppl.* **59**, 488s–496s.

Damage Tolerant Design Handbook (1973). MCIC HB-01 and MCIC He-01S, Metals and Ceramics Information Center, Battelle–Columbus Laboratories, Columbus, Ohio.

Dowling, N. E. and Begley, J. A. (1976). Fatigue crack growth during gross plasticity and the *J*-integral. In *Mechanisms of Crack Growth*, ASTM STP 590. American Society for Testing and Materials, Philadelphia, Pennsylvania, 82–103.

Experimental Techniques in Fracture Mechanics (1973) (ed. A. S. Kobayashi). Iowa State University Press and Society of Experimental Stress Analysis, Ames, Iowa.

Forman, R. G., Kearney, V. E., and Engle, R. M. (1967). Numerical analysis of crack propagation in cyclic-loaded structures. *J. Basic Eng.* **89**, 459–464.

Fowlkes, C. W. (1974). Fracture toughness tests of a rigid polyurethane foam. *Int. J. Fract.* **10** (1), 99–108.

Fowlkes, C. W. and Tobler, R. L. (1976). Fracture testing and results for a Ti–6Al–4V alloy at liquid helium temperatures. *Eng. Fract. Mech.* **8**, 487–500.

Goode, R. J. (1972). Identification of fracture plane orientation. *Mater. Res. Stand.* **12**, 31–32.

Griffith, A. A. (1920). The phenomena of rupture and flow in solids. *Philos. Trans. Soc. Lond. A.* **221**, 163–198.

Hahn, G. T., Hoaglund, R. G., and Rosenfield, A. R. (1971). The variation of K_{Ic} with temperature and loading rate. *Metall. Trans.* **2**, 537–541.

Harrod, D. L., Hengstenberg, T. F., and Manjoine, M. J. (1969). Fracture toughness of beryllium. *J. Mater.* **4**, 618–632.

Hartwig, G. and Wuchner, F. (1975). Low temperature mechanical testing machine. *Rev. Sci. Instrum.* **46** (4), 481–485.

Hayes, D. J. (1970). Some applications of elastic plastic analysis to fracture mechanics. Ph.D. thesis, University of London, London, England.

Hickerson, F. P. (1973). *Elements of Experimental Fracture Mechanics*. SLA–73–0424, Sandia Laboratories, Albuquerque, New Mexico.

Hiltner, A. and Baer, E. (1974). Mechanical properties of polymers at cryogenic temperatures. *Polymer* **15**, 805–813.

Horiuchi, T., Shimada, M., Fukutsuka, T., and Tokuda, S. (1977). Design and construction of an apparatus for testing materials at cryogenic temperature. In *Proceedings of the Fifth International Cryogenic Engineering Conference*. IPC Science and Technology Press, Guildford, England, 465–468.

Hutchinson, J. W., Needleman, A., and Shih, C. F. (1978). Fully plastic crack problems in bending and tension. In *Fracture Mechanics* (eds. N. Perrone, H. Liebowitz, D. Mulville, and W. Pilkey). University Press of Virginia, Charlottesville, Virginia, 515–523.

Irwin, G. R. (1957). Analysis of stresses and strains near the end of a crack traversing a plate. *J. of Appl. Mech.* **24**, 361–364.

Irwin, G. R. (1960a). Plastic zone near a crack and fracture toughness. In *Proceedings, 7th Sagamore Ordnance Materials Research Conference*, Syracuse University Press, Syracuse, New York, IV63–IV78.

Irwin, G. R. (1960b). Fracture mode transition of a crack traversing a plate. *J. Basic Eng.* **82**, 417–425.

Irwin, G. R. and Kies, J. A. (1954). Critical energy rate analysis of fracture strength. *Weld. J. (N.Y.)* **33**, 193s–198s.

James, L. A. (1976). Fatigue crack propagation in austenitic stainless steels. *At. Energy Rev.* **14**, 37–86.

Jin, S., Morris, J. W., Jr., and Zackay, V. F. (1975). Grain refinement through thermal cycling in an Fe–Ni–Ti cryogenic alloy. *Metall. Trans.* **6A**, 141–149.

Jolley, W. (1968). Influence of a 3.28 pct nickel addition on the yield and fracture behavior of alpha iron. *Trans. Metall. Soc. AIME* **242**, 306–314.

Kaufman, J. G. and Holt, M. (1965). *Fracture Characteristics of Aluminum Alloys*. ARL Technical Paper No. 18, Aluminum Company of America, Pittsburgh, Pennsylvania.

Kaufman, J. G., Nelson, F. G., and Wygonik, R. H. (1974). In *Fatigue and Fracture Toughness — Cryogenic Behavior*. ASTM STP 566, American Society for Testing and Materials, Philadelphia, Pennsylvania, 125–158.

Kirchner, H. P., Gruver, R. M., and Sotter, W. A.

(1976). Characteristics of flaws at fracture origins and fracture-stress–flaw-size relations in various ceramics. *Mater. Sci. Eng.* **22**, 147–156.

Kneifel, B. (1979). Fracture properties of epoxy resins at low temperatures. In *Nonmetallic Materials and Composites at Low Temperatures*. Plenum Press, New York, 123–129.

Knott, J. F. (1973). *Fundamentals of Fracture Mechanics*. Halsted Press, John Wiley & Sons, New York.

Kossowsky, R. (1977). Microstructure of Inconel X-750 materials for cryogenic applications. In *Advances in Cryogenic Engineering*, Vol. 22. Plenum Press, New York, 59–68.

Krafft, J. M., Sullivan, A. M., and Boyle, R. W. (1961). Effect of dimensions on fast fracture instability of notched sheets. In *Proceedings, Crack Propagation Symposium, Vol. I*. College of Aeronautics, Cranfield, England, 8–28.

Lake, R. L., (1975). Toughness variations through the thickness of thick 5083-O aluminum alloy plate. In *Properties of Materials for LNG Tankage*. ASTM STP 579, American Society for Testing and Materials, Philadelphia, Pennsylvania, 176–189.

Landes, J. D., and Begley, J. A. (1974). Test results from *J*-integral studies — an attempt to develop a *J*-integral test procedure. In *Fracture Analysis*. ASTM STP 560, American Society for Testing and Materials, Philadelphia, Pennsylvania, 170–186.

Lawn, B. R. and Wilshaw, T. R. (1975). *Fracture of Brittle Solids*. Cambridge University Press, London.

Liebowitz, H. (1974). *Fracture Mechanics of Aircraft Structures*. AGARDograph No. 176, AD–777–641, NATO Advisory Group for Aerospace Research and Development.*

Liu, A. F., Ehret, R. M., and Malcahy, E. J. (1974). *Fracture Control Methods for Space Vehicles*, Vols. 1–3. NASA, CR-134596, CR-134597, and CR-134598.*

Logsdon, W. A. (1977). Cryogenic fracture mechanics properties of several manufacturing process/heat treatment combinations of Inconel X-750. In *Advances in Cryogenic Engineering*, Vol. 22. Plenum Press, New York, 47–58.

Logsdon, W. A., Wells, J. M., and Kossowsky, R. (1976). Fracture mechanics properties of austenitic stainless steels for advanced cryogenic applications. In *Proceedings of the Second International Conference on Mechanical Behavior of Materials*. American Society for Metals, Metals Park, Ohio, 1283–1289.

Marshall, G. P. and Williams, J. B. (1973). The correlation of fracture data for PMMA. *J. Mater. Sci.* **8**, 138–140.

Martin, G. C. and Gerberich, W. W. (1976). Temperature effects on fatigue crack growth in polycarbonate. *J. Mater. Sci.* **11**, 231–238.

McCabe, D. E. and Heyer, R. H. (1973). *R*-curve determination using a crack-line-wedge-loaded (CLWL) specimen. In *Fracture Toughness Evaluation by* R-*curve Methods*. ASTM STP 527, American Society for Testing and Materials, Philadelphia, Pennsylvania, 17–35.

McHenry, H. I. (1971). A compliance method for crack growth studies at elevated temperatures. *J. Mater.* **6**, 862–873.

McHenry, H. I. and Reed, R. P. (1977). Fracture behavior of the heat affected zone in 5% Ni steel weldments. *Weld. J.* **56**, 1045–1125.

Methods for Crack Opening Displacement (*COD*) *Testing* (1979). BS 5762:1979, British Standards Institution, London, England.

NASA Space Vehicle Design Criteria (*Structures*) (1970). Fracture control of metallic pressure vessels. NASA SP-8040.*

Novikov, N. V. (1977). Mechanical property measurement techniques of structural materials at cryogenic temperatures. In *Advances in Cryogenic Engineering*, Vol. 22, Plenum Press, New York, 113–118.

Paris, P. C. (1964). The fracture mechanics approach to fatigue. In *Fatigue—An Interdisciplinary Approach*. Syracuse University Press, Syracuse, New York, 107–133.

Paris, P. C. (1977). Fracture mechanics in the elastic–plastic regime. In *Flaw Growth and Fracture*. ASTM STP 631, American Society for Testing and Materials, Philadelphia, Pennsylvania, 3–27.

Paris, P. C. and Sih, G. C. (1965). Stress analysis of cracks. In *Fracture Toughness Testing and Its Applications*. ASTM STP 381, American Society for Testing and Materials, Philadelphia, Pennsylvania, 30–81.

Paris, P. C., Tada, H., Zahoor, A., and Ernst, H. (1977). The theory of instability of the tearing mode of elastic–plastic crack growth. In *Elastic–Plastic Fracture* (eds. J. D. Landes, J. A. Begley, and G. A. Clarke). ASTM STP 668, American Society for Testing and Materials, Philadelphia, Pennsylvania, 5–36.

Porter, L. F. (1970). Metallurgical factors affecting the toughness of steels. Paper presented at the ASM Conference on Fracture Control, Philadelphia, Pennsylvania. Available from the author, U.S. Steel, Pittsburgh, Pennsylvania.

Porter, L. F. and Dabkowski, D. S. (1970). Grain size control by thermal cycling. In *Proceedings of the Sixteenth Sagamore Army Materials Research Conference*. Syracuse University Press, Syracuse, New York, 133–161.

Read, D. T., McHenry, H. I., Steinmeyer, P. A., and Thomas, R. D., Jr. (1980). Metallurgical factors affecting the toughness of 316L SMA weldments at cryogenic temperatures. *Weld. J.* **59**, 104s–113s.

Read, D. T. and Reed, R. P. (1978). Toughness, fatigue crack growth and tensile properties of three nitrogen-strengthened stainless steels at cryogenic temperatures. In *The Metal Science of Stainless Steels* (eds. E. W. Collings and H. W. King). Metallurgical Society of AIME, Denver, Colorado, 92–121.

Reed, R. P. (1962). A cryostat for tensile tests in the temperature range 300° to 4°K. In *Advances in Cryogenic Engineering*, Vol. 7. Plenum Press, New York, 448–454.

Reifsnider, K. L. (1978). Mechanics of failure of composite materials. In *Fracture Mechanics* (eds. N. Perrone, H. Liebowitz, D. Mulville, and W. Pilkey). University of Virginia, Charlottesville, Virginia, 317–332.

Rice, J. R. (1968). A path independent integral and approximate analysis of strain concentration by

notches and cracks. *J. Appl. Mech.* **35**, 379–386.
Rolfe, S. T. (1970). *Fracture Criteria for Structural Steels*. Paper presented at the American Society for Metals Conference on Fracture Control, Philadelphia, Pennsylvania. Available from the author, University of Kansas, Lawrence, Kansas.
Rolfe, S. T. and Barsom, J. M. (1977). *Fracture and Fatigue Control in Structures*. Prentice-Hall, Inc., Englewood Cliffs, New Jersey.
Sarno, D. A., Bruner, J. P. and Kampschaefer, G. E. (1974). Fracture toughness of 5% nickel steel weldments. *Weld. J.* **53**, 486.
Shoemaker, A. K., and Rolfe, S. T. (1971). The static and dynamic low temperature crack-toughness performance of seven structural steels. *Eng. Fract. Mech.* **2**, 319–339.
Sih, G. C. (1973). *Handbook of Stress Intensity Factors for Researchers and Engineers*. Institute of Fracture and Solid Mechanics, Lehigh University, Bethlehem, Pennsylvania.
Srawley, J. E. and Brown, W. F., Jr. (1965). Fracture toughness testing methods. In *Fracture Toughness Testing and Its Applications*, ASTM STP 381, American Society for Testing and Materials, Philadelphia, Pennsylvania, 133–196.
Stonesifer, F. R. and Armstrong, R. W. (1977). Effect of prior austenite grain size on the fracture toughness properties of A 533 B steel. In *Fracture 1977, Vol. 2.* (ed. D. M. R. Taplin). University of Waterloo Press, Waterloo, Ontario, Canada, 1–6.
Sullivan, A. M., Freed, C. N., and Stoop, J. S. (1973). Comparison of *R*-curves determined from different specimen types. In *Fracture Toughness Evaluation by* R-*curve Methods*. ASTM STP No. 527, American Society for Testing and Materials, Philadelphia, Pennsylvania, 85–104.
Szumakowski, E. R. and Reid, H. F. (1978). Cryogenic toughness of SMA austenitic stainless steel weld metals: part 1—role of ferrite. *Weld. J.* **57**, 325s–333s.
Tada, H., Paris, P. C., and Irwin, G. R. (1973). *The Stress Analysis of Cracks Handbook*. Del Research Corp., Hellertown, Pennsylvania.
Tenge, P. and Solli, O. (1973). Fracture mechanics in the design of large spherical tanks for ship transport of LNG. *Norw. Maritime Res.* **1** (2), 1–17.
Tetelman, A. S. and McEvily, A. J., Jr. (1967). *Fracture of Structural Materials*. John Wiley & Sons, New York.
Thomas, W. E. and Schwendtner, A. H. (1972). LNG carriers: the current state of the art. *Oceanol. Intern.* **7**, 1A.
Tobler, R. L. (1978). Low temperature fracture behavior of a Ti–6Al–4V alloy and its electron beam welds. In *Toughness and Fracture Behavior of Titanium*. ASTM STP 651, American Society for Testing and Materials, Philadelphia, Pennsylvania, 267–294.
Tobler, R. L., McHenry, H. I., and Reed, R. P. (1978). Fracture mechanics parameters for an iron–13% chromium–19% manganese stainless steel and its welds at cryogenic temperatures. In *Advances in Cryogenic Engineering*, Vol. 24. Plenum Press, New York, 560–572.
Tobler, R. L., Mikesell, R. P., Durcholz, R. L., and Reed, R. P. (1975). Low temperature fracture behavior of iron–nickel alloy steels. In *Properties of Materials for LNG Tankage*. ASTM STP 579, American Society for Testing and Materials, Philadelphia, Pennsylvania, 261–287.
Tobler, R. L. and Reed, R. P. (1977a). Fracture mechanics parameters for a 5083-O aluminum alloy at low temperatures. *J. Eng. Mater. Technol.* **99**, 306–312.
Tobler, R. L. and Reed, R. P. (1977b). Fatigue crack growth rates of structural alloys at 4 K. In *Advances in Cryogenic Engineering*, Vol. 22. Plenum Press, New York, 35–46.
Tobler, R. L. and Reed, R. P. (1979). Fatigue crack growth resistance of structural alloys at cryogenic temperatures. In *Advances in Cryogenic Engineering*, Vol. 24. Plenum Press, New York, 82–90.
Turner, C. E. and Burdekin, F. M. (1974). Review of current status of yielding fracture mechanics. *At. Energy Rev.* **12**, 439–503.
Wei, R. P., (1970). Some aspects of environment-enhanced fatigue crack growth. *Eng. Fract. Mech.* **1**, 633–651.
Wei, R. P. and Ritter, D. L. (1972). The influence of temperature in crack growth in a mill-annealed Ti–6Al–4V alloy. *J. Mater.* **7**, 240–247.
Wells, A. A. (1964). Notched bar tests, fracture mechanics and the brittle strength of welded structures. *Weld. World* **2** (4), 198–219.
Wells, J. M., Kossowsky, R., Logsdon, W. A., and Daniel, M. R. (1976). Structural materials for cryogenic applications. In *Materials Research for Superconducting Machinery V* (eds. R. P. Reed, H. M. Ledbetter, and E. C. van Reuth). ADA030170, National Bureau of Standards, Boulder, Colorado, 1–69.
Westergaard, H. M. (1939). Bearing pressures and cracks. *J. Appl. Mech.* **61**, A49–53.
Zackay, V. F., and Parker, E. R. (1976). The changing role of metastable austenite in the design of alloys. In *Annual Review of Materials Science, Vol. 6.* (ed. R. A. Huggins). Annual Reviews, Palo Alto, California, 139–155.

---

*Available from NTIS, Springfield, Virginia.

9 Martensitic Phase Transformations

R. P. REED

9.1 INTRODUCTION

In the late 1800s, the German metallurgist Martens identified a new crystalline phase in steel, which he associated with steel hardening. In 1895, this phase was labeled *martensite* in honor of his research. Since then, the role of martensite in influencing the strength of ferrous alloys has been continuously studied.

Martensitic transformation initially meant the transformation from face-centered-cubic (f.c.c.) austenite to body-centered-cubic (b.c.c.) or body-centered-tetragonal (b.c.t.) martensite in steels. Gradually, however, metallurgists began to describe newly discovered nonferrous transformations as martensitic in nature, and the term martensitic evolved to describe a class of structural transformations. Cohen, Olson, and Clapp (1979), who have recently presented a thorough discussion of the characterization of martensitic transformation, state "A martensitic transformation is a lattice-distortive, virtually diffusionless structural change having a dominant deviatoric component and associated shape change such that strain energy dominates the kinetics and morphology during the transformation." Thus, a change of atomic structure that involves no atomic diffusion and exhibits a change of shape with an attendant undistorted plane is considered a *martensitic phase transformation*. In such transformations, the associated strain energy has significant influence on the final product shape and size and the transformation time and temperature characteristics.

Nucleation restricts the martensitic transformation in all ferrous alloys and in some nonferrous systems. After nucleation, growth to a restricted product morphology usually proceeds very quickly; the rate may approach the speed of sound. Except in certain nonferrous alloys, additional transformation is achieved by the nucleation of new crystals, not by the growth of the existing crystals.

Martensitic transformations occur at very low temperatures in such diverse materials as stainless steels, alkali metals, and solidified gases; therefore, atomic diffusion plays no essential role. Atom transfer from parent to product structures proceeds by shear-like, coordinated motion of many atoms, resulting in a shape change. This shape change is easily observed with conventional microscopy techniques.

A consequence of the shear-like, coordinated atomic motion is that there are easily defined geometrical relationships between parent and product phases. An orientation relationship exists between all product-phase and parent-phase crystal structures. The interface plane, separating the parent and product phase, is conventionally called the *habit plane* and is usually unique for a given alloy system. Phenomenological crystallographic theories have been reasonably successful in predicting and explaining the interrelationships of these geometrical parameters, but they are not completely satisfactory. Another consequence of the atom-transfer mode of a martensitic transformation is that the identical atom concentration of the parent phase is preserved in the product phase; there is no composition gradient between the two phases. Finally, locally large stresses, which may exceed the yield strength of the parent material, may result from the shape deformation of the product phase. This is particularly apparent in ferrous-alloy transformations, where extensive slip, faulting, or in special cases, deformation twinning may be induced in the parent phase and where the generation of point defects in the product phase may be about 0.5 at.%.

The study of martensitic transformations has become an extremely large subject. The reference list includes outstanding general books, review papers, and conference proceedings on martensitic transformations. Especially, the reader is referred to Reed and Breedis (1966) for a thorough compilation (702 references) of literature prior to 1966; to Bilby and Christian (1961), Christian (1970), Wayman (1964, 1968), and Lieberman (1970) for excellent descriptions of martensite crystallography; and to Roitburd and Kurdjumov (1979), Entwisle (1971), Christian (1965), and Kaufman and Cohen (1958) for excellent reviews of martensite kinetics. Two reviews have emphasized low temperatures: Barrett (1976) and Reed and Breedis (1966). General books on martensitic transformations include *Martensite Fundamentals and Technology* (1970) and Christian (1976). There have been excellent international conference proceedings on general martensitic research: *Physical Properties of Martensite and Bainite* (1965), *Mechanism of Phase Transformations in Crystalline Solids* (1969), *Shape Memory Effect in Alloys* (1975), *New Aspects of Martensitic Transformations* (1976), and *Phase Transformations* (1970). The International Conference on Martensitic Transformations (ICOMAT) held its third conference in 1979. These ICOMAT conferences bring most of the principal research scientists together every third year. Their proceedings (e.g., *Proceedings of the International Conference on Martensitic Transformation*, ICOMAT, 1979) report current research.

This chapter concentrates on very low-temperature martensitic transformations, which are of great concern for cryogenic applications and research. The principal transformation characteristics are reviewed and then elaborated. The material classes or alloy systems that exhibit martensitic transformations at very low temperatures are discussed in greater detail. In particular, the martensitic transformations and their effects in austenitic stainless steels, iron–nickel alloys, practical superconductors, alkali metals, solidified gases, and polymers are discussed.

9.2 CHARACTERISTICS

All martensitic transformations do not have identical traits, but certain general characteristics are the result of a coordinated shear-type mechanism of atom transfer.

The driving force for a phase transformation is the difference in the Helmholtz free energies of the parent and product phases. The equilibrium temperature, T_0, is defined as the temperature where the free energies of the two phases are equal, as illustrated in Fig. 9.1. Below T_0, the martensitic transformation starts. The temperature of transformation start is labeled T_{ms} in this chapter; this nomenclature is a compromise between the metallurgical expression, M_s, and the physics label, T_m. Cooling below T_0 without transformation is termed *undercooling*, and associated with undercooling is an increase in the free-energy difference of the two phases. The degree of undercooling below T_0 with attendant increase of free-energy difference that is required to initiate the transformation depends, essentially, on the magnitude of the shape change and on the ability of both parent and product phases to accommodate the shape change of the martensitic shear transformation. In Fig. 9.1, a range of possible T_{ms} values is illustrated. Characteristics such as flow strengths, defect formation energies, and elastic constants affect the ease of transformation by contributing to the strain energies associated with the parent and product phases. Materials with small differences of strain energy between parent and product phases have rela-

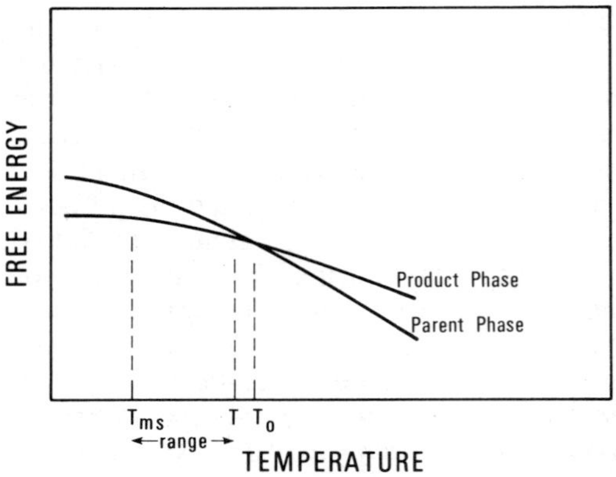

Figure 9.1 Free-energy-vs.-temperature schematic for phase transformation. The equilibrium temperature is T_0; the martensite start temperature is T_{ms}.

tively small differences between T_0 and T_{ms}; materials with larger strain energy differences require more driving energy and start transforming at relatively lower temperatures. Estimates of ferrous-alloy free-energy differences at T_{ms} range from 400 to 2600 J·mol^{-1}; some nonferrous transformation driving forces are estimated at about 40 J·mol^{-1} (Christian, 1979).

During transformation, the free energy is converted into heat and stored energy. The evolved heat can be measured and the transformation enthalpy estimated (e.g., Lee, Millman, MacDougall, Keown, and Argent, 1977). In ferrous alloys the measured enthalpies range from about 1200 to 3200 J·mol^{-1} and depend on product morphology and internal defect structure (stored energy). For instance, Lee et al. (1977) found that considerably less heat is evolved during lath (dislocation) martensite formation than plate (twinned) martensite formation. They attributed the stored energy to transformation dislocations and twins and pointed out that twins are expected to store much less energy (~100 J·mol^{-1}) compared with dislocations (10^{17} dislocations ≅ 1500 J·mol^{-1}). Christian (1979) discusses three major contributions to the stored energy: (a) coherency strains, (b) interphase interface, and (c) shape-change accommodation defects. The accommodation defects include dislocations and twins.

Large concentrations of vacancies or vacancy clusters have also been reported after a martensitic transformation (Reed, Clark, and Schramm, 1971). Their estimated energy of formation (between 0.016 to 0.18 J, 0.1 to 1.1 eV) is high enough to make a substantial contribution to the stored energy. It is not clear whether the vacancies contribute to the accommodation strain; they may be caused by the production or motion of accommodation dislocations or twins or by the advancing interface dislocations.

9.2.1 Nucleation

At discrete positions within the parent material, martensite nucleates and grows quickly to an equilibrium size and morphology. Transformation initiates on cooling below the temperature at which the product and parent phases are in thermodynamic equilibrium. The transformation initiation sites (nuclei) are not random. Calculations indicate that very sizable energies are required for homogeneous nucleation (Christian, 1969), and these could not be supplied by thermal fluctuations. Therefore, it is possible that heterogeneous nucleation begins at sites supplying a lower activation energy barrier. In principle, lattice defects, including free surfaces (Clapp, 1973), and their associated strain fields are capable of being such sites. Olson and Cohen (1976) proposed a model for stacking-fault nuclei in ferrous alloys. Stacking faults, or the presence of a very small amount of hexagonal close-packed (h.c.p.) structure within an f.c.c. structure, were originally proposed as nucleation sites in ferrous alloys by Venables (1962).

Olson and Cohen (1981) recently expanded their suggestions concerning heterogeneous nucleation. In essence they proposed that martensite nucleates via the dissociation of interphase dislocations. These interphase dislocations may be of two types and serve two functions:

a. Coherency dislocations. Homogeneous deformation to convert one crystal lattice into another produces a fully coherent interface; it is referred to as *lattice deformation*. Dislocations producing lattice deformation have Burgers vectors with climb and glide components, and their motion is usually confined to their loop plane.

b. Anticoherency dislocations. The macroscopic shape change of the lattice deformation can be altered to produce less strain energy by inhomogeneous deformation, or lattice-invariant, deformation. Dislocations required to produce these deformations have been referred to as *misfit* dislocations, but Olson and Cohen (1981) suggested the term *anticoherent* dislocations since their function is to disrupt the interface coherency. An appropriate array at special sites of coherent and anticoherent dislocations dissociates under the driving force of a negative stacking-fault energy.

Suzuki and Ledbetter (1982) suggested another approach to the understanding of martensite nucleation. They calculated the required energy, or barrier energy, for sodium from a particular homogeneous deformation reaction path (the Bain mechanism) by using an Ashcroft empty-core pseudopotential. Kelly (1979) used a similar pseudopotential model and calculation, but a different transformation deformation mechanism, to calculate the energy differences of the structures in sodium. This methodology may prove to be very useful in quantitative energy calculations, in selection of most probable transformation shear mechanisms, and in the understanding of martensite nucleation.

9.2.2 Temperature dependence

Temperature plays a critical role in martensitic transformations. The new phase begins to form at T_{ms}, but unlike most other transformations, product growth proceeds very rapidly to a final product size and morphology. Additional transformation is usually induced only by lowering the temperature.

The temperature at which the martensitic phase begins to transform back to the parent structure, T_{as}, is considerably higher than T_{ms} in ferrous alloys (≥ 600 K) but may be only a few degrees higher, or even less than T_{ms}, in some nonferrous alloys systems (Au–Cd, Ag–Cd, and Au–Zn).

Factors that influence the defect density, defect configurations, or parent-material flow stress affect T_{ms}. Cold work, reduction of grain size, irradiation, and repeated transformation cycling usually decrease T_{ms}. Relaxation of constraints, such as in thin foils and at surfaces, and the introduction of specific defects or defect arrays may result in increased T_{ms}. Hydrostatic pressure decreases T_{ms} in those alloy systems in which there is a volume expansion on transformation, for example, the ferrous alloys. Magnetic fields raise T_{ms} in ferrous alloys.

9.2.3 Time dependence

Martensitic transformations usually proceed by the nucleation of many martensite crystals. The growth of each crystal to its final product shape and size is extremely rapid; rates approaching the velocity of sound have been measured in ferrous alloys (Bunshah and Mehl, 1953).

From the standpoint of kinetics (Entwisle, 1971), there are basically four types of transformations: Three types involve multiple nucleation sources and irreversible, rapid growth. The fourth type, thermoelastic, usually has fewer nucleation sources and reversible, slow growth; it is discussed in § 9.2.4.

a. In many Fe–C alloy steels, the transformation begins at T_{ms}, and a well-defined amount of martensite forms as a function of temperature during cooling (Fig. 9.2a). The T_{ms} is essentially independent of the rate of cooling, at least to rates about 50 000 $K{\cdot}s^{-1}$; however, the rate of reaction is dependent on the rate of cooling—higher cooling rates increase the reaction rate. This type of kinetic behavior commonly has been called *athermal transformation* and usually occurs above room temperature. Within the time normally required for a specimen to attain thermal equilibrium, the transformation progresses and stops, and subsequent holding at a given temperature results in no further transformation. These characteristics have led Magee (1970) to conclude that the transformation is always dynamically stabilized and that interstitial carbon and nitrogen, which segregate to potential nucleation sites, act as stabilizers, inhibiting any subsequent martensite formation. Indeed, compared with direct cooling, holding at intermediate temperatures (usually below T_{ms}) requires cooling to lower temperatures to obtain equivalent martensite concentrations.

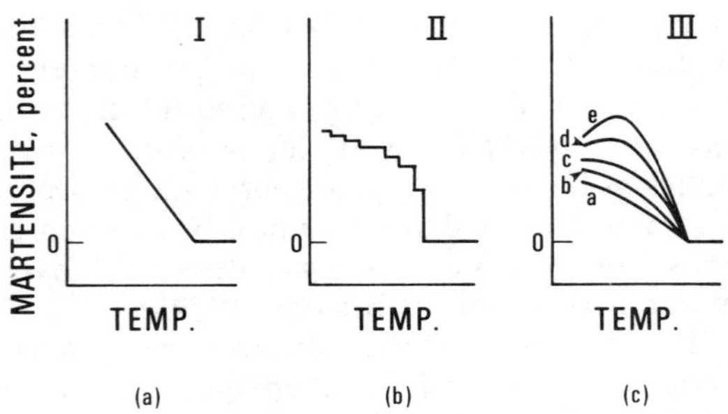

Figure 9.2 Three types of martensitic transformation kinetics on cooling. Type I is "athermal," Type II is "burst," and Type III is isothermal martensitic transformation. In (c), the curves show the effects of progressively decreasing cooling rates, from a to e.

b. In some systems, especially Fe–Ni and Fe–Ni–C alloys, a large increment of transformation will suddenly occur in one burst on cooling (Fig. 9.2b), thus the term *burst transformation*. Normally, there is a distinctly audible click associated with the transformation. The transformation occurs via the nucleation of many crystals, which grow very quickly. One or several nucleation events are sufficient to trigger others, and the transformation assumes a very rapid autocatalytic sequence. This is not unexpected since the burst transformation in iron-alloy systems produces extensive slip with limited twinning or faulting and high residual stresses remain in the vicinity of the austenite–martensite plate interface.

c. Many systems exhibit what is called *isothermal transformation* kinetics (Fig. 9.2c). Again, the martensite crystal growth is very rapid, but the kinetics are controlled by the nucleation event, which is time dependent. Typical C-curve behavior is exhibited when temperature-vs.-transformation time is plotted for constant amounts of martensite. New transformation usually occurs by the formation of new crystals, not by the growth of existing crystals. As illustrated in Fig. 9.2c, the amount of transformation depends on the cooling rate; lower cooling rates result in more martensite. Furthermore, at very low temperatures insufficient thermal energy is available to incite nucleation and less martensite forms.

9.2.4 Stress dependence

The driving force for any phase transformation is the difference in free energies of the phases. Mechanical stress may be considered as assisting this driving force in martensitic transformations because, as the transformation proceeds, the martensite shape change adds a directional increment of strain, thus relieving the applied stress field. The reviews by Warlimont and Delaey (1974) and Delaey, Krishnan, Tas, and Warlimont (1974) have excellent descriptions of stress effects in nonferrous alloy transformations.

The application of stress produces two types of transformations: (a) *Stress-assisted* transformation—the applied stress assists the transformation from existing nucleation sites of preferred crystallographic variants of the product phase. This product is the same morphological type as that which forms during cooling. (b) *Strain-induced* transformation—the plastic deformation of the parent material creates new nucleation sites. The product usually forms as very small crystals along active slip planes and in preferred orientations.

Applied stress induces martensitic transformation above T_{ms}; the highest temperature at which martensite is formed under stress is called T_{md}. The application of stress below T_{ms} assists in completion of the transformation that began during cooling. However, in the range of temperatures above T_{ms} and below T_{md} in some alloys (such as the Fe–Cr–Ni austenitic stainless steels), the strain-induced transformation may actually be suppressed below about 80 K.

Cold work may affect subsequent transformation during cooling below T_{ms} in two ways. Small amounts of plastic deformation in some ferrous alloys evidently provide additional nucleation sites and slightly enhance martensite formation below T_{ms}. Continued cold working, however, increases the parent austenite flow strength, which lowers both T_{ms} and the amount of transformation.

In some alloy systems, the growth of the martensite plate is continuous with decreasing temperature. This type of transformation is called *thermoelastic* and is characterized by low nucleation barriers and by the lack of plastic accommodation strains in the parent phase (thus, rather small deformation-shape strains). High elastic constants of the parent phase promote thermoelastic behavior. In thermoelastic martensite, with complete interface phase mobility and the lack of plastic accommodation, elastic stresses are essentially stored in the product phase and can only be released by the reverse transformation. Thus, when the mechanical forces build up sufficiently to balance the chemical driving force, growth stops. In thermoelastic martensite systems there is little temperature hysteresis between T_{ms} and T_{as}. In fact in some systems, T_{as} is lower than T_{ms}; reversion occurs prematurely by assistance of the stored elastic energy.

In some systems there is complete reversibility of individual product crystal growth as a function of applied stress, analogous to thermoelastic transformation behavior as a function of temperature. Martensites that grow reversibly with respect to applied stress are labeled *pseudoelastic* martensites. Here the strain produced in response to the applied stress is supplied by the martensitic transformation.

Martensitic transformations in the absence of plastic accommodation and stabilization effects are usually completely reversible and repeatable. When thermoelastic or pseudoelastic martensites are deformed (in the product phase) and warmed above T_{as}, the reverse transformation results in a shape change from the deformed product phase to the original parent phase shape. This shape change is called the *shape-memory effect*; alloy compounds and systems that exhibit this effect include Ti–Ni, Au–Cd, Cu–Al–Ni, Cu–Zn, Cu–Sn, Fe–Pt, and In–Tl. With such materials it is possible to produce a heat engine. The spring-back deformation on heating is larger than the low-temperature deformation because there is an elastic stored energy contribution to the reverse transformation. Preliminary engine efficiencies are estimated at between 5 and 20% (Wayman, 1975), indicating some promise for their use in energy conversion. This subject has attracted considerable attention: a special symposium on shape-memory effects has been held (*Shape Memory Effect in Alloys*, 1975).

When a martensitic transformation acts as an integral part of a deformation mode, the stress–strain characteristics of the material are altered. Particularly, there is strain associated with the transformation that exceeds that produced from normal dislocation interactions in the parent material. Therefore, flow strengths are typically less than those achieved in the absence of the transformation at higher temperature or under stabilized conditions. In some cases excessive elongation exceeding 100% has been reported.

9.2.5 Morphology and crystallography

The shear type of atom transfer in martensitic transformations produces a macroscopic shape change and results in definite crystallographic relationships between the parent and product phases and a well-defined morphology of the martensite crystals. The shape change is identifiable using optical microscopy (Fig. 9.3a) and, sometimes, by visual inspection of the material surface (Fig. 9.3b). To produce the macroscopic martensitic shape change illustrated in Fig. 9.3b for austenitic stainless steels, the austenite grain size must be very large. The surface appearance of the shape change is conditioned by the morphology of the martensitic product phase.

In ferrous alloys, the product is usually in the shape of plates, laths, or sheets. The typical plate-like ferrous-alloy morphology is illustrated in Fig. 9.4. Figure 9.4a shows the plate morphology of the product in an Fe–33Ni alloy.*

Fig. 9.4b is a micrograph of the martensitic structure of a sample of the Tishomingo meteorite (Fe–32Ni–1Co; $T_{ms} = 200$ K). Very large (≤10 cm) martensite plates are observed, which presumably formed in large single-crystal regions during slow cooling in space.

Nonferrous product morphologies usually are either plate-like or have a single, advancing-front interface. Except in the case of the single-interface transformation, it is usual for many product crystals to form within each parent grain.

Typical habit plane normals are listed in Table 9.1. On cooling, many variants are usually present, but the number of variants is reduced if the

*Unless otherwise specified, alloy elements are given in wt.%.

(a) (b)

Figure 9.3 (a) Shape deformation of martensitic transformation in austenitic stainless steel (Fe–Cr–Ni alloy, AISI 304L) on cooling to 76 K. (b) Shape deformation of martensitic transformation in steel (AISI 304) flange previously used in service at 76 K.

(a)

(b)

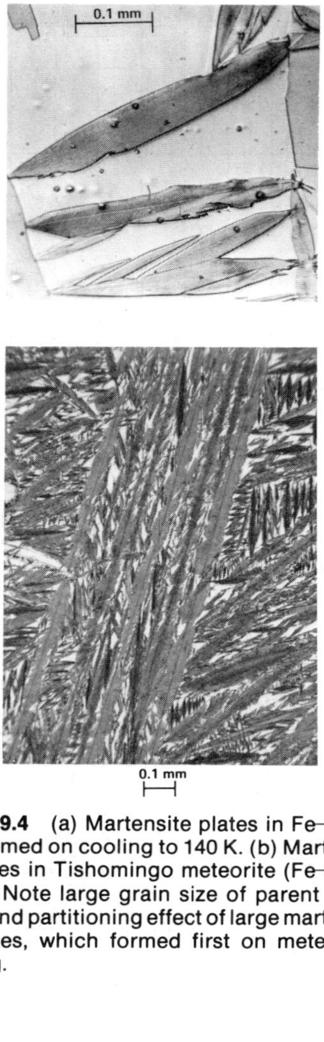

Figure 9.4 (a) Martensite plates in Fe–33Ni that formed on cooling to 140 K. (b) Martensite plates in Tishomingo meteorite (Fe–32Ni –1Co). Note large grain size of parent austenite and partitioning effect of large martensite plates, which formed first on meteorite cooling.

transformation is induced by applied stress. The variants represent those that reduce the stress field by the transformation shape deformation.

Regardless of the number of habit-plane variants, there is a unique orientation relationship between the parent and product crystal structures. Typical orientation relationships are also presented in Table 9.1.

Either slip, twinning, faulting-lattice invariant deformations, or a combination of them is necessary to produce the martensitic structure. Dislocations, twins, or stacking faults producing the transformation deformation are easily observable in the martensite by using transmission electron microscopy. An example of the occurrence of both slip (dislocation) and twinning in an Fe–30Ni alloy is shown in Fig. 9.5. It is experimentally difficult to confirm all of the lattice invariant deformation mechanisms or even to distinguish between defects produced from transformation processes and those produced from shape accommodation changes. For discussion and characterization of the morphology and crystallography of nonferrous alloys, see Warlimont and Delaey (1974); for ferrous alloys, see Wayman (1968) and Dunne and Wayman (1971). A large amount of crystallographic information is given in Reed and Breedis (1966).

The shape change that results from the formation of specific product crystals can also be measured. Measurements of surface tilt and displacement changes are made from prescribed existing surface markings, such as scratches or pits. These data are valuable for assessing phenomenological crystallographic theories.

Generally, the theories of Wechsler, Leiberman, and Read (1953), Bowles and Mackenzie (1954), and Bullough and Bilby (1956) have been

Table 9.1 Typical crystallographic and morphological parameters of common martensitic transformations.

Alloy System	Habit Plane[a]	Orientation Relationship[b]		Parent/Product Structure	Morphology
Fe–Ni	{259}	$(111)_p \parallel (011)_m$,	$[1\bar{2}1]_p \parallel [0\bar{1}1]_m$,	f.c.c./b.c.c.	Plates
Fe–Mn	{111}	$(111)_p \parallel (00.1)_m$,	$[1\bar{1}0]_p \parallel [11.0]_m$,	f.c.c./h.c.p.	Sheets
Fe–Cr–Ni	{225}, {112}	$(111)_p \parallel (101)_m$,	$[1\bar{1}0]_p \parallel [1\bar{1}1]_m$,	f.c.c./b.c.c.	Laths
	{111}	$(111)_p \parallel (00.1)_m$,	$[1\bar{1}0]_p \parallel [11.0]_m$,	f.c.c./h.c.p.	Sheets
Au–Cd	{133}	$(100)_p \parallel (1.00)_m$,	$[\bar{1}11]_p \parallel [\bar{1}\bar{1}0]_m$,	b.c.c./h.c.p.	Single interface
Cu–Al	{122}, {133}	$(110)_p \parallel (00.1)_m$,	$[1\bar{1}1]_p \parallel [11.0]_m$,	b.c.c./h.c.p.	Plates
Ti–Cr	{334}	$(111)_p \parallel (00.1)_m$,	$(110)_p \parallel (11.0)_m$,	b.c.c./h.c.p.	Plates
Cu–Sn	{233}, {133}	$(001)_p \cong 2° \ (0.11)_m$,	$[010]_p \cong 6° \ [1.00]_m$, and $[2\bar{1}0]_p \parallel [1.\bar{1}1]_m$	b.c.c./h.c.p.	Plates

[a]With respect to parent phase.
[b]*p* refers to parent phase: *m* refers to martensitic phase.

Figure 9.5 Appearance of both dislocation substructures and twins in b.c.c. martensite in Fe–30Ni.

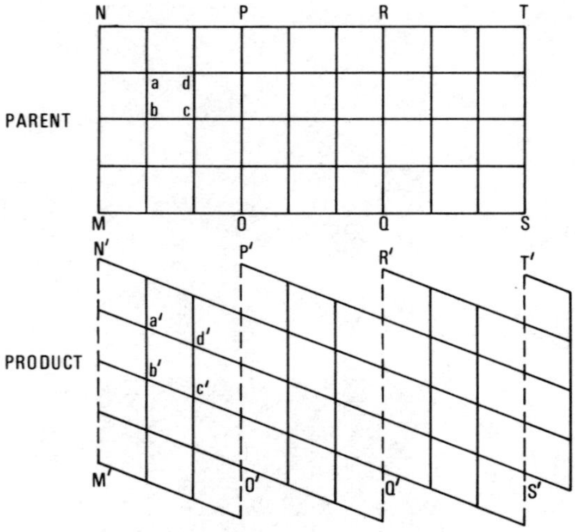

Figure 9.6 Schematic of parent phase and product martensite, illustrating homogeneous atom shear.

successful in predicting the crystallographic aspects of martensitic transformations from knowledge of the lattice parameters and crystal structures of both phases. The total macroscopic shape change, P_1, is composed of three components: a rigid-body rotation, R; a pure deformation or lattice strain, B; and the lattice-invariant or shear strain, P_2. The combined rigid-body rotation and the pure deformation make up the lattice deformation. All components can be represented as matrices and expressed as

$$P_1 = RBP_2 \tag{9.1}$$

where at minimum strain energy, P_1 is an invariant plane strain and the invariant plane is the habit plane. The solutions to Eq. 9.1 are extensively discussed by Wayman (1968) and Christian (1970).

An example of a parent lattice and a product lattice that have undergone slip to produce a macroscopic shape change is shown in Fig. 9.6. Here, the lattice strain, B, consists of a homogeneous shear of atoms that changes the shape of the basic unit cell from atom positions abcd to a′b′c′d′ and changes the volume. The lattice invariant deformation produces an overall macroscopic shape change and a microscopically incoherent interface, but leaves the unit cell geometrically unchanged. Either slip, twinning, or faulting produces this deformation (slip along M′N′, O′P′, Q′R′, and S′T′ is shown in the example). The interface or habit plane can be identified as N′P′R′T′; this line is macroscopically straight with microscopic discontinuities. In general, the habit plane is irrational; the invariant lines M′N′, O′P′, Q′R′, and S′T′ are neither normal to the habit nor contained in the habit plane, and the Burgers vector of the dislocations producing the slip is not normal to the habit plane.

Although Eq. 9.1 has four solutions for the habit plane, the number is reduced to a unique habit plane either by the degeneracy of the solution variants or by the appropriate selection of the lattice invariant deformation. Accurate experimental studies have shown that there is usually not a unique habit plane but a range of habit planes (2 to 5° on a stereographic triangle). Bowles and Mackenzie (1954) included in their crystallographic approach a dilational parameter, which allows for small (~1%) uniform expansion or contraction. Multiple lattice-invariant shear systems may, in theory, produce habit plane variations in at least two directions, and there is ample evidence in ferrous alloy systems that more than one transformation slip system, twinning system, or both are operative. Multiple shears have been proposed by Acton and Bevis (1969/70) and Ross and Crocker (1970) to account for specific habit planes in ferrous alloys.

9.3 AUSTENITIC STAINLESS STEELS

The AISI 300 series stainless steels, especially alloys 304, 310, and 316, are used extensively for cryogenic structures. The chemistry and characteristics of these Fe–Cr–Ni austenitic alloys are described in Chapter 11, Structural Alloys. These alloys have a high elastic modulus, high toughness and ductility, low thermal and electrical conductivity, and good weldability. Because of these characteristics, they are used more than any other alloy class for structures for service at 20 K and below. But their use is complicated by the metastability of the austenitic structure of most alloys of this Fe–Cr–Ni alloy series. The metastability leads to martensitic transformation during cooling, from applied stress, or during plastic deformation. The martensitic transformation is a significant design consideration in applications requiring fracture control planning, close dimensional tolerances, the absence of a ferromagnetic phase, and high toughness of weldment, heat-affected zone, and base metal. Consequently, research has been conducted to characterize the transformations and their effects on mechanical and physical properties and in-service performance.

9.3.1 Austenite stability

Either on cooling, under applied elastic stresses, or during plastic deformation, the austenite phase (γ) of Fe–Cr–Ni alloys may transform martensitically to b.c.c. (α') and h.c.p. (ϵ) phases. The alloy composition affects the chemical free-energy difference between the two phases.

The difference in free energy, F, between the two phases γ and α' is:

$$\Delta F^{\alpha'\to\gamma} = F_T^{\gamma} - F_T^{\alpha'} \qquad (9.2)$$

where F_T^{γ} and $F_T^{\alpha'}$ are the total free energies of the respective phases. The free energies of mixing, ΔF_m^{γ} and $\Delta F_m^{\alpha'}$, of the two phases may be expressed in terms of the total free energies as:

$$\Delta F_m^{\gamma} = F_T^{\gamma} - \Sigma\, X_i\, F_i^0 \quad \text{and} \qquad (9.3)$$

$$\Delta F_m^{\alpha'} = F_T^{\alpha'} - \Sigma\, X_i F_i^0 \qquad (9.4)$$

where X_i is the atom fraction of a component, i, and F_i^0 is the free energy of the pure component, i, in the standard state. The total free energy of each phase is a sum of the partial molar free energies of each component in an ideal solution of γ or α':

$$F_T^{\gamma} = \Sigma\, X_i \bar{F}_i^{\gamma} \text{ and } F_T^{\alpha'} = \Sigma\, X_i \bar{F}_i^{\alpha'} \qquad (9.5)$$

The difference in the free energy of mixing for the two phases is:

$$\Delta F_m^{\gamma} - \Delta F_m^{\alpha'} = F_T^{\gamma} - \Sigma\, X_i F_i^0 - F_T^{\alpha'} + \Sigma\, X_i F_i^0 \qquad (9.6)$$

and

$$\Delta F_m^{\gamma} - \Delta F_m^{\alpha'} = F_T^{\gamma} - F_T^{\alpha'} = \Delta F^{\alpha'\to\gamma} \qquad (9.7)$$

Therefore, from Eqs. 9.5 and 9.7,

$$\Delta F^{\alpha'\to\gamma} = \Sigma\, X_i(\bar{F}_i^{\gamma} - \bar{F}_i^{\alpha'}) = \Sigma\, X_i\,(\Delta \bar{F}_i^{\alpha'\to\gamma}) \qquad (9.8)$$

That is, the free-energy difference between the two phases can be expressed as the sum of the partial molar free-energy differences of each component.

Breedis and Kaufman (1971) have derived an expression $\Delta F^{\alpha'\to\gamma}$ for the Fe–Cr–Ni ternary system:

$$\begin{aligned}\Delta F_{\text{Fe,Cr,Ni}}^{\alpha'\to\gamma} &= (1 - X_{\text{Cr}} - X_{\text{Ni}})\Delta \bar{F}_{\text{Fe}}^{\alpha'\to\gamma} \\ &+ X_{\text{Cr}}\Delta \bar{F}_{\text{Cr}}^{\alpha'\to\gamma} + X_{\text{Ni}}\Delta \bar{F}_{\text{Ni}}^{\alpha'\to\gamma} \\ &+ X_{\text{Cr}}(1 - X_{\text{Cr}} - X_{\text{Ni}}) \\ &\times (B - A)_{\text{Fe–Cr}} \\ &+ X_{\text{Ni}}(1 - X_{\text{Cr}} - X_{\text{Ni}}) \\ &\times (B - A)_{\text{Fe–Ni}} \\ &+ X_{\text{Cr}}X_{\text{Ni}}\,(B - A)_{\text{Cr–Ni}}\end{aligned} \qquad (9.9)$$

Here X_{Cr} and X_{Ni} are the atom fractions, and A and B are interaction parameters. These parameters are described as

$$A = \frac{F_{\text{exc}}^{\alpha'}}{X(1-X)} \text{ and } B = \frac{F_{\text{exc}}^{\gamma}}{X(1-X)} \qquad (9.10)$$

where F_{exc} is the excess free energy of a binary system. $\Delta F_{\text{Cr}}^{\alpha'\to\gamma}$, $\Delta F_{\text{Ni}}^{\alpha'\to\gamma}$, and $\Delta F_{\text{Fe}}^{\alpha'\to\gamma}$ represent the partial molar free-energy difference between α' and γ for the pure components chromium, nickel, and iron, respectively.

Kaufman (1959) derived free-energy relations for the Fe–Cr binary system and combined these with the previous data for Fe–Ni binary alloys (Kaufman and Cohen, 1956) to calculate $\Delta F^{\alpha'\to\gamma}$ for stainless steels. On the basis of regular solution and for the temperature range of about 100 to 900 K,

$$\Delta F^{\alpha\rightarrow\gamma}_{\mathrm{Fe,Cr,Ni}} = \Delta F^{\gamma} - \Delta F^{\alpha'}$$
$$= (1 - X_{\mathrm{Cr}} - X_{\mathrm{Ni}})(5036 - 11\times10^{-3}T^2) + 6.5\times10^{-6}T^3 + X_{\mathrm{Cr}}(1927 + 4.2T) + X_{\mathrm{Ni}}(-15500 + 2.97\times10^{-3}T^2 + 1.64\times10^{-6}T^3) + X_{\mathrm{Cr}}(1 - X_{\mathrm{Cr}} - X_{\mathrm{Ni}})(-11730 + 3.1T) + X_{\mathrm{Ni}}(1 - X_{\mathrm{Cr}} - X_{\mathrm{Ni}})\times[15080 + 2.4T(1 - \ln T)]\ \mathrm{J\cdot mol^{-1}} \quad (9.11)$$

For calculations of T_{ms} of stainless steels, Kaufman used the term

$$X_{\mathrm{C+N}}(-2346 + 0.75T)\ \mathrm{J\cdot mol^{-1}} \quad (9.12)$$

to account for carbon and nitrogen; the nominal manganese content of about 1.5 wt.% was added to the nickel content. Values of $\Delta F^{\alpha'\rightarrow\gamma}$ ranging from about 2000 J·mol^{-1} (475 cal·mol^{-1}) for T_{ms} = 100 K to 1260 J·mol^{-1} (300 cal·mol^{-1}) for T_{ms} = 550 K were calculated for typical stainless steel compositions using Eq. 9.9. Later calculations by Breedis and Kaufman (1971) and Rao, Russell, and Winchell (1967) predict lower values of $\Delta F^{\alpha'\rightarrow\gamma}_{\mathrm{Ni}}$ and would, similarly, result in lower values of $\Delta F^{\alpha'\rightarrow\gamma}$ for the Fe–Cr–Ni ternary system.

An equation analogous to 9.11 can be developed for the free-energy difference between the ϵ and γ phases in Fe–Cr–Ni alloys. Values for $\Delta F^{\epsilon\rightarrow\gamma}_{\mathrm{Cr}}$ and $\Delta F^{\epsilon\rightarrow\gamma}_{\mathrm{Ni}}$ were estimated by Kaufman (1967) from enthalpy and entropy differences due to structure. Interaction parameter estimates were made by Breedis and Kaufman (1971) for Fe–Cr, Fe–Ni, and Cr–Ni binaries. The following expression is derived for the free-energy difference between ϵ and γ phases in Fe–Cr–Ni alloys:

$$\Delta F^{\epsilon\rightarrow\gamma}_{\mathrm{Fe,Cr,Ni}} = F^{\gamma} - F^{\epsilon}$$
$$= X_{\mathrm{Fe}}\Delta F^{\epsilon\rightarrow\gamma}_{\mathrm{Fe}} + X_{\mathrm{Cr}}(1095 + 0.628T) - X_{\mathrm{Ni}}(1047 + 1.26T) + 2095(X^2_{\mathrm{Cr}} - X_{\mathrm{Cr}} + X^2_{\mathrm{Ni}} - X_{\mathrm{Ni}})\ \mathrm{J\cdot mol^{-1}} \quad (9.13)$$

The partial molar free-energy difference ($\Delta F^{\epsilon\rightarrow\gamma}_{\mathrm{Fe}}$) can be calculated from the graphical presentation by Breedis and Kaufman (1971) of $\Delta F^{\alpha\rightarrow\gamma}_{\mathrm{Fe}}$ and $\Delta F^{\alpha\rightarrow\epsilon}_{\mathrm{Fe}}$. This term becomes positive at temperatures lower than 390 K and equals about 300 J·mol^{-1} at 300 K. Using Eq. 9.14 and the $\Delta F^{\epsilon\rightarrow\gamma}_{\mathrm{Fe}}$ data plotted in Fig. 9.7, the temperature dependence of $\Delta F^{\epsilon\rightarrow\gamma}_{\mathrm{Fe,Cr,Ni}}$ can be calculated and is plotted in Fig. 9.7. This indicates that $\Delta F^{\epsilon\rightarrow\gamma}_{\mathrm{Fe,Cr,Ni}}$ becomes positive near 325 K. Actually, ϵ transformation has been observed on cooling below room temperature (e.g., Reed, 1962). Alloy additions such as carbon, nitrogen, and manganese move the $\Delta F^{\epsilon\rightarrow\gamma}_{\mathrm{Fe,Cr,Ni}}$ curve to lower temperatures.

Expressions relating the stability of the austenite and the temperatures of transformation during cooling (T_{ms}) or during deformation (T_{md}) have been developed empirically and are summarized in Table 9.2. Eichelman and Hull (1953), working with austenitic alloys with 10 to 18% Cr, 6 to 12% Ni, 0.6 to 5% Mn, 0.3 to 2.6% Si, 0.004 to 0.129% C, and 0.01 to 0.06% N, established that all of the above alloying elements stabilize the austenite and thus lower T_{ms}. Monkman et al. (1957), using a larger number of specimens with ranges of 5 to 13% Ni, 11 to 19% Cr, and 0.035 to 0.126% C+N, produced a sim-

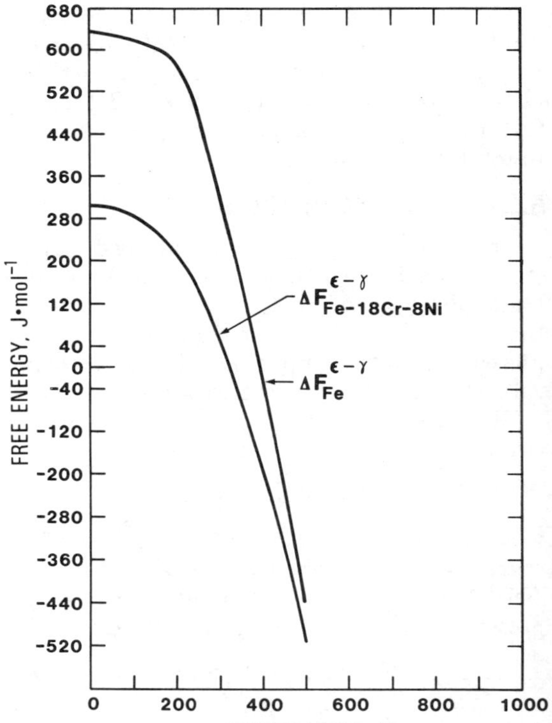

Figure 9.7 Temperature dependence of free energy difference of ϵ and γ structures of iron and Fe–18Cr–8Ni alloy.

Table 9.2 Temperature equivalents for calculation of stability parameters of austenitic steels.

Investigator (Year)	Temperature Equivalent									Comments, Composition Range (wt.%)
	Base	Cr	Ni	Mn	Si	C	N	Mo	Other	
f.c.c.→b.c.c. (T_{ms}), cooling[a]										
Eichelman, Hull (1953)	1578	−41.7	−61.1	−33.3	−27.8	−1670	−1670			21 alloys: 10–18Cr, 6–12 Ni, 0.6–5Mn, 0.3–26Si, 0.004–0.12C, 0.01–0.06N
Monkman, Cuff, Grant (1957)	1455	−36.7	−56.7			−1460	−1460			49 alloys: 11–19Cr, 5–13Ni, 0.035–0.0176(C+N)
Hammond (1963)	1105	−29	−39					−36		16 alloys: 0–12Cr, 4–8Ni, 0.03C, 2–6Mo, 0–15Co, 1–2Ti
Andrews (1965)	273	−12.1	−17.7	−30.4		−423		−7.5		184 alloys from previous studies not in this table. Notice different composition ranges. 0–4.6Cr, 0–5.0Ni, 0.04–4.9Mn, 0.1–1.9Si, 0.11–0.6C, 0–5.4Mo
Hull (1973)	1755	−47	−59	−54	−37	−2390	−3720	−56	−180 (Ti), −14 (Co)	59Ni = average of Eichelmann, Hull (1953) and Monkman et al. (1957), 29 alloys: 12–24Cr, 0–22Ni, 0–20Mn, 0–4Si, 0–0.1C, 0–0.15N, 0–6Mo, Co, 0–2Ti
f.c.c.→b.c.c. (T_{md}), deformation[a]										
Angel (1954)	686	−14	−9.5	−8.1	−9.2	−46.2	−46.2	−18.5		30% tension, 50% α′
Hull (1973)	1655	−23	−59	−41	−20	−777	−315	−24	−12 (Co)	50% compression, 60 alloys: 12–24Cr, 0–22Ni, 0–20Mn, 0–4Si, 0–0.1C, 0–0.15N, 0–6Mo, Co
Williams, Williams, Capellaro (1976)	686	−6	−25	−16	+21	−222	−222	−11		45% compression, 2.5% α′, 25 alloys: 12–25Cr, 9–20Ni, 1–2Mn, 0.1–0.6Si, 0.04–0.25C, 0.01–0.1N, 0.6–2.8Mo
Magnetic transition (paramagnetic→antiferromagnetic, Néel temperature)										
Warnes, King (1976)	90	−1.25	−2.75	+7.75	−14			−5.5		64 commercial and specially prepared alloys: 5–25Cr, 5–32Ni, 0–16Mn, 0–24Si, 0.3Mo
Enthalpies of formation[b]										
Lee, Millman, MacDougall, Keown, Argent (1977)	−8168	208	181	1087		3186			1137[c]	48 alloys: 0–18Cr, 0–30Ni, 0–3Mn, 0–0.05Si, 0.01–0.86C, 0.02–0.2Mo
f.c.c.→h.c.p. (stacking fault energy at room temperature)[d]										
Dulieu, Nutting (1964)	0		5	14		34		1	36 (Ti), 32 (Cu), −0.55 (Co)	Electron microscopy
Schramm, Reed (1975)	−530	7	62	32				93		7 alloys, x-ray, literature data-regression analysis.
Rhodes, Thompson (1977)	12	6	14	177	−47					Modified Schramm–Reed approach
Brofman, Ansell (1978)	167	9	21			260				

[a] Numbers in table represent K/wt.%.
[b] Numbers in table represent $J \cdot mol^{-1} \cdot wt.\%$.
[c] Represents stored energy from enthalpy measurements, lath, and plate α′.
[d] Numbers in table represent 10 $mJ/m^2 \cdot wt.\%$ (10 $ergs/cm^2 \cdot wt.\%$).

ilar analysis and concluded that only to a first approximation was T_{ms} linearly dependent on composition. The dependence of T_{ms} on carbon and nitrogen concentration seemed to be influenced by chromium and nickel concentration.

Hull (1973) completed another study assessing the effects of nickel, chromium, manganese, carbon, nitrogen, silicon, and cobalt (aluminum, thallium, vanadium, and tungsten sometimes were added) on α' formation during low-temperature cooling-and-deformation experiments and on ferrite retention after cooling to room temperature from the melt. Hull's study reconfirmed that all the above elements suppress low-temperature α' formation during either cooling or deformation, but that chromium, molybdenum, silicon, vanadium, tungsten, niobium, tantalum, and aluminum additions promote the formation or retention of the high-temperature b.c.c. delta-ferrite phase. To predict the effect of elemental additions on T_{ms}, Hull assumed that the nickel contribution was the average of the Monkman et al. (1957) and Eichelman and Hull (1953) results and then compared all the effects of alloying additions with that of nickel. The empirical relationships have been analyzed by King and Larbalestier (1980), who point out some expected deviations in the dependence of T_{ms} on nickel and chromium when the relationships are used for prediction of the composition range of AISI 300 series stainless steels.

The formulation of Andrews (1965) applies to lower chromium and nickel concentrations, thus higher T_{ms}, and his results are included in Table 9.2 to point out the disparity between the high- and low-temperature empirical results. Hammond (1963) was the only investigator to include molybdenum.

In all these studies, specimens of various composition were prepared and cooled, and their T_{ms} values were measured. Regression analyses were then used to obtain the empirical relation of the dependence of T_{ms} on alloy concentration.

Similar experiments have been conducted to assess the influence of cold work (either tensile or compressive) on T_{md}. Their results are also contained in Table 9.2. Note that the experimental definition of T_{md} varied in each study: Angel (1954) used 50% α' at 30% tensile elongation; Williams et al. (1976) used 2.5% α' at 45% compressive elongation; and Hull (1973) used the minimum detection level (probably about 1% α') at 50% compressive elongation.

In reviewing the current status of the predictability of Fe–Cr–Ni stability, Collings and King (1979) caution that the Breedis and Kaufman (1971) free-energy expression for T_0 near 0 K does not have the same chromium and nickel coefficients as the Eichelman and Hull (1953) and Monkman et al. (1957) T_{ms} expressions. The latter expressions are based predominantly on chromium and nickel concentrations less than those required for T_0 near 0 K ($T_{ms} \simeq 300$ K). This raises the question of the usefulness of these T_{ms} predictive expressions for the normal 18Cr–8Ni stainless steels.

Collings and King (1979) also point out that in low-interstitial-austenitic stainless steels (those steels that are most easily welded), there is a higher tendency for isothermal transformation (in the 100 to 200 K range), and these predictive equations are not useful in the estimation of the occurrence of time-dependent α' transformation (or the separation between athermal and isothermal transformation).

Schramm and Reed (1975) used x-ray peak-shift measurements, coupled with previous stacking-fault energy measurements, to estimate the dependence of the stacking-fault energy on chemical composition. This dependence should correspond to the dependence of the $\gamma \rightarrow \epsilon$ transformation on composition, because a stacking fault represents a local planar area of ϵ. Rhodes and Thompson (1977) suggested that the stacking-fault energy values of the Schramm and Reed (1975) analysis were too large, considering additional weak-beam electron-microscopy data. Weak-beam electron microscopy measurements by Bampton, Jones, and Loretto (1978) confirm that the Schramm and Reed (1975) least-square analysis produced stacking-fault energies that were excessively large and indicated a possible data spread of about ±25% from measurements of individual nodes. Brofman and Ansell (1978) added the dependence on carbon. The stacking-fault energy composition dependences are also contained in Table 9.2. The stacking-fault energy is expected to decrease at lower temperatures, and the temperature dependence should follow the form of the $\Delta F^{\gamma \rightarrow \epsilon}_{\mathrm{Fe,Cr,Ni}}$ curve illustrated in Fig. 9.7.

In all empirical relations, chromium is shown to increase the stability of austenite relative to either ϵ or α' (Table 9.2). Although it is not clear at first glance, Eq. 9.9 (intended for use below 900 K) predicts that chromium will act to suppress the $\gamma \rightarrow \alpha'$ transformation. At higher temperatures, chromium has the opposite effect and stabilizes the γ structure (Kaufman, 1959). However, Eq. 9.13 predicts that chromium additions act to increase $\Delta F^{\epsilon \rightarrow \gamma}_{Fe,Cr,Ni}$, thus promoting ϵ transformation. This contrasts with the regression analyses results of the dependence of the stacking-fault energy on composition (Table 9.2). Additional study is needed to decide whether to amend the free-energy expressions and whether chromium alloyed with manganese, silicon, carbon, and nitrogen is a stabilizing influence on austenite with respect to hexagonal close-packed structures.

Warnes and King (1976) measured the Néel temperature (paramagnetic-to-antiferromagnetic transition temperature) for a series of Fe–Cr–Ni alloys. Regression analysis of their data has resulted in an expression relating the Néel temperature to composition (Table 9.2). The magnetic term was considered to be a relatively minor contribution to the free energy of the Fe–Cr–Ni alloy structures (e.g., Breedis and Kaufman, 1971). It is of interest, however, that both chromium and nickel act, in the approximately same ratio, to suppress the antiferromagnetic phase and to stabilize the austenite structure.

Calorimetric measurements of the heat evolved during martensite formation were made by Lee et al. (1977) for a series of Fe–Cr–Ni alloys. For an Fe–18Cr–8Ni alloy transformed at temperatures ranging from 156 to 293 K, the transformation enthalpy change, ΔH, was measured, and the average value was found to be 1920 J·mol^{-1}. No temperature dependence of ΔH was observed. Results of the regression analysis of all alloys for elemental contributions to the enthalpy change are listed in Table 9.2. There is a strong similarity between these results and the effect of the contributions of the elements on T_{ms}.

9.3.2 Structure

Face-centered cubic austenite (γ) transforms on cooling to two martensitic products: one b.c.c. (α'), the other h.c.p. (ϵ). The h.c.p. phase is associated with extended stacking faults and forms as thin sheets on (111) austenite planes. In most austenitic stainless steels, the α' product forms as laths. Breedis (1964) reported a lath α' morphology for compositions ranging from Fe–19Cr–11Ni to Fe–10Cr–16Ni; at lower chromium and higher nickel concentrations the α' morphology changed to a plate-like structure.

From Reed (1962), the α' lath-like structure is parallel to $<110>_\gamma$, with $(225)_\gamma$, $(112)_\gamma$, or both habit planes (Kelly, 1965). The laths are restricted within $\{111\}_\gamma$ bands, and usually three sets of habit planes form within the band. The schematic in Fig. 9.8 is a typical representation of α' laths within a $\{111\}_\gamma$ band. For a given $<110>_\gamma$ lath direction, two possible habit planes are possible; for example, for $[1\bar{1}0]_\gamma$ $(111)_\gamma$, both $(\bar{2}\bar{2}5)_\gamma$ and $(225)_\gamma$ are possible habit planes. The habit plane observed is the one that makes the largest angle to the confining $(111)_\gamma$ band. Many adjacent laths appear to be twin related (Breedis, 1964). Optical photomicrographs of bands of α' martensite, taken in sequence after polishing and etching, are shown in Fig. 9.9. Reed (1962) notes that during thermal cycling or cooling to a lower temperature, new bands of lath martensite continued to form and morphological variations were observed. Some lath directions deviated from $<110>$, and the habit plane deviated from $\{225\}$. Whereas most laths were confined by $\{111\}$ bands, some α' was restricted by $\{211\}$ bands. Therefore, continued transformation adds considerably to the complexity of the α' structure.

Transformation during plastic deformation initially results in α' at intersections of active $\{111\}$ deformation bands. The α' assumes the shape of laths along $<110>$ common to the two active $\{111\}$ systems. Figure 9.10 illustrates α' formation at slip-plane intersections. Evidence of dislocation pileups at α' laths is observable. Also, more than one lath forms at each transformation site, and in contrast to the α' that forms on cooling (Figs. 9.8 and 9.9), all laths have the same habit-plane variant. Finally, at several sites, the lath transformation appears ir-

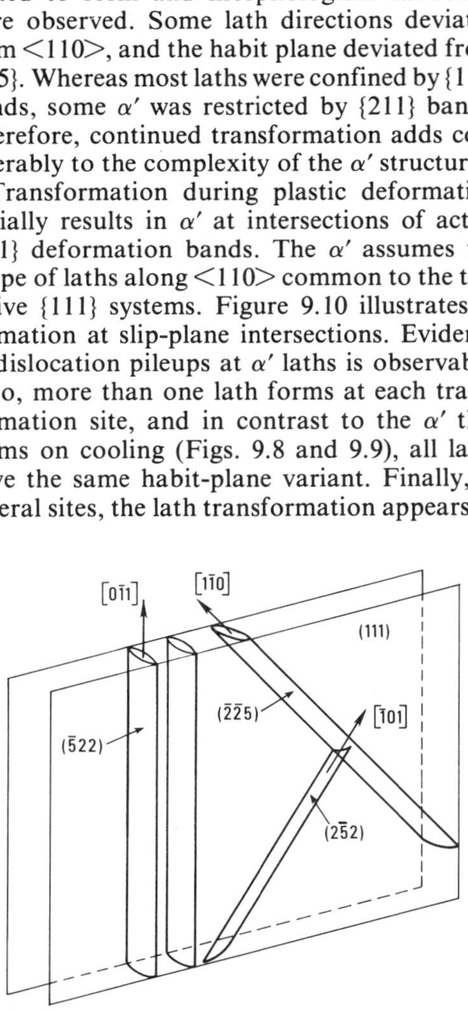

Figure 9.8 Schematic representation of α' laths within a $\{111\}$ sheet in Fe–18Cr–8Ni.

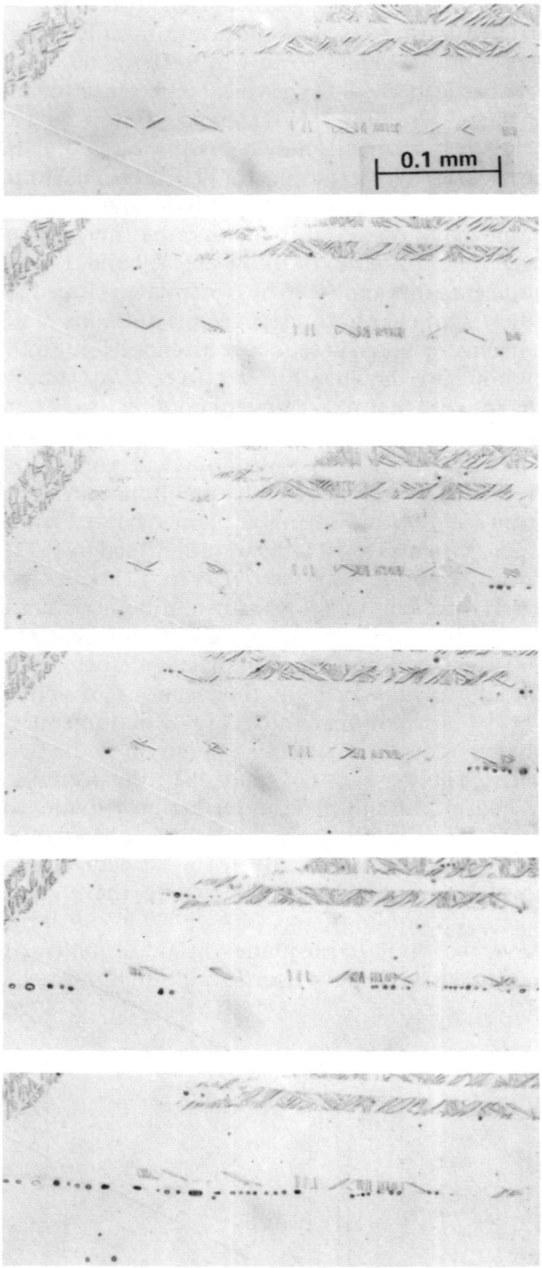

Figure 9.9 Optical photomicrographs of sheets of α' martensite in Fe–18Cr–8Ni taken after sequential electropolishing and etching steps. Sequence begins at top.

regular or incomplete; there is a lack of a well-defined austenite–martensite interface, and the interaction of transformed material with the dislocations in the immediate vicinity is very com-

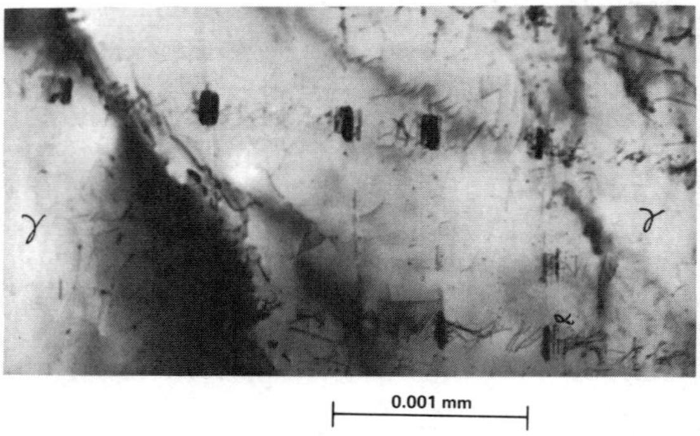

Figure 9.10 Fe–18Cr–8Ni austenite strained 0.01 in tension with two active slip systems on {111}. Formation of α' along $\langle 1\bar{1}0 \rangle$ intersections of slip planes.

plex. Growth of the α' nucleus produces a more distinct and crystallographically regular lath-like product phase morphology.

There is a considerable amount of accommodation deformation in the austenite after transformation during cooling; it is particularly large within the {111} band containing the α' laths. This leads one to suspect that lattice deformation is minimized normal to the {111} band that contains the three $\langle 110 \rangle$ directions and that the lattice-invariant deformation in these $\langle 110 \rangle$ directions may be very low. The $\{111\}_\gamma$ bands contain either ϵ martensite or a large amount of stacking faults (Kelly, 1965; Breedis, 1964).

The orientation relationships between the γ, ϵ, and α' phases are (Kelly, 1965):

$$(111)_\gamma \parallel (0001)_\epsilon \parallel (101)_{\alpha'}$$

$$[1\bar{1}0]_\gamma \parallel [1\bar{2}10]_\epsilon \parallel [11\bar{1}]_{\alpha'}$$

These relationships are apparently retained regardless of the manner of ϵ or α' formation.

The internal defect structure of the α' phase in Fe–Cr–Ni steels consists predominantly of dislocations. Wirth and Bickerstaffe (1974) have reported the presence of some twins in α' following increased chromium additions (with 8 wt.% Ni) or increased cooling rates. As chromium is replaced by nickel, Breedis (1964) reports that cellular, irregular distributions of dislocations typical of α' laths change to planar, regular arrays typical of α' plates. Also, as nickel replaces chromium, the amount of ϵ transformation decreases, the sharpness of the h.c.p. reflections decreases, and the f.c.c. twin reflections become more diffuse (Breedis, 1964).

Many metallurgists have been concerned with

the role of the ϵ transformation: does ϵ act as a precursor transformation ($\gamma \to \epsilon \to \alpha'$) or is the ϵ an accommodation effect ($\gamma \to \alpha'$, $\gamma \to \epsilon$)? Both effects have been observed. The stainless steels with low stacking-fault energy tend to form ϵ, and the α' tends to form from ϵ. Higher stacking-fault-energy alloys require transformation stresses for observable ϵ formation.

9.3.3 Effects of temperature

The amount of α' transformation is dependent on temperature, and the kinetics generally follow the form shown in Fig. 9.2. After cooling below T_{ms}, holding at constant temperature produces a C curve when temperature is plotted against holding time for constant α' product. Suppression of α' transformation usually occurs below about 100 K (Lagneborg, 1964). In most of the Fe–18Cr–8Ni (AISI 304L) alloys that have been examined and found to exhibit α' transformation, cooling to 76 K produces a maximum amount of α' (usually 5 to 10%). Cooling rapidly to 4 K in some cases suppresses α' transformation entirely.

Reversion of ϵ, formed either from deformation (Reed, 1962) or cooling to γ (Abrassart, 1973), begins when the alloy is heated to about 410 K. Reversion of α', formed either by cold working (~70% α') or cooling (~7% α'), begins at a higher temperature, about 770 K (Reed, 1962). Partial decomposition of the α' on holding at lower temperatures has been observed.

9.3.4 Effects of austenite deformation

Deformation of austenite at temperatures several hundred degrees above room temperature tends to produce a cellular dislocation substructure. At lower temperatures (about one hundred degrees above room temperature) more planar defects, stacking faults, and twins are produced. At still lower temperatures, ϵ martensite, α' martensite, or both form. The exact temperatures depend primarily on the stacking-fault energy. The lower the stacking-fault energy, the higher the deformation temperatures for formation of planar dislocations, stacking faults, and ϵ martensite. Since the stacking-fault energy is very dependent on temperature (Abrassart, 1973; Bampton et al., 1978), deformation at lower temperatures results in an increase of planar defects.

The formation of planar defects tends to enhance subsequent martensite formation at lower temperatures. In contrast, a cellular dislocation structure suppresses subsequent transformation on cooling below T_{ms} or during deformation below T_{md}.

Presumably cellular structures only influence the martensitic transformation by raising the strength of the austenitic matrix. Planar defects, with reduced cross-slip, produce high local strain fields, which may serve as nucleation sites. Evidence of austenitic prestrain effects on transformation at lower temperatures confirms the very delicate balance between austenite strengthening and planar-dislocation-induced stress fields.

Small amounts of tensile plastic deformation (0 to 10%) at room temperature stimulate transformation to α' on cooling to temperatures below T_{ms} (Reed, 1962; Lagneborg, 1964; Breedis, 1965). Added deformation steadily suppresses transformation. Breedis (1965) has reported that the α' transformation is enhanced in alloys that have a low stacking-fault energy; planar dislocation arrays, as opposed to cellular dislocation substructures, result in more transformation during cooling. Strife, Carr, and Ansell (1977) have found that prestrain above T_{md} reduces T_{ms}, may enhance subsequent transformations, and at larger strains, usually suppresses transformation.

There is ample evidence that deformation-induced defects play a major role in the nucleation of both ϵ and α'. Brooks, Loretto, and Smallman (1979) concluded that the nucleus of the ϵ martensite is a single stacking fault and that α' nucleates from faulted regions in dislocation pileups. Others (Suzuki, Kojima, Suzuki, Hashimoto, and Ichihara, 1977; Mangonon and Thomas, 1970; Lagneborg, 1964) have observed that the formation of strain-induced α' initially occurs at intersections of two active {111} slip systems (see Fig. 9.10 and associated discussion in § 9.3.5).

9.3.5 Effects on austenite deformation

Below the equilibrium temperature, T_0, martensite tends to form during (a) cooling, (b) application of elastic stress, and (c) plastic deformation of the austenite. Transformation during cooling and during application of elastic stress (stress-assisted transformation) results in similar martensite morphology. But, as described earlier, strain-induced martensite forms at slip-band intersections. During plastic deformation of metastable Fe–Cr–Ni austenite at low temperatures, two strain-induced martensitic phases form, h.c.p. ϵ and b.c.c. α'.

The AISI 300 series stainless steels used in cryogenic applications range from metastable to stable austenites. Alloys, such as 304, containing 18% Cr and 8% Ni, are metastable, whereas 310, containing 26% Cr and 20% Ni is stable with respect to martensitic transformation. The stress–strain behavior and temperature dependence of the flow strength of these alloys differ and depend on the austenite stability. In the more unstable alloys, both ϵ and α' martensite are formed; in the slightly metastable alloys (e.g., 316), only α' is formed; and in the stable alloys, neither α' nor ϵ is formed during deformation to fracture at any temperature.

Figures 9.11 and 9.12 assist in explaining the role of martensite transformations in austenite deformation. In a metastable alloy at low temperatures (4 to 200 K), three distinct stages are present. A stable alloy only exhibits dislocation work hardening and has "well-behaved" stress–strain characteristics.

Stage I (Fig. 9.11) of metastable austenite stress–strain behavior represents the microstrain and early macrostrain behavior that includes the 0.2% offset yield strength. The formation of α' is not thought to occur in this range; perhaps stacking-fault clusters, or the ϵ phase, or both contribute to low-temperature deformation in this range. The stacking-fault energy is reduced at low temperatures and apparently becomes low enough to promote ϵ martensite. With x-ray analysis to detect ϵ formation during early deformation and creep at low temperatures, Mirzagev, Goykhanberg, Shteynberg, and Rushchin (1973) related the log of volume fraction of ϵ over the volume fraction of $\gamma + \alpha'$ to the log of the plastic deformation for an Fe–18Cr–13Ni–0.02C steel; other x-ray analysis measurements on similar steels have not confirmed this ϵ-strain correspondence (Guntner and Reed, 1962; Reed and Guntner, 1964).

The temperature dependence of metastable and stable Fe–Cr–Ni alloys in Stage I is illustrated in Fig. 9.12. There are three regions of the temperature dependence of the flow strength for metastable alloys: (a) the high-temperature range ($>T_1$) that exhibits normal temperature dependence with no detectable martensitic products; (b) the middle region ($T_2 < T < T_1$) in which the flow strength decreases with decreasing temperature; and (c) the low-temperature range ($<T_2$) where flow strength again increases at lower temperatures. The temperature dependence of the flow strength of stable alloys is normal, increasing monotonically with decreasing temperature.

In Stage I, α' has not been detected. Magnetic torsion-balance measurements (Reed and Guntner, 1964) do not sense a permeability change until Stage II is reached. Suzuki et al. (1977) have also used magnetic measurements to characterize the stress level at which α' begins to form as a

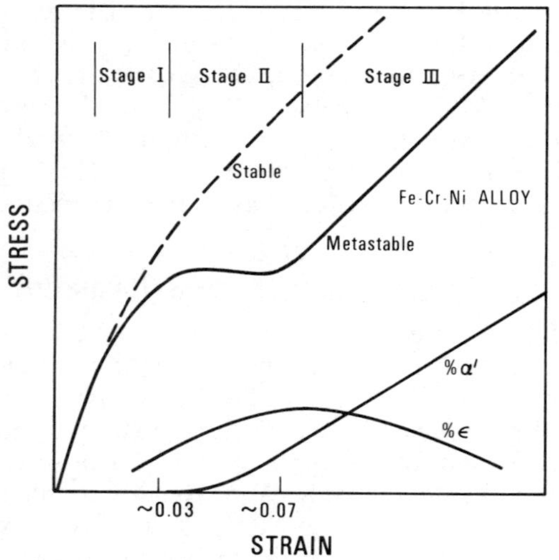

Figure 9.11 Typical stress-vs.-strain curves of stable and metastable austenitic Fe–Cr–Ni alloys at low temperatures and associated relative amounts of strain-induced ϵ and α' that form in the metastable alloy.

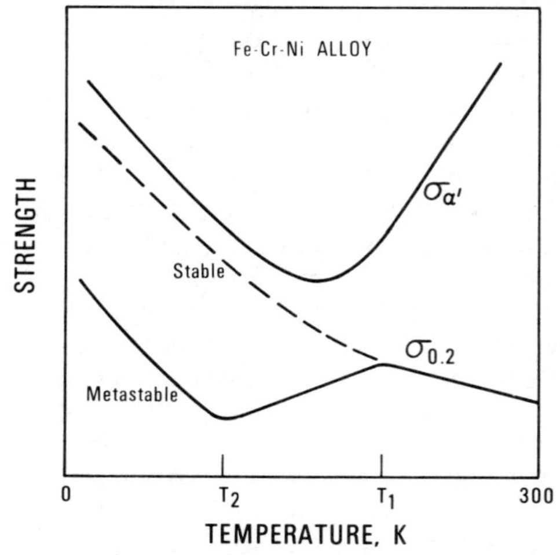

Figure 9.12 Temperature dependence of tensile flow strength at 0.2% offset, $\sigma_{0.2}$, and estimated strength at which α' begins to form, $\sigma_{\alpha'}$, of stable and metastable Fe–Cr–Ni steels (Suzuki et al., 1977).

function of temperature; their results are also included in Fig. 9.12. The stress level at which α' forms is considerably above the yield strength. At first, the study of Olson and Azrin (1978) appears to conflict with these premises. They worked with a less stable Fe–9Cr–8Ni alloy (a TRIP steel) and obtained an excellent correlation between the temperature dependence of the yield strength and the flow strength at which α' was first detected by using ac permeability measurements. It is likely, however, that martensite formation in their alloy was stress assisted and that ϵ did not form.

Stress-assisted martensite formation or the formation of α' from preexisting nucleation sites is best illustrated by the study of Breedis and Robertson (1963). They measured the temperature dependence of a metastable single-crystal Fe–16Cr–12Ni and of a stable single-crystal alloy. The critical resolved shear stress of the metastable alloy fell to zero at the T_{ms} of α' (240 K). They detected the presence of both ϵ and α' in the deformed crystals at temperatures above T_{ms}. The critical resolved shear stress of the stable alloy increased continuously with decreasing temperature. The T_{ms} values of commercially available alloys are normally below 0 K, and therefore, the flow strengths of these alloys are not expected to approach zero.

Stage II, consisting of an "easy glide" range is best correlated with the formation of α' laths at cross-slip intersections. Suzuki et al. (1977) have proposed that such α' laths, with the long $\langle 110 \rangle$ direction representing the intersection of two active slip systems, act as windows to assist cross slip. Stage II is most prevalent in the temperature range 76 to 200 K. In this temperature region, the range of strain at which the stress remains relatively constant is large, usually about 0.03 to 0.05.

In Stage III, the rate of work hardening increases to a constant, which is maintained over a relatively large amount of plastic deformation (20 to 40%). The α' volume percentage is linearly related to plastic deformation in Stage III. The nature of the transition in the roles of α' formation — from promoting earlier glide in Stage II to being associated with a linear, high rate of work hardening in Stage III — is not clear. Perhaps in Stage III essentially all active cross-slip sites have been transformed to α', and subsequent α' formation occurs within active glide bands. Such α' formation would not be expected to promote stacking-fault glide. There is a hint that this sequence may occur: the x-ray data of Reed and Guntner (1964) presented schematically in Fig. 9.10 indicate that a maximum value of the ϵ volume concentration occurs at about the transition from Stage II to Stage III. There is apparently no ϵ formation during Stage III; the amount of ϵ decreases with strain, presumably transforming to α' (Reed and Guntner, 1964).

Stages I and II decrease in significance in more stable austenites, but Stage III is retained. For instance, AISI 316 exhibits neither Stage I nor Stage II, but does begin to transform to α at strains of about 0.02 below 190 K. No ϵ martensite has been detected in this alloy.

In this discussion of the influence of martensitic products on plastic deformation, the results of current research indicate that the schematic representations of Figs. 9.10 and 9.11 are time dependent. Further research is necessary to sort out the creep and strain-rate effects and the isothermal characteristics of the strain-induced martensitic transformations.

This discussion has emphasized the roles of the strain-induced martensites. Martensite that forms during cooling below T_{ms} contributes differently. For many years α' martensite has been known and used to strengthen steel, and it may be used in a similar way to strengthen Fe–Cr–Ni alloys (e.g., Mangonon and Thomas, 1970). With severe rolling to induce a high percentage of α', room-temperature yield strengths of about 1.4 GPa (2×10^5 psi) may be achieved at the expense of significant ductility and (probably) toughness.

The complex martensitic transformation characteristics in Fe–Cr–Ni austenitic commercial alloys are summarized in Table 9.3. Stress-assisted martensite has been reported in single crystals (Breedis, 1964) and in Fe–9Cr–8Ni (Olson and Azrin, 1978), but has not been observed in Fe–18Cr–8Ni polycrystalline commercial grades. Martensite formed by the application of elastic stress should result in morphology similar to that observed from cooling. The strain-induced transformations have been subdivided into three stages, each contributing distinctly to austenite deformation. The major distinction between the stages is the degree of work hardening. Strain-induced α' (Stage III) is distinguished from stress-assisted or thermally induced martensite by the location of transformation: strain-induced α' forms at slip-band intersections with only one habit-plane variant; stress-assisted or thermally induced martensite forms in $\{111\}$

Table 9.3 Summary of martensitic transformation characteristics in commercial AISI 300 series steels.

Transformation	Characteristics: Crystal Structure	Morphology	Location	Amount (%)	Comments
Cooling	h.c.p.	{111} sheets		0–5	
	b.c.c.	<1$\bar{1}$0> laths, 3 habit plane variants	Within {111} sheets	0–10	Isothermal (more at 76 than 4 K)
Stress assisted	h.c.p.	{111} sheets			
	b.c.c.	<1$\bar{1}$0> laths			Not observed in Fe–18Cr–8Ni or more stable alloys
Strain induced					
Stage I	h.c.p.?	{111} sheets		0–5	Work hardening
Stage II	h.c.p.	{111} sheets		0–20	Reduces in Stage III
	b.c.c.	<1$\bar{1}$0> laths, 1 habit plane variant	Slip-plane intersections	0–10	No work hardening
Stage III	b.c.c.		Within {111} sheets	10–80	% linear with strain; strong work hardening

sheets with three habit-plane variants. This review emphasizes one aspect that is not conventionally recognized. It has been thought that stress-assisted transformation lowers the flow strength, and that strain-induced transformation increases the flow strength. The strain-induced martensite at first promotes no work hardening and thus decreases the flow strength in Stage II, and then in Stage III, it increases the flow strength. This leads one to suspect that the transformation mechanism changes from Stage II to III.

9.3.6 Influence on performance

Several characteristics of the austenite-to-b.c.c.-martensite transformation in stainless steels cause problems in many cryogenic applications.

The martensitic b.c.c. product has a larger specific volume (+1.7%) than the parent austenite; therefore, these steels expand during transformation. The martensitic product forms as individual crystals, and the associated transformation shear and volume expansion results in localized regions of disruption. For example, in Fig. 9.3b, α' formation near or at the surface of the flange caused localized surface upheavals. For service that requires close tolerances, such as valves or bearings, these local surface fluctuations are disastrous.

Welding of austenitic steels tends to produce chemical segregation in the heat-affected zone. Nitrides (in high-nitrogen steels) and carbides tend to form, especially in steels containing chromium. The chemically depleted austenitic solid solution has a greater tendency for martensitic transformation on cooling to cryogenic temperatures. Transformation under such circumstances is particularly critical, since it is restricted to the weld heat-affected zone and, therefore, tends to produce high local stresses from the transformation volume expansion and local shear. In a few applications of thin-gauge sheet, through-thickness cracks in the heat-affected zone of the welds or spot welds have resulted.

The transformation at cryogenic temperatures is unpredictable from room-temperature characteristics. The transformation occurs after application of applied stress or local plastic deformation at low temperatures much more readily than at room temperature. Prevention of in-service transformation depends on proper alloy selection. A good rule to follow is that the more alloying elements one uses (and can afford), the more stable the austenitic stainless steel will be. Therefore, to a first approximation, AISI 310 is stable, AISI 316 has limited strain-induced martensite, and AISI 304 is the least stable. In AISI 304 within the chemical specification limits, martensite may form on cooling and on application of stress and certainly will form with plastic deformation.

Another concern, mostly in design of superconducting magnets, is that the b.c.c. martensitic product is ferromagnetic, whereas the parent austenite is paramagnetic. A simple rule is that each percent of b.c.c. martensite results in a permeability increase of 0.01. Therefore, if time-dependent field changes corresponding to material changes of the order of 0.10 are significant, then alloy selection is important.

Finally, a common design fear has been that martensitic transformation in austenitic stainless steels adversely affects toughness. Experience suggests that this is partly true. Strain-induced martensitic transformation in austenitic stainless steels tends to produce higher toughness. Comparison of the low-temperature toughness of AISI alloy grades 304, 316, 310 (see Chapter 11) indicates that toughness is inversely related to austenite stability (and austenite tensile yield strength). But, if the austenite is sufficiently unstable such that martensite forms on cooling (e.g., in the AISI 400 series steels and in AISI 301), then the martensite is less tough.

To summarize: In the more stable austenitic stainless steel grades, martensite does not form during cooling and forms only during deformation. In these alloys, with predominantly austenite plastic deformation, martensitic transformation produces greater toughness. In metastable stainless steels in which the base structure is predominantly martensite prior to the application of stress and, therefore, in which plastic deformation must take place in the martensitic phase, the martensite has a deleterious effect on toughness.

9.4 ALKALI METALS

Lithium and sodium have been observed to undergo martensitic transformations at low temperatures, whereas in potassium, rubidium, and cesium, no transformation has been reported. Although neither lithium nor sodium has a potential for extensive use at low temperatures, they have both been studied rather extensively because their crystal and electronic structures are simple, permitting easier fundamental calculations of free energies. The characterization of their martensitic transformation has not been previously reviewed.

9.4.1 Sodium

On cooling to 26 to 36 K, the b.c.c. structure of sodium begins to transform to a probably heavily faulted h.c.p. structure with the lattice parameter a equal to 0.3767 nm and c equal to 0.6154 nm at 5 K (Barrett, 1956a). The c/a ratio is about 1.633, the ideal value within experimental accuracy. Apparently both larger grain size and fewer internal or surface defects raise the T_{ms} slightly (Barrett, 1955/56; Hull and Rosenberg, 1960). Severely broadened x-ray diffraction lines from the product phase indicate that sodium contains many stacking faults. By comparing the broadened lines and their angular positions for a faulted h.c.p. or faulted f.c.c. structure, Barrett (1956a) argued that the faulted structure was h.c.p. However, the probable ideal c/a ratio and the extreme weakness of the (004) h.c.p. diffraction line led Stedman (1976) to suggest the possibility that the product phase has an f.c.c. structure.

The transformation first appears on a specimen surface as fine lines or planes, which broaden (Hull and Rosenberg, 1959). As illustrated in Fig. 9.13, many habit-plane traces are observable. Assuming that the surface appearance is representative of the interior, Hull and Rosenberg (1959) estimate that, for a specimen cooled to 4 K (Fig. 9.13), the amount of product

Figure 9.13 Photomicrograph of sodium surface after cooling to about 6 K. Surface is estimated to contain about 40% martensite. (Courtesy of H. M. Rosenberg.)

phase is about 45%, considerably higher than the 5% predicted by Barrett (1956a) from integrated x-ray peak intensities and Lorenz polarization and structure-factor estimations. Only a wide disparity between the amounts of martensitic product formed on the surface and within the specimen could explain the different estimates.

Martensitic transformation in sodium is accompanied by a small expansion. As measured by Basinski and Verdini (1959), the volume change of a specimen cooled to 4 K was 1.45×10^{-3}; if one assumes, following Hull and Rosenberg (1959), that 45% of the specimen transformed, then

$$(V_{\text{product}} - V_{\text{parent}})/V_{\text{parent}} \cong +3 \times 10^{-3} \tag{9.14}$$

Reversion to the b.c.c. structure in sodium is apparently retarded by specimen cold work. Ultrasonic attenuation studies suggest that reversion begins above 50 K on warming previously annealed specimens that had been cooled to 4 K, but that reversion begins above 110 K in specimens that were strained at 77 K to induce h.c.p. martensite (Stern, Natale, and Radnick, 1966; Verdini, 1962). Barrett (1956a) reports that reversion following cold work at 4 K begins when heating to 60 to 75 K and that the temperature interval to complete the transformation from h.c.p. to b.c.c. is increased when the h.c.p. is strain induced. Electrical resistivity (Dugdale and Gugan, 1962) and optical microscopy observations (Hull and Rosenberg, 1960) place the reverse transformation temperature, in the absence of cold work, at about 50 K and completion of the h.c.p.-to-b.c.c. transformation near 70 K.

Low-temperature measurements (Filby and Martin, 1963) indicate that the faulted h.c.p.-martensitic-phase specific heat at 4.0 K is estimated to be 0.037 $J \cdot K^{-1} \cdot g\text{-at.}^{-1}$ (the Debye temperature, $\theta_D = 158.8$ K), lower than the b.c.c.-parent-phase specific heat of 0.041 $J \cdot K^{-1} \cdot g\text{-at.}^{-1}$ ($\theta_D = 152.0$ K); to obtain these values, 50% transformation on cooling to 3 K was assumed. Spontaneous specimen heating ($\sim$0.21 $J \cdot mol^{-1}$) was observed during warming in the temperature range 5 to 15 K after cooling to 3 K; presumably this may be attributed to small additional martensitic transformation subsequent to limited defect annealing (Gugan, 1966).

The low-temperature electrical resistivity of sodium was summarized by Dugdale and Gugan (1962), who measured its pressure dependence between 2 and 300 K. The results of these measurements are most easily explained if the ratio of resistivities of the two phases is independent of pressure. The intrinsic resistivity of the h.c.p. phase is less than the b.c.c. phase; cooling below T_{ms} reduces the resistance. Deformation of sodium at 20 K produces little change in resistivity; the increase in resistivity from defect generation is offset by the reduction due to the b.c.c.-to-h.c.p. transformation (Hull and Rosenberg, 1959). Warming to 80 K and recooling to 20 K, after strain-induced transformation at 20 K, increases the electrical resistivity at 20 K; this effect is caused by h.c.p.-to-b.c.c. reversion during warming, overwhelming the resistivity decrease from defect annealing.

Inelastic neutron-scattering measurements (Dolling, Powell, and Martel, 1968; Stedman, 1976) on sodium between 36 and 90 K show usual variations in normal-mode temperature dependence, frequency shifts, or peak broadening, indicating there was no phonon instability in the sodium preceding the martensitic transformation.

Calculations of b.c.c. and h.c.p. Helmholtz free energies in sodium have been performed by using local pseudopotential theory (Straub and Wallace, 1971) and Heine–Abarenkov model potential with quasi-harmonic phonon-frequency distributions (Pynn and Ebbsjo, 1971). Both are successful in predicting the stability of the b.c.c. structure at higher temperatures and the h.c.p. structure at lower temperatures.

Tensile stress–strain curves of sodium have been measured at 195, 80, 20, and 4 K by Hull and Rosenberg (1959). The 4-K stress–strain relationship is included in Fig. 9.14, along with the 4-K curves for lithium and potassium.

At temperatures of 80 K and below, however, sodium work-hardens more than potassium. Nevertheless, the temperature dependence of the yield strength of both elements from 195 to 4 K is similar, indicating that the martensitic transformation in sodium is strain induced, forming during plastic deformation and contributing to specimen strain hardening. Indications are that the T_{md} of sodium is above 80 K. Notice, in Fig. 9.14, that there are no multiple-yielding phenomena in the stress–strain behavior of sodium or potassium at 4 K.

9.4.2 Lithium

Barrett (1947) first identified the martensitic transformation in lithium. On cooling, b.c.c. lithium undergoes a martensitic transformation at

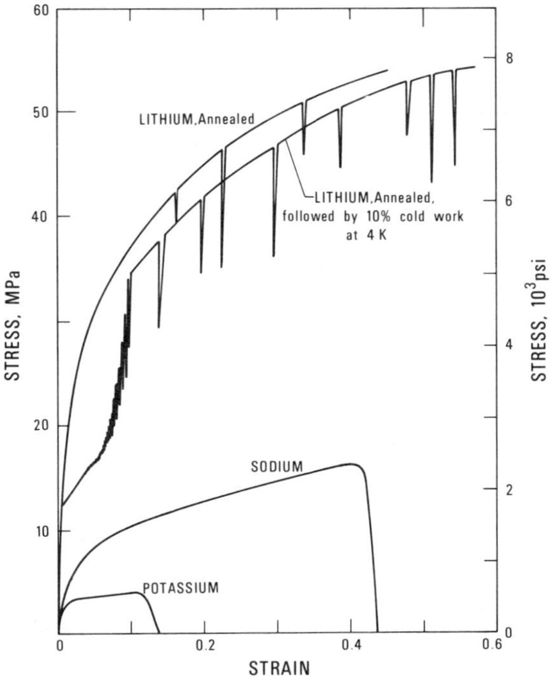

Figure 9.14 Stress–strain curves at 4 K of annealed polycrystalline sodium, potassium, lithium, and cold-worked lithium (Reed, 1977; Hull and Rosenberg, 1959).

68 to 78 K to a predominantly faulted h.c.p. structure (Barrett, 1956b). Snyder (1966) reported a slight effect of the lithium isotope on T_{ms}; lithium-6 has an average T_{ms} of 68.6 ± 2 K and lithium-7, an average T_{ms} of 72.7 ± 2 K. Plastic deformation at low temperatures produces primarily a heavily faulted f.c.c. martensitic product with both b.c.c. and h.c.p. structures transforming to f.c.c. Reversion of f.c.c. begins near 95 K and extends to about 170 K. In one x-ray investigation, Alexopoulos and Shaw (1961) detected the presence of the (111) f.c.c. diffraction line after cooling annealed, fine-grained, high-purity material. Results of x-ray studies suggest that about 50% h.c.p. martensite forms on cooling annealed lithium to 50 K (Barrett and Trautz, 1948), and optical microscopy at 20 K indicates about 90% transformation of the lithium surface with no subsequent transformation on cooling to 4 K (Hull and Rosenberg, 1960). Basinski and Verdini (1959) suggested that 50% transforms on cooling to 4 K and that a 100% f.c.c. phase would be expected to be produced by 60% compression at about 50 K. X-ray measurements indicate that about 50% f.c.c. phase is produced by 60% compression at 78 K.

From optical microscopy observations of traces remaining on specimens that had previously been cooled to 77 K and then warmed to room temperature, Bowles (1951) concluded that the habit plane of the h.c.p. product was approximately {441}. Optical microscopy of transformed lithium specimen surfaces by Hull and Rosenberg (1960) and Gindin, Lazarev, and Starodubov (1961) indicates a close similarity to the sodium martensite morphology (Fig. 9.13).

In contrast to sodium, the h.c.p. martensitic transformation in lithium produces a volume contraction. Basinski and Verdini (1959) calculated

$$(V_{\text{product}} - V_{\text{parent}})/V_{\text{product}} \cong -7 \times 10^{-4} \tag{9.15}$$

from liquid-displacement measurements and from estimates of 50% transformation on cooling to 4 K. From optical metallographic observations of the product phase, which closely resembles the sodium h.c.p. phase (Fig. 9.13), Hull and Rosenberg (1960) estimated that about 90% transformation had occurred in their specimens at 20 K.

The electrical resistivity experiments of Dugdale and Gugan (1961) provide evidence of the complex effects of a martensitic transformation and the associated production of defects on physical property measurements and on the temperature dependences of the measured physical properties. Measurements during cooling first indicate that the resistivity of Li–0.21 to 0.95 at.% Mg alloys decreases as a consequence of the spontaneous transformation, which begins at T_{ms}. Pure lithium specimens usually exhibit a slight resistivity decrease following transformation. Subsequent cooling results in an additional decrease of resistivity, but at a lower rate of change than in untransformed material. Therefore, the residual resistivity of the h.c.p. phase is larger than the residual resistivity of the b.c.c. phase, but at higher temperatures the resistivity of the martensitic phase is less. A typical resistivity-vs.-temperature curve during cooling and warming and a schematic of the deduced apparent resistivities of the two phases are illustrated in Fig. 9.15. The possible higher residual resistivity at low temperatures of the h.c.p. phase (estimated 10% higher than the b.c.c.) is consistent with the expected larger defect density found in martensitic structures.

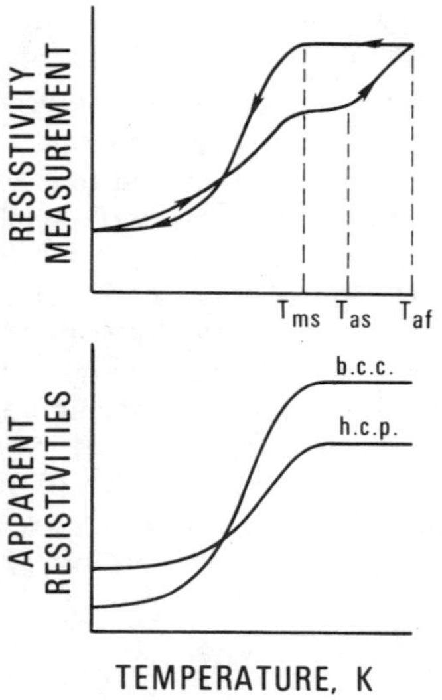

Figure 9.15 Typical resistivity-vs.-temperature curves for lithium and Li–Mg and apparent residual resistivities of b.c.c. and h.c.p. phases as a function of temperature (Dugdale and Gugan, 1961). T_{af} is the temperature of complete reversion to the b.c.c. phase.

Hedgcock (1960) reported that the h.c.p. phase exhibits a greater magnetic susceptibility than the b.c.c. phase, suggesting that the band structure changes during transformation. An increase in magnetic susceptibility corresponds to an increased electron effective mass, an increased density of states, and therefore, an increased electron-scattering contribution to the resistivity. Measurements on deformed specimens containing f.c.c. martensite suggest that the f.c.c. low-temperature residual resistivity is about 10% larger than that of the h.c.p.

Electrical resistivities at 4 K as a function of specimen strain for potassium, sodium, and lithium are presented in Table 9.4 (Hull and Rosenberg, 1959; Dugdale and Gugan, 1960, 1961; Basinski, Dugdale, and Gugan, 1959). The dependence of the change of resistivity per unit strain ($\Delta\rho/\Delta\epsilon$) is greatest for those materials that have strain-induced martensitic transformations. The intrinsic resistivity [$\rho_0(\epsilon)$] of strained lithium at 4 K is essentially composed of three contributions; the as-cooled, intrinsic resistivity (ρ_0), the resistivity contribution from deformation- and transformation-induced defects ($\Delta\rho_0^d$), and the resistivity contribution from the martensitic structure ($\Delta\rho_0^m$). This is

$$\rho_0(\epsilon) = \rho_0 + \Delta\rho_0^d + \Delta\rho_0^m \tag{9.16}$$

Table 9.4 Electrical resistivity at 4 K.

Metal	ρ_0 (nΩ·cm)	$\frac{\Delta\rho_0}{\Delta\epsilon}$, $\epsilon = 0.10$ (nΩ·cm)	Source
K	9.3	20	Basinski et al. (1959)
Na	2.9	28	Dugdale, Gugan (1960)
Na	1.3	21	Hull, Rosenberg (1959)
Li	15.0	235	Dugdale, Gugan (1961)
Li	87.0	240	Hull, Rosenberg (1959)
Li–0.07 Mg	82.8	335	Dugdale, Gugan (1961)
Li–0.35 Mg	466	505	Dugdale, Gugan (1961)
Li–0.73 Mg	969	515	Dugdale, Gugan (1961)
Li–0.95 Mg	1290	550	Dugdale, Gugan (1961)

In Table 9.4, ρ_0 and ($\Delta\rho_0^d + \Delta\rho_0^m$) are listed. The intrinsic resistivity contribution is composed of two parts, that of the b.c.c. phase and that of the h.c.p. martensitic phase that has formed on cooling to 4 K. It is apparent that larger $\Delta\rho_0$ values are obtained for larger intrinsic resistivities. Presumably, as a discussion later in this section indicates, increasing the magnesium concentration reduces the stability of the b.c.c. structure and results in higher martensitic transformation per unit strain. This contributes to an increase of resistivity at 4 K. The data also suggest an increase of $\Delta\rho_0^d$ as ρ_0 increases, since the increase of $\Delta\rho_0$ is considerably larger than one would expect from increased f.c.c. martensite formation (a maximum of 10% has been previously suggested). The added $\Delta\rho_0^d$ may result from an increase of deformation- or transformation-induced stacking faults or twins.

Electrical resistance measurements (Gugan, 1963) show that the T_{ms} of lithium is only slightly dependent on hydrostatic pressure, increasing by 0.25 K per 100 MPa (1000 atm).

Magnetic susceptibility results (Hedgcock, 1960) suggest that there is an increase in susceptibility at 80 K of approximately 24% in transforming from the b.c.c. to the h.c.p. phase.

Specific heat measurements by Martin (1960a, 1960b) provide excellent information on reversion of the faulted h.c.p.-to-b.c.c. transformation. Figure 9.16 presents Martin's data on the specific heat of lithium and Li–0.95 Mg on cooling to 20 K and then warming. During warming,

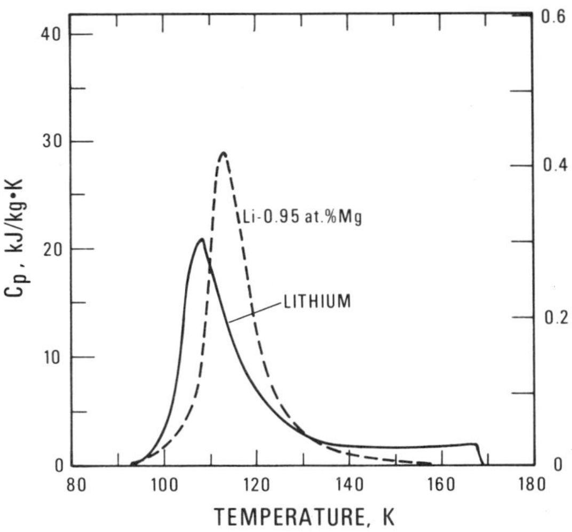

Figure 9.16 Specific heat, C_p, variations when cooling to 20 K and subsequently warming to higher temperatures. Total heat content of anomalous ΔC_p $(C_p dT)$ is 39.3 kJ/kg for lithium and 38.9 kJ/kg for Li–0.95 at.% Mg. (Martin 1960a, 1960b.)

the h.c.p.-to-b.c.c. reversion is identified by an absorption of energy, contrasting with defect recombination and recrystallization, which emit energy. Therefore, as both phase reversion and defect recombination are likely to be occurring simultaneously, the shape of the specific heat-vs.-temperature curves of Fig. 9.16 probably represents the sum of both processes. From Fig. 9.16 it is clear that the T_{as} temperature is near 90 K and that the reversion is completed near 170 K. The reversion process is time dependent for both lithium and the Li–Mg alloy; variations of the C_p–T curve are produced by changing the warming rate or by holding the specimen at a given temperature between 90 and 170 K. As Martin (1960a) suggested, it is likely that a spectrum of activation energies controls the higher temperature portion of the reversion process, especially in pure lithium. Assuming that 50% h.c.p. phase forms on cooling to 55 K, Martin (1960a, 1960b) deduced a Debye temperature at 60 K for the h.c.p. phase of 390 K and for the b.c.c. parent phase of 371 K. He estimated the heat of transformation from h.c.p. to b.c.c. to be about 59 J·g-at.$^{-1}$.

Using ultrasonic-pulse techniques, Day and Ruoff (1974) and Felice, Trivisonno, and Schuele (1977) measured the elastic constants and the temperature and pressure derivations of lithium-6 and lithium-7 from 85 to 300 K. Results of lithium measurements were similar to those of sodium measurements; no evidence for bulk elastic instability was found. All the elastic constants had negative temperature and positive pressure derivatives. Below T_{ms}, modulus data are complex. Snyder (1968), using modified vibroscope detection equipment, found a general decrease in Young's modulus below T_{ms} on cooling and, on warming, an anomalous dip in Young's modulus followed by a general decrease until T_{as} was reached. Considerable specimen-to-specimen variation was detected.

Perhaps the most extensive calculation of the structural stabilities of lithium and associated properties related to its phase transformations was attempted by Bajpai, Oho, Ohno, and Toya (1975). Calculations of electrostatic energy by Ewald's method, core repulsion by the Born–Mayer potential and Hartree–Fock potential and nonlocal pseudopotential electronic contributions led to predictions of zero-point energy, lattice parameters, Debye temperature, cohesive energy, form factors, phonon dispersion relations, and free energies of the b.c.c., h.c.p., and f.c.c. structures as a function of temperature. For these calculations the total free energy was taken as the sum of the cohesive energy at 0 K and the phonon zero-point energy, combined with the temperature-dependent phonon and conduction-electron free energy. The resulting equilibrium temperature estimations of 147 K for b.c.c.–h.c.p. and 74 K for b.c.c.–f.c.c. transitions are remarkably close to experimental results.

Tensile properties to 4 K were measured by Hull and Rosenberg (1959) and Reed (1977), and the 4-K stress–strain curves for lithium, sodium, and potassium are shown in Fig. 9.14. The phase transformations contribute considerably to the low-temperature strength of lithium. The large load drops apparently result from the phase transformation or from large-scale twinning. The load drops are too irregular to be caused by adiabatic specimen heating, although there is evidence for discontinuous yielding from adiabatic heating in lithium at 1.5 K (Gindin, Lazarev, Starodubov, 1960). The elastic limits of the three alkali metals are very close, but the strain-induced phase transformation to h.c.p. increases the work-hardening rate of sodium, compared with that of potassium, and the strain-induced transformation to the f.c.c. structure considerably increases the work hardening rate in lithium.

A comparison of flow strengths of lithium and

potassium at various tensile strains is presented vs. temperature in Fig. 9.17, using the data of Hands and Rosenberg (1969), Hull and Rosenberg (1959), and Reed (1977). Above 77 K there are only moderate differences between lithium and potassium flow strengths. At 77 and 20 K there are large differences between the flow strengths, presumably caused by the martensitic transformation. The increased temperature dependence at low temperatures (to 20 K) apparently reflects the increased martensitic transformation per unit strain, but may be at least partially accounted for if the flow strength dependence on temperature of a mixed-phase (b.c.c., h.c.p., f.c.c.) structure is significantly greater than the b.c.c. temperature dependence. The constant temperature dependence of the flow strength difference between 20 and 4 K results from the increased temperature dependence of the potassium flow strength (Hands and Rosenberg, 1969); below 20 K, work-hardening rates in potassium increase rapidly with reduced temperature from either dislocation–impurity interactions, dislocation–dislocation interactions, or an increase in the Peierls stress.

Assuming that no martensitic transformation occurs at 295 or 195 K, the temperature dependence of the flow strength is linear to near 90 K, and the influence of the increased martensitic-transformation work-hardening rate is linearly dependent on temperature below T_{md}. The latter two of these assumptions are drawn schematically in Fig. 9.17. Then the two regions, extrapolated linearly, intersect at about 93 K (see Fig. 9.17). This temperature range is only slightly lower than the T_{md} reported by Barrett and Trautz (1948), indicating that the assumption of linear dependence of the flow strength below T_{md} is reasonable. This linear dependence is possibly related to a linear increase of percent martensite below T_{md} as a function of temperature at constant strain. Barrett and Trautz reported that the percentage of f.c.c. martensite forming at constant deformation is linearly dependent on the deformation temperature.

Creep experiments of Gindin, Godzhagev, Lazareva, and Starodubov (1966) to 77 K indicate that for imposed stresses above 4 MPa, the transformation impedes the logarithmic steady-state creep stage.

In the Li–Mg alloy series, measurements have been made of transformation temperatures, elastic constants, C_{ij}, and lattice parameters; they are summarized in Fig. 9.18. Transformation temper-

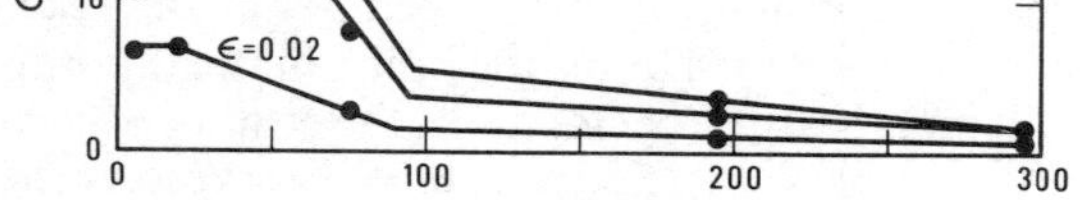

Figure 9.17 The differences between the flow strengths, σ_ϵ, of lithium and potassium at tensile strains of 0.02, 0.05, and 0.10 and temperatures 295, 195, 77, 20, and 4 K (Hands and Rosenberg, 1969; Hull and Rosenberg, 1959; Reed, 1977).

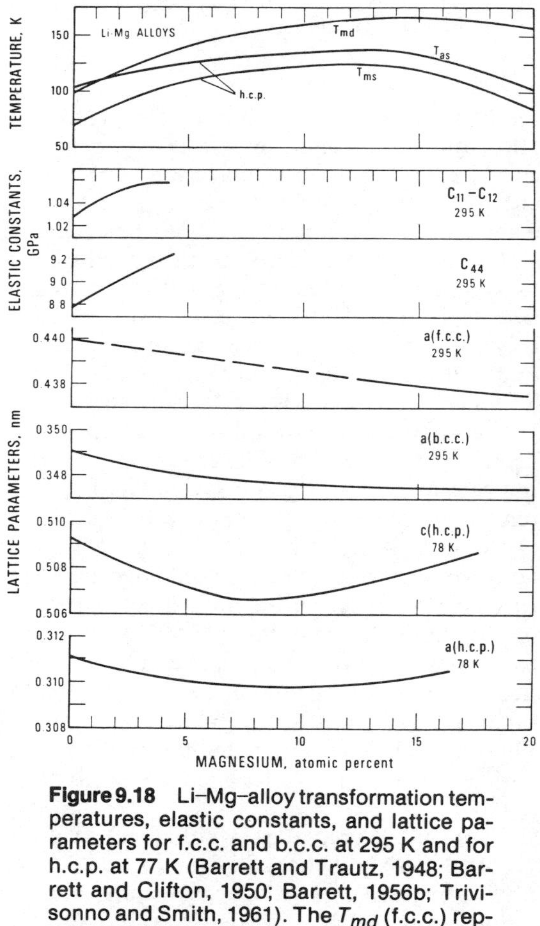

Figure 9.18 Li–Mg-alloy transformation temperatures, elastic constants, and lattice parameters for f.c.c. and b.c.c. at 295 K and for h.c.p. at 77 K (Barrett and Trautz, 1948; Barrett and Clifton, 1950; Barrett, 1956b; Trivisonno and Smith, 1961). The T_{md} (f.c.c.) represents detection of f.c.c. phase after 60% elongation.

atures are a maximum near 10 at.% Mg, and there is little temperature hysteresis between T_{ms} and T_{as}. Complete reversion to b.c.c. requires raising the temperature some 70 K above T_{as}. The h.c.p. phase retains an almost perfect 1.63 c/a ratio, within experimental scatter, throughout the alloy series. At 24 at.% Mg no h.c.p. structure has been detected, indicating complete suppression of the spontaneous transformation for magnesium concentration exceeding approximately 20%.

The elastic constants determined by the ultrasonic pulse-echo technique indicate that the high elastic anisotropy ($2C_{44}/C_{11} - C_{12}$) of lithium continues to increase slightly with magnesium addition, reaching a value at 5 at.% Mg, about 2% greater than that of pure lithium (Fig. 9.18). This trend coincides with an increase in T_{ms} in the same alloy range, supporting the argument that high elastic anisotropy relates to structural instability since high anisotropy indicates a low $\{110\} \langle 1\bar{1}0 \rangle$ shear resistance. Indeed, Masson (1962) has advanced this case for the Li–Mg alloy series.

9.5 SOLIDIFIED GASES

Gases that exist at room temperature solidify into relatively weakly bound solids at low temperatures. Binding forces consist of the van der Waals or London forces plus electric-pole interactions. Although the simple Lennard–Jones type of spherically symmetric potential is sufficient for prediction of many physical properties, this type of potential is not successful with respect to phase stability because it predicts the preference for h.c.p. crystal structures instead of f.c.c. crystal structures.

In fact, as Niebel and Venables (1976) point out, any two-body potential will predict the wrong sign for the small energy (E) differences ($E_{\text{f.c.c.}} - E_{\text{h.c.p.}}/E_{\text{f.c.c.}} \cong 1.1 \times 10^{-4}$), about 0.01% of total energy differences between the two crystal structures. They review the many-body long- and short-range forces and zero-point energy approaches and suggest a new many-body effect: consideration of s- and p-type outer orbitals results in small, but significant, differences in the van der Waals energy of the two structures. Klein and Venables have edited a two-volume collection of excellent reviews of lattice dynamics and experimental characterization of rare gas solids (*Rare Gas Solids*, 1977). Silvera (1980) has recently presented a very comprehensive review of the properties and theory of solid molecular hydrogens.

The x-ray diffraction experimental research leading to the suggestion that certain transformations in solidified gases are martensitic was done predominantly by co-workers Meyer and Barrett, who have both written reviews (Meyer, 1969; Barrett, 1969). Obviously, experimental verification of the transformation mode is difficult, and evidence has been restricted primarily to the influence of plastic deformation on transformation kinetics and to the occurrence of an orientation relationship between parent- and product-phase structures.

9.5.1 Hydrogen

The high-temperature hydrogen and deuterium phases freeze into an h.c.p.-like crystal structure at 13.96 K and 18.72 K, respectively, and undergo transformation with martensitic characteristics to an f.c.c.-like structure on cooling near 2 K. Far-infrared-spectroscopy and neutron-diffraction studies imply that slightly lower crystal symmetries are more likely than the h.c.p. and f.c.c. structures reported from x-ray diffraction work. The structure determination is difficult: diffraction is weak and diffuse from excessive beam penetration, and there is evidence that the crystal structure may be influenced by the substrate.

Specific heat anomalies have been detected between 1.5 and 1.7 K (Hill and Ricketson, 1954; Ahlers and Orttung, 1964), presumably from the transformation. Volume changes ($\Delta V/V$), on temperature cycling through the transformation temperature, have been measured for a hydrogen orthorhombic concentration of 73% at 0.5% (cooling) and 0.08% (warming) (Jarvis, Ramm, and Meyer, 1967). Here

$$\Delta V/V = (V_{\text{h.c.p.}} - V_{\text{f.c.c.}})/V_{\text{f.c.c.}} \tag{9.17}$$

and is calculated from the relation

$$\Delta V/V = -k_t \Delta P - \alpha_p \Delta T \tag{9.18}$$

where ΔP and ΔT are the pressure and temperature changes at constant volume, k_t is the isothermal compressibility, and α_p is the volume coefficient of thermal expansion. Deformation induces the f.c.c.-like structure near 3 K. Incomplete transformation after cooling or deformation usually results. Repeated thermal cycling through the apparent T_{ms} temperature produces complete transformation to the f.c.c. product. Ramm, Meyer, and Mills (1970) have measured volume changes for deuterium.

Terry (1965) prepared an extensive bibliography of solid-hydrogen crystallographic studies and McCarty, Hord, and Roder (1981) include discussions of the structure of hydrogen in their extensive review of engineering data of hydrogen.

9.5.2 Oxygen

The high-temperature, solid-oxygen-phase, cubic structure (γ) melts at 54.36 K and remains stable on cooling to 43.8 K, where a rhombohedral phase is formed. On continued cooling, the rhombohedral (β) phase martensitically transforms to a monoclinic (α) structure, beginning at 23.9 K. The $\beta \rightarrow \alpha$ transformation is described as martensitic by Barrett (1969) because there is an orientation relationship between the $(001)_\alpha$ and $(111)_\beta$, and this relationship remains after repeated thermal cycles through the transformation temperature. Meyer (1969) points out that the $\beta \rightarrow \alpha$ transformation is not completed at the transformation temperature, but is spread out over a range of temperatures.

Roder (1978) has extensively reviewed the solid transformation in solid oxygen with respect to density. He concludes that the molar volume decreases by 0.13 ± 0.11 $cm^3 \cdot mol^{-1}$ on transformation from β (20.95 ± 0.11 $cm^3 \cdot mol^{-1}$) to α (20.82 ± 0.22 $cm^3 \cdot mol^{-1}$). Compare this with a molar volume change of 1.18 ± 0.02 $cm^3 \cdot mol^{-1}$ for the transition from γ to β structures.

9.5.3 Argon-nitrogen and argon-oxygen

Argon melts at 83.81 K and, on cooling, has a predominantly f.c.c. structure. With additions of nitrogen or oxygen, an h.c.p. structure becomes more stable (Fig. 9.19). The phase boundaries drawn in Fig. 9.19 represent the temperature division between the high-temperature h.c.p. structure and the low-temperature martensitic f.c.c. structure for Ar–O_2 and Ar–N_2 mixtures. Above approximately 10% of nitrogen or oxygen mixed with argon, the f.c.c. phase must be induced by plastic deformation. Phase boundaries in Fig. 9.19 are determined from x-ray diffraction measurements of the transformation during plastic deformation. The f.c.c.-to-h.c.p. transformation is thermally induced at the higher temperatures at concentrations less than about 10% nitrogen or oxygen.

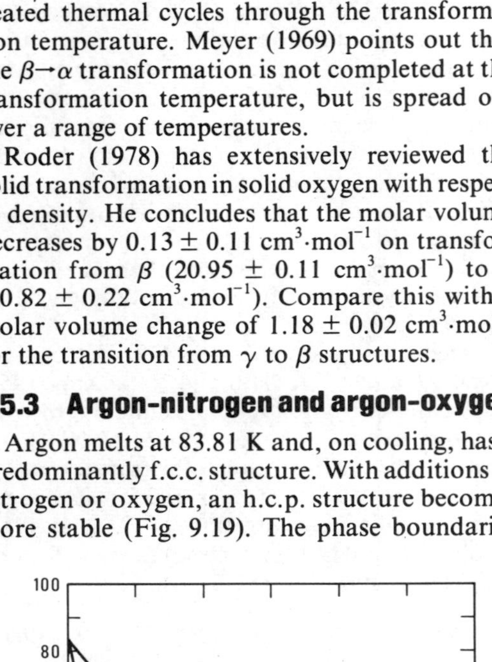

Figure 9.19 Temperature ranges of f.c.c. and h.c.p. phase stability, experimentally determined, of Ar–N_2 and Ar–O_2 mixtures (Barrett and Meyer, 1965; Barrett, Meyer, and Wasserman, 1966).

9.6 SUPERCONDUCTORS

The desire to increase the critical temperature, the critical current, and the critical fields of superconductors has led to research of complex metallurgical treatments to provide an optimized structure. Most practical superconductors, especially those with A15 structure, undergo a martensitic transformation. Martensitic transformations naturally influence the superconducting behavior and are thought to improve it. This section reviews the martensitic transformation in Ti–Nb alloys and in Nb_3Sn.

9.6.1 Titanium-niobium

A martensitic product forms from the high-temperature b.c.c. (β) phase of titanium-base niobium alloys to about 25 at.% Nb (40 wt.% Nb). Brown, Clark, Eastabrook, and Jepson (1964); Duwez (1953); Sato, Hukai, and Huang (1960); and Jepson, Brown, and Gray (1970) have measured T_{ms} to 20 at.% Nb, and these data are compiled in Fig. 9.20. Baker (1971) and Obst et al. (1980) have studied Ti–25 at.% Nb and both report that $T_{ms} = 448$ K. Hickman (1969) studied very large-grained strips of Ti–25 at.% Nb rapidly quenched from about 1500 K. If one allows for the usual suppression at higher alloy concentrations and low temperatures, it is likely that the martensitic product does not form during cooling to any temperature in compositions exceeding about 28 at.% Nb. Thus, no spontaneous martensitic transformations are expected from cooling commercial Nb–Ti superconductors [Ti–30 to 50 at.% Nb (45 to 65 wt.% Nb)]. There are no firm data on T_{md}, the strain-induced martensitic transformation temperatures. Examination of microstructures and x-ray analysis of tensile specimens deformed at 4 K of Ti–30, 39, and 50 at.% Nb reveals no evidence of a retained martensitic structure at room temperature (Read, 1978). Pulse-echo elastic constant measurements

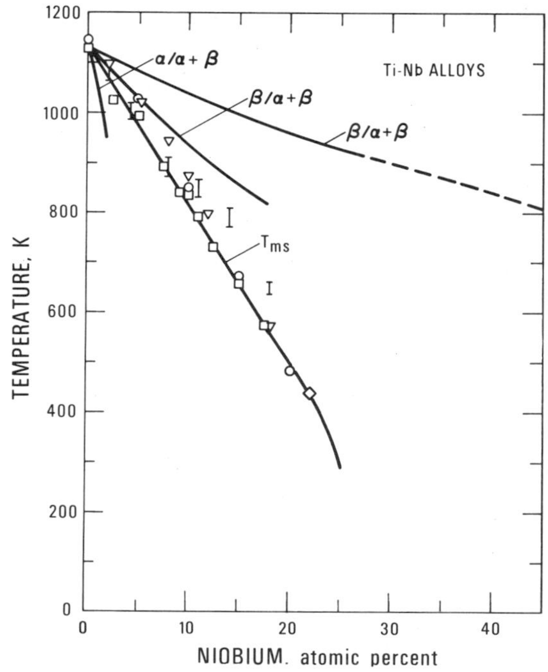

Figure 9.20 Phase boundaries and experimentally determined T_{ms} for the Ti–Nb alloy system. Phase boundary data from Hansen, Kamen, Kessler, and McPherson (1951); Jepson, Brown, and Gray (1970); Salter, Raymond, Boom, and Vetrano (1966). T_{ms} data from Brown et al. (1964); Duwez (1953); Hickman (1969); Sato et al. (1960); Jepson et al. (1970); Obst, Pattanayak, and Hochstuhl (1980).

from 4 to 300 K (Ledbetter, 1978) reveal no hysteretic effects. Pseudoelastic martensite (plate initiation and growth under applied load and reversion during removal of the load) has been suggested by Koch and Easton (1977) to explain some stress–strain "anelastic" effects observed during load cycling at 4 K. Accurate strain measurements at 4 K are very difficult, however, and to account for the observations, twinning has not been discounted (Read, 1978). Strain-induced, probably pseudoelastic, martensite has been reported in Ti–22 at.% Nb alloy at 4 K by Obst et al. (1980). A mobile interface under applied stress was reported, suggesting pseudoelasticity.

The martensitic product forms in an h.c.p. structure (α') in concentrations not exceeding about 7 at.% Nb and in an orthorhombic structure (α'') at higher niobium compositions. The room-temperature lattice parameters of each structure are presented in Fig. 9.21. The transition from hexagonal to orthorhombic is gradual, with the a and b axes contracting and expanding, respectively, at higher niobium compositions. The c axis reaches a maximum at about 10 at.% Nb and rapidly decreases at higher niobium concentrations. At concentrations above 20 at.% Nb, the high-temperature b.c.c. phase is prevalent and the room-temperature lattice parameters of this β phase are also included in Fig. 9.21. Hatt and Rivlin (1968) found the orientation relationship between β and α'' of a Ti–20.7 at.% Nb alloy to be $[001]_\beta$ 2° from $[100]_\beta$ 2° from $[010]_{\alpha''}$ and $[110]_\beta$ parallel to $[001]_{\alpha''}$.

The martensitic transformation in Ti–Ta alloys is remarkably similar to the Ti–Nb transformation. The T_{ms} values of the two-alloy systems are identical (Duwez, 1953). Bywater and Christian (1972) measured the lattice parameters of the martensitic phases; the characteristics of these parameters are the same as those of the Ti–Nb parameters (Fig. 9.21). The c axis reaches a maximum near the composition where the h.c.p. structure changes to orthorhombic. The composition for change of crystal structure in Ti–Ta alloys is about Ti–7 at.% Ta, identical to that of Ti–Nb. It is likely, therefore, that the crystallographic study of Ti–Ta martensites by Bywater and Christian (1972) may be assumed applicable to Ti–Nb alloy martensites. In compositions ranging from Ti–1.6 to 21.0 at.% Ta, plate-like morphologies were observed with no evidence of massive (lath or packet) martensite, unlike those suggested by Williams (1973) as typical in low-solute concentrations. Transformation twins were identified in alloys up to 10 at.% Ta, on $\{\bar{1}10\}_{\alpha'}$ or $\{\bar{1}11\}_{\alpha'}$; in higher tantalum compositions, no transformation twins were observed. Measurements suggest a $\{445\}_{\beta'}$ or $\{334\}_\beta$ habit plane; $\{334\}_\beta$ habit planes have been observed for other titanium-base alloys (Blackburn, 1970; Williams, 1973). The orthorhombic martensitic phase in Ti–Ta alloys is described by Bywater and Christian (1972) as fully coherent, having an irrational habit plane and no lattice-invariant deformations. To produce this, either a uniaxial strain or uniform dilatation at the interface is proposed.

The formation of the martensitic product in Ti–Nb alloys is dependent on cooling rate. Jepson et al. (1970) identified the effect of cooling rate on transformation temperature for a wide variety of Ti–Nb alloys, and their data are replotted in Fig. 9.22. Low cooling rates result in the nucleation and growth of α (Ti) or α'' (Ti–Nb alloys) from β. High cooling rates produce a martensitic transformation resulting in discontinuous suppression of the β-transformation temper-

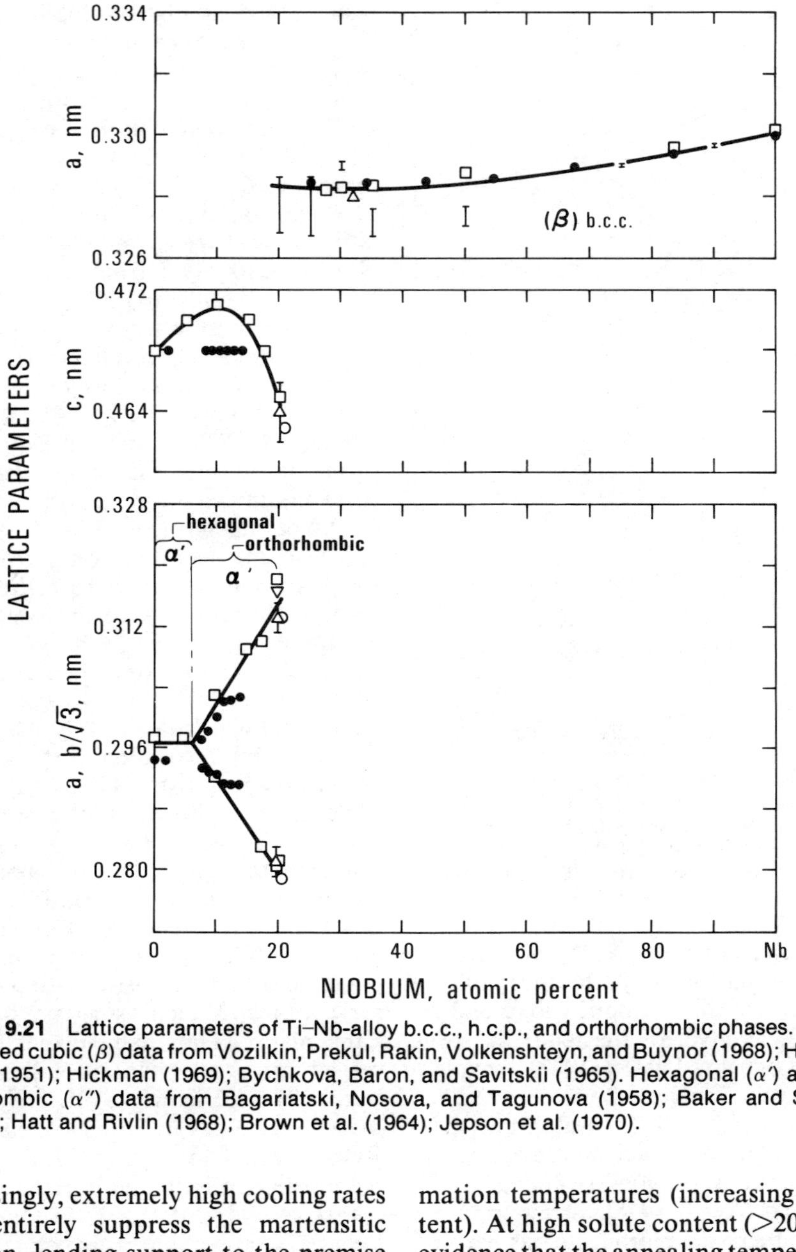

Figure 9.21 Lattice parameters of Ti–Nb-alloy b.c.c., h.c.p., and orthorhombic phases. Body-centered cubic (β) data from Vozilkin, Prekul, Rakin, Volkenshteyn, and Buynor (1968); Hansen et al. (1951); Hickman (1969); Bychkova, Baron, and Savitskii (1965). Hexagonal (α') and orthorhombic (α'') data from Bagariatski, Nosova, and Tagunova (1958); Baker and Sutton (1969); Hatt and Rivlin (1968); Brown et al. (1964); Jepson et al. (1970).

ature. Surprisingly, extremely high cooling rates retard and entirely suppress the martensitic transformation, lending support to the premise that the martensitic transformation is thermally activated. Indeed, the transformation characteristics of Ti–Nb alloys can be schematically portrayed as a C curve, if transformation temperature is plotted vs. time for transformation initiation. Also, Fig. 9.22 clearly indicates that as the temperature for the $\beta \rightarrow \alpha$ transformation decreases at higher niobium content, the ΔT decrement between α' and α'' suppression also decreases consistently with decreasing transformation temperatures (increasing niobium content). At high solute content (>20 at.%), there is evidence that the annealing temperature influences the transformation to α''; higher annealing temperatures suppress the transformation. Blackburn (1970) and Williams (1973) have provided the most recent reviews of martensitic transformations in titanium-base alloys.

Shape-memory effects have been found in Ti–22 at.% Nb by Baker (1971). As-quenched specimens plastically strained 1% at room-temperature completely recovered on heating above the temperature of reversion to the β

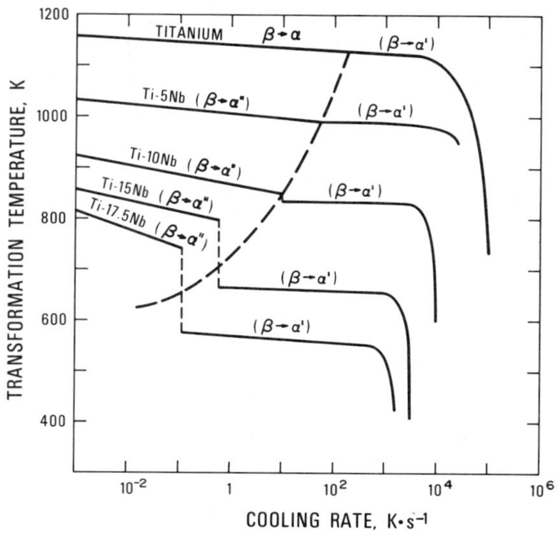

Figure 9.22 Cooling rate vs. transformation temperature for a series of Ti–Nb binary alloys (Jepson et al., 1970).

phase (457 K). A smaller percentage of recovery effects was observed for larger strains and in the presence of the ω phase.

Commercial superconductors of Ti–Nb alloys that are currently manufactured may not undergo a martensitic transformation; at present this is not clear. But, the lower-niobium-content commercial superconductor alloys do exhibit an *omega* transformation at low temperatures, which has many martensitic transformation characteristics.

Room-temperature-hardness (Bagariatski et al., 1958), magnetic-susceptibility (Collings and Smith, 1976), and elastic-property (Albert and Pfeiffer, 1972; Ledbetter, 1978; Fedotov and Belousov, 1963) measurements on quenched alloys suggest the presence of the ω phase. Recent low-temperature measurements of ultrasonic pulse and attenuation (Ledbetter, 1978) and electrical resistivity (Read, 1978) on titanium-base alloys containing up to 30 at.% Nb imply the existence of a reversible ω phase transformation at temperatures as low as 76 K. The formation of ω by precipitation in Ti–Nb alloys in the temperature range 500 to 800 K has been well documented (for a review: Sass, 1972).

The ω phase has been examined by Brammer and Rhodes (1967) using transmission electron microscopy in a Ti–22 at.% Nb alloy. They demonstrated that the ω structure is hexagonal ($c/a = 0.613$), forms as ellipsoids on $\{111\}$ with major axes along $[111]_\beta$ and $[0001]_\omega$, and has the following orientation relationship to β: $(111)_\beta$ parallel to $(0001)_\omega$, $(1\bar{1}0)_\beta$ parallel to $(11\bar{2}0)_\omega$, and $(110)_\beta$ parallel to $(10\bar{1}1)_\omega$. Each pair of ω precipitate exhibits a twin relationship across a $(011)_\beta$ plane.

In lower alloy, titanium-base systems the low-temperature ω phase forms as 1.0- to 1.5-nm-diameter particles in rows oriented along $\{111\}_\beta$. Owing to diffraction, ω reflections become more diffuse for larger solute contents, and the ω particles become more difficult to resolve (for Ti–Nb alloys, see Balcerzak and Sass, 1972). The low-temperature $\beta \rightarrow \omega$ transformation proceeds continuously during lowering of temperature and is completely reversible with temperature. Some researchers think the ω phase is a result of softening, leading to instability of the $2/3\langle 111\rangle$ longitudinal lattice vibrational mode (Sass, 1972; deFontaine, 1970; Delaey, Van Paemel, and Struyve, 1972). Such softening may result in decreased C_{44} and $C_{11} - C_{12}$ elastic constants (deFontaine, 1970). Sass (1972) demonstrated that in Nb–Zr alloys the β-phase C_{44} decreases prior to and throughout the $\beta \rightarrow \omega$ transformation. The hexagonal ω structure may be produced from the body-centered-cubic β structure by $\langle 111\rangle_\beta$ displacements (opposite each other) of adjacent, close-packed $\{111\}_\beta$ atom planes.

In alloys containing higher concentrations of niobium, the elastic property measurements of Albert and Pfeiffer (1972) suggest less ω transformation. They observed little change of Young's modulus in Ti–26 and 34 at.% Nb alloys after aging at 623 K, following extensive cold work.

Since the ω phase can be generated by linear displacements of atoms along the close-packed direction, this transformation should not be expected to have the same macroscopic crystallographic and morphological characteristics as conventional martensitic transformations having coordinated atom shear in three dimensions and possible multiple lattice displacements. Wayman (1973) discussed the general distinctions between ω and martensitic transformations.

The presence of the martensitic α' and α'' phases and the ω phase apparently improves the superconducting properties. In Ti–Nb alloys, the presence of ω phase formed during aging in the temperature range 500 to 750 K results in an increase of the critical current density, J_c (Baker and Sutton, 1969). The martensitic transformations also produce higher critical current densities as well as higher critical temperatures, T_c

(Baker and Sutton, 1969; Collings and Ho, 1969). The effect of athermal ω transformation on superconducting behavior appears to be uncertain, but there is ample evidence of the enhancement of J_c and T_c with increased ω precipitation (e.g., Lohberg, Heller, and Zwicker, 1973). The effects of metallurgical variables, such as phase transformations, are more significant at lower magnetic fields.

9.6.2 Nb_3Sn

Superconductors with the highest critical temperatures and fields are the A15-type intermetallic compounds. Many of the same A15-type structures that have good superconducting properties also are structurally unstable at temperatures near their critical superconducting temperatures and exhibit martensitic transformations. One of these is Nb_3Sn. Led by the desire to obtain higher T_c superconductors, the studies of martensitic transformations, the relation of structural instability to superconductivity, and precursor effects to both structural instability and superconductivity have progressed to a higher degree in these A15 compounds than in other alloy systems. Particularly, the ultrasonic and neutron-scattering studies have provided further insight. Many excellent, recent reviews exist for A15-structure low-temperature instabilities, including Testardi (1973, 1975, 1977), Izyumov and Kurmaev (1974), and Weger and Goldberg (1973).

The A15-type compounds represented schematically in Fig. 9.23 have a transition-metal atom, such as vanadium or niobium, in the shaded positions; these represent three linear, orthogonal chains. Usually nontransition metal atoms, such as aluminum, tin, and gallium, occupy the b.c.c. sites of the structure. The room-temperature A15 structure of Nb_3Sn transforms martensitically to a tetragonal structure on cooling to the temperature range 42 to 51 K. Table 9.5 lists the T_{ms} temperatures measured in various investigations. Temperatures of the start of the reversion to the cubic phase (T_{as}) on warming have been reported in most cases to be nearly identical to T_{ms}. Neutron-scattering measurements of Shirane and Axe (1971) have established the type and magnitude of the linear-chain niobium-atom displacements responsible for the structural change (Fig. 9.23). At 4 K the niobium atoms were found to exhibit Γ_{12} mode displacements (within the space group D^9_{4h}) of magnitude 0.315 pm to produce the tetragonal structure. The high-temperature cubic lattice parameter (Fig. 9.24) abruptly splits on cooling to T_{ms}, followed by gradual change that increases the c/a ratio to 1.0062 at 4 K.

There is only a very limited temperature range (~1 K) in which both phases coexist (Vieland, 1972). Onset of superconductivity at T_c coincides with the leveling-off of the a and c parameters, and in other A15-type superconductors the onset of superconductivity has been found to be associated with increased lattice stability. The volume change from the transformation between 4 and 50 K is less than 0.05% (Mailfert et al., 1969).

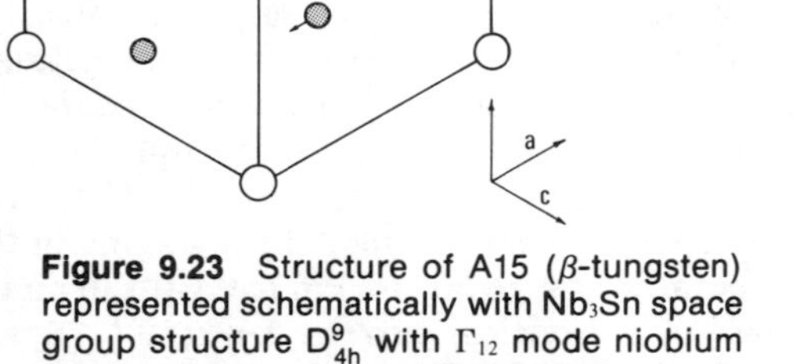

Figure 9.23 Structure of A15 (β-tungsten) represented schematically with Nb_3Sn space group structure D^9_{4h} with Γ_{12} mode niobium displacements. Arrows indicate Γ_{12} mode niobium shifts to form martensitic tetragonal structure.

Table 9.5 Nb_3Sn martensitic transformation temperatures (T_{ms}).

Source	T_{ms} (K)	Comments, Methodology
Vieland (1972)	49	Observable shift in (400) peak intensity position
	47	Observable (400) line splitting
King (1969)	42	X-ray
Chu (1974)	42.5	No pressure, x-ray and neutron scattering
	44.0	0.05 Pa pressure
	45.0	0.10 Pa pressure
	46.0	0.15 Pa pressure
Mailfert, Batterman, Hanak (1967)	<0	As-grown crystals (a_0 = 0.5295 nm)
	43	As-grown crystals, annealed 20–100 h at 1273 K (a_0 = 0.5290 nm)
Shirane, Axe (1971)	45	Neutron scattering, (300) tetragonal peak
Vieland, Wicklund (1969)	44, 51	Specific heat measurements

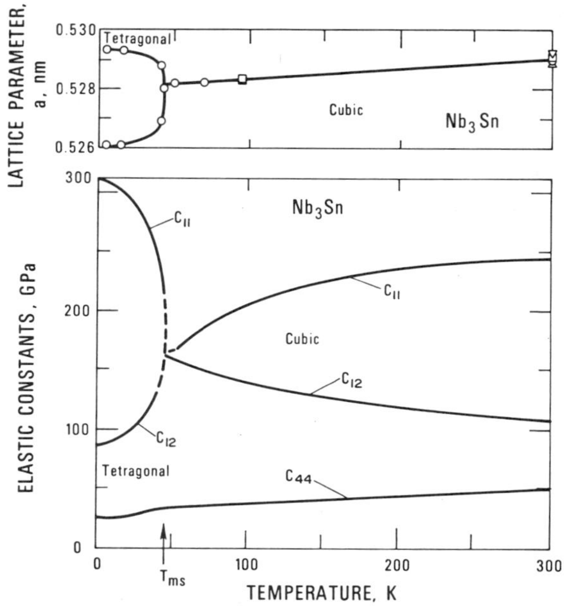

Figure 9.24 Lattice parameters and elastic constants of Nb_3Sn structures as a function of temperature. Lattice parameter data: o — Mailfert, Batterman, and Hanak (1969); □ — Vieland (1972); △ — Reed, Gatos, LaFleur, and Roddy (1969). Elastic constant data: Rehald (1968); Vieland, Cohen, and Rehwald (1971).

Optical (Wanagel and Batterman, 1970) and x-ray (Vieland, 1972) observations of martensitic product morphology reveal domain lamellae along <100> directions. These bands appear better defined and narrower at lower temperatures. Bands in two <100> directions are usually present, except near T_{ms} where only one set is prominent. No single interface (a demarcation between parent and product phases) was observed to progress across a crystal surface. The domains are about 0.2-mm wide, have alternate surface tilts of approximately 0.1°, and are twin related to their nearest neighbors.

Applied stress influences the martensitic transformation of Nb_3Sn. Chu (1974) found that the application of hydrostatic pressure linearly increases T_{ms} and decreases T_c of Nb_3Sn; conversely, the application of pressure to V_3Si linearly decreases T_{ms} and increases T_c. Applied tensile stresses also lower the T_c of Nb_3Sn slightly.

The effects of compressive stress have been studied by Hoard, Scanlon, Smith, and Farrell (1980). Their measurements imply that the application of compression, via cooled indium and copper thin-film specimen holders, acts to increase T_{ms} and to increase the amount of martensitic transformation. Their results suggest that Nb_3Sn may not transform at zero strain and that a compressive strain of about 0.018 is sufficient to induce total transformation. This possible critical dependence on strain may explain the divergence of T_{ms} measurements (Table 9.5) and the disparity in detection of the transformation reported by some investigators.

Compounds with A15 structures have in common high T_c, the tendency to transform martensitically, and elastic instability. As illustrated for Nb_3Sn in Fig. 9.24, $C_{11} - C_{12}$ approaches zero at T_{ms}. Below T_{ms} the C_{44} elastic constant begins to decrease discontinuously. The decrease of the elastic parameter, $(C_{11} - C_{12})/2$, is associated with softening of the shear mode in a [110] direction with [$1\bar{1}0$] polarization. Keller and Hanak (1967) reported ultrasonic measurements for an Nb_3Sn crystal that did not transform; in this crystal, $C_{11} - C_{12}$ approached zero at about 35 K, yet no martensitic transformation was detected. For A15 structures, probably lattice softening is a necessary, but not sufficient, condition for structural transformation.

Precursor effects related to the martensitic transformation have been identified. Attenuation increases strongly below approximately 50 K and remains high below this temperature (Rehwald, Rayl, Cohen, and Cody, 1972). Perhaps the clearest and most descriptive (with adequate interpretation) example of a precursor effect was detected by the precise neutron-scattering experiments of Axe and Shirane (1973). From examination of the soft phonon branch $[\delta\,\delta\,0]T_1$, where $\delta \lesssim 0.1$, phonon line-shape changes were detected on lowering the temperature toward T_{ms}. A "central" component in the neutron-scattering spectrum evolved; this central peak was not detected at 80 K, was easily detected at 60 K, and was very prominent and dominating at 46 K. In Fig. 9.25, the maximum of the central peak is plotted vs. temperature; also included is the (300) Bragg reflection from neutron scattering. The intensity of this reflection is proportional to the percentage of tetragonal phase. Axe and Shirane (1973) suggested that the emergence of the central peak arises from low-frequency relaxation of thermal phonons, slightly reducing the elastic stiffness within a collision-dominated phonon system. Whether these effects may be enhanced near lattice defects is not clear. From x-ray diffraction measurements, a small precursor tetragonal distortion was discovered (Vie-

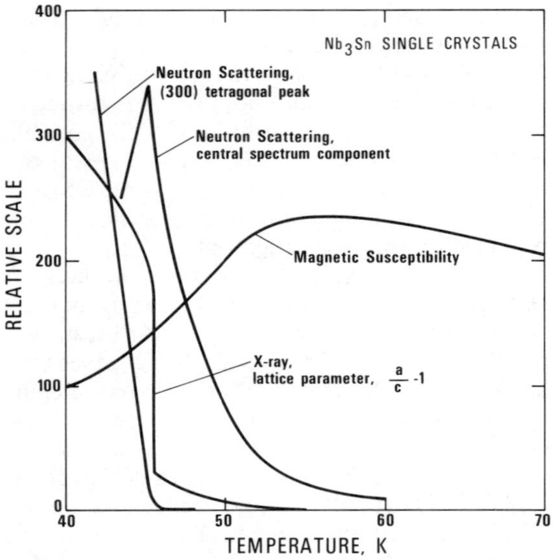

Figure 9.25 Neutron-scattering (Axe and Shirane, 1973), magnetic-susceptibility (Rehwald et al., 1972), and x-ray diffraction (Vieland et al., 1971) measurements on two Nb_3Sn single crystals.

land et al., 1971) and is included in Fig. 9.25. Magnetic susceptibility-vs.-temperature measurements indicate a maximum at a higher temperature (~56 K) for an annealed single crystal exhibiting the cubic-to-tetragonal martensitic transformation at 45 K (Rehwald et al., 1972). All the measurements described here were performed on two single crystals from the same source; the neutron scattering on one and the magnetic susceptibility and x-ray on the other. Of course, the gradual $C_{11} - C_{12}$ mode softening (Fig. 9.24), beginning above room temperature, must be regarded as the significant precursor to the transformation. It is clear that phonon and electron–phonon interactions, beginning well above T_{ms}, produce transformation precursor effects in Nb_3Sn.

Defects in the A15 structure, especially vacancies, interstitials, and antisites (substitution of A atoms on B sites, or vice versa) have been examined recently to determine their affect on T_c and T_{ms}. Although many interesting results have been obtained, the gradual structural collapse, or softening, is still considered to be the most important "defect" (Testardi, 1977, 1978).

There may be a second-phase transformation in Nb_3Sn. There is strong evidence that both the cubic and tetragonal phases have T_c temperatures near 17.5 K, with the T_c of the cubic phase slightly higher than that of the tetragonal phase. Two peaks in specific heat, spaced between 2 and 3 K in the temperature range 14.7 to 18.1 K, were observed by Viswanathan, Luo, and Vieland (1974) and Viswanathan and Johnston (1973), but only one peak at 17.78 ± 0.02 K was reported by Chu and Vieland (1974). The specimens of Viswanathan et al. (1974) and Viswanathan and Johnston (1973) were constrained by copper plating; the specimens of Chu and Vieland (1974) were unconstrained. Other experiments in which specimens were constrained on cooling through T_{ms} and below by selecting substrates of different thermal contraction (Pupp, Sattler, and Saur, 1974) and by uniaxial compressive stress (McEvoy, 1971) also indicate suppression or the appearance of abnormally low critical superconducting temperatures (~16.5 K). In other experiments on similar specimens in the absence of stress, the normal T_c of about 17.5 K was preserved. If applied stress and constraint either suppress the cubic-to-tetragonal transformation and result in another structural transformation or promote a transformation from tetragonal to a new structure, the C_p and T_c measurements can be explained. This possibility was suggested by Chu and Vieland (1974) to explain their specific heat data.

Specific heat measurements by Vieland and Wickland (1969) indicated no detectable latent heat during the martensitic transformation. Anomalous low-temperature specific heat, Knight shift, magnetic susceptibility, and electrical resistivity data have been reported by a number of investigators; their results are summarized and discussed by Testardi (1973, 1977, 1978), Izyumov and Kurmaev (1974), and Weger and Goldberg (1973).

Many theoretical approaches have been taken to explain low-temperature behavior of A15 structures. In the A15 structure, the close-packed directions <100> are composed of transition metal atoms in linear, orthogonal chains (Fig. 9.23). It is assumed that a wide band of *s* electrons from both elements of the structure exists, that the nontransition element contributes *p* electrons, and that a narrow band of *d* electrons is present.

To account for the high magnetic susceptibility and specific heat (linear term, in temperature, extrapolated to 0 K) and for the anomalous temperature dependence of the Knight shift and magnetic susceptibility of A15 structures, Clogston and Jaccarino (1961) first proposed a high density of electron states close to the Fermi level for these structures. Labbé and Friedel (1966) proposed that the A15 structure could be represented as an assembly of one-dimensional, noninteracting chains, and used a tight-binding calculation to describe the *d* electrons. Density of states were derived from this model and the strong temperature dependences of magnetic susceptibility and the Knight shift were explainable by placing the Fermi level near the band edge. Cohen, Cody, and Halloran (1967) replaced the steep density-of-states dependence on energy with a simple step function, retained the noninteracting chain concept, placed the Fermi level near the step, and predicted many of the A15 properties. Refinements of these first models have been reviewed by Izyumov and Kurmaev (1974), Bhatt (1977), and Testardi (1973, 1975, 1977, 1978).

Mattheiss's (1975) APW-LCAO (augmented plane-wave linear combination of atomic orbitals) calculations of A15 band structures indicate that there is sufficient interchain coupling to destroy the *d*-band character, not permitting the threefold degeneracy assumed by the Labbé–Friedel (1966) and the Cohen et al. (1967) models. Gor'kov and Dovokhov (1976) included interchain reactions in their earlier model, which assumed noninteracting *d* chains with the Fermi energy near the *X* point of the Brillouin zone (center of face of cubic zone). Previous models had assumed the Fermi level close to the bottom of the *d* band (at Γ, center of cubic zone). Refinements of the Gor'kov models have been discussed and reviewed by Testardi (1977, 1978), Izyumov and Kurmaev (1974), and Bhatt and McMillan (1976).

More recently Bhatt (1977) developed a two-band tight-binding model from Mattheiss's (1975) density-of-states calculations. Nearest neighbor (interchain) and next-nearest neighbor (interchain) interactions and electron–phonon coupling were considered. Both bands have sharp density-of-states peaks. As in earlier models, a Peierls gap occurs in one band and triplet degeneracy splitting in one dimension (Jahn–Teller, second-order effect) occurs in both bands.

Using Landau theory, Bhatt (1977) showed that all three approaches (Labbé–Friedel, Gor'kov, and Bhatt) qualitatively describe most features of A15 martensitic transformations, but quantitatively, the more recent Bhatt (1977) model is the most successful. Theory can relate to elastic, heat-capacity, susceptibility, phonon-dispersion, crystal-structure, and thermal-expansion experimental data. Theory relates increased phonon softening to increased phonon–electron coupling, which increases T_c. The martensitic transformation produces a decrease in the electron density of states, resulting in a decrease of T_c. McMillan (1977) discusses the relationships between superconductivity and martensitic transformations, pointing out that on approaching T_{ms} the associated lattice softening increases the electron–phonon coupling constant, raising T_c. Below T_{ms}, lattice stiffening contributes to reduced T_c.

9.6.3 Other A15-structure superconductors

In addition to Nb_3Sn, A15 compounds such as V_3Ga, V_3Si, and $Nb_3Ge_{0.3}Al_{0.7}$ are high-T_c superconductors and have martensitic transformations where T_{ms} is larger than T_c. The reported transformation characteristics of these structures are presented in a review by Testardi (1973) and, generally, show traits similar to those of Nb_3Sn.

9.7 IRON-NICKEL ALLOYS

Iron–nickel alloys play a vital role in cryogenic structures. They are used extensively for containment and transportation of cryogenic fluids. Particularly the alloys Fe–3 wt.% Ni, Fe–5 wt.% Ni, and Fe–9 wt.% Ni are used for containment of the cryogenic fluids propane, natural gas, oxygen, and nitrogen. Research is currently in progress for development of a ferritic Fe–Ni alloy suitable for service at 4 K (liquid-helium containment). The Fe–35 at.% Ni (36 wt.%) alloy (Invar) is used for applications that require extreme dimensional stability over the cryogenic temperature range; it has low thermal contraction on cooling to 4 K from room temperature.

The Fe–Ni alloys that undergo martensitic transformations range to 34 at.% Ni (35 wt.%). From Fe–35 at.% Ni to pure nickel, the alloys remain austenitic at all temperatures. Chapter 11, Structural Alloys, contains a discussion of the uses of these alloys.

A large amount of research has been devoted to the study of martensites in Fe–Ni alloys. This section summarizes only selected aspects of that research.

9.7.1 Crystallography and morphology

The study of martensite formation in Fe–Ni alloys has been in progress for many years. Extensive studies of plate-martensite crystallography and morphology in the composition range Fe–29 to 34 at.% Ni have been presented by Patterson and Wayman (1966) and by Reed (1967). The lower nickel range (Fe–0 to 20 at.% Ni) has been investigated by Wilson (1965). Dubrov (1966), Swanson and Parr (1964), Goodenow and Hehemann (1965), and Yeo (1962) have studied primarily kinetic characteristics, such as T_{ms} and the effects of rate of cooling. In the lower nickel range (0 to 20 at.% Ni), the martensitic product forms above room temperature in a morphology that has been called massive or cellular. In the intermediate nickel range (20 to 30 at.% Ni), the martensitic product degenerates from a plate-like shape as the nickel content decreases. As many as four different product shapes may be present: (a) surface martensite, (b) plate martensite, (c) cellular or massive martensite, and (d) angle martensite.

Cellular (massive) martensite. This type of martensite forms predominantly in low-nickel alloys but is observed in alloys having up to about 27 at.% Ni.

Martensite that is characterized as having a small, irregular (as viewed in optical microscopy), block-type microstructure forms in alloys containing less than about 27 at.% Ni and has been termed *cellular* (Speich and Swann, 1965) or *massive* (Owen and Wilson, 1965; Owen, Wilson, and Bell, 1965) martensite. The appearance of typical surface and internal microstructures is illustrated in Fig. 9.26, which shows a sequence of microstructures of the same region at successively deeper levels. It is typical that more than four planar traces are represented by the transformed martensite on a polished surface within one parent austenite grain. With continued polishing, however, the number of etched traces reduces to four or less; these traces represent $(111)_\gamma$ planes. The additional traces noticed on or near the surface are those associated with surface martensite.

The etched $\{111\}_\gamma$ traces do not represent habit planes; instead, they represent boundaries of a domain of a large number of similarly oriented, much smaller crystals. The smaller crystals have been resolved by Speich and Swann (1965), Wilson (1965), and Goodenow and Hehemann (1965) with electron microscopy. Individually, these are plate-like or disc-like in shape, approximately 0.3-μm thick. As first proposed by Wilson (1965), these small plates cluster in sheets or, as the transformation proceeds, domains. Speich and Swann (1965) reported that many of the small crystals are twin related. Thus, cellular (massive) martensite forms as very small, plate-like crystals that transform in a cooperative fashion within a given domain. The domain boundaries have $\{111\}_\gamma$ traces.

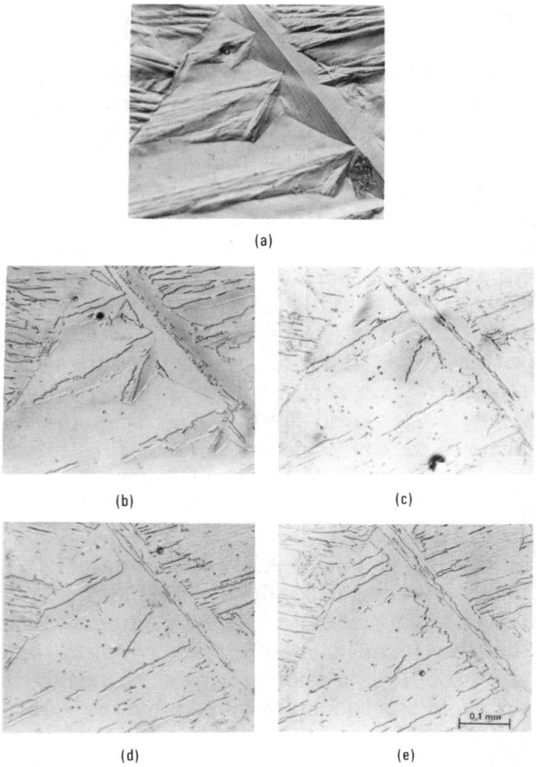

Figure 9.26 Sequence of change in microstructure of massive martensite in Fe–25 at.% Ni alloy. Transformation on polished surface is shown in (a). Sequential photos (b–e) show the etched appearance during the process of electrolytic polishing. In (e), only $(111)_\gamma$ etched traces remain.

There have been many kinetic studies that have established fairly definitive T_{ms} data (see discussion later in this section) and have documented the effects of austenitizing temperature, holding time at temperature, and the rate of cooling on T_{ms}. Greater rates of cooling lower the transformation temperature. There are discrepancies, however, when attempts have been made to identify the existence of two or more general types of transformation products, such as:

a. Bainite and martensite (Goodenow and Hehemann, 1965; Goodenow, Matas, and Hehemann, 1963).

b. Massive and martensite (Dubrov, 1966).

c. Massive, massive martensite, and martensite (Owen, Wilson, and Bell, 1965; Gilbert and Owen, 1962; Owen and Wilson, 1965).

d. Massive and martensite (Swanson and Parr, 1964; Bibby and Parr, 1964).

e. Bainite and martensite (Steven and Haynes, 1956).

The investigations have not been consistent in either terminology or experimental details pertinent to each identified transformation process. One can summarize the findings as follows: below about 8 at.% Ni, two products have been identified; one is formed via a martensitic process, the other is not. Above 8 at.% Ni two types of martensitic (or bainitic) products have been identified; these occur at different temperatures, and the higher temperature product forms in a time-dependent manner. All products are influenced by annealing treatments and rate of cooling.

Plate martensite. Plate martensite forms in iron alloys containing about 27 to 34 at.% Ni (Reed, 1967; Patterson and Wayman, 1966). As the nickel content decreases in this range, the martensite plates become more fragmented and contain fewer internal twins, and their habit planes tend to move toward $(225)_\gamma$ from a position in the stereographic triangle near the $(011)_\gamma$–$(111)_\gamma$ line.

All plates have a central planar region containing an extremely high density of twins; this region is called a *midrib* and can be clearly seen in Fig. 9.27. A plate in Fe–28 at.% Ni exhibiting a highly fragmented interface is also shown in this figure and may be contrasted with the smooth interfaces of plates of the Fe–33 wt.% Ni alloy shown in Fig. 9.4.

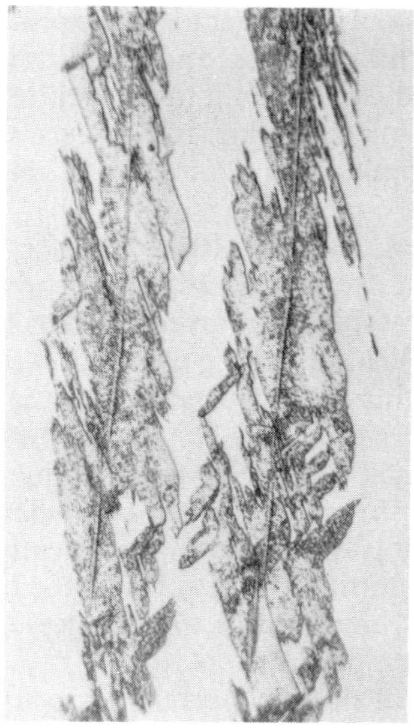

Figure 9.27 Plate martensite in Fe–28 at.% Ni alloy.

Figure 9.28 Angle martensite in Fe–29 wt.% Ni alloy, forming in association with plate martensite.

Angle martensite. Martensite that forms as two laths, intersecting with a straight interface, is called *angle* martensite. The length of the laths varies from very short to almost as long as the austenite grain diameter.

The occurrence of angle martensite was first identified by Klostermann and Burgers (1964) and later described by Klostermann (1965) as having the shape of laths (parallel to $\langle 210 \rangle_\gamma$ and having $\{1\bar{2}1\}_\gamma$ habit planes). This type of martensite is distinctive in that two laths are always associated (twin related) with each other and intersect with an angle profile across a $\{001\}_\gamma$ plane.

Figure 9.28 illustrates the formation of angle martensite in the vicinity of plates. The laths of the angle martensite are long and lie approximately parallel to the plates, and their intersecting obtuse angle usually faces outward from the plates. In the higher nickel alloys (28–29 at.% Ni), angle martensite tends to form only very close to the plates, but in Fe–27 at.% Ni, if the cooling rate is sufficient to suppress surface and cellular martensite, angle martensite forms abundantly throughout the austenite grains. This type of martensite has been observed in Fe–Ni alloys ranging from 26 to 30 at.% Ni.

Surface martensite. Surface martensite forms in the temperature region above the T_{ms} of the other types of martensite during and after electropolishing, during plastic deformation, and during aging in the temperature range 210 to 300 K. In Fe–27 at.% Ni, the entire surface may be transformed after electropolishing and room-temperature aging for about one hour. The amount of transformation is critically dependent on the condition of the surface, the holding time at a given temperature, and to a lesser extent, the holding temperature.

A typical surface martensite crystal (shown in Fig. 9.29) appears as a tapered needle. The crystals extend 5 to 10 μm below the surface; immediately below the surface they tend to fragment and assume an irregular appearance. Figure 9.29b reveals very fine striations within the individual crystals. These striations represent continuous internal twinning on $(112)_{\alpha'}$ planes; in Fig. 9.29b three distinct traces of striations are visible. As Klostermann and Burgers (1964) and Yeo (1964) first reported, surface martensite is characterized by rapid growth in the longitudinal direction and slow growth in the lateral direction on one side, with apparently no growth on the other side. Close examination of the lateral sides (example in Fig. 9.29) substantiates this. One side is very planar and the other side is irregular. The irregular side resembles an interface that has moved, but interaction with obstacles has hindered its movement.

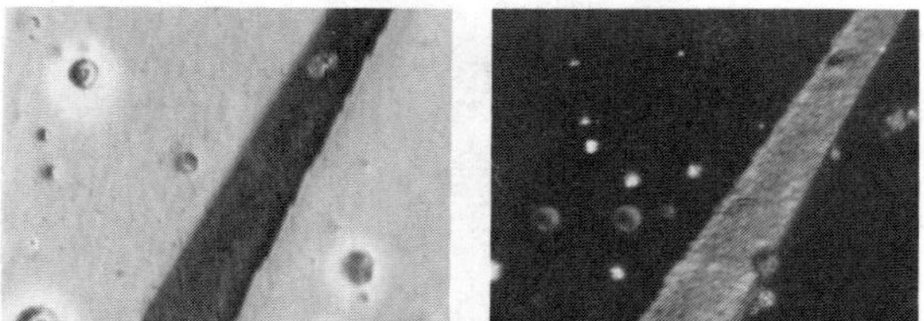

(a) (b)

Figure 9.29 Long martensite needles revealing fine striations representing internal twinning in Fe–29 wt.% Ni alloy. In (b) three different planar traces are visible. Note the irregularity of one side of each crystal.

In Fig. 9.30, a grain of Fe–27 at.% Ni that contains extensive surface martensite is analyzed. In this grain, the two polished and etched sides are approximately parallel (about 15° deviation) to the $(\bar{1}\bar{1}1)_\gamma$ and $(\bar{1}11)_\gamma$ planes. The traces of the individual martensite crystals on each side of the grain correspond to the $\{11\bar{2}\}_\gamma$ habit planes and, approximately, to the $<1\bar{1}0>$ directions. On each surface three different $\{11\bar{2}\}_\gamma$ planar traces (three $<1\bar{1}0>_\gamma$ directions) are apparent. The $\{11\bar{2}\}_\gamma$ habit planes are those that form a 90° angle to the $\{111\}_\gamma$ plane most nearly parallel to the surface [in Fig. 9.30, either $(\bar{1}\bar{1}1)_\gamma$ or $(\bar{1}11)_\gamma$]. The $<1\bar{1}0>_\gamma$ directions are those, naturally, that are common to both the individual $\{11\bar{2}\}_\gamma$ plane and to the $\{111\}_\gamma$ plane nearly parallel to the surface.

When two or three different $\{111\}_\gamma$ planes have nearly equivalent angles with the surface, then the $\{11\bar{2}\}_\gamma$ planar traces present are those asso-

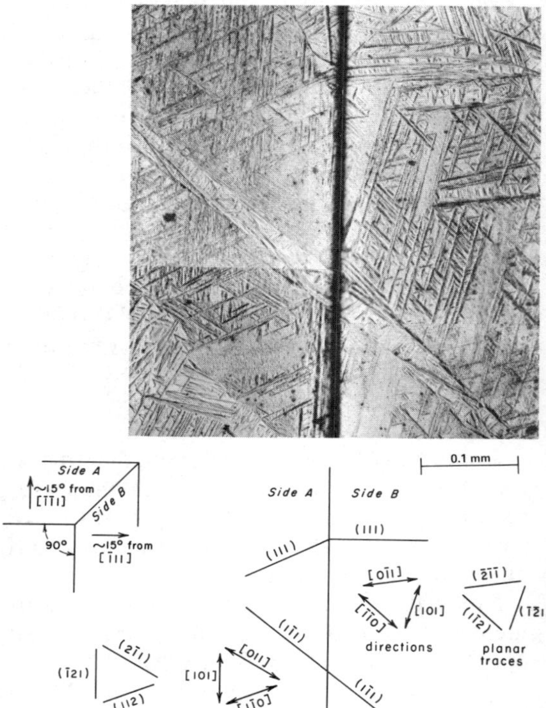

Figure 9.30 Crystallography of surface martensite. Two sides of the grain at right angles to each other are shown for an Fe–28 wt.% Ni alloy. Traces of martensite crystals and their respective f.c.c. directions and planar traces are indicated.

ciated with each respective $\{111\}_\gamma$ plane. As many as eight different $\{11\bar{2}\}_\gamma$ traces have been observed. When many $\{112\}_\gamma$ variants form, the crystals tend to form individually, and are similar to those in Fig. 9.29; however, if the surface of the grain is nearly parallel to a $\{111\}_\gamma$ plane, the surface crystals are less tapered, tend to form in a cooperative parallel fashion, and are similar to those in Fig. 9.30. Observation of many grains revealed that if a grain has a $(111)_\gamma$ plane nearly parallel to the surface plane, the grain exhibits a relatively large amount of surface transformation.

9.7.2 Transformation temperatures

Research concerning temperatures at which martensitic products begin to transform on cooling in Fe–Ni alloys has probably received greater emphasis than any other aspect of the study of martensitic transformations. Yet, there is still doubt regarding the details and interpretations of these published reports. Figure 9.31 records most of the available T_{ms} data for Fe–Ni alloys.

In the range of less than about 10% Ni equiaxed ferrite, cellular (massive) and, probably, surface martensites form. Equiaxed ferrite is characterized by the absence of significant surface upheavals. From past research, it is not possible to distinguish between the T_{ms} of surface and cellular martensite. For Fe–25 to 27 at.% Ni, surface martensite forms at temperatures 0 to 25 K above the T_{ms} for cellular martensite provided that the rate of cooling is low. As previously discussed, the formation of surface martensite may be partially suppressed by rapid quenching. In high-nickel alloys, there is no observable effect of cooling on the T_{ms} of plate martensite.

9.7.3 Effects on mechanical properties

Rather thorough research has been performed on a series of Fe–Ni–C alloys by Bolling and Richman (1970) and Richman and Bolling (1971). They have identified two relationships between deformation and transformation: (a) Deformation of austenite at temperatures above T_{ms} (253 to 371 K) reduces (usually) the subsequent spontaneous transformation temperature (labeled $T_{ms(d)}$ here) and (b) Application of tensile stress to austenite in the vicinity of T_{ms} induces martensite and results in lower flow strengths. The highest temperature at which the stress-assisted transformation is initiated is referred to here as $T_{m\sigma}$. In Fig. 9.32 the relationship between T_{ms}, $T_{ms(d)}$, and $T_{m\sigma}$ as a function of tensile strain or tensile prestrain is shown for alloys with the composition range Fe–13 to 31 at.% Ni–0.5 to 4.3 at.% C (0.11 to 0.96 wt.% C). Quite obviously, tensile prestrain usually lowers the transformation temperature, and applied stress produces martensite at temperatures higher than T_{ms}. Neither effect is very sensitive to the amount of strain. Apparently deformation of austenite in these alloys at 298 and 371 K only increased the austenite flow strength and did not create new nucleation sites or increase the tendency for existing nucleation sites to transform.

Richman and Bolling (1971) also examined the effects of changes of strain rate ($\dot{\epsilon}$) on flow strength (σ_f) and found that the lower nickel alloys (15 to 23 at.%) behaved differently than the higher nickel alloys (27 to 32 at.%). A schematic of these results is shown in Fig. 9.33. The lower nickel alloys exhibited a negative change of flow strength with respect to the change of strain rate, $\ell n\ \sigma_f / \ell n\ \dot{\epsilon}$, at temperatures above T_{ms}. The rate of reduction of flow strength with respect to strain rate was a minimum at a temperature between T_{ms} and $T_{m\sigma}$. In the higher nickel alloys, the strain-rate dependence remained positive. Whereas Richman and Bolling argue in favor of the distinction of the magnetic state (lower nickel alloys are paramagnetic, higher nickel alloys are ferromagnetic) contributing to the difference of performance, the change of stress-assisted martensite morphology from surface and massive (cellular) to plate may also relate to the change of deformation–transformation behavior.

In Fe–Ni alloys, martensite contributes to strength primarily by introducing a greater defect structure. Speich and Swann (1965) measured the tensile yield strength for a series of Fe–Ni alloys. They identified the following contributions to the flow strength of Fe–Ni alloys and assumed that they were additive:

$$\sigma_f = \sigma_{Fe} + \sigma_{Ni} + \sigma_s \tag{9.9}$$

where σ_{Fe} is the flow strength of ferritic iron, σ_{Ni} is the solid-solution-strengthening contribution of adding nickel, and σ_s is the martensitic phase contribution. Roberts and Owen (1967) measured at room temperature the (0.2%) flow strength of martensitic iron (quenched from 1200 K) as 360 MPa (53×10^3 psi). They measured the yield strength of the ferritic phase of the same iron as 140 MPa (20×10^3 psi). Thus the contribution of the martensitic substructure (σ_s) for their iron is 220 MPa (33×10^3 psi). In Fig. 9.34 the data of Speich and Swann for ferrite and the

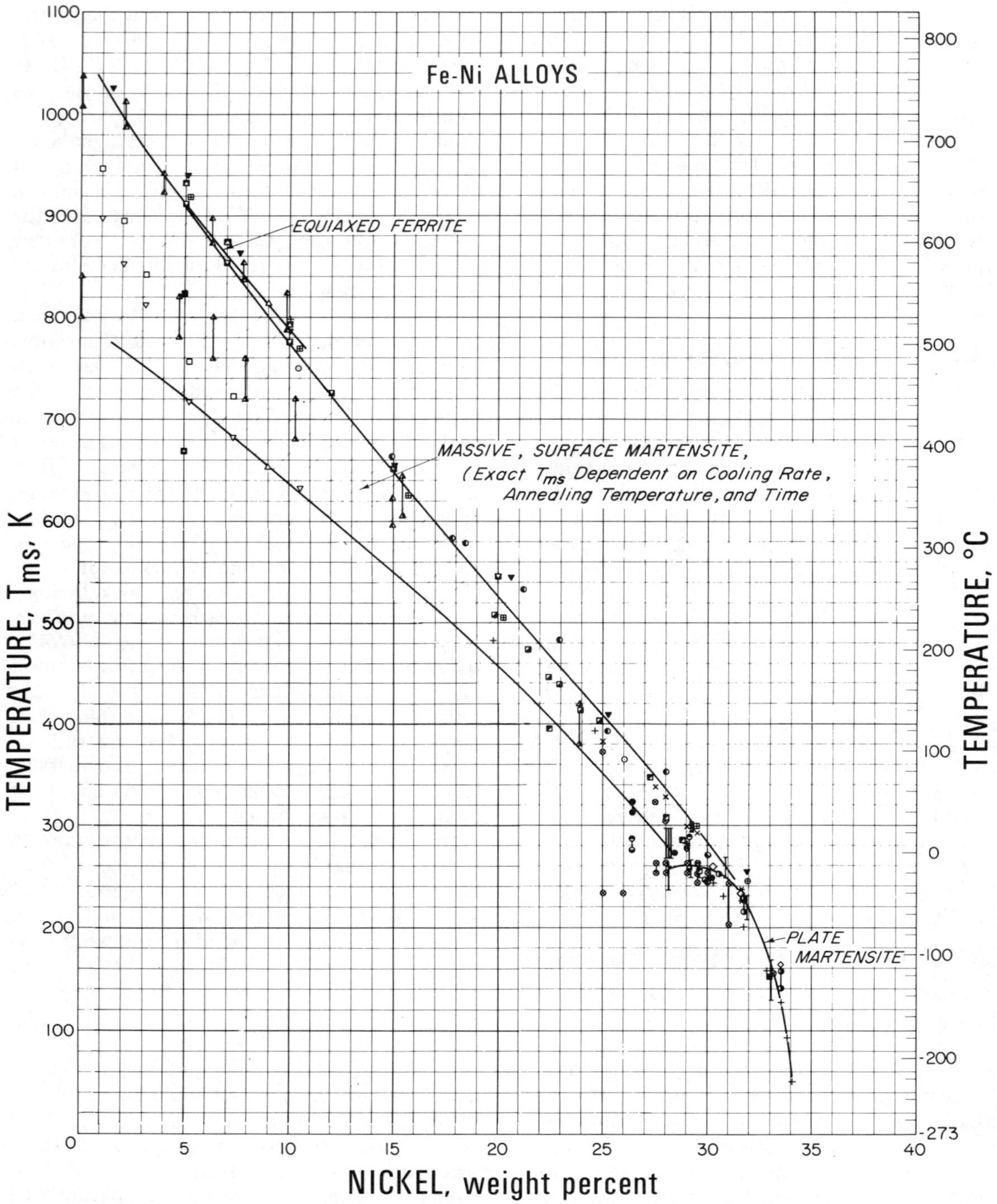

Figure 9.31 Transformation temperature, T_{ms}, vs. composition for Fe–Ni alloys. Data are from: Averbach and Cohen (1962), Bando (1964), Bibby and Parr (1964), Bokros and Parker (1963), Clark (1959); Dubrov (1966), Gilbert and Owen (1962), Goldman and Robertson (1965), Goodenow and Hehemann (1965), Honma (1957 a-c), Jones and Pumphrey (1949), Kaufman and Cohen (1956), Kaufman et al. (1961), Krauss and Cohen (1962), Krauss and Pitsch (1965), Nagakura et al. (1963), Owen et al. (1965), Porter and Dienes (1959), Reed (1980), Scheil and Saftig (1960), Steven and Haynes (1956), Swanson and Parr (1964), Yeo (1964), and Zackay, Justusson, and Schmatz (1962).

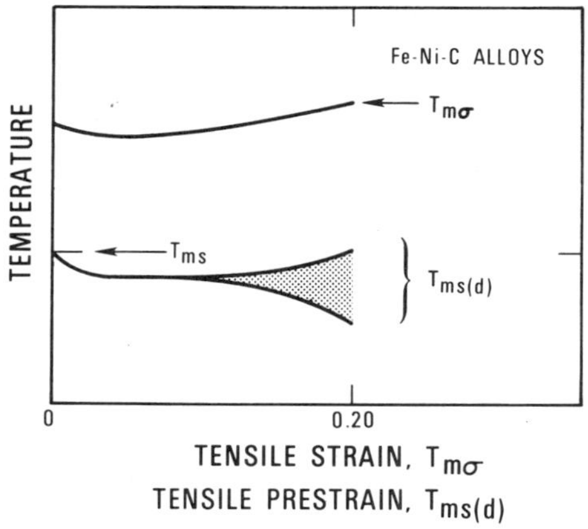

Figure 9.32 The effects of tensile prestrain at 298 and 371 K on the subsequent transformation-to-martensite temperature, $T_{ms(d)}$, and on the influence of tensile stress on the temperature at which the flow strength decreases with continued loading, $T_{m\sigma}$ (Richman and Bolling, 1971). Alloy content ranges from Fe–13 to 31 at.% Ni and 0.5 to 4.3 at.% C.

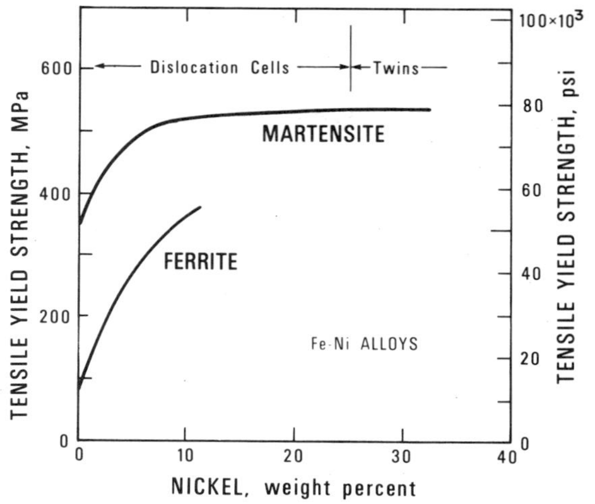

Figure 9.34 The room-temperature tensile-yield strength (0.2% offset) of a series of Fe–Ni alloys containing a martensitic or ferritic structure (Speich and Swann, 1965; Roberts and Owen, 1967).

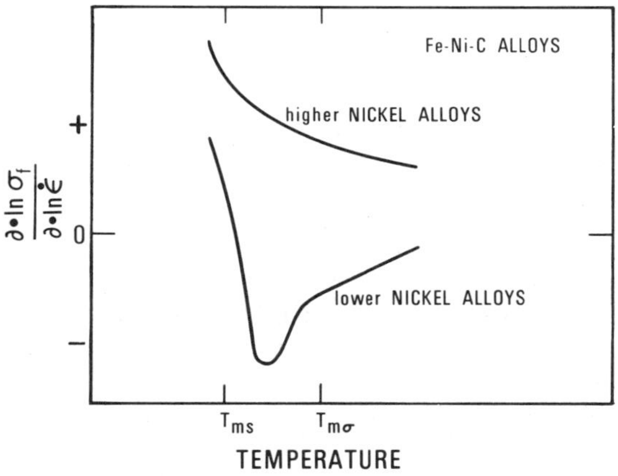

Figure 9.33 Schematic of temperature dependence of negative strain-rate sensitivity of selected (paramagnetic) Fe–Ni–C alloys compared with T_{ms} and $T_{m\sigma}$. Alloy content ranges from 15.9 to 23.6 wt.% Ni and 0.29 to 0.96 wt.% C (Richman and Bolling, 1971).

data of Speich and Swann (>10 wt.%) and Roberts and Owen for martensite are presented. Here, when both sets of data are compared, the value of σ_s for iron is larger, about 280 MPa (40×10^3 psi). Comparison of the ferrite and martensite flow strengths to about 10% Ni leads to the conclusion that, to a first approximation, both increase in the same manner with increase of nickel. Therefore, to the first order, σ_s is independent of nickel content.

The solid-solution-strengthing contribution of nickel reaches a plateau below 10% Ni at a value of slightly less than 200 MPa (30×10^3 psi). Thus the contributions from martensite and nickel to the flow strength of Fe–Ni alloys are approximately equal. The defect structure of the martensite phase typically consists of dense dislocation cells for nickel contents less than about 25% Ni and transformation twins in higher nickel alloys. The transition from dislocations to twins in the internal structure does not affect the tensile flow strength. The defect structure of the ferrite consists typically of random dislocations.

9.7.4 Lattice parameters

The lattice parameters of the f.c.c. austenite in the concentration range Fe–26 to 34 at.% Ni and of the b.c.c. ferrite and martensite from pure iron to 34 at.% Ni were measured by Reed and Schramm (1969). Their data are presented in Fig. 9.35a and b. There are three distinct ranges in the compositional dependence of the lattice parameters (a_0) of the b.c.c. structure. In the region where ferrite is dominant (0 to 6 at.% Ni), a_0 increases. In the range of nickel (5 to 25 at.%) where cellular martensite is prevalent, the lattice parameter remains quite insensitive to concentration. Finally, in the high-nickel region, a_0 decreases abruptly. As described earlier (§ 9.7), plate martensite containing internal transformation twins is the prevailing type of martensite in this range. Therefore, it appears that the changes

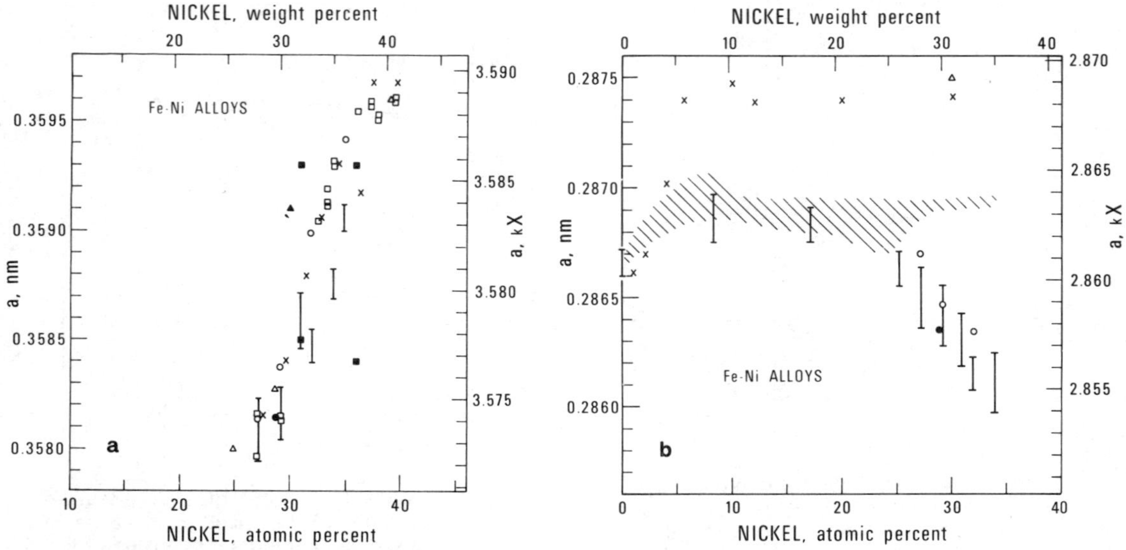

Figure 9.35 The room-temperature lattice parameters of the (a) f.c.c. and (b) b.c.c. structures in Fe–Ni alloys, both bulk and powder specimens. Data in (a) from Reed and Schramm (1969); Breedis and Wayman (1962); Owen, Yates, and Sully (1937); Jette and Foote (1936); Bradley, Jay, and Taylor (1937); Goldman and Wagner (1963); Ananthanarayanan and Peavler (1961). Data in (b) from Reed and Schramm (1969); Breedis and Wayman (1962); Goldman and Wagner (1963); Roberts and Owen (1967).

of the lattice parameter reflect the changes of the type of martensitic morphology and internal defect structure.

It proves interesting to calculate the volume change in Fe–26 at.% Ni and to compare it with that in Fe–34 at.% Ni. Assume that the temperature dependence of the lattice parameter of both phases is approximately equal; this permits the use of the room-temperature data and comparison of the two alloys with differing T_{ms} (see Fig. 9.31). The volume expansion during the martensitic transformation of Fe–26 at.% Ni is about 2.5%, but the expansion of Fe–34 at.% Ni is only 1.1%. This strong change results because the f.c.c. a_0 is rapidly increasing, and the b.c.c. a_0 is rapidly decreasing with increasing nickel content. It is unusual since the T_{ms} of Fe–34 at.% Ni is near 60 K, whereas the T_{ms} of Fe–26 at.% Ni is much higher, near 270 K. The lattice parameter of the cellular martensite remains constant at lower nickel contents. The increasing volume difference with lower nickel content may provide a clue to the reason for the change of martensite type and morphology.

9.8 POLYMERS

When polyethylene crystals with an orthorhombic structure are stressed at room temperature, they exhibit deformation mechanisms that include shear normal to the chain axis. Well-documented mechanisms include (110) twinning, (100) [010] slip, and a martensitic transformation to a monoclinic structure (Young and Bowden, 1974; Allan, Cuellin, and Bevis, 1973). Since no low-temperature martensitic transformations have come to our attention, it is not appropriate at this time for extended discussion of this topic. However, there is a strong probability for discovery of additional deformation-induced martensitic transformation in polymer crystals at low temperatures. Keith (1973) has provided the most current review of phase transformations in high polymers.

9.9 REFERENCES

9.9.1. Books, reviews, recent conferences

Barrett, C. S. (1976). Transformations at low temperatures. *Trans. Jap. Inst. Met.* **17**, 465–475.

Bilby, B. A. and Christian, J. W. (1961). The crystallography of martensitic transformations. *J. Iron Steel Inst.* **197**, 122–131.

Christian, J. W. (1965). *The Theory of Transformations in Metals and Alloys.* Pergamon Press, New York.

Christian, J. W. (1970). Basic crystallography and kinetics. In *Martensite Fundamentals and Technology* (ed. E. R. Petty). Longman Group, London, 11–41.

Christian, J. W. (1976). *The Theory of Transformations in Metals and Alloys.* Pergamon Press, New York.

Cohen, M., Olson, G. B., and Clapp, P. C. (1979). On the classification of displacive phase transformations. In *Proceedings of the International Conference on Martensitic Transformations, ICOMAT 1979.* Department of Materials Science and Engineering, M.I.T., Cambridge, Massachusetts (1979), 1–11.

Entwisle, A. R. (1971). The kinetics of martensite formation in steel. *Metall. Trans.* **2**, 2395–2407.

Kaufman, L. and Cohen, M. (1958). Thermodynamics and kinetics of martensitic transformations. In *Progress in Metal Physics*, Vol. 7 (eds. B. Chalmers and R. King). Pergamon Press, New York, 165–246.

Lieberman, D. S. (1970). Crystal geometry and mechanisms of phase transformations in crystalline solids. In *Phase Transformations.* American Society for Metals, Metals Park, Ohio, 1–58.

Martensite Fundamentals and Technology (1970) (ed. E. R. Petty). Longman Group, London.

Mechanism of Phase Transformations in Crystalline Solids (1969). Monograph and Report Series No. 33, Institute of Metals, London, England.

New Aspects of Martensitic Transformations (1976) (ed. H. Suzuki). Japan Institute of Metals, Tokyo, Japan.

Phase Transformations (1970). American Society for Metals, Metals Park, Ohio.

Physical Properties of Martensite and Bainite (1965). The Iron and Steel Institute Special Report 93. Percy Lund, Hampshires & Co., London.

Proceedings of the International Conference on Martensitic Transformation, ICOMAT-1979 (1979) (ed. W. S. Owen). Department of Materials Science and Engineering, M.I.T., Cambridge, Massachusetts.

Reed, R. P. and Breedis, J. F. (1966). Low-Temperature phase transformations. In *Behavior of Materials at Cryogenic Temperatures.* ASTM STP No. 387, American Society for Testing and Materials, Philadelphia, Pennsylvania, 60–132.

Roitburd, A. L. and Kurdjumov, G. V. (1979). The nature of martensitic transformations. *Mater. Sci. Eng.* **39**, 141–167.

Shape Memory Effect in Alloys (1975) (ed. J. Perkins). Plenum Press, New York.

Wayman, C. M. (1964). *Introduction to the Crystallography of Martensitic Transformations.* Macmillan Co., New York.

Wayman, C. M. (1968). The crystallography of martensitic transformations in alloys of iron. In *Advances in Materials Research,* Vol. 3 (ed. H. Herman). Interscience, New York, 147–304.

9.9.2. Characteristics

Acton, A. F. and Bevis, M. (1969/70). A generalized martensite crystallography theory. *Mater. Sci. Eng.* **5**, 19–29.

Bowles, J. W. and Mackenzie, J. K. (1954). The crystallography of martensite transformations I, II. *Acta Metall.* **2**, 129–147, 224–234.

Bullough, R. and Bilby, B. A. (1956). Continuous distributions of dislocations: surface dislocations and the crystallography of martensitic transformations. *Proc. Phys. Soc. B* **69**, 1276–1286.

Bunshah, R. F. and Mehl, R. F. (1953). Rate of propagation of martensite. *Trans. AIME* **197**, 1251–1258.

Christian, J. W. (1969). The nucleation of mechanical twins and martensite. In *The Physics of Strength and Plasticity* (ed. A. S. Argon). M.I.T. Press, Cambridge, Massachusetts, 85–95.

Christian, J. W. (1970). Basic crystallography and kinetics. In *Martensite Fundamentals and Technology* (ed. E. R. Petty). Longman Group, London, 11–41.

Christian, J. W. (1979). Thermodynamics and kinetics of martensite. In *Proceedings of the International Conference on Martensitic Transformation, ICOMAT 1979.* Department of Materials Science and Engineering, M.I.T., Cambridge, Massachusetts, 220–234.

Clapp, P. C. (1973). A localized soft mode theory for martensitic transformations. *Phys. Status Solidi b* **57**, 561–569.

Delaey, R., Krishnan, R. V., Tas, H., and Warlimont, H. (1974). Thermo-elasticity, pseudoelasticity and the memory effects associated with martensitic transformations. *J. Mater. Sci.* **9**, 1521–1555.

Dunne, D. P. and Wayman, C. M. (1971). The crystallography of ferrous martensites. *Metall. Trans.* **2**, 2327–2341.

Entwisle, A. R. (1971). The kinetics of martensite formation in steel. *Metall. Trans.* **2**, 2395–2407.

Kelly, M. J. (1979). Energetics of the martensite phase transition in sodium. *J. Phys. F* **9**, 1921–1938.

Lee, B., Millman, S., MacDougall, I. L., Keown, S. R., and Argent, B. B. (1977). Enthalpy of the martensitic transformation in steels containing nickel and chromium. *Met. Sci. J.* **11**, 261–271.

Magee, C. L. (1970). The nucleation of martensite. In *Phase Transformations.* American Society for Metals, Metals Park, Ohio, 115–156.

Olson, G. B. and Cohen, M. (1976). A general mechanism of martensitic nucleation. Parts I–III. *Metall. Trans.* **7A**, 1897–1923.

Olson, G. B. and Cohen, M. (1981). A perspective on martensitic nucleation. In *Annual Review of Materials Science*, Vol. 11 (eds. R. A. Huggins, R. H. Bube, and D. A. Vermilyea). Annual Reviews Inc., Palo Alto, California, 1–30.

Reed, R. P. and Breedis, J. F. (1966). Low-temperature phase transformations. In *Behavior of Materials at Cryogenic Temperatures.* ASTM STP No. 307, American Society for Testing and Materials, Philadelphia, Pennsylvania, 60–132.

Reed, R. P., Clark, A. F., and Schramm, R. E. (1971). Defect annealing (4 to 295 K) after martensitic phase transformation in an Fe–29Ni alloy. *Scr. Metall.* **5**, 485–488.

Ross, N. H. D. and Crocker, A. G. (1970). A generalized theory of martensite crystallography and its application to transformations in steel. *Acta Metall.* **18**, 405–418.

Shape Memory Effect in Alloys (1975) (ed. J. Perkins). Plenum Press, New York.

Suzuki, T. and Ledbetter, H. M. (1983). Barrier energy for the bcc–fcc martensitic transition in sodium. Accepted for publication in *Philos. Mag.*

Venables, J. (1962). The martensitic transformation in

stainless steels. *Philos. Mag.* **7**, 35–44.

Warlimont, H. and Delaey, L. (1974). Martensitic transformations in copper, silver, and gold-based alloys. In *Progress in Materials Science,* Vol. 18 (eds. B. Chalmers, J. W. Christian, and T. B. Massalski). Pergamon Press, New York.

Wayman, C. M. (1968). The crystallography of martensitic transformations in alloys of iron. In *Advances in Materials Research,* Vol. 3 (ed. H. Herman). Interscience, New York, 147–304.

Wayman, C. M. (1975). Deformation, mechanisms and other characterisitics of shape memory alloys. In *Shape Memory Effect in Alloys* (ed. J. Perkins). Plenum Press, New York, 1–27.

Wechsler, M. S., Lieberman, D. S., and Read, T. A. (1953). On the theory of the formation of martensite. *Trans. AIME* **197**, 1503–1515.

9.9.3. Austenitic stainless steels

Abrassart, F. (1973). Stress-induced $\gamma \to \alpha'$ martensitic transformation in two carbon stainless steels. Application to trip steels. *Metall. Trans.* **4**, 2205–2216.

Andrews, K. W. (1965). Empirical formulae for the calculation of some transformation temperatures. *J. Iron Steel Inst.* **203**, 721–727.

Angel, T. (1954). Formation of martensite in austenitic stainless steels. *J. Iron Steel Inst.* **177**, 165–174.

Bampton, C. C., Jones, L. P., and Loretto, M. H. (1978). Stacking fault energy measurements in some austenitic stainless steels. *Acta Metall.* **26**, 39–51.

Breedis, J. F. (1964). Martensitic transformations in iron–chromium–nickel alloys. *Trans. AIME* **230**, 1583–1596.

Breedis, J. F. (1965). Influence of dislocation substructure on the martensitic transformation in stainless steel. *Acta Metall.* **13**, 2359–2371.

Breedis, J. F. and Kaufman, L. (1971). Formation of hcp and bcc phases in austenitic iron alloys. *Metall. Trans.* **2**, 2359–2371.

Breedis, J. F. and Robertson, W. D. (1963). Martensitic transformation and plastic deformation in iron alloy single crystals. *Acta Metall.* **11**, 547–559.

Brofman, P. J. and Ansell, G. S. (1978). On the effect of carbon on the stacking fault energy of austenitic stainless steels. *Metall. Trans. A* **9A**, 879–880.

Brooks, J. W., Loretto, M. H., and Smallman, R. E. (1979). Direct observations of martensite nuclei in stainless steel. *Acta Metall.* **27**, 1839–1847.

Collings, E. W. and King, H. W. (1979). Austenitic stainless steel as magnetic transition metal alloys. In *The Metal Science of Stainless Steels* (eds. E. W. Collings and H. W. King). Metallurgical Society of the AIME, New York, 1–21.

Dulieu, D. and Nutting, J. (1964). Stacking fault energy of Fe–Cr–Ni alloys. In *Metallurgical Developments in High Alloy Steels,* Special Report 86, Iron and Steel Institute, London, 140–145.

Eichelman, G. H. and Hull, F. C. (1953). The effect of composition on the temperature of spontaneous transformation of austenite to martensite in 18-8-type stainless steel. *Trans. Am. Soc. Met.* **45**, 77–104.

Guntner, C. J. and Reed, R. P. (1962). The effect of experimental variables including the martensitic transformation on the low temperature mechanical properties of austenitic stainless steels. *ASM Trans. Q.* **55**, 399–419.

Hammond, C. M. (1963). The development of new high-strength stainless steels. In *Advances in the Technology of Stainless Steels and Related Alloys.* ASTM STP No. 369, American Society for Testing and Materials, Philadelphia, Pennsylvania, 47–53.

Hull, F. C. (1973). Delta ferrite and martensite formation in stainless steels. *Weld. J. Res. Suppl.* **52**, 193-s–203-s.

Kaufman, L. (1959). The free energy changes attending the martensitic transformation in the iron–chromium and iron–chromium–nickel systems. *Trans. AIME* **215**, 218–223.

Kaufman, L. (1967). The lattice stability of the transition metals. In *Phase Stability in Metals and Alloys* (eds. P. S. Rudman, J. Stringer, and R. I. Jaffee). McGraw–Hill, New York, 125–150.

Kaufman, L. and Cohen, M. (1956). The martensitic transformation in the iron–nickel system. *Trans. AIME* **206**, 1393–1401.

Kelly, P. M. (1965). The martensite transformation in steels with low stacking fault energy. *Acta Metall.* **13**, 635–646.

King, H. W. and Larbalestier, D. C. (1981). Austenitic stainless steels at cryogenic temperatures: The compositional dependence of the M_s. *Cryogenics* **21**, 521–524.

Lagneborg, R. (1964). The martensite transformation in 18%Cr–8%Ni steels. *Acta Metall.* **12**, 823–843.

Lee, S., Millman, S., MacDougall, I. L., Keown, P., and Argent, B. B. (1977). Enthalpy of the martensitic transformation in steels containing nickel and chromium. *Met. Sci.* **11**, 261–271.

Mangonon, P. L. and Thomas, G. (1970). Structure and properties of thermal-mechanically treated 304 stainless steel. *Metall. Trans.* **1**, 1587–1594.

Mirzagev, D. A., Goykhanberg, Y. N., Shteynberg, M. M., and Rushchin, S. V. (1973). The elastoplastic strain effects in alloys with low stacking fault energy at temperatures above M_s. *Fiz. Met. Metalloved.* **35**, 1206–1212.

Monkman, F. C., Cuff, F. B., and Grant, N. J. (1957). Computation of M_s for stainless steels. *Met. Prog.* **71**, 94–96.

Olson, G. B. and Azrin, M. (1978). Transformation behavior of TRIP steels. *Metall. Trans. A* **9A**, 713–721.

Rao, M. M., Russell, R. J., and Winchell, P. G. (1967). A correlation of thermodynamic variables for iron-rich iron–nickel–carbon alloys. *Trans. AIME* **239**, 634–642.

Reed, R. P. (1962). The spontaneous martensitic transformations in 18%Cr, 8%Ni steels. *Acta Metall.* **10**, 865–877.

Reed, R. P. and Guntner, C. J. (1964). Stress-induced martensitic transformations in 18Cr–8Ni steel. *Trans. AIME* **230**, 1713–1720.

Rhodes, C. G. and Thompson, A. W. (1977). The composition dependence of stacking fault energy in austenitic stainless steels. *Metall. Trans.* **8A**, 1901–1906.

Schramm, R. E. and Reed, R. P. (1975). Stacking fault energies of seven commercial austenitic stainless steels. *Metall. Trans.* **6A**, 1345–1351.

Strife, J. R., Carr, M. J., and Ansell, G. S. (1977). The effect of austenite prestrain above the M_s temperature on the martensitic transformation in Fe–Ni–Cr–C alloys. *Metall. Trans.* **8A**, 1471–1484.

Suzuki, T., Kojima, H. Suzuki, K., Hashimoto, T., and Ichihara, M. (1977). An experimental study on the martensite nucleation and growth in 18/8 stainless steel. *Acta Metall.* **25**, 1151–1162.

Warnes, L. A. A. and King, H. W. (1976). The low temperature magnetic properties of austenitic Fe–Cr–Ni alloys in the prediction of Néel temperatures and maximum susceptibilities. *Cryogenics* **16**, 659–667.

Williams, I., Williams, R. G., and Capellaro, R. C. (1976). Stability of austenitic stainless steels between 4 K and 373 K. In *Proceedings of the Sixth International Cryogenic Engineering Conference.* IPC Science and Technology Press, Guildford, Surrey, England, 337–341.

Wirth, A. and Bickerstaffe, J. (1974). The morphology and substructure of martensite in maraging steels. *Metall. Trans.* **5**, 799–808.

9.9.4. Alkali metals

Alexopoulos, K. D. and Shaw, C. H. (1961). *Low-Temperature Spontaneous Transformations in Fine-Grained Alkali Metals.* AD 258132.* Technical Report 809-3, Ohio State University Research Foundation, Columbus, Ohio.

Barrett, C. S. (1947). A low temperature transformation in lithium. *Phys. Rev.* **72**, 245–250.

Barrett, C. S. (1955/56). Metallographic study of sodium, potassium, rubidium, and cesium after cooling to 1.2 K. *J. Inst. Met.* **84**, 43–46.

Barrett, C. S. (1956a). X-ray study of the alkali metals at low temperatures. *Acta Crystallogr.* **9**, 671–677.

Barrett, C. S. (1956b). Crystal structures of lithium–magnesium alloys at low temperatures. *Acta Metall.* **4**, 528–531.

Barrett, C. S. and Clifton, D. F. (1950). Transformation characteristics of a lithium–magnesium alloy. *Trans. AIME* **188**, 1329–1332.

Barrett, C. S. and Trautz, O. R. (1948). Low temperature transformations in lithium and lithium–magnesium alloys. *Trans. AIME* **175**, 579–601.

Bajpai, R. P., Oho, M., Ohno, Y., and Toya, T. (1975). Self-consistent theory of the martensitic phase transformation in metallic lithium. *Phys. Rev. B* **12**, 2194–2211.

Basinski, Z. S., Dugdale, J. S., and Gugan, D. (1959). The strain dependence of the residual resistivity of potassium. *Philos. Mag.* **4**, 880–883.

Basinski, Z. S. and Verdini, L. (1959). The change of volume produced by martensitic transformation in lithium and sodium. *Philos. Mag.* **4**, 1311–1315.

Bowles, J. (1951). Metallographic study of the martensite transformation in lithium. *Trans. AIME* **191**, 44–46.

Day, J. P. and Ruoff, A. L. (1974). The variation of the elastic constants of lithium with temperature and pressure. *Phys. Status Solidi a* **25**, 205–213.

Dolling, G., Powell, B. M., and Martel, P. (1968). Nature of the martensitic transformation in sodium. *Can. J. Phys.* **46**, 1727–1729.

Dugdale, J. S. and Gugan, D. (1960). The electrical resistivity of the two phases of sodium at low temperatures. *Proc. R. Soc. A* **254**, 184–204.

Dugdale, J. S. and Gugan, D. (1961). The effect of the martensitic transformation on the electrical resistance of lithium and dilute lithium–magnesium alloys. *Cryogenics* **2**, 103–104.

Dugdale, J. S. and Gugan, D. (1962). The effect of pressure on the electrical resistance of lithium, sodium and potassium at low temperatures. *Proc. R. Soc. A* **270**, 186–211.

Felice, R. A., Trivisonno, J. and Schuele, D. E. (1977). Temperature and pressure dependence of the single-crystal elastic constants of ^{6}Li and natural lithium. *Phys. Rev. B* **16**, 5173–5184.

Filby, J. D. and Martin, D. L. (1963). The specific heats below 30 K of lithium metal of various isotopic compositions and of sodium metal. *Proc. R. Soc. A* **276**, 187–203.

Gindin, I. A., Godzhagev, V. M., Lazareva, M. B. and Starodubov, Y. D. (1966). Low-temperature creep of lithium in the polymorphous transformation range. *Phys. Met. Metalloved.* **21**(4), 600–607.

Gindin, I. A., Lazarev, B. G., and Starodubov, Y. D. (1960). Peculiarities in the mechanical properties of lithium, related to low temperature polymorphous transformations. *Phys. Met. Metalloved.* **10**(3), 153–160.

Gindin, I. A., Lazarev, B. G., and Starodubov, Y. D. (1961). Low temperature metallography of lithium. *Phys. Met. Metalloved.* **11**(1), 46–51.

Gugan, D. (1963). The effect of pressure on the electrical resistance of lithium. *Can. J. Phys.* **41**, 1381–1384.

Gugan, D. (1966). The annealing of deformed sodium. *Philos. Mag.* **13**, 533–540.

Hands, B. A. and Rosenberg, H. M. (1969). The low temperature yielding and flow of potassium and its alloys. *Acta. Metall.* **17**, 455–461.

Hedgcock, F. T. (1960). Magnetic susceptibility of the two phases of lithium at low temperatures. *Phys. Rev. Lett.* **5**, 420–422.

Hull, D. and Rosenberg, H. M. (1959). The deformation of lithium, sodium, and potassium at low temperatures: tensile and resistivity experiments. *Philos. Mag.* **4**, 303–315.

Hull, D. and Rosenberg, H. M. (1960). Metallography at liquid helium temperatures. *Cryogenics* **1**, 27–32.

Martin, D. L. (1960a). The specific heat of lithium from 20 to 300° K: the martensitic transformation. *Proc. Phys. Soc. A* **254**, 444–454.

Martin, D. L. (1960b). The specific heat of a lithium–magnesium alloy. The martensitic transformation. *Can. J. Phys.* **38**, 25–31.

Masson, D. B. (1962). Remarks on the elastic constants and temperatures of martensitic transformation in the Li–Mg alloys. *Acta Metall.* **10**, 986–987.

Pynn, R. and Ebbsjo, I. (1971). A study of the effect of temperature on the stable crystal structures of potassium and sodium. *J. Phys. F* **1**, 744–752.

Reed, R. P. (1977). Low temperature tensile properties of high-purity lithium. Unpublished data. National Bureau of Standards, Boulder, Colorado.

Snyder, D. D. (1966). *Isotope Effect on Phase Transition Temperatures for Lithium.* Rept. No. C00-972-9 presented to American Physical Society, New York, USAEC contract AT(11-1)-972. Andrews

University, Berrien Springs, Michigan.

Snyder, D. D. (1968). *Mechanical Properties of Separated Metallic Isotopes*. Final Report to USAEC, Contract AT(11-1)-972. Andrews University, Berrien Springs, Michigan.

Stedman, R. (1976). Observations on the martensitic transformation in sodium. *J. Phys. F* **6**, 2239–2246.

Stern, R., Natale, G. G., and Radnick, I. (1966). Ultrasonic attenuation in sodium at low temperatures. *J. Phys. Chem. Solids* **27**, 9–18.

Straub, G. K. and Wallace, D. C. (1971). Study of the martensitic phase transition in sodium. *Phys. Rev. B* **3**, 1234–1239.

Trivisonno, J. and Smith, C. S. (1961). Elastic constants of lithium–magnesium alloys. *Acta Metall.* **9**, 1064–1071.

Verdini, L. (1962). Ultrasonic study of the martensitic transformation in lithium and sodium at low temperatures. In *Proceedings 3rd International Congress on Acoustics*. Elsevier Publishing Co., Amsterdam, 480–484.

9.9.5. Solidified gases

Ahlers, G. and Orttung, W. H. (1964). Anomaly in the heat capacity of solid hydrogen at small molar volumes. *Phys. Rev. A* **133**, 1642–1650.

Barrett, C. S. (1969). Transformations in solidified gases. In *The Mechanism of Phase Transformations in Solids,* Institute of Metals, London, 313–318.

Barrett, C. S. and Meyer, L. (1965). Argon–nitrogen phase diagram. *J. Chem. Phys.* **42**, 107–112.

Barrett, C. S., Meyer, L., and Wasserman, J. (1966). Argon–oxygen phase diagram. *J. Chem. Phys.* **44**, 998–1000.

Hill, R. W. and Ricketson, B. W. A. (1954). An anomaly in the specific heat of solid hydrogen. *Philos. Mag.* **45**, 277–282.

Jarvis, J., Ramm, D., and Meyer, H. (1967). Density changes in solid H_2. *Phys. Rev. Lett.* **18**, 119–121.

McCarty, R. D., Hord, J, and Roder, H. M. (1981). *Selected Properties of Hydrogen.* NBS Monograph 168, U. S. Government Printing Office, Washington, D.C.

Meyer, L. (1969). Phase transitions in van der Waals lattices. *Adv. Chem. Phys.* **16**, 343–387.

Nieble, K. F. and Venables, J. A. (1976). The crystal structure problem In *Rare Gas Solids,* Vol. 1 (eds. M. L. Klein and J. A. Venables). Academic Press, New York, 558–589.

Ramm, D., Meyer, H., and Mills, R. L. (1970). Measurement of UdP/dT and related properties in solidified gases, III, Solid D2. *Phys. Rev. B* **1**(6), 2763–2767.

Rare Gas Solids (1977) (eds. M. L. Klein and J. A. Venables). Academic Press, New York.

Roder, H. M. (1978). The molar volume (density) of solid oxygen in equilibrium with vapor. *J. Phys. Chem. Ref. Data* **7**, 949–957.

Silvera, I. F. (1980). The solid molecular hydrogens in the condensed phase: Fundamentals and static properties. *Rev. Modern Phys.* **52**, 393–452.

Terry, J. E. (1965). *A Bibliography on Solid Hydrogen Crystallography.* Redstone Scientific Information Center, Report No. AD464176.*

9.9.6. Superconductors

Albert, H. and Pfeiffer, I. (1972). Über die Temperaturabhängigkeit des elastizitäts Modals von niobtitan Legierungen. *Z. Metallkd.* **63**, 126–131.

Axe, J. D. and Shirane, G. (1973). Inelastic-neutron scattering study of acoustic phonons in Nb_3Sn. *Phys. Rev. B* **8**, 1965–1977.

Bagariatski, A., Nosova, G. I., and Tagunova, T. V. (1958). Factors in the formation of metastable phases in titanium-base alloys. *Sov. Phys.–Dokl.* **3**, 1014–1018.

Baker, C. (1971). The shape-memory effect in a titanium–35 wt% niobium alloy. *Metal Sci. J.* **5**, 92–100.

Baker, C. and Sutton, J. (1969). Correlation of superconducting and metallurgical properties of a Ti–20 at.% Nb alloy. *Philos. Mag.* **19**, 1223–1255.

Balcerzak, A. T. and Sass, S. L. (1972). The formation of the ω phase in Ti–Nb alloys. *Metall. Trans.* **3**, 1601–1605.

Bhatt, R. N. (1977). Microscopic theory of the martensitic transition in A-15 compounds based on a three-dimensional band structure. *Phys. Rev. B* **16**, 1915–1932.

Bhatt, R. N. and McMillan, W. L. (1976). Landau theory of the martensitic transition in A-15 compounds. *Phys. Rev. B* **14**, 1007–1027.

Blackburn, M. S. (1970). Some aspects of phase transformations in titanium alloys. In *The Science, Technology and Application of Titanium* (eds. R. I. Jaffee and N. E. Promisel). Pergamon Press, New York, 633–643.

Brammer, W. G. and Rhodes, C. G. (1967). Determination of omega phase morphology in Ti–35%Nb by transmission electron microscopy. *Philos. Mag.* **16**, 477–486.

Brown, A. R. G., Clark, D., Eastabrook, J., and Jepson, K. S. (1964). The titanium–niobium system. *Nature* **201**, 914–915.

Bychkova, M. I., Baron, V. V., and Savitskii, E. M. (1965). Effect of heat treatment on the superconducting properties of Nb–Ti alloys. In *Metallography and Metal Physics of Superconductors.* Izdateltsvo, Nauka, Moscow, 53–60.

Bywater, K. A. and Christian, J. W. (1972). Martensitic transformations in titanium–tantalum alloys, *Philos. Mag.* **25**, 1249–1273.

Chu, C. W. (1974). Pressure-enhanced lattice transformation in Nb_3Sn single crystals. *Phys. Rev. Lett.* **33**, 1283–1286.

Chu, C. W. and Vieland, L. J. (1974). The superconducting transition temperature and its high-pressure behavior of tetragonal Nb_3Sn. *J. Low Temp. Phys.* **17**, 25–29.

Clogston, A. M. and Jaccarino, V. (1961). Susceptibilities and negative Knight shifts of intermetallic compounds. *Phys. Rev.* **121**, 1357–1362.

Cohen, R. W., Cody, G. D., and Halloran, J. J. (1967). Effect of Fermi-level motion on normal-state properties of β-tungsten superconductors. *Phys. Rev. Lett.* **19**, 840–844.

Collings, E. W. and Ho, J. C. (1969). Enhancement of superconducting transition temperatures in martensitic Ti–Mo alloys. *Phys. Lett. A* **29**, 306–307.

Collings, E. W. and Smith, R. D. (1976). The magnetic

susceptibility of titanium–niobium alloys. *J. Less-Common Met.* **48**, 187–198.

deFontaine, D. (1970). Mechanical instabilities in the b.c.c. lattice and the beta to omega phase transformation. *Acta Metall.* **18**, 275–279.

Delaey, L., Van Paemel, J., and Struyve, T. (1972). The relations between the premartensitic instability of the β-phase and the martensite structures and the shape-memory effect in the noble metal base alloys. *Scr. Metall.* **6**, 507–518.

Duwez, P. (1953). The martensitic transformation temperature in titanium binary alloys. *Trans. Am. Soc. Met.* **45**, 934–939.

Fedotov, S. G. and Belousov, O. K. (1963). The elastic properties of alloys of titanium with molybdenum, vanadium and niobium. *Sov. Phys.–Dokl.* **8**, 496–498.

Gor'kov, L. P. and Dovokhov, O. N. (1976). On the theory of the structural properties of A-15 type materials, *J. Low Temp. Phys.* **22**, 1–25.

Hansen, M., Kamen, E. L., Kessler, H. D., and McPherson, D. J. (1951). Systems titanium–molybdenum and titanium–columbium. *Trans. AIME* **191**, 881–888.

Hatt, B. A. and Rivlin, V. G. (1968). Phase transformations in superconducting Ti–Nb alloys. *Brit. J. Appl. Phys.* **1**, 1145–1149.

Hickman, B. S. (1969). Omega phase precipitation in alloys of titanium with transition metals. *Trans. AIME* **245**, 1329–1336.

Hoard, R. W., Scanlon, R. M., Smith, G. S., and Farrell, C. L. (1980). *The Effect of Strain on the Martensitic Phase Transition in Superconducting Nb_3Sn.* Preprint UCRL-84211, Lawrence Livermore National Laboratory, Livermore, California.

Izyumov, Yu. A. and Kurmaev. Z. Z. (1974). Physical properties and electronic structure of superconducting compounds with the β-tungsten structure. *Sov. Phys. Usp.* **17**, 356–380.

Jepson, K. S., Brown, A. R. G., and Gray, J. A. (1970). The effect of cooling rate on the beta transformation in titanium–niobium and titanium–aluminum alloys. In *The Science, Technology, and Application of Titanium* (eds. R. I. Jaffee and N. E. Promisel). Pergamon Press, New York, 667–690.

Keller, K. R. and Hanak, J. S. (1967). Ultrasonic measurements in single-crystal Nb_3Sn. *Phys. Rev.* **154**, 628–632.

King, H. W. (1969). A note on the dependence of the low-temperature martensitic transformation in Nb_3Sn on composition. In *The Mechanisms of Phase Transformations in Crystalline Solids*. Monograph 33, Institute of Metals, London, 196–197.

Koch, C. C. and Easton, D. S. (1977). A review of mechanical behavior and stress effects in hard superconductors. *Cryogenics* **17**, 391–413.

Labbé, J. and Friedel, J. (1966). Stabilité des modes de distorsion périodiques d'une chaine linéaire d'atomes de transition dans une structure cristalline du type V_3Si. *J. Phys. Radium* **27**, 708–716.

Ledbetter, H. M. (1978). Unpublished data on elastic constants of titanium–niobium alloys. National Bureau of Standards, Boulder, Colorado.

Lohberg, R., Heller, W., and Zwicker, U. (1973). Influence of pinning forces of titanium–niobium base alloys on the critical current up to 10 tesla. In *Titanium Science and Technology* (eds. R. I. Jaffee and H. M. Burte). Plenum Press, New York, 860–870.

McEvoy, J. P. (1971). Effect of uniaxial stress on the superconducting transition temperature of monocrystalline Nb_3Sn. *Physica* **55**, 540–544.

McMillan, W. L. (1977). Superconductivity and martensitic transformations in A-15 compounds. In *Electron–Phonon Interactions and Phase Transitions* (ed. T. Riste). Plenum Press, New York, 194–199.

Mailfert, R., Batterman, B. W., and Hanak, J. J. (1967). Low temperature structural transformation in Nb_3Sn. *Phys. Lett. A* **24**, 315–316.

Mailfert, R., Batterman, B. W., and Hanak, J. J. (1969). Observations related to the order of the low temperature structural transformation in V_3Si and Nb_3Sn. *Phys. Status Solidi* **32**, K67–K69.

Mattheiss, L. F. (1975). APW-LCAO band model for A15 compounds. *Phys. Rev. B* **12**, 2161–2180.

Obst, B., Pattanayak, D., and Hochstuhl, P. (1980). Structural effects in the superconductor NbTi65. *J. Low Temp. Phys.* **41**, 595–609.

Pupp, W. A., Sattler, W. W., and Saur, E. J. (1974). Superconductivity of thin Nb_3Sn films under tensile stress. *J. Low Temp. Phys.* **14**, 1–13.

Read, D. T. (1978). Metallurgical effects in niobium–titanium alloys. *Cryogenics* **18**, 579–584.

Reed, T. B., Gatos, A. C., LaFleur, W. J., and Roddy, J. T. (1969). Superconducting behavior of some β-tungsten structure niobium compounds and their alloys. In *Metallurgy of Advanced Electronic Materials*, Vol. 19. Interscience, New York, 71–86.

Rehwald, W. (1968). Lattice softening and stiffening of single crystal niobium stannide at low temperatures. *Phys. Lett. A* **27**, 287–288.

Rehwald, W., Rayl, M., Cohen, R. W., and Cody, G. D. (1972). Elastic moduli and magnetic susceptibility of monocrystalline Nb_3Sn. *Phys. Rev. B* **6**, 363–371.

Salter, L. C., Raymond, J., Boom, R. W., and Vetrano, J. B. (1966). Investigation of current degradation phenomenon in superconducting solenoids. NASA Report N66 23800 to Marshall Space Flight Center, Huntsville, Alabama, from Atomics International, Canoga Park, California.

Sass, S. L. (1972). The structure and decomposition of Zr and Ti b.c.c. solid solutions. *J. Less-Common Met.* **28**, 157–173.

Sato, T., Hukai, S., and Huang, Y.-C. (1960). The M_s points of binary titanium alloys. *J. Aust. Inst. Met.* **5**, 149–153.

Shirane, G. and Axe, J. D. (1971). Neutron scattering study of the lattice-dynamical phase transition in Nb_3Sn. *Phys. Rev. B* **4**, 2957–2963.

Testardi, L. R. (1973). Elastic behavior and structural instability of high-temperature A-15 structure superconductors. In *Physical Acoustics*, Vol. X (eds. W. P. Mason and R. N. Thurston). Academic Press, New York, 193–296.

Testardi, L. R. (1975). Structural instability and superconductivity in A-15 compounds. *Rev. Mod. Phys.* **47**, 637–648.

Testardi, L. R. (1977). Structural instability of A-15 superconductors. In *Physical Acoustics*, Vol. XIII (eds. W. P. Mason and R. N. Thurston). Academic Press, New York, 29–47.

Testardi, L. R. (1978). Structural phase transitions and superconductivity in A-15 compounds. In *Electron–Phonon Interactions and Phase Transitions* (ed. T. Riste). Plenum Press, New York, 181–193.

Vieland, L. J. (1972). X-ray investigation of the lattice transformation in single crystal Nb_3Sn. *J. Phys. Chem. Solids* **33**, 581–588.

Vieland, L. J., Cohen, R. W., and Rehwald, W. (1971). Evidence for a first-order structural transformation in Mo_3Sn. *Phys. Rev. Lett.* **26**, 373–376.

Vieland, L. J. and Wicklund, A. W. (1969). Specific heat and the lattice transformation in Nb_3Sn. *Solid State Commun.* **7**, 37–39.

Viswanathan, R. and Johnston, D. C. (1973). Stabilization of the cubic A-15 phase of Nb_3Sn. *Mater. Res. Bull.* **8**, 589–592.

Viswanathan, R., Luo, H. L., and Vieland, L. J. (1974). Low-temperature heat capacity of Nb_3Sn. In *Low Temperature Physics LT-13*. Plenum Press, New York, 472–474.

Vozilkin, V. A., Prekul, A. F., Rakin, V. G., Volkenshteyn, N. V., and Buynor, N. N. (1968). Structure dependence of the superconductive properties of the alloy of titanium with 47 wt.% niobium. *Phys. Met. Metalloved.* **26**, 77–84.

Wanagel, J. and Batterman, B. W. (1970). Crystallographic study of the low-temperature phase transformation in V_3Si and Nb_3Sn. *J. Appl. Phys.* **41**, 3610–3614.

Wayman, C. M. (1973). Phase transformations in "less-common" materials. *Metall. Trans.* **4**, 2781–2787.

Weger, M. and Goldberg, I. B. (1973). Some lattice and electronic properties of the β-tungstens. In *Solid State Physics*, Vol. 28 (eds. H. Ehrenreich, F. Seitz, and D. Turnbull). Academic Press, New York, 1–177.

Williams, J. C. (1973). Critical review kinetics and phase transformations. In *Titanium Science and Technology* (eds. R. I. Jaffee and H. M. Burte). Plenum Press, New York, 1433–1494.

9.9.7 Iron-nickel alloys

Ananthanarayanan, N. I. and Peavler, R. J. (1961). A new reversible solid-state transformation in iron–nickel alloys in the Invar range of compositions. *Nature* **192**, 962–963.

Averbach, B. L. and Cohen, M. (1962). *The Role of Fine Structure in the Mechanical Behavior of Metals*. Quarterly Report No. 11, AD 285459, M.I.T., Cambridge, Massachusetts.

Bando, Y. (1964). Characteristics of phase transformations in metallic fine particles (martensitic transformation of Fe–Ni alloys, and ordering of CuAu and Cu_3Au alloys). *J. Inst. Met.* **5**, 135–141.

Bibby, M. J. and Parr, J. G. (1964). The martensitic transformation in pure iron. *J. Iron Steel Inst.* **203**, 100–104.

Bokros, J. C. and Parker, E. R. (1963). The mechanism of the martensite burst transformation in Fe–Ni single crystals. *Acta Metall.* **11**, 1291–1301.

Bolling, G. F. and Richman, R. H. (1970). The plastic deformation–transformation of paramagnetic f.c.c. Fe–Ni–C alloys. *Acta Metall.* **18**, 673–681.

Bradley, A. J., Jay, A. H., and Taylor, A. (1937). The lattice spacing of iron–nickel alloys. *Philos. Mag.* **23**, 545–557.

Breedis, J. F. and Wayman, C. M. (1962). The martensitic transformation in Fe–31 wt. pct. Ni. *Trans. Metall. Soc. AIME* **224**, 1128–1133.

Clark, C. A. (1959). The dependence of M_s temperature on nickel content in nickel-alloy steels. *J. Iron Steel Inst.* **193**, 11–12.

Dubrov, V. A. (1966). High-temperature metallographic study of the kinetics of the γ–α transformation in iron–nickel alloys. *Fiz. Met. Metalloved.* **21**(4), 551–554.

Gilbert, A. and Owen, W. S. (1962). Diffusionless transformation in iron–nickel, iron–chromium and iron–silicon alloys. *Acta Metall.* **10**, 45–54.

Goldman, A. J. and Robertson, W. D. (1965). A correlation of elastic moduli, supercooling and heat evolved in the martensitic transformation in iron alloys. *Acta Metall.* **13**, 391–401.

Goldman, A. J. and Wagner, C. N. J. (1963). Faulting in deformed austenite and martensite. *Acta Metall.* **11**, 405–413.

Goodenow, R. H. and Hehemann, R. F. (1965). Transformations in iron and Fe–9 pct Ni alloys. *Trans. AIME* **233**, 1777–1785.

Goodenow, R. H., Matas, S. J., and Hehemann, R. F. (1963). Growth kinetics and the mechanism of the bainite transformation. *Trans. AIME* **227**, 651–660.

Honma, T. (1957a). Studies on martensite transformation in iron–nickel alloys (2nd report). Effects of surface and interface on "Schiebung" transformation. *Nihon Kinzuoku Gakkai-shi* **21**, 122–125.

Honma, T. (1957b). Studies on martensite transformation in iron–nickel alloys (3rd report). Coexisting range of Schiebung and Umklapp transformation. *Nihon Kinzuoku Gakkai-shi* **21**, 126–128.

Honma, T. (1957c). Studies on martensite transformation in iron–nickel alloys (4th report). Burst phenomenon of Umklapp process. *Nihon Kinzuoku Gakkai-shi* **21**, 263–267.

Jette, E. R. and Foote, F. (1936). X-ray investigation of iron–nickel alloys. *Trans. AIME* **120**, 259–265.

Jones, F. W. and Pumphrey, W. I. (1949). Free energy and metastable states in the iron–nickel and iron–manganese systems. *J. Iron Steel Inst.* **163**, 121–131.

Kaufman, L. and Cohen, M. (1956). The martensitic transformation in the iron–nickel system. *Trans. AIME* **206**, 1393–1401.

Kaufman, L., Leyenaar, A., and Harvey, J. S. (1961). The effect of hydrostatic pressure on the f.c.c. $\rightleftarrows$ b.c.c. reactions in iron-base alloys. In *Progress in Very High Pressure Research*. John Wiley & Sons, New York, 90–108.

Klostermann, J. A. (1965). Discussion 2. In *Physical Properties of Martensite and Bainite*. Special Report 93, Iron and Steel Institute, London, 43–47.

Klostermann, J. A. and Burgers, W. G. (1964). Surface martensite in iron–nickel. *Acta Metall.* **12**, 355–360.

Krauss, G. and Cohen, M. (1962). Strengthening and annealing of austenite formed by the reverse martensitic transformation. *Trans. AIME* **224**, 1212–1221.

Krauss, G. and Pitsch, W. (1965). The fine structure

and habit planes of martensite in a Fe–33 wt pct Ni single crystal. *Trans. AIME* **233**, 919–926.

Nagakura, S., Kikuchi, N., Kaneko, Y., and Oketani, S. (1963). Study of martensitic transformations in evaporated iron–nickel alloy films by electron microscopy and diffraction. *Jap. J. Appl. Phys.* **2**, 201–205.

Owen, E. A., Yates, E. L., and Sully, A. H. (1937). An x-ray investigation of pure iron–nickel alloys. Part 4: The variation of lattice-parameter with composition. *Proc. Phys. Soc. (London)* **49**, 315–322.

Owen, W. S. and Wilson, E. A. (1965). A note on massive structures. In *Physical Properties of Martensite and Bainite*. Special Report 93, Iron and Steel Institute, London, 53–57.

Owen, W. S., Wilson, E. A., and Bell, T. (1965). The structure and properties of quenched iron alloys. In *High Strength Materials*. John Wiley & Sons, New York, 167–212.

Patterson, R. L. and Wayman, C. M. (1966). The crystallography and growth of partially twinned martensite plates in Fe–Ni alloys. *Acta Metall.* **14**, 347–369.

Porter, L. F. and Dienes, G. J. (1959). Effect of neutron irradiation on the martensite transformation in iron–nickel alloys. *Trans. AIME* **215**, 854–863.

Reed, R. P. (1967). The plate-like martensite transformation in FeNi alloys. *Acta Metall.* **15**, 1287–1296.

Reed, R. P. (1980). National Bureau of Standards, Boulder, Colorado, unpublished data.

Reed, R. P. and Schramm, R. E. (1969). Lattice parameters of martensite and austenite in Fe–Ni alloys. *J. Appl. Phys.* **40**, 3453–3458.

Richman, R. H. and Bolling, G. F. (1971). Deformation and martensitic transformation. *Metall. Trans.* **2**, 2451–2462.

Roberts, M. J. and Owen, W. S. (1967). The strength of martensitic iron–nickel alloys. *Trans. Am. Soc. Met.* 60, 687–692.

Scheil, E. and Saftig, E. (1960). Measurement of heat of transformation of martensite formation for iron–nickel alloys by a dry ice (solid carbon dioxide) calorimeter. (In German.) *Arch. Eisenhuettenw.* **31**, 623–632.

Speich, G. R. and Swann, P. R. (1965). Yield strength and transformation substructure of quenched iron–nickel alloys. *J. Iron Steel Inst.* **203**, 480–485.

Steven, W. and Haynes, A. G. (1956). The temperature of formation of martensite and bainite in low-alloy steels; some effects of chemical composition. *J. Iron Steel Inst.* **183**, 349–359.

Swanson, W. D. and Parr, J. G. (1964). Transformations in iron–nickel alloys. *J. Iron Steel Inst.* **202**, 104–106.

Wilson, E. A. (1965). Diffusionless transformations in iron alloys. Ph.D. Thesis, University of Liverpool, Liverpool, England.

Yeo, R. B. G. (1962). Isothermal martensite transformation in iron-base alloys of low carbon content. *Trans. AIME* **224**, 1222–1227.

Yeo, R. B. G. (1964). Growth of martensite in an iron–28.8% nickel alloy. *Trans. Am. Soc. Met.* **57**, 48–61.

Zackay, V. F., Justusson, M. W., and Schmatz, D. J. (1962). In *Strengthening Mechanisms in Solids*, American Society for Metals, Metals Park, Ohio, 179–216.

9.9.8 Polymers

Allan, P., Cuellin, E. B., and Bevis, M. (1973). Stress-induced twinning and phase transformations in polyethylene single crystals. *Philos. Mag.* **27**, 127–145.

Keith, H. D. (1973). Phase transformations in high polymers. *Metall. Trans.* **4**, 2747–2754.

Young, R. J. and Bowden, P. B. (1974). Twinning and martensitic transformations in oriented high-density polyethylene. *Philos. Mag.* **34**, 1061–1073.

10 Compatibility of Materials with Cryogens

J. C. MOULDER and J. G. HUST

10.1 INTRODUCTION

In the broadest possible sense, the compatibility of two substances concerns their mutual tolerance. Compatible materials have the ability to combine or remain in contact with no undesirable effects. Clearly, such a broad definition encompasses a host of phenomena; in this chapter discussion is limited to those problems that arise from chemical or physical interactions between liquefied gases and the common materials used in their production, storage, transportation, distribution, and use.

Most cryogens are relatively inert and present no special hazards due to material incompatibility in normal use. The exceptions to this rule are oxygen, fluorine, and, to a lesser extent, hydrogen. Oxygen and fluorine are among the most powerful oxidizers known; both elements will react with most organic and inorganic substances that are not already fully oxidized. Even metals, excepting gold and platinum, will burn in oxygen or fluorine, and the high heats of combustion for most metals can lead to extensive damage in a metal fire.

The many hazards associated with the reactivity of materials in oxygen and fluorine have led to considerable research on the compatibility of materials with these cryogens. Numerous tests have been devised to screen and rank materials for use in liquid oxygen or liquid fluorine systems, and literally hundreds of materials have been tested. Compatibility tests generally fall into one of two categories: tests that determine the *sensitivity* of a material to ignition and combustion and tests that measure the *intensity* of a reaction once a material is ignited. Most compatibility research has been concerned with materials for liquid oxygen service, since more oxygen is used than all other atmospheric gases combined. Fluorine is less commonly used because of its toxicity and the special handling problems that toxicity engenders. But fluorine has several unique characteristics that make its use as an oxidizer for rocket propellants attractive, and a significant amount of research has been stimulated by this interest.

Liquid hydrogen, although it is highly flammable, poses less hazard from incompatibility than oxygen or fluorine. Some plastics do act as oxidizers for hydrogen and should be avoided, but hydrogen will not react exothermically with most structural materials under ordinary conditions of pressure and temperature. Hydrogen embrittlement of metals is the most serious compatibility problem with this cryogen. Strictly speaking, very little diffusion of hydrogen into metals occurs at 20 K, the boiling point of liquid hydrogen. But if long-term liquid-hydrogen service with thermal cycling is contemplated, the accumulated exposure of metal structures to hydrogen gas at room temperature and their susceptibility to hydrogen embrittlement must be considered. Hydrogen embrittlement is discussed briefly in § 10.4.

Although safety is the principal motivation for studying the compatibility of materials, material incompatibility is only one of many hazards associated with cryogenic fluids. The reader interested in a fuller account of cryogenic safety should consult one of the texts available on this subject (Zabetakis, 1967, or *Cryogenic Safety, A*

Summary Report of the Cryogenic Safety Conference, 1959).

10.2 COMPATIBILITY OF MATERIALS WITH LIQUID OXYGEN

10.2.1 Metal ignition and combustion

Metals are perhaps the single most important class of materials used in handling liquid oxygen. Their chemical stability when in contact with oxygen, high strength, resistance to thermal shock, large heat capacity, and high thermal conductivity all contribute to making metals the best available structural materials for liquid oxygen systems. Because the ignition and combustion of metals are heterogeneous processes, the experimental study of these phenomena is unique in several respects.

Experiments on metal combustion date back to at least 1782, when von Ingenn-Hausz plunged heated spirals of iron and steel into oxygen and observed that they ignited and burned, producing a great deal of heat and light (Grosse and Conway, 1958). Although two centuries have passed, metal combustion has been seldom studied and is poorly understood, as previous reviewers have noted (Clark and Hust, 1974; Markstein, 1963). It is only during the past twenty years that significant progress has been made toward understanding the complex, interrelated chemical and physical processes that occur when a metal burns.

Some insight into the complexities of metal combustion may be gained by examining Table 10.1, which lists the factors affecting metal combustion. Although the list is fairly inclusive, it is by no means exhaustive. In addition to the usual environmental factors that affect a chemical reaction (temperature, pressure, concentration of oxygen), there are a number of factors that depend on the particular configuration of the metal (size, shape, and history, for example) and a large number of material properties that affect the ignition and combustion processes. In most cases the relevant material properties for both the metal and its oxides are required. Where these properties vary with temperature, the variation must be known. In some cases, such as viscosity and surface tension, the properties change not only with temperature but with the concentration of oxygen as well.

Thermodynamics of metal-oxygen reactions. None of the metallic elements are chemically inert, although their relative activities vary markedly, depending on their position in the periodic table. In particular, with the exception of gold, no metal or alloy is stable in air at room

Table 10.1 Factors affecting metal combustion.

Material Properties[a]	Environmental Factors	Configuration
Atomic weight (m)	Physical state of oxidizer	History of metal
Molecular weight (mo)	Temperature	Condition of surface
Thermal conductivity (m + mo)	Pressure	Oxide film thickness
Thermal diffusivity (m + mo)	Concentration of oxygen	Specific surface area (cm^2/g)
Heat capacity (m + mo)	Concentration of diluent	Total mass of metal
Heats of phase transitions (m + mo)	Nature of diluent	Presence of other metals (eutectic formation or thermite reaction possible)
Heat of melting (m + mo)	Gas or liquid velocity	Presence of contaminants
Heat of vaporization (m + mo)	Reynolds number	
Heat of formation (mo)	Heat and mass transfer coefficients	
Melting point (m + mo)		
Boiling point (m + mo)		
Ignition temperature (m)		
Vapor pressure (m + mo)		
Viscosity (m + mo)		
Surface tension (m + mo)		
Solubility of oxygen in metal (m)		
Diffusivity of oxygen (m + mo)		
Diffusivity of metal (m + mo)		
Emissivity (m + mo)		
Composition of metal (m)		
Oxidation rate laws for metal (m)		
Dissociation pressure of oxide (mo)		

[a]m = metal; mo = metal oxide.

temperature. All other metals and alloys will form oxides; many form nitrides as well.

The force that drives metal oxidation is the free energy change associated with the formation of a metal oxide. The metal oxides are, as a rule, highly stable: they have a lower free energy than the free metal. Consequently, metal oxidation is a thermodynamically favored process.

The overall chemical equation for the reaction between a metal and oxygen to form the metal oxide may be written

$$aM + (b/2)O_2 = M_aO_b \qquad (10.1)$$

This is a reversible reaction; the direction in which it proceeds depends upon temperature and the concentration of oxygen present in the gas phase. If the partial pressure of oxygen is greater than the dissociation pressure of the oxide at the ambient temperature, the metal will oxidize. The metal will remain unaffected and any metal oxide will decompose if the ambient oxygen partial pressure is lower than the dissociation pressure. The oxides of the noble metals (gold, silver, copper) have dissociation pressures that are greater than or equal to ambient oxygen pressures at temperatures below their melting point. Thus, these metals can be heated in air without oxidizing excessively. They are the only pure metals found natively. The base metals have much lower dissociation pressures and will oxidize under most conditions found terrestrially. Titanium oxide, for example, has a dissociation pressure of 10 pPa (10^{-16} atm) at 1873 K (1600° C), the melting point of titanium, which helps to explain the remarkable affinity of titanium for oxygen.

Although the chemical driving force for metal oxidation is the standard free energy of formation, $\Delta G°$, the heat liberated during the chemical reaction is often the determining factor in sustaining and propagating a reaction. This important thermodynamic quantity, the standard heat of formation, $\Delta H°$, may be calculated from the standard free energy and the entropy change, $\Delta S°$, by use of the well-known Gibbs–Helmholtz equation,

$$\Delta H° = \Delta G° + T\Delta S° \qquad (10.2)$$

or obtained from readily available tabulations (Stull and Prophet, 1971). There are also several excellent textbooks that discuss the chemical thermodynamics of metal–oxygen reactions and give examples of thermophysical calculations, for example, Kubaschewski and Evans (1958); Darken and Gurry (1953); and Lewis, Randall, Pitzer, and Brewer (1961).

Metal combustion mechanisms. It must be emphasized at the outset that there is no *fundamental* difference between oxidation and combustion, only one of degree. Chemically and thermodynamically, metallic oxidation is really just slow burning: slow enough that the metal can dissipate the heat generated to its surroundings, thus preventing a temperature rise. Ignition is a thermal instability that occurs when the heat generation rate from a chemical reaction exceeds the ability of the metal to dissipate it, thus leading to a thermal runaway. The steady-state combustion that follows ignition is a new equilibrium state — one in which the heat generation rate, which is now governed by a new set of rates and mechanisms, is once again just balanced by the heat lost to the surroundings. The critical difference between oxidation and combustion, and it is an important one, is the rate at which the chemical reaction proceeds. According to Kofstad (1966), during the linear, nonprotective oxidation of titanium at 1268 K (995° C), the rate of oxidation is 0.002 mg $O_2/cm^2 \cdot s$. The corresponding rate during steady-state combustion at 3250 K (2977° C) is 100 mg $O_2/cm^2 \cdot s$ (Strobridge, Moulder, and Clark, 1979). The combustion rate is thus 50 000 times faster than the oxidation rate.

Many of the distinctive features of metal combustion are a direct result of the stability of metal oxides. Their stability implies that the heats of formation for metal oxides are large, and so a great deal of energy may be released during metal combustion. But the stability of metal oxides also means that the products of combustion, as well as the fuel, are present in the reaction zone as condensed (solid or liquid) phases, which leads to several important consequences.

In the first place, very high temperatures are attained during metal combustion. During combustion of an ordinary hydrocarbon, the flame temperature is limited by thermal dissociation of CO_2. Dissociation of CO_2 becomes appreciable at temperatures of about 1700 K (1427° C) and effectively limits hydrocarbon flame temperatures to less than 2500 K (2227° C). Although most metal oxides decompose partially at elevated temperatures, the dissociation pressure of the base metal oxides remains very low at quite high temperatures. Thus, the upper limit on the temperature of a metal flame is the boiling point of the metal oxide. For metal oxides that evaporate nonstoichiometrically, the boiling point must be defined in terms of the equilibrium of the condensed oxide with its gas-phase decomposition products at the ambient pressure. The adiabatic combustion temperatures for several metals burning in oxygen at 0.1 MPa (1 atm) are listed in Table 10.2 (Grosse and Conway, 1958). It is obvious that flame temperatures in excess of 3000 K (2727° C) are quite common in metal combustion.

The presence of metal oxides in the reaction zone also means that the condensed-phase products of the reaction play a role in determining the rates and mechanisms of the combustion process. This is perhaps best illustrated in the criteria proposed by Glassman (1959) to predict the combustion mode for metals. If the boiling point of the metal oxide is higher than that of the metal, then the flame temperature will exceed the boiling point of the metal, the metal will evaporate, and combustion will occur in the vapor phase. On the other hand, if the boiling point of the metal oxide is lower than that of the metal, the flame temperature cannot reach the metal boiling point and surface combustion will result. Because vapor-phase combustion can be suppressed, either by the formation of a protective oxide layer or by reduction of the flame temperature below the metal boiling point by radiation

Table 10.2 Adiabatic combustion temperatures in oxygen at ambient pressure (Grosse and Conway, 1958).

Metal	Adiabatic Combustion Temperature (K)
Al	3800
B	2900
Ba	3000
Be	4300
Ca	3800
Cr	3300
Fe	3000
Hf	4800
Mg	3350
Mn	3400
Mo	3000
Pb	1800
Sn	2700
Sr	3500
Th	4700
Ti	3300
W	3200
Zn	2200
Zr	4800

losses, these criteria must be modified slightly. Thus, the boiling point of a metal oxide exceeding that of the metal is a necessary but not sufficient condition for vapor-phase combustion, while the opposite case of the boiling point of a metal exceeding that of its oxide is a sufficient but not necessary condition for surface combustion.

On the basis of Glassman's criteria, titanium, iron, steels, nickel, zirconium, and hafnium are predicted to burn on the surface, and magnesium, calcium, strontium, barium, and aluminum are predicted to burn in the vapor phase.

The presence of metal oxides in the combustion zone has a strong influence on the propagation of metal fires. Owing to the refractory nature of metal oxides, most of the heat of reaction will reside in the enthalpy of the products. In a surface-burning metal, the molten oxide products will collect on the surface where their heat content aids in melting more metal. This molten mixture of metal and oxide with a high heat content can then be spread to other parts of a structure by mechanical or aerodynamic forces and thus aid in spreading a fire rapidly.

Two other characteristics of metal combustion deserve mention: the very high radiant emission of burning metals and the frequent occurrence of fragmentation and explosions during combustion. The highly luminous character of metal flames is, of course, a direct result of their high temperatures; at temperatures in excess of 3000 K (2727° C), a significant fraction of the thermal radiation falls within visible wavelengths. Even metals that burn in the vapor phase have highly luminous flames owing to the presence of large numbers of condensed oxide particles that give the flames a high emissivity. This characteristic feature of metal flames is used to good advantage in one of the practical applications of metal combustion: the photographic flashbulb.

The violent fragmentation and explosions that often accompany metal combustion are not as readily understood. Explosions are most often observed during the combustion of small particles ($m < 1$ mg) of reactive metals like zirconium, titanium, tantalum, and aluminum. Small particles of these metals will explode shortly after being ignited in an oxidizing atmosphere. Nelson (1965) discovered that for zirconium droplets to burn explosively, nitrogen had to be present, either in the metal or in the oxidizing atmosphere. Presumably, similar considerations apply to the other metals. When larger quantities of metal are burned ($m > 100$ mg), they no longer explode violently, but rather "boil" vigorously, ejecting a continuous spray of small molten droplets that continue to burn and may, themselves, explode. This behavior is thought to be analogous to the "carbon boil" that occurs during the decarburization of iron with oxygen. In this case, however, the driving force for bubble formation is the supersaturation of the liquid iron with carbon monoxide.

10.2.2 Ignition sources

There are many possible ignition sources in cryogenic systems. Often the initiating event in accidents cannot be determined because of the severity of the ensuing fire. But in many accidents, particularly those in the space program, subsequent investigations have succeeded in pinpointing the cause. Below is a list of ignition sources that are known to have contributed to failures.

Impact. Heat generated by an impact can ignite materials in liquid or gaseous oxygen. A wide variety of circumstances can lead to an impact event: foreign objects left in an oxygen system during assembly, mechanical failure of internal parts, or even a tool dropped by a worker. Impact of debris, such as rust particles or welding slag from inadequate cleaning, can cause ignition at bends in high velocity tubing. In addition to causing heating, the impact of a foreign body may remove the protective oxide coatings from metals, exposing a fresh metal surface to the oxygen.

Friction. Heating caused by one solid rubbing against another can also cause ignition. This situation can arise in a number of diverse circumstances, from closing a valve too fast to mechanical failures, such as a bearing failure leading to rubbing of a rotating part against its casing. Friction, like mechanical impact, can remove the protective oxide films from metal parts, which is an additional hazard. Severe rubbing between metal parts can also produce small metal particles that are easily ignited (see paragraph below on "Sparks").

Adiabatic heating. Heating brought about by rapid compression of a gas can also cause ignition. This hazard is most severe in high-pressure gaseous oxygen service, but high gas pressures also can evolve in liquid oxygen systems from liquid boil-off. The problem arises most often where high-pressure gaseous oxygen is suddenly admitted to a low-pressure line that is closed at the other end. The maximum theoretical gas temperatures that can result from rapid pressurization may be estimated from the formula:

$$T_f / T_i = (P_f / P_i)^{(n-1)/n} \qquad (10.3)$$

where T_i and T_f are the initial and final gas temperatures, P_i and P_f are the initial and final pressures, and n is the ratio of specific heat at constant pressure to the specific heat at constant volume. Table 10.3 gives the maximum theoretical gas temperature attained by adiabatic compression of oxygen, initially at 0.1 MPa (1 atm) and 293 K (20° C), to various final pressures.

Fracture. Fracture of a metal part can also serve as an ignition source. Titanium and zirconium are known to ignite spontaneously in oxygen at room temperatures and below, following exposure of a fresh surface by breaking in tension (Littman, Church, and Kinderman, 1961). Apparently, breaking the specimen in torsion does not have the same effect. Whether ignition results from the release of kinetic energy, the exposure of a fresh, unoxidized metallic surface to oxygen, or both is unclear.

Electrical arc. An electrical discharge, caused by either accumulation of static electricity or arcing from powered electrical circuits, can have sufficient energy to ignite materials in oxygen. A great deal of caution must be used to prevent buildup of static electricity in liquid-oxygen systems, especially where high flow rates are used. Powered electrical circuits should be avoided if at all possible in oxygen environments. If they are absolutely essential, then careful insulating and grounding are mandatory.

Table 10.3 Heating of oxygen by adiabatic compression; initial conditions are 0.1 MPa and 293 K.

P_{final} (MPa)	T_{final} (K)
0.69 (100 psi)	506
5.5 (800 psi)	922
8.3 (1200 psi)	1036
13.8 (2000 psi)	1199
27.6 (4000 psi)	1455
34.5 (5000 psi)	1550

Gas resonance. In high-pressure-gas manifolds, frequently the gas flow path crosses one or more tubes that are closed off at a distant point. This configuration can result in spontaneous fluid dynamic oscillations in the cavity thus formed. Experiments on this phenomenon have shown that temperatures in excess of 1250 K may be reached at the base of such a resonant tube and that aluminum (in the form of powders and fibers) can be ignited at pressures less than 8.3 MPa (82 atm) (Phillips, 1975). The T-shaped configuration used in these experiments is common to many flow systems, and spontaneous oscillations must be considered a hazard.

Internal flexing. Seals and gaskets in both liquid and gaseous oxygen systems sometimes extend into the flow path of the fluid through faulty design or assembly. In such a configuration, the seal or gasket can be driven into oscillation by the flowing gas or liquid. Heating caused by this flexing can be sufficient to ignite seal or gasket materials, which generally have lower ignition temperatures than metal components.

Contaminant ignition. Hydrocarbon contaminants like acetylene are commonly found in air separation plants and can collect in liquid-oxygen systems. These contaminants have extremely low ignition temperatures in gaseous or liquid oxygen and can be ignited by the slightest shock. The problem is made worse by the fact that these contaminants settle to the bottom of the tank and can collect over several cycles of refilling. The energy released by the ignition of a contaminant can readily ignite other, more ignition-resistant materials, such as metals. For this reason, *complete* cleanliness of liquid or gaseous oxygen systems is required. Even concentrations on the order of several parts per million are hazardous because of the tendency of hydrocarbons to collect over long periods of time.

Abrasion. Solid particulates are another contaminant found in oxygen systems. Sand, rust, weld slag, and alumina from gas dryers are all found to varying degrees in gaseous oxygen systems. Although traps and filters are used to remove particulates from the flow stream, such measures are rarely 100% effective. Particulates thus become entrained in the flow and in many instances have caused ignition of oxygen piping through impact and abrasion. As expected, bends in oxygen piping are the usual sites of ignition. For this reason, most oxygen producers limit the gas velocity in oxygen piping and some use sections of ignition-resistant nickel-alloy pipe as firebreaks along straight-pipe sections, as well as at bends and corners. The gas velocity limits adopted vary from one organization to another and depend on the pipe material, operating pressure, and temperature. Typical velocity limits for carbon-steel pipe are listed in Table 10.4 (Schmidt and Forney, 1975).

Puncture. Puncture of a vessel by small objects, such as micrometeorites or bullets, is another ignition source. Here, ignition is initiated by the combined effects of the kinetic energy of the projectile and the exposure of a fresh metal surface to oxygen.

Sparks. Rubbing of moving parts or the abrasive action of hard particulates like sand or alumina in a flowing stream can remove particles of metal from internal surfaces. These small particles, by virtue of their large surface-to-volume ratios, can self-heat by oxidation to incandescence, as do the sparks formed during grinding (Komanduri and Shaw, 1975). If the particles formed are small enough, they may self-heat to the ignition temperature; this property of small metal particles is termed *pyrophoricity* (Bahn, 1965). Whether particles formed by abrasion self-ignite or only heat to incandescence, they are a serious ignition hazard if they can contact other materials with low ignition temperatures.

Shock waves. There are many possible sources of shock waves in cryogenic systems,

Table 10.4 Typical velocity limits for entrained particulates in oxygen with carbon-steel pipe (Schmidt and Forney, 1975).

Pressure (MPa)	Temperature (K)	Maximum Flow Velocity (m/s)	(ft/s)
1	448	46	150
1	423	53	175
1	393	61	200
2	448	23	75
2	423	26	85
2	393	30	100
2	338	40	130
4	393	15	50
4	373	18	60
4	338	20	65
6	373	9	30
6	339	10	35

from cavitation to "water hammer" (Schmidt & Forney, 1975). Large cryogenic systems, such as those associated with space vehicles, are particularly susceptible to shock waves from these sources and from additional causes, such as engine ignition or rough combustion, explosive bolts or valves, explosions of leaking gaseous propellants, and aerodynamic shocks from supersonic flight. Military cryogenic systems may be subject to the additional hazard of shock waves from hostile weapons fire. When a shock wave impinges on a surface immersed in liquid oxygen, it is partially reflected by and partially transmitted through the interface. The value of the peak pressure and temperature at the interface depends on the ratio of the acoustic impedances of the fluid and solid; for this reason the pressure and temperature at the interface can actually exceed the maximum pressure and temperature of the propagating wave. The acoustic energy incident on the interface is partially converted to heat in localized regions, which can quickly grow and initiate a reaction if the heat is not rapidly dissipated.

10.2.3 Compatibility test methods

Compatibility tests may be readily divided into two classes: tests that measure the *sensitivity* of a material to reaction with oxygen and tests that measure the *intensity* of the reaction. Occasionally the two methods are combined in a single test. Sensitivity tests are usually conducted by bringing the material to be tested into contact with liquid or gaseous oxygen and exposing it to an ignition source of known energy. The ignition source may be any of those discussed in § 10.2.2 as possible accidental ignition sources. The information obtained from a single, isolated sensitivity test is necessarily limited to whether or not a reaction occurred in a given specimen at a given energy for a particular ignition source. Because of this limitation, a statistical approach must be employed to determine the relative compatibility of various materials.

Sensitivity tests do not take into account the severity of the reactions initiated, which vary from explosive detonation to mild charring of the specimen. Clearly, this can be a decisive factor in determining the severity of the fire that ensues from an accidental ignition and should be a factor in determining the suitability of materials for use in oxygen systems. Thus, tests that measure the *intensity* of reactions initiated in oxygen are important for a complete characterization of material compatibility. Intensity tests include combustion calorimetry and measurements of burning rates, flame propagation rates, pressure, and light emission.

Compatibility testing is highly empirical in nature; at present no single theory is capable of predicting or explaining the host of phenomena that are observed. Perhaps this is only to be expected, given the enormous number of variables that must be taken into account and the wide variability in the types of materials and their physical properties. Any significant advance in the state of the art of compatibility research awaits improvements in our understanding of the fundamentals of ignition and combustion and more complete data on the thermophysical properties of materials of technological importance.

Apart from the difficulties engendered by the lack of a solid theoretical framework for compatibility research, there are a number of problems in the inherent accuracy and reproducibility of the test methods themselves and inconsistencies in the ranking of materials by different tests. For this reason it is quite important to use test methods and ignition sources reflecting the actual hazards that will be encountered in service to select a compatible material. For example, a gasket would not ordinarily be exposed to impact in an oxygen system, but could conceivably be subject to internal flexing or adiabatic compression. Thus, one should employ a compatibility test involving one of these two ignition sources to rank candidate materials.

Other difficulties with compatibility testing are the influence of specimen size and shape, unforeseen effects arising from incompatible combinations of materials, and many other design variables that have not been quantitatively assessed. Configurational testing appears to be the only means to assess such hidden variables, and in any critical application configurational tests must certainly be undertaken. Beyond this, a common sense approach to the start-up of any new device or apparatus that will be exposed to gaseous or liquid oxygen will go a long way towards assuring safe operation. No rapid pressurizations or decompressions should be attempted until the system has "burned in" at lower pressures. Start-up procedures and practical design guidelines are discussed in more detail later (§ 10.2.6).

A test to measure the compatibility of a material with oxygen basically consists of placing the material in oxygen in the presence of an ignition source and observing the sensitivity or intensity of the reaction. The tests are conducted under different conditions, such as different temperatures, pressures, or ignition energies, to determine any dependence on these variables. The kinds of tests possible are limited primarily by the number of possible ignition sources; however, most tests are quite realistic in simulating working conditions and common ignition sources. These tests are described below under the following, somewhat arbitrary, categories: impact, acoustic, thermal, electric arc, abrasion, and fracture. Also discussed are environmental and configuration tests that involve various forms of ignition in simulation tests.

There are many experiments designed to measure specific thermodynamic properties such as heat of combustion, thermal conductivity, specific heat, or melting point. Strictly speaking, these are not compatibility tests, but do provide a basis for understanding the results of compatibility tests.

Impact tests. One of the most probable ignition mechanisms is by an impact in the presence of oxygen, whether it is caused by a wrench dropped in an oxygen spill, a piece of weld slag impinging on a bend in an oxygen pipeline, or simply closing a valve. The prevalence of impact-initiated reactions has led to the development of a number of impact tests.

The most common impact test is the mechanical impact test described by Lucas and Riehl (1960). The original impact tester was based on a device first developed by personnel of the Army Ballistic Missile Agency at Redstone Arsenal to test the impact sensitivity of high explosives. This test has been used extensively and has resulted in an enormous amount of data (Key and Riehl, 1964; Key, 1964, 1966, 1968; Jamison, 1971; *Nonmetallic Materials Design Guidelines and Test Data Handbook,* 1977). It is one of only two ASTM standard tests for oxygen compatibility *(Annual Book of ASTM Standards,* 1970). The test is conducted by imparting a known amount of energy from a falling plummet to a striker pin that is in contact with the specimen in the presence of liquid or gaseous oxygen; any reaction is noted. The test is relatively simple and readily adaptable to variable pressure and temperature (a typical mechanical impact tester is shown in Fig. 10.1). Compatibility by this ASTM test is defined as no reactions out of 20 impacts with the plummet impact energy adjusted to 100 J. Key and Riehl (1964) indicated that this go–no-go test has been remarkably successful in rejecting incompatible materials. However, the arguments described in the next paragraph suggest that poor repeatability inherent in the test may lead to the rejection of some compatible as well as incompatible materials. Blackstone and Ku (1971) suggest that this test has also passed some unacceptable materials. The mechanical impact test procedure has been criticized by Blackstone (1970); Blackstone and Ku (1971); Burmeister, Loser, and Sneegas (1967); Jamison (1971); Reynales (1958, 1961); and Staph, Baber, Ku, and Weatherford (1962).

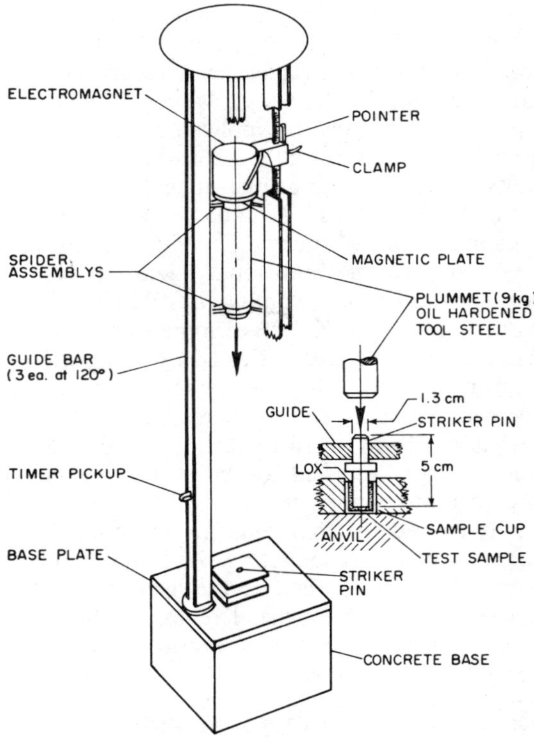

Figure 10.1 A common version of the liquid-oxygen impact sensitivity tester (Raniere, Hood, and Marker, 1972).

The criticism essentially suggests that the statistical procedure of the test is at fault rather than the test apparatus per se. Schwinghamer (1971) has refuted some of this criticism. A different statistical approach has been recommended by Jamison (1971) as a consequence of the criticism. Instead of accepting a material if it does not react in 20 tests at 100 J, one repeats the test 20 times at different energies to find the energy at which 50% reactions are obtained. This energy (determined from the drop height of the falling weight) is used to indicate the relative impact sensitivity of the material. This method is referred to as the Bruceton or "up-and-down" method of mechanical impact testing. According to Jamison (1971) the "up-and-down" method of impact testing, as an alternative to the go–no-go test, is both efficient and repeatable. That the Bruceton method is more repeatable than the go–no-go test is intuitively reasonable; however, the go–no-go test is less time-consuming and therefore is considered useful as a screening test if the occasional rejection of an acceptable material is permissible. Blackstone and Ku (1971) and Jamison (1971) give historical accounts of the development from the first impact tester described by Lucas and Riehl (1960) to current models designed for operation at high pressures (Raniere, Hood, and Marker, 1972).

As a result of recent attempts to improve the mechanical impact test, it has been suggested that material selection should be based on reaction intensity testing as well as reaction sensitivity testing. Jamison (1971) has pointed out that the improved repeatability of the Bruceton method of impact sensitivity testing shows that materials are not significantly different in impact sensitivity. However, materials differ widely in the intensity of reaction once ignited: some of them react violently and are totally consumed, whereas others are only slightly charred.

The particle impact test is used much less frequently than the mechanical impact test; however, its importance should not be neglected, especially in high-pressure-gas systems. This test is performed by impacting a test specimen with particles in a flowing oxygen stream and recording any resulting reaction (Nihart and Smith, 1964; Stradling, 1971). The test is conducted at various temperatures, pressures, flow rates (velocities), and particle sizes. The particle impact test is sometimes classified as an abrasion test because of the abrasive action of a stream of particles.

Acoustic tests. In acoustic tests, acoustic energy is imparted to the specimen in a manner that simulates actual events inside typical liquid or gaseous oxygen systems. Acoustic tests include the pneumatic impact test (or adiabatic compression test), explosive shock tests, and resonance tube ignition. The pneumatic impact test is designed to simulate the adiabatic compression resulting from rapidly opening a high-pressure valve (Pippen and Stradling, 1971). It is usually performed in just that way, with a small specimen-containing test chamber suddenly opened to a high-pressure reservoir. The adiabatic compression test uses the highest pressures of all the compatibility tests, except for a variation of this test that uses explosively generated shock waves. Pneumatic impact tests study material ignition in the compressively heated oxygen gas (see § 10.2.2). Oxygen compressed to 71 MPa (700 atm) can attain a temperature of 1900 K, which is sufficiently high to ignite almost any structural material. Whether ignition actually occurs is determined by the temperature of the material and not just the temperature of the compressed gas. Thus, the transient thermal characteristics of the system are important. The thermal diffusivity, heat capacity, and other factors, such as the existence of sharp edges, determine the peak temperature of the material. For these reasons an adiabatic compression test is most meaningful if performed on a system rather than on a specimen of material.

One recently developed test method that is related to the pneumatic impact test is the resonance tube test developed by Phillips (1975) and used by him to study ignition of aluminum powders and fibers. In this test a sonic flow of oxygen is directed into a closed resonance tube containing the material to be tested. Acoustic oscillation within the cavity heats the gas and specimen to temperatures in excess of 800 K. The test is conducted at varying pressures and flow rates and any resulting ignition or explosion is noted.

Very high dynamic pressures may be achieved in explosive shock tests, where the specimen is exposed to the shock wave created by setting off a small explosive charge nearby. Transient pressures in excess of 800 MPa (8000 atm) can be generated at the specimen–liquid-oxygen interface by this means. The simplest procedure (Riehl, Key, and Gayle, 1963) is to place a thin strip of the material to be tested in a stainless steel cup filled with liquid oxygen. A small explosive charge with a detonator is then taped to the outside of the cup and detonated; the amount of charge necessary to produce reactions 50% of the time is taken to be a measure of the shock sensitivity of the material. The reproducibility of results using this technique is about 20 to 30%.

Another technique, described by Bell, English, Murata, Pohl, and Tiner (1971), employs a shock horn to transmit the explosively generated shock wave to an insulated reaction cavity containing the specimen and liquid oxidizer. The intensity of the incident shock wave, which is measured by strain gages mounted on the shock horn, can be controlled by the size of the explosive charge and by placing various numbers of plastic cards between the charge and the end of the shock horn. This technique has the advantage of avoiding extraneous effects that are sometimes caused by exploding a charge in close proximity to the specimen, such as direct impingement of metal particles and chemical reaction between the oxidizer and gases from the explosive.

Thermal tests. Thermal tests are those in which the ignition source is primarily thermal in nature. Some tests in this category involve uniform heating of the specimen (and sometimes the environment as well), as in the oxygen-bomb test, the pot test, and the hot-wire test. Other tests in this category involve exposing the specimen and any vapors evolved to an intense thermal energy source such as a flame or an ignition pill made of a combustible solid, as in promoted ignition tests or the moving wire test.

The oxygen-bomb test, described by Nihart and Smith (1964) and Guter (1967) is performed as follows: a specimen is placed in an oxygen-filled bomb. The temperature of the contents of the bomb is gradually raised until ignition occurs, as indicated by a rapid rise in temperature, visible flash, or audible explosion. The temperature at which ignition occurs is the *autoignition temperature*. The pressure in the chamber during the controlled temperature rise either remains constant or is allowed to rise with increasing temperature. In some cases the specimen is exposed to oxygen only after equilibrium has been reached at each temperature; the specimen is changed after each such measurement. The oxygen-bomb test can be used to determine gaseous oxygen compatibility of metals or nonmetals in the solid, liquid, or gaseous states.

The pot test described by Guter (1967) is similar to the bomb test except the pot test is done at standard pressure and with flowing oxygen. It frequently is intended to simulate actual conditions of use, and with it one can assess flow effects.

The hot-wire test is conducted by heating a metal wire in a gaseous oxygen atmosphere (Grosse and Conway, 1958). An electrical current is passed through the wire, heating it by the Joule effect. The temperature is increased with increasing current, resulting eventually in the wire either igniting or melting. This test can only be used on metals that ignite below their melting temperatures; its principal advantage is its simplicity. Because of this simplicity, electrical resistance heating is frequently used to study the ignition of metals in several other forms besides wires, including foils, ribbons, and tubes (Dean and Thompson, 1961; Mellor, 1967; Laurendeau and Glassman, 1971). In one variation of the hot-wire test, the specimen is heated in a vacuum or inert gas atmosphere before oxygen is admitted to the system. This procedure prevents the buildup of a thick oxide film on the specimen during heating and may result in lower observed ignition temperatures.

Promoted ignition tests are performed by burning an ignition-promoting material in the presence of a specimen (Nihart and Smith, 1964). Many materials have been used for promoters; however, the best materials are those that do not significantly affect the composition of the specimen surface as the promoter is burned. Contamination of the surface of the specimen by the promoter can change the ignition characteristics of the specimen and thus produce erroneous results. The amount of specimen consumed by the burning of a fixed amount and type of promoter or the amount of promoter required to consume the entire specimen is used as a measure of its resistance to combustion.

The moving-wire test is a newly developed method for studying the ignition and combustion behavior of polymers (Fristrom, Hunter, and Grunfelder, 1976; Hunter, Schacke, Grunfelder, and Fristrom, 1976). In this test the polymer is coated on a supporting metal or fiberglass strand and then moved at a controlled speed through a heat source, usually a flame. The minimum surface temperature at ignition may be determined in this way with a reproducibility of better than 5%. The technique may also be combined with mass spectrometry to determine the qualitative chemical composition of the pyrolysis products and the sequence of their evolution as a function of surface temperature (Schacke, Hunter, Grunfelder, and Fristrom, 1977).

Electric arc tests. Many oxygen accidents have involved electrical ignition sources, such as arcing from electronic equipment. The flash and fire point test, used primarily on organic materials, is performed to simulate these conditions by using an electric arc to ignite a specimen in an oxygen-filled chamber (Pippen and Stradling, 1971). The temperature of the chamber is raised and the electric spark is generated periodically near the specimen. The temperature at which a momentary ignition occurs is the *flash point*; the temperature at which sustained burning occurs is the *fire point*. At high oxygen pressures the flash and fire points coincide. The relationship of flash and fire points to autoignition temperatures (see oxygen-bomb test described under Thermal tests) is not known. Intuitively, one would expect the autoignition temperature to be equal to or greater than the flash and fire points; the experimental data indicate that at high pressures there is little difference between these temperatures.

Abrasion tests. Galling caused by two pieces of metal rubbing was suggested as one of the possible causes of an oxygen-tank-truck accident, and there have been a few abrasion tests to study such phenomena (Jackson, Boyd, and Miller, 1963; Bauer, Wegener, and Windgassen, 1970; Bauer, Klein, Wegener, and Windgassen, 1971; Riehl et al., 1963). Also, the abrasive effect of fine particles in a flowing oxygen stream has been studied (Nihart and Smith, 1964; Stradling, 1971; Bates, Wren, Monroe, and Pears, 1976).

Fracture tests. A highly probable yet seldom observed accident source is fracture. The exposure of a fresh clean surface of certain metals to oxygen can cause a violent reaction. Various methods have been used to study this phenomenon, from straight tensile tests to puncture by bullets or other projectiles (Littman et al., 1961; Jackson et al., 1963; Riehl et al., 1963). Even the effect of the projectile size and material has been considered.

Configuration tests. The final test of oxygen compatibility for any material is whether it will ignite in its final configuration, complete with adjacent materials and possible contaminants. Because of our lack of understanding of the ignition and combusion of metals, most critical oxygen systems also undergo an environmental test. Artificial ignition of wire harnesses or clothing, hyperbaric testing of breathing apparatus, and overloading liquid-oxygen pumps are all examples of this test category (Schmidt and Forney, 1975). These tests will always be indispensable to safety.

10.2.4 Relative compatibility of metals

Virtually every metal will react with oxygen at some temperature and pressure. Of all the metals only gold, silver, and platinum are not expected to support combustion by virtue of their low heats of combustion and the high dissociation pressures of their oxides at modest temperatures. All of the remaining metals will ignite and burn in the presence of oxygen, yet some are consistently more difficult than others to ignite or to keep burning. It is the goal of compatibility testing to quantify this rather qualitative empirical observation; ideally, one would hope to take the results of all the various compatibility tests and compile a list of acceptable materials, ranked according to their relative compatibility with oxygen. Unfortunately, the ranking one arrives at varies according to the test method used and the environmental conditions during the test.

A more practical approach is to compile rankings for each different test method; the relative position of a given material will then be different for different tests. To choose a material for a particular application, one must use the results of test methods that come closest to reproducing actual conditions and the most probable ignition sources. McKinley (1971) has used this approach with the results of eleven different test methods, ranging from electrical ignition of wire specimens to configurational tests of rubbing in liquid-oxygen pumps. To illustrate the degree of variability that can result from different test methods, aluminum ranked first in the compatibility for electrical ignition of wires, but in a promoted ignition test it ranked last.

Although it is not possible to unequivocally rank all the metals and alloys for compatibility with oxygen, it is possible to discuss the general behavior for groups of materials in the presence of oxygen to provide some guidance where there is a wide range of choice. Some general guidelines on the compatibility of important classes of alloys are described below. Relevant thermophysical properties of the base metals and a few selected alloys are given in Table 10.5 (Hust and Clark, 1973). The dependence of metal burning rates on thickness is illustrated for several metals in Fig. 10.2 (Schmidt and Forney, 1975).

Nickel and copper alloys. Nickel and copper alloys are generally the most compatible metals available for oxygen use. The high-nickel-

Table 10.5 Thermal data on metals and alloys (Hust and Clark, 1973).

Materials	Ignition Temperature (K)	Melting Temperature (K)	Boiling Temperature (K)	Melting Temperature of Oxide(s) (K)	Boiling Temperature of Oxide(s) (K)	Thermal Conductivity at 300 K (W/cm·K)	Thermal Diffusivity at 300 K (cm^2·s)	Heat of Combustion (J/g)	Oxide Volume Ratio
Elements									
Aluminum	>1000	933	2720	2323	3800	2.2	0.91	31 000	~1.3
Beryllium	~2500	1356	2750	2820	~4300	1.6	0.44	68 000	0.7[a]
Copper	1300	1360	2855	1503–1600	—	4.0	1.15	2 400	1.71
Gold	does not ignite	1336	2980	—	—	3.1	1.24	8	—
Iron	1200	1812	3160	1693–1870	—	0.8	0.23	7 000[a]	2.23
Magnesium	900	923	1390	3075	3350	1.6	0.90	25 000	0.81
Manganese	—	1517	2314	2058	3400	0.23	0.063	7 000	1.25
Molybdenum	1023	2890	5100	1070–2880	1530	1.4	0.63	6 100	~3.3
Nickel	1730	1730	3110	2230	—	0.6	0.15	4 000	1.64
Silver	does not ignite	1233	2450	—	—	4.0	1.61	140	—
Tantalum	1550	3270	5700	2073	—	0.6	0.26	5 600	2.54
Tin	1140	505	2960	1353–2200	—	0.6	0.39	2 400	1.32
Titanium	~1600	1950–2070	3550	1920–2400	3000–3300	0.18	0.076	16 000[a]	1.73
Zinc	1170	693	1181	2250	—	1.25	0.46	5 300	1.44
Zirconium	1670	2125	4650	2960	—	0.43	0.23	12 000	1.56
Copper Alloys									
Berylco 25	1133–1144	1144–1255				0.24–0.38	~0.38	—	
Brass 70/30	1273	1183				1.26	0.39	3 600	
Nickel Alloys									
Inconel	1620	1670				0.11	0.040	4 700	
Hastelloy	1600	1640				0.11	0.040	5 000	
Monel	1520	1600				0.25	0.066	3 400	
Iron Alloys									
Cr steel	1420	1670				0.30	0.085	7 700	
Stainless Steels (300–400)	1400–1670	1670				0.15	0.043	8 000	
Carbon steel	1370	1780				0.80	0.23	7 500[a]	
Titanium Alloys									
Ti-6A1-4V	~1600	1803–1908				0.066	0.027	16 000	
Ti-8Mn	1610	1775				0.11	0.051	16 000[a]	

[a]Approximated.

content alloys and nickel–copper alloys are difficult to ignite and even more difficult to keep burning. This behavior is probably due to low heats of combustion and high thermal diffusivity. Nickel–copper alloys are being used more and more for safety in high-pressure applications in spite of the added cost.

Stainless steels. Those stainless steels with high nickel content are generally better for use with oxygen. Austenitic steels are better than martensitic. Low-carbon and low-alloy steels behave more like iron, which is more readily ignited and burns faster but occasionally will quench itself with slag. The stainless steels, as a class, have relatively high ignition temperatures and some tests show that they quench if they do ignite. With their excellent mechanical properties the stainless steels are frequently the best compromise between cost and safety.

Aluminum alloys. Aluminum alloys are widely used in oxygen systems and have an impressive safety record. Several accidents have involved aluminum burning but the cause of the accident has usually been ignition of another material. Aluminum, however, may be classified with the other light metals, which have relatively low ignition temperatures, high heats of combustion, and violent fragmentation when they do burn. Aluminum alloys will frequently melt before igniting and can flow away from the heat source, which would hinder any reaction. It could stop the reaction, but may cause hot,

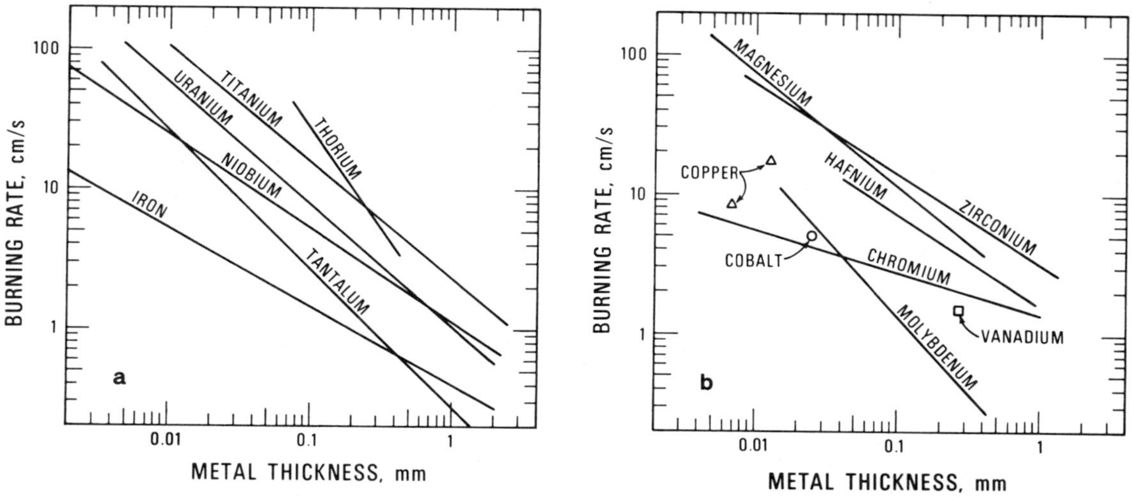

Figure 10.2 Horizontal burning rate for several metals in pure oxygen at atmospheric pressure (Schmidt and Forney, 1975).

molten aluminum to flow to another part of the system. The low cost and weight per unit strength of aluminum alloys, however, have led to their common use in oxygen systems.

Titanium alloys. These light, strong alloys are commonly used in aerospace applications but are very reactive with oxygen and are seldom used with it. A fresh surface of titanium will react violently and its alloys pass few of the compatibility tests. These facts coupled with a relatively low ignition temperature and high heat of combustion virtually exclude it from use in oxygen systems.

Other metals. Nearly all other metals suffer from low ignition temperatures and ready combustion once ignited. Basically, all of the light alkali metals, the transition metals (except copper, nickel, and perhaps cobalt), the rare earths, and the transuranium elements have shown undesirable behavior with oxygen. Notable exceptions are the noble metals, which are virtually inert but limited by their high cost and low strength.

10.2.5 Relative compatibility of nonmetals

There are many nonmetallic compounds that are compatible with oxygen: namely, those compounds that are already fully oxidized (or highly fluorinated). For example, most ceramics and glasses (including fiberglass) are fully compatible, although their physical and mechanical properties often make these materials unsuitable for structural applications. As a practical matter, oxygen systems often require nonmetals for gaskets, seals, insulation, and lubricants. For these applications a highly fluorinated polymeric material is usually chosen.

Polytetrafluoroethylene (TFE) and polychlorotrifluoroethylene (CTFE) are the solid polymers most widely used for gaskets and seals; oils and greases based on TFE and CTFE are also available. Yet even these highly fluorinated materials pose compatibility problems. Their ignition temperatures are significantly lower than most metals, and their low thermal conductivity and heat capacity make them easier to ignite. Compared with metals, these materials are highly combustible. The minimum ignition temperatures for a number of nonmetals are shown as a function of pressure in Fig. 10.3 (Schmidt and Forney, 1975). Note that TFE and CTFE have the highest reported ignition temperatures; silicone rubber has the lowest.

To offer some guidance in choosing compatible nonmetals, materials that are generally acceptable for liquid-oxygen systems are listed below by application. A fuller account of oxygen-systems design may be found in Schmidt and Forney (1975); detailed accounts of compatibility testing for nonmetals are given by Guter (1967) and Nihart and Smith (1964).

Seals, gaskets, valves, O-rings. One of the many available formulations of TFE or CTFE is the preferred material for secondary seals, gaskets, and valve seats. These materials may also be

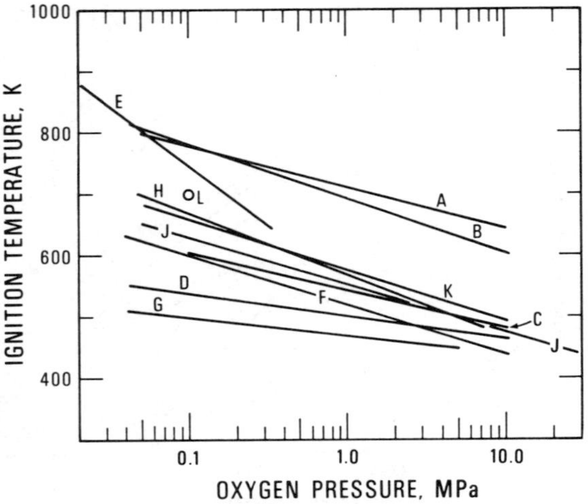

Figure 10.3 Minimum ignition temperatures for several nonmetals as a function of oxygen pressure (Schmidt and Forney, 1975). Materials: A — polytetrafluoroethylene, B — polychlorotrifluoroethylene, C — polyamide, D — polychlorobutadiene, E — polymethylmethacrylate, F — fiberglass-filled polytetrafluoroethylene, G — polyethylene, H — polyvinylchloride, J — vinylidene fluoride-hexafluoropropylene copolymer, K — polysiloxane, L — fluorinated ethylene propylene.

obtained with various fillers, such as glass fibers, nickel powder, or molybdenum disulphide. Fluorinated ethylene propylene copolymer (FEP), vinylidene fluoride, and copolymer of vinylidene fluoride and hexafluoropropylene have also been used effectively. For high-pressure or high-flow-rate applications the seal or valve seat must be recessed in a high-thermal-conductivity metal to reduce the possibility of ignition.

Lubricants, oils, and greases. There are several TFE lubricants and CTFE greases and oils available for use in oxygen systems. Such lubricants should be used sparingly in low-pressure, low-flow applications and not used at all in high-pressure or high-flow systems. Tricresylphosphate has been used successfully in vacuum pumps for oxygen.

Insulations. Since insulating materials will not usually be in direct contact with oxygen, any hazard due to incompatibility is less severe. Yet the possibility of oxygen leaks into the insulation space of a vessel make it highly desirable to use nonflammable insulation. Most conventional insulations are not readily ignitable, but they will sustain combustion once ignited. For this reason, one of the insulations based on sintered silica or volcanic glasses seems preferable. Fiberglass insulations are also nonflammable, of course. Schmidt and Forney (1975) report that up to 20% aluminum powder may be added to powdered insulation before combustion may be sustained, but this figure probably depends on aluminum-particle diameter and composition of the insulation. For this reason, any form of insulation using metal powder should be subjected to proof testing.

10.2.6 Design guidelines

To assist in choosing a safe material, some practical considerations and guidelines are outlined below. Obviously, one should first select materials that physically fulfill the mechanical requirements of the component in question. These materials can then be analyzed to select the most compatible material. This selection process may be based upon the material ranking given by a particular compatibility test or it may be based upon the consensus of the material-ranking recommendations found in the literature, as is frequently advocated. However, both of these strategies suffer from two serious drawbacks: (a) no consideration is given to the unique demands of a particular application and (b) the material-ranking recommendations based upon the results of individual investigations often are highly contradictory, owing to the specialized nature of most tests and the biased interpretation of the results. A case in point is the wide disparity in the ranking of aluminum and stainless steel reported by various investigators (Dean and Thompson, 1961; Nihart and Smith, 1964; McKinley, 1971).

The material selection process should be accomplished by matching the particular demands of the equipment component with the materials best satisfying these demands. Material selection on this basis assumes the existence of several types of compatibility and physical properties data for these materials. For example, if a component is likely to be impacted but not likely to be in a high-temperature environment, major consideration should be given to materials with a low impact sensitivity before considering their ranking according to ignition temperature. The importance of equipment design should not be ignored. For example, the introduction of slow-opening valves and heat sinks will reduce the probability of ignition by adiabatic compression.

Another way of analyzing relative compatibility is to define a compatibility index that is dependent on the physical properties of materials. Variable weighting factors for each property would be based on the particular application. For example, the compatibility index could be defined as being proportional to ignition temperature, thermal diffusivity, impact energy necessary to react, and mass, and inversely proportional to the heat of combustion, operating temperature and pressure, and flow velocity. There would have to be several weighting factors for different systems to evaluate potential ignition sources, risks, exposure to oxygen, damage effects, and other factors. It cannot be overstressed that such a compatibility index is arbitrary, but may well suffice for comparing different materials for a given specific application.

The following guidelines, listed according to priority, can be followed when selecting materials for oxygen use in the presence of ignition sources (of course any considerations must include the effects of possible contaminants in the system):

a. Eliminate ignition — select the material that is least likely to ignite under the operational conditions.

b. Prevent continued reaction — select a material that tends to quench the reaction after ignition.

c. Reduce the rate of reaction — select materials that react as slowly as possible after ignition.

The first guideline is most likely to be satisfied by selecting the material with highest ignition temperature, but it is also affected by the presence of oxide coatings or the possibility of intermaterial reactions. For example, it has been reported and discussed by Key (1964, 1968) and Keeping (1971) that chlorofluoro compounds may react strongly with aluminum in the presence of oxygen. Iron rust, according to Keeping (1971), significantly lowers the ignition resistance of aluminum in oxygen. Materials with a high thermal diffusivity are better in dynamic systems since localized heating in them results in less pronounced hot spots than in materials with low diffusivity. In a static system, high thermal conductivity materials are more desirable for essentially the same reason. Materials whose melting point is higher than the ignition temperature should be free of sharp edges. The presence of sharp edges may result in hot spots for ignition. If the ignition temperature is sufficiently above the melting point, sharp edges will probably be smoothed by melting before ignition occurs. This has been reported to be the case for aluminum by Keeping (1971). At high pressures the prevention of ignition should be the principal goal because any reaction will probably be catastrophically destructive.

The second guideline, selecting a material that tends to quench itself, is influenced by several factors. The heat of combustion is the most important. If it is sufficiently low, the heat may be conducted away rapidly enough to quench the burning. The reaction products may also interfere with continued oxidation. For example, slag buildup has been reported to have a quenching effect in iron.

The materials portion of the third guideline, select slow-burning materials, is affected by essentially the same factors. However, one can also utilize good design to slow down the reaction. For example, one might insert a nickel–copper-alloy pipe section in a predominantly stainless steel pipe. The nickel alloy section, having a lower reaction intensity, would either quench or slow a reaction propagating along the pipe. The shape and size of components at strategic points in a system can be significant in controlling a reaction. Thermal anchoring of potentially reactive components to a cold heat sink can also be useful in curtailing a burn. A fast-closing valve to shut down the oxygen supply in the event of ignition may prevent extensive damage, unless the initial burn occurs with explosive violence. Limiting the oxygen availability by restricting oxygen flow where possible is desirable for control purposes.

10.3. COMPATIBILITY OF MATERIALS WITH LIQUID FLUORINE

Fluorine is the most powerful oxidizer among the elements; it will react with nearly all substances, both organic and inorganic. The few exceptions to this rule are the light inert gases and fluorinated compounds already in their highest oxidation state. So high is the oxidation potential of fluorine that some substances ignite spontaneously upon contacting it; in particular, carbon, water, hydrogen, and most hydrogen-bearing carbon compounds exhibit this behavior to some degree. This property, termed hypergolicity, can be quite useful in rocket propulsion, where it allows multiple restarts without any ignition system.

Interest in the use of fluorine as an oxidizer for

rocket propellants first led to experimental studies of the compatibility of materials with fluorine and fluorine-oxygen mixtures (Schmidt, 1967).* Several early experiments with fluorine as an oxidizer for rocket fuels experienced failures involving the combustion of the fluorine container materials. As is often the case with liquid-oxygen accidents, the damage was usually so extensive that the exact cause of failure could not be determined. However, the research and compatibility testing that followed these early failures led to successful long-term operations of liquid-fluorine facilities as well as numerous test firings of rocket engines burning fluorine and various rocket fuels. Subsequent research and operational experience have shown that most of the early failures can be attributed to contamination of the fluorine system. Much of this research was conducted by NASA's Lewis Research Center and is summarized in Schmidt's book (1967), which is the most complete reference available on handling liquid fluorine.

Despite its promise as a rocket propellant, fluorine never advanced past the test-firing stage, largely owing to the special handling that it requires. Of paramount concern is the toxicity of fluorine and hydrogen fluoride (HF), the major combustion by-product when fluorine burns with hydrogen-bearing fuels. The toxicity of fluorine compels an even greater concern with material compatibility, since a breach of the container could be extremely hazardous over a much larger area than would be affected by the blast wave that might accompany a sudden failure. For this reason it is imperative that those considering the use of fluorine (either gas or liquid) thoroughly acquaint themselves with the personnel safety precautions that must be taken, such as those outlined in Schmidt (1967).

Above all, the use of fluorine requires scrupulous attention to the cleanliness of the system and to devising and following operating procedures that will preclude contamination by water or by hydrocarbons. Of the two, water is much more difficult to exclude, especially when working with liquid fluorine. A small leak can allow atmospheric gases containing water vapor to condense and freeze inside a liquid fluorine system, where the ice may come into contact with the fluorine and explode. "The major cause of fluorine system failures," according to Schmidt, "has been water, in the form of moisture, in the atmosphere or contained in the pressurant or purge gases and inadvertently introduced into a liquid fluorine or FLOX system." Even very small concentrations of water, which pose no ignition hazard, may cause excessive corrosion of metal parts by the formation of HF.

*Mixtures of liquid fluorine and liquid oxygen, often called FLOX, have compatibility characteristics similar to liquid fluorine. Concentrations for FLOX mixtures are expressed as percent fluorine by weight.

10.3.1 Research on materials compatibility with liquid fluorine

Compared with the diversity of test methods and the number of test results that are available for assessing compatibility with oxygen, the data base on fluorine compatibility is rather meager. To a large extent this reflects the more limited use of fluorine for industrial and technological applications as compared with oxygen. Although most of the possible ignition sources for liquid-oxygen systems apply as well to liquid-fluorine systems, only a few of the many test methods developed to study oxygen compatibility have been applied to fluorine: notably, ignition-temperature measurements, the mechanical-impact test method, and rupture of metals under fluorine to expose a clean metal surface. There have also been many studies of fluoride-film formation on metals. By far the most common test method has been simple exposure to liquid or gaseous fluorine — at various temperatures, pressures, and flow rates. Fluorine is so reactive that many materials exhibit incompatibility (either excessive corrosion or spontaneous combustion or detonation) under these conditions.

Sources for data on compatibility of materials with fluorine are relatively few in number. Aside from Schmidt's excellent book, which is the main source of data for this section, the *Design Handbook for Liquid Fluorine Ground Handling Equipment* (1965) contains a great deal of useful information.

10.3.2 Compatibility of metals with liquid fluorine

Just as some metals exposed to oxygen form a tough, protective oxide film, metals exposed to fluorine usually form a compact, protective fluoride film a short time after being exposed to fluorine. For some metals, such as aluminum, the fluoride film is so tightly bonded to the metal that it is considered to be in the surface of the metal rather than on it. If an oxide film is already present on aluminum, fluorine will replace it.

Thermodynamically, this can be viewed as a consequence of the higher oxidation potential of fluorine; AlF_3 has a lower free energy than Al_2O_3 and is therefore more stable thermodynamically. The actual mechanism for the displacement of the oxide film, while unknown, is thought to be either the diffusion of fluorine through the oxide film followed by reaction with the metal, or by direct exchange of oxygen, or by some combination of the two. The growth of fluoride films on aluminum substrates exposed to gaseous fluorine at various temperatures is shown in Fig. 10.4 (Kleinberg and Tompkins, 1962). It is interesting to note that at the highest temperature of exposure (357 K) the limiting value of film thickness was smallest. No explanation was given for this effect.

The formation of fluoride films on a number of other structural metals at 193 K is shown in Fig. 10.5 (Kleinberg and Tompkins, 1962). In general, it can be seen that aluminum, brass, nickel, copper, and Monel reach a limiting film thickness in shorter times than the steels; also the fluoride film thicknesses for these metals are only 1/2 to 1/10 the values for fluoride films on steels.

It is often stated or implied in the literature that the formation of a stable, tenacious fluoride film protects the metal from further corrosion when it is exposed to liquid fluorine. Yet there have been some experiments that seem to dispute this claim. Sterner and Singleton (1960) exposed copper and magnesium-alloy specimens to liquid fluorine while they were being brushed by stainless steel bristles with sufficient force to disrupt any fluoride film. The test results showed that corrosion did not increase during abrasion, suggesting that fluoride film protection does not control corrosion for these metals in liquid fluorine.

Metals may be divided into three classes on the basis of the characteristics of their fluoride films: (a) metals that form tough, protective fluoride films, such as aluminum, copper, nickel, magnesium, and Monel; (b) metals that are less stable and form thicker films at a greater rate than those metals in (a), such as iron, mild steels, and titanium; (c) metals that do not form a protective film, such as lead, tin, and chromium, which forms a volatile tetravalent fluoride above 423 K. For critical applications rhodium, palladium, and platinum may be used since they are practically unaffected by exposure to fluorine or hydrogen fluoride at room temperature.

As the temperature of exposure to fluorine increases, the point is reached where the fluorination rate begins to increase with time of exposure. At these higher temperatures the rate of chemical attack by fluorine may become more severe. The maximum temperatures at which various metals

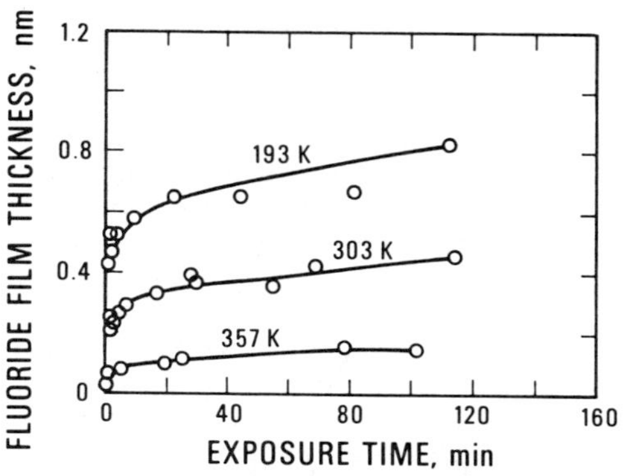

Figure 10.4 Variation of fluoride film thickness on aluminum with exposure time and fluorine temperature (Kleinberg and Tompkins, 1962).

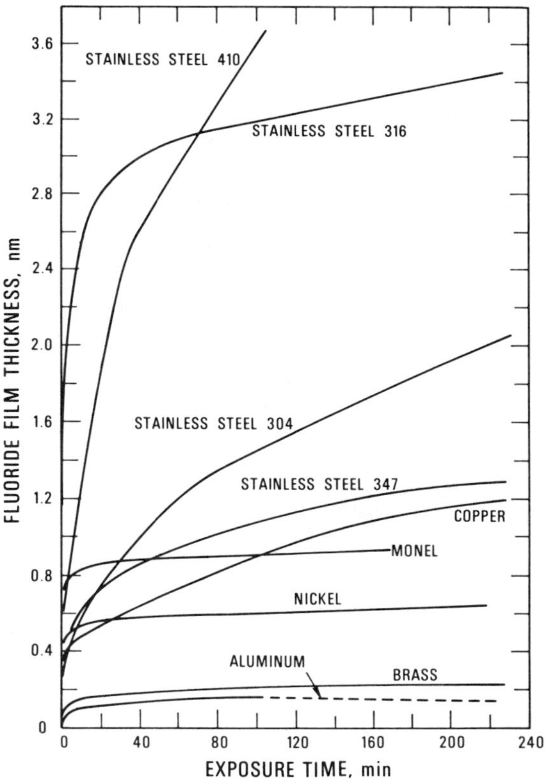

Figure 10.5 Variation of fluoride film thickness on metal powders exposed to fluorine gas at 193 K (Kleinberg and Tompkins, 1962).

can be exposed to fluorine and still remain practically unaffected are included in Table 10.6 (Schmidt, 1967).

It must be remembered that any moisture present in the system can form hydrogen fluoride and destroy the protective nature of the fluoride film on metals. Thus, one may wish to consider the resistance of a metal to corrosion by hydrogen fluoride as well as fluorine.

If heated past the point where its fluoride film is protective, a metal will ignite in fluorine, just as it will in oxygen, when the heat generation from the fluorination reaction exceeds the rate of heat dissipation by radiation, convection, and conduction. In general, the ignition temperatures of metals in gaseous fluorine are lower than the ignition temperatures in oxygen, as would be expected for a more reactive element such as fluorine. Ignition temperatures for several common metals of construction are collected in Tables 10.7 and 10.8 (Schmidt, 1967). In the former experiment, specimens were brought to temperature while exposed to the fluorine gas. In the latter experiment, the specimens were brought to temperature in an evacuated bomb and then the fluorine was admitted. Specimens suddenly exposed to fluorine while at temperature ignited after a variable induction period. Plots of induction time vs. temperature then yielded the activation energies for the reaction.

Although ignition temperatures for metals are lower in fluorine than in oxygen, they are still sufficiently high that most metals of construction

Table 10.6 Resistance of materials to chemical attack by fluorine (Schmidt, 1967).

Material	Maximum Temperature Material Remains Unaffected (K)	Temperature for Various Degrees of Chemical Attack		
		Slight (K)	Moderate (K)	Severe (K)
Carbon steel	644	—	—	772
Stainless steel 30%	—	—	—	478–533
Chromium plate	423	—	—	—
Monel	—	—	—	772–866
Inconel	672	823	—	1023
Copper	—	772–866	—	972
Aluminum	728–772	—	866	—
Nickel	866	—	972–1023	—
Soda glass	423	—	—	—
Borosilicate	473	—	—	—
Silica	523	—	—	—

Table 10.7 Metal ignition temperatures in fluorine.[a]

Metal	Wire Diameter (mm)	Average Ignition Temperature (K)	Maximum Deviation from Average Temperature (%)
Aluminum	0.25	>melting	—
	0.41	point	
Copper	0.31	965	8.0
Iron	0.36	945	0.8
Molybdenum	0.38	478	8.3
Nickel	0.20	1435	6.0
	0.39		
Tungsten	0.39	556	18.0
Monel	0.25	669	12.0
Stainless steel	0.51	954	13.0

[a]Wire specimens were resistively heated in a gaseous fluorine atmosphere until ignition occurred. Temperatures were calculated from resistivity–temperature data (Schmidt, 1967).

may be exposed to static or flowing liquid fluorine without danger of ignition as long as no external energy is added to the system. Such a perturbation as might arise from an impact or a rub could easily raise the temperature of the metal to the ignition point, however.

The susceptibility of titanium and aluminum to ignition by mechanical impact in liquid fluorine is illustrated in Table 10.9. In this experiment, a mechanical impact tester was modified for use with liquid fluorine (Sterner and Singleton, 1960) and a number of tests run with several configurations of strikers. Aluminum suffered only one reaction in 12 tests. Titanium, known to be incompatible in oxygen systems, exhibited a much higher reaction frequency. The surprising aspect of these results is that ignition never became general in tests with liquid fluorine, whereas during tests conducted in the same apparatus using liquid oxygen for comparison, one titanium specimen did sustain essentially complete combustion. Indeed, for titanium in liquid oxygen, it is well known that general ignition and combustion to completion can be expected to occur (see § 10.2.4)

Another interesting result relating to the relative severity of combustion in oxygen and fluorine came about in experiments on tensile rupture of titanium in liquid fluorine. Although titanium readily ignited and burned when pulled to failure in tension in liquid oxygen (Littman et al., 1961), the same experiment performed in liquid fluorine never resulted in ignition (Richards and Hanson, 1963). There appears to be no ready explanation for this behavior.

Table 10.8 Ignition delay times for metals in fluorine.[a]

Metal	Wire Diameter (mm)	Maximum Wire Temperature (K)	Ignition Delay (s)	Ignition Temperature Range (K)	Activation Energy (J/mol)
Copper	0.30	1178	0.8	962–974	39.5
		1125	1.0		
		1083	0.6		
		1040	0.8		
		974	1.2		
		962	No ignition		
Iron	0.36	1003	1.0	891–917	16.3
		949	1.6		
		921	2.0		
		917	2.2		
		891	No ignition		
Nickel	0.38	1630	0.6	1526–1539	—
		1579	1.2		
		1539	0.6		
		1526	No ignition		

[a]Wire specimens were brought to temperature in an evacuated bomb. Fluorine gas was then admitted to the chamber and the time delay until ignition recorded (Schmidt, 1967).

A list of the metals that are compatible with liquid fluorine at various pressures is included in Table 10.10, as recommended in the *Design Handbook for Liquid Fluorine Ground Handling Equipment* (1965). For atmospheric pressure, Monel, nickel, stainless steel, aluminum and its alloys, copper, magnesium, brass, and even mild steel are all considered acceptable materials. As the pressure climbs above 4 MPa (40 atm), however, the choice narrows considerably; for the most reliable service up to 8 MPa (80 atm) only Monel and nickel are recommended.

10.3.3 Compatibility of nonmetals with liquid fluorine

There are very few nonmetallic materials that can be used safely with liquid fluorine. The exceptions include all fully fluorinated compounds, certain ceramics and cermets, although these are usually too brittle for structural applications, and some of the more highly fluorinated polymers, such as polytetrafluoroethylene (TFE) and polychlorotrifluoroethylene (CTFE). Very severe restrictions must be placed on the use of these partially fluorinated polymers with liquid fluorine. Fluorine is capable of breaking the carbon–carbon bond in polymer molecules until the final, thermodynamically stable, end-product of the reaction between carbon and fluorine is reached: carbon tetrafluoride, a relatively nontoxic substance.

Many situations require the use of a soft-seal material such as TFE, and research has shown that under limited flow and pressure, TFE can be exposed to liquid fluorine without reacting. It is almost mandatory, however, to protect the TFE by surrounding it with enough metal in close proximity to give a heat-sink effect and to minimize the surface area of the polymer exposed to liquid fluorine or possible debris impact.

There are no known lubricants, sealants, greases, pastes, or oils suitable for use with liquid fluorine.

Most solid nonmetals have very low ignition temperatures in fluorine relative to metals. In fact, it is usually sufficient to increase the pressure to make most nonmetals ignite spontaneously. Thus, a useful test for screening materials is simply static exposure to liquid or gaseous fluorine. If several materials always ignite in pure fluorine, one can dilute the fluorine with oxygen to determine the limiting concentration of fluorine necessary to cause a reaction. The results of such a test to rank a number of common nonmetals for static compatibility in FLOX is shown in Fig. 10.6 (Russell, Schmidt, and Gordon, 1966). Nearly one-half the materials ignite merely by exposure to liquid fluorine.

To discriminate further among the materials that pass a static compatibility test, one can expose the materials to a dynamic flow of FLOX. When this is done for varying concentrations of FLOX and several different materials, each

Table 10.9 Summary of impact ignition results for metals in liquid fluorine (Sterner and Singleton, 1960).

Specimen Material	Striker Material	Striker Configuration	Number of Tests	Impact Energy (J)	Degree of Ignition: Moderate	Slight	None
Titanium	Monel	Flat, smooth	5	3.53	—	—	5
		Flat, rough	2		1	—	1
		Hemispherical	2		1	—	1
		Chisel point	2		—	2[a]	—
		Conical point	3		—	3[a]	—
Titanium	Titanium	Flat, smooth	2	54.23	1	1	—
			2	74.57	1	1	—
			4	78.64	—	1	3
		Flat, grooved	2	74.57	1	—	1
			2	78.64	2	—	—
		Conical point	3	3.53	—	1	2
			3	8.13	2	1	—
			2	12.20	1	1	—
			1	47.45	1	—	—
			2	67.79	1	1	—
			2	74.57	1	1	—
			2	78.64	—	1	1
			1	82.70	—	—	1
		Hollow point	4	8.13	—	1	3
			4	13.56	1	1	2
			1	51.52	1	—	—
Aluminum	Stainless steel	Flat, grooved	8	78.64	1	—	7
		Conical point	2	40.67	—	—	2
Aluminum	Aluminum	Conical point	2	78.64	—	—	2
Titanium	Titanium	Flat, smooth	1	74.57[b]	—	1	—
			1	81.35[b]	—	1	—
			8	88.13[b]	—	—	8
		Flat, rough	3	88.13[b]	—	1	2
		Flat, grooved	4	88.13[b]	2	—	2
		Conical point	9	88.13[b]	5[c]	4	—

[a]Ignition indicated by disappearance of striker point. No evidence of specimen igniting.
[b]Test performed in liquid oxygen.
[c]One specimen ignited and burned almost completely.

material is found to exhibit a unique reactivity limit, as illustrated in Fig. 10.7 (Russell et al., 1966) for several polymers. When conditions are above and to the right of the reactivity limit, the materials are observed to react. Below and to the left of the limit the materials do not react.

Similar tests could be performed on different batches or formulations of a material to select the most compatible. Russell et al. (1966) were able to show the effects of crystallinity on dynamic reactivity limits for KEL-F-81, as shown in Fig. 10.8. A higher degree of crystallinity raises the reactivity limit significantly.

Although research has shown certain fluorinated polymers can be used with liquid fluorine or FLOX, a great deal of caution is needed to assure safe operation since the minimum ignition energy for these compounds is relatively low. Most experts recommend the removal of any partially fluorinated polymer that can be replaced with a metal gasket or seal.

10.4 HYDROGEN EMBRITTLEMENT OF STRUCTURAL MATERIALS

In the past, low-temperature use of hydrogen has been generally restricted to aerospace applications and to the research laboratory. More recently, with the advent of fossil fuel shortages, considerable attention is being directed toward using hydrogen as an alternate portable and storable energy carrier. If the most ambitious plans for such use are realized, enormous quantities of hydrogen would be handled, transported, and stored in both the liquid and gaseous states. The somewhat limited experience of industry in handling huge quantities of hydrogen, necessar-

Table 10.10 Metals acceptable for liquid fluorine service as a function of pressure (*Design Handbook for Liquid Fluorine Ground Handling Equipment*, 1965).

Maximum Pressure MPa (psig)	Acceptable Metals
8.4 (1200)	Monel, nickel
7.0 (1000)	Monel, nickel
5.6 (800)	Monel, nickel
4.2 (800)	Monel, nickel, stainless steels 304, 316, 321, 347
2.9 (400)	Monel, nickel, stainless steel, aluminum alloys
1.5 (200)	Monel, nickel, stainless steel, aluminum alloys, copper
0.1 (0)	Monel, nickel, stainless steel, aluminum alloys; Copper, aluminum, magnesium, brass; Mild steel.

ily at high pressure, suggests that the existing technologies and systems for handling other liquid and gaseous fuels are not directly applicable to a hydrogen-based fuel economy. Ignorance of the degrading effects of hydrogen in metals can lead to serious consequences either in the laboratory or in domestic and industrial applications.

Embrittlement is the most severe technological problem encountered in the containment of gaseous hydrogen. Hydrogen readily diffuses into most engineering materials and its presence, particularly in metals, can cause significant degradation of the mechanical properties through embrittlement. A vivid example is the failure of several 34-MPa (5000-psi), 37-m^3 (1300-ft^3) hydrogen gas vessels (Jewett, Walter, Chandler, and Frohmberg, 1971). One of these vessels developed a 130-cm-long crack at a weld seam. Three similar hydrogen-containing vessels failed, while

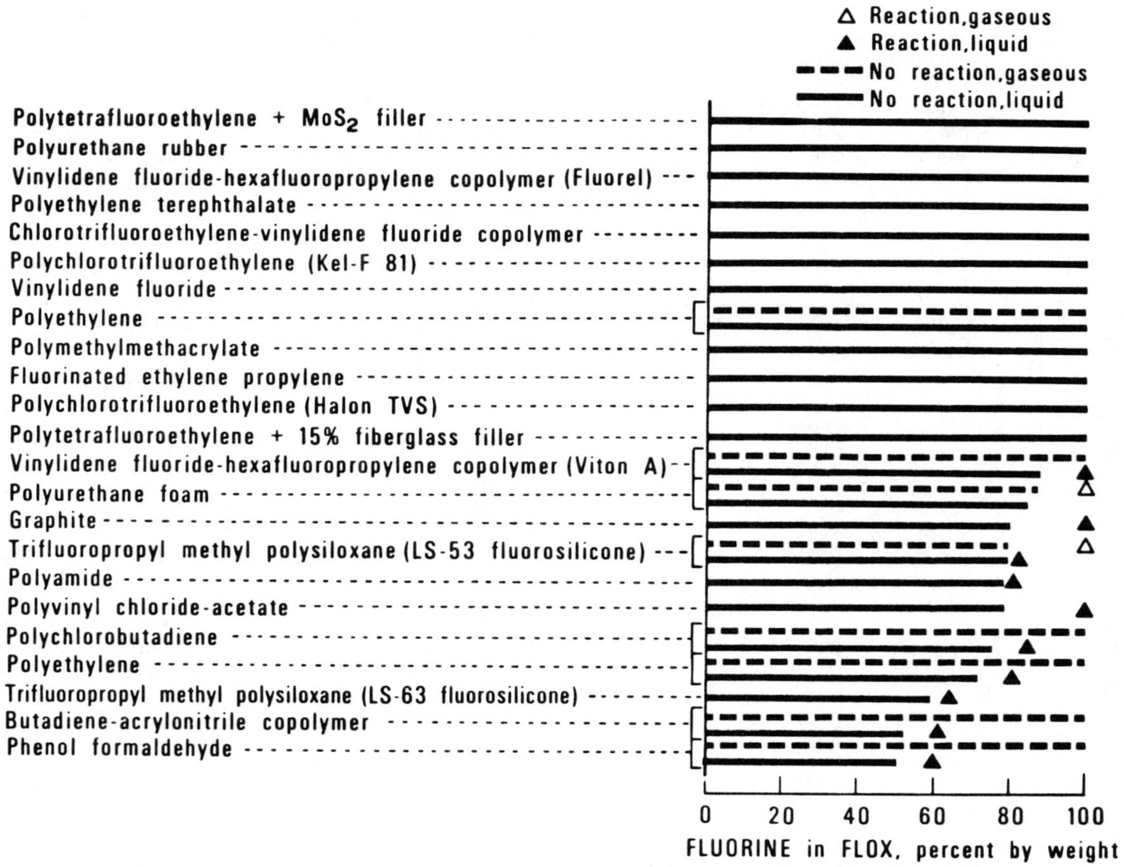

Figure 10.6 Relative compatibility of nonmetallic solids in static FLOX mixtures at ambient pressure; $T = 78$ K for liquid FLOX, $T \cong 270$ K for gaseous FLOX (Russell, Schmidt, and Gordon, 1966).

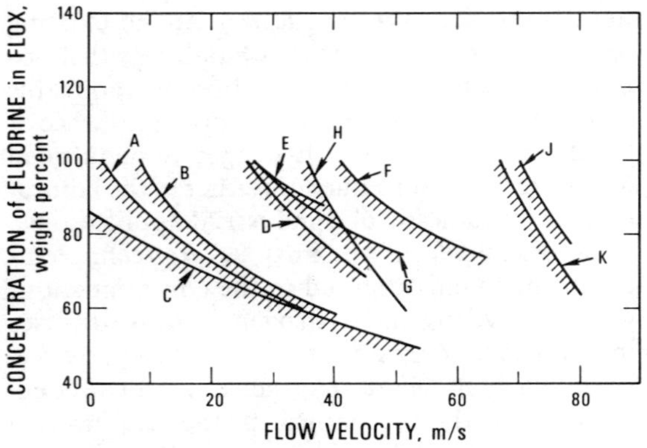

Figure 10.7 Reactivity limits for nonmetallic solids exposed to flowing liquid FLOX (77 K) (Russell, Schmidt, and Gordon, 1966). Materials react with fluorine when environmental conditions are above and to the right of shaded lines. Materials: A — vinylidene fluoride, B — chlorotrifluoroethylene-vinylidene fluoride copolymer, C — vinylidene fluoride-hexafluoropropylene copolymer, D — fluorinated ethylene propylene, E — molybdenum disulphide-filled polytetrafluoroethylene, F — polychlorotrifluoroethylene (Plaskon 2400), G — polychlorotrifluoroethylene (Halon TVS), H — polychlorotrifluoroethylene (Kel-F 81), J — polytetrafluoroethylene (Halon TFE), K — polytetrafluoroethylene (Teflon, MIL-P-19460).

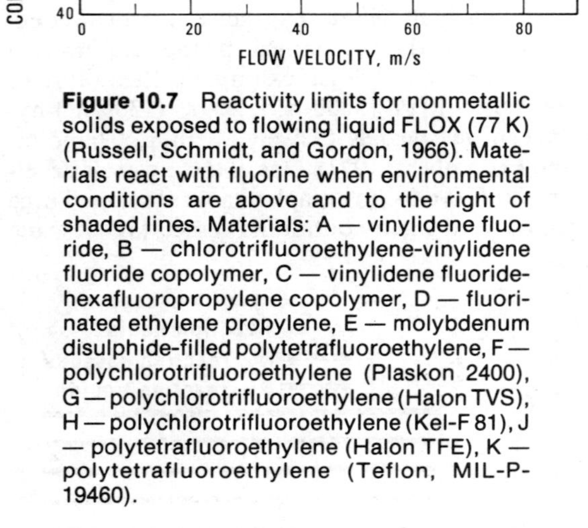

Figure 10.8 Effect of degree of crystallization upon reactivity limits for chlorotrifluoroethylene exposed to flowing liquid FLOX (Russell, Schmidt, and Gordon, 1966). Material reacts with fluorine when conditions are above and to the right of shaded lines.

several identical nitrogen-containing vessels were operated without incident for several years. Such failures can be avoided through the use of more compatible materials or more sophisticated vessel design, but this may impose significant economic restrictions in large-scale commercial facilities. Jewett et al. (1971); Voth, Arnett, and Millhiser (1970); Voth (1972); and Ordin (1974) describe many failures caused by material embrittlement in vessels and gauges. Failures have occurred at as little as one-tenth design pressure.

10.4.1 Types of embrittlement

Although it has been known for about a hundred years that hydrogen can seriously degrade the mechanical properties of metals, the exact mechanisms by which this occurs are not yet clearly understood. It is now believed that three types of hydrogen embrittlement exist: (a) hydrogen-reaction embrittlement, (b) internal-hydrogen embrittlement, and (c) hydrogen-environment embrittlement.

Hydrogen-reaction embrittlement occurs as a result of chemical reaction between hydrogen and the base metal or some alloying element of the metal. Metals such as titanium, zirconium, niobium, and tantalum irreversibly form embrittling hydride phases. The formation of high-pressure water vapor in copper voids as well as the formation of methane in steel voids during decarburization are other examples of hydrogen-reaction embrittlement. Hydrogen-reaction embrittlement is not a low-temperature phenomenon and, therefore, is not discussed further.

Both internal- and environmental-hydrogen embrittlement are due to hydrogen atoms dissolving in the metal. The most extensively studied area of internal-embrittlement is that caused by electrolytic charging, such as in plating, cleaning, or pickling baths. Internal-hydrogen and hydrogen-reaction embrittlement problems are often encountered in metal and petrochemical processing facilities. Any hydrogen-containing chemical solution is capable of producing internal-hydrogen embrittlement.

Hydrogen-environment embrittlement signifies the degradation of mechanical properties that occurs owing to the presence of hydrogen at the surface of the metal. This is also referred to as external-hydrogen embrittlement. Both internal- and external-hydrogen embrittlement are reversible so long as the hydrogen is removed before crack formation commences. Internal hydrogen can be removed by high-temperature heat treatment. Hydrogen-environment embrittlement, although probably closely related to internal-hydrogen embrittlement, is unique to low-temperature hydrogen systems and is the primary subject of this section.

10.4.2 Hydrogen-environment embrittlement

Several reviews and compilations of data on hydrogen-environment embrittlement exist (Buzzard and Cleaver, 1951; Van Ness and Dodge, 1955; Perlmutter and Dodge, 1956; Hofmann and Rauls, 1965; Jewett et al., 1971; Nelson, Tetelmon, and Williams, 1972; Louthan, Caskey, Donovan, and Rawl, 1972; Walter and Chandler, 1973; Louthan and Caskey, 1976; Thompson, 1977; and Thompson and Bernstein, 1977). It is clear from these reviews that there is disagreement regarding the exact mechanism operative in hydrogen-environment embrittlement. One argument leads to the conclusion that hydrogen-environment embrittlement is distinct from internal-hydrogen embrittlement, and other arguments lead to the conclusion that both are manifestations of the same mechanism. The main argument for distinctly different mechanisms is that only internal embrittlement causes delayed failure (failure at constant stress below the normal yield strength of the metal after a time which may vary from hours to days). Embrittlement due to hydrogen environment is immediate, and no increase in embrittlement is seen with exposure times as long as 100 days (Jewett et al., 1971; Walter and Chandler, 1973). Although the behaviors of the two types of embrittlement are quite similar, caution should be exercised in applying the results of one type of study to the other. It should be noted that early internal-embrittlement studies on Inconel 718 indicated no susceptibility to embrittlement, but recent hydrogen-environment studies showed Inconel 718 to be severely embrittled (Jewett et al., 1971).

The degree of hydrogen-environment embrittlement depends on the base metal and its alloying constituents, the microstructure of the metal, strain rate, the presence of surface imperfections (such as cracks, notches, or pits), residual stress, the yield strength of the metal, impurities present in the hydrogen environment, and probably other parameters not yet discovered. It has often been suggested that, at least in the case of steels, which are the most thoroughly investigated metal class, alloys with an f.c.c. structure are relatively immune to embrittlement whereas alloys with a b.c.c. structure are severely embrittled. It has been shown more recently that the crystal structure is less significant than the atomic structure. Louthan et al. (1972) have shown a close relationship between hydrogen solubility, and thus the degree of embrittlement, and the density of states at the Fermi surface.

Jewett et al. (1971) have shown that the degree of hydrogen-environment embrittlement is approximately proportional to the square root of pressure until saturation occurs, which is generally at a pressure of several tens of megapascals (several thousand pounds per square inch). Internal-hydrogen embrittlement generally peaks near ambient temperatures and falls off both at high temperatures and at low temperatures. No hydrogen-environment embrittlement is expected at liquid-hydrogen temperature with common structural materials, but embrittlement can occur in a liquid-hydrogen vessel if the material surrounding the ullage space approaches ambient temperature. At higher temperatures, such as those encountered in engines, burners, and hydrogen-production equipment, both hydrogen-environment and reaction embrittlement problems are likely (Jewett et al., 1971). In addition, at high temperatures and pressures, the containment material becomes saturated with hydrogen. If cooldown of the equipment is not controlled to allow the internal hydrogen to escape by permeation, classical internal-hydrogen embrittlement may result at ambient temperature. Troiano (1974) indicates that degassing by cyclic temperature control to remove hydrogen in refining operations is standard procedure.

Rate of strain is an important parameter in the degree of embrittlement exhibited. Low strain rates promote maximum embrittlement. Smooth surfaces are least susceptible to embrittlement. Metals conditioned to high strength levels are more susceptible to embrittlement than their lower strength counterparts. This effect is especially important in weldments or localized hard spots. Table 10.11 illustrates the relative degree of embrittlement of metal classes. The relative scale used is the same as that defined by Jewett et al. (1971). Table 10.12, taken from the same reference, illustrates more specifically the embrittling effects on metals as indicated by notched and unnotched strength ratios and unnotched ductility.

Hydrogen embrittlement does not significantly affect elastic properties; those properties most

Table 10.11 Relative degree of hydrogen-environment embrittlement for classes of metals (Jewett, Walter, Chandler, and Frohmberg, 1971).

Degree of Embrittlement	Materials	Characterization
Extreme	High-strength steels Nickel-base alloys	Large decrease in notch strength and notched and unnotched ductility. Some decrease in unnotched strength. Propagation of surface cracks.
Severe	Ductile, lower-strength steels Pure nickel Titanium alloys	Considerable reduction of notch strength and unnotched and notched ductility. No reduction of unnotched strength. Propagation of surface cracks.
Slight	Metastable 300 series stainless steels Beryllium–copper Pure titanium	Small decrease in notched strength and unnotched ductility. Failure of unnotched specimens from within.
Negligible	Aluminum alloys Stable austenitic stainless steels Copper	Essentially unembrittled with no surface cracks.

affected are tensile ductility, notch strength, fatigue, crack behavior, and creep. The purity of the hydrogen environment can have profound effects on the degree of embrittlement. Small amounts of oxygen, in some instances, have totally destroyed the embrittling capability of hydrogen. Small additions of SO_2, CO, and CS_2 are also effective in inhibiting hydrogen embrittlement. Jewett et al. (1971) and Fidelle, Deloron, Rows, and Rapin (1969) investigated the effects of metallic and oxide coatings to inhibit embrittlement. Jewett et al. (1971) feel that an effective coating must have low permeability, be nonporous, adhere well to the substrate, and be ductile or self-healing. They have found that electroplated coatings of copper, gold, and silver are effective barriers.

10.4.3 Selection of materials for hydrogen service

A good rule of thumb in selecting materials when the question of hydrogen embrittlement is involved is to suspect all metals until they are proven compatible. Materials that have proven reasonably immune to embrittlement are aluminum alloys, copper alloys, and some austenitic stainless steels. Generally, the higher strength alloys suffer severe embrittlement effects. Particular attention must be given to any welds in hydrogen systems, especially if the parent metal is only marginally acceptable.

For high pressure systems, above 14 MPa (2000 psi), very few materials can be recommended. Those materials that may be usable, such as the precipitation-strengthened A-286 steels, often are not economically feasible to use.

Considerable research is in progress to certify existing materials and to develop new, more suitable materials. It is expected that a much clearer picture will emerge from this current effort. However, for the present, utmost caution must be urged in materials selection, especially for high-pressure systems.

10.5 REFERENCES

Annual Book of ASTM Standards (1970), American Society for Testing and Materials, Philadelphia, Pennsylvania, 692–713.

Bahn, G. S. (1965). On pyrophoricity of metals, and of fine metal powders in particular. *Pyrodynamics* **3**, 29–41.

Bates, C. E., Wren, J. E., Monroe, R., and Pears, C. D. (1979). Ignition and combustion of ferrous metals in high pressure, high velocity, gaseous oxygen. *J. Mater. Energy Sys.* **1**, 61–76.

Bauer, H., Wegener, W., and Windgassen, K. F. (1970). Fire tests on centrifugal pumps for liquid oxygen. *Cryogenics* **10**, 241–248.

Bauer, H., Klein, G. K., Wegener, W., and Windgassen, K. F. (1971). Fire tests on centrifugal pumps handling liquid oxygen — part 2. *Cryogenics* **11**, 469–476.

Bell, L. E., English, W. D., Murata, K. K., Pohl, S. W., and Tiner, N. A. (1971). A shock sensitivity tester for liquid propellant structural materials systems. *Mater. Res. Stand.* **11**, 16–21.

Blackstone, W. R. (1970). *Development of an Improved Gaseous Oxygen Impact Test System*. Report No. 70-32875, Southwest Research Institute, San Antonio, Texas.*

Blackstone, W. R. and Ku, P. M. (1971). An assessment of impact test techniques for determining the fire or explosion hazards of materials exposed to liquid oxygen. *Mater. Res. Stand.* **11**, 30–35.

Burmeister, L. C., Loser, J. B., and Sneegas, E. C. (1967). *Advanced Valve Technology*. Report No. NASA SP-5019, National Aeronautics and Space Administration, Scientific and Technical Information Office, Washington, D.C.*

Buzzard, R. W. and Cleaves, H. E. (1951). *Hydrogen Embrittlement of Steel: Review*. NBS Circular 511, National Bureau of Standards, Washington, D.C.

Clark, A. F. and Hust, J. G. (1974). A review of the

Table 10.12 Embrittlement of metals by 68.9 MPa (10 000 psi) hydrogen at ambient temperature (Jewett, Walter, Chandler, and Frohmberg, 1971).

Material	Strength Ratio, H_2/He		Unnotched Ductility			
			Elongation (%)		RA[a] (%)	
	Notched (K_t = 8.4)	Unnotched	He	H_2	He	H_2
18 Ni–250 MAR	0.12	0.68	8.2	0.2	55	2.5
410 SS	0.22	0.79	15	1.3	60	12
1042 QT	0.22	—	—	—	—	—
17–PH SS	0.23	0.92	17	1.7	45	2.5
Fe–9Ni–4Co–0.2C	0.24	0.86	15	0.5	67	15
H–11	0.25	0.57	8.8	0	30	0
Rene 41	0.27	0.84	21	4.3	29	11
Electroformed Ni[b]	0.31	—	—	—	—	—
4140	0.40	0.96	14	2.6	48	9
Inconel 718	0.46	0.93	17	1.5	26	1
440 C	0.50	0.40	—	—	3.2	0
Ti–6Al–4V (STA)	0.58	—	—	—	—	—
430 F	0.68		22	14	64	37
Nickel 270	0.70		56	52	89	67
A-515	0.73		42	29	67	35
HY-100	0.73		20	18	76	63
A-372 Class IV	0.74		20	10	53	18
1042 Normalized	0.75				59	27
A-533-B	0.78				66	33
Ti–6Al–4V (annealed)	0.79		—	—	—	—
AISI 1020	0.79				68	45
HY-80	0.80				70	60
Ti–5Al–2.5SnELI	0.81				45	39
ARMCO Iron	0.86				83	50
304 ELC SS	0.87				78	71
305 SS	0.89				78	75
Be–Cu Alloy 25	0.93				72	71
310 SS	0.93				64	62
Titanium	0.95				61	61
A-286	0.97				44	43
7075-T73 Al Alloy	0.98				37	35
Incoloy 903[b]	1.00				50	47
316 SS	1.00				72	75
OFHC Copper	1.00				94	94
NARloy-2[c]	1.10				24	22
6061-T6 Al Alloy	1.10				61	66
1100-O Al	1.40				93	93

[a]RA = reduction in area
[b]Tested in 48.3 MPa (7000 psi) H_2
[c]Rockwell International Corporation Trademark; tested in 40 MPa (5800 psi) H_2

compatibility of structural materials with oxygen. *AIAA J.* **12**, 441–454.

Cryogenic Safety, A Summary Report of the Cryogenic Safety Conference (1959). Air Products, Inc., Allentown, Pennsylvania.

Darken, L. S. and Gurry, R. W. (1953). *Physical Chemistry of Metals*, McGraw-Hill, New York.

Dean, L. E. and Thompson, W. K. (1961). Ignition characteristics of metals and alloys. *J. Am. Rocket Soc.* **31**, 917–922.

Design Handbook for Liquid Fluorine Ground Handling Equipment (1965) (ed. R. E. Anderson). Report No. AFRPL-TR-65-133, Air Force Rocket Propulsion Laboratory, Edwards AFB, California.*

Fidelle, J. P., Deloron, J. M., Rows, C., and Rapid, M. (1969). *Influence of Surface Treatments and Coatings on the Embrittlement of High-Strength Steels by Hydrogen under Pressure*. Report No. ORNL-TR-2058, Oak Ridge National Laboratory, Oak Ridge, Tennessee.*

Fristrom, R. M., Hunter, L. W., and Grunfelder, C. (1976). Exploratory studies in polymer ablation, ignition and extinction by the moving wire technique. *Combust. Flame* **27**, 33–49.

Glassman, I. (1959). *Metal Combustion Processes*. American Rocket Society Preprint No. 938–59.

Available from author, c/o Princeton University, Princeton, New Jersey.
Grosse, A. V. and Conway, J. B. (1958). Combustion of metals in oxygen. *Ind. Eng. Chem.* **3**, 663–672.
Guter, M. (1967). *Ignition in High Pressure Oxygen.* Report No. AD 648–612, RR/D 1312, British Oxygen Co. Ltd., London, England.*
Hofmann, W. and Rawls, W. (1965). Ductility of steel under the influence of external high pressure hydrogen, *Weld. J.* **44**, 225–230.
Hunter, L. W., Schacke, H., Grunfelder, C., and Fristrom, R. M. (1976). Surface temperature measurements in the moving wire technique. *Combust. Sci. Technol.* **15**, 41–48.
Hust, J. G. and Clark, A. F. (1973). A survey of compatibility of materials with high pressure oxygen service. *Cryogenics* **13**, 325–335.
Jackson, J. D., Boyd, W. K., and Miller, P. D. (1963). *Reactivity of Metals with Liquid and Gaseous Oxygen.* DMIC Memo 163, Defense Metals Information Center, Battelle Memorial Institute, Columbus, Ohio.*
Jamison, H. H. (1971). Development of a gaseous oxygen impact test method. *Mat. Res. Stand.* **11**, 22–27.
Jewett, R. P., Walter, R. J., Chandler, W. T., and Frohmberg, R. P. (1971). *Hydrogen-Environment Embrittlement of Metals.* Contract NAS8–19(C); NASA CR 2163, Rocketdyne, Division of Rockwell International.*
Keeping, W. O. (1971). *Compatibility of Materials with Oxygen.* Compressed Gas Association — Oxygen Compressors and Pumps Symposium, Atlanta, Georgia.
Key, C. F. (1964). *Compatibility of Materials with Liquid Oxygen II.* Report No. NASA TM X-53052, National Aeronautics and Space Administration, Marshall Space Flight Center, Huntsville, Alabama.*
Key, C. F. (1966). *Compatibility of Materials with Liquid Oxygen III.* Report No. NASA TM X-53533, National Aeronautics and Space Administration, Marshall Space Flight Center, Huntsville, Alabama.*
Key, C. F. (1968). *Compatibility of Materials with Liquid Oxygen IV.* Report No. NASA TM X-53773, National Aeronautics and Space Administration, Marshall Space Flight Center, Huntsville, Alabama.*
Key, C. F. and Riehl, W. A. (1964). *Compatibility of Materials with Liquid Oxygen.* Report No. NASA TM X-985, National Aeronautics and Space Administration, Marshall Space Flight Center, Huntsville, Alabama.*
Kleinberg, S. and Tompkins, J. F. (1962). *The Compatibility of Various Metals with Liquid Fluorine.* Report No. ASD-TDR-62-250, Air Products and Chemicals, Inc., Allentown, Pennsylvania.*
Kofstad, P. (1966). *High Temperature Oxidation of Metals.* John Wiley & Sons, New York.
Komanduri, R. and Shaw, M. C. (1975). Formation of spherical particles in grinding. *Philos. Mag.* **32**, 711–724.
Kubaschewski, O. and Evans, E. L. (1958). *Metallurgical Thermochemistry*, 2nd edition. Pergamon, London.
Laurendeau, N. M. and Glassman, I. (1971). Ignition temperatures of metals in oxygen atmospheres. *Combust. Sci. Technol.* **3**, 77–82.
Lewis, G. N., Randall, M., Pitzer, K. S., and Brewer, L. (1961). *Thermodynamics*, 2nd edition. McGraw-Hill, New York.
Littman, F. E., Church, F. M., and Kinderman, E. M. (1961). A study of metal ignitions. I. The spontaneous ignition of titanium. II. The spontaneous ignition of zirconium. *J. Less-Common Met.* **3**, 367–397.
Louthan, M. R., Jr. and Caskey, G. R., Jr. (1976). Hydrogen transport and embrittlement in structural metals. *Int. J. Hydrogen Energy* **1**, 291–305.
Louthan, M. R., Jr., Caskey, G. R., Jr., Donovan, J. R., and Rawl, D. E., Jr. (1972). Hydrogen embrittlement in metals. *Mater. Sci. Eng.* **10**, 557–568.
Lucas, W. R. and Riehl, W. A. (1960). An instrument for determination of impact sensitivity of materials in contact with liquid oxygen. *Amer. Soc. Test. Mater. Bull.* **244**, 29–34.
McKinley, C. (1971). *Experimental Ignition and Combustion of Metals.* Compressed Gas Association — Oxygen Compressors and Pumps Symposium, Atlanta, Georgia. Copies available from author, care of Air Products and Chemicals, Inc., Box 538, Allentown, Pennsylvania.
Markstein, G. H. (1963). Combustion of metals. *AIAA J.* **1**, 550–562.
Mellor, A. M. (1967). *Heterogeneous Ignition of Metals: Model and Experiment.* Report No. 816, Aerospace and Mechanical Sciences, Princeton University, Princeton, New Jersey.
Nelson, H. G., Tetelman, A. S., and Williams, D. P. (1972). The kinetic and dynamic aspects of corrosion fatigue in a gaseous hydrogen environment. *Corr. Fatigue: Chem. Mech. Microstruct.,* **NACE-2**, 359–365.
Nelson, L. S. (1965). Explosion of burning zirconium droplets caused by nitrogen. *Science* **148**, 1594–1595.
Nihart, G. H. and Smith, C. P. (1964). *Compatibility of Materials with 7500 psi Oxygen*, Report No. DDC-AD 608 260, AMRL-TDR-64–76, Union Carbide Corp., Tonawanda, New York.*
Nonmetallic Materials Design Guidelines and Test Data Handbook (1977). Report No. NASA JSC 02681, Revision H, National Aeronautics and Space Administration, Johnson Space Center, Houston, Texas.*
Ordin, P. M. (1974). *Review of Hydrogen Accidents and Incidents in NASA Operations.* NASA Tech. Memo. X-71565.*
Perlmutter, D. D. and Dodge, B. F. (1956). Effects of hydrogen on properties of metals. *Ind. Eng. Chem.* **48**, 885–893.
Phillips, B. R. (1975). Resonance tube ignition of metals. Ph.D. thesis, University of Toledo, Toledo, Ohio.
Pippen, D. L. and Stradling, J. S. (1971). Techniques for determination of flash and fire points and impact sensitivity of materials in a gaseous oxygen environment. *Mater. Res. Stand.* **11**, 35–44.
Raniere, F. D., Hood, G. A., and Marker, H. E. (1972). *High-Pressure Oxygen Impact Tester.* Report No. R-8962, Rocketdyne, Canoga Park, California.
Reynales, C. H. (1958). *Compatibility of Materials*

with Oxygen. Report No. D81–444, Douglas Aircraft Co. Inc., Long Beach, California.

Reynales, C. H. (1961). Selection of lubricants and thread compounds for oxygen missile systems. In *Advances in Cryogenic Engineering*, Vol. 6. Plenum Press, New York, 117–129.

Richards, H. T. and Hanson, M. P. (1963). *Influence of Fluorine Environment on Mechanical Properties of Several Sheet Alloys*. Report No. NASA TN D-1706, National Aeronautics and Space Administration, Washington, D.C.*

Riehl, W. A., Key, C. F., and Gayle, J. B. (1963). *Reactivity of Titanium with Oxygen*, Report No. NASA TR-R-180, Marshall Space Flight Center, Huntsville, Alabama.*

Russell, L. M., Schmidt, H. W., and Gordon, L. H. (1966). *Compatibility of Polymeric Materials with Fluorine and Fluorine-Oxygen Mixtures*. Report No. NASA TN D-3392, National Aeronautics and Space Administration, Washington, D.C.*

Schacke, H., Hunter, L. W., Grunfelder, C., and Fristrom, R. M. (1977). Combustion of polyvinylchloride studied by the moving wire technique. In *Sixteenth Symposium (International) on Combustion*. The Combustion Institute, Pittsburgh, Pennsylvania, 1317–1326.

Schmidt, H. W. (1967). *Handling and Use of Fluorine and Fluorine–Oxygen Mixtures in Rocket Systems*. NASA Special Publication 3037, Scientific and Technical Information Office, National Aeronautics and Space Administration, Washington, D.C.*

Schmidt, H. W. and Forney, D. E. (1975). *ASRDI Oxygen Technology Survey, Vol. IX: Oxygen Systems Engineering Review*. NASA Special Publication 3090, Scientific and Technical Information Office, National Aeronautics and Space Administration, Washington, D.C.*

Schwinghamer, R. J. (1971). *Impact Sensitivity of Materials in Contact with Liquid and Gaseous Oxygen at High Pressure*. Compressed Gas Association Oxygen Compressors and Pumps Symposium, Atlanta, Georgia. Copies available from the author, c/o National Aeronautics and Space Administration, Marshall Space Flight Center, Huntsville, Alabama.

Staph, H. E., Baber, B. B., Ku, P. M., and Weatherford, W. D. (1962). *Survey of Lubricant Compatibility Test Methods for Missile Systems*. Report No. ASD-TDR 62–190, Southwest Research Institute, San Antonio, Texas.*

Sterner, C. J. and Singleton, H. H. (1960). *The Compatibility of Various Metals and Carbon with Liquid Fluorine*. Report No. TR 60–436, Wright Air Development Division, Wright–Patterson AFB, Ohio.*

Stradling, J. S. (1971). *Ignition by Abrasives*. Report No. NASA TP-WSTF-193, National Aeronautics and Space Administration, White Sands Test Facility, White Sands, New Mexico.*

Strobridge, T. R., Moulder, J. C., and Clark, A. F. (1979). *Titanium Combustion in Turbine Engines*. Report No. FAA-RD-79–51, Federal Aviation Administration, Systems Research and Development Service, Washington, D.C.*

Stull, D. R. and Prophet, H. (1971). *JANAF Thermochemical Tables*, 2nd edition, NSRDS-NBS 37. Available from Superintendent of Documents, U.S. Government Printing Office, Washington, D.C. (Catalog No. C 13.48:37).

Thompson, A. W. (1977). Structural materials use in a hydrogen energy economy. *Int. J. Hydrogen Energy* **2**, 299–307.

Thompson, A. W. and Bernstein, I. M. (1977). Selection of structural materials for hydrogen pipelines and storage vessels. *Int. J. Hydrogen Energy* **2**, 163–173.

Troiano, A. R. (1974). Case Western Reserve University, Cleveland, Ohio, private communication.

Van Ness, H. C. and Dodge, B. F. (1955). Effects of hydrogen at high pressures on the mechanical properties of metals. *Chem. Eng. Prog.* **51**, 266–271.

Voth, R. O. (1972). Safety of hydrogen pressure gauges. In *Advances in Cryogenic Engineering*, Vol. 17 (ed. K. D. Timmerhaus). Plenum Press, New York, 182–187.

Voth, R. O., Arnett, R. W., and Millhiser, D. R. (1970). *Cryogenic Valves, Connectors, and Pressure Gauges*. Unpublished report, National Bureau of Standards, Boulder, Colorado.

Walter, R. J. and Chandler, W. T. (1973). *Influence of Gaseous Hydrogen on Metals — Final Report*. Contract NAS8–19(C); NASA CR 124410, Rocketdyne, Division of Rockwell International.*

Zabetakis, M. G. (1967). *Safety with Cryogenic Fluids*. Plenum Press, New York.

*Available from NTIS, Springfield, Virginia.

11 Structural Alloys

H. I. McHENRY

11.1 INTRODUCTION

Many alloys have been successfully used for a wide range of structures and devices that operate at low temperatures, most commonly in superconducting magnets and machinery and for structures and equipment used in the storage, handling, and transportation of liquefied gases.

This chapter discusses the structural alloys being used for cryogenic applications in commercially significant quantities. It emphasizes the practical considerations involved in the material selection process and provides the information necessary to make preliminary selections of alloys most suitable for the intended cryogenic application. The final selection requires more information about cost, availability, and prior experience as well as a more detailed evaluation of material properties than is contained herein. Only representative data on the materials of interest are given. The fundamental significance of the properties mentioned is discussed in other chapters.

The chapter is divided into four sections; each provides general information on a class or group of alloys, their representative mechanical and physical properties, and their fabrication characteristics.

11.2 AUSTENITIC STAINLESS STEELS

The austenitic stainless steels are Fe–Cr alloys with sufficient (8 to 24%) nickel and manganese (and sometimes nitrogen) to stabilize the austenitic phase. Chromium (16 to 26%)* provides corrosion resistance. The most typical austenitic stainless steels are variations of the 18Cr–9Ni alloy. Although none of the commercial alloys were specifically developed for low-temperature service, several of them have been widely used in a variety of cryogenic applications. Their popularity stems from their retention of excellent mechanical properties, particularly toughness, at low temperatures coupled with ready availability, ease of fabrication, and good service experience. Their disadvantages are they are more expensive and have lower yield strength than ferritic Fe–Ni steels and aluminum alloys, and their machinability is poorer than that of aluminum alloys.

The austenitic stainless steels have a wide range of mechanical properties, depending on their chemical composition and thermal–mechanical processing. Three general classes of these alloys are useful for cryogenic applications:

a. *The AISI 300-series stainless steels* in the annealed condition, particularly types 304 and 304L, are the most commonly used alloys. Consequently, they have the most service experience and coverage in design codes. These grades have moderate strength and excellent toughness, and they are readily available in a variety of product forms. Casting alloys equivalent to the commonly used 300-series grades are used to manufacture complex shapes, such as valve bodies.

b. *The nitrogen-strengthened grades* are remarkably stronger at cryogenic temperatures and retain most of the advantages of the 300-series grades, except for availability, service experience, and code coverage.

c. *The cold-rolled sheets of the AISI 300-series*

*Chemical composition is in weight percent throughout this chapter, unless otherwise specified.

grades have been used as high-strength sheet materials in specialized applications, such as liquid-fueled missile cases.

In addition to the three main classes, there are precipitation-hardened austenitic grades, for example the A286 alloy (ASTM A 453), which offer promise as high-strength alloys for heavier-gauge applications, but their usage to date has been limited to nuts and bolts. Precipitation-hardened alloys must be selected carefully because these alloys are generally martensitic or semiaustenitic and have low toughness at cryogenic temperatures. The fully austenitic precipitation-hardened alloys are the preferred selection.

11.2.1 Metallurgy

Before considering the specific alloys, it is useful to review the basic metallurgical factors that contribute to the behavior and, hence, proper use of the austenitic stainless steels: composition, crystal structure, phase stability, and microstructure.

Most of the stainless steels used for cryogenic applications are austenitic Fe–18Cr alloys with nickel contents in excess of 8% and carbon content less than 0.1%. The austenitic phase has the face-centered cubic (f.c.c.) structure. The f.c.c. lattice is the optimum atomic arrangement to ease plastic deformation, and thus the f.c.c. alloys have excellent ductility at room temperature and remain ductile at low temperatures. A schematic diagram of the Fe–Cr–Ni phase diagram for alloys containing 18% Cr is shown in Fig. 11.1 (Castro and deCadenet, 1975). The ferrite phase that forms upon solidification is a body-centered cubic (b.c.c.) phase referred to as δ-ferrite. The austenite phase (γ) forms on cooling either from the liquid (at high nickel contents) or from the δ-ferrite. The α-ferrite phase (also b.c.c.) is martensite and forms from the austenite phase upon cooling below the martensite start temperature, T_{ms}, or upon deformation below the strain-induced martensite start temperature, T_{md}, where $T_{md} > T_{ms}$. The martensitic transformation is accompanied by a volume change (α is larger than γ) and by an increase in magnetic permeability (α is ferromagnetic).

The relative extent of the austenite and ferrite regions can be altered by alloying. Nickel, manganese, carbon, and nitrogen extend the austenite range, and chromium, molybdenum, and silicon extend the ferrite range. For the 18% Cr case shown in Fig. 11.1, alloys with low nickel content are ferritic, alloys with less than 8% Ni contain austenite and ferrite, and alloys with more than 8% Ni are fully austenitic at room temperature. However, either plastic deformation or further cooling may cause transformation to martensite. Each of the alloying elements increases the austenite stability (i.e., resistance to martensite formation), but only manganese does this without significantly altering the ferrite and austenite ranges. In addition, manganese increases the solubility of nitrogen in austenite, thereby permitting further increases in austenite stability. Thus, manganese is commonly used along with nitrogen as a substitute for nickel in austenitic stainless steels.

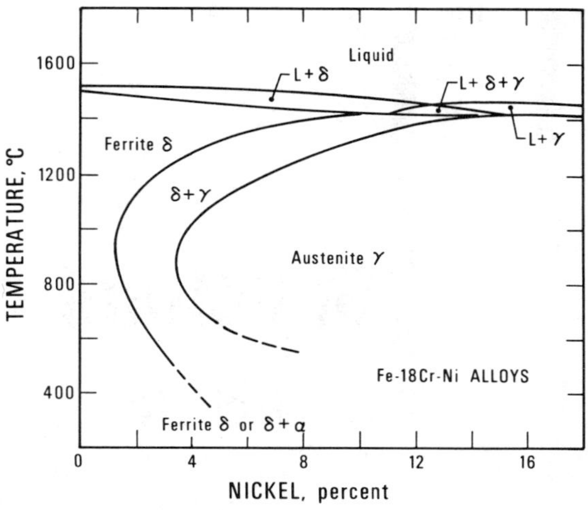

Figure 11.1 Pseudobinary phase diagram of Fe–Cr–Ni alloys with 18% Cr (Castro and de Cadenet, 1975).

The Schaeffler (1949) constitution diagram for stainless steel weld metals (Fig. 11.2) is used to estimate the ferrite content in as-deposited weld metal. Here, the effects of the various alloying elements are accounted for by use of nickel and chromium equivalents as the axes, where:

$$Ni_{eq} = \%Ni + 30 \times \%C + 0.5 \times \%Mn \quad (11.1)$$

and

$$Cr_{eq} = \%Cr + \%Mo + 1.5 \times \%Si + 0.5 \times \%Nb \quad (11.2)$$

DeLong, Ostrom, and Szumachowski (1956) have modified the Schaeffler diagram to account for nitrogen simply by adding the %N to the %C when calculating nickel content. Although the percent of ferrite can be altered by dilution ef-

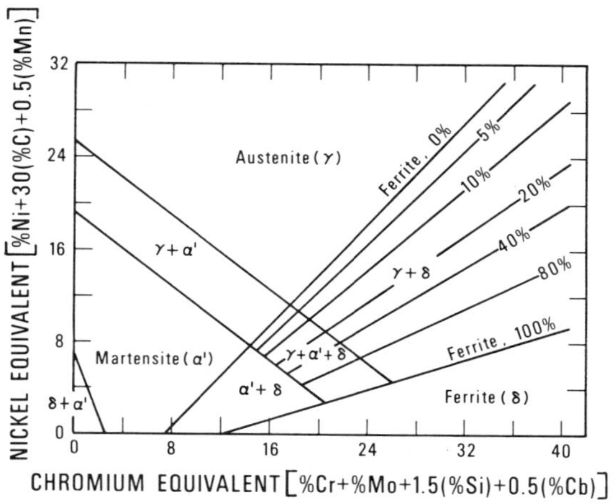

Figure 11.2 Schaeffler diagram for determining the approximate amount of δ-ferrite in the microstructure of austenitic stainless steel welds (Schaeffler, 1949).

fects, cooling rate, and other variables, it is quite useful for estimating ferrite contents in weld metals and, to a lesser extent, in castings.

The precipitation of chromium carbides near grain boundaries occurs in austenitic stainless steels in the temperature range of 500 to 800° C. This phenomenon, known as *sensitization*, causes depletion of chromium near the grain boundaries and can lead to localized intergranular corrosion. For cryogenic service, corrosion is frequently not a principal concern. However, the grain-boundary precipitation may cause a reduction in toughness, and consequently, precautions sometimes are taken to avoid it. The most common precaution is to use grades with carbon contents less than 0.03% when a potentially sensitizing thermal treatment is anticipated. Alternatively, alloys with normal (0.08% max.) carbon content and strong carbide formers such as niobium, titanium, and tantalum may be used; these grades are termed *stabilized*. The niobium, titanium, and tantalum carbides form in preference to chromium carbides and remain stable at elevated temperatures, thereby eliminating the risk of sensitization.

11.2.2 Annealed AISI 300-series grades

The designations and compositions of the most commonly used grades of the AISI 300 series are summarized in Table 11.1. These grades have moderate strength, excellent toughness, and good fabrication characteristics, and they are readily available in a variety of product forms. The ASTM specifications for the various product forms are listed in Table 11.2. Several 300-series grades not listed in Table 11.1 are covered by ASTM specifications and can meet the *ASME Boiler and Pressure Vessel Code* (1977) requirements for cryogenic vessels. These grades are not described in this chapter because they offer no particular advantage for cryogenic service, and therefore, their service experience and property data at low temperatures are limited.

The alloys listed in Table 11.1 can be compared with AISI-304, the basic 19Cr–9Ni stainless steel. Type 316 contains 2 to 3% Mo to improve resistance to pitting in chloride solutions, a property that led to its selection for cargo piping systems aboard liquefied-natural-gas (LNG) ships. Types 304L and 316L are low-carbon modifications of 304 and 316, respectively; low carbon is desirable to avoid sensitization, the grain-boundary precipitation of chromium carbides. Types 321 (alloyed with titanium) and 347

Table 11.1 Compositions of austenitic stainless steels.

AISI	Composition (%)			
Type No.	Cr	Ni	C, max.	Other
301	16–18	6–8	0.15	
302	17–19	8–10	0.15	
304	18–20	8–12	0.08	
304L	18–20	8–12	0.03	
305	17–19	10.5–13	0.12	
309	22–24	12–15	0.20	
310	24–26	19–22	0.25	1.5 Si, max.
310S	24–26	19–22	0.08	1.5 Si, max.
316	16–18	10–14	0.08	2–3 Mo
316L	16–18	10–14	0.03	2–3 Mo
321	17–19	9–12	0.08	(5 × %C) Ti, min.
347	17–19	9–13	0.08	(10 × %C) Nb+Ta, min.

Table 11.2 ASTM specifications for various product forms of austenitic stainless steels.

Product Form	ASTM Specification
Plates, sheet, strip	A240
Forgings	A473
Bars and sections	A479
Seamless tubes	A213
Welded tubes	A249
Seamless and welded tubes	A269
Small diameter tubing	A632
Bolts, screws, studs	A320
Nuts	A194

(alloyed with niobium and tantalum) are also sensitization-resistant grades. Type 310S is a 25Cr–20Ni alloy that is a low-carbon version (0.08% C max.) of AISI type 310 (0.25% C max.). The high-alloy content provides austenite stability, and thus 310S is useful where dimensional stability and nonmagnetic behavior are essential.

The properties of the 300-series austenitic stainless steels have been compiled in cryogenic handbooks (*LNG Materials and Fluids*, 1978; *Handbook on Materials for Superconducting Machinery*, 1977), in a review article by Brickner and Defilippi (1977), and in suppliers publications (*Materials for Cryogenic Service*, 1974; *Low Temperature and Cryogenic Steels*, 1966). In this section, the general trends in strength, toughness, and fatigue resistance are summarized as a function of temperature. Use the data presented here and the data in the handbooks with care because there may be significant heat-to-heat variations in the mechanical properties of the 300-series stainless steels at low temperatures, particularly in the yield strength and toughness. The principal causes of these variations are interestitial content (carbon and nitrogen) and mechanical deformation history, which controls the resulting grain size. The nitrogen content, even within the specification limit of 0.08 wt.%, is particularly significant because higher nitrogen causes substantial increases in yield strength.

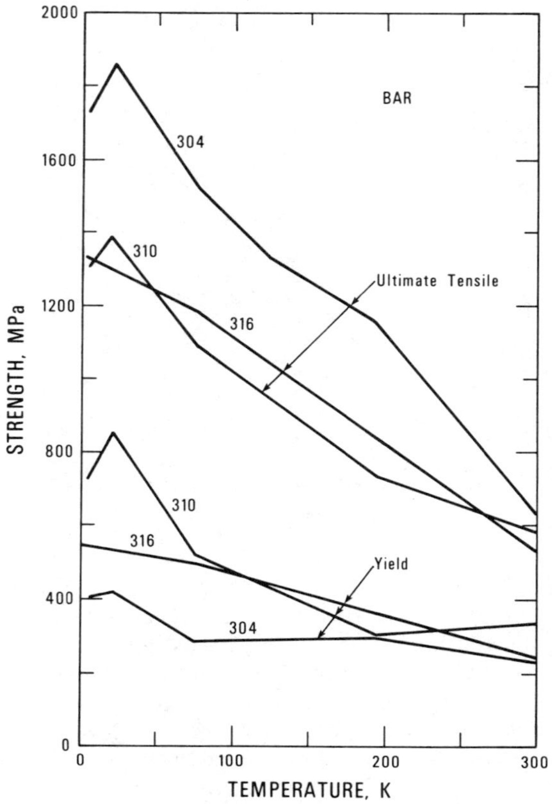

Figure 11.3 Tensile and yield strengths of three austenitic stainless steels — AISI types 304, 310, and 316 — at temperatures between 4 K and 300 K (*Handbook on Materials for Superconducting Machinery*, 1977).

Tensile properties. The ultimate tensile and yield strengths of AISI grades 304, 310, and 316 are compared in Fig. 11.3 for temperatures ranging from 4 to 300 K (*Handbook on Materials for Superconducting Machinery*, 1977). Notice that type 304 has the lowest yield strength and the highest ultimate strength at all temperatures. The increased ultimate tensile strength of type 304 is due to the martensitic transformation, which occurs more readily in 304 than in 316 or 310. Since the transformation is strain induced, the yield strength is not influenced by martensite, but the ultimate strength is increased.

The ultimate tensile and yield strengths of AISI grades 304 and 304L are compared in Fig. 11.4 (*Handbook on Materials for Superconducting Machinery*, 1977). Type 304 exhibits slightly more increase in strength than type 304L as temperature is reduced. The loss of strength associated with reduced carbon content is considered to be characteristic of all the stainless steels (e.g., 304L vs. 304, 316L vs. 316, and 310S vs. 310) and is attributed to the strengthening effect of carbon.

The ultimate tensile and yield strengths of AISI grades 304, 321, and 347 are compared in Fig. 11.5 (*Handbook on Materials for Superconducting Machinery*, 1977). The results, which are essentially the same, indicate that the small additions of titanium in type 321 and niobium plus tantalum in type 347 do not influence the mechanical behavior of these alloys at temperatures between 76 and 300 K.

The ductility of the annealed 300-series stainless steels is generally excellent at cryogenic temperatures. Elongation and reduction of area tend to drop with decreasing temperatures, but values generally exceed 30%.

Toughness. The austenitic stainless steels retain excellent toughness at cryogenic tempera-

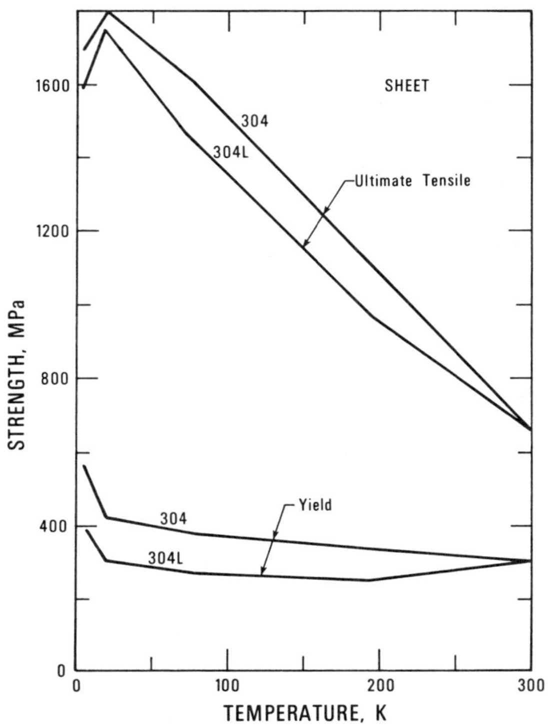

Figure 11.4 Tensile and yield strengths of two austenitic stainless steels — AISI types 304 and 304L — at temperatures between 4 and 300 K (*Handbook on Materials for Superconducting Machinery,* 1977).

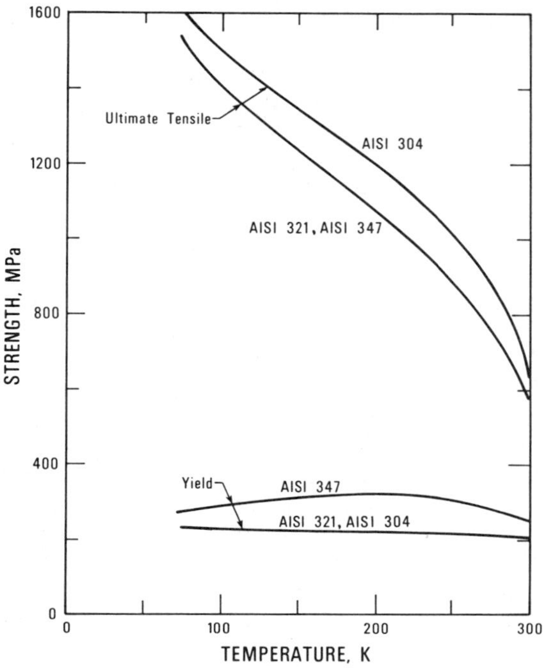

Figure 11.5 Tensile and yield strengths of three austenitic stainless steels — AISI types 304, 321, and 347 — at temperatures between 4 and 300 K (*Handbook on Materials for Superconducting Machinery,* 1977).

tures. The Charpy V-notch impact toughness does decrease with temperature (*Handbook on Materials for Superconducting Machinery*, 1977), but in all cases, it still far exceeds the 20-J (15-ft·lbf) requirement often used in codes and standards. The Charpy-toughness–vs.–temperature curves are characteristic for f.c.c. alloys; that is, ductile-to-brittle transition does not occur. The minimum temperature for Charpy testing in the United States is 20 K. A test facility for conducting the tests at 4 K is being used in the Soviet Union; however, it is likely that adiabatic heating limits the usefulness of data obtained below 20 K. Low-strain-rate tests using notched tensile specimens indicate that notch sensitivity does not develop at 4 K. Data for several alloys and product forms tested at 4 K are presented in Fig. 11.6 (*Handbook on Materials for Superconducting Machinery*, 1977). These data indicate that the notch-to-unnotched ratio of tensile strengths (NTS/TS) is approximately equal to 1, and the ratio of notched tensile strength to unnotched yield strength (NTS/YS) generally exceeds 2.

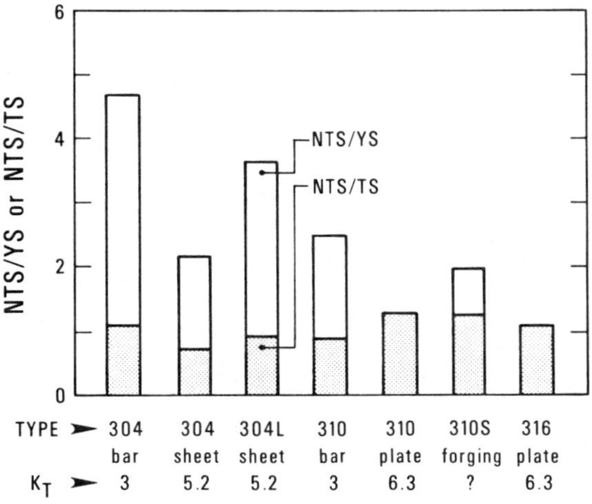

Figure 11.6 Notched tensile properties of five austenitic stainless steels — AISI types 304, 304L, 310, 310S, and 316 — at 4 K. K_T is the stress concentration factor of the notch (*Handbook on Materials for Superconducting Machinery,* 1977).

J-integral methods (see Chapter 8) have been used to measure the fracture toughness of AISI grades 310 and 316 at temperatures to 4 K. As shown in Fig. 11.7, the toughness, $K_{Ic}(J)$, at cryogenic temperatures exceeds the toughness at room temperature (*LNG Materials and Fluids*, 1978). The ratio of toughness to yield strength $[K_{Ic}(J)/\sigma_{ys}]$ is sufficiently high to ensure gross ductile deformation prior to fracture.

Fatigue. Strain–cycling fatigue properties at 295, 76, and 4 K were measured for AISI grades 304L and 310 by Nachtigall (1975) and for grades 304L and 316 by Shepic and Schwartzberg (1978). The results for 304L, which were essentially the same in both investigations, are shown in Fig. 11.8. The fatigue resistances of 310 and 316 were superior to that of 304L, particularly in low-cycle fatigue (less than 10^4 cycles). For each alloy, the fatigue resistance at low temperatures was superior to the fatigue resistance at room temperature, except at the highest strain ranges where failure occurred in less than 10^3 cycles.

The fatigue crack growth behavior (see Chapter 8 for test methods) of AISI grades 304, 304L, 310S, and 316 has been determined at 295, 76, and 4 K by Tobler and Reed (1978). The data for 304L are shown in Fig. 11.9. Notice that the growth rates are essentially the same at 76 and 4 K; similar behavior was observed in each of the other alloys. Also, notice that the growth rates at 76 and 4 K are less than those at room temper-

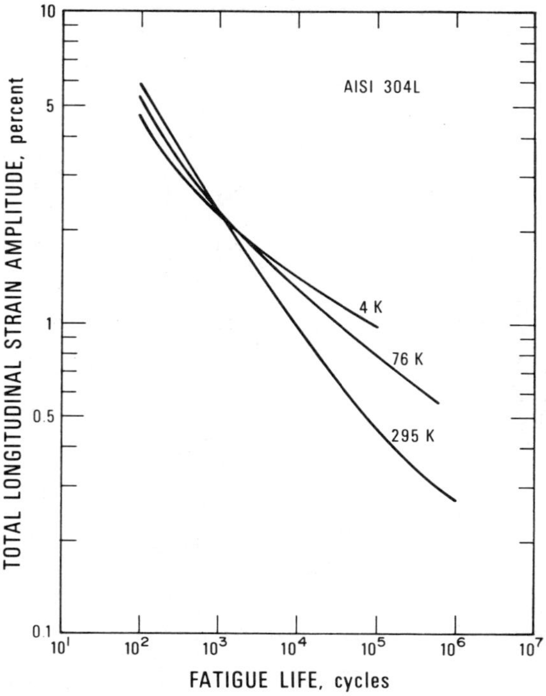

Figure 11.8 Strain–cycling fatigue curves for AISI type 304L austenitic stainless steel at 295, 76, and 4 K (Nachtigall, 1975).

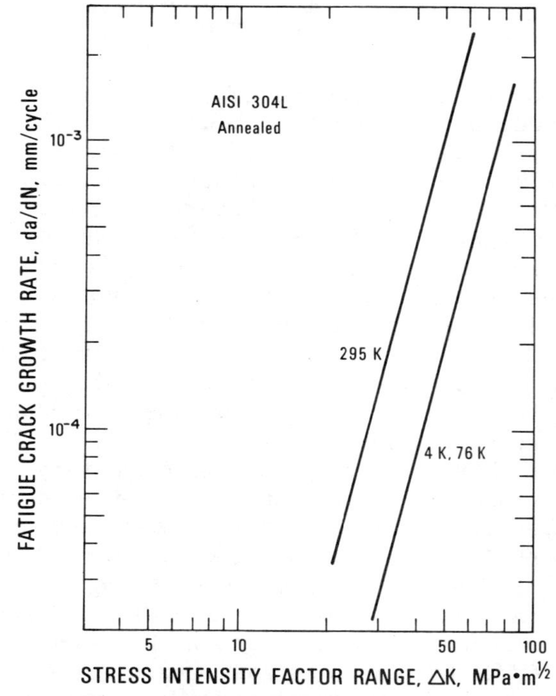

Figure 11.9 Fatigue crack growth rates of AISI type 304L austenitic stainless steel at 295, 76, and 4 K (Tobler and Reed, 1978).

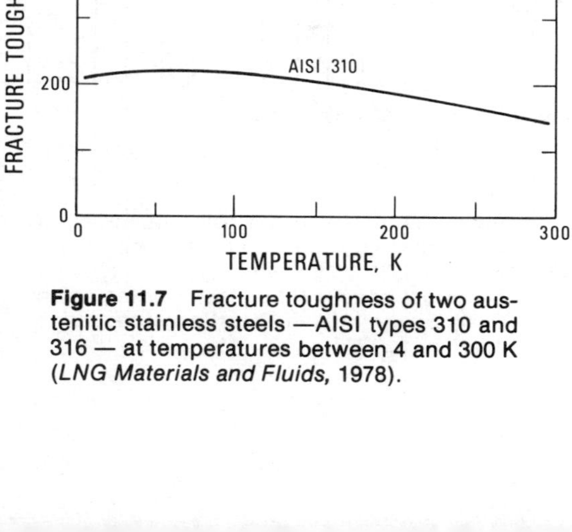

Figure 11.7 Fracture toughness of two austenitic stainless steels —AISI types 310 and 316 — at temperatures between 4 and 300 K (*LNG Materials and Fluids*, 1978).

ature. The best-fit lines through the 304L data at 295 K and at 76 and 4 K form the approximate scatter bands for the data on 304, 310S, and 316 at 295, 76, and 4 K. Thus, it can be concluded that alloy content and temperature have minimal influence on the fatigue crack growth behavior of the 300-series stainless steels.

Physical properties. The physical properties of the 300-series stainless steels at 295, 77, and 4 K are summarized in Table 11.3. The thermal properties have been taken from the best-fit lines used in *LNG Materials and Fluids* (1978), and the original sources are given for the elastic properties (Ledbetter et al., 1975), electrical resistivity (Clark et al., 1970), and magnetic permeability (Efferson and Leonard, 1976).

11.2.3 Annealed nitrogen-strengthened grades

Nitrogen significantly increases the yield strength of austenitic stainless steels, particularly at cryogenic temperatures. The influence of nitrogen on the low-temperature strength and ductility of AISI 304 (Sanderson and Llewellyn, 1969) is shown in Fig. 11.10. Notice that the 77-K yield strength of 304N (AISI 304 with deliberate additions of nitrogen) is approximately three times greater than the yield strength of 304, their ultimate strengths are approximately equal, and the elongation of the 304N is superior at temperatures below 220 K. This remarkable improvement in mechanical properties occurs in many nitrogen-strengthened grades.

The designations and compositions of the commercially available grades suitable for cryogenic service are summarized in Table 11.4. They are divided into two basic types: the Cr–Ni–N and the Cr–Ni–Mn–N alloys. The Cr–Ni–N alloys are 300-series stainless steels with deliberate additions of nitrogen. In Europe, nitrogen is deliberately added to several of the common Cr–Ni 300-series stainless steels. The equivalent AISI designations for these European grades are 304LN, 316LN, and 347N. In United States specifications, the nitrogen range is 0.10 to 0.16% for 304N and 316N. The corresponding European specifications permit nitrogen levels to 0.25%. The manganese in Cr–Ni–Mn–N alloys provides austenite stability and can thus be used to replace part of the nickel content. Manganese also increases the solubility of nitrogen in austenite and thus permits higher nitrogen levels; 0.4% N (max.) is typical.

Nickel-free Cr–Mn–N stainless steels have been evaluated for cryogenic service with mixed results. An 18Cr–15Mn alloy with high nitrogen contents (0.38 and 0.54%) exhibits a ductile-to-brittle transition between room temperature and 76 K (Defilippi, Brickner, and Gilbert, 1969). A Soviet 13Cr–19Mn alloy with 0.16% N exhibited excellent toughness at 76 K and reasonably good toughness at 4 K (Tobler, McHenry, and Reed, 1978).

Table 11.3 Physical properties of austenitic stainless steels.

Stainless Steel and Temperature	Density[a] ($g \cdot cm^{-3}$)	Young's Modulus[a] (GPa)	Shear Modulus[a] (GPa)	Poisson's Ratio[a]	Thermal Conductivity[b] ($W \cdot m^{-1} \cdot K^{-1}$)	Thermal Expansion[b] (mean) ($K^{-1} \cdot 10^{-6}$)	Specific Heat[b] ($J \cdot kg^{-1} \cdot K^{-1}$)	Electrical Resisitivity[c] ($\mu\Omega \cdot cm$)	Magnetic Permeability[d] (initial)
AISI 304									
295 K	7.86	200	77.3	0.290	14.7	15.8	480	70.4	1.02
77 K		214	83.8	0.278	7.9	13.0	—	51.4	—
4 K		210	82.0	0.279	0.28	10.2	1.9	49.6	1.09
AISI 310									
295 K	7.85	191	73.0	0.305	11.5	15.8	480	87.3	1.003
77 K		205	79.3	0.295	5.9	13.0	180	72.4	—
4 K		207	79.9	0.292	0.24	10.2	2.2	68.5	1.10
AISI 316									
295 K	7.97	195	75.2	0.294	14.7	15.8	480	75.0	1.003
77 K		209	81.6	0.283	7.9	13.0	190	56.6	—
4 K		208	81.0	0.282	0.28	10.2	1.9	53.9	1.02

[a]Ledbetter, Weston, and Naimon (1975).
[b]*LNG Materials and Fluids* (1978).
[c]Clark, Childs, and Wallace (1970).
[d]Efferson and Leonard (1976).

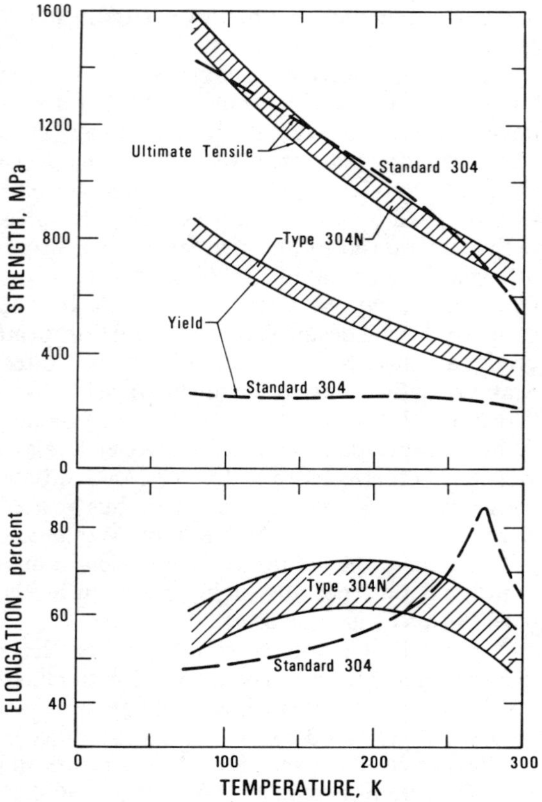

Figure 11.10 Tensile properties of two austenitic stainless steels — AISI types 304N (with 0.2% N) and 304 — at temperatures between 77 and 300 K (Sanderson and Llewellyn, 1969).

Tensile properties. The yield and ultimate strengths of the Cr–Ni–N 300-series stainless steels are summarized in Fig. 11.11. The solid lines and the broken lines represent the data of Randak, Wessling, Bock, Steinmaurer, and Faust (1971) for 304LN and 316LN, respectively. The various symbols depict the data for seven other investigations (McHenry and Reed, 1980), each on one of the following alloys: 304LN, 304N, and 316LN. As is the case for 304 and 316 (Fig. 11.3), alloys 304N and 304LN have approximately 20% greater ultimate strength at cryogenic temperatures than alloy 316LN. All three alloys have essentially the same yield strength at and below room temperature. There is relatively little scatter in the yield strength data, considering that data on 17 heats from a variety of sources are plotted. The nitrogen contents of the 17 heats range from 0.09 to 0.17%.

With the exception of type AISI 201, the tensile properties of each alloy in Table 11.4 have been evaluated at cryogenic temperatures. The 21Cr–6Ni–9Mn alloy has been most thoroughly evaluated. Its tensile and yield strength data are summarized in Fig. 11.12 (Tobler and Reed, 1979). This alloy typically has about 25% greater yield strength than the Cr–Ni–N alloys. Representative tensile data on each of the nitrogen-strengthened Cr–Ni–Mn alloys are summarized in Table 11.5. Since yield strength increases and ductility decreases with decreasing temperature, fracture resistance must be considered when selecting these alloys for service at temperatures below 77 K.

Toughness. The nitrogen-strengthened stainless steels have excellent toughness at room

Table 11.4 Compositions of nitrogen-strengthened austenitic stainless steels.

Type	Nominal Composition (%)[a]					
	Cr	Ni	Mn	C, max.	N	Other
AISI 201	16–18	3.5–5.5	5.5–7.5	0.15	0.25, max.	
AISI 202	17–19	4–6	7.5–10	0.15	0.25, max.	
AISI 304N	18–20	8–10.5	2.0, max.	0.08	0.10–0.16	
304LN	18–20	8–12	2.0, max.	0.03	0.10–0.16	
AISI 316N	16–18	10–14	2.0, max.	0.08	0.10–0.16	
AISI 316LN	16–18	10–14	2.0, max.	0.03	0.10–0.16	
ASTM XM-10[b]	19–21.5	5.5–7.5	8–10	0.08	0.15–0.40	
ASTM XM-11[b]	19–21.5	5.5–7.5	8–10	0.04	0.15–0.40	
ASTM XM-14[b]	17–19	5–6	14–16	0.12	0.35–0.50	
ASTM XM-19[b]	20.5–23.5	11.5–13.5	4–6	0.06	0.20–0.40	0.10–0.30 Cb
ASTM XM-29[b]	17–19	2.25–3.75	11.5–14.5	0.08	0.20–0.40	0.10–0.30 V 1.5–3.0 Mo

[a]For each grade: Si = 1.00, max.; P = 0.060, max.; S = 0.030, max.
[b]ASTM A412.

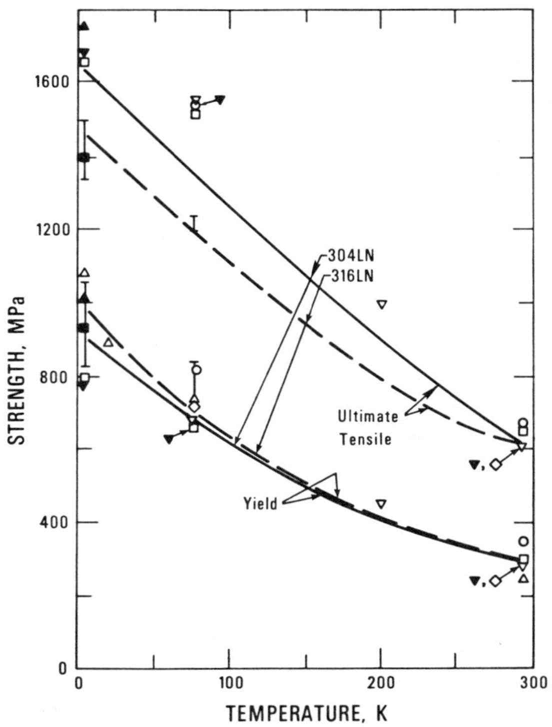

Figure 11.11 Tensile and yield strengths of nitrogen-strengthened 300-series stainless steels at temperatures between 4 and 300 K (McHenry and Reed, 1980). ▽, □, ○ — 304N; △, ▼, ▲ — 304LN; ■, ◇, I — 316LN.

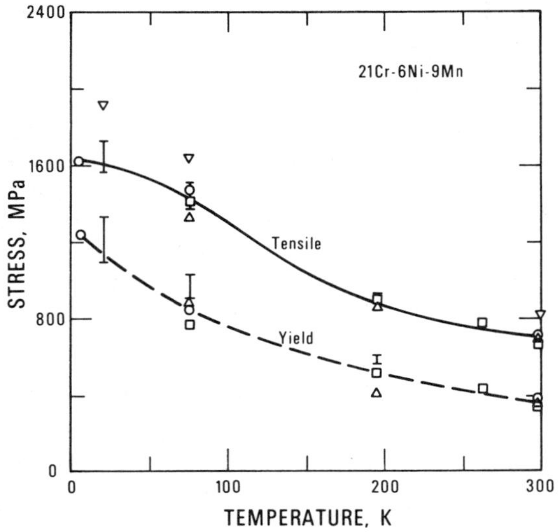

Figure 11.12 Tensile and yield strengths of 21Cr–6Ni–9Mn stainless steel at temperatures between 4 and 300 K (Tobler and Reed, 1979).

temperature. However, as shown in Fig. 11.13, significant toughness losses generally occur as the temperature is reduced (Read and Reed, 1979). At 4 K, the loss in toughness is proportional to the increases in yield strength, as shown in Fig. 11.14.

Read and Reed (1979) observed significant variations in the toughness of a single piece of 21Cr–12Ni–5Mn alloy. As-received mill-annealed material having an intergranular microconstituent had a plane strain fracture toughness at 4 K of 111 $MPa \cdot m^{1/2}$. A high-temperature (1177°C, 1.5-h) anneal dissolved the microconstituent and increased the toughness to 176 $MPa \cdot m^{1/2}$. Thus, certain nitrogen-strengthened grades can be used at 4 K, but one must be careful that the applicable melting practice, product form, and heat treatment for the alloy selected result in satisfactory toughness at 4 K.

Fatigue. The strain-cycling fatigue behavior of 21Cr–6Ni–9Mn has been measured at 295, 77, and 4 K by Shepic and Schwartzberg (1978). The results, presented in Fig. 11.15, exhibit crossover behavior similar to that of the 300-series stainless steels; that is, at high-cycle lives (greater than 10^4 cycles), the fatigue strength improves with decreasing temperature, and vice versa at low-cycle lives. A comparison of the results at 4 K (shown in Fig. 11.16) indicates that the 21Cr–6Ni–9Mn alloy has fatigue strength better than 304L but not as good as 316 (Shepic and Schwartzberg, 1978). Thus, the significantly higher strength of the nitrogen-strengthened grade does not result in a comparable improvement in fatigue life.

The fatigue crack growth behavior of several nitrogen-strengthened grades has been studied at cryogenic temperatures by Read and Reed (1979). The results at 4 K depend largely on the relative austenite stability of the alloys. For the least stable alloy evaluated, 18Cr–3Ni–13Mn, the fatigue crack growth rates are 6 times greater at 76 K than at room temperature, and even greater at 4 K. For the most stable alloy evaluated, 21Cr–12Ni–5Mn, the fatigue crack growth rates fall within the same scatter band at 295, 76, and 4 K. The fatigue behavior of types 304N and 21Cr–6Ni–9Mn is intermediate: the growth rates are essentially the same at 295 and 76 K, but increase 2- to 3-fold at 4 K. For comparison purposes, recall that the 300-series stainless steels with normal nitrogen levels generally have good fatigue resistance at cryogenic temperatures.

Tobler and Reed (1982) studied the influence of interstitial carbon and nitrogen on the fatigue

Table 11.5 Tensile properties of nitrogen-strengthened Cr-Ni-Mn stainless steels.

Alloy	Grade ASTM A412	Temperature (K)	Yield Strength (MPa)	Tensile Strength (MPa)	Elongation (%)	Reduction of Area (%)	References
18Cr–5Ni–9Mn N=0.08	202	300	324	703	74		Gunter and Reed (1961)
		76	594	1520	33		
		20	772	1160	12		
	202	300	305	634	81		Brickner and Defilippi (1977)
		77	674	1138	21[a]	17[a]	
21 Cr–6Ni–9Mn[b] N=0.28	XM-11	295	353	701	61	78	Tobler and Reed (1979)
		76	899	1470	43	37	
		4	1240	1630	16	40	
18Cr–5Ni–15Mn N=0.36	XM-14	295	432	805	59	69	Brickner and Defilippi (1977)
		77	955	1644	33	54	
21Cr–12Ni–5Mn	XM-19	295	426	766	45	69	Read and Reed (1981)
		76	916	1510	76	62	
		4	1330	1850	16	38	
18Cr–3Ni–13Mn N=0.37	XM-29	295	440	796	56	53	Read and Reed (1981)
		76	1140	1520	18	24	
		4	1540	1810	4	26	

[a]Broke at gauge mark.
[b]See also Fig. 11.13.

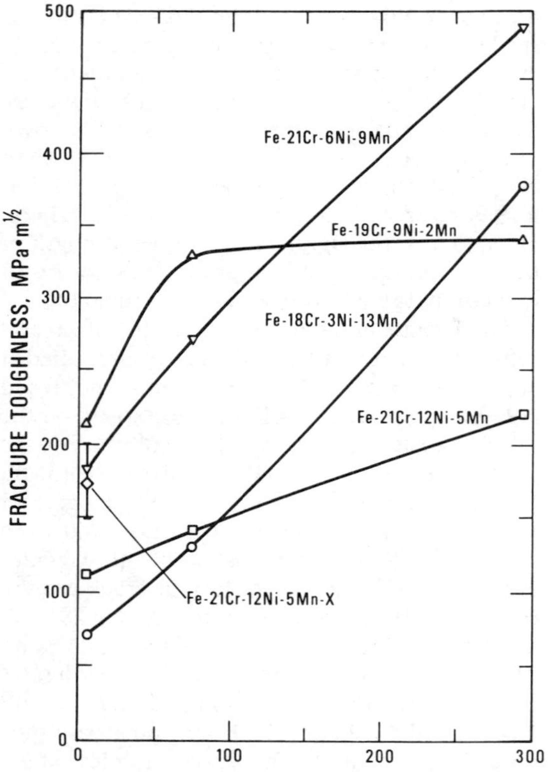

Figure 11.13 Fracture toughness of five nitrogen-strengthened austenitic stainless steels at temperatures between 4 and 300 K (Read and Reed, 1979).

crack growth resistance of nine stainless steels with AISI 304 chemistry, but with carbon contents ranging from 0.03 to 0.09% and nitrogen contents ranging from 0.04 to 0.64%. The results (shown in Fig. 11.17) indicate that alloys with greater than 0.13% C+N (alloys 3 to 9) have greater 4-K crack growth rates than those of lower interstitial alloys and, to a lesser extent, those at room temperature.

Physical properties. The physical properties of the 300-series nitrogen-strengthened stainless steels are similar to those of the 300-series alloys (summarized in Table 11.3). In the Cr–Ni–Mn grades, anomalies have been observed by Ledbetter (1977) in the elastic behavior and by Ledbetter and Collings (1979) in some of the magnetic transitions and specific heat. In general, the nitrogen contributes to austenite stability and thus reduces the magnetic permeability induced in some alloys by deformation at low temperatures.

11.2.4 High-strength cold-rolled sheet

Most austenitic stainless steels are available in the cold-worked condition as sheet, strip, rod, and wire. Heavier product forms, such as plate and bar, are not usually cold-worked owing to the nonuniformity of deformation through the thickness. Cold-rolled sheet has been extensively used at cryogenic temperatures, primarily for

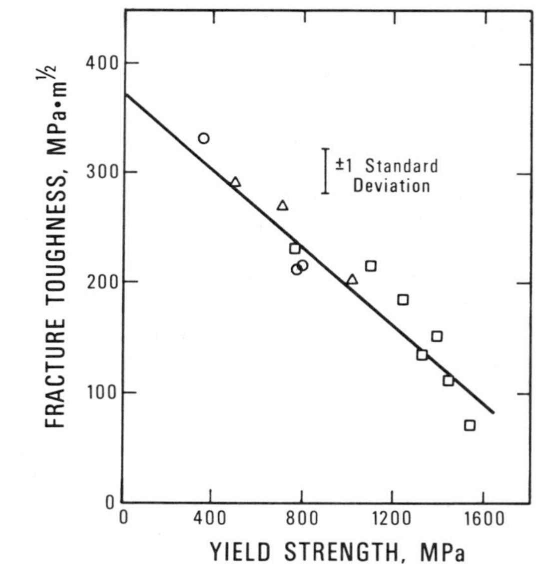

Figure 11.14 Trend line for fracture toughness vs. yield strength at 4 K for austenitic stainless steels (Read and Reed, 1981). ○ — 304 (N); △ — 316 (N); □ — other austenitic stainless steels.

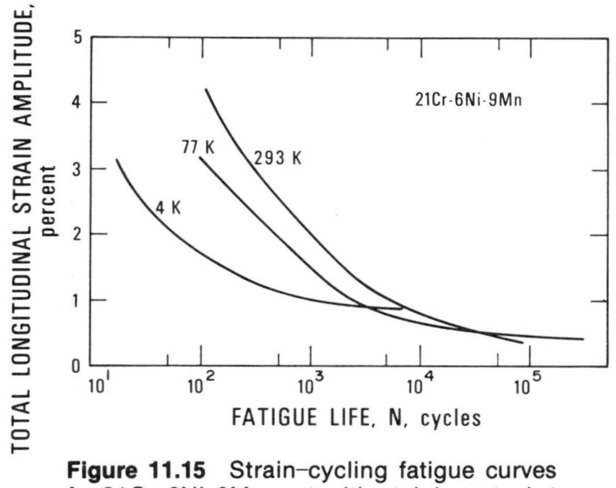

Figure 11.15 Strain–cycling fatigue curves for 21Cr–6Ni–9Mn austenitic stainless steel at room temperature, 76, and 4 K (Shepic and Schwartzberg, 1978).

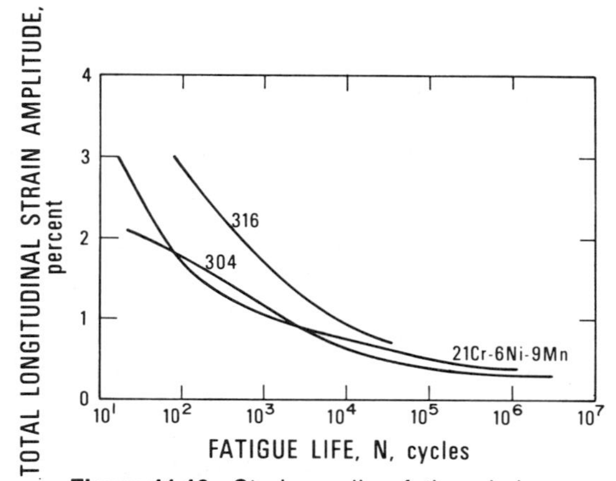

Figure 11.16 Strain–cycling fatigue behavior of three austenitic stainless steels — AISI types 304 and 316 and a 21Cr–6Ni–9Mn alloy — at 4 K (Shepic and Schwartzberg, 1978).

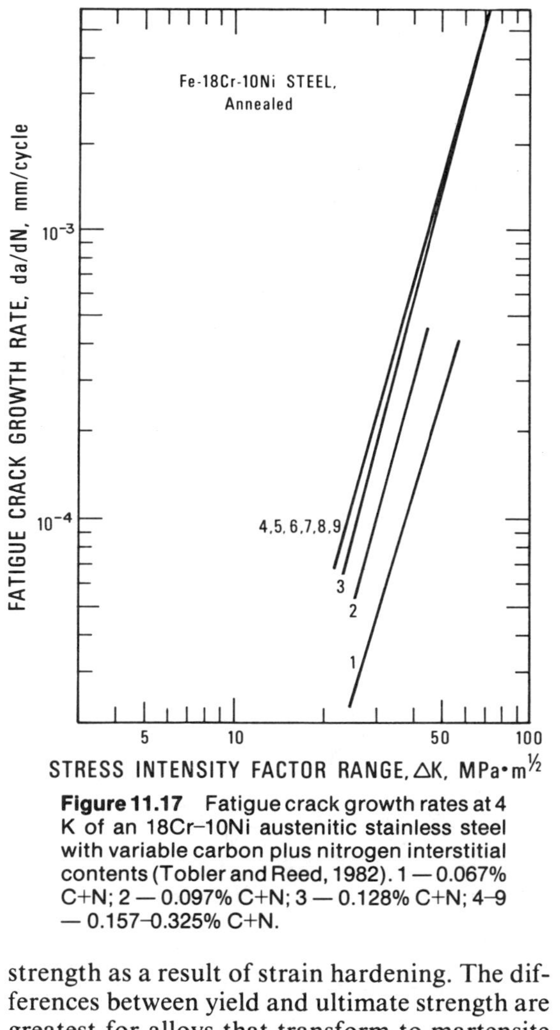

Figure 11.17 Fatigue crack growth rates at 4 K of an 18Cr–10Ni austenitic stainless steel with variable carbon plus nitrogen interstitial contents (Tobler and Reed, 1982). 1 — 0.067% C+N; 2 — 0.097% C+N; 3 — 0.128% C+N; 4–9 — 0.157–0.325% C+N.

missile cases operating at 20 K (liquid hydrogen) or 90 K (liquid oxygen). AISI grades 301, 302, 304L, and 310 have been evaluated for these applications (Watson and Christian, 1961a, 1961b).

Cold working increases the strength of stainless steels by strain hardening and, in metastable alloys, by the strain-induced martensitic transformation. Both of these effects are apparent in the tensile properties of annealed stainless steels. Recall from Figs. 11.3, 11.4, and 11.5 that the ultimate strength is much greater than the yield strength as a result of strain hardening. The differences between yield and ultimate strength are greatest for alloys that transform to martensite during the test; examples are AISI grades 304, 321, and 347. The strengthening of interest here is the increase in yield strength, which is caused by the cold-working operations, not by the tensile test. Yield strength increases are assumed to be similar for all the AISI 300-series grades.

Tensile properties. Cold rolling causes remarkable increases in yield strength, somewhat less of an increase in ultimate strength, and a severe reduction of ductility. The influence of cold working on the yield strength and elongation of AISI 310 at room temperature and 20 K is shown in Fig. 11.18 (Christian et al., 1964). At 20 K, the yield strength increases 2.4-fold as the cold work increases from 0 to 92%; the corresponding increase in ultimate strength is 1.6-fold. The ductility at 20 K decreases linearly with the percent of cold work, but remains 18% at 60% cold work, the extra-full-hard condition. In contrast, the room-temperature ductility is much lower at all levels of cold work and is only 3% at 60% cold work. Increased ductility at cryogenic temperatures is characteristic of austenitic stainless steels and has been used to advantage in forming operations, referred to as *cryoforming* (Schwartzberg and Kiefer, 1975).

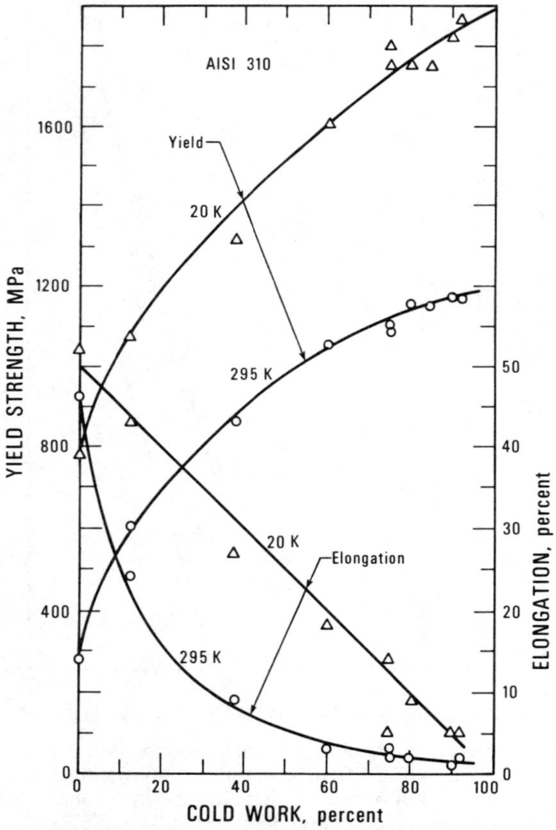

Figure 11.18 Tensile yield strength and elongation of AISI type 310 austenitic stainless steel at room temperature and 20 K as a function of percent cold work (Christian, Gruner, and Girton, 1964).

Metastable alloys, such as AISI grades 301 and 302, exhibit even greater increases in yield strength as a function of cold work owing to the strain-induced martensitic transformation. At 60% cold work, the yield strengths at 20 K decrease with increased austenite stability as follows: AISI 301 — σ_{ys} = 1940 MPa, AISI 302 — σ_{ys} = 1720 MPa, and AISI 310 — σ_{ys} = 1600 MPa. The mechanical properties of AISI 301 in the 60% cold-worked condition are summarized in Fig. 11.19 (Watson and Christian, 1961a). The drop in elongation that occurs between 77 and 20 K is accompanied by an increase in notch sensitivity and limits the usefulness of this alloy to temperatures 77 K and higher.

Toughness. Notched tensile tests are usually used to evaluate the toughness of high-strength sheet alloys. Charpy impact tests and plane strain fracture toughness tests have specimen-

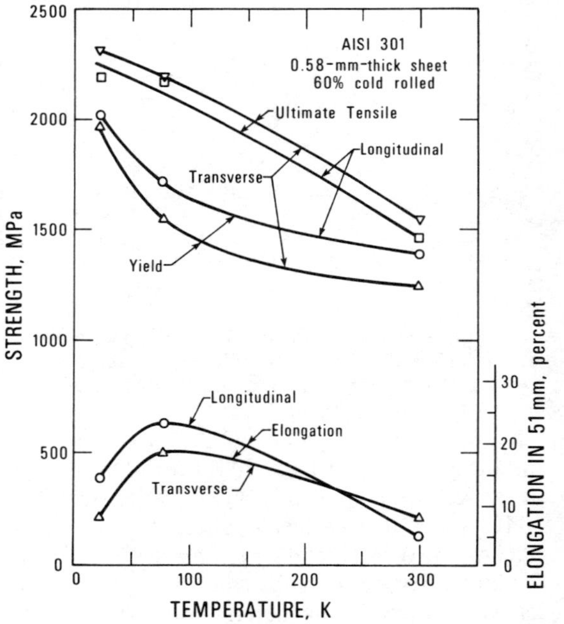

Figure 11.19 Tensile properties of AISI type 301 austenitic stainless steel, 60% cold rolled, at temperatures between 20 and 300 K (Watson and Christian, 1961a).

thickness requirements that limit their applicability to sheet products. The ratio of tensile strengths in notched and unnotched specimens is used as the index of toughness. For stable alloys, such as AISI 310, the notched (K_T = 6.3)-to-unnotched tensile ratio for room and cryogenic temperatures is equal to 1 ±0.1 for cold reductions to 92%.* Such is not the case for metastable alloys. For AISI 301, the notched (K_T = 6.3)-to-unnotched tensile ratio varies from 0.4 to 1.1, depending on the percent of cold work, test temperature, and specimen orientation, as shown in Fig. 11.20. Watson and Christian (1961a) relate this behavior to the martensitic transformation. The maxima in the notched-to-unnotched tensile ratios in Fig. 11.20 at 50 to 65% cold reduction were attributed to the stabilizing influence (during cryogenic testing) of cold working up to about 60%; cold reductions greater than 60% caused enough martensite to form during rolling to reduce notch strength at all test temperatures. The reduced toughness in the transverse specimens was attributed to the preferred orientation of the martensite rather than to the effects of inclusions.

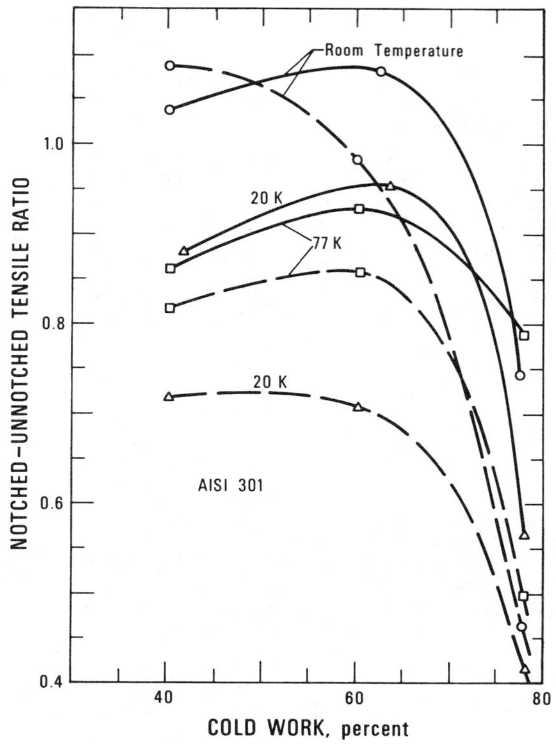

Figure 11.20 Notched-unnotched tensile ratio (K_T = 6.3) as a function of cold work for AISI type 301 austenitic stainless steel (Watson and Christian, 1961a). Solid line — longitudinal; broken line — transverse.

Fatigue. The cryogenic fatigue behavior of AISI 301 sheet, cold rolled approximately 60%, is shown in Fig. 11.21 (Favor et al., 1961). A comparison of these data with similar data on annealed AISI 347, also obtained by Favor et al. (1961), indicates that AISI 301 with a yield strength of 1530 MPa at 295 K has much better fatigue strength at room temperature than annealed AISI 347 with a yield strength of 255 MPa at 295 K. At cryogenic temperatures, however, the differences in fatigue strength are marginal; the annealed AISI 347 has better fatigue strength at fatigue lives less than 10^5 cycles and cold-rolled AISI 301 has better fatigue strength at fatigue lives greater than 10^5 cycles. At 20 K, the fatigue strength of AISI 347 is superior to that of AISI 301 at all cyclic lives. Thus, in fatigue critical applications at cryogenic temperatures, the increased yield strength of cold-rolled 301 stainless steels is of marginal value.

Notched fatigue (K_T = 3.1) tests were conducted on the same materials by Gideon, Favor, Koppenhafer, Grover, and McClure (1962). Their results further confirm that the high strength of cold-rolled AISI 301 stainless steel is not beneficial in fatigue critical applications at cryogenic temperatures. The notched fatigue behavior exhibited the same trends apparent in Fig. 11.21, that is, improved fatigue strength (but to a lesser extent) at room temperature and crossover behavior at approximately 10^5 cycles at 77 K.

The applicability of the fatigue test results on AISI 301 to other cold-rolled austenitic stainless steels has not been established. However, the trends discussed should serve as a warning to those who wish to use cold-rolled sheet in highly stressed, fatigue critical applications: Improved tensile strength does not necessarily result in improved fatigue strength, particularly at cryogenic temperatures.

11.2.5 Fabrication characteristics

The widespread use of austenitic stainless steels by a variety of industries attests to their generally satisfactory fabricability. The fabrication characteristics discussed below should re-

*K_T is the ratio of the maximum stress at the root of the notch to the nominal stress.

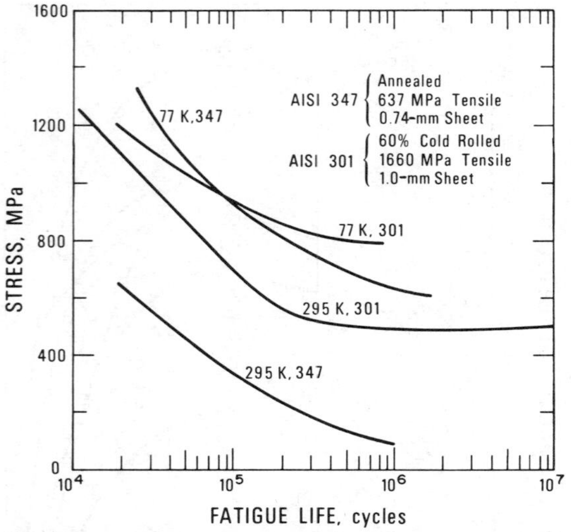

Figure 11.21 Flexural fatigue ($R = -1$) strength of two austenitic stainless steels — annealed AISI 347 and 60% cold-rolled AISI 301 — at room temperature and 77 K (Favor, Gideon, Grover, Hayes, and McClure, 1961).

ceive special consideration for cryogenic applications.

Welding. The austenitic stainless steels, with and without nitrogen strengthening, are readily weldable by all of the common welding processes, providing the appropriate consumables and procedures are used. The service experience with welded assemblies of the alloys has been satisfactory at cryogenic temperatures. The American Welding Society (AWS) designation system, which is similar to the AISI system for wrought products, is generally used to classify stainless steel filler metals. The AWS designations and chemical compositions of the filler metals commonly used for cryogenic applications are summarized in Table 11.6.

The strength of stainless steel welds at 4 K generally exceeds the corresponding base metal strength, but toughness is usually significantly lower. Four phenomena that affect the strength and toughness of the as-deposited weld metal take on added significance at cryogenic temperatures: sensitization, ferrite content, nitrogen pickup, and oxide inclusions. In addition, heat treatment generally has a stronger influence on the weld metal than on the base metal.

Sensitization, the grain-boundary precipitation of chromium carbides, reduces weld-metal toughness at low temperatures (Read, McHenry, Steinmeyer, and Thomas, 1980). Consequently, weld metals with very low carbon, such as 308L (0.04% C, max.) and 316L (0.03% C, max.) are commonly used for cryogenic applications. Sensitization is most extreme if the weldment is given prolonged exposure in the temperature range 500 to 800° C. In multipass weldments, the underlying weld beads may receive sufficient exposure to these temperatures to cause significant toughness reductions in stainless steels with greater than 0.03% C. Subsequent annealing of the welds at temperatures greater than 950° C dissolves the carbides and improves the toughness.

Table 11.6 Composition limits of selected filler metals for welding austenitic stainless steels.

AWS Type	Composition (%) Cr	Ni	C, max.	Other
Covered Electrodes[a]				
E308	18–21	9–11	0.08	
E308L	18–21	9–11	0.04	
E310	25–28	20–22.5	0.20	0.75 Si, max.
E16-8-2	14.5–16.5	7.5–9.5	0.10	0.50 Si, max.; 1–2 Mo
E316	17–20	11–14	0.08	2–2.5 Mo max.; 1–2 Mo
E316L	17–20	11–14	0.04	
E330	14–17	33–37	0.25	
Rod and Bare Electrodes[b]				
ER308	19.5–22	9–11	0.08	
ER308L	19.5–22	9–11	0.03	
ER310	25–28	20–22.5	0.08–0.15	
ER310	18–20	11–14	0.08	2–3 Mo
ER316	18–20	11–14	0.03	2–3 Mo

[a]For covered electrodes: Mn = 2.5, max.; Si = 0.9, max.; P = 0.04, max.; S = 0.03, max.
[b]For rod and bare electrodes: Mn = 1.0–2.5; Si = 0.25–60; P = 0.03, max.; S = 0.03, max.

The ferrite content of as-deposited weld metals can be estimated by using the Schaeffler constitution diagram. To avoid microfissuring, the weld metal chemistry is usually balanced to provide 4 to 10% ferrite. However, for cryogenic service, there is concern that the ferrite, which is a b.c.c. phase and therefore brittle at low temperatures, may embrittle the weld metal. Szumachowski and Reid (1978) have shown that ferrite reduces toughness at 77 K for a wide range of stainless steel weld metals. Read et al. (1980) have evaluated the effect of ferrite content on the fracture toughness of shielded metal-arc welds of AWS 316 and 316L; the results at 76 K and 4 K are shown in Fig. 11.22. Elmer, McHenry, and Whipple (1981) have evaluated the strength and toughness of several fully austenitic weld metals at 4 K. Their results for 20Cr–16Ni–7Mn–2.8Mo–2.1W, 17Cr–16Ni–4Mn–2.2Mo, 18Cr–20Ni–6Mn–0.27Nb, and 16Cr–16Ni–4Mn–1.5Mo indicate that strength and toughness combinations comparable to the base metal properties are possible.

Nitrogen increases the yield strength and decreases the toughness of stainless steel weld metals. When conventional (low-nitrogen) filler metals are used, nitrogen variations of 0.04 to 0.13% may occur, depending on the amount of air that enters the arc-shielding medium. The greatest variations in nitrogen pickup are associated with gas metal-arc (GMA) welding, where the nitrogen content ranges from 0.04 to 0.13%, depending on the type of gas coverage, the electrode extension, and the transfer mechanism (short-circuit is better than spray transfer). In shielded metal-arc (SMA) welding, smaller variations are attributed to electrode coatings; lime coatings generally give better coverage and less nitrogen pickup than titania coatings.

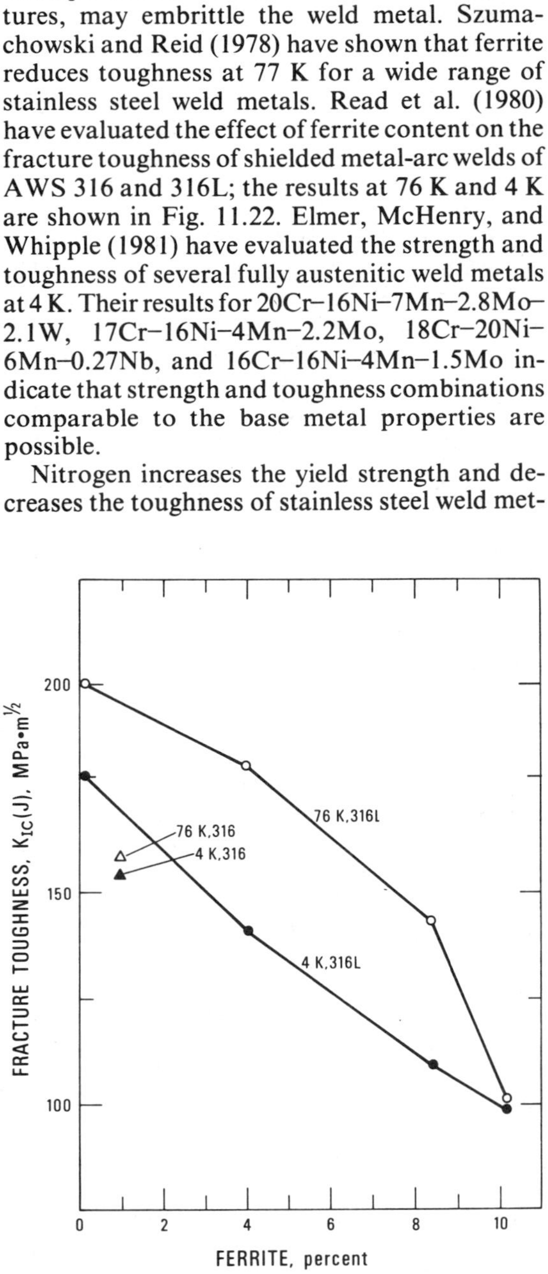

Figure 11.22 Fracture toughness of AWS 316 and 316L weld metals at 76 and 4 K as a function of ferrite content (Read et al., 1980).

The influence of oxygen content on the toughness of weld metals at cryogenic temperatures has not yet been systematically studied. However, it is well known that oxide inclusions form sites for the initiation of microvoids, and that the ductile fracture resistance decreases as the number of microvoids increases. Some evidence on the influence of oxide inclusions comes from a study by Whipple and Kotecki (1981), in which a series of AWS 316L welds were produced using three different welding processes: GMA, gas tungsten arc (GTA), and submerged arc (SA). The toughness at 4 K increased with decreasing inclusion content: SA weld — 103 MPa·m$^{1/2}$, GMA weld — 161 MPa·m$^{1/2}$, and GTA weld — 181 MPa·m$^{1/2}$. Other observations in the literature reveal the importance of inclusions. For example, Szumachowski and Reid (1978) found SMA deposits of 316L with a basic coating (lime coating, -15 electrode) generally had superior toughness to those of 316L-16 deposits with a rutile coating; basic coatings are known to have lower oxygen content and oxide inclusions than rutile coatings (Lancaster, 1965).

In summary, weld metal toughness can be improved by using low-carbon filler metals to avoid sensitization, by balancing weld-metal chemistry to avoid ferrite, by using low-nitrogen consumables and welding practices, and by using inert-gas-shielded welding processes or basic fluxes to reduce oxide inclusions. However, significant variations in the toughness still arise in ferrite-free welds containing moderate nitrogen contents (0.004 to 0.007%) that are welded by the same process. As shown in Fig. 11.23, toughness varies as a function of grain width, a feature of the weld deposit controlled by the heat input and cooling rates (Whipple et al., 1981). To minimize these toughness variations, it is imperative to qualify the welding materials and procedures

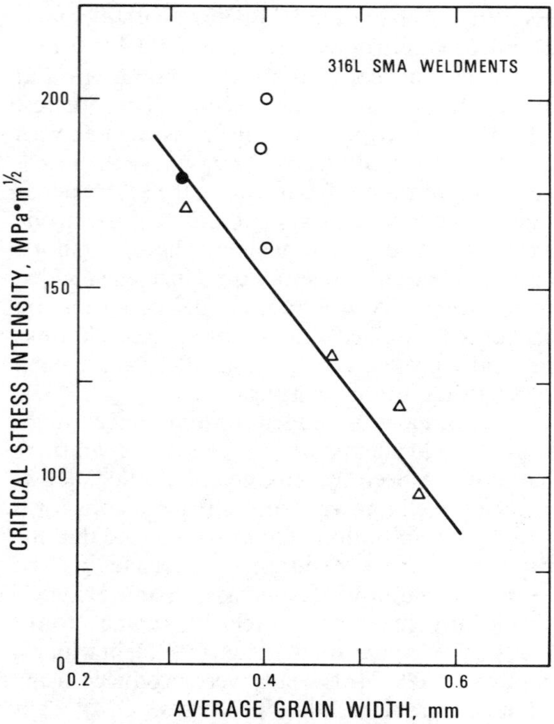

Figure 11.23 Fracture toughness of fully austenitic weld metals at 4 K as a function of grain width. ● — Read et al., 1980; △ — LCP; ○ — MFTF. LCP and MFTF refer to weld procedure qualification welds prepared for the Large Coil Project, Oak Ridge National Laboratory and the Magnetic Fusion Test Facility, Lawrence Livermore Laboratory (Whipple, McHenry, and Read, 1981).

used for construction and to ensure that the qualified procedures are used during fabrication.

Heat treatment of stainless steel welds has been studied by Bennett and Dillon (1966) for AWS types 308, 308L, and 310. Heat treating or slow cooling in the sensitization range caused significant toughness loss at 77 K in types 308 and 310, but did not affect 308L. Stress relieving at 840°C followed by furnace cooling caused about a 50% Charpy toughness loss in each alloy (20 to 32 J in as-welded and 8 to 20 J in stress-relieved alloys). Stress relieving at 950°C followed by furnace cooling caused a 50% toughness loss in 310, no change in 308L, and mixed results (±40% toughness change) in 308. Annealing at 1070°C followed by water quenching improved the toughness of 308 and 308L, but decreased the toughness of 310. The beneficial effect of a full anneal on toughness at 77 K has also been reported by Krivobok and Thomas (1950), DeLong (1974), and Mazandarany, Parker, Koenig, and Read (1980). Mazandarany et al. found that postweld annealing at 1040°C improved the 4-K $K_{Ic}(J)$ properties of 316L SMA weld deposits.

Castings. Austenitic stainless steel castings are used for cryogenic equipment having heavy sections or complex geometries that make fabrication from wrought products impractical. Typical examples are valve bodies and piston castings for liquefaction equipment. Several large castings have been used at temperatures down to 2 K for helium bubble chamber bodies. The Alloy Castings Institute (ACI) designation system is generally used for stainless steel castings. The ACI types, the equivalent AISI types for wrought products, and the chemical compositions of the casting alloys commonly used for cryogenic applications are summarized in Table 11.7.

A duplex microstructure of austenite and ferrite is typical of most of the stainless steel casting alloys. The ferrite content of the first six alloys listed in Table 11.7 may range from 0 to 40%; this is due to chemistry variations within the allowable ranges. The other three alloys (25Cr–20Ni, 20Cr–29Ni, and 16Cr–20Ni–10Mn) are fully austenitic. Ferrite content influences the strength and toughness and thus contributes to the variability observed in the mechanical properties of the castings.

The strength and ductility of five casting alloys are shown in Fig. 11.24 (Hall, 1962). The cast 304 alloy is fully austenitic, and the other alloys contain up to 15% ferrite. Notice that the cast 304 has the lowest yield strength, and the molybdenum-bearing alloys, CF 8M and CF 3M, have the highest yield strengths. The niobium-bearing alloy, CF 8C, has significantly lower ductility at cryogenic temperatures.

The influence of ferrite content, nitrogen, and sensitization on the toughness of stainless steel castings is similar to their influence on weldments. Whipple and McHenry (1982) evaluated the influences of δ-ferrite and interstitial nitrogen on the strength and toughness of stainless steel castings at 4 K using nine CF8M (nominally a 19Cr–9Ni–2Mo alloy) castings with controlled chemistries. The chromium and nickel contents were varied to obtain two series of alloys: one series of five alloys had 0.05% N and δ-ferrite contents ranging from 0 to 28.5%, the other series of five alloys had 9 ± 1% δ-ferrite and 0.02 to

Table 11.7 Compositions of austenitic stainless steel castings.

ACI Type	Nearest AISI Type	Composition[a] (%) Cr	Ni	C, max.	Other
CF-3	304L	17–21	8–21	0.03	
CF-8	304	18–21	8–11	0.08	
CF-3M	316L	17–21	9–13	0.03	2–3 Mo, 1.5 Si max.
CF-8M	316	18–21	9–12	0.08	2–3 Mo, 1.5 Si max.
CF-8C	347	18–21	9–12	0.08	(8 × %C) Nb, min. or (10%C) Nb + Ta, min.
CF-20	302	18–21	8–11	0.20	
CK-20	310	23–27	19–22	0.20	
CN-7M	—	18–22	27–31	0.07	2–3 Mo, 1.5 Si max. 3–4 Cu

[a]For each grade: Mn = 1.5, max.; P = 0.04, max.; Si = 2.0, max.; S = 0.04, max.

0.20% N. As shown in Fig. 11.25, an increase in either δ-ferrite or nitrogen content increases the yield strength and decreases ductility at 4 K. Fracture toughness decreases with increasing δ-ferrite up to a δ-ferrite content of 17% and then remains constant (Fig. 11.26). The influence of nitrogen on fracture toughness was not clearly established because the toughness at the 9% ferrite level was already a low value, 126 to 148 MPa·m$^{1/2}$ at nitrogen contents of 0.02 to 0.1%. The toughness was substantially lower at higher nitrogen contents: 74 MPa·m$^{1/2}$ at 0.14% N and 94 MPa·m$^{1/2}$ at 0.20% N. Juppenlatz (1952) reported the influence of sensitization on CF8 castings tested at 89 K. The Charpy impact toughness

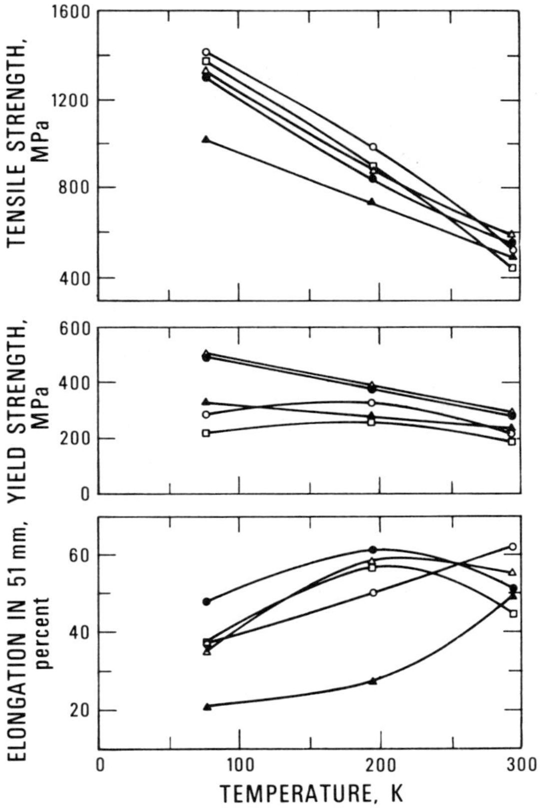

Figure 11.24 Tensile properties of five cast stainless steels at temperatures between 77 and 300 K (Hall, 1962). ○ — CF8, □ — cast 304, △ — CF8M, ● — CF3M, ▲ — CF8C.

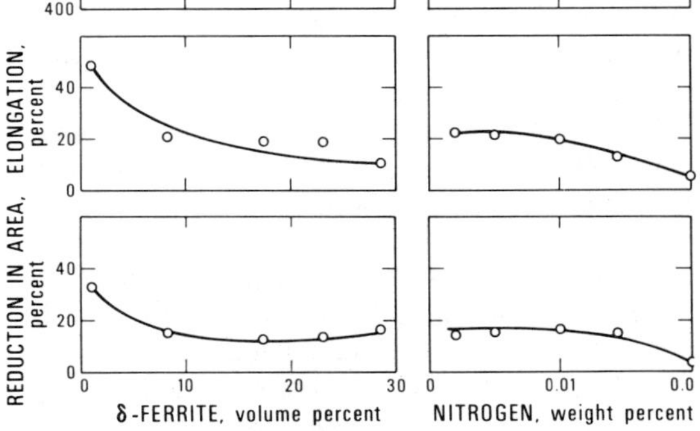

Figure 11.25 The effect of δ-ferrite and nitrogen content on the tensile properties at 4 K of CF8M austenitic stainless steel castings. The castings in the δ-ferrite study contained 0.05% N and the castings in the nitrogen study contained 9 ± 1% δ-ferrite (Whipple and McHenry, 1982).

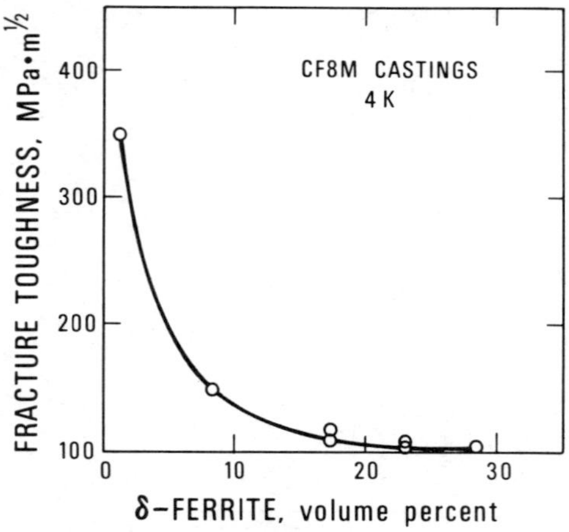

Figure 11.26 The effect of δ-ferrite on the fracture toughness at 4 K of CF8M austenitic stainless steel castings containing 0.05% N (Whipple and McHenry, 1982).

dropped from 53 J to less than 20 J for CF8 sensitized for 3 min at 675° C and to less than 10 J when sensitized for 48 h at 675° C.

Machining, forming, and forging. Annealed austenitic stainless steels are more difficult to machine than carbon steels or aluminum alloys because of their high work-hardening rates and low thermal conductivity. These properties cause poor finishes, heat buildup, and excessive tool wear. As a result, it is necessary to use heavier feeds, lower speeds, better quality tool bits, more rigid machines, and more power.

The austenitic stainless steels have excellent formability and can be forged into shapes suitable for many cryogenic applications, but they are more difficult to forge than most common structural alloys because of their high strength at elevated temperatures. Within the family of austenitic stainless steels, the forgeability of types 304 and 316 is rated good.

Inspection. During fabrication of cryogenic systems, the stainless steel material and components are likely to be inspected using conventional methods. Heavy-section stainless steel welds cannot be effectively inspected for the presence of cracks or crack-like defects. Radiography is of limited use because of its inherent inability to reliably detect cracks. Ultrasonic methods are not effective because stainless steel welds have a highly textured grain structure that is very anisotropic with respect to acoustic wave velocities. Consequently, the signals are cluttered and defect indications are masked. Standards are needed to ensure that the defect sizes that can be reliably detected do not adversely affect structural integrity.

11.3 NICKEL STEELS

The nickel steels are Fe–Ni alloys that contain relatively low carbon (<0.13%), some manganese (<1.5%), and sometimes molybdenum and chromium. The alloys have a predominantly b.c.c. crystal structure and consequently undergo a ductile-to-brittle transition as temperature is reduced. The transition temperature decreases with increasing nickel content and may be further reduced by heat treatment. The toughness transition curves obtained by Tobler, Mikesell, and Reed (1979) for 3.5, 5, and 9% Ni steels are shown in Fig. 11.27.

The commercially available steels suitable for cryogenic service (<173 K) are 3.5, 5, 5.5, 8, and 9% Ni steels. Their specifications, minimum service temperatures, and compositions are summarized in Table 11.8. These steels are normally used in plate form but can be made available as forgings, sections, bars, and pipe. They are commonly used for tanks to store and transport liquefied gases. The 3.5% Ni steel is used for containing liquid ethylene at 169 K and other liquefied gases at high temperatures, such as ethane,

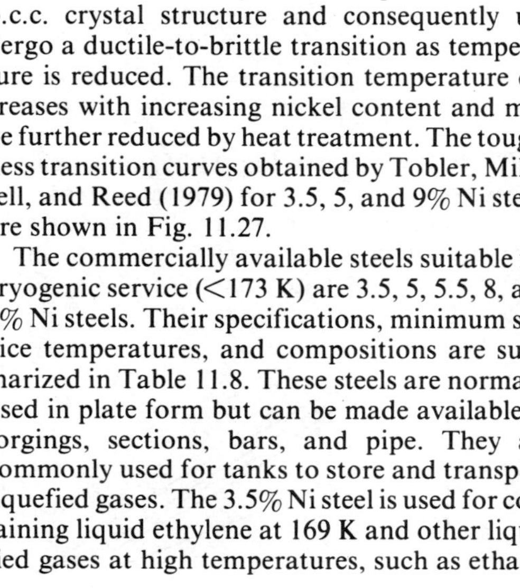

Figure 11.27 Fracture toughness of 3.5, 5, and 9% N steels at temperatures between 4 and 300 K (Tobler et al., 1976).

Table 11.8 Compositions of the nickel steels.

Alloy	Minimum Service Temp. (K)	ASTM Specification	Composition (%) C, max.	Mn	P, max.	S, max.	Si	Ni	Mo	Cr
3.5 Ni	173	A203 Grade D	0.17	0.7, max.	0.035	0.040	0.15–0.30	3.25–3.75	—	—
3.5 Ni	173	A203 Grade E	0.20	0.7, max.	0.035	0.040	0.15–0.30	3.25–3.75	—	—
5 Ni	102	0.13	0.30–0.60	0.025	0.025	0.20–0.35	4.75–5.25	0.20–0.35	—	—
5.5 Ni	77		0.13	0.90–1.50	0.030	0.030	0.15–0.30	5.0–6.0	0.10–0.30	0.10–1.00
8 Ni	102	A553 Type II	0.13	0.90, max.	0.035	0.040	0.15–0.30	7.5–8.5	—	—
9 Ni	77	A553 Type I	0.13	0.90, max.	0.035	0.040	0.15–0.30	8.5–9.5	—	—

acetylene, carbon dioxide, and propane. The 9% Ni steel is used for lower temperature applications involving containment of LNG at 111 K, liquid oxygen at 90 K, and liquid nitrogen at 77 K. The grades with 5, 5.5, and 8% Ni are relatively new alloys developed to provide a less expensive steel for LNG tankage than the 9% Ni steel. Their use to date is limited in comparison with 9% Ni steel.

Two experimental alloys are being developed for service below 77 K, particularly for liquid helium service at 4 K. These alloys are the 12% Ni family developed by Jin, Hwang, and Morris (1975) and the 13% Ni alloy developed by Ishikawa and Maruyama (1978). The 12% Ni alloy is completely ferritic and contains 0.25% Ti to reduce the interstitial content. The ductile-to-brittle transition is suppressed below 4 K by reducing the grain size by thermal cycling, a processing treatment that has limited the commercial development of the 12% Ni alloy. The 13% Ni alloy is also completely ferritic and contains 2 to 4% Mo and 1 to 2% Cr. In addition, niobium, titanium, or vanadium is added to serve as a getter for interstitials. Extremely pure components are used to prepare the alloy to reduce to the interstitial content, a requirement that has limited the commercial development of the 13% Ni alloy.

11.3.1 Metallurgy and heat treatment

Nickel in steel improves low-temperature toughness in three ways. (a) In the classical heat-treating sense, nickel improves hardenability. That is, nickel suppresses the formation of proeutectoid ferrite and promotes transformation to bainite or martensite. These fine-grain transformation products tend to lower the ductile-to-brittle transition temperatures. (b) Nickel also increases the ease of cross-slip at reduced temperatures, permitting plastic deformation to take place in preference to cleavage, and thereby improving low-temperature toughness. (c) Finally, nickel tends to stabilize the austenite phase; above about 7% Ni in pure iron, both austenite and ferrite are stable.

The toughness of the nickel steels can be improved by reducing the carbon, sulfur, and phosphorus contents. Carbides in the ferritic phase (martensite or bainite) serve as initiation sites for fracture by brittle cleavage. Thus, reducing the carbon content lowers the ductile-to-brittle transition temperature. Sulfide inclusions serve as initiation sites for ductile tearing, and thus, reducing the sulfur content increases energy required for ductile fracture, that is, raises the upper shelf toughness. Phosphorus tends to segregate to grain boundaries and causes temper embrittlement when the steel is cooled slowly in the vicinity of 500° C. Watanabe and Matsubara (1982) observed that 9% Ni steels with less than 0.005% P had markedly improved toughness at 77 K, as shown in Fig. 11.28.

Modern steel-making practices have made available 9% Ni steels with less than 0.05% C, 0.005% P, and 0.002% S. Despite the low carbon content, the steels meet the strength requirements of ASTM A553. These steels have improved fracture properties, particularly in the temperature range 77 to 111 K (Watanabe and Matsubara, 1982).

The heat treatment of Fe–Ni alloys is best understood by considering the Fe–Ni phase diagram, a portion of which is shown in Fig. 11.29. The 9% Ni steel is heat-treated in two ways: double normalize and temper (NNT) per ASTM

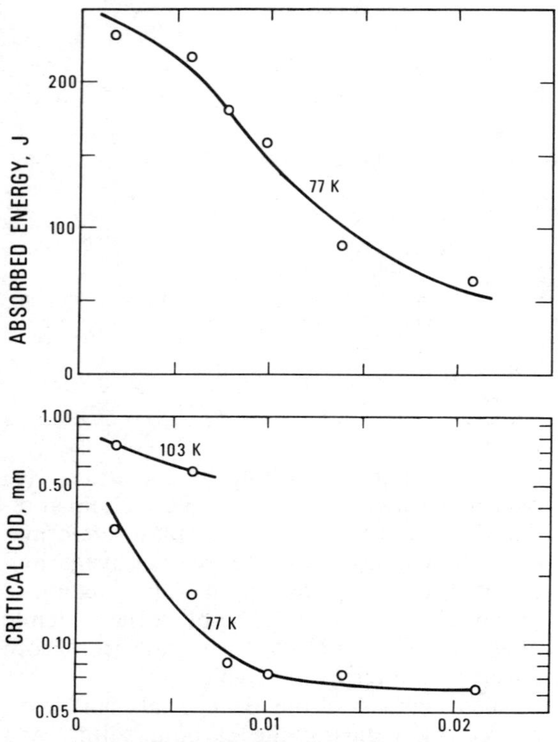

Figure 11.28 The effect of phosphorus on the toughness of 9% Ni steel (Watanabe and Matsubara, 1982).

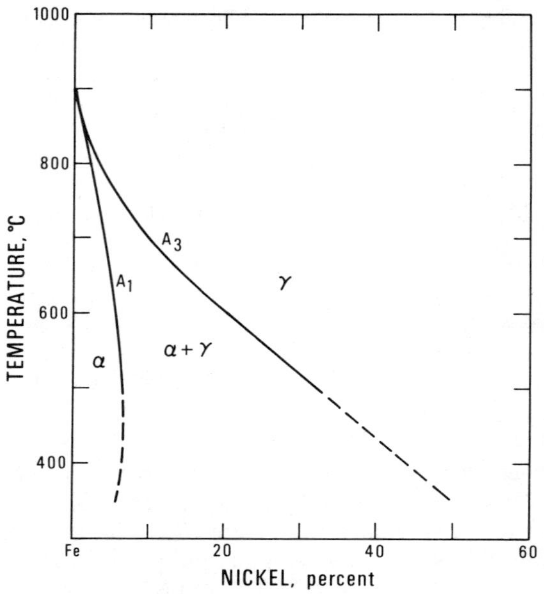

Figure 11.29 The solid-state transformation in the iron-rich portion of the Fe–Ni phase diagram (*Metals Handbook*, 1973). Austenite is labeled γ; ferrite is labeled α.

A353 or quench and temper (QT) per ASTM A553. In the NNT treatments, the first normalizing produces a bainite plus martensite microstructure. The second normalizing produces a finer grained austenite which, upon cooling, produces a finer bainite and martensite microstructure. Tempering in the two-phase field forms 5 to 10% austenite and tempers the remaining ferrite through the normal carbide-precipitation processes. The austenite dissolves carbon and other interstitials from the ferrite, increasing the ferrite toughness and the austenite stability. Tempering at a higher temperature would create more austenite, but it would be less stable and transform on cooling. In the QT treatments, the steel is austenitized, quenched in water to form a predominantly martensitic microstructure, and then tempered to form 5 to 10% austenite, which is stable on cooling. In both heat treatments, forced-air or water-quench cooling following tempering is required to avoid temper embrittlement.

A special three-step procedure is used to heat-treat 5% Ni steel per ASTM A645 and to heat-treat the 5.5% Ni Japanese grade. In both procedures, the steel is austenitized and quenched in water to form a predominantly martensitic microstructure. The second step, referred to as temperizing, involves heating to just below the A_3 temperature to produce a lamellar pattern of ferrite plus austenite and then water quenching to form alternate grains of ferrite and martensite. The final step involves heating the lamellar structure to just above the A_1 temperature to nucleate 5 to 10% austenite at the many phase interfaces where it can dissolve enough carbon from the matrix to become stabilized for 77 K service. As with 9% Ni steel, rapid cooling following tempering is required to avoid temper embrittlement.

11.3.2 Properties of the nickel steels

The properties of the nickel steels at room and cryogenic temperatures have been the subject of many reviews and data compilations in recent years. Pense and Stout (1975) reviewed the properties of the United States grades: 2.25, 3.5, 5, 8, and 9% Ni steels. Research and testing on the nickel steels in Japan has been summarized by the Subcommittee on Cryogenic Steels of the Japan Pressure Vessel Research Committee *(Cryogenic Applications of Ferrous Materials in Ja-*

pan, 1979). The properties of the grades most commonly used in Europe (3.5, 5, and 9% Ni), along with the applicable European specifications, have been compiled by International Nickel Limited *(Materials for Cryogenic Service,* 1972). Note that the 5% Ni steel commonly used in Europe is in the normalized (or quenched)-and-tempered condition and is only suitable for service down to 153 K.

Tensile properties. The tensile properties of 3.5% Ni steel, ASTM A 203 Grade D, have been evaluated by Heuttich, Pense, and Stout (1971) at temperatures in the range 116 to 295 K. Tests were conducted on 51-mm-thick plate from a single heat in three conditions: as-rolled (AR), normalized (N), and normalized followed by quench and tempering (NQT). The normalized plate, standard for A203D, had yield strengths of 369 MPa at room temperature and 455 MPa at 172 K, the minimum design temperature. The yield strengths were slightly lower in the AR plate and about 15% higher in the NQT plate. Plate in all three conditions retained good ductility (33 to 42%) in smooth-bar tensile tests at temperatures down to 116 K; however, the nil-ductility transition temperatures measured by the ASTM E 208 drop-weight test were 227, 194, and 183 K in the AR, N, and NQT conditions, respectively. Tobler et al. (1979) conducted tensile tests on 3.5% Ni steel in the QT condition at temperatures ranging from 4 to 300 K and found that the steel had 35% elongation and 58% reduction of area at 4 K. However, the fracture toughness (as shown in Fig. 11.27) dropped sharply at temperatures below 172 K.

The yield and ultimate strengths of the 5 to 9% Ni steels are compared in Fig. 11.30 for temperatures ranging from 76 to 300 K. The curves shown are for typical behavior as reported in the *LNG Materials and Fluids* (1978) for 5 and 9% Ni and by Sakai et al. (1976) for 5.5% Ni. The temperature dependences of the yield and ultimate strengths are similar for all the alloys. Typical elongation values for each of the 5 to 9% Ni steels at room and cryogenic temperatures are 25 to 35%.

Toughness. The toughness requirements for 3.5% Ni steel are 20.4 J (15 ft·lbf) at 172 K. In the normalized plate (5l-mm thick) evaluated by Heuttich et al. (1971), the 20.4-J transition temperature was 161 K. Results of the fracture toughness tests indicate that at 172 K, 3.5% Ni

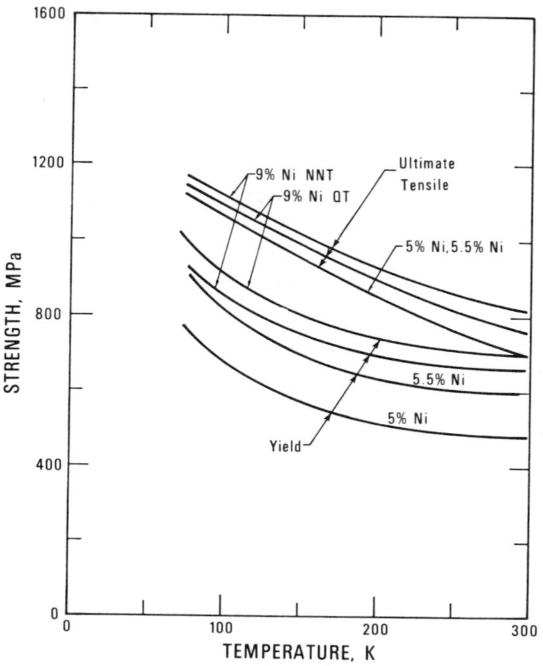

Figure 11.30 Yield and tensile strength of 5, 5.5, and 9% Ni steels at temperatures between 77 and 300 K (*LNG Materials and Fluids,* 1978); Sakai, Osaki, Moriyama, and Yano, 1976).

steel (N and NQT heat treatments) has excellent fracture toughness at low loading rates, but exhibits brittle behavior under dynamic (impact) loading conditions.

The 5 to 9% Ni steels retain satisfactory toughness at temperatures down to their intended minimum service temperature, 103 K for 5 and 8% Ni and 77 K for 5.5 and 9% Ni. The fracture toughness values at cryogenic temperatures for the 3.5, 5, and 9% Ni steels shown in Fig. 11.27 indicate that the steels do have ductile-to-brittle transitions, but the rapid drop in toughness with decreasing temperatures occurs at temperatures below the minimum service temperature.

The toughness of 5 to 9% Ni steels required by the ASTM Standards and the ASME code is a lateral expansion of 0.38 mm measured opposite the notch in a Charpy V-notch (CVN) impact test. This requirement is approximately equivalent to 35 J (25 ft·lbf) of energy absorption in a CVN test. The relationship between CVN-energy absorption and lateral expansion in 9% Ni steel tested at 77 K is shown in Fig. 11.31. In section thicknesses up to 50 mm, 9% Ni steel readily ex-

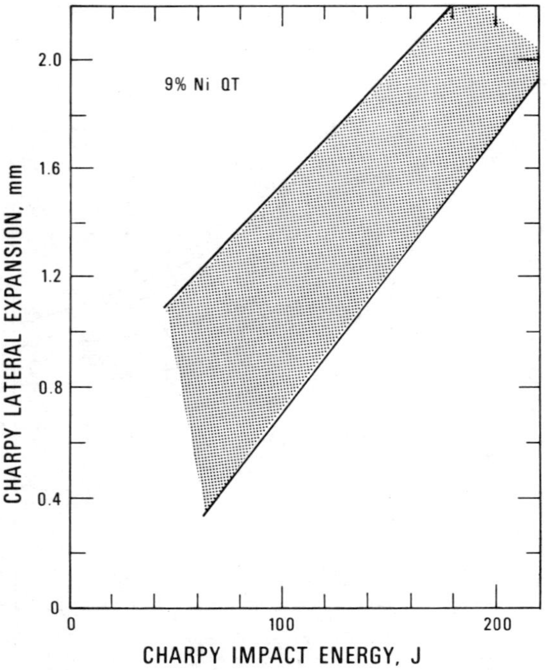

Figure 11.31 Relationship between Charpy V-notch impact energy absorption and lateral expansion in 9% Ni steel at 77 K (Pense and Stout, 1975).

ceeds the 0.38-mm lateral expansion requirement at 77 K. Typical lateral expansion values at 77 K for 9% Ni steel in the QT condition (A553) range from 0.64 to 2.0 mm; slightly lower values are typical in the NNT condition (A353). Toughnesses measured by lateral expansion for 5% Ni steel (A645) at 103 K normally range from 0.5 to 1.0 mm.

Extensive toughness testing has been conducted by Sakai et al. (1976) to compare the fracture resistance of 5.5 and 9% Ni steels. Single heats of each alloy were rolled into plates of three thicknesses: 8, 15, and 30 mm. The CVN-impact-test results indicate that the two alloys have similar toughnesses over the temperature range 77 to 300 K.

The dynamic fracture toughness of the 5, 5.5, and 9% Ni steels has been determined by Murayama, Pense, and Stout (1976). The results, shown in Fig. 11.32 indicate that loading rate does not significantly affect toughness, except in the case of 5% Ni steel, which had nearly double the static toughness when loaded dynamically. Tests conducted by Tenge, Karlsen, and Mauritzon (1975) also indicate that the toughness of 9% Ni steel, both A353 and A553, is not affected by dynamic loading.

The crack-arrest capabilities of 9% Ni steel have been widely studied in Japan. Miyakoski, Ishikura, Suzuki, and Tanaka (1981) conclude from numerous test results using a variety of specimens that 9% Ni steel possesses sufficient toughness to arrest a brittle crack running from a welded joint into the base plate. Kawaguchi, Tsukamoto, Maruyama, Bessyo, and Kato (1982) demonstrated that 9% Ni steel plate (QT) can arrest a 1500-mm-long brittle crack at LNG temperature. Nakano, Suzuki, Kamada, and Hirose (1982) observed that 9% Ni steel plates in the QT condition could arrest a crack at 103 K at stresses as high as 490 MPa; the test plates had CVN-impact-toughness values at 77 K ranging from 78 to 250 J. However, similar plates in the NNT condition did not arrest cracks at 111 K at stresses as low as 196 MPa; the test plates had CVN-impact-toughness values at 77 K ranging from 52 to 118 J. These results clearly indicate the advisability of using 9% Ni steel in the QT condition rather than the NNT condition.

Fatigue. The results of flexural fatigue tests on 9% Ni (A353 and A553), shown in Fig. 11.33, indicate that its fatigue life at cryogenic temperatures is superior to that at room temperature. Similar results have been obtained for 5% Ni steel at 103 K and room temperature (*Metallurgical Properties of Armco Cryonic 5 Steel*, 1971).

Tobler, Mikesell, Durcholz, and Reed (1975) conducted fatigue crack growth rate studies on 5, 5.5, and 9% Ni steel. Their results indicate that the crack growth rates are approximately the same at 111 and 295 K, but accelerated growth rates occur at 77 K in the 5 and 5.5% Ni steels and at 4 K in the 9% Ni steel. The results for the 9% Ni steel are shown in Fig. 11.34.

Physical properties. The physical properties of the nickel steels are summarized in Table 11.9. Thermal properties are used in the design of cryogenic tanks to determine thermal stresses, boil-off rates, and heat extraction during loading.

11.3.3 Fabrication characteristics

The nickel steels are generally supplied as plates that are formed and welded into tanks for the storage and transportation of LNG. Thus, the fabrication characteristics are limited to welding and forming. The nickel steels are machinable by the conventional processes. Heavy

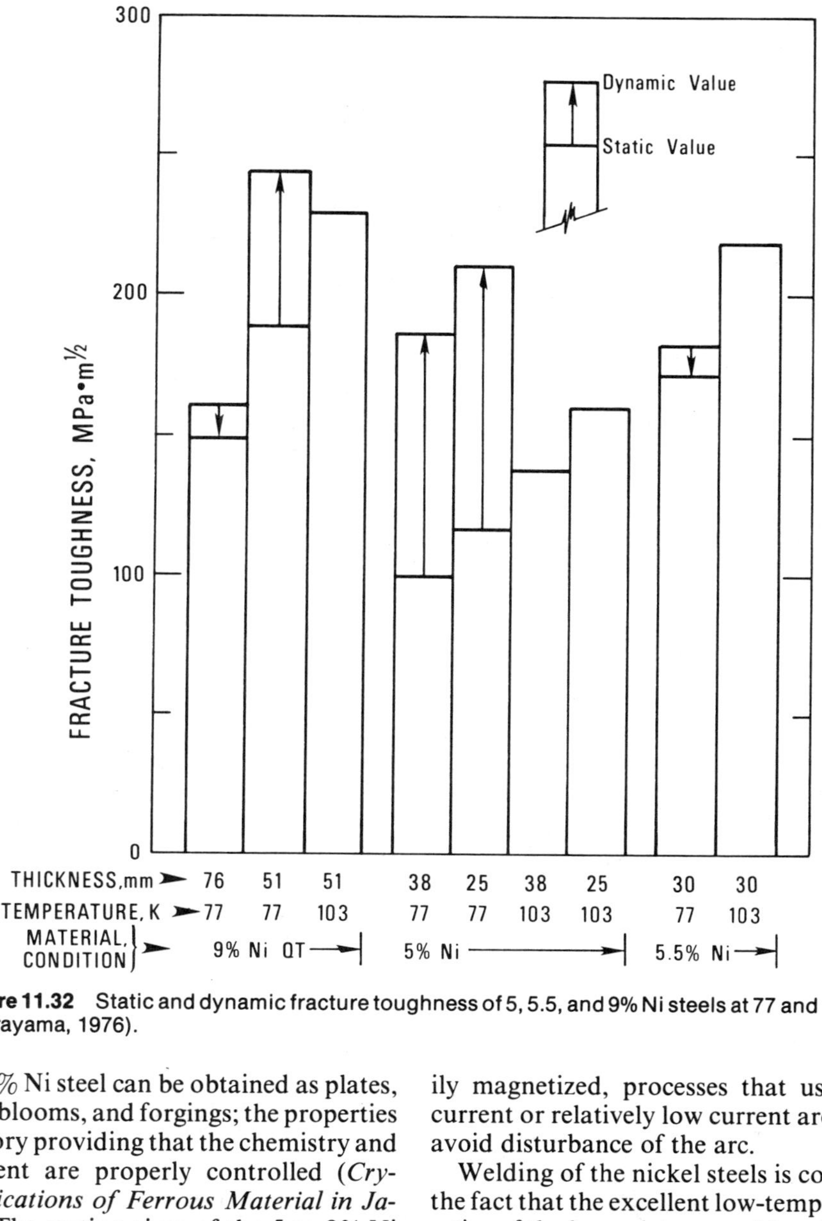

Figure 11.32 Static and dynamic fracture toughness of 5, 5.5, and 9% Ni steels at 77 and 103 K (Murayama, 1976).

sections of 9% Ni steel can be obtained as plates, bars, billets, blooms, and forgings; the properties are satisfactory providing that the chemistry and heat treatment are properly controlled (*Cryogenic Applications of Ferrous Material in Japan*, 1979). The section sizes of the 5 to 8% Ni steels are generally limited to 38 mm or less to ensure adequate heat treatment. The nickel steels are not used for castings because the coarse-grained structure tends to have low toughness.

Welding. The nickel steels have similar welding characteristics and can be welded by most processes. For construction of large tanks, the most common processes are GMA, SMA, and SA. Preheat and postweld stress relief are normally not required. Since the nickel steels are easily magnetized, processes that use alternating current or relatively low current are preferred to avoid disturbance of the arc.

Welding of the nickel steels is complicated by the fact that the excellent low-temperature properties of the base plate are achieved in large part by heat treatment. Thus, the welding procedure and filler metals must not degrade the HAZ toughness below specification requirements. In addition, the filler metal must have satisfactory toughness in the as-deposited condition because postweld heat treatments are generally impractical.

Ideally, ferritic filler metals with nearly matching chemistry would be used. Several ferritic filler metals have been developed (Mahin, Morris, and Watanabe, 1980), but industry practice

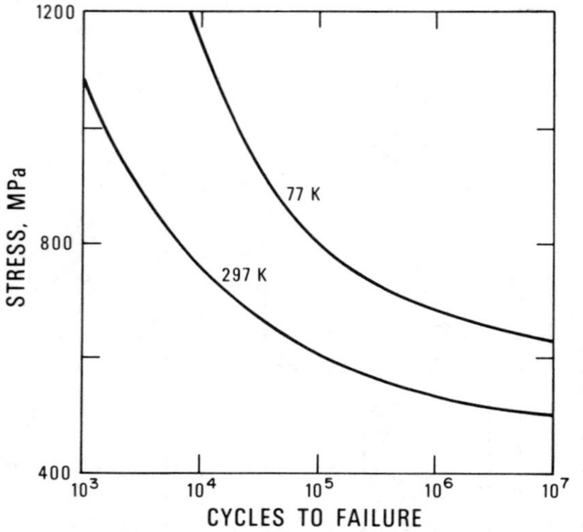

Figure 11.33 Flexural fatigue ($R = -1$) strength of 9% Ni steel at 77 and 297 K (*Materials for Cryogenic Service,* 1972).

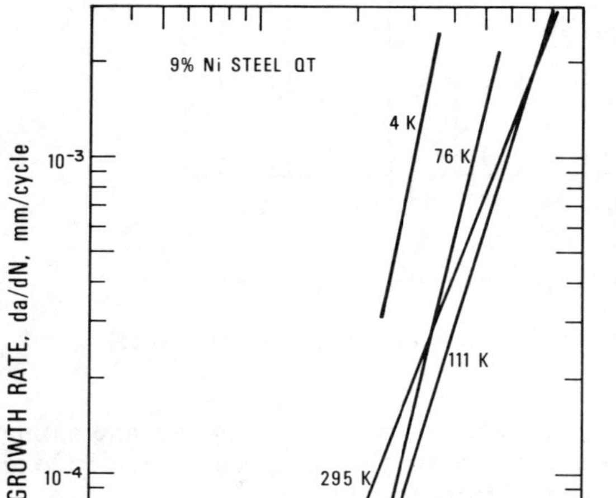

Figure 11.34 Fatigue crack growth rates for 9% Ni steel at 4, 76, 111, and 295 K (Tobler et al., 1975).

favors use of austenitic filler metals to ensure satisfactory toughness in the as-deposited weld. The most common filler metals are nickel-base alloys with approximately 60 to 70% Ni, 15 to 20% Cr, and varying amounts of molybdenum, manganese, iron, niobium, titanium, and tungsten. Compositions of commercial electrodes and wires are summarized in Table 11.10. Selection of a specific weld metal depends on the welding process used and the position (i.e., flat, vertical, overhead) requirements. Austenitic stainless steel filler metals are generally not used because of a relatively large (approximately 50%) mismatch in thermal expansion.

The tensile properties of the nickel-base weld metals are generally lower than the base-metal properties at both room and cryogenic temperatures. As shown in Table 11.11, the yield strengths range between 340 and 540 MPa at room temperature; all values are less than the minimum yield strength (585 MPa) of 5.5 or 9% Ni steel. Similarly, at 77 or 111 K, the weld deposit has significantly lower strength than the base metal. Consequently, the code allowable design stresses are based on the weld-metal strength. Blake et al. (1973) have shown that the high-nickel weld deposits have large variabilities in tensile properties depending on plate thickness, welding position, heat input, and carbon content, as shown for SMA welds in Fig. 11.35. Welland (1978) attributed the strength variations to anisotropy related to the solidification direction; in other words, a given weld will have significantly different strength depending on the direction of test.

The Charpy impact toughnesses of the high-nickel weld deposits are generally well above specification requirements, as typified by the data in Table 11.11. Relatively few fracture toughness tests have been conducted on weld deposits because there is little concern about the fully austenitic deposits becoming brittle at low temperatures. The fracture toughness tests of Carter and Harrison (1974) on AWS ENiCrMo-6 welds indicated tough behavior with crack opening displacement (COD) values at 119 K ranging from 0.29 to 0.42 mm.

The fracture toughness of the HAZ of nickel-steel weldments tends to be lower than that of the base metal or weld. The minimum toughness values (within 95% confidence levels) for the HAZ of 9% Ni steel weldments at 77 K reported by Pense and Stout (1975) were 130 MPa·m$^{1/2}$ for double-normalized plate (A353). The minimum toughness for the HAZ of 5% Ni steel (A645)

Table 11.9 Physical properties of the nickel steels.

Type of Steel and Temperature	Density[a] ($g \cdot cm^{-3}$)	Young's Modulus[a] (GPa)	Shear Modulus[a] (GPa)	Poisson's Ratio[a]	Thermal Conductivity[b] ($W \cdot m^{-1} \cdot K^{-1}$)	Thermal Expansion[b] (mean)	Specific Heat[b] ($J \cdot kg^{-1} \cdot K^{-1}$)
3.5 Ni							
295 K	7.86	204	79.1	0.282	35	11.9	450
172 K		210	81.9	0.281	29	10.2	350
5 Ni							
295 K	7.82	198	77.0	0.283	32	11.9	450
111 K		208	81.2	0.277	20	9.4	250
76 K		209	81.6	0.277	16	8.8	150
9 Ni							
295 K	7.84	195	73.8	0.286	28	11.9	450
111 K		204	77.5	0.281	18	9.4	250
76 K		205	77.9	0.280	13	8.8	150

[a]Weston, Naimon, and Ledbetter (1975).
[b]*LNG Materials and Fluids* (1978).

Table 11.10 Composition limits of selected filler metals for welding the 5-9% Ni steels.

AWS Designation	Composition Limit (%)										
	C	Mn	Fe	P	S	Si	Ni	Cr	Nb+Ta	Mo	Other[a]
Covered Electrodes											
ENiCrFe-2	0.10	1.0–3.5	12.0	0.03	0.02	0.75	62, min.	13–17	0.5–3.0	0.5–2.5	
ENiCrFe-3	0.10	5.0–9.0	10.0	0.03	0.015	1.0	59, min.	13–17	1.0–2.5	—	
ENiCrFe-4	0.20	1.0–3.5	12.0	0.03	0.02	1.0	60, min.	13–17	1.0–3.5	1.0–3.5	
ENiCrMo-3	0.10	1.0	7.0	0.03	0.02	0.75	55, min.	20–23	3.15–4.15	8–10	
ENiCrMo-6	0.10	2.0–4.0	10.0	0.03	0.02	1.0	55, min.	12–17	0.5–2.0	5–9	1.0–2.0 W
Bare Rod and Electrodes											
ERNiCr-3	0.10	2.5–3.5	3.0	0.03	0.015	0.50	67, min.	18–22	2.0–3.0	—	0.75 Ti
ERNiCrFe-6	0.08	2.0–2.7	8.0	0.03	0.015	0.35	67, min.	14–17	—	—	2.5–3.5 Ti
ERNiCrMo-3	0.10	0.5	5.0	0.02	0.015	0.50	58, min.	20–23	3.15–4.15	8–10	0.40 Al, 0.40 Ti

[a]Total of unlisted elements = 0.5 max.

Table 11.11 Mechanical properties of selected filler metals for welding 5-9% Ni steels (Tharby, Heath, and Flannery, 1973).

Filler Metal AWS Designation	Process	Strength at Room Temperature[a] Yield (MPa)	Ultimate (MPa)	CVN Energy at 77 K (J)
ENiCrFe-2	SMA	370	620	68–81
ENiCrFe-3	SMA	340	590	68–81
ENiCrFe-4	SMA	455	745	60
ENiCrMo-3	SMA	510	790	64
ENiCrMo-6	SMA	430	660	34
ERNiCr-3	SAW	379	668	54
ERNiCr-6	GMA	448	724	75
ERNiCrMo-3	GMA	545	820	40
	SAW	455	765	44

[a]All weld metal tensile tests.

weldments observed by McHenry and Reed (1977) were 103 $MPa \cdot m^{1/2}$ at 111 K and 67 $MPa \cdot m^{1/2}$ at 76 K. These low values are generally associated with a brittle pop-in fracture. In wide-plate tests with notches located in the HAZ, fracture tends to propagate through the weld metal after initiating at much higher COD values than expected from small-scale tests (*Studies on 9% Ni Steel Plates for LNG Tanks*, 1981). The stress levels at fracture are comparable to the base-metal yield strength, that is, well above the weld metal yield strength. These tests provide some hope that the brittle pop-in fracture is a laboratory phenomenon and should not occur in practice.

Forming. The nickel steels may be hot or cold formed. If hot forming is performed, the

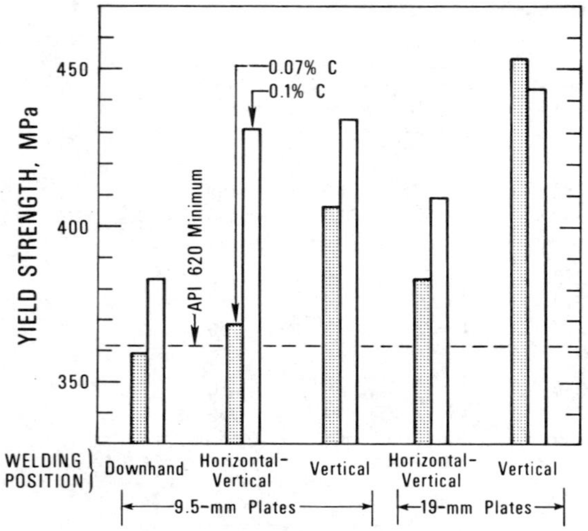

Figure 11.35 Yield strength of high-nickel weld deposits in 9% Ni steel for two plate thicknesses, three welding positions, and two carbon contents (Blake, Rowntree, and Phelps, 1973).

steel must subsequently be heat-treated in accordance with the relevant specification. If the hot-forming temperature is properly selected, the first normalizing treatment can be performed at the same time. The extent of cold forming is limited to 3% strain (ASME, 1979) unless the plate is stress relieved after forming. This rule is intended to avoid strain aging, which can elevate the yield strength and decrease notch toughness.

Inspection. Inspection of nickel steel weldments is complicated by the mismatch in the physical and chemical properties of the base metal and the weld metal. The radiographic densities of the high-nickel weld metals are different, and consequently, separate exposures for the weld region and the adjacent base metal may be required to properly inspect the joint. For ultrasonic inspection, complications are caused by the difference in crystal structure of the weld and base metal and the crystalline texture of the weld metal. Thus conventional ultrasonic techniques are not suitable for reliably detecting weld defects in nickel-steel weldments.

11.4 ALUMINUM ALLOYS

Aluminum alloys have an f.c.c. crystal structure and, like most f.c.c. alloys, retain their strength, ductility, and toughness at cryogenic temperatures. Other advantages include low density, nonmagnetic behavior, and stable microstructure. Excellent machining, forming, and forging characteristics favor their use in complex parts for aerospace systems. The disadvantages of aluminum alloys are low strength in weldments, and for some applications, low elastic modulus, high thermal expansion, and high thermal conductivity. Field weldability restrictions limit their use in large structures, such as land-based storage tanks.

Aluminum alloys have been used for many structural applications in cryogenic systems, including rocket propulsion tanks for liquid oxygen at 90 K and liquid hydrogen at 20 K, shipboard storage tanks for LNG at 111 K, and structural supports for superconducting magnets operating in liquid helium at 4 K. A notable example is the external tank system for the space shuttle, which includes a tank for liquid oxygen and a tank for liquid hydrogen. The cylindrical, 8-m-diameter tanks are integrally stiffened, fusion-welded assemblies made of a heat-treated Al–Cu alloy, 2219-T87. The tanks propel the space shuttle into orbit and then separate from the vehicle; they are not recovered. Their principal design requirements are strength-to-weight ratio, toughness at cryogenic temperatures, weldability, machinability, and cost — a combination of requirements that make aluminum alloys a clear choice.

Aluminum alloys suitable for cryogenic service include most of the alloys that are solid-solution-strengthened with up to 6% Mg, 1.5% Mn, 0.35% Cr, or a combination thereof, and some of the heat-treatable alloys that are precipitation-hardened with up to 6.5% Cu, or with magnesium and up to 1% Si or 8% Zn. The aluminum alloys commonly used for cryogenic applications include: the 2.5 to 6% Mg (5000-series) alloys, alloys with 0.6 to 1.0% Mg and 0.6 to 1.0% Si (6000-series), and the 4 to 6.5% Cu (2000-series) alloys. The preferred alloys are: (a) Al–4.5Mg (5083), a weldable alloy generally used in the annealed condition; (b) Al–1.0Mg–0.6Si (6061), a versatile alloy available in a variety of product forms and heat treatments, and (c) Al–6Cu (2219), a high-strength alloy used in the precipitation-hardened condition (e.g., T-87, T-851). The composition ranges for the preferred alloys are summarized in Table 11.12.

Several other aluminum alloys have excellent properties at 4 K and may be used for special applications. Commercially pure aluminum (alloy 1100) and the 3003 alloy have excellent formability. Alloy 3003 has been widely used for cry-

Table 11.12 Compositions of aluminum alloys suitable for cryogenic service.

Type	Composition (%)								
	Si, max.	Fe, max.	Cu	Mn	Mg	Cr	Zn, max.	Ti	Others
1100	1.0 Si (+Fe)		0.05–0.20	0.05, max.	—	—	0.10	—	99.00 Al, min.
2219	0.2	0.3	5.8–6.8	0.20–0.40	0.02, max.	—	0.10	0.02–0.10	0.05–15 V; 0.10–0.25 Zr
3003	0.6	0.7	0.05–0.20	1.0–1.5	—	—	0.10	—	
5083	0.4	0.4	0.1, max.	0.40–1.0	4.0–4.9	0.05–0.25	0.25	0.15, max.	
6061	0.4–0.8	0.7	0.15–0.40	0.15, max.	0.08–1.2	0.04–0.35	0.25	0.15, max.	
7005	0.35	0.40	0.10	0.2–0.7	1.0–1.8	0.06–0.20	4.0–5.0	0.01–0.06	0.08–0.20 Zr

ogenic applications and is the best choice when brazing is used for joining. High-strength alloys, such as 2014-T651 and 2021-T851, have been used for rocket propulsion tanks, but recently 2219-T87 has been preferred, as exemplified by the space shuttle tanks. Alloy 7005 has the best combination of strength and toughness of the 7000-series alloys, but its availability is limited to extruded products, such as rod, tubes, and shapes.

The ASTM specifications for aluminum mill products of alloys 5083, 2219, and 6061 are summarized in Table 11.13. For a more complete listing of specifications, refer to *Aluminum Standards and Data* (1981).

Table 11.13 ASTM specifications and availability of aluminum mill products.

Product Form	ASTM Specification	Available Alloys 5083	6061	2219
Sheet and plate	B209	X	X	X
Wire, rod, bar; rolled or cold finished	B211		X	X
Wire, rod, bar, shapes, tube	B221	X	X	X
Structural shapes	B308		X	X
Pipe	B241		X	
Forgings and forging stock	B247	X	X	X
Structural pipe and tube; extruded	B249		X	

11.4.1 Metallurgy and heat treatment

The aluminum alloys used for cryogenic service are of two types: solid-solution-strengthened alloys and precipitation-hardened alloys. Solid-solution strengthening is generally provided by magnesium (5000-series alloys) or manganese (3000-series alloys). These alloys can be used in the annealed condition or further strengthened by cold working. Precipitation-strengthened alloys contain an alloying element, such as copper (2000 series), silicon (6000 series), or zinc (7000 series), that is soluble in aluminum at elevated temperatures. The element is dissolved at elevated temperature by a solution treatment, retained in solution by quenching from the solution-treating temperature, and then precipitated into the aluminum-rich matrix. Precipitation hardening can occur at room temperature, referred to as *natural aging*, or elevated temperature, *artificial aging*.

Aluminum alloys are used in a variety of conditions, referred to as *tempers*. The tempers are designated by a letter followed by one to three digits. The letter is the basic temper designation: O stands for annealed: H, strain hardened; and T, thermally treated. For the strain-hardened condition, the first digit indicates the hardening operations: H1 designates strain hardening only; H2, strain hardened and partially annealed; and H3, strain hardened and stabilized. The second digit indicates the degree of strain hardening. For example, H18 is the full-hard condition; it is strain-hardened by a cold reduction of 75%. The H12, H14, and H16 conditions are the 1/4-, 1/2-, and 3/4-hard conditions, respectively. The third digit is sometimes used for specific variations. For the thermally treated tempers, the first digit indicates the sequence of basic treatments; for example, T6 designates solution heat-treated and then artificially aged, and T8 designates solution heat-treated, cold-worked, and then artificially aged. Subsequent digits indicate specific variations; for instance, Tx51 indicates stress relief by stretching, as in T651 or T851. The reader is referred to *Aluminum Standards and Data* (1981) for a complete description of the temper designation system.

11.4.2 Properties of aluminum alloys

The low-temperature properties of aluminum alloys have been compiled in cryogenic handbooks (*Handbook on Materials for Superconducting Machinery*, 1977; *LNG Materials and Fluids*, 1978), in a review article by Kaufman and Wanderer (1967), and in suppliers' publications (*Alcoa Aluminum, The Cryogenic Metal*, 1971). In this section, the general trends in strength, toughness, and fatigue resistance are summarized, particularly the data for room temperature and 4 K. The data were taken on relatively few lots of material, but generally the lot-to-lot variations in the mechanical properties of aluminum alloys are relatively small compared with those of stainless steel.

Tensile properties. The mechanical properties of many aluminum alloys have been determined by Kaufman, Bogardus, and Wanderer (1968) and by Kaufman and Wanderer (1971). In these papers, the tensile and notched-tensile properties at room temperature and some intermediate temperatures are reported for 25 different alloys and tempers. The data for alloys 5083-O 5083-H321, 6061-T651, 2219-T851, 7005-T5351, and A356-T61 are summarized in Table 11.14. Notice that for the preferred alloys (5083, 6061, and 2219), the tensile strengths and, to a lesser extent, the yield strengths increase with decreasing temperature, and the ductility values are not significantly changed. The strength and elongation trends are similar for 7005-T5351, but the reduction of area decreased from 43% at room temperature to 22% at 4 K. Similar decreases (about 50%) in reduction of area were observed in all nine of the 7000-series alloys for tempers evaluated.

Aluminum alloys having a wide range of strength levels are suitable for cryogenic applications. The yield strengths of commercially pure (CP) aluminum (types 1100-O and 1099-H14) and the preferred alloys are shown as a function of temperature in Fig. 11.36 (Kaufman et al., 1968). The yield strengths of each alloy gradually increase with decreasing temperature. Strength increases in the following order: annealed CP (1100-O), cold-worked CP (1099-H14), solution-strengthened Al–Mg alloy (5083-O), cold-worked Al–Mg alloy (5083-H321), precipitation-hardened Al–Mg–Si alloy (6061-T651), and precipi-

Table 11.14 Tensile properties of aluminum alloys at room temperature and 4 K (Kaufman et al., 1968).

Alloy and Temper	Plate Thickness (mm)	Orientation	Temperature (K)	Tensile Strength, TS (MPa)	Yield Strength, YS (MPa)	Elongation in 4D (%)	Reduction of area (%)	Notch Tensile Strength, NTS (MPa)	NTS/TS	NTS/YS
5083–O	25	L	RT	322	141	19.5	26	372	1.16	2.65
			77	434	158	32	33	420	0.97	2.65
			4	557	178	32	33	429	0.77	2.42
5083–H321	25	L	RT	335	235	15	23	421	1.26	1.80
			77	455	274	31.5	33	485	1.06	1.77
			4	591	279	29	33	508	0.86	1.82
6061–T651	25	L	RT	309	291	16.5	50	477	1.54	1.64
			77	402	337	23	48	575	1.43	1.71
			4	483	379	25.5	42	619	1.28	1.63
		T	RT	309	278	15.2	42	467	1.51	1.68
			77	405	321	20.5	39	555	1.37	1.73
			4	485	363	23	33	601	1.24	1.66
2219–T851	25	L	RT	466	371	11	27	547	1.12	1.48
			77	568	440	13.8	30	651	1.15	1.48
			4	659	484	15	26	703	1.06	1.48
		T	RT	457	353	10.2	22	531	1.16	1.50
			77	575	462	14	28	630	1.09	1.37
			4	674	511	15	23	690	1.03	1.37
7005–T5351	38[a]	L	RT	427	379	15	43	594	1.39	1.59
			77	578	465	17	27	683	1.18	1.47
			4	672	521	17	22	737	1.09	1.41
A356–T61	19	Casting	RT	287	208	8.8	10	354	1.23	1.70
			77	356	262	7.1	9	495	1.09	1.50
			4	356	262	4	4	412	1.15	1.57

[a]Extrusion.

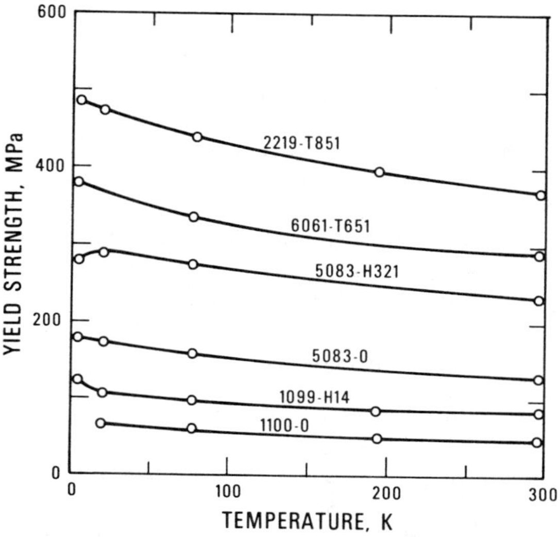

Figure 11.36 Yield strength of six aluminum alloys at temperatures between 4 and 300 K (Kaufman et al., 1968).

tation-hardened Al–Cu alloy (2219-T851). Still higher yield strengths are available in several other alloys including 2014-T651, 2024-T851, and 7005-T5351.

Toughness. Aluminum alloys generally retain excellent toughness at cryogenic temperatures. A useful measure of aluminum alloy toughness at 4 K is the notch–yield ratio, NTS/YS (Kaufman and Johnson, 1963). Ratios substantially greater than one (e.g., 1.5), indicate that the alloys retain the ability to plastically deform and resist crack initiation in the presence of sharp stress concentrations. As shown in Table 11.14, the notch–yield ratios of the preferred alloys tend to be constant over the temperature range 4 K to room temperature. The 5083 and 6061 alloys have notch–yield ratios greater than 1.6 at 77 and 4 K and the high-strength 2219-T851 and 7005-T5351 alloys have notch–yield ratios of about 1.4. Significantly lower ratios (1.0 to 1.25) occur in other high-strength alloys, such as 2014-T651 and 2024-T851, and in most 7000 series alloys (7005 is an exception), where the ratio is sometimes below 0.4 at cryogenic temperatures (Hanson, Stickley, and Richards, 1961). The relationships between notch–yield ratio and tensile yield strength for 24 alloy-and-temper combinations, as reported by Kaufman and Wanderer (1971), are shown in Fig. 11.37.

The fracture toughness of aluminum alloys tends to increase with decreasing temperature, as

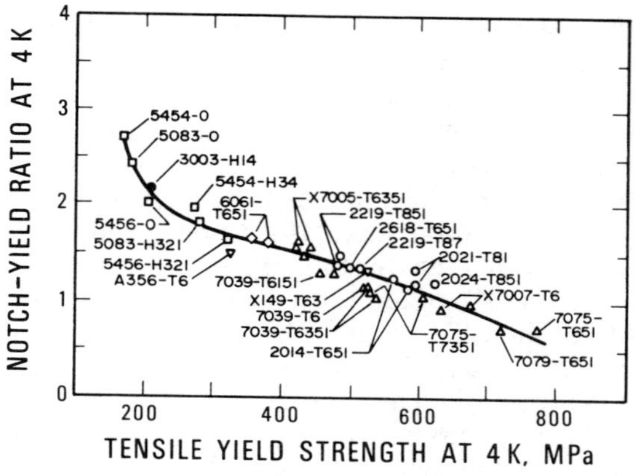

Figure 11.37 Notch-yield ratio vs. tensile yield strength for aluminum alloys at 4 K (Kaufman and Wanderer, 1971). ○ — 2xxx alloys; ● — 3xxx alloys; □ — 5xxx alloys; ◇ — 6xxx alloys; △ — 7xxx alloys; ▽ — casting alloys.

shown by Hall (1971) for 2219-87 at temperatures between 295 and 20 K and by Tobler and Read (1977) for 5083-O at temperatures between 295 and 4 K. At a given temperature, the toughness of aluminum alloys tends to be lower for higher strength alloys. The plane strain fracture toughness at 77 K of 12 alloy-and-temper combinations is shown as a function of yield strength in Fig. 11.38 (McHenry, 1977); note that the preferred alloys have the highest toughness values.

Numerous fracture tests have been conducted to verify the toughness of 5083 for use in LNG containment vessels for ships. Kaufman, Nelson, and Wygonick (1974) and Kaufman and Kelsey (1975) demonstrated that fracture in 200-mm-

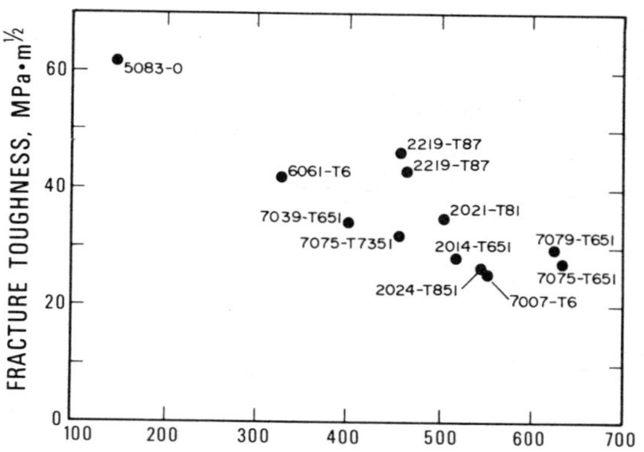

Figure 11.38 Fracture toughness of aluminum alloys at 77 K (McHenry, 1977).

thick plates was ductile and estimated the fracture toughness to be 44 to 66 MPa·m$^{1/2}$ at 77 K, depending on the crack orientation. Argy, Paris, and Shaw (1975) and Tobler and Reed (1977) used *J*-integral methods and reported the $K_{Ic}(J)$ to be 42 to 63 MPa·m$^{1/2}$ at 111 K, depending on crack orientation. Lake (1975) used dynamic tear tests to verify that the toughness at LNG temperature (111 K) was uniform through the thickness of 200-mm-thick plates. The fracture strength of numerous double-edge-notched and surface-notched tensile panels tested by Ikeda, Miyagi, Aoki, and Shirakura (1978) consistently exceeded the yield strength by 20 to 30% at temperatures ranging from 77 to 295 K; the results for a series of 400-mm-wide plates with thicknesses from 25 to 80 mm are shown in Fig. 11.39.

Limited fracture toughness test data are available on aluminum alloys at 4 K. Tobler and Reed (1977) measured the fracture toughness of 5083-O (4.3-cm-thick plate) using *J*-integral procedures and found that toughness increased with decreasing temperature. The $K_{Ic}(J)$ was 47 MPa·m$^{1/2}$ at 4 K in the transverse (TL) orientation. Shepic and Fiftal (1979) measured the toughness of 2219-T87 (3.8-cm-thick plate) using ASTM E399 plane strain fracture toughness testing procedures. The K_{Ic} for 2219-T87 at 4 K was 45.5 MPa·m$^{1/2}$ in the TL, 47.7 MPa·m$^{1/2}$ in the longitudinal (LT), and 28.6 MPa·m$^{1/2}$ in the through-thickness (SL) orientations.

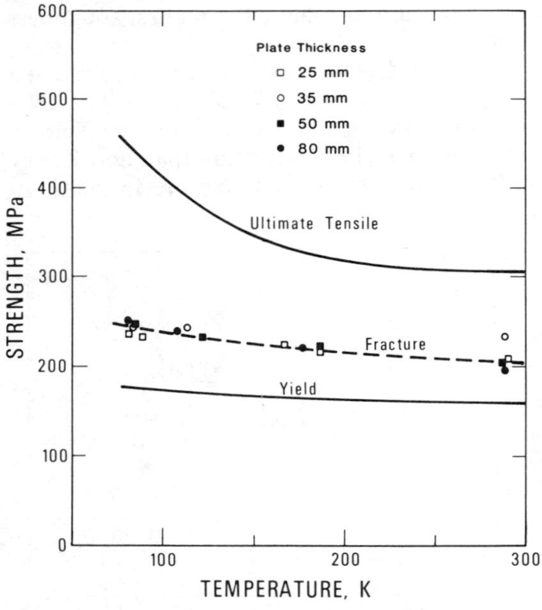

Figure 11.39 Effect of plate thickness on the fracture strength of 5083-O plate at temperatures between 77 and 295 K (Ikeda et al., 1978).

Fatigue. Constant-load amplitude fatigue tests on many aluminum alloys indicate that the fatigue strength increases with decreasing temperature (*Handbook on Materials for Superconducting Machinery*, 1977; Schwartzberg, Osgood, Keys, and Kiefer, 1964). For example, the fatigue strengths at 2·10^5 cycles of four Al–Mg (5000-series) alloys increased from an average of 276 MPa at room temperature to 317 MPa at 77 K (Kaufman and Nelson, 1962).

Strain–cycling fatigue tests have been conducted by Nachtigall (1975) on 2219-T851 and 2014-T6 alloys at room temperature, 77 K, and 4 K. For a given strain level, the fatigue lives of both alloys were greater at 4 K than at the higher temperatures for endurances greater than about 500 cycles. The results for 2219-T851 are shown in Fig. 11.40. The method of universal slopes (Manson, 1965) gave reasonable predictions of fatigue behavior based on the tensile properties at the test temperature.

The fatigue crack growth rates of aluminum alloys tend to be less at cryogenic temperatures than at room temperature. Kaufman and Kelsey (1975) showed that the greater rates at room temperature occur in moist air (90% humidity). Tobler and Reed (1977) and McHenry, Naranjo, Read, and Reed (1978) observed that the growth rates in two Al–Mg alloys at temperatures from 4 to 111 K are independent of temperature, but

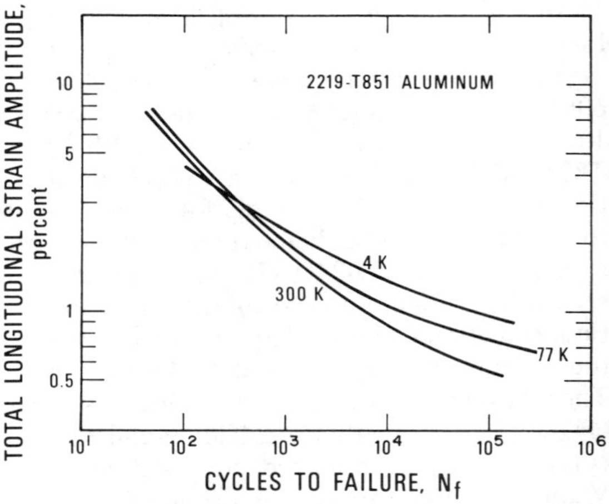

Figure 11.40 Strain–cycling fatigue curves for aluminum alloy 2219-T851 at 4, 77, and 300 K (Nachtigall, 1975).

greater growth rates occur at room temperature under uncontrolled humidity conditions.

Physical properties. The physical properties of the preferred aluminum alloys are summarized in Table 11.15. The thermal properties have been taken from the best-fit lines used in the *LNG Materials and Fluids* (1978) and from the compilation on thermal conductivity by Childs et al. (1973). Original sources are given for the elastic properties (Naiman et al., 1975; Read and Ledbetter, 1977) and electrical resistivity (Clark et al., 1970).

11.4.3 Fabrication characteristics

The relative ease of fabricating large aluminum structures is the principal advantage of aluminum over stainless steels. In addition, the light weight of aluminum structures facilitates assembly and erection of large structures and vessels. Fabrication characteristics that should receive special consideration for cryogenic applications are briefly reviewed in this section.

Welding. Aluminum alloys are readily weldable by the gas metal-arc and gas tungsten-arc processes. The compositions of aluminum welding rods and electrodes for these processes are summarized in Table 11.16. However, only the annealed 5083 alloy retains full strength in the as-welded condition. The cold-worked and precipitation-hardened alloys are weakened locally by welding. Part of the strength of precipitation-hardened alloys can be restored by a postweld heat treatment. The weld joint efficiencies (the ratio of the weld strength to unwelded-plate strength), of several aluminum alloys at low temperatures are shown in Fig. 11.41 (Kaufman, 1977).

The tensile properties of aluminum alloy weldments were determined by Nelson, Kaufman, and Wanderer (1969), and the results for selected alloys at room temperature, 77 K, and 4 K are summarized in Table 11.17. Notice that part of the strength obtained by cold working (5083-H321) and by precipitation hardening (6061-T6 and 2219-T851) is retained by the welded joint. This strength increase is not reflected in design-code allowable stresses (*ASME Boiler and Pressure Vessel Code*, 1977), which assume as-welded aluminum alloys are in the fully annealed condition. The tensile properties of 6061-T6 and 2219-T62 welds that were subsequently solution treated and aged indicate that substantial strengthening can be achieved by postweld heat treatments; however, for most cryogenic structures, postweld heat treatments are not considered practical. The low ductility of 2219 weldments without postweld heat treatment is attributed to localized deformation in the heat-affected zone. Both Ta-

Table 11.15 Physical properties of aluminum alloys.

Alloy and Temperature	Specific Gravity (g·cm^{-3})	Young's Modulus (GPa)	Shear Modulus (GPa)	Poisson's Ratio	Thermal Conductivity (W·m^{-1}·K^{-1})	Thermal Expansion (mean) (K^{-1}·10^{-6})	Specific Heat (J·kg^{-1}·K^{-1})	Electrical Resistivity (μΩ·cm)
Annealed 5083								
295 K	2.66	71.5[a]	26.8[a]	0.333[a]	120[b]	23[b]	900[b]	5.66[c]
77 K		80.2[a]	30.4[a]	0.320[a]	55[b]	18.1[b]	340[b]	3.32[c]
4 K		80.9[a]	30.7[a]	0.318[a]	3.3[b]	14.1[b]	0.28[b]	3.03[c]
Precipitation-hardened 6061								
295 K	2.70	70.1[b]	26.4[b]	0.338[b]		23[b]	900[c]	3.94[c]
77 K		77.2[b]	29.1[b]	0.328[b]		18.1[b]	340[c]	1.66[c]
4 K		77.7[b]	29.2[b]	0.327[b]		14.1[b]	0.28[c]	1.38[c]
Precipitation-hardened 2219								
295 K	2.83	77.4[d]	29.1[d]	0.330[d]	120[e]	23[f]	900[f]	5.7[b]
77 K		85.1[d]	32.3[d]	0.319[d]	56[e]	18.1[f]	340[f]	
4 K		85.7[d]	32.5[d]	0.318[d]	3[e]	14.1[f]	0.28[f]	2.9[b]

[a]Naimon, Ledbetter, and Weston (1975).
[b]*LNG Materials and Fluids* (1978).
[c]Clark et al. (1970).
[d]Read and Ledbetter (1977).
[e]Childs, Ericks, and Powell (1973).
[f]Assumed to be the same as that of 5083 and 6061.

Table 11.16 Composition limits (%) of selected filler metals for welding aluminum alloys.

Type	Cu	Mg	Mg	Cr	Si	Fe	Zn	Ti	Be	Other	Suggested for Welding
1100	0.20	—	0.05	—	Si+Fe=1.0	max.				99.0–99.5 Al	100, 3003
2319	5.6–6.8	0.02	0.20–0.40	—	0.20	0.30	0.10	0.10–0.20	0.0008	0.10-0.25 Zr	2219
4043	0.03	0.05	0.05	—	4.5–6.0	0.80	0.10	0.20	0.0008	0.05–0.15 V	1100, 6061
5183	0.10	4.3–5.2	0.50–1.0	0.50–0.25	0.40	0.40	0.25	0.15	0.0008		5083
5356	0.10	4.5–5.5	0.05–0.20	0.05–0.20	Si+Fe=0.5	max.	0.10	0.06–0.20	0.0008		5083, 6061
5556	0.10	4.7–5.5	0.5–1.0	0.5–0.20	Si+Fe=0.4	max.	0.25	0.05–0.20	0.0008		5083, 6061

ble 11.17 and Fig. 11.41 indicate that the joint efficiencies of the 5000-series weld are approximately 100% at 77 K but drop to about 80% at 4 K.

A great amount of mechanical property data has been obtained on weldments on 5083 plate welded with 5183 filler metal. Bogardus and Malcolm (1975) analyzed data on 5083/5183 weldments between 77 K and room temperature from several sources and observed that the strengths of the weld metal, fusion line, and HAZ consistently exceeded the minimum strengths (yield and tensile) of the parent metal for the same thickness range and temperature. Thus, they concluded that the maximum allowable design stresses developed for 5083-O sheet and plate are also acceptable for 5083-O sheet and plate welded with 5183 filler metal. Kelsey, Mueller, Elmer, and McHenry (1982) evaluated a series of weldments that were prepared by five different companies using procedures of their choice to weld 51-mm-thick plates with 5183 filler metal using the GMAW process. Tensile and notch tensile properties measured at room temperature and 4 K are similar despite the fact that there were considerable variations in the welding procedures. Fracture toughness and fatigue crack growth rates of the welds were measured at 4 K and found to be similar to base metal properties. It was concluded that the mechanical properties at 4 K of the 5083/5183 welds do not depend significantly on GMAW parameters and procedures. Kelsey, Wygonik, and Tenge (1975) conducted fatigue tests on large (up to 900-mm wide by 135-mm thick) plates loaded to simulate service stresses in spherical LNG cargo tanks. Fracture surface observations and comparisons with a similar test on an unwelded panel indicated that the fatigue cracking is essentially unaffected by the presence of the weld.

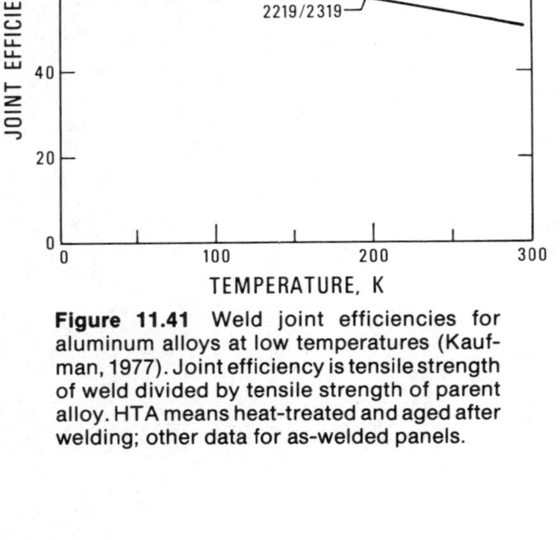

Figure 11.41 Weld joint efficiencies for aluminum alloys at low temperatures (Kaufman, 1977). Joint efficiency is tensile strength of weld divided by tensile strength of parent alloy. HTA means heat-treated and aged after welding; other data for as-welded panels.

Castings. Aluminum alloy castings can be produced by a variety of processes and generally are satisfactory for nonstructural applications at low temperatures. The *ASME Boiler and Pressure Vessel Code* (1977) permits use of Al–5Si (Type 43) and Al–7Si–0.3Mg (Type 356-T6) for nonwelded structural applications. Nelson et al. (1969) evaluated the 4-K tensile properties of several combinations of castings welded to wrought alloys; in general, superior properties were obtained when the properties of the cast and wrought components were nearly equal.

Machining, forming, and forging. Aluminum alloys, particularly the heat-treated grades, have excellent machinability. Machined and bolted assemblies of high-strength aluminum alloys can replace more costly welded stainless steel assemblies. When using steel bolts, one must consider that the thermal contraction of aluminum alloys in cooling to 4 K exceeds that of steel.

Table 11.17 Tensile properties of aluminum alloy weldments at room and cryogenic temperatures (Nelson et al., 1969).

Alloy and Temper	Plate Thickness (mm)	Filler Alloy	Post-weld Thermal Treatment	Temperature (K)	Tensile Strength, TS (MPa)	Joint Yield Strength, JYS (MPa)	Elongation in 4D (%)	Reduction of Area (%)	Joint Efficiency (%)	Location of Failure[a]	Notch Tensile Strength, NTS (MPa)	NTS/TS	NTS/JYS
5083-O	25	5183	No	RT	293	138	21.5	—	100	WM	308	1.05	2.22
				77	401	154	19	20	99	WM	345	0.86	2.24
				4	381	174	27	37	69	BM	371	0.98	2.14
5083-H321	25	5183	No	RT	305	179	14	39	96	HAZ	376	1.23	2.10
				77	446	216	19	23	100	WM	430	0.96	1.98
				4	455	246	9	14	80	WM	405	0.89	1.65
5083-H321	25	5356	No	RT	286	167	13.5	47	90	WM	371	1.30	2.22
				77	426	200	15.5	33	97	WM	418	0.98	2.08
				4	455	235	9	17	80	HAZ	398	0.87	1.69
5083-H321	25	5556	No	RT	306	176	14	36	97	WM	370	1.21	2.10
				77	450	211	20.5	26	100	WM	417	0.93	1.98
				4	474	238	13	17	83	WM	399	0.84	1.68
6061-T6	25	4043	No	RT	214	144	6	19	69	HAZ	234	1.10	1.63
				77	303	178	5.5	12	75	WM	273	0.90	1.54
				4	338	259	4.5	9	63	WM	275	0.81	1.06
			Yes	RT	298	247	11	44	96	BM	396	1.31	1.57
				77	395	291	16.5	12	97	WM	446	1.13	1.53
				4	452	309	15	16	84	WM	463	1.02	1.50
6061-T6	25	5356	No	RT	225	156	13.5	31	73	WM	323	1.44	2.07
				77	324	188	13.5	39	80	BM	373	1.15	1.98
				4	398	243	13.5	24	74	WM	367	0.92	1.50
			Yes	RT	279	202	9.5	33	90	BM	—	—	—
				77	393	234	20	29	97	BM	457	1.16	1.96
				4	476	307	19	24	89	WM	419	0.88	1.37
2219-T851	32	2319	No	RT	225	185	2	5	50	HAZ	280	1.24	1.52
				77	356	193	3.5	10	62	HAZ	334	0.94	1.73
				4	411	277	2.5	10	62	HAZ	363	0.88	1.31
2219-T62		2319	Yes	RT	395	277	7.5	7	99	HAZ	439	1.11	1.58
				77	475	321	5.5	6	94	HAZ	515	1.09	1.61
				4	496	355	3.5	5	80	HAZ	570	1.15	1.61

[a]Location of failures: WM - Weld metal; BM - Base metal; HAZ - Heat-affected zone.

Aluminum alloys have excellent formability in the annealed or the solution-treated conditions. Precipitation-hardened alloys can be formed in the solution-treated condition and then aged to the desired temper.

Aluminum alloys have the best forgeability of all structural alloys because of their very low strength at forging temperatures of 370 to 510°C. The size limit on aluminum forgings is significantly greater than that on stainless steels because less pressure is required.

Inspection. Aluminum alloys and their weldments can be readily inspected by conventional methods: dye penetration, ultrasonics, and radiography. Ikeda et al. (1978) observed that flaw detectability in heavy section plates of 5083 using ultrasonic inspection was improved by refinement of the microstructure. Tenge, Solli, and Forli (1975) recommend the use of ultrasonic testing for inspecting LNG tanks, because ultrasonic testing can detect planar defects, such as lack of fusion, whereas radiography cannot. Further, Tenge et al. recommend a calibration procedure that requires increased sensitivity over that required by the *ASME Boiler and Pressure Vessel Code* (1977) to ensure flaw detection in 5083 welds.

11.5 OTHER METALS AND ALLOYS

From a tonnage standpoint, nearly all of the structural alloys used for cryogenic applications are austenitic stainless steels, nickel steels, or aluminum alloys. However, certain important applications require the use of copper, nickel, titanium, and their alloys. Copper is used in applications where its excellent electrical and thermal conductivities are needed. Nickel-base superalloys are preferred for applications such as superconducting motors, where strengths at both room and cryogenic temperatures are important. Titanium alloys are sometimes used in aerospace applications because of their excellent strength-to-weight ratios.

11.5.1 Copper and copper alloys

Copper and copper alloys are not often used in structural applications, but they are so commonly used at cryogenic temperatures that their properties are briefly discussed. Copper is used in commercially pure form in applications where thermal or electrical conductivity is of primary importance. It can be alloyed with zinc to form the brasses, with tin to form the bronzes, with nickel to form the cupronickels, and with other elements or combinations of elements to form numerous other alloys. Copper has an f.c.c. structure, and most of the brasses, bronzes, and cupronickels are also single-phase f.c.c. alloys. Thus, toughness and ductility are retained at cryogenic temperatures, and strengthening mechanisms are limited to cold working and grain refinement.

Copper is used as the stabilizer for superconductors, an application where thermal and electrical conductivities are of prime importance. It is also commonly used for Dewar vessels and as tubing for transfer lines and heat exchangers. Copper alloys have greater strength, but their thermal and electrical conductivities are significantly lower than that of pure copper. There are many uses for copper alloys in experimental setups and special devices, such as valve components, because of their excellent fabrication characteristics.

The mechanical properties of copper and copper alloys have been compiled by Reed and Mikesell (1967a). A systematic evaluation of 15 copper alloys, including brasses, bronzes, and commercially pure coppers, has also been published by Reed and Mikesell (1967b). Tensile, notch tensile, and impact properties were tested at 295, 195, 76, 20, and 4 K. The alloys, with the exception of Ni–Al–bronze casting, had improved properties and similar or lower impact properties at low temperatures. For nine of the alloys, the yield and tensile strengths increased with decreasing temperature to 20 K, but the yield and tensile strengths were lower at 4 K than at 20 K.

The yield strength of copper can be increased significantly by cold work. The influence of cold work on the yield strength of oxygen-free, phosphorus-deoxidized copper is shown in Fig. 11.42 (Reed, 1982). Note that the increase in yield strength reaches a maximum at 25% cold work. At a deformation of 25%, the yield strength at 4 K is 440 MPa, an 8-fold increase over the annealed value.

The influence of alloy content on the strength of Cu–Zn and Cu–Ni alloys has also been evaluated by Reed (1982). As shown in Fig. 11.43, the nickel is a more potent strengthener than zinc. At about 20% Ni, a maximum is reached in the yield strength (180 MPa) at room temperature. At 76 K, the maximum in yield strength (230 MPa) occurs at about 30% Ni. The yield strength at 4 K increases monotonically with nickel content over the range 0 to 30% Ni, reaching a value of 280 MPa at 30% Ni. The yield strength of Cu–Zn increases monotonically with zinc content over the entire range of solid solubility (30% Zn). At 30% Zn, the yield strengths at 295, 76, and 4 K are 80, 130, and 160 MPa, respectively.

11.5.2 Nickel-base superalloys

Nickel is commonly alloyed with copper, chromium, iron, molybdenum, or a combination of these elements, to provide alloys with superior

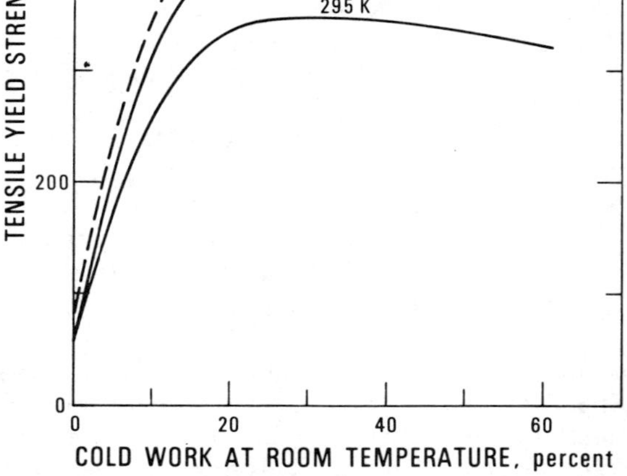

Figure 11.42 Yield strength of copper at 4, 76, and 295 K as a function of percent cold work at room temperature (Reed, 1982).

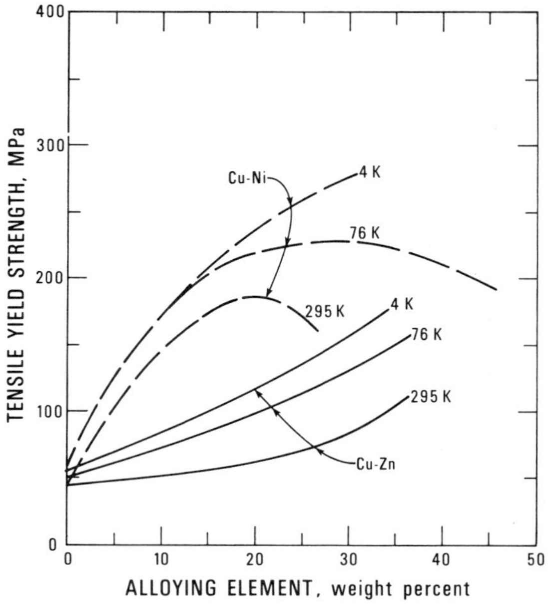

Figure 11.43 Yield strength of Cu–Ni and Cu–Zn alloys at 4, 76, and 295 K as a function of percent alloying element (Reed, 1982).

corrosion and oxidation resistance. These f.c.c. alloys tend to retain satisfactory strength and ductility at cryogenic temperatures. However, the nickel alloys are not usually used at cryogenic temperatures because they are more expensive than competitive materials, such as the stainless steels. The nickel-base superalloys are a potential exception because they have high strength at room temperature and in liquid helium. The need for material with such properties in the rotors for superconducting generators inspired the evaluation of superalloys for liquid-helium service by Wells, Kossowsky, Logsdon, and Daniel (1976).

Properties. The mechanical properties of three nickel-base superalloys have been evaluated for liquid-helium service: Inconel X-750, Inconel 718, and Inconel 706. They are Ni–Cr–Fe alloys that are hardened by a γ' precipitate of Ni_3 (niobium, titanium, and aluminum); the nominal compositions are given in Table 11.18.

The tensile properties, fracture toughness, and fatigue crack growth rates of Inconel X-750 have been evaluated by Logsdon (1977). The test materials were produced by six different processing sequences involving melting, hot working, and heat treatment. For the three vacuum-melted heats that were given the normal heat treatment (solution treat and double age), the yield strength at 4 K ranged from 866 to 1193 MPa, and the fracture toughness at 4 K ranged from 76.1 to 238 MPa·m$^{1/2}$. The large variations in properties were attributed to microstructural differences (Kossowsky, 1977). The fatigue crack growth rates also were sensitive to processing; however, in all cases, the growth rates for a given material were less at 4 K than at room temperature.

The tensile properties, fracture toughness, and fatigue crack growth rates of a single heat of Inconel 718 have been evaluated at room and cryogenic temperatures by Tobler (1976a). All test specimens were taken from a single piece of 7.7-cm-square-bar stock and tested in the solution-treated and double-aged condition. All the properties tended to improve with decreasing temperature. Between room temperature and 4 K, the yield strength increased from 1172 to 1408 MPa, the ultimate strength from 1404 to 1816 MPa, the ductility from 15 to 21% elongation, and the fracture toughness from 96 to 112 MPa·m$^{1/2}$. The fatigue crack growth rates were essentially constant at the four test temperatures, 295, 195, 76, and 4 K. Significantly lower values of fracture toughness (75.5 MPa·m$^{1/2}$ at 4 K and 61.5 MPa·m$^{1/2}$ at room temperature) were reported for Inconel 718 by Logsdon, Wells, and Kossowsky (1980); again, the need for microstructural control in the nickel-base superalloys is emphasized.

The tensile properties, fracture toughness, and fatigue crack growth rates of Inconel 706 have been evaluated at room and cryogenic temperatures by Logsdon et al. (1980). Two heats were evaluated: one was vacuum-induction melted (VIM) followed by vacuum-arc remelting (VAR), and the other was VIM followed by electric-furnace remelting (EFR). Both heats had similar microstructures and properties. The Inconel 706 alloy had strengths and toughnesses that were superior to those of Inconel X-750 and 718. The properties at 4 K were 1650 MPa ultimate strength, 1250 MPa yield strength, 30% elongation, and 149 MPa·m$^{1/2}$ fracture toughness.

11.5.3 Titanium alloys

Titanium alloys are classified according to their crystalline structure. Pure titanium solidifies as b.c.c. crystals and transforms to hexagonal close-packed (h.c.p.) at 885° C. Alloying can be used to change the transformation temperature or to eliminate the transformation entirely. Commercially pure titanium and h.c.p. alloys,

Table 11.18 Compositions (%) of nickel-base superalloys.

Designation	Ni	C	Mn	Fe	S	Si	Cu	Cr	Al	Ti	Nb (+Ta)	Mo
Inconel X-750	73	0.04	0.5	7.0	0.005	0.2	0.2	15.5	0.7	2.5	1.0	—
Inconel 718	52.5	0.04	0.2	18.5	0.008	0.2	0.2	19.0	0.5	0.9	5.1	3.0
Inconel 706	41.5	0.03	0.2	40.0	0.008	0.2	0.2	16.0	0.2	1.8	2.9	—

such as Ti–5Al–2.5Sn, are known as the α alloys. Aluminum and interstitial elements (such as oxygen, hydrogen, nitrogen, and carbon) stabilize the α phase. The b.c.c. alloys (such as Ti–13V–11Cr–3Al) are known as β alloys. Chromium, molybdenum, iron, manganese, and vanadium stabilize the β phase. Many of the most common alloys, such as Ti–6Al–4V and Ti–6Al–6V–2Sn, are predominantly α alloys with some β phase. The β and α–β alloys can be heat-treated to room-temperature yield strengths of 1100 MPa and higher. However, the β alloys and heat-treated α–β alloys are brittle at low temperatures and, thus, not suitable for cryogenic applications.

The use of titanium alloys at cryogenic temperatures has been limited to aerospace applications where strength-to-weight ratio is the primary concern. Commercially pure titanium, the Ti–5Al–2.5Sn α alloy, and the Ti–6Al–4V α–β alloy have been used. To improve the ductility and toughness at low temperatures, extra-low-interstitial (ELI) grades are specified, and the alloys are used in the annealed condition. Even with these improvements, titanium alloys have very low fracture-toughness-to-yield-strength ratios at cryogenic temperatures and, consequently, are susceptible to brittle fracture. In addition, titanium alloys are relatively expensive and difficult to fabricate. Thus, they are not considered candidates for applications where satisfactory performance can be obtained with austenitic stainless steels, nickel steels, or aluminum alloys.

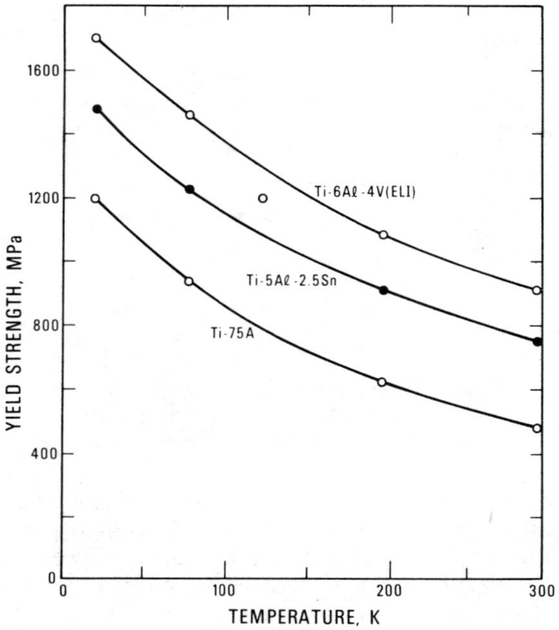

Figure 11.44 Yield strengths of three titanium alloys — Ti–6Al–4V, Ti–5Al–2.5Sn, and Ti–75A (commercial purity) — at temperatures between 4 and 300 K (*Handbook on Materials for Superconducting Machinery,* 1977).

Properties. The strengths of titanium alloys are substantially higher at cryogenic temperatures than at room temperature. The yield-strength–vs.–temperature curves for three alloys in sheet form are shown in Fig. 11.44. Note that a 2-fold increase in yield strength occurs between room temperature and 20 K. Similar trends are observed for ultimate tensile strength and for other product forms, such as plates, bars, and forgings. The ductility decreases slightly with decreasing temperatures down to about 77 K and more noticeably at lower temperatures. The strength increases and ductility decreases are slightly lower for the ELI grades than for the normal-interstitial-level grades owing to the contribution of interstitial hardening.

The low fracture toughness of titanium alloys limits their usefulness for cryogenic applications. This is well recognized for the β alloys and the α–β alloys in the heat-treated condition. The most commonly used alloys, Ti–5Al–2.5Sn and Ti–6Al–4V, are also susceptible to brittle fracture at cryogenic temperatures. Consider, for example, the data of Tobler (1976b) and Fowlkes and Tobler (1976) for Ti–6Al–4V; a curve of fracture toughness vs. temperature summarizing these data is shown in Fig. 11.45. Note that the toughness of the ELI grade is substantially better than that of the normal-interstitial-level grade,

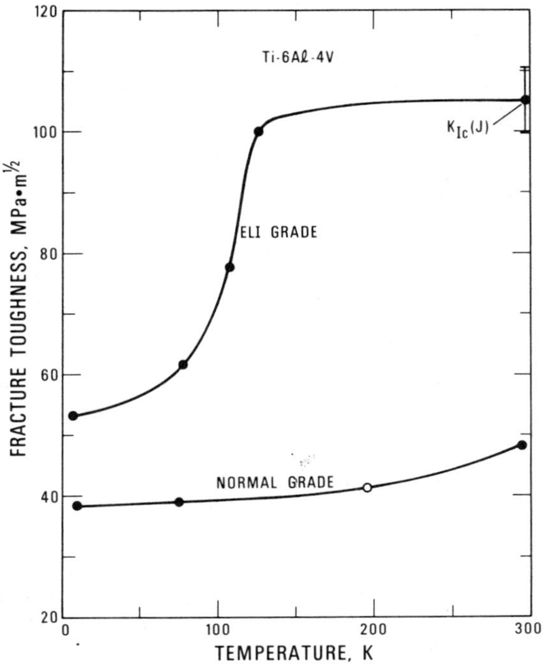

Figure 11.45 Temperature dependence of fracture toughness for Ti–6Al–4V alloys (Fowlkes and Tobler, 1976). ● — K_{Ic}; ○ — K_Q.

but the toughness of the ELI grade drops abruptly at temperatures below 125 K. The combination of reduced toughness and increased yield strength increases the risk of brittle fracture. This can be shown by comparing the specimen thicknesses, B, as required by ASTM E399, to ensure plane strain behavior, where

$$B = 2.5\,(K_{Ic}/\sigma_{ys})^2 \tag{11.3}$$

For Ti–6Al–4V (ELI) using Tobler's (1976b) data, B is 40.6 mm at room temperature, 4.4 mm at 77 K, and 1.3 mm at 4 K. Given that a material is susceptible to brittle fracture when plane strain conditions exist, these thickness limits clearly indicate that the use of Ti–6Al–4V at temperatures of 77 K and below should be restricted. Use should be limited to thin-gauge applications or critical applications where the advantages of titanium alloys offset the expense of increased care in design, analysis, fabrication, and inspection.

The fracture toughness of Ti–5Al–2.5Sn (ELI) tends to be slightly higher than that of Ti–6Al–4V (ELI), as shown by the data reviewed by Campbell (1974). Low-toughness values have been reported for Ti–5Al–2.5Sn (ELI) plates. Carman and Katlin (1968) reported an average K_{Ic} for through-thickness cracks to be 55 MPa·m$^{1/2}$ at 77 K and 59 MPa·m$^{1/2}$ at 20 K, and Terada, Nakai, and Takamatsu (1980) reported a K_{Ic} of 45 MPa·m$^{1/2}$ at 4 K. In addition, even lower K_{Ic} values have been reported for the normal-interstitial-level plates reported by Carman and Katlin (1968) and by Van Stone, Low, and Shannon (1977). Thus, the use of Ti–5Al–2.5Sn in specific cryogenic applications should be carefully evaluated.

11.6 SUMMARY

Many commercially available alloys are suitable for structural applications in a wide variety of cryogenic systems. The most commonly used alloys are the austenitic stainless steels, the nickel steels, and aluminum alloys. For these alloy classes, there are specific alloys with sufficient strength, ductility, and toughness at cryogenic temperatures that can be fabricated into large and complex structures. The preferred alloys for service to 4 K are the austenitic stainless steels (304, 310S, and 316), the nitrogen-strengthened austenitic stainless steels (304LN and 316LN), and the aluminum alloys (5083, 6061, and 2219). For service to 77 K, 9% Ni steel can be included in the list of preferred alloys.

11.7 REFERENCES

Alcoa Aluminum, The Cryogenic Metal (1971). Aluminum Company of America, Pittsburgh.

Aluminum Standards and Data (1981). Aluminum Association, Washington, D.C.

Argy, G., Paris, P. C., and Shaw, F. (1975). Fatigue crack growth and fracture toughness of 5083-O aluminum alloy. In *Properties of Materials for Liquefied Natural Gas Tankage.* ASTM STP 579, American Society for Testing and Materials, Philadelphia, 96–130.

ASME Boiler and Pressure Vessel Code (1977). Section VIII, Pressure vessels. American Society of Mechanical Engineers, New York.

Bennett, F. W. and Dillon, C. P. (1966). Impact strength of austenitic stainless steel welds at −320F —effects of composition, ferrite content, and heat treatment. *J. Basic Eng.* **88**, 33–36.

Blake, M. A. W., Rowntree, G., and Phelps, B. (1973). The variation in strength of a weld metal for 9%Ni steel. In *Welding Low Temperature Containment Plant*. The Welding Institute, Cambridge, England, 1–9.

Bogardus, K. O. and Malcolm, R. C. (1975). Design stresses for aluminum alloy 5083-O and 5183 welds at cryogenic temperatures. In *Properties of Materials for Liquefied Natural Gas Tankage.* ASTM STP 579, American Society for Testing and Materials, Philadelphia, 190–204.

Brickner, K. G. and Defilippi, J. D. (1977). Mechanical properties of stainless steels at cryogenic temperatures and at room temperature. In *Handbook of Stainless Steels*, McGraw–Hill, New York,20-1–20-39.

Campbell, J. E. (1974). Fracture toughness of high-strength alloys at low temperature — a review. In *Fatigue and Fracture Toughness — Cryogenic Behavior*. ASTM STP 556, American Society for Testing and Materials, Philadelphia, 3–25.

Carman, C. M. and Katlin, J. M. (1968). Plane strain fracture toughness and mechanical properties of 5Al–2.5Sn ELI and commercial titanium alloys at room and cryogenic temperatures. In *Applications Related Phenomena in Titanium Alloys*. ASTM STP 432, American Society for Testing and Materials, Philadelphia, 124–144.

Carter, W. P. and Harrison, J. D. (1973). The use of 9% nickel steel for LNG application. In *Welding Low Temperature Containment Plant*. The Welding Institute, Cambridge, England, 62–85.

Castro, R. and deCadenet, J. J. (1975). *Welding Metallurgy of Stainless and Heat-Resisting Steels*. Cambridge University Press, London.

Childs, G. E., Ericks, L. J., and Powell, R. L. (1973). *Thermal Conductivity of Solids at Room Temperature and Below*. NBS Monograph 131, National Bureau of Standards, Boulder, Colorado.

Christian, J. L., Gruner, J. D., and Girton, L. D. (1964). The effects of cold rolling on the mechanical properties of type 310 stainless steel at room and cryogenic temperatures. *Trans. Am. Soc. Met.* **57**, 199–207.

Clark, A. F., Childs, G. E., and Wallace, G. H. (1970). Electrical resistivity of some engineering alloys at low temperatures. *Cryogenics* **10**, 295–305.

Cryogenic Application of Ferrous Materials in Japan (1979). Japan Pressure Vessel Research Committee Report, The Iron and Steel Institute of Japan, Tokyo.

Defilippi, J. D., Brickner, K. G., and Gilbert, E. M. (1969). Ductile-to-brittle transition in austenitic stainless steels. *Trans. AIME* **245**, 2141–2148.

DeLong, W. T. (1974). Ferrite in austenitic stainless steel weld metals. *Weld. J.* **53**, 273s–286s.

DeLong, W. T., Ostrom, G. A., and Szumachowski, E. R. (1956). Measurement and calculation of ferrite in stainless steel weld metal. *Weld. J.* **35**, 526s.

Efferson, K. R. and Leonard, W. J. (1976). *Magnetic Properties of Some Structural Materials Used in Cryogenic Applications*. ORNL-4150, Oak Ridge National Laboratory, Oak Ridge, Tennessee, 126.

Elmer, J. W., McHenry, H. I., and Whipple, T. A. (1981). Strength and toughness of fully austenitic stainless steel filler metals at 4K. In *Materials Studies for Magnetic Fusion Energy Applications at Low Temperatures — IV*. NBSIR 81-1645, National Bureau of Standards, Boulder, Colorado, 289–302.

Favor, R. J., Gideon, D. N., Grover, H. J., Hayes, J. E., and McClure, G. M. (1961). *Investigation of Fatigue Behavior of Certain Alloys in the Temperature Range of Room Temperature to −423° F.* Report WADD TR 61–132, Battelle, Columbus, Ohio.

Fowlkes, C. W. and Tobler, R. L. (1976). Fracture testing and results for a Ti–6Al–4V alloy at liquid helium temperature. *Eng. Fract. Mech.* **8**, 487–500.

Gideon, D. N., Favor, R. J., Koppenhafer, A., Grover, H. J., and McClure, M. (1962). *Investigation of Notched Fatigue Behavior of Certain Alloys in the Temperature Range of Room Temperature to −423° F.* ASD-TDR-62-351, Battelle, Columbus, Ohio.

Gunter, C. J., and Reed, R. P. (1961). Mechanical properties of four austenitic stainless steels at temperatures between 300 and 20 K. In *Advances in Cryogenic Engineering*, Vol. 6. Plenum Press, New York, 565–576.

Hall, E. R. (1962). Tensile and impact properties of cast stainless steels at cryogenic temperatures. In *Evaluation of Metallic Materials in Design for Low-Temperature Service*. ASTM STP 302, American Society for Testing and Materials, Philadelphia, 85–93.

Hall, L. R. (1971). Influence of specimen design in plane strain fracture toughness testing. In *Fracture Toughness Testing at Cryogenic Temperatures*. ASTM STP 496, American Society for Testing and Materials, Philadelphia, 40–60.

Handbook on Materials for Superconducting Machinery (1977). MCIC-HB-04. Battelle, Columbus, Ohio

Hanson, M. P., Stickley, G. W., and Richards, H. T. (1961). Sharp-notch behavior of some high-strength sheet aluminum alloys and welded joints at 75, −320, and −423 F. In *Low Temperature Properties of High-Strength Aircraft and Missile Materials*. ASTM STP 287, American Society for Testing and Materials, Philadelphia, 3–15.

Huettich, N. J., Pense, A. W., and Stout, R. D. (1971). *The Toughness of 2¼% and 3½% Nickel Steels at Cryogenic Temperatures*. WRC Bulletin 165, Welding Research Council, New York.

Ikeda, M., Miyagi, Y., Aoki, M., and Shirakura, T. (1978). Fracture strength of thick 5083-O aluminum alloy for LNG storage tanks. In *Advances in Cryogenic Engineering*, Vol. 24. Plenum Press, New York, 166–174.

Ishikawa, K. and Maruyama, N. (1978). Fracture and strength of MIG welded Fe–13%Ni–3%Mo alloy for cryogenic service. *Cryogenics* **18**, 585–589.

Jin, S., Hwang, S. K., and Morris, J. W., Jr. (1975). The effect of grain size and retained austenite on the ductile–brittle transition of a titanium iron alloy. *Metall. Trans.* **6A**, 1721–1726.

Juppenlatz, J. W. (1952). Austenitic cast stainless steels good for low temperature applications — BUT. *Iron Age* **170**, 147–149.

Kaufman, J. G. (1977). Aluminum alloys for arctic service. In *Materials Engineering in the Arctic*. American Society for Metals, Metals Park, Ohio, 240–247.

Kaufman, J. G., Bogardus, K. O. and Wanderer, E. T. (1968). Tensile properties and notch toughness of aluminum alloys at −452° F in liquid helium. In *Advances in Cryogenic Engineering*, Vol. 13, Plenum Press, New York, 294–308.

Kaufman, J. G. and Johnson, E. W. (1963). Notch sensitivity of aluminum alloy sheet and plate at −320° F based upon notch–yield ratio. In *Advances in Cryogenic Engineering*, Vol. 8. Plenum Press, New York, 678–685.

Kaufman, J. G. and Kelsey, R. A. (1975). Fracture toughness and fatigue properties of 5083-O plate and 5183 welds in liquefied natural gas applications. In *Properties of Materials for Liquefied Natural Gas Tankage*. ASTM STP 579, American Society for Testing and Materials, Philadelphia, 138–158.

Kaufman, J. G. and Nelson, F. G. (1962). Cryogenic temperatures up fatigue strengths of Al–Mg alloys. *Space/Aeronaut.* **38**, 91–96.

Kaufman, J. G., Nelson, F. G., and Wygonik, R. H. (1974). Large-scale fracture toughness tests of thick 5083-O plate and 5183 welded panels at room temperature, −260 and −320° F. In *Fatigue and Fracture Toughness — Cryogenic Behavior*. ASTM STP 556, American Society for Testing and Materials, Philadelphia, 125–158.

Kaufman, J. G. and Wanderer, E. T. (1967). Aluminum alloys for cryogenic temperatures. In *Aluminum Vol. II. Design and Application*. American Society for Metals, Metals Park, Ohio, 297.

Kaufman, J. G. and Wanderer, E. T. (1971). Tensile properties and notch toughness of some 7xxx alloys at −452° F. In *Advances in Cryogenic Engineering*, Vol. 16. Plenum Press, New York, 27–36.

Kawaguchi, Y., Tsukamoto, M., Maruyama, K., Bessyo, K., and Kato, Y. (1982). Crack arrestability of Ni steel plates for low temperature service to the extremely long brittle crack. In *Fracture Toughness Testing — Methods, Interpretation and Application*. The Welding Institute, Cambridge, England, paper 6.

Kelsey, R. A., Mueller, L. N., Elmer, J. W., and McHenry, H. I. (1982). The mechanical properties of welds in aluminum alloy 5083 at 4K. In *The First International Aluminum Welding Conference*. Welding Research Council, New York, 89–105.

Kelsey, R. A., Wygonik, R. H., and Tenge, P. (1975). Crack growth and fracture of thick 5083-O plate under liquefied natural gas ship spectrum loading. In *Properties of Materials for Liquefied Natural Gas Tankage*. ASTM STP 579, American Society for Testing and Materials, Philadelphia, 44–79.

Krivobok, V. N. and Thomas, R. D., Jr. (1950). Impact tests of welded austenitic stainless steels. *Weld. J.* **29**, 493s–495s.

Kossowsky, R. (1977). Microstructures of Inconel X750 for cryogenic structural applications. In *Advances in Cryogenic Engineering*, Vol. 22. Plenum Press, New York, 59–67.

Lake, R. L. (1975). Toughness variations through the thickness of thick 5083-O aluminum alloy plate. In *Properties of Materials for Liquefied Natural Gas Tankage*. ASTM STP 579, American Society for Testing and Materials, Philadelphia, 176–189.

Lancaster, J. F. (1965). *The Metallurgy of Welding Brazing and Soldering*. American Elsevier, New York.

Ledbetter, H. M. (1977). Anomalous low-temperature elastic behavior of a nitrogen-strengthened chromium–manganese stainless steel. *Mater. Sci. Eng.* **29**, 255–260.

Ledbetter, H. M. and Collings, E. W. (1979). Low-temperature magnetic and elastic-constant anomalies in three manganese stainless steels. In *Metal Science of Stainless Steels*. American Institute of Mining, Metallurgical, and Petroleum Engineers, New York.

Ledbetter, H. M., Weston, W. F. and Naiman, E. R. (1975). Low temperature elastic properties of four austenitic stainless steels. *J. Appl. Phys.* **46**, 3855–3860.

LNG Materials and Fluids (1978) (ed. D. B. Mann). National Bureau of Standards, Boulder, Colorado.

Logsdon, W. A. (1977). Cryogenic fracture mechanics properties of several manufacturing process/heat treatment combinations of Inconel X750. In *Advances in Cryogenic Engineering*, Vol. 22. Plenum Press, New York, 47–58.

Logsdon, W. A., Wells, J. M., and Kossowsky, R. (1980). Cryogenic fracture toughness and fatigue crack-growth rate properties of Inconel 706 base metal and gas tungsten-arc weldments. In *Advances in Cryogenic Engineering*, Vol. 26. Plenum Press, New York, 137–150.

Low Temperature and Cryogenic Steels (1966). Materials Manual, U.S. Steel Corp., Pittsburgh.

Mahin, K. W., Morris, J. W., Jr., and Watanabe, I. (1980). A review of the development of ferritic consumables for the welding of 9% nickel steel: research in the United States and Japan. In *Advances in Cryogenic Engineering*, Vol. 26. Plenum Press, New York, 187–199.

Manson, S. S. (1965). Fatigue: A complex subject — some simple approximations. *Exp. Mech.* **5**, 193–226.

Materials for Cryogenic Service (1972). Engineering properties of 3½, 5, and 9 percent nickel steels and 36 percent nickel–iron alloy. International Nickel Limited, London.

Materials for Cryogenic Service (1974). Engineering properties of austenitic stainless steels. International Nickel Limited, London.

Mazandarany, F. N., Parker, D. M., Koenig, R. F., and Read, D. T. (1980). A nitrogen-strengthened austenitic stainless steel for cryogenic magnet structures. In *Advances in Cryogenic Engineering*, Vol. 26. Plenum Press, New York, 158–170.

McHenry, H. I. (1977). Fracture mechanics and its application to cryogenic structures. In *Advances in Cryogenic Engineering*, Vol. 22. Plenum Press, New York, 9–26.

McHenry, H. I., Naranjo, S. E., Read, D. T., and Reed, R. P. (1978). Low temperature fracture properties of a USSR aluminum–6% magnesium alloy. In *Advances in Cryogenic Engineering*, Vol. 24. Plenum Press, New York, 519–528.

McHenry, H. I. and Reed, R. P. (1977). Fracture behavior of the heat-affected zone in 5%Ni steel weldments. *Weld. J.* **56**, 104s–112s.

McHenry, H. I. and Reed, R. P. (1980). Structural alloys for superconducting magnets in fusion energy systems. *Nucl. Eng. Des.* **58**, 219–236.

Metallurgical Properties of Armco Cryonic 5 Steel (1971). Armco Steel Corp., Middletown, Ohio, 91.

Metals Handbook (1973). Vol. 8, *Metallography, Structures and Phase Diagrams*. American Society for Metals, Metals Park, Ohio, 304.

Miyakoshi, H., Ishikura, N., Suzuki, T., and Tanaka, K. (1981). Experience in Japanese gas industry on nine percent nickel steel LNG storage tanks. Paper 81-T-37, American Gas Association, Chicago, T-155–T-166.

Murayama, N., Pense, A. W., and Stout, R. D. (1976). The fracture toughness of cryogenic steels. In *Ad-*

vances in Cryogenic Engineering*, Vol. 22. Plenum Press, New York, 27–34.

Nachtigall, A. J. (1975). Strain cycling fatigue behavior of ten structural metals tested in liquid helium, liquid nitrogen, and ambient air. In *Properties of Materials for Liquefied Natural Gas Tankage*. ASTM STP 579, American Society for Testing and Materials, Philadelphia, 378–396.

Naimon, E. R., Ledbetter, H. M., and Weston, W. F. (1975). Low temperature elastic properties of four wrought and annealed aluminum alloys. *J. Mater. Sci.* **10**, 1309–1316.

Nakano, Y., Suzuki, S., Kamada, A. and Hirose, K. (1982). Crack initiation and arrest characteristics of 9%Ni steels with various Charpy V-notch energy values. In *Advances in Cryogenic Engineering*, Vol. 28. Plenum Press, New York, 127–136.

Nelson, F. G., Kaufman, J. G., and Wanderer, E. T. (1969). Tensile properties and notch toughness of groove welds in wrought and cast aluminum alloys at cryogenic temperature. In *Advances in Cryogenic Engineering*, Vol. 14. Plenum Press, New York, 71–82.

Pense, A. W. and Stout, R. D. (1975). *Fracture Toughness and Related Characteristics of Cryogenic Nickel Steels*. WRC Bulletin 205, Welding Research Council, New York.

Randak, V. A., Wessling, W., Bock, H. E., Steinmaurer, H. and Faust, L. (1971). *Stahl Eisen* **91**, 1255.

Read, D. T. and Ledbetter, H. M. (1977). Temperature dependences of the elastic constants of precipitation-hardened aluminum alloys 2014 and 2219. *J. Eng. Mater. Technol.* **99**, 181–184.

Read, D. T., McHenry, H. I., Steinmeyer, P. A., and Thomas, R. D., Jr. (1980). Metallurgical factors affecting the toughness of 316L SMA weldments at cryogenic temperatures. *Weld. J.* **59**, 104s–113s.

Read, D. T. and Reed, R. P. (1979). Toughness, fatigue crack growth and tensile properties of three nitrogen-strengthened stainless steels at cryogenic temperatures. In *The Metal Science of Stainless Steel.* American Society for Metals, Metals Park, Ohio, 92–121.

Read, D. T. and Reed, R. P. (1981). Fracture and strength properties of selected austenitic stainless steels at cryogenic temperatures. *Cryogenics* **21**, 415–417.

Reed, R. P. (1982). National Bureau of Standards, Boulder, Colorado, private communication.

Reed, R. P. and Mikesell, R. P. (1967a). *Low Temperature Mechanical Properties of Copper and Selected Copper Alloys.* NBS Monograph 101, National Bureau of Standards, Boulder, Colorado.

Reed, R. P. and Mikesell R. P. (1967b). Low temperature (295 to 4K) mechanical properties of selected copper alloys. *J. Mater.* **2**, 370–392.

Sakai, T., Osaki, K., Moriyama, K., and Yano, S. (1976). Application of 5.5% nickel steel in LNG storage tanks. In *Gastech 76.* Gastech Exhibitions Limited, Herts, England.

Sanderson, G. P. and Llewellyn, D. T. (1969). Mechanical properties of standard austenitic stainless steels in the temperature range −196 to +800C. *J. Iron Steel Inst. London* **207**, 1129–1146.

Schaeffler, A. L. (1949). Constitution diagram for stainless steel weld metal. *Met. Prog.* **56**, 680.

Schwartzberg, F. R. and Kiefer, T. F. (1975). *Properties of Cryogenically Worked Metals.* NASA contractor report No. NASA-CR-134757, National Aeronautics and Space Administration, Washington, D.C.

Schwartzberg, F. R., Osgood, S. H., Keys, R. D., and Kiefer, T. F. (1964). *Cryogenic Materials Data Handbook.* Technical Documentary Report No. ML-TDR-64-280, Wright Patterson Air Force Base, Ohio.

Shepic, J. A. and Fiftal, C. F. (1979). Testing of 2219-T87 aluminum alloy at 4K. In *Materials Studies for Magnetic Fusion Energy Applications at Low Temperatures — II.* NBSIR 79-1609, National Bureau of Standards, Boulder, Colorado, 17–31.

Shepic, J. A. and Schwartzberg, F. R. (1978). Fatigue testing of stainless steels. In *Materials Studies for Magnetic Fusion Energy Applications at Low Temperatures — I.* NBSIR 78-884, National Bureau of Standards, Boulder, Colorado, 15–68.

Studies on 9%Ni Steel Plates for LNG Tanks. (1981). Kawasaki Steel Corporation, Tokyo.

Szumachowski, E. R. and Reid, H. F. (1978). Cryogenic toughness of SMA austenitic stainless steel weld metals: Part 1 — Role of ferrite. *Weld. J.* **57**, 325s–333s.

Tenge, P., Karlsen, A., and Mauritzon, B. (1975). Static and dynamic fracture toughness of 9 percent nickel steel at −196C. *Norw. Maritime Res.* **3**, 11–19.

Tenge, P., Solli, O., and Forli, O. (1975). Significance of defects in liquefied natural gas tanks in ships. In *Properties of Materials for Liquefied Natural Gas Tankage.* ASTM STP 579, American Society for Testing and Materials, Philadelphia, 10–43.

Terada, H., Nakai, E., and Takamatsu (1980). Fracture toughness and fatigue behavior of Ti–6Al–4V and Ti–5Al–2.5Sn at cryogenic temperature. In *Titanium 80 Science and Technology.* Plenum Press, New York, 1607–1615.

Tharby, R. H., Heath, D. J., and Flannery, J. W. (1973). Welding 9% nickel steel — A review of current practices. In *Welding Low Temperature Containment Plant.* The Welding Institute, Cambridge, England, 158–171.

Tobler, R. L. (1976a). Low temperature effects on the fracture behavior of a nickel base superalloy. *Cryogenics* **16**, 669–674.

Tobler, R. L. (1976b). Fatigue crack growth and *J*-integral fracture parameters of Ti–6Al–4V at ambient and cryogenic temperatures. In *Cracks and Fracture.* ASTM STP 601, American Society for Testing and Materials, Philadelphia, 346–370.

Tobler, R. L., McHenry, H. I., and Reed, R. P. (1978). Fracture mechanics parameters for an iron–13% chromium–19% manganese stainless steel and its welds at cryogenic temperatures. In *Advances in Cryogenic Engineering,* Vol 24. Plenum Press, New York, 560–572.

Tobler, R. L., Mikesell, R. P., Durcholz, R. L., and Reed, R. P. (1975). Low temperature fracture behavior of iron–nickel alloy steels. In *Properties of Materials for Liquefied Natural Gas Tankage.* ASTM STP 579, American Society for Testing and Materials, Philadelphia, 261–287.

Tobler, R. L., Mikesell, R. P., and Reed, R. P. (1979). Cryogenic effects on the fracture mechanics parameters of ferritic nickel alloy steels. In *Fracture Mechanics.* ASTM STP 677, American Society for Testing and Materials, Philadelphia, 85–105.

Tobler, R. L. and Reed, R. P. (1977). Fracture mechanics parameters for a 5083-O aluminum alloy at low temperatures. *J. Eng. Mater. Technol.* **99**, 306–312.

Tobler, R. L. and Reed, R. P. (1978). Fatigue crack growth resistance of structural alloys at cryogenic temperatures. In *Advances in Cryogenic Engineering,* Vol 24. Plenum Press, New York, 82–90.

Tobler, R. L. and Reed, R. P. (1979). Tensile and fracture behavior of a nitrogen-strengthened, chromium–nickel–manganese stainless steel at cryogenic temperatures. In *Elastic-Plastic Fracture.* ASTM STP 668, American Society for Testing and Materials, Philadelphia, 85–105.

Tobler, R. L. and Reed, R. P. (1982). Interstitial carbon and nitrogen effects on the cryogenic fatigue crack growth of AISI 304 type stainless steels. Presented at the 15th National Symposium on Fracture Mechanics, College Park, Maryland, American Society for Testing and Materials, Philadelphia.

Van Stone, R. H., Low, J. R., and Shannon, J. L., Jr. (1977). *Investigation of the Fracture Mechanisms in Ti–5Al–2.5Sn at Cryogenic Temperatures.* NASA report CR 135120, National Aeronautics and Space Administration, Washington, D.C.

Watanabe, I. and Matsubara, H. (1982). Pop-ins and short arrested brittle cracks in weld-HAZ of 9%Ni steel. In *Fracture Toughness Testing — Methods, Interpretation and Application.* The Welding Institute, Cambridge, England, paper 34.

Watson, J. F. and Christian, J. L. (1961a). Low temperature properties of cold-rolled AISI types 301, 302, 304 ELC, and 310 stainless steel sheet. In *Low-Temperature Properties of High-Strength Aircraft and Missile Materials.* ASTM STP 287, American Society for Testing and Materials, Philadelphia, 170–193.

Watson, J. F. and Christian, J. L. (1961b). Mechanical properties of high-strength 301 stainless steel sheet at 70, −320, and −423°F in the base metal and welded joint configuration. In *Low-Temperature Properties of High Strength Aircraft and Missile Materials.* ASTM STP 287, American Society for Testing and Materials, Philadelphia, 136–149.

Welland, W. G. (1978). Producing strong welds in 9% nickel steel. *Weld. J.* **57**, 263s–271s.

Wells, J. M., Kossowsky, R., Logsdon, W. A., and Daniel, M. R. (1976). Structural materials for cryogenic applications. In *Materials Research for Superconducting Machinery — VI.* ADA 036919, National Bureau of Standards, Boulder, Colorado.

Weston, W. F., Naimon, E. R., and Ledbetter, H. M. (1975). Low temperature elastic properties of aluminum 5083-O and four ferritic nickel steels. In *Properties of Materials for Liquefied Natural Gas Tankage.* ASTM STP 579, American Society for Testing and Materials, Philadelphia, 397–420.

Whipple, T. A. and Kotecki, D. J. (1981). Weld process study for 316L stainless steel weld metal for liquid helium service. In *Materials Studies for Magnetic Fusion Energy Applications at Low-Temperatures — IV.* NBSIR 81-1645, National Bureau of Standards, Boulder, Colorado, 303–321.

Whipple, T. A. and McHenry, H. I. (1982). The mechanical properties of stainless steel castings at 4 K. In *Austenitic Stainless Steels at Low Temperatures,* Plenum Press, New York.

Whipple, T. A., McHenry, H. I., and Read, D. T. (1981). Fracture behavior of ferrite-free stainless steel welds in liquid helium. *Weld. J.* **60**, 72s–78s.

12 Composites

M. B. KASEN

12.1 INTRODUCTION

Effective use of composites in cryogenic applications requires familiarity with the advantages and disadvantages of available composite systems compared with conventional metals, which composites may replace. The terminology of composite systems must be understood along with approaches to design and fabrication that differ in several important aspects from those of metals technology. It is necessary to have an understanding of useful composite systems and their properties — in particular, the effects of cryogenic temperatures. An appreciation of the present data base is needed along with an understanding of the special problems encountered in testing composite materials.

This chapter emphasizes filamentary-reinforced composites because they are the most widely used composite materials. The small amount of information available on concrete aggregate composites is discussed in §12.8.

12.1.1 Why composites?

The current and future applications of composites in cryogenic systems were discussed in the proceedings of conferences held in Munich, Germany in 1978 (*Nonmetallic Materials and Composites at Low Temperatures,* 1979) and in Geneva, Switzerland in 1980 (*Fundamentals and Applications of Nonmetallic Materials at Low Temperatures,* 1982). Interest in composite materials reflects the increasingly stringent demands made on materials in modern cryogenic technology, of which superconducting machinery is an example. Undoubtedly, the first generation of superconducting motors and generators will be dependent almost entirely on metals technology. It is highly probable, however, that succeeding generations of such equipment will capitalize on composite materials technology for reasons of increased reliability, reduced weight, and increased efficiency. These advantages are a result of the higher specific strengths and moduli of composites coupled with a wider range of thermal and electrical properties than are obtainable with any conventional metal or alloy. In particular, the high strength of the polymer-matrix composites combined with very low thermal conductivity will be advantageous in minimizing heat leaks into critical low-temperature components. A similar case can be made for airborne magnetohydrodynamic (MHD) power generators, where weight is a primary problem, and the cryogenic components of magnetic fusion energy (MFE) systems. The material performance demands of the latter and the quantity of materials required in practical power plants are so great that their ultimate feasibility is questionable unless large quantities of high-performance composite materials are available for their construction.

The attractiveness of composites for such applications is illustrated in terms of the critical property ratios or "figures of merit" in Fig. 12.1. A low ratio of thermal conductivity to modulus, λ/E, or strength, λ/σ, reduces refrigeration costs whenever the structural components must be subject to a temperature gradient. High ratios of modulus or strength to density, E/ρ or σ/ρ, reduce inertial loads in rotating machinery. Uniaxial boron–epoxy or glass–epoxy composites have

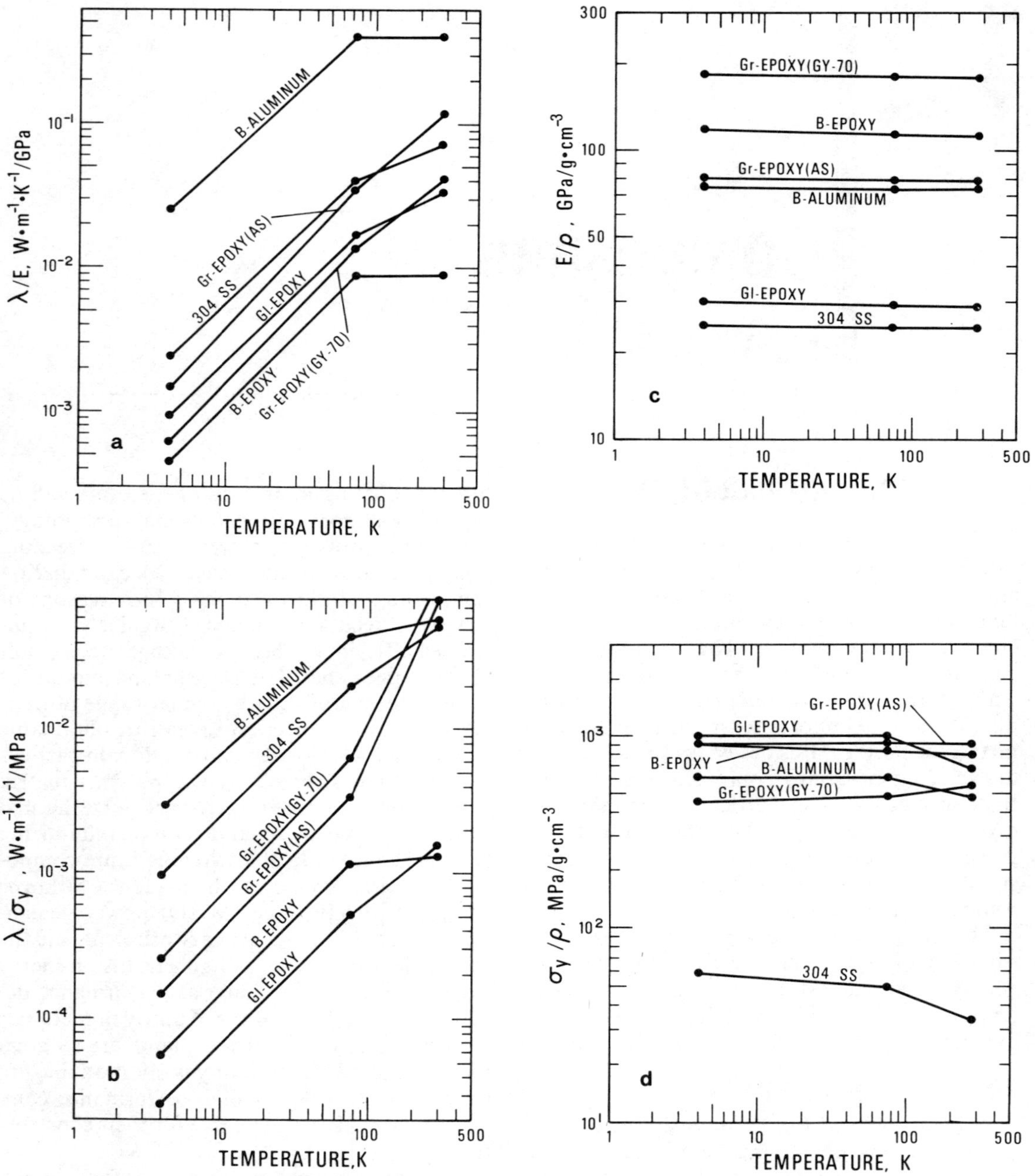

a. Ratio of thermal conductivity, λ, to Young's modulus, E.
b. Ratio of thermal conductivity, λ, to yield strength, σ_y.
c. Ratio of Young's modulus, E, to density, ρ.
d. Ratio of tensile yield strength, σ_y, to density, ρ.

Figure 12.1 Temperature dependence of critical property ratios for uniaxial composite laminates and annealed stainless steel.

1.5 to 2 orders of magnitude advantage over annealed AISI 304 stainless steel in the λ/σ ratio, and boron–epoxy has about an order of magnitude advantage in the λ/E ratio. A modest gain of 20% is expected in the E/ρ ratio when uniaxial glass–epoxy is substituted for stainless steel; nevertheless, improvements of 200 to 400% are predicted when the higher-modulus boron or graphite fibers are used. By substituting either glass, graphite, or boron-reinforced composites for stainless steel, it is possible to increase the σ/ρ ratio by an order of magnitude. These data compare the most favorable orientation of each composite with welded stainless steel. Nevertheless, the comparisons illustrate the attractiveness of composite systems for cryogenic applications.

12.1.2 Composite systems and terminology

Composite terminology is complex because of interacting diverse technologies (polymers, textiles, and metals) and because composites are inherently complex materials. In the most general sense, a composite is a material having overall properties that are some average of the properties of several distinct material components, one of which is contiguous and forms a matrix. Laminated metals, such as clad aluminum, might fall within this definition, but such laminates are a natural development of conventional metals technology and are not considered here.

Composite systems currently used in cryogenic applications are illustrated in the chart of Fig. 12.2. The first major category distinguishes between fiber-reinforced, particulate-reinforced, and aggregate-reinforced composites. Fiber-reinforced composites are constructed by bonding together individual plies (single layers of resin-impregnated reinforcement) to form a composite laminate or by continuous winding of the reinforcing fiber over a mandrel to form a filament-wound structure. A third construction method, in which continuous or discontinuous fibers of random orientation reinforce molded plastic structures, has aroused interest because of its lower cost. Conventional sand-and-gravel concretes are the only aggregate composites widely used for cryogenic structures. Particulate-reinforced composites are exemplified by filled-epoxy systems used to reduce cracking of insulators in superconducting magnets.

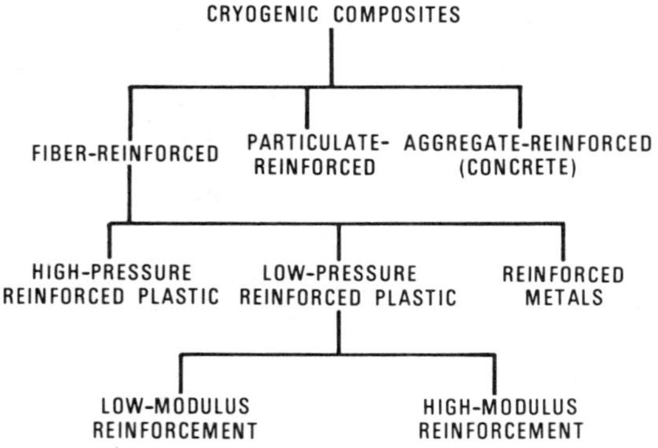

Figure 12.2 Cryogenic structural composite system flow chart.

Fiber-reinforced composites are further categorized into low- and high-pressure plastic types (which refers to the pressure used in curing the composites) and reinforced metals, a distinct group.

The pressure required to produce high-pressure reinforced plastic laminates (about 8 MPa, 1100 psi) requires that they be produced as bulk mill products, such as plates, sheets, rods, and angles. Parts are usually fabricated from these laminates by conventional machining. High-pressure laminates are frequently referred to as "industrial laminates." They are widely known by nonspecific trade names that identify all the industrial laminates produced by a company (e.g., Micarta, Phenolite, Textolite). The designations of the National Electrical Manufacturers Association (NEMA) and the American Society for Testing and Materials (ASTM) are also frequently used in referring to high-pressure laminates. Designations such as G-5, G-10, and G-11 are familiar examples of the latter. Thousands of metric tons of these relatively inexpensive industrial laminates are produced each year in the United States.

Low-pressure reinforced plastic laminates (0 to 2.8 MPa, 0 to 400 psi) are produced annually to about the same tonnage as high-pressure laminates. The low-pressure requirement permits *in situ* fabrication of completed parts from the basic composite components: For example, structures may be built up by filament winding or from an intermediate product called a *prepreg.* The latter is a tape containing the fiber and a partially cured (B-staged) polymer–resin matrix. Parts built up from the prepreg are cured under heat and pressure. Advantages are convenience and a wide choice in the selection of resin formulation and fiber content.

Reinforced metals form a distinct class because the fabrication procedures and cryogenic applications of these materials differ significantly from procedures used with laminated plastics. Reinforced-metal laminates are formed from the basic materials or from a tape product consisting of a uniaxial layer of fiber in the metal matrix.

Low-pressure laminates are reinforced by high- or low-modulus fibers, with the modulus of glass (about 69 GPa, 10^7 psi) the dividing line. This reflects a historical development: glass-reinforced composites were the first to be developed and widely used. Low-modulus fibers are generally cellulose (cotton, linen) and glass. High-modulus fibers are graphite, boron, alumina, organic aramids, and occasionally, metals. These fibers allow composites to compete with steel in stiffness-controlled engineering structures. Composites reinforced with high-modulus fibers are frequently called *advanced composites*. High-modulus fibers are not normally used in industrial laminates because of their cost and the limited number of available product forms. All metal-matrix composites are reinforced with high-modulus fibers.

Terminology such as glass-fiber-reinforced plastic and boron-fiber-reinforced plastic generally describe low-pressure laminates. Abbreviations such as GFRP and BFRP are frequently used. As one becomes more specific, the necessity to distinguish the various matrix types arises, and terms such as glass–epoxy or graphite–polyimide are used. Still more refinement is possible by referring to a particular reinforcement/matrix combination, for example, S glass/E-787 epoxy.

Glass-fiber monofilaments are sometimes used to reinforce structures, such as pressure vessels, but more frequently the reinforcement is in the form of yarn (a twisted bundle of filaments) or roving (a parallel bundle of filaments). Roving is used in unidirectional prepreg tape, whereas yarn is used in the weaving of fabrics. A large variety of fabrics are available, but only a few are widely used in composites. Fabrics may be of a balanced weave, containing an equal density of glass in both the warp and fill directions, or an unbalanced weave, having a higher density of glass in the warp direction. The warp direction is that in which the fabric emerges from the weaving operation; the fill is the direction orthogonal to the warp. The fabric type affects the composite properties. MIL-HDBK-17A (1971) provides a description of the fabric weaves most commonly used in composite laminates.

Successive layers of uniaxial fibers can be oriented in appropriate directions to tailor-make the properties of low-pressure composite laminates to the requirements of a particular application. A terminology to deal with this complexity is given in each volume of the *Advanced Composites Design Guide* (1973). The orientation of sequential layers throughout a composite sheet is defined with reference to a 0° direction determined by the first layer (see Fig. 12.3a). For example $[0/\pm45/0]_T$ defines a four-layer composite sheet in which the second layer is oriented 45° clockwise from the first, the third layer is oriented 45° counterclockwise from the first, and the fourth layer is a duplicate of the first. This laminate construction is illustrated in Fig. 12.3b. The subscript T indicates that the total laminate has been designated; a subscript S would indicate that half of a symmetrical laminate has been defined. A laminate is said to be symmetrical if there is symmetry about a midplane; a laminate is said to be balanced when all plies at angles other than 0° or 90° occur in pairs of opposite sign. The latter is sometimes called *quasi-symmetrical* and coded with a subscript Q. The unbalanced, quasi-symmetrical laminate construction of Fig. 12.3b may therefore be coded either $[0/\pm45/0]_T$ or $[45/0]_Q$. An understanding of this terminology is very valuable, because properties of laminates are strongly dependent on reinforcement orientation.

A unidirectional laminate or any balanced or symmetric laminate is orthotropic in nature, having three mutually perpendicular planes of elastic symmetry. For planar applications with loading coincident with a principal axis, this material can be characterized by four of the five elastic constants of the orthotropic class of materials. When loading does not coincide with a principal axis or if the laminate is unbalanced, the material is anisotropic, requiring six elastic constants for complete characterization. Should the laminate be neither symmetric nor balanced, coupling of the shear and normal stresses requires eighteen elastic constants for characterization. The latter, called a *coupled laminate*, is generally avoided.

The terminology used in discussing the mechanical and thermal properties of composite laminates is similar to that of metals technology with the exception of shear properties. The shear property variously called *intralaminar, in-plane,* or *longitudinal* shear, is essentially the same for laminates as for metals. Here, values reflect a pure shear stress resolved onto a 45° shear plane with respect to the tensile axis without significant compressive or tensile forces normal to the plane. But another type of laminate shear does not exist in metals. This is *interlaminar* or *horizontal* shear and refers to the shear strength between adjacent plies in a layered laminate. Intralaminar or longitudinal shear is a basic material property. Interlaminar or horizontal shear is a property strongly affected by the quality of a particular laminate.

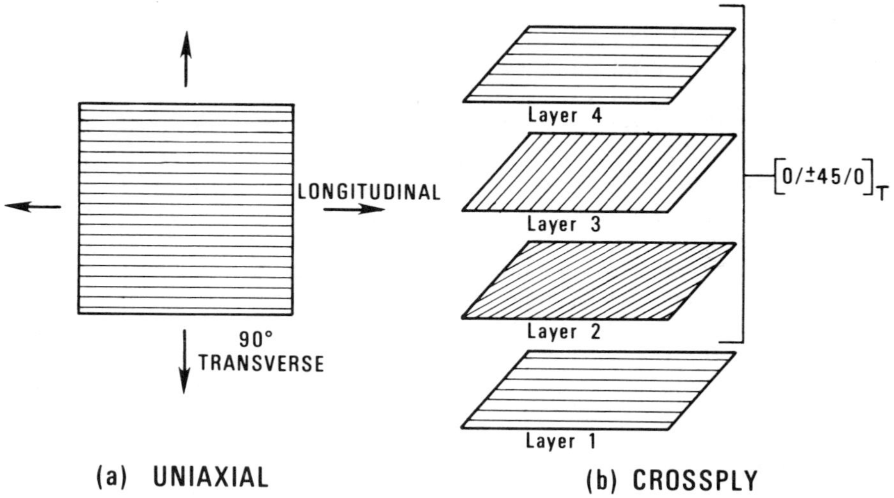

Figure 12.3 The conventional notation for uniaxial (a) and crossply (b) laminates. The crossply laminate is designated $[0/\pm 45/0]_T$.

The glossary and nomenclature sections of MIL-HDBK-17A (1971) and of the *Advanced Composites Design Guide* (1973) are helpful in clarifying terminology associated with low-pressure laminates. Additional details of the terminology associated with high-pressure laminates are provided in several texts (*Laminated Plastics,* 1958; *Handbook on Fiberglass and Advanced Plastic Composites,* 1969; *SPI Handbook of Technology and Engineering of Reinforced Plastics Composites,* 1973).

12.2 THE COMPOSITE APPROACH

12.2.1 Design and fabrication

Because composites are relatively new cryogenic structural materials, their initial use has been restricted to selective replacement of specific metal components in structures designed to be fabricated from metals. This conservative approach is necessary to build confidence in the use of composites, but the full benefit of composite technology will be achieved only when structures are designed from the beginning to obtain the maximum advantages offered by composites.

The basic approach to designing and fabricating with high-pressure industrial laminates resembles that for metals. Except in special cases where mill products can be directly utilized, final shapes must be produced by machining of bulk material. With the exception of some glass-reinforced grades, industrial laminates are easier to machine than most metals. Design and fabrication experience is extensive for these materials at normal temperatures. High-pressure laminates are the least expensive of the composite materials and are used most often for parts that do not see demanding service. Such laminates are often treated as homogeneous materials, an assumption that is justified only in the least critical applications.

The approach to designing and fabricating with low-pressure laminate composites for cryogenic applications is basically the same as that used at room and elevated temperature in the aerospace industry. Procedures for analysis, design, and fabrication with low-pressure laminates are comprehensively addressed in MIL-HDBK-17A (1971) and in *Advanced Composites Design Guide* (1973), Vols. I–III. Tailoring composite materials to meet specific property requirements of a component introduces new and often complex parameters into design. The primary problem is predicting laminate properties from laminate construction. A simple model assumes that laminate properties are a summation of the contribution made by each constituent multiplied by the volume fraction of each constituent — the *rule of mixtures.* This rule works reasonably well where the laminate property of interest is a function of the average property of the constituents. For example, the uniaxial longitudinal Young's modulus can be quite accurately predicted as a function of fiber density on this basis. But the rule does not work well where the laminate property is functionally related to the variability of a constituent property. For example, the uniaxial longitudinal tensile strength of a laminate is not predictable by the rule of mixtures because the fibers do not all break at the same stress. Here, the fracture strength is strongly affected by the lower bound of the fiber strength distribution. Furthermore, the coefficient of variation of fiber strength increases as fiber density increases.

A more sophisticated model relates fundamentally the principal strength and elastic properties of a composite laminate to the properties and structure of the basic components and interfaces of the laminate. This approach, called *micromechanics,* is analogous to establishing the relationship between properties and microstructure in metals technology. It has received considerable attention and is considered by many to be crucial to the development of a mature composites materials science. The complexity of composite materials, however, restricts the applicability of this approach to relatively simple laminates. More practical use is made of the macromechanical approach to laminate design. This approach assumes that uniaxial laminates are the basic, repeatable building blocks of the total laminate and that the strength and stiffness of complex crossply laminates, such as the one illustrated in Fig. 12.3b, are predictable from knowledge of pertinent strength, modulus, and shear properties of the uniaxial laminate from which they are constructed (Fig. 12.3a). Measured values of longitudinal and transverse uniaxial laminate strength, modulus, and Poisson's ratio plus inplane shear strength and modulus of the uniaxial laminate are combined with macromechanical laminate theory to produce a computer model of the stacking sequence required to produce the desired laminate strength and stiffness. Conversely, the computer model may be used to predict the strength and stiffness properties of existing crossply laminates. The macromechanical approach is widely used in the aerospace industry, usually as part of an iterative process of laminate design. Computer programs for such analyses are described in *Advanced Composites Design Guide* (1973), Vol. II.

Low-pressure laminate parts fabricated to the final configuration must be designed so that they can be manufactured with practical tooling. Successful use of low-pressure laminate composites requires close interaction between the materials scientist, the designer, and the fabricator. A quality assurance program must encompass all aspects of design and fabrication. Quality of both the starting materials and of the final cured laminate must be assured. Quality assurance programs are discussed in detail in *Advanced Composites Design Guide* (1973), Vol. III.

A main concern for low-temperature structures is the effect of cryogenic temperatures on materials properties. This is discussed in § 12.4.

Composite materials appear to be more expensive than metals. This is particularly true for composites formed with fibers of the highest elastic modulus. But other factors must be considered in assessing the overall life-cycle cost of a component. For example, almost all of the materials purchased for low-pressure laminate fabrication are utilized, whereas a substantial part of the material purchased for metal fabrication may wind up as machining scrap depending on the complexity of the component being manufactured. In the aerospace industry, it is not unusual for the ratio of material purchased to material used in the component to equal 1.05:1 for composite construction compared with 10:1 for metal construction. The use of composites can often reduce fabrication costs by eliminating machining and forming operations that require sophisticated equipment and highly skilled personnel. The higher stiffness-to-weight and strength-to-weight ratios offered by the high-modulus composites, compared with metals, may require less composite material to build a dependable structure. Composites often offer economies when only a few parts are required and where the cost of conventional metal tooling would impose an inordinate cost per part on metal construction. Analysis by experienced personnel is required to determine if composite fabrication of a particular component can be cost-effective.

Joining of composites or of composites to metals presents one of the major difficulties in designing with low-pressure laminates. Significant progress has been made in the aerospace industry in solving the problems of such joints for airframes. The *Advanced Composites Design Guide* (1973), Vol. I and MIL-HDBK-17A (1971) contain reviews of typical joint designs, the problem areas, and solutions for typical problems. The effect of cryogenic temperatures on the performance of various joint systems is reviewed in § 12.5.

12.2.2 Composite failure

Many composites are elastic all the way to failure, with maximum elongations of 0.5 to 1%

in the case of high-modulus fiber composites or 2 to 5% in the case of glass-reinforced composites. This is often disturbing to those accustomed to working with ductile metals. But low ductility is not necessarily a severe problem, since most structures, including those used at cryogenic temperatures, are designed to be stressed only within their elastic limits. Plastic deformation often constitutes failure. This can severely restrict the useful stress range of a metal. For example, annealed or welded AISI 304 stainless steel yields at an elastic tensile strain of about 0.2% at 4 K, although its elongation to fracture at that temperature may be as high as 30%. Conversely, a uniaxial boron–epoxy composite is elastic up to about 0.8% strain at 4 K. The corresponding limiting stress range is about 413 MPa (6×10^4 psi) for the stainless steel and about 1800 MPa (26×10^4 psi) for the boron–epoxy. A similar result is obtained with graphite–epoxy composites. Careful attention must be given to stress analysis of composite laminate structures, however, because the lack of ductility prevents redistribution of local overloads. Nevertheless, even with a substantial safety factor, the load-carrying capability of composites can equal or exceed that of metals.

The mechanism of fatigue failure in composites differs from that of metals in that polymer-matrix composites can "wear out" under cyclic loading whenever the applied strain exceeds the strain capability of the fiber or matrix. Initiation and propagation of a single crack is not normally observed. Wearout is manifested by a decrease in elastic modulus and an increase in damping owing to accumulation of internal damage, but the ultimate strength of the composite is often unaffected. Changes in damping, therefore, can serve as an early warning of structural damage in some applications. Reinforced metal composites can be cyclically strained in excess of the initial matrix yield without incurring wearout, provided that the fatigue limit of the metal matrix is not exceeded, that is, provided that the matrix work-hardens sufficiently to prevent additional damage accumulation.

Wearout occurs at lower stresses in low-modulus fiber laminates and is most evident in crossply laminates. Glass-reinforced composites, therefore, have relatively poor fatigue performance compared with composites made with high-modulus fibers. Wearout failure is usually initiated by debonding of the fiber-matrix interface, followed by crack propagation along the interface and through the matrix. Final failure is caused by fiber rupture.

Fatigue failure modes of composite materials are undergoing intensive study. Until analytical techniques become better developed, it is recommended that verification tests be conducted on components subject to cyclic stresses.

12.3 MATERIAL OVERVIEW

12.3.1 High-pressure reinforced plastic laminates

Table 12.1 lists the industrial, high-pressure laminates of greatest interest to the cryogenic industry. The largest current cryogenic use of industrial laminates is electrical and thermal insulation. Millions of kilograms are being used to support and insulate liquefied natural gas (LNG) tanks, both within ships and in shore-based storage facilities. NEMA/ASTM LE, G-10 and G-11 type laminates have been used successfully as electrical insulators for superconductors in non-radiation magnetic environments. Hundreds of metric tons of these laminates are proposed for use as both electrical and thermal insulators in future magnetic fusion energy reactors.

High-pressure industrial laminates are most commonly reinforced with woven fabric. The fabric weave is typically unbalanced, producing different properties in the warp and fill directions. For example, NEMA/ASTM G-10 products typically have a 43:32 warp-to-fill ratio, resulting in about a 30% difference in strength in the two directions.

The NEMA, ASTM, and MIL specifications (NEMA Publication L1-1; ASTM D709-72, Part 40; MIL-P and LP) establish room-temperature laminate performance requirements with emphasis on electrical insulating qualities. These requirements may be met by laminates produced by a variety of components and manufacturing techniques. This results in significant cryogenic temperature mechanical property variability among commercial laminates of the same NEMA/ASTM designation. The variability reflects differences in the type of glass, glass content, weave of the glass fabric, epoxy system used for the matrix, curve temperatures, and pressures. Furthermore, the highly competitive nature of the high-pressure laminating industry, the proprietary nature of laminate production, and production economics result in periodic changes in materials produced to a given NEMA/ASTM designation by any one company.

Table 12.1 Industrial high-pressure laminates of primary interest for cryogenic applications.

Commercial Standard	MIL-P and LP Designation	Description
NEMA/ASTM LE	MIL-P-15035, type FBE	Cellulose fabric, fine weave, phenolic resin
NEMA/ASTM C	MIL-P-15035, type FBM	Cellulose fabric, medium weave, phenolic resin
NEMA/ASTM G-5	—	Glass fabric, melamine resin
NEMA/ASTM G-10	MIL-P-18177, type GEE	Glass fabric, epoxy resin
NEMA/ASTM G-11	MIL-P-18177, type GEB	Glass fabric, high-temperature epoxy resin
None	None	Glass mat, epoxy resin
W1, W2, EH57, EH67	LP 5094	Beechwood veneer, phenolic impregnated

Problems associated with the unknown variability of cryogenic properties of high-pressure industrial laminates have recently received attention because of their potential in the Large Coil Project and the demands made on these materials in the large superconducting magnets required in the development of magnetic fusion energy (MFE) and magnetohydrodynamic (MHD) technologies. NEMA G-10 and G-11 industrial laminates are conventionally used as electrical insulators and structural support for superconducting elements in superconducting magnets. But large-coil magnets required for MFE and MHD reactors must have predictable mechanical, thermal, electrical, and radiation-resistant properties at 4 K.

As a first step in meeting this need, the National Bureau of Standards, in cooperation with NEMA and the United States laminating industry, has arranged for the commercial availability of G-10 and G-11 products manufactured to a common component and manufacturing specification. These products, designated G-10CR and G-11CR, are currently available from several United States manufacturers. The mechanical, thermal, and electrical properties of these laminates have been characterized at 4 K by the National Bureau of Standards and other laboratories. Their performance under neutron and gamma radiation has also been assessed. The data obtained in this program are discussed in § 12.4.2. Efforts are underway to extend this generic approach to other classes of laminates of interest to the cryogenic industry.

The glass-mat reinforced epoxy laminate listed in Table 12.1 reflects a recent development by a leading fiberglass manufacturer in its search for a suitable material to be used as the primary barrier in containing LNG. This type of laminate is substantially less expensive than fabric-reinforced types, and as discussed in § 12.4.2, the random-fiber reinforcement provides some desirable properties.

12.3.2 Low-pressure reinforced plastics

Low-modulus fibers. Low-modulus, low-pressure plastic laminates are almost always reinforced with glass fiber. A variety of special-purpose glasses are available; however, types E and S are most widely used for cryogenic applications. Type R is a glass of European manufacture having properties intermediate between types E and S. Compositions of each of these types are given in Table 12.2. The primary difference between types E and S is the 40% higher strength and the 18% higher modulus of virgin (untreated) S glass over virgin E glass. Table 12.3 provides a summary of their properties.

Table 12.2 Typical compositions (wt. %) of glass used as reinforcement for cryogenic laminates.

Element	Glass Designation E	S-901/S-2[a]	R[b]
SiO_2	54.4	65.0	60.0
Al_2O_3	14.4	25.0	25.0
CaO	17.5		9.0
MgO	4.5	10.0	6.0
B_2O_3	8.0		
$Na_2O + K_2O$	0.5		
Fe_2O_3	0.4		
F_2	0.3		

[a]Owens/Corning Corporation, Toledo, Ohio.
[b]Vetroflex Division, Saint-Gobain Industries, Neuilly-sur-Seine, France.

Table 12.3 Low-pressure laminate reinforcing materials.

Type	Typical 300-K Properties			
	Tensile Strength (MPa)	Tensile Strength (10^3 psi)	Elastic Modulus (GPa)	Elastic Modulus (10^6 psi)
Low Modulus				
E Glass	3450	500	72	10.5
S Glass	4825	700	85	12.4
High Modulus				
Kevlar 49 (polyaramid)	2760	400	130	19
Boron	2760	400	380	55
Graphite:				
Type I	1720–2410	250–350	310–520	45–75
Type II	2410	350	260	38
Type III	2070–2410	300–350	190–220	28–32
FP Alumina	1380	200	380	55

The lime–alumina–borosilicate E glass constitutes by far the largest production of continuous glass filaments. E glass was developed primarily for use in electrical insulation; its low content of alkali metals gives high values of volume and surface resistivity. The large industrial consumption of this type of glass has kept its price significantly below that of S glass.

The silica–alumina–magnesia S glass was developed to provide higher strength and modulus than E glass in filament-wound structures, particularly at elevated temperatures. Designations such as S-994 and S-901 denote special glass sizings applied at the time of manufacture. Glasses carrying these designations are rigidly controlled by military specifications and are the most expensive. They are not available in woven fabric, which restricts their use to filament-wound structures or to laminates built up from uniaxial B-staged tape. A lower-priced variant, type S-2, has the same composition as type S, reportedly has equivalent strength and modulus and is available in woven fabric as well as roving for filament-winding and prepreg tapes (Loud, 1979). Type S-2 is subjected to the manufacturers' quality control rather than to military specifications for traceability.

High-modulus fibers. The high-modulus fibers listed in Table 12.3 are of most interest for cryogenic structural applications. Typical fiber properties illustrate the important differences among the various types.

Aramid fiber is a proprietary organic fiber produced by the E. I. DuPont Corporation under the tradename Kevlar. The variant called Kevlar 49 is most useful in composite laminates. Aramid fiber offers a better tensile modulus than does glass at a relatively low cost compared with other high-modulus fibers. But its primary advantage is low fiber density and high fiber toughness. For equal fiber fraction, the density of an aramid-reinforced laminate is about half that for glass reinforcement. Disadvantages are a comparatively low compressive strength, low strength transverse to the fiber, and a large negative coefficient of thermal contraction in the fiber direction. This fiber has been used successfully in prototype cryogenic components.

Boron fibers are produced by vapor deposition of boron on a very fine tungsten wire substrate. Filaments of 0.10-mm (4-mil) diameter were initially produced; however, 0.14-mm (5.6-mil) fibers are now preferred and 0.20-mm (8.0-mil) fibers are being introduced. Large-diameter fibers offer some cost advantage in laminate production, although the thickness of the resulting plies and limitation on the radius of curvature of such filaments can be a disadvantage. The smaller volume fraction of tungsten lowers the overall composite density when the large-diameter fibers are used. Boron fiber coated with a thin layer of silicon carbide (produced under the trade name Borsic) is reported to have improved interfacial binding with certain metal-matrix materials.

Boron fibers were the first high-modulus fibers intensively studied and composite technology utilizing them has been highly developed. Available data indicate that the excellent room-temperature static and dynamic properties of boron-reinforced composites are maintained at cryogenic temperatures with the least variability of any composite system (Schramm and Kasen, 1977a, 1977b). The uniaxial longitudinal compressive strength of boron–epoxy is particularly impressive, frequently exceeding 3450 MPa (5×10^5 psi) at 4 K. A disadvantage is the brittle nature of the fibers, which limits the radius over which a composite can be formed. The high-neutron cross section may also be a disadvantage in certain radiation environments. But the main disadvantage is high cost, which has served as a stimulus for the development of high-modulus graphite fibers. Boron-reinforced laminates currently appear to offer the best combination of mechanical properties and reliability for cryogenic applications.

Graphite fibers are produced by graphitization (pyrolysis) of an organic precursor fiber, the most common of which are rayon and polyacrylonitrile (PAN). The properties of the fiber can be varied by controlling the extent of graphitization and hence the fiber crystallinity. Higher moduli and lower strengths are associated with higher crystallinity. This complexity has led to a proposed categorizing of graphite fibers into types I, II, and III, as indicated in Table 12.3. The primary distinction is the elastic modulus of the fiber, which can range from 190 GPa (28×10^6 psi) to over 520 GPa (75×10^6 psi). The lower-modulus fibers are less expensive, their interactions with various matrices are better understood, and the properties of composite laminates made with them are less variable. The very high modulus fibers represent the forefront of graphite-fiber development.

Composites made with the highest modulus fibers can develop room-temperature uniaxial tensile elastic moduli in excess of 310 GPa (45×10^6 psi) with tensile strengths on the order of 690 MPa (10^5 psi). Laminates made with the highest strength graphite fibers can have tensile strengths in excess of 1380 MPa (2×10^5 psi) and moduli on the order of 138 GPa (2×10^7 psi) (*Advanced Composites Design Guide*, 1973, Vol. IV). The fine filaments permit graphite-fiber composites to conform to small radii. The fibers are available as roving and as woven fabric. A slightly negative coefficient of thermal contraction in the fiber direction of high-modulus graphite fibers permits construction of crossply laminates having almost a zero overall coefficient of thermal expansion on cooling to cryogenic temperatures. This permits construction of very dimensionally stable mirror mounts and antennas for use in outer space. Disadvantages of graphite fibers are a low strength transverse to the fibers and a cost that is relatively high compared with glass, although significantly lower than that for boron. The mechanical properties of graphite–epoxy laminates are also more variable than those of boron. Recent developments in producing graphite fibers from pitch precursors offer the promise of substantially reduced fiber cost at some performance penalty.

Development of graphite fibers initially proceeded at a very rapid rate confronting the user with an abundance of fiber choices, many of which were not well characterized and many of which were shortly replaced by newer types. The industry has not matured; nevertheless, one should carefully consider the existing data base for any fiber type selected for a particular application.

Polymer matrices. Epoxy resins are most widely used as matrices for low-pressure cryogenic structural laminates and are also used for some high-pressure laminates. The effect of cryogenic temperatures on the performance of glass-reinforced composites utilizing various types of resin materials was examined in the early 1960s (Kasen, 1975a, 1975b). It was concluded that epoxies provided higher composite strengths (although not necessarily higher elastic moduli) at cryogenic temperatures than did polyester, polyurethane, Teflon, phenyl silane, polyimide, silicone, phenolic, or polybenzimidazole resins. Although epoxy systems will continue to be favored for critical applications, experience has shown that less expensive polyester matrix laminates are suitable for some noncritical applications (Khim and Ng, 1980; Wang, Kim, Kim, Cheng, Hoffman, and Gonczy, 1980).

An epoxy resin is any polymer containing one or more epoxide organic structural groups. Performance requirements and classifications of epoxy resins for aerospace use are given in specification MIL-R-9300, *Resin, Epoxy, Low-Pressure Laminating*. Combining a resin (or blend of resins) with a curing agent and possibly with accelerators or flexibilizers produces an epoxy–resin system. General discussions of epoxy systems suitable for composite applications may be found in MIL-HDBK-17A (1971) and in composite materials textbooks (e.g., *Handbook on Fiberglass and Advanced Plastics Composites*, 1969).

Mechanical properties of the matrix are more important for glass-reinforced laminates than for high-modulus laminates because the low modulus of the glass permits appreciable strain to be transferred to the matrix at relatively low stresses. Given the high strength of glass, the total elongation at fracture of glass fibers exceeds that of most resin systems even at room temperature. Cooling to cryogenic temperatures strengthens the glass, increasing its elongation at failure. But cooling embrittles the epoxy matrix, reducing its strain capability. Matrix damage therefore occurs at relatively low levels of strain (20 to 30% of ultimate strain) in glass-reinforced laminates stressed at 4 K. Adding a plasticizer to the epoxy resin system can increase the strain capability of the resin at low temperature (Soffer and Molho, 1967). But experience has shown that laminates using highly flexibilized resin matrices have cryogenic performances inferior to those obtained with laminates having fully reacted, nonflexibilized epoxy matrices. Apparently, the additional residual stress introduced into the laminate by the higher thermal contraction of the flexibilized system negates any benefit from the flexibilizer (Hartwig, 1979; Kreibich, Lohse, and Schmid, 1979). Furthermore, the very low strength of highly flexibilized resin systems at room temperature seriously reduces laminate strength and creep resistance. Very highly flexibilized systems, exemplified by the "Resin 2" formulation in Table 12.4, should therefore be avoided.

There is some evidence that partially flexibilized epoxy–resin systems can provide good cryogenic laminate performance without penalizing room-temperature performance. An example is the Epon 815/Versamid 140 formulation given in Table 12.4 (Lewis and Bush, 1967; Lewis, Bush, and Creedon, 1966). This system has been successfully used for cryogenic Dewar construction (Vogen, Clawson, and Morrison, 1975), but extensive property characterization data do not exist.

Most of the data on the performance of glass-reinforced laminates at cryogenic temperature have been obtained with laminates using fully reacted resin matrices typified by the "Polaris" formulation in Table 12.4. But there is no confirmation that this formulation is inherently superior to other fully reacted systems.

Given the present state of the art, a resin system for cryogenic laminate construction should be selected preferably on the basis of its suitability for the type of fabrication required rather than on the basis of literature data. This is particularly true for laminates reinforced with high-modulus fibers. Resin viscosity, pot life, and adaptability to pre-preg operations must be considered. Literature data can be misleading because the cryogenic performance of a laminate is strongly affected by the quality of laminate construction. Low void content and thorough fiber wetting are essential for maximum performance. Whenever the application is critical, it is wise to confirm the cryogenic performance of laminates produced under the same conditions as the final product.

Research on resin systems for cryogenic composite use has been summarized by Hartwig (1979) and by Kreibich et al. (1979). One approach to maintaining elevated-temperature strength while inducing low-temperature ductility and minimizing thermal contraction is to insert cyclo-aliphatic and phenolic segments into the epoxy crosslink network to reduce the activation energy for low-temperature relaxation mechanisms, such as bond rotation or stretching. Another approach is to insert segments that tend to crystallize into the epoxy network. Success in this work would significantly decrease the notch sensitivity of the laminate matrix and reduce the variability in cryogenic composite properties.

Hybrid laminates. Hybrid laminates contain a combination of reinforcements to achieve desired mechanical or physical properties within a laminate or to reduce the overall laminate cost. For example, adding glass or aramid plies as outer layers in a graphite–epoxy laminate increases the abrasion resistance, increases laminate toughness, and reduces the consumption of expensive graphite fiber. A hybrid laminate offers additional flexibility in tailoring composite materials to perform specific tasks. Available data on this developing technology are meager.

12.3.3 Reinforced metals

The high specific moduli (σ/ρ) and strengths (E/ρ) of fiber-reinforced metals contribute to structural efficiency where weight or inertial forces are important. Graphite-reinforced aluminum, copper, and lead; boron-reinforced alumi-

Table 12.4 Candidate resin systems for low-pressure laminates.

Resin Identification	Type	System Formulation	Concentration[a]
NASA Resin 2	Highly flexibilized	Epon 828/DSA/EMPOL 1040/BDMA[b,c,d]	100/115.9/20/1
Polaris[e]	Fully reacted	Epon 828/Epon 1031/NMA/BDMA[f,d]	50/50/90/0.55
- - - -	Partially flexibilized	Epon 815/Versamid 140[g]	66/33

[a]Parts by weight.
[b]DSA = dodecenyl succinic anhydride.
[c]EMPOL 1040 = proprietary flexibilizer.
[d]BDMA = benzyldimethylamine.
[e]Commercial designations U.S. Polymeric E-787 or Shell 58-68R.
[f]NMA = nadic methyl anhydride.
[g]Versamid 140 = proprietary flexibilizer.

num and titanium; silicon-carbide-reinforced titanium and aluminum; alumina-reinforced aluminum; and beryllium-reinforced titanium have been studied. Boron-reinforced aluminum is a highly developed, commercial product. It is the only metal-matrix composite for which cryogenic property data have been obtained (Kasen, 1975c; Schramm and Kasen, 1977a, 1977b). Most of the other materials are in the developmental stage, although graphite–aluminum is expected to become a practical commercial product in the near future. Boron fiber coated with a thin layer of silicon carbide for a diffusion barrier is known by the trade name Borsic. Limited available data indicate no advantage in using this fiber at cryogenic temperatures, reflecting the absence of diffusion-related problems.

Figures 12.1c and 12.1d illustrate the favorable E/ρ and σ/ρ ratios of boron–aluminum laminates at cryogenic temperatures compared with conventional AISI 304 stainless steel. These data compare laminate properties obtained in the fiber direction with those of annealed or welded stainless; comparing crossply properties of the laminates with those of hardened stainless steel would decrease the illustrated differences. These data show no advantage for boron–aluminum over boron–epoxy in terms of the property ratios, but there may be some cryogenic applications requiring a strong, stiff, high-electrical-conductivity laminate. An example might be the eddy-current shield of the rotor in superconducting machinery. High thermal conductivity of the metal matrix limits boron–aluminum to components whose thermal gradients are minimal or to applications where a high thermal conductivity is desired.

The ductile 6061-aluminum-alloy matrix commonly used in boron–aluminum laminates provides about 7 times higher strength and about 4 times higher modulus transverse to the boron fiber than can be obtained with an epoxy matrix. This simplifies design by reducing the probability that spurious transverse stresses will cause laminate failure. On the other hand, the ductility of the matrix combined with residual stresses from the laminating process can result in a nonlinear stress–strain relationship whenever a boron–aluminum laminate is loaded other than in a uniaxial direction. A crossplied boron–aluminum laminate or a uniaxial laminate stressed other than along a principal laminate axis may, therefore, have a negligible elastic region. This prevents reliable prediction of the stress–strain curve of crossply laminates from macromechanical composite theory. This factor tends to restrict the aerospace use of boron–aluminum to simple uniaxially loaded structures. The nonlinearity problem can be somewhat ameliorated by reducing the residual stresses in a boron–6061-aluminum laminate with a T-6 heat treatment.

Boron–aluminum composites are available in a wide range of sheet thicknesses (including monolayers) produced by diffusion bonding or in continuous monolayer tape (sometimes called *green tape*). The latter is produced by plasma-spraying aluminum onto boron fibers that are aligned on an aluminum backing sheet. The tape is laminated into the desired component shape and consolidated by diffusion bonding or low-temperature braze bonding, depending on the nature of the backing sheet. A discussion of manufacturing methods for boron–aluminum composites is provided in *Advanced Composites Design Guide* (1973), Vol. III.

Superconducting composites (superconducting filaments with a metal stabilizing matrix) logically fall within the defined scope of metal-matrix composites. An understanding of the factors affecting the mechanical properties of metal-matrix structural composites can be helpful in providing the solid-state physicist with an understanding of the behavior of superconductors in a magnet. For example, the amount of strain such a composite can undergo before the conductor goes normal may be related to the nonlinearity problem mentioned above. This is discussed further in Chapter 13.

12.3.4 Selectively reinforced structures

Advanced composites can be bonded to selected areas of alloys to increase directionally their strength, stiffness, or fatigue resistance. Stiffening of aluminum or titanium beams by overlaying or coring with graphite or boron–epoxy is a typical example. Selective reinforcement could be used to reduce the overall thermal conductivity of such cryogenic components as the rotor drive shaft of a superconducting motor or generator.

12.4 EFFECT OF CRYOGENIC TEMPERATURES

12.4.1 General

Previous reviews of the cryogenic properties of fiber-reinforced composites (Kasen, 1975a, 1975b, 1975c) identified a number of thermal and mechanical property trends for the room-temperature to cryogenic-temperature range for a variety of composite materials. A global review of available data showed that cooling generally had a beneficial effect on mechanical properties. No instance of serious property degradation was reported. Cotton-, glass-, graphite-, aramid- and boron-reinforced polymers, and boron-reinforced aluminum therefore appeared to have potential for cryogenic service.

It was equally evident that a primary obstacle to expanded use of composites was the lack of a systematic data base relating cryogenic properties to well-characterized composite materials. Available data often reported significantly differing temperature dependencies for mechanical properties of a given type of composite material. This reflected a diversity of data sources, each having different program objectives and using different experimental techniques. Property data below liquid-nitrogen temperature (77 K) were limited to 20 K. Almost no data were available at liquid-helium temperature (4 K). Some efforts to improve cryogenic composite performance by flexibilizing the resin system were reported, but results were inconclusive.

As indicated in a recent review (Kasen, 1981), substantial progress has been made in experimental methodology. More evidence associating variability in cryogenic strength properties with embrittlement of matrix resins at cryogenic temperatures has been obtained. Efforts to develop matrices having improved low-temperature toughness continue (*Fundamentals and Applications of Nonmetallic Materials at Low Temperatures*, 1982); however, the solution remains elusive.

Current efforts to understand composite behavior at cryogenic temperatures have been mostly stimulated by the need to develop superconducting power generation systems as alternatives or improvements to conventional systems (Kasen, 1978, 1979). In particular, the emerging magnetic fusion technology is placing increasing demands on the performance of composite materials that serve as electrical insulators and primary supports for magnet windings. It has also stimulated interest in the potential of advanced, high-modulus composites as contributors to lighter weight and increased structural and thermal efficiency.

12.4.2 Woven-fabric and random-mat composites

Polymer-matrix composites reinforced with woven-fabric or random-mat fibers are the least expensive of the laminate systems when they are produced by industrial high pressure laminating methods. Their low cost, reasonable strength properties and low thermal and electrical conductivities have made such materials very popular for cryogenic structures.

Glass-fabric-reinforced epoxy composites continue to dominate cryogenic interest. The best overall performance is obtained from laminates using conventional, fully reacted epoxy resin systems. Laboratory work and practical experience have provided evidence that highly flexibilized epoxy systems do not improve the cryogenic performance of glass-fabric-reinforced laminates. Furthermore, it is becoming clear that such resin systems are a distinct detriment to their room-temperature performance. The failure of highly flexibilized systems is mainly due to their high thermal contraction, which creates a high level of residual stress in the cooled laminate, negating the inherent residual strain capability of the matrix at cryogenic temperatures.

A substantial effort has been directed toward reducing the cryogenic property variability of NEMA G-10 and G-11 types of glass-fabric-reinforced high-pressure industrial laminates that are produced in large quantities by several United States laminating firms. Such products are fabricated to meet normal temperature electrical insulation performance criteria established by NEMA. Manufacturers are given considerable leeway in the selection of specific laminate components and in manufacturing techniques. Proprietary NEMA-grade laminates produced by the various firms, therefore, display substantial variability in those mechanical and thermal properties of greatest interest to the designers of large superconducting magnets and other sophisticated cryogenic apparatus.

Recognizing this problem, a cooperative program was undertaken in 1976 between the National Bureau of Standards, the NEMA organization and the United States laminating industry to establish cryogenic grades of NEMA G-10 and G-11 glass-fabric-reinforced epoxy products. The resulting laminates are designated G-10CR and G-11CR. These CR-grade products are high-quality, state-of-the-art products, differing from proprietary products in that they are produced to closely controlled, uniform specifications covering both component materials and manufacturing techniques. The CR products meet existing NEMA, MIL, and Federal specifications for conventional G-10 and G-11 products. The G-10CR is a heat-activated, amine-catalyzed, bisphenol A solid-type epoxy–resin laminate reinforced with continuous-filament E glass having a silane finish. G-11CR is an aromatic-amine hardened bisphenol A liquid-type epoxy–resin laminate reinforced in the same manner. Such laminates are currently in commercial production and have been incorporated into several prototype superconducting magnets.

The mechanical, thermal, and electrical properties of G-10CR and G-11CR have been characterized at the National Bureau of Standards and other United States laboratories (Kasen, MacDonald, Beekman, and Schramm, 1980). The effect of neutron and gamma irradiation at 4 K on the mechanical and electrical performance of the materials has been studied at Oak Ridge National Laboratory (Kernohan, Coltman, and Long, 1979; Coltman, Klabunde, Kernohan, and Long, 1979).

Similar concern for the proper selection of laminate materials for use in superconducting magnet technology has led to a substantial body of cryogenic composite characterization work on a series of European industrial laminates by the European Organization for Nuclear Research (CERN) in Geneva, Switzerland (Dahlerup-Petersen and Perrot, 1979). Mechanical and thermal property data were obtained from 275 to 4 K on E-glass-fabric-reinforced bisphenol A epoxy, epoxy–novolac, polyimide, phenolic, and silicone laminates plus a pitch-base graphite/bisphenol A and an aramid/bisphenol A laminate. The careful documentation of the studied materials and the sophistication of data acquisition and reduction methods were outstanding in this program.

A substantial contribution to understanding the effect of cryogenic temperatures on the performance of glass-mat-reinforced epoxy laminates has been made by a leading fiberglass manufacturer. Development of a practical nonmetallic containment system for LNG required a careful study of the effect of cryogenic temperatures on the integrity of the laminate serving as the primary LNG barrier. This program generated an extensive mechanical and thermal property data base to 76 K for both discontinuous and continuous strand glass-mat/epoxy laminates. Performance of the laminate systems under severely constrained thermal shock was also thoroughly investigated. Mechanical and thermal properties of these laminates have been summarized in handbook form (*LNG Materials and Fluids*, 1980) along with the properties of other nonmetallic laminates of interest to LNG technology.

The most recent data on the temperature dependence of the significant mechanical, thermal, and electrical properties of fabric and mat-reinforced polymer composites are summarized in Figures 12.4 through 12.13 and in Table 12.5. Data obtained at temperatures defined by the boiling points of cryogens (76 and 4 K) are presented in bar graph form to indicate lack of knowledge of the intervening temperature dependence. The cryogenic temperature values are taken as the average of the data at 76 and 4 K except where there is evidence of a statistically significant property change from 76 to 4 K. Data obtained over closely spaced temperature intervals are presented as continuous curves.

Strength properties. The maximum tensile strengths in fabric-reinforced laminates correspond to the warp direction of the fabric where glass content is the highest. The generally lower glass content in the fill direction results in lower strength. The G-10CR and G-11CR laminates have a 42:32 warp-to-fill ratio typical of the style of glass fabric used in high-pressure industrial laminates manufactured in the United States. The lowest strength is developed in a 45° orientation with respect to the warp and fill.

Conversely, a laminate reinforced with continuous or discontinuous randomly oriented glass mat has essentially isotropic properties in the edgewise direction, that is, in the plane normal to the direction of pressing during laminate cure. But this isotropy is obtained at the expense of reduced laminate strength and stiffness com-

Table 12.5 Electrical properties of G-10CR and G-11CR.[a]

Temperature (K)	Thickness (cm)	Volume Resistivity (Ω·cm)	Voltage Stress at Breakdown (kV/mm)
	G-10 CR		
295	0.030		48.4
	0.036	8.9×10^{15}	45.7
	0.051	9.3×10^{14}	
77	0.051	1.5×10^{17}	
4	0.036	4.0×10^{17}	48.4
	0.051	4.1×10^{17}	
	G-11 CR		
295	0.030	7.3×10^{15}	
	0.051	1.3×10^{15}	
77	0.051	6.1×10^{16}	
4	0.030	2.0×10^{17}	48.0
	0.051	2.0×10^{17}	

[a]Kasen, MacDonald, et al. (1980).

pared with the maximum values obtainable from fabric-reinforced laminates.

Recent data on the effect of cryogenic temperatures on the tensile, compressive, and interlaminar shear strengths of fabric and mat-reinforced polymer composites are summarized in Figures 12.4 through 12.6.

Figure 12.4 summarizes ultimate tensile strength data. Except as noted, all data are for the warp direction of fabric reinforcement and for the edgewise direction of mat reinforcement. In part, the data compare reported strengths of a series of E-glass-fabric-reinforced bisphenol A epoxy laminates having varying glass contents. Conventional bisphenol A epoxy–resins (diglycidyl ether of bisphenol A) and epoxy–novolacs (glycidyl ethers of novolacs) are most widely used in laminate fabrication (MIL-HDBK-17A, 1971). The composition of E glass is given in Table 12.2.

These data are compared with strengths reported for continuous and chopped-mat E-glass-reinforced bisphenol A epoxy products, for E-glass-fabric-reinforced polyimide, and for polyester and melamine laminates. Available data on the cryogenic performance of epoxy laminates reinforced with aramid fiber fabric and with a pitch-base graphite fabric are also documented along with the small amount of data available for cotton-fabric-reinforced phenolic (phenolformaldehyde) laminates.

Highest cryogenic tensile strengths were reported for the NEMA G-11CR laminate containing 67 to 72 wt.% glass. G-10CR, having 64 to 68 wt.% glass, developed slightly lower strength. A commercial laminate of European manufacture containing 74.8 wt.% glass developed properties similar to the CR grades, whereas other commercial laminates having glass contents of 62.5 to 66.1 wt.% developed slightly lower cryogenic strengths (bar 4, Fig. 12.4). These data illustrate

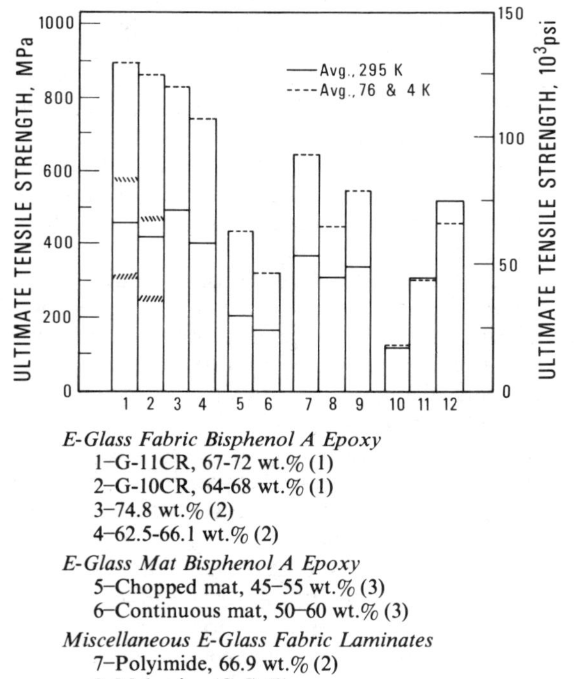

E-Glass Fabric Bisphenol A Epoxy
1–G-11CR, 67-72 wt.% (1)
2–G-10CR, 64-68 wt.% (1)
3–74.8 wt.% (2)
4–62.5-66.1 wt.% (2)

E-Glass Mat Bisphenol A Epoxy
5–Chopped mat, 45–55 wt.% (3)
6–Continuous mat, 50–60 wt.% (3)

Miscellaneous E-Glass Fabric Laminates
7–Polyimide, 66.9 wt.% (2)
8–Melamine (G-5) (3)
9–Polyester (3)

Other Laminates
10–Cotton fabric/phenolic (GE) (3)
11–Pitch-base graphite fabric/Bis A epoxy, 61.8 wt.% (2)
12–Aramid fabric/Bis A epoxy, 61.5 wt.% (2)

All data are for the warp direction except for G-10CR and G-11CR, where fill direction strengths are indicated by //// for 295 K and \\\\ for 76 and 4 K. Wt.% = fiber weight fraction. Refs.: (1) Kasen, McDonald, et al. (1980); (2) Dahlerup-Petersen and Perrot (1979); (3) *LNG Materials and Fluids* (1980).

Figure 12.4 Temperature dependence of ultimate tensile strength of fabric-reinforced polymer-matrix laminates.

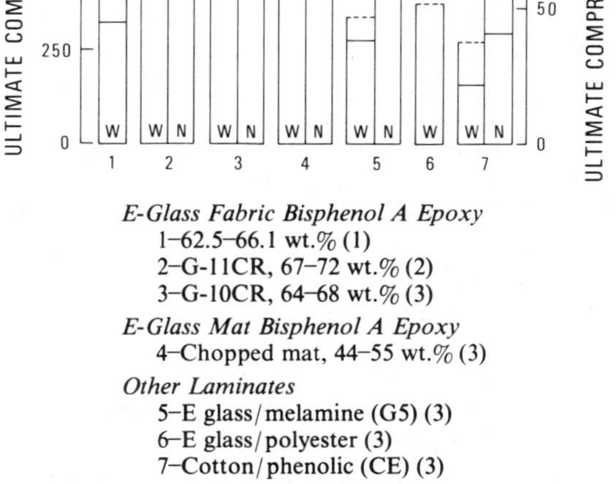

E-Glass Fabric Bisphenol A Epoxy
1–62.5–66.1 wt.% (1)
2–G-11CR, 67–72 wt.% (2)
3–G-10CR, 64–68 wt.% (3)

E-Glass Mat Bisphenol A Epoxy
4–Chopped mat, 44–55 wt.% (3)

Other Laminates
5–E glass/melamine (G5) (3)
6–E glass/polyester (3)
7–Cotton/phenolic (CE) (3)

Wt.% = fiber weight fraction. Refs.: (1) Dahlerup-Petersen and Perrot (1979); (2) Kasen, McDonald, et al. (1980); (3) *LNG Materials and Fluids* (1980).

Figure 12.5 Temperature dependence of ultimate compressive strength of fabric and mat-reinforced polymer-matrix laminates.

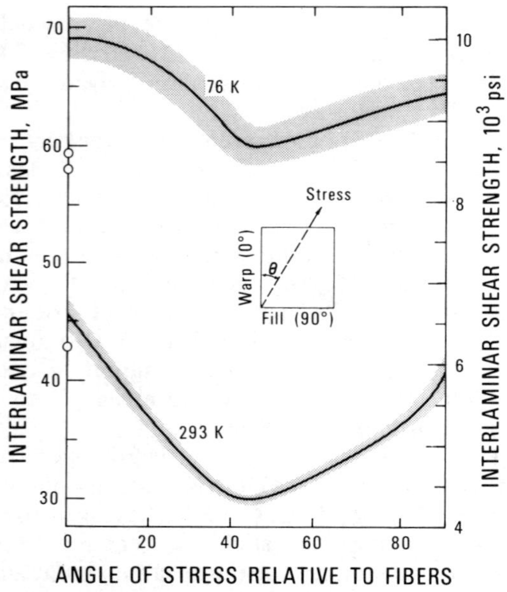

Figure 12.6 Angular dependence of interlaminar shear strength of G-10CR glass-fabric epoxy laminate at 293 and 76 K. Shaded areas reflect 2-sided, 95% confidence intervals. From Becker and Erez (1980) using the modified ASTM D 2733-70 method. NBS data for G-11CR are indicated by circles (Kasen, McDonald, et al., 1980).

the importance of specifying glass content or buying laminates to a specified range of glass content whenever maximum laminate performance is desired.

Figure 12.4 indicates that the tensile strengths in the fill direction of G-10CR and G-11CR are 35 to 70% of that in the warp direction. A similar strength anisotropy can be expected for the other fabric-reinforced laminates depending on the warp-to-fill ratio of the fabric.

The tensile strengths developed by the mat-reinforced E-glass/epoxy products (bars 5 and 6, Fig. 12.4) are substantially lower than those developed for the warp direction of fabric-reinforced E-glass/epoxy products, but the strength difference is less when compared with those of the fill direction of the fabric products. It is expected that the strength of the mat products would exceed that of the fabric products in a 45° direction from warp and fill.

Figure 12.4 indicates that a commercial polyimide-matrix laminate (Silirite 64 400) develops 7 to 10% lower average tensile strength than does bisphenol A epoxy laminates having similar type and content of glass reinforcement (bar 7, Fig. 12.4).

The data presented for glass-fabric-reinforced melamine (bar 8, Fig. 12.4) are typical values reported by the manufacturer (*LNG Materials and Fluids*, 1980). This type of laminate is commercially produced as NEMA G-5. Data presented for glass-fabric-reinforced polyester (bar 9, Fig. 12.4) reflect average data summarized in the literature for a variety of such laminates.

Glass-reinforced laminates all strengthen substantially on cooling. These data suggest that the average tensile strength of epoxy-matrix laminates will increase 45 to 55% over the room-temperature values, whereas the polyimide, melamine, and polyester laminates increase about 35 to 40%.

Despite its fairly wide use, little is known of the effect of cryogenic temperatures on the strength of cotton-reinforced phenolic laminates. Data appearing in Fig. 12.4 reflect one manufacturer's published data on its proprietary NEMA type CE product (*LNG Materials and Fluids*, 1980). Strengths obtained from other manufacturers' products or from the NEMA type LE cotton/phenolic product may differ substantially. Nevertheless, there is no doubt that the strength of the cotton/phenolic product is inferior to that of the glass-reinforced products.

Data on the pitch-base graphite fabric and on aramid fabric-reinforced epoxy laminates summarized in Fig. 12.4 represent the first cryogenic data obtained on these materials. The aramid laminate displays a significant average strength reduction on cooling — possibly a manifestation of internal stresses created by the very large negative coefficient of thermal expansion of this fiber type. The comparatively low strength of the graphite-fabric-reinforced product reflects the low strength of the inexpensive pitch-based fiber used in its construction. A comparable laminate reinforced with graphite fabric using fiber derived from the more expensive polyacrylonitrile (PAN) precursors should develop about twice the tensile strength of the pitch-based product.

The compressive strength data presented in Fig. 12.5 reflect both the warp direction and normal direction where such data are available. The direction designated "normal" is normal to the plane of reinforcement for fabric-reinforced laminates and is parallel to the pressing direction during laminate cure for mat-reinforced laminates.

These data suggest that cryogenic applications requiring compressive strength in the normal direction may be well served by some of the less expensive fabric-reinforced laminates, such as the melamine-matrix NEMA G-5 (bar 4, Fig. 12.5) or by mat-reinforced laminates (bar 3, Fig. 12.5). These laminates, the G-11CR grade of glass/epoxy and the NEMA CE cotton/phenolic products develop higher average compressive strength in the normal direction than in the direction of reinforcement. The NEMA G-10CR product is the only laminate for which the average warp direction compressive strength is reported to exceed that of the normal direction.

The data for G-11CR, G-10CR, and chopped-mat glass/epoxy (bars 1, 2, and 3, Fig. 12.5) represent systematic studies of well-characterized materials. Data for NEMA G-5 and CE (bars 4 and 6, Fig. 12.5) are manufacturers' typical data. Glass/polyester data reflect the average of literature values.

Data obtained at the National Bureau of Standards (Kasen, MacDonald, et al., 1980) indicate that testing in the fill direction of the fabric reinforcement reduces the average ultimate compressive strength by about 30% for the G-10CR product and by about 24% for the G-11CR product.

Interlaminar shear strength, the resistance to shear between the layers of a layered laminate, has become an important design property in large superconducting magnets where glass-fabric-reinforced composites are used for electrical insulation and for structural support of superconducting windings. The interlaminar shear strength of the selected materials must be adequate to resist electromagnetic forces within the magnet. The guillotine test method described by ASTM method D2733-70 is the most appropriate. However, the application of this method, as described in ASTM standards, has been found to yield extremely variable results. In response to the needs of the MFE research efforts, the National Bureau of Standards and the M.I.T. Francis Bitter National Magnet Laboratory have refined the test method, which is described in § 12.7.4.

The temperature and orientation dependence of interlaminar shear in a G-10CR glass-fabric-reinforced epoxy industrial laminate appearing in Fig. 12.6 has been reported by Becker and Erez (1980) of the Frances Bitter National Magnet Laboratory. Data determined at the National Bureau of Standards for the warp (0°) direction are in agreement with these data except for those in the warp direction of G-11CR (noted by circles in Fig. 12.6), which indicate a somewhat lower average interlaminar shear strength for this material.

Data on the average change in laminate strength properties on cooling are useful in comparing cryogenic and room-temperature performance, but such data can be misleading from a design standpoint. The relevant data are the lower bounds of performance — not the average. It is frequently observed that the variability in strength properties increases with increases in average strength on cooling. The variability appears to be less for the lower strength fabric and mat-reinforced laminates than for the higher performance uniaxial laminates discussed in a later section; nevertheless, it is a factor to consider in all laminate types. Variability arises intrinsically owing to embrittling and increased notch sensitivity of the polymer matrix in a cooled laminate. It is, therefore, a larger factor in properties such as shear and compression, which are most dependent on matrix properties. Variability is a strong function of the quality of laminate manufacture. In particular, high void contents and resin-rich areas must be avoided. Least variability is found in laminates fabricated under the close control exercised by large industrial laminating firms, particularly when controlled-specification products, such as the CR grades, are produced. Low-pressure laminating techniques such as vacuum bagging and autoclaving, used to fabricate components directly to complex shapes, can be expected to produce laminates having an inherently larger variability.

Elastic properties. Recent data on the effect of cryogenic temperatures on Young's modulus are presented in Fig. 12.7 for the warp direction of fabric laminates and for the edgewise direction of mat-reinforced laminates. Where data are available, a distinction is made between the change observed between 295 and 76 K and that observed between 76 and 4 K. In contrast to strength data, the moduli data show a monotonically increasing stiffness with decreasing temperature. Specific identification is given for each of the materials.

Figure 12.7 summarizes the first cryogenic information on pitch-base fabric graphite/epoxy laminates and illustrates the stiffness superiority of this type of laminate over conventional woven fabric materials. The high modulus is a direct reflection of the high modulus of the graphite fiber. On the other hand, the data indicate that the much higher modulus of aramid fiber (124 GPa, 18×10^6 psi) compared with that of E glass (72 GPa, 10.5×10^6 psi) is not reflected in the moduli of woven-fabric laminates. The aramid fabric laminates produced an average Young's modu-

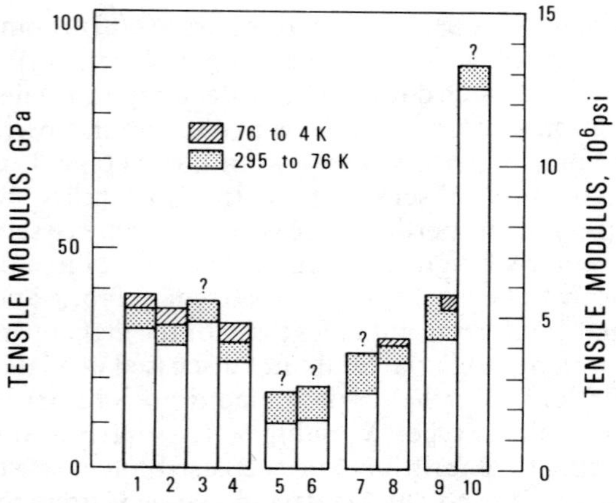

E-Glass Fabric Bisphenol A Epoxy
1–G-11CR, 67–72 wt.% (1)
2–G-10CR, 64–78 wt.% (1)
3–74.8 wt.% (2)
4–62.5–66.1 wt.% (2)
E-Glass Mat Bisphenol A Epoxy
5–Chopped mat, 45–55 wt.% (3)
6–Continuous mat, 50–60 wt.% (3)
Miscellaneous E-Glass Fabric Laminates
7–Polyester (3)
8–Polyimide, 66.9 wt.% (2)
Other Laminates
9–Aramid/epoxy, 61.5 wt.% (2)
10–Pitch-base graphite/epoxy, 61.8 wt.% (2)

Warp direction of fabric or edgewise direction of mat. Wt.% = fiber weight fraction. Refs.: (1) Kasen, McDonald, et al. (1980); (2) Dahlerup-Petersen and Perrot (1979); (3) *LNG Materials and Fluids* (1980).

Figure 12.7 Temperature dependence of tensile moduli of fabric- and mat-reinforced polymer-matrix laminates.

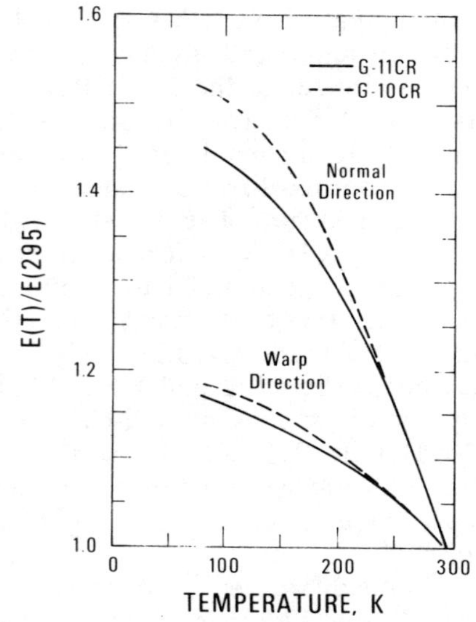

Figure 12.8 Temperature dependence of two principal Young's moduli for G-11CR and G-10CR glass-fabric-reinforced epoxy laminates by the resonance method (Ledbetter, 1979a).

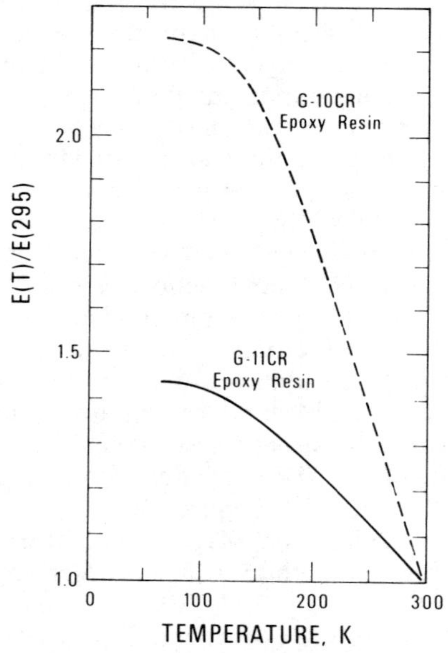

Figure 12.9 Temperature dependence of the modulus of the bisphenol A epoxy systems used in G-10CR and G-11CR glass-fabric-reinforced epoxy laminates by the resonance method (Ledbetter, 1979a).

lus very similar to that of the E-glass laminates for equivalent fiber mass fraction.

Bars 1 through 4 of Fig. 12.7 show that Young's modulus is directly related to glass content. This again illustrates the significant effect of glass content on laminate performance. The type of matrix has much less effect because its modulus is typically an order of magnitude lower than that of the glass reinforcement. These data indicate that the average modulus of glass-fabric-reinforced polymer laminates will increase about 20% upon cooling to 4 K.

Data presented in Fig. 12.7 were obtained either from strain-gauged specimens or from extensometer data. The elastic properties of G-10CR and G-11CR materials by dynamic resonance methods have also been studied at the National Bureau of Standards (Ledbetter, 1979a, 1979b). As illustrated in Figs. 12.8 and 12.9, this method can generate a continuous temperature-modulus curve for either the laminate or the resin used in the laminate construction. Results thus far obtained by the dynamic method agree reasonably well with those obtained by conventional techniques, but the dynamic method has much improved sensitivity. For example, the large difference in the temperature dependence of Young's moduli of the G-10CR and G-11CR resin systems illustrated in Fig. 12.9 is apparent in the differing temperature dependencies noted for the G-10CR and G-11CR warp direction laminate moduli in Fig. 12.8, but it is not apparent in data obtained from strain-gauged specimens. Figure 12.8 illustrates that the temperature dependence of Young's modulus for the G-10CR laminate is substantially greater in the direction normal to the reinforcement than in the warp direction.

Data presented in Fig. 12.7 indicate that a polyimide-matrix laminate (bar 8) will have a Young's modulus very similar to that of an epoxy matrix laminate of similar glass content (bar 4). The epoxy matrix laminate data in Fig. 12.7 indicate a higher Young's modulus than was reported for such laminates in an earlier review (Kasen, 1975a) because the earlier data reflected literature averages and did not distinguish between moduli in the warp and fill directions. The anisotropy of reinforcement results in anisotropy of Young's modulus, as illustrated for G-10CR in Fig. 12.10. These data suggest that average moduli values reported for the E-glass-fabric bisphenol A epoxy laminates in Fig. 12.7 would decrease about 40% if the moduli were

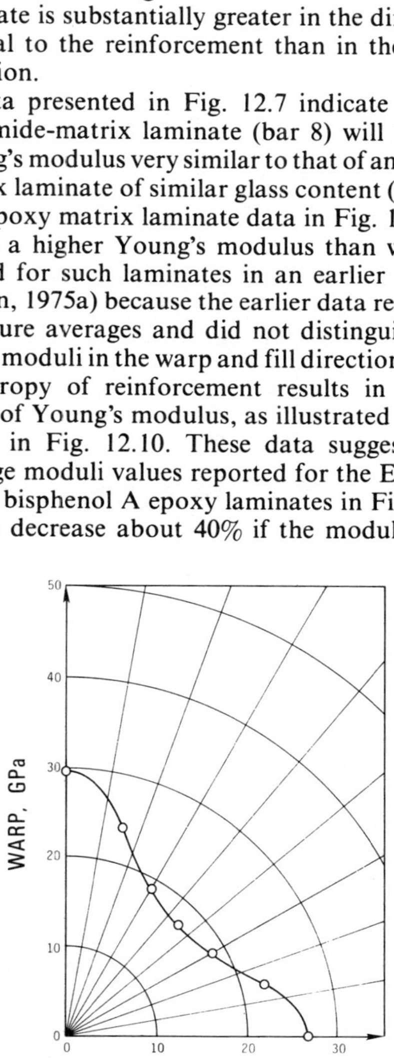

Figure 12.10 Orientation dependence of room-temperature Young's modulus of G-10CR glass-fabric/epoxy laminate by the resonance method (Ledbetter, 1979a).

evaluated in the 45° direction relative to the warp and fill of the fabric.

The Young's moduli of mat-reinforced laminates (bars 5 and 6, Fig. 12.7) are substantially lower than those of fabric-reinforced laminates when the latter are tested in the warp direction. However, the difference decreases if the aforementioned orientation effect is considered. The Young's modulus of a mat-reinforced product is essentially isotropic in the plane normal to the pressing direction during laminate cure (edgewise direction).

Elastic properties are intrinsically less variable than strength properties.

Thermal properties. Recent data on the temperature dependence of thermal expansion and thermal conductivity of woven fabric and of random mat laminates are summarized in Figs. 12.11 through 12.13.

Figure 12.11 reviews the first cryogenic thermal property data reported for graphite- and aramid-fabric-reinforced epoxy laminates. The negative coefficients of thermal expansion of graphite and aramid fibers combine with the positive coefficient of the epoxy resin to yield laminates with a high degree of dimensional stability in the warp direction (bars 10 and 11). The aramid laminate is more stable than the graphite laminate of similar fiber content owing to the higher negative coefficient of thermal expansion of the aramid fiber. The stability of these products would be lower in the fill direction for an unbalanced-weave fabric and would be still lower in the direction normal to the reinforcement.

The thermal contraction of epoxy resins on cooling is about an order of magnitude greater than for glass (Corruccini and Gniewek, 1961). Bar 12 in Fig. 12.11 summarizes the contraction for a CY 205 + HY 905 + DY 040 + DY 061 resin system. The data indicate that the contraction of a glass-fabric-reinforced laminate is about 60% of the contraction of the unreinforced resin in the normal direction and about 20% of the resin contraction in the warp direction. Contractions of the aramid and graphite-reinforced laminates in the warp direction are only about 5% of that of the unreinforced resin. This difference is reflected in residual laminate stress. The comparatively small dimensional changes observed on cooling from 76 to 4 K indicate that most of the residual stress created by cooling occurs in the 295 to 76 K temperature range. This stress is

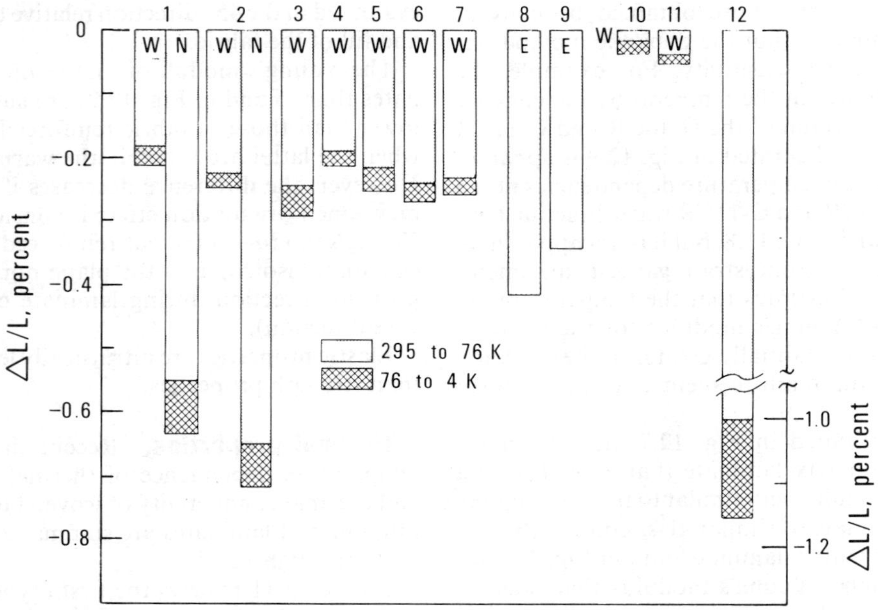

E-Glass Fabric Bisphenol A Epoxy
1–G-11CR, 67–72 wt.% (1)
2–G-10CR, 64–68 wt.% (1)
3–64.4 wt.% (2)
4–66.1 wt.% (2)
5–67 wt.% (2)
6–62.5 wt.% (2)
7–74.8 wt.% (2)

E-Glass Mat Bisphenol A Epoxy
8–Chopped mat, 45–55 wt.% (3)
9–Continuous mat, 50–60 wt.% (3)

Other Laminates
10–Aramid/epoxy, 61.5 wt.% (2)
11–Pitch-base graphite/epoxy, 61.8 wt.% (2)
12–Unreinforced epoxy (2)

For fabric laminates, W = warp direction, N = normal direction. For mat laminates, E = edgewise direction. Wt.% = fiber weight fraction. Refs.: (1) Kasen, McDonald, et al. (1980); (2) Dahlerup-Petersen and Perrot (1979); (3) *LNG Materials and Fluids* (1980).

Figure 12.11 Temperature dependence of thermal contraction for fabric- and mat-reinforced laminates.

added to that created during cooling the laminates to 295 K from the elevated cure temperature. The low modulus of epoxy (about 0.1 that of glass) results in most of the stress being resolved onto the matrix.

The glass content would, therefore, be expected to play a significant role in establishing contraction in the fiber direction. But such a correlation is not evident in Figure 12.11.

Thermal contraction of mat-reinforced laminates in the edgewise direction is substantially higher than that for fabric-reinforced laminates in the warp direction. This reflects the random fiber orientation and lower glass content of the mat product. But contraction of the mat products is isotropic in the edgewise direction, whereas that of the fabric product is anisotropic. Thermal contraction of fabric-reinforced laminates in a 45° direction relative to warp and fill would probably be greater than that of the mat-reinforced laminates in the normal direction.

Figures 12.12 and 12.13 summarize temperature dependence data recently reported for thermal conductivity of a variety of industrial laminates.

Figure 12.12 compares the reported conductivity for a series of E-glass-reinforced bisphenol A laminates and for two unreinforced bisphenol A epoxy resins. Bar 3 reflects the average conductivity of the resin systems used in CR grades. The G-10CR resin is an amine-catalyzed solid system; the G-11CR resin is an aromatic-amine hardened, liquid system. Nevertheless, their conductivities agreed within ±5%. Bar 7 reflects a resin system typical of that used in the laminates studied by Dahlerup-Petersen and Perrot (1979).

The room-temperature conductivity of the CR-grade laminates reported by Kasen, Mac-

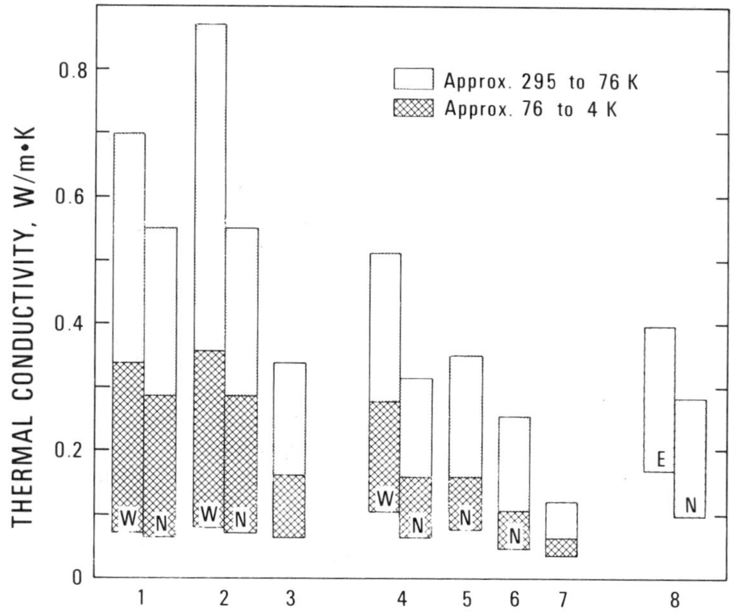

E-Glass Fabric Bisphenol A Epoxies
1–G-11CR, 67–72 wt.% (1)
2–G-10CR, 64–68 wt.% (1)
3–Bisphenol A epoxies (2)
4–64.5 wt.% (3)
5–62-74 wt.% (3)
6–67.5 wt.% (3)
7–Bisphenol A epoxy (3)

E-Glass Mat Bisphenol A Epoxies
8–Continuous mat, 50–60 wt.% (4)

Data are illustrated for fabric reinforcement (W = warp, N = normal) and mat reinforcement (E = edgewise, N = normal). Wt.% = fiber weight fraction. Refs.: (1) Kasen, McDonald et al. (1980); (2) Hust (1980); (3) Dahlerup-Petersen and Perrot (1979); (4) *LNG Materials and Fluids* (1980).

Figure 12.12 Temperature dependence of thermal conductivity for a series of E-glass-reinforced bisphenol A epoxy laminates and for unreinforced bisphenol A epoxy.

Donald, et al. (1980) at the National Bureau of Standards are significantly higher than that reported by Dahlerup-Petersen and Perrot (1979) at CERN for similar laminates. Measurements at the National Bureau of Standards on the European laminate (represented by bar 4) yielded a 45% higher conductivity at room temperature than was measured at CERN. This indicates that the difference is in the experiments rather than in the materials. Previously published literature averages for thermal conductivity of glass-reinforced laminates (Kasen, 1975a) agree with the CERN data as do the data reported for the E-glass mat laminate in Fig. 12.12. The normal conductivity range reported for unreinforced epoxy resins (MIL-HDBK-17A, 1971) is intermediate between the data of the National Bureau of Standards and CERN. Consequently, the procedures used at the National Bureau of Standards were reexamined to rule out the possibility of measurement errors owing to interface resistance, radial heat losses, incorrect calibration constants, or incorrect thermocouple calibration. The source of the discrepancy is being investigated further.

These discrepancies illustrate the difficulty of obtaining reliable data on the thermal conductivity of composite laminates, particularly in the higher temperature portion of the cryogenic range. They also illustrate the desirability of a standard reference material having conductivity in the range of composite laminates that could be used in instrument calibration.

Figure 12.13 compares data on the temperature dependence of thermal conductivity for E-glass-reinforced silicone and polyimide laminates, aramid and graphite-fabric-reinforced bisphenol A epoxy laminates and a cotton-reinforced phenolic laminate. The comparatively high conductivity of the graphite-reinforced laminate reflects the high fiber conductivity. Conductivity of the E-glass-reinforced polyimide

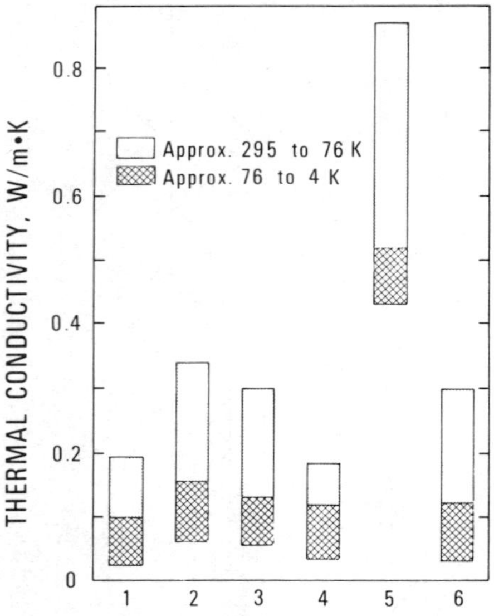

E-Glass Fabric Laminates
1–Silicone matrix, 70 wt.%
2–Polyimide matrix, 66.9 wt.%
3–Polyimide matrix, 58.9 wt.%
Miscellaneous Bisphenol A Laminates
4–Aramid fabric, 61.5 wt.%
5–Pitch-base graphite, 61.8 wt.%
Phenolic Laminates
6–65 wt.%

Figure 12.13 Temperature dependence of thermal conductivity in the direction normal to the reinforcement (flatwise direction) for a series of silicon-, polyimide-, epoxy-, and phenolic-matrix laminates (Dahlerup-Petersen and Perrot, 1979). Wt.% = fiber weight fraction.

laminates appears similar to that reported by Dahlerup-Petersen and Perrot (1979) for the epoxy-matrix laminates in Fig. 12.12. This suggests that the conductivity and temperature dependence of conductivity are similar for polyimide and epoxy resin systems.

Normal-direction conductivities of glass-reinforced silicone and aramid-reinforced epoxy are substantially lower than those of glass-reinforced silicone, and those of aramid-reinforced epoxy are substantially lower than those of glass-reinforced laminates of equal fiber mass fraction. Normal-direction conductivity of cotton-reinforced phenolic laminate (phenolformaldehyde resin) is similar to that of glass-reinforced epoxy.

The author is unaware of any substantial contribution to the characterization of specific heat properties of fabric- or mat-reinforced laminates within the cryogenic temperature range since the earlier review (Kasen, 1975a). Data of Heberlein (1976) indicate that the specific heat of E-glass laminates might be substantially lower than that of S-glass laminates, but these data did not extend below 230 K.

Electrical properties. The effect of cryogenic temperatures on the volume resistivity and electrical strength of G-10CR and G-11CR glass fabric-reinforced epoxy laminates is summarized in Table 12.5. These are the first data obtained on these parameters at 4 K. Although they are not extensive, the data indicate that the electrical strength will be independent of both temperature and material. The effect of temperature on the volume resistivity is similar for the two materials, increasing 1.5 to 2 orders of magnitude on cooling from 295 to 4 K.

12.4.3 Uniaxial fiber-reinforced composites

Composites reinforced with uniaxial glass, graphite, boron or aramid fibers continue to have limited and specialized applications in cryogenic technology. The aerospace industry is using graphite-reinforced laminates for dimensionally stable mirrors and reflectors (Kasen, 1979). The relatively small structures already flown have functioned well in the temperature range down to 100 K. The primary concern is the predictability of dimensional stability in the very large reflectors presently being designed for use in outer space.

The desire to reduce the weight of superconducting machinery and superconducting magnetohydrodynamic power generators has provided the primary impetus for continued study of the performance of advanced composites at cryogenic temperatures. Most of the related research has been carried out at the National Bureau of Standards with the assistance of United States industry (Schramm and Kasen, 1977a, 1977b) and at CERN (Dahlerup-Petersen and Perrot, 1979).

Uniaxial glass, graphite, boron, and aramid fibers all appear to be acceptable reinforcements for cryogenic laminates (Kasen, 1975a, 1975b). A wide variety of polymer matrices can be used, although epoxy systems are most common. A primary problem is the large variability in the mechanical property data reported in the literature. It is difficult to distinguish inherent material variability from experimental variability. Experimental problems are now under better control, although much remains to be done. Improved test methods have established that the earlier indications of strength degradation in graphite-reinforced epoxy laminates on cooling were incorrect. Further work has shown that the use of a highly flexibilized resin system is counterproductive in the advanced composites as well as in the glass-reinforced products. Conventional, fully reacted resin systems remain the best current choice for high-modulus fiber laminates intended for cryogenic use.

Recent mechanical property characterization studies on advanced composites have produced data required for prediction of the properties of complex angleply laminates using macromechanical laminate theory (*Advanced Composites Design Guide*, 1973; Waddoups, 1968). These data, obtained from uniaxial laminates, consist of longitudinal and transverse uniaxial strength, modulus, and Poissons' ratio plus in-plane shear strength and modulus. Laminates reinforced with S glass; high-, medium- and low-modulus graphite; boron and aramid fibers have now been characterized in this manner. Results of these studies have been published in reports (Kasen, Schramm, and Beck, 1980; Dahlerup-Petersen and Perrot, 1979) and in the open literature (Schramm and Kasen, 1977a, 1977b).

Recent data on the temperature dependence of the significant mechanical and thermal properties of uniaxial fiber-reinforced composites are summarized in Figs. 12.14 through 12.20. Bar graphs are again used to illustrate trends of strength changes from room temperature to the cryogenic temperature range for data obtained at fixed cryogenic boiling points. The average of data reported at 76 and 4 K are plotted except where the data show a statistically significant change in a property between these temperatures. Data obtained at closely spaced temperature intervals are presented as continuous curves.

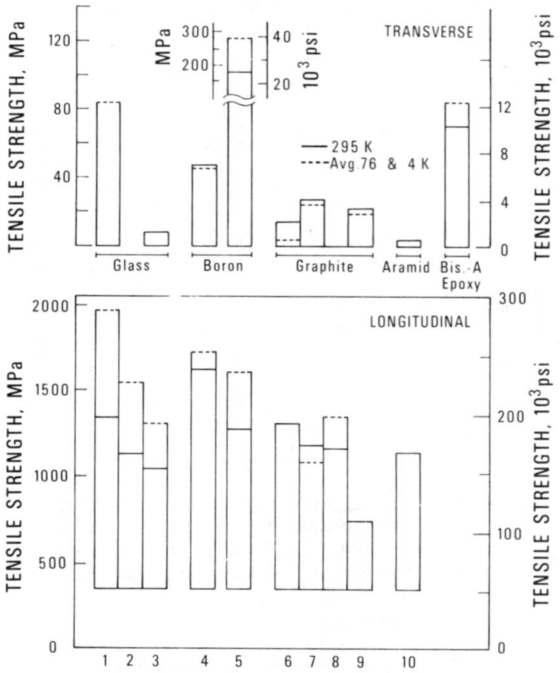

Glass Reinforcement
1–S-901/Resin 2, 66 vol.% (1)
2–R/bisphenol A, 63 vol.% (2)
3–E/bisphenol A, 50.6 vol.% (2)

Boron Reinforcement
4–140 μm (5.6 mil) B/2387 epoxy, 52 vol.% (1)
5–140 μm (5.6 mil) B/6061 aluminum, 47 vol.% (1)

Graphite Reinforcement
6–AS/Resin 2, 64 vol.% (1)
7–HM-S/934, 62 vol.% (3)
8–HM-S/bisphenol A, 55 vol.% (2)
9–GY-70/934, 65 vol.% (3)

Aramid Reinforcement
10–Kevlar 49/934, 67 vol.% (3)

Vol.% = fiber volume fraction. Refs.: (1) Schramm and Kasen (1977a); (2) Dahlerup-Petersen and Perrot (1979); (3) Kasen, Schramm, and Beck (1980). Bisphenol A epoxy: CY 205 (2).

Figure 12.14 Temperature dependence of the uniaxial longitudinal and transverse tensile strengths of glass-, boron-, graphite-, and aramid-reinforced epoxy laminates.

Strength properties. The uniaxial longitudinal and transverse tensile strengths of glass-, boron-, graphite- and aramid-reinforced epoxy laminates are summarized in Fig. 12.14.

Data for epoxy matrix laminates uniaxially reinforced with types S, E, and R glass appear on this figure. Typical glass compositions are given in Table 12.2.

Type-S glass was developed during the early 1960s as a high-strength glass for specialized aerospace purposes. The 901 designation refers to a special solvent-sizing system used with this high-cost, military-specification product. The use of S glass is confined to filament-wound or uniaxial prepreg tape products. Most of the cryogenic data on uniaxial glass-reinforced epoxy laminates have been generated with type-S glass. Potential users should also consider the recently introduced commercial grade of S glass produced under the trade designation S-2, which is

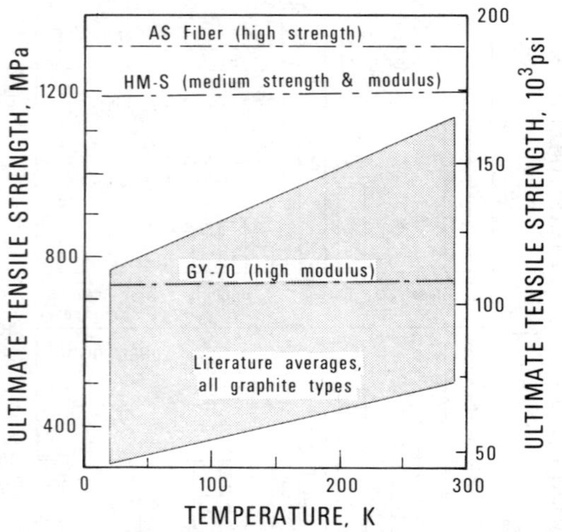

Figure 12.15 Comparison between the combined literature data (Kasen, 1975b) on the temperature dependence of uniaxial ultimate tensile strength of graphite-reinforced laminates (shaded area) and data obtained by improved test techniques (Schramm and Kasen, 1977a, 1977b; Kasen, Schramm, and Beck, 1980; Dahlerup-Petersen and Perrot, 1979).

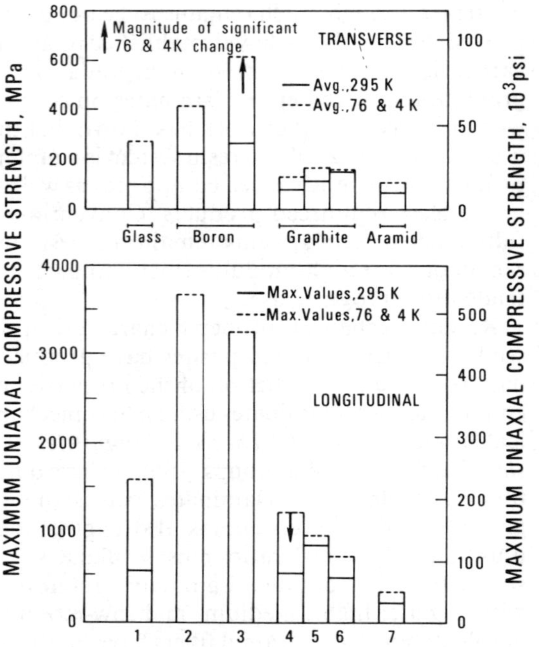

Glass Reinforcement
1–S-901/Resin 2, 66 vol.% (1)
Boron Reinforcement
2–140 μm (5.6 mil) B/2387 epoxy, 52 vol.% (1)
3–140 μm (5.6 mil) B/6061 aluminum, 47 vol.% (1)
Graphite Reinforcement
4–AS/Resin 2, 64 vol.% (1)
5–HM-S/934, 62 vol.% (1)
6–GY-70/934, 65 vol.% (2)
Aramid Reinforcement
7–Kevlar 49/934, 67 vol.% (2)

Plotted values are maximum values for the longitudinal direction and average values for the transverse direction. Vol.% = fiber volume fraction. Refs.: (1) Schramm and Kasen (1977a); (2) Kasen, Schramm, and Beck (1980).

Figure 12.16 Temperature dependence of the uniaxial longitudinal and transverse compressive strengths of glass-, boron-, graphite-, and aramid-reinforced laminates.

available in woven fabrics as well as uniaxial fiber. For reportedly comparable properties, type S-2 costs much less than other types of S glass (Loud, 1979). The author knows of no cryogenic data for laminates fabricated from S-2 glass.

Type E is the standard electrical insulating grade of glass, most often used in fabric-reinforced industrial laminates but also available in uniaxial form. It is substantially less expensive than types S or S-2. Type R is a glass of European manufacture.

The uniaxial tensile strength of the highly flexibilized Resin 2 laminate (see Table 12.4) S-901 glass averages approximately 30% higher than that of the E-glass/bisphenol A laminate at room temperature and about 50% higher at cryogenic temperature. This difference is consistent with published data on the comparative strengths of the two glass types (Loud, 1979). The R-glass/bisphenol A laminate developed uniaxial tensile strengths intermediate between that of the S-901 and E-glass laminates.

Uniaxial longitudinal tensile properties are expected to be fiber dominated. But the literature provides evidence that the average strength of the S-901 laminate has been substantially lowered by the use of the highly flexibilized Resin-2–epoxy formulation. Handbook data indicate that the room-temperature tensile strength of this laminate is about 30% lower than that expected from an S-901 glass-reinforced laminate of equal glass content in a conventional, fully reacted epoxy matrix (MIL-HDBK-17A, 1971). Cooling increases the strength of the S-901/Resin 2 laminate, but only to a value expected for a conventional-matrix laminate at room temperature. A similar discrepancy between the strengths of S-901-glass-reinforced laminates involving the two types of matrices was documented in an

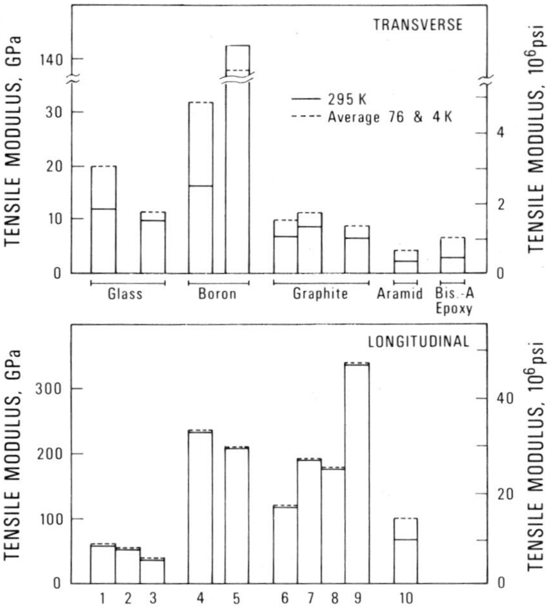

Glass Reinforcement
1–S-901/Resin 2, 66 vol.% (1)
2–R/bisphenol A, 63 vol.% (2)
3–E/bisphenol A, 50.6 vol.% (2)

Boron Reinforcement
4–140 μm (5.6 mil) B/2387 epoxy, 52 vol.% (1)
5–140 μm (5.6 mil) B/6061 aluminum, 47 vol.% (1)

Graphite Reinforcement
6–AS/Resin 2, 64 vol.% (1)
7–HM-S/934, 62 vol.% (3)
8–HM-S/bisphenol A, 55 vol.% (2)
9–GY-70/934, 65 vol.% (3)

Aramid Reinforcement
10–Kevlar 49.934, 67 vol.% (3)

Vol.% = fiber volume fraction. Refs.: (1) Schramm and Kasen (1977a); (2) Dahlerup-Petersen and Perrot (1979); (3) Kasen, Schramm, and Beck (1980).

Figure 12.17 Temperature dependence of uniaxial longitudinal and transverse tensile moduli of glass-, boron-, graphite-, and aramid-reinforced laminates.

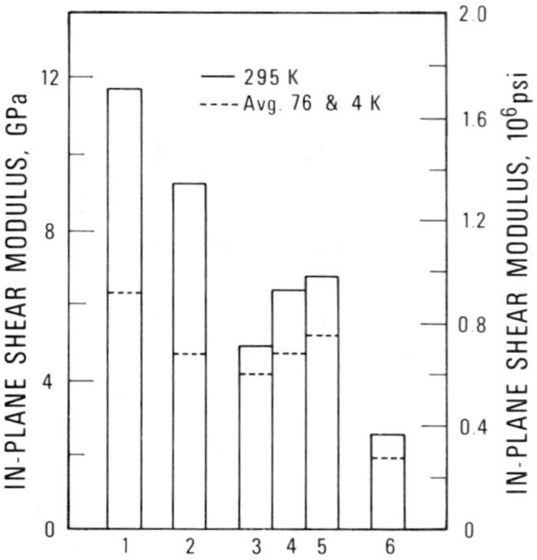

1–S-901/Resin 2, 66 vol.% (1)
2–140 μm (5.6 mil) boron/2387 epoxy, 52 vol.% (1)
3–AS graphite/Resin 2, 64 vol.% (1)
4–HM-S graphite/934 epoxy, 62 vol.% (2)
5–GY-70 graphite/934 epoxy, 65 vol.% (2)
6–Kevlar 49 aramid/934 epoxy 67 vol.% (2)

Figure 12.18 Temperature dependence of in-plane shear modulus, G_{12}, of uniaxial glass-, boron-, graphite-, and aramid-reinforced epoxy laminates. Vol.% = fiber volume fraction. Refs.: (1) Schramm and Kasen (1977a); (2) Kasen, Schramm, and Beck (1980).

earlier review (Kasen, 1975c), but the implication was not recognized at that time.

The useful strength of glass-reinforced laminates at cryogenic temperatures will be much lower than the ultimate strength whenever damage accumulation in the matrix is unacceptable. Uniaxial glass-reinforced laminates have failure strains approaching 3% at 4 K; however, the embrittled matrix will typically fail at less than 1% strain — about 20 to 30% of the ultimate fracture stress. The stress level corresponding to the onset of matrix damage may therefore be lower at cryogenic temperatures than at room temperature. This factor makes type S less functionally attractive compared with type E glass for applications where the increased damping, the slightly decreased modulus, and the dimensional instability accompanying matrix damage cannot be tolerated. Similar functional strength limitations do not exist for laminates reinforced with high-modulus fibers (graphite or boron) for which the ultimate strain to failure is less than 1% at all temperatures.

Figure 12.14 indicates that the graphite-reinforced laminates are much better cryogenic materials than was indicated by prior data. The earlier data suggested that cooling would lower the tensile properties of graphite-reinforced epoxies as a group. Correlation between laminate and fiber properties was not observed and a large scatter in temperature dependency of strength properties was reported. Figure 12.15 shows that this behavior is not confirmed by the more recent studies. It now appears that the effect of cryogenic temperatures on the average uniaxial longitudinal tensile properties is minimal for these materials.

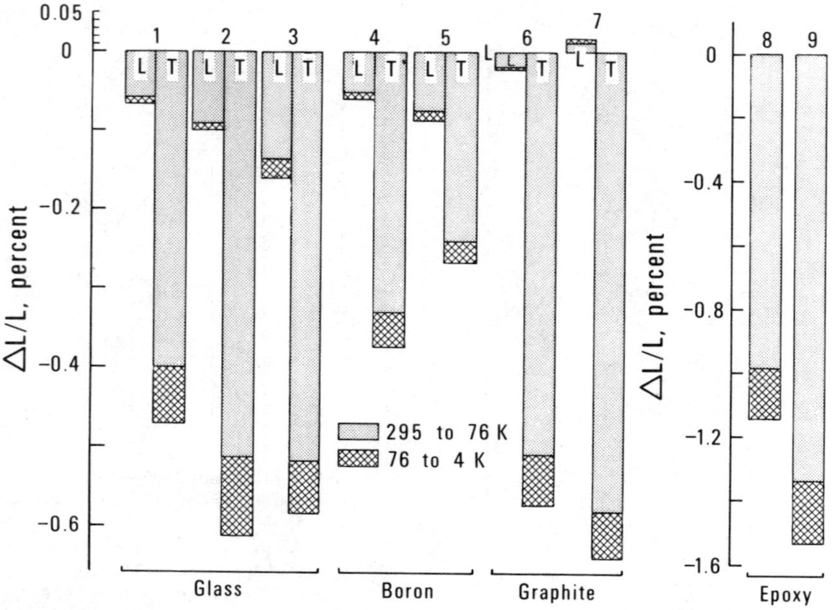

Glass Reinforcement
1–S-901/Resin 2, 66 vol.% (1)
2–R/bisphenol A, 63 vol.% (2)
3–E/bisphenol A, 50.6 vol.% (2)

Boron Reinforcement
4–140 μm (5.6 mil) B/2387 epoxy, 52 vol.% (1,3)
5–140 μm (5.6 mil) B/6061 aluminum, 47 vol.% (1,3)

Graphite Reinforcement
6–AS/Resin 2, 64 vol.% (1,3)
7–HM-S/bisphenol A, 55 vol.% (2)

Epoxy Resins
8–Bisphenol A (2)
9–Resin 2 (4)

Epoxy matrices are a conventional bisphenol A (Bis A) or a flexibilized formulation (Resin 2) (Soffer and Molho, 1967). Vol.% = fiber volume fraction. Refs.: (1) Schramm and Kasen (1977b); (2) Dahlerup-Petersen and Perrot (1979); (3) Schramm and Kasen (1977a); (4) Hust (1980).

Figure 12.19 Temperature dependence of thermal expansion of uniaxial glass-, boron-, and graphite-reinforced laminates in the longitudinal (L) and transverse (T) directions.

Properties of the matrix and of the matrix-fiber bond strongly influence the uniaxial transverse tensile strength. Thus, the very high transverse strength illustrated for boron-reinforced aluminum in Fig. 12.14 reflects the strength of the 6061-aluminum-alloy matrix. The low transverse strength observed for the AS/Resin-2–graphite laminate at 295 K reflects the weakness of this highly flexibilized resin at higher temperatures, whereas the significant decrease in strength on cooling is attributable to a high state of internal stress arising from the high thermal contraction of this resin system combined with the negative coefficient of thermal expansion of the graphite fiber. The average transverse strength of an unflexibilized, conventional, high-strength graphite laminate averages 48 MPa (7×10^3 psi) at 4 K, over 300% higher than that developed by the flexibilized-matrix laminate (Kasen, Schramm, and Beck, 1980).

The temperature dependence of the tensile strength of the unreinforced bisphenol A epoxy illustrated in Fig. 12.14 should be typical of that developed by conventional, fully reacted epoxy–resin systems. Absolute values will differ, however, depending on the formulation of the resin system.

Recent data on the temperature dependence of the uniaxial longitudinal tensile strength of aramid-fiber reinforced epoxy appearing in Fig. 12.14 is in agreement with prior literature (Kasen, 1975b). These data and the newly acquired data on the uniaxial transverse tensile strength appear to be temperature independent.

The aramid laminate data in Fig. 12.14 reflect average properties of a conventional laminate system. The large negative coefficient of thermal expansion of aramid fiber has led in the past to the belief that the cryogenic performance would be improved if a highly flexibilized epoxy resin matrix were used to accommodate the differential contraction. But studies of the performance of an

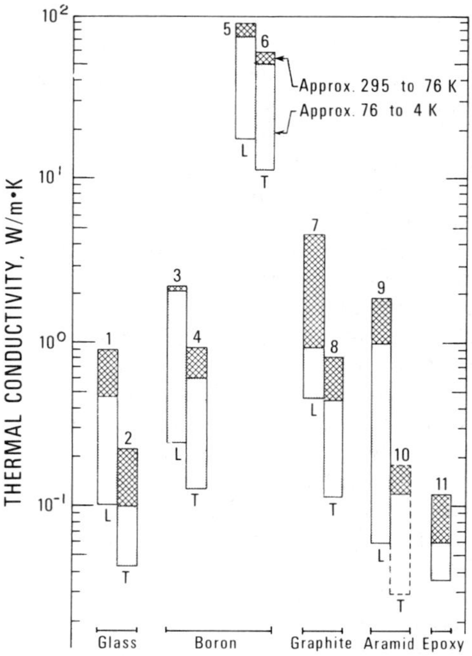

1–S glass/epoxy, 65 vol.% (1)
2–S glass/epoxy, 63 vol.% (2)
3 & 4–140 μm (5.6 mil) boron/epoxy, 52 vol.% (3)
5 & 6–140 μm (5.6 mil) boron/aluminum, 47 vol.% (3)
7–Graphite/epoxy (3)
8–Graphite/epoxy (2)
9–Aramid/epoxy (4)
10–Aramid/epoxy (2)
11–Bisphenol A epoxy (2)

Vol.% = fiber volume fraction. Refs.: (1) Hust and Arvidson (1978); (2) Dahlerup-Petersen and Perrot (1979); (3) Hust (1976); (4) Hust (1975).

Figure 12.20 Temperature dependence of thermal conductivity of uniaxial laminates. L = longitudinal, T = transverse. Transverse conductivity of the aramid laminate assumed from fabric-reinforced data.

aramid-fiber laminate using the Resin 2 system (see Table 12.4) have shown that its use is counterproductive (Kasen, Schramm, and Beck 1980). The transverse tensile strength of this combination (0.3 MPa, 42 psi at 295 K and 0.1 MPa, 17 psi at 76 K) rendered the laminate almost useless for structural applications at any temperature. Again, it is believed to be the weakness of the resin at 295 K and the development of a high state of internal stress at cryogenic temperatures that destroyed the performance of the laminate.

Reliable compressive strength properties are difficult to obtain in composite materials. A primary difficulty is the strong dependence of measured values on specimen alignment and on proper specimen support. The large surface-to-volume ratio of typical specimens accentuates the significance of surface flaws, particularly at low temperatures where the matrix becomes embrittled. Uniaxial longitudinal strengths have a distribution function that is highly skewed toward low values. The magnitude of the scatter increases as the compressive strength increases. Average values of compression data obtained from laboratory tests therefore tend to be systematically biased toward lower values than the true intrinisic material strength. The maximum compressive strength values obtained in a laboratory test series, therefore, may more accurately reflect the true compressive strengths of bulk high-strength composites than does the mean of the distribution. Uniaxial transverse compressive strength data show comparatively little scatter and a relatively normal distribution. Here, the distribution mean is perhaps the most meaningful indication of compressive strength.

Figure 12.16 summarizes recent data (Schramm and Kasen 1977a, 1977b; Kasen, Schramm, and Beck, 1980) on the effect of cryogenic temperatures on compressive strengths. A comparison is made between the uniaxial longitudinal and transverse compressive strengths of epoxy laminates reinforced with glass, boron, graphite, and aramid fibers and aluminum reinforced with boron. All data were obtained by an unsupported-beam test method.

The intrinsic uniaxial longitudinal compressive strength of typical (0.14-mm, 5.6-mil diameter) boron-reinforced laminates is on the order of 3450 MPa (500×10^3 psi) at cryogenic temperatures. Differences in strength of epoxy and aluminum-matrix products are attributable to the noted differences in fiber content. These data are at variance with prior data (Kasen, 1975b), which had indicated a negligible temperature dependence of compressive strength for boron-reinforced aluminum. Recent data also indicate a much higher compressive strength than previously reported in the literature for unsupported-beam tests. The data in Fig. 12.16 approach calculated compressive strengths based on sandwich-beam specimens (*Advanced Composite Design Guide*, 1973). Comparative room-temperature data were not available for these boron-reinforced materials.

Boron-reinforced materials are clearly the

strongest laminates in the transverse as well as in the longitudinal compression. Cooling from 76 to 4 K improves the transverse strength of the aluminum-matrix material, but has little effect on the epoxy-matrix laminates. This reflects the substantial (20 to 40%) increase in strength of the 6061-aluminum matrix between these two temperatures.

An S-901-glass laminate develops about half the compressive strength of a boron laminate, followed in strength by the graphites. Aramid laminate has the lowest compressive strength. The highly flexibilized Resin-2–epoxy matrix appears to have reduced the compressive properties of the S-901 and type-AS-graphite laminates compared with similar laminates using conventional epoxy–resin matrices. For example, room-temperature values reported for the S-901/Resin 2 laminate in Fig. 12.16 are about 20% lower in the longitudinal direction and about 50% lower in the transverse direction than the value reported for a similar laminate using a conventional resin system in MIL-HDBK-17A (1971). Cooling substantially improves the compressive strength of the S-901/Resin 2 laminate at cryogenic temperatures, but improvement in the transverse direction is only sufficient to return the laminate to the strength expected from a conventional laminate at room temperature. The highly flexibilized epoxy–resin also adversely affects the compressive strengths of the laminate reinforced with type-AS-graphite fiber. Despite a high inherent fiber strength, the longitudinal compressive strength of this laminate is reduced to values no higher than the value obtained with the much lower strength GY-70 fiber laminate. Furthermore, a significant reduction in longitudinal compressive strength is noted for the AS/Resin 2 laminate on cooling from 76 to 4 K — a phenomenon not observed in the other laminates. A deterioration in compressive strength is also evident in the transverse direction for this laminate.

The comparatively low longitudinal compressive strength of the aramid fiber laminate reflects low compressive fiber properties. The strength is low both in comparison with the other types of laminates and as a percentage of its uniaxial tensile strength. The difference is not as noticeable in the transverse-compression test mode where strength is more influenced by matrix properties.

Elastic properties. Figure 12.17 summarizes data on the temperature dependence of Young's modulus in tension for uniaxial longitudinal and transverse directions of glass-, boron-, graphite-, and aramid-reinforced epoxy laminates.

Only the aramid-reinforced epoxy laminate displayed a significant temperature dependence in the longitudinal direction. Present values for the aramid laminate are significantly lower than values reported for this type of laminate in the earlier literature review (Kasen, 1975b). The substantial modulus increase reported to occur in aramid-reinforced epoxy laminates below 76 K (Hoggatt, 1971) was not observed.

The longitudinal tensile moduli of glass-reinforced laminates are dominated by the moduli of the glass, which is about an order of magnitude higher than that of the conventional bisphenol A epoxy system illustrated in Fig. 12.17. The longitudinal modulus of a glass-reinforced laminate is, therefore, strongly dependent on the volume fraction of glass in the laminate. It is also influenced by the type of glass. The modulus of type E glass is 10 to 20% lower than that of the S type (Loud, 1979). The high transverse modulus reported for the S-901 glass laminate suggests that the type of glass and the finish on the glass have a strong influence on modulus in this direction.

The S-901 glass laminate displays a large percentage increase in transverse modulus on cooling. This reflects hardening of the highly flexibilized Resin 2 epoxy matrix in this material. However, a comparison of the present S-901 laminate data with handbook values (MIL-HDBK-17A, 1971) shows that the Resin 2 laminate must be cooled to cryogenic temperatures before developing a modulus equal to that produced at room temperature by a conventional S-glass laminate using a nonflexibilized, fully reacted epoxy–resin system.

The boron-reinforced laminate provides an excellent combination of longitudinal and transverse tensile moduli. Doubling of the transverse modulus of the boron/epoxy material is the only significant temperature effect.

The Young's modulus in tension of the AS/Resin 2 laminate is 20 to 25% lower in both the longitudinal and transverse direction than expected for a laminate reinforced with this type of fiber in a conventional, fully reacted epoxy matrix (*Advanced Composite Design Guide,* 1973).

The temperature dependence of in-plane shear moduli summarized on Fig. 12.18 represents the first cryogenic data obtained on this parameter. These data, obtained by the ±45° tensile coupon method (Sims and Halpin, 1974), indicate a substantial increase of average shear modulus for all types of laminates studied. In-plane shear modulus is strongly influenced by the modulus of the matrix. Values reported for the S-901 glass and type AS-graphite laminates in Fig. 12.16 are, therefore, lower than would be expected in conventional laminates owing to the low room-temperature shear strength of the highly flexibilized Resin-2–epoxy formulation used in their fabrication.

Thermal properties. Data on the effect of cryogenic temperatures on dimensional stability and thermal conductivity of uniaxial laminates are summarized in Figs. 12.19 and 12.20.

Figure 12.19 summarizes data on the effect of temperature on dimensional stability of boron-reinforced aluminum and well-characterized epoxy-matrix laminates reinforced with glass, boron, and graphite. Data are also provided on the stability of a typical, fully reacted bisphenol A epoxy–resin system and on the highly flexibilized Resin 2 epoxy–resin system used with several of the laminates.

The thermal contraction of the laminates is much less than that of the matrix materials. Particularly in the fiber (longitudinal) direction, the reinforcement properties dominate the contraction because the reinforcement modulus is much higher than that of the matrix. The longitudinal modulus is, therefore, dependent on the fiber volume fraction as well as on the type of fiber, with the matrix playing a minor role.

Thermal stability in the transverse direction of uniaxial laminates is influenced by both the matrix and reinforcement properties. A further analysis of the data shows that the transverse contraction of the glass- and graphite-reinforced bisphenol A laminates (fiber types R, E, HM-S) was about 52% of that expected for the unreinforced bisphenol A resin, whereas the glass- and graphite-reinforced Resin 2 laminates (S-901, AS) contracted about 35% of that expected from unreinforced Resin 2. The lesser contribution of the Resin 2 matrix to the laminate contraction is due to the low modulus of this highly flexibilized resin at temperatures above the cryogenic range. Comparable values for the boron/epoxy and boron/aluminum laminates are 32% and 63%, respectively. Here, the comparatively low transverse contraction of the epoxy laminate is due to the very high modulus of the boron, compared with that of the epoxy. The aluminum contributes more to determining the contraction of the laminate because its modulus is much higher than that of epoxy.

The temperature dependences of thermal conductivity for uniaxial laminates in the longitudinal and transverse direction are summarized in Fig. 12.20. Data for an unreinforced conventional bisphenol A epoxy is included for comparison. All of the epoxy-matrix laminates used a conventional epoxy system. The boron/aluminum laminate has a 6061 aluminum alloy matrix.

Uniaxial laminates are most frequently used as thermal standoffs in cryogenic technology because of the low conductivity of the epoxy-matrix products combined with excellent strength properties. Glass-reinforced epoxy composites are favored because they provide lower conductivity in the fiber direction than do boron-, graphite-, or aramid-reinforced composites. Boron- and aramid-reinforced laminates appear to have about the same conductivity in the longitudinal direction — approximately 20% higher than that of the glass laminate. The graphite-reinforced epoxy laminate has a longitudinal conductivity about 45% higher than that for glass reinforcement. The very high conductivity of the aluminum-matrix laminate limits its use to applications having minimum thermal gradients or to specialized applications where a high thermal conductivity is desired.

Transverse conductivities are all lower than in the longitudinal direction regardless of the relative conductivities of the fiber and matrix. This reflects the thermal resistance of the fiber-matrix interface in the transverse direction.

The aramid-reinforced epoxy laminate displays the largest change in thermal conductivity on cooling. The transverse conductivity data plotted for this material in Fig. 12.20 were measured in the direction normal to the reinforcement of a fabric-reinforced laminate. It is, therefore, only an approximation to the transverse conductivity of a uniaxial laminate. But the approximation should be quite good for similar fiber contents in the two types of laminates.

12.5 COMPOSITE JOINTS IN CRYOGENIC STRUCTURES

12.5.1 General

Replacement of metals with composite materials for structural and nonstructural applications in superconducting MHD and rotating superconducting generator systems will require composite-to-composite and composite-to-metal joints. This can result in a complex load-bearing configuration creating various levels of stress and strain discontinuity within a structure. Added complexities will arise from thermal contraction of mating parts in joints intended for cryogenic service.

This section provides an introduction to the various joining techniques developed for composite materials. Data on the performance of joints at cryogenic temperature are comparatively rare, but an examination of successful techniques used at room temperature will provide a basis for assessing the added problems arising at very low temperatures.

12.5.2 Joining techniques—room temperature experience

There are two broad classes of joining techniques applicable to composite structures and a third that is a combination of the two. Mechanical joints are characterized by a hole and a fastener, such as a rivet, pin, or bolt. Bonded joints are characterized by the use of adhesives or diffusion bonding, in the case of metal-matrix composites. A combination of bonding and bolting can be effectively used for some applications. Selection of a joint for a particular application is usually a compromise between strength requirements, reliability, flexibility in component disassembly, and cost.

Mechanically fastened joints. Typical bolted-joint configurations are illustrated in Fig. 12.21. All are applicable to composites as well as to metals. Bolting is most often used because it offers flexibility in joining materials having diverse elastic and thermal properties. Bolted joints provide high resistance to complex loading and are relatively inexpensive. Mechanically pinned joints are rarely used to join sheets of composites.

A disadvantage of mechanically fastened composite joints is the degradation of the laminate strength adjacent to the holes or cutouts required for the bolts or pins. The 50 to 80% reduction in useful laminate strength usually requires local reinforcement of the cutout region with metal inserts, doublers, or local ply buildups. The highest possible bolted joint strength for a graphite–epoxy composite is reported to be on the order of 40 to 50% of the basic laminate strength (Hart-Smith, 1976), but this applies only to joints of the type illustrated in Fig. 12.21. The mechanical

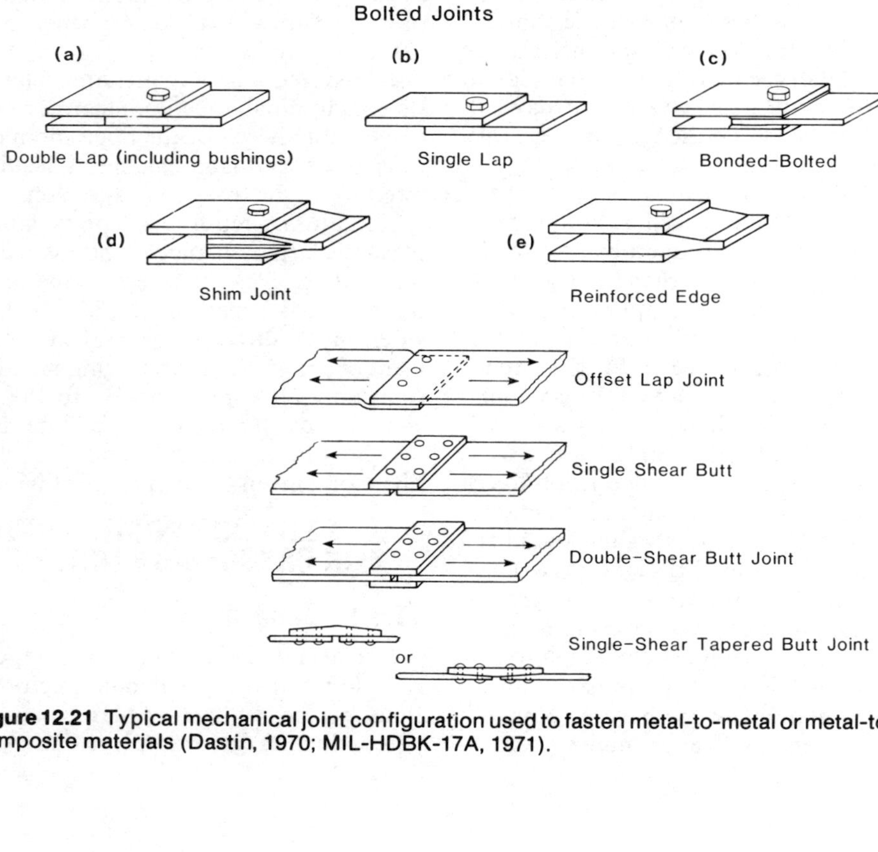

Figure 12.21 Typical mechanical joint configuration used to fasten metal-to-metal or metal-to-composite materials (Dastin, 1970; MIL-HDBK-17A, 1971).

gripping system developed for testing composites at cryogenic temperatures (see § 12.7.2) develops 100% mechanical efficiency.

The static failure modes for composite joints illustrated in Fig. 12.22 are similar to those of metals, but the lack of ductility in graphite- or boron-reinforced epoxy laminates affects the failure path. Boron-reinforced laminates tend to produce a linear elastic type of failure, whereas graphite-reinforced laminates often show substantial deformation around the bolt hole (true bearing failure) as well as tension and shear failure. Details of the failure mode are strongly dependent on the stacking sequence in the laminate.

The failure mode for a given laminate is also affected by the joint geometry reflected in the design parameters. The significant factors (illustrated in Fig. 12.23) are edge distance, fastener pitch, joint width, and laminate thickness. The type of fastener is also influential in determining joint strength and failure mode. For graphite-reinforced laminates subject to bearing failure, joints made with standard bolts are found to be significantly stronger than flush-pinned joints (Hart-Smith, 1976).

Wherever possible, mechanical joints should be designed to fail in bearing rather than in tension or shear, thereby reducing the probability of catastrophic failure. This may be difficult to achieve at cryogenic temperatures in polymer-matrix laminates. The detailed considerations needed in designing an effective bolted joint and a simplified procedure for undertaking the design have been addressed by Waszczak and Cruse (1975).

Corrosion is not a problem at cryogenic temperatures, but may occur during component storage at room temperature. The possibility of galvanic corrosion between metal fasteners and the fibers in graphite-reinforced laminates may have to be considered.

Adhesive-bonded joints. The use of organic adhesives for bonding advanced composites during fabrication of aerospace and automotive structures is increasing, concomitant with an increase in understanding of the factors affecting the performance of such joints. Adhesive bonding is now a feasible joining method for demanding applications.

Adhesive-bonded joints have lower stress concentrations and improved fatigue resistance. Weight savings can be significant. Mechanical

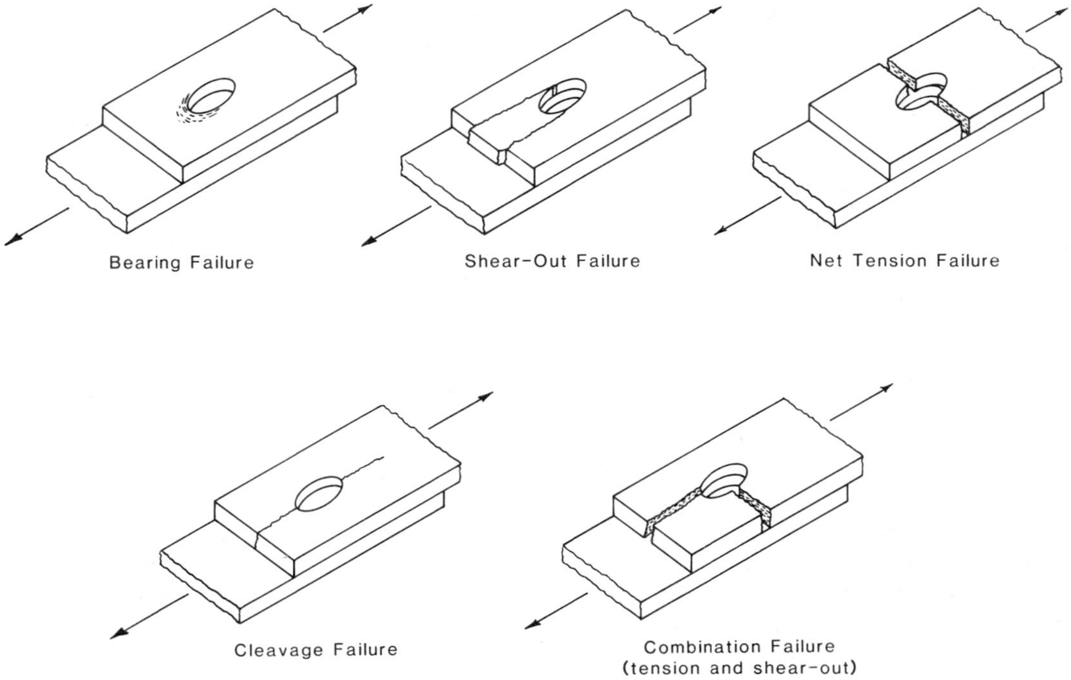

Figure 12.22 Typical failure modes of bolted or pinned mechanical joints between advanced composite laminates (*Advanced Composites Design Guide*, 1973; Hart-Smith, 1976).

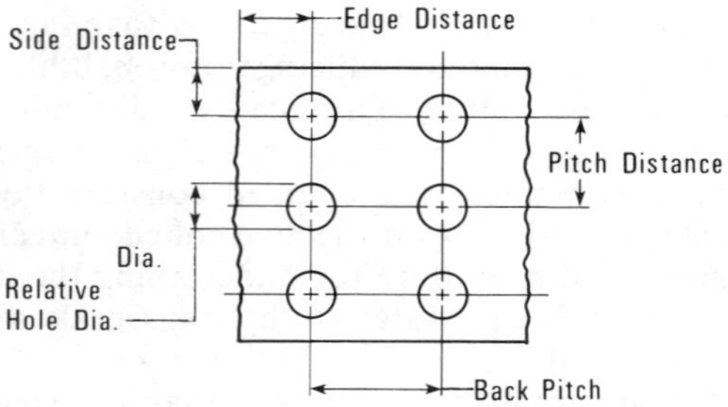

Figure 12.23 Significant parameters affecting the design of bolted joints in advanced composite materials (MIL-HDBK-17A, 1971).

efficiencies can approach 100%. Leakproof joints can be made. Corrosion problems are reduced and cost savings may be realized over bolted joints in large structures.

On the other hand, adhesive-bonded joints are more susceptible to failure by peeling than are bolted joints. They are difficult to inspect by nondestructive methods, although significant progress has been made (Alers and Elsley, 1977). Disassembly is difficult if not impossible. Differences in moduli of the mating components (adherends) can accentuate shear loads and significantly reduce the strength of the bonded interface. Thermal expansion mismatch can create similar problems in joints subjected to varying temperatures.

Major parameters affecting adhesive-bonded joint design and the failure modes of such joints at room temperature have been addressed in several publications (Dastin, 1970; MIL-HDBK-17A, 1971; *Advanced Composite Design Guide*, 1973; Hart-Smith, 1974). Primary considerations are shear loading parallel to the adherend surface and tension stresses resolved normal to such surfaces. Shear loading determines the efficiency of the joint in tension. Stresses normal to the bond interface can cause failure in a peeling or cleavage mode. The specifics of the failure mode are dependent on the relative magnitudes of the tension, shear, and bending loads imposed on the joint. These change with the size of the joint in a particular configuration.

Principal design parameters are joint width, length of overlap (interfacial area), and proper balance of modulus, Poisson's ratio, and thermal expansion of the adherend and adhesive. The strength of the interfacial bond is dependent on the care taken in preparing the adherend surface as well as on the proper choice of adhesive. Thin adhesive layers may provide better impact strength. Thin layers are much more resistant to cracking under thermal shock.

Typical bonded-composite-joint configurations are illustrated in Fig. 12.24. The elastic-plastic mechanical efficiency of a joint (joint strength expressed as a percentage of adherend strength) is determined by the proportion of the applied load transferred into the adherends at stress rupture. The elastic efficiency is determined by the applied load at the yield point of the adherends.

Hart-Smith (1974) has reviewed factors that must be considered in design and analysis of various types of adhesive-bonded joints. Stress and strain singularities across the joint are dominating factors in determining mechanical efficiency. For example, eccentricity of the load path creates stress concentrations that prevent simple lap joints from ever developing strengths equal to that of the adherends. This type of joint is therefore restricted to relatively minor structural applications. Conversely, symmetrical double-lap joints can be designed with 100% efficiency up to some determinable adherend thickness. The limit is dependent on the adherend and adhesive properties, specifically, on their ability to resist failure by peeling.

Scarf joints offer the greatest possible structural efficiency because the scarf angle can, in principle, always be made small enough to accommodate any required load transfer. Scarf joints are not inherently thickness limited. Their fabrication costs are comparatively high, however, limiting their application to thicknesses or designs inappropriate for double-lap joints. Precise alignment requirements also pose problems in many structures. Scarf joints are most effective when the composite is the type reinforced with extremely small fibers such as graphite, glass, or aramid. Scarf joints do not work well with boron-reinforced composites because of the difficulty of machining small scarf angles without breaking the ends of the boron fibers.

Increasing thermal mismatch between metal and composite adherends is a significant factor to consider in joints intended for cryogenic service. Fortunately, the data of Hart-Smith (1974) indicate that thermal mismatch or decreasing adhesive ductility has a severe effect only for intermediate overlaps in a scarf joint. Their effect can be neglected for very long overlaps.

Stepped-lap joints are widely used in aerospace technology. Joint efficiencies of 100% can be obtained, fit-up precision requirements are

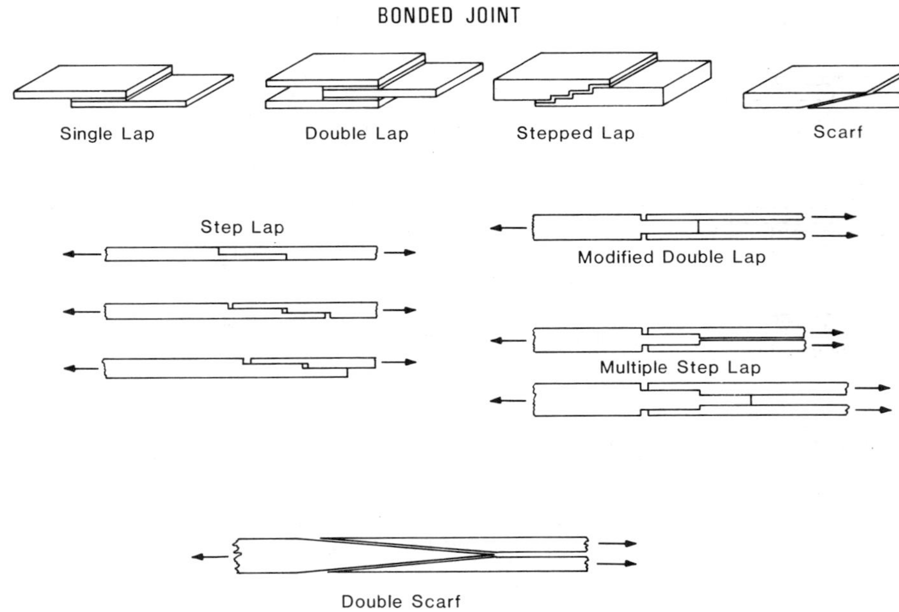

Figure 12.24 Typical adhesive-bonded joint configurations used with cured advanced composite laminates. Not shown are multiple-step lap joints in which the adhesive bond is cocured with the laminate to make a contiguous joint (*Advanced Composites Design Guide*, 1973; MIL-HDBK-17A, 1971).

less stringent than for scarf joints, and problems owing to breakage of the ends of boron fibers do not exist. Stepped lap joints are efficient only up to a defined thickness because the conditions leading to failure at each step are the same as for a simple lap joint. Computer programs encompassing both elastic and elastic-plastic failure modes have been developed to aid in joint design and analysis (Hart-Smith, 1973). There is some evidence that embrittling of the adhesive in a cold environment can drastically reduce both the elastic and elastic-plastic strength of this type of joint.

Load-bearing organic adhesive-bonded joints typically fail in a linear-elastic, brittle manner. The fracture mechanics approach to assessing the performance of structural adhesives has been developed (Ripling, Mostovoy, and Corten, 1971), and the results have been applied to furthering the understanding of adhesive-bonded joint failure under complex loading (Mulville, Hanston, and Mast, 1978; Bascom, Cottington, and Timmins, 1976). Assessment of the reliability of adhesive-bonded joints under both static and fatigue conditions has also received considerable attention (Wolff and Lemon, 1972; Dauksys, 1973).

Figure 12.25 summarizes the general room-temperature performance expected from various types of typical adhesive-bonded lap joints of precured laminates.

Combined bolted and bonded joints. Figures 12.26 and 12.27 illustrate that combined adhesive bonding and mechanical fastening can significantly improve the performance of composite-to-metal joints over that of bonding or bolting alone. This result is not always observed, however, since the fiber orientation in the laminate plays a dominant role.

12.5.3 Cryogenic experience

General considerations. Commercial structural adhesives of interest to cryogenic technology include epoxies, epoxy copolymers with nylon, phenolics, silicones, or polyamides, polyurethanes, polyimides, and polyheterocyclics. They may be obtained as unsupported calendared or cast films, as resin-impregnated (B-staged) carrier cloth, or in paste or putty form.

The reviews by Hillig (1975a, 1975b) indicate that epoxies, epoxy copolymers with nylon, and polyurethanes provide the best overall cryogenic adhesive performance. Nevertheless, the selection of the appropriate adhesive depends on how

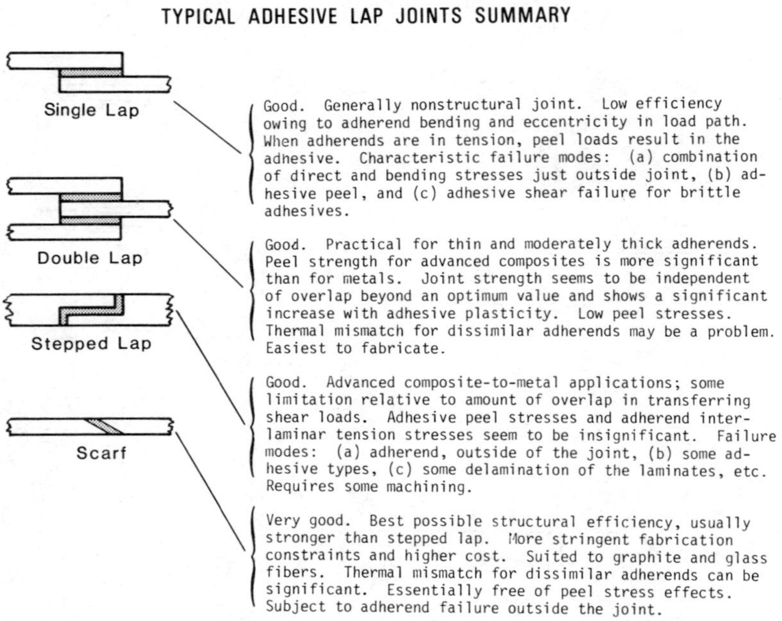

Figure 12.25 Summary of performance characteristics of typical adhesive-bonded joints between advanced composite laminates at room temperature (Hart-Smith, 1974).

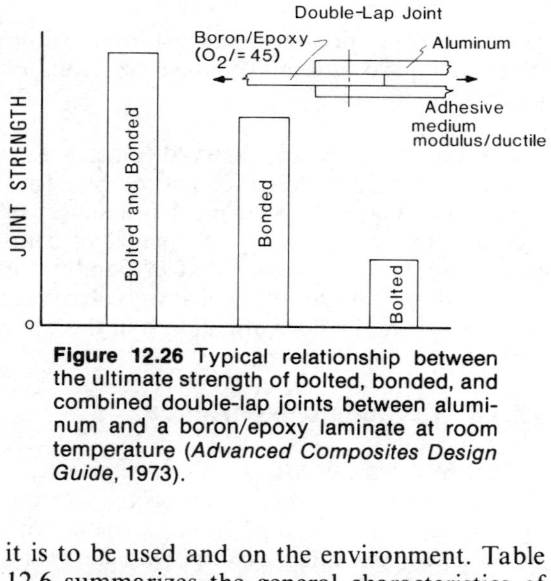

Figure 12.26 Typical relationship between the ultimate strength of bolted, bonded, and combined double-lap joints between aluminum and a boron/epoxy laminate at room temperature (*Advanced Composites Design Guide*, 1973).

it is to be used and on the environment. Table 12.6 summarizes the general characteristics of the various adhesive systems with metal adherends.

General factors considered in selecting an adhesive are initial and long-term strength, chemical inertness with adherends, low volatility, low degassing, and ease of fabricability. Specific applications introduce special considerations. For example, a specified performance is required under electromagnetic radiation or under neutron and gamma irradiation at cryogenic temperatures in magnetic fusion energy technology.

Among the conventional properties, the adhesive modulus and coefficient of thermal expansion are of particular importance for cryogenic structures. A high-modulus adhesive will perform well under tensile or shear loading, but it will be less tolerant to peeling loads, shock, and vibration. Flexibilized adhesives may provide increased fatigue life, resistance to impact loading, and resistance to thermal shock at room temperature, but strength and creep resistance will be lower than for fully reacted systems. Highly flexibilized adhesives should be used with great caution at cryogenic temperatures. Experience with using such formulations for laminate matrices indicates that the much higher thermal contractions of such systems, compared with unflexibilized resins, can induce sufficient residual stress to seriously affect cryogenic joint performance (Kasen, Schramm, and Beck, 1980).

At room temperature, failure in a composite-to-metal joint is generally found to occur in the adhesive film (cohesive failure) when a high-modulus adhesive is used. A low-modulus adhesive permits more load transfer into the laminate, tending to induce interlaminate failure in the composite near the joint interface. At cryogenic

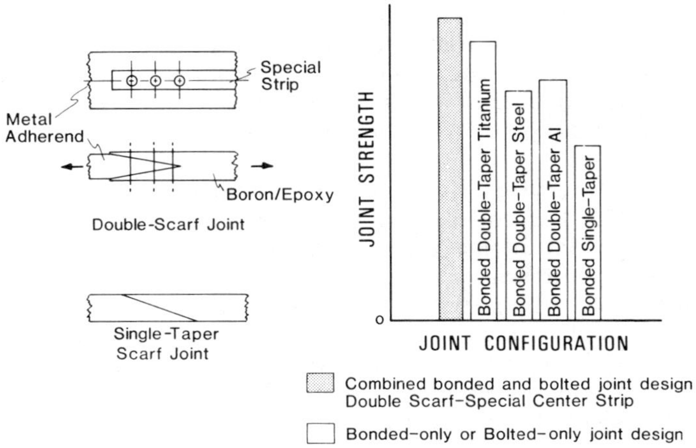

Figure 12.27 Typical relationship between the ultimate strength of single- and double-taper scarf joints at room temperature (*Advanced Composites Design Guide*, 1973)

Table 12.6 General characteristics of adhesive systems with metal adherends.[a]

System	Characteristics
Epoxies	Easy to use, relatively insensitive to bond-line thickness, stronger than epoxy-nylon at elevated temperatures but weaker at cryogenic temperatures.
Epoxy-Nylons	Unusually high shear strength, high toughness but degrades in humid environments.
Polyurethanes	Higher shear strengths than epoxy-nylons but more difficult to process.
Polyimides and Polyheterocyclics	Very good elevated-temperature adhesion and good overall performance.

[a]Hillig (1975a, 1975b).

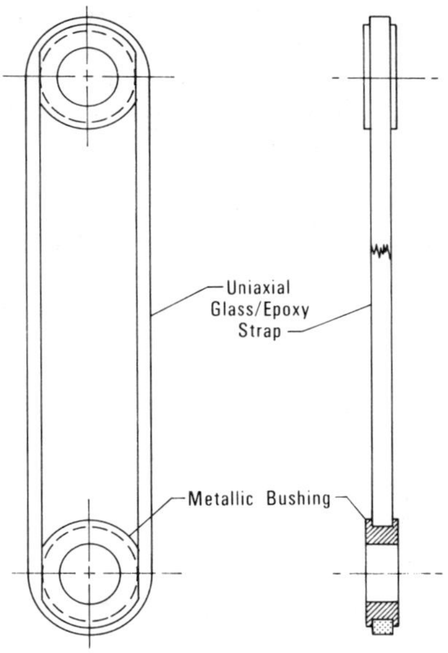

Figure 12.28 Typical design of a uniaxial glass/epoxy tensile strap for thermal isolation of cryogenic equipment.

temperatures, all adhesives have strain capability equivalent to high-modulus types. The expected failure mode is, therefore, peeling or cleavage regardless of the adhesive type.

Dependence of adhesive performance on the composition and geometry of the joint (particularly on thickness of the bond) requires evaluation of performance in simulated joint testing.

Mechanically fastened joints. Pinned racetrack-shaped mechanical joints, such as those illustrated in Fig. 12.28 are being successfully used to isolate thermally helium Dewars and superconducting magnets from the high-temperature environment in aerospace applications. The straps are always stressed in tension. Small straps are typically uniaxially filament-wound with type S-901 glass roving. The tension–tension fatigue life of such straps has been found to be an order of magnitude greater at 4 K than at 295 K (Tobler and Read, 1976). But damage accumulation (fatigue cracks and fiber delamination) was detected at the observed points of tangency between the composite and the busing at about 25% of the fatigue life, regardless of temperature. Static tests also indicated the onset of damage at about 20% of the ultimate joint strength in samples tested at 4 K. This is believed to correspond to the strain limit of the epoxy matrix at that temperature. No fatigue limit was detected for this type of joint up to 10^7 cycles at 295 K.

Niemann, Gonczy, Hoffman, Mataya, Smelzer, and Young (1980) have reported initial test results of similar designs for thermal standoffs in larger superconducting applications. Cost factors dictated construction with a type 143 glass fabric. Initial results indicated an average stress concentration factor of 1.6 at the pin. This type of thermal standoff joint must operate satisfactorily for long times under a tension preload. Therefore, creep performance at 295 K was evaluated. The first designs had an unacceptably high creep rate attributable to the selection of a highly flexibilized epoxy resin for the construction. As discussed in § 12.4, this type of resin should be used with great caution, since both experience and laboratory data show its properties to be inferior to fully reacted systems at both room and cryogenic temperatures.

Adhesive-bonded joints. The majority of research on the effect of cryogenic temperatures on the performance of adhesive-bonded joints has concentrated on metal-to-metal configurations. Mauri (1968) has reviewed the progress until 1968. Hillig (1975a, 1975b) published an updated review of the same subject. These publications show that excellent progress has been made in the development of adhesives tolerant of cryogenic temperatures. Figure 12.29 illustrates the strength of 2024 aluminum bonded by various classes of adhesives.

As of 1980, the only systematic study of the effect of cryogenic temperatures on the performance of metal-to-composite joints is that of Hillig (1975a, 1975b). This study evaluated the effect of composite laminating procedure and cure procedure on the lap and short-beam shear strength of S-glass/epoxy laminates bonded to 6061 aluminum. The bonds were cocured with the laminate. The "Polaris" type epoxy resin (Epon 828/Epon 1031/NBA/BDNA in proportions 50/50/90/0.55 by weight) therefore served as the joint adhesive as well as the laminate matrix. The surface preparation of the 6061 adherends was held constant.

Figures 12.30 and 12.31 show that cooling to 4 K increased the mean strengths, but scatter in the data negated significant differences. This in-

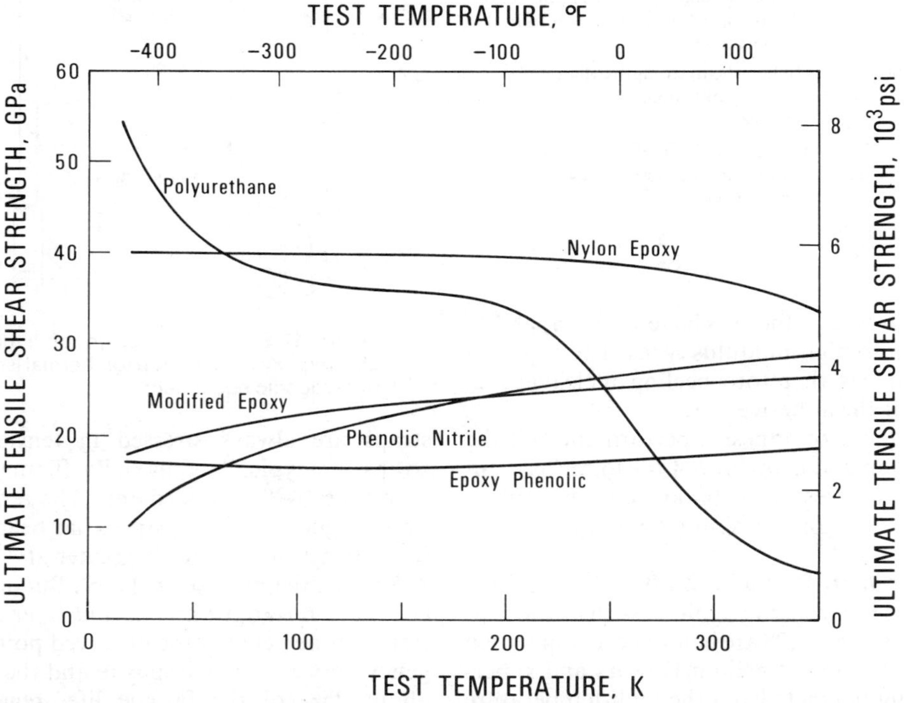

Figure 12.29 Effect of cryogenic temperature on the tensile shear strength of various classes of structural adhesives. Aluminum adherends (Mauri, 1968).

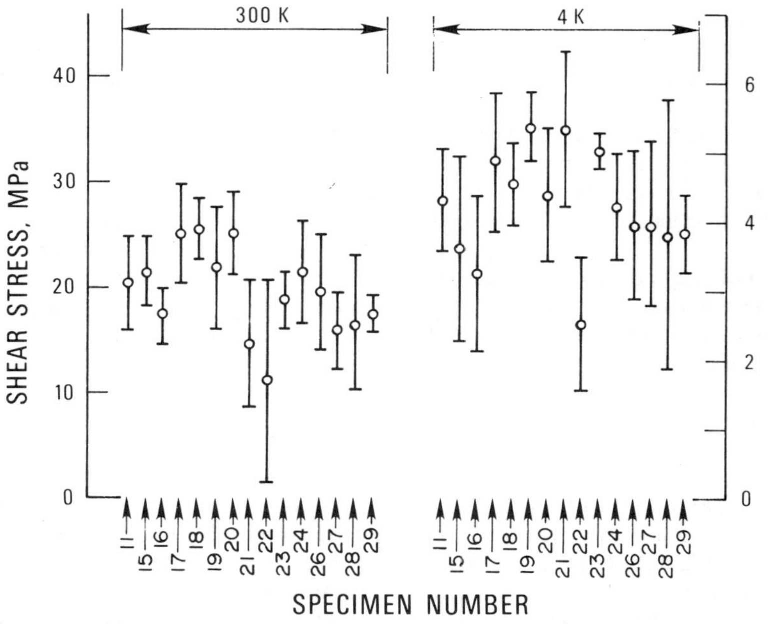

Figure 12.30 Lap shear stress at 300 and 4 K for 6061 aluminum bonded onto uniaxial S-90l glass laminates during laminate cure. The matrix resin provided the adhesive. Specimen numbers define various surface treatments. Range shown is the standard deviation (Hillig, 1975a, 1975b).

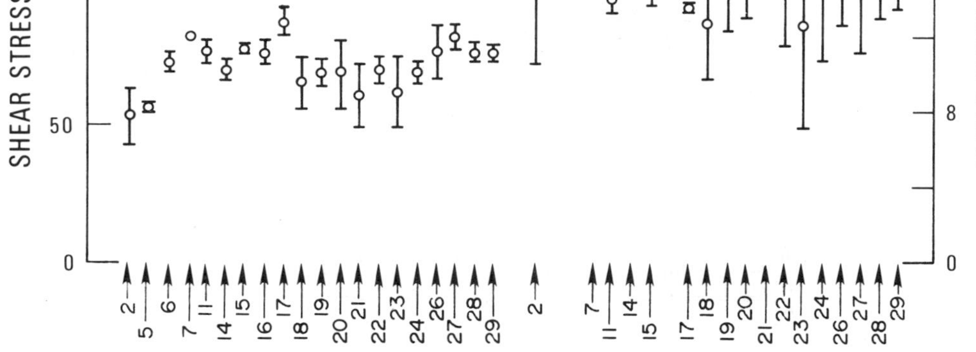

Figure 12.31 Short-beam shear stress at 300 and 4 K for 6061 aluminum bonded onto uniaxial S-901 glass laminates during laminate cure. The matrix resin provided the adhesive. Specimen numbers define various surface treatments. Range shown is the standard deviation (Hillig, 1975a, 1975b).

crease in scatter on cooling follows the general pattern observed in the other laminate strength parameters evaluated under this contract. The greater variability at cryogenic temperatures is believed to reflect the decrease in toughness and increase in notch sensitivity of conventional, fully reacted epoxy–resin systems when cooled substantially below their glass transition temperature.

These data provide added evidence that the greatest benefit to improved cryogenic performance of polymer-matrix laminates will result from development of a more strain-tolerant cryogenic resin.

Barclay, Bell, Brogren, and Strayer (1975) have provided a rare insight into the performance of an adhesive-bonded joint between an aramid fiber laminate and 6Al–4V titanium after thermal cycling to cryogenic temperatures. The experimental design (illustrated in Fig. 12.32) actually falls within the combined bonded and mechanically fastened category, because the loading pin passes through the composite laminate. The fiberglass/epoxy surface layers bind the uniaxial aramid fibers together and provide transverse strength to the strap. It is always wise to add some crossply reinforcement to uniaxial laminates.

The bond between the alumina-blasted and silane-rinsed titanium and the aramid laminate was made with a proprietary adhesive. Both end-fitting joint areas were thermally shocked to 76 K and proof tested in tension at room temperature to 16.01 kN (3600 lbf) to determine the elongation. The specimen was tested to failure at 21.59 kN (4850 lbf). These data met the performance objectives for the intended pressure vessel application.

12.5.4 Summary

The existing data base for the design and performance of composite-to-composite and composite-to-metal joints at room and elevated temperatures provides a basis for extending the technology to cryogenic temperatures. Most efforts thus far have been devoted to the development of relatively simple pinned mechanical joints or equally simple bolted flange joints as

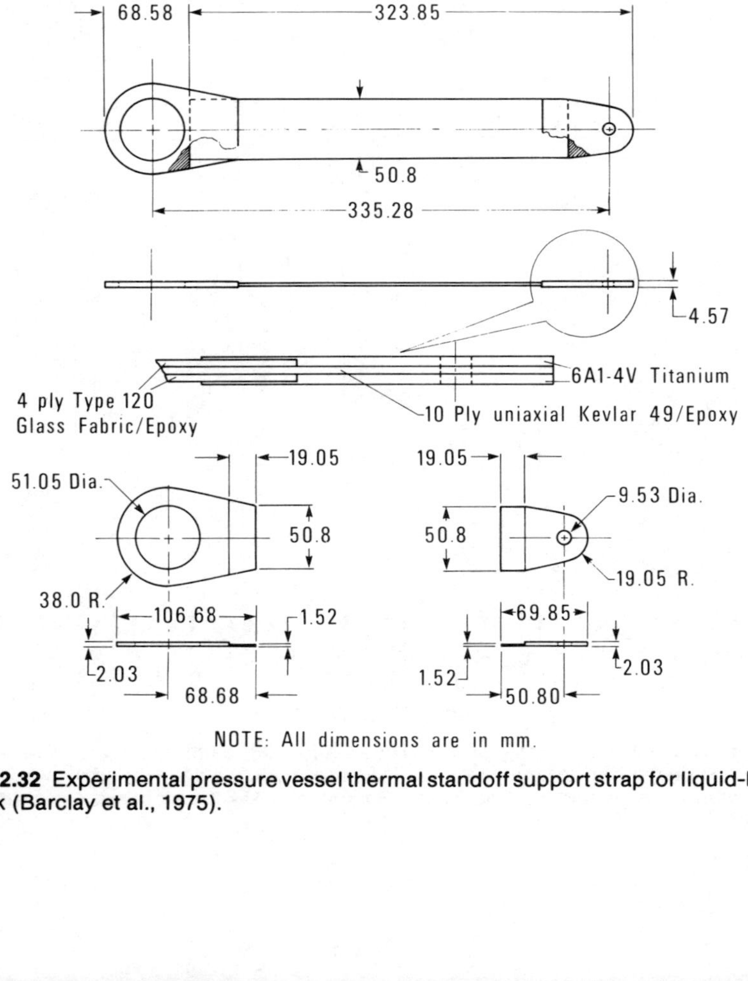

Figure 12.32 Experimental pressure vessel thermal standoff support strap for liquid-hydrogen fuel tank (Barclay et al., 1975).

required for thermal isolation members, but existing data and the results of a few studies suggest that adhesive-bonded joints, perhaps mechanically reinforced, can be designed to provide excellent cryogenic performance. The primary problems of thermal stresses and of adhesive embrittlement are more severe at cryogenic temperatures, but they are extensions of problems already recognized and addressed in existing technology.

The requirements that will be made on the performance of joints in composite components of MHD, rotating cryogenic machinery, and other high technology cryogenic systems can only be met by evaluating the essential performance elements of joints carefully prepared from well-characterized adherends and adhesives. The overall approach should have the same rationale as the development of standard grades of cryogenic composite laminates. It is first necessary to provide the designers and fabricators with basic joint systems having defined cryogenic performance and a minimum of performance variability. These systems will then form the basis for systematic future development of improved systems reflecting proven deficiencies and the need for specific functional improvement.

12.6 RADIATION EFFECTS

The resistance of nonmetallic materials and composites to damage by neutron and gamma irradiation at 4 K is an important consideration in selecting insulating and structural support materials for superconducting magnets of magnetic fusion energy systems. Current designs anticipate that glass-reinforced polymers and other nonmetallics could be exposed to lifetime dose rates equivalent to about 10^{10} rads in some critical areas (Coltman et al., 1979). The ability of conventional organic-matrix materials to withstand such doses is a significant concern.

A recent review (Brown, 1980) has noted that a number of studies have been conducted on the effect of irradiation on the room-temperature properties of nonmetallic insulators, but few studies have extended into the cryogenic range. In addition to degrading mechanical, thermal, and electrical properties, the low thermal conductivities of the nonmetallics may also result in nuclear heating, placing an added burden on the refrigeration system.

Extrapolating from room-temperature experience suggests that epoxy systems cured with aromatic amines should have better resistance than systems cured with dicyandiamide, polyimides should be better than epoxies, and filled resin systems should be superior to unfilled resins. But confirmation is lacking. Brown (1980) concluded that the complexities of the processes involved did not permit a theoretical estimate to be made of the effect of cryogenic radiation on the various properties of nonmetallic insulators — an experimental data base must be acquired. Because the radiation resistance of a polymer is strongly affected by the exact molecular structure, such experiments must be carried on with materials characterized on the molecular level.

There is a fallback position from organic insulators, should they prove to be entirely inadequate. Inorganic insulators, such as metal oxides or glasses, are known to be relatively unaffected by fast neutron irradiation (Brown, 1980), but brittleness and cost of fabrication have limited their application. Such materials will play an increasing role if organic insulators prove inadequate under service condition.

Researchers at Oak Ridge National Laboratory have recently screened the effect of 4-K gamma and neutron irradiation on the stability of eight types of organic insulation (Kernohan et al., 1979). The volume resistivities of a bisphenol A epoxy, an inorganic-filled epoxy, an aromatic-polyimide sheet material, a polyvinyl formal wire coating, and an FR-5 type glass–epoxy laminate were found to be very little affected by a dose of 2×10^8 rads (γ). The effectiveness of a filled epoxy and a B-staged glass–epoxy adhesive for joining Cu–Be lap-shear tensile specimens was also relatively unaffected, as was the flexural strength of a filled epoxy, a G-10 laminate, and an FR-5 laminate.

When dosages were increased to higher levels, however, the mechanical and electrical properties of glass-reinforced epoxy laminates and of filled epoxy resins were found to be significantly deteriorated (Coltman et al., 1979). Data from this study, summarized on Figs. 12.33, 12.34, and 12.35, show residual flexural and compressive strengths of G-10CR and G-11CR laminates reduced to 10 to 15% of original values after exposure to 2.4×10^9 rads (γ) and to 6 to 10% after exposure to 1.0×10^{10} rads (γ). These dosages were accompanied by fast neutron dosages of 2.2×10^{20} n/m^2 and 8.7×10^{20} n/m^2, respectively. The strength retention of the aromatic-amine cured G-11CR product appeared better than that of the dicyandiamide-cured G-10CR after the lower dose, but no difference was noted after the

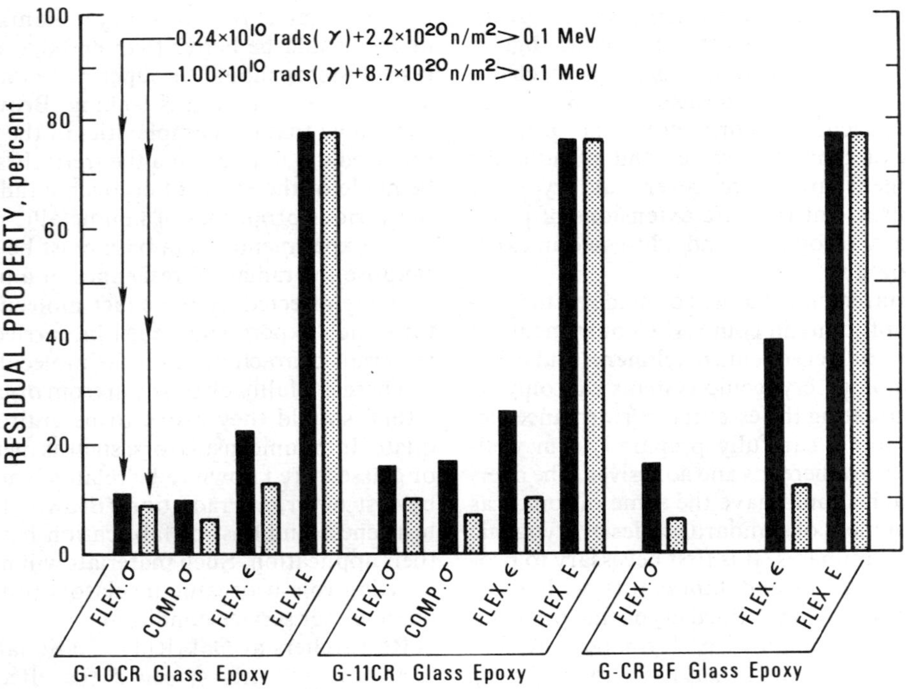

Figure 12.33 Percentage of original mechanical properties retained by G-10CR, G-11CR, and G-10CR containing boron-free glass cryogenic-grade glass-fabric-reinforced industrial laminates after two levels of irradiation at 4 K. Tested at 78 K after warm-up to 307 K. (Coltman et al., 1979). Flex. σ = flexural strength; Flex. ϵ = flexural strain; Comp. σ = compressive strength; Flex. E = flexural modulus.

higher dose. A variant of the G-10CR product containing boron-free E-glass reinforcement performed much like the G-11CR laminate.

The relatively high fraction of initial flexural modulus retained by the laminates indicates that the glass reinforcement did not sustain major damage. The low strain to failure indicates that deterioration of the matrix — and possibly of the fiber-matrix bond — accounts for most of the property degradation.

The mechanical properties of the filled resins were relatively unaffected by the lower level of radiation, but were substantially degraded at the higher level. Coltman et al. (1979) concluded that these materials were essentially at the end of their useful life after the higher level of radiation. Despite the larger percentage of strength reduction, the high initial strength of the glass/epoxy laminates permitted retention of sufficient residual strength (100 MPa, 15×10^3 psi flexural strength; 55 MPa, 8×10^3 psi compressive strength) for some magnet applications.

Except for one of the filled epoxies, no significant dimensional changes were observed as a result of the irradiation.

Warming of the specimens to room temperature after 4-K irradiation simulated periodic warming of the magnet structures throughout their lifetime and incorporated additional damage anticipated from defect recombination at higher temperatures — the Wigner energy.

It is apparent that much work remains to be done before reliable radiation-resistant insulators are developed. Current programs are examining the type of molecular damage incurred during irradiation of epoxy-matrix laminates and testing is being extended to include glass-reinforced polymide-matrix laminates.

12.7 TESTING LAMINATES AT CRYOGENIC TEMPERATURES

12.7.1 General

The mechanical properties of a laminate are functions of the quality of the components from which the laminate is fabricated. Test materials should be well characterized. The reinforcement, the resin system, and the method of fabrication should be documented. The laminate fiber content (ASTM D 3171-73) and the void content

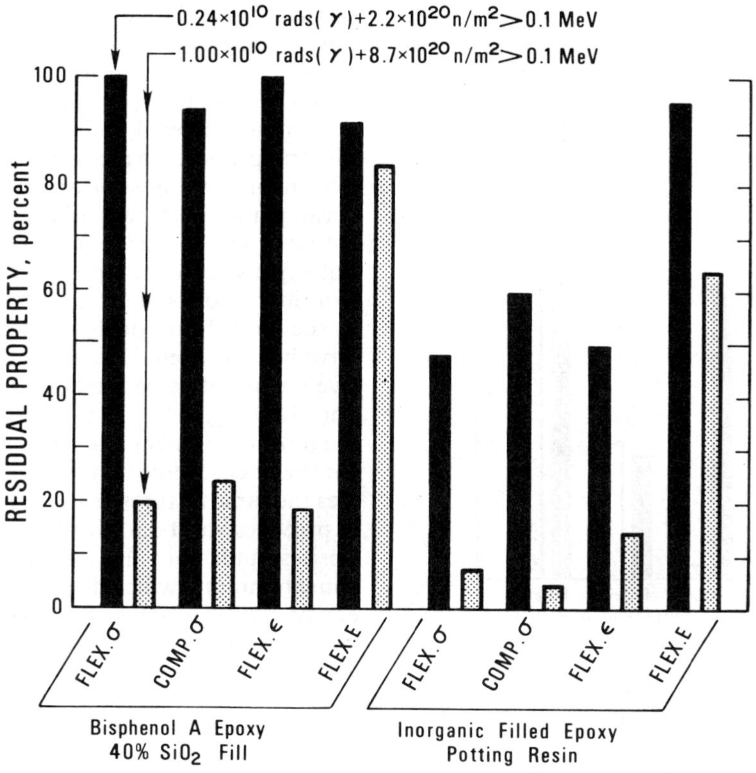

Figure 12.34 Percentage of original mechanical properties retained by two types of filled epoxies after two levels of irradiation at 4 K. Tested at 78 K after warm-up to 307 K. (Coltman et al., 1979). Flex. σ = flexural strength; Flex. ϵ = flexural strain; Comp. σ = compressive strength; Flex. E = flexural modulus.

(ASTM D 2734-70) should be established. This characterization should accompany all published data. ASTM committees are actively working to establish standard test methods; however, those that have been published must frequently be modified for cryogenic temperatures.

Obtaining valid mechanical property data on laminates is more difficult than for metals, particularly at cryogenic temperatures. A primary problem is transferring an applied load uniformly to the specimen. Improper gripping procedures produce stress concentrations, which cause premature specimen failure. It is easy to obtain data that tell a good deal about the efficiency of the test fixture, but not much about the properties of the laminate.

The high cost of testing at cryogenic temperatures requires a very high success rate accomplished with fixtures designed to conserve expensive cryogenic liquids. The cryostat required for static and dynamic mechanical testing of composite laminates is basically the same as that developed for metals (Reed, 1962), but having a greater overall length to accommodate composite specimens (Kasen, 1974). Such specimens have to be much longer than metal specimens because of their high anisotropy. Reinforcement of the cryostat barrel, which connects onto the crosshead adapter plate, is needed to withstand the shock load, which is transferred to the barrel when the large amount of elastic energy stored in the nonductile composite specimen is suddenly released at failure.

Useful static mechanical data include tensile and compressive strengths and moduli in the principal in-plane fiber directions and the shear strength. Compressive strengths should also be obtained on bulk industrial laminates in the thickness (reinforcement normal) direction. The nature of the specific tests depends on whether the laminate is reinforced with fabric or with uniaxial fiber laminae. The usual procedure is to determine the strength and modulus only in the

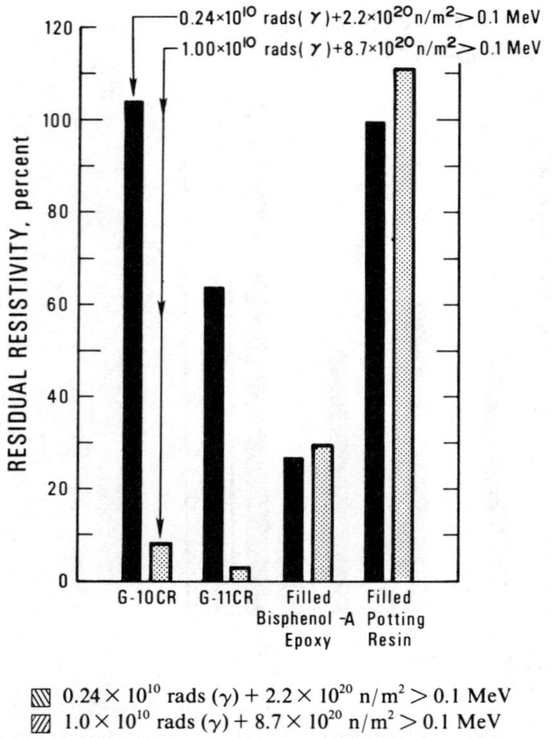

Figure 12.35 Percentage of original resistivity retained by G-10CR and G-11CR glass-epoxy industrial laminates and by two filled epoxies after two levels of irradiation at 5 K at room temperature (Coltman et al., 1979).

warp direction of fabric, since that direction usually has the maximum fiber density. Properties in the fill direction should be tested for critical structures unless they can be adequately predicted from the warp properties. For fabric-reinforced laminates, the shear parameter of interest is the strength between adjacent fabric layers, called *interlaminar* or *horizontal* shear.

Uniaxial fiber laminae may be combined in many crossply orientations to provide widely varying laminate properties. Key uniaxial lamina properties can subsequently be used to predict complex crossply properties by application of the macromechanical composite theory as discussed in § 12.2.1. Key properties are uniaxial longitudinal and transverse tensile and compressive strengths and moduli, plus the in-plane shear strength and modulus. Three such shears may be defined in a bulk, uniaxial composite laminate. Most important are the two longitudinal shears represented by shear between parallel fibers in the plane of the fibers. For example, taking the x-axis parallel to the fibers, longitudinal shear strengths may be defined as σ_{yx} or σ_{zx}; in thin laminates, however, only one such shear is of consequence. The third shear, which is transverse or cross fiber, is usually not an important design parameter. Laminate in-plane shear is a difficult property to measure accurately, because extraneous compressive or tensile stresses on the shear plane must be avoided. The torque tube (torsion) method used for metals is applicable, but expensive, for composite laminate materials. For this reason, somewhat simpler methods such as the "rail shear" or the "± 45° laminate" tests have been devised. Discussions of the comparative virtues of these methods are found in Sims and Halpin (1974). The ± 45° laminate method is the most convenient and least expensive method for use at cryogenic temperatures because it utilizes the same fixtures as those for tensile testing. It provides valid data for polymer-matrix composites but is not valid for materials such as boron–aluminum, which display significant nonlinearity between stress and strain in crossply configurations.

The strong dependence of composite laminate properties on orientation requires that alignment of test specimens meets the criteria established by ASTM Specification D 3039-71T. The procedures given in this specification for stabilizing laminate moisture content should also be followed when testing polymer-matrix materials.

12.7.2 Static and cyclic tensile testing

Flat-sided coupons 1.3-cm wide work well for longitudinal tensile testing of uniaxial laminates, whereas coupons 2.5-cm wide work well for crossply or uniaxial transverse mechanical property testing. Reinforcing the grip regions with tabs is unnecessary when a proper gripping system is used. A 28.0-cm-long coupon gripped over a 6.4-cm distance on each end leaves an adequate gauge length. Tapered or "waisted" specimens are generally avoided for laminates built up from successive layers of uniaxial laminae, since the low in-plane shear strength of such composites often causes premature failure by longitudinal splitting of the specimen. Specimens having a gradual 3- to 4-degree taper on each side to a constant gauge section of about 5.4 cm have been successfully used with glass-fabric-reinforced laminates (Dastin, 1969) and with certain types of angle-plied composites where longitudinal splitting is a minimal problem owing to the particular layup, for example, with a [0/±45/0] boron–epoxy laminate. If longitudinal splitting can be avoided, waisting can be a great convenience, because thicker laminates can be tested in machines of limited load capability and problems due to slippage in the tensile grips are minimized. It is very desirable to limit the applied tensile load as much as possible consistent with obtaining valid data, because the recoil energy released at fracture of a laminate of large cross-sectional area can damage the load train and cryostat.

A successful gripping system must transfer the tensile load systematically into the fibers of the laminate and must provide shear strength at the specimen-grip interface in excess of the ultimate strength of the laminate. It is also desirable to minimize the mass of the grips to minimize the consumption of liquid helium when testing at 4 K. Testing uniaxially reinforced laminates in the fiber direction develops the maximum laminate strength and places the most demands on the system.

Work in the author's laboratory has shown that the gripping principle illustrated in Fig. 12.36 works well for static tensile testing of uniaxial glass or high-modulus laminates up to a 6-ply thickness and loads of about 17.8 kN. A piece of 100-mesh stainless steel screen of 0.1-mm wire is wrapped around each end of the specimen for a distance of about 64 mm. The specimen ends are then sandwiched between two grips consisting of 1.6-mm thick sheets of 6Al–4V titanium alloy that has been cross-serrated with 10 teeth per cm on the sides contacting the specimen. Finally, two 4.8-mm-thick stainless steel pressure plates are bolted onto the assembly using a set of 8-32 machine screws. The tensile load is transferred only through the titanium part of the grip, which

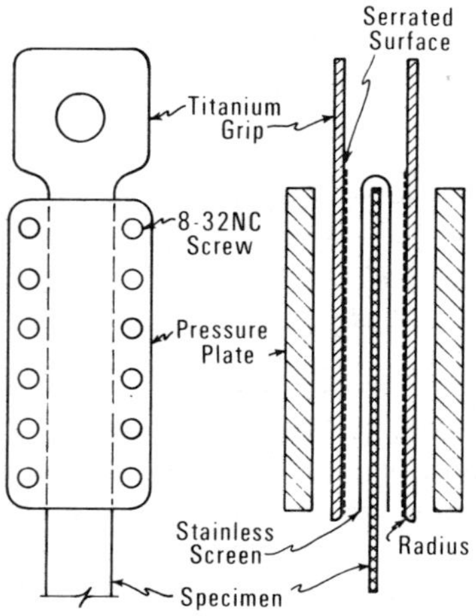

Figure 12.36 Tensile grip assembly for testing laminates at cryogenic temperature.

has an internal radius on the specimen end to minimize the stress concentration on the specimen where it leaves the grip. This stress concentration is minimized further by decreasing the torque on the machine screws systematically from the specimen ends toward the gauge section. When properly done, loading to failure will allow the specimen to slip a small amount within the grip for a distance about equal to the first two sets of screws from the gauge end, tapering off to no slippage farther into the grip. This allows the Poisson's contraction to be spread over a distance along the specimen. For example, a torque of 4.0 N·m (35 in·lbf) might be applied to the four screws closest to the specimen end, 3.7 N·m (33 in·lbf) to the fifth screw from the end, and 3.5 N·m (31 in·lbf) to the screws closest to the gauge length. The grips and specimens must be mated in an alignment fixture prior to insertion into the cryostat. Separate grips are required for differing specimen widths.

As laminate thickness increases, the additional tensile load at failure will eventually cause grip slippage in a fixture, such as the one illustrated in Fig. 12.36. Slippage will also occur when such a fixture is used for fatigue tests, unless added precautions are taken. The low-toughness titanium grips are also likely to fail under high cyclic loading. For such demanding service, the titanium grip plates may be replaced with 6.4-mm-thick Inconel X-750 plates, serrated as before, with the thickness of the stainless steel pressure plates increased to 6.4 mm. The 8-32 machine screws would then be replaced with 10-32 screws. For the most demanding fatigue tests, the uniformity of pressure may be increased by bolting directly through the grip and specimen, in addition to using side screws. An arrangement proven satisfactory in the author's laboratory is illustrated in Fig. 12.37. Here, a grip designed for a 2.5-cm-wide specimen has been modified by adding a 6.4-mm-diameter bolt through the system. Where a waisted specimen is used, as in this illustration, the 10-32 side screws are uniformly torqued to 8 N·m (72 in·lbf) each, and the center bolt is torqued to 16 N·m (144 in·lbf).

These examples of gripping techniques illustrate a typical solution. Many successful variants are possible.

Strain gauges may be used to monitor longitudinal and transverse specimen strain of laminates without problems, provided that the procedures described in the chapter on mechanical properties are followed. The only change is to use low curing temperatures for the adhesives

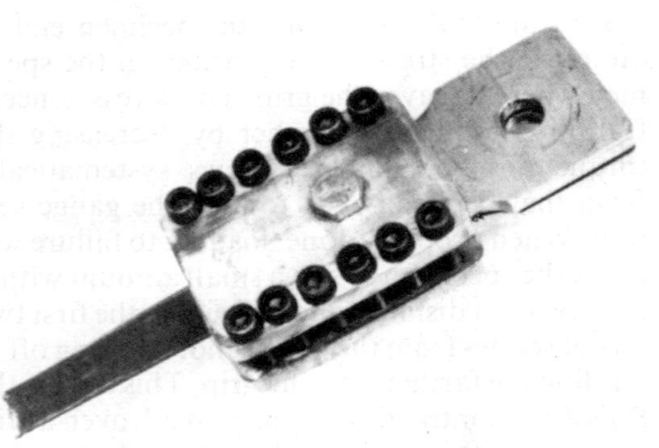

Figure 12.37 Grip system for tensile fatigue testing of laminates at cryogenic temperatures. Example is a waisted specimen used with $[0/\pm45/0]_S$, 140-μm (5.6-mil) boron–epoxy.

used to bond the gauges. Clip-on extensometers may be used, but the hard, brittle surface of polymer-matrix composites at low temperatures can cause slippage of the contacts. The cost of 4-K testing is high enough to justify the comparatively slight additional cost of bonded-on gauges whenever there is the slightest doubt as to the reliability of clip-on extensometers.

12.7.3 Static and cyclic compression testing

The large uniaxial compressive strengths of some composites require special test fixtures. A uniaxial boron–epoxy specimen loaded compressively in the fiber direction at 4 K will sustain a stress of about 2760 MPa (4×10^5 psi) before fracture, with some specimens exceeding 3450 MPa (5×10^5 psi). The test fixture, therefore, must be able to transfer the required compressive loads into the specimen while maintaining specimen alignment; no significant plastic deformation of any part of the grip system can be tolerated during this time. The fixture and grip system must be able to withstand the shock recoil load imposed upon it when specimens undergoing no plastic deformation fail suddenly at high loads. For example, a boron–epoxy specimen of only 0.26-cm^2 cross-sectional area will suddenly release about 71 kN of load into the system at failure. For cryogenic testing, there are the additional requirements that the mass of the fixture be minimized and that the fixture be usable in the cryostat designed for tensile use.

The fixture illustrated in Fig. 12.38 has been developed in the author's laboratory to meet these requirements. A similar design has been used by Dahlerup-Petersen and Perrot (1979). Pull-rod separation is converted into compression by means of interlocking yokes. The specimen with its end caps is inserted into compression blocks and alignment is maintained by a sleeve that slips around the blocks. The compression blocks are anodized to minimize friction. Additionally, MoS_2 lubricant sprayed on the fixture during assembly acts both as a release agent for the specimen end caps and as a lubricant between the compression blocks and alignment tube. When properly constructed, extraneous frictional forces are about 5 to 10 N. Split compression blocks joined by stainless steel bolts facilitate removal of the end caps after testing. Fixture dimensions and construction materials have been detailed in the literature (Schramm and Kasen, 1976). Helium consumption with this fixture is typically 5 to 6 ℓ per static compression test at 4 K.

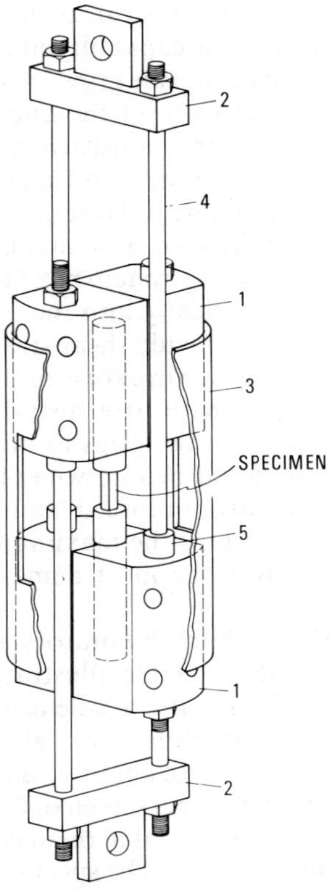

Figure 12.38 Fixture for static and fatigue compression testing at cryogenic temperatures. 1) split aluminum compression blocks, 2) stainless steel yokes, 3) aluminum alignment sleeve, 4) titanium rods, 5) lock collar.

There is no consensus on what constitutes an

acceptable failure mode in a laminate compression test. Experience has shown that a 45° shear failure is associated with the highest compressive strength. This type of failure can be reliably obtained with specimens of the configuration shown in Fig. 12.39. Specimens of square cross section, epoxied into 6061-T6 aluminum end caps, are satisfactory for most laminate types and orientations. The very high compressive strengths of uniaxial longitudinal boron-reinforced laminates require the use of round specimens epoxied into stainless steel end caps. Both types of specimens should be diamond-ground to final dimensions. Materials of low compressive strength, such as glass-cloth-reinforced polymers, can be successfully tested without bonding, and specimen widths can be increased up to 6.4 mm, provided that the aspect ratio of the unsupported section remains over 3:1.

A longitudinally split tube having an inner diameter matching the outer diameter of the end caps is useful in aligning and supporting the specimens during room-temperature cure of the epoxy used to bond the specimens into the end caps. The type of epoxy does not appear to be critical.

Compressive Young's moduli obtained from strain gauges attached to the specimen surface are not very reliable, often yielding unrealistically high moduli, possibly because of extraneous strains arising from barreling of the specimen during the test. Furthermore, it is questionable whether such data are required, because the author's and others' experiences indicate that the compressive moduli of composite laminates are very close to the tensile moduli in both the longitudinal and transverse directions.

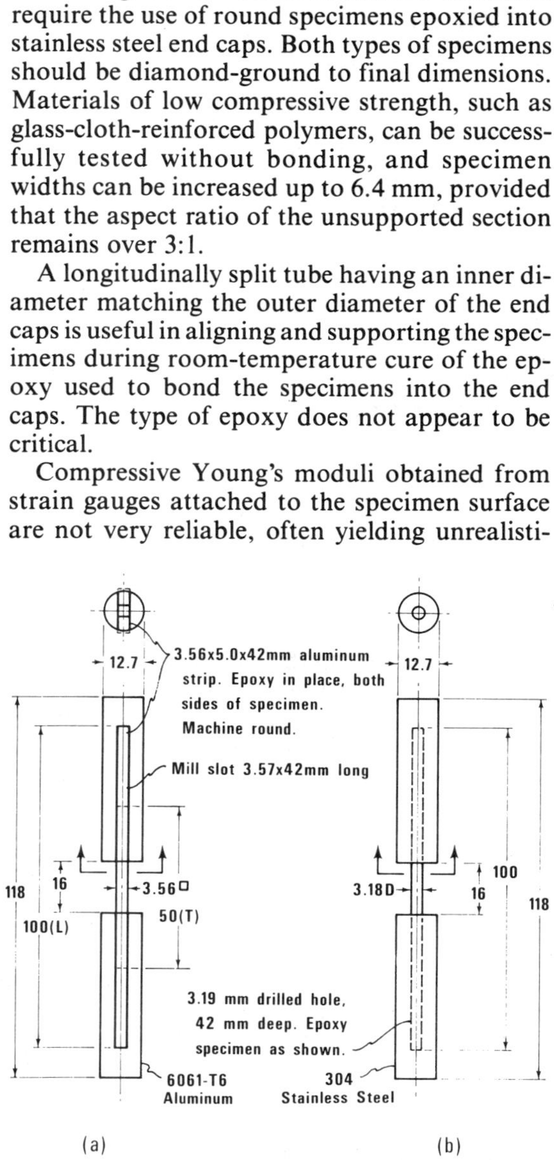

Figure 12.39 Specimen configuration used for compression testing of laminates at cryogenic temperature; (a) square specimen; (b) round specimen.

The dynamic method of determining the elastic constants of composite materials discussed in Chapter I offers an interesting alternative to the mechanical method. Programs to compare results with both methods are presently underway.

12.7.4 Interlaminar shear testing

Interlaminar shear, the resistance to failure in shear between the layers of a layered composite, is of particular interest in the design of large superconducting magnets. Thick glass–epoxy laminates of the NEMA G-10 type are frequently used as electrical insulation and structural support for magnet windings, in which case the laminates must withstand shear loads arising from magnet stresses. Some initial data on the temperature dependence of interlaminar shear in the G-10CR and G-11CR high-pressure industrial laminates appear in Fig. 12.6.

The most widely used method for determining interlaminar shear utilizes the double-offset notch "guillotine" tensile specimen illustrated in Fig. 12.40. This specimen has nonuniform shear in the test section (highest values of shear exist at the base of the notches), failure can be induced by peeling owing to unsymmetrical loading, and the results obtained are highly influenced by the distance between the notches and specimen thickness. Nevertheless, it is currently the best available method of obtaining design data. The short-beam method of determining apparent interlaminar shear (ASTM D 2344-72) cannot be used for this purpose because the state of internal stress is continuously changing throughout the test.

Experience has shown that the interlaminar shear strength obtained from a specimen, such as illustrated in Fig. 12.40, will increase systematically up to a limiting value as side support is increased. For example, a constant value is attained for screw torque of about 9 N·m (8 in·lbf) for a support fixture of the type illustrated in Fig. 12.41. A similar effect is obtained by increasing the specimen thickness. Work at Francis Bitter National Magnet Laboratory (Becker and Erez, 1980) has shown that results equal to that of

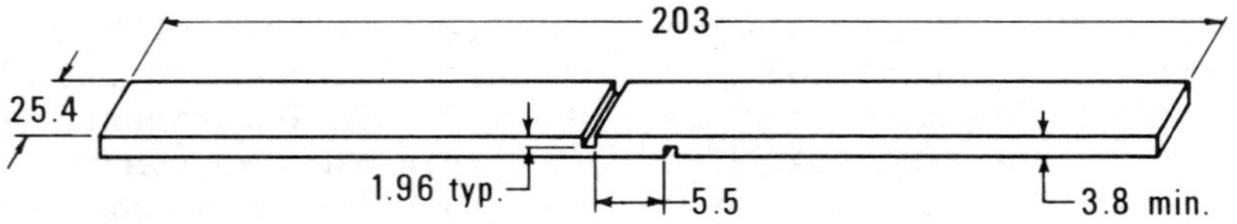

Figure 12.40 Double-offset notch "guillotine" tensile specimen for determination of interlaminar shear strength. Dimensions are in mm.

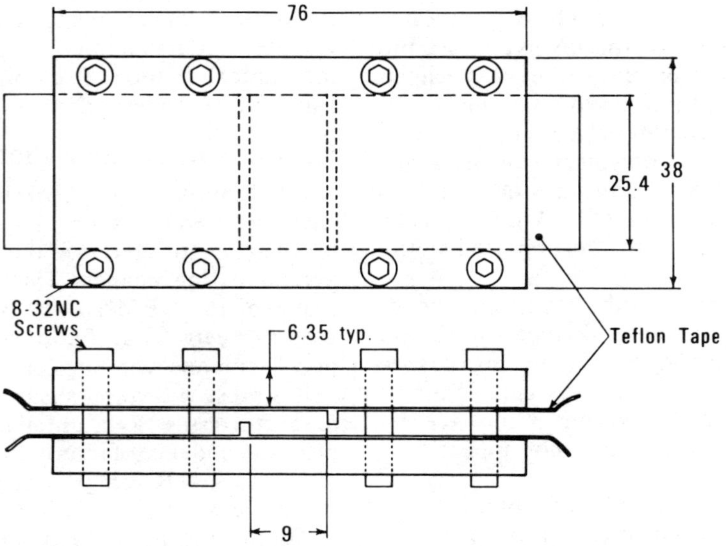

Figure 12.41 Side supports used with specimen illustrated in Fig. 12.40. Dimensions are in mm.

properly side-supported specimens are obtained with a 12.5-mm-thick specimen. These data appear in Fig. 12.6.

Despite its obvious faults, the tensile "guillotine" method, modified as above, remains the best available procedure for generating interlaminar shear strength data.

12.7.5 Secondary property tests

Secondary property tests are relatively inexpensive, simple tests that subject the composite laminate to combined stresses, that is, to simultaneous tension, compression, and shear. Such tests are useful quality control tools and are convenient for materials screening. The tests are classified as secondary because they do not produce engineering data.

Flexure tests. A flexure test is conducted by three-point loading a bar-shaped section of laminate. The ratio of support span to specimen depth is determined by the objective of the test. For flexural strength and modulus, the ratio is adjusted to develop the maximum tensile stress on the convex side of the specimen (ASTM D 790). Reducing the ratio causes the specimen to fail in shear on the internal layers, producing *short-beam* or *apparent horizontal* shear data (ASTM D 2344). There is no established correlation between flexural strength and tensile strength or between short-beam shear strength and the horizontal shear strength determined in tension.

Flexure tests are easily conducted at cryogenic temperatures. Specimens can be miniaturized and tested with dip-probe apparatus, reducing costs and facilitating testing in difficult environments, such as combined cryogenic temperatures and radiation.

Split-disc tensile tests. This test, described in ASTM D 2290, is also known as the NOL-ring test (Naval Ordinance Laboratory). It is most widely used as a quality control or screening test for composites fabricated by filament-winding processes. The specimen, in the form of a ring, is placed over a split circular disc, which is then separated to rupture the ring. This method gives only an apparent tensile strength as a result of the bending moment, which is imposed on that portion of the ring adjacent to the split. The method is adaptable to cryogenic temperatures, although the bulk of the fixture and its width (typically 146 mm) make it less desirable for testing at liquid-helium temperature.

12.8 CONCRETE AGGREGATES

Concrete is presently used as the primary structural material for land-based storage of LNG; facilities for storing some three million barrels have been constructed. Because the tensile strength of concrete is substantially lower than its compressive strength, concrete structures are either pretensioned or posttensioned in most applications. Both internal and external insulation is used. In some cases, the concrete is in contact with the LNG. The initial skepticism about the practicality of using concrete as a structural material at LNG temperatures has currently given way to optimism as practical experience has been gained. A recent review by Turner (1979) provides an overview of materials and design for such applications.

Low-temperature property data on concrete are relatively scarce. The work of Monfore and Lentz (1962) indicates that the Young's modulus of moist concretes will probably increase on cooling: up to a 50% increase was reported for a conventional sand-and-gravel concrete cooled to 115 K. Dried concretes display a much lower modulus increase. Compressive strength and splitting strength increase dramatically. The splitting strength, S, is calculated from the formula $S = 2P/\pi d\ell$, where P is the compressive load at rupture and d and ℓ are the diameter and length of the test cylinder. At 170 K, moist sand-and-gravel concretes may have 300% higher compressive strengths than at room temperature, as illustrated in Fig. 12.42; the splitting strength has been found to increase by over 200% on cooling to 210 K. The strengths may decline slightly on cooling to still lower temperatures. The extent of the strength increases is believed related to the evaporable moisture content, since increases in strength for dried concrete are substantially less than for moist concrete. The low-temperature strength depends on a number of factors, including cement content, water-to-cement ratio, type of aggregate, the age of the concrete, and particularly, on the moisture content and its condition in the concrete.

The thermal contraction of concretes on cooling also depends on the evaporable moisture content. This moisture has a high content of dissolved minerals and therefore freezes over an extended temperature range. As cooling continues below 273 K, the ice already formed contracts, while the newly formed ice provides a counteracting expansion, which minimizes the overall thermal contraction. This effect accounts for the maxima observed in the strength (Fig. 12.42) and the minimum observed in thermal conductivity on cooling (Fig. 12.43). It appears that the average coefficient of thermal contraction from 300 to 115 K is of the order 8×10^{-6} K^{-1} for a moist sand-and-gravel concrete.

As shown in Fig. 12.43, the thermal conductivity of a mortar sample was found to decrease on cooling from 300 to about 200 K, after which it increased to somewhat above that at 300 K (Eakin et al., 1963). The step function change at 273 K reflects initial freezing of the entrained moisture. Monfore and Lentz (1962) report the average conductivity of a moist sand-and-gravel concrete to be about 3.3 $W{\cdot}m^{-1}K^{-1}$ between 300 and 140 K.

The above data pertain to concretes made with air-entraining portland cement, having air contents varying from 5.3 to 7.8%. Some entrained air is considered essential for concretes subjected to subfreezing temperatures. Richard, Dobogai, Gerhardt, and Young (1975) have suggested that a still higher proportion of stable air voids, such as are obtained in cellular concretes, would be beneficial. Here, the air is introduced into the cement-water slurry in the form of a preformed stabilized protein foam. Limited data indicate that a cellular concrete having a dry density of 705 $kg{\cdot}m^{-3}$ will undergo about a 100% increase in compressive strength on cooling to 77 K, accompanied by about a 23% decrease in failure strain. Lesser strength increases are expected for higher-density products. The thermal conductivity of the 705 $kg{\cdot}m^{-3}$ product reportedly decreases slightly on cooling to 77 K, averaging about 0.15 $W{\cdot}m^{-1}K^{-1}$. This is more than an order of magnitude less conductivity than is obtained with conventional sand-and-gravel concrete. The absolute compressive strength of the cellular concrete appears to be about 50% that of conventional concrete at cryogenic temperatures.

Concretes, and perhaps other aggregates, will

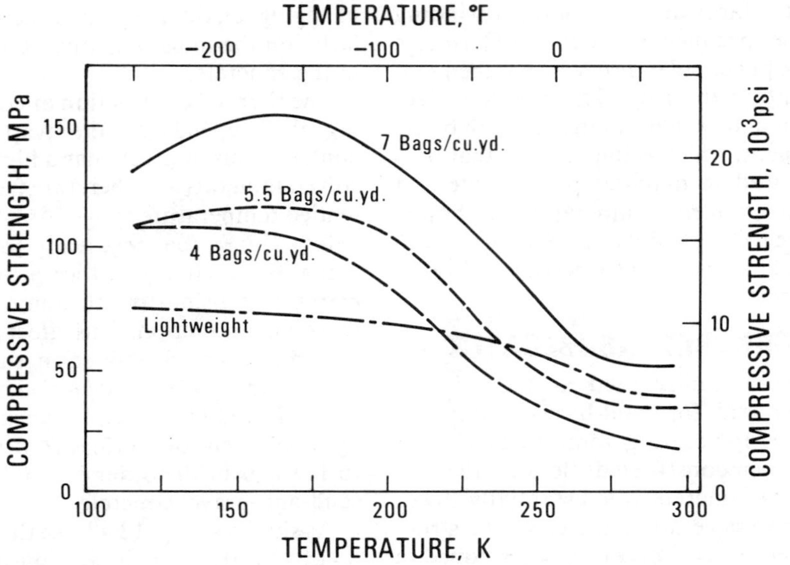

Figure 12.42 Effect of low temperatures on the strength of moist concrete using various cement contents (bags per cubic yard gravel; 1 cu yd = 0.765 m³) and for lightweight, expanded shale concrete (Monfore and Lentz, 1962).

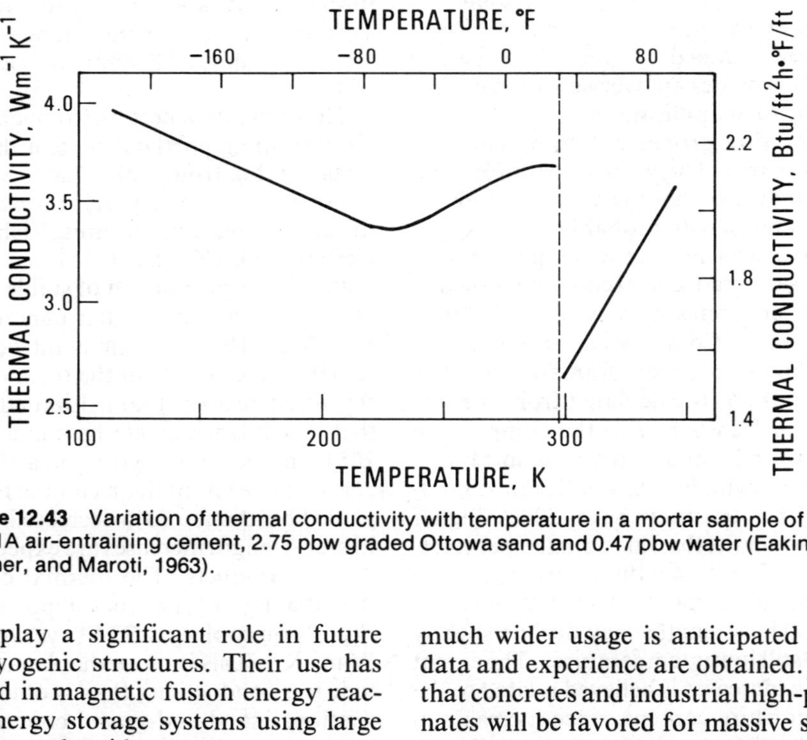

Figure 12.43 Variation of thermal conductivity with temperature in a mortar sample of 1 pbw type 1A air-entraining cement, 2.75 pbw graded Ottowa sand and 0.47 pbw water (Eakin, Bair, Closner, and Maroti, 1963).

undoubtedly play a significant role in future large-scale cryogenic structures. Their use has been suggested in magnetic fusion energy reactors and in energy storage systems using large superconducting solenoid magnets.

12.9 SUMMARY

Composite laminates and aggregates already play a significant role in cryogenic technology. A much wider usage is anticipated as additional data and experience are obtained. It is probable that concretes and industrial high-pressure laminates will be favored for massive structures owing to favorable cost factors. Industrial laminates and low-pressure glass-reinforced laminates will continue to serve intermediate-sized structures where their nonmagnetic character and insulating qualities offer advantages over metallic structures. The ability of high-modulus composites to match or to exceed the strength and moduli of stainless steel suggests that their usage will substantially increase as their cryogenic performance becomes better understood and as alloying elements become scarce and more expensive. In the interim, high-modulus composites will be used selectively for critical components where a nonmetallic, strong, stiff, low-thermal-conductivity material is required to perform specific tasks.

Conventional, off-the-shelf composite laminates and concretes developed for room-temperature service are acceptable engineering materials at cryogenic temperatures. But it is evident that the following problems are technical barriers to expanded usage:

a. Embrittlement of polymer matrices at low temperatures. Although the average values of static mechanical and elastic properties of composites improve on cooling, the low notch toughness of existing polymer systems at low temperatures limits the extent to which such improvements can be effectively utilized and increases the variability of cryogenic performance.

b. Inadequate data bases and performance experiences. Data on the effect of cryogenic temperatures on the performance of concretes is particularly deficient.

c. Lack of standardization. Composite materials must be selected from a plethora of proprietary products, among which a comparison of cost and performance is difficult to establish. A system establishing lower performance bounds with coded composite classes would greatly simplify the materials selection process at all temperatures.

Fundamentally, cryogenic composite technology as well as the general industry would benefit from:

a. Development of a workable fracture mechanics model for predicting the fitness-for-purpose of polymer-matrix composite structures.

b. Correlation of the exact molecular and interfacial structures and damage accrual on that level with the life-cycle performance of composite materials.

12.10 REFERENCES

Advanced Composites Design Guide, Vol. I-V (1973). Air Force Materials Laboratory, Wright-Patterson Air Force Base, Ohio.

Alers, G. A. and Elskey, R. K. (1977). Measurement of metal-to-adhesive bond quality using digital signal analysis. In *Ultrasonic Symposium Proceedings*. IEEE, New York, 35–40.

Barclay, D. L., Bell, J. E., Brogren, E. W., and Strayer, J. W. (1975). *Lightweight Vacuum Jacket for Cryogenic Insulation*. NASA CR-134759 (Final), Vol. I, Boeing Aerospace Corporation, Seattle, Washington.

Bascom, W. D., Cottington, R. L., and Timmins, C. O. (1976). Fracture design criteria for structural adhesive bonding — promise and problems. *Naval Eng. J.* **88** (4), 73–85.

Becker, H. and Erez, E. (1980). A study of interlaminar shear strength at cryogenic temperatures. In *Advances in Cryogenic Engineering*, Vol. 26. Plenum Press, New York, 259–267.

Brown, B. (1980). Low temperature radiation effects in superconducting magnets. *J. Nucl. Mater.* **97**, 1.

Coltman, R. R., Klabunde, C. E., Kernohan, R. H., and Long, C. J. (1979). *Radiation Effects on Organic Insulators for Superconducting Magnets*. ORNL/TM-7077, Oak Ridge National Laboratory, Oak Ridge, Tennessee.

Corruccini, R. J. and Gniewek, J. J. (1961). *Thermal Expansion of Technical Solids at Low Temperatures; A Compilation from the Literature*. NBS Monograph No. 29, National Bureau of Standards, Boulder, Colorado.

Dahlerup-Petersen, K. and Perrot, A. (1979). *Properties of Organic Composite Materials at Cryogenic Temperatures*. ISR-BOM 179–39. European Organization for Nuclear Research (CERN), Geneva, Switzerland.

Dastin, S. J. (1969). *Determination of Principal Properties of "E" Fiberglass High Temperature Epoxy Laminates for Aircraft*. DAA21-68-C-0404, Grumman Aircraft Engineering Corporation, Bethpage, Long Island, New York.

Dastin, S. J. (1970). *Joint Designed Fiberglass Reinforced Composites*. Paper No. 70-DE-56, American Society of Mechanical Engineers, New York.

Dauksys, R. J. (1973). Research and development of aerospace adhesive bonded systems and concepts. *SAMPE Q.* **5** (1), 1–12.

Eakin, B. E., Bair, W. G., Closner, J. J., and Maroti, R. (1963). *Below Ground Storage of Liquefied Natural Gas in Prestressed Concrete Tanks*. Institute of Gas Technical Report No. 8, American Gas Association, New York.

Fundamentals and Applications of Nonmetallic Materials at Low Temperatures (1982), (eds. G. Hartwig and D. Evans). Plenum Press, New York.

Handbook on Fiberglass and Advanced Plastics Composites (1969) (ed. G. Lubin). Van Nostrand Reinhold, New York.

Hart-Smith, L. J. (1973). *Adhesive-Bonded Scarf and Stepped-Lap Joints*. NASA CR-11237.*

Hart-Smith, L. J. (1974). Advances in the analysis and design of adhesive bonded joints in composite aerospace structures. *SAMPE J.* **19**, 722–737.

Hart-Smith, L. J. (1976). *Bolted Joints in Graphite Epoxy Composites*. NASA CR-144899.*

Hartwig, G. (1979). Properties of high polymers. In *Nonmetallic Materials and Composites at Low Temperatures* (eds. A. F. Clark, R. P. Reed, and G. Hartwig). Plenum Press, New York, 33–50.

Heberlein, D. C. (1976). *Heat Capacities of E-Glass, Cross-Ply; S-Glass, Cross-Ply; and S-Glass, Unidirectional Epoxy Composites*. Report 2199, U.S. Army Mobility Equipment Research and Development Command, Fort Belvoir, Virginia.

Hillig, W. B. (1975a). Composite structures materials investigation at cryogenic temperatures. In *Materials Research for Superconducting Machinery — III*. NBS-ARPA Semiannual Technical Report, ADA012365, 1–38.*

Hillig, W. B. (1975b). Composite structures materials investigation at cryogenic temperatures. In *Materials Research for Superconducting Machinery — IV*. NBS-ARPA Semiannual Technical Report, ADA019230, 1–18.*

Hoggatt, J. T. (1971). *Development of Cryogenic PRD 49-1 Filament-Wound Tanks*. NASA CR-120835.*

Hust, J. G. (1975). Low-temperature thermal conductivity of two fiber-epoxy composites. *Cryogenics* **15**, 126–128.

Hust, J. G. (1976). Thermal conductivity. In *Materials Research for Superconducting Machinery — VI*. NBS—ARPA Semiannual Technical Report, ADA036919, 305–319.*

Hust, J. G. (1980). National Bureau of Standards, Boulder, Colorado, personal communication.

Hust, J. G. and Arvidson, J. M. (1978). *Thermal Conductivity of Glass Fiber/Epoxy Composite Support Bands for Cryogenic Dewars*. Report 275.03-78-2, National Bureau of Standards, Boulder, Colorado.

Kasen, M. B. (1974). Advanced composites. In *Materials Research for Superconducting Machinery — I*. NBS-ARPA Semiannual Technical Report, AD780596 101–180.*

Kasen, M. B. (1975a). Properties of filamentary-reinforced composites at cryogenic temperatures. In *Composite Reliability*. ASTM STP 580, American Society for Testing and Materials, Philadelphia, Pennsylvania, 586–611.

Kasen, M. B. (1975b). Mechanical and thermal properties of filamentary-reinforced structural composites at cryogenic temperatures — 1: Glass-reinforced composites. *Cryogenics* **15**, 327–349.

Kasen, M. B. (1975c). Mechanical and thermal properties of filamentary-reinforced structural composites at cryogenic temperatures — 2: Advanced composites. *Cryogenics* **15**, 701–722.

Kasen, M. B. (1978). Composite laminate applications in magnetic fusion energy magnet systems. In *Proceedings 1978 International Conference on Composite Materials* (eds. B. Noton, R. Signorelli, K. Street, and L. Phillips). AIME, New York, 1493–1507.

Kasen, M. B. (1979). Cryogenic applications of composite technology in the U.S.A. In *Nonmetallic Materials and Composites at Low Temperatures* (eds. A. F. Clark, R. P. Reed, and G. Hartwig). Plenum Press, New York, 317–337.

Kasen, M. B. (1981). Cryogenic properties of filamentary-reinforced composites: An update. *Cryogenics* **21** (6), 323–340.

Kasen, M. B., MacDonald, G. R., Beekman, D. H., and Schramm, R. E. (1980). Mechanical, electrical, and thermal characterization of G-10CR and G-11CR glass cloth/epoxy laminates between room temperature and 4 K. In *Advances in Cryogenic Engineering*, Vol. 26. Plenum Press, New York, 235–244.

Kasen, M. B., Schramm, R. E., and Beck, B. (1980). *Advanced Structures Development for Superconducting Magnets*. Draft report submitted to the U.S. Air Force Aero Propulsion Laboratory, Wright-Patterson Air Force Base, Ohio.

Kasen, M. B., Schramm, R. E., and Read, D. T. (1977). Fatigue of composites at cryogenic temperatures. In *Fatigue of Filamentary Composite Materials*. ASTM STP 636, American Society for Testing and Materials, Philadelphia, Pennsylvania, 141–151.

Kernohan, R. H., Coltman, R. R., and Long, C. J. (1979). *Radiation Effects on Organic Insulators for Superconducting Magnets*. ORNL/TM-7077, Oak Ridge National Laboratory, Oak Ridge, Tennessee.

Khim, Z. G. and Ng, M. W. (1980). Testing of fiberglass-reinforced polyester composites. In *Advances in Cryogenic Engineering*, Vol. 26. Plenum Press, New York, 280–285.

Kreibich, U. T., Lohse, F., and Schmid, R. (1979). Polymers in low-temperature technology. In *Nonmetallic Materials and Composites at Low Temperatures* (eds. A. F. Clark, R. P. Reed, and G. Hartwig). Plenum Press, New York, 1–32.

Laminated Plastics (1958) (ed. D. J. Duffin). Reinhold, New York.

Ledbetter, H. M. (1979a). *Dynamic Elastic Modulus and Internal Friction in G-10CR and G-11CR Fiberglass-Cloth–Epoxy Laminates*. NBSIR 79–1609, National Bureau of Standards, Boulder, Colorado.

Ledbetter, H. M. (1979b). Dynamic elastic modulus and internal friction in fibrous composites. In *Nonmetallic Materials and Composites at Low Temperatures* (eds. A. F. Clark, R. P. Reed, and G. Hartwig). Plenum Press, New York, 267–281.

Lewis, A. and Bush, G. E. (1967). *Improved Cryogenic Resin Glass Filament-Wound Composites*. NASA CR-72163 (Final).*

Lewis, A., Bush, G. E., and Creedon, J. (1966). *Improved Cryogenic Resin/Glass-Filament-Wound Composites*. NASA CR-54867.*

LNG Materials and Fluids — Users Manual of Property Data in Graphic Format. (1980). Supplement to the first edition. National Bureau of Standards, Boulder, Colorado.

Loud, S. N. (1979). Advanced composites applications of S-2 glass fiber. In *Proceedings, 34th Annual Technical Conference*, Section 20-F. Society of the Plastics Industry, New York, 1–8.

Mauri, R. E. (1968). Adhesives. In *Space Materials Handbook*, third edition (eds. J. B. Rittenhouse and J. B. Singletary). National Aeronautics and Space Administration, Washington, D.C.

MIL-HDBK-17A (1971). *Reinforced Plastics, Plastics for Aerospace Vehicles*, Part 1. U.S. Government Printing Office, Washington, D.C.

Monfore, G. E. and Lentz, A. E. (1962). *Physical Properties of Concrete at Very Low Temperatures*. Bulletin 145, Portland Cement Association, Skokie, Illinois.

Mulville, D. R., Hanston, D. L., and Mast, P. W. (1978). Developing failure criteria for adhesive joints under complex loading. *J. Eng. Mater. Technol.* **100**, 25–31.

Niemann, R. C. Gonczy, J. D., Hoffman, J. A., Mataya, K. F., Smelser, P., and Young, W. C. (1980). Ultimate strength, low-stress creep characteristics and thermal intercept methods for an epoxy fiberglass tension member support. In *Advances in Cryogenic Engineering*, Vol. 26. Plenum Press, New York, 300–308.

Nonmetallic Materials and Composites at Low Temperatures (1979) (eds. A. F. Clark, R. P. Reed, and G. Hartwig). Plenum Press, New York.

Reed, R. P. (1962). A cryostat for tensile tests in the temperature range of 300 to 4 degrees K. In *Advances in Cryogenic Engineering*, Vol. 7. Plenum Press, New York, 448–454.

Richard, T. G., Dobogai, J. A., Gerhardt, T. D., and Young, W. C. (1975). Cellular concrete — a potential load-bearing insulation for cryogenic applications. *IEEE Trans. Magn.* MAG-11, 500–503.

Ripling, E. J., Mostovoy, S., and Corten, H. T. (1971). Fracture mechanics: a tool for evaluating structural adhesives. *J. Adhes.* **3**, 107–123.

Schramm, R. E. and Kasen, M. B. (1976). Advanced composites. In *Materials Research for Superconducting Machinery — V* (eds. R. P. Reed, H. M. Ledbetter, and E. C. VanReuth). NBS–ARPA Semi-annual Technical Report, ADA019230.*

Schramm, R. E. and Kasen, M. B. (1977a). Cryogenic mechanical properties of boron-, graphite-, and glass-reinforced composites. *Mater. Sci. Eng.* **30**, 197–204.

Schramm, R. E. and Kasen, M. B. (1977b). Static tensile properties of boron–aluminum and boron–epoxy composites at cryogenic temperatures. In *Advances in Cryogenic Engineering*, Vol. 22. Plenum Press, New York, 205–213.

Sims, D. F. and Halpin, J. C. (1944). Methods for determining the elastic and viscoelastic response of composite materials. In *Composite Materials: Testing and Design*. ASTM STP 546, American Society for Testing and Materials, Philadelphia, Pennsylvania, 46–66.

Soffer, L. M. and Molho, R. (1967). Cryogenic resins for glass-filament-wound composites. NASA CR-72114 (Final).*

SPI Handbook of Technology and Engineering of Reinforced Plastics Composites (1973) (ed. J. G. Mohr), Van Nostrand Reinhold, New York.

Tobler, R. L. and Read, D. T. (1976). Fatigue resistance of uniaxial S-glass/epoxy composite at room and liquid helium temperatures. *J. Compos. Mater.* **10**, 32–43.

Turner, F. H. (1979). *Concrete and Cryogenics*. Cement and Concrete Association, Wexham Springs, Slough, England.

Vogen, W. V., Clawson, B., and Morrison, H. G. (1975). A one-meter diameter, ac superconducting coil and fiberglass cryostat for an electromagnetic geophysical exploration system. In *Advances in Cryogenic Engineering*, Vol. 21. Plenum Press, New York 51–59.

Waddoups, M. E. (1968). Characterization and design of composite materials. In *Composite Materials Workshop*. Technomic Publishing Co., Stanford, Connecticut, 254–308.

Wang, S.-T., Kim, S.-H., Kim, N. S., Cheng, R. S., Hoffman, J., and Gonczy, J. (1980). Low-temperature measurement of the thermal and mechanical properties of phenolic laminate, the pultruded polyester fiberglass and A & B epoxy putty. In *Advances in Cryogenic Engineering*, Vol. 26. Plenum Press, New York, 286–294.

Waszczak, J. P. and Cruse, T. A. (1975). *A Synthesis Procedure for Mechanically Fastened Joints in Advanced Composite Materials*. AFML-TR-73-145, Vol. II, General Dynamics/Convair, San Diego, California.

Wolff, R. and Lemon, G. (1972). *Reliability Prediction for Adhesive Bonds*. AFML-TR-72-121, General Dynamics/Convair, Fort Worth, Texas.

---

*Available from NTIS, Springfield, Virginia.

13 Superconductors

J. W. EKIN

13.1 INTRODUCTION

For half a century after the discovery of superconductors in 1911 (Onnes, 1911), they were little more than an idle laboratory curiosity. The early superconductors could be used only at low currents and low magnetic fields. Practical superconductors were not made until the late 1960s and are still under development. It is possible now to manufacture wires that can carry extremely high electric current densities (0.1 $MA \cdot cm^{-2}$) in the presence of high magnetic fields (10 T and greater). Superconductors do this with very little power dissipation; this is the main reason for their usefulness in manufacturing high-field electromagnets. A 15-T, 5-cm bore superconductive magnet, for example, requires only a few hundred watts to operate (including refrigeration), whereas an equivalent water-cooled copper magnet would require about five megawatts of power for steady-state operation. Furthermore, such a superconducting magnet and its power supply might weigh about five hundred newtons (a hundred pounds) and occupy about a cubic meter of space, whereas the equivalent copper magnet and its current generators would weigh many tons and require an entire building to house. The savings in power, size, and weight can be substantial and is the main motivating force for the use of superconducting magnets in a number of devices.

13.1.1 Applications

Superconducting composites are used extensively today in many applications (a) where high magnetic fields must be generated economically or (b) where very low magnetic fields must be detected. The first category includes large-scale superconductive electric motors and generators, superconductive magnets for controlled thermonuclear fusion, magnets for nuclear magnetic resonance, magnets for high energy physics research, and magnets to focus pi mesons for cancer treatment. Examples of the second category are magnetometers for detection of geothermal energy sources and high-speed superconducting computers.

Applications of superconductors generally center around two main properties: their resistance to electric current conduction is negligible (i.e., they are "super" conductors) and they exhibit quantum phenomena at very low magnetic flux levels — on the order of 1 fWb (10^{-15} Wb). Most of the magnet applications listed above rely on the first property, the microelectronic applications on the second.

13.1.2 Scope

This chapter is intended as an introductory guide to interpreting handbook data on *practical*, high-current, superconducting materials, principally for magnet applications. It is not intended for advanced design. Rather, it presents an overview of the properties and operational limits of superconductive materials, as well as techniques used to fabricate practical superconducting wires. The literature on this subject is extensive and sometimes complex, but in actual practice the design considerations are surprisingly few and straightforward.

The zero resistance property of superconducting materials is a quantum mechanical effect that occurs in these materials only at low temperatures. When cooled to a low enough temperature, typically less than about 20 K, the electrons in superconductive materials condense into a low energy state wherein they can carry electric charge without resistance (the microscopic theory of this phenomenon is treated in the suggested reading cited in Appendix A). The resistance does not gradually decrease at low temperature; rather, it suddenly drops to zero at a specific temperature for each superconductor, called its *critical temperature*. For the phenomenon of superconductivity to occur, there is a limit also on the amount of magnetic field a superconductor can be subjected to, called its *critical field*, and the amount of electrical current it can carry, called its *critical current*.

Practical superconducting wires for magnet applications do not consist of a single large strand of superconductive material, but rather a *composite* structure consisting of many (10^2 to 10^5) fine filaments of superconductor embedded in normal-metal matrix to give the composite stability and limit ac losses. A cross section of a typical composite is shown in Fig. 13.1.

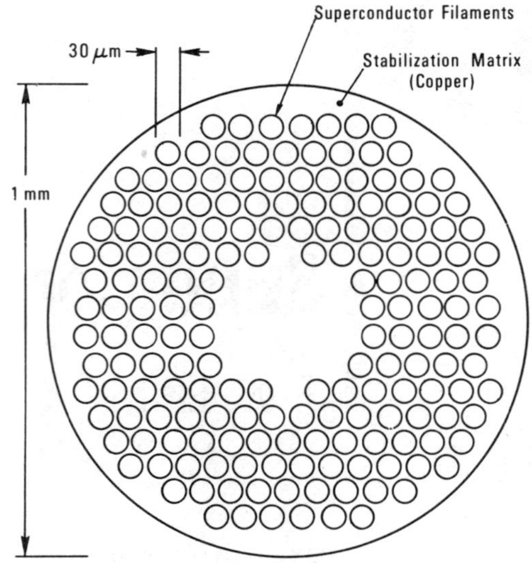

Figure 13.1 Cross-sectional view of a typical Nb–Ti, copper-stabilized superconducting composite.

The primary characteristics of such a composite wire may be broken down into two groups:

a. Properties that affect selection of the superconducting material:
- critical temperature, T_c (§ 13.2)
- critical magnetic field, H_c (§ 13.3)
- critical current density, J_c (§ 13.4)

b. Properties that affect geometrical design of the composite as a whole:
- stability (§ 13.6)
- ac loss (§ 13.7)
- mechanical characteristics (§ 13.8).

In selection of the type of superconducting material used for a particular application, critical temperature and critical field play a major role. These two properties are intrinsic to the material itself and fortunately do not change much among different samples of the same material. Critical current, on the other hand, is an extrinsic property and depends more on details of the metallurgical treatments used in the fabrication of the wire. It, consequently, is more variable and usually has to be measured for each individual spool of manufactured wire.

Stability and ac loss considerations dominate the geometrical design parameters of the superconducting composite as a whole, parameters such as the amount of stabilizing matrix material needed, the diameter of the superconducting filaments, and the filament twist pitch. Mechanical properties of the superconducting composite determine the amount of structural support needed to contain the magnetic forces developed when the superconductor winding is energized.

An introduction of each of these six properties is given in the indicated sections, along with a brief description in § 13.5 of manufacturing techniques for producing practical superconducting wires. Usually there are only one or two main considerations for each property; more detailed information is presented in appendixes at the end of the chapter. Each section also contains typical data on practical materials and a brief treatment of the experimental methods used to measure the property. These six properties are tied together in § 13.9, which summarizes their effect on superconductor material selection and the geometrical design of superconducting composites. Section 13.9 also contains a reference guide to where each composite-design parameter is treated in the chapter. A suggested reading list for more detailed information in each of these areas is provided at the end.

13.2 CRITICAL TEMPERATURE

The temperature marking the onset of superconductivity is called the *critical temperature* of the material, T_c. It is characterized by a sudden drop in the resistivity of the material from a relatively high value (~ 10 μΩ·cm) to an immeasurably low value (< 1 fΩ·cm). (By comparison, the resistivity of the purest copper is about 1 nΩ·cm, so for all practical purposes, the measured resistivity is zero.) As shown by the solid line in Fig. 13.2, the transition is typically sharp in pure materials, with a width less than 0.01 K; therefore the T_c of these materials is well defined. Most practical materials are inhomogeneous, however, and their transition is more diffuse. In this case, T_c is usually defined as the point where the resistivity of the material decreases to half its normal value (see the dashed line in Fig. 13.2).

13.2.1 Critical temperature values for practical materials

Mercury was the first superconductor to be discovered. In 1911, Onnes measured its critical temperature to be about 4 K. Since that time, 35 superconducting elements and more than a thousand alloys and compounds have been found. Compounds with the A15 or beta-tungsten crystal structure are of particular interest because they have the highest known superconducting transition temperatures. At present, Nb_3Ge holds the record with a T_c of 23.2 K (Gavaler, 1973). Appendix A briefly discusses the main empirical rules that determine high-critical-temperature materials.

Critical temperatures of some of the superconducting elements and the more practical superconducting alloys and compounds are listed in Tables 13.1 and 13.2. The values given for T_c in Tables 13.1 and 13.2 may vary by several tenths of a kelvin, depending on the individual investigator, method of material formation, and definition of T_c.

In contrast with some of the other superconductor properties, the critical temperature is an *intrinsic* property of the superconducting material, determined mainly by chemical composition and crystal structure. In compound materials, such as Nb_3Sn and V_3Ga, T_c is somewhat affected by stoichiometric composition (Appendix B) and the method of formation. In alloy materials such as Nb–Ti, however, T_c is relatively unaffected by metallurgical treatments and does not vary by more than a few tenths of a kelvin from specimen to specimen.

13.2.2 Limits for critical temperature

The effort to reach higher critical temperatures is now focused on third- and fourth-element additions to well-known high-T_c superconductors. Materials with T_c values higher than 25 to 30 K will probably not be found in the near

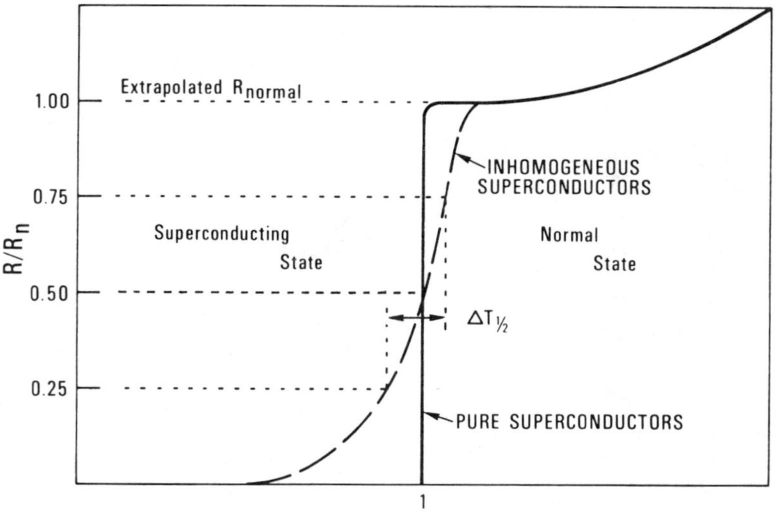

Figure 13.2 The superconducting resistance transition at T_C. Solid curve — resistive transition at the critical temperature of pure materials. Dashed curve — broad transition in inhomogeneous materials. T_C is taken as the temperature at which the resistivity of the material decreases to half its normal state value. The half-width, $\Delta T_{1/2}$, shown in the figure gives a measure of the transition width for inhomogeneous materials.

Table 13.1 Superconducting elements.

Element	T_c (K)	Type	B_c (0 K)[a] (T)	$\mu_0 H_{c2}$(4.2 K)[a] (T)
Nb	9.25	2	0.2060	0.2680
Ga(Δ)	7.8	2	0.0815	
Tc	7.8	2	0.1410	0.1450
Pb	7.19	1	0.0803	
Ga(γ)	7.0	2	0.0950	
La(β)	6.10		0.1600	
Ga(β)	6.	1	0.0560	
V	5.40	1	0.1408	
La(α)	4.88	1	0.0800	
Ta	4.47	1	0.0829	
Hg(α)	4.15	1	0.0411	
Hg(β)	3.95		0.0339	
Sn	3.72	1	0.0305	
In	3.41	1	0.0281	
Tl	2.38	1	0.0178	
Re	1.70	1	0.0200	
Pa	1.4			
Th	1.38	1	0.0160	
Al	1.18	1	0.0105	
Ga	1.08	1	0.0059	
Mo	0.92	1	0.0096	
Zn	0.85	1	0.0054	
Os	0.66	1	0.0070	
Zr	0.61	1	0.0047	
Cd	0.52	1	0.0028	
Ru	0.49	1	0.0069	
Ti	0.40	1	0.0056	
Hf	0.13			
Ir	0.11	1	0.0016	
Lu	0.1			
Be	0.026			
W	0.015	1	0.0001	

[a]The magnetic field, H, has been expressed in terms of the magnetic induction in free space, $B = \mu_0 H$, in order to present the values in the more common S.I. unit of tesla (T).

Table 13.2 Practical high-field superconducting alloys and compounds.

Material	Crystal Structure	T_c (K)	$\mu_0 H_{c2}$(4.2 K)[a] (T)
Nb–Ti	A2	10.2	12
Nb–Zr	A2	10.8	11
V_3Ga	A15	16.5	22
V_3Si	A15	17.0	
Nb_3Sn	A15	18.3	22
Nb_3Al	A15	18.9	32
Nb_3(Al,B,Be)	A15	20.0	
Nb_3Ga	A15	20.3	34
Nb_3(Al,Ge)	A15	20.5	41
Nb_3Ge	A15	23	30
NbCN	B1	16	25
V_2(Hf,Zr)	C15	10.1	24
$PbMo_6S_8$		14	45

[a]The magnetic field, H, has been expressed in terms of the magnetic induction in free space, $B = \mu_0 H$, in order to present the values in the more common S.I. unit of tesla (T).

future. Some increase may be expected, but it seems unlikely that a T_c above 25 to 30 K will be attained for superconductivity based on the phonon mechanism as it is currently understood (Appendix A).

At temperatures above 20 K, however, modest increases in T_c will have impact because of refrigeration costs and stabilization requirements. Practical superconducting devices usually operate at the boiling temperature of liquid helium, 4.2 K. A superconductor having a T_c significantly above 20 K would allow operation of magnets at temperatures considerably above 4.2 K, resulting in a significant savings in refrigeration. The critical current and critical field also scale with T_c. Thus, even for operation at 4.2 K, higher T_c materials would offer increased operating limits and enhanced stability.

It should be noted in this connection, however, that in all superconductors now known with T_c values higher than 20 K, there are major problems with their mechanical properties. They are extremely brittle, difficult to fabricate, and show significant electrical degradation at strains as low as 0.2%. Thus, the problem of increasing the ultimate T_c is coupled with the equally important problem of obtaining high-T_c materials that have better mechanical properties.

As a result of manufacturing problems, practical superconducting wires today are made from materials with T_c less than 20 K and must operate with liquid helium as the refrigerant, usually near 4.2 K. This low operating temperature is important to keep in mind when considering the temperature-dependent quantities, upper critical field, H_{c2}, and critical current density, J_c, described in § 13.4 and § 13.5.

13.2.3 Measurement of critical temperature

There are three basic methods of determining the critical temperature: measuring the change in the resistance, magnetic permeability, or heat capacity of the material. The first two are the most common. The resistance measurement is performed in zero magnetic field and the measuring current is kept low, typically less than 100 mA. If the material is fabricated in wire form, a four-contact method can be used, as shown in Fig. 13.3a. In pure samples containing few metallurgical defects, the transition is sharp and changes little with current level. Inhomogeneous samples, however, have broad transitions as shown in Fig. 13.2. A further complication is that

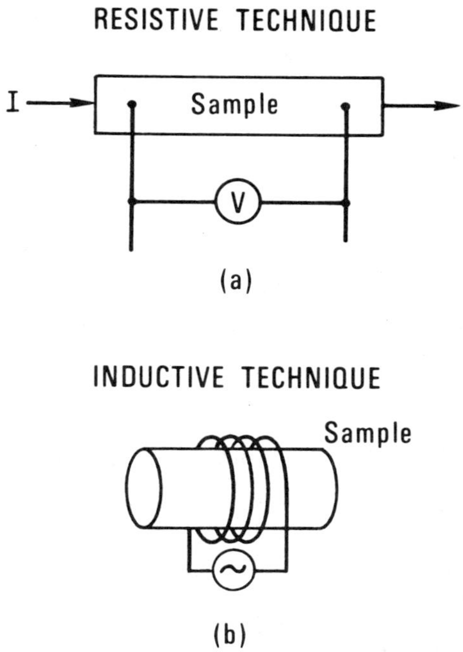

Figure 13.3 Schematic of experimental techniques for determining T_C through measurement of the resistance (a) and magnetic permeability (b).

the midpoint of the transition usually shifts with the amount of current applied to the sample. This current-dependent shift is the primary problem with the resistance method. When the current is very small, the highest T_c component of an inhomogeneous material is able to carry all the current, and a high T_c is observed. When high measuring currents are used, the lower T_c components of the material are required to carry some of the current, and the observed T_c reflects the decreased T_c of these components. Thus, because of this current-dependent shift in the transition, the resistance method is best suited to homogeneous specimens.

The second method is a probeless technique that relies on the large decrease in magnetic permeability of the material as it passes into the superconducting state (more on this in § 13.3). A coil is wound around the sample (see Fig. 13.3b), and a change in the self-inductance with temperature marks T_c. The advantage of this method is that it requires no direct electrical contacts with the sample, and consequently, it may be used with samples that are small, irregular, or in powder form. Again, the disadvantage is that the highest T_c component of the specimen can shield lower T_c components, thus dominating the measurement.

The third method utilizes the discontinuity in the specific heat, C, of a superconductor at T_c (see Fig. 13.4). Since C is a thermodynamic quantity, the entire material participates equally in affecting C, and hence, the width of the specific heat discontinuity can be used as an accurate indication of the spread of T_c values throughout the superconducting material (e.g., Matthias, Geballe, Corenzwit, Andres, Hull, Ho, Phillips, and Wohlleben, 1966; Collings and Ho, 1970). In a pure specimen, all three methods agree to within 10^{-3} K (Soulen and Colwell, 1971). The critical temperature is perhaps the best defined and easiest to measure of all superconductor parameters.

13.3 CRITICAL MAGNETIC FIELD

A second fundamentally important characteristic of superconductive materials (in addition to their very low resistance) is the effect of superconductivity on a magnetic field in the interior of the material. In 1933, Meissner and Ochsenfeld (1933) discovered that a superconductor cooled from room temperature in the presence of a magnetic field expelled the magnetic flux penetrating the material at T_c (see Fig. 13.5).

The vanishing of the magnetic field at the interior of a superconductor is brought about by super currents in the superconductor. The magnitude and distribution of these currents are such as to create an opposing interior field which cancels the applied one. There is a limit to this shielding effect, however. If the applied field becomes

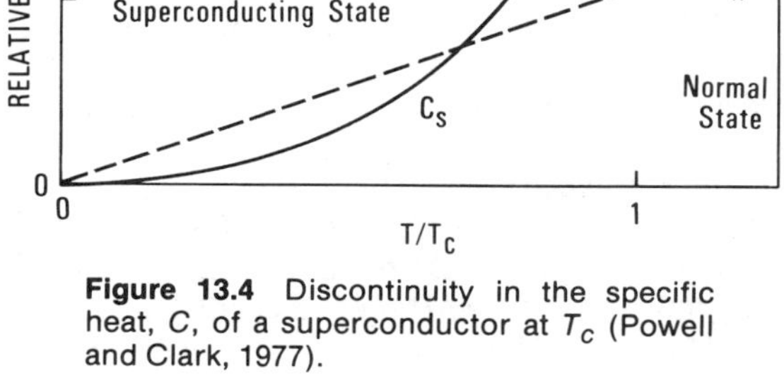

Figure 13.4 Discontinuity in the specific heat, C, of a superconductor at T_C (Powell and Clark, 1977).

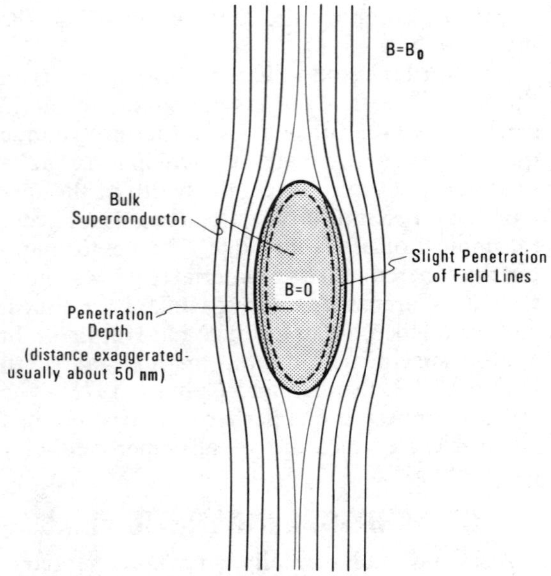

Figure 13.5 Expulsion of flux from the interior of an ideal superconductor. Dashed line shows the field penetration depth, λ, at the surface (not to scale) (Powell and Clark, 1977).

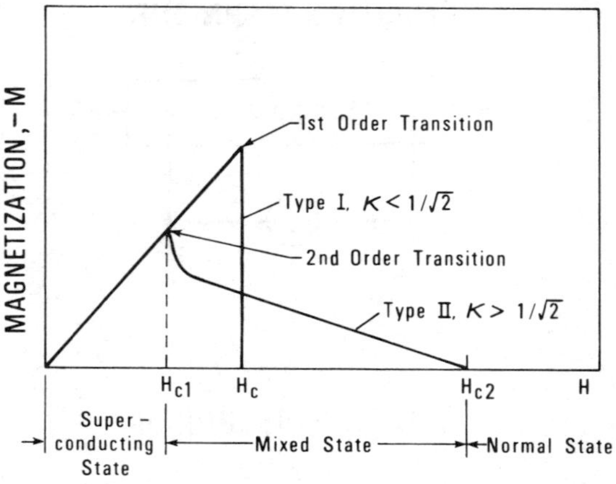

Figure 13.6 Magnetization curves on Type I and Type II superconductors, showing mixed state region for Type II materials.

too large, the shielding breaks down; the applied field fully penetrates the material, and the material reverts to a normal conductor. This is called the critical magnetic field of the superconductor. This effect is not quite so simple in some materials, leading to a division of superconductors into two types.

13.3.1 Type I and Type II superconductors

Superconductors are divided into two groups, Type I and Type II, depending on the way in which magnetic flux penetrates the material. The difference is very important for practical applications. If a magnetic field is applied parallel to a long cylinder of Type I "ideal" superconductor, the field is completely excluded from the interior of the material at all fields less than the critical field. Once the critical field is reached, however, complete flux penetration occurs. This may be seen in the magnetization curve for Type I superconductors shown in Fig. 13.6. Using the usual relation between applied field, H, magnetization, M, and the flux density, B: $B = \mu_0 (H + M)$, it is seen in Fig. 13.6 that the material is perfectly diamagnetic (i.e., $H = -M$) up to a critical field designated H_c. At this field the sample is no longer superconducting and the magnetization abruptly approaches its (small) normal-state value.

On the other hand, in Type II materials, flux penetrates the material well below H_c at a magnetic field designated H_{c1}. However, the material remains superconducting, showing zero resistance up to a much higher field, designated H_{c2}; for this reason these materials are important for practical applications (e.g., compare H_{c2} values of practical superconductors in Table 13.2 with H_c values of the elements in Table 13.1). As the applied magnetic field is increased above H_{c1}, the flux penetrates the material in the form of small flux tubes parallel to the field, shown schematically in Fig. 13.7. The core of each tube is normal (nonsuperconducting), but the material surrounding each flux tube remains superconducting. As the applied field is increased, the number of flux tubes penetrating the material increases, until eventually a field is reached where they completely overlap and their combined field is equal to the applied field, i.e., complete field penetration occurs and the material is no longer superconducting. The situation is illustrated by the magnetization curve in Fig. 13.6 labeled "Type II."

In the "mixed" state region between H_{c1} and H_{c2}, where Type II materials are occupied by an array of flux tubes, each flux tube consists of a normal core containing exactly one quantum of magnetic flux, ϕ_0. This universal constant is given by:

$$\phi_0 = hc/2e = 2.07 \text{ fWb} \tag{13.1}$$

where h is Boltzmann's constant, c is the speed of

light, and e is the charge of an electron. The flux tubes (also called flux lines or vortices) are quite small, typically 0.01 to 0.1 μm in diameter. In the absence of pinning (see § 13.4), they arrange themselves in a triangular array or flux lattice, as shown in Fig. 13.7. The presence of this flux lattice in Type II superconductors is important to an understanding of the critical current and is treated in more detail in the next section. [Type I superconductors can also exhibit an "intermediate" state, but this is not significant for practical superconductors. An extensive discussion of the intermediate state may be found in an article by Livingston and DeSorbo (1969).]

To summarize, Type I superconductors are characterized by a single low critical field, H_c, whereas Type II superconductors are characterized by two critical fields, H_{c1} and H_{c2}. The H_c values for Type I superconductors are typically less than 1 T, but Type II superconductors remain superconducting up to H_{c2}, which can be as high as 40 T or more. This makes them immensely more suitable for magnet applications. The *upper critical field*, H_{c2}, is the practical parameter from a design standpoint. Practical magnets can be safely operated at fields up to about 0.7 H_{c2}. Thus the "ideal" Type I materials may be ideal from a theoretical viewpoint, but Type II materials are far more practical. They typically have higher T_c values and remain superconducting up to much higher field levels than Type I materials.

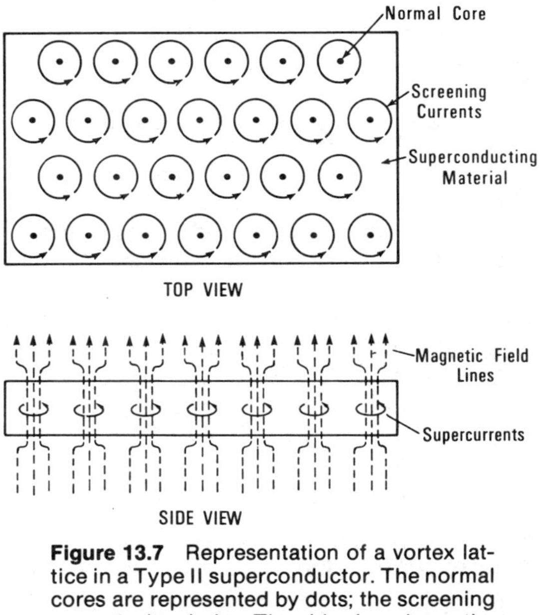

Figure 13.7 Representation of a vortex lattice in a Type II superconductor. The normal cores are represented by dots; the screening currents, by circles. The side view shows the distortion of an applied uniform magnetic field as it is channeled through each vortex core by the screening currents, leaving most of the superconducting material free of field.

13.3.2 Upper critical field values for practical materials

A listing of H_{c2} values for practical superconducting alloys and compounds is presented in Table 13.2. Like the critical temperature, the critical field is an *intrinsic* property of the superconducting material, determined by crystal structure, chemical composition, and atomic-scale characteristics rather than by metallurgical treatments (cold work and heat treatments) or other extrinsic properties. Values of H_c and H_{c2} are given for the elements in Table 13.1.*

From a historical standpoint, it was the rather recent development of the Type II superconductor Nb_3Sn by Kunzler (1961) that aroused interest in superconductors for practical magnet applications (up to ~12 T). Prior to that time only low-critical-field, low-current superconductors had been known. Soon afterward two other practical Type II superconductors were found, Nb–Zr and Nb–Ti. They have H_{c2} values lower than that of Nb_3Sn, but unlike Nb_3Sn they are ductile alloy materials rather than brittle compounds, which makes them much easier to use where moderately high magnetic fields (≤8 T) are required. Virtually all practical superconducting magnets produced at this time are made from either Nb–Ti or Nb_3Sn, depending on the magnetic field required. The very high values of H_{c2} for several other materials, such as V_3Ga, Nb_3(Al–Ge), or Nb_3Ge, qualify these superconductors also for high-field magnet applications. Practical fabrication methods for these materials are now under development.

13.3.3 Temperature dependence of critical field and upper critical field

The critical field values given in Tables 13.1 and 13.2 pertain only to the low temperatures indicated. As the temperature of a superconductor is raised toward T_c, the observed critical field de-

*For completeness, a third critical field, H_{c3}, must also be mentioned, although it is not very important in most applications. It arises when a field is applied parallel to the surface of a superconductor. Under these conditions, a very thin (< 0.1 μm) superconducting sheath can exist up to a field H_{c3} = 1.695 H_{c2}. The total current carried by the surface sheath is small and does not have much practical significance.

creases until at T_c it becomes zero. The approximate dependence of both H_c and H_{c2} on temperature is given by the relation:

$$H_c(T) = H_c(0)[1 - (T/T_c)^2] \quad (13.2)$$

Curves of T_c vs. H_{c2} for several high-field superconductors are given in Fig. 13.8. These curves define the limits of existence of superconductivity for each material.

13.3.4 Limits for upper critical field

The value of H_{c2} depends on several microscopic variables — κ, T_c, and spin-orbit scattering — described in Appendixes C, D, and E, respectively. In these appendixes it is shown that the highest critical fields are obtained in very dirty Type II superconductors, where: (a) the Ginzberg–Landau parameter, κ, is large, (b) T_c is high, and (c) there is large spin-orbit scattering. Thus far, the highest values of $H_{c2}(0)$ have been measured in the Chevrel phase superconductors,

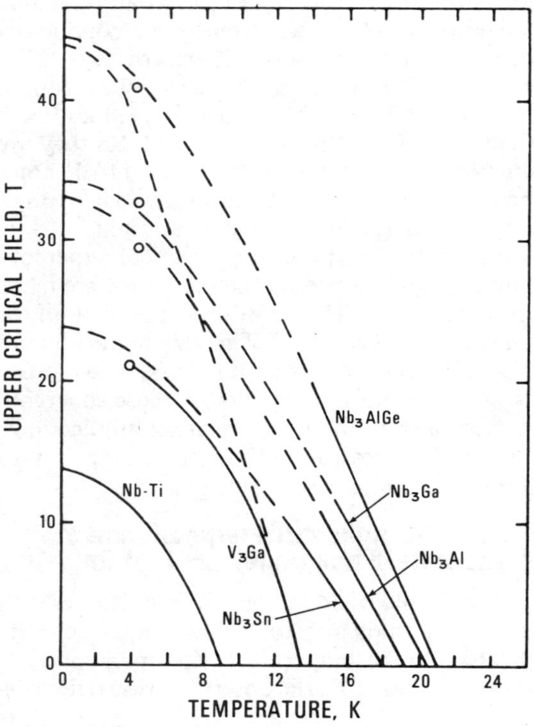

Figure 13.8 Upper critical field, H_{c2}, vs. T for several high-field superconductors. The data points at 4.2 K above about 20 T were obtained with pulsed fields. The solid lines correspond to data obtained in dc fields and the dashed lines are calculated assuming no paramagnetic limiting (Fietz and Rosner, 1974).

such as $PbMo_6S_8$, where values of $H_{c2}(0)$ in excess of 60 T have been found. Although these materials do not have exceptionally high T_c values (~14 K), they have high values of κ (see Eq. D.3 in Appendix D) and strong spin-orbit scattering to cancel the paramagnetic effect (Appendix E). (In comparison, $H_{c2}(0)$ for the highest T_c superconductor, Nb_3Ge, has been measured to be only about 36 T.)

13.3.5 Measurement of critical field and upper critical field

Critical field (for Type I superconductors) and upper critical field (for Type II superconductors) are usually determined from measurements of the superconducting magnetization curves. As shown schematically in Fig. 13.6, H_c or H_{c2} is simply the field at which the magnetization becomes zero.

The samples for such magnetization measurements are usually bulk material or bundles of short wire pieces. The magnetization is detected as shown in Fig. 13.3b by measuring the change in inductance of a coil wrapped around the material. This may be accomplished either by a mutual-inductance, self-inductance, or moving-sample technique. Calibration of the various systems is nontrivial, but is usually accomplished by assuming the specimen is perfectly diamagnetic at low fields or by using a standard reference specimen with known magnetization characteristics.

13.4 CRITICAL CURRENT

In addition to critical temperature and field, there is an upper limit to the current a superconducting material can carry called the *critical current*, I_c. If I_c is divided by the cross-sectional area of the conductor, one obtains a corresponding critical current density, J_c. The critical current is a strong function of both the applied field and temperature. The J_c, H_{c2}, and T_c are all related, as shown in Fig. 13.9. In the region between the origin and the critical J–H–T surface, the material is superconducting; outside this surface the material is normal (nonsuperconducting).

13.4.1 Critical current in Type I superconductors

In Type I superconductors there is a well-defined, intrinsic relationship among the three parameters. The current in a superconducting wire generates a magnetic self-field at the surface of the wire. The critical current is simply the cur-

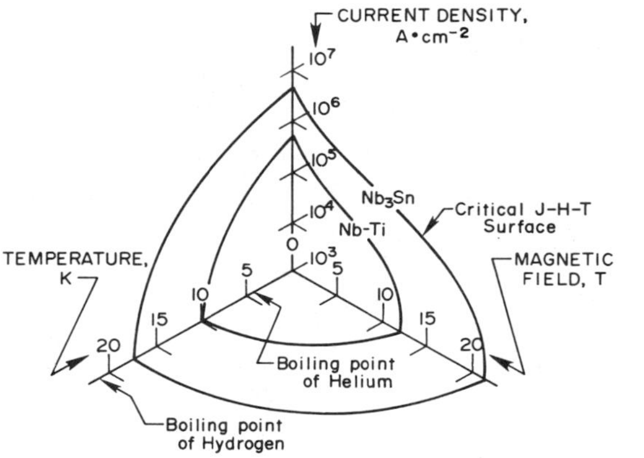

Figure 13.9 Interrelation of J_c, H_{c2}, and T_c for Nb_3Sn and Nb–Ti (W. A. Fietz, 1982). In the region between the origin and the critical *J–H–T* surface, the material is superconducting. Beyond this critical surface, the material is in the normal state. The current density is the average over the noncopper area.

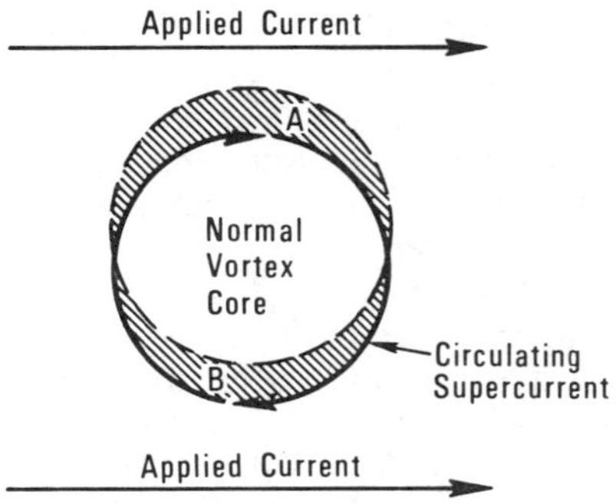

Figure 13.10 Diagram of a vortex, showing its motion perpendicular to an externally applied current. The plane of the figure is assumed to be superconducting material; the vortex core is normal (nonsuperconducting).

rent at which this self-field at the surface equals the critical field, H_c. This is known as the Silsbee rule (Silsbee, 1916). The dependence of the critical current on applied external magnetic field also arises naturally from this rule. As external field is applied to the superconductor, the critical current is reduced, because a self-field equal to the difference between the critical field and the applied field destroys superconductivity.

13.4.2 Critical current in Type II superconductors

In Type II superconductors, the situation is not so straightforward. The upper critical field, H_{c2}, is usually so high that another effect limits the current long before the self-field reaches H_{c2}. This is the phenomenon of flux motion. When a current density, J, is applied to a Type II superconductor in the mixed state, the flux tubes discussed in the previous section (§ 13.3, Fig. 13.7) experience a Lorentz force, F_L:

$$\vec{F}_L = \vec{J} \times \vec{\phi}_0 \qquad (13.3)$$

This force tries to move the flux lines at right angles to both the applied current and field.

On a microscopic scale, this can be qualitatively explained by reference to the schematic picture of a vortex shown in Fig. 13.10. Each flux tube, or vortex, consists of a normal core containing one quantum of flux, ϕ_0, surrounded by a circulating current that screens the magnetic field in the core from the rest of the superconducting material. When an externally applied current is added to the circulating supercurrents of a vortex, the combination of the two currents exceeds the critical current of the superconducting material in region A and decreases the current below the critical value in region B. Thus, region A, originally superconducting, is driven normal, and region B, originally part of the normal core, reverts to the superconducting state. The net result is that the vortex moves upward in Fig. 13.10. Although the explanation is simplified, this process is the basic microscopic origin of the Lorentz force, which tends to move vortices at right angles to the applied field and applied transport current.

The connection between the Lorentz force and the critical current is that flux-line movement under the action of the Lorentz force dissipates energy, which heats the superconductor. When vortices move with velocity, $\vec{v}$, they induce an electric field, $\vec{E}$, in the superconductor along the direction of current:

$$\vec{E} = n\vec{v} \times \vec{\phi}_0 = \vec{v} \times \vec{B} \qquad (13.4)$$

where n is the number of flux lines per unit area ($n\,\phi_0 = B$). This results in a power dissipation per unit volume of $\vec{J}\cdot\vec{E}$. If the power dissipation is large enough (as it is at the high current densities obtained in practical superconductors), the heat generated will raise the temperature of the superconductor above T_c, driving it into the normal state. Thus, *dissipative flux-line movement* causes the critical current density in Type II materials to be less than that given by the Silsbee rule.

Fortunately, this motion can be stopped by introducing metallurgical defects (e.g., voids, grain boundaries, dirt) into the material. These defects "pin" the vortices in place, arresting their movement. Practical superconducting materials have very high pinning forces, which prevent vortex motion, even when the Lorentz force acting on the vortices is very high. Thus, pinning defects enable superconductors to carry very high current densities. This is the main difference between the early "clean" (but impractical) superconductors and the "dirty" (but practical) superconductors available today.

The key factor in determining the limit of current that can be carried by Type II superconductors is the maximum pinning strength of the material that keeps the flux lines from moving. This critical current, I_c, is just the current at which the Lorentz force acting on the flux tubes equals the maximum pinning force holding them in place:

$$\vec{J}_c \times \vec{\phi}_0 = -\vec{f}_p \tag{13.5}$$

where f_p is the maximum pinning force per unit length of flux tube. Multiplying both sides of Eq. 13.5 by n, the number of flux lines per unit area, an equivalent expression is obtained for J_c in terms of the maximum pinning force, F_p, per unit volume of the superconductor:

$$\vec{J}_c \times \vec{B} = -\vec{F}_p \tag{13.6}$$

The J_c of a Type II superconductor is thus determined by its maximum pinning strength, F_p.

13.4.3 Limits for critical current density in Type II materials

The maximum pinning force is strongly influenced by heat treatments, cold work, and chemical purity (i.e., metallurgical factors). Thus, unlike T_c and H_{c2}, J_c in Type II materials is not an intrinsic atomic-scale property, but an extrinsic property dependent on the metallurgical history of the material. Its value cannot be predicted with any certainty, and it must be determined for each lot of commercially prepared wire. In fact, J_c can be made to vary anywhere from about 100 $A \cdot cm^{-2}$ to 100 $kA \cdot cm^{-2}$ in the same material (at the same temperature and field), just by changing the metallurgical treatment given the specimen.

An example of the interaction of a flux-line lattice with defects in a superconductor is shown in Fig. 13.11. Notice that the flux lines are pinned, in this case, by dislocation lines and align themselves accordingly. Note also that between dislocation lines where the pinning force is small, the flux lines arrange themselves in a triangular array or lattice. This is due to their self-interaction. Thus, flux lines not at a pinning site can be held in place by their interaction with flux lines that actually are pinned.

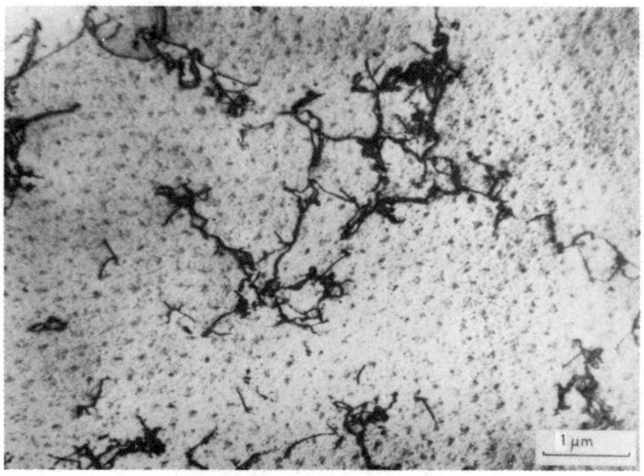

Figure 13.11 Electron micrograph of a thin film of niobium. Position of flux lines is indicated by decoration with iron particles. The underlying dislocation structure can be seen to have interfered with the flux line lattice [from C. P. Herring (Dew-Hughes, 1974)].

The need for strong pinning to obtain high J_c requires that most practical superconductors be cold-worked, impure materials. As a result, they are usually metallurgically hard, and hence, the term *hard* superconductors is sometimes used. A discussion of the pinning force and the factors that affect it is given in Appendix F.

13.4.4 Critical current density values for practical materials

The J_c–H characteristics for three commercial superconductors are shown in Fig. 13.12. The curves represent optimum values for J_c obtained in optimally cold-worked, heat-treated wire specimens. (The reasons behind the field dependence of J_c at very low and very high fields are discussed in Appendix F.)

Curves, such as those in Fig. 13.12, are needed for each wire in setting the operating currents for magnet design. Usually a magnet load line in the high-field region of the magnet is drawn on the I_c–B plot, as shown in Fig. 13.13. The magnet load line starts at the origin and has a slope determined by the relationship between I and B for the particular magnet geometry. Critical current for the magnet is determined from the intersection of the load line and the I_c–B characteristic of the wire. Depending on the conservatism of the

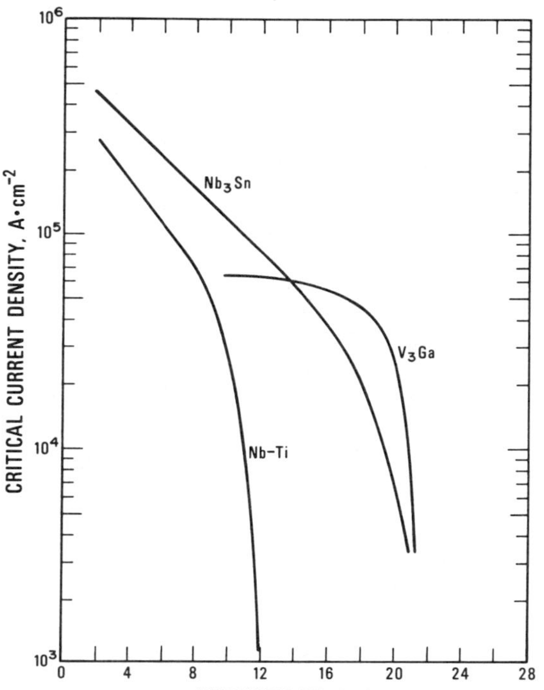

Figure 13.12 Critical current density, J_C, vs. magnetic field, *H*, for several high-field superconductors.

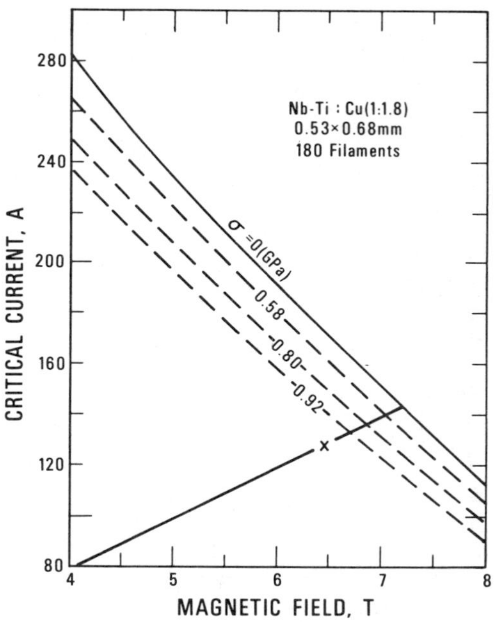

Figure 13.13 Critical current vs. field characteristic of a Nb–Ti conductor, showing magnet load line and the choice of a typical operating current point, *X*. Dashed curves indicate the degradation of the I_C–B characteristic due to stress (discussed in § 13.8).

design and expected stress levels, a magnet operating current is then chosen at some point below this critical current (point *X* in Fig. 13.13). (In large magnets where the wire may be subjected to high stress, the I_c–B characteristic may be degraded substantially, as shown in Fig. 13.13, affecting the choice of operating current. This is discussed in § 13.8.)

In most cases, superconducting magnets operate in boiling liquid helium at 4.2 K, and thus usually only the I_c–B characteristic at 4.2 K is needed. However, for elevated temperature operation and for stability calculations (see § 13.6), data on the temperature variation of J_c are needed. Very few data of this type are available, but in general, J_c has been observed to have a temperature dependence roughly given by:

$$J_c \simeq J_0(1 - T/T_c) \qquad (13.7)$$

where J_0 is the critical current density at 0 K. Data for Nb–Ti, Nb_3Sn, and V_3Ga are shown in Figs. 13.14, 13.15, and 13.16.

Much work remains to be done in systematically evaluating the effects of temperature, stress, fatigue, and other metallurgical properties on J_c for commercial materials. Virtually no J_c data are available for noncommercial materials because of the lack of specimens suitable for critical current density measurements.

13.4.5 Measurement of critical current

Critical current is usually determined by measuring the voltage–current characteristic of a short wire specimen placed in an external magnetic field. For magnet applications, the field is usually oriented perpendicular to the wire, since I_c is smallest for this orientation. Three geometries are usually used to accomplish this; they are shown in Fig. 13.17. The first two methods employ a simple solenoidal magnet to obtain the external field and use a four-terminal method for introducing current and detecting voltage. In the first method (Fig. 13.17a), the wire specimen is bent into a hairpin geometry and the voltage is detected across the short section perpendicular to the applied field, between points 1 and 2. With this method, the detection length is limited to the bore diameter of the magnet. In the second method (Fig. 13.17b), the specimen is spirally coiled around a cylindrical support such that the entire length of wire is nearly perpendicular to the ap-

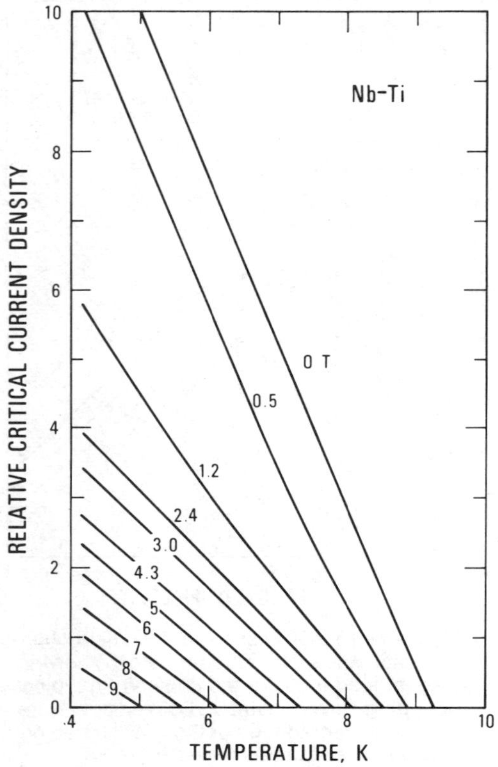

Figure 13.14 Relative critical current density vs. temperature for Nb–Ti wire. Each curve corresponds to a constant value of the magnetic field (Hampshire, Sutton, and Taylor, 1969).

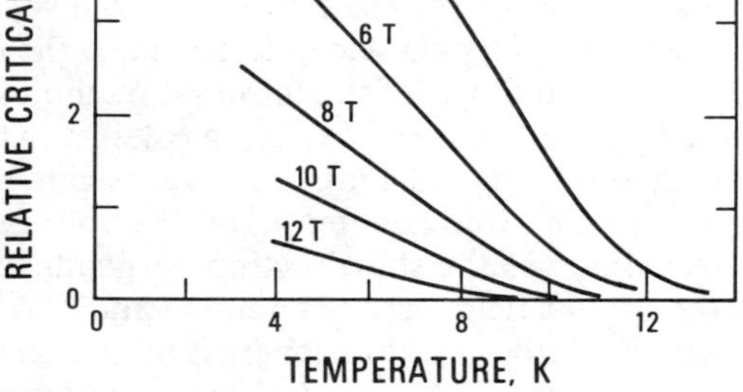

Figure 13.15 Relative critical current density vs. temperature at constant magnetic field for Nb_3Sn wire conductors (Iwasa and Montgomery, 1975).

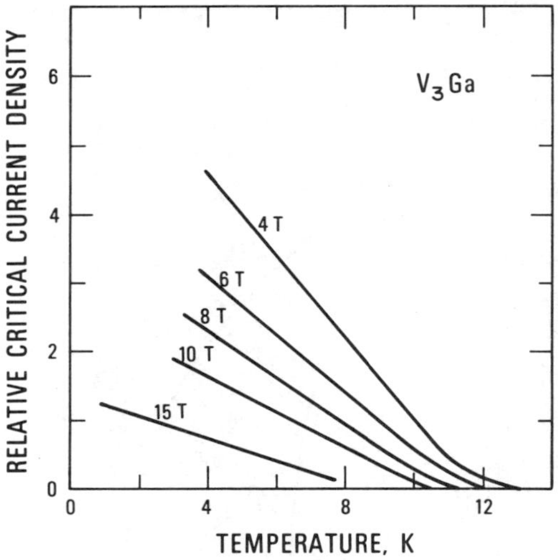

Figure 13.16 Relative critical current density vs. temperature at constant magnetic field for V_3Ga wire (Iwasa and Montgomery, 1975).

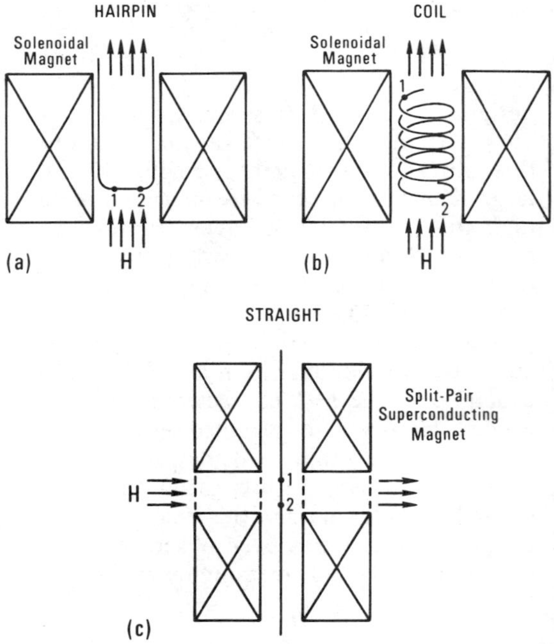

Figure 13.17 Experimental arrangements for determining short sample critical currents. Using a simple solenoidal magnet: (a) hairpin sample geometry and (b) coil sample geometry. Using a split-pair magnet: (c) long, straight geometry. Voltage contacts are attached at points 1 and 2.

plied field. With this method, a much longer length is available so that lower resistivities can be detected with the same voltage detection capability. The third method utilizes a split-pair magnet to obtain the external field. The advantage here is that brittle, straight specimens can be tested without bending them into a tight arc, which damages the superconductor (see "Bending damage" below) and the geometry is well suited to application of strain to the specimen for strain vs. critical current density studies.

The V–I characteristic is measured as a function of applied field, and I_c determined according to one of several criteria:

a. The current at which the effective resistivity reaches a given level, typically 1 to 100 pΩ·cm (referred to as ρ_c).

b. The current at which the electric field across the test section reaches a given level, typically 1 to 10 μV·cm^{-1} (referred to as E_c).

c. The current at which thermal runaway sets in and the wire is driven into the normal state.

The three criteria are indicated in Fig. 13.18. Of the three, the first two are more desirable because they are functions of the wire specimen itself. The third is a strong function of the type of cooling used, the cooled surface area, and the surface condition of the specimen. The criterion used to determined critical current should be specified because, even with the first two criteria, there can be a significant difference in I_c values obtained for specimens that have a gradual rise in the V–I curve like that shown in Fig. 13.18. This usually arises from damaged or nonuniform filaments as discussed in Appendix G. A thorough check list of the common problems encountered in measuring critical current follows.

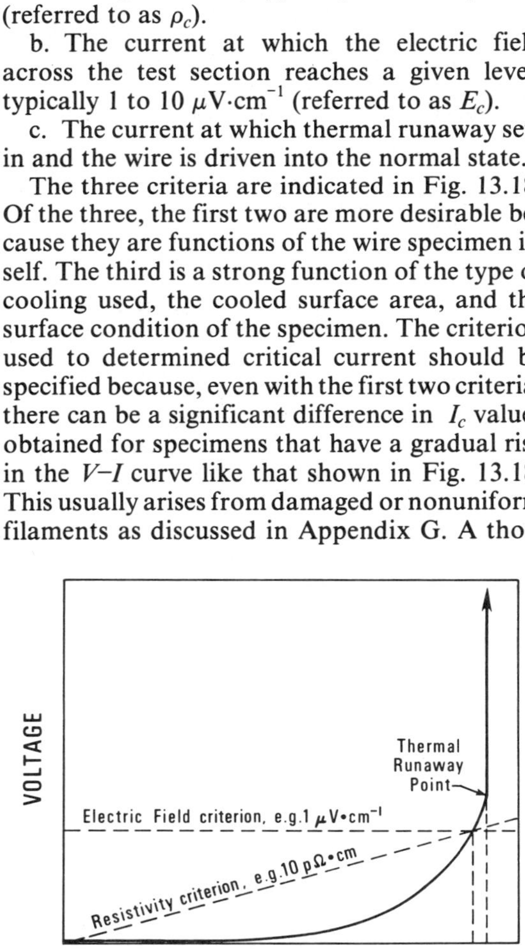

Figure 13.18 The current–voltage characteristic of a practical superconductor with transition, electric field, and resistivity criteria for critical current (Clark and Ekin, 1977).

13.4.6 Critical current measurement problems and their identification

Several experimental factors can significantly affect the measurement of critical current. They are refrigeration, resistive heating, wire movement, bending damage, voltage detection, and current transfer.

Refrigeration. If the sample is poorly cooled, thermal runaway may occur at current levels below where criteria a or b are met. Because of this problem, I_c is usually measured with the wire specimen directly immersed in liquid helium at 4.2 K. Measurements at higher temperatures are more difficult to make, but can be obtained by thermal anchoring to a controlled heat sink or immersing the specimen in liquid hydrogen. Measurements at temperatures down to 2 K can be obtained by using a pumped liquid-helium bath.

Resistive heating. Heating in the current leads and joints can also lead to premature thermal runaway. Usually, helium-vapor-cooled leads and superconducting bus bars are used to keep the resistance low. When the specimen is soldered to the current bus bar, the soldered section must be made long enough to avoid resistive heating at the joint (typically 10 to 40 wire diameters, depending on the superconductor).

Wire movement. When a wire specimen carries current at right angles to the applied magnetic field, the resultant Lorentz force can be substantial. This is a result of the flux pinning force being transmitted to the wire through its crystal structure. If the wire moves under the action of this force and the movement is opposed by friction or by impact against the support, the wire will be locally heated. A movement of only 10 μm can raise the temperature of a typical superconducting wire enough to initiate thermal runaway. Therefore, provision must be made to anchor the wire mechanically and prevent movement at right angles to the applied current and magnetic field.

Bending damage. The bending diameter of the first two test fixtures discussed above must be large enough that bending strain does not damage the specimen. For Nb–Ti, this is usually no problem; it can be severely strained with only about a 5% to 10% decrease in the critical current. For the brittle compound superconductors (e.g., Nb_3Sn, V_3Ga), it is quite a problem. As discussed in § 13.8 they show significant degradation of the critical current at a strain of only 0.2%. Typically, the bend diameter of the test fixture must be greater than about 300 wire diameters to avoid damage in brittle superconductors.

Voltage detection. A voltage detection capability in the microvolt range is usually needed to see the onset of resistivity as I_c is approached. A microvolt sensitivity is also needed to detect I_c at resistivities below those levels required by Joule heating requirements of most applications (typically 10 pΩ·cm or less). In coil specimens (see Fig. 13.17b) where a longer length of wire is available, the voltage detection requirements are less severe, provided there is some means to null induced voltages, such as counterwinding the voltage lead with the coil specimen.

Current transfer. When current is introduced into a multifilamentary composite, such as the one shown in Fig. 13.1, it must first traverse the normal matrix materials before penetrating to the inner superconducting filaments. This generates a voltage, which can be substantial near the current contacts if the matrix resistivity is high. Its effects can be avoided by leaving enough sample length between the current contact and the first voltage tap for the current distribution to equalize among the filaments. The sample length needed depends on the desired critical current resistivity criterion, ρ_c; the lower ρ_c is, the more sample length that is required. The needed minimum sample length, $x_{\min}$, may be approximately determined from the following relation (Ekin, 1978b):

$$x_{\min} = (0.1/\eta)^{1/2} (\rho_m/\rho_c)^{1/2} D \qquad (13.8)$$

where D is the diameter of the wire specimen, ρ_c is the desired level of resistivity criterion, ρ_m is the matrix resistivity transverse to the wire specimen, and η is an index of the rate of increase in resistivity as I_c is approached [$\rho \,\alpha\, (I/I_c)^\eta$, where typically $\eta \simeq 30$ to 50 for Nb–Ti, $\eta \simeq 20$ to 30 for multifilamentary Nb_3Sn and $\eta \simeq 60$ for Nb_3Sn tape]. This equation has been evaluated for typical Nb–Ti and Nb_3Sn multifilamentary wires, and the results are shown in Fig. 13.19. Thus, for example, in measuring I_c for a Nb_3Sn tape, the sample length between the current contacts and the voltage taps must be about 60 wire diameters to detect I_c at a resistivity level of a pΩ·cm.

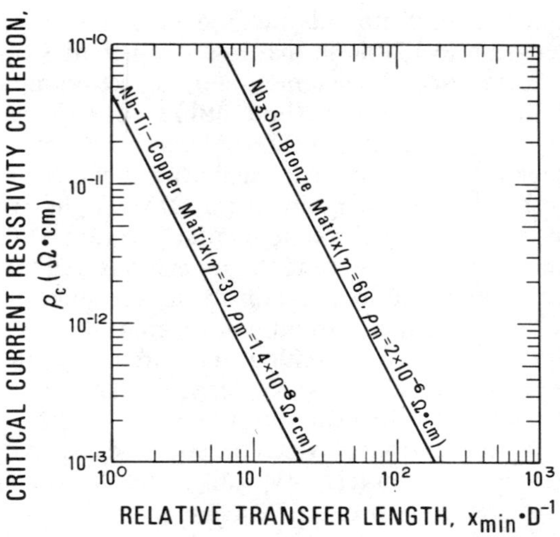

Figure 13.19 Current-transfer length $x_{\min}$ for typical Nb–Ti and Nb_3Sn multifilamentary superconductors, where D is the wire diameter and ρ_c is the critical current resistivity criterion (usually determined by the apparatus detection limit). The approximate current-transfer length for other values of η and ρ_m may be obtained from Eq. 13.8 (Ekin, 1978b).

Identification. Fortunately, each of these experimental problems can usually be recognized from the shape of the measured V–I characteristic. The ideal V–I characteristic is shown in curve A of Fig. 13.20. The current through the specimen increases until I_c is approached, whereupon the voltage starts to increase rapidly. The V–I curve is *reversible*. That is, on lowering the

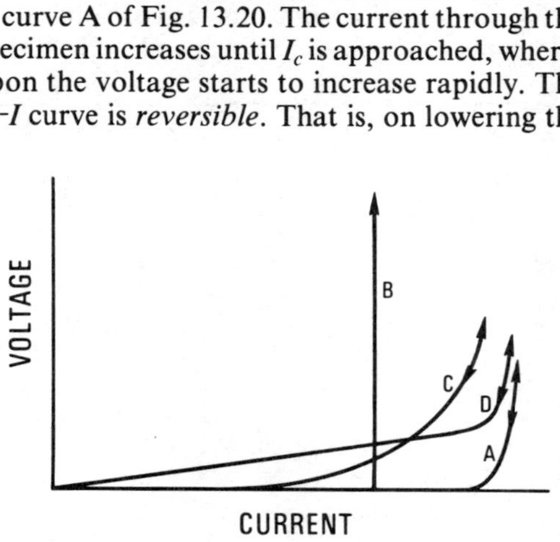

Figure 13.20 Characteristic shapes of V–I curves: A — ideal characteristic, B — premature thermal runaway, C — damaged sample, and D — inadequate current-transfer length between current contact and voltage taps.

current, the voltage retraces its original path. Curve B on the other hand, indicates the type of V–I characteristic obtained if wire movement or inadequate refrigeration causes premature thermal runaway. The voltage precipitously and *irreversibly* increases. The pen flies off the X–Y recorder, and unless the current is reduced quickly or shunted, the specimen may vaporize. Curve B is also the shape that is obtained when the voltage detection capability is inadequate. In this case, the true critical current is reached, but there is no way to tell with certainty if the quench point is the true critical current. The only sure way is to have enough sensitivity to see the reversible voltage increase. Curve C shows the broad flux–flow region that occurs when a superconductor has been damaged. Curve D shows the approximate V–I characteristic that results from current transfer.

13.5 COMPOSITE FABRICATION

Thus far we have been discussing properties of the superconducting material itself: the critical temperature, field, and current. However, practical superconducting conductors are fabricated as a twisted multifilamentary composite with substantial amounts of normal metal added. Only in such form does the wire have adequate stability, ac loss, and mechanical properties for practical application. These properties are a function of the entire composite structure, not just the superconducting component. Before discussing each of these properties in detail in § 13.6, § 13.7, and § 13.8, the processes for fabricating multifilamentary composites are considered in this section.

The various components of a typical superconducting composite strand are illustrated in Fig. 13.21. The strand is made up of many superconducting filaments embedded in a high-conductivity normal-metal matrix, such as copper or aluminum (Fig. 13.1). As many as ten thousand or more superconducting filaments may be combined into a single strand, and typically these are twisted with a pitch of about one turn per cm.

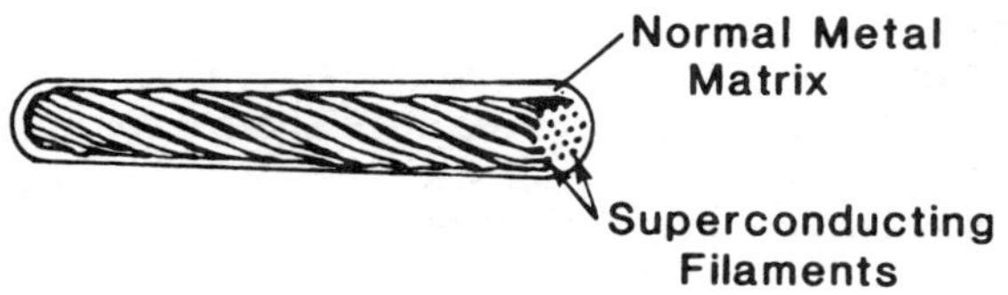

Figure 13.21 Superconducting composite strand.

The superconductor filament diameter is usually limited to less than 100 μm (as discussed in § 13.6.2 and § 13.6.3), and the overall diameter of the strand is usually limited to 1 or 2 mm (as discussed in § 13.6.2). Larger conductors may be formed by cabling or braiding together several strands, as shown in Fig. 13.22. Superconducting strands such as that in Fig. 13.21 may also be combined with stainless steel strands or other structural material to form reinforced cables (see Fig. 13.23).

Perhaps more than any other single development, the fabrication of superconducting wire in multifilamentary form has been responsible for the recent widespread use of superconducting magnets. Originally, superconducting wire was manufactured only as a single, large strand of

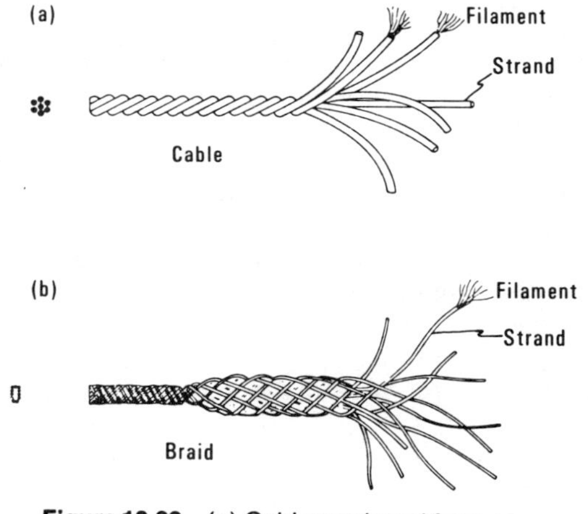

Figure 13.22 (a) Cable produced from composite strands. (b) Braid.

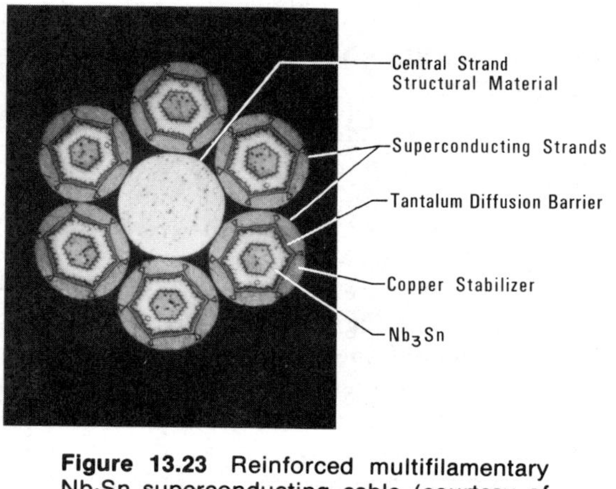

Figure 13.23 Reinforced multifilamentary Nb_3Sn superconducting cable (courtesy of Intermagnetics General Corporation).

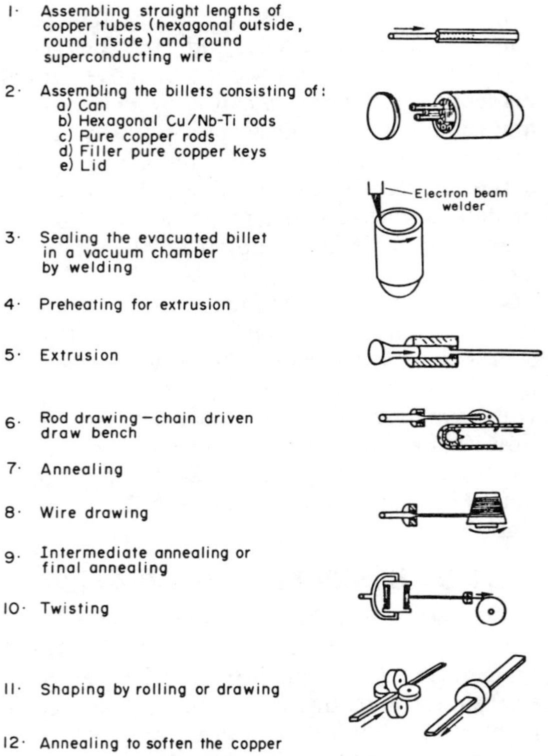

Figure 13.24 Fabrication steps of multicore Nb–Ti filament conductors (Hillman, 1981).

superconductor surrounded by a small amount of copper cladding. In such form, superconducting wire was unstable and exhibited high ac losses. The magnets produced from such wire were unreliable and rarely performed as expected, no matter how good the superconducting material from which the wire was made.

The multifilamentary composite structure now in widespread use is designed to perform two functions not found in the early wire: (a) stabilization against magnetic and mechanical disturbances is accomplished by finely subdividing the superconductor into 1- to 100-μm-diameter filaments and surrounding them with substantial amounts of high conductivity copper or aluminum (discussed in § 13.6); (b) ac losses are reduced primarily by twisting and transposing the superconducting filaments and keeping the filament diameter small (discussed in § 13.7). In some cases strength and strain tolerance are increased by adding high-strength, reinforced strands such as stainless steel to the composite (discussed in § 13.8).

There are two different types of fabrication procedures used to manufacture composite wires, depending on whether the superconductor is a ductile alloy or a brittle compound. Ductile superconducting alloys, such as Nb–Ti, are comparatively easy to fabricate into composites using conventional wire drawing procedures. Brittle superconducting compounds, such as Nb_3Sn, are not so straightforward to produce in multifilamentary form, and unconventional procedures or special heat treatments must be used. Processes for manufacturing each type are described below. In both cases, however, the ultimate aim is a composite structure having the two essential features listed above.

13.5.1 Ductile alloy superconducting composites

This group includes superconducting binary alloys, such as Nb–Zr (25 to 33 at.% Zr) and Nb–Ti (55 to 80 at.% Ti), or three-component alloys, such as Nb–Ti–Zr, or Nb–Ta–Ti. By far, the most popular is the Nb–Ti system.

Typically, the wire starts out as a billet similar to that shown in Fig. 13.24. The Nb–Ti rods in Fig. 13.24 are obtained from larger ingots by hot extrusion, interrupted occasionally for ~1/2-h high-temperature (~800° C) vacuum anneals to remove work hardening. Each rod is inserted into a copper tube having a hexagonal outer shape. A number of these are then nested in a large copper can, the can is evacuated, and a lid welded on the end. The entire assembly shown in step 3 of Fig. 13.24 is then reduced, initially by hot extrusion, followed by conventional wire drawing. Anneals are carried out as before to remove work hardening. The last of the high-temperature anneals is given before the final drawing operation so that the wire receives an additional (~90+%) reduction in cross-sectional area after this anneal. This ensures that there are sufficient cold-work dislocations in the superconducting filaments for adequate flux pinning (see § 13.4). After drawing, the wire is given an additional low-temperature (~400° C) anneal to form a strong-pinning dislocation cell structure and to enhance the critical current. The conductor is then twisted at a pitch of about one twist per centimeter to reduce ac losses. The twisted composite wire is then usually passed through a set of rollers to give it a rectangular cross section for a compact winding structure. A final low-temperature heat treatment is used to raise the conductivity of the copper stabilizer. To produce conductors with a very large number of filaments, the composite can be given a hexagonal-shaped cross section, cut into lengths, stacked in a copper tube, and the entire procedure repeated to produce composites with up to ten thousand filaments or more. (Additional details may be found in Hillman and Hauck, 1972; Whitcomb and Dew-Hughes, 1973; or Hillman, 1981.)

The fabrication procedure is carried out so that the final diameter of the superconducting filaments is about 1 to 50 μm. This is the optimum range, because at larger diameters, the wire is magnetically unstable, and at smaller diameters, irregularities introduced during fabrication reach a significant fraction of the filament size, which leads to necking and reduced critical current.

13.5.2 Brittle compound superconductors

The brittleness of Nb_3Sn, V_3Ga, and other compound superconductors prevents them from being processed from ingots (as described in § 13.5.1 for the ductile-alloy superconductors). One of three other methods is used instead: (a) the compound is deposited directly on a continuous substrate tape by vapor deposition, (b) the compound is deposited on a substrate tape by plasma-plating techniques, or (c) the constituent elements of the compound are fabricated into their final configuration as ductile alloys and then reacted by a diffusion process to form the brittle compound. The direct deposition techniques (a and b) are generally limited to ribbon or tape geometries. The diffusion technique (c), on the other hand, may be used with either tape conductors or multifilamentary wire conductors and is the most popular process for production of commercial Nb_3Sn conductor.

Tape conductors

a. Chemical vapor deposition: In the chemical vapor deposition process for Nb_3Sn (Hanak, Strater, and Cullen, 1964) a substrate ribbon of Hastelloy, stainless steel, or niobium, or other structural material is passed through a reaction chamber containing gaseous chlorides of niobium and tin and hydrogen gas. The tape is resistively heated to $\sim$1000°C, at which temperature the chlorides are reduced by the hydrogen, and Nb_3Sn is deposited on the ribbon. The thickness of the superconducting layer is controlled by the rate at which the ribbon passes through the chamber and is typically about 1-μm thick. Copper stabilizer strips can be soldered on each side of the ribbon to form a sandwich. Using a stainless steel substrate gives added mechanical strength. Different variations of this process are discussed by Echarri and Spadoni (1971).

b. Plasma plating: In the plasma plating technique for Nb_3Sn (Haid and Fietz, 1969), an inert gas carrier propels powders of niobium and tin at high velocity through an electric arc. The arc ionizes and heats the gas, which raises the powders above their melting point. This plasma stream is then sprayed onto a substrate ribbon (e.g., copper, stainless steel), leaving a thin coating of Nb–Sn mixture on the surface. The ribbon is then given a final heat treatment, which forms the Nb_3Sn layer.

c. Diffusion process: The diffusion processes discussed here are perhaps the easiest methods for forming composites of brittle superconductors. In its simplest form, the diffusion technique for Nb_3Sn consists of forming ductile niobium into a tape or ribbon and coating this with tin, usually by directly dipping the ribbon in a molten tin bath. The ribbon is then heated to between 900 and 1000°C in a reaction chamber at which temperature the tin diffuses into the niobium and reacts to form Nb_3Sn (further discussion of the original process may be found in Benz, 1968). An example of a tape conductor made in this way is shown in Fig. 13.25. Similarly, V_3Ga ribbons may be formed by passing a vanadium substrate through a molten gallium bath.

The tape geometry has two disadvantages arising from the relatively large width of the conductor. First, its stability decreases significantly in perpendicular fields, such as those in the fringing fields near the outer ends of a magnet coil. Secondly, the ac losses are high, which necessitates relatively long magnet charging time (1/2 to 1 h). The development of multifilamentary wires has proceeded for these reasons.

Multifilamentary conductors

a. Bronze diffusion technique: Perhaps the most popular method for producing multifila-

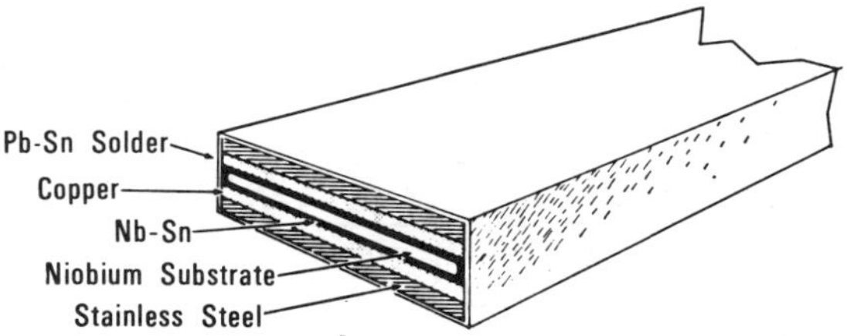

Figure 13.25 Cross-sectional view of a Nb_3Sn tape conductor (Benz, 1968).

mentary Nb_3Sn conductors is the bronze diffusion technique (Kaufmann and Pickett, 1971). Fabrication is similar to that for multifilamentary Nb–Ti, except the starting billet consists of niobium rods in a Cu–Sn bronze matrix (rather than a pure copper or aluminum matrix). The billet is processed into wire, exactly as for Nb-Ti multifilamentary conductors, except that when final size is reached (i.e., filament diameters of about 1 to 10 μm), the entire conductor is given a heat treatment at 700° C for several hours. This reacts the tin in the Cu–Sn bronze with the niobium filaments, forming a thin Nb_3Sn layer at the niobium–bronze interface. The resulting structure looks like the example in Fig. 13.26. Copper stabilization material is usually coprocessed with the bronze–niobium composite, as shown in the example in Fig. 13.26. A thin tantalum diffusion barrier separates the pure copper from the bronze to prevent tin from thermally diffusing into the pure copper during heat treatment and lowering its conductivity (and hence, the stability of the wire, § 13.6).

Variations of this technique are also being developed for other materials. For example, multifilamentary V_3Ga (Suenaga and Sampson, 1971; Tachikawa, Yoshida, and Rinderer, 1972; Howe, Francavilla, and Gubser, 1977) and V_3Si (Tachikawa et al., 1972; Suenaga and Sampson, 1972) are produced by coprocessing vanadium rods in Cu–Ga or Cu–Si alloy matrices.

b. External diffusion: Bronze-alloy matrices work-harden more rapidly than pure copper, and so an alternative process has also been developed wherein a thin, pure copper matrix is used

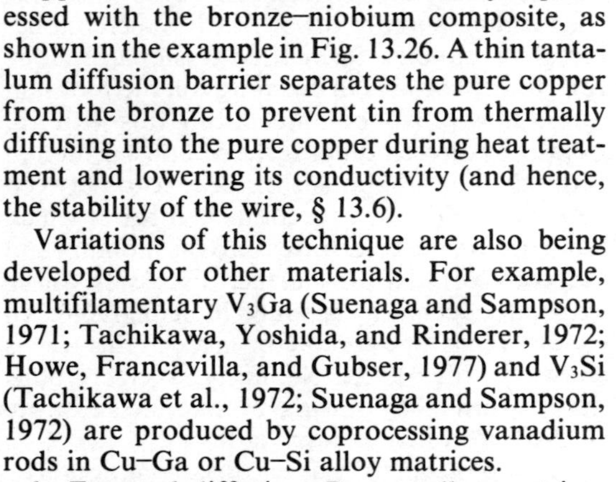

Figure 13.26 Example of multifilamentary Nb_3Sn conductor formed by the bronze technique.

instead of a bronze matrix. This eliminates the need for so many annealing heat treatments during the extrusion and drawing operations. After fabrication to final size, the wire is coated with tin and given a heat treatment that diffuses the tin through the copper matrix to react with the niobium filaments. This process also results in a higher overall current density, since a large amount of matrix material is not needed to carry the tin necessary for producing the Nb_3Sn. The disadvantages of the external diffusion technique are: (a) for highly stabilized wires, extra stabilization material must be added after the reaction heat treatment and (b) the strand size is limited to about 0.5 mm to keep diffusion times reasonably short. As a result, the bronze diffusion technique described above has been the most popular.

In addition to fabrication complexity, a problem with all the compound superconducting composites is their susceptibility to mechanical damage during handling, which further accounts for the popularity of ductile Nb–Ti conductors in spite of their restriction to lower magnetic fields ($\lesssim$ 8 T).

13.6 STABILITY

13.6.1 Types of stabilization

Perhaps the single most important function of the composite structure is to give the conductor stability. A superconductor carrying a large amount of current is inherently unstable. Any kind of magnetic or mechanical disturbance can locally heat the superconductor and drive the entire conductor into the normal state through a thermal runaway process. This liberates large amounts of heat, enough to vaporize the superconductor at practical current densities.

The origin of the thermal runaway process in a superconducting magnet (sometimes called a "quench") is fairly simple in concept: It begins when a magnetic or mechanical disturbance heats a small region of the superconductor so that it becomes resistive. Current passing through this resistive region then generates additional Joule heat. This, in turn, raises the temperature of more material, leading to more Joule heating until eventually the entire superconductor winding is driven normal. The composite structure prevents the thermal runaway process from being self-sustaining. There are two common methods for accomplishing this: flux-jump stabilization and cryostatic stabilization. The choice of method depends on the expected disturbance size. A summary of the characteristics of the two types of stability is given in Table 13.3 and discussed immediately below. Following this introduction, details of the criteria for obtaining each type of stabilization are then given in § 13.6.2 and § 13.6.3

Flux-jump stabilization is predominantly used in small (≲30-cm diameter) epoxy-filled magnets. It stabilizes the wire against magnetic instabilities (flux jumps) that might occur in response to small disturbances by (a) finely dividing the superconductor into filaments, typically 100 μm or less in diameter, and (b) embedding them in high-conductivity normal metal, such as copper or aluminum (e.g., Fig. 13.21).

The purpose of the fine subdivision is to reduce the potential energy from the excluded magnetic field, which is a source of heating. The small filament diameter also improves the cooling of the superconductor by giving it a high surface-to-volume ratio.

The purpose of the copper matrix material (see Fig. 13.1) is to conduct heat away from the superconducting filaments to the helium bath and to lower the rate of propagation of magnetic disturbances. The amount of copper required to do this is not great, typically 1 to 3 times the volume of the superconductor material.

Several theoretical criteria have been proposed for determining how fine the superconducting filaments must be and how much copper must be added to achieve stabilization against flux jumps. These are discussed in the next section (§ 13.6.2) and include the adiabatic (or enthalpy) stability criterion and the dynamic stability criterion. In practice, the dynamic stability criterion is usually used. Typically, it stipulates a filament size ranging from about 10 μm to 100 μm, depending on the type of superconductor material, the conductivity of the matrix material, and the critical current density.

The main advantage of flux-jump stabilization is that only a relatively small amount of matrix material is required. This allows a high overall current density to be achieved, typically 10 to 100 kA·cm^{-2} in most magnet applications. The chief disadvantage of flux-jump stabilization is that the superconductor can tolerate only relatively small disturbances confined to small groups of individual filaments (typically 0.1 mJ or less). The conductor is not stable against larger disturbances, which collectively affect many filaments in the wire. For example, if a flux-jump-stabilized wire in a magnet moves under the action of the Lorentz force, in most cases enough frictional heating will occur to initiate thermal runaway and the magnet will quench. For this reason, flux-jump-stabilized magnets are usually epoxy impregnated to prevent wire movement.

Table 13.3 Types of stabilization.

	Flux-Jump Stabilization	Cryogenic Stabilization: Full	Cryogenic Stabilization: Limited
Basic defining principle	Stabilization against magnetic instability (flux jump)	Heat-flow balance between composite and cryogen	Same, except based on maximum expected disturbance
Types of criteria for determining this type of stability	1) Adiabatic (enthalpy) 2) Dynamic (usually used in practice)	1) α stability parameter ($\alpha \lesssim 1$) 2) equal area theorem ($\alpha \lesssim 2$) 3) transient ($\alpha \lesssim 0.5$) (time requirement)	MPZ
Typical magnet size	Small (bore diam. usually less than ~30 cm; potted)	Large (bore diam. usually greater than ~50 cm)	Medium
Overall critical current density	0.1–1 GA·m^{-2}	<0.1 GA·cm^{-2}	0.1–1 GA·m^{-2}
Copper: superconductor ratio	1–3	5–30	1–30 (depending on expected disturbance size)
Typical mechanical disturbance tolerance	≲0.1 mJ or 10 kJ·m^{-3}	Arbitrarily large	As specified
Filament size	1–100 μm	<100 μm	<100 μm
Strand diameter	≲1 mm		

This works only if the magnetic stress is small enough that the epoxy resin will not crack when the magnet is energized (a sudden wire movement of only 10 μm can initiate thermal runaway). Thus, flux-jump stabilization is usually confined to small magnets where the stress on the wire is typically less than 100 $MN{\cdot}m^{-2}$ and the stored energy is small enough that the magnet can be easily protected in the event of a quench.

Cryogenic stabilization is predominantly used in large- and medium-sized magnets. With cryogenic stabilization, the basis for flux-jumping behavior is simply ignored, and the superconductor is paralleled with enough normal conducting material to carry the current should the superconductor be driven normal. In the event of a disturbance, the large amount of normal material can momentarily handle the current, while the liquid-helium refrigerant cools the superconductor back down below T_c. In effect, a cryogenically stabilized magnet is a liquid-helium-cooled normal magnet with a superconducting shunt to prevent losses in steady-state operation.

Cryogenic stabilization may be either full or limited. With *full* cryostabilization, enough normal conductor (stabilizer) is added so that the superconductor will recover from arbitrarily large disturbances that would drive the entire conductor normal. This eliminates the need for resin impregnation, since even wire motion is not a problem. In practice, the magnet can tolerate just about any kind of disturbance expected during normal operation. The chief disadvantage is that a lot of high-conductivity copper or aluminum is required to achieve full cryogenic stabilization, typically 10 to 20 times the amount of superconductor. This limits the overall current density to less than 10 $kA{\cdot}cm^{-2}$ in most magnet applications as well as increases the size and cost. But in large magnets with stored energies as high as 1 GJ, these considerations are not as serious as the consequences of thermal runaway.

With *limited* cryostabilization, on the other hand, only enough normal conducting matrix material is added to stabilize the superconductor against disturbances of a specified limited size. That is, the amount of stabilizer is tailored to the maximum disturbance size actually expected in practice. In many applications, the disturbances are limited in size, and only a portion of the superconductor is driven normal when they occur (not an arbitrarily large region as considered in full cryogenic stabilization). The problem is knowing what disturbances to expect in any given system. Little is known about this at present. Nevertheless, if the maximum disturbance size can be reasonably well predicted, limited cryostabilization provides a powerful means for constructing economic medium-sized magnets, tailor-designed with only enough normal metal to handle the expected disturbances. This eliminates the overdesign encountered with full cryogenic stabilization, optimizes the overall current density, and minimizes size and cost.

Several theoretical criteria have been developed for determining how much normal matrix material is needed to achieve cryogenic stability. For full cryostabilization, these include the α-stability-parameter criterion, the equal-area theorem, and the transient-stability criterion (all discussed in § 13.6.3). Criteria for limited cryostabilization are still evolving, but probably the most developed is the minimum-propagating-zone (MPZ) criterion (discussed in § 13.6.4). Basically, the principle on which all these criteria are based involves adding enough high conductivity copper or aluminum so that there is a balance between the ohmic heating in the normal conductor at operating current and the rate of cooling to the helium bath. The amount needed depends primarily on the resistivity of the matrix, the operating current, the amount of surface area cooled by cryogen, and whether the cryostabilization is full or limited.

The theoretical understanding of the dynamic processes involved in cryogenic stabilization is still evolving along with the criteria used to prescribe when it is achieved. For a quick approximate calculation, though, perhaps the easiest and simplest criterion to use is the α-stability-parameter criterion, as explained in § 13.6.3.

13.6.2 Criteria for flux-jump stabilization

In flux-jump stability, the main goal is to prevent flux jumps in the superconducting filaments from being triggered by small disturbances. Before describing the dynamic criterion for determining flux-jump stabilization, first consider what a flux jump is. It is a magnetic instability that can be diagrammed as shown in Fig. 13.27. Heat, $\triangle Q$, initially introduced by flux motion or some mechanical disturbance, locally raises the temperature of the superconductor, $\triangle T$, which lowers the critical current density, $-\triangle J_c$. This, in turn, increases the flux penetration, $\triangle\phi$ (because the shielding current densities are lowered), which produces more heat, perpetuating the cycle.

This cycle is usually broken by limiting the

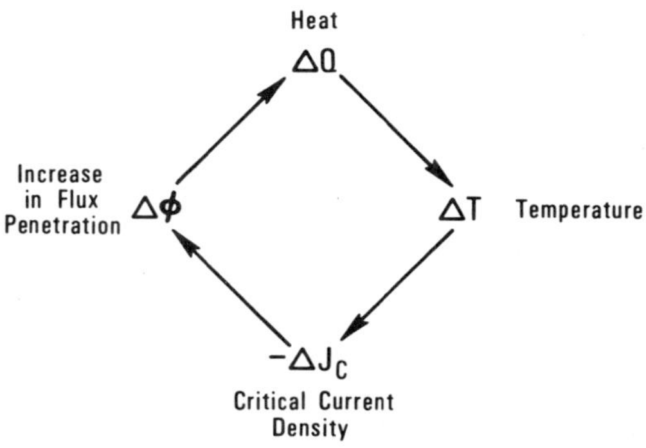

Figure 13.27 Schematic diagram of a flux jump (Wilson, 1976).

temperature rise resulting from the heat input. This is accomplished either by making the filament size so small that the small amount of heat generated can be absorbed by the heat capacity of the superconducting material without producing a large rise in temperature (adiabatic stability) or by conducting the heat away from the superconducting filament very quickly before it can cause a large temperature increase (dynamic stability). In practice, however, one does not usually consider an isolated filament in the manner prescribed by the adiabatic stability criterion, and consequently, the dynamic stability criterion is usually used for designing practical multifilamentary composites. The dynamic stability criterion is discussed below. Treatments of the adiabatic stability criterion may be found in Wilson, Walters, Lewin, and Smith (1970); Wilson (1970); or Koyama (1973), along with the details of other less-used criteria for obtaining flux-jump stability.

The dynamic stability criterion specifies a limit on filament size that ensures adequate cooling for preventing flux jumps. This is necessary because superconducting materials have a very low thermal conductivity, and considerable heat buildup can occur within a filament before it can be dissipated to the helium bath. By limiting the size of a filament the heat generated by magnetic flux motion is limited to a small volume and can be conducted quickly to the liquid helium. The dynamic stability criterion demands that the filament diameter, d, should not exceed a limiting dimension (Wilson et al., 1970):

$$d \lesssim \frac{8 T_0 k (1-f)^{1/2}}{\rho f J_c^2} \tag{13.9}$$

where T_0 is the temperature rise required to drive the superconductor normal $[T_0 = J_c/(\partial J_c/\partial T)]$, k is the thermal conductivity of the superconductor, f is the volume fraction of superconductor, ρ is the resistivity of the matrix, and J_c is the critical current density of the superconducting material alone. Also, in Eq. 13.9, the copper is assumed to be so well cooled that its temperature does not change.

The approximate maximum filament diameters for typical composites of Nb–Ti and Nb_3Sn may be evaluated using Eq. 13.9 and are shown as a function of magnetic field in Fig. 13.28. Here it is assumed for Nb–Ti in copper that $f = 0.4$, $k = 1$ mW·cm^{-1}K^{-1}, $\rho = 20$ nΩ·cm, and values of $J_c(H)$ and $T_0(H)$ have been obtained from Figs. 13.12 and 13.14, respectively. For Nb_3Sn in bronze, it is assumed that $f = 0.2$, $k = 1$ mW·cm^{-1}K^{-1}, $\rho = 2$ μΩ·cm, and values of $J_c(H)$ and $T_0(H)$ have been obtained from Figs. 13.12 and 13.15. Note that the filament size requirement for Nb_3Sn in bronze is much more stringent than for Nb–Ti in copper, because the resistivity of the bronze matrix material is much lower than that of copper. For either type of superconductor, the size requirement relaxes at high fields because J_c in the denominator of Eq. 13.9 decreases with increasing magnetic field (see Fig. 13.12).

In addition to the limitation of *filament* size,

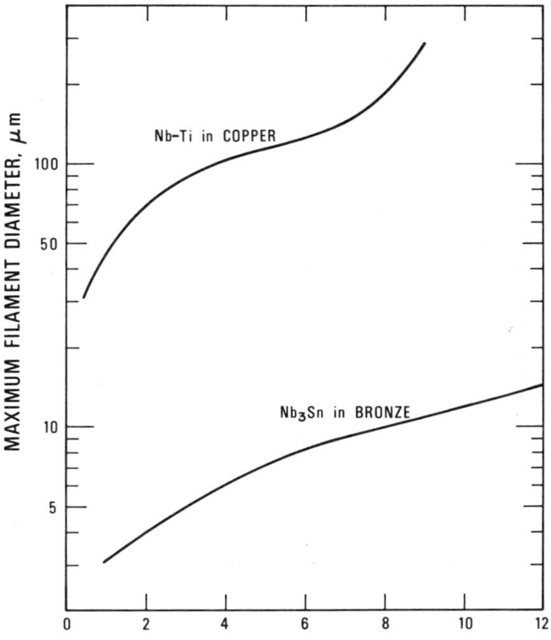

Figure 13.28 Filament diameter limits for achieving flux-jump stability in Nb–Ti and Nb_3Sn conductors, calculated using the dynamic stability criterion (Eq. 13.9).

there is also a limitation on the *overall* size of a composite strand for it to satisfy the dynamic stability criterion. This is necessary to prevent flux jumping in the self-field of the composite itself, which increases with filament diameter. The net result is that flux-jump stabilized composites are generally limited to an overall diameter less than about 2 mm (Wilson, 1976). Larger conductors may be fabricated by transposing cables or braids of smaller wires, as shown in Fig. 13.22.

13.6.3 Criteria for cryogenic stabilization

In cryogenic stability, the mechanism by which the superconducting filaments are driven normal is simply ignored. The important point is whether a normal region already formed will recover to the superconducting state. How much well-cooled normal matrix material must be placed in parallel with the superconductor to accomplish this? The answer depends on whether we know the maximum disturbance size on which to base the design. If this is known, limited cryostability criteria provide a powerful means for prescribing the minimum amount of normal material to do the job. If this is not known, or the risk of a quench cannot be tolerated, then full cryogenic stabilization is needed, and usually a relatively large amount of normal material is required.

Full cryogenic stabilization. For full cryostabilization, the α-stability-parameter criterion (Stekly and Zar, 1965; Laverick, 1965) is an approximate (but easily calculated) means for determining this. It simply says that the normal region will recover if the rate of Joule heating in the normal matrix material is less than the rate of heat dissipation to the liquid helium.

Consider a conductor of cross-section A containing a fraction, f, of superconductor and a fraction, $1-f$, of parallel normal matrix material. If the superconductor were driven normal by some disturbance, the ratio of Joule heating in the matrix to heat dissipation into the helium bath would be:

$$\alpha \equiv \frac{f^2 J_c^2 \rho}{(1-f)\dot{Q}} \frac{A}{P} \qquad (13.10)$$

where ρ is the resistivity of the matrix material, $\dot{Q}$ is the heat flux rate per unit area into the helium bath, P is the cooled perimeter of the conductor, and J is the current density in the superconducting material alone. The α-stability-parameter criterion says that the normal region will recover if α is less than unity.

The heat transfer rate, $\dot{Q}$, for film boiling is typically 0.1 W·cm^{-2}; for nucleate boiling it is 1 W·cm^{-2}. Usually a more conservative (lower) value of $\dot{Q}$ is used to design large magnets. The matrix material is usually copper because of its high conductivity. Aluminum also is starting to be used because of its lower magnetoresistance and lower cost for purified material. A comparison of the magnetoresistivity of copper and aluminum is shown in Fig. 13.29.

From Eq. 13.10, the approximate amount of matrix material required for cryogenic stabilization is determined. Results for Nb–Ti in copper are given in Table 13.4 where Eq. 13.10 has been evaluated using typical values of $\dot{Q}=0.3$ W·cm^{-2}, $\rho = 20$ nΩ·cm, and J is taken to be equal to the full critical current density of the superconductor, with a value $J_c = 0.1$ MA·cm^{-2}. Note in Table 13.4 that the required ratio of copper-to-superconductor is significantly greater than the ratio required for flux-jump stabilization (except for very small wire sizes).

The steady-state α criterion described by Eq. 13.10 does not take into account the temperature

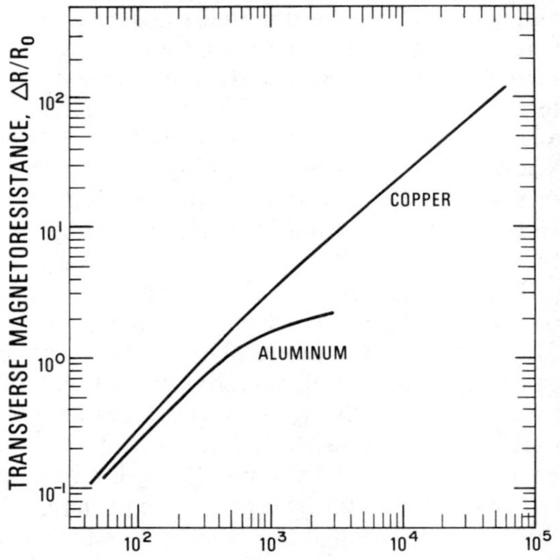

Figure 13.29 Comparison of magnetoresistance of copper (Fickett, 1972) and aluminum (Ekin, 1975b). Copper stabilizer used in commercial conductors typically has an R_0 in the range 10 to 25 nΩ·cm [R(293 K) = 1.7 μΩ·cm], whereas a typical R_0 value for aluminum stabilizer is in the range 3 to 10 nΩ·cm [R(293 K) = 2.8 μΩ·cm].

Table 13.4 Copper-to-superconductor ratio required for cryogenic stabilization.[a]

Wire Diameter (mm)	Copper-to-Superconductor Ratio	Critical Current (A)
0.5	2.4	57
1	3.6	170
2	5.3	499
5	8.6	2 040
10	12	5 890
20	18	16 600

[a]The conservative case of a round wire and film boiling has been assumed, taking $\dot{Q} = 0.3\ \text{W·cm}^{-2}$, $\rho = 20\ \text{n}\Omega\text{·cm}$ and $J_c = 0.1\ \text{MA·cm}^2$ in Eq. 13.10.

dependence of the heat-transfer coefficient, conduction between adjacent turns in a magnet, and thermal conduction along the wire. These may be taken into consideration in a more exact treatment by utilizing the so-called equal area theorem described by Maddock, James, and Norris (1969). The net result of such a detailed analysis is that, usually, somewhat less copper is actually needed than prescribed by the simple formula in Eq. 13.10.

In addition to these approximations, Eq. 13.10 also involves two implicit assumptions. First, the temperature is assumed to be uniform across the composite. Because superconducting materials have such a poor thermal conductivity, this may not be true unless the superconducting filament size is limited to a radius less than r_c, where r_c is given by Stekly (1968):

$$r_c = [k(T_c - T_h)/\rho T_c J_n]^{1/2} \qquad (13.11)$$

where k is the thermal conductivity of the superconductor, ρ is the resistivity of the matrix, J_n is the current density in the normal conductor when it carries the critical current, and T_h is the temperature of the liquid helium bath. Typically for Nb–Ti this requires the filaments to be less than about 100 μm in diameter (which, again, is comparable to that required for dynamic stabilization, § 13.6.2). So, with cryogenic stability, there is no significant relaxation of the small-filament-size requirement, even though there is considerably more normal matrix material than with flux-jump stability.

Second, recovery must be sufficiently fast that the Joule heat generated in the matrix during recovery does not vaporize all the liquid helium in a cooling channel adjacent to the conductor. This is particularly important in a force-cooled closed magnet structure with restricted cooling channels. For this situation, an additional time limit on the recovery process is needed, such as the transient-stability criterion suggested by Wilson (1976). This modifies the α-stability-parameter criterion by placing an additional requirement on the *time* it takes the normal zone to recover. It states that the total Joule heat dissipated in the normal zone during recovery should be no greater than the energy comprising the initial disturbance. This requires more copper than specified by Eq. 13.10 and in the simplest case corresponds to making $\alpha \leq 0.5$. This additional requirement ensures quick recovery, and little cryogen is required to be in contact with the conductor.

Limited cryogenic stabilization. Criteria in this category depend on a knowledge of the maximum disturbance size expected in a specific system design. If this is not known, this type of stabilization cannot be used. If it is known only approximately, however, limited-cryogenic-stabilization criteria are still very powerful design tools for optimizing current density in any given system. Criteria for limited cryostabilization are very general in scope and actually include full cryogenic stabilization as a special case where the design disturbance size is taken to be arbitrarily large.

Perhaps the most-developed criterion for limited cryostabilization is the *minimum propagating zone* (MPZ) criterion (Martinelli and Wipf, 1972; Wilson and Iwasa, 1978). It is based on the concept of the MPZ, that is, the smallest normal region of composite material that must be formed in the presence of a transport current for thermal runaway to be initiated. To understand the concept, consider a hypothetical magnet consisting of copper and superconductor, with an isotropic thermal conductivity, k, in all directions. Assume that a spherical region of radius r within the magnet has been driven normal (heated to T_c) by a transient mechanical disturbance (Fig. 13.30). The material outside r is still assumed to be superconducting.

Current passing through the normal region will generate Joule heat at a rate proportional to the volume of the region. For example, for a spherical normal zone, the heat generation rate is $(4/3)\pi r^3 J^2 \rho$, where J is the current density and ρ is the resistivity of the composite material in the normal state (essentially that of the copper). Heat will flow out of the normal region at a rate given by $4\pi rk(T_c - T_h)$ where k is the thermal conductivity of the superconducting composite

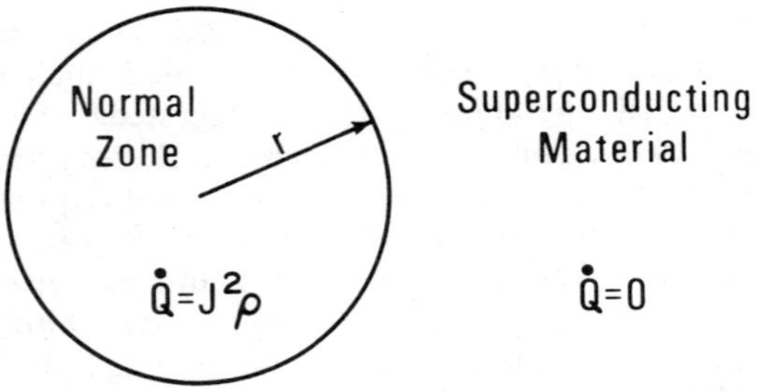

Figure 13.30 Illustration of the MPZ concept. A spherical region of radius *r* in a superconducting material is assumed to have been driven normal by some disturbance. The MPZ is just that size of normal zone where the heat transport out of the zone boundary equals the rate of Joule heating within the zone. Adiabatic disturbances that produce normal zones larger than the MPZ will initiate thermal runaway; smaller disturbances will not.

and $(T_c - T_h)$ is the temperature difference between the interior and far away from the normal zone (the critical temperature minus the helium-bath temperature). The important point is that the heat generation rate is proportional to the *volume* of the zone, whereas the heat flux out of the zone is proportional to the *radius*. Thus, for small normal zones with a large radius-to-volume ratio, the cooling rate will exceed the heating rate and the zone will shrink in size. For large zones, however, heating will exceed cooling and the zone will rapidly grow in size, leading to thermal runaway. Somewhere in between, the heating and cooling rates will be exactly equal. This defines the MPZ. For a spherical zone, this occurs when:

$$(4/3)\pi r^3 J^2 \rho = 4\pi r k\,(T_c - T_h) \tag{13.12}$$

Solving for r, one finds in this simple case that the MPZ is a spherical zone whose radius is:

$$r_{\text{crit}} = [3k\,(T_c - T_h)/(J^2\rho)]^{1/2} \tag{13.13}$$

The system will be stable against sudden mechanical disturbances that do not release enough energy for a sphere of material of radius r_{crit} to be adiabatically heated to a temperature greater than T_c.

The MPZ concept is relatively recent (Martinelli and Wipf, 1972) and still evolving (Wilson and Iwasa, 1978). For example, the thermal conductivity is not isotropic, as has been assumed here. Heat conduction between adjacent turns in a superconducting magnet is not as great as conduction along a wire. Thus, in actual practice, the MPZ is a solid ellipse rather than a sphere. Nevertheless, the concept is the same. Disturbances that form normal regions larger than the MPZ will initiate thermal runaway, smaller disturbances will not. It is simply a balance between heat generation within the volume of the zone vs. heat transfer out of the zone. The volume-to-radius ratio increases as the zone size increases, and therefore, large normal zones are unstable.

Very close in concept to the MPZ is the *minimum propagating current*. It is the current at which the normal region, once formed, just begins to grow and leads to thermal runaway (Cornish, 1966; Martinelli and Wipf, 1972). In essence, Eq. 13.13 is simply inverted. For a given disturbance size (i.e., given r_{crit} in Eq. 13.13), there is a J_{crit} above which the system is unstable. Unlike the other criteria discussed in § 13.6, the minimum propagating current is a directly measurable quantity and can be used as a design criterion in superconducting coils. For any given disturbance size, the corresponding minimum propagating current can be empirically measured (see § 13.6.4). Magnet operation below the minimum propagating current is stable. Unfortunately, little data are available at present on just what disturbance sizes to expect in different magnet types.

Other criteria for limited cryostabilization are also being developed. The MPZ criterion, for example, applies to the case of heating from adiabatic disturbances. When a steady-state heat release is also present (e.g., from continuous ac losses) the criterion must be modified. A generalized criterion has recently been developed for treating a simplified form of this situation (Wipf, 1978), but the concept is still in the evolutionary stage.

13.6.4 Measurement of stabilization

For the most part, stability is a theoretical property, calculated from the geometry of the composite, heat transfer rates, the measured critical parameters of the superconducting material (T_c, H_c, and J_c), and the measured thermal and electrical resistivities of the superconductor and matrix material (e.g., Eqs. 13.9, 13.10, or 13.13). However, as noted in § 13.6.3, empirical measurement of the minimum propagating current as a function of disturbance size is possible. This can be used as a design criterion in superconducting coils where the disturbance tolerance can be specified.

The sample geometry for measuring the minimum propagating current (Cornish, 1966;

Miller, Dresner, and Lue, 1978; Wilson and Iwasa, 1978) is similar to that for critical-current measurements. A composite wire coil is inserted into the bore of a simple solenoidal magnet (see Fig. 13.31). The coil is held mechanically immobile. Access of liquid helium to its surface or interior should simulate what will occur in actual application. A normal region in the conductor is formed by pulsing a heater co-wound with the coil. Potential taps attached to the superconducting coil are used to sense when the conductor is driven normal. The heater is pulsed periodically as the current through the specimen is increased. At low specimen current, no voltage is detected. Eventually, as the current is increased, the minimum propagated current will be reached and the normal zone will propagate, which drives the entire superconducting coil normal. To simulate localized mechanical disturbances, the heater should be no larger than the estimated size of the MPZ, and to simulate adiabatic conditions (which is the case for most mechanical disturbances), the duration of the heater pulse should be kept short, typically less than ~1 ms.

With essentially the same technique, it is also possible to measure the maximum recovery current. This is a particularly useful experimental measurement for determining when full cryostability is obtained in magnets where the superconductor windings are surrounded by liquid helium and thermally isolated from each other. The experimental setup is as described above [see, e.g., Cornish, Zbasnik, Leber, Hirzel, Johnston, and Rosdahl (1979)], except that the heater is pulsed with an arbitrarily large pulse, enough to drive a length of superconductor normal. The heater is pulsed periodically while the current through the specimen is increased. At low specimen current, voltage pulses momentarily appear and then disappear, indicating that recovery occurred after the heater initiated a normal zone. Eventually, as the current is increased above the maximum recovery current, the normal zone formed by the heater will no longer recover. The heater length is not a factor so long as neighboring superconductor turns are thermally isolated from each other by liquid helium. However, the duration of the heater pulse should be kept short (~1 ms) to simulate practical disturbance conditions.

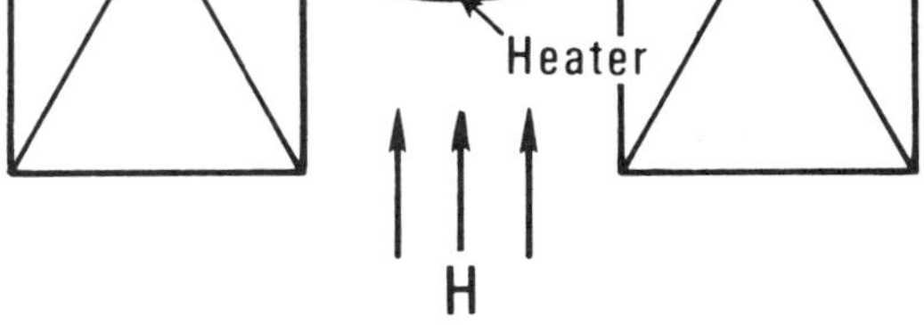

Figure 13.31 Experimental arrangement for measuring the minimum propagating current of a composite ring sample.

13.7 AC LOSSES

The requirements for stability discussed in the previous section work fine for dc applications, but further design requirements must be met for ac applications. When an ac current is passed through a superconductor, substantial losses can occur in the matrix as well as within the superconducting filaments themselves. These losses must be minimized or they will heat the superconductor above its critical temperature.

There are several different origins of ac losses:

a. Eddy current loss in the matrix
b. Self-field loss
c. Hysteretic loss
d. Surface loss

Each source of loss is minimized in a different way. Eddy current loss can be made small by twisting the wire. Self-field loss can be made small by transposing the wire. For practical superconductors, surface loss becomes important only at rf or microwave frequencies. This leaves magnetic hysteretic loss as the prime source of ac loss in the frequency domain from dc through most of the audio frequency range.

The important determining factor for hysteretic loss is the filament size. The smaller the filament diameter, the smaller the hysteretic loss per unit volume. The problem is that usually the filaments cannot be fabricated uniformly in sizes below about 1 μm in diameter. This sets a lower practical limit on the magnitude of the hysteretic loss.

A brief introduction to each type of loss is given in the following sections, along with expressions for the loss magnitude. A practical procedure for designing a superconducting composite for ac applications is to determine the hysteretic loss from an expression like Eq. 13.16

below, using the smallest filament diameter that can be practically fabricated. Then the eddy current loss and self-field loss are made negligible in comparison by using a twist pitch that is short enough (Eq. 13.14) and a strand diameter that is small enough (Eq. 13.17). A general summary of the types of ac losses and their characteristics is given in Table 13.5. More detailed discussion of the origin of each loss may be found in Wilson et al. (1970), Wipf (1978), Hancox (1966a; 1966b), Brechna (1973), or London (1963).

13.7.1 Eddy current loss

In a multifilamentary composite, the superconducting filaments are connected together electrically by the matrix material. Thus a rapidly changing perpendicular magnetic field, $\dot{H}$, will induce enough voltage to drive shielding currents across the resistive matrix from one filament to another. This is shown in Fig. 13.32a for the simple case of two filaments. The shielding current induced by the changing magnetic field can generate significant Joule heating in the matrix material.

For any given $\dot{H}$, there is a critical length of conductor that allows all of the current in a filament to cross the matrix. This is dependent on the rate of change of magnetic field, $\mu_0 \dot{H}$ $(T{\cdot}s^{-1})$, and the transverse resistivity of the matrix material, ρ ($\Omega{\cdot}$m). The critical length, ℓ_c (m), required for this complete transition of current is given by Morgan (1970):

$$\ell_c^2 \simeq 2\frac{\rho f^{1/2} J_c d}{\mu_0 \dot{H}} \frac{D}{(D+d)} \qquad (13.14)$$

where d (m) is the superconductor filament diameter, D (m) is the wire diameter, J_c (A/cm^2) is the critical current density of the superconductor, and f is the volume fraction of superconductor in the composite. The important factor is $\dot{H}$, since ρ, J_c, and the geometrical factors are usu-

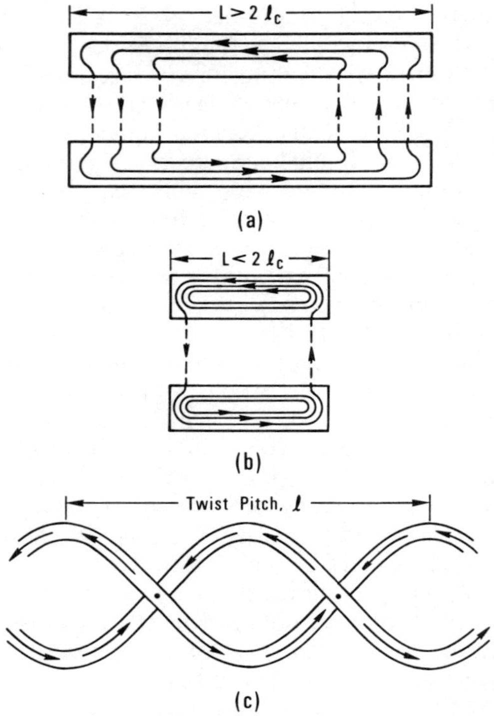

Figure 13.32 Diagram of eddy currents induced in a two-filament composite. (a) longer than the critical length, ℓ_c, (b) shorter than the critical length, ℓ_c, and (c) twisted with a pitch less than ℓ_c (Iwasa and Montgomery, 1975).

Table 13.5 Types of ac losses.

	Hysteretic	Eddy Current	Self-field	Surface
Location	Induced currents within SC filament	Induced currents between SC filaments due to external field changes	Induced currents between SC filaments due to changes in the transport current	Superconductor surface
Technique for reducing loss	Make filament diameter small (Eq. 13.16)	Twist filaments (Eq. 13.14)	Keep strand size less than about 2 mm (Eq. 13.17); transpose strands within cable	Smooth surface
Relative magnitude	Dominant loss at frequencies less than rf due to ~1 μm lower limit on filament size	Can usually be made negligible compared with hysteretic loss	Can usually be made small (<10%) compared with hysteretic loss	Dominant loss at radio frequencies and above

ally determined by material properties or stability requirements. Typically, $\mu_0\dot{H}$ ranges from 10^{-4} to 1 T/s, for which ℓ_c ranges from about 0.3 m to 3 mm (assuming typical values of J_c = 1 GA·m^{-2}, d = 50 μm, ρ = 20 nΩ·cm, f = 0.3, and $D >> d$). For conductor lengths, L, greater than $2\ell_c$, the entire current will cross the matrix at each end of the conductor (Fig. 13.32a). For $L << 2\ell_c$, only a fraction $(L/2\ell_c)^2$ of the current will cross the matrix at each end (although some of the current still forms closed current loops within each separate filament, Fig. 13.32b, which is the source of filament hysteretic loss discussed in § 13.7.3).

Eddy current losses create a problem because ℓ_c is much shorter than practical conductor lengths, L. The solution is to twist the filaments, as shown in Fig. 13.32c. The voltage between filaments then reverses periodically along the conductor, twice for each complete rotation of the filaments. The eddy currents in the conductor are equivalent to those in a conductor whose length is only half a twist pitch. Thus, if the 360° twist pitch, ℓ, is much less than $4\ell_c$, only a fraction $(\ell/4\ell_c)^2$ of the current will be able to cross the matrix. This effectively decouples the filaments and significantly reduces the eddy current loss in the resistive matrix. Typically, multifilamentary composites are given a 360° twist pitch, ℓ, of 2 cm or less.

Significant eddy current loss can also occur when copper is added to the circumference of a composite conductor for stabilization. The eddy current loss per unit volume increases with the size of the coupled volume. To minimize the eddy current loss in such stabilizer sheathing, the copper can be subdivided with sheets or fins of low conductivity Cu–Ni, as shown in Fig. 13.33.

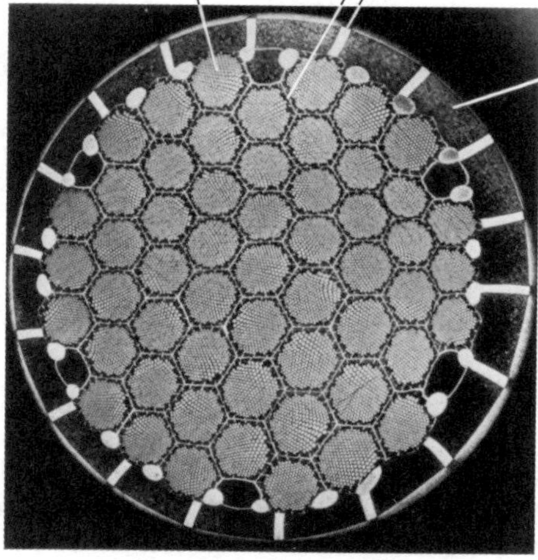

Figure 13.33 Multifilamentary superconductor with cupronickel (white areas) barriers to prevent coupling currents from flowing around filament clusters and the outer copper stabilizing sheath. Copper areas are dark and the Nb–Ti filaments are gray [courtesy of Imperial Metal Industries (Brechna, 1973)].

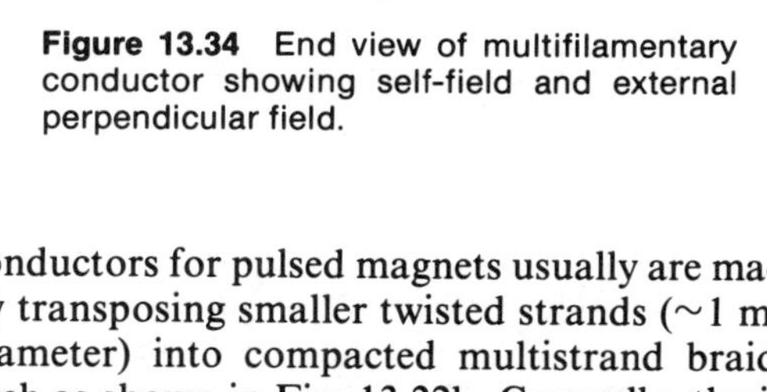

Figure 13.34 End view of multifilamentary conductor showing self-field and external perpendicular field.

13.7.2 Self-field loss

The self-field due to transport currents in the wire is not compensated by simple twisting, as shown in Fig. 13.34. The self-field does not see any effective twist, since the radial positions of the filaments are unchanged by twisting. Thus, the self-field (as opposed to the external field) acts over the entire length of conductor to couple the filaments, again leading to significant ac losses in any conductor of practical length.

The problem may be solved by transposing the superconducting filaments so that each occupies a different radial position in the conductor. Since it is not practical to manufacture a cable in which each filament is fully transposed, high-current conductors for pulsed magnets usually are made by transposing smaller twisted strands (~1 mm diameter) into compacted multistrand braids, such as shown in Fig. 13.22b. Generally, the individual strands are insulated because manufacturing limitations result in the transposition pitch being longer than the critical length given by Eq. 13.14. (Occasional shorts between strands are not a problem, however, so long as they do not occur too frequently within one critical length.)

Transposing strands eliminates most of the self-field loss, but there remains the self-field loss within each strand. This remaining loss is given by Brechna (1973):

$$W_{sf} \cong \mu_0 \frac{D^2 \overline{J_c^2} V}{192} \left(\frac{\Delta I}{I_c}\right)^3 \tag{13.15}$$

where W_{sf} (J/cycle) is the self-field loss per cycle, D (m) is the strand diameter, V (m^3) is the total volume of the composite, J_c (A·m^{-2}) is the overall critical current density of the strand, I_c is the critical current, and ΔI is the change in transport current in one cycle (in Eq. 13.15 it is assumed that $\Delta I << 2I_c$, which is the case for most applications). Also, the field dependence of J_c has been neglected, but this can be taken into consideration by taking $\overline{J_c}$ at a field that is averaged over a complete cycle and also over the coil winding, that is, at a field equal to about one-fourth the maximum field. Note that the self-field loss is proportional to the square of the diameter of the strand, D. In the next section, the size of D will be determined to make the self-field loss negligible compared with hysteretic loss.

13.7.3 Hysteretic loss

As shown in Fig. 13.32b, twisting reduces only the matrix eddy currents; a changing external field will still induce closed current loops within each separate filament. This generates loss because any movement of flux, created by a time-varying external field or transport current, generates heat in the normal vortex cores (see § 13.4.2).

The hysteretic loss associated with the shielding currents in each superconducting filament is given approximately by Brechna (1973):

$$W_{hl} \cong \frac{\mu_0 \overline{J_c} d V}{2} \Delta H_{\text{ext}} \tag{13.16}$$

where W_{hl} (J/cycle) is the hysteretic loss per cycle, d (m) is the superconducting filament diameter, V (m^3) is the total volume of the composite, J_c (A·m^{-2}) is the overall current density of the strand, and ΔH_{ext} (A·m^{-1}) is the peak-to-peak change in external field in one cycle. In Eq. 13.16, complete flux penetration has been assumed, which is a good assumption because practical multifilamentary composites usually have filament diameters in the range 1 to 100 μm (as discussed in § 13.6). Also, $\overline{J_c}(H)$ is an averaged value taken at one quarter of the peak magnetic field. Note that the hysteretic loss is directly proportional to the filament diameter, d.

From Eqs. 13.15 and 13.16, the size of the strand diameter, D, that will make the self-field loss negligible compared with the hysteretic loss can be determined. Taking the ratio of these two expressions:

$$\frac{W_{sf}}{W_{hl}} \simeq \frac{D^2}{d} \frac{\overline{J_c} (\Delta I/I_c)^3}{96 \ \Delta H_{\text{ext}}} \tag{13.17}$$

As an example, for a magnet with $\Delta H_{\text{ext}} = 2$ MA·m^{-1} [i.e., $\Delta(\mu_0 H_{\text{ext}}) = 5$ T], $J_c = 1$ GA·m^{-2} [where $\overline{J_c}(H)$ is taken at one quarter of the maximum field], and $\Delta I/I_c = 0.5$, it is found that $W_{sf}/W_{hl} \simeq 0.65 \ D^2/d$ (where D and d are in m). Thus, if the filament diameter is 10 μm (to meet stability requirements), and the self-field loss is to be less than 10% of the hysteretic loss, then the strand diameter, D, must be limited to about 1.2 mm. Note that this is consistent with limitations placed on the strand diameter by stability requirements (either cryogenic stability or dynamic stability; see § 13.6). Thus, strand diameters in most practical composites are limited to about 1 mm, both for reasons of stability and self-field loss. The approximate magnitude of the hysteretic loss in this example is found from Eq. 13.16 to be about 10 kJ·m^{-3} per cycle.

13.7.4 Measurement of ac losses

Calorimetric method. This is the least convenient, but also least error-prone method for determining the total loss under steady-state conditions. As shown in Fig. 13.35, the sample is simply placed in a calorimeter filled with helium and the increase in boil-off rate is measured with a gas flow meter when the superconductor is energized from an ac current source.

The sensitivity of the method in this configuration is moderate, about 1 mW. This can be increased by measuring pressure pulses instead of the rate of helium boil-off (Iwasa, Montgomery, and Polzer, 1970). In either case, the boil-off rate or pressure pulses are calibrated using a known energy source, such as a heater placed in the calorimeter.

The main disadvantages of this method are its moderate sensitivity and the long time usually required for the system to reach equilibrium conditions. However, the results represent a total loss figure that is not susceptible to large errors, such as can occur in the electric methods discussed below. Also, it is possible to measure the induced losses from both time-varying external magnetic fields as well as ac transport currents.

Magnetization method. As mentioned above, flux movement within the superconductor is ultimately the source of hysteretic loss. The

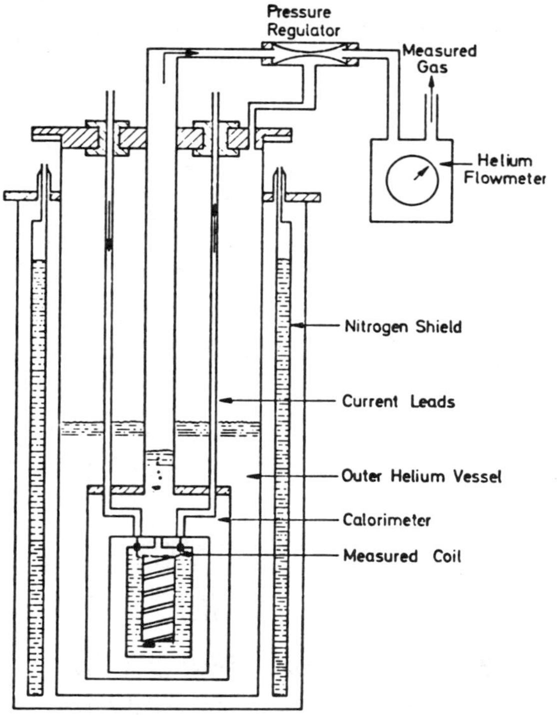

Figure 13.35 Arrangement of the calorimetric helium boil-off method to measure ac losses (Brechna, 1973).

loss per cycle can be quantitatively measured from the *area* under a full cycle of the magnetization curve, such as that shown in Fig. 13.36.

A method for measuring the magnetization as a function of magnetic field is shown in Fig. 13.37 (Brechna, 1973). The specimen (1) is placed in one of two identical coils (2 and 3) in the bore of a solenoid. When the specimen is not superconducting, the voltage produced across coils 2 and 3 by a varying external field is the same. However, when the specimen is superconducting, it excludes magnetic flux from coil 2. The difference voltage, U_d, is proportional to the time rate of change of the magnetization of the specimen. The curve in Fig. 13.36 is obtained by integrating the difference voltage and plotting it as a function of the applied magnetic field. The field is obtained by integrating the output of a pickup coil also placed in the solenoid bore. The area of this curve gives the hysteretic loss directly.

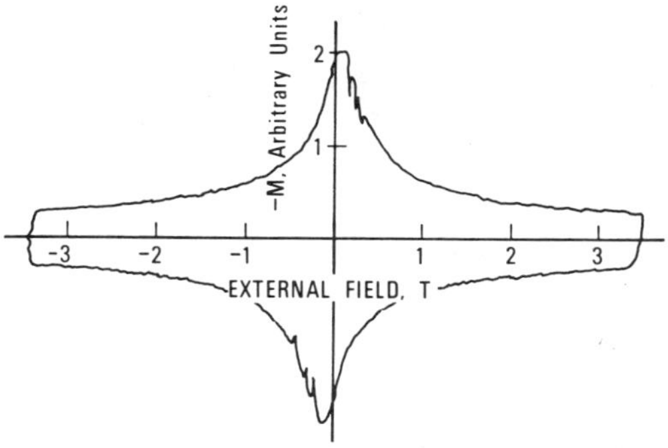

Figure 13.36 Magnetization loop traced for Nb–50 wt.% Ti (Wilson et al., 1970).

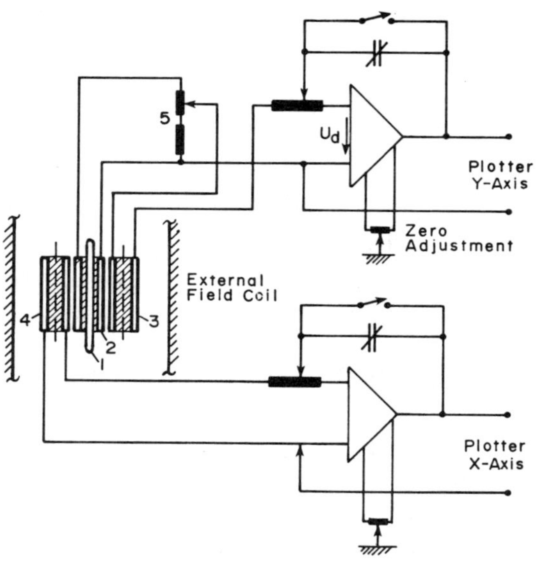

Figure 13.37 Schematic diagram of ac loss measuring system [from W. A. Fietz (Brechna, 1973)].

A modification on this technique has been described by Sekula (1971), and is shown in Fig. 13.38. In this method the difference voltage is detected with a phase-sensitive detector (PSD), using a signal proportional to the applied magnetic field as reference. The two signals are properly phased by adjusting for zero output at applied fields below H_{c1}, where there is no flux penetration and consequently, no loss. The PSD output is proportional to the in-phase product of the two signals and gives the specimen loss directly.

In principle, any of the magnetization methods can be performed under a relatively slowly varying field, since the loss per cycle is independent of frequency. The first method (Fig. 13.37) is particularly well suited for such measurement. The PSD method, however, is best suited to frequencies above several hertz. Also, the shape of the magnetization curve contains additional information about flux-jump activities and critical current, which is not available in the PSD method.

Magnetization methods, as well as the Joule-

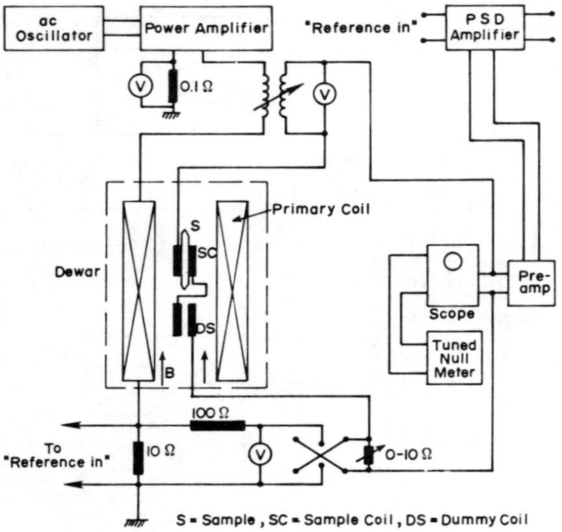

Figure 13.38 Schematic diagram of a PSD apparatus to measure ac losses (Sekula, 1971).

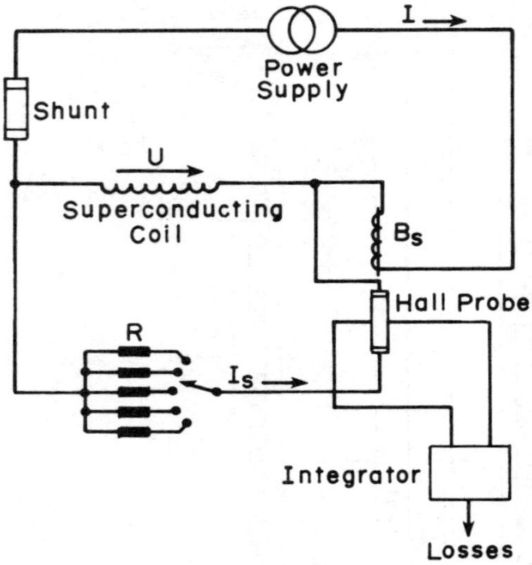

Figure 13.39 Schematic arrangements of ac loss measurement electrical method using a Hall multiplier [courtesy of Lawrence Berkeley Laboratory (Brechna, 1973)].

heating method described below, permit loss determinations to be made on small specimens, which is particularly useful where only a minimal volume of superconducting material may be available. To obtain enough sensitivity in the calorimetric method, however, several cubic centimeters of superconducting material are required.

The chief disadvantage of the magnetization methods is that the two signals must be very carefully matched and phased, since any small mismatch results in a large percentage error in the difference voltage. Also, magnetization techniques are best adapted to those cases involving time-varying external field rather than time-varying transport current.

Joule-heating method. In this method, the product of the current and voltage across the specimen is measured over one cycle and integrated to give the loss. The multiplication operation can be performed using a Hall probe, as shown in Fig. 13.39, or digitally using a computer. Sensitivity of the method ranges from 0.25% to about 5%, depending on the particular scheme used. This method is subject to the same large phasing errors described above for the magnetization techniques. By its nature, the Joule-heating method is suited to situations involving time-varying transport currents, rather than varying external fields.

13.8 MECHANICAL PROPERTIES

The final factor that enters into the design of practical superconductors is their mechanical properties. In most applications the superconductor is subjected to stress that can reach substantial levels. The three sources of stress are:

a. Fabrication stress. In magnet construction, the superconducting wire is subjected both to bending stress as it is wound into the magnet coil and to uniaxial stress from pretensioning of the wire during the winding operation (see Fig. 13.40).

b. Thermal-contraction stress. As the wire is cooled from heat-treatment temperature to liquid-helium temperature, different materials

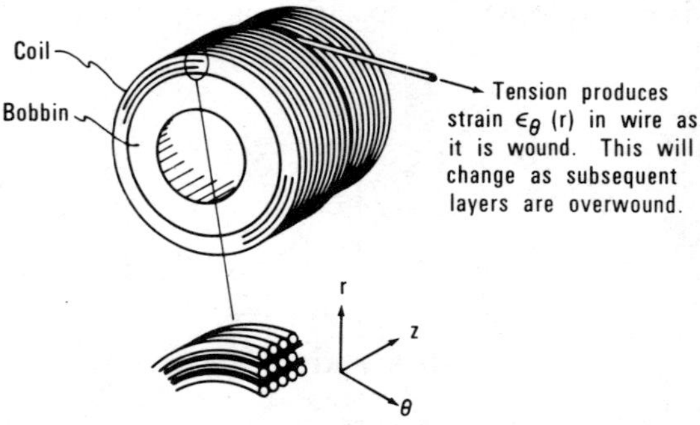

Figure 13.40 Illustration of bending strain and pretensioning uniaxial stress incurred during magnet winding (Arp, 1977).

within the composite wire and magnet structure contract at different rates. Because of this mismatch in thermal contraction rates, significant stress on the superconducting material can be generated during cooldown.

c. Magnetic stress. When a superconducting winding is energized, the Lorentz force (§ 13.4), acting on the superconductor can be quite large, particularly in large magnets. For example, the magnetic hoop stress on each wire in a solenoidal magnet (see Fig. 13.41) is given by:

$$\sigma = JBR \tag{13.18}$$

where J is the transport current density, B is the magnetic field strength, and R is the radius of the winding. In small research magnets, this is generally not a problem, where typically $\sigma \leq 0.1$ GPa. But in large magnets, where the magnet bore can be several meters in diameter, the magnetic hoop stress can approach the ultimate strength of conventional multifilamentary superconductors, 1 GPa or more.

Of the three sources of stress, the two that have the potential to cause the greatest degradation of the superconductor are magnetic stress and bending stress during fabrication.

Strain introduced into multifilamentary com-

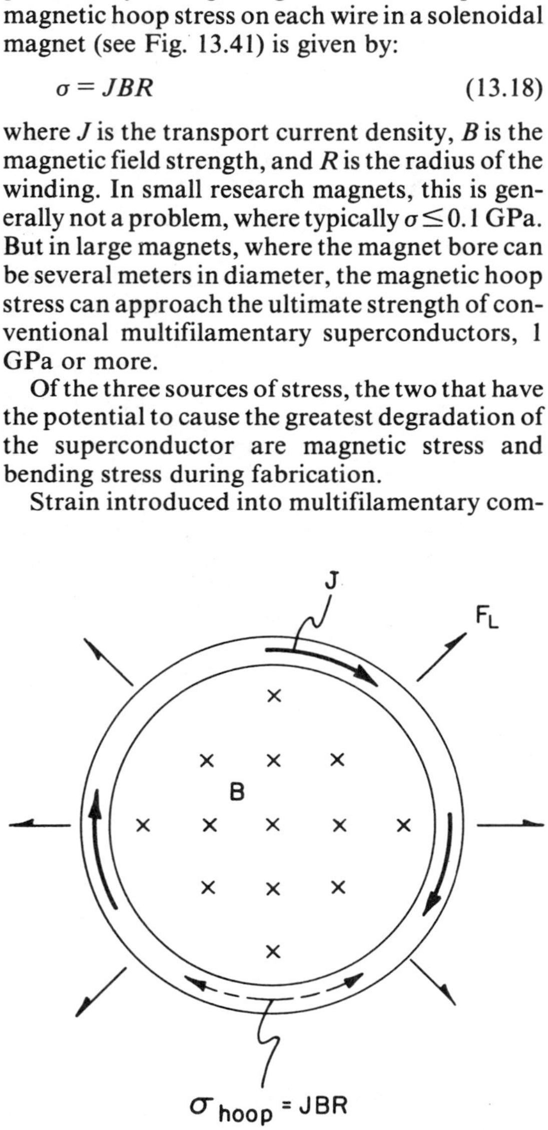

Figure 13.41 Representation of a current-carrying loop showing the transport current density, J, the associated magnetic field, B, and the resultant hoop stress, σ_{hoop}, generated by the Lorentz force, F_L, acting on the winding (Ekin, 1981b).

posites by these various sources has three important consequences:

a. Critical current degradation. The critical current of compound superconductors, such as Nb_3Sn and V_3Ga, can be lowered (reversibly) more than 30% by a static strain as low as 0.4% at 12 T; the effect becomes more severe at high fields (§ 13.8.1).

b. Matrix degradation. The stability of the matrix material in both the ductile and brittle superconductors can be more than halved by cyclic strains as low as 0.4%; the purer the material, the lower its strain tolerance to fatigue (§ 13.8.2).

c. Training. Serrated yielding of the superconducting material, fracture of magnet potting material, or wire movement can generate enough local heat to initiate thermal runaway. This results in a process known as magnet training (described in § 13.8.3).

13.8.1 Critical current degradation

Bending strain. Figure 13.42 shows an example of the effect of bending strain on the critical current of a Nb_3Sn superconductor. When a superconducting composite of thickness d is bent around a radius, R, the maximum strain along the outside of the bend is tensile and equal to $d/2R$. The maximum compressive strain along the inside is compressive and of the same absolute magnitude (see Fig. 13.43). Since the superconducting filaments are usually twisted within the strand, each filament sees alternate tensile and compressive strain. The smaller the bending diameter, the more severe the strain acting on the filaments and the greater the decrease in critical current. The effect is much greater in brittle superconductors, such as Nb_3Sn (Fig. 13.42), than for ductile superconductors, such as Nb–Ti. In fact, multifilamentary Nb–Ti composites can be bent into a hairpin shape, with little critical current degradation.

Uniaxial strain. Uniaxial strain has a similar effect on the critical current, but the degradation is generally more severe than bending strain since all the filaments are uniformly affected (not just those away from the neutral axis as with bending strain). The effects of magnetic hoop stress, for example, can be measured directly by subjecting a short section of superconducting wire to uniaxial strain. The net result of such tests show that, as with bending strain, the effect on ductile multifilamentary composites is rela-

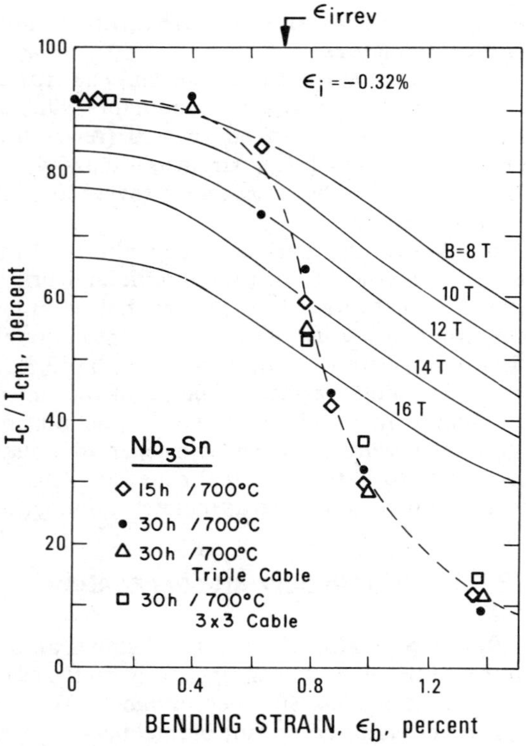

Figure 13.42 Dependence of critical current upon bend diameter for multifilamentary Nb_3Sn single-strand and cabled conductor (dashed curve) (Heyne, Hackworth, Singh, Young, and Sanger, 1979). The strain at which filament breakage occurs is indicated by ϵ_{irrev}. Solid curves indicate the calculated effect of bending strain without filament breakage (Ekin, 1980b).

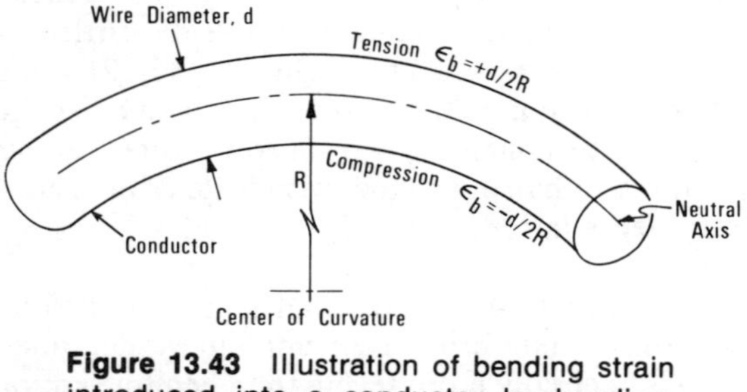

Figure 13.43 Illustration of bending strain introduced into a conductor by bending. Bending strain is zero along the neutral axis (Ekin, 1981b).

tively small, whereas for the brittle multifilamentary composites one must be extremely careful not to exceed the rather restrictive strain limits.

The effect of uniaxial strain on the critical current of *ductile* materials is shown for multifilamentary Nb–Ti at low field (7 T) by the dashed curve in Fig. 13.44. The results are plotted in normalized form with the critical current, I_c, divided by its maximum (strain-free) value, I_{cm}, and plotted as a function of the intrinsic strain, ϵ_0, experienced by the superconducting material in the composite (negative values indicate compression; positive values, tension). When the uniaxial strain is limited to less than about 0.4% tension, the I_c degradation is negligible; at higher strain the decrease in I_c is small but measurable ($\Delta I_c / I_c \simeq 5\%$ at $\epsilon_0 = 1\%$). Generally, this is not a problem in most magnet designs.

The effect of uniaxial strain on the critical field of *brittle* multifilamentary composites is shown also for Nb_3Sn in Fig. 13.44. In contrast to the dashed curve for Nb–Ti, the effect is quite large. If the superconducting filaments are subjected to either a compressive or tensile strain of only 0.4%, the critical field can be degraded by about 15%. The effect becomes even more severe at high fields. At 12 T, for example, the same strain causes a degradation of more than 30%. In designing high-field magnets, this presents the problem of measuring J_c as a function of two variables, field and strain.

Strain scaling law — design curves. Fortunately, the dependence of I_c / I_{cm} on both field

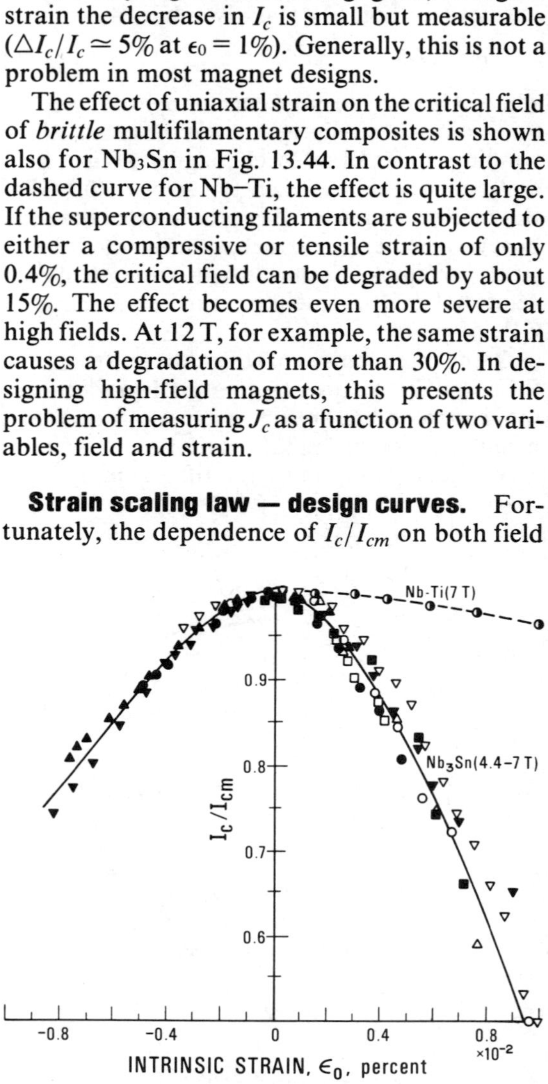

Figure 13.44 Effect of uniaxial strain on the critical current of both ductile (Nb–Ti) and brittle (Nb_3Sn) superconductors at low magnetic fields (Ekin, 1978a). ▲ — Rupp (1977); ○, △, □, ▽ — Ekin (1976; 1978a); ◑ — Ekin, Fickett, and Clark (1976); ■ — Easton and Schwall (1976).

and strain can be described quite accurately by a nearly universal relationship known as the strain scaling law (Ekin, 1980a) from which is obtained:

$$\frac{I_c}{I_{cm}} = \left[\frac{B^*_{c2}(\epsilon)}{B^*_{c2m}}\right]^{n-p} \left[\frac{1 - B/B^*_{c2}(\epsilon)}{1 - B/B^*_{c2m}}\right]^q \qquad (13.19)$$

Here B^*_{c2} is the (strain-dependent) bulk average upper critical field and is expressed by $B^*_{c2}(\epsilon) = B^*_{c2m}[1 - a|\epsilon_0|^u]$. Its dependence on strain is nearly universal for commercial multifilamentary superconductors and is shown in Fig. 13.45, where the results have been normalized by B^*_{c2m}, the maximum (strain-free) value of the upper critical field. The parameters n, p, and q are scaling parameters, which remain nearly constant for a given material and are given in Table 13.6. As an example, values of I_c/I_{cm} obtained from this relationship are plotted as a function of magnetic field and uniaxial strain for Nb_3Sn in Fig. 13.46. Note that the effects of strain shown in Fig. 13.46 are much more pronounced at higher fields, leading to extreme strain sensitivity as B approaches B^*_{c2}.

The relationship given by Eq. 13.19 and shown in Fig. 13.46 has been found to be accurate within a couple percent for multifilamentary Nb_3Sn wires with optimized critical current densities; i.e., essentially all commercial Nb_3Sn multifilamentary wires manufactured for practical applications. Equation 13.19 also holds for V_3Ga and Nb–Ti, but with a different set of the scaling parameters (given in Table 13.6).

Use of the curves — design criterion. The curves in Fig. 13.46 give the relative critical current as a function of the intrinsic strain, ϵ_0, experienced by the superconducting reaction layer. This is not to be confused with the overall strain applied to the composite, ϵ. The difference arises because of differential thermal contraction between superconductor and matrix. In most commercial wires the superconducting reaction layer is initially in a state of compression, while the bronze matrix is in tension. This results from the process used to fabricate multifilamentary Nb_3Sn; after the high-temperature (700° C) reaction heat treatment (see § 13.5), the bronze matrix contracts more than the Nb_3Sn compound on cooling. This leaves the Nb_3Sn in a state of compression with an initial compressive strain, $\epsilon_i < 0$; i.e., on the left side of the peak in Fig. 13.46. When tensile strain is applied to the composite (e.g., from winding tension and magnetic hoop stress), the first effect is to counteract the initial compressive strain in the Nb_3Sn reaction layer, $\epsilon_0 \equiv \epsilon_i + \epsilon$; that is, ϵ_0 becomes less negative and the critical current at first rises toward its peak value

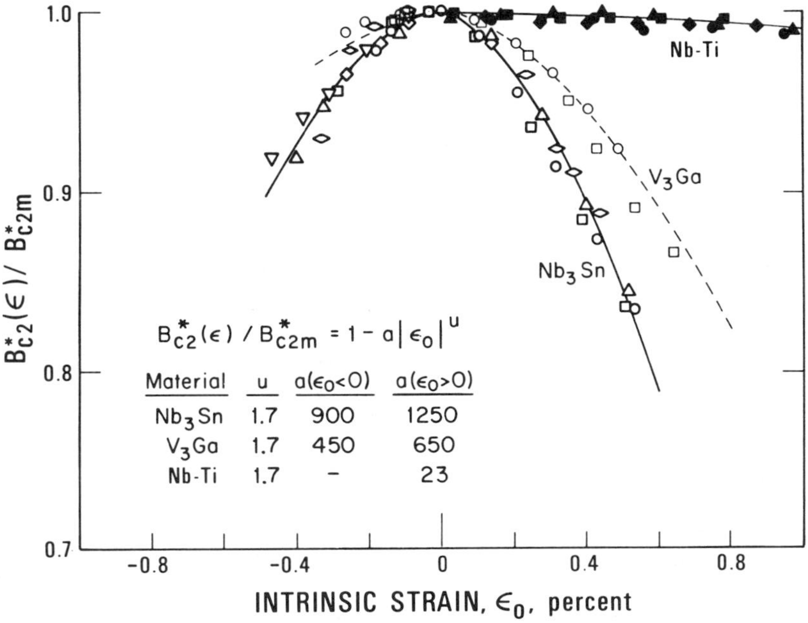

Figure 13.45 Upper critical field, B^*_{c2}, as a function of intrinsic strain in commercial multifilamentary Nb_3Sn, V_3Ga, and Nb–Ti. B^*_{c2} has been normalized by its maximum (strain-free) value, B^*_{c2m} (Ekin, 1980a; 1981a).

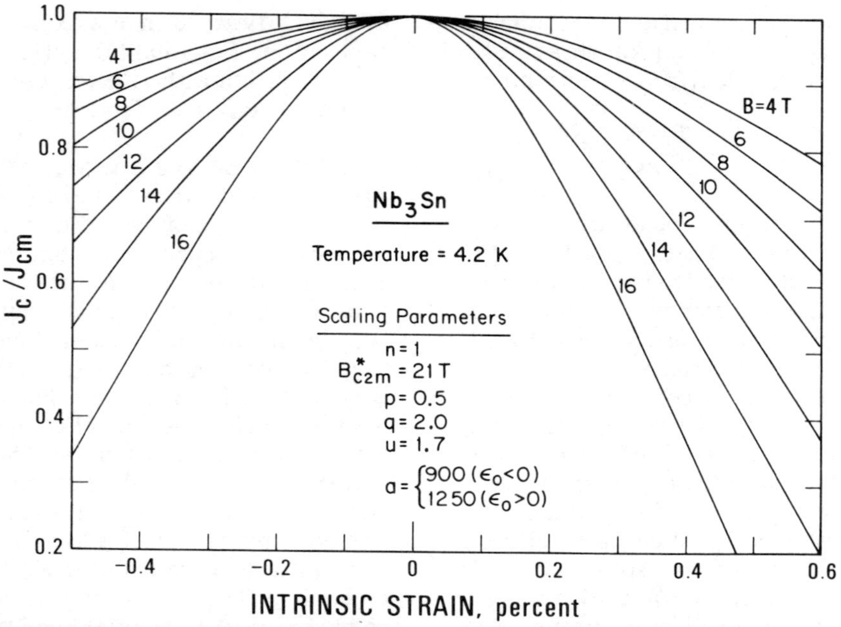

Figure 13.46 Relative critical current density, J_c/J_{cm}, as a function of intrinsic strain, ϵ_0, for different magnetic fields, evaluated using Eq. 13.19 and the typical set of scaling parameters for multifilamentary Nb_3Sn shown in the figure (Ekin, 1980a).

Table 13.6 Scaling parameters for use in Eq. 13.19 to determine the strain and field dependence of the critical current of Nb_3Sn, V_3Ga, and Nb–Ti (Ekin, 1980a; 1981a).

Material	n	p	q	B^*_{c2m}(T)	u	$a(\epsilon_0<0)$	$a(\epsilon_0>0)$
Nb_3Sn	1	0.5	2	21	1.7	900	1250
V_3Ga	1.3	0.4	1.0	21	1.7	450	650
Nb–Ti	~7	0.8	1.0	11	1.7	—	23

in Fig. 13.46 (Buehler and Levinstein, 1965; Rupp, 1977). If too much tensile strain is applied, eventually the Nb_3Sn itself experiences tension, and the critical current moves over the peak in Fig. 13.46 and starts to decrease on the right (tensile) side.

To determine ϵ_0 for use in Fig. 13.46, ϵ_i must be measured for each type of superconductive wire. This is most easily done by a single measurement of J_c vs. ϵ at a given low field. The magnitude of ϵ_i is then just equal to the strain where the J_c–ϵ curve reaches a maximum. For commercial superconductors, ϵ_i is typically in the range −0.5 to 0%. Occasionally, the initial prestrain will be positive, that is, tensile. This occurs if the compressive effects of the bronze matrix are removed by overbending or damaging the wire. This is easy to do because the bronze matrix material has a relatively low yield stress. Consequently, one must be very careful to avoid plastic deformation of the bronze during wire handling and magnet winding.

To utilize Fig. 13.46 in magnet design, the uniaxial strain design criterion can be employed. It states that the combined tensile strain from all sources of wire stress (i.e., winding prestress, ϵ_w, thermal contraction of other magnet materials, ϵ_t, and magnetic hoop stress, ϵ_m) should be approximately equal and opposite to the initial compressive prestrain of the wire, that is, $\epsilon_0 = \epsilon_i + \epsilon_w + \epsilon_t + \epsilon_m \simeq 0$ (Ekin, 1976). In this case, at full operating field the combined applied strain leaves the Nb_3Sn in a state of near-zero strain, approximately at the peak of the curve in Fig. 13.46 (or even slightly on the compressive side of maximum, since the penalty for going too far into tension is more severe than that for compression because eventually the filaments fracture). This strain balance need only be obtained in the critical region of the magnet where the field is highest.

From this criterion, mechanical strain limits may be determined for the design of large magnets. In the case of brittle superconductors (Nb_3Sn, V_3Ga), the resulting strain limits are more stringent than typical mechanical design criteria (such as yield strength limits). For example, if a 5% J_c-degradation tolerance is assumed in applying the uniaxial strain criterion, it is seen from the width of the I_c/I_{cm}-vs.-ϵ_0 peak in Fig. 13.46 that at 12 T and 4.2 K, ϵ_0 for Nb_3Sn must be limited to the range $-0.16\% \lesssim \epsilon_0 \lesssim 0.13\%$. For comparison, the elastic strain limit for most stainless steel structural materials is over 0.2%. This strain window for ϵ_0 becomes significantly more stringent as the field is increased, becoming about two-thirds the 12-T width at 16 T, for example. The strain window is shown for Nb_3Sn as a function of magnetic field in Fig. 13.47.

Structural reinforcement. Typical stress-strain relations for Nb–Ti and Nb_3Sn multifilamentary composites and their components are shown in Figs. 13.48 and 13.49. Note that for both types of composites, the matrix material plastically yields at relatively low stress. Thus, in most high-stress applications the stabilizing matrix material is not much help in reinforcing the composites. When additional strength is needed, this is usually accomplished either by surrounding the entire superconducting solenoid with external reinforcement, such as a stainless steel jacket, or by using internal stainless steel strands or tape co-wound with the superconducting strands to obtain distributed reinforcement (shown for a multifilamentary Nb_3Sn cable in Fig. 13.23).

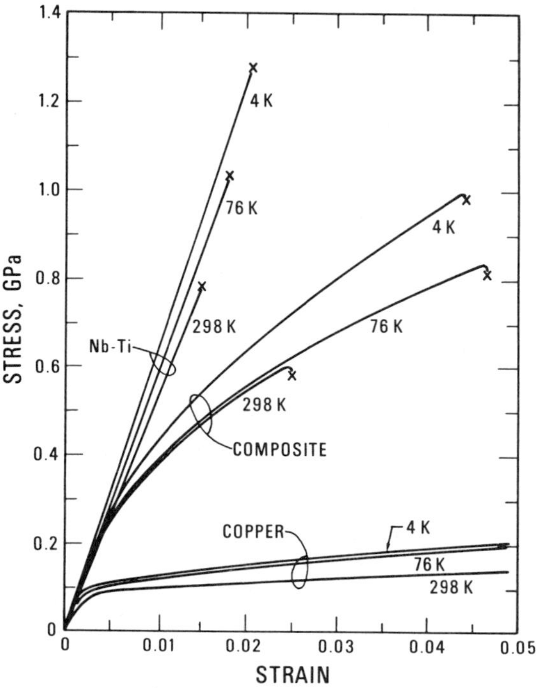

Figure 13.48 Stress–strain characteristic for a typical multifilamentary Nb–Ti conductor and its components, containing 36 vol.% Nb–Ti and 64 vol.% Cu (Reed, Mikesell, and Clark, 1977).

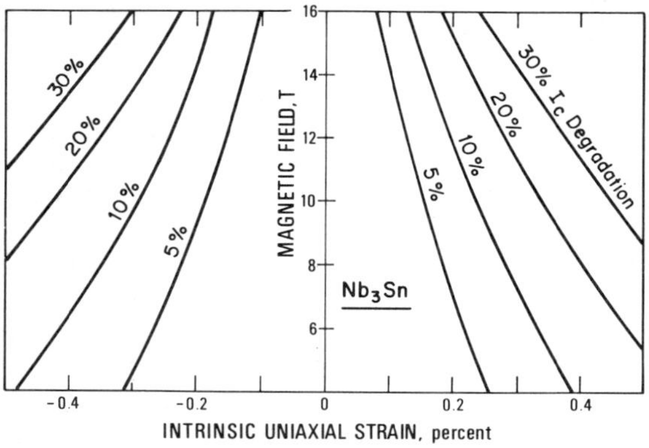

Figure 13.47 Uniaxial strain window as a function of field for multifilamentary Nb_3Sn for several J_c-degradation criteria. Note the progressive restriction on the intrinsic strain as the design field is increased (Ekin, 1981b).

13.8.2 Matrix degradation

In large pulsed-magnet applications, the cyclic strain generated by the magnetic stress can significantly lower the conductivity of the normal metal matrix through *fatigue*. The practical consequence of such a loss of conductivity in the matrix is a degradation of the stability of the wire, as discussed in § 13.6. Observations of copper-stabilized Nb–Ti and Nb_3Sn (Ekin, 1978a) indicate that although cyclic strain produces damage accumulation in the matrix material, it does not affect the superconducting material (mainly because the superconducting material behaves nearly elastically up to strains well beyond where the matrix yields; see Figs. 13.48 and 13.49).

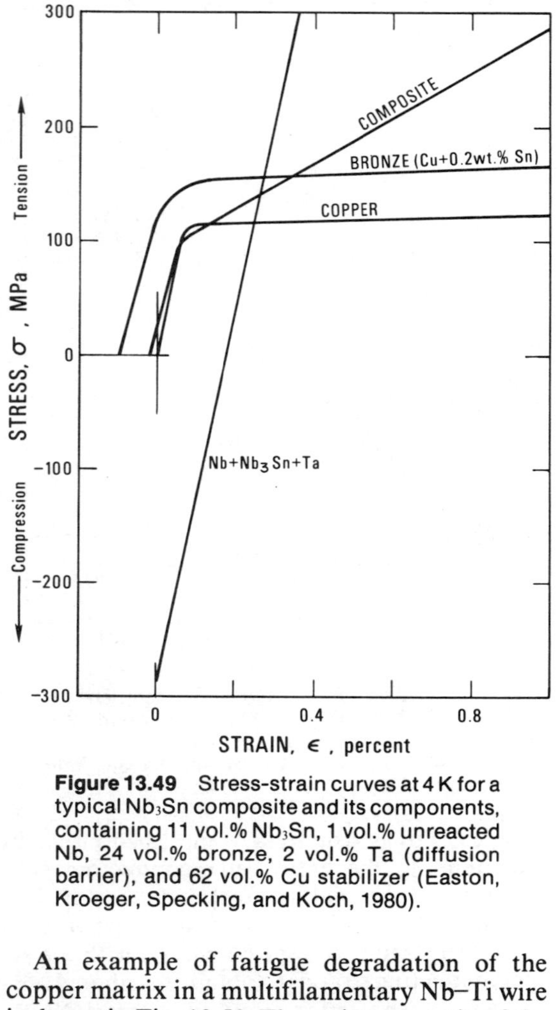

Figure 13.49 Stress-strain curves at 4 K for a typical Nb_3Sn composite and its components, containing 11 vol.% Nb_3Sn, 1 vol.% unreacted Nb, 24 vol.% bronze, 2 vol.% Ta (diffusion barrier), and 62 vol.% Cu stabilizer (Easton, Kroeger, Specking, and Koch, 1980).

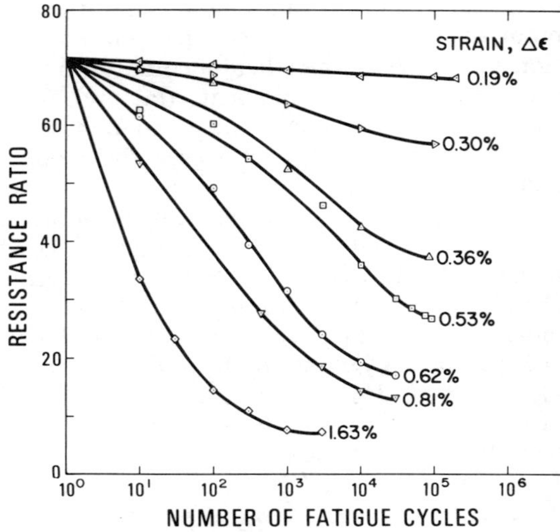

Figure 13.50 Degradation of the residual resistance ratio (R_{295K}/R_{9K}) of the copper matrix as a function of the number of fatigue cycles in a multifilamentary Nb–Ti(36 vol.%)–Cu(64 vol.%) conductor. The peak-to-peak cyclic strain amplitude is given by $\Delta\epsilon$.

An example of fatigue degradation of the copper matrix in a multifilamentary Nb–Ti wire is shown in Fig. 13.50. The resistance ratio of the copper matrix, R_{295K}/R_{4K}, is plotted as a function of the number of fatigue cycles, N, when subjected to a cyclic strain amplitude, $\Delta\epsilon$. The resistivity ratio is decreased by a factor of 2, for example, when cycled 1000 times at a cyclic strain of 0.62%; this results in a significant loss in stability (both flux-jump and cryogenic stability).

From a practical standpoint, the important number in Fig. 13.50 is the threshold strain below which there is no fatigue damage to the matrix. Designing a magnet to operate at an internal cyclic strain amplitude below this value will result in no loss of stability, no matter how many strain cycles it experiences. For the particular composite tested in Fig. 13.50, this occurs at about 0.19%. This threshold strain is approximately equal to twice the strain at which plastic yielding begins in copper (Ekin, 1981b), and therefore, can be predicted from stress–strain relations of the matrix material, such as that shown in Fig. 13.48 at 4 K.

13.8.3 Training

When a flux-jump–stabilized superconducting magnet or other device is initially energized, it often does not reach full operating field on the first try. Instead, thermal runaway occurs at a field below the design field. Usually subsequent attempts result in the thermal runaway point occurring at higher and higher fields until eventually the operating field is reached. This process is known as *training*. Cryogenically stabilized magnets usually do not train.

Experimental evidence indicates that this process is simply a relief of stress concentration points in the superconducting winding. Usually this occurs through localized fracture of the epoxy in potted magnets or frictional heating from wire movement. The first time the superconducting wire experiences stress is when the magnet is initially charged. If localized regions of the winding are under extreme stress concentration, this is the time they will move, yield, or fracture. In most cases, this liberates enough energy to cause a flux-jump–stabilized magnet to quench. Once the stress concentration is relieved, the magnet is ready to withstand higher stress.

Because of the brittle nature of epoxies at low

temperature, the strain limit for impregnated magnets is usually taken to be about 0.2% or less, in order to avoid severe training from fracture of the epoxy.

13.8.4 Measurement of mechanical properties

Stress-strain relations. Stress–strain characteristics, such as those shown in Figs. 13.48 and 13.49, can be obtained on short specimens using an apparatus shown schematically in Fig. 13.51. Tensile load is applied to the specimen by anchoring the bottom to a load frame and applying tension to the top with a mechanical loading machine. The design of the load frame is usually a compromise between having enough cross-sectional area not to buckle in compression, but not so much that heat conduction down the load frame generates excessive helium loss. As shown in Fig. 13.51, a typical load frame consists of a reaction cylinder surrounding the specimen and a concentric pull rod. It is usually made of stainless steel or titanium because of their high strength-to-thermal conductivity ratios.

Wire specimens can be anchored with a friction grip, such as the one shown schematically in Fig. 13.51, in which the wire is wrapped around a pulley and then inserted through a hole drilled in the pulley. Or it can be clamped between two serrated steel plates, although here one must be careful to avoid stress concentration at the point where the wire exits the grip. This can usually be accomplished by tapering the clamping pressure. If the wire is small enough (less than ~0.5 mm in diameter), enough mechanical strength can be obtained by simply soldering it in a hole drilled in a copper grip. In each case, the grip is preferably connected to the load train with a universal joint. This ensures self-alignment of the wire with the load train to avoid stress concentration from bending where the wire exits the grip.

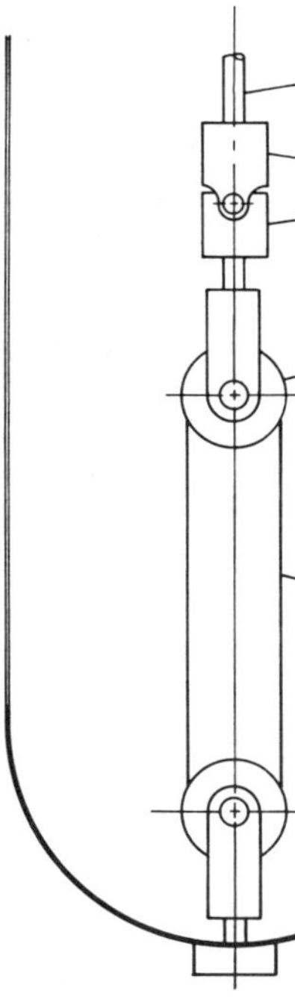

Figure 13.51 Loading assembly for wire specimens (Reed et al., 1977).

Bend tests. To test the critical current degradation resulting from bending strain, short specimens are wound on cylindrical formers having a range of diameters. The setup is exactly as described in Fig. 13.17a or 13.17b for J_c measurement, except that the former diameters range to very small values, resulting in bend degradation of J_c. The J_c may then be determined as a function of bend diameter, as shown in Fig. 13.42.

Uniaxial strain degradation of critical current density. To simulate hoop-stress degradation, uniaxial strain may be applied to a short superconducting wire specimen and its critical current measured as a function of the applied strain. For this purpose, a radial-access superconducting magnet enables application of a magnetic field perpendicular to a straight specimen, as shown schematically in Fig. 13.52. In this manner, $J_c(B,\epsilon)$ characteristics, such as those in Figs. 13.44 and 13.46, can be determined.

A problem with such measurements is that the specimen must be permitted to move uniaxially for mechanical loading, while being constrained from perpendicular movement caused by the Lorentz force. This is usually accomplished by reinforcing the specimen with a close-fitting support along its entire length. With care, specimen movement can be constrained to the point where premature thermal runaway is not initiated during measurement of the critical current.

Fatigue degradation tests can be performed with the same apparatus, except that static loading is replaced with cyclic loading, usually from a hydraulic feedback system. Data on fatigue degradation of the matrix, such as shown in Fig. 13.50, can be determined by periodically stopping the fatigue process and raising the tempera-

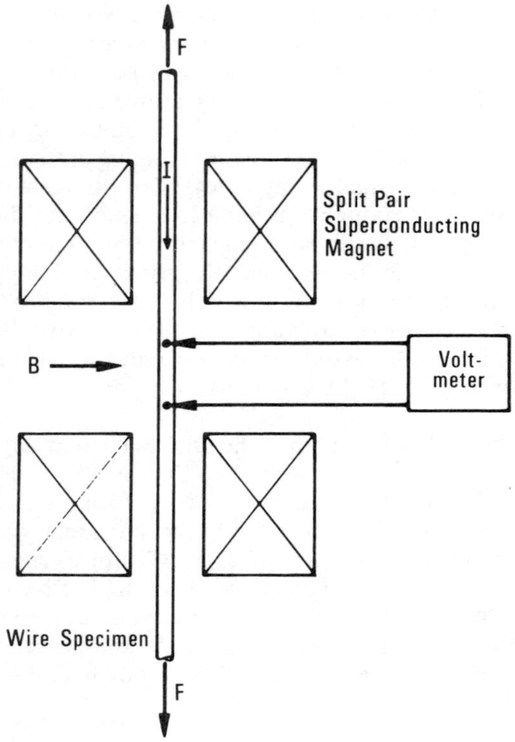

Figure 13.52 Schematic diagram of a critical current vs. strain apparatus (Ekin, 1978a).

ture of the composite just above T_c to measure the matrix resistivity. A convenient method for raising the temperature is to enclose the specimen in a small copper tube with a heater wound around it (Luhman and Suenaga, 1976). The tube has tight-fitting cotton plugs at either end so that when heated, the liquid helium within the tube is vaporized. This leaves a helium-gas insulating layer around the specimen so that the temperature can be elevated and controlled.

13.9 SUPERCONDUCTING MATERIALS SELECTION AND COMPOSITE DESIGN — A SUMMARY AND REFERENCE GUIDE

This section contains the net result of the six properties discussed in this chapter — how they affect superconductor selection and the geometrical design of superconducting composites. Unlike previous sections, this is not an introduction but a summary of the superconducting materials actually used in practical applications, the reasons for their choice, and the parameters governing the geometry of superconducting composites.

13.9.1 Commercial superconducting materials

There are basically only three superconducting materials that are being used commercially to any great extent in practical superconducting magnets: Nb–Ti, Nb_3Sn, and V_3Ga. Of these three, by far the most popular are the ductile Nb–Ti alloys; they are used in practically all applications up to magnetic fields of about 8 T (when operating with liquid helium as a refrigerant, i.e., 4.2 K). The principal reason for their popularity is their ease of fabrication (§ 13.5.1) and their superior mechanical properties [they show practically no critical current degradation from either bending or hoop stress (§ 13.8.1)].

For low-field applications ($\lesssim 3$ T) a Nb–Ti alloy containing a high-titanium composition is usually chosen, such as Nb–55 wt.% Ti. With this alloy, the critical current can be enhanced at low field using metallurgical heat treatments. For most applications at higher fields ($3\ \text{T} \lesssim H \lesssim 8$ T), however, a slightly lower titanium concentration, Nb–47 wt.% Ti, has been found to give optimum performance (reasons for this are given in Appendix F).

For applications involving fields higher than 8 T or operation at temperatures higher than 4.2 K, the brittle superconducting compounds Nb_3Sn or V_3Ga must be used. Nb_3Sn has a critical temperature of about 18 K and has a practical high-field limit of about 15 T (see Fig. 13.12). V_3Ga has a critical temperature of only about 15 K, but has a significantly higher critical current at fields above about 15 T. The practical high-field limit for V_3Ga is about 18 T (see Fig. 13.12).

Both of these superconductors have the A15-type crystal structures and are extremely brittle. One must be very careful not to overbend the wire anywhere along its length during fabrication and handling. Also, in magnet applications, enough external support or distributed reinforcement must be provided to limit the total strain on the superconductor to less than about 0.2% to prevent critical current degradation (see § 13.8.1).

Several new superconducting materials are in the developmental stages. None have primary commercial roles at this time, but they may have commercial impact in the future, and so they are mentioned briefly:

a. To alleviate the problem of mechanical degradation, superconductors are now under development having the C15-type crystal structure Laves phase (Inoue, Kuroda, and Tachikawa,

1979) and the B1 crystal structure (Gavaler, Janocko, Patterson, and Jones, 1971). These new superconductors may offer the same favorable high-temperature and high-field properties as Nb_3Sn and V_3Ga, but with considerably less strain sensitivity.

b. For operation at higher temperatures, Nb_3Ge with a critical temperature of 23 K may eventually become available in practical wire form. At the present time, however, it is fabricated only as tape specimens of limited length for laboratory testing (Gavaler, 1973; Maley, Newkirk, Thompson, and Valencia, 1981).

c. For very high-field operation, the Chevrel phase superconductors, such as $PbMo_6S_8$, are under development. These brittle materials have an $H_{c2}(0\ K)$ of ~60 T, but have been produced only in very small quantities. Considerable development is needed before it will be possible to fabricate long practical lengths.

13.9.2 Matrix materials

In practical superconducting composites, the superconductor is divided into fine filaments and embedded in a high conductivity matrix material to give the composite stability (e.g., Fig. 13.1). At present copper is the only matrix material that is used commercially to any extent. A relatively pure grade of copper is needed, usually the "oxygen-free" type. The residual resistance ratio (R_{293K}/R_{4K}) of the copper can be used as an index of its purity. Typically, the "oxygen-free" type has a ratio in the range 100 to 300, depending on the particular lot and manufacturer. Residual resistance ratios (RRRs) greater than 300 are probably not worth the additional cost of purification, since after fabrication, the copper is left in a partial cold-worked condition that typically lowers its RRR to the range 70 to 150, depending on the final anneal. The exception to this would be additional matrix material that is bonded or cabled with the superconducting composite after it is fabricated. Even here, though, the copper is usually used in a 1/4-hard condition for added strength, which lowers its RRR to about 100.

Another matrix material that is under development is pure aluminum. The advantages of aluminum are that it has a lower magnetoresistance than pure copper (Fig. 13.29), requires less refrigeration to cool (lower heat capacity), has a lower radiation cross section, and costs less. The primary problem is that the mismatch in hardness between pure aluminum and Nb–Ti or niobium makes it difficult to fabricate it into a composite. However, pure aluminum can be cabled and soldered with a superconducting composite *after* fabrication for added stability. In this case, the only concern is to prevent electrochemical corrosion caused by the mismatch between the electrochemical potential of aluminum and most low-melting-temperature solders.

13.9.3 Summary of principal composite design parameters

The main parameters that go into the design of a superconducting composite are summarized in Table 13.7 along with typical values, the determining equations, and the sections where they are discussed. Each of these parameters is discussed briefly below:

Volume fraction of superconductor. The volume fraction of composite that is superconducting is represented by f. In Fig. 13.1, for example, about one-third of the composite is occupied by superconducting material (i.e., $f \simeq 0.33$). It is also common practice to express this parameter in terms of the copper-to-superconductor ratio, which is just the ratio of the volume fraction of copper stabilizer to superconductor. For a Nb–Ti/Cu superconductor, for example, the ratio is given by $(1-f)/f$. The practical importance of f, as discussed in § 13.6, is that it has a great affect on the stability of the composite. To achieve flux-jump stability, for example, f must be limited to values in the range 0.2 to 0.6. For full cryogenic stability, f is limited to much smaller values. In both cases, the values of f are determined primarily by the electrical resistivity of the stabilizing matrix material, ρ. The smaller ρ is, the larger f can be without loss of stability. This is discussed in the stability section, § 13.6.

The volume fraction, f, is also limited to smaller values in Nb_3Sn–bronze superconductors than in Nb–Ti/Cu superconductors, not just by stability considerations, but also by the reaction kinetics in forming Nb_3Sn (see § 13.5). Typically the bronze reaction process requires a (Nb_3Sn): (bronze + unreacted Nb) ratio of less than about 0.3 to ensure stoichiometric formation of the Nb_3Sn compound.

Filament diameter. The diameter of an individual superconducting filament, d, is affected by three considerations: (a) fabrication limits, (b) stability requirements (either flux-jump stability or cryogenic stability), and (c) ac loss requirements. From stability and ac loss considerations,

Table 13.7 Composite superconductor design factors.

Parameter	Symbol	Design Consideration	Typical Values	Determining Equation	Chapter Section
Volume fraction of superconductor	f	Flux-jump stability	$f \simeq 0.2-0.6$	9	13.6.2
		Cryogenic stability	$f \simeq 0.2$	10	13.6.3
		Compressive prestress (brittle superconductors)			13.8.1
Filament diameter	d	Flux-jump stability	$d \simeq 1-100\ \mu m$	9	13.6.2
		Cryogenic stability	$d \simeq 10-100\ \mu m$	11	13.6.3
		Hysteretic loss	$d \simeq 1-100\ \mu m$	16	13.7.3
Strand diameter	D	Flux-jump stability	$D \lesssim 2$ mm		13.6.2
		Self-field loss	$D \lesssim 2$ mm	17	13.7.2 & 13.7.3
Twist pitch	ℓ	Eddy current loss	$\ell \simeq 1$ cm	14	13.7.1
Reinforcement		Enough to limit strain to about 0.2%			13.8.1–13.8.3

d should be made as small as possible, but, as discussed in the section on composite fabrication (§ 13.5), the practical lower limit on the size of d is determined mainly by difficulties in fabricating uniform filaments of diameter less than about 1 μm.

Probably the most restrictive requirement on the size of d is imposed by flux-jump stability considerations, especially in Nb_3Sn–bronze superconductors. To achieve flux-jump stability, usually d must be less than ~10 μm for Nb_3Sn–bronze superconductors. This is primarily determined by the factors entering Eq. 13.9 (discussed in § 13.6 on stability).

For cryogenic stability, the main requirement on d is that it not be so great that large temperature gradients can exist between the filament cores and the matrix material. According to the Stekly requirement (Eq. 13.11) this usually requires d to be less than about 100 μm for Nb–Ti/Cu composites, and less than about 10 μm for Nb_3Sn–bronze composites.

If the superconductor will also be subjected to a time-varying magnetic field, there is additional reason to keep d small because the accompanying hysteretic ac loss scales directly with d (Eq. 13.16). Typically, for pulsed magnet applications, the losses are high enough that it is worth the added fabrication difficulties to make d on the order of 1 μm.

Strand diameter. In flux-jump stabilized wires, the overall strand diameter, D, generally must not exceed 2 mm to prevent flux-jumping in the self-field of the composite. Furthermore, in applications involving time-varying transport currents,the self-field ac losses can become significant if the strand size is not limited. As discussed in the ac losses section (§ 13.7), the ratio of self-field to hysteretic losses scales as D^2/d. Typically for a filament diameter of 10 μm, this requires D to be less than about 1 mm if the self-field loss is to be limited to 10% of the hysteretic loss. If a larger conductor cross section is needed, it can be obtained by cabling several strands together.

Twist pitch Eddy current losses arising from a changing external field can be significantly reduced by twisting the composite strand with a twist pitch much less than the critical length given by Eq. 13.14 (discussed in the section on ac losses, § 13.7). The important factor in Eq. 13.14 is the time derivative of the magnetic field, $\dot{H}$, since the other factors are usually determined by material properties and stability requirements. For a typical $\mu_0 \dot{H}$ of 10^{-4} to 1 $T \cdot s^{-1}$, a twist pitch on the order of 1 cm usually reduces the eddy current losses to an acceptable level.

Reinforcement. Generally, tensile strain in the superconductor must be limited to less than 0.2% for a variety of reasons. In pulsed Nb–Ti magnets, cyclic strains greater than this significantly degrade the matrix conductivity and, hence, the stability of the magnet. In potted Nb–Ti magnets, the likelihood of the epoxy cracking in stress-concentration regions increases drastically for average strains greater than 0.2%, which leads to training problems. In Nb_3Sn magnets, a tensile strain greater than 0.2% can lead to severe critical-current degradation (Fig.

13.46). These considerations are discussed in more detail in the section on mechanical properties (§ 13.8).

To limit strain in large magnet systems, some type of added reinforcement is required since the stress from the magnetic force often can exceed the ultimate strength of the composite itself. This is usually accomplished by using an outer structural shell made from stainless steel or structural composite. Such an external shell, however, provides only indirect support for the inner windings. Load must be transmitted through the thickness of the magnet, which may have a low effective modulus. This can result in inadequate support of the innermost turns where it is needed most. Distributed reinforcement can be used to remedy such situations and provide adequate support throughout the winding. Such reinforcement can be obtained by co-winding stainless steel bands with the superconducting wire or adding structural strands directly to the superconducting composite during fabrication (e.g., Fig. 13.25).

13.9.4 Suggested reading

For additional information in each of these areas several general texts and articles deserve special mention. They are listed below along with the specific properties on which they are particularly helpful.

a. General: Fietz and Rosner (1974), Dew-Hughes (1974), and *Superconductor Materials Science: Metallurgy, Fabrication, and Applications* (1981).

b. Critical temperature, upper critical field, and critical current density: Tinkham (1975); de Gennes (1966); *Superconductivity* (1969); and Saint-James, Sarna, and Thomas (1969).

c. Critical current density: Campbell and Evetts (1972).

d. Stability and ac losses: Wilson et al. (1970).

e. Stability: Wipf (1979).

f. AC losses: Brechna (1973) — Chapter 4.

g. Stress vs. strain characteristics: Koch and Easton (1977).

h. Stress vs. strain, critical current density vs. strain, bending strain, fatigue, and training: Ekin (1981b).

13.10 APPENDIXES

Appendix A. Critical temperature and the occurrence of superconductivity

This appendix contains a brief treatment of some of the theoretical and empirical rules describing the occurrence of superconductivity. Although two or three possible mechanisms for superconductivity have been discussed (Little, 1964; Parmenter, 1967; Ginzburg, 1968; Heeger, 1972), it now appears that the only practical mechanism has its origin in the phonon mechanism as described by the BCS theory (Bardeen, Cooper, and Schrieffer, 1957). In the BCS theory, the critical temperature, T_c, is given by:

$$T_c \sim \theta_D \exp\{-1/[N(E_F)\Lambda]\} \qquad \text{(A.1)}$$

where θ_D is the Debye temperature, $N(E_F)$ is the electronic density of states at the Fermi surface, and Λ is an electron–phonon interaction constant. [Good introductory accounts of the BCS theory may be found in Rose-Innes and Rhoderick (1978), Lynton (1962), and Cooper (1960); a more detailed treatment is given in Tinkham (1975).]

In practice, the quantity $N(E_F)$ tends to govern T_c and explains qualitatively why metals, alloys, and compounds with high values of $N(E_F)$ have high critical temperatures. This is evident in Matthias's empirical rules (Matthias, 1957) describing the occurrence of superconductivity among elements. He observed (see Fig. A.1) that the critical temperatures of different elements in the periodic table formed a pattern with peaks occurring at specific numbers of valence electrons per atom. These values are now known to correspond to maxima in $N(E_F)$, consistent with the BCS result of Eq. A.1.

Among compounds, Matthias also pointed out that certain crystal structures tend to favor the formation of high-T_c superconductors. The highest values of T_c are found in the A15 and B1 crystal structures. The effect of crystal structure is not very well understood. However, it again

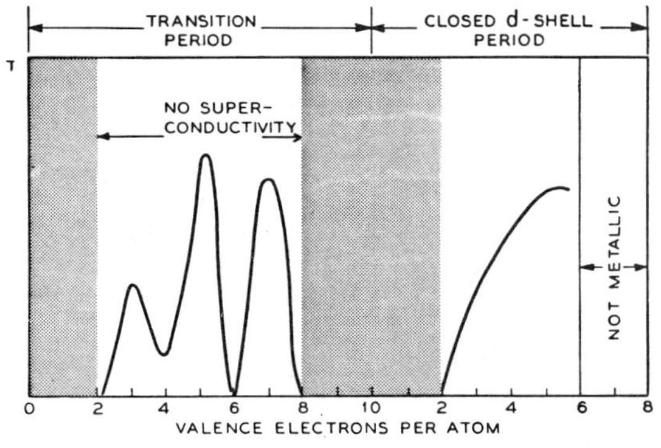

Figure A.1 Critical temperature of the superconductive elements as a function of the valence electrons per atom (Matthias, 1957).

appears that the primary role of crystal structure is connected with $N(E_F)$, the high T_c crystal structures allowing the formation of compounds with favorable values of the number of valence electrons per atom (see Fig. A.2).

Appendix B. Critical temperature and stoichiometry

Although T_c is relatively stable in most superconductive elements and alloys, it is a very sensitive function of stoichiometry in the high-T_c A15 compounds. Deviations from the ideal composition can cause a decrease in T_c by as much as 2 to 3 K per atom percent deviation from stoichiometry (Hanak, Cody, Cooper, and Rayl, 1962; Arrhenius, Corenzwit, Fitsgerald, Hull, Luo, Matthias, and Zachariasen, 1968; Blaugher, Hein, Cox, and Waterstrat, 1969). The case of the Nb–Ge system is a good example. It normally combines at a composition corresponding to Nb_4Ge, with a T_c of 6 K (Matthias, Geballe, and Compton, 1963). The use of rapid cooling techniques, however, produces a disordered Nb_3Ge compound whose T_c is 17 K (Matthias, Geballe, Willens, Corenzwit, and Hull, 1965). Films of Nb_3Ge prepared by a high-pressure dc sputtering process, however, yield T_c values up to 23 K (Gavaler, 1973).

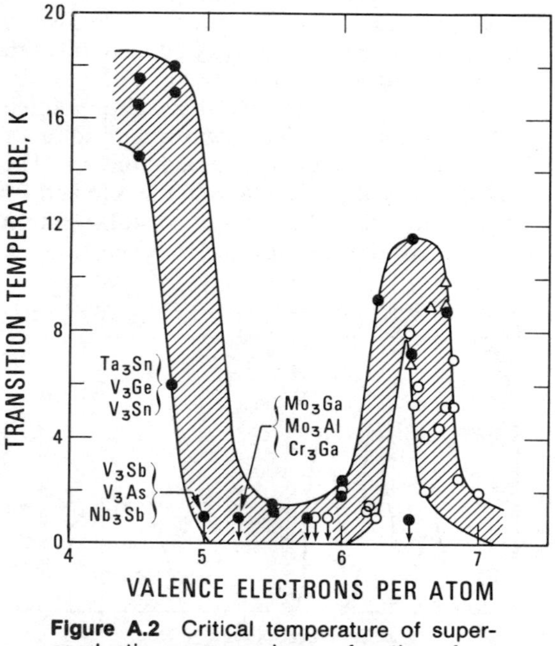

Figure A.2 Critical temperature of superconductive compounds as a function of valence electrons per atom (Blaugher and Hulm, 1961). o — sigma phase, ● — beta tungsten, △ — alpha manganese.

In summary, high T_c values in the A15 compounds are obtained by preparing the material with the ideal stoichiometric ratio. This is straightforward for V_3Ga, V_3Si, and Nb_3Sn, which are all stable at this stoichiometric composition. Unfortunately, the highest T_c compounds, Nb_3Al, Nb_3Ga, and Nb_3Ge (see Table 13.2), all have more stable phases at lower niobium compositions (e.g., Nb_2Al and Nb_5Ge_3). This makes their preparation more difficult, and normally they are produced with an excess of niobium to promote formation of the high T_c stoichiometric composition.

Appendix C. Type I and Type II superconductivity

In this section three parameters characterizing Type I and Type II superconductors are discussed along with their interrelations. They are the penetration depth, λ, the coherence length, ξ, and their ratio, κ.

Penetration depth. The penetration of a magnetic field into a Type I superconductor has already been illustrated in Fig. 13.5. The strength of the field decreases exponentially with distance, x, from the surface as:

$$H = H_0 \exp(-x/\lambda) \qquad \text{(C.1)}$$

where H_0 is the field at the surface. A similar situation exists around the normal core of each flux tube, or vortex, in a Type II superconductor. The penetration depth is temperature dependent, becoming infinite at T_c. Quantitatively, λ in a pure superconductor may be determined near T_c from the BCS relation:

$$\lambda(T)\mid_{T\to T_c} = 2^{-1/2}\lambda_L(0)[1-(T/T_c)]^{-1/2} \qquad \text{(C.2)}$$

where $\lambda_L(0)$ is a constant for each material, known as the London penetration depth:

$$\lambda_L(0) = [m^*/N(E_F)e^2]^{1/2} \qquad \text{(C.3)}$$

Here m^* is the effective mass of the superconducting electron pairs and e is their charge. The density of electronic states at the Fermi surface, $N(E_F)$, can be obtained from a measurement of the electronic specific heat of the material [$C = \gamma T$, $\gamma = (1/3)\pi^2 k_B^2 N(E_F)$; k_B = Boltzmann's constant; see Chapter 2]. Values of λ_L for several superconductors are given in Table C.1.

Coherence length. The boundary between superconducting and normal regions in a superconducting material is not perfectly sharp, but

Table C.1 BCS coherence length, ξ_0, and London penetration depth, λ_L, for several elements (Bardeen and Schrieffer, 1961).

Material	ξ_0 (nm)	λ_L(nm)
Aluminum	1600	15.7
Tin	230	35.5
Lead	83	37.0

has a finite width known as the coherence length, ξ. The normal region at the core of each flux tube in a Type II material is the minimum size possible, that is, it is approximately one coherence length in diameter. The coherence length has the same temperature dependence as λ near T_c. At T_c both ξ and λ become infinite; the normal vortex cores expand to fill the entire material and field penetration is complete. In a pure superconducting material, ξ can be obtained near T_c using the BCS relation:

$$\xi(T)|_{T \to T_c} = 0.74\, \xi_0[1 - (T/T_c)]^{-1/2} \quad \text{(C.4)}$$

where ξ_0 is a constant for each material, known as the BCS coherence length:

$$\xi_0 = 0.18\hbar v_F / k_B T_c \quad \text{(C.5)}$$

Here, $\hbar$ is Planck's constant and v_F is the Fermi velocity. Values of ξ_0 for several pure superconductors are given in Table C.1.

Kappa. The Ginzburg–Landau (1950) parameter, κ, is defined as the ratio of λ to ξ:

$$\kappa \equiv \lambda / \xi \quad \text{(C.6)}$$

This parameter provides a precise distinction between Type I and Type II superconductors. Depending on the value of κ in relation to $1/\sqrt{2}$, the material is either Type I or Type II:

$\kappa < 1/\sqrt{2}$ in Type I superconductors
$\kappa > 1/\sqrt{2}$ in Type II superconductors.

Dirty superconductors. The above formulas for λ and ξ are changed by the addition of impurities and defects to a superconductor. Such material defects shorten the electronic mean free path, ℓ, shorten ξ, and lengthen λ. In the dirty limit where $\ell << \xi_0$, Eqs. C.2 and C.4 become:

$$\lambda(T)|_{T \to T_c} = 0.615\lambda_L(0) (\xi_0/\ell)^{1/2} [1 - (T/T_c)]^{-1/2} \quad \text{(C.7)}$$

$$\xi(T)|_{T \to T_c} = 0.85(\xi_0 \ell)^{1/2}[1 - (T/T_c)]^{-1/2} \quad \text{(C.8)}$$

Although strictly valid only near T_c, these equations for the dirty limit are approximately correct at low temperatures as well. A comparison of Eqs. C.7 and C.8 with Eqs. C.2 and C.4 shows the temperature dependence for λ and ξ near T_c is not changed by the addition of defects. Note, however, that λ is scaled up by the factor $(\xi_0/\ell)^{1/2}$ and ξ scaled down by the factor $(\ell/\xi_0)^{1/2}$ compared with the values for the pure material (Eqs. C.2 and C.4). This will increase κ. In general, κ is given by:

$$\kappa|_{T \to T_c} = 0.96\lambda_L/\xi_0 + 0.7\lambda_L/\ell \quad \text{(C.9a)}$$

The first term is just the ratio of Eqs. C.2 and C.4, the second term, the ratio of Eqs. C.7 and C.8. Alternatively, κ may be expressed in terms of the resistivity, ρ (Ω·cm), and electronic specific heat, γ (J·cm^{-3}·K^{-2}), of the material:

$$\kappa|_{T \to T_c} = 0.96\, \lambda_L/\xi_0 + 2.38\, \rho\gamma^{1/2} \quad \text{(C.9b)}$$

The quantitative increase in κ with decreasing ℓ can be seen explicitly from Eqs. C.9a and C.9b. A pure Type I superconductor (with $\kappa < 1/\sqrt{2}$) can even have its κ increase to the point where it becomes a Type II superconductor (with $\kappa > 1/\sqrt{2}$) if it is made dirty (resistive) enough. Most practical superconductors are, in fact, dirty superconductors with small values of ℓ and ξ.

Appendix D. Energy gap and the relationship between thermodynamic critical field and upper critical field

Below T_c, the free energy of a superconductor is lower than that of the normal state by an amount $\Delta(T)$. This energy gap is responsible for the formation of the resistance-free superconducting state and explains why the superconducting charge carriers (Cooper pairs) cannot be scattered by impurities or other defects unless an energy greater than $\Delta(T)$ is involved. In fact, T_c is just the temperature at which the average phonon energy, $k_B T$, is approximately equal to $\Delta(T=0)$, where k_B is Boltzmann's constant. The exact relation according to BCS theory is:

$$k_B T_c = 0.57\, \Delta(0) \quad \text{(D.1)}$$

This also explains the thermodynamic critical field, H_c. The exclusion of a field, H, from a superconductor raises the free energy by an amount $\mu_0 H^2/2$ per unit volume. If H becomes large enough, the energy cost, $\mu_0 H^2/2$, can exceed the energy gain, Δ, and the superconductor will revert to the normal state. The critical field is the field at which the two terms balance. The exact relation according to BCS theory is:

$$\mu_0 H_c^2(0)/2 = (1/2)\, N(E_F)\, \Delta(0)^2 \qquad (D.2)$$

where $N(E_F)$ is the density of electronic states at the Fermi surface. At fields near H_{c2} or at temperatures near T_c, H_{c2} is related to H_c by Abrikosov's result (Abrikosov, 1957):

$$H_{c2} = \sqrt{2}\,\kappa\, H_c \qquad (D.3)$$

Thus, we can see from Eqs. D.1, D.2, and D.3 that a high value of H_{c2} requires both a high value of T_c (to make H_c large) and a high value of κ. Since κ is increased by the introduction of defects (see Appendix C), alloying tends to increase H_{c2}.

Appendix E. Paramagnetic limit and upper critical field

Upper critical field at absolute zero temperature can be determined from measurements of dH_{c2}/dT near T_c:

$$H_{c2}(0) = -0.69\, T_c dH_{c2}/dT\,|_{T\to T_c} \quad (\ell << \xi_0) \qquad (E.1)$$

This general relation for dirty superconductors neglects, however, paramagnetic reductions in Δ, which can vary substantially from material to material. The high-temperature, high-field superconductors usually have a *paramagnetic* susceptibility. This can significantly lower the observed H_{c2} from that which would theoretically be expected from Eq. E.1. This is evident in the difference between the solid experimental curves and the dashed theoretical curve for V_3Ga in Fig. 13.8. The paramagnetic effect is fortunately moderated somewhat by spin-orbit scattering, which brings about an accompanying reduction in the free energy with H. (A more detailed discussion may be found in the articles by Werthamer, Helfand, and Hohenberg, 1966; the review in Dew-Hughes, 1974; and Serin, 1969.)

The net result of Appendixes C through E is that the highest critical fields are obtained in very dirty Type II superconductors (where κ is large) with high values of T_c and where there is large spin-orbit scattering.

Appendix F. Critical current density and the pinning force

The problem of obtaining high critical currents in the practical Type II superconductors is simply one of producing the maximum pinning forces in the material. This is accomplished by the introduction of metallurgical defects. The various types of defects may be classified into four groups according to the number of defect dimensions that are large compared with the flux-line lattice spacing (see, e.g., Campbell, Evetts, and Dew-Hughes, 1968). The group classifications are point, line, surface, and volume defects.

a. Point defects. Examples include voids, small particles of a second phase and small clusters of other types of defects. Voids may be introduced by sintering, second-phase precipitates may be formed by heat treatment, and clusters may be created by radiation damage.

b. Line defects. Dislocations are the main type of line defect. These are introduced into ductile superconductors by cold work. This may be followed by a low-temperature heat treatment, which rearranges the dislocations into a more effective cellular structure. It is also possible that such a heat treatment also causes the dislocation cell walls to be decorated with precipitates that enhance their effectiveness as pinning sites. Figure 13.11 is an example of flux-line pinning by dislocation structures.

c. Surface defects. Examples include grain boundaries, martensitic boundaries, and twin or polymorphic transformations. The more defect surface area, the higher the pinning force.

d. Volume defects. Large voids or precipitates are the main examples. These are generally not very effective pinning sites; the pinning mainly occurs only at the defect surface.

In niobium alloys, such as Nb–Ti and Nb–Zr, the primary pinning sites are line defects: dislocation cell structures formed by cold work and low-temperature heat treatment. In the A15 compounds, such as Nb_3Sn, pinning is due mainly to surface defects: grain boundaries and martensite precipitates.

Size dependence. The smaller the defect size, the greater the area of defect interface per unit volume and the greater the pinning force. However, when the defect size becomes smaller than the vortex core size, the core can span several defects at once, averaging out their effect and drastically reducing the pinning force. Optimum pinning occurs in between, when the defect size is comparable to the vortex-core size, that is, about ξ in size.

As discussed in Appendix C, ξ is very small in most commercial superconductors, usually 5 nm or less at low temperatures. It is difficult to introduce defects this small. Thus, for most practical purposes, the more finely divided the defect size, the better.

Nevertheless, in some materials, where ξ is not so small, it is quite possible to reduce the average defect size to less than ξ (at all temperatures below T_c). When this occurs, the defects (although numerous) become ineffective as pinning sites. A reduction in the pinning force by as much as 3 to 4 orders of magnitude has been observed when this situation occurs (Ekin, 1975a). Again, the important point is that optimum pinning occurs when the defect size is comparable to the flux-line core size, that is, ξ. The size scale of ξ is set by the value of $\sqrt{\xi_0 \ell}$ in Eq. C.8.

Magnetic field dependence. J_c has a field dependence that varies from material to material, but the characteristic shapes are usually similar (see Fig. 13.12). This dependence may be obtained from the field dependence of the pinning force, F_p, through Eq. 13.6. At zero field where the number of flux tubes is reduced to zero, F_p also approaches zero such that $J_c(H=0)$ is finite (see Fig. F.1). As H is increased, F_p rises to a maximum. In the region of the maximum F_p, F_p is nearly constant, and J_c has an approximate inverse dependence on H, as seen from Eq. 13.6. Finally, as H_{c2} is approached, ξ becomes infinite and each vortex core spans many defects. As explained above, this rapidly reduces the pinning force and critical current to zero.

The existence of a peak in the pinning force, as shown in Fig. F.1, is a universal feature of hard high-field superconductors. Although the position of the peak and its height are sensitive to metallurgical treatments, it has been found by Fietz and Webb (1969) that F_p approximately obeys a general scaling law:

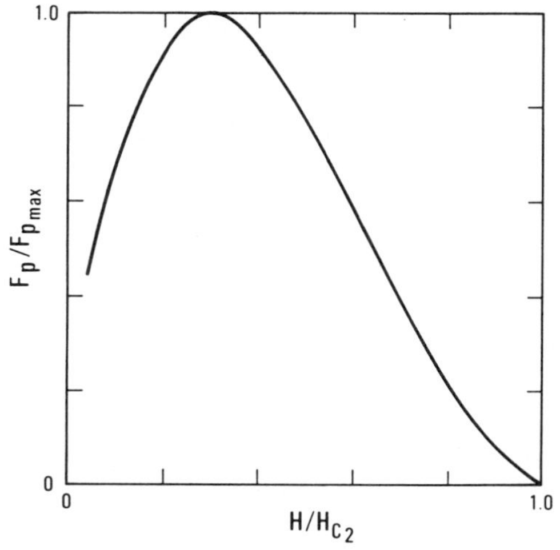

Figure F.1 Typical variation of the pinning force with magnetic field.

$$F_p = [H_{c2}(T)]^{2.5} f(h) \tag{F.1}$$

where $f(h)$ is a function of just the reduced magnetic field, $h = H/H_{c2}$, shown in Fig. F.1 [$f(h)$ can usually be represented by the form $f(h) \propto h^p(1-h)^q$, where p is typically about 0.5 and q ranges from about 1 to 3 depending on the material]. Thus, the temperature dependence of F_p (and hence J_c) is approximately predictable through Eq. F.1 from only a knowledge of $H_{c2}(T)$ for a given material.

The behavior of F_p and J_c at low and high fields is qualitatively different (Kramer, 1973). At low fields (below the peak in F_p), the onset of flux motion is determined by breakage of pins between individual flux tubes and pinning sites. At high fields (above the peak in F_p), the onset of flux motion occurs by synchronous shear of the flux tube lattice around individual pins too strong to be broken. As a result, metallurgical treatments mainly affect the low-field J_c, leaving the high-field characteristic unchanged. The latter is determined mainly by the shear modulus of the flux tube lattice, an intrinsic property of the material. Thus, in Fig. 13.12, one cannot hope to increase J_c very much at high fields by increasing the number, density, or type of defects in the material. Because of their unchanging *intrinsic* nature, the high-field J_c characteristics play a primary role in the selection of superconducting materials for specific magnet applications.

Appendix G. Critical current density and flux motion

The critical current density is usually determined from the V–I characteristics of a superconducting wire in a transverse magnetic field. The current at which the voltage rapidly rises is taken to be the critical current. However, it is not always clear what this current should be, because often the transition to the normal state is not sharp. In some cases, J_c can vary by 50% or more, depending upon how it is defined (Clark and Ekin, 1977). The broadened transition is due to the occurrence of flux motion in the material, which induces voltage at current densities below J_c.

In weak pinning materials, flux motion begins at current densities well below the thermal runaway point, giving an extremely broad transition. If the pinning is very weak, the V–I characteristic can even have a linear region, similar to that shown in Fig. G.1b. In this special case, the slope of the linear flux–flow region is an intrinsic material property called the *flux–flow resistivity,*

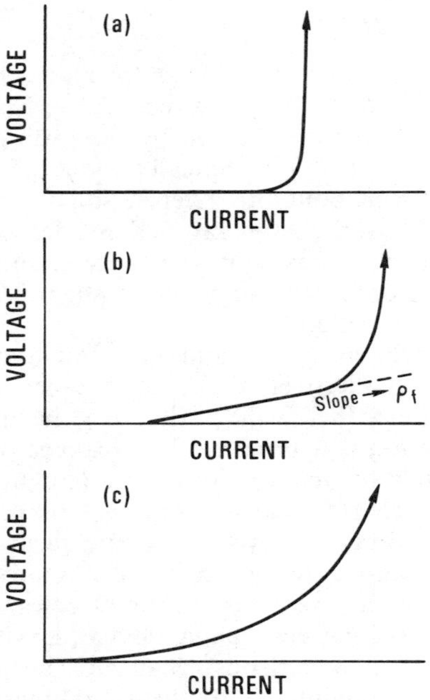

Figure G.1 Voltage–current characteristics for: (a) strong-pinning materials, (b) weak-pinning materials showing the superconducting flux flow region below J_C, and (c) damaged strong-pinning materials showing the nonlinear take-off region below J_C.

ρ_f, and is related to the normal state resistivity, ρ_n, by the approximate relation ($T << T_c$):

$$\rho_f = \rho_n(H/H_{c2}) \tag{G.1}$$

In practical materials where the pinning force is large, such an extensive linear region at low current densities is usually not observed. However, if the material is damaged, by excessive bending for example, the V–I characteristic can have a broad *nonlinear* take-off region, such as is shown in Fig. G.1c. This usually occurs in the brittle high-field superconductors where conductor damage can occur at very low strains, 0.2% to 0.6% (§ 13.8). The voltage occurring at current densities below J_c is due to localized flux motion at the damage sites in the material. In heavily damaged superconducting wire, the ohmic heating can be prohibitively large for practical magnet operation.

All *multifilamentary* superconducting wires, damaged or not, have some resistivity due to irregularities in the superconductor filament diameter. Localized flux motion occurs at the necked-down regions of each superconducting filament, giving rise to small voltages at these points. Although the accompanying resistivity is usually very small (0.1 pΩ·cm or less), the voltage developed along a long length of wire can become significant. The V–I characteristics thus serve a dual function: they determine J_c for a given material, and indicate any excessive wire inhomogeneity or damage that might lead to unacceptably high Joule heating.

13.11 REFERENCES

Abrikosov, A. A. (1957). On the magnetic properties of superconductors of the second type. *Zh. Eksp. Teor. Fiz.* **32**, 1442–1452 [*Soviet Phys. — JETP* **5**, 1174–1182 (1957)].

Arp, V. D. (1977). Stresses in superconducting solenoids. *J. Appl. Phys.* **48**, 2026–2036.

Arrhenius, G., Corenzwit, E., Fitsgerald, R., Hull, G. W., Jr., Luo, H. L., Matthias, B. T., and Zachariasen, W. H. (1968). Superconductivity of Nb_3(Al, Ge) above 20.5 degrees K. *Proc. Nat. Acad. Sci.* **61**, 621–628.

Bardeen, J., Cooper, L. N., and Schrieffer, J. R. (1957). Theory of superconductivity. *Phys. Rev.* **108**, 1175–1204.

Bardeen, J. and Schrieffer, J. R. (1961). Recent developments in superconductivity. In *Progress in Low Temperature Physics*, Vol. 3 (ed. C. J. Gorter). North-Holland, Amsterdam, 170–287.

Benz, M. G. (1968). Mechanical and electrical properties of diffusion-processed Nb_3Sn–copper stainless steel composite superconductors. *J. Appl. Phys.* **39**, 2533–2537.

Blaugher, R. D. and Hulm, J. K. (1961). Superconductivity in the σ and α–Mn structures. *Phys. Chem. Solids* **19**, 134–138.

Blaugher, R. D., Hein, R. E., Cox, J. E., and Waterstrat, R. M. (1969). Atomic ordering and superconductivity in A-15 compounds. *J. Low Temp. Phys.* **1**, 539–561.

Brechna, H. (1973). *Superconducting Magnet Systems*. Springer-Verlag, New York.

Buehler, E. and Levinstein, H. J. (1965). Effect of tensile stress on the transition temperature and current carrying capacity of Nb_3Sn. *J. Appl. Phys.* **36**, 3856–3860.

Campbell, A. M. and Evetts, J. E. (1972). *Critical Currents in Superconductors*. Barnes & Noble, New York.

Campbell, A. M., Evetts, J. E., and Dew-Hughes, D. (1968). Pinning of flux vortices in Type II superconductors. *Philos. Mag.* **18**, 313–343.

Clark, A. F. and Ekin, J. W. (1977). Defining critical current. *IEEE Trans. Magn.* **MAG-13**, 38–40.

Collings, E. W. and Ho, J. C. (1970). Influence of microstructure on the superconductivity of a dilute Ti–Mo alloy. *Phys. Rev. B* **1**, 4289–4294.

Cooper, L. N. (1960). Theory of superconductivity. *Am. J. Phys.* **28**, 91–101.

Cornish, D. N. (1966). Superconducting coils using composite stranded cables. *J. Sci. Instr.* **43**(1), 16–20.

Cornish, D. N., Zbasnik, J. P., Leber, R. L., Hirzel, D. G., Johnston, J. E., and Rosdahl, A. R. (1979).

MFTF coil construction and performance. *IEEE Trans. Magn.* **MAG-15**, 530–533.

de Gennes, P. G. (1966). *Superconductivity of Metals and Alloys*. Benjamin, New York.

Dew-Hughes, D. (1974). Practical superconducting materials. In *Superconducting Machines and Devices* (eds. S. Foner and B. B. Schwartz). Plenum Press, New York, 87–137.

Easton, D. S., Kroeger, D. M., Specking, W., and Koch, C. C. (1980). A prediction of the stress state in Nb_3Sn superconducting composites. *J. Appl. Phys.* **51**, 2748–2757.

Easton, D. S. and Schwall, R. E. (1976). Performance of multifilamentary Nb_3Sn under mechanical load. *Appl. Phys. Lett.* **29**, 312–321.

Echarri, A. and Spadoni, M. (1971). Superconducting Nb_3Sn. A review. *Cryogenics* **11**, 274–284.

Ekin, J. W. (1975a). Critical currents in granular superconductors. *Phys. Rev. B* **12**, 2676–2681.

Ekin, J. W. (1975b). Unpublished data. National Bureau of Standards, Boulder, Colorado.

Ekin, J. W. (1976). Effect of stress on the critical current of Nb_3Sn multifilamentary composite wire. *Appl. Phys. Lett.* **29**, 216–219.

Ekin, J. W. (1977). Mechanisms for critical-current degradation in NbTi and Nb_3Sn multifilamentary wires. *IEEE Trans. Magn.* **MAG-13**, 127–130.

Ekin, J. W. (1978a). Fatigue and stress effects in NbTi and Nb_3Sn multifilamentary superconductors. In *Advances in Cryogenic Engineering*, Vol. 24. Plenum Press, New York. 306–316.

Ekin, J. W. (1978b). Current transfer in multifilamentary superconductors: Part I — theory. *J. Appl. Phys.* **49**(6), 3406–3409.

Ekin, J. W. (1980a). Strain scaling law for flux pinning in practical superconductors: Part I: basic relationship and application to Nb_3Sn conductors. *Cryogenics* **20**, 611–624.

Ekin, J. W. (1980b). Strain scaling law and the prediction of uniaxial and bending strain effects in multifilamentary superconductors. In *Filamentary A15 Superconductors* (eds. M. Suenaga and A. F. Clark). Plenum Press, New York, 187–203.

Ekin, J. W. (1981a). Strain scaling law for flux pinning in NbTi, Nb_3Sn, Nb–Hf/Cu–Sn–Ga, V_3Ga, and Nb_3Ge. *IEEE Trans. Magn.* **MAG-17**, 658–661.

Ekin, J. W. (1981b). Mechanical properties and strain effects in superconductors. In *Superconducting Materials Science: Metallurgy Fabrication and Applications* (eds. S. Foner and B. B. Schwartz). Plenum Press, New York, 455–510.

Ekin, J. W., Clark, A. F., and Ho, J. C. (1978). Current transfer in multifilamentary superconductors: Part II — experimental results. *J. Appl. Phys.* **49**(6), 3410–3412.

Ekin, J. W., Fickett, F. R., and Clark, A. F. (1976). Effect of stress on the critical current of NbTi multifilamentary composite wire. In *Advances in Cryogenic Engineering*, Vol. 22. Plenum Press, New York, 449–452.

Fickett, F. R. (1972). Magnetoresistivity of copper and aluminum at cryogenic temperatures. In *Proceedings of the 4th International Conference on Magnet Technology*. A.E.C. CONF. 720903, NTIS, Springfield, Virginia, 539–541.

Fietz, W. A. (1982). Oak Ridge National Laboratory, Oak Ridge, Tennessee. Personal communication.

Fietz, W. A. and Rosner, C. H. (1974). Advances in superconductive magnets and materials. *IEEE Trans. Magn.* **MAG-10**, 239–259.

Fietz, W. A. and Webb, W. W. (1969). Hysteresis in superconducting alloys — temperature and field dependence of dislocation pinning in niobium alloys. *Phys. Rev.* **178**, 657–667.

Gavaler, J. R. (1973). Superconductivity in Nb–Ge films above 22 K. *Appl. Phys. Lett.* **23**, 480–482.

Gavaler, J. R., Janocko, M. A., Patterson, A., and Jones, C. K. (1971). Very high critical current and field characteristics of niobium nitride thin films. *J. Appl. Phys.* **42**, 54–57.

Ginzburg, V. L. (1968). The problem of high temperature superconductivity. *Contemp. Phys.* **9**, 355–374.

Ginzburg, V. L. and Landau, L. D. (1950). On the theory of superconductivity. *Zh. Eksp. Teor. Fiz.* **20**, 1064–1082.

Haid, D. A. and Fietz, W. A. (1969). *Development of Manufacturing Methods for the Production of Superconductive Devices*. Technical Report AFML-TR-69-170, Air Force Materials Lab., AFSC, Wright-Patterson Air Force Base, Ohio.

Hampshire, R., Sutton, J., and Taylor, M. T. (1969). Effect of temperature on the critical current density of Nb–44 wt. percent Ti alloy. Program preprint, IIR — British Cryogenics Council Conference on Low Temperatures and Electric Power, Royal Society, London, England (Mar. 24–26, 1969), 69–76.

Hanak, J. J., Cody, G. D., Cooper, J. L., and Rayl, M. (1962). The effect of composition on the superconducting properties of vapour-deposited niobium stannide. In *Proceedings of the Eighth International Conference on Low Temperature Physics* (ed. R. O. Davies). Butterworths, London, 353–354.

Hanak, J. J., Strater, K., and Cullen, G. W. (1964). Preparation and properties of vapor-deposited niobium stannide. *RCA Rev.* **XXV**, 342–365.

Hancox, R. (1966a). Calculation of ac losses in a Type II superconductor. *Proc. IEEE* **113**, 1221–1228.

Hancox, R. (1966b). *The Paramagnetic Limit in Type II Superconductors*. Culham Lab. Report No. CLM-P121, Culham Laboratory, Culham, England.

Heeger, A. (1972). Magnetic properties of conducting organic salts. In *AIP Conference Proceedings No. 10*, Part 2 (eds. C. D. Graham, Jr. and J. J. Rhyne). American Institute of Physics, New York, 1476–1492.

Heyne, C. J., Hackworth, D. T., Singh, S. K., Young, J. L., and Sanger, P. A. (1979). Westinghouse design of a forced flow Nb_3Sn test coil for the large coil program. In *Proceedings of the Eighth Symposium on Engineering Problems of Fusion Research*. IEEE, New York, 1148–1153.

Hillman, H. and Hauck, D. (1972). Relationship between defect structure, flux-line lattice and metallurgical treatment in high field superconductor NbTi. In *Proceedings of the 1972 Applied Superconductivity Conference*. IEEE, New York, 429–433.

Hillman, H. (1981). Fabrication technology of superconducting material. In *Superconducting Materials Science: Metallurgy, Fabrication, and Applications* (eds. S. Foner and B. B. Schwartz). Plenum Press,

New York, 275–388.
Howe, D. G., Francavilla, T. L., and Gubser, D. U. (1977). Critical current properties of multifilament V_3Ga wire at temperatures above 4.2 kelvin. *IEEE Trans. Magn.* **MAG-13**, 815–817.
Inoue, K., Kuroda, T., and Tachikawa, K. (1979). Some improvements in Laves phase high-field superconductors. *IEEE Trans. Magn.* **MAG-15**, 635–638.
Iwasa, Y. and Montgomery, B. (1975). High-field superconducting magnets. In *Applied Superconductivity*, Vol. 2 (ed. V. L. Newhouse). Academic Press, New York, 387–487.
Iwasa, Y., Montgomery, D. B., and Polzer, H. W. (1970). Energy release by flux jumps in superconductors. *J. Appl. Phys.* **14**, 2476–2481.
Kaufmann, A. R. and Pickett, J. J. (1971). Multifilament Nb_3Sn superconducting wire. *J. Appl. Phys.* **42**, 58.
Koch, C. C. and Easton, D. S. (1977). A review of mechanical behavior and stress effects in hard superconductors. *Cryogenics* **17**, 391–413.
Koyama, K. (1973). Magnetic instability in hard superconductors. *J. Appl. Phys.* **44**, 5531–5538.
Kramer, E. J. (1973). Scaling laws for flux pinning in hard superconductors. *J. Appl. Phys.* **44**, 1360–1370.
Kunzler, J. E. (1961). Superconductivity in Nb_3Sn at high current in a magnetic field of 88 K gauss. *Phys. Rev. Lett.* **6**, 89–91.
Laverick, C. (1965). The performance of large superconducting coils. In *Advances in Cryogenic Engineering*. Vol. II. Plenum Press, New York, 659–667.
Little, W. A. (1964). Possibility of synthesizing an organic superconductor. *Phys. Rev.* **134**(6A), 1416–1424.
Livingston, J. D. and DeSorbo, W. (1969). The intermediate state in Type I superconductors. In *Superconductivity*, Vol. 2 (ed. R. D. Parks). Marcel Dekker, New York, 1235–1281.
London, H. (1963). Alternating current losses in superconductors of the second kind. *Phys. Lett.* **6**, 162–165.
Luhman, T. and Suenaga, M. (1976). Brookhaven National Laboratory, Upton, New York. Private communication.
Lynton, E. A. (1962). *Superconductivity*. Methuen, London.
Maddock, B. J., James, G. B., and Norris, W. T. (1969). Superconductive composites: heat transfer and steady state stabilization. *Cryogenics,* **9**, 261–273.
Maley, M. P., Newkirk, L. R., Thompson, J. D., and Valencia, F. A. (1981). Development of long continuous Nb_3Ge tape conductors with high current density at elevated temperature and field. *IEEE Trans. Magn.* **MAG-17**, 533–540.
Martinelli, A. P. and Wipf, S. L. (1972). Investigation of cryogenic stability and reliability of operation of Nb_3Sn coils in helium gas environment. In *Proceedings of the 1972 Applied Superconductivity Conference*. IEEE Publ. No. 72CH0682-5-TABSC, IEEE, New York, 331–340.
Matthias, B. T. (1957). Superconductivity in the periodic system. In *Progress in Low Temperature Physics*, Vol. 2 (ed. C. J. Gorter). North Holland, Amsterdam, 138–150.
Matthias, B. T., Geballe, T. H., and Compton, V. B. (1963). Superconductivity. *Rev. Mod. Phys.* **35**, 1–22.
Matthias, B. T., Geballe, T. H., Willens, R. H., Corenzwit, E., and Hull, G. W., Jr. (1965). Superconductivity of Nb_3Ge. *Phys. Rev A.* **139**, 1501–1503.
Matthias, B. T., Geballe, T. H., Corenzwit, E., Andres, K., Hull, G. W., Jr., Ho, J. C., Phillips, N. E., and Wohlleben, D. W. (1966). Superconductivity of beta-uranium. *Science* **151**, 985–986.
Meissner, W. and Ochsenfeld, R. (1933). Ein neuer Effekt bei Eintritt der Supraleitfahigkeit (A new effect at the superconducting transition). *Naturwissenschaften* **21**, 787–788.
Miller, J. R., Dresner, L., and Lue, J. W. (1978). Investigation of current transfer in built-up superconductors. In *Proceedings of the 7th Symposium on Engineering Problems of Fusion Research*, IEEE Nuclear Plasma Science Society, 1282–1287.
Morgan, G. H. (1970). Theoretical behavior of twisted multicore superconducting wire in a time-varying uniform magnetic field. *J. Appl. Phys.* **41**, 3673–3679.
Onnes, H. K. (1911). The resistance of pure mercury at helium temperature. *Leiden Commun.* No. 120b.
Parmenter, R. H. (1967). Isopin formulation of the theory of a granular superconductor. *Phys. Rev.* **154**, 353–368.
Powell, R. L. and Clark, A. F. (1977). Definitions of terms for practical superconductors: 1. Fundamental states and flux phenomena. *Cryogenics* **17**, 697–701.
Powell, R. L. and Clark, A. F. (1978). Definitions of terms for practical superconductors: 2. Critical parameters. *Cryogenics* **18**, 137–141.
Reed, R. P., Mikesell, R. P., and Clark, A. F. (1977). The low temperature tensile behavior of copper-stabilized niobium–titanium superconductor wire. In *Advances in Cryogenic Engineering*, Vol. 22. Plenum Press, New York, 463–471.
Rose-Innes, A. C. and Rhoderick, E. H. (1978). *Introduction to Superconductivity*. Pergamon, Oxford.
Rupp, G. (1977). Improvement of the critical current of multifilamentary Nb_3Sn conductors under tensile stress. *IEEE Trans. Magn.* **MAG-13**, 1565–1567.
Saint-James, D., Sarna, G., and Thomas, E. J. (1969). *Type II Superconductivity*. Pergamon, Oxford.
Sekula, S. T. (1971). Low frequency electromagnetic losses at 4.2 K of neutron-irradiated Nb. *J. Appl. Phys.* **42**, 16–21.
Serin, B. (1969). Type II superconductors: experiments. In *Superconductivity* (ed. R. D. Parks). Marcel Dekker, New York, 947–948.
Silsbee, F. B. (1916). Electric conduction in metals at low temperatures. *J. Wash. Acad. Sci.* **6**, 597–602.
Smith, P. F., Wilson, M. N., Walters, C. R., and Lewin, J. D. (1969). Rutherford Laboratory Preprint RP 1A 43, Rutherford Laboratory, Chilton, Didcot, England.
Soulen, R. J., Jr. and Colwell, J. H. (1971). The equivalence of the superconducting transition temperature of pure indium as determined by electrical resistance, magnetic susceptibility, and heat-capacity measurements. *J. Low Temp. Phys.* **5**(3), 325–333.
Stekly, Z. J. J., Thome, R., and Strauss, B. (1968). Principles of stability in cooled superconducting

magnets. In *Proceedings of the 1968 Summer Study on Superconducting Devices and Accelerators*. BNL 50155(C-55), Brookhaven National Laboratory, Upton, New York, 1743–1764.

Stekly, Z. J. J. and Zar, J. L. (1965). Stable superconducting coils. *IEEE Trans. Nucl. Sci.* **23**, 367–372.

Suenaga, M. and Sampson, W. B. (1971). Superconducting properties of multifilamentary V_3Ga wires. *Appl. Phys. Lett.* **18**, 584–586.

Suenaga, M. and Sampson, W. B. (1972). Superconducting properties of multifilamentary V_3Si wires. In *Proceedings of the Applied Superconductivity Conference*. IEEE, New York, 481–485.

Superconductor Materials Science: Metallurgy, Fabrication, and Application (1981) (eds. S. Foner and B. B. Schwartz). Plenum Press, New York.

Superconductivity (1969) (ed. R. D. Parks). Marcel Dekker, New York.

Tachikawa, K., Yoshida, Y., and Rinderer, L. (1972). Studies in the formation of V_3Ga and V_3Si superconducting compounds by a new diffusion process. *J. Mater. Sci.* **7**, 1154–1160.

Tinkham, M. (1975). *Introduction to Superconductivity*, Chapter 2. McGraw-Hill, New York.

Werthamer, N. R., Helfand, E., and Hohenberg, P. C. (1966). Temperature and purity dependence of the superconducting critical field, H_{c2}. III. Electron spin and spin-orbit effects. *Phys. Rev.* **147**, 295–302.

Whitcomb, M. J. and Dew Hughes, D. (1973). Superconductivity of heat-treated Nb–65at.%Ti alloy. *J. Mater. Sci.* **8**, 1383–1400.

Wilson, M. N. (1976). Stabilization of superconductors for use in magnets. *IEEE Trans. Magn.* **MAG-13**, 440–446.

Wilson, M. N. and Iwasa, Y. (1978). Stability of superconductors against localized disturbances of limited magnitude. *Cryogenics* **18**, 17–25.

Wilson, M. N., Walters, C. R., Lewin, J. D., and Smith, P. F. (1970). Experimental and theoretical studies of filamentary superconducting composites. *J. Appl. Phys.* **3**, 1515–1585.

Wipf, S. L. (1978). *On Stability and Degradation of Superconducting Current-Carrying Devices*. LASL report LA-7275, Los Alamos Scientific Laboratory, Los Alamos, New Mexico.

Wipf, S. L. (1979). Stability of the superconductive operating mode in high current-density devices. *IEEE Trans. Magn.* **MAG-15**, 379–382.

ACKNOWLEDGMENTS

The author is pleased to thank the reviewers of the manuscript for their many excellent suggestions; the detailed reading of the final text by W. A. Fietz, J. V. Minervini, A. F. Clark, and V. D. Arp is especially appreciated. The partial support of the Naval Ship Research and Development Center for the preparation of the manuscript, and the speed and skill of V. Grove in its typing are also gratefully acknowledged. More generally, the author thanks numerous colleagues and collaborators for their contributions and lively discussions over the years that have made the study of superconductivity the fascinating subject that it is.

14 Temperature, Strain, and Magnetic Field Measurements

L. L. SPARKS

14.1 INTRODUCTION

The measurement of certain parameters at low temperatures is critical to both the development and operation of sophisticated, low-temperature machinery and processes. Three parameters have been chosen for discussion in this chapter: temperature, T, strain, ϵ, and magnetic field strength, H. These are the most critical variables in the experimental determination of the properties of materials at low temperatures. In addition to these three, pressure, fluid flow, fluid density, and liquid level measurement are important in monitoring processes and in system control. Information on the measurement of these quantities in cryogenic apparatuses may be found in the following review papers: (a) pressure — Hayakawa (1972) and Arvidson and Brennan (1975), (b) fluid flow — Mann (1974), and (c) fluid density and level measurement — Roder (1974).

The following discussion stresses the measurement of temperature because it is the primary variable in nearly all low-temperature material properties. Successful measurement of temperature involves not only utilizing the optimum sensor for a given situation, but also the correct techniques associated with low-temperature instrumentation. Section 14.2, Techniques for Low Temperatures, contains information on methods and auxiliary (but necessary) materials. Areas of frequent concern, such as thermal contact, heat leak, thermal anchoring, thermal conductivity of greases, insulators, lead wires, ground loops, and feedthroughs are reviewed in this section. The Temperature Measurement Section (§ 14.3) includes an overview and historical development of temperature scales because the practical use of all thermometers is associated with some approximation of the thermodynamic temperature scale.

Most temperature measurements in science and industry are made with resistance thermometers, thermoelectric thermometers, or filled systems. The principles involved and the performance of thermometers in each of these categories are presented along with the effect of magnetic fields. A short section (§ 14.3.5) is devoted to other types of temperature measuring devices; further details on these less commonly used devices are found in the references.

The large-scale applications of superconductivity as well as the increased use of materials at cryogenic temperatures has created a demand for data on the mechanical properties of materials at low temperatures. Strain is a primary variable in determining these properties. The instrumentation used to monitor the operation of superconducting solenoids and other cryogenic structures includes strain measuring devices.

Use of commercially available resistance-type strain gauges at low temperatures requires special precautions to account for the effect of temperature. The characteristics of this type of sensor are stressed in § 14.4. The discussion on techniques (§ 14.2) applies to the use of strain gauges as well as to thermometers.

The final topic is the measurement of magnetic field strength, H. The magnitude of H affects not only the efficiency of motors, generators, plasma containment systems, and magnetohydrodynamic systems, but also the electronic transport properties of materials. The thermal and electrical conductivities of metallic components are functions of H, and systems must be designed to take this dependence into account. Since the transport properties are utilized in most measurements of T and ϵ, H must be known if the output of T and ϵ sensors is to be correctly interpreted. Measurement devices utilizing a variety of principles are used to make magnetic field determinations (§ 14.5). These principles include nuclear magnetic resonance, induced voltage in a rotating coil or a fixed coil with integrator, magnetoresistance, and the Hall effect.

The purpose of this chapter is to present information necessary to understand the most widely used devices for measuring T, ϵ, and H. The materials used in the sensors and the characteristics of measuring devices are presented whenever it is felt that knowledge in these areas is needed to determine the parameters in question.

14.2 TECHNIQUES FOR LOW TEMPERATURES

There are several difficulties that must be overcome when making cryogenic measurements. The particular situation may require (a) good thermal, electrical, or mechanical contact between the sensor and the specimen, (b) bringing instrument wires through areas with large temperature gradients, and (c) interfacing low-voltage signals with ambient temperature junctions and switches so that spurious voltages are minimized.

14.2.1 Transducer-specimen contact

Perhaps the most difficult instrumentation problem in cryogenic measurement systems is that of establishing good thermal or mechanical contact between the sensor and the specimen. Ideally, thermal contact is such that the sensor is at the temperature that the specimen would be if the sensor did not exist. In reality, energy is transferred both from the specimen to the sensor and vice versa. The practical solution is to make the thermal contact as good as possible and then evaluate the effect of energy flow into the specimen–sensor system.

The two primary sources of thermal energy into the specimen–sensor system are heat transferred along connecting wires and power generated by the active element of the sensor (radio frequency pickup and radiant energy must also be considered in specific circumstances). Consider first the heat transfer via the connecting wires. Thermal anchoring (tempering) of the wires to a heat sink whose temperature is near that of the specimen allows the energy to dissipate before it can affect the measured parameter. The additional requirement of electrical isolation through the thermal contact is frequently present. Papers by Anderson (1969), Hust (1970), and Kopp and Slack (1971) discuss thermal anchoring to solids, and Allen, Bradford, and Crabtree (1970) discuss thermal anchoring in liquids. Hust develops a rather general approach for determining the required tempering length for wires or rods in a vacuum. The required length is a function of wire material and size, insulation, temperatures encountered, and the allowable temperature difference between the wire and the heat sink. Table 14.1, condensed from Hust's paper, presents the tempering lengths required to reduce the wire–heat sink temperature difference to 1 mK. The configuration used to develop Table 14.1 is shown in Fig. 14.1.

Heat transfer calculations require knowledge of the thermal conductivity of wires, electrical insulators, greases, and adhesives. Values for common wires are shown in Fig. 14.2. These values were obtained as follows: Au–0.03 at.% Fe and Au–0.07 at.% Fe were estimated from data by Berman, Brock, and Huntley (1964), composition change, and Nordheim's rule; TP (copper) and TN (constantan), from Powell, Rogers, and Roder (1960); JP (iron), from Hust, Powell, and Weitzel (1970); Au–2.1 at.% Co from Powell, Bunch, and Gibson (1960) with these data extrapolated to 300 K; and KP and KN, from data on other nickel alloys (White and Woods, 1958).

Thermal conductivities of greases frequently

Table 14.1 Lengths of copper and constantan wire that must be thermally anchored to a heat sink at temperature T_s in order to bring the temperature of the wire to within 1 mK of T_s.

Material	T_s (K)	T_1 (K)	Tempering Length, ℓ, for Various Wire Gauges[a] (cm) 40 AWG[b]	35 AWG	30 AWG	24 AWG
Copper	4	20	9	18	35	82
	20	78	15	25	45	95
	78	273	6	11	22	44
Constantan	4	20	0	1	1	3
	20	78	1	2	4	7
	78	273	1	2	4	8

[a]The conditions pertain to wires in an insulating vacuum environment (Hust, 1970). Thermal conductivity, λ, of the adhesive is 0.01, 0.02, and 0.05 $W \cdot m^{-1} \cdot K^{-1}$ at 4, 20, and 78 K, respectively.
[b]American wire gauge.

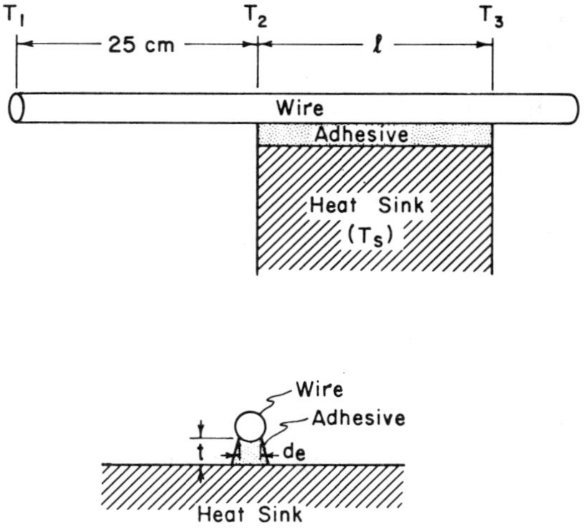

Figure 14.1 Thermal anchoring of wire to a heat sink through an adhesive.

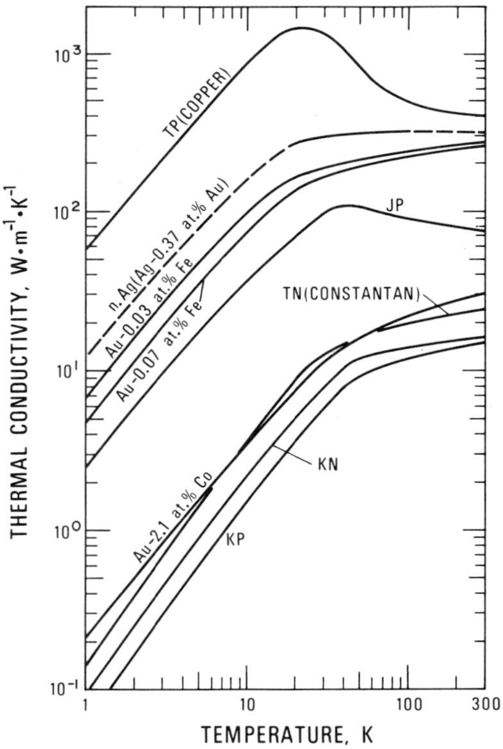

Figure 14.2 Thermal conductivity vs. temperature for frequently used thermocouple materials. The alloys corresponding to the standard thermocouple types (i.e., KP, KN, TP, TN, and JP) are discussed in 14.3.3.

used as contact agents in low-temperature thermometry are given for various temperature ranges by Denner (1969), Kreitman (1969), Kreitman and Callahan (1970), and Kreitman, Ashworth, and Rechowicz (1972). Denner also includes the thermal conductivities of several adhesives, electrical insulators, and solders; Ashworth, Johnson, Hsiung, and Kreitman (1973) give the thermal conductivity of nylon. McTaggert and Slack (1969) have determined the thermal conductivity of G.E. 7031 varnish. The thermal conductivities of Apiezon N and G.E. 7031 are tabulated in Table 14.2.

The second source of energy into a specimen–sensor system is Joule heating in the wire leads and the sensor. For resistive elements, the power generated is I^2R where I is electrical current and R is electrical resistance. A temperature gradient is established between the active element of the sensor and the specimen because of protective encapsulation. The magnitude of the gradient depends upon the particular construction of the device. Very little energy is required to increase the temperature of small specimens because heat capacity decreases with temperature; for example, the ratio of heat capacity at 300 K to that at 20 K is 18 for platinum, 50 for copper, and 76 for nickel. The Peltier heating or cooling associated with the flow of current between dissimilar metals is usually negligibly small.

14.2.2 Signal maintenance

One must be careful not to alter the signal from the sensor. This is particularly important in cryogenic systems where power dissipation and sensitivity considerations lead to relatively low-level signals. Spurious voltages can be introduced by wire-to-wire junctions, electrical feedthroughs, ground current circuits, and switching.

The general principle to remember when con-

Table 14.2 Thermal conductivity of a commercial varnish and a contact grease at several cryogenic temperatures.

Temperature (K)	Thermal Conductivity ($W \cdot m^{-1} \cdot K^{-1}$) G.E. 7031 Varnish	Apiezon N
10	0.08	
30	0.13	
100	0.24	0.15
150		0.20
200	0.35	0.21
250		0.24
300	0.44	0.24

necting electrical wires is to make as few junctions as possible and, where they must be made, to make them as isothermal as possible. This is based on one of the principles of thermoelectric thermometry — the law of intermediate materials. This law states that there will be no thermally generated voltage when dissimilar metals are joined unless the joint exists in a temperature gradient. The magnitude of the spurious voltage is dependent on the temperature gradient across the junction and the absolute thermopower of the materials involved. Room-temperature junctions seldom involve large temperature gradients; therefore, the spurious voltages are typically on the order of a few microvolts; junctions made in cryogenic systems, however, may involve large temperature gradients and produce much larger spurious voltages. Solder junctions fall into this category, since the wire–solder–wire combinations form a dissimilar metals joint even when the two wires are the same. When copper-to-copper junctions are made, a special solder ($\sim$70% Cd + $\sim$30% Sn), which nearly matches copper thermoelectrically at room temperature, is recommended. This type of solder has been used extensively at low temperatures with no observed deterioration; some caution must be taken, however, since Sinclair, Terbeek, and Malone (1970) did detect some breakdown when this type of solder was cycled to cryogenic temperatures. Junctions can also be made by spot welding, either with an electrical discharge welder or with a small torch. In some apparatus, junctions are made by simply twisting the leads together. Tools that may be used to form twisted-wire junctions are commercially available. The machine-formed, twisted junctions were intended for use in ambient temperature operations, but appear to function well in cryogenic environments as well (Rubin, 1979). All of these methods of forming junctions can be used successfully if done with care.

To transmit the transducer signal to the measuring instrument it is frequently necessary to bring wires through system walls where conditions of temperature and presssure change drastically in very short distances. There are two basic types of electrical feedthroughs used in cryogenic systems: continuous wire and junction. Figure 14.3a shows the essential design of two commercially available units. When large temperature differences exist across the feedthrough, the continuous wire variety is preferable since the junction type introduces the dissimilar metal situation discussed above. The specifications of materials, sizes, wire capacity, and pressure rating vary considerably in commercial feedthroughs so that most design criteria can be satisfied. Figure 14.3b shows one of the many "potting" methods that may be used for homemade feedthroughs. The cup-and-reservoir approach is particularly simple and effective as an ambient temperature seal. If black sealing wax is used in the reservoir, the seal is easy to remake when the wires are replaced. Use of the continuous wire type eliminates the necessity of a junction, or at least allows the junction to be made at ambient temperature where large gradients do not exist. Similarly, any disconnect plugs should be isothermal, and the number kept to a minimum.

Switches must be of the appropriate quality to prevent degradation of the signal-to-noise ratio. For manual operation, low-resistance rotary switches are completely satisfactory. Switches of this type are commercially available with contact resistances of less than 1 mΩ and thermally generated voltages of less than 0.1 μV/K. For automatic switching operations, enclosed gold-plated

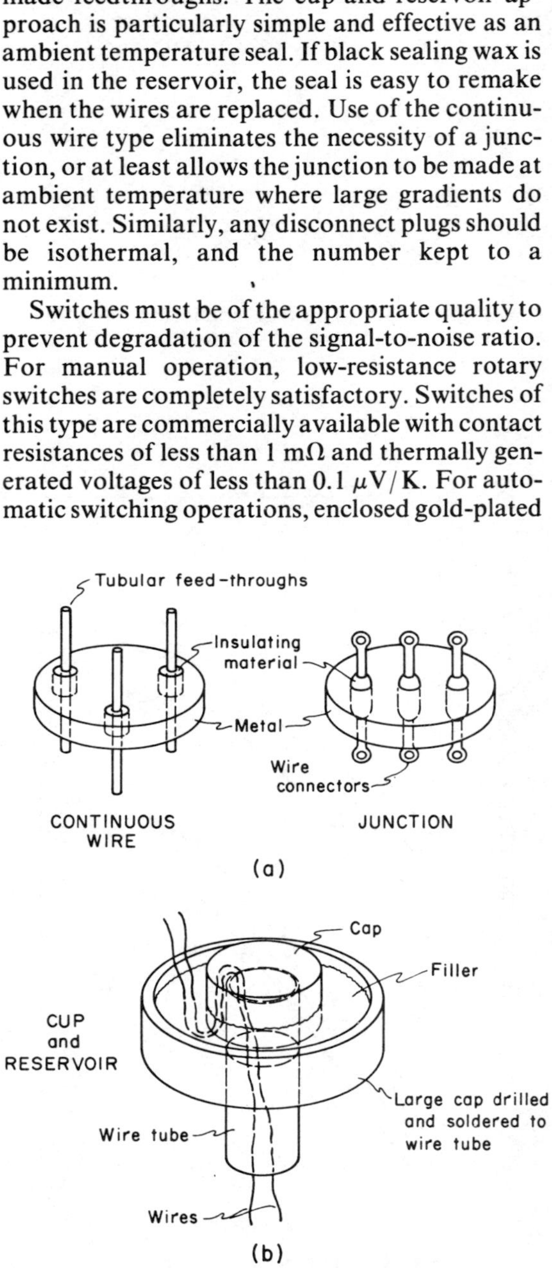

Figure 14.3 Schematic of electrical feedthroughs frequently used in low-temperature cryostats.

switches or reed-type relays (Campbell, 1979) may be used. Thermally generated offsets of less than 1 μV are possible with commercially available reed relays. Although switches generally operate at ambient temperatures, they should be thermally isolated since thermal voltages may be generated at the junctions of the wires and switch components.

When dealing with low-level voltages, one must be careful to minimize the effect of unwanted electrical signals superimposed on the sensor output. These undesirable electrical voltages are commonly referred to as *noise.* Although proper circuit design and construction can generally reduce the noise content of the signal to acceptable levels, noise pickup from external sources can seldom be completely eliminated. Noise from external sources may be introduced into a circuit via (a) electrical conduction along a conductor, which picks up noise by passing through a noisy environment; (b) common impedance coupling caused by current from two separate circuits flowing through a common impedance (this situation frequently occurs when a common ground exists or when a single power supply is used jointly by two or more circuits); and (c) electromagnetic radiation that is caused by movement of charge along a conductor, by radar, and by radio and television broadcasts. Conductor motion in the presence of a magnetic field also creates a noise voltage. Stray magnetic fields exist in most instrumentation environments. Sensible use of adhesives and cable tie-downs generally reduces this source of noise adequately. Other sources of noise include galvanic action between dissimilar metals in contact, electrolytic action, thermal noise, shot noise, and contact noise. The latter three sources are intrinsic; they can be computed but cannot, in general, be eliminated.

The omnipresent electromagnetic radiation is the most significant source of noise, and most noise suppression efforts for low-voltage circuits try to reduce its effect. The effect of an electromagnetic field on an electrical circuit varies with the characteristics of the field. The characteristics depend primarily on the source itself if the receiver is nearby (separation of $d \lesssim \lambda/2\pi$) and on the medium through which the radiation passes when $d \gtrsim \lambda/2\pi$. The case where $d \lesssim \lambda/2\pi$ (near field) is most germane to this discussion since $d \approx 46$ m when the frequency is less than 1 MHz and $d \approx 1600$ m when the frequency is 30 kHz. The magnetic and electrical components of near-field radiation interact with circuit components in different ways, and different techniques must be employed to reduce their effect. Grounding and shielding are the primary methods. Usually conductor wires are shielded by encasing them in a thin braid or wrap of conducting material. The thickness of the shield is limited by the practical need to keep the composite bundle flexible. This type of shielding does not significantly reduce the noise level by signal attenuation, as is the case when a source or receiver is encased in a relatively thick-walled metallic container. Both types of shielding have applications in reducing noise in instrument systems.

The shielding effectiveness, SE, of a conducting plate (Ott, 1976) imposed between a noise source and receiver is given in units of dB by

$$\mathrm{SE} = \mathrm{AL} + \mathrm{RL} + \mathrm{TSCF} \qquad (14.1)$$

where AL is the absorption loss, RL is the reflective loss, and TSCF is a thin-shield correction factor. The absorption term is given by

$$\mathrm{AL} = 8.69\,(t/\delta) \qquad (14.2)$$

for both magnetic and electric fields. The parameters in Eq. 14.2 are shield thickness, t, and skin depth, δ, the depth that a wave must travel into a material in order to be reduced to $1/e$ of its original value. The skin depth in cm is

$$\delta = 6.6/[f\,\mu_r \sigma_r]^{1/2} \qquad (14.3)$$

where f is the wave frequency and μ_r and σ_r are the relative permeability and relative conductivity of the shield material, respectively. Table 14.3 lists the frequency-dependent skin depths for three materials often used as shields. Note that for 60-Hz noise the skin depths for copper and aluminum are approximately 10 times larger

Table 14.3 Skin depth as a function of frequency for three shield materials.[a]

	Skin Depth (cm)		
Frequency	Copper	Aluminum	Steel
60 Hz	0.85	1.09	0.09
100 Hz	0.66	0.85	0.07
1 kHz	0.21	0.27	0.02
10 kHz	0.07	0.08	0.008
100 kHz	0.02	0.03	0.002
1 MHz	0.008	0.008	0.0008
10 MHz	0.002	0.002	0.0002

[a] The relative permeabilities are 1, 1, and 1000 and the relative conductivities are 1.00, 0.61, and 0.1 for copper, aluminum, and steel, respectively (Ott, 1976).

than that for steel; the attenuation term given by Eq. 14.2 is inversely proportional to the skin depth. Absorption is the most effective shielding mechanism when dealing with magnetic fields.

The second term in Eq. 14.1 is due to reflection of waves: the principal reflection of electric fields occurs at the first interface and is given in dB units by

$$\mathrm{RL} = 168 - 10 \log (\mu_r f / \sigma_r) \tag{14.4}$$

A material with high conductivity and low permeability results in a larger RL and concomitant increased shield efficiency than a material of low conductivity and high permeability. This reflective term represents the most effective mechanism for shielding against electric fields. The reflective term relating to magnetic field radiation requires that the source-to-receiver distance be known. This information is generally not available. The reflective term for magnetic field radiations is small compared with the absorption component and is usually approximated by zero. The final mechanism, represented by term TSCF in Eq. 14.1, describes the effect of multiple internal reflections of the magnetic field. It is given in dB units by

$$\mathrm{TSCF} = 20 \log (1 - e^{-2t/\delta}) \tag{14.5}$$

Table 14.4 summarizes the shielding effectiveness for magnetic and nonmagnetic materials in an electromagnetic environment.

Two types of noise coupling must be considered when dealing with the shielding of conducting wires: electric field (capacitive) and magnetic field (inductive). These modes of coupling require different solutions and are treated separately in the following discussion. The noise voltage, V_N, superimposed on the voltage of one conductor because of a voltage, V_1, on a second conductor is

$$V_N = j\omega R C_{1-2} V_1 \tag{14.6}$$

where $\omega = 2\pi$ frequency, R is the resistance to ground of the receiver wire, and C_{1-2} is the capacitance between the two wires. Since the source voltage, V_1, and frequency, ω, are not generally adjustable, the resistance to ground and capacitance between wires must be reduced to decrease V_N. The capacitance is reduced by physically separating the two wires. When the wire separation to wire diameter ratio is 40, about 80% of the decoupling is accomplished. If the receiving conductor is completely shielded from the source conductor by imposing a grounded shield around the receiver, $C_{1-2} = 0$, and $V_N = 0$. Practically, C_{1-2} is seldom reduced to zero owing to exposed ends of the receiver wire or holes in the shield. The noise voltage imposed on a conductor due to inductive or magnetic coupling is given by

$$V_N = j\omega BA \cos\theta \tag{14.7}$$

where B is the magnetic flux density, A is the area of the closed receiver loop, and θ is the angle between the source and receiving conductors. Reduction of this noise is more difficult than for electric fields. It can be shown that the shielding used so effectively in guarding against electric fields has no effect whatsoever on coupling due to magnetic field. The most effective way of dealing with inductive pickup is to reduce the enclosed area, A, of the receiver loop. The B term in Eq. 14.7, flux density, can also be reduced by

Table 14.4 Shielding effectiveness of magnetic and nonmagnetic materials as a function of frequency.

		Attenuation (dB)			
			Reflection		
Material	Frequency (kHz)	Absorption,[a] All Fields	Magnetic	Electric	Plane Wave[b]
Magnetic	<1	0–30	0–10	>90	>90
($\mu_r = 1000$,	1–10	30–90	0–30	>90	>90
$\sigma_r = 0.1$)	10–100	>90	10–30	>90	60–90
	>100	>90	10–60	60–90	30–90
Nonmagnetic	<1	0–10	10–30	>90	>90
($\mu_r = 1$,	1–10	0–10	30–60	>90	>90
$\sigma_r = 1$)	10–100	10–30	30–60	>90	>90
	>100	30–90	60–90	>90	>90

[a]The absorption is based on a 0.08-cm thickness (Ott, 1976).
[b]Plane wave refers to electromagnetic radiation with an impedance of 377 Ω in a vacuum (Ott, 1976).

physically separating the source and receiver and by twisting the leads of the source if it is made up of a two-wire system (as opposed to a single conductor with a ground return).

This effectiveness of cable shielding and safe operation of electrical systems depends strongly on how the various components are grounded. The following discussion pertains to low-frequency radiation environments, $f \lesssim 10$ MHz; higher frequencies require multiple grounds and are discussed by Ott (1976). Low-frequency systems require a single ground point for both power and signal components. Two methods are used to connect the various system components to this single ground: (a) series connection, which consists of a single wire to ground with the various circuit grounds being attached to this wire and (b) parallel connection, in which individual circuit ground returns are connected directly to the system ground. Conceptually, the parallel connection is better than the series connection; the series connection establishes small voltage differences between individual circuit ground-return points. Practical circuit layout usually requires a combination of the parallel and series connections. It is very important to separate the ground returns of the low-voltage circuits from those of the power circuits.

A final topic must be included here to complete this cursory review of signal maintenance: connection of the wire shield to the amplifier input. High-gain amplifiers are normally protected from electric fields by encasement in metal boxes. To eliminate capacitive coupling between the input and output, the amplifier shield must be connected to the amplifier common terminal. Two situations are frequently encountered: (a) ungrounded sensor and a grounded amplifier and (b) grounded sensor and an ungrounded amplifier. Note that only one circuit ground is allowed in either case to prevent the flow of ground currents. When the sensor is not grounded, the input wire shield should be connected to the amplifier common terminal. The input wire shield should be connected to the grounded terminal of the sensor (and not to the amplifier common) in the case of a grounded sensor.

Ott (1976) discusses other techniques of noise reduction that may be used in conjunction with the methods discussed above, for example, impedance balancing, filtering, and use of isolation transformers. Morrison (1970) discusses noise reduction techniques in general and the use of differential amplifiers to eliminate ground loops (Morrison, 1979).

14.3 TEMPERATURE MEASUREMENT

The concept of temperature is based on the cyclic operation of a reversible heat engine. Temperature, then, is established by the laws of thermodynamics and is independent of the properties of the working medium. It follows that temperature is proportional to the heat absorbed or evolved along the isotherms of a reversible Carnot cycle (Wensel, 1941). Unfortunately, this basic approach is of little use in actually establishing an absolute temperature scale. Other fundamental thermometers (theoretically independent of the working medium), such as optical pyrometers, velocity of sound thermometers, and constant-volume gas thermometers, have been utilized to approximate the thermodynamic scale. The approximation involved in the latter two systems is that of real gas properties becoming ideal gas properties as the density approaches zero. Temperature, as determined on the ideal gas scale, can be shown to be proportional to temperature on the absolute thermodynamic scale by using thermodynamics and kinetic theory (Pippard, 1957). Historically, constant-volume gas thermometry has provided a practical link between empirical thermometers (thermometers whose characteristics depend on real materials) and the thermodynamic temperature scale. In a practical sense, however, the use of the fundamental thermometers is limited to special situations. Only the national laboratories and a few universities have the capability of using these instruments at a state of the art level. Their primary use has been to transfer the best estimate of the thermodynamic scale to a few precision empirical thermometers. These transfer standards are, in turn, used to calibrate the many working thermometers by the comparison method.

The general types of temperature sensors to be considered here are resistance (§ 14.3.2), thermoelectric (§ 14.3.3) and filled systems (§ 14.3.4). Other methods of cryogenic temperature measurement are mentioned in § 14.3.5. The effect of magnetic fields on all types of transducers is discussed in § 14.3.6. The approximate ranges of use for the thermometers to be discussed are shown in Fig. 14.4.

14.3.1 Temperature scales

As instrumentation and techniques improve, temperature scales become better approxima-

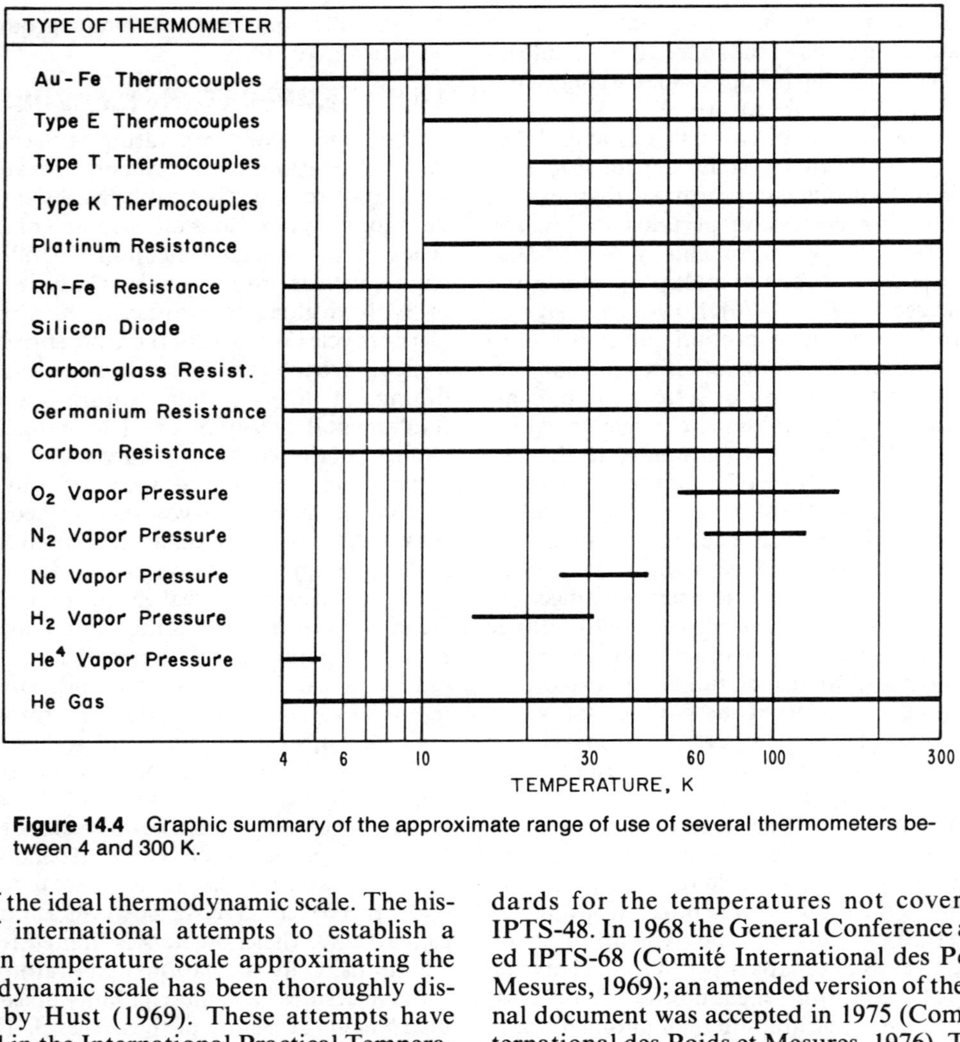

Figure 14.4 Graphic summary of the approximate range of use of several thermometers between 4 and 300 K.

tions of the ideal thermodynamic scale. The history of international attempts to establish a common temperature scale approximating the thermodynamic scale has been thoroughly discussed by Hust (1969). These attempts have resulted in the International Practical Temperature Scale (IPTS-68) and the 1976 Provisional Scale (EPT-76), and indicate that the General Conference Weights and Measures may adopt a new scale in 1987.

In 1927, the General Conference adopted the first internationally accepted temperature scale for temperatures above 83 K (Burgess, 1928). Minor revisions to the scale, including making the normal boiling point (n.b.p.) of liquid oxygen the lower limit, were made in 1948. There was a text revision in 1960 (Stimson, 1961); the revision added the word *practical* to the official name for the first time. The scale was thus represented by IPTS-48 rather than ITS-48. No internationally accepted scale existed for temperatures below the n.b.p. of oxygen. Several laboratories developed and maintained their own standards for the temperatures not covered by IPTS-48. In 1968 the General Conference adopted IPTS-68 (Comité International des Poids et Mesures, 1969); an amended version of the original document was accepted in 1975 (Comité International des Poids et Mesures, 1976). The extension of the IPTS scale to 13.81 K was of major importance. The fixed points for IPTS-68 are compared with those of previous scales in Table 14.5.

Before IPTS-68 became available, several laboratories established their own scales for temperatures below 90 K. One of the most widely used in the United States was established by Hoge and Brickwedde (1939) of the National Bureau of Standards. They compared platinum resistance thermometers to a helium gas thermometer in the range 10 to 90 K. They used 90.19 K as the best value for the n.b.p. of oxygen; this scale was designated NBS-39. In 1955 the scale was changed by assuming a value of 90.18 K for the n.b.p. of oxygen. The resulting scale was used by the National Bureau of Standards as the stan-

Table 14.5 Primary fixed points of IPTS-68 and comparative values from previous international temperature scales.

Fixed Point	ITS-27 (°C)	ITS-48 (°C)	IPTS-48 (°C)	IPTS-68 (°C)	IPTS-68 (K)	IPTS-68 Uncertainty (K)
Equilibrium between the solid, liquid, and vapor phases of equilibrium hydrogen (triple point of equilibrium hydrogen)				−259.34	13.81	0.01
Equilibrium between the liquid and vapor phases of equilibrium hydrogen at a pressure of 33 330.6 N/m² (25/76 standard atmosphere)				−256.108	17.042	0.01
Equilibrium between the liquid and vapor phases of equilibrium hydrogen (boiling point of equilibrium hydrogen)				−252.87	20.28	0.01
Equilibrium between the liquid and vapor phases of neon (boiling point of neon)				−246.048	27.102	0.01
Equilibrium between the solid, liquid, and vapor phases of oxygen (triple point of oxygen)				−218.789	54.361	0.01
Equilibrium between the liquid and vapor phases of oxygen (boiling point of oxygen)	−182.97	−182.970	−182.97	−182.962	90.188	0.01
Equilibrium between the solid and liquid phases of water (freezing point of water)	0.000	0				
Equilibrium between the solid, liquid, and vapor phases of water (triple point of water)			0.01	0.01	273.16	Exact
Equilibrium between the liquid and vapor phases of water (boiling point of water)	100.000	100	100	100	373.15	0.005
Equilibrium between the solid and liquid phases of zinc (freezing point of zinc)				419.58	692.73	0.03
Equilibrium between the liquid and vapor phases of sulfur (boiling point of sulfur)	444.60	444.600	444.6			
Equilibrium between the solid and liquid phases of silver (freezing point of silver)	960.5	960.8	960.8	961.93	1235.08	0.2
Equilibrium between the solid and liquid phases of gold (freezing point of gold)	1063	1063.0	1063	1064.43	1337.58	0.2

dard in this range until 1968. Barber (1962) established a similar scale at the National Physical Laboratory in England. As was the case with the NBS-39 scale, the fundamental thermometer was a constant-volume gas thermometer and the transfer standards were platinum resistance thermometers. The range of this scale, NPL-61, was 10 to 90 K, and the n.b.p. of oxygen was taken to be 90.18 K. Moessen, Aston, and Ascah (1962) of Pennsylvania State University established the 10- to 90-K scale known as PSU-54. Constant-volume helium gas thermometry was again used to calibrate platinum resistance thermometer transfer standards. The Soviet (USSR) national scale for the temperature range 10 to 90 K, PRMI-54, was established using 90.19 K as the n.b.p. of oxygen (Borovick-Romanov, Strelkov, Orlova, and Astrov, 1962). Table 14.6 compares these 10- to 90-K temperature scales with IPTS-68. The values in this table may be used to base "pre-IPTS-68 data" on IPTS-68 for temperatures below 90 K. Above 90 K, temperature based on IPTS-48 may be converted to IPTS-68 by using the differences given in Table 14.7. Both Tables 14.6 and 14.7 give the deviations in the form $T_{68} - T_X = \Delta T$, where T_X represents any one of the 10- to 90-K scales in Table 14.6 and IPTS-48 in Table 14.7. Douglas (1969) uses three analytical expressions to represent $T_{68} - T_{48}$ and $d(T_{68} - T_{48})/dT_{68}$ between 90 and 10 000 K.

A temperature scale between 2 and 20 K, based on velocity of sound measurements, was established by Cataland and Plumb (1965a, 1965b) and Plumb and Cataland (1966). This scale is officially known as NBS P 2-20 (1965). The international provisional scale, EPT-76, (Durieux,

Table 14.6 Temperature differences, $\Delta T = T_{68} - T_x$, are given where T_{68} represents temperatures on the IPTS-68 temperature scale, and T_x represents NBS-55, NPL-61, PRMI-54, or PSU-54 temperature scales. Temperature range for the scales represented by T_x is 14 to 90 K (Bedford, Durieux, Muijlwijk, and Barber, 1969).

	ΔT (mK)					ΔT (mK)			
IPTS-68 (K)	IPTS-68 minus NBS-55	IPTS-68 minus NPL-61	IPTS-68 minus PRMI-54	IPTS-68 minus PSU-54	IPTS-68 (K)	IPTS-68 minus NBS-55	IPTS-68 minus NPL-61	IPTS-68 minus PRMI-54	IPTS-68 minus PSU-54
14	−2.3	−1.4	−47.4	9.9	53	10.4	−4.7	−3.5	20.7
15	2.0	2.8	−36.9	16.4	54	8.9	−4.1	−3.4	20.6
16	5.1	1.3	−28.0	19.3	55	7.1	−3.2	−3.4	20.3
17	7.1	0.1	−21.0	21.4	56	5.2	−2.0	−3.6	19.9
18	8.4	4.4	−16.5	22.4	57	3.4	−0.5	−4.3	19.0
19	8.8	9.9	−14.2	22.9	58	1.7	1.1	−5.3	18.3
20	9.0	9.7	−12.9	23.3	59	0.3	2.8	−6.5	18.1
21	8.9	7.3	−11.9	23.5	60	−0.8	4.3	−7.9	18.0
22	8.8	5.3	−10.8	23.7	61	−1.4	5.9	−9.3	18.0
23	8.6	4.3	−9.7	24.0	62	−1.5	7.2	−10.6	18.0
24	8.3	3.7	−9.1	24.6	63	−1.2	8.1	−12.0	18.0
25	7.8	4.1	−8.6	24.6	64	−0.7	8.8	−13.1	18.1
26	7.3	4.5	−8.2	24.1	65	−0.1	9.2	−14.0	18.8
27	7.1	5.0	−7.8	23.5	66	0.5	9.1	−14.5	18.9
28	7.1	5.4	−7.4	23.4	67	0.9	8.8	−15.0	19.6
29	7.3	5.3	−7.2	23.6	68	1.1	8.2	−15.2	20.4
30	7.6	4.9	−7.3	24.0	69	0.9	7.3	−15.2	21.7
31	8.0	4.0	−7.5	24.5	70	0.3	6.3	−15.0	22.6
32	8.4	2.9	−7.6	25.3	71	−0.6	5.2	−15.0	22.8
33	9.0	1.6	−7.6	27.0	72	−1.7	4.0	−14.8	22.5
34	9.9	0.5	−7.6	28.5	73	−3.0	3.0	−14.7	22.0
35	11.0	−0.5	−7.5	28.8	74	−4.3	2.2	−14.7	21.2
36	12.2	−1.1	−7.3	28.8	75	−5.6	1.6	−14.6	20.9
37	13.3	−1.5	−7.3	28.8	76	−6.8	1.3	−14.5	21.7
38	14.2	−1.5	−7.3	28.8	77	−7.8	1.3	−14.4	22.6
39	14.9	−1.3	−7.0	28.7	78	−8.6	1.7	−14.3	23.3
40	15.4	−0.9	−7.0	28.6	79	−9.0	2.4	−14.4	23.9
41	15.7	−0.5	−7.0	28.1	80	−9.0	3.4	−14.5	24.4
42	15.8	−0.3	−6.9	27.5	81	−8.6	4.6	−14.7	25.2
43	15.7	−0.3	−6.9	27.0	82	−7.7	5.9	−14.9	26.2
44	15.5	−0.6	−6.9	26.7	83	−6.4	7.3	−15.0	27.2
45	15.1	−1.1	−6.7	26.7	84	−4.9	8.5	−14.9	28.9
46	14.8	−1.7	−6.6	26.4	85	−2.9	9.5	−14.8	30.0
47	14.5	−2.4	−6.1	25.7	86	−0.5	10.2	−14.5	31.0
48	14.2	−3.2	−5.8	24.7	87	2.2	10.5	−14.0	32.1
49	13.7	−4.0	−5.3	23.2	88	4.9	10.5	−13.4	33.0
50	13.1	−4.6	−4.8	21.4	89	7.4	10.1	−13.0	34.2
51	12.4	−4.9	−4.2	20.7	90	9.6	9.3	−12.7	35.5
52	11.6	−4.8	−3.8	20.6	91	11.1	7.9	−12.7	36.3

Astrov, Kemp, and Swensen, 1979; Bureau International des Poids et Mesures, 1979) covers the temperature range from 0.5 to 30 K. Two vapor-pressure scales, 1958 ^{4}He and 1962 ^{3}He, were also widely used for temperatures below 5 K. The differences between EPT-76 and the vapor-pressure scales are given in Table 14.8, and IPTS-68, NBS P 2-20, and EPT-76 are compared in Table 14.9.

14.3.2 Resistance thermometers

The variation of the electrical resistivity of certain materials as a function of temperature is the basis for one of the major approaches to determining temperature. Both metals and nonmetals are used as the resistive elements in thermometers of this type.

Metals are characterized by a positive slope of the resistance-vs.-temperature characteristic (dR/dT), whereas most nonmetals are characterized by a negative dR/dT. Both types of thermometers have been widely used for measuring cryogenic temperatures, and both types have typical strong and weak points. Metals exhibit decreasing sensitivity below certain temperatures, which ordinarily limits their use to the range T

Table 14.7 Temperature differences, $\Delta T = T_{68} - T_{48}$, are given, where T_{68} represents temperatures on the IPTS-68 temperature scale and T_{48} represents temperatures on the IPTS-48 temperature scale. Differences are given for the range −180 to 4000° C (Comité International des Poids et Mesures, 1969).

IPTS-68 (° C)	0	−10	−20	−30	−40	−50	−60	−70	−80	−90	−100
−100	0.022	0.013	0.003	−0.006	−0.013	−0.013	−0.005	0.007	0.012		
− 0	0.000	0.006	0.012	0.018	0.024	0.029	0.032	0.034	0.033	0.029	0.022
IPTS-68 (° C)	**0**	**10**	**20**	**30**	**40**	**50**	**60**	**70**	**80**	**90**	**100**
0	0.000	−0.004	−0.007	−0.009	−0.010	−0.010	−0.010	−0.008	−0.006	−0.003	0.000
100	0.000	0.004	0.007	0.012	0.016	0.020	0.025	0.029	0.034	0.038	0.043
200	0.043	0.047	0.051	0.054	0.058	0.061	0.064	0.067	0.069	0.071	0.073
300	0.073	0.074	0.075	0.076	0.077	0.077	0.077	0.077	0.077	0.076	0.076
400	0.076	0.075	0.075	0.075	0.074	0.074	0.074	0.075	0.076	0.077	0.079
500	0.079	0.082	0.085	0.089	0.094	0.100	0.108	0.116	0.126	0.137	0.150
600	0.150	0.165	0.182	0.200	0.23	0.25	0.28	0.31	0.34	0.36	0.39
700	0.39	0.42	0.45	0.47	0.50	0.53	0.56	0.58	0.61	0.64	0.67
800	0.67	0.70	0.72	0.75	0.78	0.81	0.84	0.87	0.89	0.92	0.95
900	0.95	0.98	1.01	1.04	1.07	1.10	1.12	1.15	1.18	1.21	1.24
1000	1.24	1.27	1.30	1.33	1.36	1.39	1.42	1.44			
IPTS-68 (° C)	**0**	**100**	**200**	**300**	**400**	**500**	**600**	**700**	**800**	**900**	**1000**
1000		1.5	1.7	1.8	2.0	2.2	2.4	2.6	2.8	3.0	3.2
2000	3.2	3.5	3.7	4.0	4.2	4.5	4.8	5.0	5.3	5.6	5.9
3000	5.9	6.2	6.5	6.9	7.2	7.5	7.9	8.2	8.6	9.0	9.3

Table 14.8 Temperature differences between the EPT-76 scale (T_{76}) and the vapor pressure scales 1958 ^{4}He (T_{58}) and 1962 ^{3}He (T_{62}). For temperatures up to 3.2 K, $T_{VP} = (T_{58} + T_{62})/2$ is used (Bureau International des Poids et Mesures, 1979).

T_{VP} (K)	$T_{VP}-T_{76}$ (mK)	T_{58} (K)	$T_{58}-T_{76}$ (mK)
0.5	−1.9		
0.6	−2.1		
0.8	−2.5		
1.0	−2.9		
1.2	−3.2		
1.4	−3.5		
1.6	−3.7		
1.8	−3.9		
2.0	−4.1		
2.2	−4.4		
2.4	−4.9		
2.6	−5.4		
2.8	−5.9		
3.0	−6.3		
3.2	−6.6	3.2	−6.6
		3.4	−6.8
		3.6	−7.0
		3.8	−7.0
		4.0	−7.1
		4.2	−7.1
		4.5	−7.1
		5.0	−7.1

Table 14.9 Temperature differences between IPTS-68 (T_{68}), NBS P 2-20 ($T_{NBS\ 2\text{-}20}$), and EPT-76 (T_{76}) (Bureau International des Poids et Mesures, 1979).

$T_{NBS\ 2\text{-}20}$ (K)	$T_{NBS\ 2\text{-}20}-T_{76}$ (mK)	T_{68} (K)	$T_{68}-T_{76}$ (mK)	T_{68} (K)	$T_{68}-T_{76}$ (mK)
2.3	2.1				
2.8	−1.1				
3.2	1.0				
4.2	2.5				
5.0	3.0				
6.0	1.7				
7.0	4.8				
8.0	2.2				
9.0	−1.5				
10.0	−2.1				
11.0	−1.0				
12.0	0.2				
13.0	−1.8				
		13.81	5.6	22.0	4.8
14.0	−2.2	14.0	4.6	22.5	4.2
		14.5	3.0	23.0	3.7
15.0	−0.6	15.0	2.0	23.5	3.2
		15.5	2.2	24.0	2.7
16.0	0.9	16.0	2.6	24.5	2.1
		16.5	3.6	25.0	1.6
17.0	1.7	17.0	4.6	25.5	1.1
		17.5	5.6	26.0	0.7
18.0	−0.9	18.0	6.5	26.5	0.3
		18.5	7.2	27.0	0.0
19.0	−0.2	19.0	7.4	27.5	0.0
		19.5	7.3	28.0	0.0
20.0	−0.8	20.0	6.9	28.5	0.0
		20.5	6.4	29.0	0.0
		21.0	5.8	29.5	0.0
		21.5	5.3	30.0	0.0

>10 K. The sensitivity of nonmetals, on the other hand, increases at lower temperatures, giving them the advantage when $T < 25$ K.

Measuring resistance. One of two general methods is usually used to determine the electrical resistance of a thermometer: potentiometric voltage determination at constant thermometer current or direct resistance balance via a bridge circuit. Both potentiometric and bridge techniques can utilize alternating current (ac) or direct current (dc). Traditionally, dc has been used for high-precision measurements of resistance, and most existing calibrations are based on dc. The availability of sophisticated lock-in amplifiers and operational amplifiers with extremely high input impedance and low output impedance is making ac measurements increasingly common. Elimination of thermal emf's, lower power dissipation, and automation are attractive features of ac instruments. Resistance measurements made with ac may be different from those made with dc. Kirby and Laubitz (1973) attribute the difference to the Peltier effect. Quantitative comparisons are very limited: Swenson and Wolfendale (1973) found $R_{dc} - R_{ac}$ to be equivalent to 109 mK at 78.6 K and less than 1 mK when $T < 40$ K for germanium resistance thermometers (GeRT) at 30 Hz. Rusby, Chattel, and Gilhen (1972) found that platinum resistance thermometer resistances measured at 375 Hz and with dc produced temperature equivalent differences of 2.8 mK at 13.8 K, 0.25 mK at 20.2 K, and the differences were undetectable above 54 K.

Potentiometric methods are generally used in conjunction with sensors such as GeRTs, which have large potential lead resistances. Loading effects caused by this inherent resistance limits the use of bridge networks unless high input impedance is introduced to reduce the current flow in these leads. Zero current is drawn through the potential leads in a balanced potentiometer circuit, so that lead resistance does not affect the measurement. The use of ac in a potentiometer circuit eliminates the need for current reversal, as discussed below for the dc measurement.

A schematic illustrating the standard four-lead dc potentiometric method is shown in Fig. 14.5a. Two voltage readings, V_s (voltage across a standard resistor, R_s) and V_x (voltage across the thermometer), are necessary to determine the thermometer resistance, R_x. Resistance of the thermometer is given by

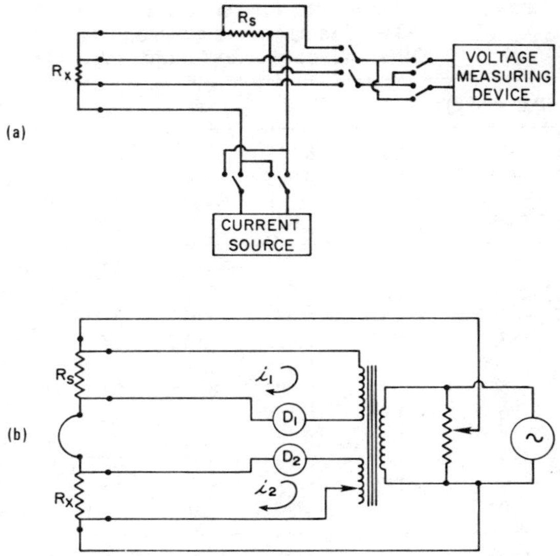

Figure 14.5 Potentiometric circuits used to determine the resistances of four-lead thermometers. (a) Direct-current circuit. The voltage measuring device may be a dc potentiometer or a digital voltmeter; the current source may be a battery or a stable dc electronic device. (b) Alternating-current circuit. D_1 and D_2 are phase-locked integrating detectors.

$$R_x = V_x R_s / V_s \tag{14.8}$$

To cancel possible spurious thermally generated voltages in the potential leads, the direction of current flow must be reversed and a similar pair of voltage readings taken. The resistance of the thermometer is then given by

$$R = 1/2\,[R_x \text{ (current forward)} + R_x \text{(current reversed)}] \tag{14.9}$$

This procedure involves four voltage measurements and, consequently, is best applied to systems with nearly constant temperatures.

A four-lead configuration utilizing an ac potentiometer (Wolfendale, 1969) is shown in Fig. 14.5b. When the currents flowing in the standard resistor and the thermometer are

$$i_s = i_x = 0 \tag{14.10}$$

then

$$R_T = R_s\,(N_2/N_1) \tag{14.11}$$

where N_1 and N_2 are voltages generated across the secondaries of the input transformer.

The second general method is to determine the

thermometer resistance using one of several possible dc or ac bridge circuits. Four of the more common circuits are shown in Fig. 14.6. The bridge shown in Fig. 14.6a is often used with two-lead thermometers and has no lead or contact resistance compensation, since both leads are in the same arm of the bridge as the unknown resistance. The measured resistance includes the resistance of the leads L_1 and L_2:

$$R = R_x + R_{L_1} + R_{L_2} \tag{14.12}$$

The circuit in Fig. 14.6b utilizes three thermometer leads; this compensates for the lead resistance only to the extent that $R_{L_1} = R_{L_3}$. These resistances appear in different arms of the bridge so that

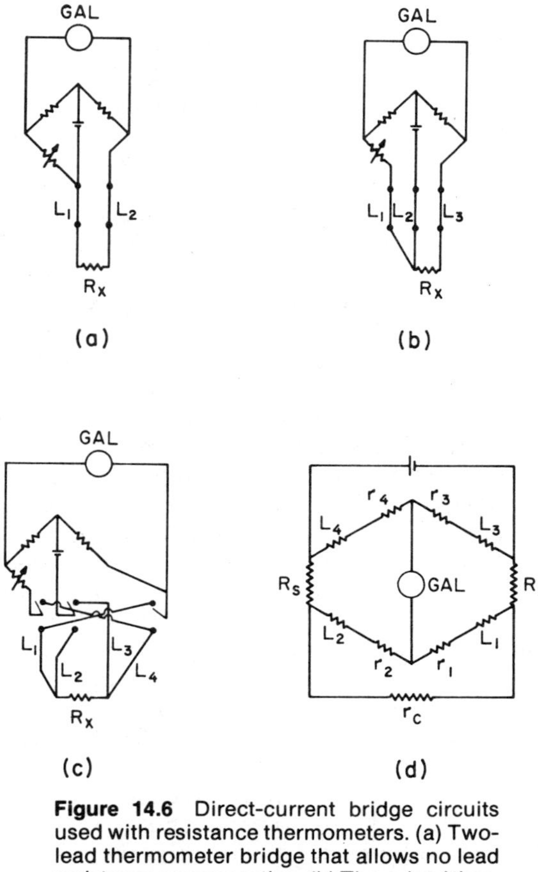

Figure 14.6 Direct-current bridge circuits used with resistance thermometers. (a) Two-lead thermometer bridge that allows no lead resistance compensation. (b) Three-lead thermometer bridge providing partial-lead resistance compensation. (c) Mueller bridge that allows full-lead resistance compensation when used with a four-lead thermometer. (d) Kelvin bridge provides full-lead resistance compensation for four-lead thermometers.

$$R = R_x + R_{L_3} - R_{L_1} \simeq R_x \tag{14.13}$$

Figure 14.6c represents the basic design of the often-used Mueller bridge. This configuration requires two readings, but does eliminate the effect of the leads when the temperature of the leads remains constant for both readings. The required pair of readings would be

$$R_1 = R_x + R_{L_1} - R_{L_4} \tag{14.14}$$

and

$$R_2 = R_x + R_{L_4} - R_{L_1} \tag{14.15}$$

Averaging R_1 and R_2 results in the cancellation of the lead resistance. A fourth bridge, Fig. 14.6d, is known as the Kelvin bridge. In principle, this circuit allows complete lead compensation. If

$$R_1 = R_{L_1} + r_1, \quad R_2 = R_{L_2} + r_2, \quad R_3 = R_{L_3} + r_3, \quad \text{and } R_4 = R_{L_4} + r_4, \tag{14.16}$$

then

$$R_x = R_s(R_3/R_4) + \{R_2 r_c/(R_1 + R_2 + r_c)\} \times (R_3/R_4 - R_1/R_2) \tag{14.17}$$

If the value of r_c and the difference ratio

$$R_3/R_4 - R_1/R_2 \tag{14.18}$$

are made small enough, the second term can be neglected and

$$R_x = R_s(R_3/R_4) \tag{14.19}$$

Harris (1952), Stout (1960), and Daneman and Mergner (1967) present further details of these and other bridge circuits.

Bridges that utilize ac have been developed for use with all types of resistance thermometers. The bridge loading that occurs when the lead resistances of the thermometers become large is eliminated by introducing operational amplifiers as voltage followers in the bridge arms containing the thermometer potential leads. The two dc bridges that allow lead resistance compensation, the Kelvin (Ekin and Wagner, 1970) and the Mueller bridges (Diamond, 1966), have been adapted to utilize ac. The lead resistance compensation is retained in the ac version of these circuits. The ac Kelvin bridge developed by Ekin and Wagner (1970) is shown in Fig. 14.7. Rubin and Golahny (1972) improved this bridge by using battery-operated operational amplifiers. The GeRT resistance equivalent of 0.5 mK at 4.2 K can be accurately measured with their modified bridge.

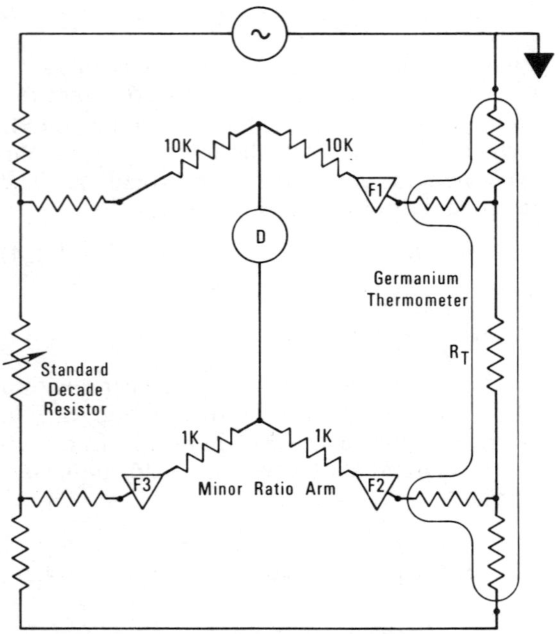

Figure 14.7 Simplified schematic of the ac Kelvin bridge. F1, F2, and F3 represent high input impedance operational amplifier voltage followers.

Temperature dependence of resistivity (metals). The resistivity of a metal, as given by Matthiessen's rule, is the sum of two terms:

$$\rho_{Total} = \rho_i(T) + \rho_0 \tag{14.20}$$

The residual resistivity, ρ_0, is assumed to be constant and is due to electron scattering by impurities and lattice defects. The intrinsic resistivity, $\rho_i(T)$, is generally an increasing function of temperature and is due to electron scattering by lattice vibrations. Lattice vibrations, and therefore temperature sensitivity, persist to lower temperatures in metals that have lower Debye temperatures. The Debye characteristic temperature is defined as

$$\theta_D = h\nu_{max}/k \tag{14.21}$$

where h is Planck's constant, k is Boltzmann's constant, and ν_{max} is the maximum lattice frequency allowable in a particular lattice. The lattice model used by Debye to arrive at this relationship is discussed in Chapter 2. The Bloch–Grüneisen relationship (discussed in Chapter 5) for the temperature-dependent resistivity is

$$\rho_i \propto T^5/\theta_D^6 \text{ when } T < \theta_D/10 \text{ and}$$
$$\rho_i \propto T/\theta_D^2 \text{ when } T > \theta_D/2 \tag{14.22}$$

In general, heavy, loosely packed materials have long minimum wavelengths, which means that ν_{max} and θ_D are low and ρ_i is high. For example, θ_D = 85, 140, 225, and 310 K for lead, indium, platinum, and copper, respectively (Gopal, 1966). Other properties being equal, the metal with the lowest θ_D would make the best thermometer element for lower temperatures.

Although platinum is the most common metal used in low-temperature thermometry, others, such as indium, copper, and Rh–0.5 at.% Fe, have been used to some extent. Important properties of a metal for use in low-temperature resistance thermometers are chemical inertness, ductility, availability of elements in a highly purified state so that ρ_0 is low, and as mentioned above, thermal sensitivity to low tempertures (θ_D as small as possible.).

Platinum resistance thermometers. Platinum resistance thermometers (PRTs) are the most accurate and reproducible empirical thermometers available over a wide range of temperatures. The IPTS-68 utilizes PRTs to define the temperature scale between the fixed points from 13.81 to 903.89 K. When state of the art thermometers, instrumentation, and techniques are used, the usable range for these thermometers can be extended down to 2 K with an accuracy of about ±10 mK (Berry, 1967a,b). Measurements in the range 15 to 725 K are made routinely with a high degree of accuracy with a number of different commercially available thermometers. At the high end of the usable temperature range, PRTs are being developed for temperatures up to 1373 K. Thermometer design and high-temperature effects are discussed by Anderson (1972), Berry (1972), Chattle (1972), Evans (1972), and Sawada and Mochizuki (1972). Berry (1980) discusses the effect of oxidation on PRTs at temperatures as low as 450 to 550° C.

PRTs are available in many configurations and can be commercially fabricated for special applications at additional cost. The following discussion divides PRTs into two somewhat arbitrary classes — laboratory and industrial. The distinction is based primarily on how the active element is supported and on its purity. The purity and strain condition are generally expressed as a ratio of resistances

$$W_T = R_T/R_{273.15} \tag{14.23}$$

where T = 373.15 K or, alternatively as the temperature coefficient of resistance

$$\alpha = (R_{373.15} - R_{273.15})/100 \cdot R_{273.15} \qquad (14.24)$$

The minimum W_T required to satisfy the IPTS-68 specifications is 1.39250. The specification for platinum for use in thermometers is not uniform throughout the world; for example, the Japanese standard JIS C 1604 requires $W_T \geqq 1.3916$, and the German DIN 43760 and British 1904:1964 require $W_T \geqq 1.3850$. This nonuniformity in thermometer materials requires that the user be aware of possible differences in $R_T/R_{273.15}$ data. In general, the absolute resistance varies from thermometer to thermometer, but the ratio $R_T/R_{273.15}$ is identical for all PRTs using strain-free platinum of a specific W_T. For instance, the resistance of a PRT with nominal $R_{273.15} = 25\ \Omega$ and $W_{373.15} = 1.385$ differs from that of a similar sensor with $W_{373.15} = 1.3926$ by approximately 0.35 Ω at 73 K. This is equivalent to 3.5 K. The resistance–temperature relationship for three grades of platinum, $W_{373.15} = 1.3850, 1.3911$, and 1.39250, are shown in Fig. 14.8. The $(1/R)dR/dT$ and dR/dT characteristics of the $W_{373.15} =$ 1.39250 PRT are shown in Fig. 14.9.

Although the resistances of laboratory-quality PRTs are extremely stable, changes do occur. IPTS-68 indicates that the ice-point resistance should remain constant to within the temperature equivalent of 0.2 mK at 273.15 K when the sensor is not exposed to temperatures above 100° C. Vibration, elevated temperatures, or even slightly rough handling can cause changes in excess of this amount. Mechanical shock induces strain into the platinum element and causes the ρ_0 term in Eq. 14.20 to increase.

Establishing the necessary temperature bath (ice or triple point of water) to within 0.5 mK or measuring resistance to an absolute accuracy of a few parts per million is not feasible in many laboratories. It is becoming increasingly common for calibration laboratories to measure the resistance of PRTs at 4.2 K (boiling point of helium) and at 273.15 K. The ratio of these resistances is extremely sensitive to changes in the residual resistance. A change in this ratio for a particular thermometer that has not been exposed to elevated temperatures indicates that the platinum resistance element has been strained. The ratio

$$\mathrm{RRR} = R_{273.15}/R_{4.2} \qquad (14.25)$$

is used to characterize the chemical and physical makeup of metals in general (Chapter 5). It is conventional in thermometry to use the inverse of RRR in the form of Eq. 14.23 with $T = 4.2$ K.

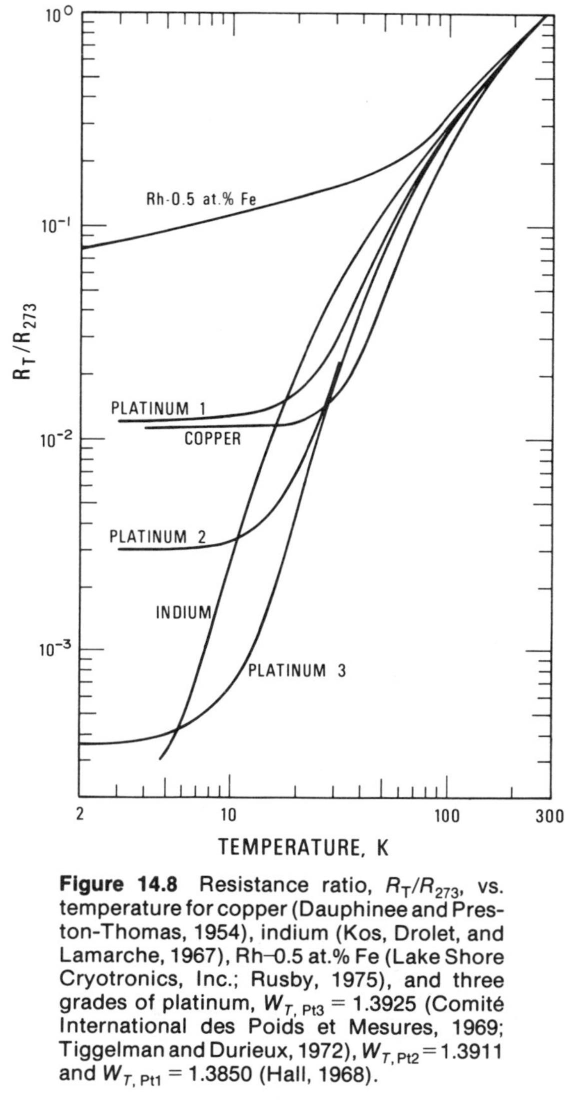

Figure 14.8 Resistance ratio, R_T/R_{273}, vs. temperature for copper (Dauphinee and Preston-Thomas, 1954), indium (Kos, Drolet, and Lamarche, 1967), Rh–0.5 at.% Fe (Lake Shore Cryotronics, Inc.; Rusby, 1975), and three grades of platinum, $W_{T,\,\mathrm{Pt3}} = 1.3925$ (Comité International des Poids et Mesures, 1969; Tiggelman and Durieux, 1972), $W_{T,\,\mathrm{Pt2}} = 1.3911$ and $W_{T,\,\mathrm{Pt1}} = 1.3850$ (Hall, 1968).

For precision thermometers of nominal $R_{273.15} = 25\ \Omega$ and $W_T = 1.3926$, the range of values for W_4 is approximately 0.00037 to 0.00073.

Meyers (1932) designed the first precision laboratory PRT (Fig. 14.10a). It consisted of a coiled helix of fine platinum wire on a mica cross support. Another Meyers design (Stimson, 1955) features a single coil of platinum on a mica support, as seen in Fig. 14.10b. Barber (1950, 1955) designed two capsule-type PRTs; the one shown in Fig. 14.10c uses a freely suspended helix of platinum in a glass support tube. All these thermometers use a protective sheath, coiling of the sensitive element to increase the length to area ratio, a suspension scheme intended to retain the

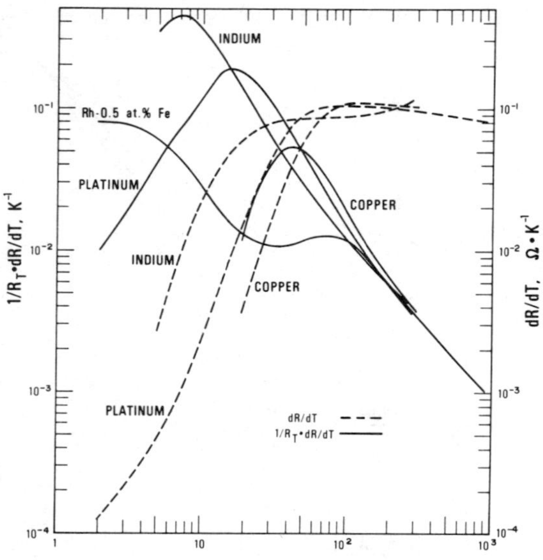

Figure 14.9 Temperature coefficient of resistance vs. temperature for copper (Dauphinee and Preston-Thomas, 1954), indium (Kos, Drolet, and Lamarche, 1967), platinum (Comité International des Poids et Mesures, 1969; Tiggelman and Durieux, 1972), and Rh- 0.5 at.% Fe (Lake Shore Cryotronics, Inc.; Rusby, 1975).

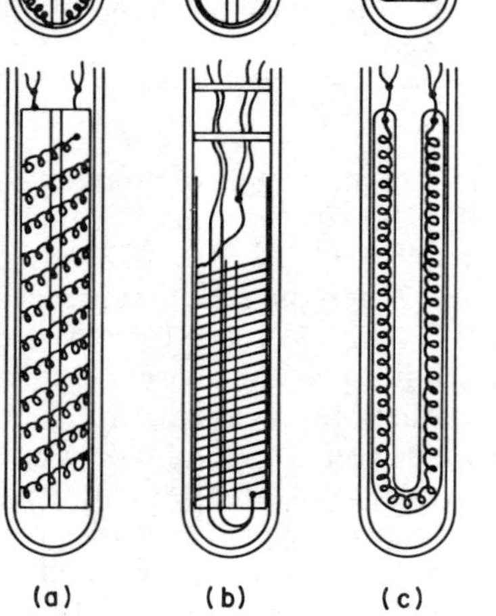

Figure 14.10 Schematic of capsule-type platimum resistance thermometers.

platinum in a strain-free state, and a filling gas to provide a thermal link between the sheath and the sensitive element. Laboratory thermometers with the essential designs discussed above are available commercially with both long-stem and capsule-types of sheaths. The strain-free state needed to meet the α requirement is achieved by annealing. The annealing temperature must be at least as high as that at which the thermometer will be used. Annealing procedures differ, depending on the amount of cold work and the purity of the platinum (Corruccini, 1951 and Berry, 1972). Annealing at temperatures in the range 740 to 773 K is generally sufficient to remove much of the cold work and all of the quenched-in resistance. PRTs are annealed at 743 to 753 K for 4 h before being calibrated at the National Bureau of Standards (Riddle, Furukawa, and Plumb, 1973). In any case, slow cooling is required, since rapid cooling from a high temperature traps lattice defects and changes the electrical properties.

Certain disadvantages associated with precision PRTs are a direct result of the necessity to make the platinum element strain free: the PRTs are susceptible to strain from vibration and moderate *g* forces (McLaren, 1957), and their active element is thermally connected to the sheath through a gas. This gaseous thermal link causes a relatively slow time response, 2 to 7 s, depending on the fill gas, the experimental conditions, and self-heating due to the thermometer current. Consequently, thermometers of this type must be handled carefully and are usually used in static or slowly changing environments where time response is not critical and where the effect of self-heating can be determined by making more than one resistance determination using different thermometer currents.

The size of the capsule-type thermometers described above is relatively large, roughly 5.7 mm in diameter, 60 mm in length, and 5 g in weight. A much smaller PRT (about 4 mm in diameter, 13 mm in length, and 1.1 g in weight) has been developed and is available commercially (Lucas, 1972). Gehring and Gerstein (1967) found that after thermally shocking these thermometers from ambient to liquid-helium temperature 20 times, the resistance at the triple point of water became stable to within about 0.2 mΩ. This is the temperature equivalent of 0.5 mK for these $R_0 \simeq 100\ \Omega$ resistors.

Industrial PRTs are available for a wide range of applications. These thermometers are, in general, less precise than the laboratory resistors just discussed. The ruggedness of these units is achieved by more rigid support of the sensitive elements. Potting or mounting procedures and differential thermal contraction between the supporting materials and the platinum thermometer element can strain the platinum and change the electrical properties. The $W_{373.15}$ of these thermometers varies from 1.3850 to 1.392.

There are two basic designs of industrial PRTs — immersion probes and surface temperature sensors. The immersion probes feature a high-purity, encapsulated platinum wire (Fig. 14.11a) or a wire securely attached to the support frame. The ice-point resistance of these thermometers is usually in the 100- to 1000-Ω range. The recommended temperature range is ~40 to 523 K and higher in cases where the platinum will not suffer contamination from the supporting material. Repeatability after thermal shocks, time response in different environments, interchangeability, and mechanical shock tolerance differ among companies and specific designs. For the most part, the specifications below represent typical values that might be used in preliminary designs; exact specifications must necessarily come from the manufacturer.

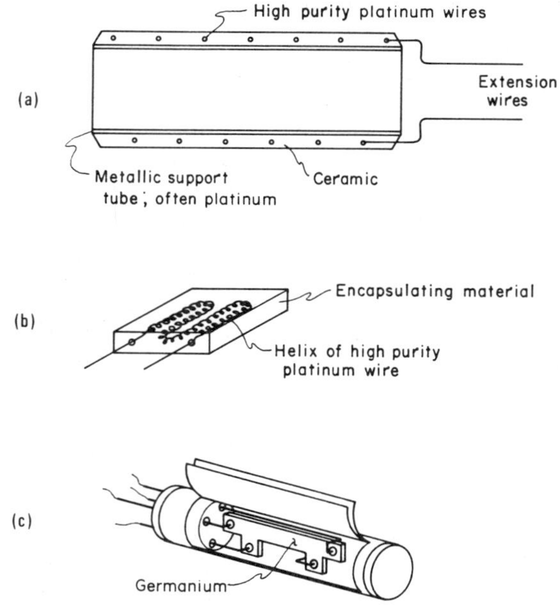

Figure 14.11 Schematic of industrial platinum resistance thermometers and a germanium resistance thermometer, (a) immersion type PRT, (b) surface type PRT, and (c) precision GeRT.

The repeatability of the typical immersion sensor is usually certified to be about ±0.1 K at the ice point after several thermal cycles to cryogenic temperatures. For most thermometers this is conservative. The time response is particularly difficult to assess in a general way because it depends critically upon the method of testing. The time response of the precision capsule-type PRT was given as 2 to 7 s, which means that the sensor reaches equilibrium with the system in this period. In the case of the industrial PRT, dynamic systems are frequently encountered. It is conventional to define the time response of an industrial thermometer to be the time it takes the sensor to reach 63.2% of the temperature of a step-function temperature change. With this definition of the time constant, a typical range of values for this type of resistor is 0.1 to 3 s.

Interchangeability is measured in terms of the errors involved when more than one thermometer is used with a single R-vs.-T relationship. It is ordinarily specified at a given temperature, that is, the resistance of the thermometers does not vary more than a specified amount at a certain temperature. Immersion-type sensors may generally be specified to have the same ice-point resistance within an equivalent of about ± 1.5 K; surface sensors show slightly worse interchangeability for particular models.

The ruggedness of industrial PRTs, that is, maintenance of specifications after severe vibration or g forces, depends strongly on the particular type of construction. An example of a ruggedness specification used by NASA (Sinclair, Terbeek, and Malone, 1968) is for a thermometer to remain within specifications after (a) 50 g's or 13-mm double amplitude (smaller of the two) from 20 to 2000 Hz for 15 min, (b) impact shock of 100 g's for a 10-ms triangular wave, and (c) velocity loading > 30.5 $m \cdot s^{-1}$ of liquid hydrogen.

The second type of industrial PRT is commonly known as a surface temperature sensor. The general design is as shown in Fig. 14.11b. These thermometers are small, typically 2.5 mm × 13 mm × 13 mm with perhaps a factor of 2 variation in any dimension. Both ceramic encapsulated and suspension mounted models are available. Their geometry is such that they make good thermal contact with surfaces of various shapes. Sensors of this type are designed to be mounted by clamping around small tubing, fitting into milled slots, clamping under bolt heads, soldering, welding, and cementing. Custom design capability is high for this general type of sensor.

The usefulness of a thermometer depends on relating a measured quantity to the temperature of the sensor. IPTS-68 assigns temperatures to specified equilibrium states between phases (fixed points) of several pure substances and provides the needed equations to interpolate between these points (Comité International des Poids et Mesures, 1969). The equations cover the range 13.81 to 903 K: below 273.15 K, temperature as a function of resistance is found from a reference function and specified deviation equations; above 273.15 K, two polynomial equations are specified. Establishing the required fixed points is extremely difficult, however, and has been done in only a few national laboratories. Detailed discussion of the interpolation formulas and fixed points may be found in the text of IPTS-68 (Comité International des Poids et Mesures, 1969). Procedures used to calibrate PRTs at the National Bureau of Standards are discussed by Riddle et al. (1973).

Methods of calibration do exist that are more consistent with the calibration capabilities found in typical low-temperature laboratories. The relationships used in conjunction with IPTS-48 (Stimson, 1961) are more simple, that is, only four fixed-point measurements are needed. The interpolating equations used between the fixed points were the Callendar

$$t(°\mathrm{C}) = (R_t - R_0)/\alpha \cdot R_0 + 0.01\partial \cdot t\,(0.01t - 1)$$
$$0 \leq t \leq 630.5\ °\mathrm{C} \tag{14.26}$$

and the Callendar–VanDusen

$$t(°\mathrm{C}) = (R_t - R_0)/\alpha \cdot R_0 + 0.01\partial \cdot t(0.01t - 1) + \beta(0.01t)^3(0.01t - 1)$$
$$-182.97 \leq t \leq 0°\mathrm{C} \tag{14.27}$$

The constants R_0, α, and ∂ in Eq. 14.26 are determined by resistance measurements at the ice point, steam point, and the sulfur point. The oxygen point resistance allows β(Eq. 14.27) to be determined.

For the very low temperature range ($2 \leq T \leq 15$ K), Kos and Lamarche (1967) found that the reduced resistance of a PRT, $W_T = R_T / R_{273.15\mathrm{K}}$, can be represented by the fifth-degree polynomial

$$W_T = \sum_{n=0}^{5} A_n\, T^n \tag{14.28}$$

to the temperature equivalent of ±10 mK. Ideally, calibrations to determine the coefficients in this equation would be done by comparison with a precision helium gas thermometer. They may also be determined by direct comparison with the vapor pressure of He^4 for $T \lesssim 4.2$ K and by comparison with calibrated germanium resistance thermometers for $T \lesssim 15$ K. Superconducting fixed points (Schooley, Soulen, and Evans, 1972) may be used to spot-check the results of the calibration and analytical representation.

In the temperature range 10 to 273 K, it is possible to determine the R-vs.-T relationship for an uncalibrated PRT by using an existing calibration for a similar resistor (ideally one with identical $W_{373.15}$), the resistance of the uncalibrated PRT at two temperatures, and Matthiessen's rule (Eq. 14.20). This procedure is based on a universal function, Z_T, discussed by Corruccini (1960)

$$Z_T = (R_T - R_1)/(R_2 - R_1) \tag{14.29}$$

This function is calculated using the existing PRT calibration: R_T is the calibration resistance at T, and R_1 and R_2 are the calibration resistances at T_1 and T_2. This approximation applies when $T_1 < T < T_2$. Since Z_T can be determined from an existing calibration table, the resistance, R_{Tx}, of the uncalibrated PRT at temperature T is given by

$$R_{Tx} = R_{1x} + Z_T(R_{2x} - R_{1x}) \tag{14.30}$$

where R_{1x} and R_{2x} are experimentally determined values of resistance at T_1 and T_2. Corruccini (1962) calibrated several capsule-type PRTs and found that the temperature differences between the values calculated from Eq. 14.30 and the NBS standard calibrations were on the order of ±10 mK when $T_1 = 20$ K and $T_2 = 80$ K. White (1959) found temperature differences on the order of ±50 mK for the capsule-type thermometers when $T_1 = 4.2$ K and $T_2 = 273$ K. Sinclair, Terbeek, and Malone (1972) applied the Z_T technique to industrial-type PRTs and found temperature differences of ±100 mK for $T_1 = 4$ K and $T_2 = 20$ K, ±20 mK for $T_1 = 20$ K and $T_2 = 77$ K, and ±50 mK for $T_1 = 77$ and $T_2 = 273$ K.

The Z_T technique can be refined (Corruccini, 1962; Besley and Kemp, 1978) to account for deviations from Matthiessen's rule. This calibration requires that the resistance of the uncalibrated PRT be determined at three temperatures and that calibrations for two similar thermometers be available. This procedure can result in errors of a few mK in the range 20 to 273 K.

Other metallic resistance thermometers. The desirable characteristics of a resistance thermometer element include either high purity or carefully controlled composition, ductility, chemical inertness, and for low temperatures ($T < 20$ K), a low Debye characteristic temperature. Platinum is the most frequently used material. Indium and copper appear to offer advantages in particular cases, but suffer from lack of development in comparison with platinum. An alloy of iron in rhodium is often used instead of nonmetallic sensors at temperatures less than 20 K and may be used to temperatures as high as 300 K.

Indium fits the low-temperature thermometer requirements rather well: it is pure ($W_{4.2} = 10^{-4}$ for indium, $\simeq 10^{-3}$ for platinum), it is extremely soft and malleable, and its θ_D is about 140 K. This low θ_D indicates that lattice vibrations in the high-purity material dominate the total resistance down to its superconducting transition temperature (~3.4 K).

Several indium thermometers have been constructed and tested by White and Woods (1957), White, Woods, and Anglin (1958), Yates and Panter (1961), Orlova, Astrov, and Medvedeva (1964), and Kos et al. (1967). Construction details and material specifications for these thermometers are included in their papers.

An important factor in determining the practical applicability of any resistance thermometer is the ability to represent the R-vs.-T characteristics analytically. White and Woods (1957) found that their data for indium were fit reasonably well by $R_T/R_{273} = A_1 + A_2T + A_3T^2$ for temperatures above 100 K; for temperatures between 100 K and 40 K, by $R_T/R_{273} = A_1 + A_2T$; and for temperatures below 10 K, by $R_T = A_1 + A_2T^5$. Orlova et al. found that this $T < 10$ K relationship yielded uncertainties of about 0.2 K at liquid-helium temperature. Caution must be observed, however, when applying these analytical representations, since Swenson (1955) found an unexplained discontinuity at 210 K in the R-vs.-T characteristic of his indium thermometer.

With sufficient development, it appears that indium thermometers could be used routinely over the temperature range 3.4 to 300 K. This would represent a significant improvement over platinum thermometers in the very low temperature range (see Fig. 14.8). Analytical representation of the resistance–temperature characteristics is a major deficiency at this time. This should be resolvable to approach the PRT level of uncertainty.

Copper is frequently used as the active element in thermometers of "in-house" construction. It offers the advantages of being cheap, ductile, and available in most laboratories. The main disadvantages are that copper oxidizes at relatively low temperatures, and compared with platinum, it is not very stable or reproducible. Dauphinee and Preston-Thomas (1954) examined eleven copper resistance thermometers, made from commercial, enameled copper wire, between the temperatures of 4.2 and 320 K and established a reference temperature–resistance relationship from these data (Figs. 14.8 and 14.9). The calibration of a thermometer at the n.b.p. of hydrogen and at the ice point enabled them to adjust the reference R_T/R_{273} values to represent the test thermometer to within ±25 mK throughout the range.

A metallic alloy has recently been developed for use in low-temperature thermometers. The resistivity of this alloy, Rh–0.5 at.% Fe, was determined by Coles (1964) while he was studying the positive temperature coefficient contribution of dilute concentrations of iron in rhodium at low temperatures. The net effect of adding this temperature coefficient to the ideal resistivity (ρ_0 in Eq. 14.20) is a monotonically decreasing resistivity with a significant temperature dependence from 273 to at least 0.1 K. Later work by Rusby (1975) has resulted in commercially available sensors built along the lines of the capsule PRTs shown in Fig. 14.10. These thermometers are produced in several sizes with overall lengths from 25 to 60 mm and diameters of 3.2 to 9 mm. Nominal ice-temperature resistances are from 20 to 100 Ω.

Although not as temperature sensitive as semiconducting sensors ($T \lesssim 20$ K) or as PRTs ($T \gtrsim 20$ K) these metallic sensors do have certain advantages. Very importantly, the simple behavior of the resistivity allows for easier calibration and analytical representation. Standard deviations of from 0.18 to 9 mK were obtained (Rusby, 1975) using polynomial fitting orders from 8 to 3 over the temperature range 2.6 to 20 K. An eleventh-order polynomial produced a ±0.2 mK fit in the expanded range 0.5 to 27 K. The long-term stability of these thermometers has been investigated by Rusby on two commercial models: calibration shifts of less than 0.1 mK per year were found for one model and 0.3 mK in 20 months for the other. The W_T and the temperature coefficient of resistance for this alloy are shown in Figs. 14.8 and 14.9, respectively.

The usable sensitivity over the entire temperature range ($T \lesssim 300$ K) offers a one-sensor capability that is extremely valuable. It appears that the chief disadvantage is lack of uniformity of resistive characteristics from thermometer to thermometer. This nonuniqueness is due to the difficulty in producing homogeneous wires from a single heat of Rh–Fe and, even more difficult, production of sufficiently identical alloys on a batch-to-batch basis. Production control related problems may be expected to diminish as experience with this Rh–Fe alloy accumulates.

Temperature dependence of resistivity (nonmetals). The resistive properties of semiconducting materials are attributed to the existence of an energy gap, E_G, between the valence band and the conduction band. The energy levels existing in a pure semiconductor are shown in Fig. 14.12a. At the absolute zero of temperature, the valence band is completely full, the conduction band is completely empty, and the crystal behaves as an insulator. As the temperature increases to a certain level, sufficient thermal energy is added to allow some electrons to jump to the conduction band. Both electrons in the conduction band and holes in the valence band contribute to the overall conduction. This temperature-dependent resistivity is known as *intrinsic* resistivity and is given by

$$\rho = A \cdot \exp(E_G/2kT) \tag{14.31}$$

where E_G represents the energy gap between the

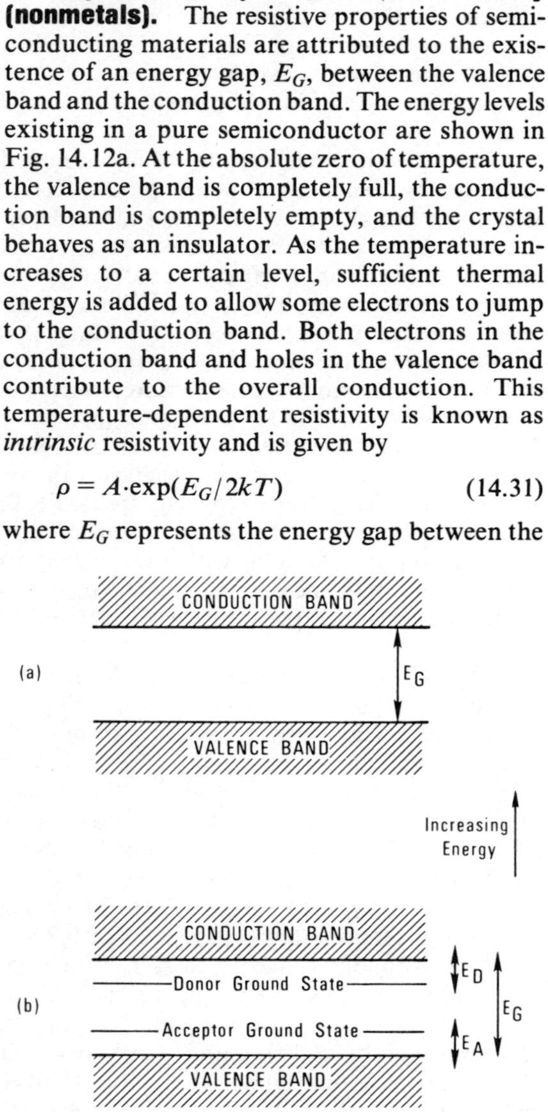

Figure 14.12 Energy levels of semiconductors.

valence and conduction bands (Fig. 14.12a). Addition of impurities such as arsenic or boron to the crystal creates energy levels as shown in Fig. 14.12b. Donor impurities give up electrons to the conduction band while acceptor impurities take electrons from the valence band. Resistivity remains a function of temperature, but the energy gaps E_D and E_A are much smaller than E_G and therefore require less thermal energy to affect the conduction properties.

When donor impurities are dominant over acceptor impurities, the conduction is primarily by electrons in the conduction band, and the resulting semiconductor is referred to as *n*-type. When acceptor impurities are dominant, conduction is primarily due to "holes" in the valence band; this type of semiconductor is known as *p*-type. The complex conduction processes involved in semiconductors make general analytical representation of R vs. T difficult. A complete discussion of the conduction properties of semiconductors is found in Kittel (1967).

Carbon resistance thermometers. In this discussion carbon is considered a semiconductor, since its general thermometric characteristics fall within this category, for example, $dR/dT < 0$. Carbon resistors are used extensively as cryogenic thermometers for temperatures below 100 K. They are inexpensive, rugged, physically small, and have a high sensitivity to temperature change at low temperatures. A very useful attribute of carbon resistors is their relatively weak magnetic field dependence (§ 14.3.6)

Carbon is used in several forms to measure temperature: common electrical circuit resistors (CRTs), colloidal suspensions of carbon painted on various substrates, vacuum-deposited carbon film, and carbon-impregnated porous glass. All exhibit the characteristic negative temperature coefficient of resistance. The primary use of carbon resistance thermometers is for temperatures at and below about 20 K. General use at higher temperatures is limited because of decreasing sensitivity and competition with metallic resistors.

The most frequently used form of the carbon thermometer is the electrical circuit resistor. The temperature dependence of R for a CRT is shown in Fig. 14.13. The CRTs are available from several commercial sources in a wide selection of absolute values and wattage ratings. Rubin (1980) and Anderson (1972) discuss the situation relating to commercial production of

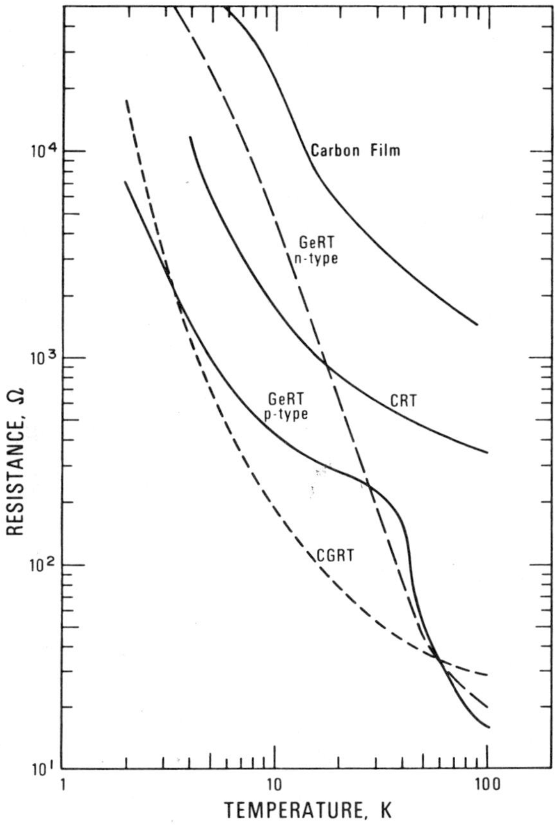

Figure 14.13 Resistance vs. temperature for a 270-Ω (ambient-temperature, 0.1-W CRT) (Sparks, 1974), a 1000-Ω (4.2-K) CGRT (Swartz, Clark, Johns, and Swartz, 1976), *n*-type and *p*-type GeRTs (Scientific Instruments, Inc.) and a vacuum-deposited carbon-film resistor (Collier et al., 1973).

resistors bearing brand names. Usually 0.1- to 1-W resistors with room-temperature resistances of 10 to 500 Ω are used.

The reproducibility of circuit carbon resistors upon thermal cycling is questioned. Since carbon exhibits graphite-like behavior in varying degrees, any lack of reproducibility could be due to the anisotropic electrical conductivity and thermal expansion observed by Friedberg (1955). Clement and Quinnell (1952) found that commercial resistors of various types were reproducible to about 0.1% or ±4 mK when cycled between room temperature and liquid-helium temperature in a vacuum environment. Plumb and Edlow (1959) and Edlow and Plumb (1962) found the short-term reproducibility to be at least this good, but did observe long-term drift when exposed to liquid-helium temperatures for extended periods of time. This drift represented an increase in resistance equivalent to about 4 mK. Johnson and Anderson (1971) also observed drift and concluded that the lack of reproducibility was intrinsic, that is, due to the active resistive element itself, rather than a thermal strain effect that would eventually saturate. Reproducibilities of the order discussed above are not possible if the CRT is exposed to moisture or is heated, for example, with a soldering iron or by baking (Herr, Terbeek, and Tiefermann, 1962). Effects of heating CRTs are confirmed by Johnson and Anderson (1971).

Clement and Quinnell found that over the temperature range 2 to 20 K, the R–T characteristics of circuit-type carbon resistors could be represented by the empirical relationship

$$\log R + D/\log R = A + B/T \tag{14.32}$$

Temperatures calculated from this equation agreed within ±0.5 K for a variety of resistors with different nominal room-temperature resistances. Schulte (1966) used this equation and coefficients $A = 4.478$, $B = 3.091$, and $D = 5.004$ to calculate $R_T/R_{296\text{K}}$ for a 0.1-W, 270-Ω CRT in the temperature range 4 to 296 K. He found the following differences between the calculated and experimental ratios: ±6.8% at 4.2 K, ±2.2% at 20.3 K, and ±0.25% at 77.4 K. These variations correspond to temperature uncertainties of ±110 mK, ±500 mK, and ±500 mK, respectively.

Other functions that may be used to represent the temperature resistance characteristics have been used by Lounasmaa (1958)

$$T + A_1 = A_2/(R_T - R_0) \tag{14.33}$$

by Hetzler and Walton (1968)

$$1/T = A_1 + A_2R + A_3R^{1/2} \tag{14.34}$$

and by Kopp and Ashworth (1972)

$$\log R = A_1 + A_2/T^P \tag{14.35}$$

Equations 14.33 and 14.34 were developed for use in specific temperature ranges: Eq. 14.33, $4 \leq T \leq 12$ K and Eq. 14.34, $0.2 \leq T \leq 3$ K. Equation 14.35 was used for $4 \lesssim T \lesssim 70$ K with uncertainties on the order of ±3 K. Ricketson (1975) found that a function of temperature, $F(T)$, could be determined that represents a particular resistor.

$$F(T) = A_1/(\log R - A_2) \tag{14.36}$$

This function, along with the resistance, determined experimentally at one or two temperatures each time the resistor was thermally cycled to ambient temperatures, was found to represent a particular CRT to within ±10 mK between 2 and 60 K.

Thin films of carbon produced by painting or by high-temperature vacuum deposition have been used as low-temperature thermometers (Star, van Dam, and van Baarle, 1969; Collier, Sparks, and Strobridge, 1973). These resistors can be extremely small with the concomitant low heat capacity and fast response time. Production of film resistors for use as thermometers is not well developed, and their reproducibility reflects this state. A well-conceived development program in this area could result in an extremely useful low-temperature thermometer.

The temperature–resistance characteristics of carbon-film thermometers show the typical negative temperature coefficient of resistance (Fig. 14.13). Star et al. and Collier et al. found that their resistors could be represented by

$$\ln T = \sum_{i=0}^{6} A_i (\ln R_i)^i \tag{14.37}$$

for temperatures up to about 80 K.

A third type of carbon resistance thermometer with a useful temperature range of 2 to 400 K was reported by Lawless (1972a). These thermometers are produced by treating porous alkali-borosilicate glass with organic liquids. The resulting composite consists of carbon filaments in a glass matrix. Their construction is similar to that of precision germanium resistance thermometers (GeRT) — the active resistor is mounted in a configuration that is as strain free as possible and encapsulated in a helium-filled metallic can. They are four-lead devices and are produced specifically for use as thermometers. The dimensions of the cylindrical, encapsulated sensor are a 3.2-mm diameter and 11-mm length. The cost of the sensors is comparable to that of GeRTs and, of course, much higher than the circuit carbon resistors discussed above.

The temperature–resistance characteristics of the carbon–glass resistance thermometers (CGRT) are typified by the negative dR/dT expected for semiconductors. The R-vs.-T behavior is smooth and monotonic (Fig. 14.13), as is the slope dR/dT. This makes calibration and analytical representation easier than for GeRTs. Lawless used Eq. 14.35 to represent three CGRTs in the ranges 2 to 5 K, 2 to 10 K, 2 to 20 K, and 2 to 40 K with a standard deviation of less than 1% of the upper temperature of each interval. This equation was found to represent the CGRTs to within ±0.5 K up to 70 K. The sensors tested by Lawless scaled well enough to allow the equation

$$R = A\,R_j^M \tag{14.38}$$

to be used within the uncertainties given above. Here R_j is the temperature-dependent resistance of a calibrated CGRT and M and A are temperature-independent terms. This allows a CGRT to be calibrated by experimentally determining $R(T)$ at only two temperatures.

Besley (1979) studied the repeatability of six CGRTs at 15, 20, and 27 K after thermally cycling the thermometers. He found that the CGRTs are better in this respect than the circuit resistors and the carbon films; they are, however, a factor of 10 less stable than GeRTs. At 20 K none of Besley's CGRTs was stable within 5 mK, and only one was stable within 10 mK.

In spite of the noted instability, CGRTs provide adequate precision for many users, particularly near 4 K, and usable sensitivity over a much broader temperature range than GeRTs. Their behavior in magnetic fields (§ 14.3.6) is also superior to that of germanium resistors.

Germanium resistance thermometers. Accurate temperature measurement in the temperature range $1.5 \leq T \leq 20$ K can be done routinely using germanium resistance thermometers (GeRTs); thermometers that allow temperatures up to 100 K to be measured with reduced sensitivity are available. Extensive development of these semiconducting devices has led to the most accurate and reproducible empirical thermometer yet devised for $T \leq 20$ K. Selective addition of various impurities (Blakemore, 1962) allows the characteristics of these thermometers to be varied to produce maximum sensitivity in particular temperature ranges. Typical temperature dependences of R for n- and p-type GeRTs are shown in Fig. 14.13.

Commercial units are four-lead devices, as shown in Fig. 14.11c. These thermometers are small (typical size is approximately 13-mm long and 3 mm in diameter) and have small masses with resulting time constants on the order of 0.1 s. The leads that actually contact the germanium crystal are very small, which causes the sensors to burn out if the measuring current is too high. Typical excitation currents are 1 μA for $T < 2$ K, 10 μA for $2 < T \leq 15$ K, 100 μA for $15 < T \leq 40$ K, and 1 mA for $40 < T < 100$ K. These thermometers are used with potentiometric circuits, as shown in Fig. 14.5. When using dc, the average voltages with forward and reverse current are needed to eliminate spurious emf's.

The reproducibility of doped, single-crystal GeRTs has been established on units from various sources by Lindenfeld (1961), Kunzler, Geballe, and Hull (1962), Edlow and Plumb (1966), and Besley and Plumb (1978). The reproducibility of the thermometers at 4 K is as good as the present ability to check them, that is, within a few tenths of a millidegree. Besley and Plumb established the reproducibility of 30 GeRTs at 20.28 K where the reduced sensitivity makes small resistance instabilities more significant. At this temperature and after 100 thermal cycles between 20 and 100 K, they found instabilities ranging from 0.1 to 20 mK with 90% of the units being stable to better than 2 mK.

The major disadvantage of GeRTs is the lack of a simple analytical representation. Each thermometer must be calibrated by comparison at many points in the range of interest if the inherent reproducibility of the thermometer is to be utilized. The fitting procedure is further complicated by the difference in R-vs.-T characteristics for n-type and p-type germanium (Fig. 14.13). Conduction in the n-type changes gradually from impurity conduction at low temperatures to free electron conduction at higher temperatures; this gradual transition results in the smooth R-vs.-T curve shown in this figure. The transition is not gradual in the p-type germanium, however, and a region of lower sensitivity appears in the R-vs.-T curve.

The most commonly used analytical forms, used with both n- and p-type thermometers, are

$$\log R = \sum_{j=0}^{m} A_j (\log T)^j \qquad (14.39)$$

and

$$\log T = \sum_{i=0}^{m} B_i (\log R)^i \qquad (14.40)$$

Both equations work well in the range $T \lesssim 15$ K, but Eq. 14.39 is superior in the range from 15 to 100 K. Application of Eq. 14.39 over the entire range from 1 to 100 K results in oscillations amounting to 0.3% of the absolute temperature. If the range is divided, $T < 20$ K and $T \gtrsim 15$ K, these oscillations are reduced to a few parts in 10^4 of temperature. The fit was usually optimized when $m = 7$ to 9 for $T < 20$ and $m = 5$ to 6 for $15 \lesssim T \lesssim 100$ K (Blakemore, Winstel, and Edwards, 1970; Blakemore, 1972). Other representations have been developed. Ahlers and Macre (1966) studied the polynomial

$$(\ln R/T)^{1/2} = \sum_{j=0}^{m} A_j (\ln R)^j \qquad (14.41)$$

and found uncertainties of 150 mK when $1.2 < T < 20$ K, but for $m = 5$ uncertainties were less than 25 mK for $15 < T < 80$ K. Equation 14.41 and four other polynomials were analyzed by Catalano, Shroyer, and English (1970). They assessed the ability of small digital computers to handle the polynomials necessary to represent the R-vs.-T characteristics of GeRTs. An alternative to polynomial fitting was proposed by Penar and Campi (1971). Their scheme involves interpolating between neighboring calibration points with the equation

$$\ln R = A/T^B \qquad (14.42)$$

This method requires only a desk calculator and reproduces calibrated data within the accuracy of the calibration. The authors do not indicate the allowable temperature interval between calibration points.

Other nonmetallic resistance thermometers. Thermistors (thermally sensitive resistors) are essentially resistors made of sintered metallic oxides. Their temperature–resistance relationship has a negative slope, much like that of carbon or germanium resistors. Features that make these devices useful thermometers include: (a) they are small, which tends to make the time response significantly less than 1 s, (b) they are high resistance units, which reduces the overall effect of lead resistances, (c) their temperature–resistance characteristics are dependent on materials and manufacturing processes, which allows thermometers to be engineered to be particularly sensitive in limited ranges of temperature, and (d) they are less sensitive to magnetic fields than several other thermometers (see § 14.3.6)

There are, of course, disadvantages. Items (b) and (c) above may be disadvantages as well as advantages: a given thermistor is not usable over a wide range of temperatures because its resistance becomes exceedingly high. To the first approximation, the temperature–resistance characteristics are represented by an equation of the form given in Eq. 14.31 (Sachse, 1962; Sachse, 1975). Reproducibilities experimentally determined by Sachse were about ±30 mK after cycling between room temperature and liquid-oxygen temperature. After 1000 cyclings, the error was on the order of a few kelvins. Schlosser and Munnings (1972) found even better reproducibility for a particular type of thermistor. Droms (1962) presents a detailed study of bridge circuits to be used with this type of resistor.

14.3.3 Thermocouples

In 1823, Seebeck found that an electrical current would flow when two dissimilar wires were joined at both ends and subjected to a temperature gradient between the two junctions. This phenomenon now bears his name — Seebeck effect. Later, Peltier showed that when electrical current flows across a junction of dissimilar metals, heat was either liberated or absorbed. The Peltier coefficient, π, is given by

$$\pi = TS \tag{14.43}$$

where T is the absolute temperature and S is the absolute thermopower. Except when one of the thermocouple materials is superconducting ($S = 0$ for materials in the superconducting state), S is actually the net thermopower of the materials making up the junction, that is, $S = S_{12} = S_1 - S_2$. Sir William Thomson analyzed the Seebeck circuit, treating it as a reversible heat engine with the Peltier voltage as the driving force. He discovered that an additional reversible effect was needed because the observed thermal voltage was not proportional to the temperature difference. The second reversible phenomenon was found to exist in a homogeneous conductor when electrical and thermal gradients were simultaneously applied. Heat, the Thomson heat, is liberated or absorbed along the length of the conductor under these conditions. This effect differs from the Peltier effect in that it occurs in a continuous, homogeneous conductor. The Thomson coefficient, μ, is defined as

$$\mu = T(dS/dT) \tag{14.44}$$

If μ is known, the absolute thermopower of a material may be computed using

$$S(T) = S(T) - S(0) = \int_0^T (\mu/T)dT \tag{14.45}$$

since the third law of thermodynamics requires that $S \rightarrow 0$ as $T \rightarrow 0$. Borelius, Keesom, Johansson, and Linde (1931) determined μ for lead using calorimetric methods. More recent discussions concerning the accuracy of these measurements are given by Pearson and Templeton (1955), Jan, Pearson, and Templeton (1958), and Roberts (1977).

Sign conventions have been adopted to determine the direction of current flow in a thermoelectric circuit. The convention is: if current flows from material A to material B at the cooler of the two junctions, then material A is thermoelectrically positive with respect to material B (Fig. 14.14a).

There are three empirical laws that govern the use of thermocouples as thermometers (Dike, 1954):

a. *Law of the homogeneous circuit.* No current will flow in a thermoelectric circuit (such as Fig. 14.14a) if material A and B are identical regardless of the magnitude of the temperature gradient or its distribution along the wires. This is presented graphically in Fig. 14.15a.

b. *Law of intermediate materials.* A third material introduced into a thermoelectric circuit (as shown in Fig. 14.14b) will have no thermoelectric effect if it exists entirely in an isothermal region. This law is shown graphically in Fig. 14.15b.

c. *Law of intermediate temperatures.* If a thermocouple pair generates a voltage V_1 when its junctions are at T_1 and T_2, and a voltage V_2 when its junctions are at T_2 and T_3, then it will generate the voltage $V_{12} = V_1 + V_2$ when its junctions are at T_1 and T_3.

A full understanding of these three empirical laws will allow successful temperature measurement with thermocouples. For instance, application of the second law ensures that a feedthrough or disconnect (being an intermediate material) will produce no thermoelectric effect if it is isothermal.

The types of thermocouples commonly used in low-temperature thermometry are the standard-

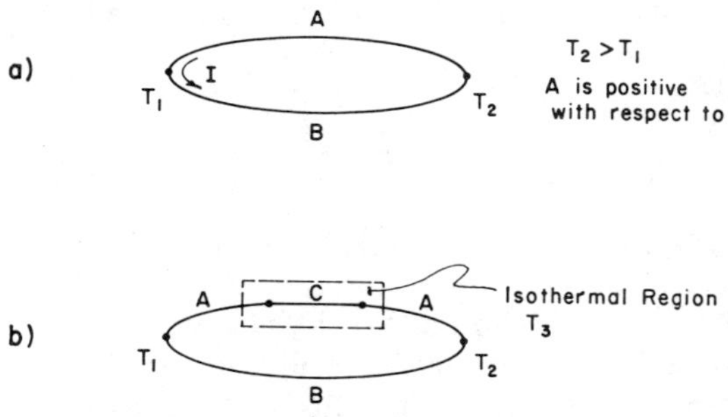

Figure 14.14 Schematic of two thermocouple circuits: (a) current flows from the positive to negative material at the cooler of the two junctions, (b) introduction of a third material, C, into an isothermal part of the thermocouple circuit has no effect on the output.

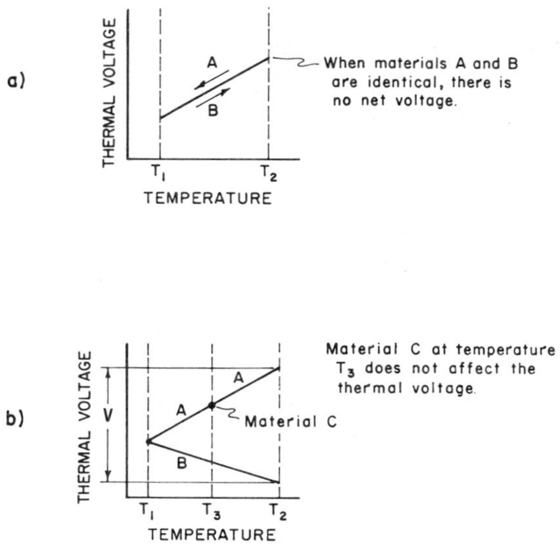

Figure 14.15 Graphic analysis of the first two laws of thermoelectricity. (a) Law of the homogeneous circuit — no net emf is developed when materials A and B are identical. (b) Law of intermediate materials — material C causes no change in the thermoelectric characteristics of the circuit because it is contained entirely in an isothermal region.

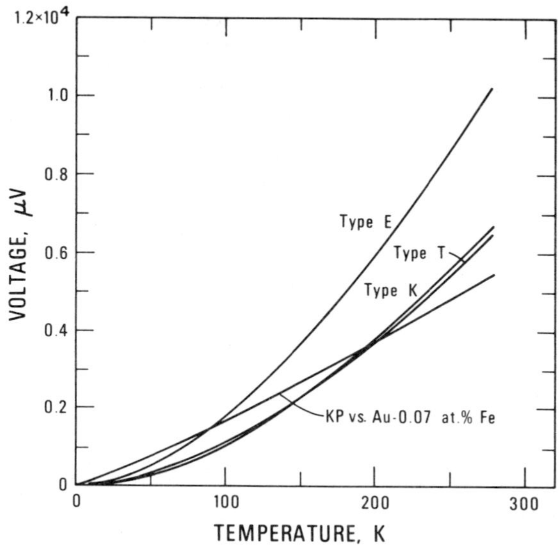

Figure 14.16 Thermoelectric voltage vs. temperature for thermocouple types E, K, T, and KP vs. Au–0.07 at.% Fe.

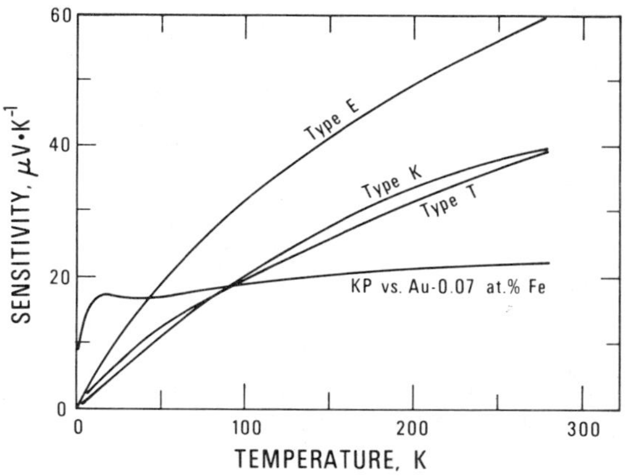

Figure 14.17 Thermoelectric sensitivity vs. temperature for thermocouple types E, K, T, and KP vs. Au–0.07 at.% Fe.

ized types — Chromel vs. constantan (type E), Chromel vs. Alumel (type K), copper vs. constantan (type T), and combinations utilizing dilute noble metal alloys as the negative thermoelements. The thermoelectric voltage and sensitivity for these combinations are shown in Figs. 14.16 and 14.17, respectively. The letter designations are those recommended by the American Society for Testing and Materials (ASTM). The letter type establishes thermoelectric properties, not exact chemical composition. The thermocouple combinations indicated by the letter designations are given in Table 14.10 along with the letter designations for the individual wires making up the thermocouples. Limits of error for the standardized thermocouple pairs are given in Table 14.11. Further discussions of the standardized materials are found in Powell, Hall, Hyink, Sparks, Burns, Scroger, and Plumb (1974).

Thermocouple calibration. Table 14.11 indicates the limits of error for the standardized thermocouple types. For more accurate temperature determination, it is necessary to determine the actual emf differences that exist between the working thermocouple and the standard data. The accuracy of a thermocouple depends not only on this working calibration, but also on the degree of inhomogeneity existing in the thermocouple wires. These thermoelectric inhomogeneities can be caused by chemical and physical imperfections, and they should be evaluated prior to calibration of the working thermocouple. Tests to evaluate these voltages serve two purposes — localized and long-range inhomogeneities in the thermocouple wires can be detected, and the magnitude of random errors due to changing temperature profiles along the wires can be determined.

Table 14.10 Material designations, temperature ranges, and single-wire component designations for standardized (ASTM) thermocouple combinations.

Thermocouple Combinations

Type Designation	Temperature Range (K)	Materials
B	273–2093	Pt–30 wt.% Rh vs. Pt–6 wt.% Rh
E	3–1273	Ni–Cr alloy vs. a Cu–Ni alloy
J	63–1473	Fe vs. a Cu–Ni alloy
K	3–1645	Ni–Cr alloy vs. Ni–Al alloy
R	223–2040	Pt–13 wt.% Rh vs. Pt
S	223–2040	Pt–10 wt.% Rh vs. Pt
T	3– 673	Cu vs. Cu–Ni alloy

Single-leg Materials

N	The negative wire in a combination
P	The positive wire in a combination
BN	Pt–nominal 6 wt.% Rh
BP	Pt–nominal 30 wt.% Rh
EN or TN	Nominally 55 wt.% Cu, 45 wt.% Ni, constantan: Cupron,[a] Advance,[b] ThermoKanthal JN[c]
EP or KP	Nominally 90 wt.% Ni, 10 wt.% Cr: Chromel,[d] Tophel,[a] ThermoKanthal KP[c]
JN	A Cu–Ni alloy similar to, but not always interchangeable with, EN and TN
JP	Nominally 99.5 wt.% Fe: ThermoKanthal JP[c]
KN	Nominally 95 wt.% Ni, 2 wt.% Al, 2 wt.% Mn, 1 wt.% Si: Alumel[d], Nial[a], T-2[b], ThermoKanthal KN[c]
RN and SN	High purity platinum
RP	Pt–13 wt.% Rh
SP	Pt–10 wt.% Rh
TP	Copper, usually electrolytic tough pitch

[a]Trademark of Wilbur B. Driver Co.
[b]Trademark of Driver Harris Co.
[c]Trademark of Kanthal Corp.
[d]Trademark of Hoskins Manufacturing Co.

The first law of thermoelectricity indicates that no thermal voltage is developed when a loop of homogeneous wire is subjected to a temperature gradient. Similarly, no voltage is generated when two identical wires are joined and the pair of wires is placed in a temperature gradient. The problem in practical thermometry is, however, that the ideal characteristics "homogeneous" and "identical" are not sufficiently approximated for real thermocouple materials. Actually, a loop of wire placed in a large temperature gradient will usually produce a thermoelectric voltage, sometimes as large as 10 μV. A thermocouple made from wires of the same nominal composition but from different spools may generate a significant thermal voltage when placed in a large temperature gradient. Readings as large as hundreds of microvolts have been observed for poorly controlled alloys (Sparks and Powell, 1972).

Test probes that may be used in cryogenic

Table 14.11 Limits of error (ASTM E230) for the standardized thermocouple types.

Thermocouple Type	Temperature Range K	°F	Limits of Error[a] Standard	Special[b]
B	1144–1978	1600–3100	±0.5%	
E	273–589	32–600	±3° F	±2.25° F
	589–1144	600–1600	±0.5%	±0.375%
J	273–550	32–530	±4° F	±2° F
	550–1033	530–1400	±0.75%	±0.375%
K	273–550	32–530	±4° F	±2° F
	550–1533	530–2300	±0.75%	±0.375%
R or S	273–811	32–1000	±2.5° F	
	811–1755	1000–2700	±0.25%	
T	89–214	−300– −75		±1.0%
	172–214	−150– −75	±2.0%	±1.0%
	198–366	−75–200	±1.5° F	±0.75° F
	366–644	200–700	±0.75%	±0.375%

[a] Where limits of error are given in percent, the percentage applies to the temperature in Fahrenheit. The limit of error in kelvins or degrees Celsius is obtained by multiplying the Fahrenheit result by 5/9.
[b] The Special designation refers to selected batches of materials whose thermoelectric characteristics closely matched the ASTM standard thermoelectric properties.

baths (Sparks and Powell, 1972) to determine short-range (length of wire < 5 m), long-range (wire from same melt, but different spools), and interlot inhomogeneities are shown in Fig. 14.18a and b. Spurious voltages tend to increase in going from the short-range to the interlot category. Spurious voltages have been determined for several of the thermocouple materials used at low temperatures, and the results are summarized in Table 14.12.

It is good practice to compare the front and back ends of a spool, or a lot of material, to determine if thermoelectric variations of the material are within acceptable limits. If the material is sufficiently homogeneous, then a single calibration will suffice for all thermocouples made from that spool. If thermoelectric differences are beyond acceptable limits, each separate part of the spool or lot must be calibrated. For relatively inhomogeneous material or extreme accuracy requirements, separate calibrations should be made for each length of wire.

The calibration procedure for a particular thermocouple depends on the application. If temperature differences are needed only to 1 or 2%, the standard ASTM calibration tables can be used (see limits of errors in Table 14.11). To obtain a more exact calibration of a particular thermocouple, the output of the thermocouple between known temperatures must be determined and used to establish the difference between the standard data and the working thermocouple.

Calibration can be done using a type 3 probe (Fig. 14.18c). This probe is used to obtain the thermoelectric emf's generated between fixed temperatures, for example, one leg in liquid helium (~ 4 K) and the other in liquid nitrogen (~ 76 K), or one leg in liquid nitrogen and the other in an ice bath (273.15 K). An alternative to the fixed-point type of calibration is measurement of the emf's that result when one leg of the thermocouple is at a fixed temperature and the other is at variable temperatures, which are determined by a calibrated thermometer. In either case, the observed voltages, V_{obs}, are compared

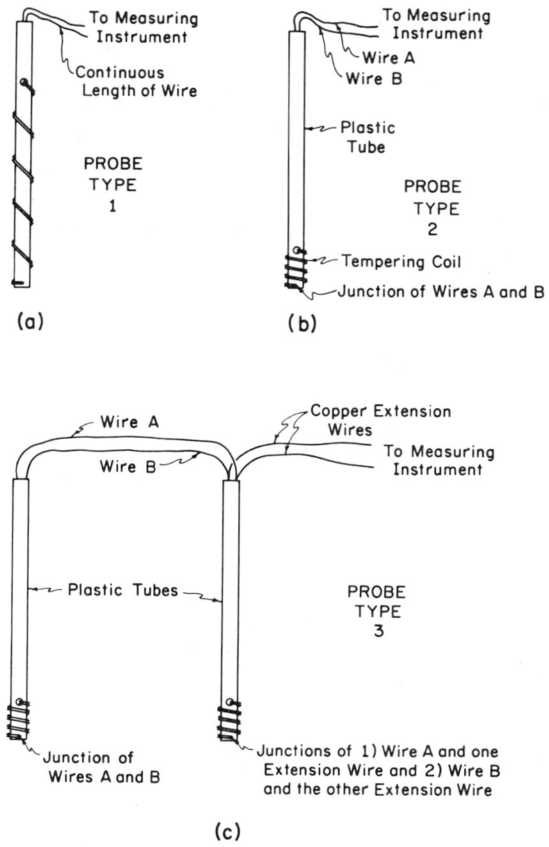

Figure 14.18 (a) Short-range inhomogeneity probe. (b) Medium-range, long-range, and interlot homogeneity probe. (c) Differential thermocouple probe.

Table 14.12 Average equilibrium values for thermocouple inhomogeneity voltages[a] in liquid nitrogen (LN_2) and liquid helium (LHe) (Sparks and Powell, 1972).

Material	Short Range LN_2 (μV)	Short Range LHe (μV)	Medium Range LN_2 (μV)	Medium Range LHe (μV)	Different Spools LN_2 (μV)	Different Spools LHe (μV)
EP or KP	1.4	1.6	2.6	2.6	23.5	26.2
JP (iron)	4.8	5.2	10.8	10.8	35.0	36.2
TP (copper)	0.3	0.9	1.2	4.2	2.3	19.2
KN	1.5	1.7	2.9	3.2	16.1	17.6
EN or TN (constantan)	2.1	2.3	5.1	5.6	22.9	26.2

[a]Each voltage represents the average maximum for several spools from the same manufacturer.

with the standard data, V_{std}, and the relationship $\Delta V = V_{std} - V_{obs}$ is calculated at each calibration temperature. A plot of ΔV vs. V_{obs} allows ΔV to be determined at any intermediate experimental voltage. The true temperature is determined by algebraically combining ΔV and V_{std} and finding the temperature in the standard data that corresponds to the adjusted voltage. For computerized analysis, the ΔV-vs.-T relationship is frequently fitted and algebraically combined with a polynomial representing the standard data. The accuracy of the working calibration depends on the number of calibration points and on the accuracy of calibration temperatures. Maximum calibration accuracy is obtained by determining the thermoelectric voltage of the thermocouple after it has been installed in the system where it will be used.

Thermocouple materials are frequently compared with standard reference materials to allow control and comparison of individual thermoelectric characteristics. The wires making up type E thermocouples, for instance, (see Table 14.10) are EP and EN. The exact compositions of EP and EN are not specified and vary slightly from manufacturer to manufacturer and even from batch to batch from a single producer. ASTM standard data pertain to the net thermoelectric characteristics of the EP-EN combination, type E. Thermoelectric comparisons of EP and EN to materials whose thermoelectric properties are constant (reference materials) allow the individual wire characterisitics to be maintained. A standard reference material (*Ag*–28 at.% Au) has been developed (Sparks and Hust, 1972) for use in comparing thermocouple materials when $T \lesssim$ 76 K. For $T \gtrsim$ 76 K, platinum may be used as the reference material (Powell, Sparks, and Hust, 1978). Standard data for the letter-designated thermocouple wires vs. these standard reference materials are found in Sparks, Powell, and Hall (1972) and Powell et al. (1974).

Type E thermocouples. The ATSM designation type E indicates a thermocouple pair consisting of a Ni–Cr alloy and a Cu–Ni alloy (the common trade names and nominal compositions for these alloys are given in Table 14.10). This type of thermocouple has the highest Seebeck coefficient, S, of the three ASTM standard thermocouple types commonly used at low temperatures, types E, K, and T. Also, both elements of this thermocouple have low thermal conductivity, reasonable homogeneity, and corrosion resistance in moist atmospheres. This type is the best thermocouple to use for temperatures down to about 40 K. There is an unfortunate ambiguity concerning the material commonly referred to as constantan (EN or TN). Iron vs. constantan (JP vs. JN) is often used for above-ambient temperature measurement. The negative element, JN, is not the same material as EN (or TN) and, generally, cannot be used interchangeably.

Type E thermocouples are recommended by the ASTM for use in the temperature range 3 to 1144 K in oxidizing or inert atmospheres. The negative wire, EN, is subject to deterioration above about 1144 K, but the thermocouple may be used up to 1273 K for short periods. The use of type E is subject to the following restrictions:

a. It should not be used in sulphurous, reducing, or alternately reducing and oxidizing atmospheres unless suitably protected with sheaths.

b. It should not be used in vacuum (at high temperatures) for extended periods of time because the chromium in the positive wire vaporizes and alters the composition of the Ni–Cr solution.

c. It should not be used in atmospheres that promote "green-rot" corrosion (those with low, but not negligible, oxygen content).

Neither thermoelement of type E thermocouples is sensitive to minor changes in composition or impurity level because both are already heavily alloyed. Also, they are not sensitive to minor differences in heat treatment (provided that the treatment complies with the restrictions mentioned above). For most general applications, they may be used with the heat treatment routinely used by the wire manufacturers. When the highest accuracy is sought, additional preparatory heat treatments may enhance their performance. Additional details of type EP (or KP) thermoelements are given by Potts and McElroy (1962).

Type K thermocouples. The ATSM designation type K indicates a thermocouple pair consisting of a Ni–Cr alloy and a Ni–Al alloy (the common trade names and nominal compositions for these alloys are given in Table 14.10). The sensitivity is only about half that of the type E combination at 20 K (4.1 $\mu V \cdot K^{-1}$ compared with 8.5 $\mu V \cdot K^{-1}$). The negative element is a bit more homogeneous than the EN element. Both materials have low thermal conductivity and are corrosion resistant in moist atmospheres.

Type K thermocouples are recommended by

the ASTM for continuous use at temperatures within the range 3 to 1533 K in inert atmospheres. Both the KP and the KN thermoelements are subject to oxidation when used in air above about 1123 K. Even so, type K thermocouples may be used at temperatures up to about 1623 K for short periods with only small changes in calibration. When oxidation occurs, a gradual increase in the thermoelectric voltage is observed. The magnitude of the change in the thermoelectric voltage depends on such factors as the temperature, time at temperature, diameter of thermoelement, and conditions of use. The thermoelectric instability of type K thermocouples in air at elevated temperatures has been studied by Burley and Acklund (1967). Restrictions on the use of type K are the same as those listed for type E thermoelements. Details on the use and behavior of type K thermocouples are given by Burley (1969, 1972).

Type T thermocouples. The ASTM designation for type T indicates a thermocouple pair consisting of Cu and Cu–Ni alloy (the common trade names and nominal composition for the alloy are given in Table 14.10). Type T is one of the older more popular combinations, and is the only one of the standardized types for which limits of error below 273.15 K have been established (Table 14.11). As mentioned in the discussion of type E, the negative element TN is thermoelectrically different than JN, even though both are called constantan.

Type T thermocouples are recommended by the ASTM for use in the temperature range from 89 to 644 K in vacuum or in oxidizing, reducing, or inert atmospheres. Later research (Sparks et al., 1972) indicates that they can be used down to about 20 K. There are, however, several reasons for not using type T at very low temperatures: (a) The sensitivity of type T at 20 K is lower than that of type E (4.6 $\mu V \cdot K^{-1}$ compared with 8.5 $\mu V \cdot K^{-1}$). (b) The thermoelectric properties of the TP element become quite dependent upon trace impurities of iron at temperatures below about 76 K. (c) The thermal conductivity of the TP element is much higher than that of the alloys in the type E and K combinations, as shown in Fig. 14.2.

Gold-iron alloy thermocouples. The increasing use of liquid hydrogen and liquid helium has created a demand for specialized thermometry below 25 K. Ordinary thermocouple combinations are only marginally acceptable because of their low sensitivity in this range. Dilute alloys of noble metals and transition metals, however, have relatively high temperature sensitivity below 25 K; Au–2.1 at.% Co is perhaps the best known of this type. Unfortunately, this alloy is a supersaturated solid solution. Powell, Caywood, and Bunch (1962) found that the thermoelectric power decreases with time when stored at room temperature. This gradual change is probably due to diffusion of cobalt atoms to grain boundaries. Dilute alloys of iron in gold are metallurgically stable and have extremely useful thermoelectric properties at very low temperatures. A differential thermocouple made with Au–0.02, 0.03, or 0.07 at.% Fe as the negative element and copper, "normal" silver (Ag–0.37 at.% Au), or KP as the positive element provides a usable sensitivity, even below 4 K.

Trace amounts of transition elements in noble metal solvents cause anomalously high thermoelectric power and a minimum in the electrical resistivity (Kondo effect) at low temperatures. Alloys of this type have been extensively studied to understand the bulk transport properties. Development of the Au–Fe alloys for use in low-temperature thermocouple thermometry began when Berman and Huntley (1963) tested Au–0.02 at.% Fe for possible use in their thermal conductivity apparatus.

Several necessary thermoelectric properties have been determined for the Au–Fe alloys: reproducibility after repeated thermal cycling (Finnemore, Ostenson, and Stromberg, 1965), behavior in a magnetic field (Berman, Brock, and Huntley, 1964), and the effect of heat treatment (Rosenbaum, 1968; 1969). It appears that sufficient information is available to establish the Au–Fe alloys as the most promising thermoelement available for use at very low temperatures.

Sparks and Powell (1973) calibrated several thermocouple combinations involving the dilute Au–Fe alloys over the temperature range 4 to 280 K. The types of wire commercially available in this country are the 0.02 and 0.07 at.% alloys of iron in gold. The annealing procedure for these alloys is rather critical, as illustrated in the comparison of sensitivity data in Fig. 14.19. The anneal used by Sparks and Powell was 623 K in air for 20 min. If the annealing temperature exceeds about 623 K in air, the iron oxidizes, resulting in a greatly reduced thermopower. Higher annealing temperatures are possible in vacuum; however, since the anneal is to relieve residual stress, the 623-K temperature is adequate. Berman and Huntley found the repeatability of the thermo-

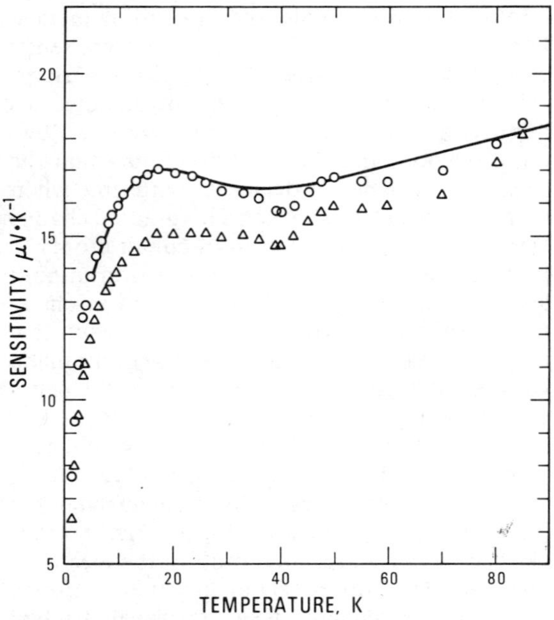

Figure 14.19 Sensitivity as a function of temperature for KP vs. Au–0.07 at.% Fe in the annealed and unannealed states. ○ — annealed (Rosenbaum, 1968); △ – hard worked (Rosenbaum, 1968); solid line – annealed (Sparks and Powell, 1973).

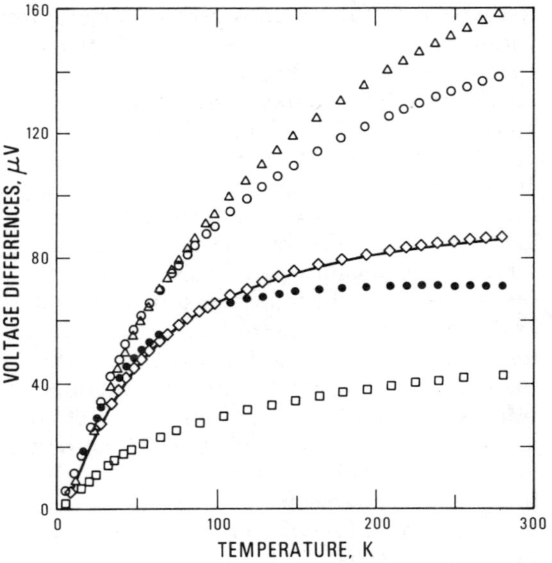

Figure 14.20 Voltage differences vs. temperature between a particular Au–0.07 at.% Fe specimen and six other Au–0.07 at.% Fe specimens.

power of a single piece of Au–Fe wire to be approximately 0.2%. The variability between wires from different manufacturers and different melts is indicated in Fig. 14.20, which shows the experimentally determined differences among seven specimens of Au–0.07 at.% Fe wire (Sparks and Powell, 1973).

Thermocouple reference data. Reference data for thermocouple types E, K, and T over the entire usable temperature range are given by Powell et al. (1974); data for KP vs. Au–0.07 at.% Fe, KP vs. Au–0.02 at.% Fe, Cu (TP) vs. Au–0.07 at.% Fe, Cu vs. Au–0.02 at.% Fe, normal silver (Ag–0.37 at.% Au) vs. Au–0.07 at.% Fe, and normal silver vs. Au–0.02 at.% Fe are given by Sparks and Powell (1973). The low-temperature tables are based on data from a thermocouple thermometry program at the National Bureau of Standards at Boulder.

The Seebeck voltage (μV) as a function of temperature (K) is represented by

$$V=\sum_{i=1}^{n} A_i T^i \qquad (14.46)$$

where n represents the number of coefficients for the particular combination and $T \leq 280$ K. The power series coefficients, A_i, used to represent these combinations, $V=f(T)$, are given in Table 14.13 with a 0 K reference temperature. The experimental range for the data is $5 \leq T \leq 280$ K. Any extension of the power series representation for $T > 280$ K is an extrapolation, and all of the uncertainties inherent in such extensions of experimental data must apply. Extension of the data to $T < 5$ K is more acceptable since the constraint $V = 0$ V when $T = 0$ K was used in the original fit. Similar coefficients for $T > 280$ K are given by Powell et al. (1974).

Coefficients representing $T = f(\text{V})$ are often needed and are also given by Powell et al. (1974) for types E, K, and T in reduced temperature ranges. These power series coefficients are given for second, third, and fourth order equations with maximum uncertainties indicated for each order and temperature range.

14.3.4 Filled systems

Filled systems refer to two types of temperature sensors: vapor pressure thermometers and constant-volume gas thermometers. Although the development of resistance and thermocouple thermometers and rapid electrical measurement systems has decreased the use of filled ther-

Table 14.13 Power series coefficients for representation of thermoelectric voltage, $V(T)$, in the range 0 to 280 K with a 0 K reference temperature.

Power Series Coefficients	T	E	K	KP vs. Au-0.07 at.% Fe
A_1	$-3.9974007864 \times 10^{-1}$	$-2.0344697205 \times 10^{-1}$	$2.4061140104 \times 10^{-1}$	6.9864426367
A_2	$2.6329515981 \times 10^{-1}$	$3.0220985715 \times 10^{-1}$	$7.3438313272 \times 10^{-2}$	$9.0607276605 \times 10^{-1}$
A_3	$-9.6491216443 \times 10^{-3}$	$-5.7844373965 \times 10^{-3}$	$1.2873437647 \times 10^{-3}$	$-4.3469694773 \times 10^{-2}$
A_4	$3.8973308068 \times 10^{-4}$	$1.7879650162 \times 10^{-4}$	$-2.2622572598 \times 10^{-5}$	$1.2468246660 \times 10^{-3}$
A_5	$-9.8186150331 \times 10^{-6}$	$-3.6597667313 \times 10^{-6}$	$2.1765238991 \times 10^{-7}$	$-2.3500537590 \times 10^{-5}$
A_6	$1.6059280063 \times 10^{-7}$	$4.9073685405 \times 10^{-8}$	$-1.3304091711 \times 10^{-9}$	$3.0837610415 \times 10^{-7}$
A_7	$-1.7932074012 \times 10^{-9}$	$-4.4751468891 \times 10^{-10}$	$5.2493539029 \times 10^{-12}$	$-2.9032251684 \times 10^{-9}$
A_8	$1.4080710479 \times 10^{-11}$	$2.8331235582 \times 10^{-12}$	$-1.2997123230 \times 10^{-14}$	$1.9881512159 \times 10^{-11}$
A_9	$-7.8671373053 \times 10^{-14}$	$-1.2476596612 \times 10^{-14}$	$1.8403309812 \times 10^{-17}$	$-9.9174829612 \times 10^{-14}$
A_{10}	$3.1144995156 \times 10^{-16}$	$3.7536769066 \times 10^{-17}$	$-1.1382797374 \times 10^{-20}$	$3.5645229362 \times 10^{-16}$
A_{11}	$-8.5433550766 \times 10^{-19}$	$-7.3627479508 \times 10^{-20}$		$-8.9864698504 \times 10^{-19}$
A_{12}	$1.5448411036 \times 10^{-21}$	$8.4898427718 \times 10^{-23}$		$1.5071673023 \times 10^{-21}$
A_{13}	$-1.6565456476 \times 10^{-24}$	$-4.3671808488 \times 10^{-26}$		$-1.5093916059 \times 10^{-24}$
A_{14}	$7.9795893156 \times 10^{-28}$			$6.8264293980 \times 10^{-28}$

mometers, they continue to offer advantages in particular situations. Both the vapor pressure and gas thermometer depend on measuring the gas pressure of a system. The pressure of an ideal gas is directly related to the thermodynamic temperature so that no calibration is required.

Vapor-pressure thermometers. Vapor-pressure thermometry utilizes the nonlinear relationship between the vapor pressure and the temperature of the saturated liquid-vapor phase of a pure gas. This type of thermometry allows high accuracy and precision, good time response, and requires no calibration. The primary disadvantage is that it can be used only between the triple point and critical point of the fill liquid. The useful ranges and approximate sensitivities are shown in Table 14.14. It is seen from this table and from Fig. 14.4 that there are gaps in the measurable ranges from 5.2 to 13.8 K and from 44.4 to 54.35 K.

The most simple and most often used vapor-pressure–temperature determination is made by measuring the pressure over a liquid surface where the liquid may or may not be confined (Fig. 14.21a). Temperature measured in this way indicates the temperature of the liquid surface and can be significantly different from temperatures existing in the bulk liquid. This is due to the effect of stratification and hydrostatic head, which is both gravity and magnetic field dependent (Sample and Rubin, 1978). Also, the purity of the fluid in a situation such as this is generally questionable. The vapor pressure represents the total pressure of the gas at the liquid surface, which is, in turn, made up of the partial pressures of all the gases present, that is, $P_{\text{total}} = P_{\text{fill fluid}} + P_{\text{residual gases}}$.

Vapor-pressure systems built specifically for temperature measurement are represented schematically in Fig. 14.21b. The accuracy of temperatures determined using a system such as this, as-

Table 14.14 The temperature and sensitivity, dP/dT, are given at the triple point, the normal boiling point, and the critical point of helium-4, equilibrium hydrogen, neon, nitrogen, and oxygen.

Substance	Triple Point Temperature (K)	Triple Point dP/dT (kPa·K^{-1})	Boiling Point Temperature (K)	Boiling Point dP/dT (kPa·K^{-1})	Critical Point Temperature (K)	Critical Point dP/dT (kPa·K^{-1})
He^4	2.177 (lambda)	10.6	4.224	93.1	5.201	114.8
H_2 (99.79% para)	13.803	4.6	20.268	32.4	32.976	133.5
Ne	24.54	16.0	27.09	30.6	44.4	334.0
N_2	63.148	2.7	77.347	11.5	126.2	141.1
O_2	54.351	0.05	90.18	10.3	154.576	182.7

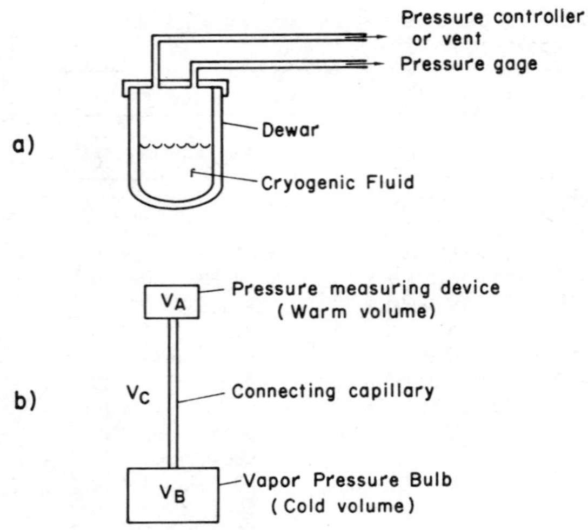

Figure 14.21 Schematic representation of two general vapor-pressure systems. (a) The ullage pressure in a cryogenic system is used to determine the cryogen temperature. (b) System specifically designed for vapor-pressure thermometry.

suming the system is properly filled, is limited by the pressure measurement accuracy, knowledge of the pressure–temperature relationship for the fill substance, and the purity of the fill substance. In most applications, the purity and pressure measurement are the principal limiting factors.

The vapor pressure of liquid helium is only negligibly affected by impurities. Only residual helium affects the vapor pressure of hydrogen. As shown by Woolley, Scott, and Brickwedde (1948), the ortho–para concentration is important in hydrogen. The vapor pressure of the nitrogen and oxygen are significantly altered by impurities; the primary impurity in nitrogen is oxygen and vice versa. Chemical preparation or high-purity cylinder gas is recommended when either is used as a fill gas. The vapor bulb system should be carefully cleaned and evacuated while being heated to remove residual gases.

The temperature indicated by a vapor-pressure measurement is the lowest temperature that contacts the thermometer system, (the vapor pressure bulb and the connecting line). Since the surface of a cryogenic liquid is usually at a lower temperature than the bulk liquid, vacuum insulation or electrical heating coils are necessary on the connecting line.

It is very easy to get erroneous readings from over-fill and fade-out. *Over-fill* is the condition that exists when enough gas has condensed to completely fill the vapor bulb and a portion of the connecting line. The temperature indicated is the temperature of the liquid–vapor interface. *Fade-out* is the condition that exists when the temperature and pressure of the system are such that there is no liquid phase; the vapor-pressure system becomes a gas thermometer in this situation. Fade-out is illustrated for temperatures greater than T_f in Fig. 14.22. This condition is particularly treacherous because there is no obvious transition, and the pressure readout does not appear much different. Careful design and filling procedures can eliminate these problems. Bulb design and filling procedures are discussed by Sparks (1974).

The response time of a vapor-pressure system depends on the rate that the system can transfer energy to and from the bath and the specific heat of the bulb. A small bulb with light, highly conductive walls requires minimum energy from the bath to alter the vapor pressure. A high ratio of outside surface area to bulb volume promotes good heat-transfer characteristics. A well-designed system in a moving liquid environment can have a response time (63.2% of a temperature step function) of less than 0.1 s.

Gas thermometers. State of the art gas thermometry falls into the category of fundamental thermometers rather than empirical thermometers and is much too demanding for common use. Only a few national laboratories

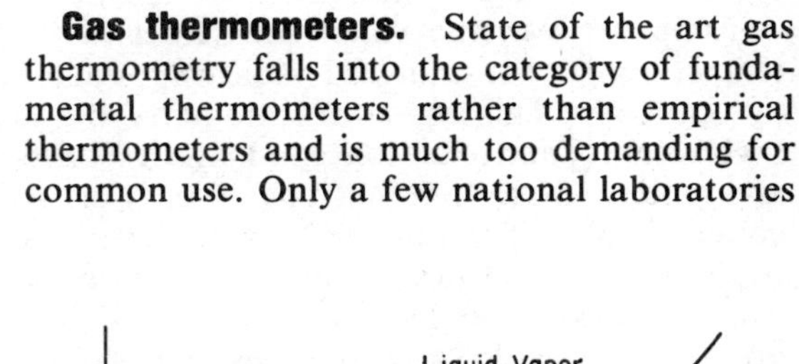

Figure 14.22 Illustration of fade-out that occurs when the vapor-pressure bulb is entirely filled with the gaseous phase of the fill substance. Critical fade-out conditions are indicated by P_f and T_f.

and universities are able to use precision gas thermometry.

Gas thermometry of a less precise nature, however, is frequently used in practical temperature measurements. Figure 14.21b, used in the discussion on vapor-pressure thermometers, may be used to illustrate gas thermometry. In this case, a fill gas must be used that will not condense at the lowest temperature to be encountered. Boyle's law states that for a constant mass of gas, $PV = RT$ or $PV/T = R$, where R is the universal gas constant. If a system, such as that shown in Fig. 14.21b, is filled to a pressure of P_f at T_f then

$$P_f(V_A + V_C)/T_f + P_f V_B/T_f = P_T V_B/T + P_T(V_A + V_C)/T_f \quad (14.47)$$

(It is assumed that the fill temperature is the same as the operating temperature of the auxiliary volume, V, i.e., $T_f = T_{\text{room}}$.) This equation may be arranged so that

$$(P_f/P_T)(V_A + V_C + V_B) = V_B T_f/T + V_A + V_C \quad (14.48)$$

If the volumes are known, the temperature, T, results directly from the measurement of pressure, P_T. If the volumes are not known, they may be determined by measuring the pressure at two calibration temperatures. Varying the ratio $(V_A + V_C)/V_B$ allows the sensitivity of the thermometer to be changed. Good low-temperature sensitivity may be obtained by making $(V_A + V_C)/V_B$ large. A high ratio means that only at very cold temperatures will an appreciable amount of gas be in the bulb; this condition also causes the thermometer to be sensitive to ambient temperature fluctuations. If $(V_A + V_C)/V_B$ is made very small, the sensitivity of the system becomes approximately constant throughout the temperature range. White (1959) and Coxon (1960) discuss corrections and precautions that must be applied to practical gas thermometer systems to achieve maximum accuracy. Inaccuracies of a few parts in a thousand are possible with reasonable care. Factors affecting the accuracy, such as shrinkage of the cold volume, capillary tube effects, and ambient temperature, are discussed by Holten (1963) along with the dynamic response of gas thermometers and methods of reducing response time. Time responses of less than 0.5 s have been achieved under optimum conditions.

Properties of fluids. The properties of the fluids used in filled thermometers are necessary to relate the measured parameter, pressure, to temperature. These data are based on various temperature scales so that conversion to the IPTS-68 temperature scale is frequently necessary (§ 14.3.1).

The saturation properties and thermodynamic properties of the He^4 are given by McCarty (1972). The lambda point is 2.177 K, the critical point is 5.201 K, and the n.b.p. is 4.224 K. Analytical representation of these vapor pressure data is given by

$$\ell n P = \sum_{i=1}^{10} A_i T^{(2-i)} \quad (14.49)$$

where the pressure is in microns and the temperature in kelvins. The coefficients for this series are given in Table 14.15. The temperatures used in this representation are $T_{58} + 0.001 + 0.002 \times T_{58}$, where T_{58} indicates temperatures on the 1958 helium vapor-pressure temperature scale (Brickwedde, van Dijk, Durieux, Clement, and Logan, 1960).

Saturation and thermodynamic properties for hydrogen are given by Roder, Weber, and Goodwin (1965). The triple point is 13.803 K, the n.b.p. is 20.268 K, and the critical point is 32.976 K. The temperatures used are based on the NBS-55 scale; these temperatures can be based on IPTS-68 by using the data in Table 14.6. The vapor-pressure–temperature relationship for hydrogen is given in three ranges by the following equations (Weber, Diller, Roder, and Goodwin, 1962). For temperatures between 13.803 and 21 K,

$$\log P = A_1 + A_2/T_{55} + A_3 T_{55} \quad (14.50)$$

between 20.268 and 29 K,

$$\log P_a = A_4 + A_5/(T_{55} + A_6) + A_7 T_{55} \quad (14.51)$$

and between 29 K and 32.976 K,

$$P = P_a + A_8(T_{55} - 29)^3 + A_9(T_{55} - 29)^5 + A_{10}(T_{55} - 29)^7 \quad (14.52)$$

Pressure is in atmospheres and temperature in kelvins. The coefficients for all three equations are given in Table 14.15. To base these functions on IPTS-68, $T_{55} = T_{68} - \Delta T$ must be used.*

---

*Values of ΔT are found in Table 14.6.

Table 14.15 Coefficients for the analytical representation of the vapor pressure–temperature relationship for oxygen, nitrogen, neon, equilibrium hydrogen, and helium-4.

	Vapor Pressure Substances				
Coefficient	Oxygen	Nitrogen	Neon	Hydrogen 99.79% Para	Helium
A_1	−62.5967185	0.8394409444×10^4	7.46116	1.772454	−3.9394635287
A_2	2.47450429	$-0.1890045259 \times 10^4$	−106.090	-4.436888×10	1.4127497598×10^2
A_3	$-4.68973315 \times 10^{-2}$	$-0.7282229165 \times 10^1$	-3.56616×10^{-2}	2.055468×10^{-2}	$-1.6407741565 \times 10^3$
A_4	$5.48202337 \times 10^{-4}$	$0.1022850966 \times 10^{-1}$	4.11092×10^{-4}	2.000620	1.1974557102×10^4
A_5	$-4.09349868 \times 10^{-6}$	$0.5556063825 \times 10^{-3}$		-5.009708×10	$-5.5283309818 \times 10^4$
A_6	$1.91471914 \times 10^{-8}$	$-0.5944544662 \times 10^{-5}$		1.0044	1.6621956504×10^5
A_7	$-5.13113688 \times 10^{-11}$	$0.2715433932 \times 10^{-7}$		1.748495×10^{-2}	$-3.2521282840 \times 10^5$
A_8	$6.02656934 \times 10^{-14}$	$-0.4879535904 \times 10^{-10}$		1.317×10^{-3}	3.9884322750×10^5
A_9		0.5095360824×10^3		-5.926×10^{-5}	$-2.7771806992 \times 10^5$
A_{10}				3.913×10^{-6}	8.3395204183×10^4

The saturation properties of neon were determined by Grilly (1962) and McCarty and Stewart (1965). The triple point is 24.54 K, the n.b.p. is 27.09 K, and the critical point is 44.40 K. The temperatures are based on the NBS-55 scale; these temperatures can be based on IPTS-68 by using the data in Table 14.6. The analytical representation is given by

$$\log P = A_1 + A_2/T_{55} + A_3 T_{55} + A_4 T_{55}^2 \quad (14.53)$$

Pressure is in torrs and temperature in kelvins. The coefficients for this equation are given in Table 14.15. In order to use this function on the IPTS-68 scale, $T_{55} = T_{68} - \Delta T$ must be used.*

Saturation and thermodynamic properties of nitrogen are given by Jacobsen, Stewart, McCarty, and Hanley (1973). The triple point is 63.148 K, the n.b.p. is 77.347 K, and the critical point, T_c, is 126.2 K. The analytical representation is given by

$$\ell n\, P = A_1/T + A_2 + A_3 T + A_4 (T_c - T)^{1.95} + A_5 T^3 + A_6 T^4 + A_7 T^5 + A_8 T^6 + A_9 \ell n(T) \quad (14.54)$$

where the pressure is in atmospheres and the temperature in kelvins. The coefficients are given in Table 14.15 and are based on IPTS-68.*

Saturation and thermodynamic properties of oxygen are given by Roder and Weber (1972). The triple point is 54.351 K, the n.b.p. is 90.18 K, and the critical point is 154.576 K. The analytical representation is given by

$$\ell n\, P = A_1 + A_2 T + A_3 T^2 + A_4 T^3 + A_5 T^4 + A_6 T^5 + A_7 T^6 + A_8 T^7 \quad (14.55)$$

where the pressure is in atmospheres and the temperature in kelvins. The coefficients for this equation are given in Table 14.15. These data are based on NBS-55 for temperatures below 91 K and IPTS-48 for temperatures above 91 K. As discussed in the hydrogen vapor-pressure paragraph, T_{55} or $T_{48} = T_{68} - \Delta T$ must be used to base the results on IPTS-68.*

14.3.5 Other thermometers

The thermometers discussed in § 14.3.2 through § 14.3.4 and diodes (discussed below) are the primary instruments used to measure low temperatures. There are, however, other phenomena and many other materials used to a lesser extent. In some cases, these thermometers may offer special characteristics that are not available in the frequently used types. The principal attributes of some of these thermometers along with additional references are included in the following discussion.

Diode thermometers. The temperature-dependent forward voltage of junction diodes has been utilized to make thermometers that have a usable temperature sensitivity in the range 1 to 400 K. Germanium, Ga–As, and silicon diodes have been studied, but only Ga–As and silicon have been developed as thermometers. The physics involved in the conduction processes are discussed by Cohen, Snow, and Tretola (1963) and by Sclar and Pollock (1972).

The temperature-dependent forward voltage is measured at constant dc current. The instruments needed to utilize diode thermometers are a dc current source with 10 to 100 μA capability

* *Values of* ΔT are found in Table 14.6.

and a high-impedance voltage-measuring device. The input impedance of the voltage-measuring device and the concomitant circuit loading must be considered when using diode thermometers. Input impedances of ~1 GΩ are sufficient for most applications; potentiometers and many digital voltmeters meet this specification. Alternatively, lower impedance voltage measuring devices may be used when the diode is calibrated and used with the same instrument.

The voltage–temperature characteristics for typical Ga–As and silicon diode thermometers are shown in Fig. 14.23 (Swartz and Swartz, 1974). As indicated in this figure, the silicon diode has a high sensitivity when $T \lesssim 30$ K, whereas the Ga–As diode becomes relatively insensitive. Some early commercial silicon diode thermometers were found to have nonmonotonic behavior in the temperature range between 11 and 16 K (Gerber and Sellmyer, 1978). It appears that this behavior occurred only in some pre-1977 diodes for which the breakdown voltage was less than 2.3 V at 4 K. Appropriate checks should be made for thermometers falling in this category. In general, the silicon diode is recommended over the Ga–As diode for all zero magnetic field measurements.

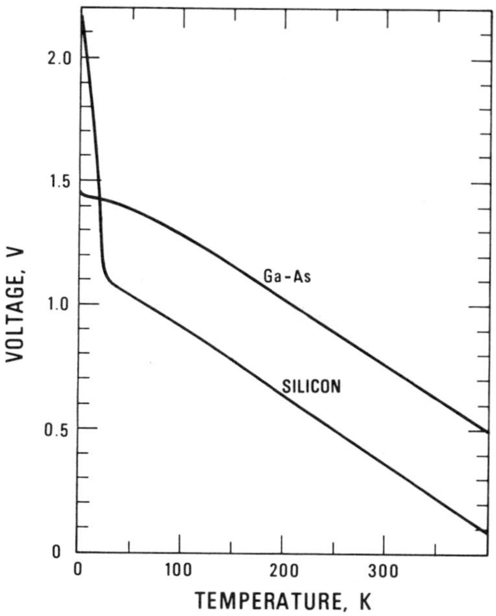

Figure 14.23 Typical forward voltages vs. temperature for silicon and Ga–As diode thermometers (Swartz and Swartz, 1974).

Miscellaneous thermometers. Pyroelectric thermometers utilize the temperature dependence of spontaneous electrical polarization that occurs in some anisotropic solids. The thermometers can be used to detect extremely small temperature changes (on the order of a few tenths μK) and rates of temperature change (on the order of hundredths of $\mu K \cdot s^{-1}$) in the temperature range 4 to 500 K. The pyroelectric phenomenon and its application to thermometry is discussed in Lang and Steckel (1965), Lang, Shaw, Rice, and Timmerhaus (1969), Lang, Rice, and Shaw (1969), and Lang (1972).

Two resonance phenomena, nuclear quadrupole resonance (NQR) and nuclear magnetic resonance (NMR), can be utilized to measure low temperatures. NQR provides usable temperature sensitivity in the range from ~20 to 500 K; accuracy of ~1 mK is possible near 77 K. Accuracies on the order of 1 mK can be achieved using an NMR thermometer at temperatures above 50 K. Utton (1967), Labrie, Infantes, and Vanier (1971), and Vanier (1972) present the principles and applications for those thermometers.

Inductance thermometers make use of the temperature dependence of the magnetic properties of core materials. When an inductor is incorporated into a tunnel diode oscillator circuit, the temperature dependence of the core is reflected in the oscillator frequency. Inductors with cores of specially treated alloys can be used in the temperature range 4 to 300 K with a resolution of better than 10 mK. Standard broadcasting techniques can be used to transmit the frequency over long distances. Details of this method of thermometry are found in Willens, Buehler, and Nesbitt (1968) and Willens (1972).

Lawless (1971) had developed glass-ceramic ($SrTiO_3$) thermometers which, although relatively new, are available commercially. The operating range for these devices is from $T < 1$ K to above room temperature. The sensitivity at 4.2 K is ~350 $pF \cdot K^{-1}$ at 5 kHz and 7.5 mV (rms). The capacitance–temperature relationship for a particular sensor is shown in Fig. 14.24. Lawless (1972b) suggests that the peak near 70 K is caused by a paramagnetic to antiferroelectric phase transition. The capacitance–temperature relationship, $T \lesssim 50$ K, is accurately represented by a simple fourth-order series expansion (Lawless and Panchyk, 1972). The temperature region below the knee in the capacitance–temperature curve is of particular importance because of the good sensitivity and essentially zero magnetic field dependence (§ 14.3.6). The field effect in

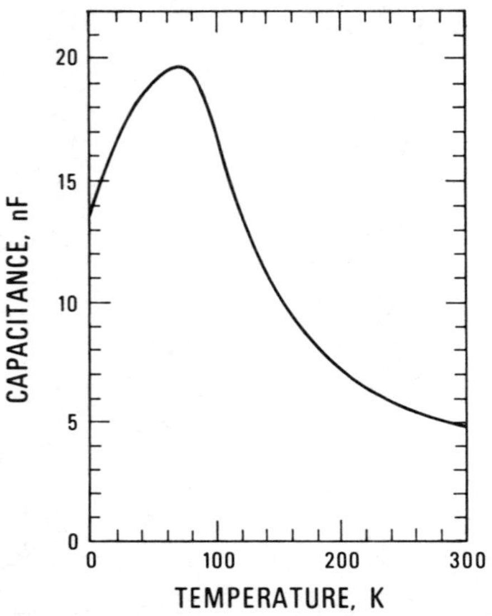

Figure 14.24 Capacitance as a function of temperature for a glass ceramic capacitance thermometer (Lawless, 1971). The measurement frequency was 5 kHz and the rms amplitude was 7.5 mV.

most other sensors is inversely related to temperature.

The capacitance thermometer is not used extensively to determine absolute temperatures, although future development may alleviate present shortcomings for this application. Lawless (1975) has determined that the aging or drift problem associated with these thermometers is due to relaxation of the glass matrix and will occur whenever this matrix is perturbed. Perturbation is caused by thermal expansion ($T > 77$ K), by electrostrictive coupling of the $SrTiO_3$ microcrystals in the glass matrix following a voltage pulse, or by crossing the ~65-K phase transition temperature. The aging brought about by any of these occurrences may cause a change in capacitance equivalent to ~15 mK over a period of 30 min. The aging phenomenon and such things as stray lead capacitances and frequency and amplitude dependence limit the use of this type of sensor as an absolute thermometer at the present time. They do not detract from the use of these thermometers as control devices in magnetic field environments. This application will be discussed further in § 14.3.6.

14.3.6 Temperature measurement in magnetic fields

With few exceptions, the instruments used to measure temperature are affected by the presence of magnetic fields. The notable exceptions to this field dependence are gas thermometers, discussed in § 14.3.4, and the $SrTiO_3$ capacitance thermometers, discussed in § 14.3.5. Vapor-pressure thermometers are nearly field independent except when the fill gas is oxygen. Sample and Rubin (1978) found that the temperature equivalent of the field-induced change in vapor pressure amounts to ~0.02, 0.03, and 0.01 mK for N_2, Ar, and He^4, respectively, at their normal boiling points and 15 T. There are temperature intervals, however, where vapor pressure techniques cannot be used (see Fig. 14.4). Constant-volume gas thermometry is difficult to the extent that it cannot satisfy the bulk of the current needs. Other alternatives are to use the common low-temperature sensors and apply corrections to account for the magnetic effect. The magnetic-field-dependent temperature errors for several types of thermometers were compiled by Sample and Rubin (1977). Table 14.16 is based on their survey. The remainder of this section supplements the information presented in this table.

Carbon and carbon-in-glass resistors in magnetic fields. Carbon circuit resistors are among the most frequently used thermometers when magnetic fields are present and $T \lesssim 40$ K. As a result, their magnetic field dependence has been widely studied and reported in the literature. The review by Sample and Rubin (1977) is an important source of references for these and the other types of sensors discussed below. In addition to the attributes that lead to the use of carbon resistors as cryogenic thermometers in zero field applications: (a) they show no orientation dependence in a magnetic field; (b) their field dependence is reproducible and essentially the same for resistors from the same company that have the same nominal room temperature resistance; and (c) their magnetoresistance is relatively small.

Neuringer and Shapira (1969) determined the relative change in resistance caused by a magnetic field

$$\Delta R / R_0 = [R(H,T) - R(0,T)] / R(0,T) \qquad (14.56)$$

for several resistors with various wattages and nominal room-temperature resistances. Figure 14.25 for a 220-Ω, 0.1-W resistor is typical of their results. Their data generally cover the temperature range $1.8 < T < 18$ K in magnetic fields up to 15.0 T. Sample, Neuringer, and

Table 14.16 Magnetic-field-dependent temperature errors for several low-temperature thermometers (Sample and Rubin, 1977).

		$\lvert \Delta T \rvert / T \times 100$ (%) Magnetic Field			
	T (K)	2.5 T	8 T	14 T	Comments
Carbon circuit resistors; 2.7, 3.9, 5.6, and 10 Ω	0.5	2–4	5–13	7–20	$\Delta R/R_0$ positive and monotonic in B and T. Good agreement with similar units.
	1.0	2–4	6–15	9–25	
	2.5	1–5	6–18	10–30	
	4.2	1–5	5–20	10–35	
Carbon circuit resistors; 47, 100, and 220 Ω	4.2	<1	5	10	$\Delta R/R_0$ positive and monotonic in B and T. Good agreement with similar units.
	10	<1	3	5	
	20	<1	1	2	
Carbon–glass resistors	2.1	0.5	1.5	4	General behavior similar to carbon circuit resistors. Negative $\Delta R/R$ when $T \gtrsim 60$ K. Useful to 300 K. Good reproducibility.
	4.2	0.5	3	6	
	15	<0.1	0.5	1.5	
	35	<0.1	0.5	1	
	77	<0.1	0.5	1.5	
Thermistors	4.2	<0.05	1	3	High sensitivity over narrow temperature range. Good reproducibility.
	10	<0.05	0.3	1	
	20	<0.05	0.1	0.5	
	40	<0.05	0.1	0.5	
	60	<0.05	0.1	0.3	
Germanium resistors	2.0	8	60	—	Not recommended except at low B owing to large, orientation-dependent temperature effect.
	4.2	5–20	30–55	60–75	
	10	4–15	25–60	60–75	
	20	3–20	15–35	50–80	
	70	3–10	15–30	25–50	
Ga–As diodes	4.2	2–3	30–50	100–250	Large orientation-dependent temperature effect. Useful in range 4.2 to 300 K.
	10	1.5–2	25–40	75–200	
	20	0.5–1	20–30	60–150	
	40	0.2–0.3	4–6	15–30	
	80	0.1–0.2	0.5–1	2–5	
Silicon diodes	4.2	75	—	—	Large orientation-dependent temperature effect.
	10	20	30	50	
	20	4	7	10	
	30	3	4	5	
	77	0.2	0.5	0.5	
Platinum resistors	20	20	100	250	Moderately orientation dependent. Suggested use only when $T \gtrsim 30$ K.
	40	<1	5	10	
	80	<0.5	1	2	
KP vs. Au–0.07 at.% Fe thermocouple	5	2	10	15	These data taken with entire thermocouple in field, cold junction at 4.2 K, errors in hot junction temperature. Handling may cause irreproducibilities.
	10	3	20	30	
	20	2	15	20	
	45	1	5	7	
	100	0.1	0.8	—	
Type E thermocouple	10	1	3	7	Useful when $T \gtrsim 10$ K. Table entires as in KP vs. Au–0.07 at.% Fe comment.
	20	<1	2	4	
	45	<1	<1	2	
$SrTiO_3$	1.5	<0.05	<0.05	≤0.05	Probably zero-field-induced temperature error. Recommended for control purposes.
	4.2	<0.05	<0.05	≤0.05	
	20	<0.05	<0.05	≤0.05	
	50	<0.05	<0.05	≤0.05	

Rubin (1974) extended the temperature to 0.5 K. Figure 14.26 shows their results for a 10-Ω, 1/8-W resistor. The data in these figures indicate that the magnetoresistance is an increasing function of field and a decreasing function of temperature. The magnetic field behavior can, however, vary dramatically depending on the manufacturer. One commonly used carbon resistor has both a negative and positive magnetoresistance (Sample et al. 1974) instead of the monotonic behavior shown in Figs. 14.25 and 14.26. Since analytical representation is a problem even in the mono-

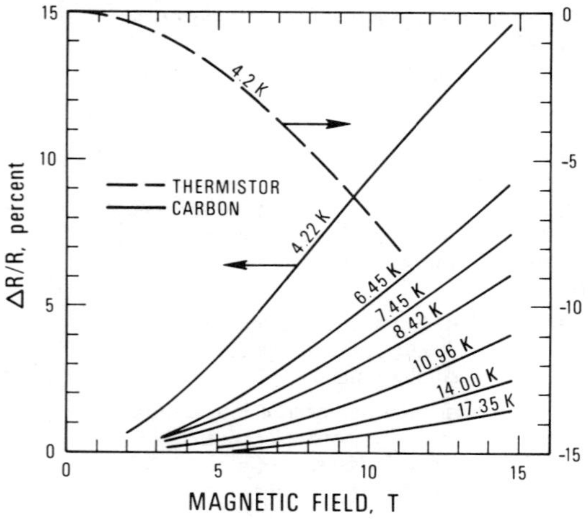

Figure 14.25 Relative change in electrical resistance as a function of magnetic field for a 220-Ω, 0.1-W carbon-circuit resistor (Neuringer and Shapira, 1969) and for a thermistor with a useful temperature range of 2.9 to 9 K (Schlosser and Munnings, 1972).

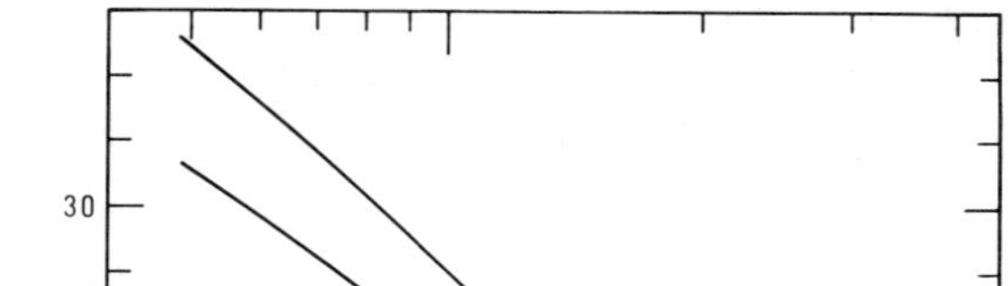

Figure 14.26 Relative change in electrical resistance as a function of temperature with magnetic field as a parameter for a 10-Ω, 1/8-W carbon-circuit resistor (Sample et al., 1974).

tonic cases, the types having double-valued behavior are not usually recommended.

The empirical formula developed by Clement and Quinnell (1952) predicts a quadratic dependence of the magnetoresistance on field. Their formula,

$$100\,(\Delta R/R_0) = (5.5 + 1.8 \log R_{300})10^{-3}T^{-3/2}H^2 \quad (14.57)$$

tends to underestimate the low-field data and overestimate the high-field data shown in Figs. 14.25 and 14.26. The relationship given in Eq. 14.32 can be used to represent the magnetoresistance of carbon resistors if the coefficients K, A, and B are assumed to be polynomial functions of the magnetic field, for example

$$B(H) = a_1 + a_2H + a_3H^2 \quad (14.58)$$

This or similar analytical representations are necessary to obtain the full benefit of computerized analysis. Swartz, Gaines, and Rubin (1975) determined the magnetic field effect on CGRTs for $T < 4.2$ K and $63 \leq T \leq 77$ K. For temperatures below ~60 K, the $\Delta R/R_0$ (Eq. 14.56) for CGRTs is similar to that for CRTs except that they are more temperature sensitive. For temperatures higher than ~60 K, the magnetoresistance becomes negative. The magnetoresistance in both ranges is described to within 0.1% by

$$R(T,H) = R(T,0) + \alpha H^2/[1 + (H/\gamma)^2] + \beta H \quad (14.59)$$

where α, β, and γ are functions of magnetic field strength and are experimentally determined. The $R(T,0)$ term must be determined by calibrating at multiple temperatures in the temperature range of interest. This term may then be represented by an empirical formula (e.g., Eqs. 14.33 through 14.37).

Platinum resistors in magnetic fields. The magnetoresistance of platinum is large, as shown in Fig. 14.27 (Neuringer, Perlman, Rubin, and Shapira, 1971). Based on a very limited sample, however, the field dependence appears to be reproducible from one unit to another. The data shown in this figure indicate that the correction required is less than 100% for $T < 30$ K and $H < 11.9$ MA·m^{-1} (15.0 T) and decreases as the temperature increases.

Unlike the carbon resistors, the magnetoresistance of platinum is a function of field direction, as shown in Fig. 14.28. The current in the platinum element is perpendicular to the mag-

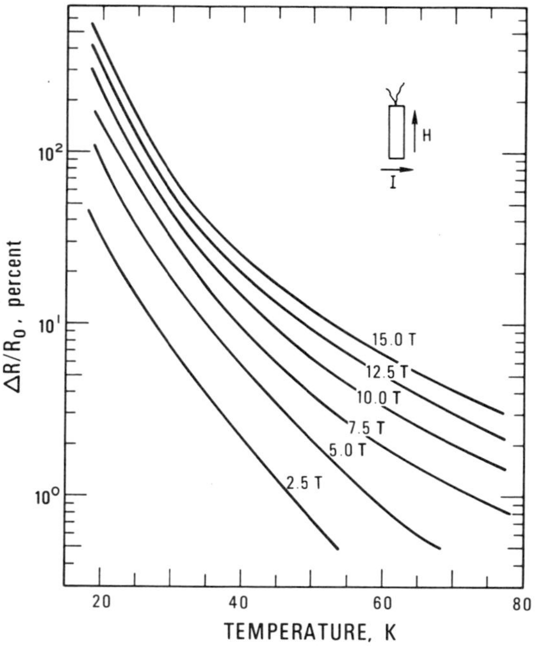

Figure 14.27 Relative change in electrical resistance as a function of temperature with magnetic field as a parameter for a platinum resistance thermometer. The magnetic field is perpendicular to the thermometer current.

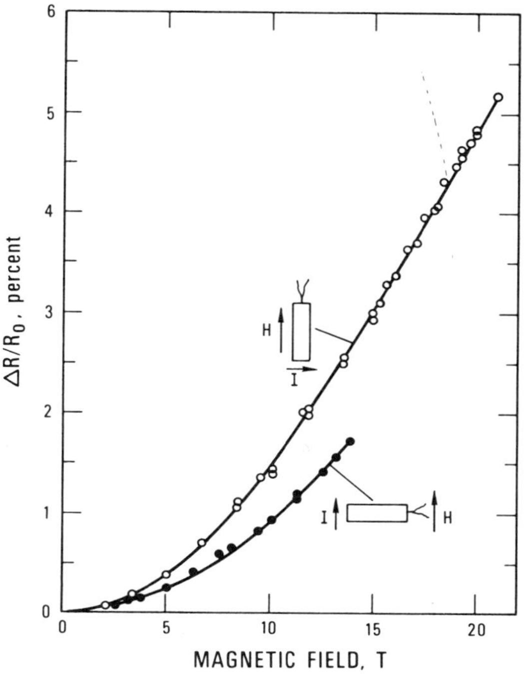

Figure 14.28 Relative change in electrical resistance as a function of magnetic field for a platinum resistance thermometer at 67.2 K with the thermometer current (I) parallel and perpendicular to the field (H).

netic field when the thermometer axis is parallel to the field and is represented by the upper curve. Thermometer current is parallel to the field in the lower curve. The magnetoresistance of germanium thermometers (Neuringer et al., 1971) changes as the thermometer is rotated (keeping the orthogonal thermometer current–magnetic field relationship). Platinum resistors do not share this trait. Despite the large, orientation-dependent field effect, platinum resistance thermometers are frequently used when $T \gtrsim 60$ K and $H = 0$ because of the lack of alternative sensors and because of their extensive use in zero magnetic field environments.

Thermocouples in magnetic fields. Thermocouples are very difficult to use as low-temperature thermometers in the presence of magnetic fields. In spite of difficulties, the size, low heat capacity, time response, and ability to measure temperature differences directly have led to some use in magnetic environments. Figure 14.29 can be used, along with the equation relating the thermoelectric voltage, V, and the thermoelectric power, S,

$$V = \int S dT \tag{14.60}$$

to illustrate some of the difficulties. Consider the situation described in Fig. 14.29a, where the unknown temperature, T_x, is in the magnetic field and the reference junctions, T_{ref}, are not. Equation 14.60 for this circuit leads to

$$V = \int_{T_{ref}}^{T_m} [S_B(0,T) - S_C(0,T)]dT + \int_{T_m}^{T_x} [S_B(H,T) - S_C(H,T)]dT \tag{14.61}$$

where S_B and S_C are the thermoelectric power of materials B and C. These thermopowers are functions of both field and temperature. The thermovoltage, V, is dependent on the temperature at which the wires leave the magnetic field. The output of a system such as this depends on both temperature and magnetic field gradients.

If, however, $T_m = T_x$,

$$V = \int_{T_{ref}}^{T_x} [S_B(0,T) - S_C(0,T)]dT \tag{14.62}$$

and the output is independent of the field. This condition has been experimentally verified by Richards, Edwards, and Legvold (1969). The ar-

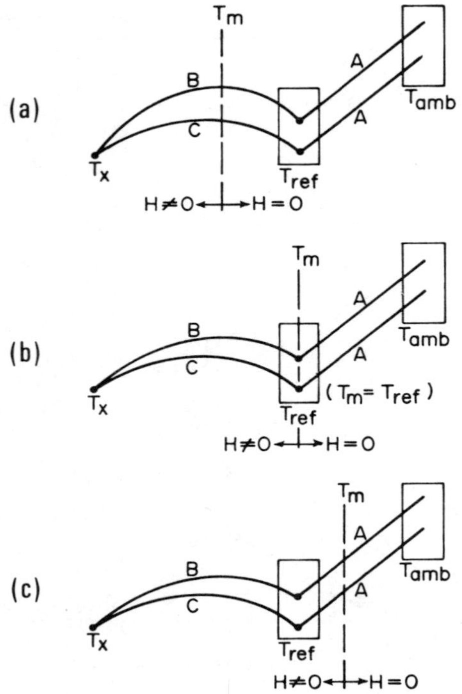

Figure 14.29 Schematic of a thermocouple circuit with the boundary of a magnetic field gradient imposed (a) between the unknown temperature, T_x, and the reference temperature, T_{ref}, (b) at T_{ref}, and (c) between T_{ref} and ambient temperature T_{amb}.

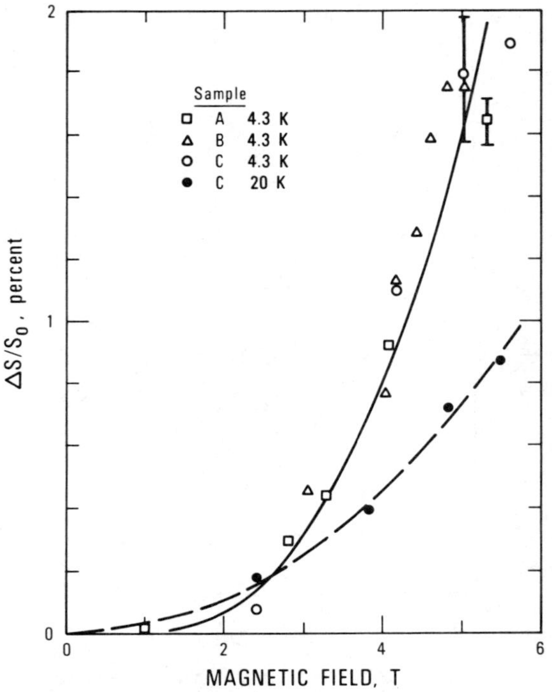

Figure 14.30 Relative change in the thermoelectric power as a function of magnetic field for a Au–0.07 at.% Fe thermocouple wire caused by aligning the wire axis parallel and perpendicular to the magnetic field.

rangements shown in Figs. 14.29 b and c are thermoelectrically equivalent and yield

$$V = \int_{T_{ref}}^{T_x} [S_B(H,T) - S_C(H,T)]dT \qquad (14.63)$$

Application of this relationship assumes that the junction at T_x and T_{ref} are in an isofield environment. Knittel (1973) has shown that the effect of a magnetic field on thermopower depends on the orientation of the wire in the field. His data (Fig. 14.30) represent the relative difference in thermopower, $\Delta S/S_{H=0}$, when half of a single Au–0.07 at.% Fe wire is parallel to a magnetic field and the other half is perpendicular.

Sample et al. (1974) and von Middendorff (1971) measured the longitudinal magnetothermoelectric effect for KP vs. Au–0.07 at.% Fe and type E (see Table 14.10 for an explanation of thermocouple wire designations). Their results are shown in Figs. 14.31 and 14.32 for KP vs. Au–0.07 at.% Fe and type E respectively. The Au–Fe combination was strongly field dependent for $T \lesssim 20$ K. Also, the field dependence for this alloy varied for specimens from the same spool, and even remounting a single wire changed the field dependence. Type E thermocouples are only marginally useful below ~10 K owing to lack of temperature sensitivity, but have a reproducible and relatively small magnetic field dependence. Chiang (1974) determined the effect of a transverse magnetic field on both of these thermocouple types.

Thermistors, diodes, and capacitors in magnetic fields. The relative magnetic field dependence, $\Delta R/R_0$, of a thermistor at 4.2 K is shown in Fig. 14.25. The magnitude of $\Delta R/R_0$ is slightly less for the thermistor than for the CRT also shown in this figure. The major disadvantage associated with the use of thermistors is the limited temperature range of usefulness of a single sensor.

The field dependence of Ga–As and silicon diodes has been investigated by Swartz and Gaines (1972) and Aldridge, Davis, and Holloway (1975), respectively. Both diodes are sensitive to field orientation, which limits their usefulness.

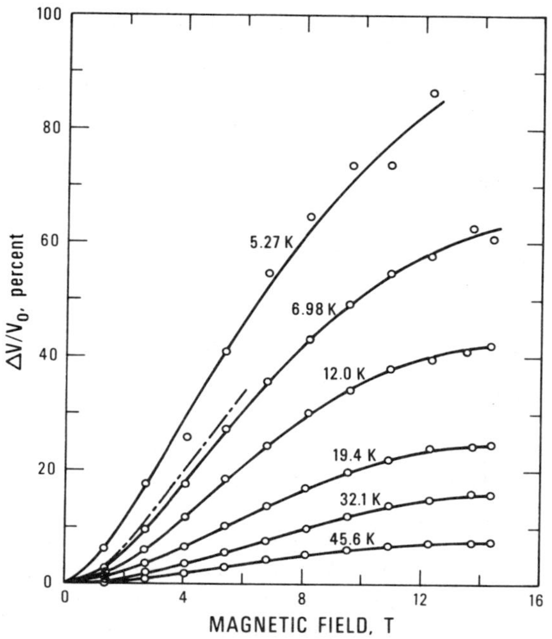

Figure 14.31 Relative change in thermoelectric voltage as a function of magnetic field with the variable junction temperature (K) as a parameter for a KP vs. Au–0.07 at.% Fe thermocouple. (Sample et al., 1974.) Data by von Middendorff (1971) has been manipulated and included (broken lines) for comparison.

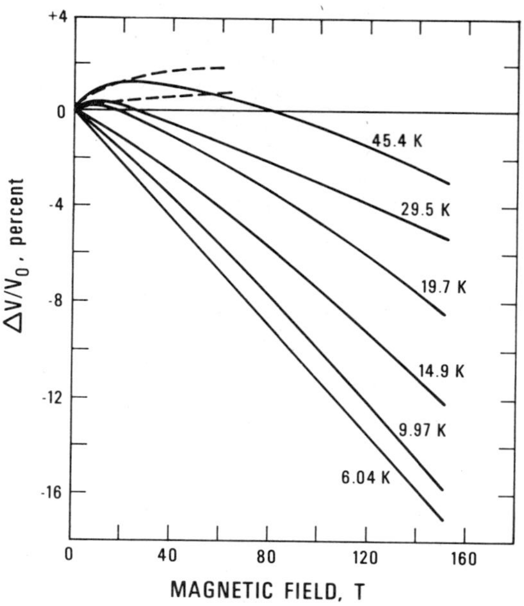

Figure 14.32 Relative change in thermoelectric voltage as a function of magnetic field with the variable junction temperature (K) as a parameter for a type E thermocouple (Sample et al., 1974). Data by von Middendorff (1971) has been manipulated and included (broken lines) for comparison.

The field-dependent temperature errors for both types are shown in Table 14.16.

The glass–ceramic capacitance thermometer that was discussed in § 14.3.5 has no measurable field dependence (Rubin and Lawless, 1971). This allows these thermometers to be used as control sensors in magnetic field environments. The method used in this application is thermal attachment of the capacitance thermometer to a calibrated thermometer, such as a GeRT or PRT. The capacitive thermometer is to be in the control circuit, and the absolute temperature is determined with the other thermometer. The desired temperature is established at zero field and is indicated by the calibrated sensor. Subsequent changes in the field do not affect the output of the capacitance thermometer, so that the zero-field temperature can be maintained in spite of eddy current heating and changes in system thermal conductivity.

14.4 STRAIN MEASUREMENT

The measurement of strain at low temperatures is in many ways similar to the measurement of temperature. The thermophysical phenomena utilized to measure strain are also used in temperature measurement, and similar techniques apply to both. On the other hand, optimization of strain-measuring devices involves minimizing the temperature coefficient of resistance, $(1/R)dR/dT$, and maximizing the effect of strain on the electrical resistivity.

Several methods are employed to measure strain: extensometers, linear variable differential transformers, measurement of piezoelectric voltage, changes in capacitance, and changes in electrical resistivity. The most widely used method involves determining the change in electrical resistance as a function of strain. The strain dependence of electrical resistivity is discussed in Chapter 5.

The resistance gauge is produced in a wide variety of sizes, patterns, materials, and forms. Although most gauges are now made by photoetching thin foils, thin wire elements are still available and vacuum-deposited thin films are being produced. A typical active element of a resistive-type gauge is shown schematically in Fig. 14.33a. Combinations of active elements are frequently used to measure multiple components of strain

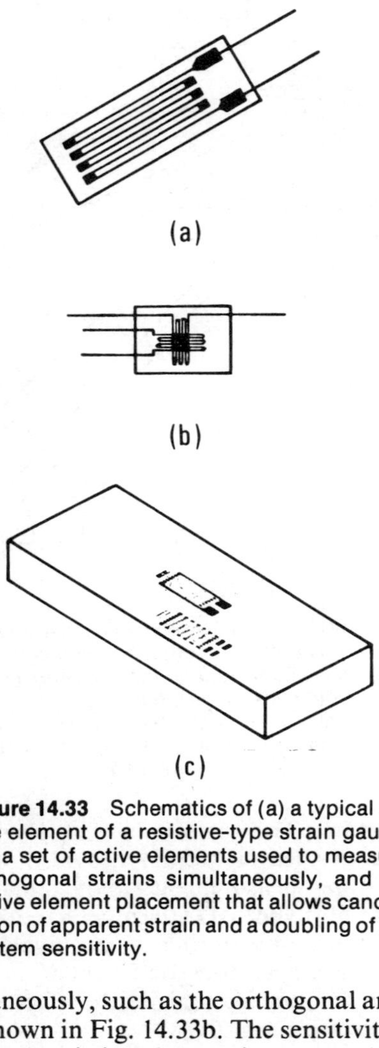

Figure 14.33 Schematics of (a) a typical active element of a resistive-type strain gauge, (b) a set of active elements used to measure orthogonal strains simultaneously, and (c) active element placement that allows cancellation of apparent strain and a doubling of the system sensitivity.

simultaneously, such as the orthogonal arrangement shown in Fig. 14.33b. The sensitivity of an individual resistive element is commonly given by the gauge factor,

$$G = \Delta R / \epsilon R \qquad (14.64)$$

where R is the electrical resistance of the gauge at a given temperature, ΔR is the change in electrical resistance due to strain, and ϵ is the strain. Metallic elements generally fall in the range $2 \lesssim G \lesssim 10$; nickel is a notable exception (see Table 14.17). Piezoresistive gauges, utilizing silicon, germanium, and gallium arsenide, have gauge factors ranging from 100 to 200. Gauge factors are affected somewhat by size, pattern, substrate, and temperature so that the numbers in Table 14.17 should be considered for comparative purposes only. More precise gauge factors are furnished with individual commercial units; for cryogenic operation, however, it is frequently necessary to

Table 14.17 Nominal gauge factors and temperature coefficients of resistance for frequently used strain gauge materials.

Material	Gauge Factor	Temperature Coefficient of Resistance ($\Omega \cdot \Omega^{-1} \cdot K^{-1}$)
Advance	2.1	10^{-4}
Chromel	2.5	
Constantan	2.0	10^{-5}
Manganin[a]	0.47	10^{-5}
Nichrome[a]	2.5	4×10^{-4}
Nickel	−12.1	6×10^{-3}
Platinum	6.0	3×10^{-3}
Carbon	20.0	-5×10^{-4}

[a]Trademark of Driver Harris Co.

determine the temperature dependence of the gauge factor experimentally.

14.4.1 Apparent strain

Although strain is one of the more common parameters used by scientists and engineers, its determination at cryogenic temperatures is by no means routine. Serious sources of error in strain gauge measurements are the temperature effects on the gauge resistance and the differential thermal contraction in the materials. Both sources of error produce a false indication of strain that can be larger than the real strain, particularly at cryogenic temperatues. The total apparent strain from a temperature difference, ΔT, can be expressed

$$\epsilon_{app} = [\gamma / G + (\alpha_S - \alpha_G)]\Delta T \qquad (14.65)$$

where γ is the temperature coefficient of resistance of the gauge and $(\alpha_S - \alpha_G)$ is the difference in thermal contraction between the substrate and the gauge. Actually, both γ and α are functions of temperature so ϵ_{app} can be a very nonlinear function of T. Figure 14.34 illustrates the apparent strain as a function of temperature for several commercially available strain gauge alloys mounted on a steel specimen.

Apparent strain can be reduced somewhat through the use of self-temperature-compensating (STC) materials, although the compensation is generally designed for the ambient temperature region. In early attempts to achieve compensation, an individual gauge contained two different alloys. The length and thermophysical properties were chosen so that both the resistivity and expansion effects tended to cancel. Contemporary STC gauges are made with a single alloy and are thermally treated to be used on very specific materials. Figure 14.35 indicates that

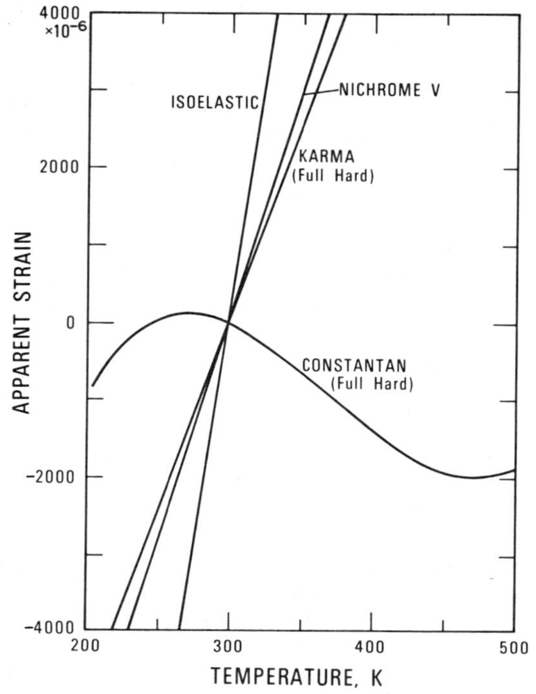

Figure 14.34 Apparent strain as a function of temperature for several commercially available strain gauges. The gauge factor used was 2.0 and the gauges were mounted on a steel specimen (*Micro-Measurements*, 1976).

the compensation achieved at both high and low temperatures leaves much to be desired. The rather limited data available at very low temperatures indicate that the apparent strain is a rapidly varying function of temperature with a minimum near 20 K. In situations where the temperature may change rapidly, such as a quench in a superconducting magnet, this apparent strain characteristic is especially troublesome.

Also, the excitation current used for measurements made in vacuum must be carefully controlled. The combination of poor heat conductivity and small heat capacity can significantly alter the temperature of the gauge.

In certain situations, the thermally induced strains are not critical. If the gauged specimen can be brought to thermal equilibrium in a zero-strain condition, the apparent strain can be electronically cancelled. Subsequent tests at this temperature will be free from apparent strain effects if the temperature profile along the lead wires remains the same as in the no-strain test and the instrument zero does not drift. In dynamic tests, the stress frequency may allow the thermal strain effects to be identified.

Use of a dummy gauge is one method of eliminating apparent strain. The thermally generated apparent strain is introduced into two adjacent arms of the measuring bridge circuit so that cancellation occurs. The mechanically induced strain remains to be measured. Physically, this is accomplished by mounting identical gauges on the

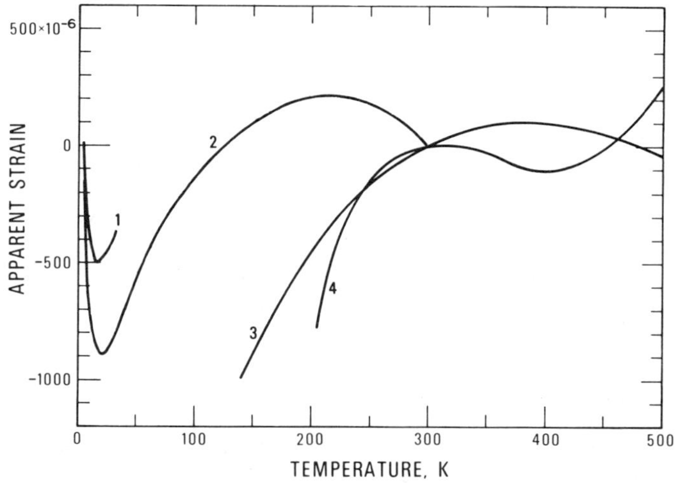

Figure 14.35 Apparent strain as a function of temperature for several STC gauges. It is difficult to make accurate strain determinations near 20 K where apparent strain is a strong function of temperature. 1 — K Alloy (Freynik, Roach, Deis, and Hirzel, 1978); 2 — K Alloy (Kaugerts, 1979); 3 — K Alloy (*Micro-Measurements*, 1976); 4 — A Alloy (*Micro-Measurements*, 1976).

specimen and on an identical piece of material that remains unstressed during the test. When the two gauged pieces are in thermal equilibrium, the bridge imbalance is due to specimen strain only. Figure 14.36a shows the active-dummy relationship in a Wheatstone bridge circuit. There are limitations to the applicability of this approach: (a) it is impossible to have perfectly matched gauges, (b) in practical situations it may not be possible to include a piece of specimen material, that is, material with the same coefficient of thermal expansion as the specimen, and (c) it is extremely difficult to maintain thermal equilibrium between the specimen and the dummy, particularly if there are transient thermal conditions in the system.

A modification of the active–dummy scheme is to mount the second or dummy gauge on the specimen so that an active–active gauge configuration is created. The relationship between the strains being monitored must be known before this method can be used. For instance, gauges may be mounted on opposite faces of thin beams in bending tests, as shown in Fig. 14.33c. In this case, both gauges C and D in Fig. 14.36a are active in the bridge circuit and the sensitivity is doubled. Of course, the requirements for gauge matching, lead equalization, and thermal equilibrium between the gauges are the same as in the active–dummy configuration.

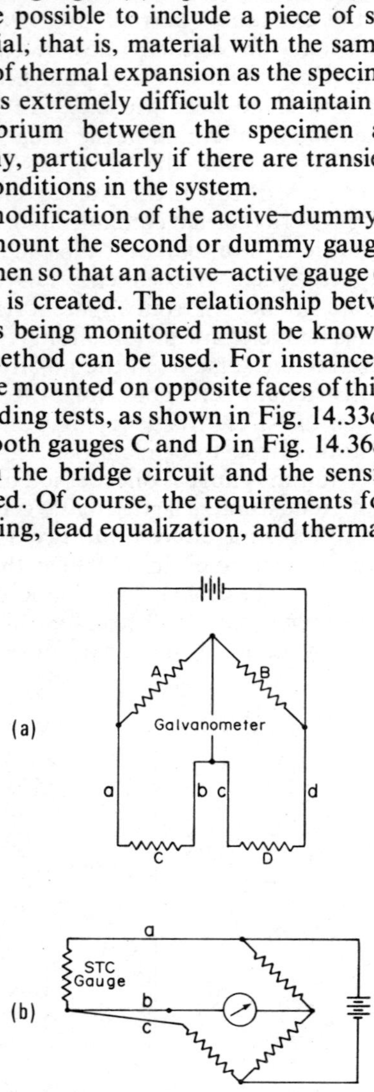

Figure 14.36 (a) Wheatstone bridge arrangement utilizing gauges C and D in either an active–dummy or an active–active relationship. (b) Wheatstone bridge arrangement utilizing a single self-temperature-compensating gauge. In this configuration the temperature dependence of the apparent strain must be known.

Very accurate strain measurements in cryogenic environments require in-place determination of apparent strain vs. temperature under zero strain conditions. This implies that both temperature and strain measurements are required in the actual tests. If the strain-vs.-temperature characteristics are known, a single gauge may be used, as shown in Fig. 14.36b. The resistance of leads a and c must be equal; differences or changes in lead resistance can cause significant errors.

14.4.2 Gauge factor

The gauge factor (Eq. 14.64) is a function of temperature, as seen in Fig. 14.37. It must be known in order to convert measured resistance changes into strain. In Fig. 14.37 the A alloy is a constantan STC material, the C alloy is an Fe–Cr–Al alloy, the K alloy is a Ni–Cr STC material similar to Karma, the N alloy is a Ni–Cr alloy similar to Nichrome V, and the D alloy is Isoelastic alloy (*Micro-Measurements*, 1976). The multiple curves shown for the K alloy indicate the effect of the various thermal treatments used to tailor the gauge to particular materials; for example, the 13 gauge would be used to measure strain on pure tin, which has an expansion coefficient of 23×10^{-6} K^{-1}.

Data from Greenough and Lee (1967a) illustrate the necessity of using gauges of the same type and from the same spool. Their results for a Cu–Ni gauge indicate a relative change in the gauge factor, $(G_T - G_{295})/G_{295} = \Delta G/G_{295}$, of 3%

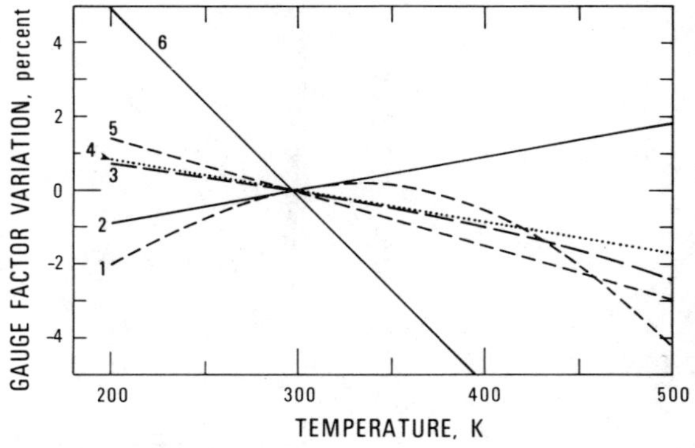

Figure 14.37 Gauge factor variation as a function of temperature for several commercially available strain gauge materials (*Micro-Measurements*, 1976). 1 — D Alloy; 2 — A Alloy; 3 — C Alloy; 4 — K(03) Alloy; 5 — K(13) Alloy; 6 — N Alloy.

at 4 K, 5.5% at 76 K, and 8% at 170 K. McClintock (1959) found values of $\Delta G/G_{295}$ for a similar, but not identical, Cu–Ni alloy to be −10% at 4 K and −4% at 76 K. These data clearly indicate that knowledge of the temperature dependence of G for a particular gauge is not sufficient to allow other gauges, which are nominally the same, to be used. Even larger variations were found when gauges made from different melts of Karma alloy were tested. Fortunately, $\Delta G/G_{295}$ for gauges made from the same reel of wire appear to have nearly identical temperature dependences (Greenough and Lee, 1967a). This allows the temperature dependence of G to be determined for a single gauge and applied to others from the same reel.

14.4.3 Strain measurements in magnetic fields

The electrical resistivity of strain gauge materials currently available is significantly affected by magnetic fields. Walstrom (1975) performed low-temperature magnetic-field-effect tests on several metal foil gauges in fields up to 6 T. He found that the apparent strain at no load (Fig. 14.38) due to the magnetic field was nearly quadratic for the 2 Ni–Cr foil gauges at 4.2 K and for the constantan foil gauge at 296 K. The Ni–Cr gauges were from different manufacturers. Walstrom also tested one of the Ni–Cr gauges under loaded conditions (upper curve, Fig. 14.38). The magnetic field dependence of the apparent strain is not affected at 4.2, 77, or 296 K, when the specimen is strained to 10^{-3}. Greenough and Lee (1967b) determined the field-induced change in electrical resistance for Ni–Cr wire gauges, which is shown in Fig. 14.39. Their data indicate that, unlike the foil gauges, the wire-wound gauges depend on field orientation.

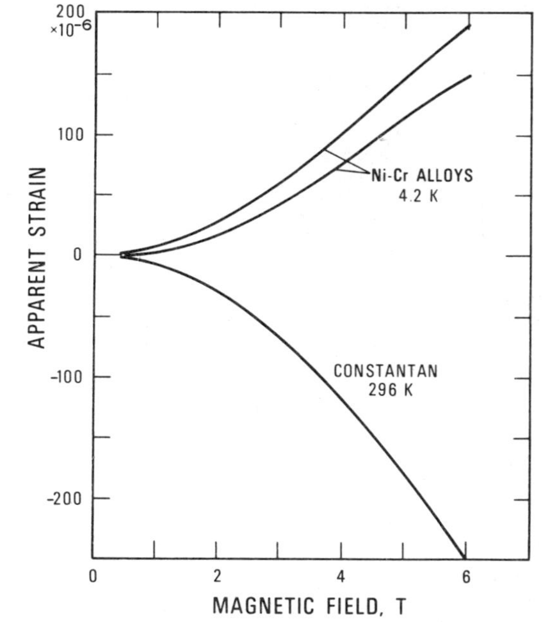

Figure 14.38 Change in apparent strain as a function of a magnetic field perpendicular to the gauges. The gauges were mounted on a steel specimen and a gauge factor of 2.05 was used to compute the apparent strain (Walstrom, 1975).

14.4.4 Other types of strain gauges

Recent developments in strain measurement involve two different solid state phenomena — piezoelectricity and piezoresistivity. Strain affects the state of bound electrons in the asymmetric piezoelectric crystals, which results in an electric potential across the crystal. The state of mobile electrons is affected by strain in the symmetric piezoresistive materials causing the crystal resistance to be proportional to the strain. From a macroscopic point of view, the piezoresistive gauges are similar to ordinary metallic gauges with very high gauge factors (100 to 200). Piezoelectricity has not been developed as much as piezoresistivity because it is difficult to maintain a steady voltage output from static or low-frequency changes in strain. The strain-induced voltage leaks off through the measurement instruments. Piezoresistive gauges, on the other hand, offer high sensitivity and are used with conventional strain measurement instruments.

The resistance change due to applied magnetic fields is very large for piezoresistive gauges. This is seen for a silicon gauge in Fig. 14.40. Even at room temperature (the effect increases as T decreases), the magnetic effect is large enough to discourage use in magnetic fields.

Low-temperature strain measurements requiring high sensitivity ($\simeq 10^{-4}$) and long range ($0.2 \lesssim \Delta\ell \lesssim 2.5$ cm) are difficult, often requiring specially designed extensometers (Warren and Reed, 1963). Brown and Lukens (1961) used the parallel-plate capacitance technique to measure increments of strain in the 10^{-6} range. White (1961) pushed the sensitivity of this type of gauge to 10^{-8} by using sophisticated equipment. Reed and Durcholz (1970) used the relationship

$$\Delta C = A_1 \epsilon \, \Delta L / 2 \, \ell n(r_2/r_1) \qquad (14.66)$$

which relates capacitance change (ΔC) to length

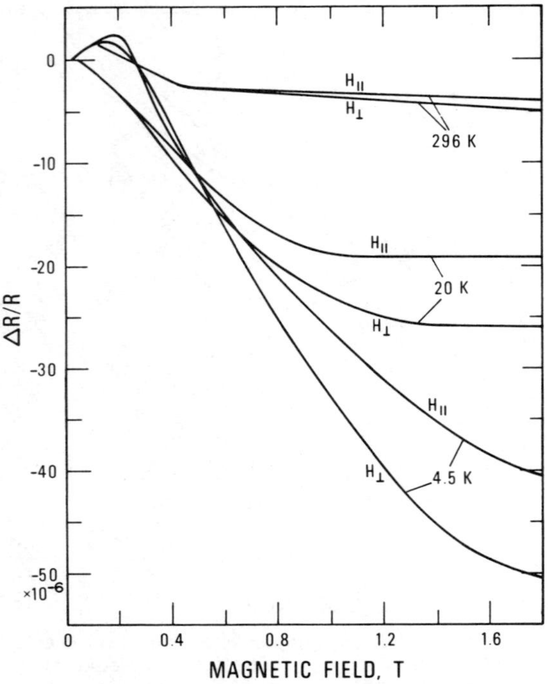

Figure 14.39 Relative change in electrical resistance as a function of magnetic field for Ni–Cr wire gauges. Longitudinal and transverse fields were applied at temperatures of 4.5, 20, and 296 K (Greenough and Lee, 1967a).

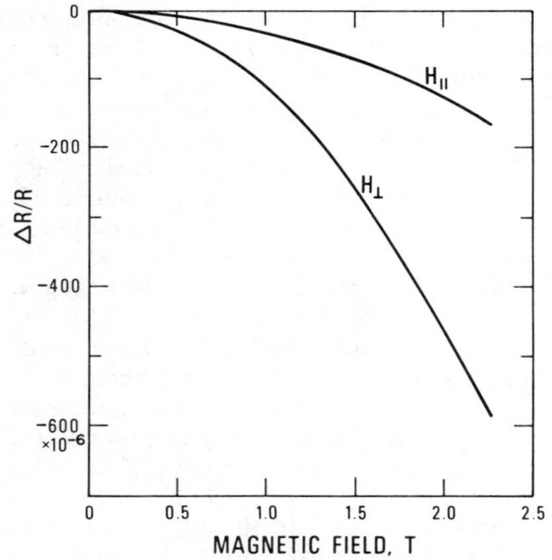

Figure 14.40 Relative change in the room-temperature electrical resistance as a function of magnetic field for silicon strain gauges. The field effect increases as the temperature decreases and is orientation dependent (Greenough and Lee, 1967a).

change (ΔL) for overlapping cylinders with radii r_1 and r_2. As Eq. 14.66 indicates, the change in capacitance is independent of length. Figure 14.41 shows a concentric cylinder device that has been used to 1.5 K with a sensitivity of 10^{-6}. Several other devices used to measure changes in length are discussed in Chapter 3.

14.5 MAGNETIC FIELD MEASUREMENT

Several methods are used to detect magnetic field strength (Watson and DiGregorio, 1965). The most widely used are nuclear magnetic resonance (NMR), Faraday coil, coil integrator, magnetoresistance, and the Hall effect. Superconducting quantum interference devices (SQUIDs) have been developed recently and are used to measure very small fields (Silver and Zimmerman, 1975). Other methods used to a lesser extent include electronic magnetic resonance (EMR), and flux gate magnetometers. In some cases these methods offer unique advantages. Important factors in selecting the appropriate sensor for a particular situation are re-

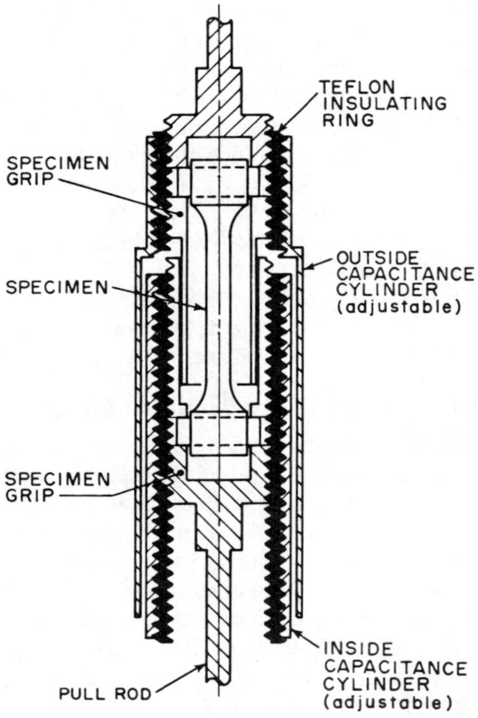

Figure 14.41 Schematic of an overlapping cylinder capacitance measuring device used to measure strain (Reed and Durcholz, 1970).

quired accuracy, available working space, field homogeneity, and temperature.

14.5.1 Nuclear magnetic resonance

Nuclei with nonzero spin quantum numbers and magnetic moments absorb electromagnetic energy at frequencies proportional to the applied magnetic field strength. Absorption occurs at the natural resonant frequencies and allows the field strength to be determined by measuring a radio frequency (0.1 to 100 MHz). These measurements can be made quite accurately. Nuclei are available that allow measurements to be made at cryogenic temperatures and over a wide range of magnetic fields. NMR techniques have traditionally been limited to relatively homogeneous and constant or slowly varying fields. Recent electronic innovations, however, have relaxed these limitations.

When the frequency supplied by an external oscillator matches the resonant frequency of the sample material (at the applied magnetic field value), energy is absorbed from the oscillator signal, and there is a small change in level of oscillation. To detect the resonant frequency, the sample nuclei are generally subjected to a small ac field superimposed on the steady-state applied field. This field is generated in small coils built into the probe itself, and its frequency (frequently 60 Hz) is in the audio range. The amplitude of the radio-frequency oscillation is applied to the y-axis of an oscilloscope, and the x-axis is swept with the signal to the ac field. When the applied radio frequency corresponds to the nuclear resonant frequency at the steady-state field, a series of oscillation level changes is displayed at the frequency of the ac magnetic field. Figure 14.42a shows an oscilloscope trace at resonance for a very homogeneous magnetic field.

The actual resonant frequency has a finite width, as indicated in Fig. 14.42a. Since the total signal is composed of the oscillations from each differentially small volume of the sample, an inhomogeneous field will cause different parts to resonate at different frequencies. Figure 14.42b and c illustrate the spreading of the line width with decreasing field homogeneity.

NMR is the most accurate method available for determining field strength. Accuracies on the order of 0.001% may be realized in very homogeneous fields. A wide variety of nuclei have been used in NMR measurements. Figure 14.43 shows the resonant frequency as a function of magnetic field for several nuclei (Andrew, 1958). The slope

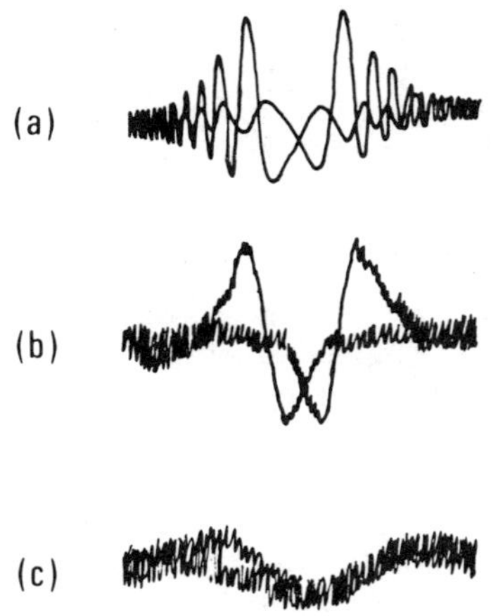

Figure 14.42 Oscilloscope trace of nuclear magnetic resonance in (a) a very homogeneous magnetic field, (b) a relatively homogeneous magnetic field, and (c) an inhomogeneous magnetic field.

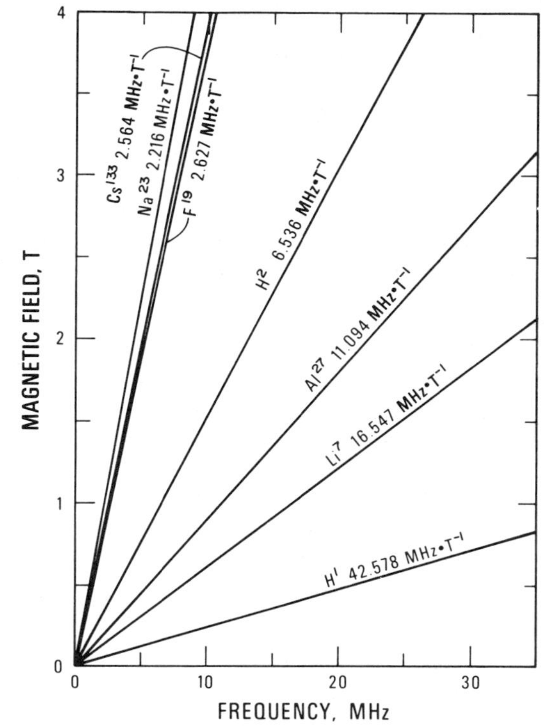

Figure 14.43 Magnetic field as a function of frequency for several nuclei frequently used in NMR detectors. The slopes of these field-dependent frequencies are unique and are known as gyromagnetic ratios.

of these curves is known as the gyromagnetic ratio and is unique for each type of nucleus. Nuclei that may be utilized at cryogenic temperatures and high magnetic fields include Al^{27}, Cs^{133}, Na^{23}, and F^{19}.

14.5.2 Faraday coil gaussmeters

Faraday coil gaussmeters operate on the principle that a voltage is generated in a coil when the magnetic flux linking the coil changes. This voltage

$$V = N \cdot 10^{-8}(d\phi / dt) \qquad (14.67)$$

is proportional to the rate of change of flux, $d\phi/dt$, and to the number of turns, N, in the coil. The relationship given in Eq. 14.67 is general and applies to either a moving coil in a stationary field or a stationary coil in a changing field. The phenomenon is widely used in two types of field measuring devices: the rotating coil gaussmeter and various coil integrator circuits.

Figure 14.44 shows schematically the basic operating parts used in the rotating devices (Lush, 1964). A sine wave voltage, V, is generated when a coil of N turns and area A rotates in a magnetic field, H. As seen in Eq. 14.68, the voltage is proportional to the field component perpendicular to the axis of the coil. The line frequency, f, is typically controlled to better than 0.1%.

$$V = NAH \cdot 10^{-8} \cdot 2\pi f \cos(2\pi f t) \qquad (14.68)$$

When the field H is not homogeneous, the generated voltage is not a perfect sine form. The average rectified voltage, however, still represents the field at the center of the coil.

The basic system shown in Fig. 14.44 can be modified to include two rotating coils with axes of rotation at right angles. Separate output from these coils allows three-dimensional mapping of the field. For extremely accurate measurements, an ac reference generator can be incorporated externally so that it rotates at the coil frequency and exactly 180° out of phase with the pickup coil. The signals from the generator and rotating coil can be compared using the null method and a precision ac potentiometer or measured directly with a digital voltmeter. The field strength can be determined with an accuracy on the order of 0.1% using this approach. The rotating coil system is used as the working standard in many laboratories.

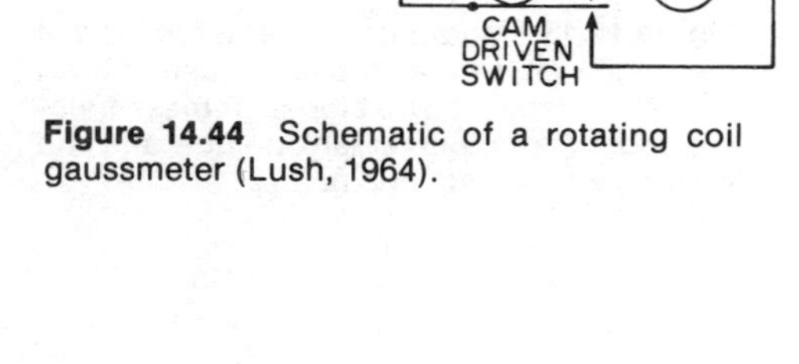

Figure 14.44 Schematic of a rotating coil gaussmeter (Lush, 1964).

The rotating coil is basically an ambient-temperature device. This limitation has been circumvented by encasing the rotating coil(s) in a nonmagnetic vacuum-insulated finger so that it can be operated at cryogenic temperatures.

Coil diameters of commercial units vary from 1.27 mm for measuring high fields (up to 20 T) to 25.4 mm for low-field measurement. The delicate coils are ordinarily used while encased in a protective tube, which increases the minimum working diameter to 3.2 mm. The low-temperature insulation further increases the probe diameter to about 22 mm.

The second application of Faraday's law to the measurement of magnetic fields also involves changes in linking flux densities due to changing fields or coil movement (rotating coils are an important subset). Other motions sometimes used include vibration, oscillation, and linear displacement. Pulsed fields or ac fields are well suited to this type of measurement.

The induced voltage, V_c, may be monitored and related to the field strength by electronic circuits, which essentially integrate the coil flux change, that is,

$$\Delta V = 1/RC\int_{t_1}^{t_2} V_c \, dt \propto B_2 - B_1 \qquad (14.69)$$

where R and C are resistive and capacitive components in the integrator circuit, B is the magnetic field intensity, and t is time (Watson and DiGregorio, 1965). Electronic circuits are available that allow adjustment of t_1 and t_2.

14.5.3 Hall effect gaussmeters

Figure 14.45a schematically illustrates the Hall effect. The Hall sample is a thin, rectangular strip of metal or semiconductor. An electric field, E_x, causes electrons to flow in the x direction, as shown in Fig. 14.45b. When a magnetic field is

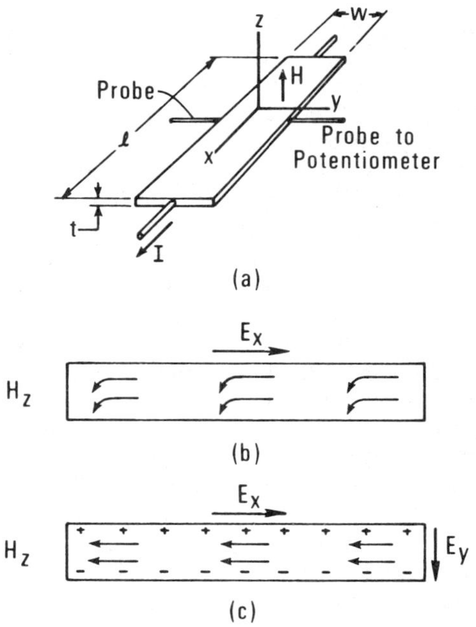

Figure 14.45 Schematics of (a) Hall effect field detection device and the relationships among current, magnetic field, and Hall voltage, (b) electron drift in the presence of the electric field, E_x, and the magnetic field H_z, out of the plane of the paper and (c) electron drift at equilibrium after the field E_y is established to balance the Lorentz force due to H_z.

applied, as indicated in Fig. 14.45c, the electrons experience a Lorentz force and deviate from the zero-field drift. The result, as indicated in Fig. 14.45c, is that an overabundance of electrons builds up on one edge of the Hall sample. This leaves a positive ion excess on the opposite edge. The concentration of electrons continues to build until an electric field, E_y, is established that exactly counteracts the effect of the magnetic field. Once this balancing field is established, the bulk electron drift returns to the zero-field mode. The magnitude of the Hall voltage, which is directly proportional to the magnetic field, H, is given by

$$E = R_H IH/t \tag{14.70}$$

where R_H is the Hall constant for the material, I is the current in the x direction (generally called the control current), and t is the specimen thickness.

Hall effect sensors, which provide very usable characteristics over the temperature range $1.5 \gtrsim T \lesssim 300$ K and in magnetic fields up to 25 T, are commercially available. These devices offer several advantages over both NMR and rotating coil magnetometers: (a) They are small — typically about 10 mm × 10 mm × 1 mm. (b) Power dissipation from self-heating is low owing to small measuring currents. Low-level current flow allows use of small electrical leads, which further limit the heat leak into the system. (c) Equipment associated with this method includes a voltage measuring device, such as a potentiometer or digital voltmeter, and a high-quality current source. More sophisticated systems using ac current sources are also used.

Three commercially available probes have been thoroughly tested at low temperatures and high fields by Rubin, Nelson, and Sample (1975) and Sample and Rubin (1976). The materials used in these probes were In–As, In–Sb, and Ga–As. These tests were conducted in fields up to 23 T and at temperatures of 300, 77, and $1.5 \lesssim T \lesssim 4.2$ K. Figure 14.46 shows a rather typical Hall voltage vs. magnetic field relationship at 4.2 K. Sample and Rubin found that when these experimental data were fitted with a straight line, differences appeared between the calculated ($V = A \cdot H + C$) and experimental voltages. Figure 14.47 shows the Hall voltage deviations from a straight line for an In–Sb probe at 4.2, 77, and 300 K. These data indicate that a change in slope of at least 2.5% is possible in the 3- to 4-T range. Temperature effects on the slope (sensitivity) are

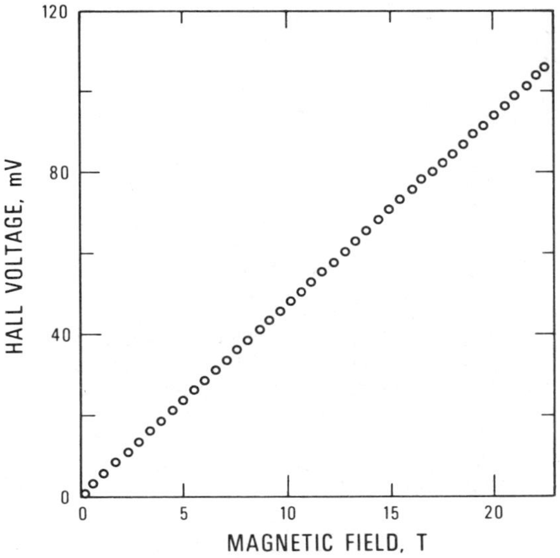

Figure 14.46 Hall voltage as a function of magnetic field at 4.2 K for a typical commercially available Hall probe (Sample and Rubin, 1976).

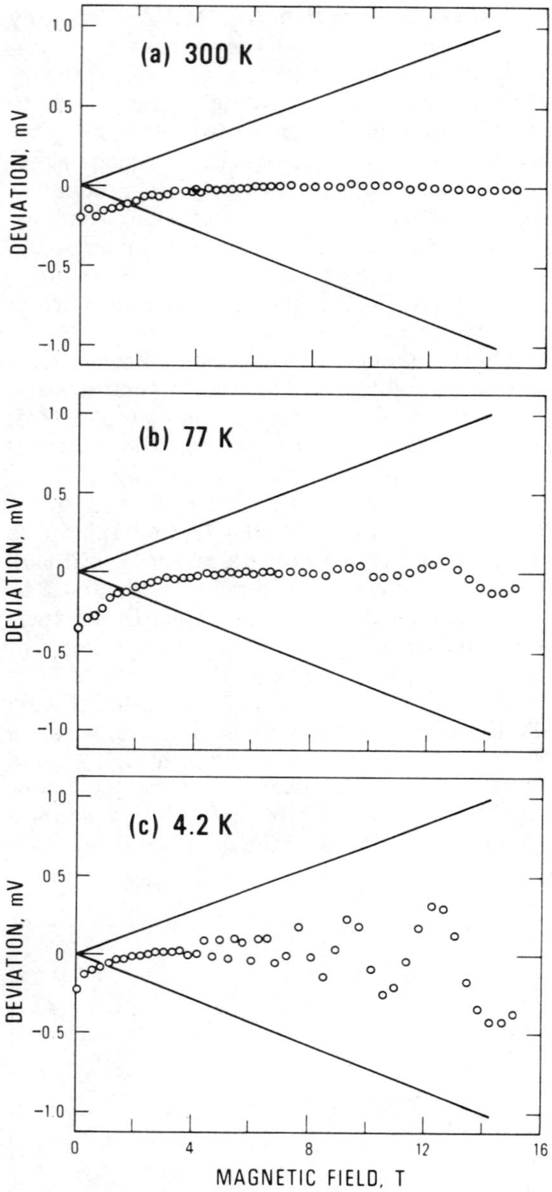

Figure 14.47 Deviations as a function of magnetic field for a In–Sb Hall probe at (a) 300 K, (b) 77 K, and (c) 4.2 K. The oscillatory behavior increases as the temperature decreases; the slope change at low fields is relatively temperature independent. The solid lines represent 1% of the total Hall voltage. (Rubin et al., 1975.)

less than 1%. The oscillatory behavior shown by these data decreases strongly as the temperature increases.

Commercially available probes suffer from varying degrees of nonreproducibility upon thermal cycling. Figure 14.48 indicates the observed change in sensitivity for three commercial probes as a result of being cycled between 4.2 K and room temperature.

Successful use of a Hall probe involves several considerations. An *in situ* calibration of a Hall probe is an extremely important step since small changes in the orientation of the probe with respect to the field can cause errors of several percent. Thermal cycling will cause the slope of the Hall voltage–magnetic field curve to change. Errors due to this change can be several percent. Both the mechanical movement and the thermal cycling effects are countered by *in situ* recalibration at a few field values. Errors caused by quantum oscillations ($H \gtrsim 6$ T) and nonlinear slope changes ($H \lesssim 4$ T) are quite reproducible, so that once they are determined in the calibration procedure, corrections may be made.

Electrical resistivity can also be used to determine magnetic field strength. It is a function of both magnetic field and temperature; this field dependence was discussed in connection with temperature (§ 14.3.6) and strain (§ 14.4.3) measurements.

Magnetoresistance is due to Lorentz forces acting on moving charges in a conductor. Under the influence of a magnetic field these charges follow a helical path. This path is longer than in the zero-field case so that, since the mean free path is constant, a charge suffers more collisions and the resistivity increases. The field sensitivity

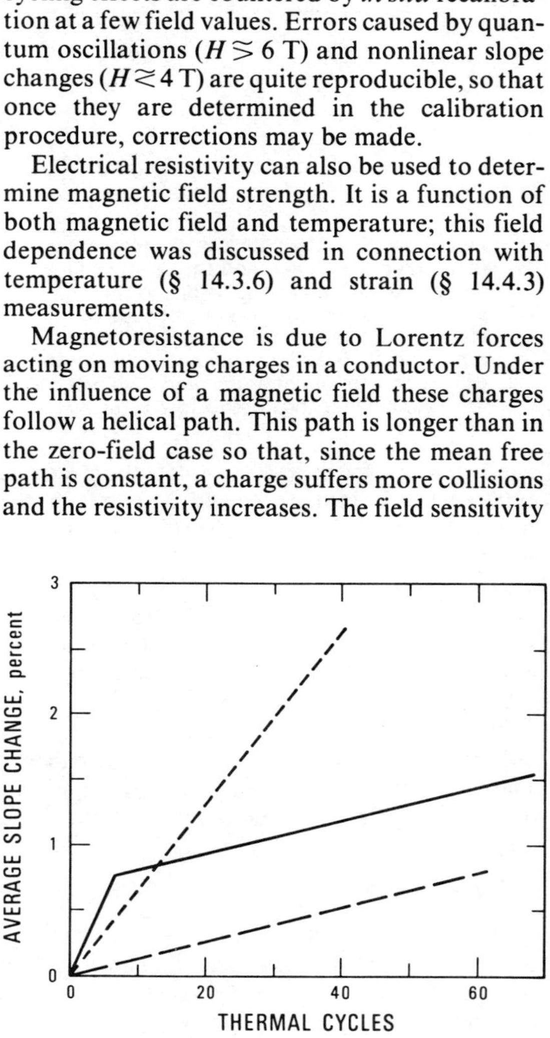

Figure 14.48 Average change in slope as a function of thermal cycling between ambient temperature and 4.2 K for three commercially available Hall probes (Sample and Rubin, 1976).

of a magnetoresistive sensor is dependent on the length-to-area ratio of the wire. The need for long, thin wires eliminates many semiconductors with potentially useful magnetoresistive coefficients because of fabrication problems. Copper and bismuth are the most commonly used materials in this type of magnetometer.

The field sensitivity of copper is adequate for use at low temperatures, down to about 4 K. The magnetoresistance of copper varies as H^2 (Hsu and Kunzler, 1963); this dependence limits the use of these sensors to fields greater than a few tenths tesla. The usable range of copper resistors can be extended to lower fields by using higher purity copper; as $R_{273\,K}/R_{4\,K}$ increases, the field sensitivity also increases.

The temperature and magnetic field dependence requires that the probe be calibrated for field effect at the same temperature that it will be used. Scott, Springford, and Stockton (1968) describe a noninductively wound copper magnetoresistor used in a superconducting solenoid. Their coil was wrapped on a former which was situated in the magnet cooling fluid (outside of the insert Dewar). This design (Fig. 14.49a) keeps the coil out of the limited working space and in a constant temperature bath. Figure

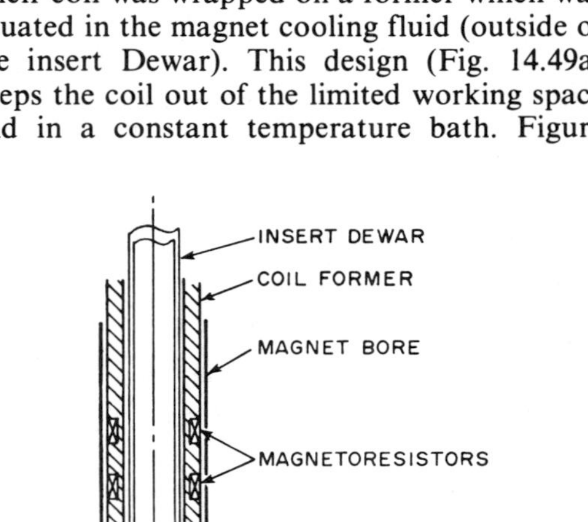

Figure 14.49 (a) Schematic cross section of the placement of copper coils in a superconducting magnet system (Scott et al., 1968). (b) Schematic of movable set of coils used to measure magnetic regions (Hsu and Kunzler, 1963).

14.49b schematically shows the set of copper coils used by Hsu and Kunzler (1963). This probe is made up of several independent copper coils so that the field in the region of each coil may be determined.

Dols, Skiff, and Watson (1958) developed a bismuth probe for field measurements. Bismuth sensors have sufficient sensitivity to be used at ambient temperature, whereas copper does not. Their system included a constant temperature-control system so that all measurements could be made near 298 K. Their work indicates that a resolution of about 0.2 mT in a field of 0.5 T is possible. A serious disadvantage of a movable bismuth probe, as opposed to a fixed coil, is the sensitivity to orientation in the field.

14.6 REFERENCES

Ahlers, G. and Macre, T. F. (1966). Temperature–resistance relation for germanium thermometers. *Rev. Sci. Instrum.* **37**, 962–963.

Aldridge, R. V., Davis, K., and Holloway, M. (1975). An investigation of the effect of a magnetic field on the forward characteristics of some silicon diodes at low temperatures. *J. Phys. D* **8**, 64–68.

Allen, L. D., Bradford, E. W., and Crabtree, R. D. (1970). Thermocouple tempering in cryogens. *Cryog. Ind. Gases* **5**, 19–20.

Anderson, A. C. (1969). The thermal grounding of electrical leads at low temperatures. *Rev. Sci. Instrum.* **40**, 1502–1503.

Anderson, A. C. (1972). Carbon resistance thermometry. In *Temperature — Its Measurement and Control in Science and Industry*, Vol. 4, Part 2. Instrument Society of America, Pittsburgh, Pennsylvania, 773–784.

Anderson, R. L. (1972). The high temperature stability of platinum resistance thermometers. In *Temperature — Its Measurement and Control in Science and Industry*, Vol. 4, Part 2. Instrument Society of America, Pittsburgh, Pennsylvania, 927–934.

Andrew, E. R. (1958). *Nuclear Magnetic Resonance.* Cambridge University Press, London.

Arvidson, J. M. and Brennan, J. A. (1975). Pressure measurement. In *ASRDI Oxygen Technology Survey*, Vol. VIII. NASA SP-3092.*

Ashworth, T., Johnson, L. R., Hsiung, C. Y., and Kreitman, M. M. (1973). Use of linear heat flow for poor conductors and its application to the thermal conductivity of nylon. *Cryogenics* **13**, 34–40.

Barber, C. R. (1950). Platinum resistance thermometers of small dimensions, *J. Sci. Instrum.* **27**, 47–49.

Barber, C. R. (1955). Platinum resistance thermometer for use at low temperatures. *J. Sci. Instrum.* **32**, 416–417.

Barber, C. R. (1962). Low-temperature scales 10 to 90° K. In *Temperature — Its Measurement and Control in Science and Industry*, Vol. 3. Part I. Reinhold Publishing Corp., New York, 345–350.

Bedford, R. E., Durieux, M., Muijlwijk, R., and Barber, C. R. (1969). Relationships between the international practical temperature scale of 1968 and the NBS-55, NPL-61, PRMI-54, and PSU-54 temperature scales in the range from 13.81 to 90.188 K. *Metrologia* **5**, 47–49.

Berman, R., Brock, J. C. F., and Huntley, D. J. (1964). Properties of gold + 0.03 percent (at.) iron thermoelements between 1 and 300 K and behavior in a magnetic field. *Cryogenics* **4**, 223–239.

Berman, R. and Huntley, D. J. (1963). Dilute gold–iron alloys as thermocouple material for low temperature heat conductivity measurements. *Cryogenics* **3**, 70–76.

Berry, R. J. (1967a). Ideal resistivity of platinum below 20° K. *Can. J. Phys.* **45**, 1693–1708.

Berry, R. J. (1967b). Platinum resistance thermometry below 10° K. *Metrologia* **3**, 53–57.

Berry, R. J. (1972). The influence of crystal defects in platinum on platinum resistance thermometry. In *Temperature — Its Measurement and Control in Science and Industry*, Vol. 4, Part 2. Instrument Society of America, Pittsburgh, Pennsylvania, 937–949.

Berry, R. J. (1980). Effect of Pt oxidation on Pt resistance thermometry. *Metrologia* **16**, 117–126.

Besley, L. M. (1979). Stability characteristics of carbon–glass resistance thermometers. *Rev. Sci. Instrum.* **50**, 1626–1628.

Besley, L. M. and Kemp, R. C. (1978). Two-point calibration of standard capsule platinum resistance thermometers for the range 13.81 K to 273.15 K. *Cryogenics* **18**, 497–500.

Besley, L. M. and Plumb, H. H. (1978). Stability of germanium resistance thermometers at 20 K. *Rev. Sci. Instrum.* **49**, 68–73.

Blakemore, J. S. (1962). Design of germanium for thermometric applications. *Rev. Sci. Instrum.* **33**, 106–112.

Blakemore, J. S., Winstel, J., and Edwards, R. V. (1970). Computer fitting of germanium thermometer characteristics. *Rev. Sci. Instrum.* **41**, 835–842.

Blakemore, J. S. (1972). Germanium resistance thermometers: resistance vs. temperature and thermal time constant characteristics. In *Temperature — Its Measurement and Control in Science and Industry*, Vol. 4, Part 2. Instrument Society of America, Pittsburgh, Pennsylvania, 827–833.

Borelius, G., Keesom, W. H., Johansson, C. H., and Linde, J. O. (1931). Measurements of thermoelectric forces of lead and tin down to the temperature of liquid helium. *Proc. K. Akad. Wet. Amsterdam* **34**, 1365–1372.

Borovick-Romanov, A. C., Strelkov, P. G., Orlova, M. P., and Astrov. D. N. (1962). The I.M.P.R. temperature scale for the 10 to 90° K region. In *Temperature — Its Measurement and Control in Science and Industry*, Vol. 3, Part 1. Reinhold Publishing Corp., New York, 113–128.

Brickwedde, F. G., van Dijk, H., Durieux, M., Clement, J. R., and Logan, J. K. (1960). *The 1958 He^4 Scale of Temperatures*. NBS Monograph 10, National Bureau of Standards, Washington, D.C.

Brown, N. and Lukens, K. F. (1961). Microstrain in polycrystalline metals. *Acta. Metall.* **9**, 106–111.

Bureau International des Poids et Mesures (1979). The provisional 0.5 K to 30 K temperature scale. *Metrologia* **15**, 65–68.

Burgess, G. K. (1928). The international temperature scale. *J. Res. Natl. Bur. Stand.* **1**, 635–640.

Burley, N. A. (1969). Solute depletion and thermo-e.m.f. drift in nickel-base thermocouple alloys. *J. Inst. Met.* **97**, 252–254.

Burley, N. A. (1972). Nicrosil and nisil: highly stable nickel-based alloys for thermocouples. In *Temperature — Its Measurement and Control in Science and Industry*, Vol. 4, Part 3. Instrument Society of America, Pittsburgh, Pennsylvania, 1677–1695.

Burley, N. A. and Acklund, R. G. (1967). The stability of thermo-emf/temperature characteristics of nickel-based thermocouples. *J. Aust. Inst. Met.* **12**, 23–31.

Campbell, B. R. (1979). Get superior performance from reed relays in switching low-level signals. *Electron. Des.* **27**, 58–61.

Cataland, G. and Plumb, H. H. (1965a). Acoustical thermometer. *Science* **150**, 155–161.

Cataland, G. and Plumb, H. (1965b). Isotherms determined by the National Bureau of Standards acoustical thermometer in the liquid helium temperature range. *J. Res. Natl. Bur. Stand.* **69A**, 531–534.

Catalano, E., Shroyer, B. L., and English, J. C. (1970). Numerical handling of germanium resistance thermometer data with small digital computers. *Rev. Sci. Instrum.* **41**, 1663–1665.

Chaing, C. K. (1974). The absolute thermopower of some low temperature thermocouple wires in high transverse magnetic fields. *Rev. Sci. Instrum.* **45**, 985–989.

Chattle, M. V. (1972). Platinum resistance thermometry up to the gold point. In *Temperature — Its Measurement and Control in Science and Industry*, Vol. 4, Part 2. Instrument Society of America, Pittsburgh, Pennsylvania, 907–918.

Clement, J. R. and Quinnell, E. H. (1952). Low temperature characteristics of carbon-composition thermometers. *Rev. Sci. Instrum.* **23**, 213–216.

Cohen, B. G., Snow, W. B., and Tretola, A. R. (1963) GaAs *p–n* junction diodes for wide range thermometry. *Rev. Sci. Instrum.* **34**, 1091–1093.

Coles, B. R. (1964). New type of low-temperature resistance anomaly in alloys. *Phys. Lett.* **8**, 243–244.

Collier, R. S., Sparks, L. L., and Strobridge, T. R. (1973). *Carbon Thin Film Thermometry*. NBS Information Report 74-355, National Bureau of Standards, Boulder, Colorado.

Comité International des Poids et Mesures (1969). International practical temperature scale of 1968. *Metrologia* **5**, 35–44.

Comité International des Poids et Mesures (1976). The international practical temperature scale of 1968, Amended edition of 1975. *Metrologia* **12**, 7–17.

Corruccini, R. J. (1951). Annealing of platinum for thermometry. *J. Res. Natl. Bur. Stand.* **47**, 94–103.

Corruccini, R. J. (1960). Interpolation of platinum resistance thermometers, 20° to 273.15° K. *Rev. Sci. Instrum.* **31**, 637–640.

Corruccini, R. J. (1962). Interpolation of platinum resistance thermometers, 10° to 273.15° K. In *Temperature — Its Measurement and Control in Science and Industry*, Vol. 3, Part 1. Reinhold Publishing Corp., New York, 329–338.

Coxon, W. F. (1960). *Temperature Measurement and*

Control. MacMillan, New York, 4–10.

Daneman, H. L. and Mergner, G. C. (1967). Precise resistance thermometry — a review. *Instrum. Technol.* **14**, 59–65.

Dauphinee, T. M. and Preston-Thomas, H. (1954). A copper resistance temperature scale. *Rev. Sci. Instrum.* **25**, 884–886.

Denner, H. (1969). Thermal conductivity of adhesives at low temperatures. *Cryogenics* **9**, 282–283.

Diamond, J. M. (1966). A Mueller bridge set for cryogenic temperature measurements. *J. Sci. Instrum.* **43**, 576–580.

Dike, P. H. (1954). *Thermoelectric Thermometry*, first edition. Leeds and Northrup, Philadelphia, Pennsylvania.

Dols, C. G., Skiff, E. W., and Watson, P. G. (1958). Temperature-regulated bismuth resistor for magnetic field measurements. *Rev. Sci. Instrum.* **29**, 349–354.

Douglas, B. T. (1969). Conversion of existing calorimetrically determined thermodynamic properties to the basis of the international practical temperature scale of 1968. *J. Res. Natl. Bur. Stand.* **73A**, 451–470.

Droms, C. R. (1962). Thermistors for temperature measurements. In *Temperature — Its Measurement and Control in Science and Industry*, Vol. 3, Part 2. Reinhold Publishing Corp., New York, 339–376.

Durieux, M., Astrov, D. N., Kemp, W. R. G., and Swenson, C. A. (1979). The deviation and development of the 1976 provisional 0.5 K to 30 K temperature scale. *Metrologia* **15**, 57–63.

Edlow, M. H. and Plumb, H. H. (1962). Resistance thermometry in the liquid helium temperature range. In *Temperature — Its Measurement and Control in Science and Industry*, Vol. 3, Part 1. Reinhold Publishing Corp., New York, 407–411.

Edlow, M. H. and Plumb, H. H. (1966). Reproducibility of germanium resistance thermometers at 4.2° K. *J. Res. Natl. Bur. Stand.* **70C**, 245–254.

Ekin, J. W. and Wagner, D. K. (1970). A simple ac bridge circuit for use in four-terminal resistance thermometry. *Rev. Sci. Instrum.* **41**, 1109–1110.

Evans, J. P. (1972). High temperature platinum resistance thermometry. In *Temperature — Its Measurement and Control in Science and Industry,* Vol. 4, Part 2. Instrument Society of America, Pittsburgh, Pennsylvania, 899–906.

Finnemore, D. K., Ostenson, J. E., and Stromberg, T. F. (1965). Secondary thermometer for the 4 to 20 K range. *Rev. Sci. Instrum.* **36**, 1369–1370.

Freynik, H. S., Roach, D. R., Deis, D. W., and Hirzel, D. G. (1978). Evaluation of metal-foil strain gauges for cryogenic application in magnetic fields. In *Advances in Cryogenic Engineering*, Vol. 24. Plenum Press, New York, 473–479.

Friedberg, S. A. (1955). Semiconductors as thermometers. In *Temperature — Its Measurement and Control in Science and Industry*, Vol. 2. Reinhold Publishing Corp., New York, 359–380.

Gehring, F. D. and Gerstein, B. C. (1967). Effect of thermal shocking on Minco model 51059 platinum resistance thermometers. *Rev. Sci. Instrum.* **38**, 280–281.

Gerber, J. A. and Sellmyer, D. J. (1978). Anomalous behavior of silicon diode thermometers. *Cryogenics* **18**, 619–620.

Gopal, E. S. R. (1966). *Specific Heats at Low Temperatures.* Plenum Press, New York, 33.

Greenough, R. D. and Lee, E. W. (1967a). Behavior of electrical resistance strain gauges at low temperatures. *Cryogenics* **7**, 10–12.

Greenough, R. D. and Lee, E. W. (1967b). Behavior of electrical resistance strain gauges at low temperatures, I. Magnetoresistance. *Cryogenics* **7**, 7–9.

Grilly, E. R. (1962). The vapor pressure of solid and liquid neon. *Cryogenics* **2**, 226–229.

Hall, L. A. (1968). *Survey of Electrical Resistivity Measurements on 16 Pure Metals in the Temperature Range 0 to 273° K.* NBS Technical Note 365, National Bureau of Standards, Boulder, Colorado.

Harris, F. K. (1952). *Electrical Measurements.* John Wiley & Sons, New York, 282–294.

Hayakawa, K. K. (1972). *Cryogenic Pressure Measurement Technology and Subjects Applied to Pressure Transducers.* Report SD72-SA-0156-1, North American Rockwell, Space Division.*

Herr, A. C., Terbeek, H. G., and Tiefermann, M. W. (1962). Suitability of carbon resistors for field measurement of temperatures in the range of 35 to 100° R. In *Temperature — Its Measurement and Control in Science and Industry*, Vol. 3, Part 2. Reinhold Publishing Corp., New York, 355–359.

Hetzler, M. C. and Walton, D. (1968). New interpolation formula for carbon resistance thermometry. *Rev. Sci. Instrum.* **39**, 1656–1657.

Hoge, H. J. and Brickwedde, F. G. (1939). Establishment of a temperature scale for the calibration of thermometers between 14° and 83° K. *J. Res. Natl. Bur. Stand.* **22**, 351–373.

Holten, D. C. (1963). *Static and Dynamic Behavior of Helium Gas Thermometers Below 77° Kelvin.* UCRL 7327, Lawrence Radiation Laboratory, University of California.

Hsu, F. S. L. and Kunzler, J. E. (1963). Magnetoresistive probe for measuring magnetic field intensity in a small space. *Rev. Sci. Instrum.* **34**, 297–298.

Hust, J. G. (1969). A compilation and historical review of temperature scale differences. *Cryogenics* **9**, 443–455.

Hust, J. G. (1970). Thermal anchoring of wires in cryogenic apparatus. *Rev. Sci. Instrum.* **41**, 622–624.

Hust, J. G., Powell, R. L., and Weitzel, D. H. (1970). Thermal conductivity standard reference materials from 4 to 300 K, I. Armco iron: including apparatus description and error analysis. *J. Res. Natl. Bur. Stand.* **74A**, 673–690.

Jacobsen, R. T., Stewart, R. B., McCarty, R. D. and Hanley, H. J. M. (1973). *Thermophysical Properties of Nitrogen from the Fusion Line to 3500 R (1944 K) for Pressures to 150,000 psia (10342 × 10^5 N/m^2).* NBS Technical Note 648, National Bureau of Standards, Boulder, Colorado.

Jan, J. P., Pearson, W. B., and Templeton, I. M. (1958). Thermoelectricity at low temperatures V. The suitability of lead as a standard reference material *Can. J. Phys.* **36**, 627–631.

Johnson, W. L. and Anderson, A. C. (1971). The stability of carbon resistance thermometers. *Rev. Sci. Instrum.* **42**, 1296–1299.

Kaugerts, J. (1979). Private communication, Brookhaven National Laboratory, Upton, Long Island,

New York.
Kirby, C. G. M. and Laubitz, M. J. (1973). The error due to the Peltier effect in direct-current measurements of resistance. *Metrologia* **9**, 103–106.
Kittel, C. (1967). *Introduction to Solid State Physics*, third edition. John Wiley & Sons, New York.
Knittel, T. (1973). Thermoelectric power dependence of a gold–iron alloy wire on the magnetic field orientation. *Cryogenics* **13**, 370–371.
Kopp, F. J. and Ashworth, T. (1972). Carbon resistors as low temperature thermometers. *Rev. Sci. Instrum.* **43**, 327–332.
Kopp, J. and Slack, G. A. (1971). Thermal contact problems in low temperature thermocouple thermometry. *Cryogenics* **11**, 22–25.
Kos, J. F., Drolet, M., and Lamarche, J. L. G. (1967). High precision indium resistance thermometers. *Can. J. Phys.* **45**, 2787–2795.
Kos, J. F. and Lamarche, J. L. G. (1967). The electrical resistivity of thermometrically pure platinum below 11° K. *Can. J. Phys.* **45**, 339–354.
Kreitman, M. M. (1969). Low temperature conductivity of several greases. *Rev. Sci. Instrum.* **40**, 1562–1565.
Kreitman, M. M., Ashworth, T., and Rechowicz, M. (1972). A correlation between thermal conductance and specific heat anomalies and the glass temperature of apiezon N and T greases. *Cryogenics* **12**, 32–34.
Kreitman, M. M. and Callahan, J. T. (1970). Thermal conductivity of apiezon N grease at liquid helium temperatures. *Cryogenics* **10**, 155–159.
Kunzler, J. E., Geballe, T. H., and Hull, G. W. (1962). Germanium resistance thermometers. *Temperature — Its Measurement and Control in Science and Industry*, Vol. 3, Part I. Reinhold Publishing Corp., New York, 391–397.
Labrie, R., Infantes, M., and Vanier, J. (1971). A practical nuclear quadrupole resonance thermometer. *Rev. Sci. Instrum.* **42**, 26–30.
Lake Shore Cryotronics, Inc. Rhodium–iron thermometer. In *Lake Shore Cryotronics, Inc. Technical Specification RFT,* Westerville, Ohio.
Lang, S. B. (1972). The pyroelectric thermometer: a sensor for measuring extremely small temperature changes or rates of temperature change. In *Temperature — Its Measurement and Control in Science and Industry,* Vol. 4, Part 2. Instrument Society of America, Pittsburgh, Pennsylvania, 1153–1167.
Lang, S. B., Rice, L. H., and Shaw, S. A. (1969). Pyroelectric effect in barium titanate ceramic. *J. Appl. Phys.* **40**, 4335–4340.
Lang, S. B., Shaw, S. A., Rice, L. H., and Timmerhaus, K. S. (1969). Pyroelectric thermometer for use at low temperatures. *Rev. Sci. Instrum.* **40**, 274–284.
Lang, S. B. and Steckel, F. (1965). Study of the ultrasensitive pyroelectric thermometer. *Rev. Sci. Instrum.* **36**, 1817–1821.
Lawless, W. N. (1971). A low temperature glass–ceramic capacitance thermometer. *Rev. Sci. Instrum.* **42**, 561–566.
Lawless, W. N. (1972a). Thermometric properties of carbon-impregnated porous glass at low temperatures. *Rev. Sci. Instrum.* **43**, 1743–1747.
Lawless, W. N. (1972b). A low-temperature, glass-ceramic capacitance thermometer. In *Temperature — Its Measurement and Control in Science and Industry,* Vol. 4, Part 2. Instrument Society of America, Pittsburgh, Pennsylvania, 1143–1151.
Lawless, W. N. (1975). Aging phenomena in a low-temperature glass-ceramic capacitance thermometer. *Rev. Sci. Instrum.* **46**, 625–628.
Lawless, W. N. and Panchyk, E. A. (1972). Thermometer equations for low-temperature glass–ceramic capacitance thermometers. *Cryogenics* **12**, 196–200.
Lindenfeld, P. (1961). Tests and comparisons of carbon and germanium thermometers. *Rev. Sci. Instrum.* **32**, 9.
Lounasmaa, O. V. (1958). A simple formula for use with carbon thermometers at low temperatures. *Philos. Mag.* **3**, 652.
Lucas, D. A. (1972). Miniature precision platinum resistance thermometers. In *Temperature — Its Measurement and Control in Science and Industry,* Vol. 4, Part 2. Instrument Society of America, Pittsburgh, Pennsylvania, 963–969.
Lush, M. J. (1964). Rotating coil gaussmeters. *Instrum. Control Syst.* **37**, 111–113.
Mann, D. B. (1974). Flow measurement instrumentation. In *ASRDI Oxygen Technology Survey,* Vol. VI. NASA SP-3084.*
McCarty, R. D. (1972). *Thermophysical Properties of Helium-4 from 2 to 1500 K with Pressures to 1000 Atmospheres.* NBS Technical Note 631, National Bureau of Standards, Boulder, Colorado.
McCarty, R. D. and Stewart, R. B. (1965). Thermodynamic properties of neon from 25 to 300° K between 0.1 and 200 atmospheres. In *Advances in Thermophysical Properties at Extreme Temperatures and Pressures.* American Society of Mechanical Engineers, New York, 84–97.
McClintock, R. M. (1959). Strain gauge calibration device for extreme temperatures. *Rev. Sci. Instrum.* **30**, 715–718.
McLaren, E. H. (1957). The freezing points of high purity metals as precision temperature standards, I. Precision measurements with standard resistance thermometers. *Can. J. Phys.* **35**, 78–90.
McTaggart, J. H. and Slack, G. A. (1969). Thermal conductivity of General Electric No. 7031 varnish. *Cryogenics* **9**, 384–385.
Meyers, C. H. (1932). Coiled filament resistance thermometers. *J. Res. Natl. Bur. Stand.* **9**, 807.
Micro-Measurements (1976). *Strain Gage Temperature Effects.* Tech. Note 128-2, Micro-Measurements, Romulus, Michigan.
Moessen, G. W., Aston, J. G., and Ascah, R. G. (1962). The Pennsylvania State University thermodynamic temperature scale below 90° K and the normal boiling points of oxygen and normal hydrogen on the thermodynamic scale. In *Temperature — Its Measurement and Control in Science and Industry,* Vol. 3, Part 1. Reinhold Publishing Corp., New York, 91–102.
Morrison, R. (1970). *Grounding and Shielding Techniques in Instrumentation.* Chapter 4. John Wiley & Sons, New York, 141–165; also in *Meas. Data* **4**, 151–165.
Morrison, R. (1979). Answers to grounding and shielding problems. *Instrum. Control Syst.* **52**, 35–38.

Neuringer, L. J., Perlman, A. J., Rubin, L. G., and Shapira, Y. (1971). Low temperature thermometry in high magnetic fields, III. germanium and platinum resistors. *Rev. Sci. Instrum.* **42**, 9–14.

Neuringer, L. J. and Shapira, Y. (1969). Low temperature thermometry in high magnetic field, I. Carbon resistors. *Rev. Sci. Instrum.* **40**, 1314–1321.

Orlova, M. P., Astrov, D. N., and Medvedeva, L. A. (1964). Indium resistance thermometers for the temperature range 3.4-300° K. *Cryogenics* **4**, 95–97.

Ott, H. W. (1976). *Noise Reduction Techniques in Electronic Systems.* John Wiley & Sons, New York.

Pearson, W. B. and Templeton, I. M. (1955). Thermoelectricity at low temperatures, III. The absolute scale of thermoelectric power: a critical discussion of the present scale at low temperatures and preliminary measurements towards its redetermination. *Proc. R. Soc.* **231A**, 534–544.

Penar, J. D. and Campi, M. (1971). Interpolation scheme for germanium resistance thermometers. *Rev. Sci. Instrum.* **42**, 528–529.

Pippard, A. B. (1957). *Elements of Classical Thermodynamics.* University Press, Cambridge, 46–48.

Plumb, H. and Cataland, G. (1966). Acoustical thermometer and the National Bureau of Standards provisional temperature scale 2-20 (1965). *Metrologia* **2**, 127–139.

Plumb, H. H. and Edlow, M. H. (1959). Constant temperature liquid helium bath and reproducibility of resistance thermometers. *Rev. Sci. Instrum.* **30**, 376–377.

Potts, J. F., Jr. and McElroy, D. L. (1962). The effects of cold working, heat treatment, and oxidation on the thermal emf of nickel-base thermoelements. In *Temperature — Its Measurement and Control in Science and Industry,* Vol. 3, Part 2. Reinhold Publishing Corp., New York, 243–263.

Powell, R. L., Bunch, M. D., and Gibson, E. F. (1960). Low-temperature transport properties of commercial metals and alloys, III. Gold–cobalt, *J. Appl. Phys.* **31**, 504–505.

Powell, R. L., Caywood, L. P., and Bunch, M. D. (1962). Low-temperature thermocouples. In *Temperature — Its Measurement and Control in Science and Industry,* Vol. 3, Part 2. Reinhold Publishing Corp., New York, 65–77.

Powell, R. L., Hall, W. J., Hyink, C. H., Sparks, L. L., Burns, G. W., Scroger, M. G., and Plumb, H. H. (1974). *Thermocouple Reference Tables Based on IPTS-68,* NBS Monograph 125, National Bureau of Standards, Boulder, Colorado.

Powell, R. L., Rogers, W. M., and Roder, H. M. (1960). Thermal conductivities of copper and copper alloys. In *Advances in Cryogenic Engineering,* Vol. 2. Plenum Press, New York, 166–171.

Powell, R. L., Sparks, L. L., and Hust, J. G. (1978). *Standard Reference Material 1967: Standard Thermocouple Material, Pt-67.* NBS Special Publication 260-56, National Bureau of Standards, Boulder, Colorado.

Reed, R. P. and Durcholz, R. L. (1970). Cryostat and strain measurement for tensile tests to 1.5 K. In *Advances in Cryogenic Engineering,* Vol. 15. Plenum Press, New York, 109–116.

Richards, D. B., Edwards, L. R., and Legvold, S. (1969). Thermocouples in magnetic fields. *J. Appl. Phys.* **40**, 3836–3837.

Ricketson, B. W. A. (1975). The 220 Ω Allen-Bradley resistor as a temperature sensor between 2 and 100 K. In *Temperature Measurement, 1975.* The Institute of Physics, London, England, 135–143.

Riddle, J. L., Furukawa, G. T., and Plumb, H. H. (1973). *Platinum Resistance Thermometry.* NBS Monograph 126, National Bureau of Standards, Washington, D.C.

Roberts, R. B. (1977). The absolute scale of thermoelectricity. *Philos. Mag.* **36**, 91–107.

Roder, H. M. (1974). Density and liquid level measurement instrumentation for the cryogenic fluids oxygen, hydrogen, and nitrogen. In *ASRDI Oxygen Technology Survey,* Vol. V. NASA SP-3083.*

Roder, H. M. and Weber, L. A. (1972). *Thermophysical Properties, Oxygen Technology Survey I.* National Aeronautical Space Administration Aerospace Safety Research and Data Institute.*

Roder, H. M., Weber, L. A., and Goodwin, R. D. (1965). *Thermodynamic and Related Properties of Parahydrogen from the Triple Point to 100° K at Pressures to 340 Atmospheres.* NBS Monograph 94, National Bureau of Standards, Boulder, Colorado.

Rosenbaum, R. I. (1968). Some properties of gold–iron thermocouple wire. *Rev. Sci. Instrum.* **39**, 890–899.

Rosenbaum, R. I. (1969). Some low temperature thermometry observations. *Rev. Sci. Instrum.* **40**, 577–583.

Rubin, L. G. (1979). Private communication, Francis Bitter National Magnet Laboratory, Massachusetts Institute of Technology, Cambridge, Massachusetts.

Rubin, L. G. (1980). Status of carbon resistors for low temperature thermometry. *Rev. Sci. Instrum.* **51**, 1007.

Rubin, L. G. and Golahny, Y. (1972). An improved ac bridge circuit for use in four-terminal resistance thermometry. *Rev. Sci. Instrum.* **43**, 1758–1762.

Rubin, L. G. and Lawless, W. N. (1971). Studies of a glass–ceramic capacitance thermometer in an intense magnetic field at low temperatures. *Rev. Sci. Instrum.* **42**, 571–573.

Rubin, L. G., Nelson, D. R., and Sample, H. H. (1975). Characterization of two commercially available Hall effect sensors for high magnetic fields and low temperatures. *Rev. Sci. Instrum.* **46**, 1624–1631.

Rusby, R. L. (1975). Resistance thermometers using rhodium–iron, 0.1 to 273 K. In *Temperature Measurement, 1975.* The Institute of Physics, London, England, 125–130.

Rusby, R. L., Chattle, M. V., and Gilhen, D. M. (1972). The calibration of resistance thermometers at low temperatures. *J. Sci. Instrum.* **5**, 1102–1105.

Sachse, H. B. (1962). Measurement of low temperatures with thermistors. In *Temperature — Its Measurement and Control in Science and Industry,* Vol. 3, Part 2. Reinhold Publishing Corp., New York, 347–353.

Sachse, H. B. (1975). *Semiconducting Temperature Sensors and Their Applications.* John Wiley & Sons, New York.

Sample, H. H., Neuringer, L. J., and Rubin, L. G. (1974). Low-temperature thermometry in high magnetic fields, III. Carbon resistors (0.5-4.2 K); thermocouples. *Rev. Sci. Instrum.* **45**, 64–73.

Sample H. H. and Rubin, L. G. (1977). Instrumentation and methods for low temperature measurements in high magnetic fields. *Cryogenics* **17**, 597–606.

Sample, H. H. and Rubin, L. G. (1976). Characterization of three commercially available Hall effect sensors for low temperatures and magnetic fields to 23 T. *IEEE Trans. Magn.* **12**, 810–812.

Sample, H. H. and Rubin, L. G. (1978). Magnetic field induced temperature changes in cryogenic liquids: N_2, Ar, and He^4. *Cryogenics* **18**, 223–229.

Sawada, S. and Mochizuki, T. (1972). Stability of 25 ohm platinum thermometer up to 1100°C. In *Temperature — Its Measurement and Control in Science and Industry,* Vol. 4, Part 2. Instrument Society of America, Pittsburgh, Pennsylvania, 919–926.

Schlosser, W. F. and Munnings, R. H. (1972). Thermistors as cryogenic thermometers. In *Temperature — Its Measurement and Control in Science and Industry,* Vol. 4. Part 2. Instrument Society of America, Pittsburgh, Pennsylvania, 795–801.

Schooley, J. F., Soulen, R. J., and Evans, J. A., Jr. (1972). *Preparation and use of superconductive fixed point devices, SRM 767.* NBS Special Publication 260–44, National Bureau of Standards, Washington, D.C.

Schulte, E. H. (1966). Carbon resistors for cryogenic temperature measurement. *Cryogenics* **6**, 321–323.

Scientific Instruments. *Germanium-Resistance Thermometers for Cryogenic Temperature Measurement.* Data sheet 1002/9170. Scientific Instruments, Lake Worth, Florida.

Sclar, N. and Pollock, D. B. (1972). On diode thermometers. *Solid-State Electron.* **15**, 473–480.

Scott, G. B., Springford, M., and Stockton, J. R. (1968). Use of a magnetoresistor to measure the magnetic field in a superconducting magnet. *J. Sci. Instrum.* **1**, 925–928.

Silver, A. H. and Zimmerman, J. E. (1975). Josephson weak-link devices. In *Applied Superconductivity,* Vol. 1. Academic Press, New York, 1–112.

Sinclair, D. H., Terbeek, H. G., and Malone, J. H. (1968). *Cryogenic Temperature Measurement Using Platinum Resistance Thermometers.* NASA Tech. Note D-4499.*

Sinclair, D. H., Terbeek, H. G., and Malone, J. H. (1970). Cryogenic temperature measurement. *Cryog. Ind. Gas.* **5**, 15–22.

Sinclair, D. H., Terbeek, H. G., and Malone, J. H. (1972). Calibration of platinum resistance thermometers. In *Temperature — Its Measurement and Control in Science and Industry,* Vol. 4, Part 2. Instrument Society of America, Pittsburgh, Pennsylvania, 983–988.

Sparks, L. L. (1974). Low temperature measurement. In *ASRDI Oxygen Technology Survey,* Vol. IV. NASA SP 3073.*

Sparks, L. L. and Hust, J. G. (1972). *Thermoelectric Voltage of Silver-28 Atomic Percent Gold Thermocouple Wire, SRM 733, Versus Common Thermocouple Materials (between Liquid Helium and Ice Fixed Points).* NBS Special Publication 260-34, National Bureau of Standards, Boulder, Colorado.

Sparks, L. L. and Powell. R. L. (1972). Method for assessing homogeneity and interchangeability of thermocouple wires. *Cryogenics* **12**, 40–43.

Sparks, L. L. and Powell, R. L. (1973). Low temperature thermocouples: KP "normal" silver, copper versus Au–0.02 at.% Fe and Au–0.07 at.% Fe. *J. Res. Natl. Bur. Stand.* **76A**, 263–283.

Sparks, L. L., Powell, R. L., and Hall, W. J. (1972). *Reference Tables for Low Temperature Thermocouples.* NBS Monograph 124, National Bureau of Standards, Boulder, Colorado.

Star, W. M., van Dam, J. E., van Baarle, C. (1969). Manufacture and precision interpolation of a film-type carbon resistance thermometer. *J. Sci. Instrum.* **2**, 257–262.

Stimson, H. F. (1955). Precision resistance thermometry. In *Temperature — Its Measurement and Control in Science and Industry,* Vol. 2. Reinhold Publishing Corp., New York, 141–168.

Stimson, H. F. (1961). *International Practical Temperature Scale of 1948 — Text Revision of 1960.* NBS Monograph 37, National Bureau of Standards, Washington, D.C.

Stout, M. B. (1960). *Basic Electrical Measurements.* Prentice-Hall, Englewood Cliffs, New Jersey, 104–126.

Swartz, J. M., Clark, C. F., Johns, D. A., and Swartz, D. L. (1976). Germanium and carbon glass resistance thermometry: a comparison of characteristics, stability and construction. In *Proceedings of the Sixth International Cryogenic Engineering Conference.* IPC Science and Technology Press, Guildford, England, 198–201.

Swartz, J. M. and Gaines, J. R. (1972). Wide range thermometry using gallium, In *Temperature — Its Measurement and Control in Science and Industry,* Vol. 4, Part 2. Instrument Society of America, Pittsburgh, Pennsylvania, 1117–1124.

Swartz, J. M., Gaines, J. R., and Rubin, L. G. (1975). Magnetoresistance of carbon–glass thermometers at liquid helium temperatures. *Rev. Sci. Instrum.* **46**, 1177–1178.

Swartz, D. L. and Swartz, J. M. (1974). Diode and resistance cryogenic thermometry: a comparison. *Cryogenics* **14**, 67–70.

Swenson, C. A. (1955). Properties of indium and thallium at low temperatures. *Phys. Rev.* **100**, 1607–1614.

Swenson, C. A. and Wolfendale, P. C. F (1973). Differences between ac and dc determinations of germanium thermometer resistances. *Rev. Sci. Instrum.* **44**, 339–341.

Tiggelman, J. L. and Durieux, M. (1972). Platinum resistance thermometry below 13.81 K. In *Temperature — Its Measurement and Control in Science and Industry,* Vol. 4, Part 2. Instrument Society of America, Pittsburgh, Pennsylvania, 849–856.

Utton, D. B. (1967). Nuclear quadrupole resonance thermometry. *Metrologia* **3**, 98–105.

Vanier, J. (1972). Nuclear magnetic resonance thermometry. In *Temperature — Its Measurement and Control in Science and Industry,* Vol. 4, Part 2. Instrument Society of America, Pittsburgh, Pennsylvania, 1197–1212.

von Middendorff, A. (1971). Thermocouples at low temperatures in high magnetic fields. *Cryogenics*

11, 318.

Walstrom, P. L. (1975). The effect of high magnetic fields on metal foil strain gauges at 4.2 K. *Cryogenics* **15**, 270–272.

Warren, K. A. and Reed, R. P. (1963). *Tensile and Impact Properties of Selected Materials from 20-300° K.* NBS Monograph 63, National Bureau of Standards Boulder, Colorado.

Watson, P. G. and DiGregorio, R. F. (1965). Techniques of magnetic-field measurement. In *Proceedings of the International Symposium on Magnet Technology.**

Weber, L. A., Diller, D. E., Roder, H. M., and Goodwin, R. D. (1962). The vapor pressure of 20° K equilibrium hydrogen. *Cryogenics* **2**, 236–238.

Wensel, H. T. (1941). Temperature. In *Temperature — Its Measurement and Control in Science and Industry.* Reinhold Publishing Corp., New York, 3–23.

White, G. K. (1959). *Experimental Techniques in Low-Temperature Physics.* University Press, Oxford, England, 91–100; 109–114.

White, G. K. (1961). Measurement of thermal expansion at low temperatures. *Cryogenics* **1**, 151–158.

White, G. K. and Woods, S. B. (1957). Indium resistance thermometer; 4 to 300° K. *Rev. Sci. Instrum.* **28**, 638–641.

White, G. K. and Woods, S. B. (1958). Thermal conductivity of solidified inert gases: argon, neon, and krypton. *Philos. Mag.* **3**, 785–797.

White, G. K., Woods, S. B., and Anglin, F. (1958). Indium resistance thermometer. *Rev. Sci. Instrum.* **29**, 181–182.

Willens, R. H. (1972). Inductance thermometry. In *Temperature — Its Measurement and Control in Science and Industry,* Vol. 4, Part 2. Instrument Society of America, Pittsburgh, Pennsylvania, 1137–1141.

Willens, R. H., Buehler, E., and Nesbitt, E. A. (1968). Inductance thermometer. *Rev. Sci. Instrum.* **39**, 194–196.

Wolfendale, P. C. F. (1969). A precise automatic ac potentiometer for low temperature resistance thermometry. *J. Sci. Instrum.* **2**, 659–660.

Woolley, H. W., Scott, R. B., and Brickwedde, F. G. (1948). Compilation of thermal properties of hydrogen in its various isotopic and ortho–para modifications. *J. Res. Natl. Bur. Stand.* **41**, 379–475.

Yates, B. and Panter, C. H. (1961). Indium resistance thermometers. *J. Sci. Instrum.* **38**, 196–197.

*Available from NTIS, Springfield, Virginia.

International System of Units

Conversion factors for the International System of Units (SI) used in this book and definitions of the SI prefixes are given below. Note that ounces and pounds are avoirdupois, Btu's are thermochemical, and pounds and pounds of force are given rather than pound-mass and poundals. See Table 6.1 for additional magnetic unit conversions and § 6.2.1 for a thorough discussion of magnetic unit systems.

$T_K = T_C + 273.15$ where T_K = temperature in kelvins
$T_K = (T_F + 459.67)/1.8$ $\quad T_C$ = temperature in degrees Celsius
$T_C = (T_F - 32)/1.8$ $\quad T_F$ = temperature in degrees Fahrenheit

* * * * *

1 A·m^{-1} = 1.257 × 10^{-2} oersteds
1 A·cm^{-2} = 6.452 A·in^{-2}
1 A·m^{-2} = 0.6452 mA·in^{-2}

1 g = 3.527 × 10^{-2} oz
1 kg = 2.205 lb
1 kg·m^{-2} = 0.2048 lb·ft^{-2}
1 g·cm^{-3} = 62.43 lb·ft^{-3}
1 g·cm^{-3} = 3.613 × 10^{-2} lb·in^{-3}
1 kg·m^{-3} = 6.243 × 10^{-2} lb·ft^{-3}
1 kg·m^{-3} = 3.613 × 10^{-5} lb·in^{-3}

1 Hz = 1 s^{-1}

1 J = 1 N·m
1 J = 0.7376 ft·lbf
1 J = 10^{7} erg
1 J = 1 W·s
1 J = 2.778 × 10^{-4} W·h
1 J = 6.241 × 10^{18} eV
1 J = 9.485 × 10^{-4} Btu
1 J·kg^{-1}·K^{-1} = 2.388 × 10^{-4} Btu·lb^{-1}·°F^{-1}
1 kJ·m^{-2} = 5.714 in-lb·in^{-2}

1 m = 3.281 ft
1 m = 39.37 in
1 m^{2}·s^{-1} = 3.875 × 10^{4} ft^{2}·h^{-1}
1 m^{2}·s^{-1} = 10.764 ft^{2}·s^{-1}

1 N = 1 kg·m·s^{-2}
1 N = 0.2248 lbf
1 N·m = 8.85 lbf·in
1 N·m = 10^{7} dyne·cm
1 N·m^{-1} = 6.852 × 10^{-2} lbf·ft^{-1}

1 Pa = 1 N·m^{-2}
1 Pa = 10 dyne·cm^{-2}
1 Pa = 0.1020 kgf·m^{-2}
1 MPa = 10 bar
1 MPa = 9.869 atm (standard)
1 MPa = 145.0 lbf·in^{-2} (psi)
1 MPa = 295.3 in Hg (at 32° F)
1 MPa·m$^{1/2}$ = 909.9 psi·in$^{1/2}$

1 T = 1 Wb·m^{-2}
1 T = 10^{4} gauss
1 T = 10^{9} gamma

(continued)

$$1\ \text{V}\cdot\text{K}^{-1} = 0.5556\ \text{V}\cdot{}^\circ\text{F}^{-1}$$
$$1\ \text{V}\cdot\text{K}^{-2} = 0.3086\ \text{V}\cdot{}^\circ\text{F}^{-2}$$

$$1\ \text{W}\cdot\text{m}^{-2} = 9.290 \times 10^{-2}\ \text{W}\cdot\text{ft}^{-2}$$
$$1\ \text{W}\cdot\text{m}^{-1}\cdot\text{K}^{-1} = 1.927 \times 10^{-3}\ \text{Btu}\cdot\text{in}\cdot\text{s}^{-1}\cdot\text{ft}^{-2}\cdot{}^\circ\text{F}^{-1}$$
$$1\ \text{W}\cdot\text{m}^{-1}\cdot\text{K}^{-1} = 6.938\ \text{Btu}\cdot\text{in}\cdot\text{h}^{-1}\cdot\text{ft}^{-2}\cdot{}^\circ\text{F}^{-1}$$
$$1\ \text{W}\cdot\text{m}^{-1}\cdot\text{K}^{-1} = 0.5782\ \text{Btu}\cdot\text{ft}\cdot\text{h}^{-1}\cdot\text{ft}^{-2}\cdot{}^\circ\text{F}^{-1}$$

$$\Omega\cdot\text{cm} = 0.3937\ \Omega\cdot\text{in}$$
$$\Omega\cdot\text{cm}^3 = 6.102 \times 10^{-2}\ \Omega\cdot\text{in}^3$$

SI PREFIXES

Factor	Prefix	Symbol	Factor	Prefix	Symbol
10^{-15}	femto	f	10^{3}	kilo	k
10^{-12}	pico	p	10^{6}	mega	M
10^{-9}	nano	n	10^{9}	giga	G
10^{-6}	micro	μ	10^{12}	tera	T
10^{-3}	milli	m	10^{15}	peta	P
10^{-2}	centi	c			

Index

Indexing of low-temperature properties and phenomena is based largely on the organization of the book into its various chapters. Materials are indexed in detail to enable the reader to find easily a specific property of a particular metal, alloy, composite or polymer.

The words "temperature" and "low temperature" are almost never used in this index, but "temperature effect" appears as a subhead when there is fairly extended general discussion in the text. The effects of low temperatures are shown in innumerable tables and graphs, all indexed by property, phenomenon and material.

The symbol (F) or (T) following a page number means that information is presented in a figure or a table on that page or the included pages. —M.R.H.

A

D

J

K

L

M